A COMPREHENSIVE GUIDE TO THE HAZARDOUS PROPERTIES OF CHEMICAL SUBSTANCES

SECOND EDITION

A COMPREHENSIVE GUIDE TO THE HAZARDOUS PROPERTIES OF CHEMICAL SUBSTANCES

SECOND EDITION

Pradyot Patnaik, Ph.D.

A John Wiley & Sons, Inc., Publication

New York • Chichester • Weinheim • Brisbane • Singapore • Toronto

For ordering and customer service, call 1-800-CALL-WILEY.

Library of Congress Cataloging-in-Publication Data:
Patnaik, Pradyot.
    A comprehensive guide to the hazardous properties of chemical
  substances / Pradyot Patnaik. — 2nd ed.
      p.   cm.
    Includes index.
    ISBN 0-471-29175-7 (cloth : alk. paper)
    1. Toxicology. 2. Chemicals — Tables. I. Title.
  RA1211.P38 1999
  615.9 — dc21   98-39972
                                                                  CIP

Printed in the United States of America.

10 9 8 7 6 5 4 3 2 1

To
Manisha and Chirag

CONTENTS

PREFACE

This book is a major revision of the first edition. Several new chapters have been added to this edition incorporating many new entries under specific classes of compounds. Some chapters have been fully rewritten or expanded to include additional compounds or update information.

Fifteen new chapters were added into this edition. Four of these are presented in Part A. They are (1) Physical Properties of Compounds and Hazardous Characteristics; (2) Target Organs and Toxicology; (3) Teratogenic Substances; and (4) Habit-Forming Addictive Substances. Chapters on toxic properties of substances, explosive characteristics of chemical substances, and cancer-causing substances have been fully revised.

In Part B, eleven new chapters were added. Azides, acetylides, and fulminates of metals have been separated into two groups. Metal azides is upgraded to a full chapter. The chapters on the pesticides and herbicides have now been expanded from two chapters in the first edition to seven chapters in this book including six structurally distinct classes of substances: organochlorine —, organophosphorus —, carbamate —, triazine —, urea —, and chlorophenoxy acid types. There is an additional chapter on their classifications, structures and chemical analyses. Also, the sulfur mustards and the sulfur esters are new additions to this text.

Metals are now grouped under the headings as reactive or toxic and thus revised from the earlier classification as RCRA metals in the last edition. Also, the toxicity and reactivity of several metals that do not fall under any Federal regulations are now discussed in these chapters. Several organometallic compounds are included under the new chapters: metal alkyls, metal alkoxides and metal carbonyls.

Three appendices are incorporated at the end. One of these is US Federal Regulations. Hazardous substances are presented as appendices in this edition. The Toxic Substance Control Act has been added and other regulatory acts have been updated for the benefit of readers. This chapter is moved from Part A in the first edition to one of the Appendices in this edition with some updated information on the Safe Drinking Water Act and Resource Conservation and Recovery Act. The Toxic Substance Control Act is a new addition to this section. The other two appendices present a complete listing of carcinogenic agents updated to 1998, one is classified by the International Agency for Research on Cancer and the other one is the listing of the US National Toxicology Program. These two appendices should be considered a continuation of Chapter VI: Cancer-Causing Chemicals in Part A of this text.

Any comments and suggestions from readers will be greatly appreciated.

Pradyot Patnaik

PREFACE TO THE FIRST EDITION

The subject of hazardous properties of chemical compounds constitutes a very wide area which includes topics of wide diversity ranging from toxicology and explosivity of a compound to its disposal or exposure limits in air. In fact, each of these topics can form a subject for a book on its own merits. There are several well-documented books on these topics. However, most of these works show important limitations. The objectives of this book, therefore, are (1) to present information on many aspects of hazardous properties of chemical substances, covering the research literature up to 1991, and (2) to correlate the hazardous properties of compounds to the functional groups, reactive sites, and other structural features in the molecules; and thus to predict or assess the hazards of a compound from its structure when there is lack of experimental data.

The hazardous properties are classified under two broad headings: health hazard and fire and explosion hazard. The former includes toxicity, corrosivity, carcinogenicity, mutagenicity, reproductive toxicity, and exposure limits. Flammability, violent and explosive reactions, incompatibility, and fire-extinguishing agents are discussed at length under fire and explosion hazard. In addition, information is provided on physical properties, uses, storage, handling, disposal/destruction, and chemical analyses. These are presented in 46 Chapters covering organics, metals and inorganics, industrial solvents, common gases, particulates, explosives, and radioactive substances. All the Chapters except those on azo dyes, industrial solvents, common toxic and flammable gases, particulates, and RCRA and priority pollutant metals are arranged in accordance with their structures and reactive functional groups. In other words, substances of same families are grouped together. A general discussion on the chemistry, toxicity, flammability, and chemical analysis is presented for each class of substances. Individual compounds of moderate to severe hazard and/or commercial importance are discussed in detail. Relationship of structures with hazardous properties are highlighted.

Individual compounds in each class of substance are discussed in most Chapters under the following features: formula; molecular weight; CAS registry number; EPA (RCRA) and DOT status; structure; synonyms; uses and exposure risk; physical properties, including color, odor, melting and boiling points, density (mg/L), solubility, and pH (if acidic or alkaline); health hazard, which includes toxic, corrosive, carcinogenic, and teratogenic properties; exposure limits set by ACGIH, OSHA, MSHA, and NIOSH; fire and explosion hazard, which includes flammability data, fire-extinguishing agents, and explosive reactions; storage and shipping; disposal/destruction, including laboratory methods for destruction and biodegradation; and chemical analyses, which include general instrumental techniques as well as specific analysis of compounds in water, soils, sediments, solid wastes, and air, and biological monitoring. Deviations from the foregoing format are noted in many Chapters for less hazardous substances or for lack of information. Also if the methods or procedures are similar, a detailed discussion is presented only for a few of the prototype compounds. The topics that are reviewed in greater detail are toxicological properties, violent reactions and explosive products, biodegradation, and chemical analysis of more hazardous substances.

This book is divided into two major parts. Part B presents a comprehensive discussion of individual substances and classes of substances, Part A highlights the four primary hazardous properties: toxicity, carcinogenicity, flammability, and explosive characteristics of chemical substances. The salient features of various federal regulations for the protection of human health and the environment are discussed in brief in Part A.

It is hoped that this book serves its purpose. Any comments, criticism, or suggestions from readers for further improvement of the book will be greatly appreciated.

ACKNOWLEDGMENTS

I would like to express my sincere thanks to Mrs. Min Yang, my colleague at the ISC Laboratory for all her help during the final phase of this manuscript preparation. I express my thanks to Mrs. Mary Ann Richardson for typing most of the manuscript in a timely manner. Also I would like to thank Mrs. Nancy Olsen, former Editor of Environmental Health and Safety of Van Nostrand Reinhold for initiating this edition and Mrs. Betty Sun, Ms. Jerilynn Caliendo, Mrs. Camille Pecoul Carter, and other staff of John Wiley for their help and advice in the preparation of this book. Finally, special words of thanks are due to my wife, Sanjukta, for her patience and support.

PART A

I

INTRODUCTION

Practically all chemical substances manifest hazardous properties in one way or another, depending on the quantities involved, system conditions, and the nature of the surrounding. The term *hazardous properties* may be broadly classified into two principal categories: namely (1) toxicity, and (2) flammability and explosivity. The term *toxicity* refers to substances that produce poisoning or adverse health effects upon acute or chronic exposure. The term may be expanded to include mutagenicity and carcinogenic potential; teratogenicity, including reproductivity and developmental toxicity; and corrosivity or irritant actions. The toxic properties of compounds may be correlated with the structures and reactive sites in the molecules, although such a classification has obvious limitations. The presence of certain functional group(s) or the possession of certain types of structural features tends to make the molecules highly toxic. Thus, both a cyanide ion (CN^-) or phosphate ester linkage (as in many organophosphorus insecticides) impart a high degree of toxicity, but of different types. The toxicity of a chemical substances depend on a variety of factors, including the electrophilic nature of the substances, steric hindrance in the molecules, adduct formations, and metabolic activations — a host of chemical and biochemical reactions ranging from simple oxidation or hydrolysis to highly complex enzyme inhibition reactions. Certain physical properties of substances also have a significant effect on the toxic actions. These include water and lipid solubility, boiling point, and vapor pressure. Substances of low molecular weights have vapor pressures and water solubilities greater that these of high-molecular-weight compounds in the same family. Such substances would therefore present a greater risk of pollution of air and inhalation hazard contamination of water, respectively. Many lipid-soluble substances, such as DDT and PCBs, are stored in the body tissues and are not excreted readily.

Animal data on the median lethal dose (LD_{50}) by various routes of administration, as well as the median lethal concentration (LC_{50}) by the inhalation route, are good indicators of the degree of toxicity of a substance. Such terms as extremely toxic, highly toxic, moderately toxic, and low or very low toxicity are often used for comparative purposes. These terms are also used to refer to the degree of toxicity of a compound. In this book, substances that exhibit acute oral LD_{50} values of <100 mg/kg in rats or mice are termed highly toxic compounds. Those that have LD_{50} values of 100–500 mg/kg and >500 mg/kg are termed moderate and low toxicants, respectively. For inhalation toxicity, LC_{50} values of <100 ppm, 100–500 ppm, and >500 ppm for a 4-hour exposure period in rats or mice refer roughly to high, moderate, and low toxicities, respectively. Substances having LD_{50} values of <10 mg/kg or LC_{50} values of <10 ppm may be considered extremely toxic. The use of such terms is very qualitative and often does not reflect human toxicity. This is because the lethal dose of a substance in one species may be quite different in another species, and entirely different in humans. Also, chronic toxicity is not taken into consideration.

In recent years there has been a notable increase in research on structure–activity relationships (SARs), also called quantitative structure–activity relationships (QSARs), used to assess the toxicity of substances for which there are few experimental data. This approach involves establishing mathematical relationships derived from computer modeling, based on known toxicity data of similar (or dissimilar) types of compounds, octanol–water partition coefficients, molar connectivity index values, and other parameters. A detailed discussion of this subject is beyond the scope of this book.

The flammable properties of substances in air include their flash point, vapor pressure, autoignition temperatures, and flammability range (i.e., their lower and upper explosive limits). Liquids that have a flash point of $<100°F$ ($37.8°C$) are termed flammable, whereas liquids that have a flash point of 100–200°F ($37.8–93.3°C$) are termed combustible. These terms are explained in detail in Sections X and XI of Part A. Many substances in their solid state at the ambient temperature are flammable too. They ignite spontaneously in air. Flammability and explosivity are very closely related terms. A violent reaction that produces flame or goes to incandescent can cause an explosion under more severe conditions. Similarly, many chemical explosions are accompanied by flame. Thus, the fire and explosion hazards of chemical compounds are closely linked. A violent reaction between two compounds depends on the energetics of the reaction and the thermodynamic properties of the reactants. Many classes of chemicals, including peroxides, ozonides, azides, acetylides, nitrides, and fulminates, undergo violent explosive reactions. By contrast, polynitroorganics detonate on heating, impact, or percussion. Another type of violent reaction involves the combining of fairly stable molecules, producing highly unstable or shock-sensitive products. Common examples include the reaction between nitric acid and glycerol and that between ethylene and ozone. It may be noted that strong oxidizers react violently with organics and other oxidizable substances. This subject is discussed further throughout the book.

Hazardous substances can be disposed of in an approved landfill disposal site, although such disposal of highly toxic compounds can lead to groundwater contamination. Incineration in a chemical incinerator equipped with an afterburner and scrubber is a widely used practice for the destruction of toxic wastes. Biodegradation of toxic compounds to nontoxic products in soils, waters, treatment plants, and bioreactors has become the subject of in-depth investigation in recent years. Other processes that are used include treatment with molten metals, wet oxidation, ultraviolet (UV) light-catalyzed oxidation using hydrogen peroxide or ozone, wet oxidation of carcinogenic amines with potassium permanganate, photodegradation, catalytic oxidation, and dechlorination or desulfurization. Many hazardous substances in small quantities can be destroyed in the laboratory by carefully treating them with various types of substances to produce stable and nontoxic end products.

Chemical analysis of hazardous substances in air, water, soil, sediment, or solid waste can best be performed by instrumental techniques involving gas chromatography (GC), high-performance liquid chromatography (HPLC), GC/mass spectrometry (MS), Fourier transform infrared spectroscopy (FTIR), and atomic absorption spectrophotometry (AA) (for the metals). GC techniques using a flame ionization detector (FID) or electron-capture detector (ECD) are widely used. Other detectors can be used for specific analyses. However, for unknown substances, identification by GC is extremely difficult. The number of pollutants listed by the U.S. Environmental Protection Agency (EPA) are only in the hundreds — in comparison with the thousands of harmful pollutants that may be present in the environment. GC/MS is the best instrumental technique for compound identification. FTIR and GC-FTIR techniques are equally effective and have been described for the analysis of many pollutants. HPLC analysis are very common for biological monitoring, as well as for several types of air and water analysis. Colorimetric, polarographic, and ion-selective electrode methods have been described for many individual compounds or types of compounds. X-ray diffraction and electron microscopy techniques may be applied for the determination of silica, asbestos, and other particulates. Metals analysis is best performed by atomic absorption spectroscopy. Aqueous samples are extracted using a suitable solvent. In the case of air analysis, the pollutants are adsorbed over charcoal, Tenax, or silica gel (depending on the nature of the substance) before analysis. Particulate material is collected using suitable membrane filters.

II

GLOSSARY

GENERAL ABBREVIATIONS

ACGIH   American Conference of Governmental Industrial Hygienists

ACS   American Chemical Society

At. no.   Atomic number

At. wt.   Atomic weight

bp   Boiling point

CAS   Chemical Abstracts Service

EPA   Environmental Protection Agency

IARC   International Agency for Research on Cancer

MCA   Manufacturing Chemists' Association

mp   Melting point

MSHA   Marine Safety and Health Administration

MW   Molecular weight

NFPA   National Fire Protection Association

NIOSH   National Institute for Occupational Safety and Health

NRC   National Research Council

OSHA   Occupational Safety and Health Administration

RCRA   Resource Conservation and Recovery Act

RTECS   *Registry of Toxic Effects of Chemical Substances* (five-volume NIOSH compilation listing the toxicity data of more than 80,000 compounds; no longer published in book format, only available in CD-ROM)

GLOSSARY OF TERMS IN TOXICITY AND FLAMMABILITY

IDLH   Immediately dangerous to life or health (term used for the purpose of respiratory selection; maximum concentration from which in the event of respirator failure one could escape within 30 minutes without experiencing irreversible health effects)

LC_{50}   Median lethal concentration 50 (calculated concentration of a chemical in air, exposure to which can cause the death of 50% of experimental animals in a specified period of time)

LD_{50}   Median lethal dose 50 (calculated dose of a chemical that is expected to cause the death of 50% of experimental animals when administered by any route other than inhalation)

LEL Lower explosive limit (in air) (also known as lower flammable limit; see Section V of Part A)

SADT Self-accelerating decomposition temperature

STEL Short-term exposure limit (maximum concentration of a substance in air to which workers can be exposed for 15 minutes for four exposure periods per day with at least 60 minutes between exposure periods)

TLV Threshold limit values (concentration of a substance in air to which workers can be exposed without adverse effect)

TWA Time-weighted average (8-hour work period per day or 40-hour work period per week)

UEL Upper explosive limit (in air) (also known as upper flammable limit; see Chapter V of Part A)

UNITS AND CONVERSION

°C centigrade or Celsius (unit of temperature)

density defined as mass per unit volume; the values given for liquids in this book are g/mL (unit omitted)

°F Fahrenheit (unit of temperature)

kg 1 kilogram (1000 g or 1,000,000 mg)

m³ cubic meter [1000 liters (L) or 1,000,000 milliliters (mL)]

mg milligram [0.001 gram (g) or 1000 micrograms (µg)]

millimole 0.001 mole

mL milliliter [0.001 liter or 1000 microliters (µL)]

mol molecular weight expressed in grams

mppcf millions of particles per cubic foot (of air) (mppcf × 35.3 = million particles per cubic meter = particles per cubic centimeter)

ng nanogram [0.001 µg or 1000 picograms (pg)]

nm nanometer (common unit for absorbance used in spectroscopy)

pH term used to refer to acidity or alkalinity of a solution; pH value less than 7 is acidic, above 7 is alkaline, and at 7 is neutral; $= -\log[H_3O^+]$ or $[H_3O^+] =$ antilog$(-pH)$ (where $[H_3O^+]$ is the concentration of hydrogen ion (or hydronium ion) in the solution)

ppb parts per billion (1 ppb = 0.001 ppm)

ppm parts per million (1 ppm = 0.0001% or 1000 ppb)

torr unit of pressure (1 torr = 1 mm of mercury; 1 atm = 760 torr)

µg microgram [0.001 mg or 1000 nanograms (ng)]

µm micrometer (0.001 meter)

Conversion of degrees Fahrenheit (F) to degrees Celsius (C), and vice versa:

$$C = \frac{5F - 160}{9}$$

$$F = \frac{9C + 160}{5}$$

Conversion of ppm concentration of gas or vapor in air to mg/m³, and vice versa:

$$1\,ppm = \frac{MW}{24.46}\ mg/m^3 \text{ at 25°C and 1 atm}$$

$$1\,mg/m^3 = \frac{24.46}{MW}\ ppm \text{ at 25°C and 1 atm}$$

where MW represents molecular weight. For example, a concentration of 200 ppm methanol in air at 25°C is equal to

$$\frac{200 \times 32}{24.46} = 261.6\ mg/m^3 \text{ methanol}$$

[The molecular weight of methanol (CH_3OH) is 32.]

The ppm concentration in air to mg/m³, and vice versa, at any temperature x(°C) may be calculated from the following relationship:

$$1 \text{ ppm} = \frac{MW \times 298}{24.46(273 + x)} \text{ mg/m}^3$$

$$1 \text{ mg} = \frac{24.46(273 + x)}{MW \times 298} \text{ ppm}$$

For example, a concentration of 200 ppm of methanol in air at 10°C is equal to

$$\frac{200 \times 32 \times 298}{24.46(273 + 10)} = 275.5 \text{ mg/m}^3$$

GLOSSARY OF TERMS IN INSTRUMENTATION AND ANALYSIS

AA Atomic absorption spectrophotometry (technique commonly used for metals analysis)

ECD Electron capture detector [used in GC primarily to analyze halogenated organics such as chlorinated pesticides and polychlorinated biphenyls (PCBs)]

FID Flame ionization detector (used in GC for general organics analysis; aqueous samples may be analyzed directly)

FPD Flame photometric detector (used in GC primarily to analyze the sulfur-containing organics)

FTIR Fourier transform infrared spectroscopy (technique useful in identifying and analyzing various unknown compounds, such as environmental pollutants)

GC Gas chromatography (most commonly used analytical technique for organics)

GC/MS Gas chromatography/mass spectrometry (technique based on chromatographic separation, followed by chemical or electron-impact ionization and identification of the mass spectra of the ionized fragments)

HECD Hall electrolytic conductivity detector (used in GC to analyze halogenated organics: generally, volatile organics)

HPLC High-performance liquid chromatography (used in a wide variety of applications in pharmaceutical and toxicological analysis)

ICP Inductively coupled plasma emission spectrophotometry (technique applied for simultaneous determination of several metals)

IR Infrared (spectroscopy)

MSD Mass selective detector

NMR Nuclear magnetic resonance

NPD Nitrogen–phosphorus detector (used in GC to analyze nitrogen- and phosphorus-containing organics)

ODS Octadecylsilane (silica-based C_{18} column used in HPLC analysis)

PID Photoionization detector (used in GC to analyze many common organics: generally, mononuclear aromatics)

SEM Scanning electron microscopy (used in the determination of asbestos, silica, and other particulates, and as a tool in toxicological studies)

SIM Selective ion monitoring [technique used to look for specific ion(s) so as to identify compounds of interest]

TCD Thermal conductivity detector (used in GC primarily to analyze common gases such as methane and CO_2)

TEM Transmittance electron microscopy (used in the determination of asbestos, silica, and other particulates, and as a tool in toxicological studies)

TLC Thin-layer chromatography

UV Ultraviolet (spectroscopy)

XRD X-ray diffraction spectroscopy (used in the determination of asbestos, silica, and other particulates, and to analyze metals and study structures of crystals)

XRF X-ray fluorescence spectroscopy (used in the determination of asbestos, silica, and other particulates, and to analyze metals and study structures of crystals)

PHYSICAL PROPERTIES OF COMPOUNDS AND HAZARDOUS CHARACTERISTICS

Certain physical properties of a substance determine its hazard and the risk of human exposure. For example, solubility of a substance in aqueous medium and lipid can indicate whether the water-soluble toxic metabolite is susceptible to excrete out from the body or high lipid solubility could result in its storage in the body fat. Similarly, water solubility of a compound should be a decisive factor in extinguishing a fire or diluting or cleaning up a spill. Also, the fate of a pollutant in the biosphere or troposphere greatly depends on certain physical properties. The risk of exposure also significantly depends on the volatility (vapor pressure) of the compound and its vapor density. Odor is another physical property of great significance. Many hazardous substances such as hydrogen sulfide or ammonia can be readily identified from their odor. Some of the physical properties that may be applied to evaluate the risk of hazard are discussed below. Also, some calculations and solved problems are illustrated in the following discussion to show their applications and conversions.

III.A VAPOR PRESSURE

Many hazardous substances, whether toxic or flammable, or both, are liquids under ambient conditions. Inhalation toxicity and the level of exposure to such liquid substances may depend on their volatility. The more volatile a substance, the greater the risk of exposure of inhaling its vapors. Similarly, the vapors of many flammable substances form explosive mixtures with air. Therefore, knowledge on volatility of a substance is essential to assess its fire, explosion, and toxic hazard and the risk of exposures, especially in confined places.

The volatility of a substance can be assessed from its vapor pressure and boiling point. The latter is defined as the temperature at which the vapor pressure of a liquid is equal to the atmospheric pressure. At any temperature, including ambient conditions, any liquid will have some kind of vapor pressure. Such vapor pressure is related to the temperature and the intermolecular forces. The higher the temperature, the higher the vapor pressure. Similarly, the weaker the intermolecular forces, the higher the vapor pressure.

Often, the experimental data in the literature on the vapor pressure of a liquid may be found at a temperature higher than that at ambient condition. Or, we may need to know the vapor pressure of a liquid at an elevated temperature, say, that of a closed confined space. The relationship between the vapor pressure and temperature can be mathematically expressed in the following equation:

$$\ln P = \frac{-\Delta H_{vap}}{R}\left(\frac{1}{T}\right) + C$$

where $\ln P$ is the natural logarithm of the vapor pressure P; ΔH_{vap} is the heat of vaporization of the liquid (heat required to vaporize a mole of molecules in the liquid state); T is the temperature in K; R is the universal gas constant (8.31 J/mol·K); and C is a constant. This is known as the *Clausius–Clapeyron* equation.

A two-point version of the above equation may be written as follows to determine the vapor pressure, boiling point, or heat of vaporization nongraphically:

$$\ln \frac{P_2}{P_1} = \frac{-\Delta H_{vap}}{R}\left(\frac{1}{T_2} - \frac{1}{T_1}\right)$$

where P_1 and P_2 are the vapor pressure of a liquid at temperatures T_1 and T_2, respectively. ΔH_{vap} values may be found in most chemical data handbooks. If ΔH_{vap} is known and the vapor pressure of a substance is known at one temperature, we can calculate the vapor pressure at any other temperature. Similarly, from the vapor pressure and ΔH_{vap}, we can calculate the boiling point of the substance. Two solved problems are illustrated below.

Problem 1. The vapor pressure of 2-bromopropane is 400 torr at 41.5°C. What would be its vapor pressure at 15°C? The heat of vaporization, ΔH_{vap} is 31.76 kJ/mol.

$$\ln \frac{P_2}{P_1} = \frac{-\Delta H_{vap}}{R}\left(\frac{1}{T_2} - \frac{1}{T_1}\right)$$

$$T_1 = 273 + 41.5 = 314.5 \text{ K}$$

$$T_2 = 273 + 15 = 285 \text{ K}$$

$$\Delta H_{vap} = 31.76 \text{ kJ/mol}$$

$$P_1 = 400 \text{ torr}$$

$$P_2 = ? \text{ (to be determined)}$$

$$R = 8.31 \text{ J/mol} \cdot \text{K}$$

$$\ln\left(\frac{P_2}{400 \text{ torr}}\right) = \left(-\frac{31.76 \times 10^3 \text{ J/mol}}{8.31 \text{ J/mol} \cdot \text{K}}\right)$$

$$\times \left(\frac{1}{285 \text{ K}} - \frac{1}{314.5 \text{ K}}\right)$$

$$= -3.824 \times 10^3 \times 0.33 \times 10^{-3}$$

$$\ln \frac{P_2}{400 \text{ torr}} = -1.262$$

$$\ln P_2 - \ln 400 = -1.262$$

$$\ln P_2 = 5.99 - 1.262 = 4.729$$

$$P_2 = 113.2 \text{ torr}$$

(Use a calculator to determine the e^x of the given natural logarithm, to obtain the number. Here, e^x of 4.729 is 113.2.) Therefore, the vapor pressure of 2-bromopropane at 15°C is 113.2 torr.

Problem 2. Determine the boiling point of 2-bromopropane from the above data. We know that the boiling point of a substance is the temperature at which its vapor pressure is 760 torr (i.e., the vapor pressure is equal to the atmospheric pressure). We can now use the same Clausius–Clapeyron equation, where P_2 is 760 torr and T_2, the boiling point. Thus,

$$\ln\left(\frac{760 \text{ torr}}{400 \text{ torr}}\right) = \left(-\frac{31.78 \times 10^3 \text{ J/mol}}{8.31 \text{ J/mol} \cdot \text{K}}\right)$$

$$\times \left(\frac{1}{T_2} - \frac{1}{314.5 \text{ K}}\right)$$

$$0.642 = (-3.824 \times 10^3)$$

$$\times \left[\frac{1}{T_2} - (3.18 \times 10^{-3})\right]$$

$$0.642 = -\frac{3.824 \times 10^3}{T_2} + 12.16$$

$$-11.518 \, T_2 = -3.824 \times 10^3 \text{ K}$$

$$T_2 = 332 \text{ K}$$

or

$$T_2 = 332 \text{ K} - 273 = 59°C$$

Thus, the boiling point of 2-bromopropane should theoretically be equal to 59°C. The actual boiling point determined experimentally is 60°C, which agrees very closely with the above calculation.

III.B VAPOR DENSITY

It is often necessary to know whether the vapors of a hazardous substance is heavier than air. For example, vapors of a flammable liquid, such as, diethyl ether are heavier than air and can present a "flashback" fire hazard. Being heavier than air, they remain longer near their sources and do not dissipate up or vent out readily into the air. Also, this may present relatively a high-level exposure and occupational health problem.

It may be noted that the vapor density is different form the gas density. While the latter is defined as the mass of a gas per unit volume, usually expressed as grams per liter (g/L) at a specific temperature, the vapor density simply indicates how many times the vapor of a substance is heavier than air. It does not have a unit and is written as a number followed by "air = 1" in parentheses.

It is calculated by dividing the molar mass of the substance with that of air. Since air is a homogeneous mixture of 78% N_2 and 20% O_2, its theoretical molar mass is equivalent to:

$$\left(\frac{28 \text{ g N}_2}{1 \text{ mol N}_2} \times \frac{0.78 \text{ mol N}_2}{1 \text{ mol air}} \right)$$
$$+ \left(\frac{32 \text{ g O}_2}{1 \text{ mol O}_2} \times \frac{0.20 \text{ mol O}_2}{1 \text{ mol air}} \right) = 29 \text{ g/mol}$$

Thus, to determine the vapor density of a compound, we divide its formula weight by 29, as in the following example.

Problem 1. Determine the vapor density of (i) diethyl ether and (ii) hydrogen cyanide.

    (i) The molecular formula of diethyl ether is $(C_2H_5)_2O$, and its formula

weight is 74. Thus, the vapor density = $(74/29) = 2.55$. That is, the vapor density 2.55 (air = 1)

    (ii) The formula weight for hydrogen cyanide, HCN is 27. Thus, the vapor density = $(27/29) = 0.93$. That is, the vapor density is 0.93 (air = 1).

III.C SOLUBILITY

Solubility of a substance in water and other solvents is a very important physical property in toxicity studies. The fate of a pollutant, its oxidation, hydrolysis, biodegradation, groundwater contamination, and overall persistence in the environment depend on the water solubility. Similarly, water solubility is a major criterion for the use of water to extinguish fire from a substance or dilute its spill.

In general, polar substances are soluble in water, while the nonpolar compounds are soluble in nonpolar solvents. Among the inorganic substances, the solubility depends on the degree of ionic bonding. All chlorides, nitrates, alkali metals, and ammonium salts are completely soluble in water. Most alkaline metal salts are also water soluble. Every substance is soluble to some extent. Even the so-called insoluble salt may exhibit slight solubility at the ppm level. The solubility of a salt in water may be readily calculated from its solubility product constant (K_{SP}) value. The higher the K_{SP}, the greater the solubility. The K_{SP} for salts may be found in any standard handbook of chemistry.

The following solved problems show how to calculate solubility of any inorganic salt in water.

Problem 1. The K_{SP} for silver chromate is 2.6×10^{-12} at 25°C. Calculate its solubility in water. First, we write the dissociation equation:

$$Ag_2CrO_4(S) \rightleftharpoons 2Ag^+(aq) + CrO_4^{2-}(aq)$$

Setting up the reaction table with $S =$ molar solubility:

Concn (M)	$Ag_2CrO_4(s)$	$\rightarrow$	$2Ag^+(aq)$	$+$	$CrO_4^{2-}(aq)$
Initial	—		0		0
Change	—		+2S		+S
Equilibrium	—		2S		S

Substituting into K_{SP} and solving for S:

$$K_{SP} = [Ag^+]^2[CrO_4^{2-}] = (2S)^2 \times S = 4S^3$$

$$= 2.6 \times 10^{-12}$$

$$S = \sqrt[3]{\frac{2.6 \times 10^{-12}}{4}} = 8.66 \times 10^{-5}$$

Thus, the molar solubility of silver chromate in water at 25°C is 8.66×10^{-5} mol/L.

Problem 2. Express the above solubility as g/L, mg/L, % and ppm.

$$\frac{8.66 \times 10^{-5} \text{ mol Ag}_2\text{CrO}_4}{1 \text{ L}} \times \frac{331.8 \text{ g Ag}_2 \text{ CrO}_4}{1 \text{ mol Ag}_2 \text{ CrO}_4}$$

$$= 2.87 \times 10^{-2} \text{ or } 0.0287 \text{ g Ag}_2 \text{ CrO}_4/\text{L}$$

$$\equiv \frac{0.0287 \text{ g Ag}_2\text{CrO}}{\text{L}} \times \frac{1000 \text{ mg}}{1 \text{ g}}$$

$$= 28.7 \text{ mg Ag}_2\text{CrO}_4/\text{L}$$

As we know, 1 L solution is equal to 1000 g (since the density of the aqueous solution is almost 1.00 g/mL). Therefore, on a (W/W) percentage scale, we should know how many grams of the salt will dissolve in 100 g (or 100 mL) solution. Thus, 0.0287 g Ag_2CrO_4 per 100-g solution. Therefore, the solubility = 0.00287%. Since mg/L is the same as ppm, therefore, the solubility of Ag_2CrO_4 in water is 28.7 ppm.

The solubility of an organic compound in water depends on the polarity and the size (diameter) of the molecule. Lower-molecular-weight alcohols, aldehydes, ketones, carboxylic acids, aliphatic amines, amides, nitriles, and many other functional groups dissolve in water. The solubility, however, decreases with an increase in the carbon chain length. The addition of double and triple bonds to the molecule increases the water solubility. Also, ring formation for a given carbon number increases the solubility. For alkanes, alkenes, and alkynes, types of hydrocarbons (all of which have very low solubility in water) branching increases the solubility; while for aromatics and cycloalkanes, there is no effect from branching. All hydrocarbons are almost insoluble in water. These substances, however, are soluble in nonpolar solvents, such as carbon tetrachloride or hexane.

III.D OCTANOL/WATER PARTITION COEFFICIENT

Toxicity of many substances, their biological uptake and lipophilic storage in the body, their bioconcentrations (movements through the food chain resulting in higher concentrations), and so forth, as well as their fate in the environment can be predicted from their octanol/water partition coefficients, P_{oct}.

The partition or distribution coefficient of a substance is the ratio of its solubility in two immiscible solvents. It is a constant without dimensions, and depends on the temperature and pressure. Thus,

$$P_{oct} = \frac{C_{octanol}}{C_{water}}$$

where $C_{octanol}$ and C_{water} are the equilibrium concentrations of the dissolved substance in octanol and water, respectively. The logarithm of the partition coefficient to the base 10 ($\log P_{oct}$) is generally used in all calculations.

Many studies have shown that the bioconcentrations or bioaccumulations of several substances in marine species follow linear relationships with their partition coefficients. The term *bioconcentration* refers to the ratio of the concentration of the substance between the muscle of the species and the surrounding water. Equations have been proposed to predict such bioconcentrations of many toxicants in fishes and other marine species. Such equations relate to $\log P_{oct}$ and are in good agreement with experimental measurements. One such equation to predict the bioconcentration of a chemical in trout muscle is

$$\log B_f = 0.542 \log P_{oct} + 0.124$$

where B_f is the bioconcentration factor.

The partition coefficient P_2 for any substance can be either experimentally measured or theoretically calculated, using the following equation:

$$\log P_2 = a \log P_{oct} + b$$

where a and b are constants that have specific values for a specific compound. The P-values for many compounds have been experimentally determined and can be found in the literature. Further discussion on the subject is beyond the scope of this text.

III.E   ODOR

Odor is probably the most sensitive of the human perceptory organs. Many chemicals can be identified at trace levels from their odor. Odor may serve as an indicator to warn of the presence of highly odorous hazardous substances. However, not all hazardous substances have strong, distinct odors. Also, not all chemicals that have strong odors are hazardous. Nevertheless, it is an immediate firsthand indicator to identify the presence of certain hazardous substances in many dangerous situations, such as spillage, gas cylinder leak, or fire.

Odors have been quantified in several different ways. The major definitions are

(1) threshold odor concentration (TOC), (2) odor recognition threshold, (3) threshold odor number (TON) and (4) odor index (OI). These terms are briefly explained below.

The TOC may be defined as the absolute perception threshold (at which the odor is barely identifiable, but too faint); while the recognition threshold (100%) means he threshold concentration of the odorant whose odor may be defined (or accepted) by 100% odor panel as being representative odor of the substance being studied. By contrast, when 50% odor panel define the odor that is representative of the substance, then the above term becomes "50% recognition threshold".

The TON is the number of times a given volume of the gaseous sample is to be diluted in clean air to bring it to the threshold odor level determined by 50% of a panel of observers. Its intensity is expressed in odor units.

The OI, is the ratio of vapor pressure to 100% odor recognition threshold. Both the terms, that is, the vapor pressure and the odor recognition threshold, are expressed in ppm. Therefore, the OI becomes a dimensionless term.

$$OI = \frac{\text{vapor pressure (ppm)}}{100\% \text{ odor recognition threshold (ppm)}}$$

where 1 atm = 1,000,000 ppm or 1 torr (or 1 mm Hg) = 1316 ppm.

Calculation of OI is shown in the following solved example.

Example. The vapor pressure of a volatile liquid at ambient temperature is 38 torr. Its odor recognition threshold as determined by 100% panel of observers is 200 ppb. Determine the OI of this substance.

$$\text{Vapor pressure} = 38 \text{ torr} \times \frac{1 \text{ atm}}{760 \text{ torr}}$$
$$\times \frac{1,000,000 \text{ ppm}}{1 \text{ atm}} = 50,000 \text{ ppm}$$

Odor recognition threshold (100%)
$$= 200 \text{ ppb} \times \frac{1 \text{ ppm}}{1000 \text{ ppb}} = 0.2 \text{ ppm}$$

TABLE III.1 Odor Recognition Concentration and Odor Index of Selected Compounds

Compound/Class	Odor Recognition Concn (100%)[a,b]	Odor Index[c]
Amines		
Methylamine	3 ppm	940,000
Ethylamine	0.8 ppm	1,450,000
Isopropylamine	1 ppm	640,000
n-Butylamine	0.3 ppm	400,000
Dimethylamine	6 ppm	280,000
Diethylamine	0.3 ppm	880,000
Diisopropylamine	0.8 ppm	110,000
Dibutylamine	0.5 ppm	5,500
Trimethylamine	4 ppm	490,000
Triethylamine	0.3 ppm	235,000
Ethanolamine	5 ppm	130
Ammonia	55 ppm	167,000
Acids, Carboxylic		
Formic acid	20 ppm	2,200
Acetic acid	2 ppm	15,000
Propionic acid	40 ppb	110,000
Butyric acid	20 ppb	50,000
Valeric acid	0.8 ppb	260,000
Alcohols		
Methanol	6000 ppm	22
Ethanol	6000 ppm	11
1-Propanol	45 ppm	480
2-Propanol	45 ppm	480
1-Butanol	5000 ppm	120
Pentanol	10 ppm	370
Octanol	2 ppb	33,000
Aldehydes		
Formaldehyde	1 ppm	5,000,000
Acetaldehyde	0.3 ppm	4,300,000
Propionaldehyde	0.08 ppm	3,900,000
Acrolein	20 ppm	19,000
Butyraldehyde	40 ppb	2,400,000
Isobutyraldehyde	300 ppb	950,000
Crotonaldehyde	200 ppb	125,000
Furfuraldehyde	200 ppb	5,300
Benzaldehyde	5 ppb	22,000
Esters		
Methylformate	2000 ppm	300
Methylacetate	200 ppm	1,100
Ethylacetate	50 ppm	1,900
n-Propylacetate	20 ppm	1,600
Isopropylacetate	30 ppm	2,100
n-Butylacetate	15 ppm	1,200
Ethylacrylate	1 ppb	38,000,000
Methylbutyrate	3 ppb	11,000,000

TABLE III.1 (Continued)

Compound/Class	Odor Recognition Concn (100%)[a,b]	Odor Index[c]
Ethers		
Diethylether	300 ppb	1,900,000
Diisopropylether	60 ppb	3,200,000
Di-*n*-butylether	500 ppb	13,500
Diphenylether	100 ppb	130
Hydrocarbons, Alkanes		
Ethane	1500 ppm	25,000
Propane	11,000 ppm	425
Butane	5,000 ppm	480
Pentane	900 ppm	570
Octane	200 ppm	100
Hydrocarbons, Alkenes		
Ethene	800 ppm	57,000
Propene	80 ppm	14,700
1-Butene	70 ppb	43,500,000
2-Butene	600 ppb	3,300,000
Isobutene	600 ppb	4,600,000
1-Pentene	2 ppb	376,000,000
Hydrocarbons, Aromatic		
Benzene	300 ppm	300
Toluene	40 ppm	720
Xylenes	0.4–20 ppm	360–18,000
Cumene	40 ppb	90,000
Ketones		
Acetone	300 ppm	720
Methylethylketone	30 ppm	3,800
Diethylketone	9 ppm	1,900
Methylisobutylketone	8 ppm	1,000
2-Pentanone	8 ppm	2,000
2-Heptanone	20 ppb	171,000
Mercaptans		
Methylmercaptan	35 ppb	53,000,000
Ethylmercaptan	2 ppb	290,000,000
Propylmercaptan	0.7 ppb	263,000,000
Butylmercaptan	0.8 ppb	49,000,000
Phenylmercaptan	0.2 ppb	940,000
Sulfides		
Hydrogen sulfide	1 ppm	17,000,000
Methylsulfide	100 ppb	2,760,000
Ethylsulfide	4 ppb	14,400,000
Butylsulfide	2 ppb	660,000
Phenylsulfide	4 ppb	14,000

Source: Verschueren, K., 1983. *Handbook of Environmental Data on Organic Chemicals*, 2nd ed. New York: Van Nostrand Reinhold.

[a]Defined by 100% odor panel.

[b]Numbers are rounded up.

[c]Measure of the detectibility of an odorant in the air under evaporative conditions, such as spills, leaks, or solvent evaporation. The numbers above are calculated and rounded up.

$$OI = \frac{50,000 \text{ ppm}}{0.2 \text{ ppm}} = 250,000$$

Table III.1 presents the 100% *odor recognition concentrations* and the *odor index* for some selected classes of compounds that are hazardous (toxic and/or flammable). It may be noted that the OI does not distinguish between a pleasant or disagreeable odor. Also, while the OI takes into account the vapor pressure of the substance and, therefore, the degree of air pollution and exposure, the recognition threshold measures the detectability of the odorant in the air.

IV

TOXIC PROPERTIES OF CHEMICAL SUBSTANCES

The toxicity of a substance and its intensity can differ widely with the nature of the compound, primarily, the nature of the functional group (also the cation or anion). In addition, it would depend on the dosage amount and the time of exposure. There are also other factors, namely, the host factors, which include the species and the strain of the subject (animal), as well as sex and age.

IV. A PATHWAY OF ENTRY: ABSORPTION AND EXCRETION

A toxicant can cause injury at the site of contact (local effects) such as skin, by damaging or destroying the tissues, or manifests its effect after it is absorbed by the specific organism of the body. The absorption of a chemical can occur through the lungs, skin, gastrointestinal tract, and several minor routes. The nature and the intensity of the toxicity of a substance would depend on its concentration in the target organs. Most toxicants are absorbed in the gastrointestinal tract upon ingestion. For example, the weak acids that are soluble in lipids and that are nonionized are susceptible to be absorbed in the stomach. On the other hand, these substances tend to exist in ionized form in

the plasma and become mobile. Similarly, the weak bases will be highly ionized in the acidic gastric juice in the stomach and therefore would not be absorbed readily in the stomach.

The route of entry of many toxicants is inhalation. Such substances include gases, vapors of volatile liquids and particulate matter. The main site of absorption is the alveoli in the lungs. This site has a large alveolar area and high blood flow, favorring absorption. The rate of absorption of such gaseous substances, however, depends on their solubility in the blood. The more soluble the substance, the greater its absorption. The particulate matter, especially the smaller particles within the range $0.01-10$ μm and the liquid aerosols, are also susceptible to absorption in the lung. While the smaller particles are likely to be deposited in the trachea, bronchi, and bronchioli, the very small particles that have a diameter of <0.01 μm can be exhaled out. By contrast, particles >10 μm are deposited in the nose and do not enter into the respiratory tract. They may enter the gastrointestinal tract with mucus and by other ways. It has been estimated that about 50% of inhaled particles are deposited in the upper respiratory tract, while about 25% are deposited in the lower respiratory tract, and 25% are exhaled.

Many chemicals can be absorbed through the skin to produce a systemic effect. Although skin is relatively impermeable and acts as a barrier, certain chemicals can diffuse through the epidermis. Such substances include both polar and nonpolar compounds. While the former may diffuse through the outer surface of the protein filaments of the hydrated stratum corneum, nonpolar solvents may find passage by dissolving and diffusing through the lipid between the protein filaments. Certain corrosive substances, such as acids and alkalis, may injure the dermis barrier and increase dermal permeability.

A chemical is distributed rapidly throughout the body, after it enters the bloodstream. The extent to which it enters into an organ depends on the blood flow, the ease of its diffusion through the local capillary wall and the cell membrane, and the affinity for the chemical.

A chemical can bind in a tissue or organ. Such binding can result in its accumulation or higher concentration in the tissue. If it is bound strongly, by irreversible covalent bonding, the effect can be significantly toxic. However, a number of substances, as well as a major portion of the dose, undergo weak noncovalent binding that is reversible. Even in such reversible binding, the bound chemical can dissociate gradually and retain the level of unbound chemical in the tissue. Many chemicals can bind to plasma protein. The liver, kidney, adipose tissue, and bone are generally the major sites of binding and storage of chemicals. While the liver and kidney manifest high capacity for binding, because of their metabolic and excretory functions, adipose tissue can store lipid-soluble compounds by dissolution in the neutral fats.

Many chlorinated pesticides and PCBs are soluble in lipid and remain stored in the adipose tissue. Metal ions, such as lead, cadmium, and strontium, exchange with calcium in the bone and deposit. Similarly, fluoride and other anions undergo ion exchange with hydroxide and accumulate in the bone. Such ion exchange takes place between ions of similar charges and radii.

After being absorbed and distributed in the organism, chemicals may be excreted, either as the parent compounds or as their metabolites and/or their conjugate adducts. The principal excretory organs are the kidney, liver, and lungs. There are also a number of minor routes of excretion.

IV. B DETOXICATION AND BIOACTIVATION

Many chemicals absorbed into various parts of the body may undergo metabolic transformation. This process, also known as *biotransformation*, occurs in the organs and tissues. The most important site is the liver. Other sites of such reactions are the lungs, stomach, intestine, kidneys, and skin. The detoxication process involves the conversion of the toxicants to their metabolites, which in many cases are less toxic than the parent compounds.

Detoxication or biotransformation of toxicants involves two types of reactions: (1) degradation, and (2) addition or conjugation reactions. Either or both types of reactions can take place. The degradation reactions are more common and include oxidation, reduction, and hydrolysis. Such reactions result in the breakdown of the toxicants. On the other hand, the conjugation process produces adducts of the toxicants or their metabolites with endogenous metabolites. All these reactions are heavily biocatalyzed by various enzyme systems in the body. The rate and nature of the reactions can vary from one species of animal to another. Also, the biotransformation can differ with the age and the sex of the animal.

Some biotransformation processes produce metabolites that are more toxic than the parent chemicals. Such reactions are also catalyzed by enzymes in a process as *bioactivation*. Thus, bioactivation increases the toxicity of a compound. Some examples of degradative and conjugation reactions of toxicants are presented below:

1. Oxidation:

Alkyl side chains of aromatics to alcohols; e.g., *n*-propylbenzene to 3-phenylpropan-1-ol, 3-phenylpropan-2-ol, 3-phenylpropan-3-ol and hexane; trimethylamine to trimethylamine oxide

2. Epoxidation:

e.g., Aldrin → dieldrin

Heptachlor → heptachlor
epoxide

3. Hydroxylation:

e.g., Aniline → phenylhydroxylamine

Naphthalene → 1-, and 2-
naphthol

4. Sulfoxidation:

e.g., Methiocarb → Methiocarb sulfone

5. Desulfuration:

e.g., Parathion → paraoxon

6. Dealkylation:

e.g., *p*-Nitroanisole → *p*-nitrophenol

7. Dehydrogenation:

e.g., Ethanol → acetaldehyde →
acetic acid

8. Reduction:

e.g., Azobenzene → aniline

Nitrobenzene → aniline

Prontosil → sulfanilamide

9. Hydrolysis:

e.g., Esters → acids + alcohols
Also, phosphates and amides undergo hydrolysis.

10. Glucuronic acid conjugation:

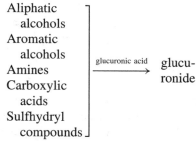

Aliphatic alcohols, Aromatic alcohols, Amines, Carboxylic acids, Sulfhydryl compounds $\xrightarrow{\text{glucuronic acid}}$ glucuronide

(The enzyme and coenzyme for the above conjugation reactions are uridine diphosphate glucuronyl transferase and uridine-5′-diphospho-α-D-glucuronic acid, respectively.)

11. Acetyl conjugation:

Primary aliphatic amine, Primary aromatic amine, Hydrazines, Sulfonamides $\xrightarrow{\text{acetylation}}$ acetyl derivatives

(The enzyme and coenzyme catalyzing above reactions are N-acetyl transferase and acetyl coenzyme A.)

12. Amino acid conjugation:

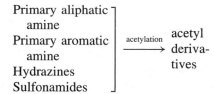

Aromatic carboxylic acids, Arylacrylic acids, Arylacetic acids $\xrightarrow[\substack{\text{glycine,} \\ \text{glutamine}}]{\substack{\alpha\text{-amino} \\ \text{acids, e.g.,}}}$ amino acid conjugates

13. Sulfate conjugation:

Aliphatic alcohols, Phenols, Aromatic amines $\xrightarrow{\text{sulfate}}$ sulfate conjugates

(The enzymes involved in the above conjugation reactions are sulfotransferases, found in the cytosolic fraction of kidney, liver and intestine; the coenzyme is 3′-phosphoadenosine 5′-phosphosulfate (PAPS).

IV. C TOXIC PROPERTIES OF CHEMICALS — OVERVIEW

Of more than 10 million compounds that have been isolated or synthesized, toxicity data have been documented for a little more than 100,000 substances — only a small fraction. Most of the information that is available is based on animal studies. Human toxicity data are known for a relatively small number of substances.

The toxic properties of chemicals may be assessed in a very general way from two relationships: the dose-response relationship and the structure–activity relationship (SAR). The toxicity depends greatly on the *dose* — the amount of substance per unit of body weight to which an organism is exposed. The effect on the organism is known as *response*. Thus, at a very low level, a highly toxic compound can be harmless. On the other hand, a substance with relatively very low toxicity can cause severe adverse effects when taken in large doses. The dose-response relationship is usually illustrated graphically. The cumulative percentage of death is plotted (on the Y-axis) against the log of the dose (on the X-axis). Normally, an S-shaped curve is obtained, the midpoint (inflection point) of which is a statistical estimate of LD_{50} (also expressed as 5 probit, or probability units).

IV. D COMMON TERMS

Some of the most common terms used in toxicity studies are defined below.

Acidosis Condition in which the pH of the blood is acidic, below the normal range. The toxicity of certain chemicals is caused by generation of an acidic metabolite or by retention of carbon dioxide.

Albinuria Presence of serum albumin in the urine.

Alkalosis Condition in which the pH of the blood is alkaline, >7.8 caused by a disturbance in the acid–base balance.

Anemia Condition in which the hemoglobin level in the blood is below normal.

Anesthetic Substance that causes dizziness, drowsiness, and headache leading to loss of consciousness and loss of response to pain.

Anorexia Lack or loss of appetite for food.

Anoxia Reduction of oxygen in the body tissues below physiologic levels.

Anticoagulant An agent that prevents blood clotting.

Antidote Substance that is administered to reverse the poisoning effects of a toxicant.

Aplasia Lack of development of an organ or tissue.

Arrhythmia Variation from the normal rhythm of the heartbeat.

Asphyxiant Substance that displaces oxygen or reduces the amount of oxygen in the air or limits the blood from transporting adequate oxygen to the vital organs of the body.

Axon Responsible for transmitting impulses in the form of sodium and potassium ion shifts from one neuron to another.

Bradycardia Slowness of the heart beat (slowing of pulse rate to <60).

Bronchoconstriction Narrowing of the air passages of the lungs.

Chloracne Acne-like eruption caused by exposure to chlorine compounds.

Cirrhosis Chronic liver disease caused by chronic ethanol poisoning. Other hepatoxicants may also cause cirrhosis.

Conjunctivitis Inflammation of the lining of the eyelids.

Corrosive Substance that causes local chemical burn and destruction of tissues on contact. The substance reacts chemically with the tissues at the point of contact.

Cyanosis Occurs due to reduced oxygenation of the blood with symptoms of discoloration of the skin (slightly blue or purple). Toxicological causes include carbon monoxide poisoning, pulmonary edema, and asphyxiant gases.

Dermatitis Symptoms of scaly and dry rashes on the skin with a chemical.

Diuretic A substance that increases urine production.

Diuresis Increased secretion of urine.

DNA Deoxyribonucleic acid; carries the genetic information of the cell.

Dose Quantity of a substance that is taken by, or administered to, an organism.

Dyspnea Shortness of breath, labored breathing, or distress in respiration that sometimes results from the toxic effects of certain chemicals.

Edema Excessive amount of interstitial fluid accumulated in subcutaneous or other body tissue.

Emesis Vomiting

Encephalopathy Clinical condition of disorganized brain functions; the signs may vary from drowsiness or confusion to seizures or coma.

Epithelium Cells covering the internal and external surfaces of the body.

Erythema Redness of the skin produced by congestion of the capillaries.

Fetotoxicity Harmful effects exhibited by a fetus, due to exposure to a toxic substance, that may result in death, reduced birthweight, or impairment of growth and physiological dysfunction.

Gastrointestinal Pertaining to the stomach or intestine.

Glomerular Pertaining to a cluster or tuft, as of blood vessels or nerve fibers.

Hallucinogen Psychedelic agent: a compound that produces changes in perception, thought, or mood without causing major disturbances in the nervous systems (autonomic). An example is LSD.

Hematuria Appearance of blood in the urine.

Hemoglobinuria Excretion of hemoglobin in the urine.

Hemolysis Breakdown of red blood cells or erythrocytes with the release of hemoglobin into the blood plasma. This results in hemoglobinuria.

Hepatitis Inflammation of the liver, often accompanied by jaundice and liver enlargement.

Hepatotoxicity Toxic effect on the liver. Chemicals that cause such adverse effects on the liver are known as *hepatotoxicants*.

Hypotensive Condition in which the blood pressure is lower than normal, caused by certain chemicals affecting cholinergic or adrenergic function.

Hypoxia Condition in which there is a lack of oxygen supply to the tissues.

Irritant Substance that induces local inflammation of normal tissues on immediate, prolonged, or repeated contact.

Lachrymator (or Lacrimator) A substance that increases the flow of tears.

Lacrimation Secretion and discharge of tears.

Methemoglobinemia Condition or symptoms resulting from the presence of methemoglobin (an oxidized form of containing iron in the ferric state). Methemoglobin cannot bind reversibly with oxygen, causing hypoxia.

Miosis (or Myosis) Contraction of the pupil.

Mitochondria Cell organelle involved in energy production in the cell utilizing oxygen.

Mutagen Substance capable of causing a heritable change in the genetic information stored in the DNA. Many chemicals, including imines, epoxides, acrolein, benzene, sulfur, and methylsulfonates, are mutagens. Mutagens can cause fertility disorders, prenatal death, hereditary diseases, and cancers

Mydriasis   Extreme dilation of the pupil.

Narcotic   Compound that produces stupor. Many opium derivatives are examples of strong narcotics. Narcotics can affect the central nervous system (CNS) and the gastrointestinal tract. The CNS effects include analgesia, euphoria, sedation, respiratory depression, and antitussive action. Chronic use of narcotics develops tolerance to the compounds and physical dependence.

Necrosis   Death of tissue, as individual or group of cells or in localized areas.

Nephritis   Inflammation of the kidney.

Neuron   Cell in the nervous system for receiving, storing, and transmitting information.

Neuropathy   Funcitnal disturbances and/or pathological changes in the peripheral nervous system.

Nephrotoxicity   Toxic effects induced by a variety of chemicals, causing disruption of the functioning of the kidney, necrosis of the proximal tubule, and renal failure.

Neurotoxicity   Toxic effects on the central or peripheral nervous system causing behavioral or neurological abnormalities.

Oliguria   Secretion of a diminished amount of urine in relation to fluid intake.

Ophthalmic   Pertaining to the eye.

Parenteral   By injection through routes other than alimentary canal, such as, subcutaneous, intramuscular, or intravenous.

Paresthesia   An abnormal sensation, as burning or prickling

Photophobia   Abnormal visual intolerance to light

Phytotoxic   Poisonous to plants, inhibiting plant growth.

Pulmonary   Pertaining to the lungs.

Pulmonary edema   Excess fluid in the lungs, caused by irritants that injure the pulmonary epithelium. Such conditions can affect oxygen uptake, causing death.

Receptor   Binding site that has a high affinity for a particular ligand. Receptors interact with biologically endogenous ligands, facilitating intracellular communication. Many chemicals interact with receptors, producing a variety of toxic effects.

Renal   Pertaining to the kidney.

Routes of entry   There are several routes by which a chemical can enter the body. The three primary routes are inhalation or breathing; ingestion through eating, drinking, or smoking; and absorption through the skin. In addition, a chemical may be administered into the body via other routes — intraperitoneal, intravenous, intramuscular, gastric lavage, renal, and ocular, as is done in animal studies.

Tachycardia   Excessively rapid heart beat.

Target organ   Part of the body: a specific organ, such as the eyes, lung, central nervous system, liver, or kidney, entered by a chemical, affecting the organ or causing injury. For example, ethanol, when ingested, affects the brain and liver.

Teratogens   Compounds that cause embryonic mortality or birth of offspring with physical, mental, or developmental defects.

Tinnitus   Ringing, buzzing, clicking, or roaring noise in the ears.

Toxic effects   The effects of toxic substances may be acute, chronic, systemic, or local. *Acute toxicity* is manifested from a single dose or one-time exposure within a short period of time, usually from a few minutes to several days. *Chronic toxicity* results from several exposures of small concentrations for a long period of time, usually over an 8-hour work shift. Certain substances cause illness after several years. *Systemic effect* is the toxic effect of a chemical at one area in the body, the chemical having entered the body at another point. When a substance affects the tissues at the point of contact or where it enters, it is terms a *local effect*.

Urticaria A vascular skin reaction causing elevated patches (wheals) and itching.

Vertigo Dizziness; an illusion of revolving around.

Vesiccant Chemical that produces blisters on contact with the skin.

The toxicity of a substance is related to its chemical properties. Thus, the biological activity of a compound may be assessed or predicted from its molecular structure. This is known as the structure–activity relationship (SAR). Their are many in-depth studies on this subject in the technical literature. Computers are used to perform complex SAR calculations.

The toxicity of many common substances or classes of compounds is as follows. Common gases such as hydrogen, nitrogen, helium, neon, argon nitrous oxide, methane, ethane, propane, butane, ethylene, and acetylene are nontoxic but simple asphyxiants. These gases can dilute the oxygen in the oxygen in the inhaled air and reduce its partial pressure in the alveoli. The results in reduction of the transfer of oxygen into venous blood. Lack of oxygen can cause death. On the other hand, carbon monoxide, hydrogen sulfide, sulfur dioxide, sulfur trioxide, nitric oxide, nitrogen dioxide, ozone, arsine, stibine, phosphine, cyanogen, ammonia, fluorine, chlorine, bromine vapor, hydrogen fluoride, hydrogen chloride, and hydrogen bromide are examples of highly toxic inorganic gaseous substances. The symptoms, severity, and toxic actions of these substances, however, are quite different. Carbon monoxide, for example, is an odorless gas that combines with hemoglobin to form carboxyhemoglobin. It is an example of a chemical asphyxiant that blocks oxygen transport and causes death even at a very low concentration. Chlorine and hydrogen halides are pungent suffocating gases, highly irritating to mucous membranes, that can cause pulmonary edema, lung injury, and respiratory arrest.

Hydrogen cyanide and metal cyanides, especially those of alkali metals, are extremely acute poisons. The cyanide ion is an inhibitor of cytochrome oxidase, blocking cellular respiration. Many metal ions, including arsenic, cadmium, lead, mercury, manganese, barium, and selenium, are highly toxic to humans. Metals such as antimony, cobalt, nickel, zinc, molybdenum, silver, copper (dust), and aluminum exhibit either low acute poisoning or chronic toxicity. The toxicity of metals also varies widely. Within the same group in the Periodic Table, for compounds with similar electronic configuration and chemical properties, the toxicity may differ significantly. For example, mercury and lead are much more severe toxicants than are zinc or tin. Although heavy metals, in general, are more toxic than the lighter ones in the same group, a reverse pattern is observed among group VA metals. Arsenic is a severe poison; antimony and bismuth exhibit much less toxicity.

The toxicity of many organics may be correlated with the functional groups on the molecules. Organophosphorus compounds, which include several insecticides as well as fatal nerve agents, are inhibitors of acetylcholinesterase, an enzyme found extensively throughout the nervous system and in many non-nervous tissues as well. The enzyme removes the neurotransmitter acetylcholine by hydrolyzing the latter to chlorine and acetate. However, its action is inhibited by phosphate esters that bind to its esteratic site. Carbamates act similarly; unlike phosphorylation of the enzyme, however, carbamylation is reversible, leading to rapid recovery of enzymatic activity. The toxic effects from carbamate poisoning thus persist for a few hours. The phosphorylated enzyme is hydrolyzed much more slowly, with inhibition lasting several days.

Most nitriles are moderate to highly toxic substances. The toxic action is attributed to the presence of the cyanide ($-CN$) group. Acrylonitrile and acetone cyanohydrin are extremely toxic, as they decompose to hydrogen cyanide in the body. The cyanide released inhibits enzymes that cause respiration in tissue. This prevents tissue cells from using oxygen.

Lower aliphatic amines are severe irritants to the skin, eyes, and mucous membranes. As basic compounds, they hydrolyze with water in body tissue and raise the pH value of exposed tissue. Contact with the eyes can result in corneal injury and loss of sight. Systemic effects on many organs in the body are manifested. This includes necrosis of the liver and kidney and hemorrhage and edema of lungs. Aromatic amines can cause anemia, cyanosis, and reticulocytosis and can be carcinogenic. Aniline is a highly toxic substance, causing methemoglobinemia in humans. Its metabolites, *p*-aminophenol and phenyl *N*-hydroxylamine, oxidize iron(II) in hemoglobin to iron(III). As a result, hemoglobin loses its ability to transport oxygen in the body. This results in cyanosis. The systemic effects from benzidine include blood hemolysis, bone marrow depression, and liver damage.

Organic compounds containing a carbonyl functional group, $>C=O$, such as aldehydes, ketones, esters, and carboxylic acids, exhibit a low order of toxicity. The lower aldehydes, such as formaldehyde, acetaldehyde, and acrolein, are severe irritants to the eyes and respiratory tract. Acrolein is a severe lachrymator. Lower carboxylic acids are corrosive substances. Formic acid is a poison, but the toxic effects arise as a result of lowering of the blood pH below normal levels and have little to do with the carbonyl group in the molecule.

Among the other oxygen-containing organic compounds, epoxides are moderately toxic; however, ethylene oxide is a severe health hazard. Lower alcohols vary in their toxicity. Ethanol is both a depressant to the central nervous system as well as an anesthetic. Methanol is highly poisonous, causing acidosis and blindness. Lower aliphatic ethers are anesthetics.

Aliphatic hydrocarbons exhibit mild toxicity but primarily an anesthetic effect. Higher fatty alcohols are nontoxic. Aromatic hydrocarbons are narcotics. Benzene exhibits both acute and chronic toxicity. The chronic effects — anemia, abnormal increase in blood lymphocytes, and decrease in the number of blood platelets — are more serious. Some of the polynuclear aromatics are carcinogens.

Substitution of hydrogen atoms in the organic structures with halogen atoms — chlorine or bromine — imparts very different toxic properties to the molecules. Organohalide compounds show toxicity that varies in symptoms and severity form one class of compounds to others. The degree of toxicity of alkyl halides is very low; rather, they exhibit anesthetic properties. Some of these compounds are human carcinogens. Chlorohydrins (which are chloroalcohols), chlorophenols, or chlorobiphenyls, are much greater health hazards than is the parent alcohol, phenol, or biphenyl. Another class of chlorinated organics, which are environmental pollutants, are the organochlorine pesticides: DDT, heptachlor, dieldrin, chlordane, lindane. The structure of these compounds differ from each other, as do their toxic properties. Some chlorinated organic compounds, including phosgene (carbonyl chloride), mustard gas [bis(2-chloroethyl)sulfide], dioxins (2,3,7,8-tetrachloro isomer), and chlorinated dibenzofurans, are known to be dangerously toxic. The former three have been used as military poisons.

Organosulfur compounds are of many types. They include compounds that contain only carbon, hydrogen, and sulfur atoms, such as mercaptans, sulfides, and disulfides; compounds containing nitrogen, such as thiourea, thiocyanates, thiazole derivatives, and dithiocarbamates; compounds containing oxygen, such as sulfoxides, sulfones, sulfonic acids and their esters, and alkyl sulfates; and chloro compounds of sulfur, the sulfur mustards. Sulfur mustards are notoriously dangerous compounds, often used as military poisons. They are blistering and severely penetrating, destroying tissues at the point of contact. Thiols or the mercaptans are nauseating, causing cyanosis at high doses. Sulfides and disulfides are allergens that also cause anemia. Thiocyanates are highly toxic, liberating hydrogen cyanide during metabolic processes. Many alkyl sulfates are also highly toxic, causing damage to the lungs, liver, and kidney. Sulfones and sulfoxides are of very low toxicity. Sulfonic acids are strong irritants.

Many inorganic and organic chemicals are irritants to the eyes, skin, and respiratory tract. Concentrated acids and bases are highly corrosive. Highly water reactive substances such as bromine pentafluoride or other interhalogen compounds can burn the skin.

Among the organic compounds, toxicity shows a general pattern of decrease with an increase in the carbon chain length. However, such a pattern is manifested above carbon 6 or 7 (C-6 or C-7) for many classes of compounds. Below carbon 6 (C-6), certain substances, such as aliphatic esters, show enhanced toxicity with an increase in the carbon chain length. Thus *n*-amyl or *n*-butyl acetates are more toxic than ethyl or *n*-propyl acetates. The first members of certain classes of compounds show the highest toxicity. Ethylene oxide formaldehyde, or formic acid are more toxic than are other members of their families. In a few instances, certain toxic properties exhibited by the first member of a group are not displayed by other members in the same group. For example, certain toxic effects shown by methanol or benzene are not manifested by other lower aliphatic alcohols or alkyl benzenes. The presence of unsaturation in the molecule enhances the toxicity. Acrolein, acrylonitrile, and vinyl chloride, for example, are more toxic than are their saturated counterparts, which contain the same number of carbon atoms. Substitution in the molecule can alter the toxic properties significantly.

Chemicals classified according to their structure and functional group, general toxicity, and individual toxic properties are discussed at length in Part B of this book. The target organs, the nature of the toxic effects, and some specific toxicants are briefly described in Section V.

V

TARGET ORGANS AND TOXICOLOGY

Toxicants can injure one or more specific organs in the body, which include the liver, kidney, respiratory tract, immune system, skin, nervous system, eye, reproductive, and cardiovascular systems. The nature of such toxicants and their effects are briefly discussed in the following sections.

V.A LIVER INJURY AND HEPATOTOXICANTS

The liver is the largest organ in the body in which the metabolism of foods, nutrients, and drugs occurs. Poisoning can commonly occur through ingestion. Most toxic substances can readily enter the body through the gastrointestinal tract. Although many toxicants undergo biotransformation or detoxification in the liver, some may be bioactivated to more toxic metabolites. Hepatocytes cells play a major role in such metabolic processes. The detoxification of toxicants is attributed to the cytochrome P-450 and other xenobiotic-metabolizing enzymes, present in the liver at high concentrations. Such enzymes can convert the toxicants into more water-soluble and less toxic metabolites, which can be readily excreted. However, such detoxification does

not always occur. The liver has many binding sites where toxicants can be activated to induce lesions and a variety of toxic effects on different organelles in the cells.

Many types of liver injury are caused by a number of biochemical reactions of toxicants or their active metabolites. Such reactions include covalent binding, lipid peroxidation, inhibition of protein synthesis, perturbation of calcium homeostasis, disturbance of biliary production, and a variety of immunologic reactions. The types of liver injury from such biochemical reactions include steatosis (fatty liver), liver necrosis, cirrhosis, cholestasis, hepatitis, and carcinogenesis. The toxicants that cause these injuries are discussed in brief.

Steatosis, also known as fatty liver, results from the accumulation of excess lipid (more than 5% lipid by weight) in the liver. Many chemicals can cause this, and the lesions can be either acute or chronic. Ethanol is a classic example, causing both acute and chronic lesions. It induces large droplets of fat in the cell. On the other hand, substances such as tetracycline and phosphorus produce many small fat droplets in the cells, causing acute lesions. Among the two common mechanisms of steatosis, one involves an increased formation of triglyceride and other lipid moieties, while the other mechanism manifests an impairment in the release of hepatic triglyceride to the plasma. The latter can occur from the inhibition of protein synthesis (protein moiety of lipoprotein) or suppression of conjugation of triglyceride with lipoproteins.

Many chemicals are known to cause *liver necrosis*, an acute injury that involves the rupture of the plasma membrane and cell death (death of hepatocytes). Among the common toxicants that manifest such effects are the halogenated hydrocarbons, such as carbon tetrachloride, chloroform, trichloroethylene, carbon tetrabromide and bromobenzene. Other hepatotoxicants include acetaminophen, phosphorus, beryllium, tannic acid, and allyl alcohol. Most of these substances produce reactive metabolites that covalently bind with unsaturated lipids and proteins present at high concentrations in subcellular membranes, inducing lipid peroxidation. Other mechanisms and biochemical changes that may be manifested in liver necrosis include disturbance of Ca^{2+} homeostatis, depletion of glutathion, mitochondrial damage, inhibition of protein synthesis, and binding to macromolecules.

Cirrhosis is a chronic liver disease, characterized by the presence of septae of collagen distributed throughout the liver, degenerative changes in parenchyma cells and single-cell necrosis. In humans, chronic ethanol ingestion can cause cirrhosis. Dietary deficiency of protein, methionine or choline can aggravate the conditions. The mechanism of cirrhosis is not well understood. Substances such as ethanol can impair mitochondria and increase the production of reactive oxygen species. Among other toxicants, carbon tetrachloride and aflatoxins may induce cirrhosis. Many chemical carcinogens, such as dioxins, can induce cirrhosis. This chronic liver disease is often induced by the chemicals that also cause necrosis and steatosis.

Hepatitis is a disease caused by viruses and chemicals producing injury to the liver cells. Such cellular damage can be initiated by a number of mechanisms, such as inhibition of enzymes by reactive metabolites, depletion of cofactors, interaction with receptors, and alternation of cell membranes. The symptoms of the disease are inflammation of the liver, jaundice, and in some cases, enlargement of the liver.

Cholestasis is another type of liver damage. It is less common than necrosis and steatosis. Certain drugs and anabolic and contraceptive steroids can induce cholestatis. These include promazine, sulfanilamide, mestranol, and estradiol.

A large number of chemicals are known to induce liver cancer in animals. These carcinogens include acetylaminofluorene, nitrosamines, PCBs, vinyl chloride, and many alkaloids.

Some common hepatotoxicants and the associated liver injury are presented in Table V.1.

V.B KIDNEY INJURY: NEPHROTOXICANTS

Urine is the primary route of excretion of most toxicants. In humans, about 150–200 L filtrate is formed per day. Only a small amount of this filtrate is excreted as urine, while about 99% of the filtered water is reabsorbed. Also, the kidney has a high volume of blood flow. As a result, many chemicals that enter the body pass into the filtrate; thus, their concentration in the kidney becomes higher than in other organs in the body. Such substances may transport across the tubular cells and are susceptible to bioactivation, forming toxic metabolites. Because of this, the kidney is another major target organ.

Many substances are known to exhibit nephrotoxicity affecting specific parts of the kidney. The mode of action of nephrotoxicants includes inhibition of oxidative phosphorylation, interaction with receptors, and disturbance of Ca^{2+} homeostasis. There are also other mechanisms. The sites of action

TABLE V.1   Common Hepatotoxicants and Liver Diseases[a]

Compounds/Compound Type	Resulting Liver Disease[b]
Acetaminophen	Necrosis
Acetylaminofluorene	Carcinogenesis
Aflatoxin	Steatosis, necrosis
Aflatoxin B_1	Carcinogenesis
4-Aminosalicylic acid	Cholestasis, viral-like hepatitis
Allyl alcohol	Necrosis
Beryllium	Necrosis
Bromobenzene	Necrosis
Carbamazepine (drug)	Cholestasis, viral-like hepatitis
Carbon tetrabromide	Necrosis
Carbon tetrachloride	Steatosis, necrosis, cirrhosis
Chloroform	Steatosis, necrosis
Chlorpromazine (drug)	Cholestatis, viral-like hepatitis
Colchicine (alkaloid)	Viral-like hepatitis
Cycasin	Carcinogenesis
Cycloheximide	Steatosis
Diazepam (drug)	Cholestasis
Dimethylbenzanthracene	Carcinogenesis
Estradiol	Cholestasis
Ethanol	Cirrhosis, steatosis
Ethionine	Steatosis
Halothane (anesthesia)	Viral-like hepatitis
Imipramine (drug)	Viral-like hepatitis, cholestasis
Isoniazid (drug)	Viral-like hepatitis
Mepazine	Cholestasis
Mestranol	Cholestasis
Nitrosamines, alkyl derivatives	Steatosis, necrosis, carcinogenesis
Papaverine (alkaloid)	Viral-like hepatitis
Phosphorus	Steatosis, necrosis
Polychlorinated biphenyls (PCBs)	Carcinogenesis
Promazine (drug)	Cholestasis
Pyrrolizidine alkaloids	Necrosis, steatosis, carcinogenesis
Puromycin	Steatosis
Safrole	Carcinogenesis
Sulfanilamide	Cholestasis
Tannic acid	Necrosis
Tetrachloroethylene	Steatosis, necrosis
Tetracycline	Steatosis
Thioacetamide	Necrosis
Trichloroethylene	Steatosis, necrosis
Urethane	Carcinogenesis
Valproic acid	Steatosis
Vinyl chloride	Carcinogenesis

[a]This is a partial listing of some common hepatotoxicants.
[b]Carcinogenesis was observed in experimental animals.

of nephrotoxicants include proximal tubules, glomeruli, distal tubules, blood vessels, interstitial tissues, and various parts of nephrons. The major nephrotoxicants are heavy metals, such as lead, mercury, cadmium, chromium and gold; certain halogenated hydrocarbons, such as chloroform, carbon tetrachloride, bromobenzene, and hexachlorobutadiene; anesthetics, such as acetaminophen and methoxyflurane; antibiotics, such as tetracycline and puromycin, and many organic compounds. The toxic actions of these and others are discussed in Part 2 of this text.

V.C IMMUNE SYSTEM: IMMUNOTOXICANTS

The immune system protects the host against viruses, bacteria, fungus, other foreign organisms and cells, as well as neoplasm. Many toxicants are known to suppress the function of the immune system. Such substances can lower the body's resistance to bacterial and viral infection. The mode of actions of immunotoxicants are wide, showing a variety of effects. These include immunosuppression, immunodysfunction, and autoimmunity.

Some immunotoxicants suppress all the specific immune functions by binding to the Ah receptors on lymphoid cells. Highly toxic dioxin (2,3,7,8-TCDD) is an example of such toxicants. However, most immunotoxicants exhibit more restricted activities. For example, some heavy metals, such as lead and mercury, suppress humoral and cell-mediated host resistance. Similarly, some carbamate pesticides depress antibody response and phagocytosis, while cyclosporine primarily impairs the B cells.

Toxicants affect the immune system in different ways. The effects are complex. Some substances suppress humoral immunity, others cell-mediated immunity, and certain toxicants stimulate specific immune functions. These effects may be classified under three major categories: immunosuppression, immunodysfunction, and autoimmunity. Immunosuppression involves all

or some specific immune functions by toxicants, as discussed earlier.

Immunodysfunction manifests allergies or hypersensitivity reactions. There are four types of allergic reactions, designated as type I to type IV. The type I hypersensitivity reactions are immediate, occurring within a few minutes of second or subsequent exposure. In such type allergy, the first exposure to an antigen produces IgE antibodies, while the second or subsequent exposure to the same antigen triggers the release of certain substances, such as histamine, heparin, and serotonin. Newly formed substances, such as prostaglandins, are released as well. These substances, released during the second or subsequent exposure to the antigen, may induce a variety of allergic responses, such as asthma, rhinitis, and uriticaria. Such type I (immediate hypersensitivity) reactions are caused by certain metals, food additives, pesticides, and therapeutic agents. Some of the common toxicants are listed in Table V.2. The type II and III allergic reactions are less common. The former may cause hemolytic anemia, leukopenia, or thrombocytopenia. The toxicants inducing such hypersensitivity reactions include many gold salts, isocyanates, sulfonamides, and phenytoin. Type III hypersensitivity reactions or Arthus reactions are also less common. These reactions involve the formation of antigen–antibody complexes. These complexes deposit in the vascular endothelium, causing damage to the blood vessels, inducing lupus erythematosus and glomerular nephritis. Certai therapeutic agents may also induce such toxic actions. The type IV reactions manifest delayed hypersensitivity, the latent period varying from 12 to 48 hours. Many toxicants, such as nickel and beryllium, causing this type of allergy to induce immediate hypersensitivity reactions. Others include formaldehyde, chromium, and certain antimicrobial agents that contain mercury.

Many autoimmune diseases, such as hemolytic anemia, glomerular nephritis, and neutropenia, are induced by certain metals,

TABLE V.2   Common Immunotoxicants and Their Toxic Effects

Toxicants	Toxic Effects
Metals	
Beryllium	Allergy (type I and IV)
Chromium	Allergy (type IV)
Gold (and salts)	Glomerular nephritis
Nickel	Allergy (type I and IV)
Mercury	Autoimmune disease (a type of glomerular nephritis), impairs humoral and cell-mediated host resistance
Lead	Impairs humoral and cell-mediated host resistance
Arsenic salts (As_2O_3, Na_3AsO_3, Na_3AsO_4)	Variety of immunodysfunction
Platinum Compounds	Allergy (type I)
Pesticides	
Dieldrin	Hemolytic anemia
Chlordane, DDT, hexachlorobenzene, and other organochlorine pesticides	Immunosuppression (impair immune functions)
Pyrethrum	Allergy (type I)
Carbaryl	Depresses antibody response and phagocytosis
Isocyanates	
Toluenediisocyanate	Allergy (type I), hypersensitivity reactions (type II)
Halogenated hydrocarbons	
(Dioxins, PCBs, Pentachlorophenol, chloroform, etc.)	Dioxin (2,3,7,8-TCDD) suppresses all specific immune functions; others have more restricted activities
Therapeutic agents	
Penicillin	Allergy (type I)
Miscellaneous drugs	
Procainamide	Lupus erythematosus
Sulfonamides, phenytoin, chloropromaxine, and others	Hypersensitivity reactions (type II)

pesticides, and a number of drugs. In such autoimmmune disorders, the immune system produces autoantibodies to endogenous antigens, causing damage to normal tissues.

V.D   RESPIRATORY TRACT

Many substances in the form of gases, vapors, solid particulate matter, or liquid droplets can adversely affect the respiratory tract in humans and animals. Such substances may either exert systemic effects or induce local effects on the respiratory tract, both. Many toxic gases are severe respiratory tract irritants and have strong pungent odors. Exposure can cause bronchial constriction and edema and may result in dyspnea and death. Such toxicants as ammonia and chlorine primarily produce local effects. By contrast, other toxic gases, such as oxides of nitrogen and ozone, can damage the cellular membranes. The toxic effects of gases, vapors, and solid particulates on the lungs and upper respiratory tract may vary widely. Such effects include local irritation, cellular damage, allergy, fibrosis, and lung cancer.

Oxidizing substances, such as ozone, can cause peroxidation of cellular membranes, leading to membrane damage and increased permeability. The edematous fluid permeates through such membranes, accumulating and blocking the airway. This results in edema. Many other toxicants may produce cellular damage and edema, however, by different mechanisms. For example, the vapors of many organic solvents distributed to various parts of the body after inhalation may recirculate back to the lung and form reactive metabolites, which may bind covalently to the macromolecules, causing cellular necrosis. This can result in edema, hemorrhage of the lungs, and death. Many toxins behave in this way.

Another type of toxic effect on the lung involves the allergic or hypersensitivity reactions. Inhalation of particulates, such as cotton dust, pollens, molds, and bacterial contaminants, could induce such allergy. The clinical symptoms are asthma, bronchoconstriction, and chronic bronchitis.

Perhaps the most serious type of lung disease is pulmonary fibrosis or pneumoconiosis. Different types of fibrosis, such as silicosis, asbestosis, and emphysema, are known. For example, silicosis, produces fibrosis and nodules in the lung, lowering lung capacity. This makes the victim more susceptible to pneumonia and other pulmonary diseases.

Chronic exposure to various types of particulates can induce these diseases. The long list of offending pollutants includes crystalline silica, asbestos, mineral fibers, dusts of aluminum, beryllium, titanium and other metals, coal dusts, kaolin, and oxides of metals. Certain gases, such as nitrogen dioxide and ozone, can also induce fibrosis (emphysema). Among the mechanisms proposed for silicosis, once the fibrogenic silica dusts are inhaled, they rupture the lysosomal membrane in the macrophages. These ruptures release the lysosomal enzymes that digest the macrophage. Silica is released back from the lysed macrophage. This process is repeated. The damaged macrophage releases factors that stimulate fibroblast and collagen formation. This process leads to the deterioration of pulmonary function. In emphysema, the elastic fibers supporting the alveoli and bronchi are damaged by the enzyme, elastase. This enzyme is released by the polymorphonuclear granulocytes during the interaction with the toxicants.

Lung cancer is probably the most lethal manifestation of toxic effect from chronic exposure to certain chemicals. Asbestos fibers, coke oven emissions, arsenic, chromates, nickel and the radioactive gas, radon, are some of the substances known to cause lung cancer. For a detailed discussion of carcinogenesis, see Part A, Section VI, while specific compounds are discussed in Part B of this text. Acute and chronic effects of some common pulmonary toxicants are presented in Table V.3.

V.E   SKIN INJURY: DERMATOTOXICANTS

Since the body is almost entirely covered by skin, the latter is a major route of entry of toxicants. Skin consists of two layers — epidermis and dermis — separated by a basement membrane. The epidermis is a relatively thin layer with a thickness of 0.1–0.2 mm. Below this lies the dermis, a thicker layer, about 2 mm thick. The dermis rests over the subcutaneous tissues.

The epidermis consists of a basal cell layer, which provides new cells to other layers. It also contains melanocytes, Langerhans cells, and lymphocytes. While the former produces pigments, the latter two types of cell are involved in immune responses. Thus, the epidermis is a protective cover of the body. The dermis is primarily composed of collagen and elastin. It has several types of cells, including fibroblasts, which synthesize fibrous proteins. In addition to epidermis, dermis, and the subcutaneous tissues, the skin has a number of other structures. These include hair follicles, sweat glands, sebaceous glands, and small blood vessels.

Dermal exposure to toxicants can produce a variety of effects, such as primary irritation, sensitization reactions, phototoxic skin reactions, photoallergy, urticarial reactions, hair loss, chloracne, and cutaneous cancer. These toxic effects and the nature of the reactions are discussed in this section.

Certain corrosive chemicals, such as strong acids, alkalies, and some organic solvents, can

TABLE V.3 Common Pulmonary Toxicants and Their Toxic Effects[a]

Toxic Substances	Acute Effects	Chronic Effects
Aluminum dust	Cough, shortness of breath	Aluminosis (interstitial fibrosis)
Ammonia	Severe respiratory tract irritation, edema, bronchitis	Chronic bronchitis
Arsenic	Bronchitis	Lung cancer, bronchitis asbestosis (pulmonary fibrosis mesothelioma), lung cancer
Asbestos	—	
Beryllium	Pulmonary edema, pneumonia	Berylliosis (pulmonary fibrosis, progressive dyspnea, interstitial granulomatosis)
Bromine	Severe respiratory tract irritant, pulmonary edema	—
Chlorine	Cough, severe lung irritation, edema, dyspnea, tracheobronchitis, bronchopneumonia	—
Chromium(VI)	Nasal irritation, bronchitis	Lung cancer pneumonconiosis (pulmonary fibrosis)
Coal dust	—	
Coke oven emissions	—	Tracheobronchial cancer
Cotton dust	Tightness in chest, wheezing, dyspnea	Byssinosis (reduced pulmonary function and chronic bronchitis)
Fluorine	Dangerously attacks mucous membranes	—
Hydrogen fluoride	Severe respiratory tract irritant, hemorrhagic pulmonary edema	
Hydrogen chloride	Severe respiratory tract irritant, spasms of the larynx, pulmonary edema	—
Isocyanates	Cough, dyspnea	Asthma, reduced pulmonary function
Kaolin		Kaolinosis (pulmonary fibrosis)
Metal carbides of W, Ti, Ta	Hyperplasia and metaplasia of bronchial epithelium	Peribronchial and perivascular fibrosis
Metal oxides		
CdO	Cough, pneumonia	Emphysema
CrO_3	Nasal irritation	Bronchitis, lung tumor
ReO, Fe_2O_3	Cough	A type of pnuemoconiosis
Nickel	Pulmonary edema	Lung cancer
Nickel tetracarbonyl	Cellular damage, edema	—
Nitrogen oxides	Severe lung irritants, pulmonary Congestion, edema	Emphysema
Ozone	Severe lung irritant, pulmonary Edema	—
Phosgene	Edema	Emphysema
Phosphine	Severe lung irritant	Bronchitis
Phosphorus trifluoride	Strong mucous membrane irritant	

TABLE V.3   (*Continued*)

Toxic Substances	Acute Effects	Chronic Effects
Silica	—	Silicosis, pneumoconiosis (pulmonary fibrosis)
Sulfur dioxide	Respiratory tract irritant, cough, tightness in chest, bronchoconstriction	
Talc		Talcosis (pulmonary fibrosis)
Tetrachloroethylene	Pulmonary edema	
Vanadium	Upper respiratory tract irritant, produces mucus	

[a]All the lung irritant gases can readily cause death. Only a partial list of a few selected pulmonary toxicants is presented.

produce skin burn or irritation at the site of contact. Such irritations occur upon first contact and, therefore, differ from the sensitization reactions. The toxic symptoms can range from hyperemia, edema, and vesiculation to ulceration. In sensitization reactions, however, the first contact shows little or no effect, while subsequent exposure to the chemical produces a more severe reaction. The induction period can vary from a few days to years. The reaction can occur after a delay of 12–48 hours from the time of second exposure. Such sensitization reactions involve the immune system. The toxicant binds to the surface of certain cells upon entering the skin. It reacts with and sensitizes T lymphocytes, which release certain compounds upon reexposure to the same toxicant. The released substances produce edema, hyperemia, and other effects.

Certain chemicals can produce skin reactions, induced by light, after their topical application to the skin or systemic administration into the body. There are two distinct types of photoreaction: phototoxicity and photoallergy. Phototoxicity is the more common photoreaction, with the appearance of sunburn and hyperpigmentation. The reaction can occur from the first exposure to the toxicant and usually requires a high dose of the substance. *Photoallergy* is a light-induced immune reaction that occurs from reexposure to the toxicant following an incubation period after the first exposure. The clinical manifestations of photoallergy are delayed papules and eczema. Both the shorter and longer wavelength UV rays can induce phototoxicity and photoallergy. Certain toxicants can produce both types of reactions.

Another type of skin reaction can occur within a short period of time, usually within 1 hour after the contact with the chemical. These are different from the sensitization reactions. Skin contact can produce urticaria or eczema in a short time. Some of these toxicants are listed in Table V.4.

Certain chemicals are known to cause skin cancer. Animal studies show that topical applications of substances, such as polynuclear aromatic hydrocarbons, can induce skin cancer. Coal tar, soots, shale oils, and many arsenical compounds can induce skin tumor.

Some chemicals manifest harmful effects on hair, sweat glands, and sebaceous glands. Many oral contraceptives and anticoagulants can cause hair loss. Topical applications of phenol solution (95%) and chloroform can block the sweat ducts, causing a disorder known as milimaria. Skin contact with many chlorinated aromatic hydrocarbons can affect the sebaceous glands, forming various skin lesions, such as chloracne. Repeated contact with oil and grease and fatty matters can increase the formation of acne. Some bromides and iodides may induce acne on the skin when administered systemically.

TABLE V.4 Toxic Effects on Skin and Dermatotoxicants

Skin Reaction	Dermatotoxicants
Primary irritation[a]	Strong acids; strong alkalies; many organic solvents; most detergents; all dehydrating agents, such as P_2O_5 and $CaCl_2$; most halides of phosphorus and sulfur, such as PCl_5 and $SOCl_2$; oxidizing substances, such as hypochlorites
Sensitization reaction	Certain metals, such as nickel and beryllium; salts of nickel and chromium; organomercuric compounds; many diamines, such as ethylenediamine and p-phenylenediamine; some pesticides such as captan; local anesthetics, such as, benzocaine; many antibiotics; certain poisonous plants, such as poison ivy, poison oak, poison sumac
Phototoxicity	Coal tar derivatives, such as, anthracene, phenanthrene; pyridine; anthraquinone dyes; sulfanilamide; aminobenzoic acid derivatives; tetracyclines; chloropromazine; many nonsteroidal anti-inflammatory drugs (NSAIDs) sulfonamides
Photoallergy	Aminobenzoic acid derivatives; chloropromazine; sulfonamides; sulfanilamide; halogenated salicylanilides
Urticarial reactions	Certain metals, such as copper and platinum; many antibiotics and local anesthetics; biogenic polymers
Cutaneous cancer	Coal tars; soots; mineral oils; shale oils; cresote oils; arsenic compounds; certain polynuclear aromatics, such as benzo[a]pyrene and heterocyclic compounds like benz[c]acridine induce skin cancer on animals
Effect on hair follicles (hair loss)	Several types of medications (e.g., antimiotic agents in chemotherapy), oral contraceptives and anticoagulants
Effect on sebaceous glands (e.g., acne, chloracne)	Chlorinated pesticides, such as chlordane; PCBs; dioxins; many chlorinated aromatics; certain bromides and iodides
Effect on sweat glands	Phenol; chloroform

[a]Many corrosive acids and alkalies can cause severe skin burn and damage the tissues

The toxic effects of chemicals on the skin and examples of some common dermatotoxicants are summarized in Table V.4.

Certain vesicants, such as sulfur mustards, nitrogen mustards, and dimethyl sulfate, can produce severe inflammation and blistering after a latent period that depends on the nature of the substance and the of exposure. Healing is slow. Some of these vesicants are discussed in Part B of this text.

V.F EYE INJURY

The eye is a delicate organ. Many gases, vapors, liquids, and solid particles can cause irritation or injury to the eye upon contact. Some toxicants exhibit a systemic effect as well. All skin irritants are also eye irritants. Corrosive substances such as strong acids and alkalies can damage the cornea. In addition, many common organic solvents, such as hexane, benzene, toluene, acetone, and methyl ethyl ketone, can dissolve fat and damage the epithelial cells in the cornea. The extent of corneal injury may range from minor destruction of tissues to opacity or perforation of the cornea and depends on such factors as the nature of the chemical, its concentration, duration of contact, the level of exposure, and the pH value.

Like the cornea, other parts of the eye, such as the iris, lens, retina, ciliary muscle, and optic nerve, can be affected by chemical agents. Many chemicals alter the transparency of the lens, producing cataracts.

Systemic exposure to substances such as 2,4-dinitrophenol, thallium, and bisulfan or to topical applications of corticosteroids can produce cataracts. Certain chemicals, such as dimethyl sulfoxide (DMSO), can alter both transparency and refraction of the lens.

The iris is susceptible to irritation, resulting in leakage of serum proteins and leukocytes from the blood vessels. Chemicals that are sympathomimetic or parasympatholytic can dilate the pupil, while parasympathomimetic and sympatholytic chemicals can constrict the pupil. Certain mydriatics, such as atropine, can cause glaucoma by dilating the pupil, blocking drainage of the aqueous humor.

Certain chemicals can induce changes or injury to the retina. The effect may vary widely, ranging from adaptation to the dark and visual adjustment to the rupture of blood vessels to hemorrhage and partial detachment of the retina. Such retinopathic chemicals may act differently, and to varying degrees, and include many iodates, 4,4-methylenedianiline, thioridazine, chloroquine, and hydroxychloroquine. Toxicants can damage specific structures of the retina, such as ganglion cells and the optic nerve, affecting central vision or peripheral vision, or both. Many organic solvents can damage the optic nerve affecting the central vision. Such substances include methanol, carbon disulfide, and nitrobenzene. Chemicals that may produce constriction of the visual field and affect the peripheral vision include carbon monoxide, pentavalent arsenic (As^{5+}) and also nitrobenzene. Certain chemicals, such as hexane, acrylamide, and methyl n-butylketone, can induce peripheral neuropathy. The effects of some common chemicals

TABLE V.5   Effect of Chemicals on the Eye

Injury/Effect	Chemicals
Destruction of corneal tissues, corneal opacification, and perforation	Concentrated mineral acids, such as H_2SO_4; strong alkalies, i.e., KOH; HF: strong dehydrating substances, i.e., P_2O_5; ammonium and phosphonium ions; many interhalogen compounds; sulfur mustards, i.e., mustard gas (extremely dangerous to eyes; can cause blindness)
Damage to corneal epithelial cells	Organic solvents, i.e., acetone, hexane, or toluene; cationic and anionic detergents
Glaucoma	Anesthetics; morphine; atroprine, and other mydriatrics
Cataract	Many alkylating agents; naphthalene; pesticides; i.e., heptachlor, diquat; thioacetamide; iodoacetic acid; thallium; galactose and other sugars (produced galactosemia in rats when fed large amounts); 2,4-dinitrophenol; 2,4,6-trinitrotoluene; 2,6,-dichloro-4-nitroaniline; dimethyl sulfoxide; chlorpromazine; corticosterois; many types of oral contraceptives; anticholinesterases
Retinopathy (affecting adaptation to the dark and visual acuity)	Many iodates; 4,4-methylenedianiline; certain polycyclic compounds, i.e., chloroquine, hydroxychloroquine
Damage to optic nerve/ganglion cells	
Affecting central vision	Methanol; carbon disulfide; thallium; disulfuram; nitrobenzene
Affecting peripheral vision	Nitrobenzene; carbon monoxide; compounds of As^{5+}; quinine
Peripheral neuropathy	Certain organic solvents, i.e., hexane, acrylamide; methyl-n-butyl ketone
Primary irritation	A large number of chemicals are eye irritants, even at trace quantities (discussed under the hazardous properties of title compounds in Part B of this text)

on the eye and severe injurious substances are briefly summarized in Table V.5.

V.G NERVOUS SYSTEM: NEUROTOXICANTS

The nervous system is a vital part of the body. Certain protective mechanisms, primarily the blood–brain barrier (BBB) and, to a lesser extent, the blood–nerve barrier (BNB) shield the nervous system from the toxic chemicals in the blood. For example, the BBB is effective against many toxins, such as those of tetanus and diphtheria. Despite such protective barriers, certain chemicals may penetrate the brain and exhibit adverse effects to different degrees. This may be partly due to the absence of the BBB in certain sites, where the cells produce hormones or act as hormonal receptors.

Substances that are highly lipophilic may readily cross the BBB and affect the brain. Such substances include many organics and organometallics that are lipid soluble. On the other hand, many ionizable inorganic salts that are hydrophilic (with no affinity toward lipids) cannot penetrate the BBB and show no effect on the central nervous system (CNS). The permeability of a compound through the BBB also depends on the molecular diameter. Larger molecules having a diameter approximately over 5 nm are impermeable into the endothelium in the brain. The neurotoxic effect of chemicals may be attributed to their damaging effects on the neurons, axons, glial cells, and vascular system. A toxicant may have a specific effect on a specific site or may affect more than one site. A few select neurotoxicants and their effects are highlighted below in Table V.6

TABLE V.6   Toxicity of Some Selected Neurotoxicants

Neurotoxicants	Effects
Aluminum	Induces encephalopathy with neurofibrillar degeneration (a probable cause of Alzheimer's disease)
Azide	Inhibits cytochrome oxidase, producing cytotoxic anoxia
Barbiturates	Induces anoxia in the brain
Botulinum toxin	Impairs the release of acetylcholine from motor nerve endings, can cause paralysis of muscles and death
Carbon monoxide	Leukoencephalopathy (sclerosis of the white matter)
Cyanide	Inhibits cytochrome oxidase, producing cytotoxic anoxia; induces cellular edema
Lead	Highly neurotoxic; affects PNS, CNS, and motor and sensory nerves; can damage endothelial cells and cause accumulation of fluids in extracellular space
Mercury compounds	Can damage endothelial cells and cause extracellular edema
Organoarsenic compounds	Edema and focal hemorrhage
Organotin compounds	Swelling and necrosis of cells; edema of myelin sheaths in CNS
Organophosphorus compounds (only certain type of compounds, such as EPN and leptophos)	Delayed neuropathy, causing paralysis of muscles
Saxitoxin	Acts by blocking the sodium channels, can cause death by respiratory failure
Tellurium	Causes edema in endoneurium
Tetrodotoxin	Acts by blocking the sodium channels, can cause death by respiratory failure

V.H CARDIOVASCULAR SYSTEM: CARDIOTOXIC CHEMICALS

The cardiovascular system consists of the heart and blood vessels. The heart is primarily composed of myocardial cells made up of mitrochondria. The blood vessels consist of arteries, arterioles, capillaries, and veins. Although the heart is not a common target organ, many chemicals can affect it, either by damaging the myocardium or by acting indirectly through the blood vessels or nervous system. Some of the general toxic effect of chemicals on the cardiovascular system are briefly discussed in this section.

Certain chemicals can depress oxygen uptake and interfere with cardiac energy metabolism. The toxicity of such substances is greatly enhanced by the deficiency of certain amino acids in the food. Cobalt is a notable example of such types of toxicant. Furthermore, certain metals, such as cobalt, can act as an antagonist to endogenous Ca^{2+}. Many metals can also reduce the available myocardial Ca^{2+} by complexing with macromolecules. Certain substances, especially some vasodilating antihypertensive drugs, such as hydralizine and diazoxide, can induce myocardial necrosis. Also, a number of adrenergic β-receptor agonists, such as isoproterenol, manifest a similar effect through a different mechanism. Such effects produce an increased transmembrane calcium influx, resulting in cardiac hypoxia.

Certain respiratory tract irritants, including a number of fluorocarbons, can produce cardiac arrhythmias. Such effects may be attributed to the reduction of coronary blood flow, depression, or contractility and sensitization of the heart to epinephrine and several other factors. Many lipid-soluble substances can depress cardiac contractility. These include organic solvents, many general anesthetics, and aminoglycoside antibiotics.

Toxic effects may also arise when chemicals interfere with nucleic acid synthesis. Such substances bind covalently to mitochondria and to nuclear DNA. Such binding interferes with the synthesis of RNA and protein. The effects are hypotension, tachycardia, and arrhythmias. If these chemicals are administered into the body for a greater length of time, they can induce degeneration and atrophy of the cardiac muscle cells. They can also produce interstitial edema and fibrosis. Among such chemicals are the antineoplastic drugs, the anthracycline-type antibiotics (e.g., doxorubicin and daunorubicin). Certain chemicals can interfere with nucleic acid synthesis by inhibiting Q enzymes or by causing peroxidation of membrane lipids. Certain vegetable oils and coloring materials can produce morphologic changes in cardiac myofibrils and heart muscles.

A number of substances are known to exhibit toxic effects on blood vessels. Their modes of actions, however, may be different. Some of these toxic mechanisms are discussed below. Certain substances or their active metabolites are susceptible to cause cross-linking of DNA in the endothelial cells, damaging the capability of these cells to repair. Such endothelial damage can lead to thrombosis, progressive hypertension, and peripheral endarteritis. The latter can produce a gangrene-type disease, known as blackfoot disease. Arsenic and its compounds are known to cause this disease. Toxicants are known to damage endothelial cells of capillaries in the brain, increasing capillary permeability and resulting in brain edema. Pulmonary edema can result.

Many toxicants promote the permeability of capillaries surrounding blood vessels such as the coronary and carotid arteries. This can lead to degenerative changes in the blood vessels. Such changes can result in degenerative diseases, such as atherosclerosis. The mechanism of action of toxicants is complex and may differ among the various compounds. Certain substances may promote narrowing of arteries, leading to heart attack and stroke. Some cardiovascular toxicants and their effects are presented in Table V.7.

TABLE V.7 Toxic Effects on the Cardiovascular System and Selected Toxicants

Toxic Effects	Examples of Toxicants
Cardiac arrhythmias	Fluorocarbons; propylene glycol; and certain tricyclic antidepressants
Cardiomyopathy	Cobalt; ethanol; allylamine; certain antihypertensive drugs, such as hydralazine; and many adrenergic β-receptors agonists, such as isoproterenol
Depression of Cardiac Contractility	General anesthetics; antibiotics of aminoglycoside types, such as streptomycin
Damage to Endothelial Cells	Arsenic; many plant toxins, such as monocrotaline
Vasoconstriction	Arsenic; nitroglycerin; ergot alkaloids
Degenerative changes (and degenerative diseases, e.g., atherosclerosis)	Allylamines; aromatic hydrocarbons; carbon disulfide; carbon monoxide
Pulmonary edema	Nitrogen dioxide; ozone; hydrogen chloride; ammonia; and many other irritant gases
Tumors of blood vessels	Vinyl chloride; thorium dioxide

TABLE V.8 Reproductive Toxicity and Selected Toxicants[a]

Toxic Effects	Examples
Adverse effects on spermatogenesis (reduced sperm count); and testicular atrophy (in male reproductive system)	Metals, such as lead and cadmium; chlorinated pesticides such as dibromochloropropane; food colors, such as Oil Yellow AB; organic solvents
Causing defect or death to spermatozoa (in male reproductive system)	Alkylating agents (attack the DNA of the cells), such as methylmethane sulfonate
Miscellaneous adverse effects on the testis (in male reproductive system)	Many chlorinated organics, such as hexachlorophene; steroid hormaones; and alkylating agents
Tumors (in male reproductive system) such as testicular tumors and prostate cancer	Certain urea pesticides, such as linuron; cadmium
Damage to oocytes (in female reproductive system)	Polycyclic aromatic hydrocarbons, such as benzo[a]pyrene; nitrogen mustards; and certain aspirin-like drugs
Adversely affect the growth and development of conceptus (lower fetal weight) (in female reproductive system)	Certain chlorinated pesticides, such as DDT; nicotine; certain lactones
Increase the weight of uterus (in female reproductive system)	Ceratin chlorinated pesticides, such as methoxychlor; estradiol

[a]A partial list of a few selected toxicants.

V.I REPRODUCTIVE SYSTEM

Many chemicals can affect the reproductive system in a variety of ways. They may act directly on the reproductive system upon reaching the target organs in sufficient amounts. Many substances, such as dibromochloropropane, a fumigant, may act indirectly via certain endocrine organs.

In the male reproductive system, toxicants can adversely affect the formation, development, and delivery of spermatozoa.

Such substances can affect spermatogenesis and cause testicular atrophy. Among these chemicals are certain metals, solvents, pesticides, and food colors. Some chemicals manifest their toxicity on the testis. These include alkylating agents, naphthenic amines, and steroid hormones. Substances such as methylmethane sulfonate can cause the spermatozoa to become defective or less mobile. Studies on experimental animals indicate that certain compounds can induce tumors in testis. A notable example is linuron, a herbicide. In the female reproduction system, toxicants can damage oocytes and block the release of ovulation; may increase the weight of the uterus; affect the development and growth of the conceptus (causing lower fetal weight); and affect the onset of menopause. Some of these toxic effects discussed and some selected chemicals manifesting these actions are highlighted in Table V.8. Specific toxicants are discussed in Part B of this text. Section VII, on Teratogenic Substances, highlights certain mechanisms.

VI

CANCER-CAUSING CHEMICALS

VI.A CONCEPT OF CARCINOGENESIS

A carcinogen is a substance capable of causing cancer in mammals. Chemicals, radiation, and certain viruses are cancer-causing agents. Most cancers are induced by many synthetic or naturally occurring chemicals, which include inorganic substances, organics, hormones, and solid-state materials. A human carcinogen is a substance that induces cancer in humans.

Any possible carcinogenic properties of chemicals are identified from (1) human epidemiological studies establishing relationships between exposure to any given chemical and cancer, and (2) laboratory studies in experimental animals. Chemical carcinogens exert their effects after prolonged exposure for years, often with a latent period of several years. For example, asbestos fibers have demonstrated long latent periods, with the first signs of cancers appearing 20–30 years after exposure. Also, most carcinogens manifest a dose-response relationships. In animal studies large doses of suspected carcinogens are administered to two or more species of test animals over a range of time. Maximum doses are administered that would

possibly produce cancer. It is generally assumed that chemicals that produce cancers in animals would do so in humans. Such assumptions cannot always be confirmed. Therefore, any extrapolation of animal experiments data to humans should be established on the basis of human epidemiological studies.

Tumor or neoplasm is heritable altered autonomous growth of tissue, composed of abnormal cells that grow more rapidly than the surrounding normal cells. Neoplasm may be of two types, benign and malignant. While the former grows much more slowly than the latter, and does not metastasize (spread) throughout the body, malignant tumor or cancer grows rapidly and spreads throughout the body. Neoplasms use the blood, oxygen, and nutrient supplies of other tissues and organs affecting the normal function of body's organs and tissues. Neoplasms that continue to grow unabated and spread, as happens in malignancies, ultimately lead to death.

The biochemical processes involved in carcinogenesis are not well understood. It is known that many mutagens are also chemical carcinogens. Therefore, the mutation of normal cells exposed to chemical carcinogens can lead to the development of cancerous cells. Mutation is a permanent change in the base sequence of the DNA involving alterations of a single gene, a block of gene, or a whole chromosome. Carcinogenic chemicals may alter Deoxyribonucleic acid (DNA) to form a type of cell that replicates itself, forming cancerous tissue. DNA a long macromolecule that carries a genetic code through which genotypic characteristics are inherited. A chemical may interact with DNA through genetic mechanism, altering the structure or number in the chromosome, causing gene mutation or duplication. Such substances are known as genotoxic carcinogens. They initiate carcinogenesis, affecting the organism at an early stage. They are also carcinogenic at subtoxic doses and are often effective after a single exposure, exhibiting a short latent period. A large number of chemical carcinogens metabolize to reactive electrophilic intermediates that interact with DNA. Thus, the substance must undergo metabolism to produce the actual cancer-causing intermediate. Such compounds, called proximate carcinogens, include aromatic amines, polynuclear aromatic hydrocarbons (PAHs), nitrosamines, nitrosoureas, azo dyes, and chlorinated hydrocarbons.

Many chemicals in their unmetabolized form react directly with DNA or other cell constituents, causing cancer. These substances are classed as primary carcinogens. Alkylating agents, ethylene oxide, peroxides, dimethyl sulfate, sulfones, sulfur mustards, and bis(chloromethyl)ether are examples of primary carcinogens.

Another class of carcinogens are the epigenetic carcinogens. These substances, which are nongenotoxic in nature, cause changes in the structure of DNA without binding to DNA and without causing cell transformation or chromosomal aberrations. Solid-state carcinogens, hormones, and cytotoxic agents are examples of this class.

Many alkylating agents — chemicals that can add alkyl groups to DNA — are carcinogens and/or mutagens. They form reactive carbonium ions, which react with electron-rich bases in DNA, such as adenine, guanine, cytosine, and thymine. This leads to either mispairing of bases or chromosome breaks, causing growth and replication of neoplastic (cancerous) cells. The mechanism of chemical carcinogenesis is discussed below in brief.

VI.B   MECHANISM OF CHEMICAL CARCINOGENESIS

The mechanism of chemical carcinogenesis involves a complex multistage process that occurs in certain distinct phases. While genotoxic carcinogens manifest their actions in three phases: (1) initiation, (2) promotion, and (3) progression, the mechanisms for nongenotoxic carcinogen can vary and are more complex. The routes of exposure, dose amount, absorption, and distribution of the chemical within the body as well as the nature of the metabolic products, play important roles in the process.

The process of *initiation* by genotoxic carcinogens occurs when the chemical reacts with the DNA molecule. An electrophilic (electron-seeking) compound can bind covalently to certain nucleophilic sites in DNA, forming a carcinogen–DNA adduct. Some binding sites in DNA are shown below:

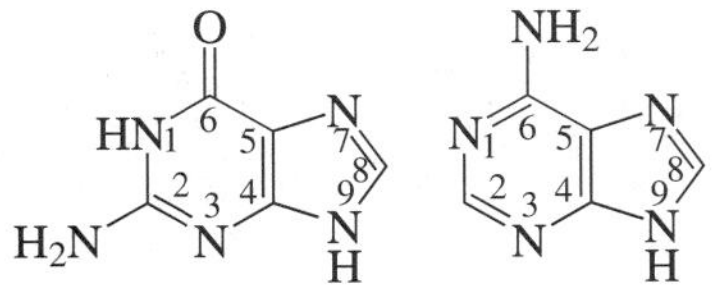

(Guanine)   (Adenine)

Binding sites for reactive carcinogens in purine base

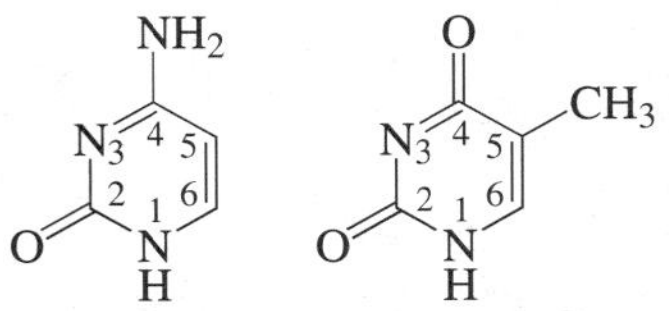

(Cytosine)   (Thymine)

Binding sites of reactive carcinogens in pyrimidine base

In these purine and pyrimidine bases, the covalent binding sites for the alkylating agents are N-7, O-6, and N-3 in guanine; N-1, N-3, and N-7 in adenine; N-1 and N-3 in cytosine; and N-3 and O-4 positions in thymine structures respectively. Among these, N^7-alkylguanine accounts for about two-thirds of all alkyl bound adducts. On the other hand, polycyclic aromatic hydrocarbons bind to the terminal $-NH_2$ group in the guanine, adenine, and cytosine rings. Experimental data also suggest that 2-acetylaminofluorene, and probably many carcinogenic azodyes, bind to the carbon atom at the C-8 position in the guanine ring. The binding sites for many cancer-causing chemicals remain unknown. Initiation leads to one or more mutations in the target cell genome. Such mutations are irreversible and dose-related and theoretically show no threshold.

Initiation is succeeded by a long phase of *promotion*. Gross tumors form during this long phase. The formation of such tumors results from the progressive clonal expansion of initiated cells. The mechanisms involve cell proliferation, tissue damage, cell death, regeneration, and immune suppression. Promotion is dose-related and partly reversible and often has a threshold. The same chemical or its metabolic products may act as both initiator or promoter, or the promoter may be an entirely different substance. Several promoters have been identified. One such promoter is 12-O-tetradecanoylphorbol-13-acetate, also known as TPA. Many other phorbol esters, similar to TPA and derived from croton oil, show similar actions. Such promoters can activate the enzyme protein kinase C, a calcium-dependent protein kinase, causing alterations in membrane function, alterations in intercellular communication, inhibition of terminal differentiation in a number of cell types, and other biochemical effects. Certain promoters may act genotoxic as well, in indirect ways. For example, TPA is not directly genotoxic and does not undergo metabolic activation, but it can release oxygen free radicals and can cause genotoxic damage in target tissues. Many other chemicals or groups have shown promoting actions in experimental animals. They include benzo(a)pyrene, anthralin, retinoic acid, and indole alkaloids, promoting skin tumor in mice; saccharin and cyclamates, promoting bladder tumor; and goitrogens, causing thyroid tumor in rats.

Progression is the final phase of tumor growth after various substances of neoplastic cells have emerged. During this stage, benign neoplasms are transformed into malign ones and the malignant cells disseminate from the primary tumors, invading and spreading to distant sites. At this stage, secondary metastatic deposits keep on forming at increasing numbers of sites, with additional mutational changes occurring.

The formation of an electrophile that would bind covalently to certain sites in the DNA is often the starting point in chemical carcinogenesis. The terms *'electrophile'* or *'nucleophile'* refer to compounds, ions, radicals, or functional groups that require or donate electrons, respectively, in chemical reactions. However, many "innocuous" substances that are not electrophiles may metabolize to form reactive electrophile species that may act as ultimate carcinogens. For example, N-hydroxylation and esterification can generate reactive carbocations and nitroenium ions. Aromatic amines, azo dyes, urethane, and similar substances undergo enzyme catalytic reactions forming reactive electrophiles. Similarly polynuclear aromatic hydrocarbons, aflatoxins, and vinyl chloride metabolize to reactive epoxides. Such metabolic activation occurs primarily in the liver and to a lesser extent in the kidneys, gastrointestinal tract, and lungs. Many alkylating substances, such as mustard gas or chloroethers, do not require metabolic activation and can bind directly to nucleophilic sites in the DNA.

When a carcinogen covalently binds to DNA, it may result in depurination, single-base substitution or single- or double-stranded breaks. Binding to alkylating agents may cause inter- or intra-strand cross-linkages. In other words, DNA is altered or damaged. A cross-linking adduct (as in the case of an alkylating agent binding to DNA) can prevent the replication of DNA or block the transcription of its own RNA and prevent the synthesis of an essential protein. In such a scenario the damage can kill the cell. In some cases, the damage may be repaired but imperfectly where the cell survives, but the genome is altered permanently. The damage may be fully repaired, with the cell recovering and restored back to its original intact state. These cases explain the origin of tumor formation and genotoxic carcinogenesis.

Not all chemical carcinogens act through genotoxic effects. Nongenotoxic carcinogens include various halogenated hydrocarbons, polychlorinated biphenyls (PCBs), dioxin, barbitals, synthetic and endogenous hormones, and many substances that are enzyme inducers. Their chemical structures are diverse, as are their biological effects and the mode of action. Studies on rats and mice indicate these substances act in the liver and induce the cytochrome P-450 mono-oxygenase system. Many hormones induce tumors in the liver and endocrine glands. Many nongenotoxic carcinogens interact with RNA and regulatory proteins and prevent the synthesis of normal proteins. They may bind to proteins that regulate cell-surface receptors for growth factors or control cell proliferation. The mechanisms and mode of actions of chemical carcinogens are not well understood.

VI.C CANCER REVIEW PANELS AND CLASSIFICATIONS OF CARCINOGENS

Several research and review panels are working on carcinogens. These groups periodically review and classify the cancer-causing potency of chemical, microbial, and physical carcinogenic agents. Unfortunately, many of their terms differ, although the meanings often overlap. Among the leading agencies are the International Agency for Research on Cancer (IARC), the National Toxicology Program (NTP) of the U.S. Department of Health and Human Services, and the Maximum Allowable Concentration Review Panel of German Research Society (DFK MAK). The classification of carcinogens and terminology used by these agencies are presented below. The known human carcinogens are listed in Table VI.1. This is not a complete list. A complete updated list of IARC and NTP carcinogenic agents, including chemicals, viruses, bacteria, physical agents, mixture, and occupational exposure conditions, is presented in Appendixes B and C.

VI.D CHEMICALS AND MIXTURES CONFIRMED AS HUMAN CARCINOGENS

Aflatoxins [1402-68-2]

Alcoholic beverages

4-Aminobiphenyl [92-67-1]

Analgesic mixture containing phenacetin

Arsenic [7440-38-2]

Arsenic compounds

Asbestos [1332-21-4]

Azathioprine [446-86-6]

Benzene [71-43-2]

Benzidine [92-87-5]

Beryllium [7440-41-7]

Beryllium compounds

Cadmium [7440-43-9]

Cadmium compounds

Chlorambucil [305-03-3]

Bis(chloromethyl)ether [542-88-1]

Chlornaphazine [494-03-1]

Chloromethyl methyl ether [107-30-2]

Chromium [7440-47-3]

Chromium(VI) compounds

Cyclosporine [79217-60-0]

Cyclophosphamide [50-18-0]

Diethylstilbesterol [56-53-1]

TABLE VI.1 Classification of Carcinogenic Agents

Agencies	Classification and Meaning	
IARC	Group 1	The agent is carcinogenic to humans — sufficient evidence established between exposure to the agent and human cancer
	Group 2A	The agent is probably carcinogenic to humans — limited evidence of carcinogenicity in humans and sufficient evidence of carcinogenicity in experimental animals
	Group 2B	The agent is possibly carcinogenic to humans — limited evidence in humans but absence of sufficient evidence in experimental animals or inadequate evidence in humans (when human data not available) but sufficient evidence in experimental animals
	Group 3	The agent is not classifiable as to its carcinogenicity to humans — when agents do not fall into any other group
	Group 4	The agent is probably not carcinogenic to humans — evidence suggesting lack of carcinogenicity in humans and experimental animals
NTP		Known to be carcinogen — sufficient evidence of carcinogenicity between the agent and human cancer
		Reasonably anticipated to be carcinogens — limited evidence of carcinogenicity from human studies, the interpretation may be credible, but alternative explanations, such as chance, bias or confounding could not adequately be excluded; or (2) sufficient evidence of incidence of malignant tumors from studies in experimental animals in multiple species or strains; or in multiple experiments with different routes of administration or using different dose levels; or to an unusual degree with regard to incidence, site or type of tumor, or age at onset
DFK MAK	Group A1	Capable of inducing malignant tumors, with human evidence
	Group A2	Carcinogenic in experimental animals only
	Group B	Suspected of having carcinogenic potential

Erionite [66733-21-9]

Ethylene oxide [75-21-8]

Melphalan [148-82-3]

Mineral oils [8007-45-2]

Mustard gas [505-60-2]

Myleran [55-98-1]

β-Naphthylamine [91-59-8]

2-Naphthylamine [91-59-8]

Nickel compounds

4-Nitrobiphenyl [92-93-3]

Radon [10043-92-2]

Shale oils [68308-34-9]

Silica [14808-60-7], crystalline (inhaled as quartz as cristobalite from occupational sources)

Soots [8001-58-9]

Talc-containing asbestiform fibers

Tamoxifen [10540-29-1]

Tar [8021-39-4]

2,3,7,8-TCDD [1746-01-6]

Thiotepa [52-24-4]

Tobacco products, smokeless

Tobacco smoke

Treosulfan [299-75-2]

Vinyl chloride [75-01-4]

Wood dust

VII

TERATOGENIC SUBSTANCES

Teratogens are chemical and physical agents that can cause birth defects and mortality among newborns, malformations, growth retardation, and functional disorders. The nature and severity of such effects may widely vary, ranging from fetal deaths and major malformation affecting the survival, growth, and fertility to delayed developmental toxicity and minor anomalies. Probably, the most dangerous teratogenic effects have been shown by thalidomide, a sedative hypnotic drug. This drug, used in the past for the relief of morning sickness, caused phocomelia, a very rare type of congenital malformation among newborn babies. Ingestion of thalidomide by pregnant mothers resulted in the births of thousands of deformed babies with shortening or absence of limbs. Although thalidomide may be the most widely known teratogen, many other substances tend to produce such severe chemical teratogenesis.

Although most teratogens are chemical in nature, ionizing radiation, nutritional imbalance, and certain infections are also known to be teratogenic to humans. A number of chemicals have been found to be teratogenic to humans. These include many common organic solvents, alcohols, pesticides, herbicides, azo dyes, heavy metals, sulfonamides, antibiotics, and certain alkaloids. Table VII.1 presents a partial list of common chemicals of industrial or pharmaceutical applications that are known human/animal teratogens. The teratogenicity of these and other substances is discussed under their titled headings in Part B of this text.

VII.A MECHANISM AND MODE OF ACTION OF TERATOGENS

After fertilization, the ovum passes through a sequence of events, characterized by cell proliferation, differentiation, migration, and organogenesis. The embryo formed undergoes a state of metamorphoses, followed by a period of fetal development before birth, passing through three distinct stages: predifferentiation, embryonic, and fetal. Of these three stages, the embryonic is the most critical, as this is when the embryo is most susceptible to the effects of teratogenic agents.

During the predifferentiation stage, which may vary from 5 to 9 days depending on the species, teratogens can either show no apparent effect on the embryo or kill the cells, causing the death of the embryo. During this

TABLE VII.1   Partial List of Known Teratogens to Humans and Animals

Classes of Substances	Examples
Metals	Lithium, selenium, strontium, lead, thallium
Organic solvents	Benzene, xylenes, carbon tetrachloride, 1,1-dichloroethane, dimethyl sulfoxide
Alcohols	Ethanol, diethylstilbestrol
Nitrosamines	N-Nitrosodimethylamine, N-nitrosodi-n-butylamine
Azo dyes	Niagara blue, Evans blue, trypan blue
Pesticides	Dinocap, phenytoin
Alkaloids	Quinine, pilocarpine, caffeine, nicotine, rauwolfia, veratrum, and vinca alkaloids
Antibiotics	Streptomycin, penicillin, tetracyclines
Sulfonamides	Sulfanilamide
Miscellaneous drugs	Antifolic drugs, antithyroid drugs, meclizine, chlorpromazine, chlorcyclizine, carbutamide
Natural toxins	Aflatoxin B_1, ergotamine, ochratoxin A
Anesthetics	Halothane
Miscellaneous substances	Thalidomide, polychlorinated biphenyls, organomercuric compounds, methimazole, bisulfan, cyclophosphamide, procymidone, trimethadione, abamectin, aminopterin, isotretinoin, valproic acid

period, cells that survive from any mildly harmful effects of the teratogenic agents compensate and form a normal embryo.

Most of the organogenesis occurs during the embryonic stage, with the cells undergoing intensive differentiation, mobilization, and organization. Studies on rodents have shown that brief exposure to a teratogen on the 10th day of gestation induced a variety of malformations in rats. These malformations included brain, eyes and heart defects with an incidence of more than 25%, and skeletal and urological defects to a minor extent. Such incidences of malformations of different organs and their susceptibilities to teratogens can vary according to the day of gestation. For example, the incidence of eye defects and urogenital defects was maximum in rats from brief pulse treatment of teratogen on the 9th and 15th days of gestation, respectively. Rat embryos show maximum susceptibility to teratogens between days 8 and 12. The embryonic stage generally ends in the 10th to the 14th day in rodents. In humans it is generally in the 14th week of the gestation period.

Teratogens may not cause morphologic defects in the fetal stage, during which growth and functional maturation occur. However, they may induce functional abnormalities, such as central nervous system deficiencies, which may not be detected at birth or shortly thereafter.

How teratogens act depends on their certain chemical properties. Thus, their mode of action in humans and animals may involve many different mechanisms. Many teratogens can interfere with replication or transcription of nucleic acid or RNA translation. Such effects can cause cell death or somatic mutation in the embryo. These may result in structural and functional defects. Therefore, the toxicants that produce reactive metabolites that react with nucleic acids can be potential teratogens. Thus, the mode of action of such teratogens is similar to that of genotoxic carcinogens. Such reactive metabolites must be able to reach the embryo for teratogenesis to occur. If such metabolites are too unstable, they may be carcinogens, but not teratogens.

Interference with nucleic acids is not the only cause of teratogenesis. Other mechanisms are involved in teratogenesis as well. These include inhibition of enzymes, deficiency of energy supply and osmolarity, lipid peroxidation, and oxidation and degradation of proteins both in maternal and in embryonic tissues. For example, 6-aminonicotinamide is a potent teratogen because it inhibits the enzymatic action of glucose-6-phosphate dehydrogenase (G6PD). Similarly, 5-fluorouracil can inhibit thymidylate synthetase and induce malformation. Certain substances can affect metabolic processes by depriving oxygen supply or restricting energy supply. Thus, gases causing hypoxia and substances that are antagonists of essential amino acids and vitamins can be teratogenic agents.

Thus, the mode of action of many teratogens is different and is not fully understood. The mechanism depends on several factors, such as bioactivation of toxicants, the stability of toxicants or their reactive metabolites to reach the embryo, and biotransformation or the detoxifying capability of the embryonic tissues.

VII.B TESTING AND EVALUATION OF TERATOGENIC EFFECTS

Teratogenic studies have generally been performed on experimental animals, primarily, rats, mice, rabbits, and hamsters. These species have a short gestation period. The chemical is administered at least at three dosage levels: one level, at which no observable ill effect should be induced, while the higher dosages should produce some maternal and/or fetal toxicity. Several animals are placed in each dose group, as well as in two control groups. Subjects of both control groups receive a chemical compound of known teratogenic activity and physiologic saline, respectively.

The chemical in the study is either incorporated into the animal feed or administered by gastric gavage. The period of testing may differ from one species to another. For rats and mice, the dosing period is 6–15 days. In general, the chemical should be administered when the embryo is most susceptible, especially throughout organogenesis.

The dosed animals are examined daily for any signs of toxic effects. The fetuses are surgically removed 1 day before the expected delivery and observed whether they are live or dead. Their number, sex, and weight are observed. Also, the number of corpora lutea, implantations, and resorptions is determined. Detailed examinations involve determination of abnormalities and external defects, such as skeletal abnormalities and visceral defects. For large animals, such as dogs or pigs, the skeletal structure of fetuses is examined by radiography. For small animals, skeletal abnormalities are determined after staining with a coloring substance, such as alizarin red. Delayed effects of teratogens may be studied by neuromotor and behavioral tests. Such tests may be used to detect the teratogenic effects on the fetal central nervous system. This includes measuring motor activity, acoustic startle, brain weight, T-maze delayed alteration, and glial fibrillary acidic protein (GFAP). Delayed effects in male animals may be manifested by altered androgenic status, such as reduced sperm count and testosterone levels.

Teratogenic effects of substances on test animals are evaluated from resorption (death of the conceptus), fetal toxicity (reduced body weight), aberrations (malformations), and minor anomalies. Statistical analysis is performed by surveying the four most important parameters; namely, the number of litters with malformed fetuses, increase in the average number of fetuses with defects per litter, the number of resorptions, and dead fetuses. Finally, the incidence of malformation (response) is plotted against doses administered. Any dose-response relationship should indicate the teratogenicity of the chemical under experimental conditions.

Teratogenic effects may be determined from a variety of in vitro tests, using either mammalian whole embryo, cells, tissues, or embryonic organs in culture or nonmammalian model systems. The results of teratogenic studies in animals do not always apply to humans. For example, thalidomide, the most potent human teratogen, has been found to be innocuous to rats and mice. Similarly, many animal teratogens are potentially harmless to humans. With the knowledge acquired so far from abundant experimental data and epidemiological studies, the teratogenic potential of chemical substances may be determined from their molecular structures and certain physicochemical properties. This should eliminate unnecessary animal experiments.

VIII

HABIT-FORMING ADDICTIVE SUBSTANCES

A number of substances of widely varying chemical structures and physical and chemical properties are known to produce compulsive physical and psychological dependence. The user feels an urge to use more of the substance to preserve the original effect. Also, when a dependent user stops taking a drug, a 'withdrawal reaction' can set in, implying a biochemical change in the body that can manifest physical and psychological symptoms. Such habit-forming chemicals may have the following common features:

1. Euphoric effect and/or respite from pain
2. Taking an amount more than intended
3. Enhanced substance dependence caused by repeat use
4. Tolerance
5. Withdrawal reaction

Chemical dependence may arise from repeat use of the substance. Not all psychological effects, are similar. Such effects may widely differ, depending on the substance. A large number of compounds may be included under the category of such habit-forming substances that have the potential to cause dependence upon excess and/or repeat intakes. Such substances include an array of chemicals that vary widely in chemical structure and properties, ranging from ethanol and caffeine to amphetamines and opiates. They include hallucinogens, narcotics, stimulants, sedatives, and hypnotics.

Such compounds may be broadly classified as

1. LSD-type hallucinogens
2. Phencyclidine hallucinogens
3. Amphetamine stimulants
4. Cocaine-type stimulants
5. Opiate narcotics
6. Cannabis
7. Barbiturate sedatives
8. Alcohols
9. Organic vapors and inhalants

Some of these compounds are discussed at length in Part B, Chapter 7 (Alkaloids). Some of the common drugs of abuse and their structural features are briefly discussed below.

VIII.A HALLUCINOGENS

Hallucinogens are psychedelic substances that produce a wide range of illusory effect, altering mood, thinking and perception. The effect can be intense, kaleidoscopic, and brilliant in colors and textures, as well as sweet, sharp, and profound in sound, taste, and smell. The effects often produce flashbacks and visual distortion, varying from pleasant to traumatic. The quality and duration of psychedelic reaction depend on the nature of the substance and amount of intake. The duration can last several hours. Common adverse effects from some psychedelic drugs involve acute panic reaction, psychotic symptoms, mental instability, and emotional crisis, which may vary with the person. Many hallucinogens do not cause physical addiction or psychological dependence. Some produce milder effect than others.

Hallucinogens may be naturally occurring or synthetic. The best known synthetic drug is lysergic acid diethylamide (LSD). The monamide of lysergic acid, however, is naturally occurring and is found in morning glory seed. The latter is a psychedelic alkaloid. LSD is probably one of the most potent hallucingogens, producing the widest range of effect, as well as profound alterations in mood and perception. The effective dose for this substance is 0.075–0.10 mg/kg, almost 10 times more potent than the corresponding monoamide. Its mode of action is not certain but probably involves the inhibition of the release of serotonin, by acting on a presynaptic serotonin receptor. Among other synthetic psychedelic drugs are diethyltryptamine (DET), dipropyltryptamine (DPT), 3,4-methylenedioxyamphetamine (MDA), 3,4-methylenedioxymethamphetamine (MDMA), and 2,5-dimethoxy-4-methylamphetamine (DOM or STP). These compounds have a tryptamine or methoxylated amphetamine structure.

MDMA, structurally related to amphetamine and mescaline, is a mild psychedelic drug. Its effects last for a shorter duration, 2–4 hours. Unlike LSD or mescaline, it does not usually evoke the psychotic reaction of flashback, the so-called "bad trip."

Another common hallucinogen of street use is phencyclidine, also known by several other names, such as, PCP, "angel dust," or "hog." This drug was originally introduced as an anesthetic is and now used in veterinary practice. The drug was withdrawn from human use because of its hallucinogenic effect. This substance is taken orally or intravenously or more commonly by mixing with tobacco, to be smoked. It produces less schizophrenia than occurs with the LSD class of hallucinogens. Symptoms are distortion of body image, detachment, and disorientation. Adverse somatic effects include numbness, sweating, hypertension, and rapid heart rate. Overdosage can lead to death. The hallucinogenic property of phencyclidie and structurally related substances may be attributed to reduced uptake of dopamine and enhance of dopaminergic transmission. Some selective hallucinogens and their structures are presented in Table VIII.1.

VIII.B STIMULANTS – GENERAL TYPE

Stimulants are therapeutic agents that stimulate the central nervous system. They are capable of exciting the CNS over the normal range. Many are natural compounds, some are synthetic. They differ in their structures, as well as in clinical and pharmacological effects. The addictive potential of these compounds is relatively much lower than that associated with narcotics or hallucinogens. Many stimulants, such as strychnine or caffeine, are alkaloids (see Chapter 7, Alkaloids).

Caffeine is the most popular psychoactive stimulant. This xanthine alkaloid occurs in coffee, tea, and cola beverages. At a modest dose of 100–200 mg, it enhances wakefulness and reduces fatigue. However, intake of higher quantities (>500 mg/day) can cause addiction. Other psychostimulant substances include deanol, pemoline, and their derivatives. These substances have limited clinical application. The structures of caffeine,

TABLE VIII.1   Common Hallucinogens and Their Structures

Hallucinogens	Structures	Comments
Lysergic acid diethylamide (LSD)		Synthetic; a highly potent and strong hallucinogen; can invoke very wide-ranging perceptual and illusory effects, including "bad trips"; effective dose 0.1 mg/kg; perceptual, psychic, and somatic effects can overlap; toxic but nonfatal
Mescaline		Naturally occurring; less potent than the LSD, but with psychedelic effects very similar to those of LSD
Dimethyltryptamine (DMT) Diethyltryptamine (DET) Dipropyltryptamine (DPT)	$R = CH_3, C_2H_5, C_3H_7$ in DMT, DET, and DPT, respectively.	DMT occurs in leaves of *Prestonia amazonica*; highly potent; the synthetic analogues DET and DPT are highly potent
2,5-dimethoxy-4-amphetamine (DOM or STP)		Mild effects, similar to MDMA
3,4-methylenedioxymethamphetamine (MDMA)		A mild hallucinogen; a common drug of abuse; generally does not evoke a "bad" trip
3.4-Methylenedioxy-amphetamine		Mild effects, similar to MDMA
Harmine		Occurs in Harmala alkaloids: low potency; also a central stimulant
Harmaline		Occurs in *Peganum harmala*; dihydro derivative of harmine; hallucinogen potency low; a central stimulant

(continued)

TABLE VIII.1 *(Continued)*

Hallucinogens	Structures	Comments
Ibogaine	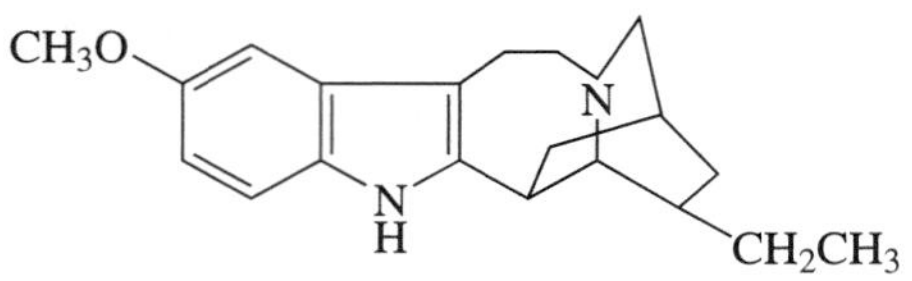	Obtained from the shrub *Tabernanthe iboga*; a powerful hallucinogen producing a wide range of effects
Psilocybin	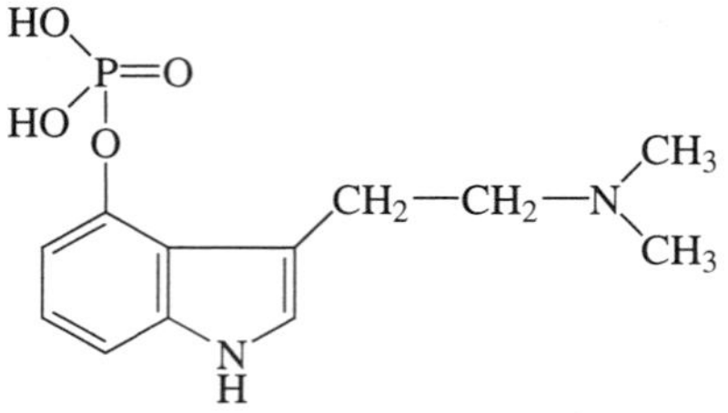 $(C_{12}H_{17}N_2O_4P)$	Highly potent hallucinogen; obtained from *Psilocybe mexicane*
Psilocin	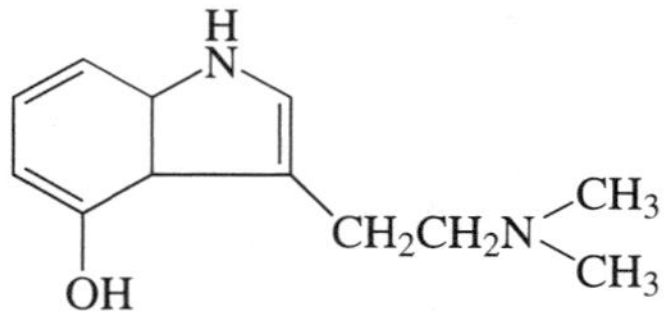 $(C_{12}H_{16}N_2O)$	Tryptamine-type hallucinogen; unstable and difficult to isolate; metabolite of psylocybin; highly potent
Phencyclidine (PCP, "angel-dust", "hog")	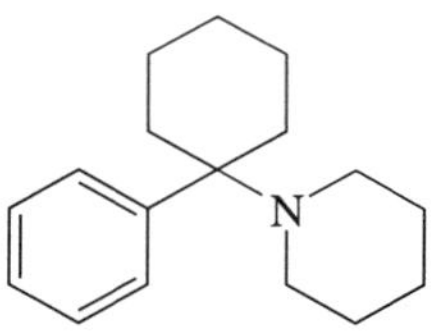 $(C_{12}H_{25}N)$	A common drug of abuse; milder than LSD; anesthetic properties

deanol, and pemoline are shown below:

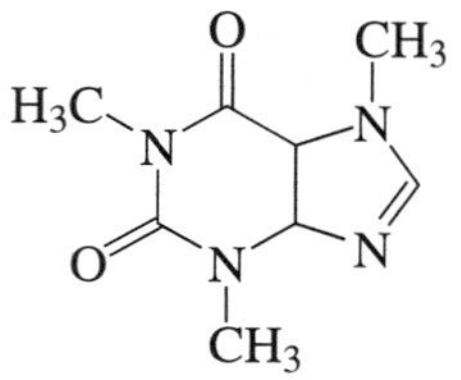

(caffeine)

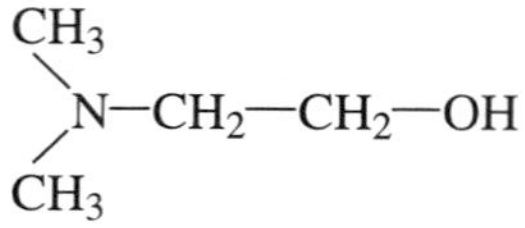

(deanol)

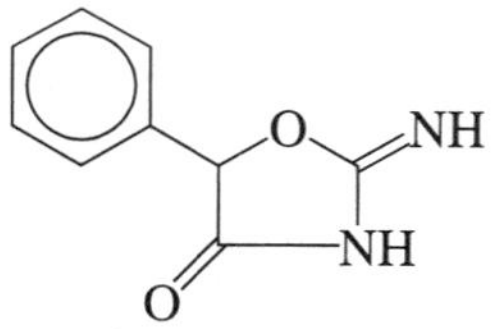

(pemoline)

Many sympathomimetics have strong excitatory effects on the central nervous system. These are mostly synthetic compounds and include amphetamines. The chemical structures and abusive potential of amphetamines are discussed separately in the following section.

VIII.C AMPHETAMINE-TYPE STIMULANTS

Amphetamines constitute a large group of sympathomimetic drugs. These substances stimulate the central nervous system. They were earlier used to control obesity and to, resist sleep (narcolepsis), and in the treatment of depression. Their clinical application, however, has decreased in recent years because of their adverse psychological effect. Amphetamines are strongly addictive compounds. Repeated use leads to dependence and the development of tolerance.

Adverse psychological effects are insomnia, restlessness, tension, confusion, irritability, hostility, panic, and psychosis. The most notable effect, is psychosis. Short-term administration of these drugs can induce psychosis even in normal people. The symptoms are restlessness, delusion, euphoria, and visual and auditory hallucinations, often similar to those associated with paranoid schizophrenia. Such effects, however, require large doses of the drug, e.g., 50–100 times the therapeutic dose. Amphetamine psychosis may disappear within a few days. Chronic use can create a high degree of tolerance toward the drug and a compulsive craving or desire for it. The user may need a much larger dose that may increase to 15 or 20 times as much to recover the original effect of euphoria. Withdrawal from amphetamine may produce headache, sweating, hunger, and stomach and muscle cramps. The psychological symptoms of withdrawal from the drug by heavy users can be severe and suicidal at times. Such symptoms include lethargy and increasing bouts of depression that can lead to a compulsive urge to start its use again.

Acute poisoning from overdose can produce pallor, fever, bluish skin, nausea, vomiting, elevated heart rate, elevated blood pressure, respiratory distress, and loss of sensory capacities. Death can occur from overdose, after high fever, convulsions, and shock.

The parent compound, amphetamine [300-62-9] is (phenylisopropyl) amine or β-amino-phenyl benzene that exists in stereoisomeric d- and l-forms. Its structure is given below:

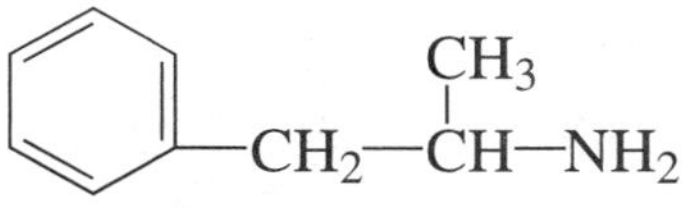

[300-62-9]

Owing to the presence of an amino group ($-NH_2$) in the molecule, amphetamine can form acid derivatives with both mineral and carboxylic acids. Aromatic ring substitution can produce a number of derivatives. Alkyl, alkoxy, alkylthio, or halogen substitution in the ring can produce a number of potent hallucinogenic substances. Similarly, alkyl substitution on the amino group can produce derivatives, such as methamphetamine, that are sympathomimetics and central stimulants. Many amphetamine derivatives may be readily synthesized in the laboratory. Some common drugs of abuse include dextroamphetamine (Dexedrine), metamphetamine (Methedrine) [537-46-2], phenmetrazine, methylphenidate (Ritaline) [113-45-1], and the racemic amphetamine sulfate (Benezedrine). The structures of a few amphetamine compounds are shown below:

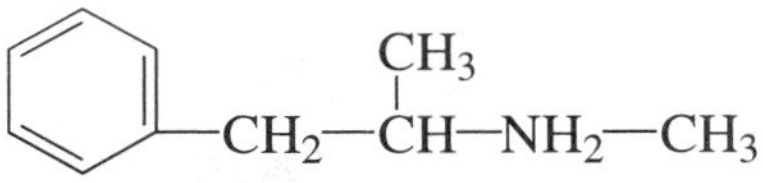

(methamphetamine)
[537-46-2]

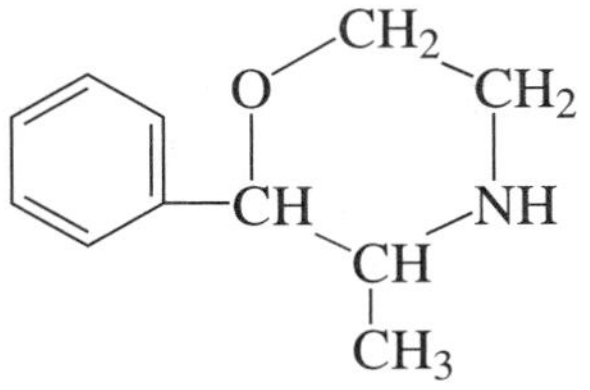

(phenmetrazine)

(methylphenydate)
[113-45-1]

(dextroamphetamine sulfate)

Methamphetamine or 1-phenyl-2-(methylamine)propane is another widely abused drug. It is also known as "speed." At higher doses, this strong CNS stimulant produces delusions and bizarre visual and auditory hallucinations. Chronic use at high doses can produce schizophrenia-like conditions. The toxic and psychological effects are similar to those of other amphetamine drugs. Its pharmaceutical uses have been discontinued.

Amphetamines are synthetic compounds, structurally related to the natural product cathinone. The latter is an alkaloid extracted from the leaves of the plant, Khat (*Catha edulis*). Chewing of fresh leaves of Khat produces effects similar to those of amphetamines. The structure of cathinone or α-aminopropiophenone is as follows:

(cathinone)

Because of the abuse potential of amphetamines, the Drug Enforcement Agency has strictly controlled the manufacture, distribution, and use of these substances in the United States.

VIII.D COCAINE

Cocaine is one of the most popular illicit drugs in the world. It is a strong CNS stimulant and is more potent than amphetamines. It belongs to the class of tropane alkaloids. For a discussion of the structures, physical and chemical properties, and toxicity, see Chapter 7 (Alkaloids). Cocaine is a natural alkaloid derived from coca leaves. Chewing of the leaves for stimulant action and to overcome fatigue was known since early times. Its anesthetic properties were discovered much later, in the nineteenth century. It has since been in use as a topical anesthetic, in diluted form.

The illicit use of cocaine increased dramatically after amphetamine drugs were banned. Street cocaine is usually adulterated with sugars and other substances. The adverse psychological effects of cocaine are similar to those of amphetamines. The effects include insomnia; weight loss; visual disturbances; hypersensitivity to light, sound, and touch; paranoid thinking; as well as a feeling of increased energy and confidence resulting in irritation and aggressiveness. It can lead to dependence far greater than that experienced with amphetamines. The euphoric effect from a single dose can last 1 or 2 hours. For this reason, users take repeat doses several times a day. Unlike heroin, cocaine-induced euphoric effects are not altered by methadone. A combination of heroin and cocaine administered intravensously (known as "speedball") can double the euphoric effect. High doses can cause death. Overdoses of cocaine alone can be fatal too. It is more toxic than amphetamines. Diazepam and propranolol (intravenous administration) are effective against cocaine overdose.

Like any other drug, common methods of cocaine intake are snuffing or snorting and subcutaneous or intravenous injection. Also, free-base cocaine may be smoked in a water pipe. Cocaine hydrochloride is converted to a more potent free base cocaine (also known as "crack") by treatment with a mild base, such as ammonia or sodium carbonate.

Among the tropane alkaloids, cocaine is the most potent and addictive drug. The euphoric effects from related compounds with tropane structures, such as *l*-hyoscyamine and its racemic isomer atropine, or scopolamine, are distinctively lower than with cocaine. Their habit-forming potential is also very low in comparison with cocaine. The structures of tropane and cocaine are presented below. The toxicity of this class of substances is described in Chapter 7 (Alkaloids).

(tropaine)

(cocaine)

Cocaine salts are also stimulants, exhibiting the same degree of habituation or addiction. These include cocaine hydrochloride, cocaine sulfate, and cocaine nitrate.

VIII.E   OPIATE NARCOTICS

Opium has been known since ancient times. It is the milky latex exudate obtained by incision of unripe capsules of the opium poppy, Papaveraceae (*Papaver somiferum* L.). The white exudate that oozes out from the seed, air dries and turns sticky and brown on standing. This brown gum, known as opium, contains about 25 alkaloids. Opium products have been used extensively as narcotic analgesics and sedatives. Many of these compounds are habit-forming substances, causing severe dependence. Because of the serious abuse problem, the cultivation, sale, and distribution of opium and its products are strictly controlled all over the world. However, illicit production of opium continues in many parts of the world.

The principal alkaloid of opium is morphine. Its concentration in opium within the range 5–20%. Other alkaloids present in opium at lower but significant concentrations are codeine, thebaine, noscapine, and papverine. Among these, thebaine and papaverine do not exhibit analgesic properties. Thebaine, however, is a precursor to many synthetic agonists, such as etorphine, and to antagonists like naloxone. The analgesic dose of morphine is about 10 mg to produce analgesia for 4–5 hours.

Morphine and its surrogates that have affinity for μ receptors affect the central nervous system, producing analgesia, euphoria, sedation, and respiratory depression. The euphoric effects produce a pleasant feeling of floating, free from anxiety and distress. People addicted to drugs, or those in pain, generally experience such euphoria. Normal people who are not in pain may sometimes experience restlessness after administration of morphine or its surrogates. Drowsiness or sedation is another common effect experienced by humans. Older people are more susceptible to sleep. The sedative effect is greater with compounds with phenanthrene-type structures. Certain animal species, on the other hand, have displayed excitation or stimulant behavior. All opiate analgesics can cause respiratory depression and decreased responsiveness to carbon dioxide. The toxic effects of opium products vary with the compounds and include miosis (constriction of the pupils), emesis, and constipation. The toxicity of these substances is not discussed in this chapter and Chapter 7.

Many opiate compounds are highly addictive. Repeated administration of therapeutic doses can develop a strong tolerance, or gradual loss in the effectiveness of the drug. Increasingly larger doses are needed to reproduce the original effects, including euphoria, analgesia, sedation, mental clouding, and dysphoria. This leads to a strong physical dependence on the drug, which then becomes necessary to prevent a characteristic abstinence or withdrawal syndrome. Such tolerance and physical dependence may be attributed to neuronal calcium that accumulates after frequent opiate intake. Calcium readily antagonizes opiate effects. When the drug is discontinued, the ability of calcium release is lost. This enhances the release of a number of neurotransmitters that can cause abstinence syndrome.

Heroin is probably the most common abused drug of the opiate class. Unlike morphine, oxycodone, and meperidine, heroin has no medicinal application. Heroin, morphine, and other opiate drugs are commonly administered intravenously, enabling the drug to reach the brain quickly, producing a "rush," followed by euphoria and tranquility. A 25-mg heroin dose can produce the effect for 3–5 hours. Because of this short duration effect, the user often administers the drug several times per day, strongly reinforcing physical dependence and tolerance.

The symptoms of opiate withdrawal can be severe. It starts 8–12 hours after the last dose. The initial symptoms are lacrimation, yawning, and sweating. This is followed by restlessness, weakness, and chills. The first phase of withdrawal may last a week. The other symptoms during this period include nausea, vomiting, muscle pain, involuntary movements, hyperthermia, and hypertension. This is soon followed by a secondary phase of abstinence, which may continue for a prolonged period of 6–7 months. The symptoms of such protracted abstinence include hypotension, bradycardia, hypothermia, mydriasis, and decreased responsiveness to carbon dioxide.

Methadone maintenance therapy is effective against heroin and other opiate dependence. Methadone, a synthetic analgesic, is pharmacologically equivalent to opiates and gradually substitutes for the abused drug. It saturates the opiate receptors. This substance is addictive too; the addictive potential, however, is lesser and milder than that associated with heroin or morphine. Among other substances that are known to be effective are clonidine, methadyl acetate, and naltrexone. Clonidine is non-narcotic and nonaddictive and is a centrally acting sympathoplegic agent. Naltrexone, on the other hand, is a narcotic antagonist.

The chemical structures of morphine, heroin, and other opiates and their derivatives are presented in Table VIII.2. Some of these compounds are discussed in Chapter 7 (Alkaloids). The structures and addictive potentials of a few selected nonopium analgesics are presented in Table VIII.3.

VIII.F   CANNABIS

Cannabinoids are the active components of the plant cannabis or hemp, *Cannabis sativa* L. Its use has been recorded for thousands of years. The drug marijuana (marihuana) comes from the leaves and the flowering tops of the plant. When the plant materials are grounded, they appear as grass clippings, for which it is known in the street as "grass." The more potent 'hashish' is obtained by alcohol extraction of the resin from the tops of the flowering plant. The most common route of administration of marijuana is smoking. It is also taken orally and intravenously.

More than 400 chemicals have been isolated from marijuana. The active components, known as cannabinoids, however, are primarily Δ^9-tetrahydrocannabinol (Δ^9-THC), Δ^8-tetrahydrocannabinol (Δ^8-THC), cannabinol (CBN), and cannabidiol (CBD). Among these, the THC and homologues are the psychoactive compounds. The THC content of cannabis varies among plants from less than 1% to more than 10%. Sensemilla, the flowering tops of female cannabis plant, unpollinated and seedless, may contain 6–15% THC. The plant extract, hashish, can contain as high as 20% THC. The psychological effects from cannabis are characterized by euphoria, uncontrollable laughter, and loss of short-term memory. Other symptoms

TABLE VIII.2   Selected Opiates and Their Derivatives

Compound	Structure/Formula	Addiction/Habituation
Morphine	 $(C_{17}H_{19}NO_3)$	High
Morphine hydrochloride	$C_{17}H_{19}NO_3 \cdot HCl$	High
Morphine N-oxide	$C_{17}H_{19}NO_4$	High
Morphosan (morphine methylbromide)	$C_{18}H_{22}BrNO_3$	High
Morphine sulfate	$(C_{17}H_{19}NO_3)_2 \cdot H_2SO_4$	High
Heroin (diacetyl morphine)	 $(C_{21}H_{23}NO_5)$	High
Heroin hydrochloride	$C_{21}H_{23}NO_5 \cdot HCl$	High
Oxymorphone	 $(C_{19}H_{19}NO_4)$	High
Oxycodone	 $(C_{18}H_{21}NO_4)$	Medium

(continued)

TABLE VIII.2 (*Continued*)

Compound	Structure/Formula	Addiction/Habituation
Codeine (methyl morphine)	$(C_{18}H_{21}NO_3)$	Medium
Codeine hydrochloride	$C_{18}H_{21}NO_3 \cdot HCl$	Medium
Codeine N-oxide	$C_{18}H_{21}NO_4$	Medium
Codeine sulfate	$(C_{18}H_{21}NO_3)_2 \cdot H_2SO_4$	Medium
Dihydromorphine	$(C_{17}H_{21}NO_3)$	High
Paramorphan	$C_{17}H_{21}NO_3 \cdot HCl$	High
Dihydromorphinone (hydromorphone)	$(C_{17}H_{19}NO_3)$	High
Hydromorphone hydrochloride (Dilaudid)	$C_{17}H_{19}NO_3 \cdot HCl$	High
Thebaine	$(C_{19}H_{21}NO_3)$	Medium

TABLE VIII.3 Selected Nonopiate Analgesic

Compound	Structure/Formula	Addiction/Habituation
Meperidine		High

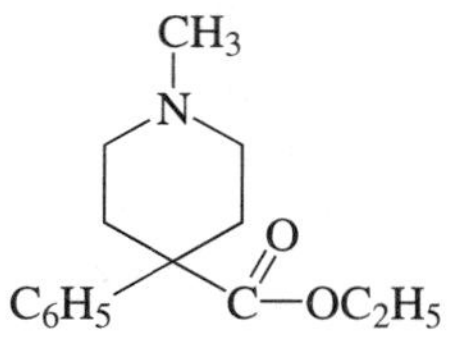

$(C_{15}H_{21}NO_2)$

| Methadone | | Medium |

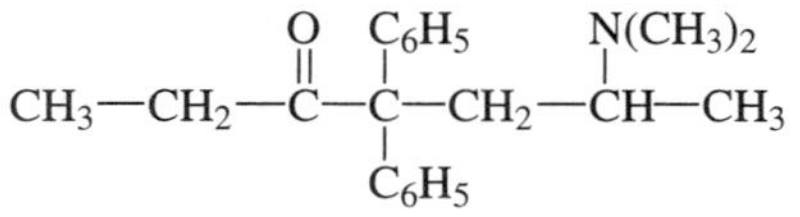

$C_{21}H_{27}NO$

| α-Prodine | | High |

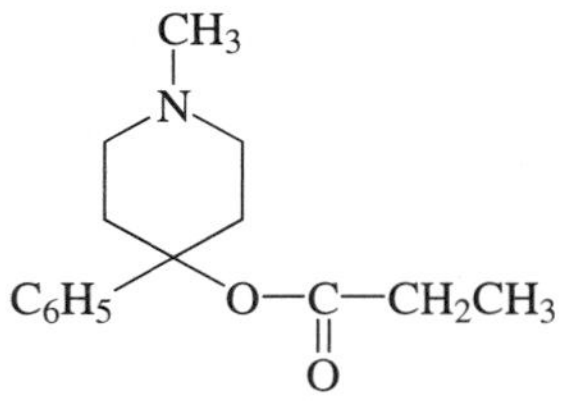

$(C_{16}H_{23}NO_2)$

| Fentanyl | | High |

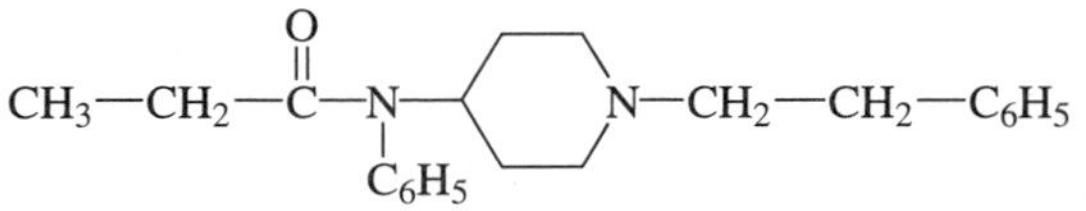

$(C_{22}H_{28}N_2O)$

include calmness, rapid flow of ideas, slow-down of time, illusions, brightness of color, and hallucinations. Higher doses may produce feelings of driving, flying, and loss of physical strength. Other psychotic reactions, such as acute mania, paranoid states, toxic delirium, or panic, have been reported from long-term marijuana use. Some people are more vulnerable than others to these mental dysfunction. The presence of a toxic metabolite of cannabis in the brain interferes with cerebral functions, causing such acute brain syndrome. This type of toxic psychosis, characterized by symptoms ranging from restlessness and confusion to fear, illusions, and hallucinations, occurs from ingestion of large doses of the drug, and rarely from smoking. Such delirium ends when the metabolite disappears from the brain. While ingestion produces a longer effect of 5–10 hours, that from smoking may last for 3–4 hours.

Cannabis differs from opiates, amphetamines, and cocaine in many respects. First, its toxicity is very low. Also, because its action is sedative, its long-term use produces lethargy, apathy, and passiveness. By contrast, many other drugs of abuse, especially the stimulants, such as cocaine and amphetamines or the LSD-type hallucinogens, induce aggressiveness and wakefulness, respectively.

Although it produces some the same hallucinogen effects as LSD-type substances, it is a far less potent drug, lacking the powerful mind-altering qualities of LSD, psilocybin, and mescaline. Probably the most significant difference is observed in the extent of tolerance developed, additive potential, and withdrawal from the drug. Tolerance develops much more slowly with marijuana than with LSE, heroin, or cocaine. The withdrawal reaction after frequent use of high doses of marijuana has been found to be mild moderate.

The chemical structures of (Δ^8-THC) and (Δ^9-THC), the psychoactive components of cannabinoids are given below. One of the psychoactive metabolites is 11-hydroxy-THC (11-OH-THC) (the −OH group attached to the C-11), which is produced in high concentration in the liver after ingestion of cannabis and at a lesser concentration in the lung when smoked. The THC is highly soluble in fat and can accumulate in fat tissue. Another major metabolite, 9-carboxy-THC, is inactive.

(Δ^9-THC)
Δ^9-tetrahydrocannabinol

(Δ^8-THC)
Δ^8-tetrahydrocannabinol

VIII.G ALCOHOL, BARBITURATES, AND OTHER SEDATIVE HYNOTICS

Sedatives are substances that reduce anxiety and exert an effect of calmness. At therapeutic doses, their effects on the depression of the central nervous system, as well as the effect on the motor and mental functions, should be minimal. By contrast, a hypnotic drug produces drowsiness and a state of sleep. The CNS depression from hypnotic effects is greater than that from sedation. Sedation progresses into hypnosis with increased dosage. Thus, the three phases of CNS effects — sedation, hypnosis, and anesthesia — are caused by the increasing the dosage amount, respectively. A further increase in the dose may produce coma and, in many cases, death.

Many sedative-hypnotics have varying degrees of addiction potential. Alcohol (ethanol) is probably the oldest and most widely abused drug. It was the first psychoactive drug in the United States to come under regulation. While its interaction with other drugs are discussed below, its toxic effects are presented in Chapter 4 (Alcohols). It is no longer being therapeutically used as a sedative-hypnotic because of the danger of abuse. Among the alcohol class of substances, many chloroalcohols have found applications as sedatives. Ethchlorvynol [113-18-8], chloral hydrate [302-17-0], and trichloroethanol are three such sedative-hypnotics. While ethchlorvynol does not induce addiction with continued administration the same is not true for the other two sedative-hypnotics.

Acute and chronic administration of alcohol can inhibit the biotransformation or detoxification of many drugs, such as barbiturates, meprobamate, and amphetamines by liver enzymes. The effect can occur in two opposite ways. Alcohol and cannabinoids effects are additive. Both are CNS depressants. Animal studies indicate that simultaneous administration of alcohol and tetrahydrocannabinol (THC), the psychoactive component of marijuana, increased the tolerance and physical dependence to alcohol. Human studies show that alcohol and THC combination enhanced the impairment of physical and mental performance only, and there is no evidence of any interaction between both drugs. With barbiturates, synergism (i.e., a combined effect greater than the algebraic sum of their individual effects) has been observed. A combination of alcohol with phenobarbital or pentobarbital can lead to increased toxicity and death. The effect with methaqualone or meprobamate is similar. By contrast, benzodiazepines potentiate the effects of alcohol to a lesser degree. The effects with opiates are biphasic. While chronic use of large quantities of alcohol inhibits the metabolism of heroin and other opiates under high alcohol level in the body, it is accelerated when there is no alcohol in the body. Chronic alcohol consumption produces an enhancement of the liver-detoxifying enzyme system. There is antagonism between alcohol and a stimulant, such as cocaine or amphetamine. The abuse pattern primarily stems from the desire by alcoholic addicts to maintain a wakeful state.

Barbiturates are derivatives of barbituric acid. These substances were once used extensively in clinical practices as sedatives-hynotics. Because of their abuse and physical dependence, their therapeutic application has diminished considerably. The sedative effects of barbiturates are similar to those of alcohol. Higher dosage can cause hypnosis and anesthesia. An intake of barbiturates 10–15 times that of the hypnotic dose can be toxic. Coma and death can result from overdose.

Barbiturates can be classified into a few types: long-acting, intermediate-acting, and short-acting drugs. Such classification of a barbiturate is based pharmacologically on its elimination half-life. The half-lives of most drugs depend on the rate of their metabolic transformation. Among the long-acting barbiturates are phenobarbital (>12 hours), intermediate-range amobarbital (6–12 hours) and short-acting secobarbital and pentobarbital sodium (3–6 hours).

Chronic use of barbiturates for insomnia or anxiety can lead to drug dependence, which can go unnoticed for months or years. The symptoms are slurred speech and incoordination. Similar to alcohol, periodic use of barbiturates to produce a "high" or euphoria can lead to the development of serious dependence. The drug is taken orally in pill or capsules form or by the intravenous route. Because of the low cost, many heroin addicts use barbiturates to get the same "rush" of pleasant and drowsy feeling. Barbiturates are often administered along with other drugs; for example, to sedate the paranoia of cocaine or amphetamine, or to alleviate alcohol intoxication or boost the effect of heroin.

Mild acute and chronic intoxication from barbiturates is similar to that associated with alcohol, including symptoms of sluggishness, slurred speech, poor memory, and difficulty in thinking. Its effect with alcohol, another CNS depressant, is additive. Death can occur from intake of barbiturates with large amounts of alcohol.

Barbiturates are strong habit-forming substances. Tolerance may develop to their sedative effects. The withdrawal symptoms are somewhat different different for short-acting and long-acting drugs, with the former producing more severe effects. Withdrawal symptoms from short-acting drugs may have onset 8–10 hours after withdrawal, producing anxiety, twitches, tremor, nausea, and vomiting and continuing for the next several hours (about 16–20 hours). Convulsions, delirium, and hallucination may follow, for the next 48 hours, especially in severe cases of addiction. In the case of long-acting barbiturates, withdrawal reactions generally manifest late, a few days after withdrawal. The syndrome is mild, and generally there is no adverse sign for the first 2 or 3 days. Convulsions may appear late. For all barbiturates and, in fact, for most sedative-hypnotics, the severity of withdrawal depends on the particular drug as well as dosage range used immediately before withdrawal. For example, while secobarbital in doses of <400 mg/day produces mild symptoms of withdrawal on discontinuation, a high dosage with in the range of 1000 mg/day can produce severe syndrome that includes anxiety, anorexia, psychosis, and convulsion. Abrupt withdrawal from short-acting barbiturates at high dosage levels can cause seizure and can be life-threatening, which can become as severe as withdrawal from opiates.

Among the sedative-hypnotics of non-barbiturate types, meprobamate [57-53-4], glutethimide [77-21-4], and methaqualone [72-44-6] are compounds that can be strongly addictive. Such substances of abuse potential can lead to physical dependence on chronic use. Benzodiazepines are another class of sedative-hypnotics. These drugs are much safer than the barbiturates and are currently being used practically in all clinical applications. The CNS effect of benzodiazepines against barbiturates is shown in the dose-response curve depicted in Figure VIII.1. As seen from the curve, benzodiazepines show deviation from linearity in the hypnosis-anesthesia region of the CNS effect. By contrast, barbiturates exhibiting linearity in the full scale of the dose-response curve can cause coma and death at high over-doses.

Benzodiazepines produce little respiratory depression. Dosages of >2000 mg produce lethargy, drowsiness, confusion, and ataxia. However, the effect of CNS depression is additive when taken with ethanol, barbiturates, or other sedative-hypnotics. Also, the euphoric effects are lower than those experienced with most other sedative-hypnotics. Tolerance, risk of dependence, or addiction is relatively low. Tolerance can still

develop upon chronic use. Withdrawal reactions can occur for certain benzodiazepines, which also depends on the amounts and the period of drug use. Withdrawal symptoms include anxieties, dysphoria, sweating, nausea and twitching of muscles. Other symptoms include irritation and an intolerance for bright lights and loud noises. Convulsions are rare. Unlike most sedative drugs of abuse, withdrawal does not produce any craving for the drug. However, withdrawal symptoms may continue for several weeks, as the drug is eliminated from the body slowly.

The structures of a few selected sedative hypnotics are shown below. The toxicity and chemical structures of all other barbiturates are discussed at length in Chapter 57.

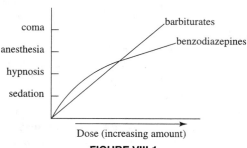

(pentobarbital)

(secobarbital)

(phenobarbital)

(chloral hydrate)
[302-17-0]

(trichloroethanol)

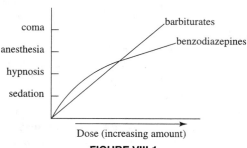

coma

anesthesia

hypnosis

sedation

barbiturates

benzodiazepines

Dose (increasing amount)

FIGURE VIII.1

(meprobamate)
[57-53-4]

(glutethimide)
[77-21-4]

(methaqualone)
[72-44-6]

VIII.H INHALANT GASES AND VAPORS

Many common gases and the vapors of organic solvents are known to produce euphoria, stimulant action, or mood alteration. Such substances differ in their chemical properties, structures, and bondings, as well as in their mode of actions. These include simple inorganic gases, such as nitrous oxide or xenon; organic solvents, such as toluene or hexane; and many anesthetics, such as ether or halothane. These compounds are classified in Table VIII.4. These inhalants are abused because of their low cost and easy availability.

Nitrous oxide (N_2O), also known as laughing gas, is one of the most common abused inhalants. It is an anesthetic gas. Inhalation can produce euphoria, dreaminess, tingling, numbness, as well as visual and auditory hallucination. Such effects arise form administering the gas diluted in oxygen, usually 30–35% mixed in oxygen. Inhalation of 100% N_2O can cause asphyxiation and death. Many volatile organic compounds that have anesthetic properties can produce exhilaration and euphoria. These include diethylether, chloroform, and halothane. High doses of these substances can produce unconsciousness.

Many common industrial solvents used in paint, glue, rubber, cement, shoe polish, degreasers, and several gasoline components, as well as a number of fluorocarbons (used as aerosol propellants) and alkyl nitrites, are known to produce a euphoric exhilarating or stimulant effect. Such compounds include volatile aliphatic and aromatic hydrocarbons, halogenated hydrocarbons, ethers, aldehydes, and ketones.

The psychological effects from most of these inhalants last a short period, 5–15

TABLE VIII.4 Inhalants of Abusive Potential

Chemical Types	Industrial Applications	Examples of Compounds
Alphatic hydrocarbons	Solvents	Hexane, pentane
Aromatic hydrocarbons	Solvents, gasoline components	Benzene, toluene
Cycloalkanes	—	Cyclopropane
Chlorinated hydrocarbons	Solvent, dry cleaning, anesthesia	Chloroform, tetrachloroethylene, halothane
Fluorocarbons	Aerosol propellants	Dichlorodifluoromethane
Ethers	Anesthesia solvent	Diethyl ether, n-amyl ether
Ketones	Solvent	Methyl ethyl ketone
Alkyl nitrites	Miscellananeous	Isobutyl nitrite locker room, rush, amyl nitrite (popper)
Inorganic gases	Anesthesia, miscellaneous	Nitrous oxide (laughing gas), xenon

minutes. Thus, the duration is much shorter and the effects milder than with the high-molecular-weight, natural alkaloids or synthetic hallucinogens having indole, tropane, phenanthrene, or morphinan-type structures. Repeat use can cause addiction or psychological dependence.

Among the inhalants that are gases at ambient conditions, cyclopropane, dichlorodifluoromethane, ethylene, xenon, and nitrous oxide have mdoerate narcotic potency to produce a certain degree of anesthesia. Some inhalants, especially many organic vapors, are toxic, producing a variety of adverse health effects, including kidney and liver damage, lung disease, bone marrow suppression, and brain impairment. Liberal use of these inhalants can be dangerous. Alkyl nitrites (amyl or isobutyl nitrites) are somewhat safte and less toxic. Inhalation can give a few minutes of "speeding" and a prolonged sexual stimulus. Other effects are dizziness, giddiness, lowered blood pressure, and rapid heart rate. However, although rare, very high doses can produce methemoglobinemia. Fluorocarbon vapors are another class of abused inhalants. These compounds can produce euphoria, disorientation, and hallucination. At high doses, ventricular arrhythmia and asphyxiation can occur. Asphyxiation can occur with all inhalants when inhaled in undiluted high quantities.

IX

FLAMMABLE AND COMBUSTIBLE PROPERTIES OF CHEMICAL SUBSTANCES

Flammable or combustible substances are those that catch fire or burn in air. Thus, the process of burning requires two types of materials: (1) a combustible substance, and (2) a supporter of combustion. A flammable substance at ambient temperature may be a solid such as white phosphorus, a liquid such as *n*-pentane or diethyl ether, or a gas such as hydrogen or acetylene. Air (oxygen) supports combustion. Many substances can burn in nitrous oxide, chlorine, and other atmospheres. We confine our discussion to substances that burn in air.

Most cases of fire hazard involve flammable liquids. A flammable liquid does not burn itself. It is the vapors from the liquid that burn. Thus, the flammability of a liquid depends on the degree to which the liquid forms flammable vapors: in other words, its vapor pressure. The molecules in the interior of the liquid are bound strongly by the molecular force of attraction, which is less for surface molecules. Evaporation occurs from the surface. The molecules that escape from the liquid make up vapor, increasing with increasing temperature. Flammability of a liquid is explained by the term *flash point*. The greater the flammability of a liquid, the lower is its

flash point. The flash point of a liquid is the lowest temperature at which the liquid releases vapor in a sufficient amount to form an ignitable mixture with air near its surface. It may be determined by standard tests. Many common chemicals have flash points below ambient temperature. Liquids that have flash points of $<100°F$ are generally termed flammable liquids, and those with flash points of $100-200°F$ are called combustible liquids. The flash point can be calculated theoretically from the boiling point and molecular structure of the compound. Equations based on additive group contributions, relative boiling points, and flame ionization properties may be applied to calculate flash points. Some of the common flammable liquids are carbon disulfide, diethyl ether, acetone, *n*-pentane, gasoline, acrolein, benzene, and methanol. In general, low-molecular-weight liquid hydrocarbons, aldehydes, ketones, ethers, glycol ethers, and many other classes of organic compounds with various functional groups are flammable (or combustible).

The flammability of a substance also depends on pressure, oxygen enrichment, and process conditions, such as mechanical agitation. Flammability, for example, increases

TABLE IX.1 Flash Point and Autoignition Temperatures of Common Flammable Liquids and Gases

Compound	Flash Point (Closed Cup) [°C (°F)]	Autoignition Temperature (°C)
Acetal	−21(−5)	230
Acetaldehyde	−39(−38)	175
Acetone	−20(−4)	465
Acetylene	Gas	305
Acrolein	−26(−15)	220
Acrylonitrile	0 (32)	481
Allylamine	−29(−20)	374
Allylbromide	−1 (30)	295
Allylchloride	−32(−25)	485
Ammonia	Gas	651
Benzene	−11 (12)	498
1,3-Butadiene	Gas	420
Butadiene monoxide	<−50 (<−58)	—
Butane	Gas	287
1-Butene	Gas	385
2-Butene	Gas	325
n-Butylamine	−12 (10)	312
sec-Butylamine	−9 (16)	—
Butyl chloride	−9 (15)	240
Butyl cyclopentane	—	250
Butyl cyclohexane	—	246
sec-Butyl mercaptan	−23(−10)	—
1,2-Butylene oxide	−22(−7)	439
Carbon disulfide	−30(−22)	94
Chloroprene	−20(−4)	—
Cyclohexane	−20(−4)	245
Cyclopentane	−29(−20)	395
Dichlorosilane	−37(−35)	36
Diethylamine	−23(−9)	312
Diethyl ether	−45(−49)	160
Dihydropyran	−18 (0)	—
2,2-Dimethylbutane	−48(−54)	405
Dimethyl ether	−2 (29)	202
1,1-Dimethylhydrazine	−15 (5)	249
Ethyl acetate	−4 (24)	426
Ethyl chloride	−50(−58)	519
Ethylene	Gas	450
Ethylene glycol monoethyl ether	39 (102)	285
Ethyl formate	−20(−4)	455
Ethyl mercaptan	−17(1)(oc)[a]	300
Ethyl nitrite	−35(−31)	90 (dec.)[b]
Fluorobenzene	−15 (5)	—
Gasoline	−43 to −38 (−45 to −36)	Varies with grade
n-Heptane	−4 (25)	204
n-Hexane	−22(−7)	223

TABLE IX.1 (*Continued*)

Compound	Flash Point (Closed Cup) [°C (°F)]	Autoignition Temperature (°C)
Hydrogen	Gas	500
Hydrogen cyanide	−18 (0)	538
Hydrogen sulfide	Gas	260
Iron pentacarbonyl	−15 (5)	—
Isoamyl alcohol	43 (109)	350
Isobutylamine	−9 (15)	378
Isobutyraldehyde	−18 (0)	196
Isoheptane	−18 (0)	—
Isooctane	−12 (10)	418
Isopentene	−57(−70) (oc)	420
Isoprene	−54(−65) (oc)	395
Isopropyl alcohol	12 (53)	399
Isopropylamine	−26 (−15) (oc)	402
Isopropyl chloride	−32 (−26)	593
Isopropyl ether	−28 (−18)	443
Kerosene	43 to 72 (100–162)	210
Methane	Gas	537
Methanol	11 (52)	385
Methyl acetate	−10 (14)	454
Methylal	−18 (0)	237
Methyl cyclohexane	−4 (25)	250
Methyl ethyl ether	−37 (−35)	190
Methyl ethyl ketone	−9 (16)	404
Methyl formate	−19(−2)	449
2-Methylfuran	−30(−22)	—
Methylhydrazine	−8 (17)	194
Methyl isobutyl ketone	18 (64)	448
Methyl isocyanate	−7 (19)	534
4-Methyl-1,3-pentadiene	−34(−30)	—
Methylpyrrolidine	−14 (7)	—
Mineral spirits	40 (104)	245
Octane	13 (56)	206
Phosphine	Gas	100
Propane	Gas	450
n-Propylamine	−37(−35)	318
Propylene	Gas	455
Propylene oxide	−37(−35)	449
n-Propyl ether	21*(70)	188
Pyridine	20 (68)	482
Pyrrole	39 (102)	—
Tetrahydrofuran	−14 (6)	321
Thiophene	−1 (30)	—
Toluene	4 (40)	480
Triethylamine	−6 (20) (oc)	249

(*continued*)

TABLE IX.1   *(Continued)*

Compound	Flash Point (Closed Cup) [°C (°F)]	Autoignition Temperature (°C)
Trimethylamine	Gas	190
Trimethylchlorosilane	−28(−18)	—
2,2,4-Trimethylpentane	−12 (10)	415
Valeraldehyde	12 (54)	222
Vinyl acetate	−8 (18)	402
Vinyl chloride	Gas	472
Vinyl ethyl ether	<−46 (<−50)	202
Vinyl isopropyl ether	−32(−26)	272
Vinyl methyl ether	Gas	287
m-Xylene	27 (81)	527
o-Xylene	32 (90)	463
p-Xylene	27 (81)	528

[a]oc, open cup.
[b]dec., decomposes.

under oxygen enrichment. Substances burn more readily and fiercely in an atmosphere that has a greater amount of oxygen (>21% by volume) than that in the air. Many petroleum products and highly flammable liquids may ignite spontaneously under such oxygen-rich environment. Also, high agitation of a flammable liquid can form a mist, generating sufficient vapor to form a flammable mixture with air.

If the vapor density (approximately the molecular weight divided by 29) of a flammable liquid at ambient temperature is greater than that of air, the liquid can pose another type of hazard — flashback fire — in the presence of a source of ignition. When the air is still, the vapors can settle or concentrate in lower areas and the trail of vapors can spread over a considerable distance. If there is a source of ignition nearby, the vapors on contact can result in a fire that can flash back to the source of the vapors.

Another flammable characteristic of compounds is autoignition (ignition) temperature. It is the minimum temperature required to cause or initiate self-sustained combustion independent of the source of heat. In other words, a substance will ignite spontaneously

when it reaches its autoignition temperature. For example, the surface of a hot plate can ignite diethyl ether, which has an autoignition temperature of 160°C. Such substances undergo oxidation at a high rate to initiate combustion. Semiempirical formulas relating to molecular structures have been derived to calculate autoignition temperature theoretically when experimental data are not available. Many substances can burn spontaneously in air and may cause a severe fire hazard if proper storage, handling, or disposal procedures are not followed. These include finely divided pyrophoric metals; alkali metals such as sodium or potassium; a large number of organometallics, including alkylaluminum and alkylzinc; organics mixed with strong oxidizers; silane; and white phosphorus. Some of the common compounds that have low autoignition temperatures (400°C) are carbon disulfide (80°C), diethyl ether (160°C), acetaldehyde (175°C), *n*-heptane (215°C), *n*-hexane (225°C), cyclohexane (245°C), *n*-pentane (260°C), ethanol (365°C), methanol (385°C), and isopropanol (399°C). Table IX.1 provides flash point and autoignition temperature data for some common flammable compounds. The

TABLE IX.2 Explosive Limits of Gases and Vapors in air

Compound	LEL (%)	UEL (%)
Acetaldehyde	3.97	57.00
Acetone	2.55	12.80
Acetylene	2.50	80.00
Allyl alcohol	2.50	18.00
Ammonia	15.50	27.00
Benzene	1.40	7.10
Butane	1.86	8.41
1-Butene	1.65	9.95
2-Butene	1.75	9.70
Carbon disulfide	1.25	50.00
Carbon monoxide	12.50	74.20
Carbonoxysulfide	11.90	28.50
Crotonaldehyde	2.12	15.50
Cyanogen	6.60	42.60
Cyclohexane	1.26	7.77
Cyclopropane	2.40	10.40
Diethyl ether	1.85	36.50
Dioxan	1.97	22.25
Divinyl ether	1.70	27.00
Ethane	3.00	12.50
Ethanol	3.28	18.95
Ethyl chloride	4.00	14.80
Ethyl formate	2.75	16.40
Ethyl nitrite	3.01	50.00
Ethylene	2.75	28.60
Ethylene oxide	3.00	80.00
Hydrogen	4.00	74.20
Hydrogen cyanide	5.60	40.00
Hydrogen sulfide	4.30	45.50
Isobutane	1.80	8.44
Methane	5.00	15.00
Methanol	6.72	36.50
Methyl amine	4.95	20.75
Methyl chloride	8.25	18.70
Methyl formate	5.05	22.70
Propane	2.12	9.35
n-Propanol	2.15	13.50
Propyl chloride	2.60	11.10
Propylene	2.00	11.10
Propylene oxide	2.00	22.00
Pyridine	1.81	12.40
Toluene	1.27	6.75
Vinyl chloride	4.00	21.70
o-Xylene	1.00	6.00

flammability data for these compounds, as well as those of other flammable and combustible substances, are presented throughout this book.

The flammability of a liquid (as vapor) or that of a gas falls between two fairly definite limits of concentrations of the vapor (or gas) in the air. These are the lower and upper explosive limits. These concentration limits in air are also known as the lower and upper flammable limits. The lower explosive limit (LEL) is the minimum concentration of the vapor (or gas) in air below which a flame is not propagated on contact with a source of ignition. There is also a maximum concentration of vapor or gas in air above which the flame does not propagate. Thus, a vapor (or gas)–air mixture below the lower explosive limit is too "lean" to burn, and the mixture above the upper explosive or flammable limit is too "rich" to burn. Within these lower and upper boundaries only, a gas or the vapor of a liquid will ignite or explode when it comes in contact with a source of ignition. It may be noted that the explosive limits vary with temperature and pressure. An increase in temperature or pressure lowers the lower limit and raises the upper limit, thus broadening the flammable or explosive concentration range of the vapor or the gas in the air. A decrease in temperature or pressure will have an opposite effect. Table IX.2 lists the explosive limits (flammable limits) of some common combustible substances in air, arranged in decreasing wideness range. The explosive limits of various substances are also presented later in the book.

EXPLOSIVE CHARACTERISTICS OF CHEMICAL SUBSTANCES

X.A THEORY OF EXPLOSION

An explosion occurs when an unstable compound undergoes a reaction that produces rapid and violent energy release. If this sudden release of energy is not dissipated rapidly, the reaction products that are predominantly gaseous or fumes may cause an exceedingly high pressure, resulting in violent rupture of containers. An explosive reaction can be spontaneous or initiated by light, heat, friction, impact, or a catalyst. Table X.1 lists examples of explosive reactions initiated by different types of initiators.

Explosions are not confined to closed systems. If the propagation rate of the gaseous products from the initiation site exceeds the velocity of sound, detonation may also occur in an open system. For example, 1 g of nitroglycerin is completely transformed into carbon dioxide, nitrogen, oxygen, and water in microseconds. Such high explosives undergo instantaneous reactions that release enormous energy which can rapidly heat the ambient temperature to 2000–3000°C. This results in an exceedingly large increase in pressure, producing shock waves that cause an explosion's shattering power. This is known as *brisance*.

The rate of detonation for high explosives can exceed 5000 meters per second (m/sec). The explosive characteristics thus depend on two factors: (1) how much energy is released, and (2) how fast the energy release occurs. The former is a thermodynamic property and can be calculated from the heats of formation. Compounds with high positive heats of formation, such as azides, fulminates, and acetylides of certain metals, are explosives (see part B, Chapter 11). An explosion can also result when the rate of an exothermic reaction is exceedingly fast. Many unsaturated organic compounds, such as acrolein or allyl alcohol, can readily polymerize. The reaction is exothermic. In the presence of a catalyst such as caustic soda, the reaction is too fast, resulting in an explosion. Compounds containing the functional groups azide, acetylide, diazo, nitroso, haloamine, peroxide, and ozonide are sensitive to shock and heat and can explode violently. Strong oxidizing substances, such as perchlorates, chlorates, permanganates, chromates, bromates, iodates, chlorites, and nitrates, can react violently with organics, acids, bases, and reducing agents. Such reactions can be explosive at elevated temperatures. Organic compounds containing nitro

TABLE X.1 Initiators of Explosive Reactions

Initiator	Examples
Light	Photochemical combination of hydrogen and chlorine to form HCl
Heat[a]	Violent explosions can occur when a primary explosive is heated (e.g., azides, acetylides, fulminates, nitrides, azo compound [diazomethane])
Shock[a]	Most primary explosives are sensitive to shock or impact (e.g., azides, fulminates, and acetylides of copper, silver, gold, and mercury; many organic peroxides and peroxy acids)
Catalyst	Base- or acid-catalyzed polymerizations (e.g., acrolein explodes with caustic potash)
Booster	All high explosives or detonators require a low explosive booster (e.g., dynamite, ammonium picrate)

[a]Normally, a chemical explosive of any type should explode when heated or subjected to mechanical impact. The terms *heat* and *shock* refer to mild to moderate heating and slight shock or jarring, respectively.

groups are highly reactive, especially in the presence of halogen substituents.

Nitro organics are a class of high explosives that are well known and widely used. This class of compounds is discussed in detail in Part B, Chapter 36. Organic peroxides and peroxy acids, which explode violently when subjected to heat or mechanical shock, are discussed in Chapters 3 and 39, respectively. Substances that are susceptible to the formation of peroxides are highlighted in Part A, Section VII.

Explosives are classified as high or low. *High explosives* undergo chemical decomposition at an exceedingly fast rate, producing intense shock waves that cause detonation. The *high explosives* often require an activating device such as a blasting cap to initiate their detonation. *Low explosives* deflagrate, bursting into flame and burning persistently. The detonation rate may be less than 250 m/sec.

Another common classification of chemical explosives is primary or secondary. *Primary explosives* are highly unstable and are sensitive to heat or shock. Secondary explosives are also unstable compounds, but their sensitivity to heat or shock is lower than that of primary explosives. A booster is used to bring about detonation. Many polynitro explosives fall under the latter class. Whether an explosive is primary or secondary, its shattering power and destructive capability can be severe.

X.B THERMODYNAMIC PROPERTIES

Substances that have positive heat of formation, ΔH_f°, are mostly unstable. The greater the value of positive ΔH_f°, the more unstable the compound. Such endothermic compounds may undergo exothermic decompositions (or reactions) releasing energy. If the energy released is sudden (i.e., the reaction is too fast), and adiabatic and its magnitude high, the reaction can proceed to explosive violence. Most compounds, however, have a negative ΔH_f°, i.e., exothermic formation reactions under standard conditions. Whether a reaction is exothermic or endothermic and how much energy would be released or absorbed can be determined from their heat of reactions. Such heat of reaction in standard state, ΔH_{rxn}°, can be calculated by different methods, including from the standard heats of formation, ΔH_f°. The ΔH_f° values can be found in most handbooks in chemistry. An example is presented below showing how to calculate the heat of reaction, ΔH_{rxn}°. It may be noted that ΔH_{rxn}° is determined at the standard states only, which are 1 atm for gases, and 1 M for liquids at 25°C. This should serve as a fairly close approximation to assess the amount of energy release in a violent reaction.

The standard heat of reaction ΔH_{rxn}° can be calculated from ΔH_f° (standard heat of formation) by applying Hess law. That is,

ΔH°_{rxn} is equal to the sum of ΔH°_{f} of the products minus the sum of the ΔH°_{f} of the reactant(s). That is,

$$\Delta H^{\circ}_{rxn} = \sum m \Delta H^{\circ}_{f} \text{ (products)}$$
$$- \sum n \Delta H^{\circ}_{f} \text{ (reactants)}$$

where the symbol $\sum$ means summation of, and m and n are the coefficients of the products and reactants in the balanced equation.

Problem X.1 Methyl nitrate, CH_3NO_3, is used as a rocket propellant. It explodes when heated to its boiling point, 64.5°C (at 1 atm). Determine the amount of energy release from the standard heat or reaction. First, we write a balance reaction for the oxidation of this compound.

$$4CH_3NO_3(l) + O_2(g) \longrightarrow 4CO_2(g)$$
$$\quad -38 \qquad\quad 0 \qquad\qquad -94.0$$
$$\qquad\qquad\qquad\qquad + 6H_2O(g) + 2N_2(g)$$
$$\qquad\qquad\qquad\qquad -57.8$$

The ΔH°_{f} for $CH_3NO_3(l)$, $CO_2(g)$ and $H_2O(g)$ are −38.0, −94.0 and −57.8 kcal/mol, respectively, while those of the elements O_2 and N_2 are 0. Thus,

$$\Delta H^{\circ}_{rxn} = \{[4 \times (-94.0) + 6$$
$$\times (-57.8)] \text{ kcal} - [2$$
$$\times (-38.0)] \text{ kcal}\} = -646.8 \text{ kcal}$$

This is a highly exothermic reaction releasing 646.8 kcal heat when 4 mol of methyl nitrate burn. This is equal to 161.7 kcal/mol or 2.1 kcal/g. Also, in addition to such heat release, the above reaction shows that 12 mol of gaseous product are formed, causing a sudden and enormous increase in pressure and, thus, an explosion.

X.C OXYGEN BALANCE

The explosive power or energy release of explosive chemicals that contain oxygen may be qualitatively assessed and predicted from their oxygen balance. It is the difference between the oxygen content of a compound and that required to oxidize the carbon, hydrogen, and other oxidizable elements in that compound to form carbon dioxide, water and other substances. Nitrogen is not considered an oxidizable element in the calculation, as it is liberated as N_2 rather than NO_x in explosive compositions. Oxygen balance can be calculated from the molecular formula of the compound and is expressed as a percentage. It can have positive, negative, or zero values. Deficiency of oxygen in the compound gives a negative balance. Surplus oxygen gives a positive balance and such substances usually behave as oxidants. On the other hand, a zero oxygen balance indicates maximum energy release. It may be noted that the concept of oxygen balance is highly empirical and there is a lack of adequate experimental data to establish any quantitative relationship. Although positive and zero oxygen balance manifest explosive characteristics, substances such as trinitrotoluene (TNT) showing negative oxygen balance (64%) are high explosives. Such substances, however, are stable at ambient conditions. Oxygen balance qualitatively measures the *unstability* of a compound. The concept should be applied in conjunction with a more rational and quantitative approach including endothermic heat of formation of the compound, its exothermic energy release in reaction, kinetics of the reaction, and the adiabatic process.

The oxygen balance may be calculated as follows:

$$\text{Oxygen balance} = \frac{(a - b)}{b} \times 100\%$$

where a is the number of oxygen atoms in the compound, and b is the number of oxygen atoms in the products formed (usually CO_2 and H_2O, but not always confined to only these two products).

Example 1 Determine the oxygen balance of trinitromethane, $(NO_2)_3CH$.

TABLE X.2   Structural Features of Selected Classes of Explosive Compounds

Classes of Substances	Structural Features	Examples
Acetylenic	$-C{\equiv}C-$	Acetylene
Acetylide	$-C{\equiv}C-$metal	Silver acetylide
Alkyl nitrate	$R-O-NO_2$	Ethyl nitrate
Alkyl nitrite	$R-O-N{=}O$	Methyl nitrite
Azide	$-N{=}N{=}N:$ or $-\overset{..}{N}-N{\equiv}N:$	Hydrazoic acid
Azo compound	$-\!\!:C-N{=}N-C-$	Azoformamide
Diazo compound	$\cdot C{=}N{=}N:$	Diazomethane
Difluoroamino compound	$-N\!\!<\!\!\genfrac{}{}{0pt}{}{F}{F}$	1,1-Difluorourea
1,2-Epoxide	$C\!\!-\!\!C$ with O	
Fulminate	$:C{=}N-O-$metal	Mercury(II) fulminate
Metal alkyl	R_n-metal	Trimethylaluminum
Nitroso compound	$C-N{=}O$ or $N-N{=}O$	1-Chloro-1-nitroso-cyclohexane
Nitrocompound (e.g., nitroalkane, nitroaromatic, nitrophenol)	$-NO_2$	2,4,6-Trinitrotoluene; nitroglycerin
Nitride	N^{3-}	Potassium nitride
Oxidizer (e.g., perchlorate, persulfate, nitrate)	ClO_4^-, ClO_3^-, $S_2O_8^{2-}$, NO_3^-, etc.	Ammonium perchlorate
Peroxide (e.g., peroxy acid, peroxy ester, hydroperoxide)	$-O-O-$	Benzoyl peroxide
Tetrazole	$-N{=}N-N{=}N-$	5-Aminotetrazole
Triazene	$C-N{=}N-N-C-$ $\mid$ H (OH, CN, etc.)	1,3-Dimethyltriazene

To determine the oxygen balance, we write a decomposition reaction of the compound, showing CO_2 and H_2O as being formed from the C and H atoms, respectively. In explosive reactions, N is mostly converted to N_2. Thus, we balance the equation for all atoms except O.

Thus, for trinitromethane, the explosive decomposition reaction would be

$$2CHN_3O_6 \longrightarrow 2CO_2 + H_2O + 3N_2$$
$$\quad 12 \qquad\qquad 4 \quad\; 1$$

The above reaction is balanced for all elements except oxygen. Now we count the number of O atoms for the reactant and the product. There are 12 O atoms on the left and 5 O atoms on the right. Thus, $a = 12$ and $b = 5$. Therefore, the oxygen balance

$$= \frac{(12 - 5)}{5} \times 100\% = +140\%$$

Example 2   Determine the oxygen balance of ethylene dinitrate, $C_2H_4N_2O_6$.

The balanced equation for the decomposition of ethylene dinitrate for oxidation of C and H is

$$C_2H_4N_2O_6 \longrightarrow 2CO_2 + 2H_2O + N_2$$
$$\quad\; 6 \qquad\qquad 4 \qquad 2$$

(C, H, and N are balanced)

$$a = 6, b = 4 + 2 = 6$$

Therefore, the oxygen balance

$$= \frac{(6 - 6)}{6} \times 100 = 0\%$$

The above compound is zero oxygen balanced.

X.D   STRUCTURAL FEATURE

The explosive characteristic of a chemical substance may be predicted from its structure, bonding or the functional group present in the molecule. For example, highly unsaturated acetylenic compounds or their metal derivatives of strained cyclic compounds such as 1,2-epoxides are highly unstable and tend to react with many compounds with explosive violence. There are many other classes of substances with a high degree of unsaturation or instability. Many such compounds (but not all) have a positive heat of formation. Some explosive compounds are very stable at ambient conditions, requiring a initator for explosion. Some substances are so unstable that they explode on a slight shock or warming. The distinct structural features in compounds that manifest myriads of violent explosive reactions under different physical and chemical conditions are outlined below. Only a few selected classes of substances are listed in Table X.2.

PEROXIDE-FORMING SUBSTANCES

Many organic and inorganic substances form peroxides when exposed to air. The peroxide-forming properties of organic compounds are governed by certain structural features of the compounds. In general, compounds that contain an α-hydrogen atom on an adjacent C atom or a C atom attached to a functional group or a double or triple bond are susceptible to forming peroxides. This is shown in the following examples.

(H) ◄— α-hydrogen atom

$$CH_3 \diagdown C-O-$$
$$CH_3 \diagup$$

(ethereal oxygen attached to isopropyl group)

$$CH_2{=}C \diagup^{Cl}_{(H)}$$

(vinyl compounds)

$$CH_3 \diagdown C{=}C-C{=}C \diagup CH_3$$
$$CH_3 \diagup \qquad \diagdown CH_3$$
(H) (H)

(dienes)

$$CH_3 \diagdown C-C{\equiv}CH$$
$$CH_3 \diagup$$
(H)

(alkylacetylene)

$$CH_3 \diagdown C-OH$$
$$CH_3 \diagup$$
(H)

(isopropanol)

$$CH_3-C{=}O$$
(H)

(acetaldehyde)

$$CH_3-\overset{\overset{O}{\|}}{C}-\overset{(H)}{C}-CH \diagup^{CH_3}_{CH_3}$$

(methyl isobutyl ketone)

In addition, hydrocarbons containing tertiary hydrogen atoms may form peroxides as follows:

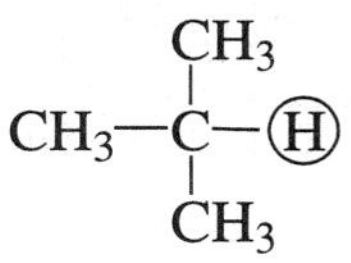
CH3
CH3—C—H
CH3

(isobutane)

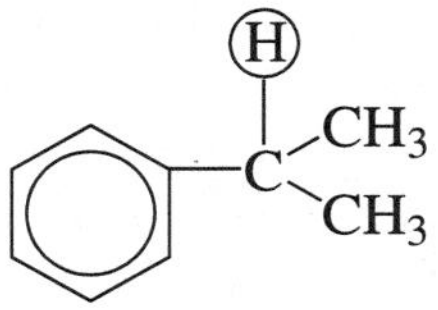
H
C
CH3
CH3

(cumene)

Thus, a large number of substances with the structural features shown above may form peroxides. However, these substances do not present a peroxide hazard to the same extent. The tendency of organic compounds to form peroxides decreases according to their structures as follows: ethers and acetals > olefins > halogenated olefins > vinyl compounds > dienes > alkynes > alkylbenzenes > isoparaffins > alkenyl esters > secondary alcohols > ketones > aldehydes > ureas and amides.

Presented below are some of the common compounds or common classes of compounds that can form peroxides on storage or on concentration (evaporation or distillation).

Ethers

Diethyl ether
Diisopropyl ether
Divinyl ether
Other aliphatic ethers
Tetrahydrofuran
1,3-Dioxane
1,4-Dioxane
1,3-Dioxolane
Tetrahydropyran
Other cyclic ethers
Ethylene glycol monomethyl ether
Ethylene glycol dimethyl ether
Other lower glycol ethers

Hydrocarbons

Butadiene
Isobutane
Cumene
Cyclohexene
Cyclopentadiene
Cyclopentene
Decalin
Diacetylene
Methylcyclopentane
Styrene
Tetralin

Vinyl compounds ($CH_2{=}CH{-}$)

Acrolein
Acrylamide
Acrylic acid
Methacrylic acid
Vinyl acetate
Vinyl acetylene
Vinyl bromide
Vinyl chloride
Vinyledene chloride
Vinyl propionate
Vinyl pyridine
Miscellaneous compounds containing vinyl group

Allylic structures ($CH_2{=}CH{-}CH$)

Allyl alcohol
Allyl amine
Allyl chloride
Allyl esters
Allyl sulfide
Miscellaneous compounds containing allylic structures

Organics with carbonyl functional group

All aldehydes (e.g., acetaldehyde, propionaldehyde.)

Some ketones (containing α-hydrogen atoms)

Inorganic compounds

Cesium metal

Metal alkoxides

Certain organometallics

Potassium amide

Potassium metal

Sodium amide

Rubidium metal

Removal of Peroxides from Solvents

Peroxides may be removed by treating the solvents with a reducing agent such as ferrous sulfate or sodium bisulfite. About 1 L of water-insoluble solvent is treated with 10 g of hydrated ferrous sulfate in 25 mL of 50% sulfuric acid in a separatory funnel and shaken vigorously for a few minutes. Other procedures for removing peroxides involve passing through a column of basic activated alumina or 4-Å molecular sieve pellets under nitrogen. Solvents containing high peroxides should be diluted with the pure solvents or dimethyl phthalate. Test papers are available to detect organic peroxides colorimetrically.

Qualitative detection may be done using sodium or potassium iodide in glacial acetic acid. The reagent in the hydroperoxide test is made by preparing a solution of 100 mg of iodide in 1 mL of glacial acetic acid. The reagent for alkyl peroxide tests may be prepared similarly, by mixing 3 g of iodide to 50 mL of glacial acetic acid and acidifying with 2 mL of 37% HCl. Low to high concentrations of peroxides in solvents produce yellow to brown colorations when 1 mL of solvent is mixed with 1 mL of the freshly prepared reagent.

Individual organic peroxides and peroxy acids are discussed in detail in Chapters 3 and 43.

PART B

1

ACIDS, CARBOXYLIC

1.1 GENERAL DISCUSSION

Carboxylic acids are weak organic acids that exhibit the following characteristics of acids: (1) undergo dissociation in aqueous solutions, forming a carboxylate ion $RCOO^-$ and H^+ (proton) or a hydronium ion, H_3O^+ (hydronium); (2) transfer protons to bases forming salts; and (3) the water-soluble acids have a sour taste. The carboxylic acids are characterized by carboxyl functional groups,

$$\overset{\overset{\displaystyle O}{\parallel}}{-C}-OH \qquad (or\ -COOH)$$

comprised of carbonyl and hydroxyl groups. The general formula for the homologous series of monocarboxylic acids is $C_nH_{2n}O_2$ or $R-COOH$, where R is an alkyl group. Dicarboxylic acids such as oxalic or succinic acid contain two $-COOH$ groups. The acid strength of carboxylic acids is much lower than those of mineral acids.

Carboxylic acids form a large number of derivatives that are very useful. The important general reactions include (1) the formation of esters when these compounds react with alcohols in the presence of an acid catalyst,

(2) reaction with alkalies to form the corresponding metal salts, (3) reduction to primary alcohols, and (4) decomposition to ketones on heating with a catalyst. Carboxylic acids react with sodium azide and sulfuric acid or with hydrazoic acid (Schmidt reaction) in an inert solvent such as chloroform to produce amines. Many aliphatic amines are strong irritants to the skin and respiratory tract (see Chapter 8). Reactions with inorganic acid chlorides may produce acyl halides ($RCOCl$), many of which are strongly corrosive, causing severe burns. With thionyl chloride, the reaction products include the toxic gases sulfur dioxide and hydrogen chloride.

$$RCOOH + SOCl \longrightarrow$$
$$RCOCl + SO_2 + HCl$$

Heat- and shock-sensitive peroxyacids (peracids) are formed when carboxylic acids react with hydrogen peroxide in the presence of methanesulfonic acid or sulfuric acid, or a strong acid cation-exchange resin:

$$RCOOH + HO-OH \xrightarrow{H_2SO_4}$$
$$\overset{\overset{\displaystyle O}{\parallel}}{RC}-O-OH + H_2O$$

The toxicity of monocarboxylic acids is moderate to low, and decreases with increase in carbon chain length. Some of the lower dicarboxylic acids exhibit moderate to high toxicity. The high-molecular-weight long-chain fatty acids are nontoxic compounds. Low-molecular-weight carboxylic acids are combustible but not flammable liquids. A fire or explosion hazard due to carboxylic acids is uncommon. However, there are cases of formic acid bottles exploding when opened after long storage. As mentioned earlier, among the most hazardous reactions is the formation of peroxy acids with hydrogen peroxide. This reaction occurs with a concentrated solution of hydrogen peroxide and in the presence of an acid catalyst. Reactions with strong oxidizing agents such as perchlorates, permanganates, chromic acid, nitric acid, and ozone may proceed to explosive violence.

Analysis

The carboxylic functional group, $-COOH$, can be identified by IR and NMR spectra. The characteristic IR absorption of saturated aliphatic acids produces strong bands at $1725-1700$ cm^{-1} and $1320-1211$ cm^{-1} due to $C-O$ stretching absorption, and a broad band of $O-H$ stretching absorption over the region $3500-2500$ cm^{-1} with various submaxima. The $C-H$ stretching absorptions occur at $2960-2850$ cm^{-1}. Individual carboxylic acids may be analyzed by a GC technique using a flame ionization detector or by GC/MS. The acid may be converted to a suitable derivative and the molecular ion may be identified as a further confirmatory test. NIOSH Method 1603 (NIOSH 1984, Suppl. 1989) describes the analysis of acetic acid in air. Between 20 and 300 L of air at a flow rate of $0.01-1.0$ L/min is passed over coconut shell charcoal. The analyte is desorbed with 1 mL of formic acid, allowed to stand for an hour, and injected into a GC equipped with an FID. Suitable GC columns for this analysis are 0.3% SP-1000 + 0.3% H$_3$PO$_4$ on Carbopack A, 0.3% Carbowax 20M + 0.1% H$_3$PO$_4$ on Carbopack C, and

Carbopack B 60/80 mesh + 3% Carbowax 20M + 0.5% H$_3$PO$_4$. This method should also be effective for the analysis of C$_3$- and C$_4$- carboxylic acids in air.

1.2 FORMIC ACID

EPA Designated Toxic Waste, RCRA Waste Number U123; DOT Label: Corrosive Material, UN 1779
Formula CH$_2$O$_2$; MW 46.03; CAS [64-18-6]
Structure:

$$H-\underset{\underset{O}{\|}}{C}-OH$$

the first member of the homologous series of carboxylic acids. The carboxyl group is attached to a hydrogen atom rather than to a carbon atom as in all other carboxylic acids. The acid strength of formic acid is about 10 times greater than that of acetic acid but less than that of strong mineral acids.

Synonyms: methanoic acid; formylic acid

Uses and Exposure Risk

Formic acid occurs in the stings of ants and bees. It is used in the manufacture of esters and salts, dyeing and finishing of textiles and papers, electroplating, treatment of leather, and coagulating rubber latex, and also as a reducing agent.

Physical Properties

Colorless liquid with a pungent, penetrating odor; boils at 100.5°C; freezes at 9.4°C; density 1.220 at 20°C; miscible in water, alcohol, and ether; pK_a 3.74 at 25°C.

Health Hazard

Formic acid is a low to moderately toxic but highly caustic compound. It is corrosive to the skin, and contact with pure liquid can cause burns on the skin and eyes.

It is more toxic than acetic acid. Formic acid is a metabolite of methanol responsible for the latter's toxicity. Thus, the acute acidosis of methanol is due to the in vivo formation of formic acid generated by the action of enzymes, alcohol dehydrogenase, and aldehyde dehydrogenase. Ingestion of formic acid can cause death.

Exposure to formic acid vapors may produce irritation of the eyes, skin, and mucous membranes, causing respiratory distress.

LD_{50} value, oral (mice): 700 mg/kg

LC_{50} value, inhalation (mice): 6200 mg/m^3/ 15 min

Exposure Limits

TLV-TWA 5 ppm (~9 mg/m^3) (ACGIH, MSHA, OSHA, and NIOSH); IDLH 100 ppm (180 mg/m^3) (NIOSH).

Fire and Explosion Hazard

Combustible liquid; flash point (open cup) 69°C (156°F), flash point of 90% solution 50°C (122°F); vapor pressure 23–33 torr at 20°C; autoignition temperature 601°C (1114°F) for anhydrous liquid and 456°C (813°F) for 90% solution. Formic acid vapors form explosive mixtures with air within the range 18–57% by volume in air.

When mixed with a concentrated solution of hydrogen peroxide, formic acid can form peroxyformic acid,

$$HC{-}OOH$$
$$\underset{O}{\overset{\|}{}}$$

which is highly sensitive to shock and heat and can explode. The reaction is catalyzed by a mineral acid catalyst. Its reaction with other strong oxidizers can be violent. Explosions arising from reaction of formic acid with furfuryl alcohol and thallium nitrate trihydrate in the presence of vanillin have been documented (NFPA 1986). Formic acid decomposes to carbon monoxide and water upon heating or when mixed with concentrated sulfuric acid. Robertson (1989) reported an explosion in which a stores clerk lost an eye when he lifted a 1-L bottle of 98–100% formic acid off the shelf. Slow decomposition to carbon monoxide and water on prolonged storage produced sufficient gas pressure, which probably ruptured the sealed glass container.

1.3 ACETIC ACID

DOT Label: Corrosive Material, UN 2789 (glacial, more than 80% acid by weight) and UN 2789 (between 25 and 80% acid by weight)

Formula $C_2H_4O_2$; MW 60.06; CAS [64-19-7]

Structure:

$$CH_3{-}\overset{\displaystyle}{\underset{\underset{O}{\|}}{C}}{-}OH$$

Synonyms: glacial acetic acid; ethanoic acid; methanecarboxylic acid; vinegar acid

Uses and Exposure Risk

Acetic acid occurs in vinegar. It is produced in the destructive distillation of wood. It finds extensive application in the chemical industry. It is used in the manufacture of cellulose acetate, acetate rayon, and various acetate and acetyl compounds; as a solvent for gums, oils, and resins; as a food preservative in printing and dyeing; and in organic synthesis.

Physical Properties

Colorless liquid with a pungent odor of vinegar; boils at 118°C; solidifies at 16.7°C; density of the liquid 1.049 at 25°C; miscible with water and most organic solvents, insoluble in carbon disulfide; weakly acid, pK_a 4.74, pH of 1.0 M, 0.1 M, and 0.01 M aqueous solutions: 2.4, 2.9, and 3.4, respectively.

Health Hazard

Glacial acetic acid is a highly corrosive liquid. Contact with the eyes can produce mild to moderate irritation in humans. Contact with the skin may produce burns. Ingestion of this acid may cause corrosion of the mouth and gastrointestinal tract. The acute toxic effects are vomiting, diarrhea, ulceration, or bleeding from intestines and circulatory collapse. Death may occur from a high dose (20–30 mL), and toxic effects in humans may be felt from ingestion of 0.1–0.2 mL. An oral LD_{50} value in rats is 3530 mg/kg (Smyth 1956).

Glacial acetic acid is toxic to humans and animals by inhalation and skin contact. In humans, exposure to 1000 ppm for a few minutes may cause eye and respiratory tract irritation. Rabbits died from 4-hour exposure to a concentration of 16,000 ppm in air.

Exposure Limits

TLV-TWA 10 ppm ($\sim$25 mg/m^3) (ACGIH, OSHA, and MSHA); TLV-STEL 15 ppm (37.5 mg/m^3) (ACGIH).

Fire and Explosion Hazard

Combustible liquid; flash point (closed cup) 39°C (103°F); vapor pressure 11 torr at 20°C; autoignition temperature 463°C (867°F) (NFPA 1986), 426°C (800°F) (Meyer 1989). The vapor of acetic acid forms explosive mixtures with air; the LEL and UEL values are 4% and 16% by volume of air, respectively. Fire-extinguishing agent: water spray, dry chemical, CO_2, or "alcohol" foam; use water to keep the fire-exposed containers cool and to flush and dilute the spill.

Acetic acid may react explosively with the fluorides of chlorine and bromine: chlorine trifluoride and bromine pentafluoride (Mellor 1946, Suppl. 1971). Explosions can result when acetic acid is mixed with strong oxidizing agents such as perchlorates, permanganates, chromium trioxide, nitric acid, ozone, and hydrogen peroxide and warmed. When warmed with ammonium nitrate, the mixture may ignite (NFPA 1986). Acetic acid may react violently with phosphorus isocyanate (Mellor 1946, Suppl. 1971). It may react violently and vigorously with potassium hydroxide and sodium hydroxide, respectively.

1.4 PROPIONIC ACID

DOT Label: Corrosive Material, UN 1848
 Formula $C_3H_6O_2$; MW 74.09; CAS [79-09-4]

Structure:

$$CH_3-CH_2-\underset{\underset{O}{\|}}{C}-OH$$

Synonyms: propanoic acid; ethylformic acid; ethanecarboxylic acid; carboxyethane

Uses and Exposure Risk

Propionic acid is used in the production of propionates used as mold inhibitors and preservatives for grains and wood chips, in the manufacture of fruit flavors and perfume bases, and as an esterifying agent.

Physical Properties

Colorless oily liquid with pungent odor; boils at 141°C; melts at −21°C; density 0.993 at 20°C; soluble in water and most organic acids.

Health Hazard

Propionic acid is a toxic and corrosive liquid. Contact with the eyes can result in eye injury. Skin contact may cause burns. Acute exposures to its vapors can cause eye redness, mild to moderate skin burns, and mild coughing (ACGIH 1986). Ingestion of high amounts of this acid may produce corrosion of the mouth and gastrointestinal tract in humans. Other symptoms include vomiting, diarrhea, ulceration, and convulsions. Oral LD_{50} value in rats is about 3500–4300 mg/kg. The LD_{50} value by skin absorption in rabbits is 500 mg/kg.

Exposure Limit

TLV-TWA 10 ppm ($\sim$30 mg/m^3) (ACGIH).

Fire and Explosion Hazard

Combustible liquid; flash point (closed cup) 54.5°C (130°F), (open cup) 58°C (136°F); autoignition temperature 465°C (870°F); vapor forms explosive mixtures in air within the range 2.9–12.1% by volume in air. Reactions with strong oxidizers can become violent, especially at elevated temperatures.

1.5 ACRYLIC ACID

EPA Designated Toxic Waste, RCRA Waste Number U008; DOT Label: Corrosive Material, UN 2218

Formula $C_3H_4O_2$; MW 72.07; CAS [79-10-7]

Structure:

$$CH_2{=}CH{-}\underset{\underset{O}{\|}}{C}{-}OH$$

an unsaturated monocarboxylic acid containing a vinyl group

Synonyms: glacial acrylic acid; propenoic acid; acroleic acid; ethylenecarboxylic acid; vinylformic acid

Uses and Exposure Risk

Acrylic acid is produced by oxidation of acrolein or hydrolysis of acrylonitrile. It is used in the manufacture of plastics; in paints, polishes, and adhesives; and as coatings for leather.

Physical Properties

Colorless liquid with an acrid odor; corrosive; boils at 141°C; solidifies at 14°C; polymerizes when exposed to air; density 1.052; miscible with water, alcohol, ether, and other organic solvents.

Health Hazard

Acrylic acid is a corrosive liquid that can cause skin burns. Spill into the eyes can damage vision. The vapors are an irritant to the eyes. The inhalation hazard is of low order. An exposure to 4000 ppm for 4 hours was lethal to rats. The oral LD_{50} values reported in the literature show wide variation. The dermal LD_{50} value in rabbits is 280 mg/kg.

Exposure Limit

TLV-TWA 10 ppm (30 mg/m^3) (ACGIH).

Fire and Explosion Hazard

Combustible liquid; flash point (closed cup) 54°C (130°F), (open cup) 68°C (155°F); vapor pressure 31 torr at 25°C; vapor density 2.5 (air = 1); autoignition temperature 360°C (774°F). Vapors of acrylic acid form explosive mixtures with air within the range 2.9–8.0% by volume in air. Fire-extinguishing agent: water spray, "alcohol" foam, dry chemical, or CO_2; use a water spray to flush and dilute the spill and to disperse the vapors.

Acrylic acid may readily polymerize at ambient temperature. Polymerization may be inhibited with 200 ppm of hydroquinone monomethyl ether (Aldrich 1990). In the presence of a catalyst or at an elevated temperature, the polymerization rate may accelerate, causing an explosion. The reactions of acrylic acid with amines, imines, and oleum are exothermic but not violent. Acrylic acid should be stored below its melting point with a trace quantity of polymerization inhibitor. Its reactions with strong oxidizing substances can be violent.

1.6 METHACRYLIC ACID

DOT Label: Corrosive Material, UN 2531

Formula $C_4H_6O_2$; MW 86.10; CAS [79-41-4]

Structure:

$$CH_2=\overset{\overset{\displaystyle CH_3}{|}}{C}-\overset{\overset{\displaystyle O}{||}}{C}-OH$$

Synonyms: 2-methylacrylic acid; 2-methyl propenoic acid

Uses and Exposure Risk

Methacrylic acid is used in the manufacture of methacrylate resins and plastics.

Physical Properties

Colorless liquid with an acrid and repulsive odor; boils at 163°C; solidifies at 16°C; density 1.015 at 20°C; soluble in water and most organic solvents; polymerizes readily.

Health Hazard

Methacrylic acid is a highly corrosive liquid. Contact with eyes can result in blindness. Skin contact may produce burns. No inhalation toxicity was observed in rats. Exposure to its vapors may produce skin and eye irritation, which can be mild to moderate. A dermal LD_{50} value in rabbits is 500 mg/kg.

Exposure Limit

TLV-TWA 20 ppm ($\sim$70 mg/m^3) (ACGIH).

Fire and Explosion Hazard

Combustible liquid; flash point (open cup) 76°C (170°F); vapor pressure <0.1 torr at 20°C. Fire-extinguishing agent: water spray, "alcohol" foam, dry chemical, or CO_2; use a water spray to dilute and flush the spill and to disperse the vapors.

Methacrylic acid polymerizes readily. The reaction is exothermic. The rate of reaction accelerates on heating, which may result in violent rupture of closed containers. The polymerization may be inhibited with a trace quantity of hydroquinone and hydroquinone monomethyl ether (Aldrich 1990). The acid may be stored safely below its melting point.

1.7 OXALIC ACID

Formula $C_2H_2O_4$; MW 90.04; CAS [144-62-7]

Structure:

$$HO-\overset{\overset{\displaystyle }{||}}{\underset{\underset{\displaystyle O}{}}{C}}-\overset{\overset{\displaystyle }{||}}{\underset{\underset{\displaystyle O}{}}{C}}-OH$$

a dicarboxylic acid

Synonyms: ethanedionic acid; ethanedioic acid

Uses and Exposure Risk

Oxalic acid occurs in the cell sap of *Oxalis* and *Rumex* species of plants as the potassium and calcium salt. It is the metabolic product of many molds (Merck 1989). There are a large number of applications of this compound, including indigo dyeing; calico printing; removal of paint, rust, and ink stains; metal polishing; bleaching leather; and manufacture of oxalates. It is also used as an analytical reagent and as a reducing agent in organic synthesis.

Physical Properties

White powder (anhydrous) or colorless crystals (dihydrate); odorless; hygroscopic; the anhydrous acid melts at 189.5°C (decomposes) and the dihydrate melts at 10.5°C; sublimes at 157°C; soluble in water, alcohol, and glycerol, insoluble in benzene, chloroform, and petroleum ether.

Health Hazard

Oxalic acid is a strong poison. The toxic symptoms from ingestion include vomiting, diarrhea, and severe gastrointestinal disorder, renal damage, shock, convulsions,

TABLE 1.1   Toxicity and Flammability of Miscellaneous Carboxylic Acids

Compound/Synonyms/ CAS No.	Formula/MW/ Structure	Toxicity	Flammability
n-Butyric acid (butanoic acid, 1-propanecarboxylic acid, propylformic acid, ethylacetic acid) [107-92-6]	$C_4H_8O_2$ 88.12 $CH_3-CH_2-CH_2-COOH$	Corrosive liquid; contact with eyes can damage vision; irritant to skin; low toxicity; ingestion of high doses may cause vomiting, diarrhea, gastrointestinal problems, ulcerations, and convulsions; LD_{50} oral (rats): 2940 mg/kg; the LD_{50} values show wide variations from species to species; DOT Label: Corrosive Material, UN 2820	Combustible liquid; flash point (closed cup) 72°C (161°F) (open cup) 76°C (170°F; autoignition temperature 443°C (830°F); vapors form explosive mixtures with air within the range 2–10% by volume in air; reactions with strong oxidizers can be violent
Valeric acid (n-pentanoic acid, butanecarboxylic acid, propylacetic acid) [109-52-4]	$C_5H_{10}O_2$ 102.15 $CH_3-CH_2-CH_2-CH_2-COOH$	Corrosive liquid; irritant to eyes and skin; low toxicity; toxic effects similar but less than those of propionic acid; LD_{50} oral (mice): 600 mg/kg; LC_{50} inhalation (mice): 4100 mg/m^3/2 hr; DOT Label: Corrosive Material, NA 1760	Combustible liquid; flash point (closed cup) 88°C (192°F), (open cup) 96°C (205°F); autoignition temperature 400°C (752°F); may react violently with strong oxidizing compounds
Malonic acid (propanedioic acid, ethanedicarboxylic acid, carboxyacetic acid, dicarboxyethane) [141-82-2]	$C_3H_4O_4$ 104.07 $HOOC-CH_2-COOH$	Skin and eye irritant; toxicity of this compound is very low; LD_{50} oral (mice): 4000 mg/kg	Noncombustible liquid
Succinic acid (ethanedicarboxylic acid) [110-15-6]	$C_4H_6O_4$ 118.10 $HOOC-CH_2-CH_2-COOH$	Irritant to eyes, application of about 1 mg in rabbit eyes caused severe irritation; no toxicity is reported	Noncombustible solid

(continued)

TABLE 1.1   *(Continued)*

Compound/Synonyms/ CAS No.	Formula/MW/ Structure	Toxicity	Flammability
Adipic acid (1,6-hexane-dioic acid,1,4-butane-dicarboxylic acid) [124-04-9]	$C_6H_{10}O_4$ 146.16 CH_2-CH_2-COOH | CH_2-CH_2-COOH	Mild eye irritant, irritation due to 20 mg over 24 hr was moderate in rabbit eyes; toxicity from oral intake was low in mice; LD_{50} oral (mice): 1900 mg/kg	Noncombustible solid; auto-ignition tempe-rature 420°C (788°F)
Benzoic acid (benzene-methanoic acid, benzene-formic acid, benzene-carboxylic acid, phenyl-carboxylic acid, carboxy-benzene) [65-85-0]	$C_7H_6O_2$ 122.13 ⬡—COOH	Mild irritant to skin and eye; low to very low toxicity in animals; toxic symptoms include somnolence, respi-ratory depression, and gastrointestinal disorder; LD_{50} oral (mice): 2000–2500 mg/kg; tested nega-tive in histidine reversion — Ames test for mutageni-city	Noncombustible solid; auto-ignition tempe-rature 570°C (1058°F)
Phthalic acid (1,2-benzene-dicarboxylic acid, *o*-dicarboxybenzene) [88-99-3]	$C_8H_6O_4$ 166.14 ⬡—COOH —COOH	Acute oral toxicity in rats was found to be very low, LD_{50} value is 7900 mg/kg; high intraperitoneal doses caused change in motor activity, muscle contraction, and cyanosis (NIOSH 1986), LD_{50} intraperitoneal (mice): 550 mg/kg	Noncombustible solid; nitration with a mixture of fuming nitric acid and sulfuric acid produces explosive products, which include phtha-loyl nitrate, nitrite, or their nitro derivatives (NFPA 1986)

and coma. Death may result from cardio-vascular collapse. The toxicity arises as oxalic acid reacts with calcium in the tissues to form calcium oxalate, thereby upsetting the calcium/potassium ratio (ACGIH 1986). Deposition of oxalates in the kidney tubules may result in kidney damage (Hodgson et al. 1988).

Oxalic acid may be absorbed into the body through skin contact. It is corrosive to the skin and eyes, producing burns. Dilute solu-tions of 10% strength may be a mild irritant to human skin. However, the inhalation toxi-city is low because of its low vapor pressure. Airborne dusts can produce eyeburn and irri-tation of the respiratory tract.

LD$_{50}$ value, oral (rats): 375 mg/kg

Exposure Limits

TLV-TWA for anhydrous acid 1 mg/m^3 (ACGIH, MSHA, and OSHA); TLV-STEL 2 mg/m^3 (ACGIH).

Hazardous Reaction Products

At high temperatures oxalic acid decomposes, producing toxic carbon monoxide and formic acid. Mixing with warm sulfuric acid may produce the same products: CO_2, CO, and formic acid. It reacts with many silver compounds, forming explosive silver oxalate (NFPA 1986). An explosion occurred when water was added to an oxalic acid/sodium chlorite mixture in a stainless steel beaker. There was also evolution of highly toxic chlorine dioxide gas (MCA 1962). Oxalic acid reacts violently with strong oxidizing substances.

1.8 MISCELLANEOUS CARBOXYLIC ACIDS

Among the monocarboxylic acids, compounds containing six carbon atoms or more are almost nontoxic, with mild or no irritant action. Similarly, the dicarboxylic acids become less toxic with an increase in alkyl chain length. However, the decrease in toxicity is very sharp immediately after oxalic acid. Aromatic acids are of low toxicity. One or two carboxyl groups attached to the benzene ring do not impart any significant toxic characteristics to the molecule.

The series of acids starting with valeric acid are noncombustible compounds. Simple aromatic acids such as benzoic and phthalic acids are noncombustible solids. As with most other classes of organic compounds, the reactions of carboxylic acids may become violent with strong oxidizing compounds. Table 1.1 presents toxicity and flammability data for a few carboxylic acids in the low-molecular-weight range.

REFERENCES

ACGIH. 1986. *Documentation of the Threshold Limit Values and Biological Exposure Indices*, 5th ed. Cincinnati, OH: American Conference of Governmental Industrial Hygienists.

Aldrich. 1990. *Aldrich Catalog*. Milwaukee, WI: Aldrich Chemical Company.

Hodgson, E., R. B. Mailman, and J. E. Chambers. 1988. *Dictionary of Toxicology*. New York: Van Nostrand Reinhold.

MCA. 1962. *Case Histories of Accidents in the Chemical Industry*. Washington, DC: Manufacturing Chemists' Association.

Mellor, J. W. 1971. *A Comprehensive Treatise on Inorganic and Theoretical Chemistry*. London: Longmans, Green & Co.

Merck. 1989. *The Merck Index*, 11th ed. Rahway, NJ: Merck & Co.

Meyer, E. 1989. *Chemistry of Hazardous Materials*, 2nd ed. Englewood Cliffs, NJ: Prentice-Hall.

NFPA. 1986. *Fire Protection Guide on Hazardous Materials*, 9th ed. Quincy, MA: National Fire Protection Association.

NIOSH. 1984. *Manual of Analytical Methods*, 3rd ed. Cincinnati, OH: National Institute for Occupational Safety and Health.

NIOSH. 1986. *Registry of Toxic Effects of Chemical Substances*, ed. D. V. Sweet. Washington, DC: U.S. Government Printing Office.

Robertson, A. V. 1989. Formic acid explosion. *Chem. Eng. News*, Nov. 13, *67*(13): 2.

Smyth, H. F. 1956. *Am. Ind. Hyg. Assoc. Q.* 17: 143.

2

ACIDS, MINERAL

2.1 GENERAL DISCUSSION

Acids are substances that dissociate into a proton (H^+), forming a hydronium ion, H_3O^+, in aqueous solutions. Thus, hydrochloric acid in water would dissociate into hydrogen and chloride ions:

$$HCl \rightleftharpoons H^+ + Cl^-$$

Water solvates the proton and the reaction may be written more specifically as follows:

$$HCl + H_2O = H_3O^+ + Cl^-$$

The acidity of an aqueous solution is expressed as

$$pH = -\log[H^+]$$

where $[H^+]$ is the concentration of hydrogen ion in the solution. Thus the lower the pH, the greater the acid strength of the solution. Mineral acids are stronger acids than carboxylic acids. The acid strengths of mineral acids also vary widely. For example, sulfuric acid is a strong acid, and phosphoric acid, is a weak acid.

The characteristic properties of acids are as follows: (1) they are sour in taste, (2) they react with a base to form salt and water, (3) they produce hydrogen when reacting with most common metals, and (4) they produce carbon dioxide when reacting with most carbonates.

Toxicity and Violent Reactions

Among the most important hazardous properties common to all mineral acids is that they are corrosive. Some are more corrosive than others. Hydrofluoric acid, for example, is extremely corrosive. It can attack glass; on human skin, contact with the acid can destroy tissues, causing severe burns. Their anhydrides or the fumes are a strong irritant to the eyes, skin, and mucous membranes. Ingestion can generally produce burning of the mouth and gastrointestinal tract.

Mineral acids are noncombustible substances. However, some are highly reactive to certain substances, causing fire and/or explosions. In this respect, mineral acids are not all similar. Few of the properties of mineral acids lead to violent reactions. These are reactions with nitrosophenol, exothermic acid–base reactions, generation of flammable hydrogen gas with most metals, and, in certain cases, formation of flammable phosphine resulting from reactions with phosphides. Often, explosion hazard arises because of the oxidizing and reducing nature of some mineral acids. Perchloric and nitric acids are strong oxidizing substances; hydriodic acid and hypophosphorous acids are examples of reducing agents. The violent reactions of these compounds are due primarily to their oxidizing or reducing strength and are independent of their acidity. Anhydrous perchloric acid, for example, is a severe explosion hazard. Contact with organic compounds, including common substances, such as wood, paper, and plastics, can result in violent explosions. Permanganic acid is also a strong oxidizer and can undergo many violent reactions. But it does not have any commercial use and is not discussed in this chapter.

Discussed in this chapter are the common mineral acids of commercial applications, which are high to moderately corrosive and undergo many violent reactions.

Cleanup, Disposal, and Destruction

Disposal of mineral acids in the laboratory may be done by dilution with water, followed by neutralization with a dilute alkali. The solution may be washed down the drain. Spent acid wastes may be contained in drums and disposed of in a waste disposal landfill. HCl and HF from waste incineration flue gas may be removed by dry scrubbing with $Ca(OH)_2$, CaO, $CaCO_3$, or $CaSO_4$ (Jons et al. 1987; Narisoko and Nanbu 1986). A removal rate of >95% has been reported for HCl.

Mandel and associates (1987) have described in a patent a method for cleanup of hazardous acidic spills. A composition containing alkaline-earth oxides, alkali-metal carbonates, highly absorptive silica or clay, less absorptive clay, hydrophobic lubricating agent, and portland cement was applied to the spill in a soft spray from a pressurized canister. Two gallons of sulfuric acid with an initial pH of 0.99 was treated with 26 lb

(11.7 kg) of composition, producing a very hard product with pH 8.85.

Analysis

The acid strength of a solution is determined from the pH measurement. Individual acid anions can be determined by wet analysis or by ion chromatography. Many anions can be analyzed by using ion-specific electrodes.

The analysis of mineral acids — sulfuric, nitric, hydrochloric, hydrobromic, hydrofluoric, and phosphoric acids — in air may be performed by NIOSH (1984) Method 7903. 3 and 100 L of air at a flow rate of 200–500 mL/min is passed through a solid sorbent tube containing washed silica gel and a glass fiber filter plug. The analytes are eluted with 10 mL of a sodium bicarbonate/sodium carbonate mixture (3.0 and 2.4 mmol strength, respectively) and analyzed by ion chromatography. Hydrofluoric acid (or hydrogen fluoride) can be analyzed by NIOSH (1984) Method 7902. A volume of 20–800 L of air at a flow rate of 1–2 L/min is passed over a 0.8-μm cellulose ester membrane and an Na_2CO_3-treated cellulose pad. HF is extracted with water and total ionic strength activity buffer (TISAB) solution and analyzed by a fluoride ion-specific electrode.

Dharmarajan and Brouwers (1987) described the use of an instrument, the atomizing trace gas monitoring system (TCG analyzer), equipped with ion-specific electrodes for measuring airborne HCl and HF. A lidar system, based on measuring backscattered light from a short pulse of laser radiation, has been reported for determining hydrogen chloride in incinerator ship plumes (Weitkamp 1985).

2.2   SULFURIC ACID

DOT Label: Corrosive Material, UN 1830, UN 1832 (for the concentrated and spent acids)

Formula H_2SO_4; MW 98.08; CAS [7664-93-9]

Structure:

$$HO-\overset{\overset{\displaystyle O}{\|}}{\underset{\underset{\displaystyle O}{\|}}{S}}-OH$$

a strong mineral acid, an aqueous solution of sulfur trioxide; both hydrogen atoms may dissociate as protons (hydronium ion, H_3O^+), producing HSO_4^- and SO_4^{2-} anions in dilute aqueous solutions

Synonyms: oil of vitriol; hydrogen sulfate

Uses and Exposure Risk

Sulfuric acid is the leading chemical in the world in terms of production and consumption. It is used in the production of phosphate fertilizers, dyes, explosives, glues, and a number of sulfates. It is also used in the purification of petroleum, cleaning of steel surfaces (metal pickling), and as a dehydrating agent. Commercially sold concentrated H_2SO_4 contains 98% acid, with the remaining water; normality 36.

Physical Properties

Colorless, odorless, viscous oily liquid; absorbs moisture from air; abstracts water from many organic substances; chars sugar and wood; bp 338°C; decomposes at 340°C to sulfur trioxide and water; anhydrous acid freezes at 10°C; 98% acid freezes at 3°C; density 1.84; infinitely soluble in water and alcohol, evolving heat.

Health Hazard

Concentrated sulfuric acid is a very corrosive liquid that can cause severe, deep burns to tissue. It can penetrate through skin and cause tissue necrosis. The effect may be similar to that of thermal burns. Contact with the eyes can cause permanent loss of vision.

Inhalation of its vapors or mist can produce severe bronchial constriction. Because the vapor pressure of sulfuric acid is negligible, <0.001 torr at 20°C, the inhalation hazard is low. However, the acid mists, having a particle size of <7 μm, may penetrate the upper respiratory tract and nasal passage. Human exposure to acid mist at a concentration of 5 mg/m³ in air produced coughing. At concentrations of <1 mg/m³ there was no irritation. Chronic exposure to sulfuric acid mist may produce bronchitis, conjunctivitis, skin lesions, and erosion of teeth. Frequent contact with dilute acid can cause dermatitis of skin.

LD_{50} value, oral (rats): 2140 mg/kg

LC_{50} value, inhalation (rats): 510 mg/m³/2 h

Exposure Limits

TLV-TWA air 1 mg/m³ (ACGIH, MSHA, and OSHA); TLV-STEL 3 mg/m³ (ACGIH).

Fire and Explosion Hazard

Noncombustible liquid, but ignites finely divided combustible substances when in contact. The acid is highly reactive. It reacts violently with water, alcohol, and many organic compounds. When diluting, the acid should be added cautiously to the water. Sulfuric acid reacts with explosive violence with a large number of chemicals. This includes metals, strong oxidizers, and many organic compounds. The reaction of sulfuric acid, even in dilute form with alkali metals, especially sodium and potassium, can be violent. The reaction with alkaline-earth metals or any other metal in powder form can also be violent. Hydrogen is liberated when sulfuric acid reacts with metals. Shock-sensitive explosives such as acetylides, azides, fulminates, and certain nitrides can become extremely hazardous in contact with sulfuric acid. It reacts with potassium permanganate to form an unstable compound, manganese heptoxide (Mn_2O_7), which detonates at 70°C. An explosion occurred at 0°C from adding the concentrated acid to an intimate mixture of potassium permanganate and potassium chloride (NFPA 1986). Ignition and/or explosion may result if concentrated acid is mixed with perchlorates. With chlorates, violent explosion occurs due to the formation of unstable chlorine dioxide, ClO_2, which is also toxic. With concentrated perchloric acid, the product is chlorine heptoxide, Cl_2O_7, resulting in a violent explosion. It catalyzes the reaction of acetic or formic acids with hydrogen peroxide, resulting in the formation of peroxy acids, which explode readily. There have been reports of explosions as well as considerable debate on the subject when concentrated sulfuric acid/hydrogen peroxide mixtures were used as cleaning agents for glassware (Bergman et al. 1990; Ericson 1990; Wnuk 1990; Matlow 1990). Such explosions have been attributed to the reaction between acetone and hydrogen peroxide, forming shock-sensitive organic peroxide, which may occur even in the absence of H_2SO_4 (Ericson 1990). Sulfuric acid may catalyze H_2O_2–organics reactions. By contrast, H_2O_2 may oxidize sulfuric acid, forming peroxydisulfuric acid, $H_2S_2O_8$, which on hydrolysis forms shock-sensitive Caro's acid, H_2SO_5 (Matlow 1990). A sulfuric acid/dichromate mixture may ignite acetone.

Violent exothermic reaction can occur when concentrated sulfuric acid is mixed with acrylonitrile, picrates, bromine pentafluoride (Mellor 1956), and chlorine trifluoride. Reactions with caustic alkalies, amines, alcohols, aldehydes, epoxides, vinyl and ally compounds, cellulose, and sugar are vigorously exothermic.

Hazardous Reaction Products

In addition to the explosive reactions of sulfuric acid discussed above and the flammable and explosive products it forms, sulfuric acid also undergoes numerous reactions in which highly toxic products are formed. Some of these products are as follows.

Sulfuric acid reacts with alkali chloride, such as sodium chloride, to form hydrogen chloride gas. It generates carbon monoxide when reacted with formic or oxalic acid or an alkali cyanide. Concentrated and dilute acids generate highly toxic hydrogen cyanide when mixed with a cyanide salt or complex. Reaction with thiocyanate yields the highly toxic and flammable gas carbonyl sulfide, COS. Reaction with sodium bromide or sodium iodide produces sulfur dioxide along with bromine or iodine vapors; and with hydrogen iodide, hydrogen sulfide is produced. Esterification with methanol may yield a highly toxic and blistering compound, dimethyl sulfate. However, this compound does not form simply by mixing H_2SO_4 and methanol but is produced on distillation. Hot and concentrated acid oxidizes copper, lead, and carbon, evolving sulfur dioxide.

2.3 OLEUM

DOT Label: Corrosive Material, UN 1831

Formula $H_2SO_4 \cdot SO_3$; MW 178.14; CAS [8014-95-7]

Composition: sulfuric acid containing free sulfur trioxide, SO_3 in concentrations within range of 10–80%

Synonyms: fuming sulfuric acid; pyrosulfuric acid; dithionic acid

Health Hazard

Oleum is an extremely corrosive substance. It emits suffocating fumes of sulfur trioxide that can cause severe lung damage. Skin contact can cause severe burns.

LC_{50} value, inhalation (rats): 347 ppm/L/h

Hazardous Reaction Products

See Section 2.2.

2.4 NITRIC ACID

DOT Label: Oxidizer and Corrosive, UN 2031

Formula HNO_3; MW 63.02; CAS [7697-37-2]

Composition: Commercial concentrated nitric acid contains about 68–70% nitric acid and 30–32% water, normality 15–16; white fuming nitric acid contains 97.5% nitric acid, less than 2% water, and less than 0.5% nitrogen oxides by weight; and red fuming nitric acid contains more than 86% acid, less than 5% water, 6–15% nitrogen oxides by weight.

Synonyms: aquafortes; acotic acid; hydrogen nitrate

Uses and Exposure Risk

Nitric acid is one of the most widely used industrial chemicals. It is employed in the production of fertilizers, explosives, dyes, synthetic fibers, and many inorganic and organic nitrates; and as a common laboratory reagent.

Physical Properties

Colorless liquid, yellow to brown, due to the presence of dissolved nitrogen dioxide, NO_2, formed as a result of slow photochemical decomposition of nitric acid catalyzed by sunlight; fumes in moist air; boils at 86°C (68% acid boils at 121.6°C); freezes at −42°C; density 1.413 (70% acid) and 1.513 (100% acid) at 20°C; infinitely miscible with water.

Health Hazard

Nitric acid is a corrosive substance causing yellow burns on the skin. It corrodes the body tissues by converting the complex proteins to a yellow substance called xanthoproteic acid (Meyer 1989). Ingestion of acid can produce burning and corrosion of the mouth and stomach. A dose of 5–10 mL can be fatal to humans.

Chronic exposure to the vapor and mist of nitric acid may produce bronchitis and chemical pneumonitis (Fairhall 1957). It emits NO_2, a highly toxic gas formed by its decomposition in the presence of light. Nitric acid is less corrosive than sulfuric acid. Its vapor and mist may erode teeth.

Exposure Limits

TLV-TWA 2 ppm ($\sim$5 mg/m^3) (ACGIH, MSHA, OSHA, and NIOSH); STEL 4 ppm ($\sim$10 mg/m^3) (ACGIH).

Fire and Explosion Hazard

Nitric acid is a noncombustible compound but is a strong oxidizer and highly reactive. It reacts explosively when its concentrated form is mixed with finely divided metals. The common metals that are not attacked by concentrated nitric acid are iron, cobalt, nickel, aluminum, and chromium. Hot concentrated nitric acid can ignite or explode when mixed with a large number of organic compounds. Common organic compounds that may explode with nitric acid are acetic acid, acetic anhydride, acetone (in the presence of sulfuric acid), ethanol, cyclohexanol, cyclohexanone, cyclobutanone, cyclopentanone, toluene (especially in the presence of sulfuric acid, water is removed, and the products, nitrocresols, may explode violently), nitrobenzene, mesitylene (when heated above 100°C), acetylene (due to the formation of trinitromethane, an explosive), indane (with sulfuric acid), triazine, thiophene, and many alkyl pyridines. An explosion occurred when a mixture of nitric acid, lactic acid, water, and hydrofluoric acid at a ratio of 5 : 5 : 2 : 1 was stored in a plastic bottle (NFPA 1986). Cellulose, fats, and glycerides may react explosively with nitric acid. Aromatic amines such as aniline, toluidine, and other alkyl anilines; furfuryl alcohol; and terpenes may ignite spontaneously with fuming nitric acid. Among the inorganic compounds that can react explosively with concentrated or fuming nitric acid are many interhalogen compounds, such as chlorine trifluoride or bromine pentafluoride; many cyanides; phosphorus trichloride; and hydrides such as phosphine, arsine, stibine, and tetraborane. Fluorine; ammonia; hydrazine; hydrazoic acid; diborane; hydrides of group VIB elements, such as hydrogen sulfide, hydrogen selenide, and hydrogen telluride; hydrogen iodide; hydrogen peroxide; phosphorus (burns with an intense white light); phosphorus halides; and thiocyanates (often explosively) ignite with nitric acid. A violent reaction occurs when pulverized carbon is mixed with nitric acid. Reactions of nitric acid should reasonably be anticipated to be vigorous to violent when it is mixed with compounds structurally similar to those mentioned above. Nitric acid undergoes exothermic hydrolysis when combined with a base such as caustic soda.

2.5   HYDROCHLORIC ACID

DOT Label: Corrosive Material, UN 1050

Formula HCl; MW 36.46; CAS [7647-01-0]

Composition: Hydrochloric acid is an aqueous solution of the gas hydrogen chloride; reagent-grade concentrated acid contains 38% HCl by mass (normality 11–12); a constant-boiling acid (an azeotrope with water) contains about 20% HCl.

Synonym: muriatic acid

Uses and Exposure Risk

Hydrochloric acid is one of the most widely used acids and a common laboratory reagent. It is used in the manufacture of chlorides, in the pickling and cleaning of metal products, as a processing agent for manufacturing various food products, as a cleaning agent, in organic synthesis, and for neutralizing alkalies.

Hydrogen chloride is a fire-effluent gas. Firefighters are frequently exposed to significant concentrations of HCl (Brandt-Rauf

et al. 1988). Large amounts of HCl are released from the oxidative thermal degradation of polyvinyl chloride (PVC)-derived fiberglass, cotton, and jute brattices in mines. At 250°C its concentration is found to be greater than 5 ppm (De Rosa and Litton 1986). The gas is absorbed by water droplets, entrapped in soot particles, causing risk of exposure of the acid to the eyes, throat, and lungs of mine workers. Stack emissions of HCl can result from burning plastic-rich wastes (e.g., hospital wastes) (Powell 1987). Emissions of 1.0–1.6 g HCl/kg waste have been reported (Allen et al. 1986).

Physical Properties

Pure acid is a colorless and fuming liquid with a pungent odor. Dissolved iron imparts slight yellow coloration (Merck 1989). Density 1.20 for 36–38% acid, 1.1 for 20% acid (at 15°C); constant-boiling 20% azeotrope boils at 108.6°C, freezes at −114°C; pH of $1\,N$ solution is 0.1, and $0.1\,N$ is 1.1.

Health Hazard

Concentrated hydrochloric acid is a corrosive substance that can cause severe burns. Spilling into the eyes can damage vision. Ingestion can produce corrosion of the mouth, gastrointestinal tract, and stomach, and diarrhea.

Hydrogen chloride is a toxic gas with a characteristic pungent odor. Inhalation can cause coughing, choking, and irritation of the mucous membranes. Exposure to concentrations at >5 ppm in air can be irritating and disagreeable to humans (Patty 1963; ACGIH 1986). A short exposure to 50 ppm may cause irritation of the throat. Workers exposed to hydrochloric acid were found to suffer from gastritis and chronic bronchitis (Fairhall 1957).

Rats exposed continuously to a hydrogen chloride atmosphere died after physical incapacitation (Crane et al. 1985). Hartzell and coworkers (1987) have studied the toxicological effects of smoke containing hydrogen chloride in fire gases. The lethality of PVC smoke was high but not entirely due to the hydrogen chloride produced. Postexposure death in rats was observed after pulmonary irritation caused by high concentration of HCl. Lethality in the presence of carbon monoxide may be additive. In another paper, Hartzell and associates (1988) reported that guinea pigs were three times as sensitive as rats to HCl exposure. HCl produced bronchoconstriction in animals and showed additive toxicity with CO at relatively high concentrations of the latter.

Exposure Limit

Ceiling limit 5 ppm ($\sim$7 mg/m^3).

Fire and Explosion Hazard

Hydrochloric acid or HCl gas is noncombustible. Contact of acid with alkali metals such as sodium or potassium can result in explosion. It reacts violently with alkaline-earth metals. Hydrogen is produced from the reaction of metals with HCl. Phosphine (flammable) is generated when a metal phosphide reacts with the acid. Its reaction with caustic alkalies is exothermic and can be vigorous to violent. Crowley and Block (1989) reported rupture of vessels resulting from the accidental introduction of HCl solution into hydrogen peroxide.

Hydrogen chloride gas can ignite carbides of metals, alkali metal silicides, and magnesium boride. Absorption of the gas over mercury sulfate above 125°C can become violent (Mellor 1946, Suppl. 1956).

Hazardous Reaction Products

Chlorine is generated when concentrated HCl reacts with oxidizing substances such as permanganates, chlorates, and dichromates:

$$2KMnO_4 + 16HCl \longrightarrow$$
$$2MnCl_2 + 2KCl + 8H_2O + 5Cl_2$$

Reactions with cyanides and sulfides generate hydrogen cyanide and hydrogen sulfide, respectively. Reaction with acetylene produces ethylene dichloride. Many chlorinated toxic compounds, such as dibenzodioxins and dibenzofurans, may be formed from HCl and phenol. However, such reactions may occur only at very high temperatures (550°C) during the incineration of wastes (Eklund et al. 1986) and not from inadvertent mixing of HCl and phenol.

2.6 HYDROFLUORIC ACID

EPA Designated Toxic Waste, RCRA Waste Number U134; DOT Label: Corrosive Material, UN 1052, UN 1790

Formula HF; MW 20.01; CAS [7664-39-3]

Composition: The acid is an aqueous solution of hydrogen fluoride gas, commercially available in varying strengths: 47, 53, and 79% concentrations.

Synonym: fluohydric acid

Uses and Exposure Risk

Hydrofluoric acid is used as a fluorinating agent, as a catalyst, and in uranium refining. It is also used for etching glass and for pickling stainless steel. Hydrogen fluoride gas is produced when an inorganic fluoride is distilled with concentrated sulfuric acid.

Physical Properties

Hydrogen fluoride is a colorless gas or a fuming liquid. It boils at 19.54°C and freezes at −83°C, and the liquid has a density 0.991 at 19.54°C and vapor pressure 76 torr. The constant-boiling liquid contains 35.55% HF by weight in water and boils at 120°C. HF is infinitely soluble in cold water, highly soluble in alcohol, and slightly soluble in ether. Anhydrous HF is strongly acidic; aqueous solution is a weak acid.

Health Hazard

Hydrofluoric acid and hydrogen fluoride gas are extremely corrosive to body tissues, causing severe burns. The acid can penetrate the skin and destroy the tissues beneath and even affect the bones. Contact with dilute acid can cause burns, which may be perceptible hours after the exposure. The healing is slow. Contact with the eyes can result in impairment of vision.

Prolonged exposure to 10–15 ppm concentrations of the gas may cause redness of skin and irritation of the nose and eyes in humans. Inhalation of high concentrations of HF may produce fluorosis and pulmonary edema. In animals, repeated exposure to HF gas within the range 20–25 ppm has produced injury to the lungs, liver, and kidneys.

LC_{50} value, inhalation (mice): 342 ppm/h

Exposure Limits

Ceiling limit 3 ppm ($\sim$2.5 mg/m^3) as F (ACGIH); TWA 3 ppm (MSHA and OSHA).

Fire and Explosion Hazard

Hydrogen fluoride and its aqueous solution, hydrofluoric acid, are noncombustible. An explosion occurred when hydrofluoric acid was mixed with nitric acid, lactic acid, and water (see Section 2.4). It may generate hydrogen with some metals. Reaction with alkali metals can be explosively violent.

Storage and Shipping

Hydrofluoric acid corrodes glass. It reacts with calcium silicate, a component of glass, forming silicon tetrafluoride:

$$CaSiO_3 + 6HF \longrightarrow CaF_2 + SiF_4 + 3H_2O$$

Among the substances it does not corrode are lead, platinum, wax, and polyethylene.

It is therefore stored in lead carboys or in waxed or polyethylene bottles. It is shipped in passivated metal drums or tank trucks or rail tank cars.

2.7 HYDROBROMIC ACID

DOT Label: Corrosive Material, UN 1048, UN 1788

Formula HBr; MW 80.92; CAS [10035-10-6]

Composition: The acid is a solution of hydrogen bromide gas in water, marketed as 40%, 48%, and 62% acid.

Uses and Exposure Risk

Hydrobromic acid is used in the manufacture of bromide, as an alkylation catalyst, and in organic synthesis.

Physical Properties

Hydrobromic acid or hydrogen bromide is a colorless and corrosive liquid or gas, respectively, with a sour taste and acrid odor. The gas liquefies at $-66.5°C$ and freezes at $-86°C$. The constant-boiling acid is an aqueous solution of 47.5% HBr, boiling at 126°C. Saturated aqueous solution contains 66% HBr at 25°C; the aqueous solutions are strongly acidic.

Health Hazard

Hydrobromic acid is a corrosive liquid. The gas is a strong irritant to the eyes, nose, and mucous membranes. In humans, exposure to 5 ppm for a few minutes can cause irritation of the nose. Irritation of the eyes and lungs may be felt at higher concentrations. The detectable odor threshold is 2 ppm.

Exposure Limits

Ceiling limit 3 ppm ($\sim$10 mg/m^3) (ACGIH); TLV-TWA 3 ppm ($\sim$10 mg/m^3) (MSHA and OSHA).

Fire and Explosion Hazard

Hydrogen bromide is a noncombustible gas. The acid or the gas in contact with common metals and in the presence of moisture produces hydrogen. Violent reactions occur with ammonia and ozone (Mellor 1946, Suppl. 1956).

2.8   HYDRIODIC ACID

DOT Label: Corrosive Material, UN 1787, UN 2197

Formula HI; MW 127.91; CAS [10034-85-2]

Composition: The acid is a solution of hydrogen iodide gas in water, available in various concentrations (57%, 47%, and 10%).

Uses and Exposure Risk

Hydriodic acid is used in the manufacture of iodides, as a reducing agent, and in disinfectants and pharmaceuticals.

Physical Properties

The acid is a colorless liquid, rapidly turning yellow or brown when exposed to light and air. The anhydrous hydriodic acid or hydrogen iodide is a colorless gas, fumes in moist air; decomposed by light; liquefies at $-35°C$; freezes at $-51°C$; extremely soluble in water, more so in cold water (900 g/100 mL at $0°C$), soluble in many organic solvents. Hydriodic acid is a strong acid (the pH of a 0.1 M solution is 1.0).

Health Hazard

Hydriodic acid is a corrosive liquid that can produce burns on contact with the skin. Contact of acid with the eyes can cause severe irritation. The gas, hydrogen iodide, is a strong irritant to the eyes, skin, and mucous membranes. No exposure limit has been set for this gas.

Fire and Explosion Hazard

Hydrogen iodide is a noncombustible gas. Reactions with sodium, potassium, and other alkali metals and with magnesium can be violent. Hydrogen iodide is a reducing agent. Therefore, its reactions with strong oxidizing agents can be vigorous to violent. It ignites when mixed with fuming nitric acid, molten potassium chlorate, or other strong oxidizing compounds. When mixed with ozone, explosion may occur.

2.9   PERCHLORIC ACID

DOT Label: Oxidizer and Corrosive Material, UN 1873 (50–72% acid strength) and UN 1802 (not >50% acid); concentrations >72% forbidden for transportation

Formula $HClO_4$; MW 100.46; CAS [7601-90-3]

Composition: Aqueous solution containing about 72% $HClO_4$ by weight is the concentrated perchloric acid.

Uses and Exposure Risk

Perchloric acid salts are used as explosives and in metal plating. They are also used as an oxidizer and as a reagent in chemical analysis. These salts are produced by distilling potassium chlorate with concentrated H_2SO_4 under reduced pressure.

Physical Properties

Anhydrous acid is a colorless and volatile liquid; very hygroscopic; density 1.768 at $22°C$; decomposes when distilled at atmospheric pressure; freezes at $-112°C$; highly soluble in water. Concentrated perchloric acid with 72% strength has density 1.705; boils at $203°C$ and freezes at $-18°C$.

Health Hazard

Concentrated perchloric acid is a highly corrosive substance that can produce burns on skin contact. It is also a severe irritant to the eyes and mucous membranes. The toxicity of this compound is moderate. The toxic symptoms from ingestion include excitement, decrease in body temperature, and distress in breathing. An oral LD_{50} value in dogs is reported as 400 mg/kg (NIOSH 1986).

Fire and Explosion Hazard

Perchloric acid is a noncombustible substance. Anhydrous acid presents a severe explosion hazard. It is unstable and decomposes explosively at ordinary temperatures or in contact with many organic compounds. It combines with water, with evolution of heat. Concentrated perchloric acid at 72% strength is stable at ambient temperature. However, it is a strong oxidizing substance. Therefore, its reactions with oxidizable compounds, especially at elevated temperatures, can become explosively violent.

Concentrated perchloric acid produces a fire or explosion, especially under hot conditions, when mixed with cellulose materials such as wood, paper, and cotton. When the acid spills absorb into wood or sawdust at room temperature, it is still dangerous, as the acid may burn or explode when it dries into anhydrous perchloric acid. The anhydrous acid explodes instantly when a drop is poured on paper or wood. Organic compounds that are oxidizable react violently with concentrated solutions of perchloric acid. This includes alcohols (as well as glycols and glycerols), ketones, aldehydes, ethers (including glycol ethers), and dialkyl sulfoxides. High temperature and/or the presence of dehydrating reagents can cause an explosion, even with dilute perchloric acid solutions. Explosions resulting from mixing perchloric acid and acetic acid in electrolytic polishing baths have been documented (NFPA 1986).

When hot, perchloric acid forms explosive salts with trivalent antimony compounds and with bismuth at 110°C. Contact with fluorine gas results in the formation of unstable fluorine perchlorate, which explodes. Explosion or ignition occurs when the concentrated acid is mixed with a reducing agent such as a metal hypophosphite, hydride, or iodide. Hydriodic acid, sodium iodide, or potassium iodide ignites with the acid. Hydrogen forms an explosive mixture with vapors of perchloric acid at 400°C or at 215°C in the presence of steel particles (NFPA 1986). Mixing concentrated perchloric acid with a dehydrating substance such as concentrated H_2SO_4, P_2O_5, anhydrous $CaCl_2$, or Na_2SO_4 may produce anhydrous acid, which can explode.

Disposal/Destruction

Perchloric acid is diluted with cold water, by slow addition, to a concentration below 5%, and then neutralized with dilute caustic soda solution. The solution is washed down the drain with an excess of water.

Storage, Spillage, and Shipping

Perchloric acid is stored in small glass bottles inside a heavy glass container padded with glass wool. It is stored well above its freezing point, which is −20°C, separated from organic compounds, reducing agents, noncombustible substances, and dehydrating agents. It is shipped in glass bottles or carboys. In case of spill, the area is flushed with plenty of water and treated with sodium bicarbonate. Contact with wood and plastic should be avoided. Any distillation must be performed in an explosion-proof hood.

2.10 PHOSPHORIC ACID

DOT Label: Corrosive Material, UN 1805
Formula H_3PO_4; MW 98.00; CAS [7664-38-2]
Structure:

$$
\begin{array}{c}
\text{OH} \\
| \\
\text{HO——P——OH} \\
\| \\
\text{O}
\end{array}
$$

Its aqueous solution of 85% strength is the acid that is commonly used.

Synonym: orthophosphoric acid

Uses and Exposure Risk

Phosphoric acid is one of the most widely used chemicals. Among its most important applications are its use in fertilizers and detergents, in pickling and rustproofing metals, as a catalyst, and as an analytical reagent.

Physical Properties

A colorless, odorless, and unstable solid (forming orthorhombic crystals), or a clear, viscous, syrupy liquid; mp 42.35°C; 88% acid crystallizes on prolonged cooling, forming hemihydrate, which melts at 29.3°C, loses water at 150°C; gradually changes to pyrophosphoric acid at 200°C and metaphosphoric acid on heating to 300°C; density 1.685 and 1.333 for 85% and 50% solutions, respectively, at 25°C; vapor pressure 0.03 torr at 20°C; very soluble in water and alcohol; pH 1.5 for a 0.1 N aqueous solution.

Health Hazard

Phosphoric acid is less corrosive and hazardous than is concentrated sulfuric or nitric acid. Its concentrated solutions are irritants to the skin and mucous membranes. The vapors (P_2O_5 fumes) can cause irritation to the throat and coughing but could be tolerated at <10 mg/m^3 (see also Section 52.4). The acute oral toxicity in rats is reported to be low, the LD$_{50}$ value being 1530 mg/kg (NIOSH 1986).

Exposure Limits

TLV-TWA 1 mg/m^3 (ACGIH, MSHA, and OSHA); TLV-STEL 3 mg/m^3 (ACGIH).

Fire and Explosion Hazard

Phosphoric acid is a noncombustible substance in both solid and liquid forms. The acid can react with metals to produce hydrogen. There is no report of any explosion reported in the published literature. Its reactions with strong caustics can be vigorous to violent (exothermic acid–base reaction).

REFERENCES

ACGIH. 1986. *Documentation of the Threshold Limit Values and Biological Exposure Indices*, 5th ed. Cincinnati, OH: American Conference of Governmental Industrial Hygienists.

Allen, R. J., G. R. Brenniman, and C. Darling. 1986. Incineration of hospital waste. *J. Air Pollution Control Assoc.* 36(7): 829–31.

Bergman, R., D. Dobbs, and T. Klaus. 1990. *Chem. Eng. News*, Apr. 23.

Brandt-Rauf, P. W., L. F. Fallon, Jr., T. Tarantini, C. Idema, and L. Andrews. 1988. Health hazards of fire fighters: exposure assessment. *Br. J. Ind. Med.* 45(9): 606–12.

Crane, C. R., D. C. Sanders, B. R. Endecott, and J. K. Abbott. 1985. Inhalation toxicology. 4. Times to incapacitation and death for rats exposed continuously to atmospheric hydrogen chloride gas. *Govt. Rep. Announce. Index (U.S.)* 85(23), Abstr. No. 552, 419; cited in *Chem. Abstr.* CA *104*(21): 181379q.

Crowley, C. J., and J. A. Block. 1989. Safety relief system design and performance in a hydrogen peroxide system. *International Symposium on Runaway Reactions*, pp. 395–424. New York: AIChE; cited in *Chem. Abstr.* CA *111*(4): 25668x.

De Rosa, M. I., and C. D. Litton. 1986. Oxidative thermal degradation of PVC-derived, fiberglass, cotton, and jute brattices, and other mine materials. A comparison of toxic gas and liquid concentrations and smoke-particle characterization. *U.S. Bureau of Mines Rep. 9058*; cited in *Chem. Abstr.* CA *106*(6): 37827h.

Dharmarajan, V., and H. J. Brouwers. 1987. Advances in continuous toxic gas analyzers for process and environmental applications. Proceedings of the *Air Pollution Control Association, 80th Annual Meeting*; cited in *Chem. Abstr.* CA *108*(16): 136891x.

Eklund, G., J. R. Pedersen, and B. Stoemberg. 1986. Phenol and hydrochloric acid at 550°C yield a large variety of chlorinated toxic compounds. *Nature 320*(6058): 155–56.

Ericson, C. V. 1990. Piranha solution explosion. *Chem. Eng. News*, Aug. 13, p. 2.

Fairhall, L. T. 1957. *Industrial Toxicology*, 2nd ed., p. 83. Baltimore: Williams & Wilkins.

Hartzell, G. E., A. F. Grand, and W. G. Switzer. 1987. Modelling of toxicological effects of fire gases. VI. Further studies on the toxicity of smoke-containing hydrogen chloride. *J. Fire Sci. 5*(6): 368–91.

Hartzell, G. E., A. F. Grand, and W. G. Switzer. 1988. Modelling of toxicological effects of fire gases. VII. Studies on evaluation of animal models in combustion toxicology. *J. Fire Sci. 6*(6): 411–31.

Jons, E. S., J. T. Moller, and K. Kragh Nielsen. 1987. Method for cleaning a hot flue gas stream from waste incineration. European Patent 23,019, July 29; cited in *Chem. Abstr. CA 107*(24): 222487f.

Mandel, F. S., J. A. Engman, W. R. Whiting, and J. A. Nicol. 1987. Novel compositions and method for control and clean-up of hazardous acidic spills. International Patent 8,706,758, Nov. 5; cited in *Chem. Abstr. CA 108*(12): 100697z.

Matlow, S. L. 1990. Mixtures of sulfuric acid and hydrogen peroxide. *Chem. Eng. News*, July 23, p. 2.

Mellor, J. W. 1946. *A Comprehensive Treatise on Inorganic and Theoretical Chemistry*. London: Longmans, Green & Co.

Merck. 1989. *The Merck Index*, 11th ed. Rahway, NJ: Merck & Co.

Meyer, E. 1989. *Chemistry of Hazardous Materials*, 2nd ed. Englewood Cliffs, NJ: Prentice Hall.

Narisoko, M., and T. Nanbu, 1986. Removing hydrogen chloride gas from waste gases from incineration of municipal wastes. European Patent 206,499, Dec. 30; cited in *Chem. Abstr. CA 106*(10): 72247r.

NFPA. 1986. *Fire Protection Guide on Hazardous Materials*, 9th ed. Quincy, MA: National Fire Protection Association.

NIOSH. 1984. *Manual of Analytical Methods*, 3rd ed. Cincinnati, OH: National Institute for Occupational Safety and Health.

NIOSH. 1986. *Registry of Toxic Effects of Chemical Substances*, ed. D. V. Sweet. Washington, DC: U.S. Government Printing Office.

Patty, F. A. 1963. *Industrial Hygiene and Toxicology*, Vol. 2, p. 851. New York: Interscience.

Powell, F. C. 1987. Air pollutant emissions from the incineration of hospital wastes: the Alberta experience. *J. Air Pollution Control Assoc. 37*(7): 836–39.

Weitkamp, C. 1985. Methods to determine hydrogen chloride in incineration ship plumes. In *Wastes Ocean*, ed. D. R. Kester, Vol. 5, pp. 91–114. New York: Wiley.

Wnuk, T. 1990. Cleaning glass funnels. *Chem. Eng. News*, June 25, p. 2.

3

ACIDS, PEROXY

3.1 GENERAL DISCUSSION

Peroxyacids are of two types: peroxycarboxylic acids, $R(CO_3H)_n$, and peroxysulfonic acids, RSO_2OOH, where R is an alkyl, aryl, cycloalkyl, or heterocyclic group and n is 1 or 2. Peroxycarboxylic acids are weaker acids than the corresponding carboxylic acids. The lower peroxyacids are volatile liquids, soluble in water. The higher acids with greater than a C_7-carbon chain are solids and insoluble in water. In the solid state these are dimeric bound by intermolecular hydrogen bonding; in the vapor and liquid states and in solutions, peroxycarboxylic acids exhibit monomeric structures (Sheppard and Mageli 1982). The $O-O$ bond in peroxy acids is weak and can cleave readily, which makes these compounds highly unstable. The low-molecular-weight peroxy acids lose oxygen readily at room temperature.

These unstable acids can decompose violently on heating and may react dangerously with organic matter and readily oxidizable compounds (see Chapter 39). Among organic peroxides, peroxyacids are the most powerful oxidizing compounds. Handling and storage of these compounds are discussed by Castrantas and Banerjee (1979), and Shanley (1972). The lower acids are also shock sensitive, but such sensitivity to shock is relatively much lower than that of some of the organic peroxides, such as acetyl peroxide or diisopropyl peroxydicarbonate.

The health hazard from peroxyacids is due primarily to their irritant actions. C_1-C_6 acids are irritants. Peroxyacetic acid can cause severe burns. The lower peroxyacids have exhibited low toxicity on laboratory test animals, but the higher acids are nontoxic and nonirritants.

Peroxyacids may be analyzed by an iodide reduction method. It is reduced by excess iodide ion and the liberated iodine is measured by thiosulfate titration. A GC pyrolytic method may be applicable to unstable peroxyacids to measure the products of decomposition. Other instrumental analytical methods may be applicable (see Chapter 43).

3.2 PEROXYFORMIC ACID

Formula CH_2O_3; MW 62.03; CAS [107-32-4]

Structure and functional group:

$$H-\overset{\overset{\text{O}}{\|}}{C}-O-O-H$$

simplest of the peroxyacids, peroxy linkage between a formyl group and hydrogen

Synonyms: performic acid; formyl hydroperoxide; permethanoic acid

Uses and Exposure Risk

Peroxyformic acid is used as an epoxidizing agent and in organic synthesis involving oxidation and hydroxylation reactions.

Physical Properties

The 90% solution of peroxyformic acid is colorless; unstable in concentrated form; soluble in water, alcohol, ether, benzene, and chloroform.

Health Hazard

Peroxyformic acid is nontoxic. It is a skin and eye irritant. Its irritant action is less severe than that of peroxyacetic acid. There are no reports on its tumorigenic properties.

Fire and Explosion Hazard

Peroxyformic acid is the only oxygen-balanced organic peroxide. Its active oxygen content (25.8%) is greater than that of other peroxy compounds. It is therefore expected to undergo extremely violent decomposition. However, in practice, this compound forms only in situ and the commercial formulations contain its aqueous or dilute solutions in organic solvents.

The solutions of peroxyformic acid are shock and heat sensitive and highly reactive and decompose violently when exposed to heat. Dilution with water increases the sensitivity to shock and heat. It is a strong oxidizer and may react violently with readily oxidizable substances. It may ignite and

explode when mixed with accelerators or flammable substances. Peroxyformic acid at 60% strength can react violently with formaldehyde, benzaldehyde, aniline, powdered aluminum, and lead dioxide (NFPA 1986). Water from a sprinkler system may be used from an explosion-resistant location to fight fire and keep the containers cool.

Spillage and Disposal/Destruction

The spilled solution of peroxyformic acid should be absorbed by vermiculite or other noncombustible absorbent and disposed of immediately. Do not use paper, wood, or spark-generating metals for sweeping and handling. It is destroyed by burning on the ground in a remote place using a long torch. Concentrated aqueous solutions in a small quantity may be diluted with copious quantities of water and flushed down the drain.

3.3 PEROXYACETIC ACID

DOT Label: Organic Peroxide, UN 2131, for <43% solution or <6% in H_2O_2; above these strengths it is DOT Forbidden

Formula $C_2H_4O_3$; MW 76.05; CAS [79-21-0]

Structure and functional group:

$$H_3C-\overset{\overset{\text{O}}{\|}}{C}-O-O-H$$

peroxide functional group is bound to acetyl group and hydrogen

Synonyms: peracetic acid; acetyl hydroperoxide; perethanoic acid

Uses and Exposure Risk

Peroxyacetic acid is used as an epoxidizing agent, for bleaching, as a germicide and fungicide, and in the synthesis of pharmaceuticals.

Physical Properties

Colorless liquid with an acrid odor; bp 105°C (40% solution in acetic acid); density at 20°C 1.150 (40% solution); readily soluble in water, alcohol, ether, and sulfuric acid.

Health Hazard

Peroxyacetic acid is a severe irritant to the skin and eyes. It can cause severe acid burns. Irritation from 1 mg was severe on rabbits' eyes. Its toxicity is low. The toxicological routes of entry to the body are inhalation, ingestion, and skin contact. The toxicity data are as follows (NIOSH 1986):

LC_{50} inhalation (rats): 450 mg/m^3

LD_{50} oral (mice): 210 mg/kg

LD_{50} oral (guinea pigs): 10 mg/kg

Its toxicity in humans should be very low, and a health hazard may arise only from its severe irritant action. Studies on mice showed that it caused skin tumors at the site of application. Its carcinogenicity on humans is not reported. No exposure limit is set for peroxyacetic acid in air.

Fire and Explosion Hazard

A very powerful oxidizing compound; highly reactive; flammable; flash point (open cup) 40.5°C (105°F); explodes when heated to 110°C (230°F); decomposes violently when exposed to heat or flame; pure compound extremely shock sensitive. A 40% solution in acetic acid is also shock sensitive. The shock sensitivity of peroxyacetic acid may increase in combination with certain organic solvents (NFPA 1986). It may explode during distillation, even under reduced pressure.

Peroxyacetic acid may react dangerously with readily oxidizable substances, causing an explosion. Ignition and/or explosion may occur when mixed with accelerators or flammable compounds. It explodes when mixed with acetic anhydride, due to the formation of highly shock-sensitive acetyl peroxide. It undergoes highly exothermic reaction with olefins.

Fire-extinguishing procedure: fight fire from a safe and explosion-resistant location. Use water from a sprinkler system or fog nozzle to extinguish the fire and dilute the spill and to keep the containers cool.

Spillage and Disposal/Destruction

Absorb the spilled material with a noncombustible absorbent such as vermiculite and place it in a metal container for immediate disposal. Do not use paper, wool, or other organic materials or spark-generating tools for sweeping and handling. Peroxyacetic acid should be disposed of in small amounts by placing it on the ground in a remote area and burning with a long torch. The containers should be washed with dilute caustic soda solution of 5–10% strength.

Storage and Shipping

Peroxyacetic acid is stored in a cool, well-ventilated place, isolated from other chemicals. It is shipped in glass or earthenware containers not exceeding 1-gallon capacity and packed inside wooden or fiberboard boxes.

3.4   PEROXYBENZOIC ACID

Formula $C_7H_6O_3$; MW 138.12; CAS [93-59-4]

Structure and functional group:

peroxy linkage between benzoyl group and hydrogen

Synonyms: perbenzoic acid; benzoyl hydroperoxide

Uses and Exposure Risk

Peroxybenzoic acid is used in organic analyses to measure the degree of unsaturation, and in making epoxides.

Physical Properties

Volatile solid with a pungent odor; sublimes; mp 41–43°C; bp 100–105°C at 15 torr (partially decomposes); slightly soluble in water but mixes readily with most organic solvents.

Health Hazard

The toxicity of peroxybenzoic acid is very low on animals. In humans it is almost nontoxic. Peroxybenzoic acid caused skin tumor in mice on prolonged contact (NIOSH 1986). Its tumorigenic action was lower than that of peroxyacetic acid. Its carcinogenicity in humans is unknown.

Fire and Explosion Hazard

The presence of the peroxide functional group renders its strong oxidizing properties similar to those of other peroxy compounds. However, this compound is not hazardous like peroxyacetic or peroxyformic acids. It is not shock sensitive. Violent decomposition may occur only at high temperatures and/or in conjunction with certain organic contaminants that are easily oxidizable. Although there is no report of its explosion, general safety measures for handling peroxy compounds should be followed.

3.5 PEROXYMONOSULFURIC ACID

Formula H_2SO_5; MW 114.09; CAS [7722-86-3]

Structure and functional group:

$$HO-\underset{\underset{O}{\parallel}}{\overset{\overset{O}{\parallel}}{S}}-O-O-H$$

peroxy linkage

Synonyms: permonosulfuric acid; persulfuric acid; sulfomonoperacid; Caro's acid

Uses and Exposure Risk

Peroxymonosulfuric acid is used as an oxidizing agent to make glycols, lactones, and esters; for making dyes; and in bleaching composition.

Physical Properties

Colorless syrupy liquid (containing sulfuric acid); corrosive; decomposes at 45°C, decomposes in cold and hot water; soluble in H_3PO_4.

Health Hazard

Peroxymonosulfuric acid is a strong irritant to the skin, eyes, and mucous membranes (Merck 1989). Toxicity data for this compound are not available.

Fire and Explosion Hazard

Peroxymonosulfuric acid is highly unstable, decomposes dangerously on heating, and evolves oxygen at room temperature. It may react violently with organic matter and readily oxidizable compounds. Violent explosions have been reported with acetone, due to the formation of acetone peroxide (Toennis 1937). It may explode when mixed with many primary and secondary alcohols, manganese dioxide, cotton, many metals in finely divided form, and aromatics such as benzene, phenol, and aniline.

3.6 MISCELLANEOUS PEROXYACIDS

See Table 3.1.

TABLE 3.1 Toxicity and Flammability of Miscellaneous Peroxyacids

Compound/ Synonyms/CAS No.	Formula/MW/Structure	Toxicity	Flammability
Peroxypropionic acid (propionyl hydroperoxide, peroxypropanoic acid) [4212-43-5]	$C_3H_6O_3$ 90.08 $H_3C-CH_2-\overset{\displaystyle O}{\overset{\|}{C}}-O-O-H$	Irritant	Can explode on heating; may decompose violently by metal ions, free radicals, and readily oxidizable compounds; may explode during distillation
Peroxybutyric acid (peroxybutanoic acid) [13122-71-9]	$C_4H_8O_3$ 104.10 $H_3C-CH_2-CH_2-\overset{\displaystyle O}{\overset{\|}{C}}-O-O-H$	Irritant	Can explode on heating; may decompose violently by metal ions, free radicals, and readily oxidizable compounds; may explode during distillation
m-Chloroperbenzoic acid (3-chloroperoxy benzoic acid, m-chlorobenzoyl hydroperoxide) [937-14-4]	$C_7H_5ClO_3$ 172.57	Causes skin tumors in mice at the site of application; carcinogenicity in humans not reported	Explodes on heating, DOT forbidden for shipping when concentration exceeds 86%

Name	Formula / MW	Structure	Toxicity	Stability / Hazard
Peroxyhexanoic acid	$C_6H_{12}O_3$ 132.16 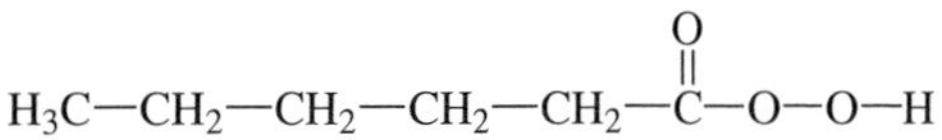	$H_3C-CH_2-CH_2-CH_2-CH_2-\overset{\displaystyle O}{\overset{\|}{C}}-O-O-H$	Nontoxic; mild irritant	May ignite and/or explode on rapid heating, stable at ambient temperature
3-Carboxyperpropionic acid [3504-13-0]	$C_4H_6O_5$ 146.10 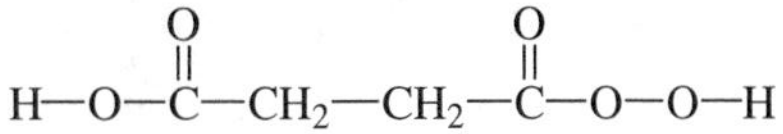	$H-O-\overset{\displaystyle O}{\overset{\|}{C}}-CH_2-CH_2-\overset{\displaystyle O}{\overset{\|}{C}}-O-O-H$	Toxicity not reported	Decomposes at 107°C; may decompose violently with readily oxidizable compound
Peroxytrichloroacetic acid	$C_2HO_3Cl_3$ 179.38 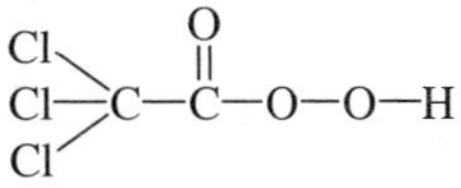	$\begin{array}{c} Cl \\ Cl-\overset{\displaystyle O}{\overset{\|}{C}}-\overset{\displaystyle O}{\overset{\|}{C}}-O-O-H \\ Cl \end{array}$	Decomposition produces dangerously toxic gases, phosgene, chlorine, HCl, and carbon monoxide	Very unstable, decomposes
Peroxycamphoric acid (percamphoric acid)			Nontoxic	Explodes when heated rapidly to 80–100°C
Peroxytrifluoroacetic acid	$C_2HF_3O_3$ 127.1 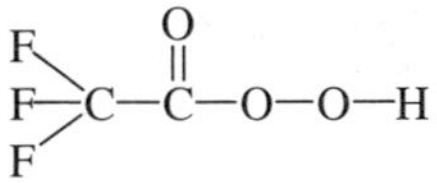	$\begin{array}{c} F \\ F-\overset{\displaystyle O}{\overset{\|}{C}}-\overset{\displaystyle O}{\overset{\|}{C}}-O-O-H \\ F \end{array}$	Nontoxic	Very powerful oxidizing agent; may react violently with organics and readily oxidizable compounds
Peroxyfuroic acid	$C_5H_4O_4$ 129.13 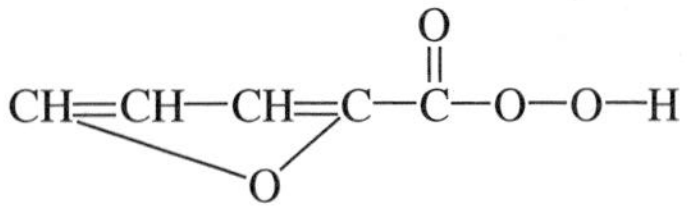	$CH=CH-CH=C-\overset{\displaystyle O}{\overset{\|}{C}}-O-O-H$ (furan ring, O)	Nontoxic	Decomposes violently at 40°C; may explode on intimate contact with finely divided metal halides

REFERENCES

Castrantas, H. M., and D. K. Banerjee. 1979. *Laboratory Handling and Storage of Peroxy Compounds*. ASTM Spec. Tech. Publ. 471. Philadelphia: American Society for Testing and Materials.

Merck. 1989. *The Merck Index*, 11th ed. Rahway, NJ: Merck & Co.

NFPA. 1986. *Fire Protection Guide on Hazardous Materials*, 9th ed. Quincy, MA: National Fire Protection Association.

NIOSH. 1986. *Registry of Toxic Effects of Chemical Substances*, ed. D. V. Sweet. Washington, DC: National Institute for Occupational Safety and Health.

Shanley, E. S. 1972. Organic peroxides: evaluation and management of hazards. In *Organic Peroxides*, ed. D. Swern, Vol. 3, pp. 341–64. New York: Wiley.

Sheppard, C. S., and O. L. Mageli. 1982. Peroxides and peroxy compounds, organic. In *Kirk-Othmer Encyclopedia of Chemical Technology*, 3rd ed. Vol. 17, pp. 27–90. New York: Wiley.

Toennis, W. 1937. *J. Am. Chem. Soc.* 59: 552–57.

4

ALCOHOLS

4.1 GENERAL DISCUSSION

Alcohols constitute a class of organic compounds containing one or more hydroxyl groups (−OH) attached to an aliphatic carbon atoms. Alcohols differ from phenols, in which the hydroxyl groups are bound to aromatic ring(s). Alcohols containing two hydroxyl groups are called diols or glycols. Higher aliphatic alcohols containing eight or more C atoms are also known as fatty alcohols.

The most characteristic reactions of alcohols are their oxidation to aldehydes and ketones, which may undergo further oxidation, producing carboxylic acids. Whereas primary alcohols oxidize to aldehyde, secondary alcohols produce ketones:

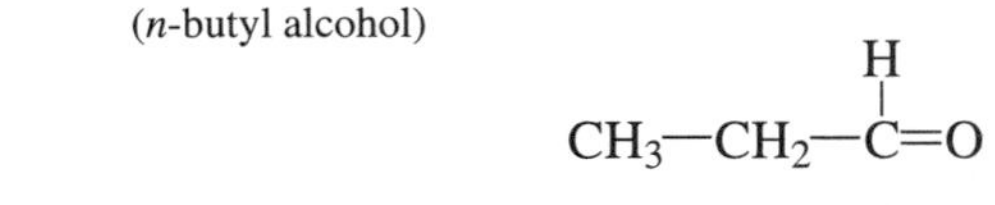

$$CH_3\!-\!CH_2\!-\!CH_2\!-\!OH \xrightarrow{\ O_2\ }$$

(*n*-butyl alcohol)

$$CH_3\!-\!CH_2\!-\!\overset{\displaystyle H}{\underset{|}{C}}\!=\!O$$

(*n*-butyraldehyde)

$$CH_3\!-\!CH_2\!-\!\overset{\displaystyle CH}{\underset{|}{CH}}\!-\!OH \xrightarrow{\ O_2\ }$$

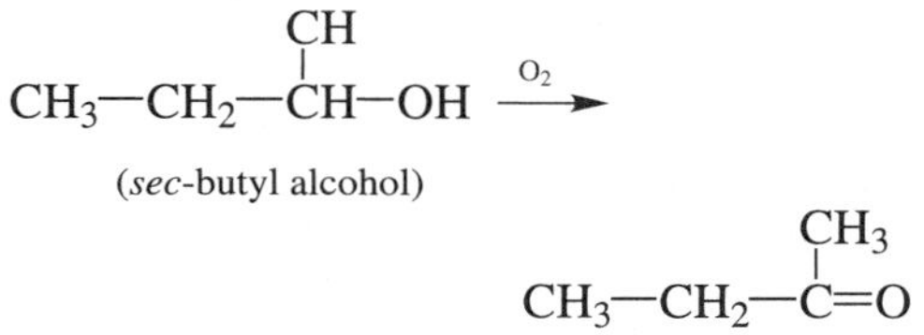

(*sec*-butyl alcohol)

$$CH_3\!-\!CH_2\!-\!\overset{\displaystyle CH_3}{\underset{|}{C}}\!=\!O$$

(methyl ethyl ketone)

Esterification is another characteristic reaction of an alcohol resulting from reaction with carboxylic acid.

Dehydration over catalytic surfaces produces ethers. Substitution of alkyl hydrogen atoms may occur in alcohols without affecting the hydroxyl functional group, resulting in formation of products such as chlorohydrins and cyanohydrins.

Alcohols are used widely in industry. The lower alcohols are excellent solvents. Most alcohols are used to synthesize a large number of substances of commercial interest.

The lower aliphatic alcohols are low to moderately toxic. Because of low vapor pressures, the acute inhalation toxicity of most alcohols is low. The vapors may be an irritant to the eyes and mucous membranes. Ingestion and absorption of the liquids through the skin can be a major health hazard. Methanol is a highly toxic substance, ingestion of which can cause blindness, acidosis, and death. It metabolizes to formic acid, which is excreted slowly. Thus, small doses may have a cumulative effect. Ethanol is relatively less toxic among the lower aliphatic alcohols. It is a central nervous system depressant, causing narcosis, and at high dosages may lead to death. Chronic intake may cause cirrhosis or liver damage. Other low aliphatic alcohols exhibit toxic actions similar to those of ethanol. Lower alcohols containing double or triple bonds exhibit a greater degree of toxicity and irritation. Fatty alcohols are almost nontoxic.

Lower alcohols are flammable or combustible liquids. The flammability decreases with increase in the carbon number. The explosive reactions of alcohols with different substances under ambient temperature are few and are listed under individual alcohols in the following sections.

4.2 METHANOL

Formula CH_3OH; MW 32.04; CAS [67-56-1]
Functional group: primary —OH, first member of the series of aliphatic alcohols
Synonyms: methyl alcohol; wood spirits; wood alcohol; carbinol

Uses and Exposure Risk

Methanol is used in the production of formaldehyde, acetic acid, methyl *tert*-butyl ether, and many chemical intermediates; as an octane improver (in oxinol); and as a possible alternative to diesel fuel; being an excellent polar solvent, it is widely used as a common laboratory chemical and as a methylating reagent.

Physical Properties

Colorless liquid with a mild odor; bp 65.15°C; mp −93.9°C; density 0.7914; miscible in water, alcohol, ether, acetone, benzene, and chloroform, immiscible with carbon disulfide.

Health Hazard

Ingestion of adulterated alcoholic beverages containing methanol has resulted in innumerable loss of human lives throughout the world. It is highly toxic, causing acidosis and blindness. The symptoms of poisoning are nausea, abdominal pain, headache, blurred vision, shortness of breath, and dizziness. In the body, methanol oxidizes to formaldehyde and formic acid — the latter could be detected in the urine, the pH of which is lowered (when poisoning is severe).

The toxicity of methanol is attributed to the metabolic products above. Ingestion in large amounts affects the brain, lungs, gastrointestinal tract, eyes, and respiratory system and can cause coma, blindness, and death. The lethal dose is reported to be 60–250 mL. The poisoning effect is prolonged and the recovery is slow, often causing permanent loss of sight.

Other exposure routes are inhalation and skin absorption. Exposure to methanol vapor to at ≤2000 ppm at regular intervals over a period of 4 weeks caused upper respiratory tract irritation and mucoid nasal discharge in rats. Such discharge was found to be a dose-related effect.

Inhalation in humans may produce headache, drowsiness, and eye irritation. Prolonged skin contact may cause dermatitis and scaling. Eye contact can cause burns and can damage vision.

First Aid/Antidote

On skin or eye contact, wash immediately with plenty of water. On inhalation of a large amount of vapor, move the person to fresh air. If breathing stops, perform artificial respiration. Hemodialysis may be the best treatment for severe poisoning. Ethanol may be partially effective as an antidote. It competes with methanol and inhibits the latter's oxidation in the body. Administration should be done carefully.

The treatment to combat severe acidosis involves oral administration of sodium bicarbonate and magnesium sulfate and of sodium bicarbonate intravenously. Gastric lavage should be performed using a 5% bicarbonate solution. Solid bicarbonate (2–6 g) may be swallowed every 2 h (Polson and Tattersall 1959). The urine should be made alkaline as rapidly as possible.

Exposure Limits

TLV-TWA (200 ppm) (ACGIH), 260 mg/m^3, 1040 mg/m^3 (800 ppm) 15 min (NIOSH); STEL 310 mg/m^3 (250 ppm); IDLH 25,000 ppm (NIOSH).

Fire and Explosion Hazard

Methanol is a highly flammable liquid; flash point (closed cup) 12°C (52°F); autoignition temperature 470°F. Methanol–air mixture is explosive; LEL and UEL values are 6% and 36% by volume in air, respectively. Methanol may catch fire or explode when in contact with flame and/or strong oxidizing substances. The flammability of a methanol–air mixture depends on temperature, pressure, and the presence of diluents. Carbon dioxide, nitrogen, argon, and water vapor prevent methanol from flaming. A 0.62 mol fraction of nitrogen added to the air at 300°C and 1.52 bar is reported to prevent flaming (Bercic and Levec 1987).

Hazardous Reaction Products

Methanol oxidizes to formaldehyde (toxic), reacts with ethylene oxide to form 2-methoxyethanol (macrocytosic), and reacts with formic acid to form methyl formate (flammable; nasal, conjunctival, and pulmonary irritant). It reacts with HBr or NaBr/H_2SO_4, forming methyl bromide (pulmonary edema, narcotic). It forms methylchlorocarbonate (methyl chloroformate), CH_3COOCl (a strong irritant), with phosgene. Esterification with salicylic acid gives methyl salicylate (toxic — nausea, acidosis, pneumonia, and convulsion). With H_2S in the presence of a catalyst, it produces methyl mercaptan, CH_3SH (toxic — nauseating and narcotic), and forms methyl acetate (respiratory tract irritant) with acetic acid. On catalytic vapor-phase oxidation in the presence of a small amount of HCl, it forms methylal (dimethoxymethane), $CH_3OCH_2OCH_3$ (toxic to the liver and kidney; flammable). Heating methanol with ammonium chloride and zinc chloride at 300°C produces methyl amine (irritant). Heating with aniline chloride under pressure may form methyl aniline (highly toxic). With ammonia it forms dimethyl amine (irritant). With aniline and H_2SO_4 under pressure, it forms dimethyl aniline (toxic); with phthalic anhydride it forms dimethyl phthalate (central nervous system depressant and mucous membrane irritant). Distillation with oleum may produce dimethyl sulfate (blistering and necrotic to skin).

Storage and Handling

Methanol should be stored in clean carbon-steel drums. Because of its flammability, tanks should be protected by a foam-type or dry chemical fire-extinguishing system. Loading and unloading should be performed under inert gas pressure in the absence of air.

Analysis

Methanol is analyzed by GC-FID. In blood it is determined by an enzymic method based on its conversion to formaldehyde by alcohol oxidase and the oxidation of formaldehyde to formic acid by the enzyme formaldehyde dehydrogenase. The presence of a toxic concentration of ethanol may give a high positive value. Methanol in air is estimated by passing 1–4 L of air over silica gel absorbent, desorbing the analyte into water, and injecting the latter into GC-FID (NIOSH 1984, Method 2000). Tenax, SP-1000, SP-2000, or 10% FFAP (polar polyethyleneglycol ester) are suitable columns. The working range is 25–900 mg/m^3 (19–690 ppm) in air. High humidity decreases the adsorption capacity of silica gel. Methanol may be analyzed by GC/MS. A primary ion of 47 was detected using an ion-trap detector. Belgsir et al. (1987) described a method for kinetic analysis of oxidation of methanol, using liquid chromatography.

4.3 ETHANOL

EPA Priority Pollutant in Solid/Hazardous Wastes; DOT Label: Flammable Liquid, UN 1170

Formula C_2H_5OH; MW 46; CAS [64-17-5]

Structure and functional group: CH_3-CH_2-OH, primary $-OH$

Synonyms: ethyl alcohol; ethyl hydroxide; grain alcohol; ethyl hydrate; rectified spirits; algrain; absolute alcohol; ethyl carbinol

Uses and Exposure Risk

Ethanol is used primarily as a solvent—an important industrial solvent for resins, lacquers, pharmaceuticals, toilet preparations, and cleaning agents; in the production of raw materials for cosmetics, perfumes, drugs, and plasticizers; as an antifreeze; as an automotive fuel additive; and from ancient times, in making beverages. Its pathway to the body system is mainly through the consumption of beverages. It is formed by the natural fermentation of corn, sugarcane, and other crops.

Physical Properties

Colorless, clear, volatile liquid; alcoholic smell and taste (when diluted); bp 78.3°C; mp −114°C; vapor pressure 43 torr at 20°C; density 0.789 at 20°C; soluble in water, acetone, ether, chloroform, and benzene; polar, dielectric constant 25.7 (at 20°C).

Health Hazard

The toxicity of ethanol is much lower than that of methanol or propanol. However, the literature on the subject is vastly greater than that of any other alcohol. This is attributable essentially to its use in alcoholic beverages. There are exhaustive reviews on alcohol toxicity and free-radical mechanisms (Nordmann et al. 1987). The health hazard arises primarily from ingestion rather than inhalation. Ingestion of a large dose, 250–500 mL, can be fatal. It affects the central nervous system. Symptoms are excitation, intoxication, stupor, hypoglycemia, and coma—the latter occurring at a blood alcohol content of 300–400 mg/L. It is reported to have a toxic effect on the thyroid gland (Hegedus et al. 1988) and to have an acute hypotensive action, reducing the systolic blood pressure in humans (Eisenhofer et al. 1987). Chronic consumption can cause cirrhosis of the liver.

Inhalation of alcohol vapors can result in irritation of the eyes and mucous membranes. This may happen at a high concentration of 5000–10,000 ppm. Exposure may result in stupor, fatigue, and sleepiness. There is no report of cirrhosis occurring from inhalation. Chronic exposure to ethanol vapors has produced brain damage in mice. The neurotoxicity increases with thiamine deficiency (Phillips 1987). Both acute and chronic doses of ethanol elevated the lipid peroxidation in rat brain. This was found to be elevated further by vitamin E deficiency, as well as its supplementation (Nadiger et al. 1988).

The toxicity of ethanol is enhanced in the presence of compounds such as barbiturates, carbon monoxide, and methyl mercury. With the latter compound, ethanol enhanced the retention of mercury in the kidney of rats and thus increased nephrotoxicity (McNeil et al. 1988). When combined with cocaine and fed to rats, increased maternal and fetal toxicity was observed (Church et al. 1988). Ethanol is reported to be synergistically toxic with caffeine (Pollard 1988) and with *n*-butanol and isoamyl alcohol. Prior ethanol consumption increased the toxicity of acetaminophen [103-90-2] in mice (Carter 1987).

Antidote

Studies on mice showed that certain organic zinc salts, such as aspartate, orotate, histidine, and acetate, alone or in conjunction with thiols, exhibited protective action against ethanol toxicity (Floersheim 1987). Corresponding Mg, Co, Zr, and Li salts showed similar action. Experimentation reduced lipid peroxides produced in the liver in chronic ethanol toxicity (Marcus et al. 1988). Vitamin E may have an antidotal action against alcohol toxicity. N-Acetylcysteine [616-91-1] is reported to combat ethanol–acetaminophen toxicity (Carter 1987).

Exposure Limit

TLV-TWA 1900 mg/m^3 (1000 ppm) (ACGIH).

Fire and Explosion Hazard

Flammable liquid; flash point (closed cup) 12.7°C (55°F); autoignition temperature 363°C (685°F); fire hazard in contact with flame or heat; forms explosive mixtures with air; LEL and UEL values 4.3% and 19.0% by volume in air, respectively.

It may explode violently when added to 90% H_2O_2 and acidified with concentrated H_2SO_4. Ethanol-concentrated H_2O_2 in the absence of concentrated H_2SO_4 can detonate by shock. It can explode with concentrated HNO_3; perchloric acid (due to the formation of ethyl perchlorate); metal perchlorate; permanganate-sulfuric acid, or permanganic acid, $HMnO_4$ (explodes with flame); potassium superoxide, KO_2; potassium *tert*-butoxide; silver nitrate, $AgNO_3$ (due to the formation of an explosive fulminate); chlorine trioxide, ClO_3; calcium hypochlorite, $Ca(OCl)_2$; mercuric nitrate (due to the formation of mercuric fulminate);

sodium hydrazide, $NaNHNH_2$; and acetyl chloride, CH_3COCl. Explosion was reported when ethanol was distilled off from vacuum after treatment with iodine and mercuric oxide. There is a report of a violent explosion when manganese perchlorate and 2,2-dimethoxypropane were refluxed with ethanol under a nitrogen atmosphere (NFPA 1986). Mixture of ethanol with chromium trioxide, chromyl chloride, and bromine pentafluoride may ignite.

Hazardous Reaction Products

It liberates hydrogen when it reacts with metal; forms acetaldehyde (toxic, flammable) on catalytic vapor-phase dehydrogenation; ethyl ether (flammable) on dehydration with H_2SO_4 or a heterogeneous catalyst such as alumina, silica, $SnCl_2$, $MnCl_2$, or $CuSO_4$; ethyl bromide (flammable, irritant) with phosphorus tribromide or HBr; ethyl chloride (flammable, narcotic) with PCl_3, thionyl chloride, or HCl in the presence of $AlCl_3$ or $ZnCl_2$; ethyl acetate (flammable, irritant) with acetic acid and mineral acid catalyst or when heated over copper oxide and thoria at 350°C; ethyl aniline (toxic) on heating at 180°C; ethyl acrylate (flammable, lacrimator) with ethylene chlorohydrin and H_2SO_4 or acetylene and CO (with a catalyst); ethylamine (flammable and strong irritant) with ammonia at 150–200°C over a catalyst such as Ni or silica–alumina in the vapor phase at 13 atm; ethyl formate (flammable, irritant, and narcotic) with formic acid; ethyl nitrite (causes methemoglobinemia, hypotension) with $NaNO_3$ and H_2SO_4 in cold or with nitrosyl chloride; ethyl phthalate (irritant and narcotic) with phthalic acid or phthalic anhydride; ethyl silicate (irritant, narcotic) with $SiCl_4$; ethyl mercaptan (narcotic and mucous membrane irritant) when reacting catalytically with H_2S; 1,2-ethanedithiol (causes nausea and headache) with thiourea and ethylene dibromide, followed by alkaline hydrolysis; and ethyl vinyl ether (flammable) on vapor-phase reaction with acetylene at 160°C with KOH.

Analysis

Ethanol may be analyzed colorimetrically by forming colored complexes with 8-hydroxy-quinoline or vanadates and measuring the absorbance at 390 nm. GC-FID is equally suitable for trace analysis. Other analytical methods include a differential refractometer technique for measuring ethanol content in wine and enzymatic methods in combination with GC for analysis in blood. It can be analyzed by GC/MS, using a purge and trap technique (U.S. EPA 1986, Method 8240, SW-846) in solids, sludges, and groundwater. Ethanol in air can be estimated by adsorbing it on charcoal, followed by desorption into CS_2 and analysis by GC-FID (NIOSH 1984, Method 1400). Lazaro and co-workers (1986) have described enzymic determination of ethanol in wine at concentrations of 20–160 ppm. The analysis was done by a single-beam spectrophotometer equipped with a diode-array detector using the alcohol dehydrogenase.

4.4   1-PROPANOL

Formula C_3H_7OH; MW 60.09; CAS [71-23-8]

Structure and functional group: $CH_3-CH_2-CH_2-OH$, primary $-OH$

Synonyms: *n*-propyl alcohol; *n*-propanol; ethyl carbinol

Uses and Exposure Risk

1-Propanol is used in making *n*-propyl acetate; and as a solvent for waxes, resins, vegetable oils, and flexographic printing ink. It is produced from the fermentation and spoilage of vegetable matter.

Physical Properties

Colorless liquid with a smell of alcohol; mp $-126.5°C$; bp $97.4°C$; density 0.8035; soluble in water, alcohol, acetone, ether, and benzene.

Health Hazard

Target organs: skin, eyes, gastrointestinal tracts, and respiratory system. Toxic routes: ingestion, inhalation, and skin contact.

LD_{50} value, oral (rats): 5400 mg/kg (NIOSH 1986)

LD_{50} value, skin (rabbits): 6700 mg/kg (NIOSH 1986)

Ingestion causes headache, drowsiness, abdominal cramps, gastrointestinal pain, ataxia, nausea, and diarrhea. Eye contact produces irritation. It may cause dermatitis on repeated skin contact. Although the toxicity of 1-propanol is low, at a high concentration it may produce a narcotic effect, as well as irritation of the eyes, nose, and throat.

Exposure Limits

TLV-TWA (200 ppm); (500 mg/m^3); STEL 250 ppm (625 mg/m^3); IDLH 4000 ppm.

Fire and Explosion Hazard

Flammable liquid; flash point (open cup) 29°C (59°F); autoignition temperature 371°C (700°F). 1-Propanol forms explosive mixtures with air within the range of 2–14 by volume in air. Violent reactions may occur when it is mixed with strong oxidizing substances.

Hazardous Reaction Products

1-Propanol reacts with various substances, forming moderate to highly toxic and/or flammable products. These are as follows: It reacts with ammonia to form propylamine (strong irritant and skin sensitizer); and on oxidation the aldehyde 1-propanal (irritant and flammable) is formed. Reaction with concentrated nitric and sulfuric acids gives *n*-propyl nitrate (explosive, flammable, and toxic). Heating with benzenesulfonic acid produces *n*-propyl ether (flammable).

Analysis

1-Propanol is analyzed by GC-FID. In air, it is adsorbed over coconut shell charcoal, which is then desorbed in a mixture of CS_2 and 2-propanol (99 : 1) and analyzed (NIOSH 1984, Method 1401). A GC column of 10% SP-1000 on Chromosorb W-HP or equivalent is suitable. The working range is 50–900 mg/m³ for 10 L of air.

4.5 2-PROPANOL

Formula C_3H_7OH; MW 60.09; CAS [67-63-0]
Structure and functional group

$$\underset{}{CH_3-\overset{\overset{\displaystyle OH}{|}}{CH}-CH_3}$$

secondary −OH
Synonyms: isopropyl alcohol; isopropanol; *sec*-propyl alcohol; dimethyl carbinol; rubbing alcohol (70% solution in water)

Uses and Exposure Risk

2-Propanol is used in the production of acetone, isopropyl halides, glycerin, and aluminum isopropoxide; employed widely as an industrial solvent for paints, polishes, and insecticides; as an antiseptic (rubbing alcohol); and in organic synthesis for introducing the isopropyl or isopropoxy group into the molecule. Being a common laboratory solvent like methanol, the exposure risks are always high; however, its toxicity is comparatively low.

Physical Properties

Colorless liquid with alcohol-like odor; bitter taste; bp 82.3°C (at 101.3 kPa); mp −89.5°C; density 0.7849 at 20°C; soluble in water, ether, acetone, benzene, and chloroform, insoluble in CS_2.

Health Hazard

Target organs: eyes, skin, and the respiratory system. Inhalation produces mild irritation in the eyes and nose. At 400- and 5500-ppm concentrations isopropanol caused deterioration of ciliary activity in guinea pigs. Some pathological changes were observed. Recovery was rapid, however. Ingestion causes drowsiness, dizziness, and nausea. A large dose may result in coma. Doses ranging from 100 to 166 mL may be fatal to humans.

Exposure Limits

TLV-TWA 980 mg/m³ (400 ppm); STEL 1225 mg/m³ (500 ppm) (ACGIH); IDLH 12,000 ppm (NIOSH).

Fire and Explosion Hazard

Flammable liquid; flash point (anhydrous grade) 17.2°C (Tag open cup) and 11.7°C (closed cup). A 91% alcohol in water (by volume) is also flammable. The LEL and UEL values are 2.02 and 7.99% by volume in air, respectively.

2-Propanol stored for a long time with large head space may explode. There is a report of such detonation during the distillation of this compound taken from a 5-year-old container, one-fourth filled.

Hazardous Reaction Products

2-Propanol undergoes reactions characteristic of a secondary −OH group. Oxidation is highly exothermic (43 kcal/mol) but occurs above 300°C, forming acetone. It reacts with acid chlorides, chlorine, and phosphorus chloride to produce chloroacetones (toxic). It loses water molecules over acid catalysts such as alumina or sulfuric acid, to form diisopropyl ether (flammable) and propylene (flammable, forms an explosive mixture with air). It forms isopropyl acetate (mucous membrane irritant, narcotic) with acetic acid. Treatment with nitrosyl chloride or with sodium nitrite/concentrated H_2SO_4 produces isopropyl nitrite (toxic — vasodilation, tachy-cardia, headache).

Storage and Handling

2-Propanol is stored in glass containers or steel drums. Baked phenolic-lined steel tanks are used for storage and shipping. Aluminum containers should not be used, as isopropanol may react to form aluminum isopropoxide.

Analysis

2-Propanol is analyzed by GC-FID. It may be determined colorimetrically by forming colored complexes with acetone, which is produced by the oxidation of 2-propanol. Analysis in air is performed by sampling the air over charcoal, desorbing the adsorbent in a CS_2/2-butanol mixture (99 : 1), and analyzing by GC-FID using Carbowax 1500 on Carbopack C or any equivalent column [i.e., 10% FFAP on Chromosorb W-AW (NIOSH 1984, Method 1400)]. High humidity reduces sampling efficiency.

4.6 ALLYL ALCOHOL

EPA-listed priority pollutant in solid waste matrices and groundwater)

Formula C_3H_5OH; MW 58.1; CAS [107-18-6]

Structure and functional group: $CH_2=CH-CH_2-OH$, primary $-OH$

Synonyms: 2-propene-1-ol; 2-propenol; vinyl carbinol

Uses and Exposure Risk

Allyl alcohol is used to produce glycerol and acrolein and other allylic compounds. It is also used in the manufacture of military poison gas. The ester derivatives are used in resins and plasticizers.

Physical Properties

Colorless liquid with a pungent smell of mustard; bp 97°C; mp −50°C; density 0.854

at 20°C; soluble in water, alcohol, ether, and chloroform.

Health Hazard

The toxicity of allyl alcohol is moderately high, affecting primarily the eyes. The other target organs are the skin and respiratory system. Inhalation causes eye irritation and tissue damage. A 25-ppm exposure level is reported to produce a severe eye irritation. It may cause a temporary lacrimatory effect, manifested by photophobia and blurred vision, for some hours after exposure. Occasional exposure of a person to allyl alcohol does not indicate chronic or cumulative toxicity. Dogterom and associates (1988) investigated the toxicity of allyl alcohol in isolated rat hepatocytes. The toxicity was independent of lipid peroxidation, and acrylate was found to be the toxic metabolite.

Ingestion of this compound may cause irritation of the intestinal tract. The oral LD_{50} value in rats is 64 mg/kg (NIOSH 1986).

Exposure Limits

TLV-TWA 5 mg/m^3 (2 ppm); STEL 10 mg/m^3 (4 ppm) (ACGIH); IDLH 150 ppm.

Fire and Explosion Hazard

Flammable liquid; flash point (closed cup) 21°C (70°F); vapor pressure 17 torr at 25°C; vapor density 2.0 (air = 1); the vapor is heavier than air and may travel a considerable distance to a source of ignition and flash back; autoignition temperature 378°C (713°F); forms explosive mixtures with air in the range 2.5–18.0% by volume in air. It is susceptible to forming peroxides on prolonged exposure to air.

Hazardous Reaction Products

Oxidation products are acrolein (toxic, flammable) and acrylic acid (strong irritant). Reactions with hydrogen halides produce strongly toxic allyl halides; allyl iodide is produced by reaction with methyl iodide and triphenyl phosphite. Dehydration occurs in the presence of cuprous chloride, $-H_2SO_4$, forming allyl ether (flammable and skin irritant). Amyl amine (toxic and irritant) is produced on reaction with ammonia or amines. Dehydrogenation produces propargyl alcohol (flammable and toxic). It reacts with carbon tetrachloride, forming di- and trichlorobutylene epoxides (explode), and an unstable product of unknown composition (explodes) when it is reacted with tri-*n*-bromomelamine. It polymerizes explosively in the presence of caustic soda.

Analysis

Allyl alcohol in air (range $1-10$ mg/m^3) is measured by flowing air over a charcoal adsorbent, eluting the analyte into a CS_2-isopropanol mixture (ratio $95:5$), and analyzing by GC-FID (NIOSH 1984, Method 1402). A suitable column is 10% SP-1000 on Supelcoport or 10% FFAP on Chromosorb W-AW. Other, equivalent columns may be used. High humidity reduces sampling capacity. In solid wastes, soils, sediments, and groundwater it may be analyzed by GC/MS (U.S. EPA 1986, Method 8240, SW-846) using a purge and trap technique or direct injection. The primary ion used to identify this compound by GC/MS is 57 and the secondary ions are 58 and 39.

4.7 2-PROPYN-1-OL

EPA-listed priority pollutant in solid and hazardous wastes

Formula C_3H_3OH; MW 56.06; CAS [107-19-7]

Structure and functional group: $HC\equiv C-CH_2-OH$, primary hydroxyl group

Synonyms: propargyl alcohol; acetylene carbinol

Uses and Exposure Risk

It is used in metal plating and pickling and as a corrosion inhibitor of mild steel in mineral acids. It also finds application in preventing the hydrogen embrittling of mild steel in acids. It is used as an intermediate for making miticide and sulfadiazine.

Physical Properties

A colorless and moderately volatile liquid with a mild odor of geranium; mp $-48°C$; bp $113.6°C$; density 0.948; soluble in water, alcohol, ether, benzene, chloroform, and acetone; insoluble in aliphatic hydrocarbon solvents.

Health Hazard

2-Propyn-1-ol is a moderately toxic substance causing depression of the CNS and irritation of the eyes and skin.

LD_{50} value, oral (rats): 70 mg/kg
LD_{50} value, oral (guinea pigs): 60 mg/kg

Exposure Limits

No exposure limit has been set for 2-propyn-1-ol. A TLV-TWA of 1 ppm ($\sim$2.3 mg/m^3) should be appropriate for this compound. This estimation is based on its similarity to allyl alcohol in chemical properties and toxic actions.

Fire and Explosion Hazard

Flammable liquid; flash point (open cup) $33°C$ ($91°F$); vapor forms explosive mixtures with air; LEL and UEL data not available. 2-Propyn-1-ol may react violently with strong oxidizers. It ignites when mixed with phosphorus pentoxide.

Hazardous Reaction Products

Because of the presence of a triple bond, an acetylenic hydrogen, and the primary hydroxyl group, this compound is highly active and undergoes reactions of substitution, hydrogenation, and oxidation types to give a wide variety of products, some of which may be toxic or flammable.

Oxidation with chromic acid produces the aldehyde propynal, $HC{\equiv}C-CHO$ (toxic and flammable). Hydrogenation gives allyl alcohol, $CH_2{=}CH-CH_2-OH$, and propenal (acrolein), $CH_2{=}CH-CHO$ (both products are toxic and flammable). Phosphorus trichloride and other chlorinating agents can substitute the $-OH$ group to form propargyl chloride (flammable). It reacts with HCl in the presence of mercury salt catalysts to form 2-chloro-2-propen-1-ol (2-chlorallyl alcohol), $CH_2{=}C(Cl)CH_2-OH$ (toxic). Halogen adds onto the double bond of 2-propyn-1-ol to form dihaloallyl alcohol, $CHX{=}CHX-CH_2OH$ (toxic).

Storage and Handling

2-Propyn-1-ol is stored in stainless steel-lined, glass-lined, or phenolic-lined tanks or drums. Unlined steel containers may be used if free of rust. Aluminum, rubber, and epoxy materials should not be used. Use protective wear when handling. Wash thoroughly with water on spillage.

Analysis

2-Propyn-1-ol is analyzed by GC-FID using a suitable column for alcohol. It can be analyzed in all types of samples, such as groundwater or soils, sediments, and sludges by GC/MS using a purge and trap technique or by direct injection (U.S. EPA 1986, Method 8240, SW-846). A column containing 1% SP-1000 on Carbopack-B is suitable. The primary ion is 55 and the secondary ions are 39, 38, and 53 (electron-impact ionization).

4.8 1-BUTANOL

Formula C_4H_9OH; MW 74.1; CAS [71-36-3]

Structure and functional group: $CH_3-CH_2-CH_2-CH_2-OH$, primary $-OH$

Synonyms: n-butanol; n-butyl alcohol; propyl carbinol

Uses and Exposure Risk

1-Butanol is used in the production of butyl acetate, butyl glycol ether, and plasticizers such as dibutyl phthalate; as a solvent in the coating industry; as a solvent for extractions of oils, drugs, and cosmetic nail products; and as an ingredient for perfumes and flavor.

1-Butanol occurs in fusel oil and as a by-product of the fermentation of alcoholic beverages such as beer or wine. It is present in beef fat, chicken broth, and nonfiltered cigarette smoke (Sherman 1979).

Physical Properties

Colorless liquid with a characteristic vinous odor; mp $-90°C$; bp $117.2°C$, density 0.810; moderately soluble in water, 7 g/100 mL at 30°C, soluble in alcohol, ether, acetone, and benzene.

Health Hazard

The toxicity of 1-butanol is lower than that of its carbon analog. Target organs are the skin, eyes, and respiratory system. Inhalation causes irritation of the eyes, nose, and throat. It was found to cause severe injury to rabbits' eyes and to penetrate the cornea upon instillation into the eyes. Chronic exposure of humans to high concentrations may cause photophobia, blurred vision, and lacrimation.

A concentration of 8000 ppm was maternally toxic to rats, causing reduced weight gain and feed intake. Teratogenicity was observed at this concentration with a slight increase in skeletal malformations (Nelson et al. 1989).

In a single acute oral dose, the LD_{50} value (rats) is 790 mg/kg; in a dermal dose the LD_{50} value (rabbits) is 4200 mg/kg.

n-Butanol is oxidized in vivo enzymatically as well as nonenzymatically and is eliminated rapidly from the body in the urine and in expired air. It inhibits the metabolism of ethanol caused by the enzyme alcohol dehydrogenase.

Based on the available data, the CIR expert panel concluded that the use of *n*-butanol as an ingredient is safe under the present practices and concentrations in cosmetic nail products (Cosmetic, Toiletry and Fragrance Association 1987a).

Exposure Limits

TLV-TWA 300 mg/m^3 (100 ppm) (NIOSH), 150 mg/m^3 (50 ppm) (ACGIH); IDLH 8000 ppm (NIOSH).

Fire and Explosion Hazard

Flammable liquid, flash point 35°C; autoignition temperature 367°C. It presents a fire hazard when exposed to heat or flame. It forms explosive mixtures with air, with LEL and UEL values of 1.4% and 11.2% by volume in air, respectively. It emits toxic fumes when heated to decomposition. It may react violently with strong oxidizers.

Hazardous Reaction Products

1-Butanol exhibits the characteristic reactions of a primary —OH group, undergoing oxidation, dehydrogenation, dehydration, and esterification. The hazardous products include butyraldehyde (narcotic and irritant) formed from oxidation with air over copper or silver catalysts at 250–350°C. It forms Butyl Cellosolve (highly toxic) with ethylene carbonate; butyl nitrite (toxic — causes headache and lowering of blood pressure) with nitrous acid, and butyl mercaptan (nauseating and narcotic) with H$_2$S over ThO. Esterifications with organic acids in the presence of mineral acid catalysts yield products such as *n*-butyl acetate (irritant, narcotic); butyl phthalate (toxic to gastrointestinal tract) and butyl propionate (skin irritant). It reacts with ammonia over alumina at 300°C, forming butyl amine (skin and mucous membrane irritant). In the presence of a Ni—Al$_2$O$_3$ catalyst, butyronitrile (highly toxic) may form. Heating with HCl and anhydrous ZnCl$_2$ may yield *n*-butyl chloride (highly flammable).

Analysis

1-Butanol is analyzed by GC-FID using a suitable column for alcohols, such as 10% SP-1000 on Chromosorb W-HP or equivalent. It is analyzed in air by charcoal adsorption, desorbing the analyte in a CS$_2$—isopropanol mixture (99 : 1) and injecting the eluant onto GC-FID (NIOSH 1984, Method 1401). Other methods, such as GC/MS, HPLC, IR, and TLC, have been employed. A fluorophotometric method using alcohol dehydrogenase may be used.

4.9   2-BUTANOL

Formula C$_4$H$_9$OH; MW 74.1; CAS [78-92-2]
Structure and functional group

$$CH_3-\underset{\underset{\displaystyle OH}{|}}{CH}-CH_2-CH_3$$

secondary —OH

Synonyms: *sec*-butyl alcohol; methyl ethyl carbinol; 2-hydroxybutane

Uses and Exposure Risk

2-Butanol is used in the production of methyl ethyl ketone and *sec*-butyl acetate, as a solvent in lacquers and alkyd enamels, in hydraulic brake fluids, in cleaning compounds, and its xanthate derivatives in ore flotation.

Physical Properties

Colorless liquid with a strong pleasant odor; mp −114.7°C; bp 99.5°C; density 0.807 at 20°C; moderately soluble in water (18 g/mL), miscible with alcohol, ether, benzene, and acetone.

Health Hazard

Exposure to 2-butanol may cause irritation of the eyes and skin. The latter effect is produced by its defatting action on skin. This toxic property is mild and similar to that of other butanol isomers. High concentration may produce narcosis. The narcotic effect is stronger than that of *n*-butanol, probably due to the higher vapor pressure of the secondary alcohol.

The toxicity is lower than that of its primary alcohol analogue.

LD$_{50}$ value, oral (rats): 6480 mg/kg

Exposure Limits

TLV-TWA 450 mg/m^3 (150 ppm) (NIOSH), 305 mg/m^3 (100 ppm) (ACGIH); IDLH 10,000 ppm.

Fire and Explosion Hazard

Flammable liquid, flash point 24.4°C (76°F); autoignition temperature 405°C (761°F); vapor forms explosive mixtures with air within the range 1.7–9.8% by volume in air at 100°C. It may react violently with strong oxidizing substances.

Analysis

2-Butanol is analyzed by GC-FID. Analytical methods are similar to those of 1-butanol (see Section 4.8).

4.10   *tert*-BUTYL ALCOHOL

Formula C$_4$H$_9$OH; MW 74.1; CAS [75-65-0] Structure and functional group:

$$H_3C-\underset{\underset{CH_3}{|}}{\overset{\overset{CH_3}{|}}{C}}-OH$$

tertiary −OH

Synonyms: 2-methyl-2-propanol; trimethyl carbinol; *tert*-butanol

Uses and Exposure Risk

tert-Butyl alcohol is used in the production of *tert*-butyl chloride, *tert*-butyl phenol, and isobutylene; in the preparation of artificial musk; and in denatured alcohols.

Physical Properties

Colorless liquid (solid); mp 25.5°C; bp 82.5°C; density 0.789 at 20°C; miscible with water, alcohol, and ether.

Health Hazard

tert-Butyl alcohol is more toxic than *sec*-butyl alcohol but less toxic than the primary alcohol. However, its narcotic action is stronger than that of *n*-butanol. Inhalation may cause drowsiness and mild irritation of the skin and eyes. Ingestion may produce headache, dizziness, and dry skin.

Acute oral LD$_{50}$ value (rats): 3500 mg/kg

Exposure Limits

TLV-TWA 300 mg/m^3 (100 ppm) (ACGIH); IDLH 8000 ppm.

Fire and Explosion Hazard

Flammable liquid, flash point 9°C (48°F); autoignition temperature 478°C (892°F); presents a fire hazard in contact with heat or flame; vapor forms explosive mixtures with air, LEL and UEL values 2.4% and 8.0% by volume of air, respectively. It may react violently with strong oxidizers.

Hazardous Reaction Products

When added to a cold aqueous solution of NaOH and passing chlorine into the mixture it forms *tert*-butyl hypochlorite,

$(CH_3)_3COCl$ (explosion hazard on exposure to strong light, heat, and rubber; eye and mucous membrane irritant). Other hazardous products are *tert*-butyl ether (flammable) on heating with dilute H_2SO_4; and *tert*-butyl acetate (toxic and irritant) on esterification with acetic acid; and di-*tert*-butyl peroxide (explodes violently) on adding to a 2 : 1 weight mixture of 78% H_2SO_4 and 50% hydrogen peroxide.

Analysis

tert-Butyl alcohol may be analyzed by GC-FID; and air analysis may be done by NIOSH Method 1400 (see Section 4.5).

4.11   ISOBUTYL ALCOHOL

EPA Priority Pollutant under Solid and Hazardous Waste Category; RCRA Hazardous Waste Number U140

Formula C_4H_9OH; MW 74.1; CAS [78-83-1]

Structure and functional group:

$$\begin{array}{c} CH_3 \\ \diagdown \\ CH_3 \diagup CH-CH_2-OH \end{array}$$

primary $-OH$

Synonyms: 2-methyl-1-propanal; isobutanol; isopropyl carbinol

Uses and Exposure Risk

Isobutanol is widely used in the production of isobutyl acetate for lacquers, isobutyl phthalate for plasticizers; as a solvent for plastics, textiles, oils, and perfumes; and as a paint remover.

Physical Properties

Colorless liquid with mild odor; mp $-108°C$, bp $18.1°C$: vapor pressure 8.8 torr at $25°C$; density 0.8018 at $20°C$; moderately soluble in water, 7.5 g/mL at $30°C$, soluble in alcohol, ether, and acetone.

Hazardous Reaction Products

Isobutanol forms isobutylamine (toxic — erythema and blistering) with NH_3; isobutyl chlorocarbonate, $(CH_3)_2CH-CH_2-COOCl$ (eye and mucous membrane irritant) with phosgene; and isobutyraldehyde (flammable) on air oxidation at $300°C$ in the presence of Cu.

Health Hazard

Inhalation causes eye and throat irritation and headache. Ingestion may cause depression of the central nervous system. It is an irritant to the skin, causing cracking. Target organs are the eyes, skin, and respiratory system.

LD_{50} value, oral (rabbits): 3750 mg/kg

Exposure Limits

TWA 300 mg/m^3 (100 ppm) NIOSH, 150 mg/m^3 (50 ppm) (ACGIH); IDLH 8000 ppm.

Fire and Explosion Hazard

Flammable liquid, flash point $27.5°C$ ($81.5°F$); autoignition temperature $415°C$ ($780°F$); vapor forms explosive mixtures with air in the range 1.2–10.9% by volume in air at $100°C$.

Analysis

Isobutanol may be analyzed by GC-FID and GC/MS techniques. Air analysis may be performed by NIOSH Method 1401. GC/MS volatile organic analysis using a purge and trap or thermal desorption technique may be suitable for analyzing isobutanol in water, soil, and solid wastes (U.S. EPA 1986, Method 8240) (see Section 4.6). Characteristic mass ions from electron-impact ionization are 41, 42, 43, and 47.

4.12 AMYL ALCOHOLS

Formula $C_5H_{11}OH$; MW 88.15

Amyl alcohols occur in several isomeric forms, most of which are flammable and slightly toxic. These isomers, with their synonyms, CAS number, and some physical properties are presented in Table 4.1.

Uses and Exposure Risk

Amyl alcohols are used as solvents for resins and gums; as diluents for lubricants, printing inks, and lacquers; and as frothers in ore flotation process. Amyl xanthates are used as ore collectors in the mining industry.

Physical Properties

Amyl alcohols are colorless liquids with the characteristic odor of alcohol. Neopentyl alcohol is a solid (mp 52°C). Amyl alcohols are slightly soluble in water but mix readily with most organic solvents. Other properties are listed in Table 4.1.

Health Hazard

Inhalation of isoamyl alcohol can cause coughing and irritation of the eyes, nose, and throat. Ingestion may produce narcosis, headache, and dizziness. Cracking of skin may result from the contact. Prolonged contact with isoamyl alcohol can cause dysuria, nausea, and diarrhea. The toxic effects of other isomers in humans are unknown. On the basis of animal studies, it appears that at high concentrations all these isomers could be narcotic and cause eye irritation.

Exposure Limits

No threshold limit values have been assigned to amyl alcohol isomers. For isoamyl alcohol, TLV-TWA is set at 360 mg/m³ (100 pm), STEL at 450 mg/m³ (125 ppm) (ACGIH), and the IDLH level is 7000 ppm (NIOSH).

Fire and Explosion Hazard

These are flammable compounds. The flash point data are presented in Table 4.1. The flammability of 2-methyl-2-butanol (*tert*-amyl alcohol) is greater than that of other isomers. The vapors of all isomers form explosive mixtures with air. The LEL and UEL values for isoamyl alcohol are 2.4% and 10.5% by volume in air, respectively. Reactions with strong oxidizing substances may be vigorous to violent.

Storage and Handling

These compounds should be stored and shipped in stainless steel, aluminum, and lined steel tanks. The tanks should be purged with dry inert gas to expel air, which may form flammable mixture. Handling of large amounts should be performed using proper protective wear and keeping away from flame.

Analysis

Amyl alcohol isomers may be analyzed by GC-FID on a column suitable for alcohols (e.g., Carbowax 1500). A capillary column should be effective in the separation of isomers. Isoamyl alcohol in air may be measured by sampling the air on charcoal, eluting the analyte with a CS_2/isopropyl alcohol mixture, and analyzing by GC-FID.

4.13 CYCLOHEXANOL

Formula $C_6H_{11}OH$; MW 100.16; CAS [108-93-0]

Structure and functional group:

the −OH group is secondary

Synonyms: cyclohexyl alcohol; hexahydrophenol; hydroxycyclohexane; hexalin; hydralin; anol; adronal; naxol

TABLE 4.1 Physical Properties of Amyl Alcohols

Isomer/Synonyms/CAS No.	Structure/Functional Group	Boiling Point (°C)	Flash Point (°C)	Density (Specific Gravity)
1-Pentanol (*n*-amyl alcohol) [71-41-0]	$CH_3-CH_2-CH_2-CH_2-CH_2-OH$ Primary—OH	137.8	58(oc), 33(cc)	0.814
2-Pentanol (active *sec*-amyl alcohol) [6032-29-7]	$\overset{\displaystyle OH}{CH_3-CH_2-CH_2-CH-CH_3}$ Secondary—OH	119.3	42(oc), 34(cc)	0.830
3-Pentanol [584-02-1]	$\overset{\displaystyle OH}{CH_3-CH_2-CH-CH_2-CH_3}$ Secondary—OH	115.6	39(oc), 41(cc)	0.822
2-Methyl-1-butanol (active amyl alcohol) [137-32-6]	$\overset{\displaystyle CH_3}{CH_3-CH_2-CH-CH_2-OH}$ Primary—OH	128.0	50(oc)	0.816
3-Methyl-1-butanol (isoamyl alcohol, isobutyl carbinol, isopentyl alcohol, fusel oil, fermentation amyl alcohol) [123-51-37]	$\overset{\displaystyle CH_3}{CH_3-CH-CH_2-CH_2-OH}$ Primary—OH	110.5	43(oc)	0.809

(continued)

123

TABLE 4.1 *(Continued)*

Isomer/Synonyms/CAS No.	Structure/Functional Group	Boiling Point (°C)	Flash Point (°C)	Density (Specific Gravity)
2-Methyl-2-butanol (*tert*-amylalcohol) [75-85-4]	CH$_3$—CH$_2$—C—CH$_3$ with CH$_3$ and OH Tertiary —OH	101.8	24(oc), 19(cc)	0.809
3-Methyl-2-butanol (*sec*-isoamyl alcohol) [598-75-4]	CH$_3$—CH—CH—CH$_3$ with CH$_3$ and OH Secondary —OH	112.0	35	0.819
2,2-Dimethyl-1-propanol (neopentyl alcohol) [75-84-3]	CH$_3$—C—CH$_2$—OH with CH$_3$ and CH$_3$ Primary —OH	113–114	37	0.812

oc, Open cup; cc, closed cup.

Uses and Exposure Risk

Cyclohexanol is used for the production of adipic acid and caprolactam for making nylon. Its phthalate derivatives are used for plasticizers. It is used as a stabilizer for soaps and detergents; as a solvent for lacquers, varnishes, and shellacs; and as a dye solvent for textiles.

Physical Properties

Colorless viscous liquid or a sticky solid with a mild smell of camphor; hygroscopic; mp 25.15°C; bp 161.1°C; density 0.924 at 20°C; solubility: moderately soluble in water, 4.3 g/100 g at 30°C, readily mixes with most organic solvents and also with waxes, resins, gums, and many oils.

Health Hazard

Cyclohexanol is moderately toxic. Target organs are the eyes, skin, and respiratory system. At high concentrations cyclohexanol absorbed through the skin may possibly injure the brain, kidney, and heart.

LD$_{50}$ value, single oral dose (rats): 2060 mg/kg

LD$_{50}$ value, single intravenous dose (mice): 270 mg/kg

Inhalation of vapors may cause irritation of the eyes, nose, and throat. However, because of its low vapor pressure (1.12 torr at 25°C), the health hazard due to inhalation is low. Ingestion can cause nausea, trembling, and gastrointestinal disturbances. Repeated skin contact may produce erythema and edema.

Exposure Limits

TLV-TWA 200 mg/m^3 (50 ppm) (ACGIH); IDLH 3500 ppm (NIOSH).

Fire and Explosion Hazard

Combustible liquid; flash point (open cup) 67.2°C (154°F); autoignition temperature 300°C (572°F); vapor forms explosive mixtures with air; LEL 2.4% by volume in air, UEL data not available.

Analysis

Cyclohexanol may be analyzed by GC-FID using Carbowax 20 M on Chromosorb support or an equivalent column. It may be analyzed colorimetrically by reaction with *p*-hydroxybenzaldehyde in sulfuric acid (absorbance maxima at 535 and 625 nm). Cyclohexanol in air (working range 15–150 mg/m^3) may be measured by sampling 10 L of air on charcoal, desorbing the adsorbent in a CS$_2$/isopropyl alcohol mixture, and analyzing the eluant by GC-FID, using the column 10% SP-1000 on Supelcoport (NIOSH 1984, Method 1402). 10% FFAP on Chromosorb may be used as an alternative column.

4.14   METHYLCYCLOHEXANOL

Formula CH$_3$C$_6$H$_{10}$OH; MW 114.19; CAS [25639-42-3]

Structure and functional group:

o-isomer   *m*-isomer   *p*-isomer

mixture of *o*-, *m*-, and *p*-isomers, −CH$_3$ group in 2-, 3-, or 4-position, −OH is secondary

Synonyms: hexahydrocresol; hexahydromethyl phenol; methyl hexalin

Uses and Exposure Risk

Methylcyclohexanol is used as a blending agent in textile soaps as an antioxidant in lubricants, and as a solvent for lacquers.

Physical Properties

Colorless viscous liquid with a weak coconut oil odor; bp 155–175°C; density 0.919–0.945; soluble in alcohol and ether.

Health Hazard

Methylcyclohexanol is mildly toxic. Inhalation may produce mild irritation of the eyes and the respiratory system, and headache. Studies in rabbits showed that this compound could cause rapid narcosis and convulsion at sublethal doses. The minimum lethal dose by oral administration was 2000 mg/kg. Severe exposure may produce narcosis in humans. Skin absorption may cause dermatitis.

Exposure Limits

TWA-TWA 235 mg/m^3 (50 ppm) (ACGIH), 470 mg/m^3 (100 ppm) (NIOSH); IDLH 10,000 ppm (NIOSH).

Fire and Explosion Hazard

Noncombustible liquid; incompatible with strong oxidizers.

Analysis

The air is sampled over charcoal and desorbed by methylene chloride, and the eluant is analyzed by GC-FID.

4.15 1,2-ETHANEDIOL

Formula $C_2H_4(OH)_2$; MW 62.07; CAS [107-21-1]

Structure: $HO-CH_2-CH_2-OH$

Synonyms: ethylene glycol; 1,2-dihydroxyethane; ethylene dihydrate; ethane-1,2-diol

Uses and Exposure Risk

Ethylene glycol is used as an antifreeze in heating and cooling systems (e.g., automobile radiators and coolant for airplane motors). It is also used in the hydraulic brake fluids; as a solvent for paints, plastics, and inks; as a softening agent for cellophane; and in the manufacture of plasticizers, elastomers, alkyd resins, and synthetic fibers and waxes.

Physical Properties

Colorless syrupy liquid; odorless; sweet taste; hygroscopic; boils at 197.6°C; freezes at −13°C; density 1.1135 at 20°C; soluble in water, alcohol, acetone, acetic acid, and pyridine, slightly soluble in ether, and insoluble in benzene, chloroform, and petroleum ether.

Health Hazard

The acute inhalation toxicity of 1,2-ethanediol is low. This is due to its low vapor pressure, 0.06 torr at 20°C. Its saturation concentration in air at 20°C is 79 ppm and at 25°C is 131 ppm (ACGIH 1986). Both concentrations exceed the ACGIH ceiling limit in air, which is 50 ppm. In humans, exposure to its mist or vapor may cause lacrimation, irritation of throat, and upper respiratory tract, headache, and a burning cough. These symptoms may be manifested from chronic exposure to about 100 ppm for 8 hours per day for several weeks.

The acute oral toxicity of 1,2-ethanediol is low to moderate. The poisoning effect, however, is much more severe from ingestion than from inhalation. Accidental ingestion of 80–120 mL of this sweet-tasting liquid can be fatal to humans. The toxic symptoms in humans may be excitement or stimulation, followed by depression of the central nervous system, nausea, vomiting, and drowsiness, which may, in the case of severe poisoning, progress to coma, respiratory failure, and death. When rats were administered sublethal doses over a long period, there was deposition of calcium oxalate in tubules, causing uremic poisoning.

LD$_{50}$ value, oral (rats): 4700 mg/kg

Ingestion of 1,2-ethanediol produced reproductive effects in animals, causing fetotoxicity, postimplantation mortality, and specific developmental abnormalities. Mutagenic tests proved negative. It tested negative to the histidine reversion–Ames test.

Exposure Limits

Ceiling limit in air for vapor and mist 50 ppm (~125 mg/m^3) (ACGIH); TWA 10 mg/m^3 (particulates) (MSHA).

Fire and Explosion Hazard

Noncombustible liquid; flash point (open cup) 115°C (240°F); autoignition temperature 398°C (748°F). The reactivity of this compound is low. Its reaction with concentrated sulfuric or nitric acids is vigorous (exothermic).

Analysis

1,2-Ethanediol may be analyzed by GC-FID using any column that is generally suitable for alcohols. HPLC and GC/MS techniques are also suitable.

4.16   1,4-BUTANEDIOL

Formula $C_4H_8(OH)_2$; MW 90.1; CAS [110-63-4]

Structure and functional group: $HO-CH_2-CH_2-CH_2-CH_2-OH$, primary $-OH$ groups

Synonyms: tetramethylene glycol; 1,4-butylene glycol; 1,4-dihydroxybutane

Uses and Exposure Risk

1,4-Butanediol is used to produce polybutylene terephthalate, a thermoplastic polyester; and in making tetrahydrofuran, butyrolactones, and polymeric plasticizers.

Physical Properties

Colorless liquid; bp 228°C; mp 20.1°C; density 1.017 at 20°C; soluble in water, alcohol,

and acetone, insoluble in aliphatic hydrocarbons and benzene.

Health Hazard

The acute toxic effects are mild. 1,4-Butanediol is less toxic than its unsaturate analogs, butenediol and the butynediol. The oral LD_{50} value in white rats and guinea pigs is ~2 mL/kg. The toxic symptoms from ingestion may include excitement, depression of the central nervous system, nausea, and drowsiness.

Fire and Explosion Hazard

Nonflammable liquid, flash point (open cup) 121°C.

Hazardous Reaction Products

With acid catalysts 1,4-butanediol cyclizes to tetrahydrofuran (toxic); forms 1,4-dichlorobutane with thionyl chloride and 1,4-dibromobutane with HBr. It forms bis(chloromethyl) ether (toxic) with formaldehyde and HCl.

4.17   2-BUTENE-1,4-DIOL

Formula $C_4H_6(OH)_2$; MW 88.1; CAS [110-64-5]

Structure and functional group: $HO-CH_2-CH=CH-CH_2-OH$, olefinic diol with two primary hydroxyl groups

Synonyms: ethylene dicarbinol; dimethylol ethylene

Uses and Exposure Risk

It is used to make agricultural chemicals and the pesticide endosulfan; and as an intermediate for making vitamin B.

Physical Properties

Colorless liquid; bp 234°C; mp 25°C (*trans*-isomer), 4°C (*cis*-isomer); density 1.070; soluble in water, alcohol, and acetone, insoluble in benzene and aliphatic hydrocarbons.

Health Hazard

2-Butene-1,4-diol is a depressant of the CNS. Inhalation toxicity is very low due to its low vapor pressure. The oral LD_{50} value in rats and guinea pigs is 1.25 mL/kg. It is a primary skin irritant.

Fire and Explosion Hazard

Noncombustible liquid; flash point (open cup) 128°C.

Hazardous Reaction Products

2-Butene-1,4-diol forms furan (narcotic) when treated with dichromate in acidic solution. Dehydration of the *cis*-isomer over acid catalysts gives 2,5-dihydrofuran (narcotic). Halogens form substitution or addition products, 4-halobutenols, or 2,3-dihalo-1,4-butanediol. These are toxic compounds. Ammonia or amine form pyrroline or its derivatives (moderately toxic).

4.18   2-BUTYNE-1,4-DIOL

Formula $C_4H_4(OH)_2$; MW 86; CAS [110-65-6]

Structure and functional group: $HO-CH_2-C\equiv C-CH_2-OH$, unsaturated diols with two primary $-OH$ groups

Synonyms: butynediol; acetylene dicarbinol; dimethylolacetylene

Uses and Exposure Risks

2-Butyne-1,4-diol is used to produce butanediol and butenediol, in metal plating and pickling baths, and in making the carbamate herbicide Barban (Carbyne).

Physical Properties

Crystalline solid; mp 58°C; bp 248°C; specific gravity 1.114 at 60°C; soluble in water, ethanol, and acetone, sparingly soluble in ether, and slightly soluble in benzene.

Health Hazard

2-Butyne-1,4-diol exhibits moderate to high toxicity in test animals. It is about 10 times more toxic than are the saturated C_4 diols. The oral LD_{50} value in rats is 0.125 mL/kg. It causes irritation to the skin.

Fire and Explosion Hazard

Noncombustible solid, flash point (open cup) 152°C. It is stable at room temperature. But the dry compound explodes in the presence of certain heavy metal salts, such as mercuric chloride. Heating with alkaline solution may result in an explosion.

Hazardous Reaction Products

Halogen addition reactions form various haloderivatives according to the conditions of temperature and catalysts. Chlorination and bromination produce mucochloric and mucobromic acids, respectively (strong irritants). HCl replaces one $-OH$ group to form 2,4-dichloro-2-buten-2-ol (toxic). In the presence of a catalyst, copper or mercury salt, the product is 3-chloro-2-buten-1,4-diol.

Storage and Handling

2-Butyne-1,4-diol can be stored in steel, aluminum, nickel, glass, epoxy, and phenolic liner containers. Rubber hose may be used for transfer. Avoid contact with heavy metal salt contaminants.

Analysis

The commercial aqueous solution is 34% minimum butynediol; determined by bromination or refractive index; analyzed by GC-FID.

4.19   2,3-NAPHTHALENEDIOL

Formula $C_{10}H_6(OH)_2$; MW 160.17; CAS [92-44-4]

Structure and functional group:

dicyclic aromatic compound with two hydroxyl groups in *ortho*-positions

Synonym: 2,3-dihydroxynaphthalene

Uses and Exposure Risk

2,3-Naphthalenediol is used in cosmetics as a component of oxidative hair dye. It is used at a concentration of <0.1%.

Physical Properties

White crystalline solid; mp 163–164°C; soluble in alcohol, ether, benzene, and acetic acid.

Health Hazard

2,3-Naphthalenediol shows low toxicity and mild irritant actions on the skin and eyes. The oral LD_{50} value for rats of a 5% solution in propylene glycol may be on the order of 675 mg/kg (calculated by the method of Weil). The intravenous LD_{50} value for mice is 56 mg/kg. At 1% concentration, it caused slight eye irritation in female albino rabbits and exhibited erythemal response in guinea pigs. At the 0.1% level (concentration in hair dye), it had no reaction on human skin. This compound was nonmutagenic in *Salmonella typhimurium* strain tests.

There is very little information in the literature on the toxicity of 2,3-naphthalenediol. Assessment by the CIR expert panel on the safety of this compound as used in cosmetics has been inconclusive (Cosmetic, Toiletry and Fragrance Association 1987b).

Analysis

2,3-Naphthalenediol can be analyzed by HPLC and GC-FID techniques.

4.20 MISCELLANEOUS ACETYLENIC ALCOHOLS

These alcohols are characterized by an acetylenic triple bond. Table 4.2 lists some of these alcohols with their physical properties, commercial applications, and toxicity data. Owing to the presence of the triple bond and labile hydroxyl group(s), these alcohols are easily susceptible to two general types of reactions: addition to double bonds and substitution of hydroxyl groups forming several toxic products.

The addition reactions are similar to those of 2-propyn-1-ol (propargyl alcohol). Hydrogenation gives alkenyl and alkyl alcohols. Oxidation produces ketones, mostly toxic. Halogens, hydrogen halides, or thionyl halides form dihaloalcohols.

Under normal conditions these compounds are stable and do not decompose. Most of these are noncombustible substances. Only 2-methyl-3-butyn-2-ol and 3-methyl-1-pentyn-3-ol are flammable liquids [flash point (open cup) 25° and 38°C, respectively].

4.21 BENZYL ALCOHOL

Formula $C_6H_5CH_2OH$; MW 108.13; CAS [100-51-6]

Structure and functional group:

first member of aromatic alcohol, primary —OH

Synonyms: phenyl methanol; phenyl carbinol; benzene carbinol; benzoyl alcohol; α-hydroxytoluene; α-toluenol

Uses and Exposure Risk

Esters of benzyl alcohol are used in making perfume, soap, flavoring, lotion, and ointment. It finds application in color photography; the pharmaceuticals industry, cosmetics,

TABLE 4.2 Properties, Uses, and Toxicity of Miscellaneous Acetylenic Alcohols[a]

Compound/CAS No.	Structure/Functional Group	Physical Properties and Solubility in water	Uses	Toxicity
2-Methyl-3-butyn-2-ol [115-19-5]	$HC{\equiv}C{-}\overset{\displaystyle OH}{\underset{\displaystyle CH_3}{C}}{-}CH_3$ Tertiary —OH	Liquid; bp 103.6°C; soluble in water, density 0.867	In making isoprene, vitamin A; metal pickling and plating	Low toxicity; oral LD_{50} value (mice): 1900 mg/kg
3-Methyl-1-pentyn-3-ol [77-75-8]	$CH_3{-}CH{-}\overset{\displaystyle OH}{\underset{\displaystyle CH_3}{C}}{-}C{\equiv}CH$ Tertiary —OH	Liquid; bp 121.4°C; soluble in water, 10 g/100 mL; density 0.872	In the production of vitamin A; metal pickling and plating	Moderately toxic; oral LD_{50} value (mice): 610 mg/kg
1-Hexyn-3-ol [105-31-7]	$CH_3{-}CH_2{-}CH_2{-}\underset{\displaystyle OH}{CH}{-}C{\equiv}CH$ Secondary —OH	Liquid; bp 142°C; sparingly soluble in water, 3.8 g/100 mL; density 0.882	Corrosion inhibitor	Moderately toxic; oral LD_{50} value (mice): 154 mg/kg

Compound	Structure	Physical properties	Uses	Toxicity
4-Ethyl-1-octyn-3-ol [5877-42-9]	$CH_3-(CH_2)_3-CH-CH-CH-OH$ with C_2H_5 and $C{\equiv}CH$ Secondary —OH	Liquid; bp 197.2°C; insoluble in water; density 0.873	Corrosion inhibitor	Low toxicity; oral LD_{50} value (mice): 1830 mg/kg
2,4,7,9-Tetramethyl-5-decyne-4,7-diol [126-86-3]	(acetylenic diol structure) Tertiary —OH	Solid; mp 37° C; bp 260°C; slightly soluble in water, 0.12 g/100 mL (at 20°C)	Antifoaming wetting agent	Low toxicity; oral LD_{50} value (mice): 4000 mg/kg

[a]These acetylenic alcohols are soluble in alcohols and ethers

131

and leather dyeing; and as an insect repellent. It occurs in natural products such as oils of jasmine and castoreum.

Physical Properties

Colorless liquid with a mild sweet odor; hygroscopic; bp 205.5°C; mp −15°C; density 1.044 at 25°C; soluble in water (4 g/100 mL), readily mixes with alcohol, ether, and chloroform.

Health Hazard

Benzyl alcohol is a low acute toxicant with a mild irritation effect on the skin. The irritation in 24 hours from the pure compound was mild on rabbit skin and moderate on pig skin. A dose of 750 μg produced severe eye irritation in rabbits. The toxicity of benzyl alcohol is of low order, the effects varying with the species. Oral intake of high concentrations of this compound produced behavioral effects in rats. The symptoms progressed from somnolence and excitement to coma. Intravenous administration in dogs produced ataxia, dyspnea, diarrhea, and hypermotility in the animals.

Adult and neonatal mice treated with benzyl alcohol exhibited behavioral change, including sedation, dyspnea, and loss of motor function. Pretreatment with pyrazole increased the toxicity of benzyl alcohol. With disulfiram the toxicity remained unchanged. The study indicated that the acute toxicity was due to the alcohol itself and not to bezaldehyde, its primary metabolite (McCloskey et al. 1986).

Exposure Limits

No exposure limit is set. Because of its low vapor pressure and low toxicity, the health hazard to humans from occupational exposure should be very low.

Fire and Explosion Hazard

Noncombustible liquid; flash point (open cup) 93°C (200°F); autoignition temperature

436°C (860°F); vapor density 3.7 (air = 1); vapor pressure 0.97 torr at 30°C.

4.22   HIGHER ALIPHATIC ALCOHOLS

These are the monohydric aliphatic alcohols containing six or more C atoms, derived naturally or by synthetic processes. The natural alcohols derived from oils, fats, and waxes are also known as fatty alcohols. Although some of these alcohols are of only low toxicity, a brief discussion is presented here because of their enormous applications. The alcohols containing 6–11 C atoms are plasticizer-range alcohols used as their ester derivatives to produce plasticizers and lubricants. The detergent-range alcohols contain 12 or more C atoms and are used in making detergents and surfactants. They are also used in cosmetics and food additive applications.

These are saturated or unsaturated alcohols containing one hydroxyl group (primary). C_6–C_{12} alcohols are sparingly soluble in water; their solubility decreases with carbon chain length. Higher alcohols are insoluble in water. All the compounds are miscible with ethanol, petroleum ether, and many other organic solvents.

Health Hazard

The toxicity of these alcohols decreases with the increase in their carbon number. n-Hexanol shows very low toxic action in mice. The oral LD_{50} value in rats is reported as 4200 mg/kg. C_6–C_{10} alcohols caused eye and skin irritation in rabbits. Cetyl alcohol may cause depression of the central nervous system. Isostearyl (C_{17}-), myristyl (C_{14}-), and behenyl (C_{22}-) alcohols are all nontoxic substances (Cosmetic, Toiletry and Fragrance Association 1988).

Inhalation of cetyl alcohol vapors (26 ppm) caused slight irritation of the eyes, nose, and throat in test animals: mice, rats, and guinea pigs. Exposure to a concentration of 2220 mg/m^3 resulted in death.

The long-chain saturated and unsaturated alcohols stearyl ($C_{18}H_{37}OH$), oleyl ($C_{18}H_{35}OH$), and octyl dodecanol ($C_{20}H_{41}OH$) at high concentrations produced minimal ocular and mild cutaneous irritation in rabbits. No irritation effect was observed on human skin. Acute oral toxicity studies in rats indicate a very low order of toxicity of undiluted stearyl alcohol and octyldodecanol and 20% oleyl alcohol.

Based on the data available, these alcohols are viewed as safe in the present practices of use and concentration in cosmetics and other applications (Cosmetic, Toiletry and Fragrance Association 1985).

REFERENCES

ACGIH. 1986. *Documentation of the Threshold Limit Values and Biological Exposure Indices*, 5th ed. Cincinnati, OH: American Conference of Governmental Industrial Hygienists.

Belgsir, E. M., H. Huser, J. M. Leger, and C. Lamy. 1987. A kinetic analysis of the oxidation of methanol at platinum-based electrodes by quantitative determination of the reaction products using liquid chromatography. *J. Electroanal. Chem. Interfacial Electrochem.* 225 (1–2): 281–86.

Bercic, G., and J. Levec. 1987. Limits of flammability of methanol–air mixtures: effect of diluent, temperature, and pressure. *Chem. Biochem. Eng. Q.* 1 (2–3): 77–82; cited in *Chem. Abstr. CA 108* (7): 55457h.

Carter, E. A. 1987. Enhanced acetaminophen toxicity associated with prior alcohol consumption in mice: prevention by *N*-acetylcysteine. *Alcohol 4* (1): 69–71.

Church, M. W., B. A. Dintcheff, and P. K. Gessner. 1988. The interactive effects of alcohol and cocaine on maternal and fetal toxicity in the Long-Evans rat. *Neurotoxicol. Teratol. 10* (4): 355–61.

Cosmetic, Toiletry and Fragrance Association. 1985. Final report on the safety assessment of stearyl alcohol, oleyl alcohol and octyl dodecanol. *J. Am. Coll. Toxicol. 4* (5): 1–25.

Cosmetic, Toiletry and Fragrance Association. 1987a. Final report on the safety assessment

of *n*-butyl alcohol. *J. Am. Coll. Toxicol. 6* (3): 403–22.

Cosmetic, Toiletry and Fragrance Association. 1987b. Final report on the safety assessment of 2,3-naphthalenediol. *J. Am. Coll. Toxicol. 6* (3): 353–56.

Cosmetic, Toiletry and Fragrance Association. 1988. Final report on safety assessment of cetearyl alcohol, cetyl alcohol, isostearyl alcohol, myristyl alcohol and behenyl alcohol. *J. Am. Coll. Toxicol. 7* (3): 359–407.

Dogterom, P., G. J. Mulder, and J. F. Nagelkerke. 1988. Allyl alcohol and acrolein toxicity in isolated rat hepatocytes is independent of lipid peroxidation. *Arch. Toxicol. Suppl. 12:* 269–73; cited in *Chem. Abstr. CA 110* (7): 52457p.

Eisenhofer, G., D. G. Lambie, and R. H. Johnson. 1987. Effects of ethanol ingestion on blood pressure reactivity. *Clin. Sci. 72* (2): 251–54.

Floersheim, G. L. 1987. Synergism of organic zinc salts and sulfhydryl compounds (thiols) in the protection of mice against acute ethanol toxicity and protective effects of various metal salts. *Agents Actions 21* (1–2): 217–22; cited in *Chem. Abstr. CA 107* (19): 170411e.

Hegedus, L., N. Rasmussen, V. Ravn, J. Kastrup, K. Krogsgaard, and J. Aldershvile. 1988. Independent effects of liver disease and chronic alcoholism on thyroid function and size: the possibility of a toxic effect of alcohol on the thyroid gland. *Metabol. Clin. Exp. 37* (3): 229–33; cited in *Chem. Abstr. CA 108* (11): 89325d.

Lazaro, F., M. D. Luque de Castro, and M. Valcarcel. 1986. Individual and simultaneous enzymic determination of ethanol and acetaldehyde in wines by flow injection analysis. *Anal. Chim. Acta 185:* 57–64.

Marcus, S. R., M. W. Chandrakala, and H. V. Nadiger. 1988. Liquid peroxidation and ethanol toxicity in rat liver: effect of vitamin E deficiency and supplementation. *Med. Sci. Res. 16* (16): 879–80.

McCloskey, S. E., J. J. Gershanik, J. J. L. Lertora, L. White, and W. J. George. 1986. Toxicity of benzyl alcohol in adult and neonatal mice. *J. Pharm. Sci. 75* (7): 702–5.

McNeil, S. I., M. K. Bhatnagar, and C. J. Turner. 1988. Combined toxicity of ethanol and methylmercury in rat. *Toxicology 53* (2–3): 345–63.

Nadiger, H. A., S. R. Marcus, and M. V. Chandra-kala. 1988. Lipid peroxidation and ethanol toxicity in rat brain: effect of vitamin E deficiency and supplementation. *Med. Sci. Res. 16*(24): 1273–74.

Nelson, B. K., W. S. Brightwell, A. Khan, J. R. Burg, and P. T. Goad. 1989. Lack of selective developmental toxicity of three butanol isomers administered by inhalation to rats. *Fundam. Appl. Toxicol. 12*(3): 469–79.

NFPA. 1986. *Fire Protection Guide on Hazardous Materials*, 9th ed. Quincy, MA: National Fire Protection Association.

NIOSH. 1984. *Manual of Analytical Methods*, 3rd ed. Cincinnati, OH: National Institute for Occupational Safety and Health.

NIOSH. 1986. *Registry of Toxic Effects of Chemical Substances*, ed. D. V. Sweet. Washington, DC: U.S. Government Printing Office.

Nordmann, R., C. Ribiere, and H. Rouach (editors). 1987. Alcohol toxicity and free radical mechanisms, *Proceedings of the First Congress of the European Society for Biomedi-cal Research on Alcoholism*, Sept. 18–19, Paris; *Adv. Biosci. 71*, papers and references therein.

Phillips, S. C. 1987. Neurotoxic interaction in alcohol-treated, thiamine deficient mice. *Acta Neuropathol. 73*(2): 171–76; cited in *Chem. Abstr. CA 107*(19): 174871c.

Pollard, M. E. 1988. Enhancement of ethanol toxicity by caffeine: evidence of synergism. *Biochem. Arch. 4*(2): 117–24; cited in *Chem. Abstr. CA 109*(7): 50107e.

Polson, C. J., and R. N. Tattersall. 1959. *Clinical Toxicology*, pp. 368–85. Philadelphia: J. B. Lippincott.

Sherman, P. D., Jr. 1979. Butyl alcohols. In *Kirk-Othmer Encyclopedia of Chemical Technology*, Vol. 4, pp. 338–45. New York: Wiley-Interscience.

U.S. EPA. 1986. *Test Methods for Evaluating Solid Waste*, 3rd ed., Vol. 1B. Washington DC: Office of Solid Waste and Emergency Response.

5

ALDEHYDES

5.1 GENERAL DISCUSSION

Aldehydes are a class of organic compounds containing the functional group $-CHO$ and are written structurally as

$$\underset{\underset{\displaystyle}{|}}{-}\overset{\displaystyle H}{\underset{\displaystyle}{C}}=O$$

Thus, the carbonyl group ($-\overset{|}{C}=O$) in an aldehyde is bound covalently to a hydrogen atom. Such a $-CHO$ group may attach itself covalently via the single valence electron left on the carbon atom to form an array of compounds with the general structure

$$R-\overset{\displaystyle H}{\underset{\displaystyle}{C}}=O$$

where R is generally an alkyl (C_nH_{2n+1}) or an aryl (aromatic) group. R could constitute any other group, such as an unsaturated alkenyl (C_nH_{2n}) or alkynyl (C_nH_{2n-2}) moiety, an alicyclic such as a cyclohexyl ($-C_6H_{11}$) radical, or a heterocyclic moiety, as in

When R is a H atom, the aldehyde thus formed will have two H atoms attached to the carbonyl group, having a structure

$$H-\overset{\displaystyle H}{\underset{\displaystyle}{C}}=O$$

This is formaldehyde, the simplest member of the aldehyde family, the most abundant and most hazardous of all the aldehydes. In the present book, only those aldehydes that present toxic, corrosive, and flammable hazards are discussed. Those presenting a mild hazard but being used commercially will be mentioned briefly.

Aldehydes are intermediate products in the conversion of primary alcohols to carboxylic acids, or vice versa:

$$\underset{\text{(ethanol)}}{CH_3CH_2OH} \underset{\text{reduction}}{\overset{\text{oxidation}}{\rightleftharpoons}} \underset{\text{(acetaldehyde)}}{CH_3CHO} \overset{\text{oxidation}}{\underset{\text{reduction}}{\rightleftharpoons}}$$

$$\underset{\text{(acetic acid)}}{CH_3COOH}$$

Toxicity

Like many other organic classes of compounds, the low-molecular-weight aldehydes are more toxic than the higher-molecular-weight aldehydes. Formaldehyde, acrolein, and acetaldehyde are the most toxic; those containing five or more carbon atoms in the alkyl chain are least toxic. The toxicity decreases with increase in the carbon chain. Thus decanal, $C_9H_{19}CHO$, is almost nontoxic. Aromatic aldehydes are less toxic than the low-molecular-weight aliphatic aldehydes. However, for heptanal and benzaldehyde, both of which contain the same number of C atoms, the latter, which is aromatic, is more toxic, which may be attributed to the greater number of reactive sites in the molecule.

The reactive site in any aldehyde is the carbonyl group, which is susceptible to oxidation and reduction processes. The toxic metabolites of aldehydes are their oxidation products, which are mostly, but not exclusively, acids, the formation of which is catalyzed by enzymes. The toxicity is attributed to the fact that whether the aldehyde or its metabolite reacts with proteins and hemoglobin-forming compounds, the biological functioning is affected. Such a reaction would depend on the presence of other active sites in the aldehyde, such as C=C unsaturation, acid–base properties, substituents, steric hindrance, and solubility. The latter two properties depend on the chain length; thus the lower aldehydes are more toxic than the higher ones. Similarly, acrolein, with three carbon atoms but containing additional reactive sites because of its carbon-carbon unsaturation, is more toxic than propionaldehyde with the same number of C atoms. As mentioned earlier, due to its reactive sites in the ring in addition to the −CHO functional group, benzaldehyde should exhibit high toxicity. But this is not so, because its solubility (octanol–water partition coefficient) is low. Thus, the toxicity of this compound is low.

Table 5.1 presents the toxicity data of substituted cinnamaldehyde, based on the intraperitoneal LD_{50} values in mice (NIOSH 1986). A chloro substitution in the β-position renders the molecule more toxic than it does with α-substitution. This may be attributed to the steric hindrance, however small, that the chlorine atom may exert, thus preventing the aldehyde group from participating in any reaction. A further chloro or bromo substitution, this time in the benzene ring at the p-position, slightly increases the toxicity. A similar effect was noted with other electron-withdrawing groups in the p-position, thus making the ring more reactive toward electron-donor nitrogen and phosphorus atoms in the protein molecules. The toxicity expected to result from a fluoro substitution, however, is anomalous to the experimental value and does not fit into the foregoing explanation. More experimental data are required to propose a detailed structure-related mechanism.

Flammability

Low-molecular-weight aldehydes are highly flammable, the flammability decreasing with increasing carbon chain length. Based on the flash point data, the flammability of some of the lower aliphatic aldehydes is as follows, in decreasing order: formaldehyde (C_1) > acetaldehyde (C_2) > propionaldehyde (C_3) > acrolein (C_3-double bond) > isobutyraldehyde (C_4) > n-butyraldehyde (C_4) > n-valeraldehyde (C_5) $\geq$ crotonaldehyde (C_4-double bond). It may be noted that for a given carbon number, saturated aldehydes exhibit higher flammability than those having olefinic double bonds. Autoignition temperatures of these aldehydes reflect the same pattern, as presented in Table 5.2. Flammability of an aldehyde decreases with substitution in the alkyl group. Acetaldehyde is extremely flammable, with flash point (closed cup) $-38°C$ and vapor pressure 750 torr, while its chloro-substituted product, chloroacetaldehyde, is a combustible liquid with a flash point of $87.8°C$ and a vapor pressure of 100 torr.

TABLE 5.1 Toxicity of Substituted Cinnamaldehyde

Compound/CAS No.	Formula/MW/Structure	Toxicity	LD$_{50}$
α-Chlorocinnamaldehyde (2-chloro-3-phenyl2-propenal) [18365-42-9]	C$_9$H$_7$ClO 166.6	Mutagen	Data not available
β-Chlorocinnamaldehyde (3-chloro-3-phenyl2-propenal) [40133-53-7]	C$_9$H$_7$ClO 166.6	Causes tremor and muscle contraction; may have adverse effect on lungs and respiratory tract	220 mg/kg: intraperitoneal (mice)
β,p-Dichlorocinnamaldehyde [14063-77-5]	C$_9$H$_6$Cl$_2$O 201.05	Causes tremor and muscle contraction; may have adverse effect on lungs and respiratory tract	170 mg/kg: intraperitoneal (mice)
β-Chloro-p-fluorocinnamaldehyde [55338-97-1]	C$_9$H$_6$ClFO 184.6	Causes tremor and muscle contraction; may have adverse effect on lungs and respiratory tract	220 mg/kg: intraperitoneal (mice)

(continued)

TABLE 5.1 (Continued)

Compound/CAS No.	Formula/MW/Structure	Toxicity	LD$_{50}$
β-Chloro-p-bromocinnamaldehyde [14063-78-6]	C$_9$H$_6$BrClO 245.5 Cl C=CH—CHO Br	Causes tremor and muscle contraction; may have adverse effect on lungs and respiratory tract	158 mg/kg: intraperitoneal (mice)
β-Chloro-p-iodocinnamaldehyde [55404-82-5]	C$_9$H$_6$ClIO 292.5 Cl C=CH—CHO I	Causes tremor and muscle contraction; may have adverse effect on lungs and respiratory tract	188 mg/kg: intraperitoneal (mice)
β-Chloro-p-methylcinnamaldehyde [40808-08-0]	C$_{10}$H$_9$ClO 180.6 Cl C=CH—CHO CH$_3$O	Causes tremor and muscle contraction; may have adverse effect on lungs and respiratory tract	236 mg/kg: intraperitoneal (mice)
β-Chloro-p-methoxycinnamaldehyde [14063-79-7]	C$_{10}$H$_9$ClO$_2$ 196.6 Cl C=CH—CHO CH$_3$O	Causes tremor and muscle contraction; may have adverse effect on lungs and respiratory tract	187 mg/kg: intraperitoneal (mice)
β-Chloro-p-nitrocinnamaldehyde [2888-10-0]	C$_9$H$_6$ClNO$_3$ 211.6 Cl C=CH—CHO O$_2$N	Causes tremor and muscle contraction; may have adverse effect on lungs and respiratory tract	168 mg/kg: intraperitoneal (mice)

TABLE 5.2   Autoignition Temperatures of Aldehydes

Aldehyde	Structure	Autoignition Temperature ($°C$)
Formaldehyde	$HCHO$	
Acetaldehyde	CH_3CHO	175
Propionaldehyde	CH_3CH_2CHO	207
Acrolein	$CH_2=CH_2CHO$	234
n-Butyraldehyde	$CH_3CH_2CH_2CHO$	218
Isobutyraldehyde	$(CH_3)_2CHCHO$	196
Crotonaldehyde	$CH_3CH=CHCHO$	232
n-Valeraldehyde	$CH_3CH_2CH_2CH_2CHO$	222

Low aromatic aldehydes, on the other hand, are combustible or nonflammable liquids. With regard to their flammability and formation of explosive mixtures with air, they present a low hazard. They exhibit relatively much higher flash points and autoignition temperatures and lower vapor pressures than do those of their aliphatic counterparts.

Disposal/Destruction

Hazardous aldehydes are burned in a chemical incinerator equipped with an afterburner and scrubber. Current ongoing research suggests that other methods may be applicable. Mention will be made throughout the book of various disposal methods. Some are described under ketones (see Chapter 32), which may as well be applicable as well to aldehydes. A laboratory method based on potassium permanganate oxidation is described under the disposal/destruction of acetaldehyde (see Section 5.3).

Analysis

The aldehyde functional groups can be tested by classical wet methods. Individual aldehydes can be analyzed by various instrumental techniques, such as GC, HPLC, GC/MS, colorimetry, polarography, and FTIR. Of these, GC, GC/MS, and HPLC are referred to here because of their versatility and wide application. Although a flame ionization detector is commonly employed in GC, a thermal conductivity detector can also be suitable, especially for lower aliphatic aldehydes. The advantage of FID is that aqueous samples can be injected straight into the column.

HPLC, GC/MS, and sometimes GC techniques require the derivatization of aldehydes. This is highly recommended for C_1 to C_4 aldehydes. The advantages are that (1) retention time is shifted (delayed), so that peaks do not coelute with the solvents; (2) volatile compounds are stabilized, so the loss due to evaporation is reduced; and (3) when free aldehydes are analyzed by GC/MS, electron-impact ionization produces ions the same as those that would result from the corresponding alcohols: therefore, the mass selective detector cannot distinguish between a low-molecular-weight aldehyde and its alcohol. This drawback is overcome by converting the aldehyde into its derivative. There are several compounds that can be used to derivatize an aldehyde. Some of these are 2,4-dinitrophenylhydrazine, semicarbazine, thiosemicarbazine, oximes, and the commercial Girard-T reagent, which is (carboxymethyl)trimethylammonium chloride hydrazide.

Trace determination of aldehydes in water in the range 1–10 ppb can be carried out effectively by an HPLC method. Takami et al. (1985) reported a method using a PTFE sampling cartridge packed with a moderately sulfonated cation-exchange resin charged with 2,4-dinitrophenylhydrazine. The water sample was passed through the resin. The 2,4-dinitrophenylhydrazone derivatives of the aldehydes were eluted with acetonitrile and analyzed by HPLC with a 3-μm ODS column. Other equivalent cartridges and columns have been used.

Derivatives of aldehydes can be analyzed by TLC. Infrared (IR) spectra can exhibit the presence of an aldehyde functional group. Due to the stretching mode of the C−H bond of the formyl group, characteristic absorption near 2720 cm^{-1} is produced. Owing to the −CO stretching mode, saturated aldehydes show absorption in the region 1740–1720 cm^{-1}, α,β-olefinic aldehydes in the region 1705–1685 cm^{-1}, and more conjugated systems in the region 1677–1664 cm^{-1}.

5.2 FORMALDEHYDE

EPA Classified Toxic Waste, RCRA Waste Number U122; DOT Label: Combustible Liquid (aqueous solutions), UN 1198, UN 2209

Formula HCHO; MW 30.03; CAS [50-00-0]

Structure and functional group:

$$\begin{array}{c} \text{H} \\ | \\ \text{H}-\text{C}=\text{O} \end{array}$$

simplest member of aliphatic aldehydes

Synonyms: methanal; methyl aldehyde; methylene oxide; methylene glycol; oxymethylene; oxymethane; paraform; morbicid; formalin (the latter two are a solution of 37% by weight)

Uses and Exposure Risk

Formaldehyde is used in the manufacture of phenolic resins, cellulose esters, artificial silk, dyes, explosives, and organic chemicals. Other uses are as a germicide, fungicide, and disinfectant; in tanning, adhesives, waterproofing fabrics, and for tonic and chrome

printing in photography; and for treating skin diseases in animals. In vitro neutralization of scorpion venom toxicity by formaldehyde has been reported (Venkateswarlu et al. 1988).

Formaldehyde constitutes about 50% of all aldehydes present in the air. It is one of the toxic effluent gases emitted from burning wood and synthetic polymeric substances such as polyethylene, nylon 6, and polyurethane foams. Firefighters have a greater risk to its exposure. Incapacitation from the toxic effluent gases is reported to occur more rapidly from the combustion of synthetic polymers than from that of natural cellulose materials.

Formaldehyde is directly emitted into the air from vehicles. It is released in trace amounts from pressed wood products such as particleboard and plywood paneling, from old "sick" buildings, and from cotton and cotton–polyester fabrics with selected cross-link finishes. Formation of formaldehyde has been observed in some frozen gadoid fish due to enzymic decomposition of the additive trimethylamine oxide (Rehbein 1985). Its concentration can build up during frozen storage of fish (Leblanc and Leblanc 1988; Reece 1985). It occurs in the upper atmosphere, cloud, and fog; it also forms in photochemical smog processes.

Health Hazard

Formaldehyde can present a moderate to severe health hazard injuring eyes, skin, and respiratory system. It is a mutagen, teratogen, and probably carcinogenic to humans. It is a severe eye irritant. An amount of 0.1 mg/day caused severe eye irritation in rabbits. In humans a 1-ppm concentration can cause burning in the eyes. Its lachrymating effect on humans can become intolerable at 10 ppm in air. Exposure to 50 mg/day caused moderate skin irritation in rabbits. Contact with formaldehyde solution or its resins can cause sensitization to dermatitis. Exposure by humans at 1–2 ppm of formaldehyde in air can exhibit the symptoms of itching eyes, burning nose, dry and sore throat, sneezing, coughing, headache, feeling thirsty, and disturbed sleep.

Inhalation of a high concentration of formaldehyde can lead to death. Animal studies indicated that exposure to 700 ppm for 2 h was fatal to mice; cats died of an 8-h exposure. Chronic exposure to a 40-ppm concentration was lethal to mice, with symptoms of dyspnea, listlessness, loss of body weight, inflammation in the nasal tissues, and pathological changes in the nose, larynx, trachea, and bronchi. In addition to this, pathological changes in ovaries and uterus were observed in female mice. Neurotoxicity studies indicate that acute low-level exposure (5–20 ppm) for 3 h/day for 2 days can result in decreased motor activity in rats associated with neurochemical changes in dopamine [51-61-6] and 5-hydroxytryptamine [50-67-9] neurons. Similar to acrolein, formaldehyde can induce nasal toxicity; short-term exposure to 6–15 ppm can cause respiratory epithelial injury—the severity related to concentration (Monteiro-Riviere and Popp 1986). The injury as detected from SEM and TEM studies was not specific to cell but to the area of exposure (Popp et al. 1986). In a subchronic (13-week) inhalation toxicity study of formaldehyde in rats, Woutersen et al. (1987) reported that the compound was hepatotoxic to rats only at >10 ppm concentration. At 20 ppm concentration, the treatment-related changes observed were stained coats, yellowing of the fur, growth retardation, and degeneration of nasal respiratory epithelium.

Subacute oral toxicity of formaldehyde and acetaldehyde in rats has been reported by Til et al. (1988). In a drinking water study a dose level of 125 mg/kg formaldehyde per day was fed to rats for 4 weeks. The symptoms noted were yellow discoloration of the fur, decreased protein and albumin levels in their blood plasma, hyperkeratosis in the forestomach, and gastritis. However, no adverse effect was noted at a dose of 25 mg/kg/day. By comparison, only a low toxicity was observed with acetaldehyde at a dose of 675 mg/kg/day.

Upreti et al. (1987) studied the mechanism of toxicity of formaldehyde in male rats by intraperitoneal injection of ^{14}C-labeled HCHO. In 72 h 41% of the dose was eliminated through expired air and another 15% in urine. A significant level of radioactivity was detected bound to subcellular microsomal fractions, DNA, RNA, protein, lipid fractions of liver, and spleen tissues. The study indicates that formaldehyde undergoes rapid absorption and distribution in the body.

Formaldehyde-induced mutation has been studied in both human lymphoblasts and *Escherichia coli* (Crosby et al. 1988). In human lymphoblasts, it induced large losses of DNA. In *E. coli*, varying concentrations of formaldehyde produced different genetic alterations.

Animal studies indicate that it can cause cancer. There is sufficient evidence of its carcinogenicity in test species resulting from its inhalation. It caused olfactory tumor. Subcutaneous dosages produced skin tumors at the sites of applications. Similar tumorigenic properties of formaldehyde are expected in humans. The evidence of its carcinogenic behavior in humans, however, is limited.

Fire and Explosion Hazard

Formaldehyde in pure gas form is extremely flammable, and the commercial aqueous solution can be moderately flammable—the gas vaporizes readily from the solution. Flammable data: bp of 37% solution 101°C (bp pure gas −19°C); flash point (closed cup) for methanol-free 37% aqueous solution with 15% methanol 50°C (122°F); autoignition temperature 300°C (572°F); fire-extinguishing agent: water spray, dry chemical, alcohol foam, or CO_2; a water spray may be used to flush and dilute the spills.

Formaldehyde forms explosive mixtures with air in the range 7–73% by volume. It explodes when heated with NO_2 at 180°C, burns with explosive violence when treated with a mixture of perchloric acid and aniline, and explodes with concentrated performic acid (peroxyformic acid),

$$H-\underset{\underset{O}{\|}}{C}-O-O-H$$

Spillage

If there is a spill, use a water spray to reduce the vapors. Use sand or other noncombustible material to absorb the spill. Injection of hydrogen peroxide is suggested as a remedial measure against contamination of groundwater from the spill (Staples 1988).

Disposal/Destruction

Formaldehyde is burned in a chemical incinerator. It may be destroyed by biodegradation at a concentration of 100–2300 ppm by an activated sludge process (Bonastre et al. 1986).

Analysis

Formaldehyde can be analyzed by several instrumental techniques, such as GC, colorimetry, polarography, and GC/MS. The GC method involves the passage of air through a solid sorbent tube containing 2-(benzylamino)ethanol on Chromosorb 102 or XAD-2. The derivative, 2-benzyloxazolidine, is desorbed with isooctane and injected into GC equipped with an FID (NIOSH 1984, Method 2502). Carbowax 20M or a fused-silica capillary column is suitable. 2-Benzyloxazolidine peak is sometimes masked under the peaks, due to the derivatizing agent or its decomposition and/or polymeric products. The isooctane solution of 2-benzyloxazolidine can readily be analyzed by GC/MS without any interference using a capillary DB-5 column (Patnaik 1989).

Alternatively, air is passed through 1-µm PTFE membrane and 1% sodium bisulfite solution. The solution is treated with chromotropic and sulfuric acid mixture. The color development due to formaldehyde is measured by a visible spectrophotometer at 580-nm absorbance (NIOSH 1984, Method 3500). In polarography analysis, a Girard-T reagent is used. Formaldehyde forms a derivative that is analyzed by sampled DC polarography (NIOSH 1984, Method 3501). Auel et al. (1987) reported a similar electrochemical analysis of industrial air using an iridium electrode backed by a gas-permeable fluorocarbon-based membrane.

Igawa et al. (1989) have reported analysis of formaldehyde and other aldehydes in cloud and fogwater samples by HPLC with a postcolumn reaction detector. The aldehydes were separated on a reversed-phase C_{18} column, derivatized with 3-methyl-2-benzothiazolinone hydrazone, and detected at 640 nm.

Fluorescence-based liquid-phase analysis for selective determination of formaldehyde and other gases is reported (Dong and Dasgupta 1987; Dasgupta 1987). In this analytical method fluorescence of 3,5-diacetyl-1,4-dihydrolutidine formed upon reaction of formaldehyde with ammonium acetate and 2,4-pentanedione is monitored with a filter fluorometer. Draeger tubes used to monitor formaldehyde concentrations in air can give excessively high results (Balmat 1986).

5.3 ACETALDEHYDE

EPA Classified Toxic Waste; RCRA Waste Number U001; DOT Label: Flammable Liquid, UN 1089

Formula CH_3CHO; MW 44.05; CAS [75-07-0]

Structure and functional group:

$$H_3C-\underset{}{\overset{\overset{\displaystyle H}{|}}{C}}=O$$

contains an aldehyde carbonyl group (−CHO)

Synonyms: ethanal; acetic aldehyde; ethyl aldehyde

Uses and Exposure Risk

Acetaldehyde is used in producing acetic acid, acetic anhydride, cellulose acetate, synthetic pyridine derivatives, pentaerythritol, terephthalic acid, and many other raw materials. Release of acetaldehyde from poly(ethylene terephthalate) (PET) bottles into carbonated mineral waters has been observed (Lorusso et al. 1985); 180 ppm was detected in samples kept for 6 months at 40°C.

Physical Properties

Colorless mobile liquid; pungent odor, fruity smell when diluted; bp 20.8°C; mp −121°C; density 0.7846 at 15°C, vapors heavier than air (vapor density 1.52); soluble in water, alcohol, ether, acetone, and benzene.

Health Hazard

Acetaldehyde is moderately toxic through inhalation and ingestion routes. Ingestion can result in conjunctivitis, central nervous system (CNS) depression, eye and skin burns, and dermatitis. Large doses can be fatal. Because of its metabolic link to ethanol, its intoxication consequences are similar to those of chronic ethanol intoxication.

Inhalation can produce irritation of the eyes, nose, and throat, and narcotic effects. High concentrations can cause headache, sore throat, and paralysis of respiratory muscles. Prolonged exposure can raise blood pressure and cause a decrease in red and white blood cells. A 4-h exposure to 1.6% acetaldehyde in air was lethal to rats (ACGIH 1986).

The functional groups $-NH_2$, $-OH$, and $-SH$ in the three-dimensional protein molecules are susceptible to $-CHO$ attack. Acetaldehyde can therefore bind to liver protein and hemoglobin to form stable adducts. Such covalent binding probably alters the biological functions of protein and hemoglobin and thus contributes to its toxicity.

Rats subjected to inhalation of acetaldehyde for 21 days showed the presence of such "bound" aldehyde adducts in their intracellular medium. A control experiment on unexposed rats, however, showed similar adducts, but at a low concentration. This could probably have formed from trace aldehyde generated from intestinal microbial fermentation of alcohols.

In a study on chronic inhalation toxicity of acetaldehyde on rats, the compound was found to effect increased mortality, growth retardation, and nasal tumors (Woutersen et al. 1986). The study indicates that acetaldehyde is both cytotoxic and carcinogenic to the nasal mucosa of rats. Investigating the toxicity of tobacco-related aldehydes in cultured human bronchial epithelial cells, Graftstrom et al. (1985) reported that acetaldehyde was weakly cytotoxic, less so than acrolein and formaldehyde.

Exposure Limits

TLV-TWA 180 mg/m³ (100 ppm) (ACGIH), 360 mg/m³ (200 ppm) (NIOSH); STEL 270 mg/m³ (150 ppm); IDLH 10,000 ppm.

Fire and Explosion Hazard

Acetaldehyde is highly flammable, flash point (closed cup) −38°C (−36.4°F) (Merck 1989); autoignition temperature 175°C; vapor pressure 750 torr at 15°C. Its low flash point and autoignition temperature, coupled with high vapor pressure, make it a dangerous fire and explosion hazard. Explosive limits of mixtures with air are 4–60% by volume. Active surfaces may ignite and detonate fuels containing acetaldehyde. It is susceptible to forming peroxide, which can catch fire or explode.

Hazardous Reaction Products

Acetaldehyde polymerizes on treatment with mineral acids such as H_2SO_4, HCl, and H_3PO_4 to form paraldehyde (toxic — causes respiratory depression and cardiovascular collapse) at ambient temperature, and metaldehyde (toxic to the intestine, kidney, and liver) at low temperature (subambient). It forms ethyl acetate (irritant, narcotic) with aluminum ethoxide catalyst; peracetic acid (explodes at 110°C, corrosive) on oxidation with or without a catalyst (cobalt salt); glyoxal (explodes with air, irritant) on oxidation with HNO_3 or selenious acid; acetaldehyde ammonia (eye and mucous membrane irritant) with NH_3 and H_2 in the presence of Ni at 50°C; chloroacetaldehyde (highly corrosive and strong irritant) with chlorine at room temperature and with chloral (strong irritant) at 80–90°C; acetyl chloride (highly corrosive, dangerous eye irritant, flammable and explosion hazard) with chlorine in a gas-phase reaction; acetyl bromide (eye irritant, violent reaction with water) on bromination; 1,1-dichloroethane (toxic — irritant) with PCl_5; and phosgene (highly toxic, fatal at high concentrations) with CCl_4 in the presence of anhydrous $AlCl_3$. It decomposes above 400°C, forming CO and methane.

Disposal/Destruction

Acetaldehyde is burned in a chemical incinerator equipped with an afterburner and scrubber. It may be disposed of in a drain (not recommended, highly volatile) in small amounts, <100 g at a time, mixed with 100 volumes of water, as it is biodegradable.

In the laboratory it may be destroyed by $KMnO_4$ oxidation (National Research Council 1983). To an aqueous solution of 0.1 mol aldehyde, excess $KMnO_4$ solution is added slowly. The mixture is refluxed until the purple color decolorizes. This is followed by addition of some more $KMnO_4$ solution and heating. It is cooled and acidified with $6\,N$ H_2SO_4. (Concentrated H_2SO_4 should not be mixed freely with $KMnO_4$ as it forms Mn_2O_7 — explosion hazard.) Sodium bisulfite is added with stirring to reduce Mn to its divalent state. When the purple color disappears and the solid MnO_2 dissolves, the mixture is washed down the drain with large amounts of water.

Analysis

Acetaldehyde is analyzed by GC or HPLC. The acetaldehyde level in blood and liver can be estimated by forming its derivative with 2,4-dinitrophenylhydrazine and analyzing by HPLC. A GC headspace method using FID is equally suitable. Other analytical methods include colorimetry using thymol blue on silica gel, and derivatizing to 2,4-dinitrophenylhydrazone or semicarbazone followed by polarography or paper chromatography. It can be tested by wet methods such as reduction of Fehling's solution and Tollens' reagent; mercurimetric oxidation; sodium bisulfite/iodometry; and argentometric titration. Its presence in air can be estimated quantitatively by converting it to its derivative, 2,4-dinitrophenylhydrazone, and analyzing by HPLC using a UV detector. Analysis in air by NIOSH (1984) Method 3507 involved bubbling 6–60 L of air through a Girard-T solution at pH 4.5 and measuring the derivative at 245 nm (HPLC/UV detector). The working range is 170–670 mg/m^3 (18–372 ppm).

Jones et al. (1985) estimated the concentration of acetaldehyde in blood from analysis in breath. The method is based on liquid–air partition coefficients of acetaldehyde determined by GC-FID. Jones et al. (1986) reported a GC-headspace method for its analysis in wine. Habboush and co-workers (1988) have reported the analysis of acetaldehyde and other low-molecular-weight aldehydes in automobile exhaust gases by GC-FID.

5.4 ACROLEIN

EPA priority pollutant; EPA Classified Acute Hazardous Waste, RCRA Waste Number P003; DOT Label: Flammable Liquid and Poison, UN 1092

Formula C_2H_3CHO; MW 56.07; CAS [107-02-8]

Structure and functional group:

$$CH_2{=}CH{-}\overset{\displaystyle H}{\underset{\displaystyle |}{C}}{=}O$$

—CHO (aldehyde carbonyl), simplest member of the class of unsaturated aldehydes

Synonyms: 2-propenal; 2-propen-1-one; allyl aldehyde; acrylaldehyde; acraldehyde; ethylene aldehyde; aqualin; biocide

Uses and Exposure Risk

Acrolein is used as an antimicrobial agent to prevent the growth of microbes against plugging and corrosion, to control the aquatic weed and algae, in slime control in paper manufacturing, as a tissue fixative, and in leather tanning.

Because of its widespread use it occurs in the environment — in air and water. After formaldehyde it is the second most abundant aldehyde, constituting 5% of total aldehydes in air. Acrolein is one of the toxic gases produced in a wood or building fire or when polyethylene or other polymer substances burn (Morikawa 1988; Morikawa and Yanai 1986). Firefighters are at greater risk of exposure to this gas.

Physical Properties

Colorless volatile liquid; bp 53°C; mp −87°C; density 0.8427 at 20°C; moderately soluble in water, readily mixes with alcohol and ether.

Health Hazard

Acrolein is a highly toxic compound that can severely damage the eyes and respiratory system and burn the skin. Ingestion can cause acute gastrointestinal pain with pulmonary congestion.

LD_{50} value, oral (mice): 40 mg/kg

Acrolein is a strong lachrymator and a nasal irritant. Direct contact of liquid in the eyes may result in permanent injury to the cornea. Inhalation can result in severe irritation of the eyes and nose. A concentration of 0.5 ppm for 12 min can cause intolerable eye irritation in humans. In rats, exposure to a concentration of 16 ppm acrolein in air for 4 h was lethal.

Acrolein can be absorbed through the skin; the spillage of liquid can cause severe chemical burns. Skin contact may lead to chronic respiratory disease and produce delayed pulmonary edema. Subcutaneous administration of acrolein produced degeneration of fatty liver and a general anesthetic effect.

LD_{50} value, subcutaneous (mice): 30 mg/kg

Based on the available data, a concentration of 68 and 55 ppb may be toxic to aquatic life in fresh and salt water, respectively (U.S. EPA 1980). A concentration as low as 21 ppb may produce chronic toxicity to freshwater aquatic life. Acrolein is reported to be more toxic to aquatic organisms than are phenol, chloro- and nitrophenols, aniline, o-xylene, and other toxic compounds (Holcombe et al. 1987). Rainbow trout, spinally transected, were exposed to an acutely toxic aqueous concentration of acrolein to monitor their respiratory–cardiovascular responses. A steady increase was recorded in their cough rate. The ventilation rate, oxygen utilization, and heart rate steadily fell throughout their period of survival.

In a study on inhalation toxicity in rats, Crane et al. (1986) observed that the exposure to 1 atm of acrolein vapors caused physical incapacitation. The animals lost the ability to walk and expired. In a study on cytotoxicity of tobacco-related aldehydes to cultured human bronchial epithelial cells, acrolein was found to be more toxic than formaldehyde (Graftstrom et al. 1985). Both compounds induced DNA damage.

Certain sulfur compounds, such as dithiothreitol [3483-12-3] and dimercaptopropanol [59-52-9], reacted with acrolein to reduce its toxicity (Dore et al. 1986). Such protection against its toxicity was observed in isolated rat hepatocytes.

Exposure Limits

TLV-TWA 0.25 mg/m^3 (0.1 ppm) (ACGIH and OSHA); STEL 0.8 mg/m^3 (0.3 ppm); IDLH 5 ppm (NIOSH).

Fire and Explosion Hazard

Highly flammable; flash point (open cup) $-18°C$ ($-0.5°F$) (Aldrich 1989), closed cup $-36°C$ ($-33°F$); vapor pressure 214 torr at 20°C; vapor density 1.93 (air = 1); vapor may travel a considerable distance to a source of ignition and flash back; autoignition temperature 234°C (453°F) (unstable); fire-extinguishing agent: "alcohol" foam, dry chemical, or CO_2; use a water spray to keep the fire-exposed containers cool, to flush and dilute the spill, and to disperse the vapor.

Vapor forms an explosive mixture with air in the range 2.8–31.0% by volume of air. It undergoes self-polymerization, liberating heat. It polymerizes at elevated temperatures. Closed containers may rupture violently as a result of polymerization. The reaction may become extremely violent in contact with alkaline substances such as caustic soda, amines, or ammonia. Contact with acid can result in polymerization, liberating heat. However, the reaction is not as violent as that with caustic soda or caustic potash.

Storage and Handling

Acrolein should be stored under an inert atmosphere below 38°C. Steel, ceramic, glass, Teflon, silicone rubber, and containers with baked phenolic coatings are suitable for storage (NFPA 1986). Copper, zinc, or materials of polyethylene or other organic coatings should not be used. Hydroquinone or 4-methoxyphenol is added to acrolein to inhibit its polymerization. It should be stored at a pH value of 5–6, adjusted with acetic acid. In case of acid or alkali contamination, a buffer solution should be added.

Disposal/Destruction

Acrolein is destroyed by controlled burning in an incinerator. It is also destroyed by biodegradation after being treated with dilute NaOH or sodium bisulfite solution. Acrolein and other toxic pollutants from industrial laundry wastewater can be removed by treatment involving lime coagulation, carbon adsorption, and ultrafiltration (Van Gils and Pirbazari 1985).

Analysis

Acrolein in water can be analyzed by a purge and trap GC method, using a flame ionization detector (U.S. EPA 1984, Method 603). A column containing Porapak QS (80/100 mesh) or Chromosorb 101 (60/80 mesh) may be suitable. Wastewater and hazardous waste samples can be analyzed by GC/MS (U.S. EPA 1984, Methods 624 and 1624 and U.S. EPA 1986 Method 8240/SW-846) on any column suitable for volatile organics. Characteristic ions are 56, 55, and 54 (electron-impact ionization).

Acrolein in air in the range 0.12–1.5 mg/m^3 can be analyzed by NIOSH (1984) Method 2501. Air is passed over a solid sorbent tube containing 2-(hydroxymethyl)piperidine on XAD-2. It is converted to 9-vinyl-1-aza-8-oxabicyclo[4.3.0]nonane, desorbed with toluene and analyzed by GC with a nitrogen specific detector. 5% SP-2401-DB on Supelcoport (100–120 mesh) is a suitable column.

5.5   PROPIONALDEHYDE

DOT Label: Flammable Liquid, UN 1275
Formula C_2H_5CHO; MW 58.08; CAS [123-38-6]
Structure and functional group:

$$H_3C-CH_2-\overset{\displaystyle H}{\underset{\displaystyle |}{C}}=O$$

reactive site: carbonyl ($-\overset{|}{C}=O$) group

Synonyms: propyl aldehyde; propionic aldehyde; propanal; methyl acetaldehyde

Uses and Exposure Risk

Propionaldehyde is used in the production of propionic acid, propionic anhydride, and many other compounds. It is formed in the oxidative deterioration of corn products, such as corn chips. It occurs in automobile exhaust gases.

Physical Properties

Colorless liquid with a fruity but suffocating odor; bp 48.8°C; freezes at −81°C; density 0.806 at 20°C; soluble in water (17%), alcohol, and ether.

Health Hazard

Propionaldehyde is a mild irritant to human skin and eyes. The irritation effect from 40 mg was severe in rabbits' eyes. The toxicity of this compound observed in test animals was low. Subcutaneous administration in rats exhibited the symptoms of general anesthetic effect, convulsion, and seizure. Inhalation toxicity was determined to be low. A concentration of 8000 ppm (19,000 mg/m^3) in air was lethal to rats.

LD$_{50}$ value, oral (rats): 1400 mg/kg

LD$_{50}$ subcutaneous (rats): 820 mg/kg

Fire and Explosion Hazard

Highly flammable; flash point (open cup) <-7°C (<19°F); vapor density 2.0 (air = 1); vapor heavier than air and can travel a considerable distance to an ignition source and flash back; autoignition temperature 207°C (405°F). Fire-extinguishing agent: "alcohol" foam, CO$_2$, or dry chemical; use a water spray to dilute the spill.

Propionaldehyde forms an explosive mixture with air in the range 2.9–17% by volume of air. Reactions with strong oxidizers and alkaline substances can be exothermic.

Disposal/Destruction

Propionaldehyde is destroyed by burning in a chemical incinerator equipped with an afterburner and scrubber. Permanganate oxidation is a suitable laboratory method of destruction (see Section 5.3).

Analysis

Propionaldehyde is analyzed by GC-FID, using Carbowax 20 or any equivalent column. It is converted into a derivative of 2,4-dinitrophenylhydrazone and analyzed by HPLC at 254 nm or by GC/MS.

5.6 CROTONALDEHYDE

EPA Classified Toxic Waste, RCRA Waste Number U053; DOT Label: Flammable Liquid, UN 1143

Formula C$_3$H$_5$CHO; MW 70.09; CAS [4170-30-3] ([123-73-9] for the *trans*-isomer, *trans*-2-butenal)

Structure and functional group:

$$H_3C-CH=CH-\overset{\overset{\textstyle H}{|}}{C}=O$$

reactive sites: olefinic double bond and the −CHO group

Synonyms: 2-butenal; propylene aldehyde; β-methyl acrolein; crotonic aldehyde

Uses and Exposure Risk

Crotonaldehyde is used in the manufacture of butyl alcohol, butyraldehyde, and in several organic synthesis.

Physical Properties

Colorless liquid turning pale yellow on contact with air or light; pungent suffocating smell; bp 102°C; freezes at −76.5°C; density 0.8531 at 20°C; mixes readily with water and most organic solvents.

Health Hazard

Crotonaldehyde causes severe irritation of the eyes, nose, lungs, and throat. Exposure to a concentration of 12 mg/m³ in air for 10 min can cause burning of the lungs and throat in humans. The symptoms of inhalation toxicity in rats were excitement, behavioral change, convulsion, and death. The same symptoms were observed when crotonaldehyde was administered subcutaneously.

LC_{50} value, inhalation (rats): 4000 mg/m³/ 30 min

LD_{50} value, subcutaneous (rats): 140 mg/kg

Crotonaldehyde is less toxic than acrolein or formaldehyde. The toxic symptoms, however, were similar to those of acrolein. The cis-isomer of crotonaldehyde is mutagenic; it caused cancer in test animals. Oral administration of 2660 mg/kg for 2 years produced tumor in liver in rats. Evidence of carcinogenicity in humans is not yet confirmed.

Exposure Limits

TLV-TWA 6 mg/m³ (2 ppm)(ACGIH); IDLH 400 ppm (NIOSH).

Fire and Explosion Hazard

Highly flammable, flash point (open cup) 12.8°C (55°F) and 53°C (127.4°F) for 93% commercial grade; vapor pressure 30 torr at 20°C; autoignition temperature 232°C (450°F); vapor density 2.4 (air = 1). It can ignite under normal temperature conditions; flashback fire risk. It forms explosive mixtures with air within the range of 2.1–15.5%

by volume. Explosive polymerization reaction can occur at high temperatures or in contact with alkaline compounds such as caustic alkalies, amines, or NH_3. It presents a peroxide hazard. There is a report of an explosion during its reaction with 1,3-butadiene under pressure. Fire-extinguishing agent: foam, dry chemical, or CO_2; in small fires, a water spray may be partially effective.

Disposal/Destruction

Crotonaldehyde can be analyzed by GC-FID, HPLC, and GC/MS after derivatizing with oxazolidine.

Analysis

Crotonaldehyde can be analyzed by GC-FID, HPLC, and GC/MS after derivatizing with oxazolidine.

5.7 n-BUTYRALDEHYDE

DOT Label: Flammable Liquid, UN 1129

Formula C_3H_7CHO; MW 72.1; CAS [123-72-8]

Structure and functional group:

$$H_3C-CH_2-CH_2-\overset{\overset{\displaystyle H}{|}}{C}=O$$

reactive site: ($-\overset{|}{C}=O$) group

Synonyms: n-butanal; butyric aldehyde

Uses and Exposure Risk

n-Butyraldehyde is used to make rubber accelerators, synthetic resins, and plasticizers; and as a solvent.

Physical Properties

Colorless liquid with a pungent odor; bp 75.7°C; mp −99°C; density 0.817 at 20°C;

soluble in alcohol, ether, and acetone, slightly soluble in water.

Health Hazard

n-Butyraldehyde is a mild skin and eye irritant. The liquid in 100% pure form produced moderate irritation on guinea pig skin. The irritation resulting from 20 mg in 24 h on rabbit eye was moderate. A higher dose could produce severe irritation.

Toxicity of *n*-butyraldehyde is very low. The effect is primarily narcotic. No toxic effect, however, was observed in mice from 2-h exposure at a concentration of 44.6 g/m^3. At a higher concentration, 174 g/m^3 for 30 minutes, it exhibited a general anesthetic effect on rats. Subcutaneous administration of a high dose, >3 g/kg, produced the same effect, affecting the kidney and bladder.

Exposure Limit

No exposure limit is set for *n*-butyraldehyde.

Fire and Explosion Hazard

Highly flammable; flash point (closed cup) $-7°C$ (19°F); vapor density 2.48 (air = 1); the vapor is heavier than air and can travel a considerable distance to a source of ignition and flash back; autoignition temperature 218°C (435°F); fire-extinguishing agent: foam, dry chemical, or CO_2; use a water spray to flush the spill and disperse the vapors.

n-Butyraldehyde forms an explosive mixture with air, with LEL and UEL values of 1.9% and 12.5% by volume of air, respectively. It generates heat when mixed with concentrated acids.

Analysis

GC, TIC, GC/MS, and HPLC techniques are employed for analysis after derivatizing *n*-butyraldehyde.

5.8 ISOBUTYRALDEHYDE

DOT Label: Flammable Liquid, UN 2045

Formula C_3H_7CHO; MW 72.1; CAS [78-84-2]

Structure and functional group:

$$\begin{array}{c} H_3C \\ \diagdown \\ CH{-}C{=}O \\ \diagup | \\ H_3C H \end{array}$$

active site: carbonyl ($-\overset{|}{C}{=}O$) group

Synonyms: isobutanal; 2-methyl-1-propanal; isobutyric aldehyde; isobutyl aldehyde; 2-methyl propionaldehyde; valine aldehyde

Uses and Exposure Risk

Isobutyraldehyde is used in the synthesis of cellulose esters, resins, and plasticizers; in the preparation of pantothenic acid and valine; and in flavors.

Physical Properties

Colorless liquid with a pungent smell; bp 64.5°C; mp $-66°C$; density 0.794; soluble in water (11 g/100 mL), ether, acetone, and chloroform.

Health Hazard

Isobutyraldehyde is a moderate skin and eye irritant; the effect may be slightly greater than that of *n*-butyraldehyde. An amount totaling 500 mg in 24 h produced severe skin irritation in rabbits; 100 mg caused moderate eye irritation.

The toxicity of isobutyraldehyde determined on test animals was very low. Exposure to 8000 ppm (23,600 mg/m^3) for 4 h was lethal to rats.

LD_{50} value, oral (rats): 2810 mg/kg

Fire and Explosion Hazard

Extremely flammable; flash point (closed cup) $-18°C$ $(-1°F)$; vapor density 2.5 (air = 1); vapors can travel some distance to an ignition source and flash back; autoignition temperature 196°C (385°F). Fire-extinguishing agent: "alcohol" foam, dry chemical, or CO_2; use a water spray to disperse the vapors and to keep fire-exposed containers cool. Vapor–air mixture is explosive. The LEL and UEL values in air are 1.6% and 10.6% by volume of air, respectively.

Disposal/Destruction

Isobutyraldehyde is burned in a chemical incinerator equipped with an afterburner and scrubber.

Analysis

Isobutyraldehyde is derivatized and analyzed by GC, GC/MS, or HPLC (see Section 5.3).

5.9 n-VALERALDEHYDE

DOT Label: Flammable Liquid, UN 2058
Formula C_4H_9CHO; MW 86.1; CAS [110-62-3]
Structure and functional group:

$$H_3C-CH_2-CH_2-\overset{\overset{\displaystyle H}{|}}{C}=O$$

reactive site: carbonyl group ($-\overset{|}{C}=O$)
Synonyms: n-pentanal; valeric aldehyde; amyl aldehyde; butyl formal

Uses and Exposure Risk

n-Valeraldehyde is used for food flavoring and in resin and rubber products.

Physical Properties

Colorless liquid, bp 103°C; freezes at $-91.5°C$; density 0.8095 at 20°C; slightly soluble in water but mixes readily with alcohol and ether.

Health Hazard

n-Valeraldehyde is a moderate skin and eye irritant. At a high concentration the irritation may be severe; 100 mg/day was severely irritating on rabbits' eyes. Pure liquid caused severe irritation to guinea pig skin. The systemic toxicity of valeraldehyde is very low.

LD_{50} value, skin (rabbits): 4857 mg/kg
LD_{50} value, oral (rats): 3200 mg/kg

Inhalation toxicity is very low. Exposure to 4000 ppm for air was lethal to rats.

Exposure Limit

TLV-TWA 175 mg/m^3 (50 ppm) (ACGIH and OSHA).

Fire and Explosion Hazard

Highly flammable; flash point (open cup) 12°C (54°F); autoignition temperature 222°C (432°F); vapor density 3 (air = 1). Vapor–air mixture is explosive. Fire-extinguishing agent: dry chemical, foam, or CO_2.

Disposal/Destruction

n-Valeraldehyde is burned in a chemical incinerator equipped with an afterburner and scrubber.

Analysis

n-Valeraldehyde can be analyzed by GC-FID using Carbowax 20M on Chromosorb or equivalent column or by GC/MS. An aldehyde functional group can be tested by wet methods (see Section 5.3). Air analysis is performed by passing air through a solid sorbent tube containing 10% 2-(hydroxymethyl)piperidine on XAD-2. The derivative valeraldehyde oxazolidine (9-butyl-1-aza-8-oxabicyclo[4.3.0]nonane) is desorbed with toluene and analyzed by GC-FID (NIOSH 1984, Suppl. 1989, Method 2536).

5.10 CHLOROACETALDEHYDE

EPA Classified Acute Hazardous Waste, RCRA Waste Number P023; DOT Label: Poison, UN 2232

Formula $ClCH_2CHO$; MW 78.5; CAS [107-20-0]

Structure and functional group:

$$ClCH_2-\overset{\overset{\displaystyle H}{|}}{C}=O$$

aldehyde carbonyl ($-\overset{|}{C}=O$) group, chloro substitution in the methyl group

Synonyms: 2-chloroacetaldehyde; 2-chloro-1-ethanal

Uses and Exposure Risk

Chloroacetaldehyde is used in the production of 2-aminothiazole.

Physical Properties

Clear colorless liquid with a pungent smell; bp 85°C; density 1.068 at 20°C; soluble in water, alcohol, and ether.

Health Hazard

Chloroacetaldehyde is a highly toxic and corrosive compound that can injure the eyes, skin, and respiratory system. Exposure to its vapor at high concentrations can produce severe irritation and impair vision. At low concentrations the vapor can cause irritation

and sore eyelids. Brief contact with 40% aqueous solution can result in skin burn and destruction of tissues. A 0.5% dilute solution can still be irritating on skin.

Inhalation of its vapor at the 5-ppm level can irritate the eyes, nose, and throat. Ingestion may result in pulmonary edema. Swallowing a concentrated solution may be fatal. The acute toxicity data are as follows:

LD_{50} value, intraperitoneal (rats): 2 mg/kg

LD_{50} value, oral (rats): 23 mg/kg

LD_{50} value, skin (rabbits): 67 mg/kg

This compound is a mutagen, testing positive in the Ames test.

Exposure Limits

Ceiling 3 mg/m^3 (1 ppm) (ACGIH); IDLH 250 ppm (NIOSH).

Fire and Explosion Hazard

Combustible; flash point (closed cup) 87.8°C (190°F); flash point of 50% aqueous solution 53°C (128°F) (at this concentration it may form insoluble hemihydrate); it forms an explosive mixture with air. Reactions with strong acids and oxidizers are exothermic.

Safety Precautions

When handling this compound, wear appropriate eye goggles, hand gloves, and protective clothing. Avoid skin contact with solution >0.1%. Wash the contaminated skin with soap and water.

Disposal/Destruction

Chloroacetaldehyde is burned in an incinerator equipped with an afterburner and scrubber.

Analysis

Chloroacetaldehyde is analyzed by GC and GC/MS techniques.

5.11 GLYOXAL

Formula $C_2H_2O_2$; MW 58.04; CAS [107-22-2]

Structure and functional group:

$$O=\overset{\overset{\displaystyle H}{|}}{C}-\overset{\overset{\displaystyle H}{|}}{C}=O$$

active sites: the two carbonyl ($-\overset{|}{C}=O$) groups

Synonyms: ethanedial; 1,2-ethanedione; biformal; biformyl; oxal; oxaldehyde; glyoxylaldehyde

Uses and Exposure Risk

Glyoxal is used in the production of textiles and glues and in organic synthesis.

Physical Properties

Yellowish liquid, in the solid form — yellow prism becoming white; vapors green; bp 50.4°C; mp 15°C; density 1.14; soluble in water (polymerizes), alcohol, and ether.

Health Hazard

Glyoxal is a skin and eye irritant; the effect may be mild to severe. Its vapors are irritating to the skin and respiratory tract. An amount of 1.8 mg caused severe irritation in rabbits' eyes. Glyoxal exhibited low toxicity in test subjects. Ingestion may cause somnolence and gastrointestinal pain.

LD_{50} value, oral (guinea pigs): 760 mg/kg

Fire and Explosion Hazard

Noncombustible; polymerizes on standing or when mixed with water; polymerization is exothermic and can become violent if uncontrolled; reactions with strong acids, bases, and oxidizers can become violent; vapor–air mixture is explosive.

Disposal/Destruction

Glyoxal is mixed with a combustible solvent and burned in a chemical incinerator equipped with an afterburner and scrubber.

Analysis

Glyoxal is derivatized and analyzed by GC, HPLC, or GC/MS.

5.12 GLUTARALDEHYDE

Formula $(CH_2)_3(CHO)_2$; MW 100.1; CAS [111-30-8]

Structure and functional group:

$$O=\overset{\overset{\displaystyle H}{|}}{C}-CH_2-CH_2-CH_2-\overset{\overset{\displaystyle H}{|}}{C}=O$$

an aliphatic dialdehyde; reactive sites are the two carbonyl ($-\overset{|}{C}=O$) groups

Synonyms: 1,5-pentanedial; 1,5-pentanedione; glutaral; glutaric dialdehyde; Cidex; Sonacide

Uses and Exposure Risk

Glutaraldehyde is used as a cold sterilizing disinfectant, as fixatives for tissues, in tanning, and in cross-linking proteins.

Physical Properties

Colorless crystals; bp 187°C (decomposes); mp −6°C; density 1.062; soluble in water, alcohol, ether, and other organic solvents.

Health Hazard

Glutaraldehyde is a strong irritant to the nose, eyes, and skin. In rabbits, 250 μg and 500 mg in 24 h produced severe irritation in the eyes and skin, respectively. The corrosive effect on human skin of 6 mg over 3 days was severe. However, the acute toxicity of glutaraldehyde by the oral and dermal routes is low to mild. Ohsumi and Kuroki (1988) determined that the symptoms of acute toxicity of this compound were less severe than those of formaldehyde. But the restraint of growth was more pronounced in mice treated with glutaraldehyde. An oral LD_{50} value of 1300 mg/kg was reported for mice. Inhalation of this compound can cause upper respiratory tract irritation, headache, and nervousness. Mice exposed at 33 ppm showed symptoms of hepatitis.

Exposure Limit

Ceiling (ACGIH) 0.8 mg/m^3 (0.2 ppm).

Fire and Explosion Hazard

Not flammable; it does not present explosion hazards.

Disposal/Destruction

Glutaraldehyde is dissolved in a combustible solvent and burned in a chemical incinerator equipped with an afterburner and scrubber.

Analysis

Glutaraldehyde is analyzed by GC-FID using any column suitable for aldehyde; other analytical techniques are GC/MS and HPLC.

5.13 BENZALDEHYDE

Formula C_6H_5CHO; MW 106.10; CAS [100-52-7]

Structure and functional group:

—CHO group attached to benzene ring, first member of aromatic aldehyde homologous series

Synonyms: benzoic aldehyde; phenyl methanal; oil of bitter almond; benzene carbaldehyde; benzene carbonal

Uses and Exposure Risk

Benzaldehyde is used as an intermediate in the production of flavoring chemicals, such as cinnamaldehyde, cinnamalalcohol, and amyl- and hexylcinnamaldehyde for perfume, soap, and food flavor; synthetic penicillin, ampicillin, and ephedrine; and as a raw material for the herbicide Avenge. It occurs in nature in the seeds of almonds, apricots, cherries, and peaches. It occurs in trace amounts in corn oil.

Physical Properties

Colorless liquid; bp 178°C; mp −26°C; density 1.046 at 20°C; readily soluble in alcohol and ether, moderately soluble in acetone and benzene, and slightly soluble in water (0.6% at 20°C).

Health Hazard

Benzaldehyde exhibited low to moderate toxicity in test animals, the poisoning effect depending on dosage. Ingestion of 50–60 mL may be fatal to humans. Oral intake of a large dose can cause tremor, gastrointestinal pain, and kidney damage. Animal experiments indicated that ingestion of this compound by guinea pigs caused tremor, bleeding from small intestine, and an increase in urine volume; in rats, ingestion resulted in somnolence and coma.

LD_{50} value, oral (guinea pigs): 1000 mg/kg
LD_{50} value, oral (rats): 1300 mg/kg

A 500-mg amount for a 24-h period resulted in moderate skin irritation in rabbits. Because of its low toxicity, high boiling point, and low vapor pressure, the health hazard to humans from exposure to benzaldehyde is very low.

Fire and Explosion Hazard

Combustible, flash point (open cup) 74°C (148°F) and (closed cup) 64.5°C; vapor pressure 0.97 torr at 26°C; autoignition temperature 192°C (377°F). Fire-extinguishing agent: water spray, dry chemical, foam, or CO_2.

Benzaldehyde forms explosive mixture with air; explosive limits are not reported. It can explode when treated with performic acid.

Storage and Handling

Benzaldehyde is stored in stainless steel, glass, Teflon, or phenolic resin-lined containers.

Analysis

Benzaldehyde is analyzed by GC-FID or GC/MS either in its pure form or as a derivative with oxime, semicarbazone, 2,4-dinitrophenylhydrazone, or thiosemicarbazone.

5.14 CINNAMALDEHYDE

Formula C_8H_7CHO; MW 132.17; CAS [104-55-2]

Structure and functional group:

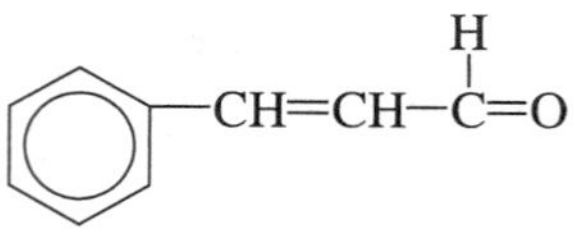

unsaturated aromatic aldehyde; active sites are the aldehyde carbonyl ($-\overset{|}{C}=O$) group, olefinic double bond in the side chain, and the positions in the benzene ring

Synonyms: cinnamal; cinnamic aldehyde; 3-phenyl-2-propenal; 3-phenyl acrolein; zimtaldehyde; cassia aldehyde

Uses and Exposure Risk

Cinnamaldehyde is used in flavor and perfumes. It occurs in cinnamon oils.

Physical Properties

Yellowish liquid with a strong smell of cinnamon; bp 253°C; mp −7.5°C; density 1.0497 at 20°C; immiscible with water, soluble in alcohol, ether, and chloroform.

Health Hazard

Cinnamaldehyde can cause moderate to severe skin irritation. Exposure to 40 mg in 48 h produced a severe irritation effect on human skin. The toxicity of this compound was low to moderate on test subjects, depending on the species and the toxic routes. However, when given by oral route in large amounts, its poisoning effect was severe. Amounts greater than 1500 mg/kg have produced a wide range of toxic effects in rats, mice, and guinea pigs. The symptoms were respiratory stimulation, somnolence, convulsion, ataxia, coma, hypermotility, and diarrhea.

LD_{50} value, oral (guinea pigs): 1150 mg/kg

Cinnamaldehyde is a mutagen. Its carcinogenic effect is not established.

Fire and Explosion Hazard

Combustible liquid; flash point (*trans*-form) 71°C (160°F). Fire-extinguishing agent: "alcohol" foam, CO_2, or dry chemical.

Disposal/Destruction

Cinnamaldehyde is mixed with a combustible solvent and burned in a chemical incinerator equipped with an afterburner and scrubber.

Analysis

A GC, GC/MS, or HPLC technique is suitable for analyzing cinnamaldehyde.

5.15 SALICYLALDEHYDE

$C_6H_4(OH)CHO$;   MW   122.13;   CAS [90-02-8]

Structure and functional group:

an aldehyde and a phenolic $-OH$ group on the benzene ring

Synonyms: *o*-hydroxybenzaldehyde; 2-form-ylphenol; salicylal

Health Hazard

Salicylaldehyde is a skin irritant; 500 mg/day caused moderate irritation to rabbit skin. It can have injurious effects on fertility. Studies on rats indicate that subcutaneous administration of salicylaldehyde in a high dose of >400 mg/kg can produce developmental abnormalities, fetal death, and postimplantation mortality.

The toxicity of this compound, however, is low. No toxic symptoms were noted.

LD_{50} value, oral (rats): 520 mg/kg

LD_{50} value, skin (rats): 600 mg/kg

Fire and Explosion Hazard

Combustible liquid; flash point 78°C (172°F); fire-extinguishing agent: "alcohol" foam. The range for its explosive limits in air is not reported.

Disposal/Destruction

Salicylaldehyde is burned in a chemical incinerator equipped with an afterburner and scrubber.

Analysis

GC-FID, GC/MS, and HPLC techniques are suitable for the analysis of salicylaldehyde.

5.16 MISCELLANEOUS ALDEHYDES

Many aldehydes other than those discussed above can be hazardous. Such substances, which contain the characteristic functional group

falling under the category of aldehyde, do not necessarily show toxicity, corrosivity, or flammability attributable exclusively to the aldehyde functional group in the molecule. For example, the toxicity may be generated from the halo substitution in the molecule. However, for the sake of convenience, these are classified under aldehydes, even if their hazardous properties are caused by the "toxic" or "reactive" sites present elsewhere in the molecule. Listed in Table 5.3 are some of the aldehydes that present moderate to severe health and/or fire hazards. Also presented in the lists are compounds of commercial use but with mild to moderate hazards. CAS registry numbers, synonyms, structure, physical properties, health, and fire hazards are documented in the table.

Safety, precautions, method of disposal, and analyses, being somewhat similar, are omitted. Burning in a chemical incinerator equipped with an afterburner and scrubber is recommended for disposal of these compounds. The standard methods for the analyses of these chemicals include GC-FID, GC/MS, HPLC, and colorimetry techniques. Other techniques, including FTIR and NMR, may be applied wherever necessary for substantial information on structure and for confirmation.

TABLE 5.3 Properties, Toxicity, and Flammability of Miscellaneous Hazardous Aldehydes

Compound/Synonyms/ CAS No./RCRA Waste No.	Formula/MW/Structure	Physical Properties	Toxicity	Flammability
Tribromoacetaldehyde (Bromal) [115-17-3]	Br_3C_2HO 280.76 (structure)	Liquid, bp 174.6°C, density 2.665, soluble in alcohol, ether, acetone	Skin irritant	Combustible, flash point 65°C (150°F)
Trichloroacetaldehyde (Chloral, Grasex, trichloroethanal), [75-87-6]; DOT UN 2075 (Poison B); RCRA U034	Cl_3C_2HO 147.38 (structure)	Liquid, bp 97.8°C, mp −57.5°C, density 1.5121, soluble in water, alcohol, ether	Mutagenic; may be mildly toxic; no toxic effect noted on rats at 600 mg/kg (intraperitoneal dose)	Noncombustible
Chloral hydrate (trichloroacetaldehyde hydrate, 2,2,2-trichloro-1,1-ethanediol, aqua-chloral, hydral, Dormal, Sontec, Lorinal, Phaldrone] [302-17-0]	$Cl_3C_2H_3O$ 165.40 (structure Cl_3C-C=$O \cdot H_2O$)	mp 57°C, bp 96.3°C, density 1.9081, soluble in water, alcohol, ether, acetone, chloroform, pyridine	General anesthetic effect; causes sleep; high dose may be lethal—180 mg/kg intravenous (dogs) LD_{50} intraperitoneal (rats): 472 mg/kg; may cause skin tumors; mutagenic, teratogen	Noncombustible
o-Anisaldehyde (2-methoxybenzaldehyde, 2-methoxybenzene-carboxaldehyde, salicylaldehyde methyl ether) [135-02-4]	$C_8H_8O_2$ 136.16 (structure, CHO, OCH_3)	Solid, mp 38°C, bp 243°C, density 1.1326, soluble in alcohol, ether, acetone, benzene, chloroform	Skin irritant; 500 mg/24 h caused moderate irritation on rabbit skin; no toxic effect noted	Noncombustible, flash point >100°C (212°F)

Name	Formula/Structure	Physical properties	Hazards	Flammability
p-Anisaldehyde (4-methoxybenzaldehyde, anisic aldehyde, Crategine, Aubepine) [123-11-5]	$C_8H_8O_2$ 136.16 CHO—benzene ring—OCH_3	Liquid, mp 0°C, bp 249.5°C, density 1.1191, soluble in water, ether, acetone, benzene, chloroform	Skin irritant; caused moderate irritation on rabbit skin; ingestion may cause somnolence	Flash point >100°C
o-Chloromethyl-*p*-anisaldehyde (2-chloromethyl-4-methoxybenzaldehyde) [73637-11-3]	$C_9H_9ClO_2$ 184.6 CHO, CH_2Cl—benzene ring—OCH_3		Skin irritant; 200 mg can produce severe skin irritation in humans	
2-Ethylhexanal (2-ethyl hexaldehyde, ethyl butyl acetaldehyde, β-propyl-α-ethylacrolein) [123-05-7]	$C_8H_{16}O$ 128.24 CH_3—CH_2—CH_2—CH_2—CH_2—CH—CHO with C_2H_5	Liquid, bp 163°C, density 0.8, soluble in alcohol, ether	Skin and eye irritant, 10 mg/24 h open caused severe skin irritation in rabbit; no toxic effect noted at 3700 mg/kg oral (mice)	Combustible; vapor density 4.4; flash point (closed cup) 44°C (112°F); auto-ignition temperature 190°C (375°F); vapor–air mixture explosive; LEL 0.85% at 93°C, UEL 7.2% at 135°C

(*continued*)

TABLE 5.3 (Continued)

Compound/Synonyms/ CAS No./RCRA Waste No.	Formula/MW/Structure	Physical Properties	Toxicity	Flammability
Methacrolein (methacrylaldehyde, 2-methacrolein, isobutenal, methacrylic aldehyde, 2-methyl propenal) [78-85-3], DOT UN 2396, flammable liquid and poison	C_4H_6O 70.1 $CH_2{=}\overset{\displaystyle CH_3}{C}{-}CHO$	Liquid, bp 68°C (154°F), density 0.8, soluble in water, alcohol, acetone, ether	Skin irritant, lachrymator; 500 mg/24 h caused severe skin irritation in rabbit; 0.05 mg/kg caused severe eye irritation; exposure to 125 ppm/4 h was lethal to rats; LD_{50} oral (rats); 111 mg/kg, mutagen	Highly flammable; flash point (open cup) 2°C (35°F); vapor density 2.4; danger of vapor propagation to ignition source and flashback fire; fire-extinguishing agent: "alcohol" foam; can react violently with oxidizers
Cumene aldehyde (α-methyl benzene acetaldehyde, α-formyl ethyl benzene, 2-phenyl propanal, α-methyl-α-toluic aldehyde, hyacinthal hydratropic aldehyde, 2-phenyl propionaldehyde) [93-53-8]	$C_9H_{10}O$ 134.2	Liquid, bp 202°C, density 1.009, soluble in alcohol	Toxicity—low; it can injure skin and hair; ingestion of large dose can result in coma	Noncombustible
2-Methoxyacetaldehyde (methoxyethanal) [10312-83-1]	$C_3H_6O_2$ 74.1 $CH_3{-}O{-}CH_2{-}\overset{\displaystyle H}{C}{=}O$	Liquid, bp 92.3°C, density 1.005, soluble in water, alcohol, ether, acetone	Eye irritant; 100 mg produced moderate irritation in rabbit eyes; low toxicity (oral and skin)	Noncombustible

Name	Formula / Structure	Properties	Toxicity	Hazard
2,2-Dichloroacetaldehyde (chloraldehyde) [79-02-7]	$Cl_2C_2H_2O$ 112.94 Cl—CH(Cl)—CH=O (H)	Liquid, bp 90°C, soluble in alcohol	Mutagen	
Pivaldehyde (trimethylacetaldehyde, trimethyl ethanal) [630-19-3]	$C_5H_{10}O$ 86.1 CH_3—C(CH_3)(CH_3)—CHO	bp 77–78°C, mp 6°C, density 0.7923, soluble in alcohol, ether	Toxicity—low to very low	Highly flammable; flash point −15°C (4°F); vapor density 2.9; propagation and flashback fire hazard
Protocatechualdehyde (3,4-dihydroxybenzaldehyde; 3,4-dihydroxybenzene-carbonal; rancinamycin IV) [139-85-5]	$C_7H_6O_3$ 138.1 CHO—C₆H₃(OH)(OH)	Solid, mp 153°C, soluble in water, alcohol, ether	Corrosive; toxic, can cause convulsive and respiratory stimulation; LD_{50} intraperitoneal (mice): 205 mg/kg	Noncombustible

(continued)

TABLE 5.3 (Continued)

Compound/Synonyms/ CAS No./RCRA Waste No.	Formula/MW/Structure	Physical Properties	Toxicity	Flammability
Cumaldehyde [*p*-isopropylbenzaldehyde, cumic aldehyde, cuminal aldehyde, isopropylbenzenecarbox- aldehyde, 4-(1-methylethyl) benzaldehyde] [122-03-2]	$C_{10}H_{12}O$ 148.2	Liquid, density 0.975, soluble in alcohol, ether	Toxicity—low; oral administration of ≈1400 mg/kg in rats produced somnolence, ulceration, bleeding from stomach: target organs: liver, gastrointestinal tract; mild irritant	Noncombustible
Acetaldol (aldol, 3-butanonal, 3-hydroxybutanal, 3-hydroxybutyraldehyde, oxybutanal) [107-89-1]. DOT UN 2839 (Poison B)	$C_4H_8O_2$ 88.1 $$CH_3-\underset{\underset{\displaystyle OH}{|}}{CH}-CH_2-CHO$$	bp 83°C, density 1.103, readily mixes with water, alcohol	Skin and eye irritant; 10 mg/24 hr caused mild irritation on rabbit skin; toxic via skin absorption; LD_{50} oral (rats): 2180 mg/kg	Combustible; flash point data not available
2-Methylbutyraldehyde (2-methyl-1-butanal, methylethylacetaldehyde) [96-17-3]	C_4H_9CHO 86.15 $$CH_3-CH_2-\underset{\underset{\displaystyle CH_3}{|}}{CH}-CHO$$	bp 92°C, density 0.803, soluble in alcohol, ether, acetone	Skin and eye irritant; exposure to 500 mg/24 hr caused severe eye irritation in rabbits; 100% in 24 h caused moderate skin irritation in guinea pigs	Combustible; flash point data not available

Name	Formula/MW	Properties	Toxicity	Fire hazard
1-Hexanal (hexaldehyde, caproic aldehyde, caproaldehyde) [66-25-1], DOT UN 1207, flammable liquid	$C_5H_{11}CHO$ 100.18 $CH_3—CH_2—CH_2—CH_2—CH_2—CHO$	Liquid, bp 128°C, mp −56°C, density 0.814, soluble in alcohol, ether, acetone, benzene	Mild skin and eye irritant; effect of 100 mg/24 h in rabbit's eye was mild	Flammable; vapor density 3.6; flash point (open cup) 32°C (90°F)
Metaldehyde II (tetramer of acetaldehyde)	$(CH_3CHO)_4$ 176.2	bp 110°C, mp 47°C, soluble in alcohol, ether, acetone, benzene	Moderately toxic; male dogs given 600 mg/kg via stomach tube show ataxia and tremor; vomiting occurred less often (Booze and Oehme 1986)	Flammable, flash point (closed cup) 36°C (97°F); fire-extinguishing agent: water
Metaldehyde (meta-acetaldehyde—oligomer of acetaldehyde between 4 and 6 molecules) [37273-91-9]	$(CH_3CHO)_{4–6}$	Sublimes at 1105°C, mp 246.2°C		
Cyclohexanecarboxaldehyde (hexahydrobenzaldehyde, cyclohexylformaldehyde) [2043-61-0]	$C_6H_{11}CHO$ 112.17	Liquid, bp 161°C, density 0.926, soluble in alcohol, ether, acetone, benzene	Exposure can cause mild to moderate eye irritation	Combustible; flash point 40°C (105°F)

(*continued*)

TABLE 5.3 (Continued)

Compound/Synonyms/ CAS No./RCRA Waste No.	Formula/MW/Structure	Physical Properties	Toxicity	Flammability
Pyruvaldehyde (pyruvic aldehyde methyl glyoxal, 2-oxopropanal, 2-ketopropionaldehyde, acetyl formaldehyde, acetyl formyl) [78-98-8]	$C_3H_4O_2$ 72.07 $$CH_3-\overset{\overset{O}{\|}}{C}-CHO$$	Yellow liquid hygroscopic, bp 72°C, density 1.0455, soluble in alcohol, ether, benzene	Eye irritant; mutagen— 1.5 mmol/L caused DNA damage in hamster ovary; toxicity—low	
Citral (3,7-dimethyl-2-6-octadienal) [5392-40-5]	$C_{10}H_{16}O$ 152.26 $$CH_3-\overset{\overset{CH_3}{\|}}{C}=CH-CH_2-CH_2-\overset{\overset{CH_3}{\|}}{C}=CH-CHO$$	Amber liquid, bp 95°C (5 mm), density 0.887	Skin irritant; 500 mg/24 h caused moderate irritation on rabbit skin; 40 mg/24 h may cause mild irritation on humans	
Malonaldehyde (malonic aldehyde, malonic dialdehyde, 1,3-propanedial, 1,3-propanedialdehyde, 1,3-propanedione) [542-78-9]	$C_3H_4O_2$ 72.07 $$O=C-CH_2-C=O$$ with H on each carbonyl		Mutagen; suspected carcinogen: the evidence, however, is inconclusive, animal studies data indicate that it may produce cancer primarily in skin and perhaps in lungs, respiratory tract, and blood; tumors may develop at the site of application	

Name	Formula / MW	Structure	Properties
α-Chlorocinnamaldehyde (2-chloro-3-phenyl-2-propenal) [18365-42-9]	C_9H_7ClO 166.6		Mutagen
β-Chlorocinnamaldehyde (3-chloro-3-phenyl-2-propenal) [40133-53-7]	C_9H_7ClO 166.6		Toxic; causes tremor and muscle contraction; may have adverse effects on lung and respiratory tract; LD_{50} intraperitoneal (mice): 220 mg/kg
p-Bromo-β-chlorocinnamaldehyde [14063-78-6]	C_9H_6BrClO 245.5		Toxic; causes tremor and muscle contraction; may have adverse effects on lung and respiratory tract; LD_{50} intraperitoneal (mice): 158 mg/kg
β-Chloro-p-fluorocinnamaldehyde [55338-97-1]	C_9H_6ClFO 184.6		Toxic; causes tremor and muscle contraction; may have adverse effects on lung and respiratory tract; LD_{50} intraperitoneal (mice): 220 mg/kg

(continued)

TABLE 5.3 *(Continued)*

Compound/Synonyms/ CAS No./RCRA Waste No.	Formula/MW/Structure	Physical Properties	Toxicity	Flammability
β-Chloro-p-iodocinnamaldehyde [55404-82-5]	C_9H_6ClIO 292.5		Toxic; causes tremor and muscle contraction; may have adverse effects on lung and respiratory tract; LD_{50} intraperitoneal (mice): 188 mg/kg	
β-Chloro-p-methoxycinnamaldehyde [14063-79-7]	$C_{10}H_9ClO_2$ 196.6		Toxic; causes tremor and muscle contraction; may have adverse effects on lung and respiratory tract; LD_{50} intraperitoneal (mice): 187 mg/kg	
β-Chloro-p-methylcinnamaldehyde [40808-08-0]	$C_{10}H_9ClO$ 180.6		Toxic; causes tremor and muscle contraction; may have adverse effects on lung and respiratory tract; LD_{50} intraperitoneal (mice): 236 mg/kg	

β-Chloro-p-nitrocinnamaldehyde
[2888-10-0]

$C_9H_6ClNO_3$
211.6

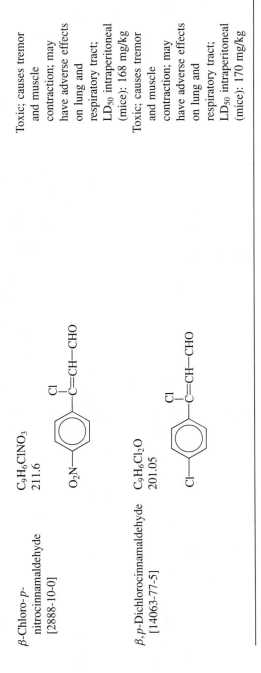

Toxic; causes tremor and muscle contraction; may have adverse effects on lung and respiratory tract; LD_{50} intraperitoneal (mice): 168 mg/kg

β,p-Dichlorocinnamaldehyde
[14063-77-5]

$C_9H_6Cl_2O$
201.05

Toxic; causes tremor and muscle contraction; may have adverse effects on lung and respiratory tract; LD_{50} intraperitoneal (mice): 170 mg/kg

REFERENCES

ACGIH. 1986. *Documentation of the Threshold Limit Values and Biological Exposure Indices*, 5th ed. Cincinnati, OH: American Conference of Governmental Industrial Hygienists.

Aldrich. 1989. *Aldrich Catalog*. Milwaukee, WI: Aldrich Chemical Company.

Auel, R. A. M., J. D. Jolson, and D. A. Stewart. 1987. Electrochemical air analysis for formaldehyde. European Patent 230, 289, July 29.

Balmat, J. L. 1986. Accuracy of formaldehyde analysis using the Draeger tube. *Am. Ind. Hyg. Assoc. J. 47*(8): 512.

Bonastre, N., C. De Mas, and C. Sola. 1986. Vavilin equation in kinetic modelling of formaldehyde biodegradation. *Biotechnol. Bioeng. 28*(4): 616–19.

Booze, T. F., and F. W. Oehme. 1986. An investigation of metaldehyde and acetaldehyde toxicities in dogs. *Fundam. Appl. Toxicol. 6*(3): 440–46.

Crane, C. R., D. C. Sanders, B. R. Endecott, and J. K. Abbott. 1986. Inhalation toxicology. VII. Times to incapacitation and death for rats exposed continuously to atmospheric acrolein vapor. *Govt. Rep. Announce. Index* (U.S.), *86*(22), Abstr. No. 648,599; cited in *Chem. Abstr. CA 106*:79971f.

Crosby, R. M., K. K. Richardson, T. R. Craft, K. B. Benforado, H. L. Liber, and T. R. Skopek. 1988. Molecular analysis of formaldehyde induced mutations in human lymphoblasts and *E. coli. Environ. Mol. Mutagen. 12*(2): 155–66.

Dasgupta, P. K. 1987. Application of continuous liquid phase fluorescence analysis for the selective determination of gases. *Proc. SPIE Int. Soc. Opt. Eng. 743*: 97–102.

Dong, S., and P. K. Dasgupta. 1987. Fast fluorometric flow injection analysis of formaldehyde in atmospheric water. *Environ. Sci. Technol. 21*(6): 581–88.

Dore, M., L. Atzori, and L. Congiu. 1986. Protection by sulfur compounds against acrolein toxicity in isolated rat hepatocytes. *IRCS Med. Sci. 14*(6): 595–96.

Graftstrom, R. C., J. C. Wiley, K. Sundqvist, and C. C. Harris. 1985. Toxicity of tobacco-related aldehydes in cultured human bronchial epithelial cells. *Air Force Aerospace Med. Res. Lab.*

Tech. Rep. ARAMRL-TR (U.S.)-84-002; cited in *Chem. Abstr. CA 104*(5): 30233m.

Habboush, A. E., S. M. Farroha, O. F. Naoum, and A. A. F. Kassir. 1988. The collection and direct analysis of low molecular weight aldehydes from automobile exhaust gases by GLC. *Int. J. Environ. Anal. Chem. 32*(2): 79–85.

Holcombe, G. W., G. L. Phipps, A. H. Sulaiman, and A. D. Hoffman. 1987. Simultaneous multiple testing: acute toxicity of 13 chemicals to 12 diverse freshwater amphibian, fish, and invertebrate families. *Arch. Environ. Contam. Toxicol. 16*(6): 697–710.

Igawa, M., W. J. Munger, and M. R. Hoffmann. 1989. Analysis of aldehydes in cloud- and fogwater samples by HPLC with a postcolumn reaction detector. *Environ. Sci. Technol. 23*(5): 556–61.

Jones, A. W., A. Sato, and O. A. Forsander. 1985. Liquid/air partition coefficients of acetaldehyde: values and limitations in estimating blood concentrations from analysis in breath. *Alcohol Clin. Exp. Res. 9*(5): 461–64; cited in *Chem. Abstr. CA 104*(3): 17224q.

Jones, J. S., G. D. Sadler, and P. E. Nelson. 1986. Acetaldehyde and accelerated storage of wine: a new rapid method for analysis. *J. Food Sci. 51*(1): 229–30.

Leblanc, E. L., and R. J. Leblanc. 1988. Effect of frozen storage temperature on free and bound formaldehyde content of cod (*Gadus morhua*) fillets. *J. Food Process. Preserv. 12*(2): 95–113.

Lorusso, S., L. Gramiccioni, S. DiMarzio, M. R. Milana, P. DiProspero, and A. Papetta. 1985. Acetaldehyde migration from poly(ethylene terephthalate) (PTE) containers. GC determination and toxicological assessment. *Ann. Chim. 75*(9–10): 403–14; cited in *Chem. Abstr. CA 104*(5): 33187k.

Merck. 1989. *The Merck Index*, 11th ed. Rahway, NJ: Merck & Co.

Monteiro-Riviere, N. A., and J. A. Popp. 1986. Ultrastructural evaluation of acute nasal toxicity in the rat respiratory epithelium in response to formaldehyde gas. *Fundam. Appl. Toxicol. 6*(2): 251–62.

Morikawa, T. 1988. Toxic gases evolution from commercial building materials under fire atmosphere in a semi-full-scale room. *J. Fire Sci. 6*(2): 86–99.

Morikawa, T., and E. Yanai. 1986. Toxic gases evolution from air-controlled fires in a semi-full scale room. *J. Fire Sci. 4*(5): 299–314.

National Research Council. 1983. *Prudent Practices for Disposal of Chemicals from Laboratories.* Washington, DC: National Academy Press.

NFPA. 1986. *Fire Protection Guide on Hazardous Materials*, 9th ed. Quincy, MA: National Fire Protection Association.

NIOSH. 1984. *Manual of Analytical Methods*, 3rd ed. Cincinnati, OH: National Institute for Occupational Safety and Health.

NIOSH. 1986. *Registry of Toxic Effects of Chemical Substances*, ed. D. V. Sweet. Washington, DC: U.S. Government Printing Office.

Ohsumi, T., and K. Kuroki. 1988. Acute toxicity of glutaraldehyde. *Kyushu Shika Gakkai Zasshi 42*(4): 470–76; cited in *Chem. Abstr. CA 109*(25): 224269b.

Popp, J. A., K. T. Morgan, J. Everitt, X. Z. Jiang, and J. T. Martin. 1986. Morphologic changes in the upper respiratory tract of rodents exposed to toxicants by inhalation. *Microbeam Anal. 21*: 581–82; cited in *Chem. Abstr. CA 105*(19): 166390p.

Reece, P. 1985. The fate of nicotinamide adenine dinucleotide in minced flesh of cod (Gadus morhua) and its association with formaldehyde production during frozen storage. *Sci. Tech Froid. 4*: 375–79.

Rehbein, H. 1985. Does formaldehyde form crosslinks between myofibrillar proteins during frozen storage of fish muscle? *Sci. Tech. Froid 4*: 93–99; cited in *Chem. Abstr. CA 106*(7): 48808d.

Staples, C. A. 1988. How clean is clean? A use of hazard assessment in ground water for evaluation of an appropriate formaldehyde spill remedial action endpoint. *ASTM Spec. Tech. Publ. 971; Aquat. Toxicol. Hazard Assess. 10*: 483–90; cited in *Chem. Abstr. CA 109*(10): 79321f.

Takami, K., K. Kuwata, S. Akiyoshi, and M. Nakamoto. 1985. Trace determination of aldehydes in water by high performance liquid chromatography. *Anal. Chem. 57*(1): 243–45.

Til, H. P., R. A. Woutersen, V. J. Feron, and J. J. Clary. 1988. Evaluation of the oral toxicity of acetaldehyde and formaldehyde in a 4-week drinking-water study in rats. *Food Chem. Toxicol. 26*(5): 447–52.

Upreti, R. K., M. Y. H. Farooqui, A. E. Ahmed, and G. A. S. Ansari. 1987. Toxicokinetics and molecular interaction of ^{14}C-formaldehyde in rats. *Arch. Environ. Contam. Toxicol. 16*(3): 263–73.

U.S. EPA. 1980. 45 *Federal Register* 79318, November 28, 1980.

U.S. EPA. 1984 (Revised 1990). Methods for organic chemical analysis of municipal and industrial wastewater. 40 CFR, Part 136, Appendix A.

U.S. EPA. 1986. *Test Methods for Evaluating Solid Waste*, 3rd ed. Washington DC: Office of Solid Waste and Emergency Response.

Van Gils, G. J., and M. Pirbazari. 1985. Pilot plant investigations for the removal of toxic pollutants from industrial laundry wastewater. In *Proceedings of the Mid-Atlantic Industrial Waste Conference on Toxic Hazardous Wastes*, ed. I. J. Kugelman, pp. 186–95. Lancaster, PA: Technomic.

Venkateswarlu, Y., B. Janakiram, and G. Rajaram Reddy. 1988. In vitro neutralization of the scorpion, *Buthus tamulus* venom toxicity. Indian *J. Physiol. Pharmacol. 32*(3): 187–94.

Woutersen, R. A., L. M. Appelman, A. Van-Garderen-Hoetmer, and V. J. Feron. 1986. Inhalation toxicity of acetaldehyde in rats. III. Carcinogenicity study. *Toxicology 41*(2): 213–31.

Woutersen, R. A., L. M. Appelman, J. W. G. M. Wilmer, H. E. Falke, and V. J. Feron. 1987. Subchronic (13-week) inhalation toxicity study of formaldehyde in rats. *J. Appl. Toxicol. 7*(1): 43–49.

6

ALKALIES

6.1 GENERAL DISCUSSION

Alkalies are water-soluble bases, mostly the hydroxides of alkali- and alkaline-earth metals. Certain carbonates and bicarbonates also exhibit basic properties, but they are weak bases. These compounds react with acids to form salts and water:

$$NaOH + HCl \longrightarrow NaCl + H_2O$$

Strong alkalies such as caustic soda and caustic potash have a wide range of commercial applications.

The health hazard from the concentrated solutions of alkalies arises from their severe corrosive actions on tissues. These compounds are bitter in taste, corrosive to skin, and are a severe irritant to the eyes. However, such corrosive properties of alkalies vary markedly from one compound to another depending on the electropositive nature of the metal. Thus, whereas caustic potash is extremely corrosive, slaked lime or calcium hydroxide is a mild irritant. The strength of the base among the alkalies decreases in the following order: rubidium hydroxide > caustic potash > caustic soda > lithium hydroxide > barium hydroxide > strontium

hydroxide > calcium hydroxide > potassium carbonate $\geq$ sodium carbonate > potassium bicarbonate $\geq$ sodium bicarbonate.

The bicarbonates of sodium and potassium are very weak bases and their concentrated solutions are not corrosive to the skin. Beryllium and magnesium hydroxides are only slightly soluble in water and are very weakly basic. While the corrosive actions of the alkalies can be predicted from their basicity strength, their toxicity is governed by the metal ions (especially for the alkaline-earth metal hydroxides). For example, while caustic soda is a stronger base, and more corrosive than barium hydroxide, the latter is toxic in a manner similar to other barium salts.

The hydroxides and carbonates of alkali- and alkaline-earth metals are noncombustible. The strong caustic alkalies, however, react exothermically with many substances, including water and concentrated acids, generating heat that can ignite flammable materials. The violent reactions of alkalies, resulting in explosions, are listed in this chapter under the individual compounds.

The hazardous alkali spills may be treated effectively by methods involving neutralization and solidification (Mandel et al. 1987). The composition contains an organic acid, clays of high and low absorbing types, respectively, and a weak water-soluble acid that may be applied to the spill in a soft spray from a pressurized canister similar to a fire extinguisher. Landfill leachate containing high NH_3 concentrations may be destroyed by biological treatment and oxidation (Ehrig 1984).

6.2 SODIUM HYDROXIDE

DOT Label: Corrosive Material, UN 1823 (for solid) and UN 1824 (for solution)

Formula NaOH; MW 39.998; CAS [1310-73-2]

Synonyms: caustic soda; sodium hydrate; soda lye; white caustic

Uses and Exposure Risks

Caustic soda is one of the most widely used chemicals. It is used to neutralize acids; to make sodium salts; to precipitate metals as their hydroxides; in petroleum refining; in the saponification of esters; in the treatment of cellulose, plastics, and rubber; and in numerous synthetic and analytical applications.

Physical Properties

White translucent crystalline solid (flakes or beads); odorless; bitter in taste; hygroscopic; mp 318°C; bp 1390°C; density 2.10 for pure solid and 1.53 for 50% aqueous solution; highly soluble in water (109 g/dL at 20°C), aqueous solutions highly alkaline (pH 13 for a 0.1 M solution).

Health Hazard

Sodium hydroxide is a highly corrosive substance that causes damage to human tissues. Its action on the skin is somewhat different from acid burns. There is no immediate pain, but it penetrates the skin. It does not coagulate protein to prevent its further penetration, and thus the caustic burn can become severe

and slow healing. Spilling of its concentrated solutions into the eyes can result in severe irritation or permanent injury.

It is toxic by ingestion as well as inhalation of its dust. Although the oral toxicity of a 5–10% solution of caustic soda was found to be low in test animals, high dosages at greater concentrations can cause vomiting, prostration, and collapse. The oral lethal dose in rabbits is 500 mg/kg (NIOSH 1986).

Sodium hydroxide dusts or aerosols are irritating to the eyes, nose, and throat. Prolonged exposure to high concentrations in air may produce ulceration of the nasal passage.

Exposure Limits

TLV-TWA air 2 mg/m³ (OSHA); ceiling 2 mg/m³ (ACGIH) and 2 mg/m³/15 min (NIOSH).

Hazardous Reactions

The hazardous reactions of sodium hydroxide may be classified under two categories: (1) exothermic reactions, which cause a noticeable rise of temperature and pressure in a closed container, and (2) violent polymerization reactions, causing explosions. Type 1 reactions include the neutralization of concentrated acids to form salts and water; highly exothermic hydrate formation when caustic soda is mixed with water; and reactions with many alcohols and substituted alcohols, such as chlorohydrins and cyanohydrins. These reactions can become violent if large amounts of reactants are mixed rapidly.

Type 2 reactions involve violent base-catalyzed polymerization of reactants causing explosions. A base such as caustic soda may catalyze the polymerizations of many organic unsaturates. Violent explosions occurred when caustic soda was mixed with acrolein, acrylonitrile, or allyl alcohol (NFPA 1986).

Sodium hydroxide reacts with trichloroethylene, forming explosive mixtures of dichloroacetylene. When heated with phosphorus pentoxide, a violent explosion can result (Mellor 1946, Suppl. 1971). Phosphorus boiled with caustic soda solution can produce phosphine, which ignites spontaneously in air. Amphoteric metals such as aluminum, zinc, and tin react with sodium hydroxide, generating hydrogen, which may form explosive mixtures in air.

Storage and Shipping

Sodium hydroxide is stored in a dry place, separated from acids, metals, and organic peroxides, and protected against moisture. It is shipped in bottles, cans, and drums.

Mild steel may be suitable as a material of construction for handling caustic soda at ambient temperature. At elevated temperatures, >60°C, corrosion may occur. Nickel and/or its alloys are most suitable for caustic handling at all temperatures and concentrations, including anhydrous molten caustic up to 480°C (Leddy et al. 1978). Polypropylene, fluorocarbon plastics, and fiberglass/vinyl ester resins are being used for many applications. Aluminum, tin, zinc, and other amphoteric metals should not be used in construction materials.

Analysis

Sodium hydroxide solutions are tested for their total alkalinity by titration with standard acid. Alkaline dusts in air may be estimated by flowing air through a 1-mm PTFE membrane filter, extracting the hydroxides with 0.01 N HCl under nitrogen, followed by acid–base titration using a pH electrode (NIOSH 1984, Method 7401). This method also measures the total alkaline dusts in air that includes caustic potash.

6.3 POTASSIUM HYDROXIDE

DOT Label: Corrosive Material, UN 1813 (solid) and UN 1814 (solution)

Formula KOH; MW 56.11; CAS [1310-58-3]

Synonyms: caustic potash; potassium hydrate; potassa

Uses and Exposure Risk

Potassium hydroxide is used in making liquid soap and potassium salts, in electroplating and lithography, in printing inks, as a mordant for wood, and finds wide applications in organic syntheses and chemical analyses.

Physical Properties

White deliquescent solid (lumps, rods, flakes, or pellets); mp 360–380°C; bp 1320°C; density 2.044; highly soluble in water, alcohol, and glycerol, slightly soluble in ether; highly alkaline, pH 13.5 for a 0.56% solution.

Health Hazard

Potassium hydroxide is extremely corrosive to tissues. Its corrosive action is greater than that of sodium hydroxide. It gelatinizes tissues to form soluble compounds that may cause deep and painful lesions (ACGIH 1986). Contact with the eyes can damage vision. Ingestion can cause severe pain in the throat, vomiting, and collapse.

LD$_{50}$ value, oral (rats): 365 mg/kg

Exposure to its dusts can cause irritation of the nose and throat.

Exposure Limit

Ceiling in air 2 mg/m^3 (ACGIH).

Hazardous Reactions

Potassium hydroxide is noncombustible. However, mixing it with water can produce sufficient heat to ignite combustible materials. Dissolution in water is exothermic due to the formation of hydrates. Neutralization with acids is exothermic, which may become violent if large amounts in high concentrations are mixed rapidly.

Contact with amphoteric metals such as aluminum, zinc, and tin generates hydrogen, which is flammable, forming explosive mixtures in air. When heated with tetrachloroethane or 1,2-dichloroethylene, potassium hydroxide generates chloroacetylene, a toxic gas that ignites spontaneously and explodes in air.

Similar to caustic soda, caustic potash can initiate base-catalyzed polymerizations of acrolein, acrylonitrile, allyl alcohol, and other organic unsaturates susceptible to polymerization, often resulting in explosions.

Most hazardous reactions of potassium hydroxide are similar to those of caustic soda and many are well documented (NFPA 1986). Explosions occurred when caustic potash was added to liquid chlorine dioxide (Mellor 1946), nitrogen trichloride, *N*-nitrosomethyl urea in *n*-butyl ether (NFPA 1986), and maleic anhydride (MCA 1960). Reaction with phosphorus yields toxic and flammable gas phosphine, which ignites spontaneously. Explosion may occur if impure tetrahydrofuran is heated or distilled in the presence of concentrated caustic potash solution.

Storage and Shipping

Caustic potash is stored in a dry place, protected against moisture and water, and isolated from acids, metals, organic peroxides, and flammable materials. Nickel, its alloys, polypropylene, or fluorocarbon plastics are suitable for use as construction materials for caustic potash handling. Aluminum, zinc, tin, or their alloys should not be used for such purposes. Potassium hydroxide is shipped in bottles, barrels, drums, and tank cars.

Disposal/Destruction

Potassium hydroxide is added slowly with stirring to water at cold temperature, neutralized with dilute acid, and flushed down the drain.

Analysis

Titrations with standard acids are commonly applied to test for the alkalinity of caustic potash solutions. Air analysis for its dusts may be performed by filtering air through a PTFE membrane filter followed by acid titration (NIOSH 1984, Method 7401; see Section 6.2).

6.4 MISCELLANEOUS ALKALIES

See Table 6.1.

TABLE 6.1   Toxicity and Hazardous Reactions of Miscellaneous Alkalies

Compound/Synonyms/ CAS No.	Formula/ MW	Toxicity	Hazardous Reactions
Lithium hydroxide (lithium hydrate) [1310-65-2]; lithium hydroxide monohydrate [1310-66-3]	LiOH 23.95 LiOH·H_2O 41.97	Highly corrosive, very irritating to skin and eyes; caustic effects similar to caustic soda but to a lesser extent; low toxicity; systemic toxicity similar to other compounds of lithium; DOT Label: Corrosive Material, UN 2679 (for anhydrous salt), UN 2680 (for hydrate)	Explosive polymerization may occur when in contact with certain unsaturate organics; strongly alkaline but less caustic to sodium hydroxide; expected to undergo many hazardous reactions similar to caustic soda

(*continued*)

TABLE 6.1 (Continued)

Compound/Synonyms/ CAS No.	Formula/ MW	Toxicity	Hazardous Reactions
Ammonium hydroxide (aqueous ammonia, aqua ammonia, ammonia solution, ammonia water) [1336-21-6]	NH_4OH 35.06	Highly caustic and corrosive; contact with eyes and skin can cause severe irritation; concentrated solutions emit ammonia, which can cause lachrymation and severe irritation of respiratory tract and acute pulmonary edema (see Section 20.1); static 24-h median lethal tests with channel catfish indicated that while ammonia was lethal with LC_{50} value of 1.45 NH_3^- N mg/L, NH_4 ion from NH_4Cl was essentially nontoxic to channel catfish at the same pH (Sheehan and Lewis 1986); moderately toxic by ingestion; LD_{50} oral (rats): 350 mg/kg; DOT Label: Corrosive Material, NA 2672 (for 12–44% ammonia content)	Reacts violently with many silver salts, such as silver nitrate, silver oxide, silver perchlorate, and permanganate, causing explosions — attributed to the formation of shock-sensitive silver nitride or silver complexes; explosion may occur when its reaction products with gold and mercury with variable compositions are heated and dried; in concentrated solution it reacts with excess of iodine to form nitrogen iodide, which may detonate on drying (Mellor 1946, Suppl. 1964); produces flame and/or explosion when mixed with fluorine; reacts violently with dimethyl sulfate (NFPA 1986); exothermic reactions occur when its concentrated solution is mixed with concentrated acids
Rubidium hydroxide (rubidium hydrate) [1310-82-3]	RbOH 102.48	Stronger base and more caustic than caustic potash; highly corrosive to tissues; acute oral toxicity (probably an effect of the cation), Rb is lower than caustic potash LD_{50} oral (rats): 586 mg/kg; DOT Label: Corrosive Material, UN 2677 (solid), UN 2678 (solution)	Exothermic reactions with concentrated acids and water; violent polymerization of certain organic unsaturates may occur when in contact with its solutions; produces flammable gases phosphine, chloroacetylene, and hydrogen on reactions with phosphorus, 1,2-dichloroethylene, and metals such as aluminum, zinc, and tin, respectively
Barium hydroxide (barium hydrate, caustic baryta) [17194-00-2]	$Ba(OH)_2$ 171.38	Highly alkaline, corrosive to skin; strong irritant to eyes; moderately toxic to test animals; toxicity	Reaction with phosphorus may yield toxic and flammable gas, phosphine; causes explosive

TABLE 6.1   *(Continued)*

Compound/Synonyms/ CAS No.	Formula/ MW	Toxicity	Hazardous Reactions
		similar to other barium compounds	decomposition of maleic anhydride; may undergo violent reactions typical to strong alkalies
Calcium hydroxide (slaked lime, calcium hydrate) [1305-62-0]	$Ca(OH)_2$ 74.10	Can cause moderate skin and eyes irritation; vapors can cause irritation to respiratory tract; no report on its acute or chronic toxicity; LD_{50} oral (mice): 7300 mg/kg; TLV-TWA 5 mg/m^3 (ACGIH)	Reacts with phosphorus to form phosphine, which ignites spontaneously in air; reacts with nitro-alkanes forming salts, which in dry state decomposes violently; causes explosive decomposition of maleic anhydride; violent base-catalyzed polymerization of acrolein and acrylo-nitrile may occur when mixed with its concentrated solutions
Potassium carbonate (dipotassium carbonate pearl ash, potash) [584-08-7]	K_2CO_3 138.21	Irritant to skin and eyes; oral toxicity low to very low; LD_{50} oral (rats): 1870 mg/kg	Exothermic reaction with concentrated sulfuric acid; and chlorine tri-fluoride; decomposed with fluorine, produces incandescence (Mellor 1946)
Sodium carbonate (dipotassium carbonate, crystol carbonate, soda ash) [497-19-8]	Na_2CO_3 105.99	Mild irritant to skin and eyes; no toxic effect from oral intake; LD_{50} oral (rats): 4100 mg/kg; breathing dusts may cause lung irritation	Explosion occurred in contact with red hot aluminum (NFPA 1986); reacts with fluorine decomposing to incande-scence (Mellor 1946); liberates heat when mixed with concentrated acids

REFERENCES

ACGIH. 1986. *Documentation of the Threshold Limit Values and Biological Exposure Indices*, 5th ed. Cincinnati, OH: American Conference of Governmental Industrial Hygienists.

Ehrig, H. J. 1984. Treatment of sanitary landfill leachate: biological treatment. *Waste Manage. Res.* 2(2): 131–52; cited in *Chem. Abstr. CA 102*(2): 11727t.

Leddy, J. J., I. C. Jones, Jr., B. S. Lowry, F. W. Spillers, R. E. Wing, and C. D. Inger. 1978. Alkali and chlorine products. In *Kirk-Othmer Encyclopedia of Chemical Technology*. New York: Wiley-Interscience.

Mandel, F. S., J. A. Engman, W. R. Whiting, and J. A. Nicol. 1987. Novel compositions and method for neutralization and solidification of hazardous alkali spills. U.S. Patent 8,706,757, Nov. 5; cited in *Chem. Abstr. CA 108*(28): 226388t.

MCA. 1960. *Case Histories of Accidents in the Chemical Industry*. Washington, D.C.: Manufacturing Chemists' Association.

Mellor, J. W. 1946. *A Comprehensive Treatise on Inorganic and Theoretical Chemistry*. London: Longmans, Green & Co.

NFPA. 1986. *Fire Protection Guide on Hazardous Materials*, 9th ed. Quincy, MA: National Fire Protection Association.

NIOSH. 1984. *Manual of Analytical Methods*, 3rd ed. Cincinnati, OH: National Institute of Occupational Safety and Health.

NIOSH. 1986. *Registry of Toxic Effects of Chemical Substances*, ed. D. V. Sweet. Washington, DC: U.S. Government Printing Office.

Sheehan, R. J., and W. M. Lewis. 1986. Influence of pH and ammonium salts on ammonia toxicity and water balance in young channel catfish. *Trans. Am. Fish. Soc. 115*(6): 891–99.

7

ALKALOIDS

7.1 GENERAL DISCUSSION

Alkaloids comprise a broad class of organic compounds, generally nitrogenous plant products that are basic in character and that exhibit marked physiological action. These substances occur throughout the plant kingdom, in about 40 classes of plants. Some alkaloids are also found in animals, insects, marine organisms, and microorganisms. The general characteristics of alkaloids are as follows. First, alkaloid molecules contain at least one atom of nitrogen, usually in a heterocyclic ring. In addition, these compounds may contain nitrogen atoms in amine or imine functional groups. The alkaloids are mostly tertiary bases. A few are secondary bases or quaternary ammonium salts. The basicity varies from strong to very weak. Second, in nature, alkaloids are generally found in the form of salts or combined with tannin. Third, most alkaloids are optically active. Fourth, being basic, alkaloids react readily with mineral or organic acids to form salts.

Within the foregoing broad definition and chemical properties, there are nearly 10,000 alkaloids, the nitrogenous natural products, with a wide diversity of structures, have been obtained from leaves, stems, roots, barks, and seeds of plants or from animals or microorganisms. Many alkaloids may be produced by synthetic routes.

Because of their marked pharmacologic activities, many alkaloids are used in medicines. Alkaloids have been in use for thousands of years in medicines and potions, as euphoric substances and hallucinogens, and as poisons. Table 7.1 lists some of the known alkaloids, their plant origin, and their skeletal ring structures.

Toxicity

All alkaloids are toxic compounds. However, the degree of toxicity varies widely. Some are extremely toxic compounds, and a good number of these are mild toxicants. The extremely toxic alkaloids include coniine, aconitine, amanitins, chaksine, saxitoxin, and surugatoxin; strychnine, nicotine, cytisine, reserpine, and emetine are examples of moderate to highly poisonous alkaloids. Atropine, pilocarpine, quinine, lobeline, ephedrine, and yohimbine are some of the known alkaloids used in medicine and are toxic at excessive doses.

Probably the most important alkaloids are some of the potent habit-forming narcotic and hallucinogenic substances extracted

TABLE 7.1 Principal Plant Alkaloids and Their Structural Heterocyclic Rings

Alkaloid Type/Source	Principal Alkaloids	Skeletal Ring/Structural Unit
Tobacco	Nicotine Anabasine	Pyridine–pyrrolidine *or* Pyridine–piperidine
Cocoa belladonna	Cocaine Atropine Scopolamine	Tropane
Opium	Morphine Codeine Thebaine Heroin	Epoxymorphinan or iminoethanophenanthrofuran
	Papaverine	Isoquinoline
Hemlock	Coniine Coniceine	Pyrimidine
Cinchona	Quinine Cinchonine Quinidine	Quinoline–pyrimidine

TABLE 7.1   (*Continued*)

Alkaloid Type/Source	Principal Alkaloids	Skeletal Ring/Structural Unit
Ergot	Lysergamide LSD Ergonovine	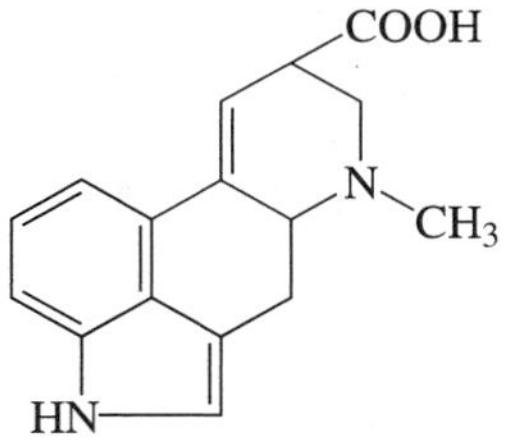 Lysergic acid
Rauwolfia	Reserpine Deserpidine Rescinnamine Yohimbine	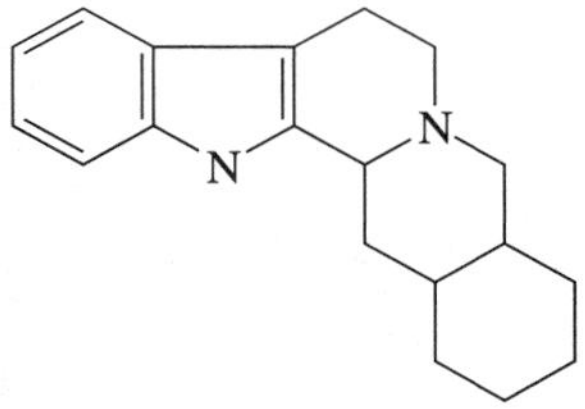 Yohimban
Strychnos	Strychnine Brucine	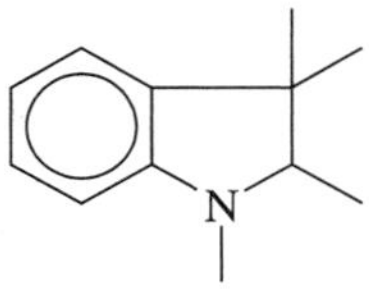 Indole unit, part of the condensed ring systems
Ipecac	Emetine Cephaeline	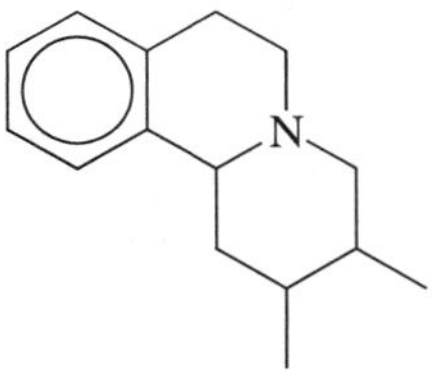 Benzoquinolizine
Lupinous	Lupinine Sparteine Cytisine	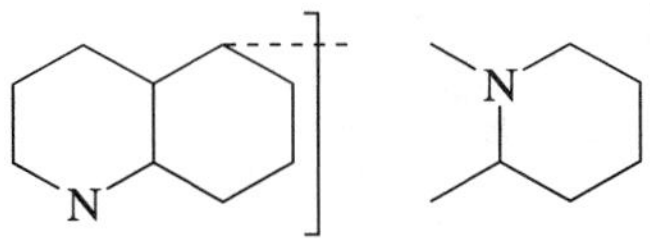 Octahydroquinolizine and/or diazocine
Cocoa	Caffeine	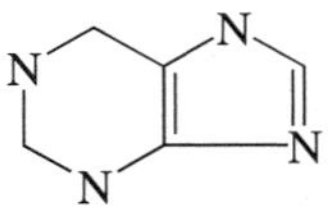 Purine

(*continued*)

TABLE 7.1 (*Continued*)

Alkaloid Type/Source	Principal Alkaloids	Skeletal Ring/Structural Unit
Autumn crocus	Colchicine	Benzoheptalene; no heterocyclic ring; the alkaloid is an amide derivative
Peyote	Mescaline	No heterocyclic ring, an amine derivative
Mushroom	Amanitins	Presence of an indole unit in a complex structure
	Muscazone Muscimol Ibotenic acid	Oxazole Isooxazole
Aconite	Aconitine Feraconitine	Five-, six-, and seven-membered fused cyclic rings; N atom is outside the ring

from coca, opium, and peyote cactus plants. These include cocaine, morphine, heroin, and mescaline. LSD, an amide derivative of lysergic acid, first synthesized three decades ago, is among the most potent psychedelic drugs. Halo derivatives of mescaline are several times as potent as mescaline.

Alkaloids in this chapter are arranged according to their plant origin and chemical structures. Compounds from the same species have similar heterocyclic structures, differing only in functional groups or substituents attached to the rings. Although the pharmacologic properties are the same among alkaloids of the same species, the degree of toxicity may differ widely. In certain cases, compounds of the same family with the same skeletal structure may show additional toxic effects of a different type. Thebaine, the methoxy derivative of morphine, is also a habit-forming substance, but its primary toxic effects are similar to those of strychnine, a convulsant poison,

rather than narcosis. Also, there are alkaloids having different structures but similar toxic actions. Lobeline, a lobelia alkaloid, which has a piperidine nucleus, shows pharmacologic action similar (but less potent) to that of nicotine, a tobacco alkaloid that has a pyridine–pyrrolidine ring system. The structure, occurrence, physical properties, and toxicity of some important alkaloids are presented in the following sections.

Analysis

Since most alkaloids are basic, they form salts with dilute mineral or organic acids. The presence of alkaloids in plants may be tested by using a metal iodide reagent such as potassium mercuric iodide or potassium bismuth iodide. A precipitate formation occurs. But these tests can give false results in the presence of proteins and coumarins. Ergot alkaloids produce blue color with Ehrlich's reagent, which is a mixture of *p*-dimethylaminobenzaldehyde in sulfuric acid. Indole alkaloids form distinct colors with ceric ammonium sulfate reagent.

Individual compounds may be analyzed by GC or HPLC techniques. Perrigo and associates (1985) have described the use of dual-column fused-silica capillary gas chromatography for drug screening. A fused-silica DB 1 capillary column using dual nitrogen–phosphorus and flame ionization detectors have enabled the analysis of 188 compounds. High reproducibility has been reported (Lora-Tamayo et al. 1986) by the capillary system in the analysis of these drugs and their metabolites, based on retention indexes and retention times relative to diazepam.

HPLC is by far the most common analytical technique. Jinno and coworkers (1988) have described rapid identification of poisoned human fluids by reversed-phase liquid chromatography. Using diode array detectors, solutes in urine and plasma can be identified to 10 ppm concentration with a high degree of certainty (Engelhardt and Koenig 1989). Computer-assisted HPLC analysis using a multiwavelength UV detection system has been applied to analyze several toxic drugs qualitatively and quantitatively to permit clinical diagnosis in cases of multiple drug ingestion (Hayashida et al. 1990; Jinno et al. 1990).

Alkaloids may be identified from their mass spectra and analyzed by GC/MS. Molecular ions and the characteristic ions of skeletal units may aid in structure elucidations.

Strychnine, used as a pesticide, is the only alkaloid having a NIOSH method for air analysis (NIOSH 1984, Method 5016). An air volume of 70–1000 L is passed over a 37-mm glass fiber and desorbed with 5 mL of mobile phase containing aqueous 1-heptane sulfonic acid and acetonitrile and analyzed by HPLC using UV detection at 254 nm. The column contains Bondapak C18 of particle size 10 nm.

TOBACCO ALKALOIDS

Toxicity and Carcinogenesis of Tobacco Components

Among the tobacco alkaloids, nicotine and anabasine are the major components. All tobacco alkaloids are toxic. The toxicity of nicotine and anabasine is discussed below under separate headings. There are also other harmful substances that are not alkaloids. These include nitrosamines, neophytadiene, solanesol, polyphenols, hydrocarbons, and steroids.

Aoyama and Tachi (1989) investigated the acute toxicity caused by chewing tobacco. Loose-leaf chewing tobacco (6 g) was extracted with artificial saliva (8 mL) for an hour and centrifuged. Ingestion of 1 mL of extract caused 100% mortality in mice. The LD_{50} value estimated was 14 g of chewing tobacco leaves per kilogram of mouse body weight.

Accidental ingestion of dried tobacco leaves caused nausea, vomiting, perspiration, and dizziness to this author (Patnaik 1990). The dose is estimated to be about 3 mg/kg, and the effect lasted about 6 hours. Such symptoms may be manifested especially after first-time intake.

Rao and Chakraborty (1988) have studied the effects of solvent extractions on the chemical composition of tobacco. Water extractions significantly reduced the levels of nicotine, reducing sugar and polyphenols. Extractions with methanol reduced the content of solanesol, neophytadiene, hydrocarbons, and steroids. The solvent extractions did not alter the structural identity of the tobacco. There were no significant cellular changes.

Certain compounds in tobacco are known to cause cancer. Among these are the *N*-nitrosamines and certain hydrocarbons. The alcohol extract of the chewing variety of tobacco tested mutagenic in a *Salmonella typhimurium* test (Shah et al. 1985). Administration of the extract in mice resulted in an increased incidence of lung and liver tumors.

The two major carcinogenic *N*-nitrosamines in tobacco are *N*′-nitrosonornicotine (NNN) [16543-55-8] and 4-(methylnitrosoamino)-1-(3-pyridyl)-1-butanone (NNK) [64091-91-4]. The structures of these compounds are as follows:

(NNN)

(NNK)

of lactating rats (La Voie et al. 1987). NNK formed in tobacco smoke has been found to induce lung, liver, and nasal cavity tumors in rats (Belinsky et al. 1986). NNK, NNN, and *N*′-nitrosoantabine are present in ppm concentrations in many tobacco products and in nanogram amounts in tobacco smoke (Nair et al. 1989). Whereas both NNN and NNK induced tumors of the lung, forestomach, and liver, they failed to induce oral tumors in mice (Padma et al. 1989). A subcutaneous cumulative dose of 50–300 mg/kg NNK developed tumors in the respiratory tract, nasal cavity, adrenal glands, pancreas, and liver in Syrian golden hamsters (Correa et al. 1990). The carcinogenic hydrocarbons found in tobacco smoke are primarily the two aromatics benz[a]pyrene and benzene.

The toxicity of tobacco smoke may be decreased by treatment with dilute potassium permanganate solution (Rosenthal 1989). Nascent oxygen from KMnO may oxidize the nicotine and tar components of a smoke stream. Zinc oxide and ferric oxide are found to reduce or eliminate the carcinogens in tobacco smoke (Hardy and Ayre 1987). Amonkar and coworkers (1989) found that hydroxychavicol, a phenolic component of betel leaf, showed protective action against the tobacco-specific carcinogens NNN and NNK. The authors suggest that this compound may reduce the risk of oral cancer

The carcinogenic hydrocarbon found in tobacco is benz[a]pyrene (see Chapter 29). The carcinogens mentioned above are present in both cigarette smoke and tobacco. Studies show that these compounds may be transferred from circulating blood into the milk with tobacco chewers. Hydroxychavicol exhibited an antimutagenic effect against the mutagenicity of *N*-nitrosamines in both the Ames *Salmonella*/microsome assay and the micronucleus test using Swiss male mice.

7.2 NICOTINE

EPA Designated Toxic Waste, RCRA Waste Number P075; DOT Label: Poison B, UN 1654

Formula $C_{10}H_{14}N_2$; MW 162.26; CAS [54-11-5]

Structure:

contains a pyridine and a pyrrolidine ring

Synonyms: 3-(1-methyl-2-pyrrolidinyl)pyridine; 1-methyl-2-(3-pyridyl)pyrrolidine; β-pyridil-α-*N*-methylpyrrolidine; black leaf

Uses and Exposure Risk

Nicotine is one of the principal constituents of tobacco. It occurs in the dried leaves of *Nicotiana tabacum* and *Nicotiana rustica* to the extent of 2–8%. Exposure risk to this alkaloid arises from smoking, chewing, or inhaling tobacco. Nicotine and its salts are used as insecticides and fumigants, in tanning, and in medicine.

Physical Properties

Colorless to pale yellow, oily liquid; turns brown on exposure to light or air; faint odor of pyridine; burning taste; hygroscopic; boils at 247°C with decomposition at 745 torr; vapor pressure 0.0425 torr at 20°C; density 1.0097 at 20°C; soluble in water and organic solvents; aqueous solution alkaline (pH of 0.05 M solution 10.2); reacts with acids to form water-soluble salts.

Health Hazard

Nicotine is a highly toxic compound. It stimulates neuromuscular junctions and nicotinic receptors, causing depression and paralysis of autonomic ganglia. Exposure routes are ingestion, absorption through skin, or inhalation (smoking or inhaling tobacco). The acute toxic symptoms in humans include nausea, vomiting, salivation, muscular weakness, twitching, and convulsions. The symptoms of confusion, hallucinations, and distorted perceptions have also been noted in humans. Death may occur from respiratory failure. The lethal dose is approximately 40 mg/kg. Chronic poisoning from occupational exposure may exhibit symptoms of vomiting and diarrhea. Fatal cases due to occupational poisoning are unusual.

Nicotine administered in animals produced toxic symptoms that include somnolence, change in motor activity, ataxia, dyspnea, tremor, and convulsions.

LD$_{50}$ values, intraperitoneal (rats): 14.5 mg/kg

LD$_{50}$ values, oral (rats): 50 mg/kg

Kramer and co-workers (1989) investigated the effect of nicotine on the accumulation of dopamine in synaptic vesicles prepared from mouse cerebral cortex or bovine striatum. It was found to be a weak inhibitor of dopamine accumulation.

The role of nicotine in tobacco carcinogenesis is not yet fully understood (see the preceding section, "Tobacco Alkaloids"). Nicotine is a precursor of N'-nitrosonornicotine [16543-55-8], which is a suspected lung carcinogen (Hoffmann et al. 1985). Nicotine-derived N-nitrosoamines contribute significantly to carcinogenesis caused by tobacco. However, there is no evidence in animals or humans of cancers caused by nicotine itself. Berger and coworkers (1987) have found a beneficial effect of perinatal nicotine administration in decreasing the tumors of the neuorogenic system induced by N-methylnitrosourea [684-93-5] in Sprague-Dawley rats. Although nicotine is noncarcinogenic, it acts as a cofactor in carcinogenesis induced by 7,12-dimethyl benz[a]anthracene in male Syrian golden hamsters (Chen and Squier 1990).

Nicotine produced teratogenic effects in test animals, causing postimplantation mortality, fetal death, and developmental abnormalities. Nicotine and its primary metabolite cotinine exhibited teratogenic potential with *Xenopus* frog embryo teratogenesis assay (Dowson et al. 1988).

Nicotine tested negative in the *Neurospora crassa*–aneuploidy and histidine reversion–Ames tests for mutagenicity.

Exposure Limit

TLV-TWA air 0.5 mg/m^3 (ACGIH, MSHA, and OSHA).

7.3 ANABASINE

Formula C$_{10}$H$_{14}$N$_2$; MW 162.24; CAS [494-52-0]

Structure:

contains a pyridine and a piperidine ring

Synonyms: 2-(3-pyridyl)piperidine; 3-(2-pi-peridyl)pyridine; neonicotine

Uses and Exposure Risk

Anabasine occurs in the tobacco species *Nicotiana glauca*, *Anabasis aphylla* L., *Chenopodiaceae*, and *Solanaceae*. It is used as an insecticide and as a metal anticorrosive agent.

Physical Properties

Colorless liquid; boils at 271°C; freezes at 9°C; soluble in water and most organic solvents.

Health Hazard

The acute toxic symptoms include increased salivation, confusion, disturbed vision, photophobia, nausea, vomiting, diarrhea, respiratory distress, and convulsions. It causes respiratory muscle stimulation similar to that caused by lobeline. An oral lethal dose in dogs is 50 mg/kg. A subcutaneous lethal dose in guinea pigs is 10 mg/kg.

7.4 LOBELINE

Formula $C_{22}H_{27}NO_2$; MW 337.47; CAS [90-69-7]

Structure:

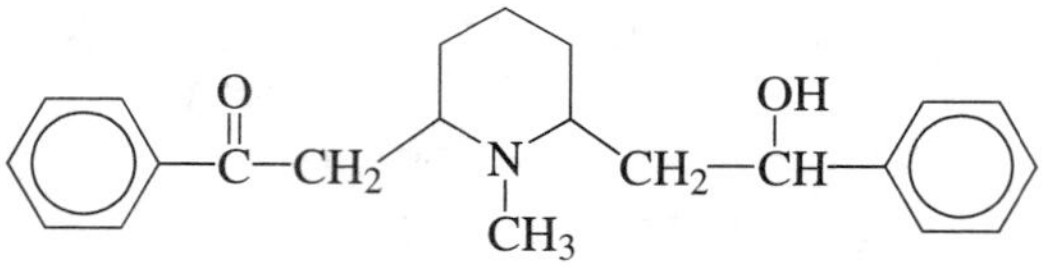

contains a piperidine ring

Synonyms: 2-(6-(2-hydroxy-2-phenylethyl)-1-methyl-2-piperidyl)-1-phenylethanone; 2-(6-β-hydroxyphenylethyl)-1-methyl-2-piperidyl)acetophenone; 8,10-diphenyllo-belionol

Uses and Exposure Risk

Lobeline is the principal lobelia alkaloid. It occurs in the seeds and herb of Indian tobacco (*Lobelia inflata* and *Lobeliaceae*). It is used as a respiratory stimulant. Its sulfate salt is used in antismoking tablets.

Physical Properties

Crystalline solid; melts at 130°C; slightly soluble in water, dissolves readily in hot alcohol, ether, chloroform, and benzene.

Health Hazard

The structure of lobeline is different from those of nicotine and anabasine. It does not have a pyridine ring, similar to the latter two alkaloids. However, its pharmacologic action is similar to but less potent than that of nicotine. Like anabasine, it is a respiratory stimulant. The toxic symptoms include increased salivation, nausea, vomiting, diarrhea, and respiratory distress.

LD_{50} value, intravenous (mice): 6.3 mg/kg

TROPANE ALKALOIDS

7.5 COCAINE

Formula $C_{17}H_{21}NO_4$; MW 303.35; CAS [50-36-2]

Structure:

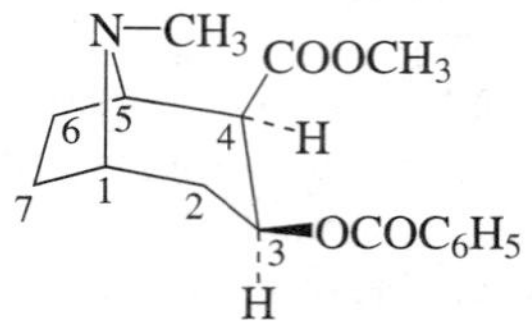

belongs to the class of tropane alkaloids

Synonyms: benzoylmethylecgonine; 2-β-carbomethoxy-3-β-benzoxytropane; 3-tropanylbenzoate-2-carboxylic acid methyl ester; 3-β-hydroxy-1-α-H,5-α-H-tropane-2-β-carboxylic acid methyl ester benzoate; l-cocaine; β-cocaine

Uses and Exposure Risk

Use of cocaine is known since early times. It occurs in the South American coca leaves. Chewing of leaves mixed with lime was a common practice among natives, who traveled great distances without experiencing fatigue (Cordell 1978). Cocaine is obtained by extraction of coca leaves. It is also prepared by methylation and benzoylation of the alkaloid, ecgonine. The dilute aqueous solutions of its hydrochloride is used as a topical anesthetic in ophthalmology. Cocaine and its derivatives are controlled substances listed in the U.S. Code of Federal Regulations (Title 21, Parts 321.1 and 1308.12, 1987).

Health Hazard

The physiologic responses from the use of cocaine in humans are euphoria and excitement, making it a habit-forming substance. Such addictive potential has also been observed in rats (Hartman 1978). High doses can produce confusion, hallucinations, delirium, convulsions, hypothermia, and respiratory failure. It is a toxicant to the cardiovascular and central nervous systems. The acute poisoning symptoms, in addition to those stated above, are nausea, vomiting, abdominal pains, and dilation of the pupils.

Cocaine may be inhaled, smoked, or injected. Low to average inhalation doses range between 20 and 150 mg. Ether refined powder is more potent than the unrefined substance. It may also be refined by other chemical treatments. The lethal doses in humans as reported in the literature are widely varying. A dose of 1–1.5 g may be fatal to humans.

LD$_{50}$ value, oral (mice): 99 mg/kg

Bozarth and Wise (1985) compared the toxicity of cocaine with that of heroin in rats, resulting from intravenous self-administration. The animals were given unlimited access to both compounds. Animals self-administering cocaine showed a loss of 47% of their body weight and deterioration of general health; 90% of the subjects died in 30 days. Whereas heroin showed stable drug self-administration that increased gradually, causing 36% mortality, cocaine showed excessive self-administration. The study showed that cocaine was much more toxic than heroin when rats were given unlimited access to intravenous drugs.

Langner and coworkers (1988) have reported arteriosclerotic lesions in rabbits resulting from repeated injection of cocaine. These data indicate that the abuse of cocaine may cause damage to the aorta, resulting in the premature onset of cardiovascular disease and its complications. Intravenous administration of doses of 1 and 2 mg/kg demonstrated dose-dependent increases in systolic, diastolic, mean arterial, and pulse pressures in pregnant and nonpregnant ewes (Woods et al. 1990). The study showed that pregnancy increased the cardiovascular toxicity caused by cocaine. In squirrel monkeys, an increase in blood pressure was noted, along with increases in dopamine, epinephrine, and nonepinephrine plasma concentrations (Nahas et al. 1988). Hoskins and coworkers (1988) investigated diabetes potentiation of cocaine toxicity in rats. The study shows that diabetic subjects are at special risk to the toxicity and lethality of this alkaloid.

Rosenkranz and Klopman (1990) have predicted the carcinogenic potential of cocaine in rodents. There is no substantial evidence of its carcinogenic action. Exposure is linked to the risk of transplacental cancer induction in the developing human fetus.

Conners and associates (1989) have investigated interactive toxicity of cocaine with phenobarbitol, morphine, and ethanol in organ-cultured human and rat liver slices.

The study indicated that cocaine combined with any of these substances showed greater toxicity than that observed with single components. A similar interactive additive effect with alcohol on maternal and fetal toxicity in Long-Evans rats has been reported (Church et al. 1988). In general, alcohol and cocaine in combination with drugs poses a greater risk to pregnancy than that of either compound alone.

Many anticonvulsant drugs have been studied for their efficacy against cocaine-induced toxicity. Pretreatment with diazepam or phenobarbitol prevented seizure and death from intoxication with cocaine in rats (Derlet and Albertson 1990a). N-Methyl-D-aspartate, valproic acid, and phenytoin showed partial protection against cocaine-induced seizures. In another paper, Derlet and associates (1990) reported that diazepam and propranolol pretreatment afforded protection against cocaine-induced death. Pretreatment with clonidine (0.25 mg/kg), prazosin (5–20 mg/kg), propranolol (8–32 mg/kg), or labetalol (40 mg/kg) protected rats against intraperitoneal LD_{86} values of cocaine (70 mg/kg) (Derlet and Albertson 1990b). These substances interact with α- or β-adrenoreceptors. A combination of these agents did not provide more protection than that provided by single agents. Trouve and Nahas (1986) reported the antidotal action of nitrendipine [39562-70-4] against cardiac toxicity and the acute lethal effects of cocaine. These investigators report that simultaneous administration of nitrendipine (0.00146 mg/kg/min) and cocaine (2 mg/kg/min) increased the survival time of rats from 73 minutes to 309 minutes and the lethal dose of the alkaloid from 146 mg/kg to 618 mg/kg.

7.6   ATROPINE

Formula $C_{17}H_{23}NO_3$; MW 289.41; CAS [51-55-8]
Structure:

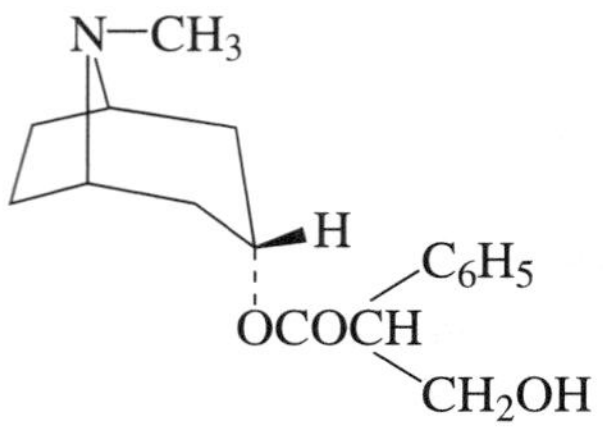

Synonyms: *dl*-hyoscyamine; α-(hydroxymethyl) benzeneacetic acid 8-methyl-8-azabicyclo [3.2.1] oct-3-yl ester; 2-phenylhydracrylic acid 3-α-tropanyl ester; *dl*-tropanyl 2-hydroxy-1-phenylpropionate, tropine tropate; *dl*-tropyltropate

Uses and Exposure Risk

Atropine is the racemic form of the alkaloid *l*-hyoscyamine. The latter is a common tropane alkaloid found in solanaceous plants, such as belladonna (*Atropa belladonna*), henbane (*Hyoscyamus niger*), and the deadly nightshade (*Datura stramonium*). During extraction, *l*-hyoscyamine is readily racemized to atropine, which does not occur naturally in more than traces.

Atropine is used medicinally as an anticholinergic and antispasmodic agent. It is applied in ophthamological treatment and is used as an antidote against organophosphorus insecticide and nerve gases (see Chapter 35). Exposure risks may arise for infants and young children from ophthalmological treatments involving atropine application.

Health Hazard

Although atropine is a known antidote against several toxic substances, its overdose can cause severe poisoning. As with other tropane alkaloids, atropine competitively blocks muscarinic acetylcholine receptor sites. This results in dilation of pupil of the eye (mydriatic effect), dryness of the mucous membranes, especially of the mouth, and inhibition of activity of sweat glands (parasympatholytic action). At toxic doses, it causes palpitation, speech disturbance, blurred vision, confusion, hallucinations, and delirium. Overdoses may produce depression of central nervous system, convulsions, and paralysis. The oral LD_{50} value in rats is within the range 600 mg/kg.

7.7 SCOPOLAMINE

Formula $C_{17}H_{21}NO_4$; MW 303.39; CAS [51-34-3]

Structure:

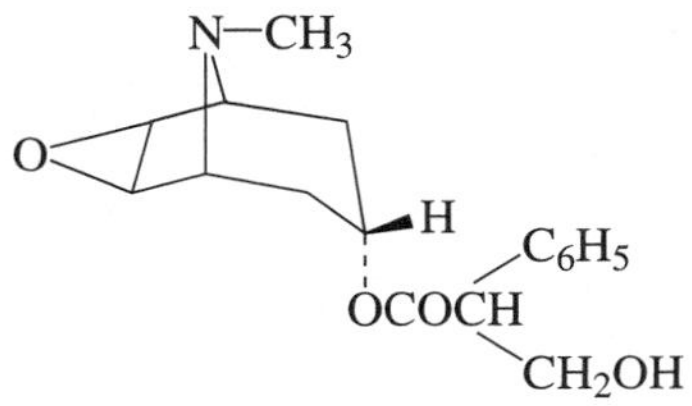

Synonyms: *l*-scopolamine; 6,7-epoxytropine tropate; scopine tropate; hyoscine; 6-*β*-7-*β*-epoxy-3-*α*-tropanyl *S*-(−)-tropate; 9-methyl-3-oxa-9-azatricycle [3.3.1.0²,⁴] non-7-yl ester

Uses and Exposure Risk

Scopolamine is found in the leaves of *Datura metel* L., *D. meteloides* L., and *D. fastuosa* var. *alba* (Cordell 1978). It is used as a sedative, a preanesthetic agent, and in the treatment of motion sickness (Merck 1989).

Physical Properties

Colorless viscous liquid; decomposes on standing; forms a crystalline monohydrate; moderately soluble in water (10 g/100 g at 20°C), dissolves readily in hot water, alcohol, ether, acetone, and methylene chloride; slightly soluble in petroleum ether, benzene, and toluene. It is readily hydrolyzed when mixed with acids and alkalies.

Health Hazard

The toxic effects are similar to atropine. The symptoms at toxic doses are dilation of the pupils, palpitation, blurred vision, irritation, confusion, distorted perceptions, hallucinations, and delirium. However, the mydriatic effect is stronger than that of many other tropane alkaloids. Scopolamine is about three and five times more active than hyocyamine and atropine, respectively, in causing dilation of the pupils. Its stimulating effect on the central nervous system, however, is weaker than that of cocaine but greater than that of atropine. The oral LD_{50} value in mice is within the range of 1200 mg/kg.

The histidine reversion−Ames test for mutagenicity gave inconclusive results.

OPIUM ALKALOIDS

7.8 MORPHINE

Formula $C_{17}H_{19}NO_3$; MW 285.33; CAS [57-27-2]

Structure:

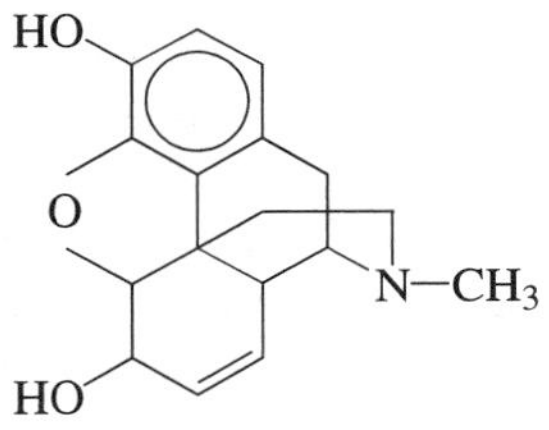

Synonyms: 7,8-didehydro-4,5-*α*-epoxy-17-methylmorphinan-3,6-*α*-diol; 4*a*,5,7*a*,8-tetrahydro-12-methyl-9*H*-9,9*c*-iminoethanophenanthro(4,5-*bcd*) furan-3,5-diol; morphium

Uses and Exposure Risk

Morphine is the most important opium alkaloid, found in the poppy plant (*Papaver somniferum* L.). It is obtained by extracting the dried latex of poppy. It is also obtained from opium, which is an air-dried milky exudate of the same plant produced by incising unripe capsules. Opium contains about 9−14% of morphine. It is used in medicine for its narcotic analgesic, anesthetic, and pain-relief actions.

Physical Properties

White crystalline or amorphous powder; obtained as short orthorhombic needles or prisms from alcohols; melts at 197°C; decomposes at 254°C; darkens on exposure to light; almost insoluble in water (0.02 g/100 g at 20°C), slightly soluble in ethanol, chloroform, ether, and benzene; dissolves in alkali solutions.

Health Hazard

Morphine is a habit-forming substance leading to strong addiction. It exhibits acute toxicity of several types and complexity. Morphine simultaneously produces depressing and stimulating actions on the central nervous system. Depression of the central nervous system results in drowsiness and sleep. Large doses may produce coma and lowering of heart rate and blood pressure. Initial doses of morphine may produce stimulant action, inducing emesis, which can cause nausea and vomiting. Subsequent doses may block emesis. The variety of effects on the central nervous system are manifested by behavioral changes ranging from euphoria and hallucinations to sedation. Repeated dosing enhances tolerance and dependence, with increasingly larger doses needed to produce the effect of euphoria. Thus any abrupt termination to morphine intake after chronic use may lead to physiological rebound (Hodgson et al. 1988). Haffmans and associates (1987) reported the combating action of phelorphan against morphine addiction in chronic morphine-dependent rats. Administration of phelorphan (158 mmol/2 mL), an inhibitor of enzymes involved in the biodegradation of enkephalins, affected the withdrawal symptoms in the addicted animals.

Severe morphine poisoning may result in respiratory failure after fainting fits, accompanied by coma and acute miosis. It exerts effects on the gastrointestinal tract, decreasing the spasmotic reflexes in the intestinal tract, thus resulting in constipation.

LD_{50} value, oral (mice): 524 mg/kg

LD_{50} value, subcutaneous (mice): 220 mg/kg

A fatal dose in humans may be about 0.5–1 g.

Girardot and Holloway (1985) have investigated the effect of cold-water immersion on analgesic responsiveness to morphine in mature rats of different ages. This study indicates that chronic stress affects the reactivity to morphine in young mature rats but not in old rats.

Studies show that when coadministered with morphine, certain substances potentiated morphine's toxicity or mortality in rats. These substances include gentamicin [1403-66-3] (Hurwitz et al. 1988), and doxorubicin (Adriamycin) [23214-92-8] (Innis et al. 1987).

Stroescu and co-workers (1986) have reported the influence of the body's electrolyte composition on the toxic effect of morphine. In mice, depletion of body sodium ion increased the toxicity, which was found to decrease by sodium loading. Glutathione depletors such as cocaine, when coadministered with an opiate such as morphine or heroin, may enhance hepatotoxicity in humans (McCartney 1989). Such a potentiation effect has been explained by the authors as being a result of depletion of endogenous glutathione, which conjugates with morphine to prevent toxic interaction with hepatic cells.

Ascorbic acid and sodium ascorbate have been reported to prevent toxicity of morphine in mice (Dunlap and Leslie 1985). Sodium ascorbate (1 g/kg) injected intraperitoneally 10 minutes before morphine (500 mg/kg) protected the animals against mortality due to respiratory depression. These investigators postulated that ascorbate antagonized the toxicity of morphine by selectively affecting the neuronal activity.

7.9 CODEINE

Formula $C_{18}H_{21}NO_3$; MW 299.36; CAS [76-57-3]

Structure:

the methyl ether of morphine

Synonyms: 7,8-didehydro-4,5-α-epoxy-3-methoxy-17-morphinan-6-α-ol; morphine-3-methyl ether; methylmorphine; morphine monomethyl ether; N-methylnor-codeine

Uses and Exposure Risk

Codeine is used in medicine for its narcotic analgesic action. It is used as a sedative in cough mixtures. Codeine occurs in opium from 0.7% to 2.5%. It is prepared from morphine by methylating the phenolic $-OH$ group of the morphine. It is also obtained by extraction of opium.

Physical Properties

White crystalline powder; obtained as mono-hydrate in the form of orthorhombic rods or tablets from water; dehydrates at 80°C; melts at 155°C; sublimes under high vacuum; slightly soluble in water (0.8 g/100 g at 20°C), readily soluble in most organic solvents; aqueous solution alkaline (pH of a saturated solution 9.8).

Health Hazard

The toxic effects due to codeine are similar but less toxic than those of morphine and other opium alkaloids. An overdose can cause respiratory failure. It is a weak depressant of the central nervous system. It also exhibits stimulant action. Toxic symptoms from high dosages may include drowsiness, sleep, tremors, excitement, and hallucinations. It may also produce gastric

pains and constipation. An oral LD_{50} value in rats is 427 mg/kg. The habit-forming effects of codeine are lower than those associated with morphine.

Nagamatsu and co-workers (1985) have reported in vitro formation of codeinone from codeine by rat or guinea pig liver homogenate. Codeinone may be a metabolic intermediate in the presence of NAD. Its acute toxicity in mice was determined to be 30 times higher than that of codeine.

7.10   THEBAINE

Formula $C_{19}H_{21}NO_3$; MW 311.37; CAS [115-37-7]

Structure:

the methyl enol ether of codeinone

Synonyms: 6,7,8,14-tetradehydro-4,5-α-epoxy-3,6-dimethoxy-17-methylmorphinan; paramorphine; morphine-3,6-dimethyl ether

Uses and Exposure Risk

Thebaine is present in opium at 0.3–1.5%. It is extracted from opium or from latex of the plant *Papaver bracteatum*, in which it is present at up to 26%. It is used to produce codeine.

Physical Properties

Orthorhombic crystals; sublimes; melts on rapid heating at 193°C; slightly soluble in water (0.068 g/100 g at 20°C), moderately soluble in hot alcohol, chloroform, pyridine, and benzene; aqueous solution alkaline (pH of saturated solution is 7.6).

Health Hazard

Although thebaine has a morphine- or codeine-like structure, its pharmacology action is somewhat different. It is a convulsant poison like strychnine, rather than a narcotic. High doses can cause ataxia and convulsion.

LD_{50} value, intraperitoneal (mice): 42 mg/kg

Thebaine is a habit-forming compound. It is a controlled substance (opiate) listed under U.S. Code of Federal Regulations (Title 21 Part 1308.12, 1985).

7.11   HEROIN

Formula $C_{21}H_{23}NO_3$; MW 369.45; CAS [561-27-3]

Structure:

a diacetyl derivative of morphine

Synonyms: 7,8-didehydro-4,5-α-epoxy-17-methylmorphinan-3,6-α-diol   diacetate; morphine   diacetate;   diacetylmorphine; acetomorphine

Uses and Exposure Risk

Heroin does not occur in opium. It is made from poppy extract. Acetylation of morphine with acetic anhydride or acetyl chloride produces heroin. Heroin is not generally used clinically. Its illicit use in the street has been rampant.

Physical Properties

Orthorhombic plates or tablets crystallized from ethyl acetate; turns pink on long exposure to air, emiting acetate odor; mp 173°C;

practically insoluble in water (0.06 g/100 g at 20°C); solubility in ether, alcohol, and chloroform is 1, 3.2, and 66 g/dL of solvent, respectively, at 20°C.

Health Hazard

Heroin is a strongly habit-forming drug. It is more potent than its parent compound, morphine, and much more potent than opium. The toxic effects are similar to those of morphine. An overdose can cause respiratory failure. It is a depressant and a stimulant of the central nervous system, showing a wide range of effects, ranging from drowsiness to distorted perceptions and hallucinations. Toxicological problems from an overdose of heroin may also lead to coma and pinpoint pupil. The toxicity of heroin may be enhanced when it is coadministered with cocaine. Administration of a narcotic antagonist such as naloxone or naltrexone may be performed for therapy. A subcutaneous lethal dose in rabbits may be 180 mg/kg. Heroin may produce harmful reproductive effects in animals and humans. It is a controlled substance (opiate) listed under the U.S. Code of Federal Regulations (Title 21, Part 329.1, 1308.11, 1987).

7.12   PAPAVERINE

Formula $C_{20}H_{21}NO_4$; MW 339.38; CAS [58-74-2]

Structure:

an opium alkaloid of benzylisoquinoline type

Synonyms: 1-((3,4-dimethoxyphenyl)me-
thyl)-6,7-dimethoxyisoquinoline; 6,7-di-
methoxy-1-veratrylisoquinoline

Uses and Exposure Risk

Papaverine occurs in opium to the extent of
0.8–1.0%, commonly associated with narco-
tine. It is used as a smooth muscle relaxant
and in medicine for its vasodilator action on
the blood vessels in the brain. It is effective
against asthma.

Physical Properties

White crystalline powder; obtained as
orthorhombic prisms from an alcohol–ether
mixture; melts at 147°C; sublimes under
vacuum; insoluble in water; soluble in
acetone, glacial acetic acid, and benzene.

Health Hazard

Papaverine is an inhibitor of cyclic nucleotide
phosphodiesterase, producing vasodilatory
effect. The acute toxic effects relative to
phenanthrene-type opium alkaloids (e.g.,
morphine, heroin) are low and the symptoms
are not the same. Papaverine is neither a narcotic nor an addictive substance. Excessive
doses may produce drowsiness, headache,
facial flushing, constipation, nausea, vomiting, and liver toxicity.

The LD_{50} data reported in the literature
show variation. An oral LD_{50} value in rats is
on the order of 400 mg/kg.

ERGOT ALKALOIDS

Ergot alkaloids are derived from the fungus
Claviceps purpurea, which grows on rye and
other grains. These alkaloids are extracted
from ergot produced parasitically on rye.
These compounds may also be produced
by partial or total synthesis. There are four
different classes of ergot alkaloids: clavine

types, lysergic acids, lysergic acid amides,
and ergot peptides. These alkaloids exhibit a
wide range of physiological actions, which
may arise from their partial agonist or
antagonist effects at amine receptor sites,
causing complex effects on cardiovascular
function, suppression of prolactin secretion,
and increase in uterine motility. Although
many of these alkaloids are highly toxic,
some are of great therapeutic value. Their
medicinal applications include treatment
of Parkinson disease, hemorrhage, and
migraine. Discussed below are individual
alkaloids of high toxicity. The acute
poisoning effects of ergot include nausea,
vomiting, diarrhea, thirst, rapid and weak
pulse, confusion, and coma. The chronic
toxicity signs are gastrointestinal disturbance
and gangrene.

Lysergic acid, produced by hydrolysis
of ergot alkaloids, is biologically inactive,
but many of its derivatives are toxic or
hallucinogens.

7.13 LSD

Formula $C_{20}H_{25}N_3O$; MW 323.48; CAS [50-
37-3]
Structure:

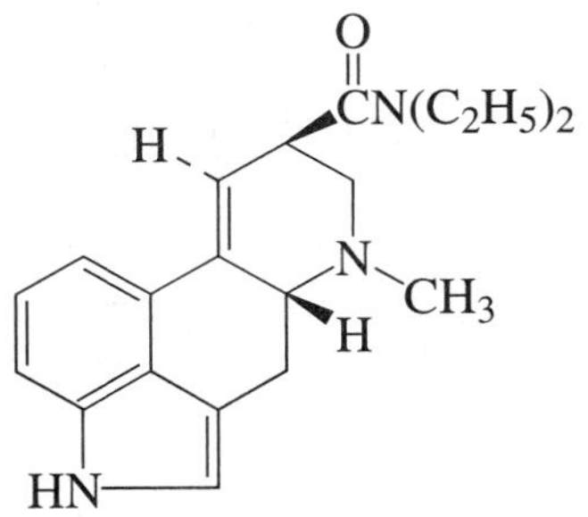

a diethylamide derivative of D-lysergic
acid.
Synonyms: 9,10-didehydro-N, N-diethyl-6-
methylergoline-8-β-carboxamide; N, N-
diethyllysergamide; D-lysergic acid diethylamide; LSD-25; lysergide; delysid

Uses and Exposure Risk

LSD is obtained by partial synthesis from D-lysergic acid. It is also produced by microbial reaction of *Claviceps paspali* over the hydroxylethylamide. It is a well-known hallucinogen and a drug of abuse, listed as a controlled substance in the U.S. Code of Federal Regulations (Title 21, Part 1308.11, 1987). It is used as an antagonist to serotonin and in the study and treatment of mental disorders.

Physical Properties

Crystalline solid; melts at 80–85°C; soluble in alcohol and pyridine.

Health Hazard

LSD is a strong psychedelic agent. The effects in human are excitement, euphoria, hallucinations, and distorted perceptions. It alters the thinking process, producing illusions and loss of contact with reality. In humans, a dose (intramuscular) of 0.7–0.9 mg/kg or an oral dose of 2.5–3.0 mg/kg may produce the effects above. Other symptoms may include nausea, vomiting, dilation of pupils, restlessness, and peripheral vasoconstriction. However, there is no reported case of overdose death. In rabbits, somnolence, ataxia, and an increase in body temperature were the symptoms noted at the LD_{50} (intravenous) doses at 0.3 mg/kg.

LD_{50} value, intravenous (mice): 46 mg/kg

LD_{50} value, subcutaneous (guinea pigs): 16 mg/kg

7.14 MISCELLANEOUS LYSERGIC ACID AMIDES WITH HALLUCINOGEN ACTIVITY

MLD 41

Formula $C_{21}H_{27}N_3O$; MW 337.51; CAS [4238-85-1]

Structure:

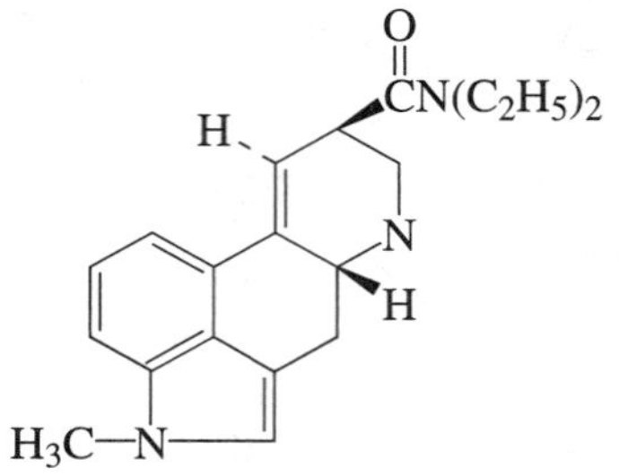

Synonyms: D-1-methyl lysergic acid diethylamide; 9,10-didehydro-N,N-diethyl-1,6-dimethylergoline-8-β-carboxamide

Toxicity

Hallucinogen. In humans an oral dose of 3 μg/kg may produce hallucinogen effects.

ALD-52

Formula $C_{22}H_{27}N_3O_2$; MW 365.52; CAS [3270-02-8]

Structure:

Synonyms: D-1-acetyl lysergic acid diethylamide; 1-acetyl-9,10-didehydro-N,N-diethyl-6-methylergoline-8-β-carboxamide

Toxicity

Hallucinogen. An oral dose of 1 μg/kg has been reported to cause hallucinogen effects (NIOSH 1986).

MLA 74

Formula $C_{19}H_{23}N_3O$; MW 309.45; CAS [7240-57-5]

Structure:

Synonyms: D-1-methyl lysergic acid mono-ethylamide; 9,10-didehydro-1,6-dimethyl-N-ethylergoline-8-β-carboxamide

Toxicity

Hallucinogen. The effects in humans may occur from an oral dose of 25 μg/kg (NIOSH 1986). The intravenous LD_{50} value in rabbits is 9 mg/kg.

LAE 32

Formula $C_{18}H_{21}N_3O$; MW 295.42; CAS [478-99-9]

Structure:

Synonyms: D-lysergic acid monoethylamide; 9,10-didehydro-N-ethyl-6-methylergoline-8-β-carboxamide

Toxicity

Hallucinogen. Psychedelic effects in humans result from an oral dose of 20 μg/kg (NIOSH 1986).

DAM 57

Formula $C_{18}H_{21}N_3O$; MW 295.42; CAS [4238-84-0],

Structure:

Synonyms: 9,10-didehydro-6-methylergo-line-8-β-carboxamide; ergine; LA

Toxicity

Hallucinogen. The toxic dose in humans may be 14 μg/kg.

7.15 ERGONOVINE

Formula $C_{19}H_{23}N_3O_2$; MW 325.45; CAS [60-79-7]

Structure:

Synonyms: 9,10-didehydro-N-(α-(hydroxy-methyl)ethyl)-6-methylergoline-8-β-car-boxamide; N-(1-(hydroxymethyl)ethyl)-D-lysergomide; D-lysergic acid-1,2-pro-panolamide; ergobasine; ergometrine

Uses and Exposure Risk

Ergonovine is a water-soluble lysergic acid derivative that occurs in ergot. It is used as an oxytocic agent. Its maleate derivative is used in the treatment of postpartum hemorrhage.

Physical Properties

Crystalline solid, obtained as fine needles or solvated crystals from benzene or ethyl acetate; melts at 162°C; solution alkaline, pK 6.8; soluble in water, alcohol, acetone, and ethyl acetate.

Health Hazard

The toxicity of ergonovine is low to moderate. The toxic symptoms include excitability, respiratory depression, cyanosis, and convulsions. The intravenous LD_{50} value in mice is 144 mg/kg.

7.16 METHYLERGONOVINE

Formula $C_{20}H_{25}N_3O_2$; MW 339.48; CAS [113-44-8]
Structure:

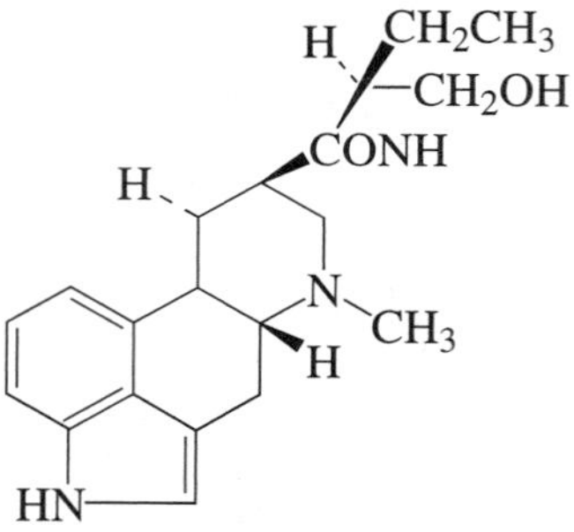

Synonyms: 9,10-didehydro-N-(α-(hydroxy-methyl)propyl)-6-methylergoline-8-β-car-boxamide; d-lysergic acid-dl-hydroxy-butylamide-2; methylergobasine; meth-ergine

Uses and Exposure Risk

This compound is a homologue of ergono-vine. It is used in the treatment of postpartum hemorrhage.

Physical Properties

Shiny crystals; melts at 172°C; moderately soluble in water, dissolves readily in alcohol and acetone.

Health Hazard

The toxic effects due to methylergonovine in humans include nausea, vomiting, dizziness, and hypertension. The oral and intravenous LD_{50} values in mice are 187 and 85 mg/kg, respectively.

CINCHONA ALKALOIDS

7.17 QUININE

Formula $C_{20}H_{24}N_2O_2$; MW 326.21; CAS [130-95-0],
Structure:

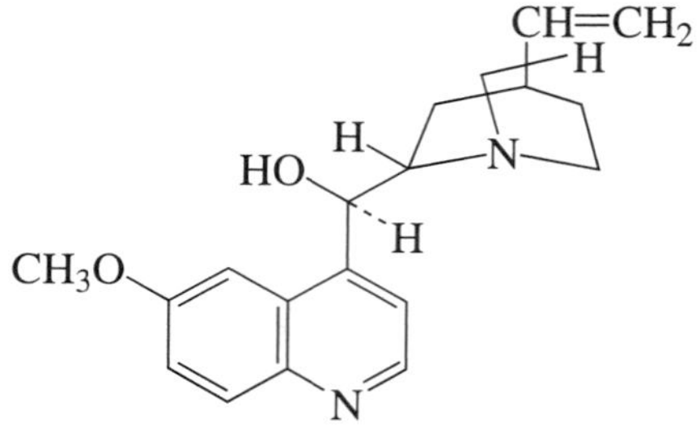

Synonyms: 6′-methoxycinchonan-9-ol; 6-methoxycinchonine; α-(6-methoxy-4-qui-nolyl)-5-vinyl-2-quinuclidine methanol; (−)-quinine

Uses and Exposure Risk

Quinine occurs in the dried stems or root barks of cinchona (*Cinchona ledgeriana* Moens). It is used in the treatment of malaria. It is also used as an analgesic and antipyretic agent.

Physical Properties

White crystalline solid; bitter in taste; melts at 177°C; pH of saturated aqueous soln. 8.8; produces blue fluorescence; slightly soluble in water (0.05 g/100 mL) at 20°C, dissolves readily in alcohols and chloroform. Quinine forms three stereoisomers: epiquinine, quinidine, and epiquinidine.

Health Hazard

The toxicity of quinine is characterized by cinchonism, a term that includes tinnitus, vomiting, diarrhea, fever, and respiratory depression. Other effects include stimulation of uterine muscle, analgesic effect, and dilation of the pupils. Severe poisoning may produce neurosensory disorders, causing clouded vision, double vision, buzzing of the ears, headache, excitability, and sometimes coma (Ferry and Vigneau 1983). Death from quinine poisoning is unusual. Massive doses may be fatal, however.

LD_{50} value, oral (guinea pigs): 1800 mg/kg

7.18   QUINIDINE

Formula: $C_{20}H_{24}N_2O_2$; MW 324.46; CAS [56-54-2]

Structure: Quinidine is a stereoisomer of quinine (see Section 7.17).

Synonyms: *β*-quinine; (+)-quinidine

Uses and Exposure Risk

Quinidine occurs in cinchona bark to about 0.25–0.3% and also in cuprea bark. It is present in quinine sulfate mother liquor. It is formed by isomerization of quinine. It is used in the prevention of certain cardiac arrhythmias.

Physical Properties

White crystalline solid; bitter taste; mp 174°C; blue fluorescence; slightly soluble in water, soluble in alcohols and chloroform.

Health Hazard

Quinidine is more potent than quinine in its action on the cardiovascular system. Overdoses may cause lowering of blood pressure. Gastric effects are lower than quinine. Toxicity is lower relative to quinine; subcutaneous lethal dose in mice is 400 mg/kg against 200 mg/kg for quinine.

7.19   CINCHONINE

Formula $C_{16}H_{11}NO_2$; MW 249.26; CAS [118-10-5]

Structure:

similar to quinine, except there is no methoxy group at position 6 in the ring

Uses and Exposure Risk

Cinchonine occurs in most varieties of cinchona bark (*Cinchona micrantha* and *Rubiaceae*). It is used as an antimalarial agent.

Physical Properties

White prismatic crystals; bitter in taste; melts at 265°C; insoluble in water, moderately soluble in boiling alcohol, slightly soluble in chloroform and ether.

Health Hazard

The toxic effects from high doses of cinchonine include tinnitus, vomiting, diarrhea, and fever. The toxic effects noted in rats were excitement, change in motor activity, and convulsions. An intraperitoneal LD_{50} value in rats is 152 mg/kg.

RAUWOLFIA ALKALOIDS

7.20   RESERPINE

Formula $C_{33}H_{40}N_2O_9$; MW 608.75; CAS [50-55-5]

Structure:

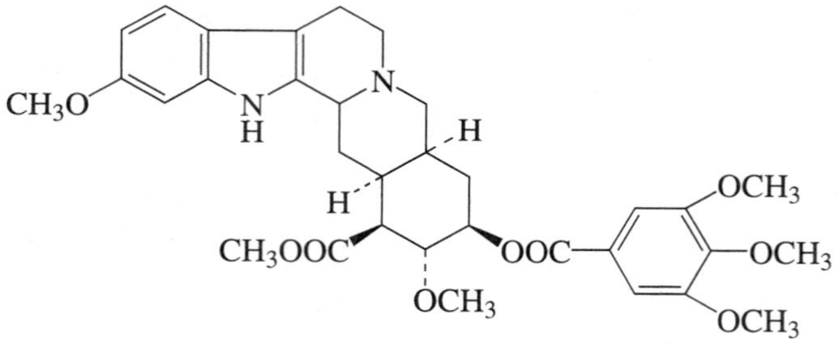

Synonyms: $(3\beta,16\beta,17\alpha,18\beta,20\alpha)$-11,17-dimethoxy-18-((3,4,5-trimethoxybenzoyl)-oxy)yohimban-16-carboxylic acid methyl ester; 3,4,5-trimethoxybenzoyl methyl reserpate; austrapine; sandril; serpasil, serpine

Uses and Exposure Risk

Reserpine occurs in the roots of *Rauwolfia serpentina* and other *Rauwolfia* species, such as *R. micrantha*, *R. canescens*, and *R. vomitoria* Hook. It is used therapeutically as an antihypertensive agent and a tranquilizer. Its use has been reduced significantly because of toxic side effects.

Physical Properties

Crystalline prismatic solid; decomposes at 264°C; weakly alkaline, pK 6.6; slightly soluble in water, dissolves readily in chloroform, glacial acetic acid, ethyl acetate, and benzene; produces fluorescence; solutions turn yellow on standing.

Health Hazard

Reserpine produces sedative, hypotensive, and tranquilizing effects. This is due to its actions of causing depletion of monoamines from presynaptic nerve terminals in central and peripheral nervous systems. The adverse side effects are drowsiness, nightmare, depression, excessive salivation, nausea, diarrhea, increased gastric secretion, abdominal cramps, and hypotension.

LD_{50} value, intraperitoneal (mice): 5 mg/kg

LD_{50} value, oral (mice): 200 mg/kg

There is some evidence of its carcinogenic actions in animals and humans. Reserpine given orally to rats caused tumors in skin, liver, and blood. In humans it may cause skin cancer. Incidence of breast cancers have been found to be high in women treated with reserpine.

7.21   RESCINNAMINE

Formula $C_{35}H_{42}N_2O_9$; MW 634.79; CAS [24815-24-5]

Structure:

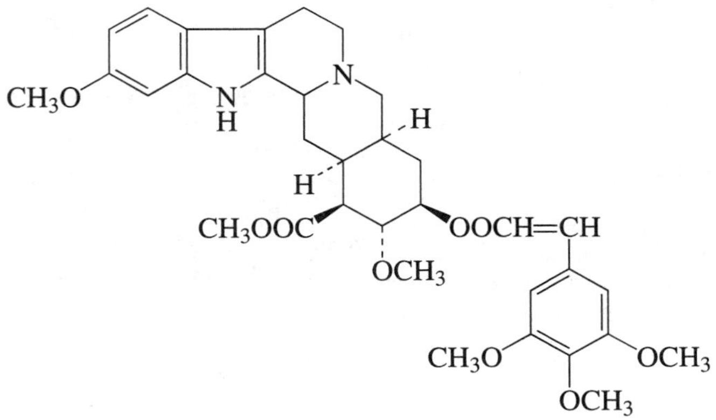

trimethoxycinnamate derivative of methyl reserpate

Synonyms: 3,4,5-trimethylcinnamoylmethyl reserpate; rescamin; anaprel

Uses and Exposure Risk

Rescinnamine occurs in the roots of a number of *Rauwolfia* species. Therapeutically, it is used as an antihypertensive agent.

Physical Properties

Crystalline solid, crystallized as needles from benzene; melts at 238°C; insoluble in water, moderately soluble in methanol, benzene, and chloroform.

Health Hazard

The pharmacological properties of rescinnamine are similar to those of reserpine, but the side effects are low. It acts as a sedative and hypotensive. A dose of 0.25 mg/day may produce the effect of sleepiness in humans. Excessive doses may produce depression, nightmare, nausea, diarrhea, and hypotension.

LD_{50} value, oral (rats): 1000 mg/kg

The carcinogenicity of rescinnamine in animals or humans is unknown.

7.22 DESERPIDINE

Formula $C_{32}H_{38}N_2O_8$; MW 578.72; CAS [131-01-1]

Structure:

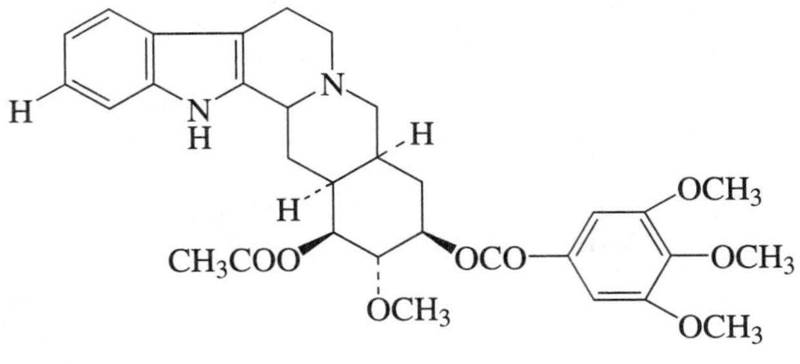

Synonyms: 17α-methoxy-18β((3,4,5-trimethoxybenzoyl)oxy)-3β,20α-yohimban-16 β-carboxylic acid methyl ester; 11-desmethoxy reserpine; canescine; tranquinil

Uses and Exposure Risk

Deserpidine is found and extracted from the roots of *Rauwolfia canescens* L., and Apocyanaceae. Therapeutically, it is used as an antihypertensive agent.

Physical Properties

Crystalline solid; exists in three crystalline forms; mp 228–232°C; soluble in chloroform and methanol, insoluble in water; p*K* 6.68 in saturated solution of 40% methanol in water.

Health Hazard

The pharmacological properties of deserpidine are similar to those of reserpine, causing sedation and tranquilization. Toxic effects from high doses include drowsiness, depression, nausea, diarrhea, abdominal cramps, and hypotension. It is less toxic than reserpine. However, the poisoning effects may be greater than those of other *Rauwolfia* alkaloids, such as rescinnamine. An oral LD_{50} value in mice is 500 mg/kg.

There is some evidence of its carcinogenic actions in animals and humans. Rats given oral doses of 54 mg/kg over 77 weeks developed blood tumor. In humans, it may produce tumors in skin and appendages. Cancer-causing properties of deserpidine, however, are not yet fully established.

7.23 YOHIMBINE

Formula $C_{21}H_{26}N_2O_3$; MW 354.49; CAS [146-48-5]

Structure:

an indole-derived alkaloid

Synonyms: (16α,17α)-17-hydroxyyohimban-16-carboxylic acid methyl ester; corynine; aphrodine; quebrachine

Uses and Exposure Risk

Yohimbine occurs in *Corinanthe johimbe* K. and Rubiaceae trees. It is also found in the roots of *Rauwolfia serpentina* L. and Apocyanaceae. Its derivatives are used therapeutically as adrenergic blocking agents.

Physical Properties

Orthorhombic needles; melts at 235°C; slightly soluble in water, soluble in alcohol, chloroform, and hot benzene.

Health Hazard

Pharmacologically, yohimbine is an adrenergic blocking agent. It exhibits hypotensive and cardiostimulant activities. Poisoning from excessive doses may become severe, causing convulsions and respiratory failure.

LD_{50} value, intraperitoneal (mice): 16 mg/kg

LD_{50} value, oral (mice): 37 mg/kg

ACONITE ALKALOIDS

7.24 ACONITINE

Formula $C_{34}H_{47}NO_{11}$; MW 645.72; CAS [302-27-2]

Structure:

Synonyms: 16-ethyl-1,16,19-trimethoxy-4-(methoxymethyl)aconitane-3,8,10,11,18-

pentol 8-acetate 10-benzoate; aconitane; acetyl benzoyl aconine

Uses and Exposure Risk

Aconitine occurs to the extent of 0.4–0.8% in dried tuberous roots of aconite or monkshood (*Aconitum napellus* L. and Ranunculaceae) found in India, North America, and Europe. It is used to produce heart arrhythmia in experimental animals and as an antipyretic agent.

Physical Properties

Crystalline solid obtained as hexagonal plates from chloroform; melts at 204°C; aqueous soln. alkaline; slightly soluble in water (0.033 g/100 mL at 20°C), moderately soluble in ethanol and ether, but dissolves readily in chloroform and benzene.

Health Hazard

Aconitine is among the most toxic alkaloids known. Toxic doses are close to therapeutic doses, and in humans as little as 2 mg may cause death (Ferry and Vigneau 1983). An oral lethal dose of 28 mg/kg has also been recorded (NIOSH 1986). The toxic symptoms at low doses may be excitement, drowsiness, and hypermotility. The first sign of poisoning from ingestion is a tingling, burning feeling on the lips, mouth, gums, and throat (Hodgson et al. 1988). This is followed by nervous disorders, anesthesia, loss of coordination, vertigo, hypersalivation, nausea, vomiting, and diarrhea. Toxic actions of aconitine are very rapid. The crystalline form of this compound is much more toxic than the amorphous aconitine. At a lethal dose, death may result from cardiorespiratory failure. Procaine may be an effective antidote against aconitine poisoning.

LD_{50} value, oral (rat): 1 mg/kg

7.25 FERACONITINE

Formula $C_{36}H_{51}NO_{12}$; MW 689.78; CAS [127-29-7]

Structure:

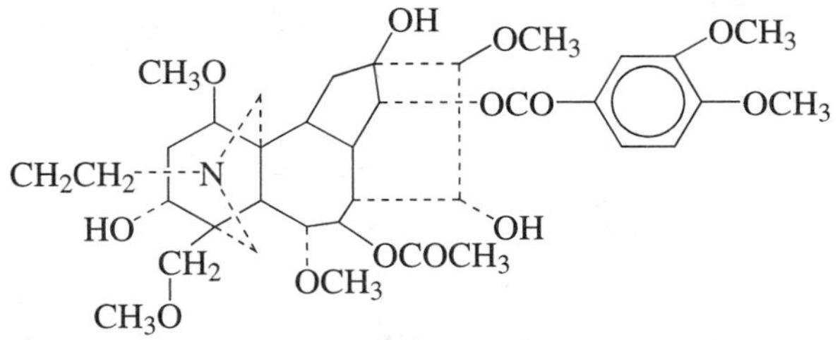

Synonyms: (1α,3α,6α,14α,16β)-20-ethyl-1, 6,16-trimethoxy-4-(methoxy ethyl)aconitane-3,8,13,14-tetrol 8-acetate 14-(3,4-dimethoxybenzoate); pseudoaconitine; nepaline; veratroylaconine

Uses and Exposure Risk

Feraconitine occurs in the tubers of Indian aconite (bish or *Aconitum ferox*), found in India and Nepal. It is used in arrow poisoning for hunting. Clinically, it is used in treating rheumatism.

Physical Properties

White crystalline, amorphous, or syrupy substance; salts are optically active; melts at 214°C; insoluble in water, soluble in alcohol, ether, and chloroform.

Health Hazard

Feraconitine is a highly toxic alkaloid. The acute toxic effects are similar to those of aconitine. The symptoms from ingestion are drowsiness, nervous disorders, weakness, loss of coordination, vertigo, nausea, vomiting, and diarrhea, progressing to bradycardia. Death may result even from small doses, due to cardiac depression. LD_{50} data for this compound are not available.

HEMLOCK ALKALOIDS

7.26 CONIINE

Formula $C_{18}H_{17}N$; MW 127.26; CAS [458-88-8]

Structure:

Synonyms: 2-propylpiperidine; β-propylpiperidine; cicutine; conicine

Uses and Exposure Risk

Coniine is present in all parts of the plant poison hemlock (*Conium maculatum*). It is the first alkaloid to be synthesized and structure determined. Because of its high toxicity, its clinical application is very limited.

Physical Properties

Colorless liquid; darkens on exposure to light and air; polymerizes; boils at 166°C; melts at −2°C; alkaline, pK_a 3.1; density 0.845 at 20°C; slightly soluble in water and chloroform, readily soluble in most organic solvents.

Health Hazard

Coniine is a highly toxic alkaloid. The toxic symptoms from ingestion are weakness, drowsiness, nausea, vomiting, muscle contraction, and labored breathing. A high dose can cause convulsions, cyanosis, asphyxia, and death. Chronic ingestion of coniine produced adverse reproductive effects and specific developmental abnormalities in cattles.

LD_{50} value, oral (mice): 100 mg/kg

7.27 γ-CONICEINE

Formula $C_8H_{15}N$; MW 125.24; CAS [1604-01-9]

Structure:

Synonyms: 2,3,4,5-tetrahydro-6-propylpyridine; 2-*n*-propyl-3,4,5,6-tetrahydropyridine

Uses and Exposure Risk

γ-Coniceine occurs in the seeds of poison hemlock (*Conium maculatum* L. or Umbelliferae). It is produced by reduction of coniine.

Physical Properties

Colorless liquid with characteristic odor similar to coniine; boils at 171°C; density 0.875 at 15°C; alkaline; slightly soluble in water, dissolves readily in organic solvents.

Health Hazard

Acute oral toxicity of γ-coniceine is greater than that of coniine. The symptoms are similar. Ingestion may cause drowsiness, nausea, vomiting, respiratory distress, convulsions, cyanosis, and at high doses, death. An oral LD_{50} value in mice is reported as 12 mg/kg (NIOSH 1986).

STRYCHNOS ALKALOIDS

7.28 STRYCHNINE

EPA Designated Toxic Waste; RCRA Waste Number P108; DOT Label: Poison B, UN 1692

Formula $C_{21}H_{22}N_2O_2$; MW 334.40; CAS [57-24-9]

Structure:

Synonym: strychnidin-10-one

Uses and Exposure Risk

Strychnine occurs in the seeds of strychnos species (*S. nux vomica* L., *S. Loganiaceae*, and *S. ignatii* Berg). The total alkaloid content in these plants is 2–3%. The composition of strychnine in these species ranges between 1% and 3%. Strychnine is widely known as a poison. Its therapeutic applications are very limited. It is used as a rodent poison.

Physical Properties

White crystalline powder; highly bitter; melts at 268°C; very slightly soluble in water (0.016 g/100 mL at 20°C), moderately soluble in boiling alcohol, dissolves readily in chloroform, pH of a saturated aqueous soln. 9.5.

Health Hazard

Strychnine is a highly toxic alkaloid. It causes hypersensitivity to sensory stimuli. It is a powerful convulsant. This results in respiratory and metabolic acidosis (Hodgson et al. 1988). Death occurs from asphyxia after a few seizures. Its convulsant actions are attributed to the antagonism of the inhibitory effects of glycine. It excites all portions of the central nervous system. It produces green-colored vision, which is an effect of sensory disorders. Toxic symptoms from continued medication with strychnine include photophobia, muscular rigidity, stiffness in joints, lassitude, and headache (von Oettingen 1952, ACGIH 1986). Ingestion of 0.1 g may be fatal to humans.

LD_{50} value, oral (mice): 2 mg/kg

Intravenous administration of diazepam is applied for the treatment of strychnine poisoning.

Exposure Limit

TLV-TWA air 0.15 mg/m^3 (ACGIH, MSHA, and OSHA)

7.29   BRUCINE

EPA Designated Toxic Waste, RCRA Waste Number P018; DOT Label: Poison B, UN 1570

Formula $C_{23}H_{26}N_2O_4$; MW 394.51; CAS [357-57-3]

Structure:

Synonyms: 2,3-dimethoxystrychnine; 10,11-dimethoxy strychnine; 2,3- dimethoxystrychnidin-10-one

Uses and Exposure Risk

Brucine occurs in the seeds of strychnos species (*Strychnos nux vomica* L. and other species). It is used for denaturing alcohols and oils; for separating racemic mixtures; and as an additive to lubricants. It is also used for colorimetric analysis of nitrate. Therapeutically, it is used as a stimulant.

Physical Properties

White crystalline solid; very bitter in taste; mp 178°C; hydrated crystals (tetra- and dihydrates lose water at 100°C; pH of saturated aqueous soln. 9.5; slightly soluble in water (0.07 g/100 mL), dissolves readily in alcohols and chloroform.

Health Hazard

The toxicity of brucine is similar to that of strychnine. It is a very poisonous alkaloid. It is a strong convulsant; excites all portions of the central nervous system. Toxic symptoms include headache, tremor, muscular rigidity, and convulsions. Death may occur from respiratory arrest after a few convulsions.

LD$_{50}$ value, subcutaneous (mice): 60 mg/kg

LUPIN ALKALOIDS

7.30   CYTISINE

Formula C$_{11}$H$_{14}$N$_2$O; MW 190.27; CAS [485-35-8]

Structure:

contains quinolizidine nucleus

Synonyms: 1,2,3,4,5,6-hexahydro-1,5-methano 8*H*-pyrido-[1,2-*a*][1,5]diazocin-8-one; sophorine; ulexine; baptitoxine

Uses and Exposure Risk

Cytisine occurs in the seeds of *Cytisus laburnum* L. and other Leguminosae.

Physical Properties

Orthorhombic prisms crystallized from acetone; melts at 152°C; aqueous soln. strongly basic; soluble in water and organic solvents.

Health Hazard

Cytisine is highly toxic to humans and animals. Ingestions may cause nausea, vomiting, and convulsions. Death may occur from respiratory failure.

LD$_{50}$ value, oral (rats): 101 mg/kg

7.31   LUPININE

Formula C$_{10}$H$_{19}$NO; MW 169.27; CAS [486-70-4]

Structure:

Synonyms: *l*-lupinine; (1*R*-trans)-octahydro-2*H*-quinolizine-1-methanol

Uses and Exposure Risk

The *l*-form of lupinine occurs in seeds and herb of *Lupinus luteous* L., Chenopodiaceae, and other lupinus species. Its clinical applications are very limited.

Physical Properties

Crystalline solid, obtained as orthorhombic prisms from acetone; melts at 69°C; bp 270°C; strongly basic; soluble in water and organic solvents.

Health Hazard

This alkaloid is moderately toxic. The toxic action, however, is lower than that of cytisine. Ingestion of high doses may produce nausea, convulsions, and respiratory failure. The lethal dose in guinea pigs by the intraperitoneal route is 28 mg/kg.

7.32   SPARTEINE

Formula C$_{15}$H$_{26}$N$_2$; MW 234.37; CAS [90-39-1]

Structure:

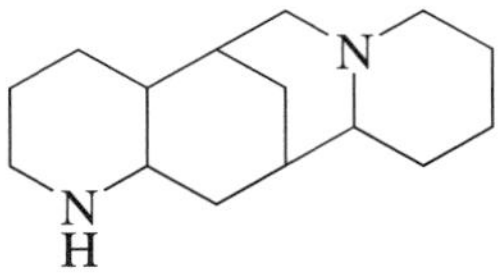

Synonyms: dodecahydro-7,14-methano-2*H*, 6*H*-dipyrido [1,2-*a*:1′,2′- *e*][1,5]diazocine; lupinidine; 6β,7α,9α,11α-pachycarpine; *l*-sparteine

Uses and Exposure Risk

Sparteine occurs in yellow and black lupin beans (*Lupinus luteus* L. and *L. niger*). It also occurs in other *Lupinus* species as well as in *Cytisus* and *Spartium* species. Therapeutically, it is used as an oxytocic agent.

Physical Properties

Viscous oily liquid; boils around 180°C; density 1.02 at 20°C; strongly basic, a 0.01 *M* soln. has pH 11.6; slightly soluble in water (0.3 g/100 mL), dissolves readily in alcohol, ether, or chloroform.

Health Hazard

Toxic actions of sparteine are similar to those of coniine but are much less toxic than those of coniine. The toxic symptoms in mice were tremor, muscle contraction, and respiratory distress. An intraperitoneal dose lethal to guinea pigs is 23 mg/kg.

LD$_{50}$ value, oral (mice): 220 mg/kg

IPECAC ALKALOIDS

7.33 EMETINE

Formula $C_{19}H_{40}N_2O_4$; MW 480.63; CAS [483-18-1]

Structure:

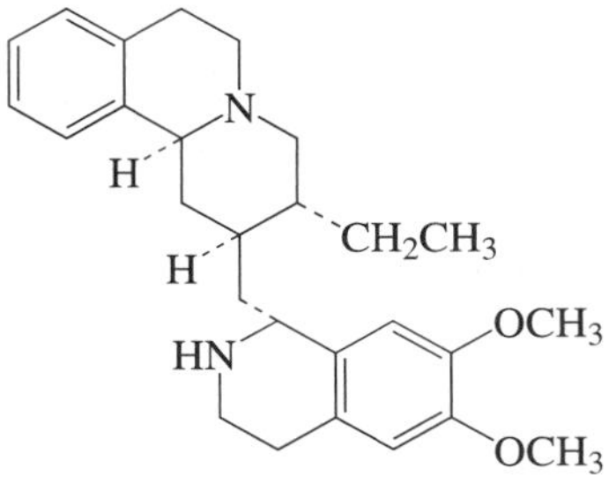

Synonyms: 3-ethyl-1,3,4,6,7,11*b*-hexahydro-9,10-dimethoxy-2-((1,2,3,4-tetrahydro-6, 7-dimethoxy-1-isoquinolyl)methyl)2*H*-benzo [a]quinolizine; 6′,7′,10,11-tetra-methoxymetan; cephaeline methyl ether

Uses and Exposure Risk

Emetine occurs in the ground roots of ipecac (*Uragoga ipecacuanha*, *Cephaelis ipecacuanha*, *Uragoga acuminata*, and Rubiaceae). Ipecacs are of two varieties, Brazilian and Cartagena. The former contains about 1.2–1.5% emetine, and the latter constitutes about 1.1–1.4% emetine. It is used as an emetic to induce vomiting for the treatment of poisoning. It is also used as an antiamebic.

Physical Properties

White amorphous powder with very bitter taste; becomes yellow when exposed to light or heat; melts at 74°C; aqueous solutions alkaline; slightly soluble in water, dissolves readily in most organic solvents.

Health Hazard

Emetine is a moderate to highly toxic alkaloid. At toxic doses it affects the digestive, neuromuscular, and cardiovascular systems, causing nausea, diarrhea, vomiting, muscle weakness, and lowering of blood pressure. The poisoning effects are cumulative and may be manifested as side effects at therapeutic doses. The oral LD$_{50}$ value in rats is reported to be 68 mg/kg (NIOSH 1986). Chronic exposure to this alkaloid and its salts have resulted in conjunctivitis, epidermal inflammation, and asthma attacks in susceptible individuals (Cordell 1978).

7.34   CEPHAELINE

Formula $C_{28}H_{38}N_2O_4$; MW 466.60; CAS [483-17-0]

Structure:

Synonyms: 7',10,11-trimethoxymetan-6'-ol; dihydropsychotrine; desmethylemetine

Uses and Exposure Risk

Cephaeline occurs in the ground roots of ipecac along with emetine. The Brazilian variety usually contains about 0.5% cephaeline. It is used as an emetic and anti-amebic reagent.

Physical Properties

Crystalline solid, obtained as needles from ether; faintly bitter in taste; melts at 115°C; insoluble in water, dissolves in organic solvents and dilute mineral acids.

Health Hazard

The toxicity of cephaeline is lower than that of emetine. The toxic effects are cumulative. Ingestion of high doses may produce hypotension, muscle weakness, and gastrointestinal problems, including nausea, vomiting, and diarrhea.

MUSHROOM ALKALOIDS

7.35   AMANITINS

Among the highly toxic species of mushrooms are the *Amanita* species. The group of amanitin alkaloids present in *Amanita phalloides* are α-, β-, and γ-amanitins and amanin. The former two types of alkaloids occur to the extent of 4–5 mg/100 g. The structure of α- and β-amanitin is as follows:

α-amanitin: $C_{39}H_{54}N_{10}O_{13}S_7$, R=NH$_2$
β-amanitin: $C_{39}H_{53}N_9O_{14}S$, R=OH

Amanitins are highly toxic alkaloids. The fatal dose for humans is about 0.1 mg/kg. The delayed toxic effects begin about 10–15 hours after ingestion. The symptoms are salivation, vomiting, diarrhea, muscular twitching, liver damage, lesions of the kidneys, convulsions, hepatic coma, and death.

α-Amanitin [21150-20-9] LD$_{50}$ value, intraperitoneally (mice): 0.1 mg/kg

β-Amanitin [21150-22-1] LD$_{50}$ value, intraperitoneally (mice): 0.5 mg/kg

γ-Amanitin [21150-23-2] LD$_{50}$ value, intraperitoneally (mice): 0.2 mg/kg

ε-Amanitin [21705-02-2] LD$_{50}$ value, intraperitoneally (mice): 0.3 mg/kg

Amanin [21150-21-0] LD$_{50}$ value, intraperitoneally (mice): 0.5 mg/kg

Certain species of poisonous mushrooms contain toxic substances that are structurally different from those mentioned above. These toxic components are characterized by the presence of oxazolidine heterocycles in their structures. These are ibotenic acid [2552-55-8], muscazone [2255-39-2], and muscimol [2763- 96-4].

7.36 IBOTENIC ACID

Formula C$_5$H$_6$N$_2$O$_4$; MW 158.11; CAS [2552-55-8]

Structure:

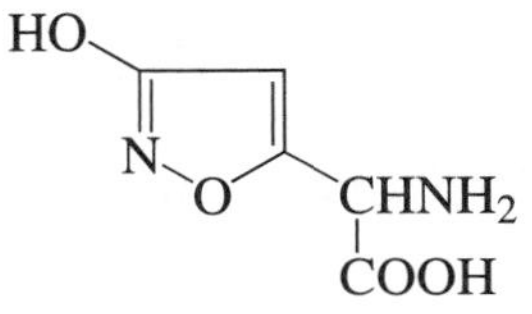

Synonyms: α-amino-2,3-dihydro-3-oxo-5-isoxazoleacetic acid; α-amino-3-hydroxy-5-isoxazoleacetic acid; amino-(3-hydroxy-isoxazolyl) acetic acid

Ibotenic acid occurs in *Amanita pantherina, Amanita muscaria* L., and Agaricaceae.

Physical Properties

Crystalline solid; forms monohydrate; mp 151°C (anhydrous), 145°C (monohydrate); soluble in water and alcohol.

Health Hazard

Ibotenic acid is a potent neurologic amino acid. It exhibits neuroexcitatory activity and causes sedative actions on spinal neurons. High doses can cause sleep, hallucinations, distorted perceptions, and nausea. In humans the symptoms above may be manifested from ingestion of 6–8 mg of ibotenic acid.

LD$_{50}$ value, oral (mice): 38 mg/kg

LD$_{50}$ value, intraperitoneal (mice): 15 mg/kg

7.37 MUSCAZONE

Formula C$_5$H$_6$N$_2$O$_4$; MW 158.11; CAS [2255-39-2]

Structure:

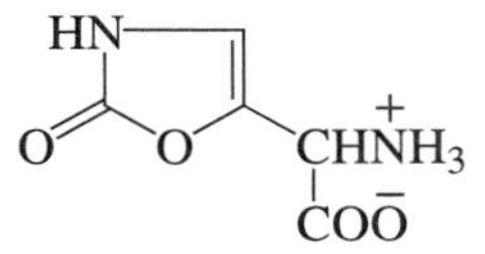

Synonyms: α-amino-2,3-dihydro-2-oxo-5-oxazoleacetic acid; α-amino-2-oxo-4-oxazoline-5-acetic acid

Muscazone is a crystalline solid, decomposes above 190°C; sparingly soluble in water.

Health Hazard

Muscazone is one of the potent neurologic toxins of *Amanita muscaria*. The toxic symptoms from high doses are similar to those of ibotenic acid. The LD$_{50}$ data for this compound are not available.

7.38 MUSCIMOL

EPA Designated Toxic Waste, RCRA Waste Number P007

Formula C$_4$H$_6$N$_2$O$_2$; MW 114.10; CAS [2763-96-4]

Structure:

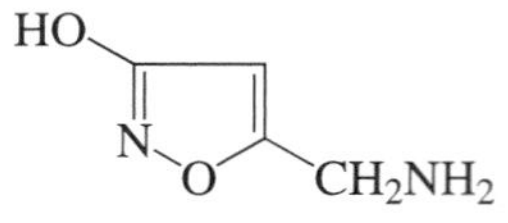

Synonyms: 5-(aminomethyl)-3(2H)-isooxazolone; 3-hydroxy-5-aminomethylisooxazole; agarin

Muscimol is found in *Amanita muscaria* L. and Agaricaceae.

Physical Properties

Crystalline solid, melts at 175°C (decomposes).

Health Hazard

Muscimol is a toxic alkaloid of poisonous mushroom species, producing neurologic action. It is a potent depressant of the central nervous system. The adverse health effects from its ingestion include sleep, hallucination, distorted perceptions, and vomiting.

LD$_{50}$ value, oral (mice): 17 mg/kg
LD$_{50}$ value, intraperitoneal (mice): 2.5 mg/kg

MISCELLANEOUS ALKALOIDS

7.39 MESCALINE

Formula C$_{11}$H$_{17}$NO$_3$; MW 211.29; CAS [54-04-6]
Structure:

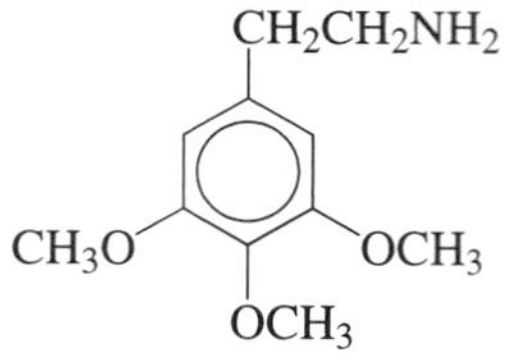

Synonyms: 3,4,5-trimethoxyphenylethylamine; 3,4,5-trimethoxybenzeneethanamine; 1-amino-2-(3,4,5-trimethoxyphenyl) ethane; mezcline

Uses and Exposure Risk

Mescaline occurs in peyote cactus, *Lophophora williamsii* (Lemaire) Coulter, grown in Mexico. It is used as a hallucinogen in Native American religious ceremonies. It is also used in medicine. It is a controlled substance (hallucinogen) and listed in the U.S. Code of Federal Regulations (Title 21, Part 1308.11, 1987).

Physical Properties

Crystalline solid; melts at 35°C; moderately soluble in water, dissolves readily in alcohol, benzene, and chloroform, slightly soluble in ether; absorbs carbon dioxide from air, forming carbonate.

Health Hazard

Mescaline is a psychedelic agent, causing hallucinations. It promotes psychosis in predisposed individuals. The mechanism of action probably involves its function as serotonin 5-HT agonist (Nichols 1986). Toxic symptoms include drowsiness, dilation of pupil, hyperreflexia, and restlessness. These symptoms progress with increasing doses, resulting in euphoria, distorted perceptions, and hallucinations. Such an illusion effect in humans may be manifested from intramuscular administration of about 150 mg of mescaline. The toxicity of this alkaloid, is otherwise, moderate, rarely causing death from overdose.

LD$_{50}$ value, intraperitoneal (mice): 315 mg/kg
LD$_{50}$ value, oral (mice): 880 mg/kg

7.40 COLCHICINE

Formula C$_{22}$H$_{25}$NO$_6$; MW 399.48; CAS [64-86-8]

Structure:

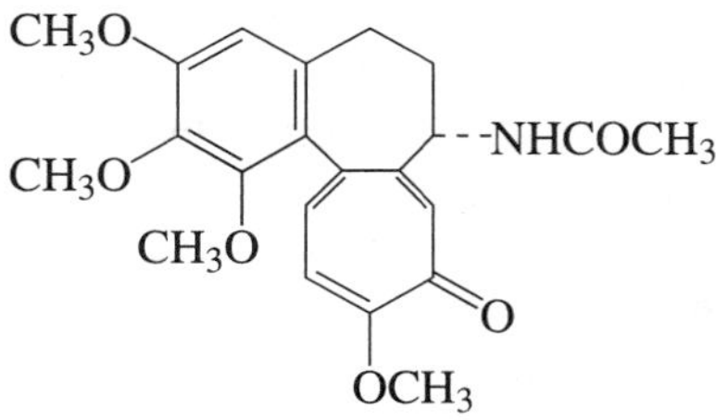

contains an amide functional group

Synonyms: *N*-(5,6,7,9-tetrahydro-1,2,3,10-tetramethoxy-9-oxobenzo(α)heptalen-7-yl) acetamide; *N*-acetyl trimethylcolchicinic acid methylether; colcin

Uses and Exposure Risk

Colchicine is present in the poisonous autumn crocus (meadow saffron). It is the major alkaloid of *Colchicum autumnale* L. and Liliaceae. It was used in poison potions in the ancient kingdom of Colchis (Greece). It is used therapeutically as an antineoplast, for the suppression of gout, and in the treatment of Mediterranean fever. It is used in plant studies for doubling chromosome groups.

Physical Properties

Pale yellow crystals or powder; darkens when exposed to light; melts at 145°C; moderately soluble in water (4.5 g/dL at 20°C), slightly soluble in ether and benzene, dissolves readily in chloroform and ethanol; aqueous solution is neutral.

Health Hazard

Colchicine is a highly toxic alkaloid. The target organs are the lungs, kidney, gastrointestinal tract, and blood. The toxic effects include drowsiness, nausea, hypermotility, diarrhea, lowering of body temperature, lowering of blood pressure, tremors, convulsions, and respiratory distress. Chronic ingestion may cause aplastic anemia and hemorrhage.

LD_{50} value, oral (mice): 5.9 mg/kg

Colchicine solution at 10,000 ppm concentration caused severe irritation when applied repeatedly to rabbits' eyes. Colchicine produced teratogenic effects in test animals. It caused fetal death, cytological changes, and developmental abnormalities in hamsters, rabbits, domestic animals, and mice. It tested positive to in vitro mammalian nonhuman micronucleus and *D. melanogaster* — nondisjunction tests for mutagenicity (U.S. EPA 1986; NIOSH 1986).

7.41   SAXITOXIN

Formula $C_{10}H_{17}N_7O_4$; MW 299.34; CAS [35523-89-8]

Structure:

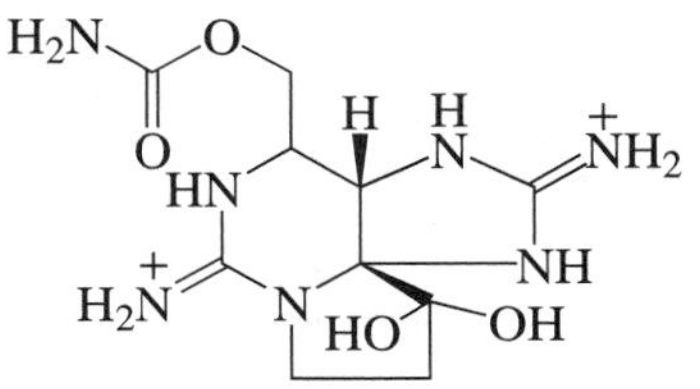

Synonyms: 2,6-diamino-4-(((aminocarbonyl) oxy)methyl)-3*a*,4,8,9-tetrahydro-(3*a*S-(3 *a*-α,4α,10a*R*))1*H*,10*H*-pyrrolo(1,2-*c*)purine-10,10-diol; mussel poison; clam poison

Uses and Exposure Risk

Saxitoxin is an alkaloid of nonplant origin. It is the neurotoxic constituent of dinoflagellates (*Gonyaulax catenella* and *G. excavata*) the so-called "red tide" found along the U.S. coast. Shellfish, clams, and scallops consume this and become extremely poisonous for human consumption.

Physical Properties

Crystalline solid; soluble in water and methanol; forms dihydrochloride with HCl.

Health Hazard

Saxitoxin is an extremely toxic substance. It binds to sodium channels and blocks nerve membrane. In humans, ingestion of this compound can produce tingling and burning in the lip, tongue, face, and the whole body within an hour. This is followed by numbness, muscular incoordination, confusion, headache, and respiratory failure. Death may occur within 12 hours.

LD_{50} value intraperitoneal (mice): 0.005 mg/kg

LD_{50} value oral (mice): 0.26 mg/kg

Intravenous administration of 1 mL of 1:2000 solution of prostigmine methylsulfate has been reported to be effective against saxitoxin poisoning (Hodgson et al. 1988).

7.42 PILOCARPINE

Formula $C_{11}H_{16}N_2O_2$; MW 208.25; CAS [92-13-7]

Structure:

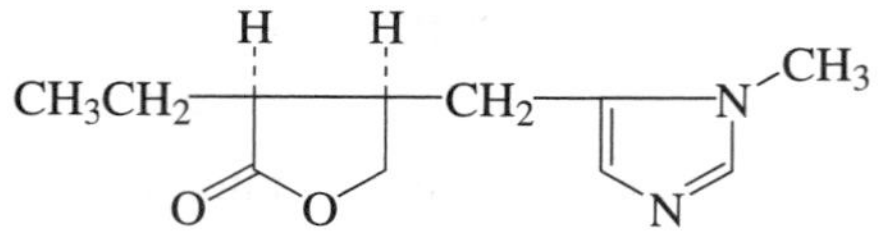

CH₃CH₂ — CH₂ — N—CH₃ (furanone imidazole structure)

Synonyms: 3-ethyldihydro-4-((1-methyl1H-imidazol-5-yl)methyl)-2(3H)-furanone; almocarpine

Uses and Exposure Risk

Pilocarpine occurs in the leaves of various species of pilocarpus. It is used as an antidote for atropine poisoning and in ophthalmology to produce contraction of the pupil.

Physical Properties

Colorless crystalline solid or an oil; melts at 34°C; dissolves in water, alcohol, and chloroform; slightly soluble in ether and benzene.

Health Hazard

Pilocarpine is a tropane alkaloid. Toxic symptoms are characterized by muscarinic effects. Toxic effects include hypersecretion of saliva, sweat, and tears; contraction of the pupils of the eyes; and gastric pain accompanied with nausea, vomiting, and diarrhea. Other symptoms are excitability, twitching, and lowering of blood pressure. High doses may lead to death due to respiratory failure. A lethal dose in humans is estimated within the range of 150–200 mg.

7.43 CAFFEINE

Formula $C_8H_{10}N_4O_2$; MW 194.22; CAS [58-08-2]

Structure:

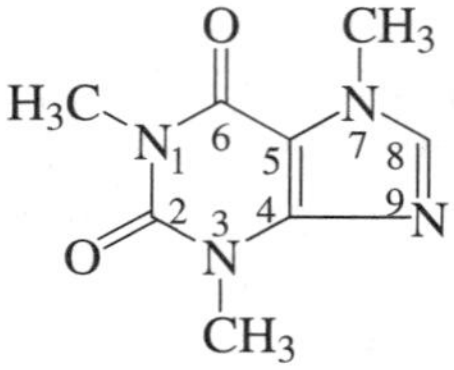

Synonyms: 3,7-dihydro-1,3,7-trimethyl-1H-purine-2,6-dione; 1,3,7- trimethylxanthine; 1,3,7-trimethyl-2,6-dioxopurine; methyl-theobromine; guaranine; theine

Uses and Exposure Risk

Caffeine is consumed in coffee, tea, cocoa, chocolate, and soft drinks. It occurs naturally in the leaves of coffee, tea, and maté and in cola nuts. It is used in medicine and found in many drugs. It is used as a cardiac stimulant.

Physical Properties

White prismatic crystals; odorless; bitter in taste; melts at 237°C; sublimes; slightly soluble in water (2.2 g/100 mL at 20°C), soluble in boiling water, chloroform, tetrahydrofuran, and ethyl acetate, slightly soluble in alcohol, ether, and benzene.

Health Hazard

Caffeine is a stimulant of the central nervous system. It eliminates fatigue and drowsiness. However, high doses cause gastrointestinal motility, restlessness, sleeplessness, nervousness, and tremor. Acute poisoning effects include nausea, vomiting, headache, excitability, tremor, and sometimes, convulsive coma. Other symptoms may be respiratory depression, muscle contraction, distorted perception, and hallucination. Ingestion of 15–20 g may be fatal to humans.

LD_{50} value, oral (mice): 127 mg/kg
LD_{50} value, oral (rabbits): 224 mg/kg

Animal studies indicate that caffeine at high doses produces adverse reproductive effects, causing developmental abnormalities. It tested negative in the histidine reversion–Ames and TRP reversion tests.

REFERENCES

ACGIH. 1986. *Documentation of the Threshold Limit Values and Biological Exposure Indices*, 5th ed. Cincinnati, OH: American Conference of Governmental Industrial Hygienists.

Amonkar, A. J., P. R. Padma, and S. V. Bhide. 1989. Protective effect of hydroxychavicol, a phenolic component of betel leaf, against the tobacco-specific carcinogens. *Mutat. Res.* 210(2): 249–53.

Aoyama, M., and N. Tachi. 1989. Acute toxicity in male DDY mice orally administered chewing tobacco extract. *Nagoya Med. J.* 34(1): 55–61; cited in *Chem. Abstr. CA 113*(17): 147117d.

Belinsky, S. A., C. M. White, J. A. Boucheron, F. C. Richardson, J. A. Swenberg, and M. Anderson. 1986. Accumulation and persistence of DNA adduct in respiratory tissue of rats following multiple administrations of the tobacco specific carcinogen, 4-(N-methyl-N-nitrosamino)-1-(3-pyridyl)-1-butanone. *Cancer Res.* 46(3): 1280–84.

Berger, M. R., E. Petru, M. Habs, and D. Schmaehl. 1987. Influence of perinatal nicotine administration on transplacental carcinogenesis in Sprague–Dawley rats by N-methylnitrosourea. *Br. J. Cancer 55*(1): 37–40.

Bozarth, M. A., and R. A. Wise. 1985. Toxicity associated with long-term intravenous heroin and cocaine self-administration in the rat. *J.A.M.A. 254*(1): 81–83.

Chen, Y. P., and C. A. Squier. 1990. Effect of nicotine on 7,12- dimethylbenz[a]anthracene carcinogenesis in hamster cheek pouch. *J. Natl. Cancer Inst.* 82(10): 861–64.

Church, M. W., B. A. Dintcheff, and P. K. Gessner. 1988. The interactive effects of alcohol and cocaine on maternal and fetal toxicity in the Long–Evans rat. *Neurotoxicol. Teratol. 10*(4): 355–61.

Conners, S., D. R. Rankin, C. L. Krumdieck, and K. Brendel. 1989. Interactive toxicity of cocaine with phenobarbital, morphine and ethanol in organ cultured human and rat liver slices. *Proc. West. Pharmacol. Soc. 32*: 205–8; cited in *Chem. Abstr. CA 111*(11): 89815a.

Cordell, G. A. 1978. Alkaloids. In *Kirk-Othmer Encyclopedia of Chemical Technology*, 3rd ed., Vol. 1, pp. 883–943. New York: Wiley-Interscience.

Correa, E., P. A. Joshi, A. Castonguay, and H. M. Schuller. 1990. The tobacco-specific nitrosamine 4-(methylnitrosamino)-1-(3-pyridyl)-1-butanone is an active transplacental carcinogen in Syrian golden hamsters. *Cancer Res. 50*(11): 3435–38.

Derlet, R. W., and T. E. Albertson. 1990a. Acute cocaine toxicity: antagonism by agents interacting with adrenoceptors. *Pharmacol. Biochem. Behav. 36*(2): 225–31.

Derlet, R. W., and T. E. Albertson. 1990b. Anticonvulsant modification of cocaine-induced toxicity in the rat. *Neuropharmacology 29*(3): 255–59.

Derlet, R. W., T. E. Albertson, and P. Rice. 1990. Antagonism of cocaine, amphetamine, and methamphetamine toxicity. *Pharmacol. Biochem. Behav. 36*(4): 745–49.

Dowson, D. A., D. J. Fort, G. J. Smith, D. L. Newell, and J. A. Bantle. 1988. *Teratog. Carcinog. Mutagen. 8*(6): 329–38.

Dunlap, C. E., III, and F. M. Leslie. 1985. Effect of ascorbate on the toxicity of morphine in mice. *Neuropharmacology 24*(8): 797–804.

Engelhardt, H., and T. Koenig. 1989. Application of diode array detectors for solute identification in toxicological analysis. *Chromatographia* 28(7–8): 341–53.

Ferry S., and C. Vigneau. 1983. Alkaloids. In *Encyclopedia of Occupational Health and Safety*, 3rd ed., Vol 1., ed. L. Parmeggiani, pp. 117–120. Geneva: International Labor Office.

Girardot, M. N., and F. A. Holloway. 1985. Chronic stress, aging and morphine analgesia: chronic stress affects the reactivity to morphine in young mature but not old rats. *J. Pharmacol. Exp. Ther.* 233(3): 545–53.

Haffmans, J., M. V. Walsum, J. C. C. Van Amsterdam, and M. R. Dzoljic. 1987. Phelorphan, an inhibitor of enzymes involved in the biodegradation of enkephalins, affected the withdrawal symptoms in chronic morphine-dependent rats. *Neuroscience* 22(1): 233–36.

Hardy, L. R., and R. G. Ayre. 1987. Use of ferric oxide and zinc oxide for decreasing the carcinogens in tobacco smoke. *Chem. Abstr. CA* 108(13): 109700c.

Hartman, D. E. 1978. Identification and assessment of human neurotoxic syndromes. In *Neuropsychological Toxicology*, pp. 217–221. New York: Pergamon.

Hayashida, M., M. Nihira, T. Watanabe, and K. Jinno. 1990. Application of a computer-assisted high-performance liquid chromatographic multi-wavelength ultraviolet detection system to simultaneous toxicological drug analyses. *J. Chromatogr.* 506: 133–43.

Hodgson, E., R. B. Mailman, and J. E. Chambers. 1988. *Dictionary of Toxicology.* New York: Van Nostrand Reinhold.

Hoffmann, D., E. J. Lavoie, and S. S. Hecht. 1985. Nicotine: a precursor for carcinogens. *Cancer Lett.* 26(1): 67–75.

Hoskins, B., C. K. Burton, and I. K. Ho. 1988. Diabetes potentiation of cocaine toxicity. *Res. Commun. Subst. Abuse* 9(2): 117–23; cited in *Chem. Abstr. CA* 109(15): 122401w.

Hurwitz, A., M. Garty, and Z. Ben-Zvi. 1988. Morphine effects on gentamicin disposition and toxicity in mice. *Toxicol. Appl. Pharmacol.* 93(3): 413–20.

Innis, J., M. Meyer, and A. Hurwitz. 1987. A novel acute toxicity resulting from the administration of morphine and adriamycin to mice. *Toxicol. Appl. Pharmacol.* 90(3): 445–53.

Jinno, K., M. Kuwajima, M. Hayashida, T. Watanabe, and T. Hondo. 1988. Automated identification of toxic substances in poisoned human fluids by retention prediction system in reversed-phase liquid chromatography. *J. Chromatogr.* 436(3): 445–53.

Jinno, K., M. Hayashida, and T. Watanabe. 1990. Computer-assisted liquid chromatography for automated qualitative and quantitative analysis of toxic drugs. *J. Chromatogr. Sci.* 28(7): 367–73.

Kramer, H. K., H. Sershen, A. Lajtha, and M. E. A. Reith. 1989. The effect of nicotine on catecholaminergic storage vesicles. *Brain Res.* 503(2): 296–98; cited in *Chem. Abstr. CA* 112(9): 72125y.

Langner, R. O., C. L. Bement, and L. E. Perry. 1988. Arteriosclerotic toxicity of cocaine. *NIDA Res. Monogr.* 88: 325–26; cited in *Chem. Abstr. CA 113*(11): 91267h.

La Voie, E. J., S. L. Stern, C. I. Choi, J. Reinhardt, and J. D. Adams. 1987. Transfer of the tobacco-specific carcinogens N′-nitrosonornicotine and 4-(methylnitrosamino)-1-(3-pyridyl)-1-butanone and benzo[*a*]pyrene into the milk of lactating rats. *Carcinogenesis* 8(3): 433–37.

Lora-Tamayo, C., M. A. Rams, and J. M. R. Chacon. 1986. Gas chromatographic data for 187 nitrogen- or phosphorus-containing drugs and metabolites of toxicological interest analyzed on methyl silicone capillary column. *J. Chromatogr.* 374(1): 73–85.

McCartney, M. A. 1989. Effect of glutathione depletion on morphine toxicity in mice. *Biochem. Pharmacol.* 38(1): 207–9.

Merck. 1989. *The Merck Index*, 11th ed. Rahway, NJ: Merck & Co.

Nagamatsu, K., T. Terao, and S. Toki. 1985. In vivo formation of codeinone from codeine by rat or guinea pig liver homogenate and its acute toxicity in mice. *Biochem. Pharmacol.* 34(17): 3143–46.

Nahas, G. G., R. Trouve, W. M. Manager, C. Vinyard, and S. R. Goldberg. 1988. Cocaine cardiac toxicity and endogenous catecholamines.

Neurol. Neurobiol. 42B: 457–62; cited in *Chem. Abstr. CA 110*(19): 166028e.

Nair, J., S. S. Pakhale, and S. V. Bhide. 1989. Carcinogenic tobacco-specific nitrosamines in Indian tobacco products. *Food Chem. Toxiol. 27*(11): 751–53.

Nichols, D. E. 1986. *CRC Handbook of CNS Agents and Local Anesthetics*, Boca Raton, FL: CRC Press.

NIOSH. 1984. *Manual of Analytical Methods*, 3rd ed. Cincinnati, OH: National Institute for Occupational Safety and Health.

NIOSH. 1986. *Registry of Toxic Effects of Chemical Substances*, ed. D. V. Sweet. Washington, DC: U.S. Government Printing Office.

Padma, P. R., V. S. Lalitha, A. J. Amonkar, and S. V. Bhide. 1989. Carcinogenicity studies on the two tobacco-specific *N*-nitrosamines, N′-nitrosonornicotine and 4-(methylnitrosamino)-1-(3-pyridyl)-1-butanone. *Carcinogenesis 10*(11): 1977–2002.

Patnaik, P. 1990. Toxic effects from accidental ingestion of dried tobacco leaves: personal experience. Unpublished data

Perrigo, B. J., H. W. Peel, and D. J. Ballantyne. 1985. Use of dual-column fused-silica capillary gas chromatography in combination with detector response factors for analytical toxicology. *J. Chromatogr. 341*(1): 81–88.

Rao, C. V. N., and M. K. Chakraborty. 1988. Influence of solvent extraction on levels of carcinogenic precursors in tobacco. *Tobacco Res. 14*(1): 69–73; cited in *Chem. Abstr. CA 110*(7): 52804z.

Rosenkranz, H. S., and G. Klopman. 1990. The carcinogenic potential of cocaine. *Cancer Lett. 52*(3): 243–46.

Rosenthal, W. 1989. Decreasing tobacco smoke toxicity with potassium permanganate. *Chem. Abstr. CA 112*(21): 195452k.

Shah, A. S., A. V. Sarode, and S. V. Bhide. 1985. Experimental studies on mutagenic and carcinogenic effects of tobacco chewing. *J. Cancer Res. Clin. Oncol. 109*(3): 203–7.

Stroescu, V., D. Manu, and I. Fulga. 1986. Experimental research of the influence of the body electrolyte composition on the toxic effect of morphine. *Rev. Roum. Morphol. Embryol. Physiol. Physiol. 23*(3): 147–49; cited in *Chem. Abstr. CA 106*(23): 188860q.

Trouve, R., and G. Nahas. 1986. Nitrendipine: an antidote to cardiac and lethal toxicity of cocaine. *Proc. Soc. Exp. Biol. Med. 183*(3): 392–97; cited in *Chem. Abstr. CA 106*(15): 114721y.

U.S. EPA. 1986. *Test Methods for Evaluating Solid Waste*, 3rd ed. Washington DC: Office of Solid Waste and Emergency Response.

von Oettingen, W. F. 1952. *Poisoning*, p. 466. New York: Hoeber.

Woods, J. R., Jr., and M. A. Plessinger. 1990. Pregnancy increases cardiovascular toxicity to cocaine. *Am. J. Obstet. Gynecol. 106*(2): 529–33.

8

AMINES, ALIPHATIC

8.1 GENERAL DISCUSSION

Amines are a class of organic compounds that contain the functional group $-NH_2$. Thus the structure of an amine is $R-NH_2$, where R, in the case of an aliphatic amine, is an alkyl group; or for aromatic amines is an aryl group. Also, R may be an alkene or alkyne group attached to the nitrogen atom of the amine. In a broad definition, compounds containing the structure $R-NH-R'$ (an imine), or $R-\underset{\underset{O}{\|}}{C}-NH_2$ (an amide), or a diamine or triamine (containing two or three amino groups, respectively) are classified as aliphatic amines. Alicyclic amines, which exhibit hazardous properties similar to those of simple alkyl amines, are also grouped under aliphatic amines in this chapter. Common amides such as formamide or acetamide are relatively nonhazardous and are not discussed. Aziridine and methyl aziridine are three-membered heterocyclic imines that exhibit toxic and flammable properties similar to those of lower aliphatic amines. These compounds are also discussed in this chapter.

The toxicity of most aliphatic amines may fall in the low to moderate category, except for ethylenimine (aziridine), which is moderate to high. Such classification of low, moderate, or high is too simplistic, based on the oral LD_{50} or inhalation LC_{50} values for rodents. The health hazard from amines arises primarily from their caustic nature. All lower aliphatic amines are severe irritants to the skin, eyes, and mucous membranes. Skin contact can cause burns, while contact with eyes can result in corneal injury and loss of sight. All these compounds have a strong to mild odor of ammonia, and their vapors produce irritation of the nose and throat.

Aliphatic amines, especially the lower ones, are highly flammable liquids, many of which have flash points below 0°C. Such flammability presents a greater threat in handling these compounds than that of their caustic and irritant nature. Being heavier than air, the vapors of these substances present the danger of flashback fire hazard when the container bottles are opened near a source of ignition.

The reactivity of aliphatic amines is low. Strongly basic, these compounds react vigorously with concentrated mineral acids. The explosion hazards from amines are very limited. They may promote base-catalyzed polymerization of many unsaturated organics.

Such reactions are exothermic and can become violent if not controlled properly. Reactions with perchloric acid, hypochlorites, and chlorine may produce unstable products that may explode.

Analysis

Aliphatic amines may be analyzed by GC using a flame ionization detector. Aqueous samples may be injected directly into the GC. A nitrogen-specific detector is more sensitive and can be used for nonaqueous solutions.

Analysis of dimethyl- and diethylamine in air may be performed in accordance to NIOSH Method 2010 (NIOSH 1984, Suppl. 1989). About 3–30 L of air at a flow rate of 10–1000 mL/min is passed through a silica gel (20/40 mesh, 150 mg front and 75 mg back) sorbent tube. The analytes are desorbed with dilute sulfuric acid in 10% aqueous methanol (3-h ultasonication) and injected into a GC equipped with a FID. A packed column, 4% Carbowax M over 0.8% KOH on Carbosieve B (60/80 mesh) was used in the NIOSH laboratory. A nitrogen-specific detector may be used for amine analysis instead of an FID, using a DB-5 fused-silica capillary column. This method may be applied for the analyses of other aliphatic amines in air and also for certain ethanolamines (NIOSH 1984, Suppl. 1989, Method 2007). Alternatively, ethanolamines may be analyzed by NIOSH Method 3509 (NIOSH 1984, Suppl. 1989). These compounds present in air (5–300 L of air) are collected over an impinger containing 15 mL of 2 mM hexanesulfonic acid and analyzed by ion chromatography.

8.2 METHYLAMINE

DOT Label: Flammable Gas (Anhydrous)/ Flammable Liquid, UN 1061 and UN 1235

Formula CH_3NH_2; MW 31.07; CAS [74-89-5]

primary amine, first member of a homologue series of aliphatic amines

Synonyms: Methanamine; monomethylamine; aminomethane

Uses and Exposure Risk

Methylamine is used in dyeing and tanning; in photographic developer, as a fuel additive, and as a rocket propellant. It is also used in organic synthesis and as a polymerization inhibitor. It occurs in certain plants, such as *Mentha aquatica*.

Physical Properties

Colorless gas with a strong odor of ammonia at high concentrations and a fishy odor at low concentrations; liquefies to a fuming liquid at −6.8°C; freezes at −93.5°C; soluble in water, alcohol, and ether.

Health Hazard

Methylamine is a strong irritant to the eyes, skin, and respiratory tract. Kinney and coworkers (1990) have studied its inhalation toxicity in rats. Rats were exposed to methylamine by nose-only inhalation for 6 hr/day, 5 days/week for 2 weeks. Exposure to 75 ppm caused mild nasal irritation; 250 ppm produced damage to respiratory mucosa of the nasal turbinates. Exposure to 750 ppm produced severe body weight losses, liver damage, and nasal degenerative changes.

Any adverse health effects in humans due to methylamine, other than its irritant action, is unknown.

LC_{50} value, inhalation (mice): 2400 mg/kg/ 2 hr

Exposure Limits

TLV-TWA 10 ppm (∼12.3 mg/m^3)(ACGIH, MSHA, and OSHA); IDLH 100 ppm (NIOSH).

Fire and Explosion Hazard

Flammable gas; the aqueous solutions are also flammable, flash point of a 30% solution 1.1°C (34°F); the gas (vapor) is heavier than air (1.1 times that of air) and can travel a considerable distance to a source of ignition and flashback; autoignition temperature 430°C (806°F); fire-extinguishing procedure: shut off the flow of gas; in case its aqueous solution catches fire, use a water spray, "alcohol" foam, dry chemical, or CO_2; use a water spray to keep fire-exposed containers cool.

Methylamine forms explosive mixtures with air; the LEL and UEL values are 4.9% and 20.7% by volume in air, respectively. Methylamine reacts explosively with mercury. Violent reactions may occur when mixed with nitrosyl perchlorate or maleic anhydride.

Storage and Shipping

Methylamine is stored in a cool, well-ventilated noncombustible area separated from possible sources of ignition and oxidizing substances and mercury. Its solutions are stored in a flammable liquid storage room or cabinet. The gas is shipped in steel cylinders or tank cars; the liquid is shipped in steel drums or tank cars.

8.3   ETHYLAMINE

DOT Label: Flammable Liquid, UN 1036

Formula $C_2H_5NH_2$; MW 45.10; CAS [75-04-7]

Structure: $CH_3-CH_2-NH_2$, primary amine

Synonyms: ethanamine; monoethylamine; aminoethane

Uses and Exposure Risk

Ethylamine is used in the manufacture of dyes and resins, as a stabilizer for rubber latex, and in organic synthesis.

Physical Properties

Colorless liquid or gas with an odor of ammonia; boils at 16.6°C; freezes at −81°C; density 0.689 at 15°C; soluble in water, alcohol, and ether; strongly alkaline.

Health Hazard

Ethylamine is a severe irritant to the eyes, skin, and respiratory system. The pure liquid or its highly concentrated solution can cause corneal damage upon contact with eyes. Skin contact can result in necrotic skin burns.

Rabbits exposed to 100 ppm ethylamine for 7 hr/day, 5 days/week for 6 weeks manifested irritation of cornea and lung, and liver and kidney damage (ACGIH 1986). A 4-hr exposure to 3000 ppm was lethal to rats. The acute oral and dermal toxicity of this compound was moderate in test animals.

LD_{50} value, oral (rats): 400 mg/kg

LD_{50} value, skin (rabbits): 390 mg/kg

Exposure Limits

TLV-TWA 10 ppm ($\sim$18 mg/m^3) (ACGIH, MSHA, and OSHA): IDLH 4000 ppm (NIOSH).

Fire and Explosion Hazard

Highly flammable liquid; flash point (closed cup) < −17°C (<0°F); vapor pressure of liquid 400 torr at 2°C; vapor density 1.55 (air = 1), vapor is heavier than air and can travel a considerable distance to a source of ignition and flash back; autoignition temperature 384°C (723°F); fire-extinguishing agent: CO_2 or dry chemical; a water spray may be used to keep fire-exposed containers cool and to dilute the spill.

Ethylamine forms explosive mixtures with air in the range 3.5–14.0% by volume in air. Reaction with calcium or sodium hypochlorites may produce explosive chloroamine. Explosive reactions may occur if mixed with nitrosyl perchlorate or with maleic anhydride (NFPA 1986). It reacts vigorously with concentrated mineral acid. It catalyzes the exothermic polymerization of acrolein.

Storage and Shipping

Ethylamine should be stored in a flammable-liquids storage room or cabinet. It should be stored away from oxidizing materials and sources of ignition. It is shipped in steel cylinders or drums.

8.4 *n*-PROPYLAMINE

EPA Classified Toxic Waste, RCRA Waste Number, U194; DOT Label: Flammable Liquid, UN 1277

Formula $C_3H_7NH_2$; MW 59.13; CAS [107-10-8]

Structure: $CH_3-CH_2-CH_2-NH_2$, a primary amine

Synonyms: 1-propylamine; mono-1-propylamine; 1-aminopropane

Uses and Exposure Risk

n-Propylamine is used as an intermediate in many organic reactions.

Physical Properties

Colorless liquid with strong ammoniacal odor; density 0.719 at 20°C; boils at 48°C; freezes at −83°C; soluble in water, alcohol, and ether; strongly alkaline.

Health Hazard

n-Propylamine is a strong irritant and a moderately toxic substance. the toxic routes are inhalation, ingestion, and absorption through the skin. Contact of the liquid on the skin can cause burns and possibly skin sensitization. Irritation in the eyes from exposure to high concentrations can be severe. It is a respiratory tract irritant, similar to ethylamine. Inhalation of 2300 ppm of *n*-propylamine for 4 hr caused labored breathing, hepatitis, and hepatocellular necrosis in rats. It is somewhat less toxic than ethylamine by oral and dermal routes.

LC_{50} value, inhalation (mice): 2500 mg/m^3/2 hr

LD_{50} value, oral (rats): 570 mg/kg

LD_{50} value, skin (rabbits): 560 mg/kg

Exposure Limit

No exposure limit has been set. Based on its similarity to ethylamine in irritation and toxicity, a TLV-TWA of 10 ppm ($\sim$24 mg/m^3) should be appropriate.

Fire and Explosion Hazard

Highly flammable liquid; flash point (closed cup) −37°C (−35°F); vapor density 2.0 (air = 1), vapor can travel a considerable distance to a source of ignition and flash back; autoignition temperature 318°C (604°F); fire-extinguishing agent: dry chemical, CO_2, or "alcohol" foam; water should be used to keep fire-exposed containers cool and to flush and dilute the spill.

n-Propylamine forms explosive mixtures in air; LEL and UEL values are 2.0% and 10.4% by volume in air, respectively. There is no report of explosion associated with this compound. It is expected to exhibit violent reactions characteristic of lower aliphatic primary amines (see Section 8.3).

8.5 ISOPROPYLAMINE

DOT Label: Flammable Liquid, UN 1221

Formula $C_3H_7NH_2$; MW 59.13; CAS [75-31-0]

Structure:

$$\begin{array}{c} CH_3 \\ \diagdown \\ CH-NH_2 \\ \diagup \\ CH_3 \end{array}$$

a secondary amine

Synonyms: 2-propylamine; *sec*-propylamine; 2-propanamine; 2-aminopropane; monoisopropylamine

Uses and Exposure Risk

Isopropylamine is used as a dehairing agent and as an intermediate in the preparation of many organics.

Physical Properties

Colorless volatile liquid with an odor of ammonia; density 0.694 at 15°C; boils at 33–34°C; freezes at −101°C; soluble in water, alcohol, and ether; strongly alkaline.

Health Hazard

Isopropylamine is a strong irritant to the eyes, skin, and respiratory system. A short exposure to 10–20 ppm can cause irritation of the nose and throat in humans (Proctur and Hughes 1978). Prolonged exposure to high concentrations may lead to pulmonary edema. Skin contact can cause dermatitis and skin burns. Exposure to 8000 ppm for 4 hours was lethal to rats.

LD$_{50}$ value, oral (mice): 2200 mg/kg

Exposure Limits

TLV-TWA 5 ppm (~12 mg/m^3) (ACGIH, MSHA, and OSHA); TLV-STEL 10 ppm (~24 mg/m^3) (ACGIH); IDLH 4000 ppm (NIOSH).

Fire and Explosion Hazard

Highly flammable liquid; flash point (closed cup) −37°C (−35°F), open cup −26°C (−15°F); vapor pressure 478 torr at 20°C; vapor density 2.0 (air = 1), the vapor is heavier than air and can travel some distance to a source of ignition and flashback; autoignition temperature 402°C (756°F); fire-extinguishing agent: dry chemical, CO$_2$, or "alcohol" foam; use water to keep fire-exposed containers cool and to flush and dilute the spill.

Isopropylamine forms explosive mixtures with air in the range 2.0–10.4% by volume in air. Vigorous reactions may occur with strong acids and oxidizers. Contact with acrolein may cause base-catalyzed polymerization, which is highly exothermic.

8.6 n-BUTYLAMINE

DOT Label: Flammable Liquid, UN 1125

Formula C$_4$H$_9$NH$_2$; MW 73.16; CAS [109-73-9]

Structure: CH$_3$−CH$_2$−CH$_2$−CH$_2$−NH$_2$, a primary amine

Synonyms: 1-butylamine; 1-butanamine; 1-aminobutane; mono-n-butylamine

Uses and Exposure Risk

n-Butylamine is used as an intermediate for various products, including dyestuffs, pharmaceuticals, rubber chemical, synthetic tanning agents, and emulsifying agents. It is used for making isocyanates for coatings.

Physical Properties

Colorless liquid with ammoniacal odor; density 0.733 at 25°C; boils at 78°C; freezes at −50°C; miscible with water, alcohol, and ether; alkaline.

Health Hazard

n-Butylamine is a severe irritant to the eyes, skin, and respiratory tract. Contact of the liquid with the skin and eyes can produce severe burns. Irritation effect on rabbits' eyes was as severe as that produced by ethylamine (ACGIH 1986). Exposure can cause irritation of the nose and throat, and at high concentrations, pulmonary edema. Scherberger and associates (1960) have reported erythema of the face and neck occurring within 3 hours after exposure to n-butylamine, along with a burning and itching sensation.

n-Butylamine is more toxic than is either n-propylamine or ethylamine. A 4-hour exposure to 3000-ppm concentration in air was lethal to rats. Toxic symptoms in animals from ingestion include increased pulse rate, labored breathing, and convulsions. Cyanosis and coma can occur at near-lethal dose.

LD$_{50}$ value, oral (rats): 366 mg/kg
LD$_{50}$ value, skin (guinea pigs): 366 mg/kg

Exposure Limits

Ceiling 5 ppm (~15 mg/m^3) (ACGIH, MSHA, and OSHA); IDLH 2000 ppm (NIOSH).

Fire and Explosion Hazard

Flammable liquid; flash point: (closed cup) −14°C (6°F) (Aldrich 1990), −12°C (10°F) (NFPA 1986; NIOSH 1984, Suppl. 1985): (open cup) −1°C (30°F) (Merck 1989), 7°C (45°F) (Scherberger et al. 1960); vapor pressure 82 torr at 20°C; vapor density 3.0 (air = 1); the vapor is heavier than air and can travel some distance to a source of ignition and flash back; autoignition temperature 312°C (594°F); fire-extinguishing agent: dry chemical, CO$_2$, or "alcohol" foam; use water to keep fire-exposed containers cool and to flush and dilute any spill.

n-Butylamine forms explosive mixtures with air within the range 1.7–9.8% by volume in air. Its reactions with strong acids or oxidizers can be vigorous. Contact with acrolein may cause base-catalyzed polymerization of the latter, which is highly exothermic. *n*-Butylamine may exhibit violent reactions, characteristic of lower aliphatic primary amines (see Section 8.3).

8.7 CYCLOHEXYLAMINE

DOT Label: Flammable Liquid, Corrosive, UN 2357
Formula C$_6$H$_{11}$NH$_2$; MW 99.20; CAS [108-91-8]

Structure:

an alicyclic amine

Synonyms: cyclohexanamine; hexahydrobenzenamine; aminocyclohexane; hexahydroaniline

Uses and Exposure Risk

Cyclohexylamine is used in the manufacture of a number of products, including plasticizers, drycleaning soaps, insecticides, and emulsifying agents. It is also used as a corrosion inhibitor and in organic synthesis.

Physical Properties

Colorless or yellowish liquid with a strong fishy, amine odor; density 0.8645 at 25°C; boils at 134.5°C; solidifies at −17.7°C; miscible with water and most organic solvents; forms an azeotropic mixture with water containing 44% cyclohexylamine, which boils at 96.5°C; strongly basic.

Health Hazard

Cyclohexylamine is a severe irritant to the eyes, skin, and respiratory passage. Skin contact can produce burns and sensitization; contact of the pure liquid or its concentrated solutions with the eyes may cause loss of vision.

The acute oral and dermal toxicity of cyclohexylamine was moderate in test subjects. The toxic effects include nausea, vomiting, and degenerative changes in the brain, liver, and kidney. Inhalation of its vapors at high concentrations may cause a narcotic effect.

LD$_{50}$ value, oral (rats): 156 mg/kg
LD$_{50}$ value, skin (rabbits): 277 mg/klg

Cyclohexylamine may be mutagenic, the test for which has so far given inconclusive results. Administration of this compound in animals produced a reproductive effect, including embryotoxicity and a reduction in male fertility. Intraperitoneal injection of the amine in rats caused a dose-dependent increase in chromosomal breaks. Roberts and coworkers (1989) studied the metabolism and testicular toxicity of cyclohexylamine (a metabolite of cyclamate) in rats and mice. Chronic dietary administration of 400 mg/kg/day for 13 weeks showed decrease in organ weigh, histological changes, and testicular atrophy in both the Wistar and DA rats, but to a widely varying extent, while mice exhibited no evidence of testicular damage.

There is no evidence of carcinogenicity in animals or humans caused by cyclohexylamine.

Exposure Limit

TLV-TWA 10 ppm ($\sim$40 mg/m^3) (ACGIH).

Fire and Explosion Hazard

Flammable liquid; flash point (open cup) 32°C (90°F); autoignition temperature 293°C (560°F); vapor density 2.4 (air = 1); the vapor is heavier than air and can travel a considerable distance to a source of ignition and flash back. Fire-extinguishing agent: dry chemical, CO_2, or "alcohol" foam; water may be used to flush and dilute any spill and to keep fire-exposed containers cool.

Cyclohexylamine vapors form explosive mixtures with air; explosive limits data are not available. Vigorous reactions may occur when the amine is mixed with strong acids or oxidizers.

8.8 DIMETHYLAMINE

EPA Classified Toxic Waste, RCRA Waste Number U092; DOT Label: Flammable Gas/Flammable Liquid (Aqueous), UN 1032, UN 1160

Formula C_2H_6NH; MW 45.10; CAS [124-40-3]

Structure: $CH_3-NH-CH_3$, a primary amine

Synonym: *N*-methylmethanamine

Uses and Exposure Risk

Dimethylamine is used in the manufacture of *N*-methylformamide, *N*-methylacetamide, and detergent soaps; in tanning; and as an accelerator in vulcanizing rubber. It is commercially sold as a compressed liquid in tubes or as a 33% aqueous solution.

Physical Properties

Colorless gas with a pungent fishy ammoniacal odor; liquefies at 7°C; freezes at −96°C; density of liquid 0.680 at 0°C; highly soluble in water, soluble in alcohol and ether; aqueous solution strongly alkaline.

Health Hazard

Dimethylamine is a strong irritant to the eyes, skin, and mucous membranes. Spill of liquid into the eyes can cause corneal damage and loss of vision. Skin contact with the liquid can produce necrosis. At sublethal concentrations, inhalation of dimethylamine produced respiratory distress, bronchitis, pneumonitis, and pulmonary edema in test animals. The acute oral toxicity was moderate, greater than for monomethylamine.

LC$_{50}$ value, inhalation (rats): 4540 ppm/6 hr
LD$_{50}$ value, oral (mice): 316 mg/kg

Buckley and coworkers (1985) have investigated the inhalation toxicity of dimethylamine in F-344 rats and B6C3F1 mice. Animals exposed to 175 ppm for 6 hr/day, 5 days/week for 12 months showed significant lesions in the nasal passages. Rats developed more extensive olfactory lesions than did mice. The study indicated that olfactory sensory cells were highly sensitive to dimethylamine. Even at a concentration of

10 ppm, the current threshold limit value, the rodents developed minor lesions from exposure.

Exposure Limits

TLV-TWA 10 ppm ($\sim$18 mg/m^3) (ACGIH, MSHA, and OSHA); IDLH 2000 ppm (NIOSH).

Fire and Explosion Hazard

Flammable gas; the gas (vapor) is heavier than air and can travel a considerable distance to a source of ignition and flash back; autoignition temperature 402°C (755°F); fire-extinguishing procedure: shut off the flow of gas; use dry chemical, CO$_2$, or "alcohol" foam to extinguish fire involving its solution; use a water spray for diluting and flushing the spill and to keep fire-exposed containers cool.

Dimethylamine forms explosive mixtures with air in the range between 2.8% and 14.4% by volume in air. Its reactions with strong acids and oxidizers can be vigorous; contact with acrolein may catalyze violent exothermic polymerization. Dimethylamine may react violently with chlorine and hypochlorites. Explosive reaction may occur in contact with mercury.

8.9 DIETHYLAMINE

DOT Label: Flammable Liquid, UN 1154

Formula C$_4$H$_{10}$NH; MW 73.16; CAS [109-89-7]

Structure: CH$_3$−CH$_2$−NH−CH$_2$−CH$_3$, a primary amine

Synonyms: *N*,*N*-diethylamine; *N*-ethylethanamine; diethanamine

Uses and Exposure Risk

Diethylamine is used as a flotation agent; in dyes and pharmaceuticals; and in resins. It also finds applications in rubber and petroleum industry.

Physical Properties

Colorless liquid with a strong ammonia smell; density 0.707 at 20°C; boils at 55°C; freezes at −50°C; soluble in water, alcohol, and most organic solvents; strongly alkaline; forms a hydrate.

Health Hazard

Diethylamine is a strong irritant to the eyes, skin, and mucous membranes. Contact of the pure liquid or its concentrated solutions with eyes can produce corneal damage. Skin contact can cause necrosis. Exposure to its vapors at 1000-ppm concentration produced lacrimation, muscle contraction, and cyanosis in mice. Repeated exposures at sublethal concentrations produced bronchitis, pneumonitis, and pulmonary edema in test animals. The acute oral toxicity was moderate.

LC$_{50}$ value inhalation (rats): 4000 ppm/4 hr
LD$_{50}$ value oral (mice): 500 mg/kg
LD$_{50}$ value skin (rabbits): 820 mg/kg

Lynch and associates (1986) investigated subchronic inhalation toxicity of diethylamine vapors in Fischer-344 rats. Rats exposed to 240 ppm for 6.5 hr/day, 5 days/week for 24 weeks developed sneezing, tearing, and reddened noses and lesions of the nasal mucosa. Animals exposed to 25 ppm did not show any of these signs. No evidence of cardiotoxicity was observed.

Exposure Limits

TLV-TWA 10 ppm ($\sim$30 mg/m^3) (ACGIH, MSHA, and OSHA), 25 ppm ($\sim$75 mg/m^3) (NIOSH); TLV-STEL 25 ppm (ACGIH); IDLH 2000 ppm (NIOSH).

Fire and Explosion Hazard

Highly flammable liquid; flash point (closed cup) < -18°C (<0°F); vapor pressure 195 torr at 20°C; vapor density 2.5 (air = 1); the vapor is heavier than air and can travel a considerable distance to a source of ignition and flash back; autoignition temperature 312°C (594°F); fire-extinguishing agent: dry chemical, CO_2, or "alcohol" foam; use a water spray to flush and dilute any spill and disperse vapors.

The vapors of diethylamine form flammable mixtures in air; LEL and UEL values are 1.8% and 10.1% by volume in air, respectively. Its reactions with strong acids and oxidizers can be vigorous to violent. It may undergo violent reactions that are typical of lower aliphatic amines.

8.10 ETHYLENIMINE

EPA Classified Acute Hazardous Waste; RCRA Waste Number P054; DOT Label: Flammable Liquid and Poison, UN 1185

Formula C_2H_4NH; MW 43.08; CAS [151-56-4]

Structure:

a three-membered heterocyclic compound, properties are similar to those of aliphatic amines

Synonyms: aziridine; azirane; azacyclopropane; aminoethylene; dimethylenimine

Uses and Exposure Risk

Ethylenimine is used in the manufacture of triethylenemelamine and other amines.

Physical Properties

Colorless liquid with a strong odor of ammonia; density 0.832 at 20°C; bp 56°C; solidifies at −74°C; infinitely soluble in water, soluble in most organic solvents; aqueous solution strongly alkaline.

Health Hazard

Ethylenimine is a highly poisonous compound and a severe irritant to the skin, eyes,

and mucous membranes. A 20–25% aqueous solution on contact with the eyes can cause corneal opacity and loss of vision. Skin contact with the pure liquid or its concentrated solution can produce severe burns and skin sensitization.

Ethylenimine is highly toxic by all routes of exposure. Inhalation of its vapors can cause eye, nose, and throat irritations and difficulty in breathing. The toxic symptoms noted from repeated exposures include chest congestion, delayed lung injury, vomiting, hemorrhage, and kidney damage. An 8-hour exposure to 100 ppm in air proved fatal to rabbits. The symptoms from acute oral toxicity in humans may include nausea, vomiting, dizziness, headache, and pulmonary edema. Chronic toxic effects may result in kidney and liver damage. The acute oral LD_{50} value in rats was 15 mg/kg. Ethylenimine is also absorbed through the skin, exhibiting poisoning effects similar to those of acute oral toxicity.

It exhibited reproductive toxicity in animals, indicating paternal effects and specific developmental abnormalities in the central nervous system, eyes, and ears. Ethylenimine is a mutagen, testing positive in the histidine reversion–Ames test as well as in the in vitro cytogenetics–human lymphocyte, *Drosophila melanogaster*–reciprocal translocation, *Saccharomyces cerevisiae* gene conversion, and other mutagenic tests. Animal studies show sufficient evidence of carcinogenicity. It may cause cancers in the lungs and liver.

Exposure Limits

TLV-TWA (skin) 0.5 ppm (~1 mg/m^3) (ACGIH, OSHA, and MSHA); Potential Human Carcinogen in the workplace (OSHA).

Fire and Explosion Hazard

Flammable liquid; flash point (closed cup) −11°C (12°F); vapor pressure 160 torr at 20°C; vapor density 1.48 (air = 1); the vapor is heavier than air and can travel a considerable distance to a source of ignition and flash back; autoignition temperature 322°C (612°F); fire-extinguishing agent: dry chemical or "alcohol" foam; firefighting should be done from an explosion-resistant area; use water to flush and dilute any spill and to keep fire-exposed containers cool. Ethylenimine may polymerize under fire conditions. The reaction is exothermic, which may cause violent rupture of the container.

Ethylenimine vapors form explosive mixtures with air within a wide range, 3.3–46.0% by volume in air. Undiluted ethylenimine can undergo violent polymerization in the presence of an acid. Reaction with chlorine or hypochlorite forms 1-chloroethyleneimine, which is an explosive compound. Its reaction with concentrated mineral acids is vigorous and with strong oxidizers is vigorous to violent.

Storage and Shipping

Ethylenimine is stored in a flammable liquid storage room or cabinet isolated from acids, and from combustible and oxidizing substances. It is shipped in metal drums or cans of ≤30-gallon capacity placed in wooden boxes, cushioned with noncombustible packing.

8.11   ETHYLENEDIAMINE

DOT Label: Corrosive, Flammable Liquid, UN 1604

Formula $C_2H_4(NH_2)_2$; MW 60.12; CAS [107-15-3]

Structure: $NH_2-CH_2-CH_2-NH_2$

Synonyms: 1,2-ethanediamine; 1,2-ethylenediamine; 1,2-diaminoethane

Uses and Exposure Risk

Ethylenediamine is used as a stabilizer for rubber latex, as an emulsifier, as an inhibitor in antifreeze solutions, and in textile lubricants. It is also used as a solvent for albumin, shellac, sulfur, and other substances.

Physical Properties

Colorless, viscous liquid with ammonia-like smell; density 0.899 at 20°C; bp 116°C; freezes at 8.5°C; soluble in water, alcohol, and benzene, slightly soluble in ether; forms hydrate with water; aqueous solution strongly alkaline; absorbs CO_2.

Health Hazard

Ethylenediamine is a severe skin irritant, producing sensitization and blistering on the skin. Pure liquid on contact with the eyes can damage vision. A 25% aqueous solution can be injurious to the eyes. Inhalation of its vapors can produce a strong irritation to the nose and respiratory tract. Such irritation in humans with symptoms of cough and distressed breathing may be noted at concentrations of >400 ppm. Repeated exposure to high concentrations of this substance in air may cause lung, liver, and kidney damage. The toxicity of this compound, however, is much less than that of ethylenimine.

The acute oral toxicity value in animals was low to moderate. An oral LD_{50} value in rats is 500 mg/kg (NIOSH 1986).

LD_{50} value, skin (rabbits): 730 mg/kg

Exposure Limits

TLV-TWA 10 ppm (~25 mg/m^3) (ACGIH, MSHA, and OSHA); IDLH 2000 ppm (NIOSH).

Fire and Explosion Hazard

Combustible liquid; flash point (closed cup) 34°C (93°F) (NFPA 1986; NIOSH 1984, Suppl. 1985; Aldrich 1990), 41°C (110°F) (Merck 1989); vapor pressure 10 torr at 20°C; autoignition temperature 385°C (725°F); the vapor forms explosive mixtures with air in the range 5.8–11.1% by volume in air. Ethylenediamine undergoes vigorous reactions when mixed with strong acids, oxidizers, and chlorinated organic compounds. An explosion occurred when ethylenediamine was added dropwise to silver perchlorate (NFPA 1986). Its perchlorate salt, ethylenediamine diperchlorate, $NH_2CH_2CH_2NH_2(ClO_4)_2$, is a powerful explosive.

8.12 MONOETHANOLAMINE

DOT Label: Corrosive Material, UN 2491

Formula $C_2H_4(OH)NH_2$; MW 61.10; CAS [141-43-5]

Structure: $HO-CH_2-CH_2-NH_2$; the amine contains a primary alcoholic $-OH$ group

Synonyms: ethanolamine; 2-aminoethanol; β-aminoethyl alcohol; 2-hydroxyethylamine; colamine

Uses and Exposure Risk

Monoethanolamine is used as a dispersing agent for agricultural chemicals, in the synthesis of surface active agents, as a softening agent for hides, and in emulsifiers, polishes, and hair solutions.

Physical Properties

Colorless, viscous liquid with a mild ammoniacal odor; hygroscopic; density 1.018 at 20°C; bp 171°C; solidifies at 10.3°C; miscible with water, alcohol, and acetone; strongly alkaline, pH 12.05 (0.1 N aqueous solution); absorbs CO_2.

Health Hazard

Monoethanolamine causes severe irritation of the eyes and mild to moderate irritation of the skin. The pure liquid caused redness and swelling when applied to rabbits' skin. The acute oral toxicity of this compound was low in animals. The toxic symptoms included somnolence, lethargy, muscle contraction,

and respiratory distress. The oral LD_{50} values showed a wide variation with species.

LD_{50} value, oral (rabbits): 1000 mg/kg

Monoethanolamine showed reproductive toxicity when administered at a dose of 850 mg/kg/day, causing 16% mortality to pregnant animals (Environmental Health Research and Testing 1987). This study also indicated that monoethanolamine reduced the number of viable litters but had no effect on litter size, the birth weight, or percentage survival of the pups.

Exposure Limits

TLV-TWA 3 ppm (~7.5 mg/m^3) (ACGIH, MSHA, and OSHA); TLV-STEL 6 ppm (~15 mg/m^3) (ACGIH); IDLH 1000 ppm (NIOSH).

Fire and Explosion Hazard

Combustible liquid; flash point (closed cup) 85°C (185°F); vapor pressure 0.4 torr at 20°C; autoignition temperature 410°C (770°F). The vapor forms explosive mixtures with air; LEL value 5.5% by volume in air, UEL data not available. Its reactions with strong oxidizers and acids can be vigorous.

8.13 DIETHANOLAMINE

Formula $C_2H_8(OH)_2NH$; MW 105.16; CAS [111-42-2]

Structure: $HO-CH_2-CH_2-NH-CH_2-CH_2-OH$; the amine contains two alcoholic $-OH$ groups

Synonyms: 2,2'-iminodiethanol; 2,2'-dihydroxydiethylamine; 2,2'-iminobisethanol; bis(2-hydroxyethyl)amine

Uses and Exposure Risk

Diethanolamine is used in the production of surface-active agents and lubricants for the textile industry; as an intermediate for rubber

TABLE 8.1 Toxicity and Flammability of Miscellaneous Aliphatic Amines

Compound/Synonyms/ CAS No.	Formula/MW/Structure	Toxicity	Flammability
Diallylamine (di-2-propeny-lamine, *N*-2-propenyl-2-propen-1-amine) [124-02-7]	$C_6H_{10}NH$ 97.18 $CH_2=CH-CH_2-NH-CH_2-CH=CH_2$	Moderately toxic by inhalation, ingestion, and skin absorption; exposure to its vapors can cause lacrimation, irritation of nose and throat, coughing, and respiratory distress; in humans irritant effect of diallylamine may be noted at 5-min exposure to 5-ppm concentration; ingestion may cause tremor, gastrointestinal disorders, and dyspnea; LC_{50} inhalation (rats): 795 ppm/8 hr; LD_{50} oral (rat): 578 mg/kg (NIOSH 1986)	Flammable liquid; flash point (closed cup) 15°C(60°F) (Aldrich 1990); vapor forms explosive mixtures with air, LEL and UEL data not available. DOT Label: Flammable Liquid, UN 2359
Diisopropylamine (*N*-(1-methy-lethyl)-2-pro-panamine, DIPA) [108-18-9]	$C_6H_{14}NH$ 101.22 (structure: CH3–CH(CH3)–NH–CH(CH3)–CH3)	Severe irritant to eyes, skin, and respiratory tract; can cause skin burn on contact; causes visual disturbance and cloudy swelling of cornea with total or partial loss of vision; acute inhalation and oral toxicity low; symptoms: irritation, nausea, and headache; exposure to concentration >2000 ppm for 1–2 hr was lethal to most test subjects; LD_{50} oral (mice): 2120 mg/kg; TLV-TWA (skin) 5 ppm (~20 mg/m^3) (ACGIH, MSHA, and OSHA), IDLH 1000 ppm (NIOSH)	Flammable liquid; flash point (open cup) −1°C (30°F); vapor pressure 60 torr at 20°C; vapor density 3.5 (air = 1), vapor heavier than air and presents flashback fire hazard in presence of an ignition source; autoignition temperature 316°C (600°F); vapor forms explosive mixtures with air in the range 1.1–7.1% by volume in air; reactions with strong acids and oxidizers can be vigorous; DOT Label: Flammable Liquid, UN 1158
Dibutylamine (*N*-butyl-1-butanamine, di-*n*-buty-lamine) [111-92-2]	$C_8H_{18}NH$ 129.8 $CH_3-CH_2-CH_2-CH_2-NH-CH_2-CH_2-CH_2-CH_3$	Strong irritant to skin, eyes, and respiratory tract; 4-hr exposure to 500 ppm was lethal to rats; acute oral toxicity— moderate, more toxic than diisopropylamine; LD_{50} oral (mice): 290 mg/kg	Combustible liquid; flash point: (closed cup) 41°C (106°F) (Aldrich 1990), 47°C (117°F) (NFPA 1986), (open cup) 57°C (135°F) (Merck 1989); vapor forms explosive mixtures with air, LEL 1.1% by volume in air, UEL not established. DOT Label: Corrosive and Flammable Liquid; UN 2248

(continued)

221

TABLE 8.1 (Continued)

Compound/Synonyms/ CAS No.	Formula/MW/Structure	Toxicity	Flammability
Diamylamine (dipentylamine, N-pentyl-1-pentanamine) [2050-92-2]	$C_{10}H_{22}NH$ 157.34 $CH_3-(CH_2)_4-NH-(CH_2)_4-CH_3$	Irritant to eyes, skin, and respiratory passage; direct contact may produce secondary burn; moderate to high toxicity by inhalation, ingestion, and absorption through skin; inhalation can produce nose and throat irritation, cough, headache, and respiratory distress; exposure to 63 ppm for 48 hr was lethal to rats; LD_{50} skin (rabbits): 350 mg/kg; DOT Label: Poison B and Flammable Liquid, UN 2841	Combustible liquid; flash point (closed cup) 51°C (124°F), (open cup) 72°C (162°F); vapor forms explosive mixtures with air, LEL and UEL data not available
Dicyclohexylamine (N,N-dicyclohexylamine, N-cyclohexyl-cyclohexanamine, dodecahydrodiphenylamine) [101-83-7]	$C_{12}H_{22}NH$ 181.36 (structure: two cyclohexyl rings joined by NH)	Irritant to skin and eyes; moderately toxic by ingestion and skin absorption; LD_{50} oral (rats): 373 mg/kg; exhibits carcinogenicity, causing liver and gastrointestinal tumors—such evidence in animals, however, inadequate, no human evidence; DOT Label: Corrosive Material, UN 2565	Noncombustible liquid
Diethylene triamine (2,2'-diaminodiethylamine, bis(2-aminoethyl)amine, 3-azapentane-1,5-diamine) [111-40-0]	$C_4H_{13}N_3$ 103.20 $NH_2-CH_2-CH_2-NH-CH_2-CH_2-NH_2$	Severe eye, skin, and respiratory tract irritant; pure liquid or its solutions above 15–20% strength can cause corneal injury; vapor or the liquid can produce sensitization of skin and respiratory tract; LD_{50} intraperitoneal (mice): 71 mg/kg; LD_{50} oral (rats) 1080 mg/kg; TLV-TWA (skin) 1 ppm (~4 mg/m³) (ACGIH, MSHA); DOT Label: Corrosive Material, UN 2079	Noncombustible liquid; autoignition temperature 399°C (750°F)

Propyleneimine (2-methylaziridine, methylethylenimine, 2-methylazacyclopropane) [75-55-8]	C_3H_6NH 57.11 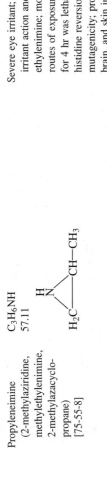	Severe eye irritant; can cause skin burns; irritant action and toxicity lower than ethylenimine; moderately toxic by all routes of exposure; inhalation of 500 ppm for 4 hr was lethal to rats; tested positive to histidine reversion — Ames test for mutagenicity; produced tumors in blood, brain, and skin in animals; A2 — suspected human carcinogen; TLV-TWA skin 2 ppm ($\sim$5 mg/m^3) (ACGIH, MSHA, and OSHA), IDLH 500 ppm (NIOSH); A2 — Suspected Human Carcinogen (ACGIH); EPA Classified Toxic Waste, RCRA Waste Number P067	Flammable liquid; flash point (closed cup) $-4°C$ ($25°F$); vapor pressure 112 torr at 20°C; vapor density 2.0 (air = 1), vapor heavier than air and can travel a considerable distance to a source of ignition and flashback; vapor forms explosive mixtures with air; reactions with chlorine, hypochlorites, and silver may form unstable products that may decompose violently; may react vigorously with strong acids and oxidizers; DOT Label: Flammable Liquid, UN 1921
1,2-Propanediamine (propylenediamine, 1,2-diaminopropane) [78-90-0]	$C_3H_{10}N_2$ 74.15 $CH_3-\overset{\overset{\displaystyle NH_2}{\displaystyle |}}{CH}-CH_2-NH_2$	Contact with the liquid can cause severe skin irritation and corneal injury; acute oral toxicity is low; more toxic when absorbed through skin; LD$_{50}$ skin (rabbits): 500 mg/kg	Flammable liquid; flash point (open cup) 33°C (92°F); autoignition temperature 416°C (780°F); vapor forms explosive mixtures with air, LEL and UEL data not available; DOT Label: Flammable Liquid, UN 2258
1,3-Propanediamine (1,3-propylenediamine, trimethylenediamine, 1,3-diaminopropane) [109-76-2]	$C_3H_{10}N_2$ 74.15 $NH_2-CH_2-CH_2-CH_2-NH_2$	Strong irritant to eyes and skin; contact with skin may cause burn; the liquid is injurious to eyes; moderately toxic by ingestion and skin absorption; toxicity is greater than 1,2-propanediamine; LD$_{50}$ oral (rats): 350 mg/kg; LD$_{50}$ skin (rabbits): 200 mg/kg	Combustible liquid; flash point (open cup) 48°C (120°F); vapor forms explosive mixtures with air; the LEL and UEL data not available

chemicals; as an emulsifier; as a humectant and softening agent; as a detergent in paints, shampoos, and other cleaners; and as an intermediate in resins and plasticizers.

Physical Properties

A viscous liquid or a deliquescent solid at room temperature; mild odor of ammonia; mp 28°C; bp 269°C; density of liquid, 1.088 at 30°C; soluble in water, alcohol, and acetone, insoluble in ether, hydrocarbons, and carbon disulfide; aqueous solution strongly alkaline, pH 11.0 (0.1 N solution).

Health Hazard

The irritant action of diethanolamine on the eyes can be severe. Direct contact of the pure liquid can impair vision. Irritation on the skin may be mild to moderate. The acute oral toxicity of this compound was low in test animals. The toxic symptoms include somnolence, excitement, and muscle contraction.

LD$_{50}$ value, oral (mice): 3300 mg/kg

The vapor pressure of diethanolamine is negligibly low (<0.01 torr at 20°C). At ordinary temperature, this compound should not cause any inhalation hazard. The mists, fumes, or vapors at high temperatures, however, can produce eye, skin, and respiratory tract irritation.

In contrast to monoethanolamine, diethanolamine administered to mice at 1125 mg/kg/day caused no change in maternal mortality, litter size, or percentage survival of the pups (Environmental Health Research and Testing 1987).

Exposure Limit

TLV-TWA 3 ppm (~13 mg/m^3) (ACGIH).

Fire and Explosion Hazard

Noncombustible liquid; flash point >93°C (>200°F); autoignition temperature 662°C (1224°F). The reactivity of this compound is low.

8.14 MISCELLANEOUS AMINES

Hazardous properties of some of the aliphatic amines (including diamines, triamines, and aziridines) are presented in Table 8.1. Amines in general, as mentioned earlier, are strongly caustic and can injure the cornea upon direct contact. Skin contact can cause burns, which may vary from severe to second degree or to mild, depending on the carbon chain length. The toxicity is low to moderate. The flammability decreases with an increase in the carbon number. The reactivity of these compounds, in general, is low. Outlined in Table 8.1 are some of the amines used commercially that are hazardous.

REFERENCES

ACGIH. 1986. *Documentation of the Threshold Limit Values and Biological Exposure Indices*, 5th ed. Cincinnati, OH: American Conference of Governmental Industrial Hygienists

Aldrich. 1990. *Catalog Handbook of Fine Chemicals*. Milwaukee, WI: Aldrich Chemical Company.

Buckley, L. A., K. T. Morgan, J. A. Swenberg, R. A. James, T. E. Hamm, Jr., and C. S. Barrow. 1985. The toxicity of dimethylamine in F-344 rats and B6C3F1 mice following a 1-year inhalation exposure. *Fundam. Appl. Toxicol.* 5(2): 341–52.

Environmental Health Research and Testing. 1987. Screening of priority chemicals for reproductive hazards. Monoethanolamine; diethanolamine; triethanolamine. *Govt. Rep. Announce. Index (U.S.) 89*(9), Abstr. No. 922, p. 787; cited in *Chem. Abstr. CA 111*(17): 148328u.

Kinney, L. A., R. Valentine, H. C. Chen, R. M. Everett, and G. L. Kennedy, Jr. 1990. Inhalation toxicology of methylamine. *Inhalation Toxicol.* 2(1): 29–35.

Lynch, D. W., W. J. Moorman, P. Stober, T. R. Lewis, and W. O. Iverson. 1986. Subchronic inhalation of diethylamine vapor in Fischer-344 rats: organ system toxicity. *Fundam. Appl. Toxicol.* 6(3): 559–65.

Merck. 1989. *The Merck Index*, 11th ed. Rahway, NJ: Merck & Co.

NFPA. 1986. *Fire Protection Guide on Hazardous Materials*, 9th ed. Quincy, MA: National Fire Protection Association.

NIOSH. 1984. *Manual of Analytical Methods*, 3rd ed. Cincinnati, OH: National Institute for Occupational Safety and Health.

NIOSH. 1986. *Registry of Toxic Effects of Chemical Substances*, ed. D. V. Sweet. Washington, DC: US Government Printing Office.

Proctor, N. H., and J. P. Hughes. 1978. *Chemical Hazards of the Workplace*, p. 303. Philadelphia: J. B. Lippincott.

Roberts, A., A. G. Renwick, G. Ford, D. M. Creasy, and I. Gaunt. 1989. The metabolism and testicular toxicity of cyclohexylamine in rats and mice during chronic dietary administration. *Toxicol. Appl. Pharmacol.* 98(2): 216–29.

Scherberger, R. F., F. A. Miller, and D. W. Fassett. 1960. *Am. Ind. Hyg. Assoc. J. 21*: 471.

9

AMINES, AROMATIC

9.1 GENERAL DISCUSSION

Aromatic amines constitute compounds containing one or more amino (or imino) groups attached to an aromatic ring. These amines are similar in many respects to aliphatic amines in reactions that are characteristics of an $-NH_2$ functional group. However, the presence of other groups in the ring, and their electrophilic or nucleophilic nature or size, often alter the course of the reactions to a significant extent. These amines are basic, but the basicity is lower to aliphatic amines. Aromatic amines are widely used in industry for making dyes, as intermediates in many chemical syntheses, and as analytical reagents.

Toxicity

The health hazard from aromatic amines may arise in two ways: (1) moderate to severe poisoning, with symptoms ranging from headache, dizziness, and ataxia to anemia, cyanosis, and reticulocytosis; and (2) cancer causing, especially cancer of the bladder. Many amines are proven or suspected human carcinogens. Notable among these

are benzidine, *o*-tolidine, dianisidine, α- and β-naphthylamines, *o*-phenylenediamine, and phenylhydrazine. Among the aromatic amines, *ortho*-isomers generally exhibit stronger carcinogenic properties than those of the *para*- and *meta*-isomers. In many instances, the *meta*-isomers are either non-carcinogenic or exhibit the least carcinogenic potential.

Unlike most aliphatic amines, the aromatic amines do not cause severe skin burn or corneal injury. The pure liquids (or solids) may, however, produce mild to moderate irritation on the skin. The toxicity of individual compounds is discussed in the following sections.

Fire and Explosion Hazard

Lower aromatic amines are combustible liquids and form explosive mixtures with air. Except for aniline and methylanilines, most amines are noncombustible liquids or solids. Amines may react violently with strong oxidizing compounds. Lower amines may form explosive chloroamines when mixed with hypochlorites. Ignition may occur when mixed with red fuming nitric acid.

Analysis

Gas chromatographic techniques involve the use of a flame ionization or nitrogen-specific detector. Direct injection of the aqueous solution can be made into the GC equipped with an FID. Aniline, benzidine, o-anisidine, o-toluidine, 2,4,5-trimethylaniline, 1,4-phenylenediamine, 1-naphthylamine, 2-naphthylamine, diphenylamine, 2,4-diaminotoluene, and 4-aminobiphenyl are some of the aromatic amines that are listed as EPA priority pollutants in solid and hazardous wastes. The analysis of these compounds (as well as some nitro- and chloroanilines) in soil, sediment, groundwater, and solid and hazardous wastes may be performed by GC/MS by EPA Method 8270 (U.S. EPA 1986) using a fused-silica capillary column. Benzidine in wastewaters may be analyzed by HPLC in accordance with EPA Method 605 (U.S. EPA 1984). The analysis of benzidine in air, water, blood, and urine is discussed separately in Section 9.14. Generally speaking, any aromatic amine may be analyzed suitably by one or more of the following techniques: GC(FID/NPD), HPLC, or GC/MS. Barek and co-workers (1985) reported picogram-level analysis of aromatic amines by an HPLC system using voltammetric detection with a carbon-fiber detector. Such a system was substantially more sensitive than was the commonly used HPLC-UV photometric detection. Urine metabolites of toxic amines are best analyzed by HPLC techniques.

Aromatic amines, such as aniline, o-toluidine, 2,4-xylidine, N,N-dimethyl-p-toluidine, and N,N-dimethylaniline in air may be analyzed by NIOSH (1984) Method 2002. About 10–30 L of air at a flow rate of 20–200 mL/min is passed over a silica gel sorbent tube, desorbed with 95% ethanol for 1 hour in an ultrasonic bath, and injected into a GC equipped with an FID. Chromosorb 103 (80/100 mesh) is suitable for the purpose. A nitrogen-specific GC detector may be used instead of an FID. Anisidines in air (25–300 L at a flow rate of 500–1000 mL/min) may be collected over a XAD-2 solid sorbent tube, desorbed with 5 mL of methanol, and analyzed by HPLC, UV detection at 254 nm (NIOSH 1984, Method 2514). A stainless steel-packed column containing Bondapak C or equivalent is suitable for the purpose. o-Tolidine and o-dianisidine in air may be analyzed by an HPLC-UV technique similar to benzidine by NIOSH (1984) Method 5013 (see Section 9.14).

Disposal/Destruction

Aromatic amines are dissolved in a combustible solvent and burned in a chemical incinerator equipped with an afterburner and scrubber. Aromatic amines in laboratory wastes may be destroyed through oxidation with potassium permanganate/sulfuric acid into nonmutagenic derivatives (Castegnaro et al. 1985; Barek 1986). One report describes the destruction of carcinogenic amines and 4-nitrobiphenyl in the laboratory wastes by treatment with $KMnO_4/H_2SO_4$, followed by horseradish peroxidase [9003-99-0] and deamination by diazotization in the presence of hypophosphorus acid (IARC 1985). Barek and associates (1985) described a method based on enzymic oxidation of the amines in solution of hydrogen peroxide and horseradish peroxidase. The solid residues are then oxidized with $KMnO_4$ in sulfuric acid medium. The authors reported greater than 99.8% destruction efficiency.

Biodegradation of aniline in sludges by *Pseudomonas putida* FW have been reported (Park et al. 1988; Patil and Shinde 1988). Kawabata and Urano (1985) reported improved biodegradability of aniline and many refractory organic compounds after wet oxidation at 150°C in the presence of Mn/Ce composite oxides.

9.2 ANILINE

EPA Designated Toxic Waste, RCRA Waste Number U012; DOT Label: Poison B, UN 1547

Formula $C_6H_5NH_2$; MW 93.14; CAS [62-53-3]

Structure:

simplest aromatic amine

Synonyms: aminobenzene; aminophen; phenylamine; benzenamine

Uses and Exposure Risk

Aniline is used in the manufacture of dyes, pharmaceuticals, varnishes, resins, photographic chemicals, perfumes, shoe blacks, herbicides, and fungicides. It is also used in vulcanizing rubber and as a solvent. It occurs in coal tar and is produced from the dry distillation of indigo. It is also produced from the biodegradation of many pesticides. Aniline is a metabolite of many toxic compounds, such as nitrobenzene, phenacetin, and phenylhydroxylamine.

Physical Properties

Colorless oily liquid with characteristic odor and burning taste; darkens when exposed to air or light; density 1.022 at 20°C; boils at 184.5°C; freezes at −6.1°C; moderately soluble in water (3.5 g/dL at 20°C), miscible with most organic solvents; weakly alkaline, pH 8.1 for 0.2 M aqueous solution.

Health Hazard

Aniline is a moderately toxic compound. The routes of exposure are inhalation, ingestion, and absorption through skin. The absorption of the liquid aniline through skin may occur at a rate of 3 mg/hr/cm^2 of skin area (ACGIH 1986, Supplement 1989); that from an aqueous solution is slower, depending on its concentration.

The toxic symptoms include headache, weakness, dizziness, ataxia, and cyanosis. Acute poisoning arises due possibly to methemoglobin formation, which may result in cyanosis. Overexposure may lead to death from respiratory paralysis. Inhalation of 250 ppm aniline in air for 4 hours was lethal to rats. The concentration of aniline in samples of rapeseed food oil that caused Spanish toxic oil syndrome was determined to be within the range of 110–1300 ppb (Hill et al. 1987). Contact of the pure liquid on the skin can produce moderate irritation, while the effect on the eyes can be severe. The oral LD$_{50}$ value in animals varied with the species. An LD$_{50}$ value in mice is 464 mg/kg. Toxicity of aniline in aqueous species was very high. An LC$_{50}$ value based on static acute toxicity was calculated to be 0.17 mg/L to daphnids (*Daphnia magna*) (Gersich and Mayes 1986).

Aniline is metabolized to aminophenols, phenylhydroxylamine, and their glucuronide and sulfate derivatives, and excreted. *p*-Aminophenol is the major metabolite in humans and is excreted in urine.

Aniline administered to rats by the oral route caused tumors in the kidney and bladder. The evidence of carcinogenicity in animals, however, is inadequate. Any cancer-causing action of aniline in humans is not known.

Exposure Limits

TLV-TWA skin 2 ppm (∼8 mg/m^3) (ACGIH), 5 ppm (∼19 mg/m^3) (MSHA, OSHA, and NIOSH); IDLH 100 ppm (NIOSH).

Fire and Explosion Hazard

Combustible liquid; flash point (closed cup) 70°C (158°F); autoignition temperature 770°C (1418°F); the vapor forms explosive mixtures with air; the LEL value is 1.3% by volume in air, UEL not reported. Aniline undergoes violent reaction when mixed with hexachloromelanamine or trichloromelanamine (NFPA 1986). Its reactions with strong oxidizing substances are violent, accompanied by flame and/or explosion. Ignition occurs with red fuming nitric acid even at a very low temperature of $-70°C$. It forms ozobenzene, a white, gelatinous explosive compound as one of the products, when mixed with ozone (Mellor 1946). It ignites with alkali metal peroxides, perchromic acid, or perchloric acid followed by formaldehyde addition. Many perchlorates form solvated adducts with organic solvents, including aniline, which explode on impact. Aniline reacts vigorously with concentrated mineral acids.

9.3 *o*-TOLUIDINE

EPA Designated Toxic Waste, RCRA Waste Number, U328; DOT Label: Poison B, UN 1708

Formula $C_7H_7NH_2$; MW 107.17; CAS [95-53-4]

Structure:

NH$_2$
CH$_3$

Synonyms: 2-toluidine; 2-methylbenzenamine; 2-methylaniline; 2-aminotoluene; 2-amino-1-methylbenzene; 1-amino-2-methylbenzene; *o*-tolylamine

Uses and Exposure Risk

o-Toluidine is used in the manufacture of various dyes, in printing textiles blue-black, and as an intermediate in rubber chemicals, pesticides, and pharmaceuticals.

Physical Properties

Colorless to light yellow liquid becoming reddish brown on exposure to air or light; weak aromatic odor; density 1.008 at 20°C; boils at 200°C; solidifies at $-16°C$; slightly soluble in water (1.5% at 20°C), very soluble in organic solvents and dilute acids.

Health Hazard

o-Toluidine is a cancer-causing compound. Its acute toxicity in test species was low to moderate. Severe poisoning may occur at high doses. It may enter the body by inhalation of its vapors, ingestion, or absorption through skin contact. The target organs are kidneys, liver, blood, cardiovascular system, skin, and eyes. The toxic symptoms include methemoglobinemia, anemia, and reticulocytosis, which are similar to aniline. This produces anoxia (lack of oxygen), cyanosis, headache, weakness, drowsiness, dizziness, increase in urine volume, and hematuria.

The pure liquid on skin contact can cause irritation and dermatitis. Contact with eyes can cause burns.

LD$_{50}$ value, oral (mice): 520 mg/kg

o-Toluidine is a suspected human carcinogen. The evidence of carcinogenicity in human is inadequate. It caused tumors in the kidney, bladder, and lungs in rats, mice, and rabbits resulting from oral and subcutaneous administration.

Exposure Limits

TLV-TWA 2 ppm ($\sim$9 mg/m^3) (ACGIH), 5 ppm ($\sim$22 mg/m^3) (MSHA, OSHA, and NIOSH); IDLH level 100 ppm (NIOSH); carcinogenicity: Suspected Human Carcinogen (ACGIH), Human Limited Evidence (IARC).

Fire and Explosion Hazard

Combustible liquid; flash point (closed cup) 85°C (185°F); autoignition temperature 482°C (900°F); the vapor forms explosive mixtures with air, the explosive limits not established. It ignites when mixed with red fuming nitric acid. Violent reaction may occur when combined with strong oxidizers, such as perchloric acid, permanganic acid, or peroxides.

9.4   *m*-TOLUIDINE

DOT Label: Poison B, UN 1708

Formula $C_7H_7NH_2$;   MW   107.17;   CAS [108-44-1]

Structure:

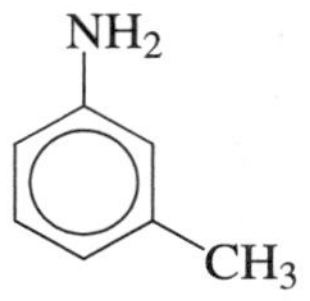

Synonyms: 3-toluidine;   3-methylbenzen-amine;   3-methylaniline;   3-aminotoluene; 3-amino-1-methylbenzene;   3-amino-phenylmethane; *m*-tolylamine

Uses and Exposure Risk

m-Toluidine is an intermediate in the manufacture of dyes and other chemicals.

Physical Properties

Light-yellow liquid; boils at 203°C; freezes at −50°C; density 0.990 at 25°C; slightly soluble in water, dissolves readily in organic solvents and dilute acids.

Health Hazard

The toxicity of *m*-toluidine is similar to that of *o*-toluidine. The exposure routes are ingestion, inhalation, and absorption through skin. The toxic effects are methemoglobinemia, anemia, and hematuria. It metabolizes to 2-amino-4-methylphenol, the major component excreted in urine. Inhalation of 40-ppm concentration for an hour produced severe poisoning in humans. The oral LD_{50} value in mice is 740 mg/kg, which is in a range comparable to its *o*- and *p*-isomers.

The pure liquid is a mild to moderate skin irritant. Its irritating effect on the eyes of rabbits was strong.

m-Toluidine did not induce any carcinogenicity in test subjects. This finding is in contrast to its *ortho*-isomer. Nor did it produce any mutagenic activity.

Exposure Limit

TLV-TWA skin 2-ppm ($\sim$9 mg/m^3) (ACGIH).

Fire and Explosion Hazard

Combustible liquid; flash point (closed cup) 85.5°C (186°F); the vapor forms explosive mixtures with air, the explosive range not determined.

m-Toluidine ignited when mixed with red fuming nitric acid. Reactions with perchloric acid, perchromic acid, alkali metal peroxides, and ozone can be violent or produce unstable products.

9.5   *p*-TOLUIDINE

EPA Designated Toxic Waste, RCRA Waste Number U353; DOT Label: Poison B, UN 1798

Formula $C_7H_7NH_2$;   MW   107.17;   CAS [106-49-0]

Structure:

Synonyms: 4-toluidine;   4-methylbenzen-amine;   4-methylaniline;   4-aminotoluene; 4-amino-1-methylbenzene; *p*-tolylamine

Uses and Exposure Risk

p-Toluidine is used as an intermediate in the manufacture of various dyes. It is also used as a reagent for lignin and nitrites.

Physical Properties

White crystalline solid; melts at 44°C; boils at 200°C; slightly soluble in water (0.74% at 20°C), dissolves readily in organic solvents and dilute acids.

Health Hazard

p-Toluidine is a mild to moderate irritant on the skin. The irritant effect on rabbits' eyes was strong. The toxic properties of p-toluidine are similar to its *ortho-* and *meta-*isomers and aniline. The clinical signs of toxicity are methemoglobinemia, anemia, and cyanosis. The major metabolite in urine after oral application in male rats was 2-amino-5-methylphenol, which was excreted along with 3.5% unchanged p-toluidine (ACGIH 1986). Exposure to 40-ppm concentration for 1 hour resulted in severe poisoning in humans.

LD$_{50}$ value, oral (mice): 330 mg/kg

ACGIH lists p-toluidine as a suspected human carcinogen. It caused tumors in liver of experimental animals. The evidence of carcinogenicity in animals or humans is not yet fully established.

Exposure Limit

TLV-TWA skin 2 ppm ($\sim$9 mg/m^3) (ACGIH); Suspected Human Carcinogen (ACGIH).

Fire and Explosion Hazard

Volatile solid, giving off flammable vapors on heating; flash point (closed cup) 86.6°C (188°F); autoignition temperature 482°C (900°F); the vapor forms explosive mixtures with air; the explosive limits not reported. It ignites when mixed with red fuming nitric acid. It reacts violently with strong oxidizing substances.

9.6   *o*-PHENYLENEDIAMINE

DOT Label: Poison B, UN 1673

Formula $C_6H_8N_2$;   MW   108.16;   CAS [95-54-5]
Structure:

Synonyms: 1,2-benzenediamine;   1,2-di-aminobenzene; 1,2-phenylenediamine; 2-aminoaniline

Uses and Exposure Risk

o-Phenylenediamine is used as an intermediate in the manufacture of dyes, pigments, and fungicides. It is also used in hair dye formulation, as a photographic developing agent, and in organic synthesis.

Physical Properties

Brownish yellow crystals; mp 103°C; bp 257°C; slightly soluble in water, dissolves readily in organic solvents.

Health Hazard

The symptoms of acute poisoning in animals were excitement, tremor, convulsions, salivation, and respiratory depression. Oral application of o-phenylenediamine in rats produced irritation of the stomach. An intraperitoneal administration of 100 μmol kg body weight in male rats caused methemoglobin formation to an extent of 10.8% (Watanabe et al. 1976).

LD$_{50}$ value, oral (mice): 331 mg/kg (Galushka et al. 1985)

o-Phenylenediamine is classified as a suspected human carcinogen by ACGIH (1989). It produced liver tumors in male rats and male and female mice (Weisburger et al. 1978). Evidence of carcinogenicity of this compound, however, is inadequate.

Exposure Limits

TLV-TWA 0.1 mg/m^3; carcinogenicity: A2-Suspected Human Carcinogen (ACGIH 1989).

Fire and Explosion Hazard

Noncombustible solid. It may react violently with strong oxidizers.

9.7 *m*-PHENYLENEDIAMINE

DOT Label: Poison B, UN 1673

Formula $C_6H_8N_2$; MW 108.16; CAS [95-54-5]

Structure:

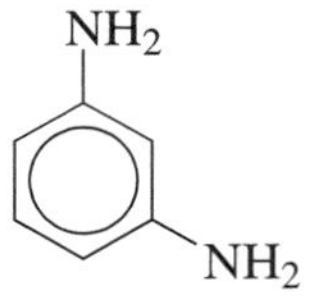

Synonyms: 1,3-benzenediamine; 1,3-diaminobenzene; 1,3-phenylenediamine; 3-aminoaniline

Uses and Exposure Risk

m-Phenylenediamine is used in the manufacture of a variety of dyes, in hair dye formulations, as a rubber curing agent, in petroleum additives, as a photographic developing agent, and as an analytical reagent.

Physical Properties

White crystalline solid; turns red when exposed to air; mp 62°C; bp 284–287°C; soluble in water, alcohol, acetone, chloroform, and dimethylformamide; slightly soluble in ether, carbon tetrachloride, and benzene.

Health Hazard

The toxic effects of *m*-phenylenediamine are similar to its *ortho-* and *para-*isomers.

The toxic symptoms were tremor, excitement, convulsions, and cyanosis, and collapse at fatal dose. Other symptoms included decreased blood pressure and pulse rate. Chronic toxic effects resulting from the oral application of an aqueous solution of this compound were increase in liver and to a lesser extent, kidney weights. No carcinogenicity was observed in test animals. The compound was excreted rapidly, unchanged.

LD$_{50}$ value, oral (rats): 650 mg/kg

Exposure Limits

TLV-TWA 0.1 mg/m^3 (ACGIH 1989); carcinogenicity: Animal Inadequate Evidence (IARC).

Fire and Explosion Hazard

Noncombustible solid. Its reactions with strong oxidizers can be violent.

9.8 *P*-PHENYLENEDIAMINE

DOT Label: Poison B, UN 1673

Formula $C_6H_8N_2$; MW 108.16; CAS [106-50-3]

Structure:

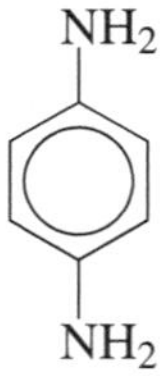

Synonyms: 1,4-benzenediamine; 1,4-diaminobenzene; 1,4-phenylenediamine; 4-aminoaniline

Uses and Exposure Risk

p-Phenylenediamine is used for dyeing hair and fur, in the manufacture of azo dyes, in accelerating vulcanization of rubber, and in antioxidants.

Physical Properties

White crystalline solid; becomes purple to black on exposure to air; melts at 145–147°C; boils at 267°C; sparingly soluble in water, dissolves readily in alcohol, ether, chloroform, and acetone.

Health Hazard

p-Phenylenediamine is a moderate to highly toxic compound; the acute, subacute and chronic toxicity of this amine is greater than that of its *ortho-* and *meta*-isomers. The acute poisoning effects in animals were manifested by lacrimation, salivation, ataxia, tremor, lowering of body temperature, increased pulse rate, and respiratory depression. An intraperitoneal dose of 10.8 mg/kg (suspended in propylene glycol) in male rats caused the formation of methemoglobin to the extent of 12.9% after 5 hours (Watanabe et al. 1976). The hydrochloride of this amine has been reported to cause edema of the head and neck in animals dosed with 120–350 mg/kg. p-Phenylenediamine in hair dye formulations produced skin irritation and mild conjunctivial inflammation in a variety of test animals (Lloyd et al. 1977). In guinea pigs, contact produced skin sensitization. Hair dyes containing p-phenylenediamine damaged vision when applied into eyes. In addition, allergic asthma and inflammation of the respiratory tract resulted from exposure to higher concentrations. Reports in early literature cite several cases of human poisoning resulting from the use of hair dyes containing p-phenylenediamine. The toxic symptoms reported were liver and spleen enlargement, vertigo, gastritis, jaundice, atrophy of liver, allergic asthma, dermatitis, cornea ulcer, burning and redness in eyes, and presbyopia (the latter effects arising from using hair dyes on the eyebrows and eye lashes).

Tests for mutagenicity in *Salmonella* microsome assays (in vitro) were negative. With metabolic activation, upon oxidation with hydrogen peroxide, most mutagenic tests showed positive results. Tests for carcinogenicity were negative, although it slightly increased the overall tumor rate in experimental animals. After oxidation with hydrogen peroxide, the amine produced tumors in the mammary glands of female rats (Rojanapo et al. 1986).

Exposure Limits

TLV-TWA 0.1 mg/m^3 (ACGIH 1989); TWA skin 0.1 mg/m^3 (MSHA and OSHA); IDLH 25 mg/m^3 (NIOSH); carcinogenicity: Animal Inadequate Evidence (IARC).

Fire and Explosion Hazard

Noncombustible solid. Reactions with strong oxidizers may become vigorous to violent.

9.9 PHENYLHYDRAZINE

DOT Label: Poison B, UN 2572
Formula $C_6H_8N_2$; MW 108.16; CAS [100-63-0]
Structure:

a hydrazino derivative of benzene
Synonyms: hydrazinobenzene; hydrazinebenzene

Uses and Exposure Risk

Phenylhydrazine is used in the manufacture of dyes and pharmaceuticals; as a stabilizer for explosive; as a reagent for aldehydes, ketones, and sugars in chemical analysis; and in organic synthesis.

Physical Properties

Pale yellow crystals or an oily liquid; darkens when exposed to light or air; mp 19.5°C bp 243.5°C (decomposes); slightly soluble in water, readily dissolves in alcohol, ether, acetone, benzene, chloroform, and dilute acids.

Health Hazard

Phenylhydrazine is a moderate to highly toxic compound and a carcinogen. Acute toxic symptoms include hematuria, changes in liver and kidney, vomiting, convulsions, and respiratory arrest. Additional symptoms are lowering of body temperature and fall in blood pressure. Chronic exposure to this compound caused hemolytic anemia and significant body weight loss in rats. An oral administration of 175 mg/kg was lethal to mice. An oral LD_{50} value in rat is 188 mg/kg.

Phenylhydrazine caused adverse reproductive effect when dosed intraperitoneally in pregnant mice. It caused jaundice, anemia, and behavioral deficit in offspring.

Oral or subcutaneous administration of phenylhydrazine or its hydrochloride produced lung and liver tumors in experimental animals. ACGIH lists this compound as a suspected human carcinogen. The evidence of carcinogenicity of this compound in human is inadequate.

Exposure Limits

TLV-TWA skin 0.1 ppm (0.44 mg/m^3) (ACGIH), 5 ppm (22 mg/m^3) (OSHA); STEL 10 ppm (44 mg/m^3) (OSHA); carcinogenicity: A2-Suspected Human Carcinogen (ACGIH), Carcinogen (NIOSH).

Fire and Explosion Hazard

Combustible liquid; flash point (closed cup) 89°C (192°F). It reacts with lead dioxide and strong oxidizing compounds vigorous to violently.

9.10   2,4-TOLUENEDIAMINE

EPA Designated Toxic Waste, RCRA Waste Number U221; DOT Label: Poison B, UN 1709

Formula $C_7H_{10}N_2$; MW 122.19; CAS [95-80-7]

Structure:

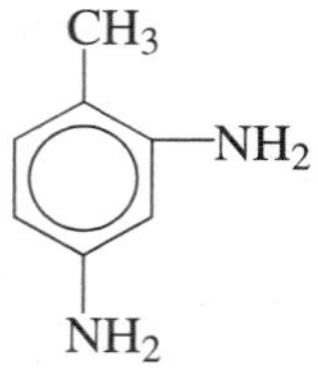

Synonyms: 2,4-diamino-1-toluene; 4-methyl-1,3-benzenediamine; 4-methyl-*m*-phenylenediamine; 3-amino-*p*-toluidine; 5-amino-*o*-toluidine; 2,4-tolylenediamine

Uses and Exposure Risk

2,4-Toluenediamine is used as an intermediate in the manufacture of dyes.

Physical Properties

Crystalline solid; melts at 98°C; boils at 284°C; slightly soluble in water, dissolves in organic solvents.

Health Hazard

2,4-Toluenediamine is a cancer-causing compound. Animal studies indicate sufficient evidence of its carcinogenicity. Oral application of this amine resulted in blood and liver cancers in rats and mice.

The acute oral toxicity was moderate to high in test animals. It produced methemoglobinemia, cyanosis, and anemia. The oral LD_{50} value in rats is 250–300 mg/kg. It is a mild skin irritant. The irritation effect on rabbits' eyes was moderate.

Thysen and associates (1985a,b) evaluated the reproductive toxicity of 2,4-toluenediamine in adult male rats. Diets containing up to 0.03% of amine fed to the animals for 10 weeks reduced the fertility and exerted an inhibitory or toxic effect on spermatogenesis. The amine affected androgen action in the male rats and induced damage to the germinal components of the testes.

2,4-Toluenediamine tested positive to the histidine reversion-Ames and *Drosophila melanogaster* sex-linked lethal tests. When administered to rats pretreated with the microsomal enzyme inducers phenobarbital [50-06-6], β-naphthoflavone [6051-87-2], or 3-methylcholanthrene [56-49-5], the urinary metabolites showed increased mutagenic activity compared with the controls (Reddy et al. 1986).

Exposure Limit

Carcinogenicity: Animal Sufficient Evidence (IARC); no TWA assigned (because it is a carcinogen, there should be no exposure to this compound).

Fire and Explosion Hazard

Noncombustible solid. It ignites when mixed with red fuming nitric acid. Reactions with strong oxidizers and hypochlorites can be violent.

9.11 1-NAPHTHYLAMINE

EPA Designated Toxic Waste, RCRA Waste Number U167; DOT Label: Poison B, UN 2077

Formula $C_{10}H_7NH_2$; MW 143.20; CAS [134-32-7]

Structure:

Synonyms: α-naphthylamine; 1-aminonaphthalene; naphthalidine

Uses and Exposure Risk

1-Naphthylamine is used in the manufacture of dyes and tonic prints.

Physical Properties

Colorless crystals turning red on exposure to air and light; disagreeable odor; mp 50°C; bp 301°C; sublimes; slightly soluble in water (0.16 g/dL at 20°C); dissolves readily in alcohol and ether

Health Hazard

1-Naphthylamine is a moderately toxic and cancer-causing substance. The toxic symptoms arising from oral intake or skin absorption of this compound include acute hemorrhagic cystitis, dyspnea, ataxia, dysuria, and hematuria. An intraperitoneal LD_{50} value in mice is 96 mg/kg. Inhalation of dusts or vapors is hazardous, showing similar symptoms. 1-Naphthylamine caused leukemia in rats. There is substantial evidence of its cancer-causing effects in animals and humans.

Exposure Limit

TLV-TWA not assigned; carcinogenicity: Human Carcinogen (skin) (MSHA and OSHA).

Fire and Explosion Hazard

Noncombustible solid. There is no report of explosion caused by this compound.

9.12 2-NAPHTHYLAMINE

EPA Designated Toxic Waste, RCRA Waste Number U168; DOT Label: Poison B, UN 1650

Formula: $C_{10}H_7NH_2$; MW 143.20; CAS [91-59-8]

Structure:

Synonyms: β-naphthylamine; 2-aminonaphthalene; 2-naphthaleneamine

Uses and Exposure Risk

2-Naphthylamine was widely used in the manufacture of dyes and in rubber. Currently, its use is curtailed because of the health hazard.

Physical Properties

White to reddish crystalline solid; mp 111–113°C; bp 306°C; soluble in hot water, alcohol, and ether.

Health Hazard

2-Naphthylamine poses a severe health hazard because of its carcinogenicity. Administration of this compound by all routes resulted in cancers in various tissues in test animals. It caused tumors in the kidney, bladder, liver, lungs, skin, and blood tissues. There is sufficient evidence that this compound causes bladder cancer in humans after a latent period of several years.

The toxicity of 2-naphthylamine is low to moderate. However, high doses can produce severe acute toxic effects. The routes of exposures are ingestion, skin contact, and inhalation of its dusts or vapors. The acute toxic symptoms are similar to those produced by 1-naphthylamine: hemorrhagic cystitis or methemoglobinemia (causing hypoxia or inadequate supply of oxygen to tissues), respiratory distress, and hematuria (blood in urine).

LD$_{50}$ value, oral (rats): 727 mg/kg

Exposure Limits

Since it is a carcinogen, there is no TLV-TWA for this compound. Recognized Carcinogen (ACGIH); Carcinogen (OSHA); Human Sufficient Evidence (IARC).

Fire and Explosion Hazard

Noncombustible solid; low reactivity. There is no report of explosion in the literature.

9.13 DIPHENYLAMINE

Formula $C_{12}H_{10}NH$; MW 169.24; CAS [122-39-4]
Structure:

Synonyms: *N*-phenylaniline; (phenylamino) benzene; *N*-phenylbenzeneamine; anilinobenzene

Uses and Exposure Risk

Diphenylamine is used in the manufacture of dyes, as a stabilizer for nitrocellulose explosives, and as an analytical reagent for colorimetric tests for nitrate and chlorate.

Physical Properties

Colorless crystalline solid with floral odor; melts at 52.8°C; boils at 302°C; insoluble in water, dissolves readily in organic solvents.

Health Hazard

Diphenylamine is much less toxic than aniline. The acute oral toxicity is low. A dose of 3000 mg/kg was lethal to rats. At a concentration of >500 ppm, a diet fed to rats for over 7 months resulted in renal cysts in animals. Its absorption through the skin and the respiratory system is lower than that of aniline. Exposure to its dusts caused changes in liver, spleen, and kidney in test animals. Industrial exposure to diphenylamine has caused tachycardia, hypertension, eczema, and bladder symptoms in workers (Fairhall 1957). Carcinogenicity of this compound is unknown. It showed an adverse reproductive effect in animals, causing developmental abnormalities in urogenital system in pregnant rats.

Exposure Limit

TLV-TWA 10 mg/m^3 (ACGIH and MSHA).

Fire and Explosion Hazard

Noncombustible solid; autoignition temperature 634°C (1173°F); low reactivity.

9.14   BENZIDINE

EPA Priority Pollutant in Wastewater and Solid and Hazardous Wastes, RCRA Toxic Waste Number U021; DOT Label: Poison B, UN 1885

Formula $C_{12}H_{12}N_2$; MW 184.26; CAS [92-87-5]

Structure:

Synonyms: 4,4'-biphenyldiamine; 4,4'-di-amino-1,1'-biphenyl; 4,4'-diphenylene-diamine; p,p'-dianiline

Uses and Exposure Risk

Benzidine was used extensively in the manufacture of dyes. Because of its cancer-causing effects in humans, its application in dyes has been curtailed. Other uses of this compound are in chemical analysis: as a reagent for the determination of hydrogen peroxide in milk and in the analysis of nicotine. Its hydrochloride is used as a reagent to analyze metals and sulfate.

Physical Properties

White crystalline solid; darkens on exposure to air or light; mp 120°C; bp 400°C; slightly soluble in cold water (0.04%); moderately soluble in hot water, alcohol, and ether.

Health Hazard

Benzidine is a known carcinogen, causing bladder cancer in humans. Numerous reports in the literature document its carcinogenicity in animals and humans. Oral or subcutaneous application of this compound in experimental animals produced tumors in liver, blood, lungs, and skin. The routes of entry into human body are primarily the inhalation of its dusts and absorption through skin.

The acute oral toxicity in animals was moderate. Ingestion can produce nausea, vomiting, kidney, and liver damage. The oral LD_{50} values in test animals were in the range 150–300 mg/kg.

Exposure Limits

Because it is a carcinogen and readily absorbed through skin, no TLV has been assigned. Exposure should be at an absolute minimum.

Recognized Human Carcinogen (ACGIH); Human Carcinogen (MSHA); Carcinogen (OSHA); Human Sufficient Evidence, Animal Sufficient Evidence (IARC).

Analysis

Benzidine in wastewater may be analyzed by HPLC and GC/MS techniques by U.S. EPA (1984) Methods 605 and 625, respectively, or an oxidation/colorimetric method using Chloramine T. Analysis in soil, solid wastes, and groundwater may be performed in accordance to U.S. EPA (1986) Method 8270 (by GC/MS technique).

Benzidine in urine may be screened by NIOSH (1984) Method 8304. Two urine samples (150 mL each) before and after 6 hours of exposure are collected. The samples are derivatized with 2,4,5-trinitrobenzene sulfonic acid and analyzed by visible absorption (at 400-nm wavelength) thin-layer chromatography. Alternatively, the samples may be analyzed by GC using an ECD (NIOSH 1984, Method 8306). Air analysis may be done by NIOSH (1984) Method 5013. About 120–300 L of air at a flow rate of 1–3 L/min is passed over a 5-μm PTFE membrane. The analyte is desorbed with water and analyzed by an HPLC-UV technique (UV set at 280 nm).

9.15 o-TOLIDINE

EPA Designated Toxic Waste, RCRA Waste Number U095

Formula $C_{14}H_{16}N_2$; MW 212.32; CAS [119-93-7]

Structure:

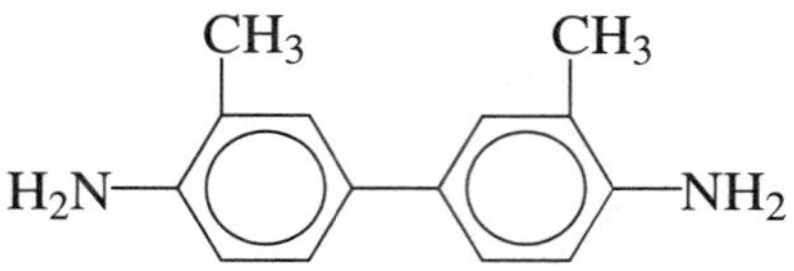

Synonyms: 3,3'-dimethylbenzidine; 4,4'-diorthotoluidine; 4,4'-diamino-3,3'-dimethylbiphenyl; 3,3'-dimethyl-4,4'-biphenyldiamine; 3,3'-dimethyl-(1,1'-biphenyl)-4,4'-diamine

Uses and Exposure Risk

o-Tolidine is used extensively in the manufacture of dyes. It is also used as a reagent for the analysis of gold and free chlorine in water.

Physical Properties

White to reddish crystals or powder; melts at 129°C; bp 200°C; slightly soluble in water, soluble in organic solvents and dilute acids.

Health Hazard

The information on toxicity and carcinogenicity of o-tolidine is very little. o-Tolidine often contains other biphenylamine contaminants. Its structure and chemical properties are similar to those of benzidine. The toxic properties are therefore expected to be similar to those of benzidine.

o-Tolidine is absorbed into body through the skin, respiratory, and gastrointestinal tract. The acute oral and dermal toxicity was moderate in experimental animals. An oral LD_{50} value in rats is within the range of 400 mg/kg.

o-Tolidine produced bladder cancer in test species. Also, cancers in other tissues were noted in some animals. ACGIH lists this substance as a suspected human carcinogen.

Exposure Limits

TLV-TWA none; carcinogenicity: A2 — Suspected Human Carcinogen (ACGIH), Animal Limited Evidence (IARC).

Fire and Explosion Hazard

Noncombustible solid; incompatible with strong oxidizers.

9.16 MISCELLANEOUS AROMATIC AMINES

Presented in Table 9.1 are toxicity data for some commercially used aromatic amines. These substances exhibit low toxicity and may cause bladder cancer (NIOSH 1986). The carcinogenicity is much weaker than benzidine or α- or β-naphthylamine. The toxicity data on many of these substances are too little. These compounds are noncombustible liquids or solids. Any explosive reactions exhibited by these amines are not known. These may, however, undergo vigorous to violent reactions if mixed with strong oxidizers.

TABLE 9.1   Toxicity of Miscellaneous Amines

Compound/Synonyms/ CAS No.	Formula/MW/ Structure	Toxicity
o-Anisidine (2-anisidine, 2-methoxy-1-aminobenzene, 2-methoxybenzenamine, o-methoxyaniline, o-anisylamine) [90-04-0]	C_7H_9NO 123.17	Causes anemia, reticulocytosis, and cyanosis on chronic exposure; absorbed through skin; other toxic routes are ingestion and inhalation of its vapors, and symptoms are headache and dizziness; LD_{50} oral (rats): 2000 mg/kg; caused bladder cancer in experimental animals, human limited evidence. DOT Label: Poison B, UN 2431; exposure limits: TLV-TWA skin 0.1 ppm (0.5 mg/m^3) (ACGIH, MSHA, OSHA); IDLH 50 mg/m^3 (NIOSH)
p-Anisidine (4-anisidine, 4-methoxy-1-aminobenzene, 4-methoxybenzenamine, p-methoxyaniline, p-anisylamine) [104-94-9]	C_7H_9NO 123.17	Exposure route: ingestion and skin absorption; target organs: kidney, liver, blood, and cardiovascular system; causes anemia, cyanosis, and changes in liver and kidney; LD_{50} oral (mice): 810 mg/kg; its carcinogenicity in humans not yet established. Exposure limit: TLV-TWA skin 0.1 ppm (0.5 mg/m^3)
5-Methyl-o-anisidine (2-methoxy-5-methyl benzenamine, p-cresidine, 2-methoxy-5-methylaniline, 3-amino-4-methoxytoluene) [120-71-8]	$C_8H_{11}NO$ 137.20	Low oral toxicity; cancer-causing compound; caused bladder cancer in experimental animals, limited evidence of carcinogenicity in humans
2-Methyl-p-anisidine (m-cresidine, 2-methyl-p-anisidine, 4-methoxy-2-methylbenzenamine) [102-50-1]	$C_8H_{11}NO$ 137.20	Oral administration of this compound caused tumors in kidney and liver; human carcinogenicity not reported; toxicity data not available

(continued)

TABLE 9.1 (*Continued*)

Compound/Synonyms/ CAS No.	Formula/MW/ Structure	Toxicity
Dianisidine (*o*-dianisidine, 3,3′-dimethoxybenzidine, 3,3′-dimethoxy-4,4′-diphenyldiamine, 4,4′-diamino-3,3′-dimethoxybiphenyl) [119-90-4]	$C_{14}H_{16}N_2O_2$ 244.32	Information on the toxicity of this compound in the literature is too scant; the toxicity profile and carcinogenic properties of this compound are expected to be similar to those of benzidine; an oral LD_{50} value in rats is in the range 2000 mg/kg; caused cancer in bladder and skin in experimental animals; EPA Designated Toxic Waste, RCRA Waste Number U091
3,3′-Dichlorobenzidine (4,4′-diamino-3,3′-dichlorodiphenyl, 3,3′-dichloro-(1,1′-biphenyl)-4,4′-diamine) [91-94-1]	$C_{12}H_{10}Cl_2N_2$ 253.14	Low oral toxicity; LD_{50} value in rats is in the range 5000 mg/kg; carcinogenic; causes bladder cancer; cancer-causing actions similar to benzidine; animal sufficient evidence and human inadequate evidence (IARC); human carcinogen (MSHA); carcinogen (OSHA); suspected carcinogen (ACGIH); no TLV assigned; EPA Designated Toxic Waste, RCRA Waste Number U073
N-Phenyl-β-naphthylamine (*N*-phenyl-2-naphthylamine, β-naphthyl phenylamine, 2-anilino naphthalene) [135-88-6]	$C_{16}H_{13}N$ 219.30	Low toxicity; metabolizes to β-naphthylamine; oral LD_{50} value in mice is in the range 1500 mg/kg; caused tumors in kidney, liver, and lungs in animals; animal limited evidence; suspected human carcinogen
N-Phenyl-α-naphthylamine (*N*-phenyl-1-naphthylamine, α-naphthyl phenylamine, 1-anilino naphthalene) [90-30-2]	$C_{16}H_{13}N$ 219.30	Low toxicity; the toxic effects and doses producing toxic symptoms are comparable to its β-isomer; oral LD_{50} value in mice is in the range 1200–1300 mg/kg; caused kidney, liver, and lung cancer in experimental animals; negative to histidine reversion — Ames test and *Saccharomyces cerevisiae* gene conversion test for mutagenicity

REFERENCES

ACGIH. 1986. *Documentation of the Threshold Limit Values and Biological Exposure Indices*, 5th ed. Cincinnati, OH: American Conference of Governmental Industrial Hygienists.

Barek, J. 1986. Destruction of carcinogens in laboratory wastes. II. Destruction of 3,3-dichlorobenzidine, 3,3-diaminobenzidine, 1-naphthylamine, 2-naphthylamine, and 2,4-diaminotoluene by permanganate. *Microchem. J* 33(1): 97–101.

Barek, J., V. Pacakova, K. Stulik, and J. Zima. 1985. Monitoring of aromatic amines by HPLC with electrochemical detection. Comparison of methods for destruction of carcinogenic aromatic amines in laboratory wastes. *Talanta* 32(4): 279–83.

Castegnaro, M., C. Malaveille, I. Brouet, J. Michelon, and J. Barek. 1985. Destruction of aromatic amines in laboratory wastes through oxidation with potassium permanganate/sulfuric acid into non-mutagenic derivatives. *Am. Ind. Hyg. Assoc. J.* 46(4): 187–91.

Galushka, A. J., A. K. Manenko, et al. 1985. Hygienic basis for the maximum allowable concentration of *o*-phenylenediamine and methylcyanocarbamate dimers in bodies of water. *Gigiena Sanit. 6*: 78; cited in *Chem. Abstr. CA 103*: 99843u.

Gersich, F. M., and M. A. Mayes. 1986. Acute toxicity tests with *Daphnia magna* Straus and *Pimephales promelas* Rafinesque in support of national pollutant discharge elimination permit requirements. *Water Res. 20*(7): 939–41.

Hill, R. H., Jr., G. D. Todd, E. M. Kilbourne, et al. 1987. Gas chromatography/mass spectrometric determination of aniline in food oils associated with the Spanish toxic oil syndrome. *Bull. Environ. Contam. Toxicol.* 39(3): 511–15.

IARC. 1985. Decontamination and destruction of carcinogens in laboratory wastes: some aromatic amines and 4-nitrobiphenyl. *IARC Sci. Publ.*, 64: 1–85; cited in *Chem. Abstr. CA 104*(1): 1978c.

Kawabata, N., and H. Urano. 1985. Improvement of biodegradability of organic compounds by wet oxidation. *Chem. Abstr. CA 105*(6): 48363j.

Lloyd, G. K., M. P. Ligget, S. R. Kynoch, and R. E. Davies. 1977. Assessment of the acute toxicity and potential irritancy of hair dye constituents. *Food Cosmet. Toxicol. 15*(6): 607.

Mellor, J. W. 1946. *A Comprehensive Treatise on Inorganic and Theoretical Chemistry*. London: Longmans, Green & Co.

NFPA. 1986. *Fire Protection Guide on Hazardous Materials*, 9th ed. Quincy, MA: National Fire Protection Association.

NIOSH. 1984. *Manual of Analytical Methods*, 3rd ed. Cincinnati, OH: National Institute for Occupational Safety and Health.

NIOSH. 1986. *Registry of Toxic Effects of Chemical Substances*, ed. D. V. Sweet. Washington, DC: U.S. Government Printing Office.

Park, Y. K., J. S. Oh, C. I. Ban, S. J. Yoon, and M. S. Choi. 1988. Biodegradation of aniline by *Pseudomonas putida* FW. *Chem. Abstr. CA 110*(12): 101065v.

Patil, S., and V. M. Shinde. 1988. Biodegradation studies of aniline and nitrobenzene in aniline plant wastewater by gas chromatography. *Environ. Sci. Technol. 22*(10): 1160–65.

Reddy, T. V., R. Ramanathan, T. Benjamin, P. H. Grantham, and E. K. Weisburger. 1986. Effect of microsomal enzyme inducers on the urinary excretion pattern of mutagenic metabolites of the carcinogen 2,4-toluenediamine. *J. Natl. Cancer Inst. 76*(2): 291–97.

Rojanapo, W., P. Jupradinum, A. Tepsuwan, et al. 1986. Carcinogenicity of an oxidation product of p-phenylenediamine. *Carcinogenesis 7*(12): 1997.

Thysen, B., S. K. Varma, and E. Bloch. 1985a. Reproductive toxicity of 2,4-toluenediamine in the rat. 1. Effect on male fertility. *J. Toxicol. Environ. Health 16*(6): 753–61.

Thysen, B., E. Bloch, and S. K. Varma. 1985b. Reproductive toxicity of 2,4-toluenediamine in the rat. 2. Spermatogenic and hormonal effects. *J. Toxicol. Environ. Health 16*(6): 763–69.

U.S. EPA. 1984. Methods for organic chemical analysis of municipal and industrial wastewater. 40 CFR, Part 136, Appendix A.

U.S. EPA. 1986. *Test Methods for Evaluating Solid Waste*, 3rd ed. Washington, DC: Office of Solid Waste and Emergency Response.

Watanabe, T., N. Ishihara, and M. Ikeda. 1976. Toxicity of and biological monitoring for 1,3-diamino-2,4,6-trinitrobenzene and other nitro-amino derivatives of benzene and chlorobenzene. *Int. Arch. Occup. Environ. Health 37*(3): 157

Weisburger, E. K., A. B. Russfield, and F. Homberger. 1978. Testing of twenty aromatic amines or derivatives for long-term toxicity or carcinogenicity. *J. Environ. Pathol. Toxicol. 2*: 325.

10

ASBESTOS

10.1 STRUCTURE AND PROPERTIES

Asbestos is a general term used to describe a variety of naturally occurring hydrated silicates that produce mineral fibers upon mechanical processing. Such fibrous silicate minerals were known from early times and were used in various products through the years. There are two varieties of asbestos: (1) serpentine and (2) amphibole minerals. The occurrence and formation of these two varieties differ widely. The chemical composition of these species are shown in Table 10.1.

The serpentine group of minerals, which include chrysotile asbestos, are almost identical in composition. The chemical composition of unit cell is $Mg_6(OH)_8Si_4O_{10}$. Chrysotiles have layered or sheeted crystal structure containing a silica sheet of $(Si_2O_5)_n$ in which silica tetrahedra point one way (Streib 1978). A layer of brucite, $Mg(OH)_2$, joins the silica tetrahedra on one side of the sheet structure. Two out of every three $-OH$ are replaced by oxygen atoms. X-ray and electron microscope studies indicate chrysotile fibers to be of hollow cylindrical form, with diameters of $0.02-0.03$ μm.

Compositions of amphibole asbestos are more complex than the serpentine group.

The structure consists of two chains based on Si_4O_{11} units separated by a group of cations. The central cation in each unit cell is attached to two hydroxyls. The metal ions that are present predominantly in amphiboles are Mg^{2+}, Fe^{2+}, Fe^{3+}, Ca^{2+}, and Na^+. Minor cation substituents are Al^{3+}, Ti^{4+}, and K^+. The ultimate diameters of amphiboles are about 0.1 μm.

Both chrysotile and amphibole fibers lose hydroxyls at elevated temperatures. Dehydroxylation and decomposition occurs at $600-1050°C$, depending on the species. Amphibole fibers show great resistance to acids, while chrysotiles may break down by acids. Under normal conditions, chrysotile is not readily attacked by caustic alkalies because of its alkaline surface. Dispersion of chrysotile fibers in reagent water produces alkaline solution, reaching a pH of around 10.3, attributed to $Mg(OH)_2$.

10.2 USES AND EXPOSURE RISK

Asbestos occurs in nature in fibrous stones, silicate minerals, rocks, open-pit quarries, and underground mines. It also occurs in burial pits and human-made sites, where it was dumped during the era before regulations.

243

TABLE 10.1   Asbestos Types and Their Chemical Compositions

Asbestos Type[a]	CAS No.	Composition
Chrysotile	[12007-29-5]	$3MgO \cdot 2SiO_2 \cdot 2H_2O$
Crocidolite	[12001-28-4]	$Na_2O \cdot Fe_2O_3 \cdot 3FeO \cdot 8SiO_2 \cdot H_2O$
Amosite	[12172-73-5]	$11FeO \cdot 3MgO \cdot 16SiO_2 \cdot 2H_2O$
Actinolite	[13768-00-8]	$2CaO \cdot 4MgO \cdot FeO \cdot 8SiO_2 \cdot H_2O$
Anthophyllite	[17068-78-9]	$7MgO \cdot 8SiO_2 \cdot H_2O$
Tremolite	[14567-73-8]	$2CaO \cdot 5MgO \cdot 8SiO_2 \cdot H_2O$

[a]Chrysotile belongs to serpentine group, all other asbestos listed are of amphibole type; chrysotile, crocidolite, and amosite are of commercial use.

There are several such dump sites in the United States that contain significant amounts of asbestos fibers. Ashes from cellulose and wood chips resulting from low combustion temperatures in certain municipal refuse incinerators have been found to contain asbestos (Patel-Mandlik et al. 1988). Because of their unique physical and chemical properties, such as noncombustibility, withstanding temperatures of over 500°C, resistance to acids, high tensile strength, and thermal and acoustic insulation, asbestos was used widely in making innumerable products for day-to-day use. Such products included textiles, cement, paper, wicks, ropes, floor and roofing tiles, water pipe, wallboard, fireproof clothing, gaskets, curtains, and so on. Such widespread use of asbestos products is currently being curtailed because of possible health hazards that may arise from exposure to asbestos fibers.

10.3   HEALTH HAZARD AND TOXICOLOGY

Health hazards associated with excessive exposure to asbestos were known as early as 1900. The asbestos-related diseases usually show a long latency period, sometimes many years after the exposure. The latency time, however, varies depending on the species of asbestos, the degree of exposure, and the type of disease. Inhalation is primarily the route of entry of asbestos particles into biogenic system. Other than the pulmonary system,

fibers may accumulate in the gastrointestinal tract from drinking water or transportation from nasal passage. Exposure to airborne dusts can cause asbestosis, mesothelioma, lung cancer, and gastrointestinal cancer.

Asbestosis is a disease characterized by lung disorder and a diffuse interstitial fibrosis that is irreversible in nature. It can lead to respiratory disability, causing death from inability of the body to obtain requisite oxygen or cardiac failure. Chest radiographs show a granular change. The sputum may contain asbestos bodies. It is a progressive disease that develops even after the exposure ceases. It may fully develop within 7 to 10 years, and death may result in a few years after that. The symptoms of asbestosis are shortness of breath, dry coughing, irritation or chest pain during coughing, and diffuse basal rales. The symptoms in advanced stages are clubbing of fingers and toes; cyanosis, which may cause deficient blood oxygenation and dark blue to purple coloration of mucous membranes and the skin; and chronic elevation of pulmonary arterial pressure, leading to respiratory insufficiency and right ventricular failure (Peters and Peters 1980).

Mesothelioma is a rare malignant tumor that can occur after a short intensive exposure to certain types of asbestos and also to many other materials (ASTM 1982). Unlike asbestosis, mesothelioma develops in a shorter latent time, and death may occur within a year of diagnosis (Proctor and

Hughes 1978). Malignant cancers (mesotheliomas) spread over the surfaces of lung, heart, and abdominal organs. Early symptoms are pleural effusion, shortness of breath, chest pain, progressive weight loss, and abdominal swelling. Among the asbestos types, chrosidolite is most hazardous, followed by amosite and chrysotile, in causing mesothelioma in humans.

Lung cancer may occur from moderate exposure to airborne asbestos fibers. The latency period may be about 15–35 years. The clinical symptoms may be irritating cough, blood-flecked sputum, chest pain, loss of weight and appetite, and clubbing of fingers. Cigarette smoking can enhance the risk of bronchogenic carcinoma in asbestos workers by many times. Harrison and Heath (1986) reported synergic property of N-nitrosoheptamethyleneimine (NHMI) and cadmium particulate toward lung carcinogenesis of crocidolite asbestos fibers in rats. The study showed that lung tumor incidence in rats receiving crociodolite, cadmium, and NHMI was significantly higher than in the groups receiving either crocidolite and NHMI together or crocidolite and cadmium together.

Toxicology

There have been extensive studies in recent years to understand the toxicity and carcinogenic properties of asbestos fibers and the mechanism of such actions. Among the factors that contribute to the health hazard from exposure to asbestos particles, the route of entry probably is most important. Asbestos fibers in food or drinking water do not seem to produce any notable adverse health effect. Delahunty and Hollander (1987) investigated the toxic effect of chronic administration of chrysotile asbestos in drinking water. Rats were given a solution of 0.5 g/L in their drinking water at a total dose of 7 mg/day for 1.5 years. Asbestos fibers were not detected in the bowels of the animals by polarized light microscopy. There was no impairment

of kidney function in the treated group. The only effect observed was a decrease in the ability of the intestine to absorb some nonmetabolizable sugar.

There was no toxic or carcinogenic response from crocidolite asbestos in male or female rats when the mineral was administered at 1% concentration in the feed for the lifetime of the animals (NTP 1988). In an earlier study it was determined that feed consumption and survival were the same for the short-range and intermediate-range-fiber-length chrysotile asbestos and controls (NTP 1985).

Carcinogenicity of asbestos fibers in rats was observed with all fibers of different types and sources, producing mesotheliomas at the injection sites (Minardi and Maltoni 1988). Jaurand and co-workers (1987) studied the carcinogenicity of asbestos and other mineral fibers of varying dimensions and the effects of acid treatment. A 20-mg dose was injected into the pleural cavity of Sprague-Dawley rats. Chrysotile fibers of mean length 3.2–1.2 μm induced mesotheliomas, the shorter fibers showing low effect. Crocidolite and amosite forms also induced mesotheliomas. While acid treatment of these two forms did not significantly modify the percentage of mesotheliomas, acid leaching of chrysotile asbestos prior to intrapleural injection reduced the carcinogenicity by 25%. Other than the dimensions of asbestos fibers, and acid leaching, carcinogenicity also depended on certain physical and chemical properties. The in vitro experiments of these investigators showed that there was no strong correlation between cytotoxicity and the carcinogenic potency; and the effects of the fibers were modulated by their size.

Mechanisms based on electron transfer and active oxygen species have been proposed to explain asbestos-induced toxicity and lung disease. Fisher et al. (1987) studied the effect of heat treatment on chrysotile asbestos toxicity. The in vitro study showed that heat treatment reduced cytotoxicity. IR spectra indicated a reduction of external hydroxyl group population, which repopulated after irradiation. There is, apparently, an electron transfer from the asbestos matrix to biological receptors. In an earlier study, Fisher and coworkers (1985) reported that irradiation of chrysotile samples heated to 400°C restored the biological activity to near-control values. X-ray diffraction pattern showed no change in the crystal structure. Brucite, present as a surface contaminant, was removed by heating.

Mossman and Marsh (1989) attributed the asbestos-induced cell damage to the production of oxygen-free radicals. This was based on their findings that cytotoxicity to hamster tracheobronchial epithelial (HTE) cells induced by longer fibers in vitro could be prevented by scavengers of superoxide, which is released by both HTE cells and alveolar macrophages after being exposed to chrysotile and crocidolite asbestos fibers. The oxygen free radicals so formed could damage the HTE cells directly. Asbestos of all types caused lipid peroxidation in mammalian cells and artificial membranes, a process catalyzed by iron (Mossman and Marsh 1989). A similar mechanism was suggested by Goodlick and Kane (1986) for crocidolite fibers. Iron present in the fibers catalyzed the formation of hydroxyl radical from superoxide anion and hydrogen peroxide formed during phagocytosis. These −OH radicals are highly reactive, causing cell injury.

Crocidolite asbestos induced greater than a 20-fold increase in the cells in anaphase with abnormalities (Hesterberg and Barret 1985). Fibers were observed in mitotic cells, interacting with chromosomes, causing missegregation of chromosomes, resulting in aneuploidy. Such an effect could induce cell transformation and cancer.

Exposure Limits

TLV-TWA air 0.1 fiber/cm^3 ($= 100,000$ fibers/m^3) of length >5 μm (NIOSH); 2 fibers/cm^3 ($= 1$ million fibers/m^3) of length >5 μm (OSHA). The TLV-TWA recommen-

ded by ACGIH for various forms of asbestos are as follows:

Amosite: 0.5 fiber/cm^3 ($= 500,000$ fibers/m^3)

Chrysotile: 2 fibers/cm^3 ($= 2$ million fibers/m^3)

Crosidolite: 0.2 fiber/cm^3 ($= 200,000$ fibers/m^3)

Other forms: 2 fibers/cm^3 ($= 2$ million fibers/m^3)

(the fibers >5 μm with an aspect ratio of $\geq 3:1$ as measured by the membrane filter method at 400 to 450× or 4-mm objective phase-contrast illumination).

10.4 ANALYSIS

Asbestos can be characterized and identified by a number of analytical techniques, which may be classified under the following five categories: (1) optical microscopy, (2) electron microscopy, (3) physical analysis, (4) chemical analysis, and (5) gravimetry.

Optical Microscopy

Membrane filter method using an optical microscope is widely used for counting asbestos fibers in air. A known volume of dust-laden air is drawn through a membrane filter. The fibers are counted by an optical microscope fitted with a 4 mm/40× objective at 450–500 magnification using transmitted light and phase-contrast techniques. Only the fibers with length >5 μm and breadth <3 μm and a ratio of length to breadth exceeding $3:1$ may be counted by this technique. Other optical microscopy techniques that may be applicable for measuring asbestos fibers are dispersion staining, interference contrast, and coupled with image analyzer.

Electron Microscopy

Scanning electron microscopy (SEM) with an x-ray analyzer and transmission electron microscopy (TEM) are the most common techniques in electron microscopy. In the former technique, samples are scanned at $600\times$ and $2000\times$ and the metal ratios are quantitated from x-ray data. Element mass ratios, cation/anion ratios, and morphology are compared with empirical data from reference standards to identify the asbestos type. Air, water, and bulk samples may be analyzed by this technique. In blind tests, the correct identifications were made for more than 94% of fibers (Sherman et al. 1989).

Both amphibole and chrysotile asbestos in air and water may be analyzed by TEM. Accuracies above 90% may be obtained on chrysotile fibers of >1 μm (Steel and Small 1985).

Physical Methods

X-ray diffraction, light scattering, light modulation, IR spectroscopy, and β-ray absorption are some of the techniques for asbestos analysis. Industrially processed asbestos fibers may be characterized by laser microprobe mass analysis (LAMMA). Organic components adsorbed on the asbestos surface at concentrations of around 1000 ppm can also be detected by this means (DeWaele and Adams 1985). DeWaele and coworkers (1986) reported identification of asbestos fibers in human lung tissue and bronchoalveolar lavage fluid by LAMMA techniques. Alterations of the chemical surface caused by fluids in the human lung were characterized.

Radial distribution of x-ray intensities diffracted by asbestos samples subjected to heat treatment can provide information on interatomic distances, coordination numbers, and expelled hydroxyl water (Datta et al. 1987) and thus the structure of asbestos. Bulk asbestos samples may also be identified by thermal analysis.

Chemical Analysis

Metals commonly found in asbestos samples are quantitatively determined by x-ray

fluorescence, atomic absorption, or neutron activation techniques. pH measurement is usually done in chrysotile form.

Gravimetry

Fibers may be estimated gravimetrically by weighing directly or by using a quartz microbalance (Health and Safety Commission 1978).

NIOSH Methods

Asbestos fibers in air may be analyzed by NIOSH (1984) Methods 7402, 7400, and 9000. Method 7402 specifies the analysis using a transmission electron microscope with electron diffraction and energy dispersive x-ray capabilities. Cellulose ester membranes 25 mm in diameter and $0.8-1.2$ μm long and a conductive cassette are used to collect the sample. The counting range in the method is $80-100$ fibers. Nonasbestiform amphiboles may interfere if the individual particles have the aspect ratio $>3:1$. This may be eliminated by quantitative zone axis electron diffraction analysis.

The revised NIOSH (1984) Method 7400 is an optical counting procedure for airborne fibers based on phase-contrast light microscopy. The analytical range is $100-1300$ fibers/mm^2 area. Method 9000 may be used to analyze chrysotile asbestos for bulk samples. The sample is ground in a liquid nitrogen-cooled mill and wet sieved with 2-propanol. The suspension is filtered through a nonfibrous filter, mixed with 2-propanol, agitated, and filtered through a 0.45-μm silver membrane filter. The chrysotile dust is analyzed by x-ray powder diffraction technique.

10.5   DISPOSAL/DESTRUCTION

In accordance with the Federal Toxic Substance Control Act, the EPA is to be reported at least 10 days before any asbestos abatement project that contains $>1\%$ by weight of friable asbestos material in an area of

>3 ft (linear) or 3 ft^2. Flooding, fire, or any other emergency projects are to be reported 48 hours after the project begins (U.S. EPA 1986).

Asbestos wastes may be solidified prior to their landfill burial. This may be achieved by a cementing process such as that using pozzolanic concrete, which contains fly ash or kiln dust mixed with lime, water, and additives (Peters and Peters 1980). Other processes for solidification include thermoplastic and polymeric processes. In the former, a binder such as paraffin, polyethylene, or bitumen is used. In the latter, polyester, polybutadiene, or polyvinyl chloride is used to trap the asbestos fibers or particles over a spongy polymeric matrix. The solidified waste should be disposed of in a licensed hazardous waste dump or disposal site.

Heasman and Baldwin (1986) reported complete destruction of chrysotile using waste acids, nitric/chromic acid, or waste sulfuric acid. The fiber-to-fluid ratio was 1 : 10, and the destruction was achieved at room temperature in 10 hours. Nitric/chromic acid was most effective, removing >80% Mg; and the presence of metal cations increased this leaching further.

REFERENCES

ACGIH. 1986. *Documentation of the Threshold Limit Values and Biological Exposure Indices*, 5th ed. Cincinnati, OH: American Conference of Governmental Industrial Hygienists.

ASTM. 1982. *Definitions for Asbestos and Other Health-Related Silicates*, ed. B. Levadie, ASTM Spec. Tech. Publ. 834. Philadelphia: American Society for Testing and Materials.

Datta, A. K., B. K. Mathur, B. K. Samantaray, and S. Bhattacharjee. 1987. Dehydration and phase transformation in chrysotile asbestos: a radial distribution analysis study. *Bull. Mater. Sci. 9*(2); 103–10.

Delahunty, T., and D. Hollander. 1987. Toxic effect on the rat small intestine of chronic administration of asbestos in drinking water. *Toxicol. Lett. 39*(2–3): 205–9.

DeWaele, J., and F. Adams. 1985. Study of asbestos by laser microprobe mass analysis (LAMMA). *NATO ASI Ser. B119:* 273–75; cited in *Chem. Abstr. CA 103*(18): 153028k.

DeWaele, J. K., F. C. Adams, P. A. Vermeire, and M. Neuberger. 1986. Laser microprobe mass analysis for the identification of asbestos fibers in lung tissue and broncho-alveolar washing fluid. *Microchim. Acta 3*(3–4): 197–213.

Fisher, G. L., K. L. McNeill, B. T. Mossman, J. Marsh, A. R. McFarland, and W. R. Hart. 1985. Investigations into the mechanisms of asbestos toxicity. *NATO ASI Ser. G3:* 31–38; cited in *Chem. Abstr. CA 104*(15): 124419d.

Fisher, G. L., B. T. Mossman, A. R. McFarland, and R. W. Hart. 1987. A possible mechanism of chrysotile asbestos toxicity. *Drug Chem. Toxicol. 10*(1–2): 109–31; cited in *Chem. Abstr. CA 107*(35): 230901b.

Goodlick, L. A., and A. B. Kane. 1986. Role of reactive oxygen metabolites in crocidolite asbestos toxicity to mouse macrophages. *Cancer Res. 46*(11): 5558–66.

Harrison, P. T. C., and J. C. Heath. 1986. Apparent synergy in lung carcinogenesis: interactions between *N*-nitrosoheptamethyleneimine, particulate cadmium and crocidolite asbestos fibers in rats. *Carcinogenesis 7*(11): 1903–8.

Health and Safety Commission, UK. 1978. *Measurement and Monitoring of Asbestos in Air*, Second report by the advisory committee on Asbestos. London: Crown Publishers.

Heasman, L., and G. Baldwin. 1986. The destruction of chrysotile asbestos using waste acids. *Waste Manage. Res. 4*(2): 215–23; cited in *Chem. Abstr. CA 105*(6): 48387v.

Hesterberg, T. W., and C. J. Barret. 1985. Induction by asbestos fibers of anaphase abnormalities: mechanism for aneuploidy induction and possibly carcinogenesis. *Carcinogenesis 6*(3): 473–75.

Jaurand, M. C., J. Fleury, G. Monchaux, M. Nebut, and J. Bignon. 1987. Pleural carcinogenic potency of mineral fibers (asbestos, attapulgite) and their cytotoxicity on cultured cells. *J. Natl. Cancer Inst. 79*(4):797–804.

Minardi, F., and C. Maltoni. 1988. Results of recent experimental research on the carcinogenicity of natural and modified asbestos. *Ann. N.Y. Acad. Sci. 534:* 754–61.

Mossman, B. T., and J. P. Marsh. 1989. Evidence supporting a role for active oxygen species in asbestos-induced toxicity and lung disease. *Environ. Health Perspect. 81:* 91–94.

NIOSH. 1984. *Manual of Analytical Methods,* 3rd ed. Cincinnati, OH: National Institute for Occupational Safety and Health.

NTP. 1985. Toxicology and carcinogenesis studies of chrysotile asbestos in F344/N rats (feed studies). *National Toxicology Program, Report NTP-TR-295; Govt. Rep. Announce. Index (U.S.) 1986 86* (12), Abstr. No. 625, p. 932.

NTP. 1988. Toxicology and carcinogenesis studies of crocidolite asbestos in F344/N rats (feed studies). *National Toxicology Program, Report NTP-TR-280; Govt. Rep. Announce. Index (U.S.) 89* (15), Abstr. No. 942, p. 454.

Patel-Mandlik, K. J., C. G. Manos, and D. J. Lisk. 1988. Analysis of municipal refuse incinerator ashes for asbestos. *Bull. Environ. Contam. Toxicol. 41* (6): 844–46.

Peters, G. A., and B. J. Peters. 1980. *Sourcebook on Asbestos Diseases.* New York: Garland STPM Press.

Proctor, N. H., and J. P. Hughes. 1978. *Chemical Hazards of the Workplace*, Philadelphia: J. B. Lippincott.

Sherman, L. R., K. T. Roberson, and T. C. Thomas. 1989. Qualitative analysis of asbestos fibers in air, water and bulk samples by scanning electron microscopy-energy dispersive x-ray analysis. *J. Pa. Acad. Sci. 63* (1): 28–33; cited in *Chem. Abstr. CA 113* (14): 125507d.

Steel, E. B., and J. A. Small. 1985. Accuracy of transmission electron microscopy for the analysis of asbestos in ambient environments. *Anal. Chem. 57* (1): 209–13.

Streib, W. C. 1978. Asbestos. In *Kirk-Othmer Encyclopedia of Chemical Technology*, 3rd ed., pp 267–83. New York: Wiley-Interscience.

U.S. EPA. 1986. Toxic substances; asbestos abatement projects. *Fed. Register* Apr. 25, *51* (80): 15722–33.

11

AZO DYES

11.1 GENERAL DISCUSSION

Azo dyes are an important class of synthetic dyes characterized by one or more azo groups $(-N=N-)$ attached to two aromatic rings. There are well over 2000 azo dyes synthesized to date. These are prepared by two basic reactions: diazotization and coupling:

$$R-NH_2 + 2HX + NaNO_2$$
$$\longrightarrow RN_2^+X^- + NaX + 2H_2O$$
$$\text{(diazonium salt)}$$

where $X = Cl^-, Br^-, NO_3^-, HSO_4^-, BF_4^-$, and so on, and R = aryl group.

In coupling reactions, the diazo compound formed above is reacted with the coupling components, such as aromatic amines, phenols, and naphthols. The reaction is as follows:

$$R-N_2X + R'H \longrightarrow R-N=N-R' + HX$$
$$\text{(azo dye)}$$

All coupling components have an active hydrogen atom bound to a carbon atom. Other coupling groups include carboxylic acids, compounds with active methylene group, and heterocyclic compounds such as pyrrole and indole. A detailed discussion is presented by Catino and Farris (1980).

Azo dyes are classified according to a '*color index*' system based on usage and chemical constituents. These are subdivided into monoazo, disazo, trisazo, and polyazo derivatives with a specific assigned range of color index number. Another classification system involves dividing dyes into dyeing classes such as acid, basic, disperse, direct, mordant, and reactive dyes. Azo compounds are used extensively as dyes to color varnishes, paper, fabrics, inks, paints, plastics, and cosmetics. They are used in color photography.

11.2 HEALTH HAZARD

The health hazards from azo dyes are presented in Table 11.1. More detailed toxicological data may be found in RTECS (NIOSH 1986). Generally, azo dyes exhibit low toxicity on test animals. Human toxicological data are scant. Some dyes exhibited mutagenic, carcinogenic, and teratogenic actions on animals. Only a few are human carcinogens. The cleaved coupling components such as phenols or aromatic amines may be toxic. These can be formed when the dyes break down under specific reaction conditions.

TABLE 11.1 Toxicity of Azo Dyes

Compound/Synonyms/CAS No.	Formula/MW/Structure	Toxicity
Acid yellow 3 [2-(2-quinolyl)-1,3-indanedione disulfonic acid disodium salt, dye quinoline yellow, Japan yellow 203, CI 47005, CI food yellow 13] [8004-92-0]	$C_{18}H_9NO_8S_2 \cdot 2Na$ 477.38 Acid dye of quinoline type	Low toxicity; LD_{50} oral (rats): 2000 mg/kg
Acid yellow 7 [2,3-dihydro-6-amino-1,3-dioxo-2-(p-tolyl)-1H-benz[d,e]isoquinoline-5-sulfonic acid monosodium salt, brilliant sulfaflavin, CI 56205] [2391-30-2]	$C_{19}H_{13}N_2O_5S \cdot Na$ 404.39 Acid dye of amino ketone type	Low toxicity; LD_{50} intravenous (mice): 110 mg/kg
Acid yellow 23 [5-hydroxy-1-(p-sulfophenyl)-4-(p-sulfophenyl)azopyrazole-3-carboxylic acid trisodium salt, trisodium 3-carboxy-5-hydroxy-1-sulfophenyl-4-sulfophenyl-azopyrazole, tartrazine, tartraphenine, tartrazol yellow, wool yellow, CI 19140] [1934-21-0]	$C_{16}H_9H_4O_9S_2 \cdot 3Na$ 534.38	Mutagenic; carcinogenicity unknown; caused birth defect and developmental abnormalities in blood and lymphatic systems in rats; low toxicity in animals; in humans, ingestion may cause paresthesias and affect the teeth and supporting structure; LD_{50} oral (mice): 12,750 mg/kg.
Acid yellow 42[4,4'-bis((4,5-dihydro-3-methyl-5-oxo-1-phenyl-1H-parazol-4-yl)azo)(1,1'-biphenyl)-2,2'-disulfonic acid disodium salt, acid yellow K, CI 22910] [6375-55-9]	$C_{32}H_{26}N_8O_8S_2 \cdot 2Na$ 760–76	Mutagenic; carcinogenicity not reported; nontoxic
Fast oil yellow 2G [(p-phenylazo)phenol, p-hydroxyazobenzene, CI solvent yellow 7, CI 1800] [1689-82-3]	$C_{12}H_{10}N_2$ 198.24	Low to moderate toxicity on mice; LD_{50} intraperitoneal (mice): 75 mg/kg; caused tumor in test animals, the evidence of its carcinogenicity inadequate
Azoxybenzene (azobenzene oxide) [495-48-7]	$C_{12}H_{10}N_2O$ 214.24	Skin irritation mild on rabbits; mutagenic; low toxicity on test animals, LD_{50} oral (rats): 620 mg/kg

(continued)

TABLE 11.1 (*Continued*)

Compound/Synonyms/CAS No.	Formula/MW/Structure	Toxicity
Metanil yellow [acidic metanil yellow, acid yellow 36, 11363 yellow, acid leather yellow R, CI 13065, 3-(4-phenylazo) benzene-sulfonic acid monosodium salt] [587-98-4]	$C_{18}H_{15}N_3O_3S \cdot Na$ 376.41	Mutagenic; carcinogenicity not reported; reproductive effect of paternal type observed in mice; low toxicity in mice, LD_{50} intraperitoneal (mice): 1000 mg/kg; toxic symptoms in *Heteropneustes fossilis* were loss of body weight, restlessness, jerky and random movement, etc. (Goel and Gupta 1985); an oral LD_{50} value of 6500 mg/kg reported for a blend of metanil yellow and orange II in male rats (Singh et al. 1988)
Dye sunset yellow [6-hydroxy-5-(*p*-sulfophenyl)azo)-2-naphthalene-sulfonic acid disodium salt, 1-*p*-sulfophenylazo-2-hydroxynaphthalene-6-sulfonate disodium salt, yellow no. 6, CI 15985] [2783-94-0]	$C_{16}H_{10}N_2O_7S_2 \cdot 2Na$ 452.38	Low toxicity in rat via intraperitoneal route, high dosages may cause symptoms of gastrointestinal pain, diarrhea, convulsions, and coma; LD_{50} intraperitoneal (rats): 3800 mg/kg
Acid orange 7 [4-((2-hydroxy-1-naphthalenyl) azo) benzenesulfonic acid monosodium salt, solar orange, mandarin G, CI 15510] [633-96-5]	$C_{16}H_{12}N_2O_4S \cdot Na$ 351.35	Oral and intratesticular administration in rats indicated teratogenic nature of this compound with paternal effect; no adverse effect in humans reported; nontoxic, nonmutagenic
Acid orange 10 [7-hydroxy-8-(phenylazo)-1,3-naphthalenedisulfonic acid disodium salt, 1-phenylazo-2-naphthol-6,8-disulfonic acid disodium salt, acid-fast orange G, CI 16230] [1936-15-8]	$C_{16}H_{10}N_2O_7S_2 \cdot 2Na$ 452.38	Mutagenic; evidence of carcinogenicity in animals inadequate; oral administration in pigs indicated teratogenic nature with paternal effect; nontoxic
Sudan orange R [1-(phenylazo)-2-naphthol, benzene-1-azo-2-naphthol, CI solvent yellow 14, oil orange, CI 12055] [842-07-0]	$C_{16}H_{12}N_2O$ 248.30	Carcinogenic to animals, caused kidney, bladder, and liver cancer in laboratory test animals; no evidence of carcinogenicity in humans

Name	Formula	Toxicity/Effects
Oil orange TX [o-tolueno-azo-β-naphthol, 1-(o-tolylazo)-2-naphthol; lacquer orange V, CI solvent orange 2, CI 12100] [2646-17-5]	$C_{17}H_{14}N_2O$ 262.33	Caused tumor in kidney, bladder, and gastrointestinal tract in mice; no evidence of cancer-causing effect in humans; low to moderate toxicity on animals — poisoning effect varying with species; 60- and 200 mg/kg doses were lethal to rabbits and dogs by intravenous route
Acid red 18 [7-hydroxy-8-(4-sulfo-1-naphthyl) azo-1,3-naphthalene disulfonic acid trisodium salt, SX purple, coccine, CI 16255] [2611-82-7]	$C_{20}H_{14}N_2O_{10}S_3 \cdot$ Na 607.51	Mutagenic; caused liver cancer in rats; test on rodents indicate mild to moderate toxicity, occasionally associated with symptoms of somnolence, convulsion, and coma; oral administration of 8 g/kg was lethal to mice, LD_{50} intravenous (rats): 100 mg/kg; toxic effects in human unknown
Acid red 14 [4-hydroxy-3,4-azodi-1-naphthalene sulfonic acid disodium salt; disodium 2-(4-sulfo-1-naphthyl-azo)-1-naphthol-4-sulfonate, chromotrope FB, azorubin, CI 14720] [3567-69-9]	$C_{20}H_{12}N_2O_7S_2 \cdot$ 2Na 502.44	Low toxicity in rodents with symptoms of somnolence and convulsion; large amounts can cause coma; toxicity in humans not reported; LD_{50} intraperitoneal (mice): 900 mg/kg; carcinogenic action in animals not fully established
Acid red 26 [3-hydroxy-4-(2,4-xylylazo)-2,7-naphthalenedisulfonic acid disodium salt, acid scarlet 2B, xylidine red, CI 16150] [3761-53-3]	$C_{18}H_{14}N_2O_7S_2 \cdot$ 2Na 480.44	Mutagenic; caused liver and gastrointestinal cancer in mouse; carcinogenicity in humans not reported; low toxicity in mice — symptom of sleepiness; LD_{50} intraperitoneal (mice): 2000 mg/kg
AF red 1 [3-hydroxy-4((2,4,5-trimethyl-phenyl)azo)-2,7-naphthalenedisulfonic acid disodium salt, CI food red 6, CI 16155] [6226-79-5]	$C_{19}H_{16}N_2O_7S_2 \cdot$ 2Na 494.47	Laboratory tests on animals showed carcinogenicity in kidney, liver, and lungs; carcinogenicity in humans not reported; toxicity not reported

(continued)

TABLE 11.1 (Continued)

Compound/Synonyms/CAS No.	Formula/MW/Structure	Toxicity
Congo red [3,3′-4,4′-biphenylenebis(azo)bis(4-amino)-1-naphthalenesulfonic acid disodium salt, sodium diphenyldizao bis(α-naphthylamine sulfonate), benzo Congo red, direct red 28, CI 22120] [573-58-0]	$C_{32}H_{24}N_6O_6S_2 \cdot 2Na$ 698.72	Mutagenic; carcinogenicity not reported; strong teratogenic action in rats, causing harmful effects on fertility and embryo, as well as specific developmental abnormalities in urogenital system, central nervous system, eyes, and ear; moderately toxic; can cause respiratory congestion and somnolence; ingestion of 8–12 g can be fatal to humans
CI reactive red 2 [5-((4,6-dichloro-s-triazin-2-yl)amino)-4-hydroxy-3-(phenylazo)2,7-naphtahlenedisulfonic acid disodium salt, ostazin brilliant red S 5B] [17804-49-8]	$C_{19}H_{10}Cl_2N_6O_7S_2 \cdot 2Na$ 615.35	Laboratory tests on animals indicate low toxicity with symptoms of excitement and aggressiveness; LD_{50} oral (mice): 5700 mg/kg
Sudan red [1-(2,4-xylylazo)-2-naphthol, 1-xylylazo-2-naphthol, oil scarlet L, CI solvent orange 7, CI 12140] [3118-97-6]	$C_{18}H_{16}N_2O$ 276.36	Carcinogenic to animals; caused tumor in kidney and bladder in mice, no evidence of carcinogenicity in humans
Acid red 37 [3-hydroxy-4-4-((4-sulfo-1-naphthyl)azo)-2,7-napthalenedisulfonic acid trisodium salt, acid amaranth 1, CI 16185], [915-67-3]	$C_{20}H_{11}N_2O_{10}S_3 \cdot 3Na$ 604.48	Caused blood and gastrointestinal tumors in rats; however, the evidence of carcinogenicity in animals is inadequate; carcinogenicity in humans not reported; harmful effect on fertility and embryo; developmental abnormalities and menstrual cycle change in rats; low toxicity — symptom of sleepiness in mice LD_{50} intraperitoneal (mice) 1000 mg/kg
Acid blue 9 [ammonium ethyl(4-(p-(ethyl(m-sulfobenzyl)amino)-α-(o-sulfophenyl)benzylidene)-2,5-cyclohexadiene-1-ylidene)(m-sulfobenzyl)hydroxide diammonium salt, α-azurine, neptune blue bra, CI 42090] [2650-18-2]	$C_{37}H_{36}N_2O_9S_3 \cdot 2NH_4$ 783.01	Mutagenic; subcutaneous application in rats caused tumors at sites of application carcinogenicity in humans not reported; no animal data on toxicity available; toxicity in humans may be moderate to severe, with symptoms of muscle contraction and shortness of breath; 2–4 mg via intravenous route may be lethal to humans
Acid blue 92 [4-((4-anilino-5-sulfo-1-naphthyl)azo-5-hydroxy-2,7-naphthalenedisulfonic acid trisodium salt, trisodium 4′-anilino-8-hydroxy-1,1′-azonaphthalene-3,6,5-trisulfonate, benzyl blue R, CI 13390] [3861-73-2]	$C_{26}H_{19}N_3O_{10}S_3 \cdot 3Na$ 698.63	Mutagenic; carcinogenicity in animals not reported; low toxicity — an intravenous administration of 450 mg/kg proved lethal to mice

Name	Formula / MW	Toxicity
Amanil blue 2BX[3,3'-((4,4'-biphenylylene)bis(azo)bis(5-amino-4-hydroxy-2,7-naphthalene disulfonic acid tetrasodium salt, sodium diphenyl-4,4'bis-azo-2',8'-amino-1'-naphthol-3',6'-disulfonate, benzanil blue 2B, paramine blue 2B, CI 22610] [2602-46-2]	$C_{32}H_{20}N_6O_{14}S_4 \cdot 4Na$ 932.78	Mutagenic; caused cancers in kidney, bladder, and liver in rodents, human carcinogenicity not reported; teratomer, caused birth defects, developmental abnormalities in eyes, ears, and central nervous system and embryo death in rats; toxicity data not available
Amanil sky blue [3,3'-((3,3'-dimethoxy-4,4'-biphenylene)bis(5-amino-4-hydroxy)-2,7-naphthalenedisulfonic acid tetrasodium salt; benzanil sky blue; direct blue 5; CI 24400] [2429-74-5]	$C_{34}H_{28}N_6O_{16}S_4 \cdot 4Na$ 996.88	Mutagenic; carcinogenicity not reported; teratogen, caused developmental abnormalities in eyes, ear, and central nervous system and showed harmful effects on embryo and fertility in rats
Amanil sky blue R [3,3'-((3,3'-dimethyl-4,4'-biphenylene)bis(azo))bis(5-amino-4-hydroxy)-2,7-naphthalenedisulfonic acid, amidine blue 4B, benzanil blue, Congo blue, direct blue 14; CI 23850] [72-57-1]	$C_{34}H_{28}N_6O_{14}S_4 \cdot 4Na$ 964.88	Mutagenic; no report of carcinogenicity; strong teratomer, causing harmful effects on fertility and embryo, and specific developmental abnormalities in central nervous system, eyes, ear, and cardiovascular and gastrointestinal systems in rats, guinea pigs, hamsters, and pigs; low toxicity— a dose of 300 mg/kg by intravenous or subcutaneous route was lethal to mice and guinea pigs
Acid green 3 [ammonium ethyl 4-(p-ethyl(m-sulfobenzyl)amino)-α-phenylbenzyidene)2,5-cyclohexadien-1-ylidene)(m-sulfobenzyl)hydroxide sodium salt, acid green B, CI 42085] [4680-78-8]	$C_{37}H_{36}N_2O_6S_2 \cdot Na$ 691.86	Mutagenic; produced blood cancer in laboratory rats; carcinogenicity in humans not reported
Acid green 25 [2,2'-(1,4-anthraquinonyl-enediimino)bis(5-methylbenzenesulfonic acid, CI 61570] [4403-90-1]	$C_{28}H_{20}N_2O_8S_2 \cdot 2Na$ 622.60	Mild irritant to skin and eyes, 500 mg/day led to mild skin and eye irritation in rabbit; toxicity not reported

(*continued*)

TABLE 11.1 *(Continued)*

Compound/Synonyms/CAS No.	Formula/MW/Structure	Toxicity
Acid green 5 [ammonium ethyl(4-(*p*-ethyl (*m*-sulfobenzyl)amino-α-(*p*-sulfophenyl) benzylidene)-2,5-cyclohexadien-1-ylidene)(*m*-sulfobenzyl) hydroxide disodium salt, acid brilliant green SF, CI 42095] [5141-20-8]	$C_{37}H_{36}N_2O_9S_3 \cdot 2Na$ 794.91	Animal studies indicate that this compound can cause cancer in blood and gastrointestinal tract; low toxicity observed in rodents; LD_{50} intravenous (mice): 700 mg/kg
Direct brown 95 [(5-((4'-((2,5-dihydroxy-4-((2-hydroxy-5-sulfophenyl)azo phenyl)azo)(1,1'-biphenyl)-4-yl)azo-2-hydroxybenzoato(2-)) copper disodium salt, amanil fast brown brilliant, benzamil supra brown brilliant, CI 30145] [16071-86-6]	$C_{31}H_{20}N_6O_9S \cdot Cu \cdot 2Na$ 762.15	May cause liver cancer; may be carcinogenic to humans
Direct black 38 [4-amino-3-((4'-((2,4-diamino-phenyl)azo)(1,1'-biphenyl)-4-yl)azo)-5-hydroxy-6-(phenylazo)-2,7-naphthalene-disulfonic acid disodium salt, amanil black GL, benzo deep black E, chlorazol black E, CI 30235] [1937-37-7]	$C_{34}H_{25}N_9O_7S_2 \cdot 2Na$ 781.78	Sufficient evidence of its carcinogenicity in rodents; can cause tumor in kidney, bladder, and liver; may cause cancer in humans
Azoic diazo component 11 (4-chloro-2-toluidine hydrochloride, 2-methyl-4-chloro-aniline hydrochloride, 5-chloro-2-amino toluene hydrochloride; 4-chloro-2-methyl benzeneamine hydrochloride, azanil red salt TRD, fast red salt TRA, CI 37085) [316593-3]	$C_7H_8ClN \cdot HCl$ 178.07	EPA Classified Toxic Waste, RCRA Waste Number U049; DOT Label: Poison UN 1579; can cause liver and vascular tumors in mice; may cause cancer in humans; low toxicity—LD_{50} intraperitoneal (rats): 560 mg/kg
Azoic diazo component 49 (2-chloro-5-trifluoromethyl) aniline, 2-amino-4-chloro-α,α,α-trifluorotoluene, 3-amino-4-chloro-benzotrifluoride, CI 37050) [121-50-6]	$C_7H_5ClF_3N$ 195.58	Low toxicity in mice, LD_{50} intraperitoneal (mice): 100 mg/kg

11.3 ANALYSIS

Identifications and analyses of azo dyes can be performed by several techniques. These are absorption spectroscopy (i.e., visible, UV, and IR spectroscopy), fluorescence analysis, and GC. Chemical testings involve reduction methods. Reducing agents such as zinc dust, stannous chloride, and sodium hydrosulfite can split the azo linkage, producing the diazonium and the coupling fragments. The cleaved products may be identified by paper chromatography. Detailed analytical methods are well documented (Rounds 1968).

Benzidine-based dyes such as Congo Red, Amanil Blue 2BX, Direct Brown 95, Direct Black 38, and *o*-tolidine- and *o*-dianisidine-based dyes in air may be analyzed by NIOSH Method 5013 (NIOSH 1984). The method involves passing 120–500 L of air through a 5-mm PTFE membrane filter, desorping the compounds with water, reducing with sodium hydrosulfite in aqueous phosphate buffer to free benzidine, *o*-tolidine, and *o*-dianisidine from the dyes. The cleaved aromatic amines are analyzed by HPLC technique using a UV detector set at 280 nm. Colorimetric methods may be applied for analysis of dyes.

11.4 DISPOSAL/DESTRUCTION

Azo dyes are disposed of by burning in a chemical incinerator equipped with an after-burner and scrubber. Aerobic biodegration tests by Pagga and Brown (1986) indicated that the dyes did not biodegrade significantly.

REFERENCES

Catino, S. C., and R. E. Farris. 1980. Azo dyes. In *Kirk-Othmer Encyclopedia of Chemical Technology*, Vol. 3, pp. 387–433. New York: John Wiley & Sons.

Goel, K. A., and K. Gupta. 1985. Acute toxicity of metanil yellow to *Heteropneustes fossilis. Indian J. Environ. Health* 27(3): 266–69.

NIOSH. 1984. *Manual of Analytical Methods*, 3rd ed. Cincinnati, OH: National Institute for Occupational Safety and Health.

NIOSH. 1986. *Registry of Toxic Effects of Chemical Substances*, ed. D. V. Sweet. Washington, DC: U.S. Government Printing Office.

Pagga, U., and D. Brown. 1986. The degradation of dyes. II. Behavior of dyes in aerobic biodegration tests. *Chemosphere* 15(4): 479–91.

Rounds, R. L. 1968. Azo dyes. In *Encyclopedia of Industrial Chemical Analysis*, ed. F. D. Snell. New York: Wiley.

Singh, R. L., S. K. Khanna, and G. B. Singh. 1988. Acute and short-term toxicity of a popular blend of metanil yellow and orange II in albino rats. *Indian J. Exp. Biol.* 26(2): 105–11.

12

CHLOROHYDRINS

12.1 GENERAL DISCUSSION

Chlorohydrins are a class of organic compounds containing one or more hydroxyl groups and one or more chlorine atoms. The structures of these compounds are as follows:

$$Cl—CH_2—CH_2—OH$$

(ethylene chlorohydrin)

$$Cl—CH_2—CH_2—CH_2—OH$$

(trimethylene chlorohydrin)

$$Cl—CH_2—\underset{\underset{OH}{|}}{CH}—CH_2—OH$$

(α,γ-dichlorohydrin)

The reactions of epichlorohydrin are similar to those of glycerol dichlorohydrins (Riesser 1979). Epichlorohydrin is often classified as a chlorohydrin. However, as it is a chloroether with an epoxide structure, it is discussed in Chapter 16.

Among the industrially important chlorohydrins are ethylene, propylene, and glycerol chlorohydrins. The methods of preparing chlorohydrins are by the addition of hypochlorous acid to olefins or by reacting HCl with epoxides or glycols.

Ethylene chlorohydrin is by far the most hazardous chlorohydrin. The higher compounds are less hazardous. The latter can cause depression of the central nervous system to varying degrees. Because of the presence of two substitution sites in the molecule, —OH and —Cl, chlorohydrins may undergo many reactions, which may be either hazardous or may yield toxic products. The toxic effects of chlorohydrins are different from their parent alcohols.

Fluoro- and bromohydrins are similar to chlorohydrins and their hazardous properties are highlighted in the foregoing section.

12.2 ETHYLENE CHLOROHYDRIN

DOT Poison B and Flammable Liquid, UN 1135

Formula C_2H_5OCl; MW 80.52; CAS [107-07-3]

Structure and functional group: $Cl—CH_2$ $—CH_2—OH$, primary —OH and a chlorosubstituted alkyl group

Synonyms: 2-chloroethanol; 2-chloroethyl alcohol; glycol monochlorohydrin; ethylene glycol chlorohydrin

Uses and Exposure Risk

Ethylene chlorohydrin is used in the manufacture of insecticides, as a solvent for cellulose esters, in treating sweet potatoes before planting (Merck 1989), and in making ethylene glycol and ethylene oxide. Exposure risks to this compound may arise when it is formed from ethylene oxide during the sterilization of grain and spices, drugs, and surgical supplies.

Physical Properties

Colorless liquid with a faint smell of ether; bp 129°C; mp −67°C; density 1.197 at 20°C; soluble in water, alcohol, and ether.

Health Hazard

Ethylene chlorohydrin is a severe acute poison. The target organs are central nervous system, cardiovascular system, kidney, liver, and gastrointestinal system. The symptoms of acute toxicity are respiratory distress, paralysis, brain damage, nausea, and vomiting. In addition, ethylene chlorohydrin can cause glutathione depletion in liver and formation of polyuria in kidney.

Ingestion of 20–25 mL can be fatal to humans. Death may occur in 48 hours. It may be more toxic by skin contact than by oral intake. Inhalation of its vapors can produce nausea, vomiting, headache, chest pain, and stupor (ACGIH 1986). At high concentrations, death may occur. Exposure to 300 ppm for 2 hours can be fatal to humans.

LD$_{50}$ value, oral (mice): 81 mg/kg

LD$_{50}$ value, intraperitoneal (mice): 97 mg/kg

Ethylene chlorohydrin is an irritant to the skin, eyes, nose, and mucous membranes. It is a confirmed mutagen by the Ames test in *Salmonella typhimurium*. It inhibits the growth of DNA-deficient bacteria. Exposure to this compound increased the chromosome aberration in the bone marrow of rats. The odor threshold is 0.4 ppm.

Exposure Limits

Ceiling limit 3 mg/m^3 (1 ppm) skin (ACGIH); TLV-TWA air 16 mg/m^3 (5 ppm) skin (OSHA); IDLH 10 ppm (NIOSH).

Fire and Explosion Hazard

Combustible; flash point (closed cup) 60°C (140°F), (open cup) 40°C (105°F); vapor pressure 5 torr at 20°C; vapor density 2.8 (air = 1); fire-extinguishing agent: "alcohol" foam; use a water spray to dilute the spill and keep fire-exposed containers cool (NFPA 1986).

Ethylene chlorohydrin vapors form an explosive mixture with air; the LEL and UEL values are 4.9% and 15.9% by volume of air, respectively. Among the hazardous reaction products are ethylene oxide, formed by internal displacement of the chlorine atom by the alkoxide ion, ethylene glycol formed by hydrolysis with sodium bicarbonate at 105°C, and ethylene cyanohydrin resulting from the reaction with alkali metal cyanides.

Disposal/Destruction

Ethylene chlorohydrin is mixed with a combustible solvent and burned in a chemical incinerator equipped with an afterburner and scrubber.

Analysis

Ethylene chlorohydrin can be analyzed by GC using FID. Air analysis involves sampling air over petroleum charcoal, desorbing the analyte with 2-propanol/CS2 mix (5% v/v), followed by GC analysis using the column FFAP on Chromosorb W-HP (NIOSH 1984, Suppl. 1985, Method 2513).

12.3 PROPYLENE β-CHLOROHYDRIN

DOT Label: Poison B and Flammable Liquid, UN 2611

Formula C$_3$H$_7$OCl; MW 94.54; CAS [78-89-7]

Structure and functional group:

$$H_3C-\underset{\underset{Cl}{|}}{CH}-CH_2-OH$$

reactions characteristic of primary $-OH$ and alkyl chlorides

Synonyms: 2-chloro-1-propanol; 2-chloropropyl alcohol

Uses and Exposure Risk

Propylene β-chlorohydrin is used in the production of propylene oxide.

Physical Properties

Colorless liquid with a pleasant smell; bp 133–134°C; density 1.103 at 20°C; soluble in water and alcohol and other organic solvents.

Health Hazard

Propylene β-chlorohydrin is a moderately toxic compound having irritant actions on the skin and eyes. Instilling 2.2 mg of liquid caused severe eye irritation in rabbits. It can cause poisoning via ingestion, inhalation, and skin absorption. Exposure to 500 ppm for 4 hours was lethal to rats. The lethal dose in dogs by oral intake was 200 mg/kg. The target organs are the central nervous system, gastrointestinal system, liver, and kidney.

LD_{50} value, skin (rabbits): 529 mg/kg

Exposure Limit

No exposure limit is set. However, based on its similar toxic properties to ethylene chlorohydrin, a TLV-TWA air 10 ppm (approximately 40 mg/m^3) is recommended.

Fire and Explosion Hazard

Combustible liquid; flash point (closed cup) 44°C (112°F); vapor density 3.2 (air = 1);

fire-extinguishing agent: "alcohol" foam; use a water spray to keep fire-exposed containers cool and to flush and dilute the spill.

The vapor may form an explosive mixture with air; the explosion limits are not reported. Dehydrochlorination in the presence of a catalyst can produce propylene oxide.

Analysis

GC/FID is the most common method of analysis. The column FFAP on Chromosorb W-HP is suitable. GC/MS may be used as an alternative method of analysis.

12.4 TRIMETHYLENE CHLOROHYDRIN

DOT Label: Poison B, UN 2849

Formula C_3H_7OCl; MW 94.55; CAS [627-30-5]

Structure and functional group: $Cl-CH_2$ $-CH_2-CH_2-OH$, characteristic reactions of primary $-OH$, Cl function undergoes the reactions of alkyl chlorides

Synonyms: 3-chloro-1-propanol; 3-chloropropyl alcohol; 1-chloro-3-hydroxypropane

Uses and Exposure Risk

Trimethylene chlorohydrin is used in organic synthesis to produce cyclopropane and trimethylene oxide.

Physical Properties

Colorless oily liquid with agreeable odor; bp 165°C; density 1.1318 at 20°C; soluble in water, alcohol, and other organic solvents.

Health Hazard

Trimethylene chlorohydrin is a moderately toxic compound. Ingestion can result in central nervous system depression, muscle contraction, gastrointestinal pain, and ulceration or stomach bleeding. High dosage can cause liver injury. The toxicity of this compound, however, is less severe than that of its isomer, propylene β-chlorohydrin.

LD$_{50}$ value, oral (mice): 2300 mg/kg

Skin contact can cause irritation.

Exposure Limit

No exposure limit is set. A TLV-TWA air value of 100 ppm (approximately 400 mg/m^3) is suggested for this compound.

Fire and Explosion Hazard

Combustible; flash point 73°C (164°F) (Aldrich 1989); vapor density 3.2 (air = 1). The vapor may form an explosive mixture with air; the LEL and UEL values are not reported. It decomposes during distillation at ordinary pressure, liberating HCl.

Disposal/Destruction

Trimethylene chlorohydrin is burned in a chemical incinerator equipped with an afterburner and scrubber.

Analysis

Trimethylene chlorohydrin can be analyzed by GC-FID.

12.5 TETRAMETHYLENE CHLOROHYDRIN

Formula C$_4$H$_9$OCl; MW 108.58; CAS [928-51-8]

Structure and functional group: Cl—CH$_2$ —CH$_2$—CH$_2$—CH$_2$—OH, exhibits reactivity of primary alcohol and alkyl halides

Synonyms: 4-chloro-1-butanol; 4-chlorobutyl alcohol

Uses and Exposure Risk

Tetramethylene chlorohydrin is used in organic synthesis.

Physical Properties

Colorless, oily liquid, bp 84.5°C at 16 torr; density 1.0083 at 20°C; soluble in water and in organic solvents.

Health Hazard

The toxicity of this compound is low. However, the acute toxic symptoms are those of ethylene and propylene chlorohydrins. Oral intake of this compound caused muscle contraction, gastrointestinal pain, ulceration, and liver injury in test animals.

LD$_{50}$ value, oral (mice): 990 mg/kg

Tetramethylene chlorohydrin caused tumors in lungs in test animals. Its carcinogenicity, however, is not yet fully established.

Fire and Explosion Hazard

Flammable; flash point 36°C (97°F); vapor density 3.7 (air = 1); the vapor forms an explosive mixture with air, range is not reported. Fire-extinguishing agent: "alcohol" foam; a water spray may be used to cool fire-exposed containers and to flush any spill. It decomposes to HCl and tetrahydrofuran on heating.

Disposal/Destruction

Tetramethylene chlorohydrin is burned in a chemical incinerator equipped with an afterburner and scrubber.

Analysis

Tetramethylene chlorohydrin can be analyzed by GC-FID using a suitable column.

12.6 GLYCEROL α-MONOCHLOROHYDRIN

DOT Label: Poison B, UN 2689

Formula C$_3$H$_7$O$_2$Cl; MW 110.55; CAS [96-24-2]

Structure and functional group:

$$Cl-CH_2-\underset{\underset{OH}{|}}{CH}-CH_2-OH$$

two primary alcoholic —OH groups that exhibit the reactions of alcohols; Cl function characterizes alkyl chloride-type reactions

Synonyms: 3-chloro-1,2-propanediol; α-monochlorohydrin; 2,3-dihydroxypropyl chloride; 3-chloro-1,2-dihydroxypropane; 3-chloropropylene glycol; 1-chloro-2,3-propanediol

Uses and Exposure Risk

It is used in the synthesis of glycerol esters, amines, and other derivatives; to lower the freezing point of dynamite; and as a rodent chemosterilant (Merck 1989).

Physical Properties

Colorless and slightly viscous liquid; faint pleasant smell; sweetish taste; bp 213°C; density 1.3204 at 20°C; soluble in water, alcohol, and ether.

Health Hazard

Glycerol α-monochlorohydrin is a highly toxic, teratogenic, and carcinogenic compound. It is also an eye irritant.

Inhalation of 125 ppm in 4 hours was fatal to rats. The lethal dose on mice via intraperitoneal route was 10 mg/kg. Low dosage can cause sleepiness, and on chronic exposure, weight loss.

LD$_{50}$ value, oral (rats): 26 mg/kg

This compound is a strong teratomer, causing severe reproductive effects. Animal studies indicated that the adverse effects were of spermatogenesis type, related to the testes, sperm duct, and Cowper's gland. These were paternal effects.

Studies on rats indicated that high exposure levels to glycerol chlorohydrin can give rise to thyroid tumors.

Fire and Explosion Hazard

Glycerol α-monochlorohydrin is noncombustible; flash point >100°C (212°F). The fire hazard is very low. It can decompose at high temperature, liberating toxic HCl gas.

Disposal/Destruction

Chemical incineration is the most appropriate method of disposal.

Analysis

Glycerol α-monochlorohydrin can be analyzed by GC-FID and GC/MS.

12.7   CHLOROBUTANOL

Formula $C_4H_7OCl_3$; MW 177.46; CAS [57-15-8]

Structure and functional group:

$$H_3C-\underset{\underset{OH}{|}}{\overset{\overset{CH_3}{|}}{C}}-C\overset{\diagup Cl}{\underset{\diagdown Cl}{-Cl}}$$

tertiary —OH group and three Cl atoms

Synonyms: 1,1,1-trichloro-2-methyl-2-propanol; trichloro-*tert*-butyl alcohol; acetone chloroform; methaform; sedaform; chloretone

Uses and Exposure Risk

Chlorobutanol is used as a plasticizer for cellulose esters, as a preservative for hypodermic solutions, and in veterinary medicine as a mild sedative and antiseptic.

Physical Properties

Crystalline solid with a camphor-like odor and taste; sublimes; mp 97°C (hemihydrate

78°C); bp 167°C; soluble in water and organic solvents.

Health Hazard

Chlorobutanol exhibited moderate toxicity in test animals. A lethal dose by oral intake in dogs and rabbits were 238 and 213 mg/kg, respectively (NIOSH 1986). The toxic symptoms are not reported. This compound is a mild skin and eye irritant and may be mutagenic.

Disposal/Destruction

A suitable method of destruction of chlorobutanol is the burning of its solution in a combustible solvent in a chemical incinerator.

Analysis

Chlorobutanol may be analyzed by GC and GC/MS. FID is the suitable detector.

12.8   ETHYLENE BROMOHYDRIN

Formula C_2H_5OBr; MW 124.98; CAS [540-51-2]

Structure and functional group: $Br-CH_2$ $-CH_2-OH$, undergoes reactions characteristic of alkyl bromide and a primary $-OH$ functional group

Synonyms: glycol bromohydrin; 2-bromoethanol

Uses and Exposure Risk

Ethylene bromohydrin is not used much for any commercial purpose. The risk of exposure to this compound arises when ethylene oxide reacts with hydrobromic acid.

Physical Properties

Colorless liquid; hygroscopic; aqueous solution having a sweet burning taste; bp 150°C (decomposes); density 1.763 at 20°C; soluble in water and organic solvents, insoluble in petroleum ether.

Health Hazard

The vapors of ethylene bromohydrin are an irritant to the eyes and mucous membranes. It is corrosive to the skin. Ingestion of this compound can produce moderate to severe toxic effects. The target organs are the central nervous system, gastrointestinal tract, and liver. The lethal dose in mice by the intraperitoneal route was 80 mg/kg.

Ethylene bromohydrin manifested carcinogenicity in test animals. It caused tumors in lungs and the gastrointestinal tract in mice from intraperitoneal (150 mg/kg/8 weeks) and oral (43 mg/kg/80 weeks) dosages, respectively. It is a mutagen, positive to the histidine reversion–Ames test.

Fire and Explosion Hazard

Combustible; flash point 40°C (105°F); vapor density 4.3 (air = 1); vapor can form an explosive mixture with air; the explosive limits are not reported. Fire-extinguishing agent: "alcohol" foam; a water spray may also be effective and should be used to cool fire-exposed containers and to dilute and flush any spill. It decomposes to hydrogen bromide (toxic) when heated above 150°C.

Disposal/Destruction

Ethylene bromohydrin is mixed with a combustible solvent and burned in a chemical incinerator.

Analysis

Ethylene bromohydrin is analyzed by GC-FID and GC/MS.

12.9   MISCELLANEOUS CHLOROHYDRINS

See Table 12.1.

TABLE 12.1 Toxicity and Flammability of Miscellaneous Chlorohydrins

Compound/Synonyms/ CAS No.	Formula/MW/Structure	Toxicity	Flammability
Propylene α-chloro-hydrin (1-Chloro-2-propanol, *sec*-propylene chlorohydrin) [127-00-4]	C_3H_7OCl 94.54 $H_3C-CH-CH_2-Cl$ $\quad\quad\;\; OH$	Low toxicity, LD_{50} value not reported; may be mutagenic; histidine reversion–Ames test inconclusive	Combustible, flash point 51°C (125°F)
Propylene fluorohydrin (fluoropropanol, 1-fluoro-3-propanol) [430-50-2]	C_3H_7OCl 78.10 $H_3C-CH-CH_2-F$ $\quad\quad\;\; OH$	Toxic via inhalation and ingestion; low toxicity; LC_{Lo} (rats): 4000 ppm/4 hr LD_{50} oral (rats): 3260 mg/kg	Combustible liquid, flash point not reported
β-Fluoropropanol [462-43-1]	$F-CH_2-CH_2-CH_2-OH$	Moderate to high toxicity; exposure to 350 mg/m^3/10 min was lethal to guinea pigs; LD_{50} intraperitoneal (mice): 47 mg/kg	Combustible liquid, flash point not reported
Ethylene fluorohydrin (2-fluoroethanol, 3-fluoroethanol) [371-62-0]	C_2H_5OF 64.07 $F-CH_2-CH-CH_3$ $\quad\quad\;\; OH$	Highly toxic; LC_{50} values varying widely with species-conflicting reports; LD_{50} subcutaneous (mice): 15 mg/kg, intraperitoneal (rats): 17.5 mg/kg; toxic effects — CNS depression and convulsion	Flammable liquid, flash point 31°C (88°F); can form explosive mixture with air; fire-extinguishing agent: "alcohol" foam or water spray
Glycerol α,β-dichloro-hydrin (1,3-dichloro-2-propanol, α-dichlo-rohydrin) [96-23-1]	$C_3H_6OCl_2$ 128.99 $Cl-CH_2-CH-CH_2-Cl$ $\quad\quad\quad\; OH$	Skin irritant; moderately toxic via inhalation, ingestion and skin absorption; exposure to 125 ppm/4 hr was lethal to rats; LD_{50} oral (rats): 110 mg/kg; mutagenicity: positive to histidine reversion–Ames test; DOT Label: Poison B UN 2750	Combustible, flash point 85°C (186°F)

TABLE 12.1 *(Continued)*

Compound/Synonyms/ CAS No.	Formula/MW/Structure	Toxicity	Flammability
Glycerol α,β-dichloro-hydrin (2,3-dichloro-1-propanol, 1,2-dichloro-3-propanol, α,β-dichlorohydrin) [616-23-9]	C_3H_6OCl 128.99 $Cl-CH_2-CH-CH_2-OH$ $\quad\quad\quad\quad\mid$ $\quad\quad\quad\quad Cl$	Skin and eye irritant, 6.8 mg caused severe eye irritation in rabbits; moderately toxic; 500 ppm/4 hr was lethal to rats; LD_{50} oral (rats): 90 mg/kg	Combustible, flash point data not available
6-Fluorohexanol [373-32-0]	$F-CH_2-CH_2-CH_2-CH_2$ $-CH_2-CH_2-OH$	Moderately toxic; LD_{50} intraperitoneal (mice): 1.2 mg/kg	Combustible, flash point data not available

REFERENCES

ACGIH. 1986. *Documentation of the Threshold Limit Values and Biological Exposure Indices*, 5th ed. Cincinnati, OH: American Conference of Governmental Industrial Hygienists.

Aldrich. 1989. *Aldrich Catalog*. Milwaukee, WI: Aldrich Chemical Company.

Merck. 1989. *The Merck Index*, 11th ed. Rahway, NJ: Merck & Co.

NFPA. 1986. *Fire Protection Guide on Hazardous Materials*, 9th ed. Quincy, MA: National Fire Protection Association.

NIOSH. 1984. *Manual of Analytical Methods*, 3rd ed. Cincinnati, OH: National Institute for Occupational Safety and Health.

NIOSH. 1986. *Registry of Toxic Effects of Chemical Substances*, ed. D. V. Sweet. Washington, DC: U.S. Government Printing Office.

Riesser, G. H. 1979. Chlorohydrins. In *Kirk-Othmer Encyclopedia of Chemical Technology*, Vol. 5, pp. 848–64. New York: Wiley-Interscience.

13

CYANIDES, ORGANIC (NITRILES)

13.1 GENERAL DISCUSSION

Organic cyanides or the nitriles are the organic derivatives of hydrogen cyanide or the cyano-substituted organic compounds. This class of compounds is characterized by CN functional group. Nitriles are highly reactive. The CN group reacts with a large number of reactants to form a wide variety of products, such as amides, amines, carboxylic acids, aldehydes, ketones, esters, thioamides, and other compounds. Most nitriles can be prepared by treating an alkali metal cyanide with an alkyl halide or a sulfonate salt (Smiley 1981). These are also formed by addition of hydrogen cyanide to unsaturated compounds; dehydration of an amide; and oxidation of olefins and aromatics in presence of ammonia. Some of the nitriles are commercially important for the production of a large number of synthetic products. Acrylonitrile and acetonitrile are two of the most widely used organic cyanides.

Toxicity

Nitriles are highly toxic compounds — some of them are as toxic as alkali metal cyanides. Acute and subchronic toxicity, teratogenicity, and biochemical mechanism studies of a series of structurally similar aliphatic nitriles indicated that although the toxicological profiles were generally the same for most of the compounds, there were unique differences, too (Johannsen and Levinska, 1986). In this respect, acetonitrile was different from the other nitriles within the homologous series. Nitriles liberated cyanide both in vivo and in vitro. Tanii and Hashimoto (1984a,b, 1985) observed a dose–cyanide liberation relationship in liver and hepatic microsomal enzyme system in mice pretreated with carbon tetrachloride. For most nitriles, the toxicity was greatly reduced by carbon tetrachloride pretreatment. By contrast, certain nitriles, exhibited an increase of toxicity as a result of such pretreatment.

The organic cyanides that demonstrated a lowering of toxicity as a result of carbon tetrachloride (CCl_4) pretreatment, however, produced the highest CN' level in the brain, $\sim 0.7–0.8$ mg/g of brain, which is comparable to KCN dosing (Tanii and Hashimoto 1984a,b). Microsomal liberation of such CN^- was found to be inhibited when microsomes were prepared from the liver of mice pretreated with CCl_4. The relationship between toxicity and the octanol–water partition coefficient P was determined to be linear: representing the equation

$$\log \left(\frac{1}{LD_{50}} - CCl_4 \right)$$
$$= -0.371 \log P - 0.152$$

There was no correlation between $\log P$ and $1/LD_{50}$ for dinitriles.

A single oral dose of organic cyanide can be absorbed rapidly, and most of it eliminated in urine in 24 hours. In a study on toxicity of 4-cyano-N,N-dimethyl aniline in mice, Logan et al. (1985) found that the residues in tissues 48 hours after dosing were 0.19, 0.10, 0.01, 0.10, and 0.02 ppm in liver, kidney, testes, fat, and blood, respectively. A major metabolite was 2-amino-5-cyanophenyl sulfate [95774-25-7].

The toxicity of individual nitriles is discussed in the following section. Miscellaneous nitriles are presented in Table 13.1.

Fire and Explosion Hazard

Because of the CN^- functional group, the nitriles exhibit high reactivity toward several classes of reactant. However, only certain reactions as outlined under individual nitriles could be violent. Its reactions with strong oxidizers and acids are exothermic and sometimes violent.

Lower aliphatic nitriles are flammable. These form explosive mixtures with air. The explosive range narrows down with an increase in carbon chain length.

13.2 ACRYLONITRILE

EPA Classified Toxic Waste, RCRA Waste Number U009; EPA Priority Pollutant; DOT Label: Flammable Liquid and Poison, UN 1093

Formula C_2H_3CN; MW 53.06; CAS [107-13-1]

Structure and functional group: $H_2C{=}CH{-}C{\equiv}N$, $-CN$ functional group and an olefinic double bond impart high reactivity to the molecule

Synonyms: vinyl cyanide; 2-propenenitrile; cyanoethylene; Acrylon; Acritet; Ventox

Uses and Exposure Risk

Acrylonitrile is used in the production of acrylic fibers, resins, and surface coating; as an intermediate in the production of pharmaceuticals and dyes; as a polymer modifier; and as a fumigant. It may occur in fire-effluent gases because of pyrolyses of polyacrylonitrile materials. Acrylonitrile was found to be released from the acrylonitrile–styrene copolymer and acrylonitrile–styrene–butadiene copolymer bottles when these bottles were filled with food-simulating solvents such as water, 4% acetic acid, 20% ethanol, and heptane and stored for 10 days to 5 months (Nakazawa et al. 1984). The release was greater with increasing temperature and was attributable to the residual acrylonitrile monomer in the polymeric materials.

Physical Properties

Clear, colorless liquid; bp 77.3°C; freezes at −83.5°C; density 0.806 at 20°C, moderately soluble in water (7.3 g/dL), readily miscible with most organic solvents.

Health Hazard

Acrylonitrile is a highly toxic compound, an irritant to the eyes and skin, mutagenic, teratogenic, and causes cancer.

Acrylonitrile is a moderate to severe acute toxicant via inhalation, oral intake, dermal absorption, and skin contact. Inhalation of this compound can cause asphyxia and headache. Firefighters exposed to acrylonitrile have reported chest pains, headache, shortness of breath, lightheadedness, coughing, and peeling of skin from their lips and hands (Donohue 1983). These symptoms were manifested a few hours after exposure and persisted for a

TABLE 13.1 Toxicity and Flammability of Miscellaneous Nitriles

Compound/Synonyms/ CAS No.	Formula/MW/Structure	Toxicity	Flammability
Allyl cyanide (vinyl acetonitrile, allyl nitrile, 3-butenenitrile) [109-75-1]	C_3H_5CN 67.1 $H_2C{=}CH{-}CH_2{-}C{\equiv}N$	Moderately toxic in rats oral intake; low toxicity when inhaled or absorbed through skin; exposure to 500 ppm for 4 hr was lethal to rats; LD_{50} oral (mice): 66.8 mg/kg; mild skin irritant	Flammable liquid flash point 23°C (75°F); vapor forms explosive mixtures with air
Acrolein cyanohydrin (vinyl glycolonitrile, 2-hydroxy-3-butenenitrile) [5809-59-6]	$C_3H_4(OH)CN$ 83.10 $H_2C{=}CH{-}CH{-}C{\equiv}N$ $\quad\quad\quad\quad\; \overset{\mid}{OH}$	Acute toxicity greater than allyl cyanide; toxic routes: inhalation, ingestion, and absorption through skin; LC_{50} inhalation (rats): 54 mg/m^3 (16 ppm)/4 hr; LD_{50} oral (rats): 65 mg/kg; LD_{50} skin (rabbits): 7.5 mg/kg; skin and eye irritant 50 mg produced severe eye irritation in rabbits	Combustible liquid
Cyanoacetic acid (malonic mononitrile) [372-09-8]	$C_2H_3O_2CN$ 85.07 $N{\equiv}C{-}CH_2{-}COOH$	Low to moderate toxicity; adverse effects on lungs, peripheral nerve, skin, and hair; oral and subcutaneous intake caused dyspnea, respiratory depression, change in motor activity, spastic paralysis, decrease in body temperature and loss of hair in test subjects; LD_{50} oral (rats): 1500 mg/kg; skin irritant	Noncombustible solid

Name (synonyms) [CAS]	Formula, MW, structure	Toxicity	Physical properties
Ethyl cyanoformate (cyanoformic acid ethyl ester) [623-49-4]	C₃H₅O₂CN 99.09 N≡C—C(=O)—OC₂H₅	Low to moderate toxicity; can cause tremor, dyspnea, and respiratory stimulation; LD₅₀ data not available	Flammable liquid flash point 24°C (76°F)
Ethyl cyanoacetate (cyanoacetic acid ethyl ester) [105-56-6]	C₄H₇O₂CN 113.13 N≡C—CH₂—C(=O)—OC₂H₅	Low to moderate toxicity in test animals; subcutaneous administration caused tremor, dyspnea, respiratory stimulation, and spastic paralysis; a dose of 1410 mg/kg was lethal to rabbits; lachrymator; DOT Label: Poison B, UN 2666, can cause lacrymation and change in motor activity; LD₅₀ oral (rats): 160 mg/kg	Noncombustible liquid; flash point >94°C (200°F)
Allyl cyanoacetate (cyanoacetic acid allyl ester, cyanoacetic acid 2-propenyl ester) [13361-32-5]	C₅H₇O₂CN 125.14 N≡C—CH₂—C(=O)—O—CH₂—CH=CH₂		Noncombustible liquid; flash point >94°C (200°F)
2-Acetoxy-3-butene nitrile (acrolein cyanohydrin acetate) [22581-05-1]	C₅H₇O₂CN 125.13 H₂C=CH—CH(C≡N)—O—C(=O)—CH₃	Skin irritant	Combustible liquid; flash point 72°C (163°F)
2-Acetoxyacrylonitrile (2-(acetoxy)-2-propenenitrile, 1-cyanovinyl acetate, acetic acid 1-cyanovinyl ester) [3061-65-2]	C₄H₅O₂CN 111.11 H₃C—C(=O)—O—C(C≡N)=CH₂	Toxic by inhalation, ingestion, and skin absorption; exposure to 560 mg/m³ (125 ppm) for 4 hr was lethal to rats; LD₅₀ oral (rats): 100 mg/kg; LD₅₀ skin (rabbits): 140 mg/kg; irritant to skin	Combustible liquid; flash point 63°C (147°F)

(continued)

TABLE 13.1 (Continued)

Compound/Synonyms/ CAS No.	Formula/MW/Structure	Toxicity	Flammability
Glyconitrile (hydroxyacetonitrile, glycolic nitrile, formaldehyde cyanohydrin; cyanomethanol) [107-16-4]	CH_3OCN 57.06 $HO-CH_2-C{\equiv}N$	Highly toxic by oral, ocular, inhalation, and skin absorption routes; exposure to 27 ppm/4 hr was lethal to mice following lacrimation and somnolence; ocular and oral intake produced tremor, convulsion, coma, and death; subcutaneous dose of 15 mg/kg was lethal to mice; LD_{50} oral (mice): 10 mg/kg; LD_{50} skin (rabbits): 5 mg/kg; eye irritant; exposure limit: 5 mg/m^3 (2.15 ppm)/15 min (NIOSH)	Combustible liquid; flash point data not available
Chloroacetonitrile (2-chloroacetonitrile; chloromethyl cyanide) [107-14-2]	CH_2ClCN 75.70 $ClCH_2-C{\equiv}N$	Toxic by inhalation, oral intake, and skin absorption; exposure to 250 ppm for 4 hr was lethal to rats; LD_{50} oral (mice): 139 mg/kg; LD_{50} skin (rabbits): 71 mg/kg; caused skin cancer in mice; no evidence of human carcinogenicity; DOT Label: Poison B, UN2268	Combustible liquid; flash point 47°C (118°F); DOT Label: Flammable Liquid, UN 2668

270

N,N-Diethylglycinonitrile (N,N-diethylamino acetonitrile) [3010-02-4]	$C_5H_{12}NCN$ 112.20 $N\equiv C-CH_2-N\overset{\textstyle C_2H_5}{\underset{\textstyle C_2H_5}{}}$	Moderately toxic by inhalation and ingestion; manifests low toxicity when absorbed through skin; LC_{50} inhalation (rats): 125 ppm/4 hr; LD_{50} oral (rats): 93.2 mg/kg; caused mild skin and severe eye irritation in rabbits	Noncombustible liquid; flash point 94°C (200°F)
Ethylene cyanohydrin (2-cyanoethanol, glycol cyanohydrin, 3-hydroxypropionitrile, 3-hydroxypropanenitrile, hydracrylonitrile) [109-78-4]	$C_2H_4(OH)CN$ 71.08 $HO-CH_2-CH_2-C\equiv N$	Mild skin and eye irritant; very low toxicity; an oral dose of 900 mg/kg was lethal to rabbit, LD_{50} oral (rats): 10,000 mg/kg	Noncombustible liquid; flash point >100°C (212°F)
Lactonitrile (2-hydroxypropionitrile) [78-97-7]	$C_2H_4(OH)CN$ 71.08 $H_3C-\overset{\textstyle }{\underset{\textstyle OH}{CH}}-C\equiv N$	Moderate to high toxicity; toxic routes inhalation, ingestion, and skin absorption; target organs: lungs, central nervous system, and peripheral nerve; symptoms observed in test animals were respiratory stimulation, dyspnea, ataxia, convulsion, lowering of body temperature, and spastic paralysis; LD_{50} oral (rats): 87 mg/kg	Combustible liquid; flash point (closed cup) 76°C (170°F)

(continued)

271

TABLE 13.1 (Continued)

Compound/Synonyms/CAS No.	Formula/MW/Structure	Toxicity	Flammability
3-Chlorolactonitrile [33965-80-9]	$C_3H_3(OH)ClCN$ 105.53 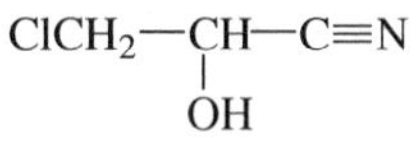 $ClCH_2-CH-C{\equiv}N$ $\quad\quad\quad\underset{OH}{\vert}$	Skin absorption and administration by intraperitoneal and intravenous routes exhibited high acute toxicity; lethal dose in mice intraperitoneally was 4 mg/kg	Noncombustible liquid; flash point >94°C (200°F)
Chloral cyanohydrin (3,3,3-trichloro lactonitrile, chlorocyanohydrin, chloral hydrocyanide) [513-96-2]	$C_2(OH)Cl_3CN$ 174.41 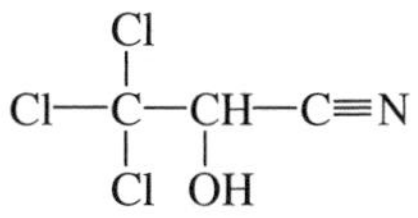 $Cl-\underset{\underset{Cl}{\vert}}{\overset{\overset{Cl}{\vert}}{C}}-\underset{OH}{\overset{\vert}{CH}}-C{\equiv}N$	Moderate to high toxicity in mice when applied subcutaneously, lethal dose 23 mg/kg; oral toxicity data not available	Noncombustible liquid
Crotononitrile (2-butenenitrile, 1-cyanopropene, 1-propenyl cyanide, β-methylacrylonitrile) [4786-20-3]	C_3H_5CN 67.10 $H_3C-CH{=}CH-C{\equiv}N$	Low to moderate toxicity in test subjects; intravenous dose of 60 mg/kg was lethal to rabbits; irritant to skin	Flammable liquid; flash point 20°C (68°F)
Cinnamonitrile (cinnamyl nitrile, 3-phenyl-2-propene nitrile) [4360-47-8]	C_8H_7CN 129.17 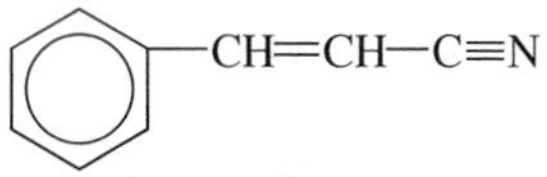 $\text{(phenyl ring)}-CH{=}CH-C{\equiv}N$	Low oral toxicity; subcutaneous dose of 130 mg/kg was lethal to guinea pigs; irritant to skin	Noncombustible liquid; flash point >94°C (200°F)

272

Benzonitrile (benzenenitrile, cyanobenzene, phenyl cyanide) [100-47-0]	C_6H_5CN 103.13 C≡N (attached to benzene ring)	Low inhalation toxicity; exposure to high concentration, however, can result in shortness of breath and toxic spasms of face and arm muscles; the acute toxic symptoms from ingestion and subcutaneous applications were tremor, convulsion, somnolence, dyspnea, and paralysis of peripheral nervous system; LD_{50} oral (rabbits): 800 mg/kg; skin irritant; DOT Label: Poison B, UN 2224	Combustible liquid; flash point 71°C (161°F); DOT Label: Combustible Liquid UN 2224
o-Chlorobenzonitrile [873-32-5]	C_6H_4ClCN 137.57 C≡N, Cl (attached to benzene ring)	Eye irritant; low to moderate acute toxicity; LD_{50} oral (rats): 435 mg/kg	Noncombustible solid
Azodiisobutyronitrile (2,2'-azobis (isobutyronitrile),2,2'-dicyano-2,2'-azopropane, Vazo-64) [78-67-1]	$C_8H_{12}N_4$ 164.24 $H_3C-\underset{\underset{C≡N}{|}}{\overset{\overset{CH_3}{|}}{C}}-N=N-\underset{\underset{C≡N}{|}}{\overset{\overset{CH_3}{|}}{C}}-CH_3$	Low oral toxicity; LD_{50} oral (mice): 700 mg/kg; mild skin and eye irritant; can decompose to highly toxic tetramethyl succinonitrile	Flammable solid; can explode if its solution in acetone is concentrated; liberates nitrogen on heating or with solvents that can cause pressure buildup and rupture of vessel; DOT Label: Flammable Solid, UN 2952

(continued)

TABLE 13.1 (*Continued*)

Compound/Synonyms/CAS No.	Formula/MW/Structure	Toxicity	Flammability
2,2′-Azobis (2,4-dimethylvaleronitrile) (Vazo-52) [4419-11-8]	$C_{14}H_{24}N_4$ 248.36 $H_3C-CH-CH_2-C-N=N-C-CH_2-CH-CH_3$ with CH_3, CH_3, CN groups and CH_3, CH_3, CN	Mild eye irritant; nontoxic	Flammable solid; liberates N_2 on heating or solvent treatment, causing pressure buildup in closed containers
Succinonitrile (butanedinitrile, 1,2-dicyanoethane, ethylene dicyanide, succinic acid dinitrile) [110-61-2]	$C_2H_4(CN)_2$ 80.10 $N\equiv C-CH_2-CH_2-C\equiv N$	Exhibited moderate toxicity in test animals from intraperitoneal, subcutaneous, and oral intakes; extent of toxicity varied with species; target organs: kidney and bladder; LD_{50} oral (mice): 129 mg/kg; teratogenic, causing fetotoxicity and developmental abnormalities in central nervous system in hamster; exposure limit: TLV-TWA 20 mg/m³ (6 ppm) (NIOSH)	Noncombustible solid
Tetramethylsuccinonitrile (tetramethylbutanedinitrile, tetramethylsuccinic acid dinitrile) [3333-52-6]	$C_6H_{12}(CN)_2$ 136.22 $N\equiv C-C-C-C\equiv N$ with H_3C, CH_3 and H_3C, CH_3	Toxic symptoms were tremor, convulsion, respiratory stimulation; nausea, vomiting, and headache; coma can result from severe exposure; exposure to 60 ppm for 3 hr was fatal to rat; LD_{50} oral (rats): 40 mg/kg; teratogen; exposure limits: TLV-TWA 3 mg/m³ (0.5 ppm) (ACGIH	Noncombustible solid

Phenylacetonitrile
(benzeneacetonitrile,
benzyl cyanide, benzyl
nitrile, (cyanomethyl)
benzene, α-cyano-
toluene, α-tolunitrile)
[140-29-4]

C_7H_7CN
117.16

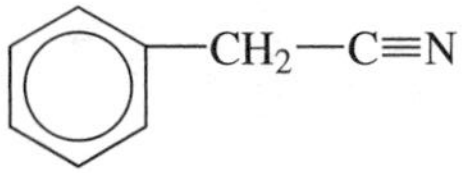

and OSHA) 6 mg/m^3/15 min
(NIOSH)

Moderate acute toxicity by
inhalation and ingestion;
toxic symptoms in test
subjects were tremor,
convulsion, excitement,
ataxia, dyspnea, and
paralysis of peripheral nerve
system similar to other
nitriles and manifested when
administered subcutaneously
and intraperitoneally; LC_{50}
inhalation (rats): 430 mg/m^3
(90 ppm)/2 hr; LD_{50} oral
(mice): 45.5 mg/kg; low
toxicity when absorbed
through skin; mild skin
irritant; DOT Label: Poison
B, UN 2470

Noncombustible liquid;
flash point >94°C
(200°F)

Diphenylacetonitrile
(diphenylmethylcyanide,
α-cyanodiphenyl-
methane, α-phenyl-
benzeneacetonitrile,
diphenatrile) [86-29-3]

$C_{13}H_{11}CN$
193.26

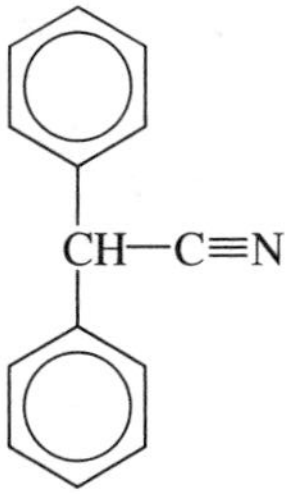

Carcinogenic; oral and
subcutaneous administration
caused cancer in lungs, liver,
and blood in mice;
carcinogenic dose 61 g/kg/78
weeks (oral); low oral
toxicity; LD_{50} oral (rats):
3500 mg/kg; CN toxicity is
not well manifested,
probably due to the
inhibition of CN reactivity
because of steric hindrance
exerted by the benzene rings

Noncombustible; flash
point >94°C (200°F)

275

few days. Inhalation of 110 ppm for 4 hours was lethal to dogs. In humans, inhalation of about 500 ppm for an hour could be dangerous. The toxicity symptoms in humans from inhaling high concentrations of acrylonitrile were somnolence, diarrhea, nausea, and vomiting (ACGIH 1986).

Ingestion and absorption of acrylonitrile through the skin exhibited similar toxic symptoms: headache, lightheadedness, sneezing, weakness, nausea, and vomiting. In humans, the symptoms were nonspecific but related to the central nervous system, respiratory tract, gastrointestinal tract, and skin. Severe intoxication can cause loss of consciousness, convulsions, respiratory arrest, and death (Buchter and Peter 1984).

Investigating the acute and subacute toxicity of acrylonitrile, Knobloch et al. (1971) reported that the compound caused congestion in all types of organs and damage to the central nervous system, lungs, liver, and kidneys. A dose of 50 mg/kg/day given intraperitoneally to adult rats for 3 weeks resulted in body weight loss, leukocytosis, and functional disturbances and degeneration of the liver and kidneys. There was also light damage to the neuronal cells of the brain stem and cortex.

LD$_{50}$ value, oral (mice): 27 mg/kg

LD$_{50}$ value, subcutaneous (mice): 34 mg/kg

The lethal effect of acrylonitrile increased in rats when coadministered with organic solvents (Gut et al. 1981), although the latter decreased the formation of cyanide. Metabolic cyanide formation was found to play only a minor role in the inhalation toxicity of acrylonitrile (Peter and Bolt 1985). This was in contrast to the toxicity of methylacrylonitrile, where the observed clinical symptoms suggest a metabolically formed cyanide.

Combination of styrene and acrylonitrile enhanced the renal toxicity of the former in male rats (Normandeau et al. 1984). The lethal effect of acrylonitrile increased with hypoxia or the condition of inadequate supply of oxygen to the tissues (Jaeger and Cote 1982).

Acrylonitrile is a mild skin irritant. It caused severe irritation in rabbits' eyes. Inhalation and oral and intraperitoneal dosages exhibited birth defects in rats and hamsters. Abnormalities in the central nervous system, as well as cytological changes and postimplantation mortality were the symptoms observed. Acrylonitrile is a mutagen. It tested positive in TRP reversion and histidine reversion–Ames tests. This compound caused cancer in test species. Inhalation and ingestion of this compound produced cancers in the brain, gastrointestinal tract, and skin in rats. An oral dose that was carcinogenic to rats was determined to be 18,000 mg/kg given over 52 weeks (NIOSH 1986). On the basis of animal studies, acrylonitrile is expected to cause cancer in humans.

Antidote

The cyanide antidote 4-dimethylaminophenol plus sodium thiosulfate showed some protective action only after oral intake. However, against inhalation and other toxic routes, the antidote above was ineffective (Appel et al. 1981). Buchter and Peter (1984) reported the effectiveness of cysteine [52-90-4] and N-acetyl cysteine [616-91-1] in combating acrylonitrile poisoning.

Rats pretreated by inhalation with sublethal concentrations of acrylonitrile became immune to subsequent lethal exposure (Cote et al. 1983). Such inhalation-induced tolerance to acrylonitrile did not protect against subsequent poisoning by cyanide.

Exposure Limits

TLV-TWA 4.35 mg/m^3 (2 ppm) (ACG1H and OSHA), 2.17 mg/m^3 (1 ppm) (NIOSH); ceiling 21.7 mg/m^3 (10 ppm)/15 min (NIOSH).

Fire and Explosion Hazard

Highly flammable; flash point (open cup) 0°C (32°F); vapor pressure 83 torr at 20°C;

vapor density 1.83 (air = 1); the vapor is heavier than air and can travel a considerable distance and flash back; dilute solutions are flammable, too: flash points (open cup) of 2% and 3% aqueous solutions were 21°C (70°F) and 12°C (54°F), respectively (NFPA 1986); autoignition temperature 481°C (898°F); fire-extinguishing agent: dry chemical, CO_2, or "alcohol" foam; use a water spray to dilute the spill and keep water-exposed containers cool.

Vapors of acrylonitrile form an explosive mixture with air in the range 3–17% by volume of air. It polymerizes violently in the presence of concentrated alkali or at elevated temperature. It undergoes photochemical polymerization, liberating heat. Violent rupture of glass containers can occur due to polymerization.

When mixed with bromine, acrylonitrile can explode. Dropwise addition of bromine to acrylonitrile without cooling the flask can also be violent. Reactions with strong oxidizers, strong bases, strong acids, ammonia, amines, and copper and its alloys can be violent. Lithium reacts exothermically with acrylonitrile.

Storage and Shipping

Acrylonitrile is stored in standard flammable liquid cabinet, isolated from alkalies and oxidizing materials. Pure compound is inhibited from polymerization by adding 35–45 ppm hydroquinone monomethyl ether (Aldrich 1996). It is shipped in tank cars, drums, and lined pails.

Disposal/Destruction

Acrylonitrile is burned in a chemical incinerator equipped with an afterburner and scrubber. Investigating the flame-mode destruction of acrylonitrile waste, Kramlich et al. (1985) reported that in a microspray reactor the destruction was complete at >602°C in an oxygen-rich atmosphere. It can be removed from sludges and wastewater by biodegradation. Stover et al. (1985) reported that

addition of powdered activated carbon to activated sludge systems was effective for removal of many toxic organic compounds, including acrylonitrile. The compounds were absorbed over activated carbon, enhancing their biodegradation because of the longer retention time of the compounds in the system. However, ozonization decreased the biodegradability of acrylonitrile.

Analysis

Among the instrumental techniques, GC and GC/MS are most suitable for its analysis. Air analysis is performed by absorbing it over coconut shell charcoal, desorbing the analyte into carbon disulfide containing 2% acetone and injecting the eluant into GC equipped with FID (NIOSH 1984, Method 1604); 20% SP-1000 on 80/100-mesh Supelcoport or any equivalent column is suitable for the purpose. Analysis of acrylonitrile in wastewaters may be performed by GC-FID using Method 603 or by GC/MS Method 624 (U.S. EPA 1984). U.S. EPA Method 8240 (1986) describes soil and hazardous wastes analysis by GC/MS technique. The primary characteristic ion is 53.

13.3 ACETONITRILE

EPA Classified Toxic Waste, RCRA Waste Number U003; DOT Label: Flammable Liquid and Poison, UN 1648

Formula CH_3CN; MW 41.05; CAS [75-05-8]

Structure and functional group: $H_3C-C{\equiv}N$, the first member of the homologous series of aliphatic nitriles, reactive site $-C{\equiv}N$, undergoes typical nitrile reactions

Synonyms: methyl cyanide; ethanenitrile; ethyl nitrile; methanecarbonitrile; cyanomethane

Uses and Exposure Risk

Acetonitrile is used as a solvent for polymers, spinning fibers, casting and molding plastics, and HPLC analyses; for extraction of butadiene and other olefins from hydrocarbon streams; in dyeing and coating textiles; and as a stabilizer for chlorinated solvents. It occurs in coal tar and forms as a by-product when acrylonitrile is made.

Physical Properties

Colorless liquid with ether-like smell; bp 81.6°C; freezes at −45.7°C; density 0.786 at 20°C; dielectric constant 38.8 (at 20°C); dipole moment 3.11 D; fully miscible with water and most organic solvents, immiscible with petroleum fractions and some saturated hydrocarbons.

Health Hazard

The toxicity of acetonitrile to human and test animals is considerably lower than that of some other nitriles. However, at high concentrations, this compound could produce severe adverse effects. The target organs are the kidney, liver, central nervous system, lungs, cardiovascular system, skin, and eyes. In humans, inhalation of its vapors can cause asphyxia, nausea, vomiting, and tightness of the chest. Such effects can probably be manifested at several hours exposure to concentration in air above 400–500 ppm. At a lower concentration of 100 ppm, only a slight adverse effect may be noted. It is excreted in the urine as cyanate. The blood cyanide concentration does not show any significant increase in cyanide at low concentrations.

The acute oral toxicity of acetonitrile is generally of low order. The toxic symptoms associated with oral intake can be gastrointestinal pain, nausea, vomiting, stupor, convulsion, and weakness. These effects may become highly marked in humans from ingestion of 40–50 mL of acetonitrile. Freeman and Hayes (1985) observed toxicological interaction between acetone and acetonitrile when administered in rats by oral dose. There was a delay in the onset of toxicity (due to acetonitrile) and

an elevation of blood cyanide concentration when the dose consisted of a mixture of acetone and acetonitrile. Acetone inhibited the cyanide formation. The toxicity of both the solvents were prevented by administering sodium thiosulfate. Sodium nitrite also provided protection against mortality from lethal concentrations (Willhite 1981). Intraperitoneal administration of acetonitrile resulted in damage to cornea, ataxia, and dyspnea in mice. It is an eye and skin irritant.

LD_{50} value, oral (mice): 269 mg/kg

LD_{50} value, intraperitoneal (mice): 175 mg/kg

Acetonitrile is a teratomer. Pregnant hamsters were exposed to this compound by inhalation, ingestion, or injection during the early stage of embryogenesis. Severe axial skeletal disorders resulted in the offspring at a high concentration of 5000–8000 ppm (inhalation) or 100–400 mg/kg (oral dose) (Willhite 1983). Teratogenic effects were attributed to the release of cyanide, which was detected in high concentrations along with thiocyanate in all tissues after an oral or intraperitoneal dose. Sodium thiosulfate-treated hamsters did not display a teratogenic response to acetonitrile.

Exposure Limits

TLV-TWA 70 mg/m³ (40 ppm) (ACGIH and OSHA); STEL 105 mg/m³ (60 ppm) (ACGIH); IDLH 4000 ppm (NIOSH).

Fire and Explosion Hazard

Flammable liquid; flash point (open cup) 5.5°C (42°F); vapor pressure 73 torr at 20°C; vapor density at 38°C 1.1 (air = 1); the vapor is heavier than air and can travel some distance to a source of ignition and flash back; ignition temperature 524°C (975°F); fire-extinguishing agent: dry chemical, CO_2, or "alcohol" foam; use a water spray to flush and dilute the spill and keep fire-exposed containers cool.

Acetonitrile vapors form an explosive mixture with air; the LEL and UEL values are 4.4% and 16.0% by volume of air, respectively. It reacts with strong oxidizers and acids, liberating heat along with pressure increase. Thus contact in a close container can result in rupture of the container. Erbium perchlorate tetrasolvated with acetonitrile when dried to disolvate exploded violently on light friction (Wolsey 1973). Neodymium perchlorate showed similar heat and shock sensitivity when dried down to lower levels of solvation (*Chemical & Engineering News*, Dec. 5, 1983). Bretherick (1990) proposed that the tendency for oxygen balance to shift toward zero for maximum energy release, with diminishing solvent content, decreased the stability of solvated metal perchlorates at lower levels of solvation. Such a zero balance for maximum exotherm should occur at 2.18 mol of acetonitrile solvated to metal perchlorate. Metals such as lithium react exothermically with acetonitrile at ambient temperature (Dey and Holmes 1979).

Disposal/Destruction

Acetonitrile is burned in a chemical incinerator equipped with an afterburner and scrubber. In the laboratory, it can be destroyed by refluxing with ethanolic potassium hydroxide or with 36% HCl for several hours.

Analysis

Acetonitrile is analyzed by GC using a FID detector on a Porapak column. Analysis in air is performed by absorption over coconut shell charcoal, followed by desorption with benzene and injection onto GC-FID (NIOSH 1984, Method 1606). Large charcoal tubes (400 mg/200 mg) are necessary for collecting acetonitrile.

13.4 PROPIONITRILE

EPA Classified Hazardous Waste, RCRA Waste Number P101; DOT Label: Flammable Liquid and Poison, UN 2404

Formula CH_3CH_2CN; MW 55.09; CAS [107-12-0]

Structure and functional group: $H_3C-CH_2 -C\equiv N$, aliphatic mononitrile; active site: $-C\equiv N$ functional group

Synonyms: ethyl cyanide; propionic nitrile; propanenitrile; cyanoethane

Uses and Exposure Risk

Propionitrile is used as a chemical intermediate. It is formed as a by-product of the electrodimerization of acrylonitrile to adiponitrile.

Physical Properties

Colorless liquid with a sweet ether-like odor; bp 97°C; mp −92°C; density 0.782 at 20°C; soluble in water (12 g/dL at 40°C), alcohol, ether, and most organic solvents.

Health Hazard

Propionitrile is a moderate to highly toxic compound, an eye irritant, and a teratomer. The toxic symptoms are similar to acetonitrile. However, the acute inhalation toxicity is greater than that associate with acetonitrile. Willhite (1981) reported a median lethal concentration of 163 ppm in male mice exposed for 60 minutes. By comparison, acetonitrile and butyronitrile exhibited median lethal concentrations of 2693 and 249 ppm, respectively.

The toxic routes are inhalation, ingestion, and absorption through skin. The target organs are kidney, liver, central nervous system, lungs, and eyes. Inhalation of 500 ppm for 4 hours was lethal to rats. When administered intraperitoneally to mice, it caused corneal damage, ataxia, and dyspnea. The acute oral toxicity of this compound was found to be moderately high in rodents.

LD$_{50}$ value, oral (mice): 36 mg/kg

Propionitrile exhibited teratogenic effects in hamsters. Intraperitoneal administration of 238 mg/kg caused cytological changes in embryo and developmental abnormalities in the central nervous system. There is no report of any cancer-causing effects of this compound in animals or humans.

Exposure Limit

TLV-TWA 14 mg/m^3 (6 ppm) (NIOSH).

Fire and Explosion Hazard

Flammable liquid; flash point (closed cup) 2°C (36°F); vapor density 1.9 (air = 1); the vapor is heavier than air and can travel a considerable distance to a source of ignition and flashback; fire-extinguishing agent: dry chemical, CO_2, or "alcohol" foam; a water spray should be used to flush and dilute the spill and to keep fire-exposed containers cool.

Propionitrile forms explosive mixtures with air; the LEL value is 3.1% by volume of air, the UEL value is not available. When heated to decomposition, it liberates toxic gases: HCN and ethylene. HCN formation can occur in contact with strong acids. It reacts exothermically with strong oxidizers.

Disposal/Destruction

Propionitrile is burned in a chemical incinerator equipped with an afterburner and scrubber. A suitable method for its destruction in the laboratory is to reflux with ethanolic KOH for several hours. Nonhazardous water-soluble cyanate can be drained down the sink with copious volume of water.

Analysis

GC-FID and GC/MS techniques are most suitable for analyzing propionitrile. Air analysis may be performed similar to acetonitrile-involving charcoal absorption, followed by desorption of the analyte with benzene and analyzing by GC.

13.5 BUTYRONITRILE

DOT Label: Flammable Liquid and Poison, UN 2411

Formula C_3H_7CN; MW 69.12; CAS [109-74-0]

Structure and functional group: $H_3C-CH_2-CH_2-C\equiv N$, aliphatic nitrile, the primary alkyl group (butyl) attached to the reactive $-C\equiv N$ functional group

Synonyms: *n*-butanenitrile; butyric acid nitrile; propyl cyanide; 1-cyanopropane

Uses and Exposure Risk

Butyronitrile is used as a chemical intermediate.

Physical Properties

Colorless liquid; bp 117.5°C; freezes at −112°C; density 0.795 at 15°C; sparingly soluble in water, mixes with alcohol, ether, and dimethylformamide.

Health Hazard

Butyronitrile showed moderate to high toxicity on test animals. It is an acute toxicant by all routes: inhalation, ingestion, and absorption through skin. The target organs are the liver, kidney, central nervous system, lungs, and sense organs, as well as the peripheral nerve. Its toxicity is on the same order as that of propionitrile; its inhalation toxicity is slightly lower than that of propionitrile, and its oral toxicity is slightly greater than that of propionitrile.

Inhalation can cause nausea, respiratory distress, and damage to liver. Willhite (1981) reported a LC$_{50}$ value of 249 ppm in mice from 1 hour exposure to its vapor. It produced ataxia, dyspnea, and corneal damage in test animals when given intraperitoneally. The toxic symptoms from subcutaneous applications are tremor, dyspnea, respiratory depression, and spastic paralysis. It is toxic only at low levels by skin absorption.

LD$_{50}$ value, oral (mice): 27.7 mg/kg

There is no report on its teratogenicity. The reproductive effect of this compound is expected to be similar to that of propionitrile.

Exposure Limit

TLV-TWA 22.5 mg/m^3 (8 ppm) (NIOSH).

Fire and Explosion Hazard

Flammable liquid; flash point (open cup) 24°C (76°F); vapor density 2.4 (air = 1); ignition temperature 501°C (935°F); fire-extinguishing agent: dry chemical, CO_2, or "alcohol" foam; use a water spray to cool fire-exposed containers.

Butyronitrile forms an explosive mixture with air; the LEL value is 1.65% by volume of air, the UEL value is not available. It reacts exothermically with strong oxidizers and acids. Liberation of HCN can occur on reactions with acids.

Disposal/Destruction

Burning in a chemical incinerator equipped with an afterburner and scrubber is the most effective way to destroy the compound. Oxidation with ethanolic–KOH can convert butyronitrile to nonhazardous cyanate.

Analysis

Analysis can be performed by GC-FID and GC/MS. Its presence in air can be detected in the same manner as for acetonitrile (NIOSH 1984, Method 1606) following successive steps of charcoal absorption, desorption with benzene, and GC analysis.

13.6 ACETONE CYANOHYDRIN

EPA Classified Acute Hazardous Waste, RCRA Waste Number P069; DOT Label: Poison B, UN 1541

Formula $C_3H_6(OH)CN$; MW 85.10; CAS [75-86-5]

Structure and functional group:

$$\begin{array}{c} CH_3 \\ | \\ H_3C-C-C\equiv N \\ | \\ OH \end{array}$$

a −CN and an alcoholic −OH functional group in the molecule

Synonyms: 2-hydroxy-2-methyl propionitrile; 2-methyl lactonitrile; α-hydroxyisobutyronitrile

Uses and Exposure Risk

Acetone cyanohydrin is used to make methyl methacrylate; in the production of pharmaceuticals, foaming agents, and insecticides; and to produce polymerization initiators.

Physical Properties

Colorless liquid; bp 95°C; freezes at −19°C; density 0.932; soluble in water and most organic solvents, insoluble in petroleum ether, hydrocarbons, and carbon disulfide.

Health Hazard

Acetone cyanohydrin is a highly poisonous compound. Although the acute toxicity is lower than that associated with hydrogen cyanide, the toxic effects are similar to those of the latter compound. The toxic routes are inhalation, ingestion, and skin contact.

Inhalation of its vapors at high concentrations can cause instantaneous loss of consciousness and death. Exposure to 63 ppm for 4 hours was lethal to rats. The oral toxicity of this compound in test animals was found to be high with LD_{50} values falling between 10 and 20 mg/kg.

LD_{50} value, oral (mice): 14 mg/kg

It is a mild irritant to the skin but moderately toxic by skin absorption. There is no report of its teratogenic and carcinogenic actions in humans or animals.

Exposure Limit

Ceiling 4 mg (1.2 ppm)/m^3/15 min (NIOSH).

Fire and Explosion Hazard

Combustible; flash point (open cup) 74°C (165°F); ignition temperature 688°C

(1270°F); the vapor forms an explosive mixture in air within the range of 2.2–12.0% by volume of air; fire-extinguishing agent: water spray, CO_2, dry chemical, or "alcohol" foam.

Acetone cyanohydrin decomposes on heating at 120°C, releasing HCN. HCN is also liberated when this compound comes in contact with bases or reacts with water. With the latter, acetone is formed along with HCN. Contact with strong acids can cause a violent reaction.

Disposal/Destruction

Acetone cyanohydrin is burned in a chemical incinerator equipped with an afterburner and scrubber.

Analysis

Gas chromatography is the most suitable technique for its analysis. An NP detector and a column consisting of OV-17 on Chromosorb are used in GC. Because acetone cyanohydrin is thermally unstable, the column is pushed up to the detector to eliminate the contact of the analyte with a hot metallic or glass surface.

Air analysis is performed using the foregoing GC techniques (NIOSH 1984, Suppl. 1985, Method 2506). The analyte is absorbed on Porapak QS and then desorbed with ethyl acetate and injected onto the GC column.

13.7   METHYLACRYLONITRILE

EPA Classified Toxic Waste, RCRA Waste Number U152

Formula C_3H_5CN; MW 67.10; CAS [126-98-7]

Structure and functional group:

$$H_2C{=}C{-}C{\equiv}N$$
$$\underset{\displaystyle CH_3}{|}$$

reactive sites are the cyanide functional group and the olefinic double bond

Synonyms: methacrylonitrile; isopropenyl-nitrile; 2-methyl-2-propenenitrile; isopropene cyanide; 2-cyanopropene

Uses and Exposure Risk

Methylacrylonitrile is used to make coating and elastomers and as an intermediate in the preparation of acids, amine, amides, and esters.

Physical Properties

Colorless liquid; bp 90°C; freezes at −35.8°C; density 0.8001 at 20°C; slightly soluble in water (2.57 g/dL water at 20°C, mixes readily with common organic solvents.

Health Hazard

Methylacrylonitrile is a moderate to severe acute toxicant. The degree of toxicity varied with toxic routes and species. Inhalation, ingestion, and skin application on test subjects produced convulsion. Exposure to high concentrations can result in asphyxia and death. The lethal concentrations varied among species from 50 to 400 ppm over a 4-hour exposure period. The clinical symptoms observed in rats suggested a toxic activity of metabolically formed cyanide (Peter and Bolt 1985). This finding was in contrast with acrylonitrile toxicity in the same species, where formation of metabolic cyanide played a minor role.

Methylacrylonitrile is a mild skin and eye irritant. However, it is readily absorbed by skin. It showed delayed skin reaction. In mice, the lethal dose from intraperitoneal administration was 15 mg/kg. The oral toxicity due to this compound was also relatively high; an LD_{50} of 11.6 mg/kg was determined in mice. There is no report of its mutagenic, teratogenic, or carcinogenic actions in animals or humans. 4-Dimethylaminophenol [616-60-3] plus sodium thiosulfate or N-acetylcystein [616-91-1] was shown to antagonize the acute toxicity of methylacrylonitrile (Peter and Bolt 1985).

Exposure Limit

TLV-TWA skin 2.7 mg/m^3 (1 ppm) (ACGIH).

Fire and Explosion Hazard

Flammable; flash point (open cup) 13°C (55°F); vapor pressure 40 torr at 12.8°C; vapor density 2.3 (air = 1); fire-extinguishing agent: dry chemical, CO_2, or "alcohol" foam; a water spray can be used to keep fire-exposed containers cool.

Vapor forms an explosive mixture with air within the range of 2–6.85 by volume of air. It may react violently, undergoing exothermic polymerization, when heated with concentrated alkali solutions. Reactions with strong oxidizers and acids can be exothermic.

Disposal/Destruction

Methylacrylonitrile is burned in a chemical incinerator equipped with an afterburner and scrubber.

Analysis

Instrumental techniques suitable to analyze methylacrylonitrile are GC and GC/MS. NPD is a suitable detector for GC.

13.8 MALONONITRILE

EPA Classified Toxic Waste, RCRA Waste Number U149; DOT Label: Poison B, UN 2647

Formula $CH_2(CN)_2$; MW 66.07; CAS [109-77-3]

Structure and functional group: N≡C—CH$_2$—C≡N, two —CN functional groups

Synonyms: cyanoacetonitrile; dicyanomethane; methylene cyanide; propanedinitrile

Uses and Exposure Risk

Malononitrile is used in organic synthesis.

Physical Properties

Colorless solid; mp 32°C; bp 218°C; density 1.191 at 20°C; soluble in water, alcohol, ether, acetone, and benzene.

Health Hazard

Malononitrile is a highly toxic compound by all toxic routes. Its acute toxicity is somewhat greater than that of the aliphatic mononitriles, propionitrile, and butyronitrile. The increased toxicity may be attributed to the greater degree of reactivity in the molecule arising from two —CN functional groups. The acute toxic symptoms in test animals have not been well documented. An intraperitoneal dose of 10 mg/kg was lethal to rats.

LD$_{50}$ value, intravenous (rabbits): 28 mg/kg
LD$_{50}$ value, oral (mice): 19 mg/kg

Malononitrile is an eye irritant. The irritation from 5 mg in 24 hours was severe in rabbits' eyes. There is no report of teratogenic and carcinogenic action in animals or humans.

Exposure Limit

TLV-TWA 8 mg/m^3 (3 ppm) (NIOSH).

Fire and Explosion Hazard

Noncombustible solid; flash point >94°C (200°F). Highly reactive; reactions with strong acids and oxidizers are exothermic. HCN liberation can occur when reacted with acids.

Disposal/Destruction

Malononitrile is dissolved in a combustible solvent and burned in a chemical incinerator equipped with an afterburner and scrubber.

Analysis

GC and GC/MS are suitable techniques for analysis. FID is used as a detector for GC.

13.9 ADIPONITRILE

DOT Label: Poison B, UN 2205

Formula $C_4H_8(CN)_2$; MW 108.16; CAS [111-69-3]

Structure and functional group: $N{\equiv}C-CH_2-CH_2-CH_2-CH_2-C{\equiv}N$, an aliphatic dinitrile

Synonyms: 1,4-dicyanobutane; hexanedinitrile; adipic acid dinitrile; tetramethylene cyanide

Uses and Exposure Risk

Adiponitrile is used in the production of hexamethylenediamine for manufacturing nylon 6,6 and to produce adipoguanamine to make amino resins.

Physical Properties

Colorless liquid; bp 295°C; mp −2.5°C, density 0.965 (at 20°C); sparingly soluble in water (8% at 20°C), moderately soluble (35%) at 100°C; readily soluble in alcohol, chloroform, and benzene, low solubility in ether, carbon disulfide, and hexane.

Health Hazard

The acute toxicity of adiponitrile is somewhat lower than that of malononitrile. It is toxic by inhalation and oral routes. Inhalation of its vapors can cause nausea, vomiting, respiratory tract irritation, and dizziness. The symptoms are similar to those of other aliphatic mono- and dinitriles. Similar poisoning effects may be manifested from ingestion of this compound. However, its toxicity is very low from skin absorption.

LC_{50} value, inhalation (rats): 1710 mg/m^3/ 4 hr

LD_{50} value, oral (mice): 172 mg/kg

There is no report of its teratogenicity or cancer-causing effects in animals or humans.

Exposure Limit

TLV-TWA 18 mg/m^3 (4 ppm) (NIOSH).

Fire and Explosion Hazard

Noncombustible; flash point >94°C (200°F); autoignition temperature 550°C (958°F). It forms explosive mixtures with air in the range 1.7–5.0% by volume in air. It undergoes typical reactions of nitriles. When mixed with strong acids, HCN evolution can occur.

Disposal/Destruction

Adiponitrile is burned in a chemical incinerator equipped with an afterburner and scrubber.

Analysis

The analytical methods are similar to those of other nitriles. GC-FID and GC/MS techniques are suitable for its analysis. Wet methods can be applied to determine the total cyanide (−CN) contents.

13.10 MISCELLANEOUS NITRILES

See Table 13.1.

REFERENCES

ACGIH. 1986. *Documentation of the Threshold Limit Values and Biological Exposure Indices*, 5th ed. Cincinnati, OH: American Conference of Governmental Industrial Hygienists.

Aldrich. 1996. *Aldrich Catalog.* Milwaukee, WI: Aldrich Chemical Co.

Appel, K. E., H. Peter, and H. M. Bolt. 1981. Effect of potential antidotes on the acute toxicity of acrylonitrile. *Int. Arch. Occup. Environ. Health* 49(2): 157–63.

Bretherick, L. 1990. Solvated metal perchlorates. *Chem. Eng. News* 68(22): 4.

Buchter, A., and H. Peter. 1984. Clinical toxicology of acrylonitrile. *G. Ital. Med. Lav.* 6(3–4): 83–86.

Cote, I. L., A. Bowers, and R. J. Jaeger. 1983. Induced tolerance to acrylonitrile toxicity by prior acrylonitrile exposure. *Res. Commun. Chem. Pathol. Pharmacol.* 42(1): 169–72.

Dey, A. N., and R. W. Holmes. 1979. Safety studies on lithium/sulfur dioxide cells. *J. Electrochem. Soc.* 126(10): 1637–44.

Donohue, M. T. 1983. Health hazard evaluation report. HETA-83-157-1373, *Govt. Rep. Announce. Index (U.S.)* 85(11): 61.

Freeman, J. J., and E. P. Hayes. 1985. Acetone potentiation of acute acetonitrile toxicity in rats. *J. Toxicol. Environ. Health* 15(5): 609–29.

Gut, I., J. Kopecky, J. Nerudova, M. Krivucova, and L. Pelech. 1981. Metabolic and toxic interactions of benzene and acrylonitrile with organic solvents. In *Proceedings of the International Conference of Industrial Environmental Xenobiotics,* ed. I. Gut, M. Cikrt, and G. L. Plaa, pp. 255–62. Berlin: Springer-Verlag.

Jaeger, R. J., and I. L. Cote. 1982. Effect of hypoxia on the acute toxicity of acrylonitrile. *Res. Commun. Chem. Pathol. Pharmacol.* 36(2): 345–48.

Johanssen, F. R., and G. J. Levinska. 1986. Relationships between toxicity and structure of aliphatic nitriles. *Fundam. Appl. Toxicol.* 7(4): 690–97.

Knobloch, K., S. Szendzikowski, T. Czajkowska, and B. Krysiak. 1971. Acute and subacute toxicity of acrylonitrile. *Med. Presse,* 22(3): 257–69.

Kramlich, J. C., M. P. Heap, W. R. Seeker, and G. S. Samuelsen. 1985. Flame-mode destruction of hazardous waste compounds. In *Proceedings of the International Symposium on Combustion,* Vol. 20, pp. 1991–99; cited in *Chem Abstr. CA* 104(4): 23772h.

Logan, C. J., D. H. Hutson, and D. Hesk. 1985. The fate of 4-cyano-*N,N*-dimethylaniline in mice: the occurrence of a novel metabolite during *N*-demethylation of and aromatic amine. *Xenobiotica* 15(5): 391–97.

Nakazawa, H., N. Jinnai, and F. Masahiko. 1984. Toxicological review of acrylonitrile with reference to migration from food-contact polymer. *Koshu Eiseiin Kenkyu Hokoku* 33(1): 35–38; cited in *Chem. Abstr. CA* 102(9): 77304j.

NFPA. 1986. *Fire Protection Guide on Hazardous Materials,* 9th. ed. Quincy, MA: National Fire Protection Association.

NIOSH. 1984. *Manual of Analytical Methods,* 3rd ed. Cincinnati, OH: National Institute for Occupational Health and Safety.

NIOSH. 1986. *Registry of Toxic Effects of Chemical Substances,* ed. D. V. Sweet. Washington DC: U.S. Government Printing Office.

Normandeau, J., S. Chakrabarti, and J. Brodeaur. 1984. Influence of simultaneous exposure to acrylonitrile and styrene on the toxicity and metabolism of styrene in rats. *Toxicol. Appl. Pharmacol.* 75(2): 346–49.

Peter, H., and H. M. Bolt. 1985. Effect of antidotes of the acute toxicity of methacrylonitrile. *Int. Arch. Occup. Environ. Health* 55(2): 175–77; cited in *Chem. Abstr. CA* 102(23): 199173m.

Smiley, R. A. 1981. Nitriles. In *Kirk-Othmer Encyclopedia of Chemical Technology,* Vol. 15, pp. 888–909. New York: Wiley-Interscience.

Stover, E. K., A. Fazel, and D. F. Kincannon. 1985. Powdered activated carbon and ozone-assisted activated sludge treatment for removal of toxic organic compounds. *Ozone Sci. Eng.* 7(3): 191–203; cited in *Chem. Abstr. CA* 104(14): 115399v.

U.S. EPA. 1984 (Revised 1990). Methods for organic chemical analysis of municipal and industrial wastewater. 40 CFR, Part 136, Appendix A.

U.S. EPA. 1986. *Test Methods for Evaluating Solid Waste,* 3rd ed. Washington DC: Office of Solid Waste and Emergency Response.

Tanii, H., and K. Hashimoto. 1984a. Structure–toxicity relationship of aliphatic nitriles. *Toxicol. Lett.* 22(2): 267–72.

Tanii, H., and K. Hashimoto. 1984b. Studies on the mechanism of acute toxicity of nitriles in mice. *Arch. Toxicol.* *55* (1): 47–54.

Tanii, H., and K. Hashimoto. 1985. Structure–toxicity relationship of mono- and dinitriles. *Pharmacochem. Libr. 8*: 73–82; cited in *Chem. Abstr. CA 103* (17): 136603b.

Willhite, C. C. 1981. Inhalation toxicology of acute exposure to aliphatic nitriles. *Clin. Toxicol. 18* (8): 991–1003.

Willhite, C. C. 1983. Development toxicology of acetonitrile in the Syrian golden hamster. *Teratology 27* (3): 313–25.

14

CYANIDES, INORGANIC

14.1 GENERAL DISCUSSION

Inorganic cyanides are the metal salts of hydrocyanic acid, HCN. Alkali and alkaline metal cyanides are characterized by ionic bonding between the metal ion and the cyanide, CN^-, and thus having crystalline structure. These are formed by the typical acid–base reaction:

$$HCN + NaOH \longrightarrow NaCN + H_2O$$

In addition to the wet method above, other commercial methods have been developed to produce cyanides from carbonates, ferrocyanides, or metals.

Hydrogen cyanide exhibits physical and chemical properties that typify both inorganic cyanide and nitrile. It is a very weak acid; the magnitude of dissociation constant is on the same order as that of amino acids. It is also a nitrile of formic acid and exhibits many reactions similar to nitriles. However, because its toxicology is similar to that of alkali cyanides, HCN is discussed in this chapter.

Toxicity

Cyanides of alkali metals are extremely toxic and the poisoning effects of sodium and potassium cyanides are well documented. Hydrogen cyanide, which is the toxic metabolite of nitriles, is more toxic than sodium or potassium cyanide. The order of toxicity in rabbits is as follows: HCN > NaCN > KCN; with LD_{50} values of 0.039, 0.103, and 0.121 mmol/kg, respectively (Ballantyne et al. 1983b).

Although the difference in the lethality of sodium and potassium cyanide is not much, such a pattern of gradual lowering of toxicity with increasing molecular weight is well manifested in general by most classes of organic compounds. Among other inorganic cyanides hazardous to health are cyanogen, cyanogen chloride, cyanogen bromide, cyanamide, barium cyanide, and other metal cyanides. In addition to being extremely toxic by ingestion or skin absorption, most metal cyanides present a serious hazard of forming extremely toxic HCN when coming into contact with acids.

The toxicity of cyanide is attributed to its ability to inhibit enzyme reactions. The action of one such enzyme, cytochrome oxidase, essential for the respiration of cells is inhibited by cyanide ions. Cytochrome oxidase is a component of the mitochondrial electron transport system. It transfers electrons from cytochrome c to oxygen, forming water, while releasing sufficient free energy to permit the formation of adenosine 5'-triphosphate (ATP). The latter is essential for normal metabolic processes. Cyanide ion forms complexes with heavy metal ions such as iron and copper to stop electron transport and thus prevent ATP formation. Several enzyme reactions have been listed that cyanide can inhibit several enzyme reactions by forming complexes.

14.2 HYDROGEN CYANIDE

EPA Classified Hazardous Waste, RCRA Waste Number P063; DOT Label: Poison A and Flammable Gas, UN 1614, 5% solution or more, UN 1613

Formula HCN; MW 27.03; CAS [74-90-8]

Structure: H—C≡N, linear triply bonded

Synonyms: hydrocyanic acid; prussic acid; formonitrile

Uses and Exposure Risk

Hydrogen cyanide is used to produce methyl methacrylate, cyanuric chloride, triazines, sodium cyanide, and chelates such as EDTA; and in fumigation. It occurs in beet sugar residues and coke oven gas. It occurs in the roots of certain plants, such as sorghum, cassava, and peach tree roots (Branson et al. 1969; Esquivel and Maravalhas 1973; Israel et al. 1973) and in trace amounts in apricot seeds (Souty et al. 1970) and tobacco smoke (Rickert et al. 1980).

Firefighters chance a great risk to the exposure to HCN, which is a known fire-effluent gas. Materials such as polyurethane foam, silk, wool, polyacrylonitrile, and nylon fibers burn to produce HCN (Sakai and Okukubo 1979; Yamamoto 1979; Morikawa 1988; Levin et al. 1987; Sumi and Tsuchiya 1976) along with CO, acrolein, CO_2, formaldehyde, and other gases. Emissions of these toxic gases take place primarily under the conditions of oxygen deficiency, and when the air supply is plentiful the emissions are decreased considerably (Hoschke et al. 1981).

Bertol et al. (1983) determined that 1 g of polyacrylonitrile generated 1500 ppm of HCN. Thus a lethal concentration of HCN could be obtained by burning 2 kg of polyacrylonitrile in an average-sized living room. Table 14.1 presents the concentration of HCN generated in fires of different polymeric materials.

Jellinek and Takada (1977) reported evolution of HCN from polyurethanes as a result

TABLE 14.1 Emission of Hydrogen Cyanide from Combustion of Polymeric Materials

Material Burned	HCN (ppm)	Reference
Polyacrylonitrile, 1 g	1,500	Bertol et al. 1983
Polyurethane foam, 10–15 g	100	Yamamoto and Yamamoto 1971
Aliphatic polyurethane[a]	1,000	Jellinek and Takada 1977
Aromatic polyurethane[a]	4,000–8,000	Jellinek and Takada 1977
Modacrylic drape[b]	64,000	Sarkos 1979
Wool[b]	42,000	Sarkos 1979
Wool carpet[b]	15,000	Sarkos 1979
Natural lawn[c]	8.2	Nishimaru and Tsuda 1982
Artificial lawn[c]	6.3	Nishimaru and Tsuda 1982

[a]At 311°C after 1 hr, 20% oxygen volume.
[b]At 600°C, sample weight 250 mg, airflow rate 2000 mL/min and test time 5 min.
[c]The oxygen level in the chamber dropped down to 14%.

of oxidative thermal degradation while no HCN evolved from pure thermal degradation. Copper inhibited HCN liberation due to the catalytic oxidation of evolved HCN. Herrington (1979) observed that the isocyanate portion of polyurethane foam volatilizes first, releasing heat, smoke, HCN, nitrogen oxides, and organic compounds. Volatilization of polyolefin portion occurs next, releasing CO and CO_2. Kishitani and Nakamura (1974) reported that the largest amount of HCN was evolved from urethane foam at 500°C, whereas with polyacrylonitrile and nylon 66, the amount of HCN increased with increasing temperature.

Physical Properties

Hydrogen cyanide is a colorless or pale blue liquid or a gas; odor of bitter almond; mp -13.4°C; bp 25.6°C; density of liquid 0.688 at 20°C and gas 0.947 (air = 1) at 31°C; very weakly acid; soluble in water and alcohol, slightly soluble in ether.

Health Hazard

Hydrogen cyanide is a dangerous acute poison by all toxic routes. Lethal effects due to inhalation of its vapor depend on its concentration in air and time of exposure. Inhalation of 270 ppm HCN in air can be fatal to humans instantly, while 135 ppm can cause death after 30 minutes (Patty 1963; ACGIH 1986). Exposure to high concentration can cause asphyxia and injure the central nervous system, cardiovascular system, liver, and kidney.

HCN is extremely toxic via ingestion, skin absorption, and ocular routes. Swallowing 50 mg can be fatal to humans. The symptoms of HCN poisoning at lethal dosage are labored breathing, shortness of breath, paralysis, unconsciousness, convulsions, and respiratory failure. At lower concentrations toxic effects are headache, nausea, and vomiting.

LD_{50} value, intravenous (mice): 0.99 mg/kg
LD_{50} value, oral (mice): 3.70 mg/kg

Investigating the relationship between pH (in the range 6.8–9.3) and the acute toxicity of HCN on fathead minnow, Broderius et al. (1977) observed that similar to H_2S, the toxicity of HCN increased at an elevated pH value. This was attributed to certain chemical changes occurring at the gill surface and possible penetration of the gill by both molecular and anionic forms.

In an acute lethal toxicity study on the influence of exposure route, Ballantyne (1983a) observed that the blood cyanide concentrations varied with the route. Concentrations in certain specific tissues varied markedly with exposure route. The blood cyanide concentration was lowest by inhalation and skin penetration. For a given exposure route, the cyanide level in blood were similar for different species. Among the most toxic cyanides, HCN was more toxic than NaCN or KCN by intramuscular and transocular routes.

Alarie and Esposito (1988) proposed a blood cyanide concentration of 1 mg/L as the fatal threshold value for HCN poisoning by inhalation. A cyanide concentration of 1.2 mg/L was measured in test animals exposed to nylon carpet smoke. The combined toxicities of fire effluent gases CO and HCN was found to be additive (Levin et al. 1988). The study indicated that the sublethal concentrations of the individual gases became lethal when combined. Furthermore, the presence of CO_2 combined with decreasing oxygen concentration enhanced the toxicity of the CO–HCN mixture (Levin et al. 1987). HCN and nitric oxide hastened the incapacitation in rats produced by carbon monoxide. Such incapacitation occurred at a carbonyl hemoglobin concentration of 42.2–49%; while for CO alone 50–55% carbonyl hemoglobin manifested the same effect (Condit et al. 1978).

Exposure Limits

Ceiling 11 mg/m^3 (10 ppm) (ACGIH), 5 mg CN/m^3/10 min (NIOSH); TWA air 11 mg/m^3 (10 ppm) skin (OSHA); IDLH 50 ppm.

Antidote

Amyl nitrate is an effective antidote against cyanide poisoning. The treatment involves artificial respiration, followed by inhalation of amyl nitrite. In severe poisoning, intravenous administration of sodium nitrite, followed by sodium thiosulfate is given. The dose is 0.3 g of $NaNO_2$ (10 mL of a 3% solution at a rate of 2–5 mL/min) and 25 g of $Na_2S_2O_3$ (50 mL of a 25% solution at a rate of 2–5 mL/min) (Hardy and Boylen 1983).

Nitrite displaces cyanide bound to methemoglobin and the released cyanide is oxidized to harmless cyanate.

Fire and Explosion Hazard

Highly flammable liquid; flash point (closed cup) −17.8°C (0°F); vapor pressure 620 torr at 20°C; vapor lighter than air and diffuses laterally and upward; vapor density 0.9 (air = 1); ignition temperature 540°C (1000°F); the vapor forms an explosive mixture with air in the range 6–41% by volume of air; fire-extinguishing agent: dry chemical, "alcohol" foam, or CO_2; use a water spray to flush the spill, disperse the vapors, and keep fire-exposed containers cool.

Hydrogen cyanide becomes unstable if stored for long periods and may explode (NFPA 1986). In the presence of 2–5% water or trace alkali it can polymerize exothermically. It can react violently when mixed with acetaldehyde and anhydrous ammonia. Violent hydrolysis can occur if an excess of strong acid such as 10% H_2SO_4 is added to confined HCN.

Storage and Shipping

Hydrogen cyanide is stored in a flammable liquids storage cabinet or a cool outside area isolated from sources of ignition and caustic alkalies. The storage temperature should be less than 32°C (90°F) and should be inhibited with 0.06–0.5% sulfuric or phosphoric acid (Jenks 1979). The containers should be discarded after about 3 months. It is shipped in steel cylinders.

Disposal/Destruction

Hydrogen cyanide can be destroyed by burning in a combustion furnace with a fuel–air mixture (Samuelson 1988). The quantity of oxygen should be less than the stoichiometric amount to decrease the formation of nitric oxide. Morley (1976) reported that HCN completely oxidized in rich hydrocarbon flame, producing partly NO.

A small amount of HCN can be burned in a hood. It can be converted to nontoxic ferrocyanide by treating with ferrous sulfate under strong basic conditions (pH 12). A convenient method is to oxidize the cyanide to cyanate, which is nontoxic. Hypochlorite solution (10%), permanganate, H_2O_2, ozone, and Cl_2/NaOH are some of the oxidizing agents. Caustic soda is added to ice-cold hydrogen cyanide solution before the dropwise addition of the oxidizing agent.

Dave et al. (1985) reported the biodegradation of toxic cyanide waste. The effluent was pretreated with hot alkali digestion and chemical coagulation followed by biological aeration. The HCN content in black gram and other seeds at a level of 1 mg/100 g flour was completely destroyed in 20-minute cooking (Dwivedi and Singh 1987).

Analysis

HCN can be tested colorimetrically. A method suitable for detecting 10 ppm HCN involves the use of test paper treated with *para*-nitrobenzaldehyde and K_2CO_3. A reddish purple stain is produced due to HCN. Its presence in plants, tissue, and toxicologic substances can be tested colorimetrically using 4-(2-pyridylazo)resorcinol-palladium in a carbonate–bicarbonate buffer solution. The intense red color changes to yellow upon the action of HCN (Carducci et al. 1972).

It can be analyzed by GC and other techniques. Air analysis is done by sampling on the filter bubbler, 0.8 μmm cellulose ester membrane, and 0.1 N KOH. The cyanide ion is analyzed by the cyanide-specific electrode (NIOSH 1984, Method 1904).

Many portable devices to measure HCN and other toxic gases have been developed (Diehl et al. 1980) in which the user is warned as to its acoustic and optic effects.

14.3 SODIUM CYANIDE

EPA Classified Toxic Waste, RCRA Waste Number P106; DOT Label: Poison, UN 1689

Formula NaCN; MW 49.02; CAS [143-33-9]

Structure and functional group: Na^+CN^-, ionic bond between Na^+ and CN^-, a face-centered-cubic (fcc) crystal lattice structure similar to that of sodium chloride

Synonyms: Cyanobrik; Cyanogran; white cyanide

Uses and Exposure Risk

Sodium cyanide is used for electroplating metals such as zinc, copper, cadmium, silver, and gold, and their alloys; for extracting gold and silver from ores; and as a fumigant and a chelating agent. It occurs in many varieties of maniocs (cassava), especially in bitter manioc.

Physical Properties

White crystalline granules or fused pieces, isomorphous to NaCl, forming mixed crystals (Jenks 1979); faint odor of bitter almond, odorless when dry; mp 564°C; density 1.62; readily soluble in water and liquid ammonia, low solubility in alcohol (6.4 g/100 g methanol at 15°C), furfural, and dimethyl formamide.

Health Hazard

Sodium cyanide is a highly poisonous compound by oral intake and by ocular and skin absorption. Accidental ingestion of a small quantity; as low as 100–150 mg could result in immediate collapse and instantaneous death in humans. At a lower dosage it can cause nausea, vomiting, hallucination, headache, and weakness. The toxicology of NaCN is the same as that of HCN. The metal cyanide forms HCN rapidly in the body, causing immediate death from a high dosage.

The lethal effect from cyanide poisoning varied with species. Investigating the acute oral toxicity of sodium cyanide in birds, Wiemeyer et al. (1986) observed that the LD_{50} values for the flesh-eating birds were lower than that for the birds that fed on plant material; vulture 4.8 mg/kg versus chicken 21 mg/kg. In a study on marine species, Pavicic and Pihlar (1983) found that at 10 ppm concentration of NaCN, invertebrates were more sensitive than fishes. In animals, the lethal dose of NaCN were in the same range by different toxic routes. A dose of 8 mg NaCN/kg resulted in ataxia, immobilization, and death in coyotes (Sterner 1979); however, the lethal time was longer, at 18 minutes.

Ballantyne (1983b) studied the acute lethal toxicity of sodium and other cyanides by ocular route. He found that cyanide instilled into the eye was absorbed across conjunctival blood vessels causing systemic toxicity and death within 3–12 minutes of the eye being contaminated. The toxicity of the cyanide did not decrease by mixing the solid with an inert powder such as kaolin.

LD_{50} value, intraperitoneal (mice): 4.3 mg/kg
LD_{50} value, oral (rats): 6.4 mg/kg

Sodium cyanide is a teratogen, causing fetus damage and developmental abnormalities in the cardiovascular system in hamsters (NIOSH 1986).

Sodium cyanide reacts with acids to form highly toxic hydrogen cyanide. There could be a slow liberation of HCN in contact with water.

Exposure Limits

TLV-TWA (measured as CN) skin 5 mg CN/m^3 (ACGIH and OSHA), 5 mg CN/m^3/ 10-min ceiling (NIOSH).

Antidote

Many antidotes have been developed to combat cyanide poisoning. Some of these are simple salts such as sodium thiosulfate and sodium nitrate, which demonstrated partial effect. Friedberg and Shukla (1975) reported the efficiency of aquacobalmin acetate [49552-79-6] as an antidote when given alone or combined with sodium thiosulfate. Aquacobalmin 100 and 200 mg/kg and thiosulfate 500 mg/kg injected separately with a 1-minute interval detoxified cyanide in guinea pigs and cats, respectively.

Atkinson et al. (1974) reported the antidote action of Kelocyanor [15137-09-4] and enzyme. The bacterial enzyme rhodanese [55073-14-8] mixed with sodium thiosulfate (40 µg and 100 mg/kg) was effective against NaCN in rabbits. Amyl nitrite is a common cyanide antidote.

Hazardous Reactions

Sodium cyanide explodes violently when heated with chlorate or nitrite at 450°C or melted with nitrite or nitrate salt. It reacts violently with nitric acid and vigorously with fluorine. It produces highly toxic HCN when mixed with acids; cyanogen when reacted with copper sulfate; cyanogen chloride, bromide, and iodide when mixed with chlorine, bromine, and iodine, respectively.

Disposal/Destruction

Small amounts of sodium cyanide in the laboratory can be oxidized by aqueous sodium hypochlorite at pH 12 to form nonhazardous cyanate.

$$NaCN + NaOCl \longrightarrow NaOCN + NaCl$$

The cyanide solution is added dropwise to a stirred hypochlorite solution maintained at 45–50°C. Commercial laundry bleach that contains NaOCl can be used. The reaction mixture along with unreacted laundry bleach can be flushed down the drain with excess water (National Research Council 1983).

Cyanide waste can be disposed in a landfill by the lab pack method: The waste is placed in metal containers and packed with an adsorbent material such as vermiculite or fuller's earth and placed in a 55-gallon steel drum for landfill disposal.

Bacterial treatment was effective in reducing the high concentration of NaCN in manioc processing wastes from 15,000 ppm to 250 ppm (Branco 1979).

Analysis

The cyanide content of a NaCN sample can be estimated by titration with 0.1 N $AgNO_3$ solution using potassium iodide or para-dimethylaminobenzalrhodamine as indicator. The selective-ion electrode method may be accurate below 0.5 ppm concentration.

Analysis of cyanide particulate in air is performed by flowing air through a 0.8-mm cellulose ester membrane, followed by extraction with 0.1 N KOH and analysis for cyanide by specific-ion electrode.

14.4 POTASSIUM CYANIDE

EPA Classified Toxic Waste, RCRA Waste No PO98; DOT Label: Poison UN 1680

Formula KCN; MW 65.12; CAS [151-50-8]

Structure: K^+ and CN^- are held by ionic bond, cubic and orthorhombic forms, and fcc crystal structure at room temperature

Synonym: kalium cyanide

Uses and Exposure Risk

Potassium cyanide is used for electrolytic refining of platinum; fine silver plating; as an electrolyte for the separation of gold, silver, and copper from platinum; and for metal coloring.

Physical Properties

White crystalline solid; deliquescent; mp 634.5°C; density 1.553 at 20°C; readily soluble in water (71.6 g/100 g), formamide, glycerol, and hydroxylamine, moderately soluble in liquid ammonia and methanol.

Health Hazard

Potassium cyanide is a dangerous poison; toxicity comparable to that of sodium cyanide. Ingestion of 100–150 mg of this compound could be fatal to humans. Similar toxicity is observed when KCN is absorbed through skin or eyes. Intake of the quantity above can cause collapse and cessation of breathing.

At lower concentrations the acute toxic symptoms are nausea, vomiting, headache, confusion, and muscle weakness. KCN administered in test animals by the intramuscular, intravenous, intraperitoneal, ocular, and oral routes exhibited LD_{50} values within the range of 3–9 mg/kg; the acute toxic effects were ataxia, respiratory stimulation, paralysis, and seizure.

Smith and Heath (1979) observed the effect of temperature on KCN toxicity of freshwater fish. When the temperature of exposure was lowered from 15°C to 5°C the toxicity of KCN to goldfish decreased by a factor of 5.

LD_{50} value, intraperitoneal (rats): 4 mg/kg
LD_{50} value, oral (rats): 10 mg/kg
LD_{50} value, oral (humans): 2.86 mg/kg

Potassium cyanide caused reproductive damage in test animals, producing harmful effects on fertility and embryo.

Exposure Limits

TLV-TWA (measured as CN) skin 5 mg CN/m3 (ACGIH and OSHA); 5 mg CN/m^3/ 10 min ceiling (NIOSH).

Antidote

Sodium thiosulfate and sodium sulfate have shown antidotal properties toward KCN toxicity. Certain compounds and enzymes may exhibit similar action (see Sections 14.2 and 14.3). Ohkawa et al. (1972) reported that *para*-aminopropiophenone [70-69-9] protected mice against lethal doses of KCN.

Hazardous Reactions

Potassium cyanide is a noncombustible solid; vapor pressure is 0 at 20°C and does not present any fire or explosion hazard at ambient temperature. It liberates dangerously toxic HCN when it comes in contact with acid:

$$KCN + HCl \longrightarrow HCN + KCl$$

Powdered metals, magnesium, calcium, beryllium, boron, and aluminum reduce KCN on heating in the absence of air, forming potassium (highly reactive):

$$2KCN + 3Mg \longrightarrow 2K + 2C + Mg_3N_2$$

It forms highly toxic cyanogen gas on reacting with copper sulfate or chloride:

$$2KCN + CuSO_4 \longrightarrow K_2SO_4 \\ + Cu + (CN)_2$$

Disposal/Destruction

Small amounts of KCN can be destroyed in the laboratory by treating with sodium hypochlorite. Laundry bleach containing about 5% NaOCl is added dropwise to cyanide solution maintained at a constant temperature of 45–50°C. Nontoxic potassium cyanate formed according to the reaction

$$KCN + NaOCl \longrightarrow KOCN + NaCl$$

is drained with excess water along with unreacted bleach (see Section 14.3).

The lab pack method for landfill disposal and bacterial degradation are suitable modes for disposal and destruction (see Section 14.3).

Wolverton and Bounds (1988) reported the removal of KCN and other toxic chemicals from contaminated water by microbial aquatic plants, torpedo grass, and southern bulrush.

Analysis

The cyanide content of a KCN sample can be determined by titration with $0.1\ N$ $AgNO_3$ solution. Potassium iodide is a suitable indicator at a high cyanide concentration of approximately 1%. *para*-Dimethylaminobenzalrhodamine is more sensitive for very low concentrations of cyanide. Among other methods, a selective-ion electrode is accurate for concentrations below 0.5 ppm.

Cyanide particulate in aerosol is analyzed by collecting it on a 0.8-μm cellulose ester membrane, extracting with $0.1\ N$ KOH solution, and analyzing by a selective-ion electrode (NIOSH 1984, Method 1904).

14.5 CALCIUM CYANIDE

EPA Classified Toxic Waste, RCRA Waste Number P021; DOT Label: Poison, UN 1575

Formula $Ca(CN)_2$; MW 92.12; CAS [592-01-8]

Structure: Two cyanide radicals held to Ca^{2+} ion by weak Coulombic forces, rhombohedric crystal

Synonyms: Calcyanid; Cyanogas; black cyanide

Uses and Exposure Risk

Calcium cyanide is used for the extraction of gold and silver from their ores, in the froth flotation of minerals, as a fumigant, and as a rodenticide.

Physical Properties

White crystalline solid; mp 640°C (decomposes); soluble in water, formamide, and hydroxylamine, slightly soluble in liquid ammonia.

Health Hazard

Calcium cyanide is a highly poisonous compound to humans, animals, and fish. The toxic routes are ingestion, skin contact, and inhalation of the dust. It forms HCN readily when it reacts with CO_2 or water. This makes it highly hazardous, more so than the alkali metal cyanides, although the LD_{50} value of $Ca(CN)_2$ is greater than the sodium or potassium cyanides.

LD_{50} value, oral (rats): 39 mg/kg

Exposure Limits

TLV-TWA (measured as CN) skin 5 mg(CN)/m^3 (ACGIH); 5 mg(CN)/m^3/10-min ceiling (NIOSH).

Hazardous Reactions

Reactions with acids, acidic salts, water, and CO_2 can release HCN.

$$Ca(CN)_2 + 2HCl \longrightarrow CaCl_2 + 2HCN$$

$$Ca(CN)_2 + 2H_2O \longrightarrow$$
$$Ca(OH)_2 + 2HCN \text{ (black polymer)}$$

HCN formation can even occur when calcium cyanide is exposed to humid atmosphere.

Storage and Shipping

Calcium cyanide is stored in tight containers free from moisture. Proper ventilation and protective equipment should be used while handling the solid or while preparing an aqueous solution. It is shipped in mild-steel or fiber drums.

Disposal/Destruction

Calcium cyanide is destroyed by oxidizing it to calcium cyanate using sodium hypochlorite or other oxidizing compounds (see Section 14.3):

$$Ca(CN)_2 + NaOCl \longrightarrow Ca(OCN)_2$$
$$+ 2NaCl$$

It may be converted to calcium ferrocyanide by treating with ferrous sulfate:

$$Ca(CN)_2 + FeSO_4 \longrightarrow CaFe(CN)_6$$
$$+ CaSO_4$$

Analysis

Calcium cyanide can be analyzed by titration with silver nitrate using potassium iodide indicator. Sulfide interfering in the reaction is removed by sodium carbonate–lead acetate solution.

14.6 CYANOGEN

EPA Classified Toxic Waste, RCRA Waste Number P031; DOT Label: Poison, UN 1026

Formula C_2N_2; MW 52.04; CAS [460-19-5]

Structure: $N{\equiv}C{-}C{\equiv}N$

Synonyms: ethanedinitrile; oxalic acid dinitrile; oxalonitrile; oxalyl cyanide; carbon nitride; dicyan

Uses and Exposure Risk

Cyanogen is used as a fumigant, as a fuel gas for welding and cutting metals, as a propellant, and in organic synthesis. It occurs in blast furnace gases.

Physical Properties

Colorless gas; almond-like smell; becoming acrid and pungent at high concentrations; density 0.87 at 20°C; mp −27.9°C; bp −21.1°C; freezes at −57°C; highly soluble in water, alcohol, and ether.

Health Hazard

Cyanogen is a highly poisonous gas having toxic symptoms similar to those of HCN. Acute exposure can result in death by asphyxia. The toxic routes are inhalation and percutaneous absorption. At sublethal concentrations the symptoms of acute toxicity are nausea, vomiting, headache, confusion, and weakness.

Rats exposed to cyanogen exhibited toxic symptoms of respiratory obstruction, lacrimation, and somnolence. Exposure to 350 ppm for 1 hour caused death to 50% of test animals. In humans, exposure to 16 ppm for 5 minutes produced irritation of eyes and nose. Toxicity of cyanogen is considerably lower than that of HCN. Lethal dose in test animals from subcutaneous injection varied between 10 and 15 mg/kg.

A subchronic toxicity study conducted on male rhesus monkeys and male albino rats exposed over a period of 6 months (6 hours/day, 5 days/week) indicated marginal toxicity of cyanogen at 25 ppm (Lewis et al. 1984). Total lung moisture content and body weights were significantly lower. The odor threshold level for cyanogen is about 250 ppm.

Exposure Limit

TLV-TWA 20 mg/m^3 (10 ppm) (ACGIH).

Antidote

Antidotes for HCN and other cyanides may be effective for cyanogen. Inhalation of amyl nitrite is recommended in case of low poisoning.

Fire and Explosion Hazard

Highly flammable, burns with a purple-tinged flame; vapor density 1.8 (air = 1); the vapor may travel a considerable distance to an ignition source and flash back; fire-extinguishing procedure: use a water spray to fight fire and keep fire-exposed containers cool; shut off the flow of gas.

Cyanogen forms an explosive mixture with air within the range of 6.6–32%. Liquid cyanogen can explode when mixed with liquid oxygen. When mixed with an acid or water or when heated to decomposition, it produces toxic fumes.

Storage and Shipping

Cyanogen is stored outside or in a detached area: cool, dry, and well ventilated, and isolated from acid, acid fumes, and water. It is shipped in high-pressure metal cylinders of <45-kg (<100-lb) water capacity.

Disposal/Destruction

Cyanogen is dissolved in a combustible liquid and burned in an incinerator.

Analysis

Cyanogen may be analyzed by GC by direct injection of air. Alternatively, air is passed through a cellulose ester membrane filter, extracted with KOH solution, and analyzed for cyanide by ion-specific electrode. It can also be analyzed by GC/MS technique.

14.7 CYANOGEN CHLORIDE

EPA Classified Acute Hazardous Waste, RCRA Waste Number PO33; DOT Label: Poison and Corrosive, UN 1589

Formula CNCl; MW 61.47; CAS [506-77-4]

Structure: Cl—C≡N

Synonyms: chlorine cyanide; chlorocyanide; chlorocyan; chlorocyanogen

Uses and Exposure Risk

Cyanogen chloride is used in organic synthesis and as a military poisonous gas.

Physical Properties

Colorless liquid or gas with pungent smell; density 1.186; mp −6°C; bp 13.8°C; highly soluble in water, alcohol, ether, and other organic solvents.

Health Hazard

Cyanogen chloride is a highly poisonous compound and a severe irritant. In humans, exposure to 1 ppm for 10 minutes caused severe irritation of eyes and nose. Irritation of respiratory tract is followed by hemorrhage of the bronchi and trachea, as well as pulmonary edema.

The toxicity of cyanogen chloride is attributed to its relatively easy decomposition to cyanide ion in an aqueous medium. The cyanide attacks the cells in the body and interferes with the cellular metabolism. Tests on rats indicated that exposure to cyanogen chloride caused lacrimation and chronic pulmonary edema and somnolence. At a high concentration of 300 ppm, death occurred to the test animals. Inhalation of 48 ppm for 30 minutes was fatal to humans. In animals, subcutaneous intakes of 5 and 15 mg/kg were lethal to both dogs and rabbits.

Chronic exposure to cyanogen chloride can cause conjunctivitis and edema of the eyelid.

Exposure Limit

Ceiling limit 0.7 mg/m^3 (0.3 ppm) (ACHIH).

Antidote

Inhalation of oxygen and amyl nitrite with artificial respiration may be effective against cyanogen chloride poisoning.

Disposal/Destruction

Lahaye et al. (1987) reported an effective method of destroying cyanogen chloride by treatment with pyridine-4-carboxylic acid adsorbed on activated carbon. The efficiency of the system is enhanced when water is preabsorbed on the carbon support.

Cyanogen chloride can be destroyed in an incinerator. It is dissolved in a combustible solvent and burned.

Analysis

Cyanogen chloride can be analyzed by GC and colorimetric techniques.

14.8 CYANOGEN BROMIDE

EPA Classified Toxic Waste, RCRA Waste Number U246; DOT Label: Poison and Corrosive UN 1889
Formula CNBr; MW 105.93; CAS [506-68-3]
Structure: $Br-C\equiv N$
Synonyms: bromine cyanide; bromocyanide; bromocyan; bromocyanogen; cyanobromide; Campilit

Uses and Exposure Risk

Cyanogen bromide is used in organic synthesis and as a reagent in bioanalysis.

Physical Properties

Crystalline solid; volatile at ordinary temperature; mp 52°C; density 2.015 at 20°C; readily soluble in water, alcohol, and ether.

Health Hazard

Cyanogen bromide is a highly toxic substance. Its toxic effects are similar to those of HCN. However, it is not as toxic as HCN. Because it volatilizes readily at ambient temperature, inhalation is the major toxic route. The toxic symptoms in humans may be nausea, headache, and chronic pulmonary edema. Exposure to 100 ppm for 10 minutes can be fatal to humans.

LC_{50} inhalation (mice): 500 mg/m³/10 min

Cyanogen bromide is an irritant.

Exposure Limit

No exposure limit is set. However, based on the exposure limits of related compounds a ceiling limit of 0.5 ppm (2 mg/m³) is recommended.

Antidote

Inhalation of oxygen and amyl nitrite may combat cyanogen bromide poisoning.

Hazardous Reactions

Cyanogen bromide may react violently with large amounts of acid. It decomposes in water, releasing toxic gases. Impure compounds decompose rapidly. Among the toxic gases produced from the reactions of cyanogen bromide are cyanogen, bromine, HCN, and HBr.

Disposal/Destruction

Cyanogen bromide can be destroyed by slowly adding to it an aqueous solution of caustic soda followed by an addition of excess of laundry bleach. Nontoxic sodium thiocyanate and unreacted bleach are drained down with large volumes of water.

Analysis

Cyanogen bromide is analyzed by colorimetric and GC techniques.

14.9 CYANOGEN IODIDE

Formula CNI; MW 152.92; CAS [506-78-5]
Structure: $I-C\equiv N$
Synonyms: iodine cyanide; iodocyanide; Jodcyan

Uses and Exposure Risk

Cyanogen bromide is used in taxidermy for preserving insects, butterflies, and so on (Merck 1989).

Physical Properties

White crystalline solid; highly pungent smell; acrid taste; mp 147°C; soluble in water, alcohol, and ether.

Health Hazard

Cyanogen bromide is a highly poisonous substance. Toxic routes are oral intake and skin absorption. Acute toxic symptoms on test animals were convulsion, paralysis, and respiratory failure. Ingestion of a 5-g amount could be fatal to humans.

LD_{Lo} value, oral (cats): 18 mg/kg
LD_{Lo} value, subcutaneous (dogs): 19 mg/kg

Cyanogen iodide is an irritant to skin.

Exposure Limit

Because cyanogen iodide is a solid with very low vapor pressure, it does not present any inhalation hazard.

Antidote

No specific antidote is reported. Sodium thiosulfate and other cyanide antidotes may be used.

Hazardous Reactions

Cyanogen iodide produces toxic gases when mixed with acids. It reacts with molten phosphorus with incandescence forming phosphorus iodide.

Disposal/Destruction

A suitable method for destroying cyanogen iodide may consist of treatment with caustic soda, followed by adding sodium hypochlorite (laundry bleach) to oxidize the cyanide to nontoxic cyanate.

Analysis

Cyanogen iodide may be analyzed by classical wet methods for cyanide and iodide tests, by colorimetry, and by GC and GC/MS.

14.10   HYDROGEN CYANATE

Formula CHNO; MW 43.03; CAS [420-05-3]
Structure: $N \equiv C - OH$
Synonym: cyanic acid

Uses and Exposure Risk

Hydrogen cyanate is used in the preparation of cyanates.

Physical Properties

Colorless liquid or gas with an acrid smell; mp −86°C; bp 23.5°C; density 1.140 at 20°C; soluble in water, alcohol, ether, benzene, and toluene.

Health Hazard

Hydrogen cyanate is a severe irritant to the eyes, skin, and mucous membranes. Exposure to this compound can cause severe lacrimation. Inhalation can produce irritation and injury to the respiratory tract. LD_{50} values are not reported.

Exposure Limit

No exposure limit has yet been set for this compound.

Fire and Explosion Hazard

Flammable; the liquid can explode when heated rapidly.

Disposal/Destruction

Hydrogen cyanate can be disposed of in the drain in small amounts. It decomposes in water forming CO_2 and NH_3.

Analysis

Hydrogen cyanate can be analyzed by measuring its decomposition products: CO_2 by GC/TCD or NH_3 by colorimetry.

14.11 CUPROUS CYANIDE

EPA Classified Hazardous Waste, RCRA Waste Number P029; DOT Label: Poison, UN 1587

Formula CuCN; MW 89.56; CAS [544-92-3]

Structure: orthorhombic or monoclinic crystals

Synonyms: copper(I) cyanide; cupricin

Uses and Exposure Risk

Cuprous cyanide is used in electroplating; as an insecticide and fungicide; and as a catalyst for polymerization.

Physical Properties

White powder, dark green orthorhombic and dark red monoclinic crystals; mp 474°C; density 2.92; slightly soluble in water and alcohol, soluble in NH_4OH and boiling dilute HCl (Merck 1989).

Health Hazard

Cuprous cyanide is a highly toxic substance. The toxic routes are inhalation of dust, ingestion, and skin contact. Toxicology and LD_{50} values for this compound are not reported. Because it is slightly soluble in water, its dissociation to cuprous and cyanide ions in the body may not be significant. The role of cyanide ion in the toxicity of cuprous cyanide is not established. The inhalation hazard, however, is attributable to copper. It is a skin irritant.

Exposure Limit

TLV-TWA 1 mg Cu/m^3 (ACGIH).

Disposal/Destruction

Cuprous cyanide can be destroyed by slowly adding its dilute alkaline solution with an excess of laundry bleach maintaining a constant temperature. After allowing it to stand for several hours and adjusting the pH value to 7 and precipitating heavy metals as sulfide, the filtrate is drained down and the precipitate is disposed for burial in a chemical landfill (Aldrich 1996).

Analysis

Cuprous cyanide is dissolved in hot dilute HCl and the solution is analyzed for copper by atomic absorption spectrophotometry. Cyanide may be analyzed by selective ion electrode.

14.12 MERCURIC CYANIDE

DOT Label: Poison, UN 1636

Formula $Hg(CN)_2$; MW 252.63; CAS [592-04-1]

Structure: tetragonal crystals

Synonym: mercury(II) cyanide

Uses and Exposure Risk

Mercuric cyanide finds veterinary application as a topical antiseptic for cats and other animals.

Physical Properties

White powder or a colorless crystal; odorless; decomposes at 320°C; density 3.996; soluble in acetone, alcohol, and hot water, slightly soluble in cold water (8%).

Health Hazard

Mercuric cyanide is a highly poisonous compound. Its components, mercury(II) and the cyanide ions, are both highly toxic. Its toxicity, however, is lower than that of sodium and potassium cyanides.

Acute toxic symptoms from oral intake of this compound in humans are hypermotility, diarrhea, nausea or vomiting, and injury to kidney and bladder. Toxic symptoms may be manifested in humans from consuming 15–20 g of this compound. Lower doses may produce somnolence. An intraperitoneal dosage of 7.5 mg/kg was fatal to rats.

LD_{50} value, oral (mice): 7.5 mg/kg

Exposure Limit

TLV-TWA 0.1 mg Hg/m^3 (skin) (ACGIH).

Disposal/Destruction

Mercuric cyanide solution is precipitated with a sulfide as HgS at pH 7. The precipitate

TABLE 14.2   Toxicity and Hazardous Reactions of Miscellaneous Inorganic Cyanides

Compound/Synonyms/ CAS No.	Formula/MW	Toxicity	Hazardous Reactions
Ammonium cyanide [12211-52-8]	NH_4CN 44.046	Highly toxic; toxic effects similar to those of NaCN and KCN; LD_{50} values not reported	Readily decomposes to very toxic mixture of NH_3 and HCN
Lithium cyanide [2408-36-8]	LiCN 32.949	Highly toxic; toxic symptoms of NaCN and KCN; LD_{50} values not reported	Produces toxic HCN with acids; heating with nitrites, nitrates, nitric acid, and chlorates can be violent
Potassium ferricyanide (tripotassium hexacyanoferrate) [13746-66-2]	$K_3Fe(CN)_6$/MW 329.27	Low toxicity; LD_{50} oral (rats): 1600 mg/kg	Decomposes by acids to toxic HCN
Cupric cyanide [copper (II) cyanide] [14763-77-0]	$Cu(CN)_2$ 115.58	Moderate to severe poison; LD_{50} intraperitoneal (rats): 50 mg/kg; DOT Label: Hazard, Poison B	Forms toxic HCN with acids
Gold cyanide (aurus cyanide) [506-65-0]	AuCN 223.02	Highly toxic; LD_{50} values not reported	Liberates toxic HCN with acids
Silver cyanide [506-64-9], RCRA Hazardous Waste No P104	AgCN 133.89	Highly toxic; LD_{50} oral (rats): 123 mg/kg severe eye irritant: 5 mg/24 hr (rabbits) mild skin irritant; DOT Label: Hazard, Poison B	Forms toxic HCN with acids
Cadmium cyanide [542-83-6]	$Cd(CN)_2$ 164.45	Very poisonous; LD_{50} values not reported	Releases HCN with dilute mineral acids; reacts exothermically with Mg
Zinc cyanide [557-83-6], RCRA Hazardous Waste No. P121	$Zn(CN)_2$ 117.41	Highly poisonous; an intraperitoneal dosage of 100 mg/kg was lethal to rats; DOT Label: Poison, UN 1713	Releases HCN with dilute mineral acids; reacts with incandescence with Mg

is disposed of in a chemical landfill. The filtrate is treated slowly with sodium hypochlorite to oxidize the cyanide to cyanate.

Analysis

Mercury is analyzed by atomic absorption spectrometry using a graphite furnace or by a mercury analyzer. Cyanide is tested by an ion-specific electrode.

14.13 MISCELLANEOUS INORGANIC CYANIDES

The hazardous properties of miscellaneous cyanides are presented in Table 14.2.

REFERENCES

ACGIH. 1986. *Documentation of the Threshold Limit Values and Biological Exposure Indices*, 5th ed. Cincinnati, OH: American Conference of Governmental Industrial Hygienists.

Alarie, Y., and F. M. Esposito. 1988. Role of small scale toxicity testing in hazard control. *Fire Mater. 13*: 122–29.

Aldrich. 1996. *Aldrich Catalog*. Milwaukee, WI: Aldrich Chemical Company.

Atkinson, E., D. A. Rutter, and K. Sargent. 1974. Enzyme antidote for experimental cyanide poisoning. *Lancet 2*(7894): 1446.

Ballantyne, B. 1983a. The influence of exposure route and species on the acute lethal toxicity and tissue concentrations of cyanide. *Dev. Toxicol. Environ. Sci. 11*: 583–86.

Ballantyne, B. 1983b. Acute systemic toxicity of cyanides by topical application to the eye. *J. Toxicol. Cutaneous Ocul. Toxicol. 2*(2–3): 119–29.

Bertol, E., F. Mari, G. Orzalesi, and I. Volpato 1983. Combustion products from various kinds of fibers: toxicological hazards from smoke exposure. *Forensic Sci. Int. 22*(2–3): 111–16.

Branco, S. M. 1979. Investigations on biological stabilization of toxic wastes from manioc processing. *Prog. Water Technol. 11*(16): 51–54.

Branson, T. A., P. L. Guss, and E. E. Ortman. 1969. Toxicity of sorghum roots to larvae of the western corn rootworm. *J. Econ. Entomol. 62*(6): 1375–78.

Broderius, S. J., L. L. Smith, Jr., and D. T. Lind. 1977. Relative toxicity of free cyanide and dissolved sulfide forms to the fathead minnow. *J. Fish Res. Board Can. 34*(12): 2323–32.

Carducci, C. N., P. Luis, and A. Mascaro. 1972. Par [4-(2-pyridylazo)resorcinol] palladium as an analytical reagent for hydrogen cyanide. Application to the detection of thiocyanate and to biological and toxicological analysis. *Microchim. Acta 3*: 339–44.

Condit, D. A., D. Malek, A. D. Cianciolo, and C. H. Hofrichter. 1978. Life hazard evaluation of flexible polyurethane foam/fabric composites typical of cushioned upholstered furniture. 1978. *J. Combust. Toxicol. 5*(4): 370–90.

Dave, Y. I., K. T. Oza, and S. A. Puranik. 1985. Hazard prevention: biodegradation of toxic cyanide waste. *Chem. Age India 36*(8): 775–77.

Diehl, W., L. Grambow, and K. H. Huneke. 1980. Toxiwarn: a portable measuring and warning device for toxic gases. *Draeger Rev. 45*: 39–52.

Dwivedi, G. K., and V. P. Singh. 1987. Occurrence of some toxic factors associated with newly evolved strains of black gram and effect of cooking on these toxic factors. *Indian J. Agric. Chem. 20*(1): 33–38.

Esquivel, T. F., and N. Maravalhas. 1973. Rapid field method for evaluating hydrocyanic toxicity of cassava root tubers. *J. Agric. Food Chem. 21*(2): 321–22.

Friedberg, K. D., and U. R. Shukla. 1975. Efficiency of aquacobalamine as an antidote in cyanide poisoning when given alone or combined with sodium thiosulfate. *Arch. Toxicol. 33*(2): 103–13.

Hardy, H. L., and G. W. Boylen, Jr. 1983. Cyanogen, hydrocyanic acid and cyanides. In *Encyclopedia of Health and Occupational Safety*, 3rd ed., ed. L. Parmeggiani. Geneva: International Labor Office.

Herrington, R. M. 1979. The rate of heat, smoke and toxic gases release from polyurethane foams. *J. Fire Flammability 10*(Oct.): 308–25.

Hoschke, B. N., J. J. Madden, J. W. Milne, and M. F. R. Mulcahy. 1981. Toxic gas emission from materials subjected to flame. *J. Combust. Toxicol. 8*(Feb.): 19–32.

Israel, D. W., J. E. Giddens, and W. W. Powell. 1973. Toxicity of peach tree roots. *Plant Soil 39*(1): 103–12.

Jellinek, H. H. G., and K. Takada. 1977. Toxic gas evolution from polymers: evolution of hydrogen cyanide from polyurethanes. *J. Polym. Sci. Polym. Chem. Ed. 15*(9): 2269–88.

Jenks, W. R. 1979. Cyanides. In *Kirk-Othmer Encyclopedia of Chemical Technology*, Vol. 7, pp. 307–34. New York: Wiley-Interscience.

Kishitani, K., and K. Nakamura. 1974. Toxicities of combustion products. *J. Fire Flammability/Combust. Toxicol. Suppl. 1*(May): 104–23.

Lahaye, J., P. Ehrburger, and R. Fangeat. 1987. Destruction of cyanogen chloride with 4-pyridinecarboxylic acid impregnated activated carbon: I. The physical chemistry of interaction. *Carbon 25*(2): 227–31.

Levin, B. C., M. Paabo, J. L. Gurman, and S. E. Harris. 1987. Effects of exposure to single or multiple combinations of the predominant toxic gases and low oxygen atmospheres produced in fires. *Fundam. Appl. Toxicol. 9*(2): 236–50.

Levin, B. C., M. Paabo, J. L. Gurnam, H. M. Clark, and M. F. Yoklavich. 1988. Further studies of the toxicological effects of different time exposures to the individual and combined fire gases: carbon monoxide, hydrogen cyanide, carbon dioxide, and reduced oxygen. *Proc. SPI Annu. Tech. 31*: 249–52.

Lewis, T. R., W. K. Anger, and R. K. Te Vault. 1984. Toxicity evaluation of subchronic exposure to cyanogen in monkeys and rats. *J. Environ. Pathol. Toxicol. Oncol. 5*(4–5): 151–63.

Merck. 1989. *The Merck Index*, 11th ed. 1989. Rahway, NJ: Merck & Co.

Morikawa, T. 1988. Toxic gases evolution from commercial building materials under fire atmosphere in a semi-full-scale room. *J. Fire Sci. 6*(2): 86–99.

Morley, C. 1976. The formation and destruction of hydrogen cyanide from atmospheric and fuel nitrogen in rich atmospheric-pressure flames. *Combust. Flame 27*(2): 189–204.

NFPA. 1986. *Fire Protection Guide on Hazardous Materials*, 9th ed. Quincy, MA: National Fire Protection Association.

National Research Council. 1983. *Prudent Practices for Disposal of Chemicals from Laboratories*. Washington, DC: National Academy Press.

NIOSH. 1984. *Manual of Analytical Methods*, 3rd ed. Cincinnati, OH: National Institute for Occupational Safety and Health.

NIOSH. 1986. *Registry of Toxic Effects of Chemical Substances*, ed. D. V. Sweet. Washington DC: U.S. Government Printing Office.

Nishimaru, Y., and Y. Tsuda. 1982. Study of toxic gas generated during combustion: in case of natural and artificial lawn. *NBS Spec. Publ. (U.S.) 639*: 104–15.

Ohkawa, H., R. Ohkawa, I. Yamamoto, and J. E. Casida. 1972. Enzymic mechanisms and toxicological significance of hydrogen cyanide liberation from various organothiocyanates and organonitriles in mice and houseflies. *Pestic. Biochem. Physiol. 2*(1): 95–112.

Patty, F. A. 1963. *Industrial Hygiene and Toxicology*, 2nd ed. New York; Interscience.

Pavicic, J., and B. Pihlar. 1983. Toxic effects of cyanides (including complex metal cyanides) on marine organisms. *Chem. Abstr. CA 101*: 164912d.

Rickert, W. S., Robinson, J. C., and J. C. Young. 1980. Estimating the hazards of "less hazardous" cigarettes: I. Tar, nicotine, carbon monoxide, acrolein, hydrogen cyanide, and total aldehyde deliveries of Canadian cigarettes. *J. Toxicol. Environ. Health 6*(2): 351–65.

Sakai, T., and A. Okukubo. 1979. Application of a test for estimating the relative toxicity of thermal decomposition products. In *Fire Retardant: Proceedings of the European Conference on Flammability and Fire Retardants*, ed. V. M. Bhatnagar, pp. 147–53. Westport, CT: Technomic.

Samuelson, O. 1988. Destruction of volatile nitrogen compounds produced during treatment of lignocellulose with nitrogen dioxide. *Res. Discl. 286*: 81.

Sarkos, C. P. 1979. Recent test results from FAA/NAFEC cabin fire safety program. *NBS Spec. Publ. (U.S.) 540*: 589–623.

Smith, M. J., and A. G. Heath. 1979. Acute toxicity of copper, zinc and cyanide to freshwater fish: effect of different temperatures. *Bull. Environ. Contam. Toxicol. 22*(1–2): 113–19.

Souty, M., F. Pascal, and R. Platon. 1970. *Fruits* *25*(7–8): 539–42.

Sterner, R. T. 1979. Effects of sodium cyanide in coyotes (*Canis latrans*): applications as predacids in livestock toxic collars. *Bull. Environ. Contam. Toxicol. 23*(1–2): 211–17.

Sumi, K., and Y. Tsuchiya. 1976. Toxicity of decomposition products: polyacrylonitrile, nylon 6, ABS. *Build. Res. Note Natl. Res. Counc. Can. Div. Build. Res. 111*: 1–4.

Wiemeyer, S. N., E. F. Hill, J. W. Carpenter, and A. J. Krynitsky. 1986. Acute oral toxicity of sodium cyanide in birds. *J. Wildl. Dis. 22*(4): 538–46.

Wolverton, B. C., and B. K. Bounds. 1988. Aquatic plants for pH adjustment and removal of toxic chemicals and dissolved minerals from water supplies. *J. Miss. Acad. Sci. 33*: 71–80.

Yamamoto, K. 1979. On the acute toxicities of the combustion products of various fibers, with special reference to blood cyanide values. *NBS Spec. Publ. (U.S.) 540*: 520–27.

Yamamoto, K., and Y. Yamamoto. 1971. Toxicity of gases released by polyurethane foams subjected to sufficiently high temperature. *Nippon Hoigaku Zasshi 25*(4): 303–14.

15

DIOXIN AND RELATED COMPOUNDS

15.1 GENERAL DISCUSSION

Polychlorinated dibenzo-*p*-dioxins are tricyclic aromatic compounds with extreme high toxicity. These compounds received high publicity for their destructive actions in Indochina. The deadly effect of Agent Orange, the forest defoliant used in Vietnam, is attributed to these polychlorinated dibenzo-*p*-dioxins occurring in phenoxy acid herbicide, 2,4,5-T. Agent Orange was a liquid mixture soluble in oil and insoluble in water with a specific gravity 1.285. It was composed of *n*-butyl esters of 2,4,5-trichlorophenoxy acetic acid (2,4,5-T), 2,4-dichlorophenoxy acetic acid (2,4-D), and 2,3,7,8-tetrachlorodibenzo-*p*-dioxin in percent ratio of 52.5 : 46.9 : 0.0037, respectively (Westing 1984).

Structure

The structure of dibenzo-*p*-dioxin is as follows:

(dibenzo-*p*-dioxin)

One to eight chlorine atoms can bind to different positions in the ring, producing a total of 75 isomers. These would contain 10 dichloro, 14 trichloro, 22 tetrachloro, 14 pentachloro, 10 hexachloro, and 2 each of monochloro and heptachloro monomers and 1 octachloro isomer of dioxin. Chlorine-rich isomers are susceptible to photochemical dechlorination. In 2,3,7,8-TCDD chlorine atoms occupy positions at 2, 3, 7, and 8 in dibenzo-*p*-dioxin.

The structures of dibenzofuran and xanthene are as follows:

(dibenzofuran)   [9(*H*)-xanthene]

Chloro substitution in the aromatic rings, similar to dioxins, can give rise to several polychlorinated dibenzofurans and their isomers.

Physical Properties of 2,3,7,8-TCDD

White crystalline solid; mp 303°C; decomposes above 700°C; vapor pressure 1.5 ×

10^{-9} torr at 25°C; insoluble in water (solubility 0.317 ppb), solubility in nonpolar organic solvents is higher (chloroform 0.037%).

15.2 USES AND EXPOSURE RISK

2,3,7,8-Dioxin, the principal toxic impurity in chlorophenoxy herbicides, has been in use in chemical warfare to destroy crops and defoliate jungles. It occurs as an impurity in trace amount in the herbicide 2,4,5-T. It is formed as a by-product in the synthesis of 2,4,5-trichlorophenol. It has been detected in distillation bottom tars from 2,4-dichlorophenol, in pentachlorophenolic wastewaters, in forest fires; and in smoke emissions from municipal incinerators. Combustion gases, soots, and ashes from explosions of chlorinated phenol plants were found to contain dioxins.

Trace dioxins have been detected in soil, sediment, vegetation, fruits, fish and mammalian tissue, and bovine milk. The major source of human exposure to dioxin is the food chain. According to an estimate, the long-term average daily intake of dioxin in humans is 0.05 ng/day (Hattemer-Frey and Travis 1987).

Dioxins have been detected in sealed dried sewage sludges even after several years of storage. The natural enzyme chloroperoxidase, commonly found in soils can biosynthesize dioxin in garden compost piles, peat bogs, and sewage containing phenols and chloride (Gribble 1998).

The 2,3,7,8-isomer, which is the most toxic dioxin, can be formed by photochemical or thermal reactions of 2,4,5-T or other phenoxy herbicides (Rappe 1978; Akermark 1978). Heating salts of 2,4,5-T at 400–450°C or the ester at 500–580°C (Ahling et al. 1977) produced this isomer to the extent of 1000 and 0.3 ppm, respectively.

Dioxins absorb onto soil and exhibit little downward mobility. It undergoes lateral erosional movement due to water and wind and partial photodegradation in nature. The half-life reported for 2,3,7,8-TCDD are somewhat varying, from 3.5 years to 5 years and 10 years (Westing 1984). The 2,3,7,8-TCDD contamination in soils is confined to the top layer only. Therefore, its poisoning of underlying groundwater may not take place unless the soil is sandy or has low organic content (U.S. EPA 1986a).

15.3 FORMATION OF DIOXINS AND DIBENZOFURANS IN INCINERATORS

Dioxins and dibenzofurans form catalytically from pentachlorophenol wastes, chlorinated phenol precursors, and from nonchlorinated compounds by the reaction of phenol with inorganic chloride ion in the low-temperature region of the incinerator plant. Karasek and Dickson (1987) determined with optimum temperature range as 250–350°C for the formation of dioxins from pentachlorophenol. Eklund et al. (1986) observed that the formation of these toxic compounds depended on the HCl concentration in the incinerator. Such formation is probably caused by phenol and HCl and may be controlled by injecting suitable alkaline sorbent along with the fuel. These compounds may also be formed from the carbon particulates in the incinerators. The concentrations of dioxins produced from pentachlorophenol were much higher than those produced from carbon (Dickson et al. 1988). In addition, carbon catalyzed the formation of dioxins from pentachlorophenol. An occurrence of surface-catalyzed reactions on the fly ash has been proposed. Similar surface phenomena involving adsorption of precursor compounds on carbon in the fly ash, as well as oxidation of carbon, were also suggested by Vogg et al. (1987) to explain the mechanism of dioxin–dibenzofuran formation in the municipal solid waste incinerator at a low-temperature region of 300°C.

Hagenmaier et al. (1987) studied the catalytic effect of fly ash on the formation and decomposition of dioxins and dibenzofurans under oxygen-deficient and oxygen-surplus conditions. While under oxygen-deficient conditions, fly ash catalyzed dechlorination or hydrogenation reaction, in the presence of surplus oxygen an increase in PCDD and PCDF was observed.

During high-temperature incineration of PCBs and polychlorobenzene wastes, no 2,3,7,8-TCDD was detected in stack emissions, slag, or wash water, while some PCDDs were detected only in ppt amounts in stack emission (Brenner et al. 1986). The levels of PCDDs and PCDFs in the incinerator flue gases were found to be slightly high when the amount of wet leaves in the municipal solid waste was higher (Marklund et al. 1986). The increase in polyvinyl chloride (PVC) content in the waste feed did not produce emissions with detectable concentrations of dioxins and dibenzofurans (Carrol 1988; Giugliano et al. 1988). There is no evidence that the amount of PVC in the wastes affects the formation of PCDDs and PCDFs.

15.4 HEALTH HAZARD

The long-term effect on humans due to dioxins from herbicide spray has been very serious. Dioxins can produce a wide range of metabolic dysfunctions, even at a very low level of nanograms to micrograms per kilogram of mammalian body weight. Laboratory tests on animals show that dioxins can cause cancer and birth defects and induce genetic damage. However, not all polychlorinated dioxins pose such a severe health hazard. It is primarily the 2,3,7,8-tetrachloroisomer, and to a lesser extent, some of the hexachloro isomers that are indeed hazardous. Few of the pentachloro dioxins are highly toxic but not carcinogenic. Monochloro, dichloro (except 2,7-dichloro isomer), trichloro, and heptachloro compounds have a much lower order of toxicity and are neither carcinogens nor teratogens. Thus it is interesting to note that chlorine substitutions on the aromatic rings of dibenzo-p-dioxin (and dibenzofuran) in certain numbers at certain positions can convert an "innocuous" molecule to an extremely "dangerous" one.

Chlorinated dibenzo-p-dioxins that present moderate to extreme health hazards are listed in Table 15.1. The toxicity data of halogenated dibenzofurans are presented in Table 15.2.

TABLE 15.1 Toxicity of Polychlorinated Dibenzo-p-dioxins

Compound/CAS No.	Formula/MW	Health Hazard
2,7-Dichloro dibenzo-p-dioxin [33857-26-0]	$C_{12}H_6Cl_2O_2$ 253.08	*Teratogen:* can cause developmental abnormalities in cardiovascular system *Carcinogen:* can cause tumor in liver and leukemia in mice *Toxicity:* moderate; 60 times less toxic than 2,3,7,8-TCDD (Sijm and Opperhuizen 1988) *Eye irritant:* 2 mg had mild effect on rabbits
1,3,6,8-Tetrachlorodibenzo-p-dioxin [33423-92-6]	$C_{12}H_4Cl_4O_2$ 321.96	*Teratogen:* shows fetotoxicity in rats (harmful to embryo)
2,3,7,8-Tetrachlorodibenzo-p-dioxin [1746-01-6]	$C_{12}H_4Cl_4O_2$ 321.96	The most harmful in the class of polychlorinated dioxins; chromosomal damage and mutagen: affects mitosis or cell division showing a powerful inhibitory effect and chromosome damage (Hay 1984); studies by Vietnamese scientists (Trung and Dieu 1984; Can et al. 1983) have established its long-lasting effect and chromosomal aberration in humans

TABLE 15.1   *(Continued)*

Compound/CAS No.	Formula/MW	Health Hazard
		Teratogen: exposure can cause birth defects, miscarriage, still births, and fetotoxicity; a dose of 12 mg/kg can cause fetal death in mice
		Carcinogen: tests on animals indicate that it can cause cancer in liver, kidney, thyroid, and lungs; mouse (oral) 1 mg/kg/2 yr: liver, thyroid, lungs; oral (rats) 328 mg/kg/78 weeks: liver, kidney; it can cause skin tumor at the site of application: 62 mg/kg/2 yr (mice); it should be considered a carcinogen to humans
		Toxicity: human exposure can cause delayed effect of neurotoxicity with symptoms of fatigue, headache, irritability, abdominal pain, blurred vision, and ataxia (Dwyer and Epstein 1984); liver damage; disorders of pancreas, cardiovascular system and urinary and respiratory tracts; disorders of fat and carbohydrate metabolism; and chloracne, an acne-like skin eruption (Oliver 1975), the symptoms subsiding after months; 0.107 mg/kg in humans may cause dermatitis and allergy after topical application on skin; LD_{50} skin (rabbits): 0.275 mg/kg LD_{50} oral (rabbits): 0.01 mg/kg; LD_{50} oral (mice): 0.114 mg/kg; toxic effect was markedly observed only in certain species of animals, such as mice, rabbits, and chickens
		Eye irritation: 2 mg on rabbit was moderately irritating
1,2,3,7,8-Pentachlorodibenzo-*p*-dioxin [40321-76-4]	$C_{12}H_3Cl_5O_2$ 356.40	*Toxicity:* high with symptoms of neurotoxicity, liver damage, and chloracne; LD_{50} oral (guinea pigs): 0.0031 mg/kg, LD_{50} oral (mice): 0.3375 mg/kg
		Carcinogen: inadequate evidence on animals
1,2,4,7,8-Pentachlorodibenzo-*p*-dioxin [58802-08-7]	$C_{12}H_3Cl_5O_2$ 356.40	*Toxicity:* moderate, lower than its 1,2,3,7,8-isomer; LD_{50} oral (guinea pigs): 1.125 mg/kg
		Carcinogen: inadequate evidence
1,2,3,6,7,8-Hexachlorodibenzo-*p*-dioxin [34465-46-8]	$C_{12}H_2Cl_6O_2$ 390.84	*Teratogen:* causes specific developmental abnormalities and fetotoxicity and can impair urogenital system
		Toxicity: highly toxic, can impair liver and neurological system; cause chloracne; LD_{50} oral (guinea pigs): 0.07 mg/kg, LD_{50} oral (rats): 1.25 mg/kg

(continued)

TABLE 15.1 (*Continued*)

Compound/CAS No.	Formula/MW	Health Hazard
		Carcinogen: inadequate evidence
		Eye irritation: moderate on rabbit eyes (2 mg/kg)
1,2,3,4,7,8-Hexachloro dibenzo-*p*-dioxin [57653-85-7]	$C_{12}H_2Cl_6O_2$ 390.84	*Toxicity:* high; LD_{50} oral (guinea pigs): 0.072 mg/kg, LD_{50} oral (mice): 0.825 mg/kg
1,2,3,7,8,9-Hexachloro dibenzo-*p*-dioxin [19408-74-3]	$C_{12}H_2Cl_6O_2$ 390.84	*Toxicity:* high; LD_{50} oral (guinea pigs): 0.06 mg/kg
1,2,3,4,6,7,8,9-Octachlorodibenzo-*p*-dioxin [3268-87-9]	$C_{12}Cl_8O_2$ 459.72	*Teratogen:* homeostasis
		Carcinogen: causes skin tumor, animal inadequate evidence
		Eye irritant: mild in rabbits (2 mg)

TABLE 15.2 Toxicity of Polyhalogenated Dibenzofurans

Compound/CAS No.	Formula/MW	Health Hazard
Dichlorodibenzofuran [43047-99-0]	$C_{12}H_6Cl_2O$ 237.08	*Toxicity:* moderate; an oral dose of 250 mg/kg can be lethal to rats
2,3,7,8-Tetrachlorodibenzofuran [51207-31-9]	$C_{12}H_4Cl_4O$ 305.96	This isomer presents severe health hazard like its dioxin counterpart
		Teratogen: can cause developmental abnormalities and fetal death, 0.25 mg/kg oral (mice)
		Toxicity: highly toxic, can injure skin and hair, cause dermatitis and anemia and damage liver; LD_{50} oral (guinea pigs): 5 µg/kg, LD_{50} oral (monkeys): 1 mg/kg
2,3,7,8-Tetrabromodibenzofuran [67733-57-7]	$C_{12}H_4Br_4O$ 483.80	*Toxicity:* highly toxic; LD_{50} oral (guinea pigs): 4.74 µg/kg, symptom of lowering body temperature
2,3,4,7,8-Pentachlorodibenzofuran [57117-31-4]	$C_{10}H_3Cl_5O$ 340.40	*Toxicity:* highly toxic; LD_{50} oral (guinea pigs): 10 µg/kg; lowering of body temperature

15.5 TOXICOLOGY

Toxicity and carcinogenicity of 2,3,7,8-TCDD in laboratory animals have been investigated extensively. Studying the acute toxicity of this compound in mink, Hochstein et al. (1988) observed that adult male minks were a highly sensitive mammalian of species to its toxicity. Single oral dose of 2.5–7.5 mg/kg caused dose-dependent decrease in feed consumption with corresponding body weight loss; mottling and discoloration of the liver and spleen; and at higher doses, there was enlargement of brain, kidneys, heart, and thyroid gland with respect to percentage of body weight. A 28-day LD_{50} value was calculated to be 4.2 mg/kg. It can bioaccumulate in the body, reacting slowly and causing mortality for several weeks after exposure. Such an effect was noted in fathead minnows exposed for 1 day to a concentration as low as 7.1 ng/L (Adams et al. 1986).

This cumulative mortality after a progressive loss of body weight was observed in yellow perch treated with a single TCDD dose of 5 mg/kg (Spitsbergen et al. 1988). The toxic symptoms were fin necrosis and petechial cutaneous hemorrhage. At higher dosages of 25 and 125 mg/kg no weight loss was observed, and the mortality occurred in a short period following the symptoms of thymic atrophy, focal myocardial necrosis, and submucosal gastric edema.

Kumar et al. (1986) reported that the acute oral toxicity of 2,3,7,8-TCDD and 2,3,7,8-TCDF in guinea pigs were greater in peanut oil than in water, with 40-day LD_{50} value for TCDD being 2.5 and 10.54 mg/kg, respectively. Studies on guinea pigs indicated cardiotoxicity of TCDD, showing evidence of increased intracellular calcium and decreased β-adrenergic responsiveness resulting from a single intraperitoneal dose of 10 mg/kg in corn oil (Canga et al. 1988).

Electron microscopic studies by Rozman et al. (1986) indicated that interscapular brown adipose tissue was a target tissue in TCDD-induced toxicity in rats. The medical diagnoses of people exposed to Agent Orange in Vietnam also showed primarily 2,3,7,8-TCDD in their adipose tissue; the highest levels were the primary carcinoma of the liver (56.7 ppt), adenoma of the prostate (55.9 ppt), and adenocarcinoma of the stomach (28.9 ppt) (Pham et al. 1988).

The major metabolites of 2,3,7,8-TCDD in mammals are 2-hydroxy-3,7,8-trichlorodibenzo-p-dioxin [82019-04-3] and 2-hydroxy-1,3,7,8-tetrachlorodibenzo-p-dioxin [82019-03-2] (Mason et al. 1987a). The former was found to be active as an inducer of the microsomal monooxygenases at dose levels of 1000–5000 mg/kg in immature male Wistar rats (Mason and Safe 1986). With respect to relative enzyme induction activities, it was three orders of magnitude less active than TCDD.

The carcinogenic action of 2,3,7,8-TCDD has been studied on Syrian golden hamsters, which otherwise, showed maximum resistance to the toxic effect of this compound (Rao et al. 1988). TCDD was fully carcinogenic in hamsters, developing squamous cell carcinomas of the skin of the facial region from a total dose of 600 mg/kg by a subcutaneous or intraperitoneal route within 52–56 weeks.

Polychlorinated dibenzofurans are high to moderately toxic compounds showing toxic actions similar to those associated with the chlorinated dioxins. The toxicity, however, is lower to 2,3,7,8-TCDD. Table 15.3 presents the toxic potency of some of the isomers as compared with TCDD. Diets containing 200 mg/kg to rats, 2,3,4,7,8-pentachlorodibenzofuran for 13 weeks, caused thymus atrophy, decreased food consumption and body weight, severe lesions in the liver, and death (Pluess et al. 1988).

The consumption of toxic rice oil causing the "Yusho" tragedy was attributed to PCDF contaminants in PCBs that mixed into the oil (Hori et al. 1986).

The teratogenic effects of pentachloro isomers with chlorine atoms at 1,2,3,7,8-hexachlorodibenzofuran were assessed in mice (Birnbaum et al. 1987). All three compounds were highly teratogenic, and the responses

TABLE 15.3 Toxic Potency of Polychlorinated Dibenzofurans as Compared with 2,3,7,8-TCDD (2,3,7,8-TCDD = 1)

Isomer	2,3,7,8-TCDD Equivalence Factor	Species	Reference
2,3,7,8-tetrachloro-	0.31,[a] 0.7[b]	Guinea pigs	Kumar et al. (1986)
1,2,3,7,8-pentachloro-	0.01	Rats	Poiger et al. (1987)
2,3,4,7,8-pentachloro-	0.4	Rats	Poiger et al. (1987)
1,2,3,4,8-pentachloro-	<0.0003	Rats	Poiger et al. (1987)
1,2,3,6,7,8-hexachloro-	0.1	Rats	Poiger et al. (1987)

[a]In peanut oil.
[b]In water.

were similar to TCDD but were only $\frac{1}{10}$ to $\frac{1}{100}$ as potent.

Polybrominated dioxins (PBDDs) are highly toxic, however, less toxic than the chloro compounds. The effects and symptoms are comparable to those of chlorinated dioxins. In rats, 2,3,7,8-bromo isomer was 3 to 10 times less toxic than 2,3,7,8-TCDD (Loeser and Ivens 1988). Mason et al. (1987b) have investigated the biological and toxic effects and structure–activity relationships of brominated and chlorinated dioxins. Halo substitutions in all four lateral positions imparted maximum toxicity in dioxins and dibenzofurans. The binding affinities of PBDDs as ligands for the 2,3,7,8-TCDD cytosolic receptor showed a decrease with a decrease in lateral substituents or an increase in nonlateral substituents. The binding affinities did not significantly alter with interchange of Br and Cl substituents. The study indicated that the toxic potencies of PBDDs decreased in the following order: 2,3,7,8-, > 1,2,3,7,8-, > 1,2,4,7,8-, > 1,3,7,8-. 1,2,4,5,7,8-Hexachloro (9H) xanthene [38178-99-3] exhibited low to moderate toxicity in guinea pigs from a single oral dose of 12.5 mg/kg (DeCaprio et al. 1987). It may be noted that the absence of the second oxygen atom in the heterocyclic ring substantially reduced the dioxin-like activity.

Antidote

The effects of cotreatment with a hypolipidemic agent, diethylhexyl phthalate were studied in F344 rats (Tomaszewski et al. 1988). This compound decreased the TCDD-induced hyperlipidemia when administered before or after TCDD. Microscopic examinations of liver showed that treatment with diethyl hexyl phthalate suppressed TCDD-induced fatty liver. The antagonizing effects were low when the treatment was initiated after the dose. Thus treatment with a hypolipidemic agent such as diethyl hexyl phthalate may protect against TCDD toxicity.

15.6 DISPOSAL/DESTRUCTION

Destruction of dioxins and dibenzofurans in contaminated soils and wastes is best achieved by high-temperature pyrolysis. Treatment of soils contaminated with polychlorinated di-benzo-p-dioxins at ~2200°C in an electrically heated pyrolyzer reduced levels of tetra-, penta-, and hexachlorinated dibenzo-p-dioxins and dibenzofurans to <0.12 ppb (Boyd et al. 1987). Waterland et al. (1987) reported destruction and removal efficiencies greater than 99.9999% for TCDD in still-bottom wastes with initial concentration of 37 ppm TCDD using a rotary kiln incineration system. Portable infrared incinerators having removal efficiencies greater than 99.9999% have been reported (Daily 1987).

Reductive pyrolysis may be an effective method for destruction of chlorinated organic wastes, including dioxins and dibenzofurans (Louw et al. 1986). The process involves passing the vaporized or finely dispersed waste and hydrogen through a tubular flow reactor heated to 600–950°C at residence times of 1–10 seconds, causing dechlorination without charring. Other techniques based on thermal destruction involved treatment of the waste combined with oxygen in a molten metal bath of iron, nickel, or copper in the presence of silicon at 1400°C (Bach and Nagel 1986); and exposure to highly concentrated solar radiation of 300 suns at >300°C (Sutton and Hunter 1988).

Low-temperature degradation of dioxins may be attained catalytically. Dioxins in flue gases from incinerators degraded at 300–450°C in the presence of platinum on a honeycomb support (Hiraoka et al. 1988). The honeycomb support prevented clogging by particulates. Removal of chlorinated dioxins from industrial wastewater using modified clays has been reported (Srinivasan and Fogler 1987). Oxidative degradation by ozone at 20–50°C, using a chlorofluoro solvent was effective in treating contaminated water (Palauschek and Scholz 1987). Rogers and

Kornel (1987) described chemical destruction of dioxins and dibenzofurans in wood preservation wastes to nondetectable levels by dehalogenation with potassium salt of polyethylene glycol at 100°C and 30 minutes.

Chlorinated dioxins can undergo photodegradation, losing their chlorine atoms. Such dechlorination takes place in the presence of a hydrogen donor such as alcohols and hydrocarbons and UV light (Crosby et al. 1971; Liberti et al. 1978); and occurring preferentially at the 2,3,7, and 8 positions (Nestrick et al. 1980).

Chlorinated dioxins and dibenzofurans are not easily biodegraded in soils. Only a few of the microorganisms have shown some ability to degrade 2,3,7,8-TCDD (Matsumura and Benezet 1973). This isomer can accumulate in plants and animals, and its uptake by maize and bean plants from soil has been studied by Facchetti et al. (1985). The molecule does not appear to break down.

15.7   SAFETY MEASURES

Use latex gloves, impervious laboratory apron or clothing, and full-face shields when handling these compounds. In the event of a spill, wipe up with absorbent paper and then clean with methylene chloride, methanol, and a detergent. The contaminated papers, clothing, and so on, should be placed into a hazardous waste drum. Contaminated glasses should be compacted in a hazardous waste trash compactor and placed in a waste disposal drum. Effluents from hoods, GC/MS, and vacuum pumps should be passed through HEPA particulate filters and charcoal.

15.8   ANALYSIS

Dioxins and dibenzofurans are analyzed by high-resolution capillary column gas chromatography, low- and high-resolution mass spectroscopy, and HPLC. A suitable GC column for the analysis of dibenzofuran is 3% SP2250. Chlorinated dioxins and dibenzofurans, however, are best analyzed by GC/MS technique. Some of the suitable chromatographic columns are DB-5, CP-Sil-88, and SP-2250. Other equivalent columns can be used. Dioxins in wastewater can be analyzed by EPA method 613 (U.S. EPA 1984). Soils, sludges, solids, and hazardous wastes can be analyzed for chlorinated dioxins and dibenzofurans by EPA Method 8280 (U.S. EPA 1986b). The GC/MS method involved the identification of ions in the selective-ion monitoring mode and measuring the isotopic ratios. The samples may be cleaned up over a carbon column to separate some of the interfering compounds. Some of the isomers, including 2,3,7,8-TCDD, can be analyzed by HPLC (Lamparski et al. 1979).

REFERENCES

Adams, W. J., G. M. DeGraeve, T. D. Sabourine, J. D. Cooney, and G. M. Mosher. 1986. Toxicity and bioconcentration of 2,3,7,8-TCDD to fathead minnows *(Pimephales promelas)*. *Chemosphere 15* (9–12): 1503–11.

Ahling, B., A. Lindskog, B. Jansson, and G. Sundstrom. 1977. Formation of polychlorinated dibenzo-*p*-dioxins and dibenzofurans during combustion of a 2,4,5-T formulation. *Chemosphere 6*: 461–68.

Akermark, B. 1978. Photochemical reactions of phenoxy acids and dioxins. *Ecol. Bull. 1978* (27): 75–81.

Bach, R. D., and C. J. Nagel. 1986. Destruction of toxic chemicals. U.S. Patent 4,574,714, Mar. 11, cited in *Chem. Abstr. CA 105* (2): 11520g.

Birnbaum, L. S., W. M. Harris, E. R. Barnhart, and R. E. Morrissey. 1987. Teratogenicity of three polychlorinated dibenzofurans in C57BL/6N mice. *Toxicol. Appl. Pharmacol. 90* (2): 206–16; cited in *Chem. Abstr. CA 107* (23): 213376s.

Boyd, J., H. D. Williams, R. W. Thomas, and T. L. Stoddart. 1987. Destruction of dioxin contamination by pyrolysis techniques. In *ACS*

Symp. Ser. 338: 299–310; cited in *Chem. Abstr. CA 107*(10): 83329r.

Brenner, K. S., I. H. Dorn, and K. Herrmann. 1986. Dioxin analysis in stack emissions, slags and the wash water circuit during high-temperature incineration of chlorine-containing industrial wastes; Part 2. *Chemosphere 15*(9–12): 1193–99.

Can, N., T. K. Anh, and L. Bisanti. 1983. Reproductive epidemiology: symposium summary. *Summary Report of a Working Group of the International Symposium on Herbicides and Defoliants in War: The Long-Term Effects on Man and Nature*, Jan. 13–20, Ho Chi Minh City, Vietnam.

Canga, L., R. Levi, and A. B. Rifkind. 1988. Heart as a target organ in 2,3,7,8-tetrachlorodibenzo-*p*-dioxin toxicity: decreased β-adrenergic responsiveness and evidence of increased intracellular calcium. *Proc. Natl. Acad. Sci. USA 85*(3): 905–9.

Carroll, W. F., Jr. 1988. PVC and incineration. *J. Vinyl Technol. 10*(2): 90–94.

Crosby, D. G., A. S. Wong, J. P. Plimmer, and E. A. Woolson. 1971. Photodecomposition of chlorinated dibenzo-*p*-dioxins. *Science 195*: 1337–38.

Daily, P. L. 1987. Performance assessment of a portable infrared incinerator: thermal destruction testing of dioxin. *ACS Symp. Ser. 338*: 311–18; cited in *Chem. Abstr. CA 107*(10): 83345t.

DeCaprio, A. P., R. Briggs, J. F. Gierthy, C. S. Kim, and R. D. Kleopfer. 1987. Acute toxicity in the guinea pig and in vitro "dioxin-like" activity of the environmental contaminant 1,2,4,5,7,8-hexachloro(9H)xanthene. *J. Toxicol. Environ. Health 20*(3): 241–48.

Dickson, L. C., D. Lenoir, and O. Hutzinger. 1988. Surface-catalyzed formation of chlorinated dibenzodioxins and dibenzofurans during incineration. *Chemosphere 19*(1–6): 277–88.

Dwyer, J. H., and S. S. Epstein. 1984. Cancer and clinical epidemiology: an overview. In *Herbicides in War: The Long Term Ecological and Human Consequences*, ed. A. H. Westing, pp. 123–127. Philadelphia: Taylor & Francis.

Eklund, G., J. R. Pedersen, and B. Stroemberg. 1986. Phenol and hydrochloric acid at 550°C

yield a large variety of chlorinated toxic compounds. *Nature 320*(6058): 155–56.

Facchetti, S., A. Balasso, C. Fichiner, G. Frare, A. Leoni, C. Mauri, and M. Vasconi. 1985. Assumption of 2378-TCDD by some plant species. Presented at the *ACS National Meeting*, Miami, FL, Apr.

Giugliano, M., S. Cernuschi, and U. Ghezzi. 1988. The emission of dioxins and related compounds from the incineration of municipal wastes with high contents of organic chlorine (PVC). *Chemosphere 19*(1–6): 407–11.

Gribble, G. W. 1998. Dioxin discoveries. *Chem. Eng. News 76*(24): 4.

Hagenmaier, H., M. Kraft, H. Brunner, and R. Haag. 1987. Catalytic effects of fly ash from waste incineration facilities on the formation and decomposition of polychlorinated dibenzo-*p*-dioxins and polychlorinated dibenzofurans. *Environ. Sci. Technol. 21*(11): 1080–84.

Hattemer-Frey, H. A., and C. C. Travis. 1987. Comparison of human exposure to dioxin from municipal waste incineration and background environmental contamination. *Chemosphere 18*(1–6): 643–49.

Hay, A. W. M. 1984. Experimental toxicology and cytogenetics. In *Herbicides in War: The Long Term Ecological and Human Consequences*, ed. A. H. Westing, pp. 161–65. Philadelphia: Taylor & Francis.

Hiraoka, M., N. Takeda, S. Okajima, T. Kasakura, and Y. Imoto. 1988. Catalytic destruction of PCDDs in flue gas. *Chemosphere 19*(1–6): 361–66.

Hochstein, J. R., R. J. Aulerich, and S. J. Bursian. 1988. Acute toxicity of 2,3,7,8-tetrachlorodibenzo-*p*-dioxin to mink. *Arch. Environ. Contam. Toxicol. 17*(1): 33–7; cited in *Chem. Abstr. CA 108*(7): 50813m.

Hori, S., H. Obana, R. Tanaka, and T. Kashimoto. 1986. *Eisei Kagaku 32*(1): 13–21; cited in *Chem. Abstr. CA 105*(13): 113803u.

Karasek, F. W., and L. C. Dickson. 1987. Model studies of polychlorinated dibenzo-*p*-dioxin formation during municipal refuse incineration. *Science 237*(4816): 754–56.

Kumar, S. N., A. Srivastava, and M. K. Srivastava. 1986. *J. Recent Adv. Appl. Sci. 1*(1): 17–21.

Lamparski, L., T. J. Nestrick, and R. H. Stehl. 1979. Determination of part-per-trillion concentrations of 2,3,7,8-tetrachlorodibenzo-*p*-dioxin in fish. *Anal. Chem. 51*: 1453–58.

Liberti, A., D. Brocco, I. Allegrini, A. Cecinato, and M. Possanzin. 1978. Solar and UV photodecomposition of 2,3,7,8-tetrachlorodibenzo-*p*-dioxin in the environment. *Sci. Total Environ. 10*: 97–104.

Loeser, E., and I. Ivens. 1988. Preliminary results of a 3 month toxicity study on rats with 2,3,7,8-tetrabromodibenzo-*p*-dioxin (2,3,7,8-TBDD). *Chemosphere 19* (1–6): 759–64.

Louw, R., J. A. Manion, and P. Mulder. 1986. Reductive pyrolysis: an effective method for destruction of chlorinated organic wastes. In *Proceedings of the International Conference on Chemicals in the Environment*, ed. J. N. Lester, R. Perry, and R. M. Sterritt, pp. 710–14. London: Selper.

Marklund, S., L. O. Kjeller, M. Hansson, M. Tysklind, C. Rappe, C. Ryan, H. Collazo, and R. Dougherty. 1986. Determination of PCDDs and PCDFs in incineration samples and pyrolytic products. In *Chlorinated Dioxins and Dibenzofurans Perspectives*, ed. C. Rappe, G. Choudhary, and L. H. Keith, pp. 79–92. Chelsea, MI: Lewis Publishers.

Mason, G., and S. Safe. 1986. Synthesis, biologic and toxic effects of the major 2,3,7,8-tetrachloro-dibenzo-*p*-dioxin metabolites in the rat. *Toxicology 41* (2): 163–69.

Mason, G., M. A. Denomme, L. Safe, and S. Safe. 1987a. Polybrominated and chlorinated dibenzo-*p*-dioxins metabolites in the rat. *Toxicology 41* (2): 153–59.

Mason, G., M. A. Denomme, L. Safe, and S. Safe. 1987b. Polybrominated and chlorinated dibenzo-*p*-dioxins: synthesis, biological and toxic effects and structure–activity relationships. *Chemosphere 16* (8–9): 1729–31.

Matsumura, F., and H. J. Benezet. 1973. Studies on the bioaccumulation and microbial degradation of 2,3,7,8-tetrachloro-*p*-dioxin. *Environ. Health Perspect. 5*: 253–58.

Nestrick, T. J., L. Lamparski, and D. I. Townsend. 1980. Identification of tetrachloro benzo-*p*-dioxin isomers at the 1 ng level by photolytic degradation and pattern recognition techniques. *Anal. Chem. 52*: 1865–75.

Oliver, R. M. 1975. Toxic effects of 2,3,7,8-tetrachlorodibenzo 1,4-dioxin in laboratory workers. *Br. J. Ind. Med. 32*: 49–53.

Palauschek, N., and B. Scholz. 1987. Destruction of polychlorinated dibenzo-*p*-dioxins and dibenzofurans in contaminated water samples using ozone. *Chemosphere 16* (8–9): 1857–63.

Pham Hoang Phiet, Trinh Kim Anh, Dan Quoc Vu, and A. Schecter. 1988. Preliminary observations on the clinical histories, polychlorinated dibenzodioxins and dibenzofuran tissue levels and 2,3,7,8-TCDD toxic equivalents of potentially dioxin-exposed patients living in the south of Vietnam. *Chemosphere 19* (1–6): 937–40.

Pluess, N., H. Poiger, C. Hohbach, M. Suter, and C. Schlatter. 1988. Subchronic toxicity of 2,3,7,8-pentachlorodibenzofuran (PeCDF) in rats. *Chemosphere 17* (6): 1099–1110.

Poiger, H., N. Pluess, and C. Schlatter. 1987. Subchronic toxicity of some chlorinated dibenzofurans in rats. *Chemosphere 18* (1–6): 265–75.

Rao, M. S., V. Subbarao, J. D. Prasad, and D. G. Scarpelli. 1988. Carcinogenicity of 2,3,7,8-tetrachlorodibenzo-*p*-dioxin in the Syrian golden hamster. *Carcinogenesis 9* (9): 1677–79.

Rappe, C. 1978. Chlorinated phenoxy acids and their dioxins: chemistry: summary. *Ecol. Bull. 1978* (27): 19–27.

Rogers, C. J., and A. Kornel. 1987. Chemical destruction of chlorinated dioxins and furans. In *Proceedings of the 2nd International Conference on New Frontiers of Hazardous Waste Management*, U.S. EPA, pp. 419–24; cited in *Chem. Abstr. CA 109* (8): 60872q.

Rozman, K., B. Strassle, and M. J. Iatropoulos. 1986. Brown adipose tissue is a target tissue in 2,3,7,8-tetrachlorodibenzo-*p*-dioxin (TCDD)-induced toxicity. *Arch. Toxicol. Suppl. 9*: 356–60; cited in *Chem. Abstr. CA 106* (9): 62534t.

Sijm, D. T. H. M., and A. Opperhuizen. 1988. Biotransformation, bioaccumulation and lethality of 2,8-dichlorodibenzo-*p*-dioxin: a proposal to explain the biotic fate and toxicity of PCDD's and PCDF's. *Chemosphere 17* (1): 83–99.

Spitsbergen, J. M., J. M. Kleeman, and R. E. Peterson. 1988. 2,3,7,8-Tetrachlorodibenzo-*p*-dioxin toxicity in yellow perch *(Perca flavescens)*. *J. Toxicol. Environ. Health 23* (3): 359–83.

Srinivasan, K. R., and S. H. Fogler. 1987. Use of modified clays for the removal and disposal of chlorinated dioxins and other priority pollutants from industrial wastewaters. *Chemosphere* *18*(1–6): 333–42.

Sutton, M. M., and E. N. Hunter. 1988. Photolytic/thermal destruction of dioxins and other toxic chloroaromatic compounds. *Chemosphere* *19*(1–6): 685–90.

Tomaszewski, K. E., C. A. Montgomery, and R. L. Melnick. 1988. Modulation of 2,3,7,8-tetrachlorodibenzo-*p*-dioxin toxicity in F344 rats by di(2-ethylhexyl) phthalate. *Chem. Biol. Interact.* *65*(3): 205–22; cited in *Chem. Abstr. CA* *109*(9): 68399m.

Trung, C. B., and V. V. Dieu. 1984. Chromosomal aberrations in humans following exposure to herbicides. In *Herbicides in War: The Long Term Ecological and Human Consequences*, ed. A. H. Westing, pp. 157–159. Philadelphia: Taylor & Francis.

U. S. EPA. 1984. Analysis for organic priority pollutants in wastewaters. *Fed. Register* Oct. 24.

U. S. EPA. 1986a. Fate of 2,3,7,8-TCDD in soil. In *The National Dioxin Study Tiers 3,5,6 and 7, Appendix B*. Final Draft Report, Apr. Washington, DC: Office of Water Regulations and Standards.

U. S. EPA. 1986b. *Test Methods for Evaluating Solid Wastes, Physical/Chemical Methods*. EPA Publ. SW-846. Washington, DC: Office of Solid Waste and Emergency Response.

Vogg, H., M. Metzger, and L. Steiglitz. 1987. Recent findings on the formation and decomposition of PCDD/PCDDF in municipal solid waste incineration. *Waste Manage. Res.* *5*(3): 285–94.

Waterland, L. R., R. W. Ross II, T. H. Backhouse, R. H. Vocque, J. W. Lee, and R. E. Mournighan. 1987. Pilot-scale incineration of a dioxin-containing material. *U.S. EPA Res. Rep., EPA/600/9-87/018F*; cited in *Chem. Abstr. CA* *109*(8): 60900x.

Westing, A. H. 1984. Herbicides in war: past and present. In *Herbicides in War: The Long Term Ecological and Human Consequences*, ed. A. H. Westing, pp. 3–23. Philadelphia: Taylor & Francis.

16

EPOXY COMPOUNDS

16.1 GENERAL DISCUSSION

Epoxides are cyclic three-membered ethers containing an oxygen atom in the ring. They are also termed oxiranes and 1,2-epoxides. The general structure of this class of compounds is as follows:

$$H_2C\overset{\diagdown O \diagup}{}CH_2-R$$

where R is an alkyl, cycloalkyl, or aryl group.

Epoxides are formed by oxidation of olefinic or aromatic double bond. The oxidizing agents for such epoxidation reactions are peracids or peroxides with or without catalysts. The epoxy compounds are widely used as intermediates to produce synthetic resins, plasticizers, cements, adhesives, and other materials.

Exposure to epoxides can cause irritation of the skin, eyes, and respiratory tract. Low-molecular-weight epoxides are strong irritants and more toxic than those of higher molecular weight. Inhalation can produce pulmonary edema and affect the lungs, central nervous system, and liver. Many epoxy compounds have been found to cause cancer in animals.

Lower epoxides are highly flammable. They polymerize readily in the presence of strong acids and active catalysts. The reaction generates heat and pressure, which may rupture closed containers. Contact with anhydrous metal halides, strong bases, and readily oxidizable substances should therefore be avoided.

Ethylene oxide and other hazardous epoxides characterized by the presence of three-membered strained oxirane ring structures are discussed individually in the following sections. Miscellaneous epoxy compounds with low toxicity and flammability and low commercial applications are discussed in Table 16.1.

16.2 ETHYLENE OXIDE

EPA Classified Toxic Waste, RCRA Waste Number U115; DOT Label: Flammable Liquid/Flammable and Poison Gas, UN 1040

Formula C_2H_4O; MW 44.06; CAS [75-21-8]

Structure and functional group:

$$H_2C\overset{\diagdown O \diagup}{}CH_2$$

TABLE 16.1 Toxicity and Flammability of Miscellaneous Epoxy Compounds

Compound/Synonyms/ CAS No.	Formula/MW/Structure	Toxicity	Flammability
Trichloropropene oxide (1,1,1-tri-chloropropane-2,3-oxide, 3,3,3-trichloropropene oxide, 1,2-epoxy-3,3,3-trichloro-propane, trichloro-methyl oxirane) [3083-23-6]	$C_3H_3Cl_3O$ 161.41	Mutation test positive; low to moderate toxicity in rats; LD_{50} intraperi-toneal (rats): 142 mg/kg	Combustible liquid
Epifluorohydrin (1,2-epoxy-3-fluoropropane) [503-09-3]	C_3H_5FO 76.08	Corrosive; moderately toxic on mice; LD_{50} intra-venous (mice): 178 mg/kg	Flammable liquid; flash point 4°C (40°F); vapor density 2.6 (air = 1); forms explosive mixture with air polymerizes generating heat and pressure, may be violent in contact with strong acids and bases, anhy-drous metal halides, potassium-*tert*-butoxide

Name (CAS)	Formula / Structure	Toxicity / Hazard	Fire hazard
Epibromohydrin (3-bromo-1,2-epoxypropane) [3132-64-7]	C_3H_5BrO 136.99; 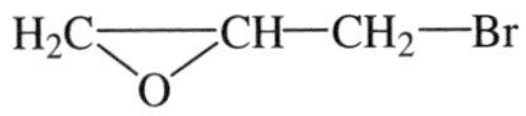 H_2C—CH—CH_2—Br (epoxide)	Corrosive; moderately toxic on mice; LD_{50} intraperitoneal (mice): 300 mg/kg, mutation test positive; carcinogenicity not reported; DOT Poison UN 2558	Combustible liquid; flash point 56°C (133°F); DOT Flammable Liquid UN 2558
1-Chloro-1,2-epoxypropane (chloropropene oxide, 2-chloro-3-methyloxirane) [21947-75-1] cis-isomer [21947-76-2] trans-isomer	C_3H_5ClO 92.53; 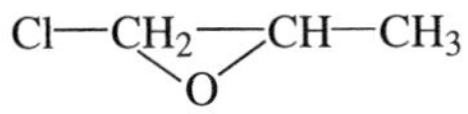 Cl—CH_2—CH—CH_3 (epoxide)	Caused skin cancer in mice at the site of application	Combustible liquid
Cyclopentene oxide (6-oxabicyclo-[3.1.0]hexane, cyclopentane oxide, cyclopentene epoxide, 1,2-epoxycyclopentane) [285-67-6]	C_5H_8O 84.13; 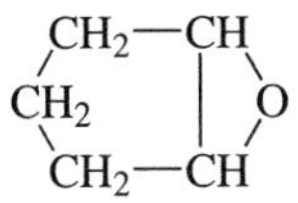 CH_2—CH, CH_2, CH_2—CH with O bridge	Irritant; may be mutagenic; carcinogenicity not reported	Flammable liquid; flash point 10°C (50°F) (Aldrich 1989)

(continued)

317

TABLE 16.1 (Continued)

Compound/Synonyms/CAS No.	Formula/MW/Structure	Toxicity	Flammability
Butadiene monoxide (3,4-epoxy-1-butene, 1,2-epoxy-but-3-ene) [930-22-3]	$C_4H_{16}O$ 70.10 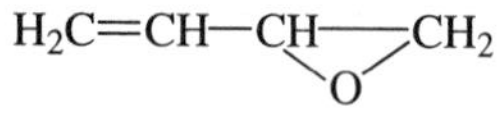 $H_2C{=}CH{-}CH{-}CH_2$ (with epoxide O)	Moderately to highly toxic on mice; 8 mg/kg (intravenous) was lethal; caused skin tumors in mice at the site of application	Highly flammable; flash point −50°C (−58°F); vapor density 2.4 (air = 1); vapor may form explosive mixture with air
D,L-Butadiene epoxide (D,L-diepoxybutane, (+−)-1,2,3,4-diepoxybutane, D,L-butadiene dioxide) [298-18-0]	$C_4H_6O_2$ 86.09 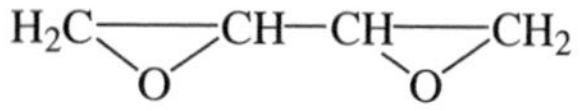 $H_2C{-}CH{-}CH{-}CH_2$ (with two epoxide O)	Eye and skin irritant, 4 hr inhalation in rats caused clouding of cornea and watering of eyes; can cause skin burns and blistering; toxic by inhalation and ingestion; produced lung congestion, labored breathing, and pulmonary irritation in test animals (Rose 1983); exposure to humans can cause irritation	Combustible liquid; flash point 45°C (114°F); reactions can be vigorous with strong acids, bases, and anhydrous metal halides

318

Name [CAS]	Formula, Mol. Wt.	Toxicity	Carcinogenicity	Flammability
L-Butadiene epoxide (L-1,2,3,4-diepoxybutane, L-diepoxybutane) [30031-64-2]	$C_4H_6O_2$ 86.09 An isomer of D,L-butadiene epoxide	and swelling of eyelids; LD_{50} oral (rats): 210 mg/kg; caused skin cancer in mice at the site of application	Positive to mutation test; intraperitoneal administration caused lung and endocrine tumors in mice; sufficient evidence of carcinogenicity in animal tests	Combustible liquid
meso-Butadiene epoxide (meso-diepoxybutane, meso-1,2,3,4-diepoxybutane) [564-00-1]	$C_4H_6O_2$ 86.09 An isomer of D,L-butadiene epoxide		Positive to mutation test; caused skin tumors in mice at the site of application; sufficient evidence of carcinogenicity in animal tests; toxic by skin absorption, 400 mg/kg was lethal to mice	Combustible liquid

(continued)

TABLE 16.1 (Continued)

Compound/Synonyms/CAS No.	Formula/MW/Structure	Toxicity	Flammability
1-Hexene epoxide (1-hexene oxide, 1,2-hexene oxide, 2-butyloxirane, 1,2-epoxyhexane) [1436-34-6]	$C_6H_{12}O$ 100.18 H_2C⟍$CH-CH_2-CH_2-CH_2-CH_3$ (epoxide O)	Toxicity not reported; no report on carcinogenicity	Flammable liquid; flash point 15°C (60°F); moisture sensitive; reacts exothermically with strong acids, bases, and anhydrous metal halides
1,2-Epoxy-5-hexene [2-(1-butenyl)oxirane] [10353-53-4]	$C_6H_{10}O$ 98.15 H_2C⟍$CH-CH-CH_2-CH_2-CH=CH_2$ (epoxide O)	No report on toxicity	Flammable liquid; flash point 15°C (60°F) (Aldrich 1989); moisture sensitive; may react vigorously with strong acids, bases, and anhydrous metal halides
1,2,3,4-Diepoxy-2-methyl butane (2-methyl-2,2'-bioxirane) [6341-85-1]	$C_5H_8O_2$ 100.13 CH_3 H_2C⟍C⟋CH⟍CH_2 (two epoxide O)	Caused blood tumors in mice; toxicity data not available	Combustible liquid
1-(2,3-Epoxy-propoxy)-2-(vinyloxy)ethane [[1,2-epoxy-3-(2-vinyloxy)ethoxy)propane; glycidyl	$C_7H_{12}O_3$ 144.19 H_2C⟍$CH-CH_2-O-CH_2-CH_2-O-CH=CH_2$ (epoxide O)	Low toxicity in mice, causing muscle weakness and altered sleeping time; LD_{50} oral	Combustible liquid

(mice): 1250 mg/kg

Diglycidyl ether [di(2,3-epoxy-propyl)ether, bis(2,3-epoxy-propyl) ether, DGE] [2238-07-5]

$C_6H_{10}O_3$
130.14

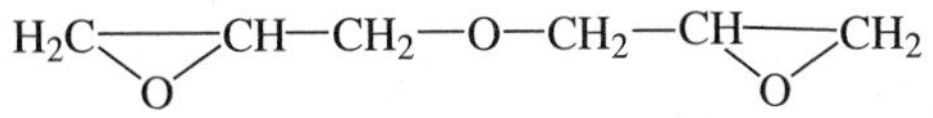

vinyloxyethyl ether; ((((2-ethenyl)oxy) ethoxy)methyl) oxirane; Vinilox] [16801-19-7]

Caused severe skin and eye irritation in test animals; in humans, acute irritation of eyes and respiratory tract can occur from exposure to 10 ppm; moderately toxic symptoms in rats: depression of central nervous system, swollen eyelids, clouding of cornea, nasal discharge, and loss in body weight; LC_{50} (mice): 30 ppm (4 hr); LD_{50} oral (mice): 150 mg/kg; caused skin cancer in mice;

(continued)

TABLE 16.1 *(Continued)*

Compound/Synonyms/ CAS No.	Formula/MW/Structure	Toxicity	Flammability
		TLV-TWA 0.5 mg/m^3 (0.1 ppm) (ACGIH), 2.5 mg/m^3 (0.5 ppm) (OSHA), 15-min ceiling 0.2 ppm (NIOSH)	
1,4-Epoxycyclohexane (7-oxabicyclo[2.2.1] heptane) [279-49-2]	C$_6$H$_{10}$O 98.15	Toxicity not reported	Flammable liquid; flash point 12°C (55°F) (Aldrich 1989)
Cyclohexene oxide (7-oxabicyclo-[4.1.0] heptane, cyclo-hexene epoxide, cyclohexane oxide, 1,2-epoxycyclo-hexane, tetramethy-lene oxirane) [286-20-4]	C$_6$H$_{10}$O 98.15	Toxic on test animals by inhalation, ingestion, and contact; LD$_{50}$ oral (rats): 1090 mg/kg, LD$_{50}$ skin (rabbits): 630 mg/kg, inhalation of 2000 ppm for 4 hr was lethal to rats, caused lung tumor in mice	Flammable liquid; flash point 27°C (81°F)

Compound	Formula / MW	Structure	Toxicity	Flammability
exo-2,3-Epoxy-norbornane (exo-2,3-norbornane oxide) [3146-39-2]	$C_7H_{10}O$ 110.17 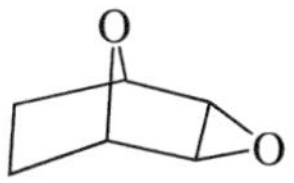		May be mutagenic; toxicity not reported	Flammable solid; flash point 10°C (50°F) (Aldrich 1989)
1,2-Epoxyoctane (octylene epoxide) [2984-50-1]	$C_8H_{16}O$ 128.24 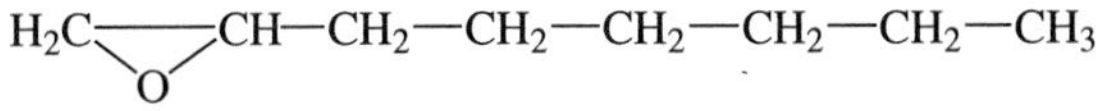	$H_2C\!\!-\!\!CH\!-\!CH_2\!-\!CH_2\!-\!CH_2\!-\!CH_2\!-\!CH_2\!-\!CH_3$ (O)	May be carcinogenic, caused blood tumors in mice	Flammable liquid; flash point 37°C (99°F) (Aldrich 1989) moisture sensitive; generates heat and pressure when mixed with strong acids and bases
4,5-Epoxyoctane [27415-21-0]	$C_8H_{16}O$ 128.24 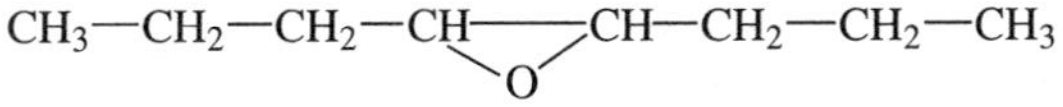	$CH_3\!-\!CH_2\!-\!CH_2\!-\!CH\!-\!CH\!-\!CH_2\!-\!CH_2\!-\!CH_3$ (O)	Moderately toxic to mice via intravenous route; LD$_{50}$ intravenous (mice): 56 mg/kg (NIOSH 1986)	Flammable liquid; flash point 38°C (101°F); exothermic reactions with strong acids and bases
1,2-Epoxy-7-octene [19600-63-6]	$C_8H_{14}O$ 126.20 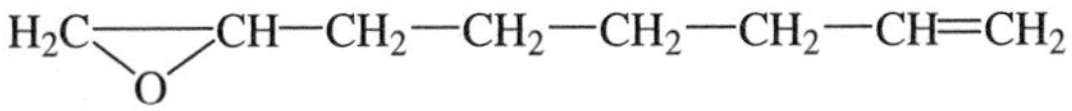	$H_2C\!\!-\!\!CH\!-\!CH_2\!-\!CH_2\!-\!CH_2\!-\!CH_2\!-\!CH\!\!=\!\!CH_2$ (O)	Mild irritant to skin and eyes	Combustible liquid; flash point 38°C (101°F)
1,2-Epoxydodecane (dodecene epoxide) [2855-19-8]	$C_{12}H_{24}O$ 184.36 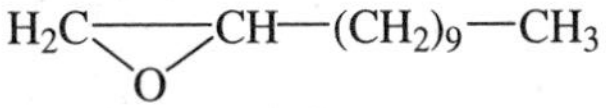	$H_2C\!\!-\!\!CH\!-\!(CH_2)_9\!-\!CH_3$ (O)	Caused blood tumors in mice; irritant to skin	Combustible liquid; flash point 44°C (112°F)

(continued)

323

TABLE 16.1 (*Continued*)

Compound/Synonyms/ CAS No.	Formula/MW/Structure	Toxicity	Flammability
1,2-Epoxyhexadecane (hexadecene epoxide)	$C_{16}H_{32}O$ 240.48 	Caused skin tumors in mice at the site of application; irritant to skin	Noncombustible liquid
n-Butyl glycidyl ether (glycidyl butyl ether, 2,3-epoxy-propyl butyl ether, 1-butoxy-2,3-epoxy-propane) [2426-08-6]	$C_7H_{14}O_2$ 130.21 	Mild skin and eye irritant; low toxicity; caused depression of central nervous system in rats; inhalation of 670 ppm was lethal; LD_{50} oral (rats): 2050 mg/kg; mutagenic effect in microbial and mammalian test systems; in humans, exposure may cause irritation and skin sensitization; TLV-TWA 135 mg/m^3 (25 ppm) (ACGIH), 270 mg/m^3 (50 ppm)	Combustible liquid; flash point 54.5°C (130°F); vapor pressure 3 torr at 20°C (4200 ppm in air)

Name (CAS)	Formula / MW	Structure	Toxicity	
1,2-Epoxy-3-(o-methoxyphenoxy)propane [2210-74-4]	$C_{10}H_{12}O_3$ 180.22 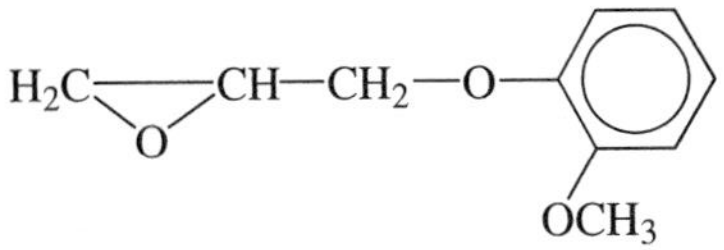		(OSHA), ceiling 30 mg/m³ (15 min) (NIOSH)	Data not available
Cresyl glycidyl ether (glycidyl methyl phenyl ether, tolyl glycidyl ether, 1,2-epoxy-3-(tolyloxy)propane, (methylphenoxy) methyloxirane) [26447-14-3]	$C_{10}H_{12}O_2$ 164.22 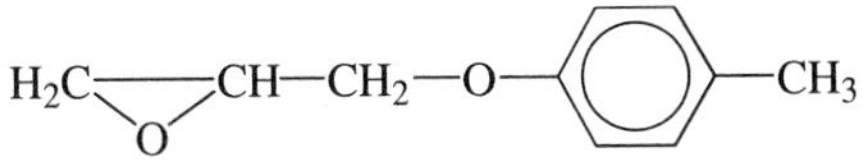		Low toxicity in mice, symptom: muscle contraction; LD_{50} subcutaneous (mice): 1355 mg/kg	Data not available
1,3-Bis(2,3-Epoxypropoxy)-2,2-dimethylpropane (neopentylglycol diglycidyl ether, Heloxy WC68) [17557-23-2]	$C_{11}H_{20}O_4$ 216.31 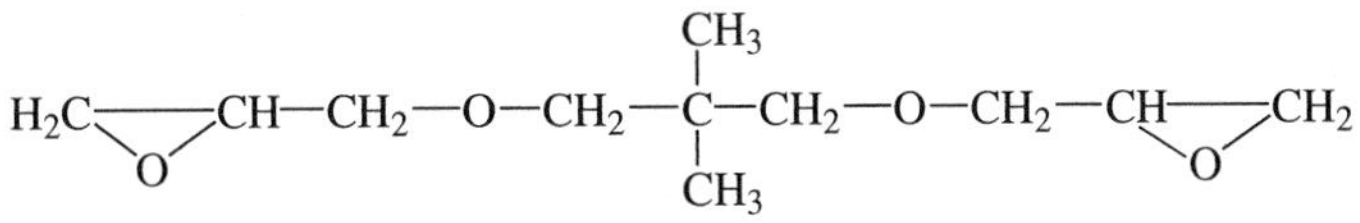		Low toxicity in test animals; toxic routes: inhalation and ingestion symptoms: somnolence, muscle weakness and excitement; LC_{50} (mice): 310 mg/m³, LD_{50} oral (mice): 1700 mg/kg Caused skin tumor in mice	Data not available

325

(continued)

TABLE 16.1 *(Continued)*

Compound/Synonyms/ CAS No.	Formula/MW/Structure	Toxicity	Flammability
2,2-Bis(4-(2,3-epoxy-propoxy)phenyl) propane [bis(4-glycidyloxyphenyl) dimethylmethane, bis(4-hydroxy-phenyl)propane diglycidyl ether. Araldite 6010, Epon 828, Epoxide A] [1675-54-3]	$C_{21}H_{24}O_4$ 340.45	Irritant: mild on rabbit skin, application of 2 mg in 24 hr caused severe irritation in rabbits' eyes; low toxicity in rats; mutation test positive; caused cancer in mice skin, lung, and liver; no report of carcinogenicity in humans	Data not available

first member of epoxide series of compounds; the ring cleaves readily to undergo a wide array of reactions, due to the strained configuration

Synonyms: oxirane; ethene oxide; dimethylene oxide; 1,2-epoxyethane

Uses and Exposure Risk

Ethylene oxide is widely used as a sterilizing agent; as a fumigant; as a propellant; in the production of explosives; in the manufacture of ethylene glycol, polyethylene oxide, glycol ethers, crown ethers, ethanolamines, and other derivatives; and in organic synthesis.

Physical Properties

Colorless gas with characteristic ether-like smell; mobile liquid at low temperature; bp 10.4°C; freezes at −111°C; density 0.8824 at 10°C; soluble in water, alcohol, ether, acetone, and benzene.

Health Hazard

Ethylene oxide is a severe irritant, as well as a toxic and carcinogenic compound. Inhalation can cause severe irritation in the eyes, respiratory tract, and skin. In humans, the delayed symptoms may be nausea, vomiting, headache, dyspnea, pulmonary edema, weakness, and drowsiness. Exposure to high concentrations can cause central nervous system depression.

Contact with an aqueous solution of ethylene oxide on skin can produce severe burns after a delay period of a few hours. It may be absorbed by plastic, leather, and rubber materials if not handled properly, and can cause severe skin irritation.

Exposure of test animals to a high concentration of ethylene oxide resulted in the watering of eyes, nasal discharge, and labored breathing. The toxic effects observed after a few days were vomiting, diarrhea, pulmonary edema, dyspnea, and convulsion, followed by death.

LC$_{50}$ value, inhalation (rats): 800 ppm (4 hr)
LC$_{50}$ value, inhalation (dogs): 960 ppm (4 hr)
LD$_{50}$ value, oral (guinea pigs): 270 mg/kg

Ethylene oxide is a teratogen, causing birth defects. Laboratory tests on animals indicated that exposure could cause fetal deaths, specific developmental abnormalities, and paternal effects related to testes and sperm ducts.

Ethylene oxide showed positive carcinogenicity in test animals. Inhalation, ingestion, and subcutaneous application over a period of time developed tumors of all kinds in rats and mice. It caused brain, liver, gastrointestinal, and blood cancers in test subjects. Although the evidence of its carcinogenicity in humans is inadequate, it is suspected to be cancer causing to humans.

Ethylene oxide is a mutagen in animals and humans. It causes chromosomal aberrations (Thiess et al. 1981), errors in DNA synthesis (Cumming et al. 1981), and alkylation of hemoglobin (Calleman et al. 1978).

Exposure Limits

TLV-TWA 1.8 mg/m^3 (1 ppm) (ACGIH), 0.18 mg/m^3 (0.1 ppm), 5 ppm/10 min (NIOSH).

Fire and Explosion Hazard

Highly flammable; flash point (closed cup) −17°C (0°F); vapor pressure 1095 torr at 20°C; vapor density 1.5 (air = 1); the vapor can travel a considerable distance to a source of ignition and flash back; ignition temperature in air 429°C (804°F), ignition temperature in the absence of air (100% ethylene oxide) 570°C (1058°F); fire-extinguishing agent: use a water spray from an explosion-resistant location to fight fire; use water to flush the spill and to keep containers cool.

Ethylene oxide forms an explosive mixture with air; the LEL and UEL values are 3% and 100% by volume, respectively. It explodes when heated in a closed vessel. It polymerizes violently when in contact with active catalyst surfaces. Rearrangement or polymerization occurs exothermically in the presence of concentrated acids and bases, alkali metals, oxides of iron and aluminum, and their anhydrous chlorides (Hess and Tilton 1950). It may explode when combined with alcohols and mercaptans or with ammonia under high pressure (NFPA 1986).

Storage and Shipping

Ethylene oxide is stored in an outside cool area below 30°C, provided with a water-spray system, and isolated from combustible materials such as acids, bases, chlorides, oxides, and metallic potassium. Protect against physical damage. Avoid inside storage. Shipping should be done in steel cylinders, drums, and insulated tank cars.

Analysis

Ethylene oxide can be analyzed by a gas chromatograph equipped with ECD or PID. Ambient air drawn directly into a syringe is injected into a Carbopack column in a GC equipped with PID (NIOSH 1984, Method 3702). In an alternative method air is passed through a solid sorbent tube containing HBr-coated petroleum charcoal, whereby ethylene oxide is converted to 2-bromoethylheptafluorobutyrate, which is then desorbed with dimethyl formamide. The eluant is injected into a Chromosorb W-HP column of a GC equipped with an ECD (NIOSH 1984, Method 1614).

16.3 PROPYLENE OXIDE

DOT Label: Flammable Liquid, UN 1280
Formula C_3H_7O; MW 58.08; CAS [75-56-9]
Structure and functional group:

$$H_2C\overset{\diagup\diagdown}{\underset{O}{}}CH_2-CH_3$$

Strained three-membered oxirane ring imparts a high degree of reactivity
Synonyms: 1,2-epoxypropane; methyl oxirane; propene oxide

Uses and Exposure Risk

Propylene oxide is used as a fumigant for foodstuffs; as a stabilizer for fuels, heating oils, and chlorinated hydrocarbons; as a fuel–air explosive in munitions; and to enhance the decay resistance of wood and particleboard (Mallari et al. 1989).

Physical Properties

Colorless liquid, exists as two optical isomers; bp 34.3°C; mp 112°C; density 0.859 at 0°C; soluble in water, alcohol, and ether.

Health Hazard

Exposure to propylene oxide vapors can cause moderate to severe irritation of the eyes, mucous membranes, and skin. Inhalation can also produce weakness and drowsiness. Symptoms of acute exposure in test animals were lachrymation, salivation, gasping, and labored breathing and discharge from nose. Harris et al. (1989) in a study on Fischer-344 female rats found no adverse effect below a 300-ppm exposure level. However, at the chronic inhalation level of 500 ppm the gain in the maternal body weight and food consumption were reduced significantly. In a similar chronic inhalation study on Wistar rats, Kuper et al. (1988) observed a decrease in the body weight and degenerative and hyperplastic change in the nasal mucosa when rats were exposed to 300 ppm of propylene oxide. The investigators have reported an increase in the incidence of malignant tumors in the mammary glands and other sites in female rats.

Contact with its dilute aqueous solutions can produce edema, blistering, and burns on the skin. It is mutagenic in the Ames test and a suspected animal carcinogen. Its carcinogenicity in humans is not established. Its odor threshold is 200 ppm.

Exposure Limit

TLV-TWA 50 mg/m^3 (20 ppm) (ACGIH), 240 mg/m^3 (100 ppm) (OSHA); IDLH 2000 ppm.

Fire and Explosion Hazard

Highly flammable; flash point $-37°C$ ($-35°F$); vapor pressure 440 torr at 20°C; vapor density 2.0 (air = 1); the vapor may travel a considerable distance to an ignition source and flash back; fire-extinguishing agent: "alcohol" foam, dry chemical, or CO$_2$; although water is ineffective in fighting fire, it may be used to disperse the vapors and to keep containers cool (NFPA 1986).

Propylene oxide evolves heat when it polymerizes, and therefore closed containers may rupture. Contact with concentrated acids and alkalies, finely divided metals, and catalyst surfaces may be violent. It forms an explosive mixture with air in the range 2.8–37.0% by volume of air.

Storage and Shipping

Propylene oxide is stored in a flammable liquid cabinet isolated from combustible and oxidizable materials. It is shipped in glass bottles and metal containers under a nitrogen atmosphere.

Disposal/Destruction

Propylene oxide is dissolved in a solvent and burned in a chemical incinerator equipped with an afterburner and scrubber. Alternatively, it is mixed in small amounts with copious quantities of water and drained down the sink.

Analysis

Propylene oxide may be analyzed by several techniques, including colorimetry, gas chromatography, mass spectroscopy, IR, and NMR. In colorimetric methods sodium periodate or 4-(4′-nitrobenzyl)pyridine are used. Air analysis is performed by passing air through charcoal cartridge or porous polymer; eluting the charcoal adsorbent into carbon disulfide; or heating the polymer to 200°C and analyzing by GC/FID.

16.4 BUTYLENE OXIDE

Formula C$_4$H$_8$O; MW 72.11; CAS [106-88-7]

Structure and functional group:

an ethyl group attached to the strained oxirane ring

Synonyms: 1,2-epoxybutane; ethyl oxirane; 1,2-butylene oxide; 1,2-butene oxide

Uses and Exposure Risk

Butylene oxide is used as a fumigant and in admixture with other compounds. It is used to stabilize fuel with respect to color and sludge formation.

Physical Properties

Colorless liquid with disagreeable odor; bp 63.3°C; density 0.837 at 17°C; soluble in acetone and ether, moderately soluble in alcohol, and slightly soluble in water.

Health Hazard

Butylene oxide is an irritant to the skin, eyes, and respiratory tract. It may cause burns on prolonged contact with skin. It exhibited low to moderate toxicity in rats via inhalation and ingestion. Prolonged exposure caused inflammatory lesions of the nasal cavity in rats.

LD$_{50}$ value, oral (rats): 500 mg/kg

Carcinogenicity of butylene oxide on rats has been reported by Dunnick (1988). In 2-year inhalation studies conducted on F344/N rats and B6C3F1 mice carcinogenic activity was observed when the level of exposure was 50 ppm.

Butylene oxide exhibits mutagenicity. Inhalation of 1000 ppm for 7 hours caused harmful effects on fertility in rabbits.

Health hazard to humans may arise primarily from its irritant actions. No exposure limit is set for this compound.

Fire and Explosion Hazard

Highly flammable; flash point (closed cup) −15°C (−5°F); ignition temperature 515°C (959°F); vapor density at 38°C 1.6 (air = 1); the vapor may travel a considerable distance to an ignition source and flash back; fire-extinguishing agent: foam, dry chemical, or CO_2; water sprays may be used to flush any spill, disperse the vapors, and keep fire-exposed containers cool (NFPA 1986).

Butylene oxide may react violently with concentrated alkalies, anhydrous chlorides of aluminum, iron, and tin, and peroxides of these metals. It forms an explosive mixture with air in the range 3.1–25.1% by volume.

Storage and Shipping

Butylene oxide is stored in a "flammable liquids" cabinet isolated from combustible and oxidizable substances. It is shipped in glass bottles and metal containers under nitrogen atmosphere.

Analysis

Butylene oxide may be analyzed by colorimetric methods using sodium periodate or by GC-FID. Analysis for butylene oxide in air may be performed by adsorbing the analyte on charcoal, eluting with carbon disulfide, and analyzing by GC.

16.5   GLYCIDALDEHYDE

EPA Classified Toxic Waste, RCRA Waste Number U126; DOT Label: Flammable Liquid and Poison UN 2622

Formula $C_3H_4O_2$; MW 72.07; CAS [765-34-4]

Structure and functional group:

$$H_2C{-}\!\!\overset{}{\underset{O}{\triangle}}\!\!{-}CH_2{-}CHO$$

an aldehyde group linked to the strained cyclic ether

Synonyms: 2,3-epoxypropionaldehyde; 2,3-epoxy-1-propanal; glycidal; epihydrine aldehyde; oxirane carboxaldehyde

Uses and Exposure Risk

Butylene oxide is used as a cross-linking agent in wool finishing, for tanning and fat liquoring of leather, and to insolubilize protein.

Physical Properties

Colorless liquid with a pungent smell; bp 112°C; freezes at −62°C; density 1.140 at 20°C; soluble in water and organic solvents.

Health Hazard

Glycidaldehyde is a severe irritant, moderately toxic, and a carcinogenic compound. Exposure to 1 ppm for 5 minutes resulted in moderate eye irritation in humans. It produced severe skin irritation with slow healing, causing pigmentation of affected areas (Rose 1983).

The symptoms of its toxicity in humans are central nervous system depression, excitement, and effects on olfactory sense organs. Such ill effects may be observed on exposure to concentrations exceeding 5 ppm.

An intravenous administration of glycidaldehyde at 20 mg/kg in rabbits caused miosis, lacrimation, and respiratory depression followed by death. In rats, 50 mg/kg, given orally, was fatal.

LD$_{50}$ value, intraperitoneal (mice): 200 mg/kg

Glycidaldehyde showed positive in mutation tests and is a proven carcinogen in animals. It caused skin cancer in mice at the site of application. There is no report of its carcinogenicity in humans.

Exposure Limit

No exposure limit is set. It is recommended that human exposure to glycidaldehyde in the work environment should not exceed 1 ppm (3 mg/m^3) concentration in air.

Fire and Explosion Hazard

Flammable; flash point (closed cup) 31°C (88°F); vapor density 2.6 (air = 1); the vapor forms an explosive mixture with air, LEL and UEL values not reported; fire-extinguishing agent: water spray, "alcohol" foam, dry chemical, or CO$_2$; use a water spray to dilute the spill and keep fire-exposed containers cool.

Its reactions with strong acids and bases and anhydrous metal halides can become violent. It can polymerize exothermically on active catalytic surfaces.

Disposal/Destruction

Glycidaldehyde is burned in a chemical incinerator equipped with an afterburner and scrubber. Drain disposal in small amounts with large volumes of water may be done.

Analysis

Glycidaldehyde may be analyzed by GC, GC/MS, and other instrumental techniques. Analysis of glycidaldehyde in air may be performed by passing air over charcoal, desorbing the analyte into CS$_2$, and injecting the eluant into GC equipped with FID.

16.6 EPICHLOROHYDRIN

EPA Classified Toxic Waste, RCRA Waste Number UO41; DOT Label: Flammable Liquid and Poison, UN 2023

Formula C$_3$H$_5$ClO; MW 92.53; CAS [106-89-8]

Structure and functional group:

$$H_2C\overset{}{\underset{O}{\diagup\!\!\diagdown}}CH_2-Cl$$

a chloromethyl group attached to the strained oxirane ring which cleaves to form chlorohydrin, thus epichlorohydrin is similar to chlorohydrins in reactions

Synonyms: chloropropylene oxide; (chloromethyl)ethylene oxide; 2-(chloromethyl) oxirane; 1,2-epoxy-3-chloropropane; 2,3-epoxypropyl chloride

Uses and Exposure Risk

Epichlorohydrin is used to make glycerol, epoxy resins, adhesive, and castings; as derivatives for producing dyes, pharmaceuticals, surfactants, and plasticizers; and as a solvent for resins, gums, paints, and varnishes.

Physical Properties

Colorless, mobile liquid with chloroform-like smell; exists as a racemic mixture of its optical isomers in equal amounts; bp 116°C; freezes at −57°C; density 1.1807 at 20°C; soluble in most organic solvents, moderately soluble in water (6.6% by weight).

Health Hazard

Epichlorohydrin is toxic, carcinogenic, and a strong irritant. Its vapors can produce irritation in the eyes, skin, and respiratory tract. Exposure to high concentration resulted in death in animals, injuring the central nervous system. The liquid can absorb through human skin, causing painful irritation of subcutaneous tissues (ACGIH 1986). The symptoms of toxicity from high dosage in test animals were paralysis of muscles and slow development of respiratory distress. Long exposures at 120 ppm for several hours resulted in lung, kidney, and liver injury in rats (Gage 1959). Ingestion by an oral route caused tremor, somnolence, and ataxia in mice (NIOSH 1986). The toxic symptoms and lethal doses varied widely with animal species. The toxic metabolite of epichlorhydrin could be α-chlorohydrin [96-24-2]; the latter was produced in vitro by rat liver microsomes (Gingell et al. 1987).

LD_{50} value, oral (guinea pigs): 280 mg/kg
LC_{50} value, (rats): 250 ppm (8 hr)

A 25 ppm concentration may be detectable by odor. Exposure at this level may cause burning of the eyes and nose in humans. Above 100 ppm even a short exposure may be hazardous to humans, causing nausea, dyspnea, lung edema, and kidney injury.

Epichlorohydrin is mutagenic and has shown carcinogenicity in test animals. It caused tumors in the lungs and nose and at gastrointestinal and endocrine sites. Exposure to this compound caused harmful reproductive effects on fertility and birth defects in mice.

Exposure Limits

TLV-TWA (skin) 8 mg/m^3 (2 ppm) (ACGIH); STEL (15 min) 19 mg/m^3 (5 ppm) (NIOSH).

Fire and Explosion Hazard

Flammable; flash point (open cup) 40°C (105°F); vapor pressure 13 torr; vapor density 3.2 (air = 1); ignition temperature 415°C; forms an explosive mixture with air in the range 3.8–21.0% by volume; fire-extinguishing agent: water spray, dry chemical, "alcohol" foam, or CO_2; use a water spray to flush or dilute the spill, disperse the vapors, and keep fire-exposed containers cool.

Epichlorohydrin can polymerize in the presence of acids, bases, and other catalysts generating heat and pressure. In a closed and confined space, explosion can occur. Its reactions with anhydrous metal halides can be violent, and with amines and imines, exothermic. Ignition may occur with potassium-*tert*-butoxide.

Storage and Shipping

Epichlorohydrin is stored in a well-ventilated, cool place isolated from combustible and oxidizable materials, all acids and bases, and anhydrous metal halides. Protect from physical damage. It is shipped in metal drums.

Disposal/Destruction

Epichlorohydrin is mixed with vermiculite and then with caustic soda and slaked lime, wrapped in paper, and burned in a chemical incinerator equipped with an afterburner and scrubber (Aldrich 1996).

Analysis

Epichlorohydrin is analyzed by GC, GC/MS, and other instrumental techniques. Epichlorohydrin in air is adsorbed over coconut charcoal, desorbed into carbon disulfide, and analyzed by GC-FID (NIOSH 1984; Method 1010).

16.7 GLYCIDOL

Formula $C_3H_6O_2$; MW 74.08; CAS [556-52-5]

Structure and functional group:

$$H_2C \overset{\diagdown}{\underset{O}{\diagup}} CH_2 - CH_2 - OH$$

an alcoholic −OH group (primary) on the oxirane side chain

Synonyms: 3-hydroxy-1,2-epoxypropane; 2,3-epoxy-1-propanol; oxyranyl methanol; epihydrin alcohol; glycidyl alcohol

Uses and Exposure Risk

Glycidol is used as a stabilizer for natural oils and vinyl polymers, as a demulsifier, and as a leveling agent for dyes.

Physical Properties

Colorless liquid; bp 166°C; freezes at −45°C; density 1.1143 at 25°C; soluble in water, alcohol, and ether.

Health Hazard

Glycidol is an eye, lung, and skin irritant. The pure compound caused severe but reversible corneal injury in rabbit eyes (ACGIH 1986). Exposure to its vapor caused irritation of lung in mice, resulting in pneumonitis. There is no evidence of any cumulative toxicity. From the limited toxicity data, it appears that the health hazard to humans from its exposure is, primarily, respiratory irritation, stimulation of the central nervous system, and depression.

LC$_{50}$ value, (mice): 450 ppm (4 hr)

LD$_{50}$ value, oral (rats): 420 mg/kg

Glycidol is mutagenic, testing positive in the histidine reversion–Ames test. There is no report of its carcinogenic action. Oral and intraperitoneal administration of glycidol in rats showed harmful effects on fertility.

Exposure Limits

TLV-TWA 75 mg/m^3 (25 ppm) (ACGIH); 150 mg/m^3 (50 ppm) (OSHA); IDLH 500 ppm (NIOSH).

Fire and Explosion Hazard

Combustible; flash point (closed cup) 72°C (162°F); vapor pressure 0.9 torr (at 25°C), at which its concentration in air is 1180 ppm; may form explosive mixtures with air, LEL and UEL values not reported; fire-extinguishing agent: water spray, "alcohol" foam, dry

chemical, or CO_2. It is incompatible with strong oxidizers and nitrates.

Disposal/Destruction

Glycidol is burned in a chemical incinerator equipped with an afterburner and scrubber.

Analysis

Analysis may be performed by GC, GC/MS, and other instrumental techniques. Air analysis for glycidol may be performed by charcoal adsorption, followed by desorption of analyte with tetrahydrofuran and injecting into GC equipped with FID.

16.8   ISOPROPYL GLYCIDYL ETHER

Formula $C_6H_{12}O_2$; MW 116.16; CAS [4016-14-2]

Structure and functional group:

an isopropyl group attached to the epoxide via ethereal linkage

Synonyms: 1,2-epoxy-3-isopropoxypropane; isopropyl epoxypropyl ether; isopropoxymethyl oxirane

Uses and Exposure Risk

Isopropyl glycidyl ether (IGE) is used to stabilize chlorinated solvents and as a viscosity reducer of epoxy resins.

Physical Properties

Colorless volatile liquid; bp 127°C; density 0.919 at 20°C; soluble in water (19%) and most organic solvents.

Health Hazard

IGE is a skin and eye irritant with low toxicity. Irritation in rabbits' eyes and skin caused by 100 mg was moderate to low. In humans, frequent skin contact may cause dermatitis.

Studies on rats indicated low toxicity of this compound with symptoms of respiratory distress. Prolonged exposure caused ocular irritation and depression of the central nervous system. An increase in hemoglobin and a decrease in peritoneal fat was observed.

LC_{50} value, inhalation (mice): 1500 ppm (4 hr)

LD_{50} value, oral (mice): 1300 mg/kg

LD_{50} value, oral (rats): 4200 mg/kg

Exposure Limits

TLV-TWA 240 mg/m^3 (50 ppm); TLV-STEL 360 mg/m^3 (75 ppm) (ACGIH); 15-minute ceiling 50 ppm (NIOSH); IDLH 1500 ppm (NIOSH).

Fire and Explosion Hazard

Flammable; flash point (closed cup) 33°C (92°F); vapor pressure 9.4 torr at 25°C; forms explosive mixtures with air, LEL and UEL values not reported; fire-extinguishing agent: use a water spray to extinguish fire.

Analysis

GC and GC/MS are suitable analytical techniques. Air analysis may be done by charcoal adsorption, desorption with CS_2, and injecting the eluant onto a GC equipped with FID.

16.9 STYRENE OXIDE

Formula C_8H_8O; MW 120.15; CAS [96-09-3]

Structure and functional group:

Synonyms: (1,2-epoxyethyl)benzene; phenylethylene oxide; phenyl oxirane; 1-phenyl-1,2-epoxyethane; styrene epoxide

Uses and Exposure Risk

Styrene oxide is used in organic synthesis.

Physical Properties

Colorless liquid; bp 194.1°C; freezes at −35.6°C; density 1.0523 at 16°C; soluble in alcohol, ether, and benzene, insoluble in water.

Health Hazard

Styrene oxide is a mild to moderate skin irritant. Irritation from 500 mg was moderate on rabbit skin. The toxicity of this compound was low on test animals. Inhalation of 500 ppm in 4 hours was lethal to rats. An in vivo and in vitro study in mice (Helman et al. 1986) indicates acute dermal toxicity, causing sublethal cell injury.

LD_{50} value, oral (mice): 1500 mg/kg

Styrene oxide, however, may present a considerable health hazard as a mutagen, teratogen, and carcinogen. The reproductive effects from inhalation observed in rats were fetotoxicity, developmental abnormalities, and effects on fertility (Sikov et al. 1986). There is sufficient evidence of its carcinogenicity in animals, producing liver, gastrointestinal tract, and skin tumors. Its cancer-causing effects on humans are unknown.

No exposure limit has been set for this compound. Its toxic and irritant effects in humans are quite low.

Fire and Explosion Hazard

Combustible liquid; flash point 79°C (175°F); vapor density 4.1 (air = 1); the vapor may form an explosive mixture with air, LEL and

UEL values not reported; fire-extinguishing agent: foam, dry chemical, or CO_2; although water is ineffective, it may be used to keep the fire-exposed containers cool.

Styrene oxide may polymerize with evolution of heat. It may react exothermically with concentrated alkalies, anhydrous catalyst surfaces, and metal peroxides; the violence of such reactions is, however, lower than low-carbon-numbered aliphatic epoxides.

Disposal/Destruction

Styrene oxide is burned in a chemical incinerator equipped with an afterburner and scrubber.

Analysis

GC-FID and GC/MS. Other instrumental techniques may be applied.

16.10 PHENYL GLYCIDYL ETHER

Formula $C_9H_{10}O_2$; MW 150.18; CAS [122-60-1]

Structure and functional group:

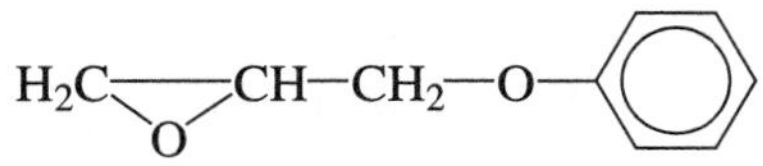

The strained oxirane ring can cleave readily, exibiting high reactivity; substitution can occur on the benzene ring attached to the epoxide through ethereal linkage.

Synonyms: 1,2-epoxy-3-phenoxypropane; 2,3-epoxypropylphenyl ether; (phenoxymethyl)oxirane; phenoxypropylene oxide

Uses and Exposure Risk

Phenyl glycidyl ether (PGE) is used as an intermediate in organic syntheses.

Physical Properties

Colorless liquid; bp 245°C; freezes at 3.5°C; density; soluble in alcohol, ether, and benzene, insoluble in water.

Health Hazard

PGE is a toxic compound exhibiting moderate irritant action and carcinogenicity in animals. Application of 0.25 mg resulted in severe eye irritation in rabbits, while 500 mg caused moderate skin irritation over a period of 24 hours. Prolonged or repeated contact can cause moderate irritation and skin sensitization in humans.

The symptoms of its toxicity in animals were depression of the central nervous system and paralysis of the respiratory tract. Prolonged exposure caused changes in the kidney, liver, thymus, and testes, and loss of hair in rats. The toxicity of this compound in humans is low and the health hazard can arise primarily from its skin-sensitization action.

LD_{50} value, oral (mice): 1400 mg/kg

DGE showed carcinogenicity in rats, causing nasal cancer.

Exposure Limits

TLV-TWA 6 mg/m^3 (1 ppm) (ACGIH); 1 ppm (15 min) (NIOSH).

Fire and Explosion Hazard

Noncombustible liquid, flash point 120°C (248°F); vapor pressure 0.01 torr at 25°C; incompatible with strong acids and bases, amines, and strong oxidizers.

Analysis

PGE can be analyzed by GC using an FID and by GC/MS. Analysis of this compound in air may be done by adsorbing onto charcoal and analyzing the CS_2 eluant by GC.

REFERENCES

ACGIH. 1986. *Documentation of the Threshold Limit Values and Biological Exposure Indices,*

5th ed. Cincinnati, OH: American Conference of Governmental Industrial Hygienists.

Aldrich. 1996. *Aldrich Catalog.* Milwaukee, WI: Aldrich Chemical Company.

Calleman, C. J., L. Ehrenbert, and B. Jannson. 1978. *J. Environ. Pathol. Toxicol. 2:* 427.

Cumming, R. B., T. R. Michaud, L. R. Lewis, and W. H. Olson. 1981. Cited in ACGIH. 1986. *Documentation of the Threshold Limit Values and Biological Exposure Indices*, 5th ed. Cincinnati, OH: American Conference of Governmental Industrial Hygienists.

Dunnick, J. 1988. Toxicology and Carcinogenesis Studies of 1,2-Epoxybutane in F344/N Rats and B6C3F1 Mice. *Nat. Toxicol. Prog. Tech. Rep. Ser. 329.* Research Triangle Park, NC: National Toxicology Program.

Gage, J. C. 1959. *Br. J. Ind. Med. 16:* 11.

Gingell, R., P. W. Beatty, H. R. Mitschke, R. L. Mueller, V. L. Sawin, and A. C. Page. 1987. Evidence that epichlorohydrin is not a toxic metabolite of 1,2-dibromo-3-chloropropane. *Xenobiotica 17*(2): 229–40.

Harris, S. B., J. L. Schardein, C. E. Ulrich, and S. A. Radlon. 1989. Inhalation developmental toxicology study of propylene oxide in Fischer 344 rats. *Fundam. Appl. Toxicol. 13*(2): 323–31.

Helman, R. G., J. W. Hall, and J. Y. Kao. 1986. Acute dermal toxicity: in vivo and in vitro comparisons in mice. *Fundam. Appl. Toxicol. 7*(1): 94–100.

Hess, L. G., and V. V. Tilton. 1950. *Ind. Eng. Chem. 42:* 1251–58.

Kuper, C. F., P. G. J. Reuzel, V. J. Feron, and H. Verschuuren. 1988. Chronic inhalation toxicity and carcinogenicity study of propylene oxide in Wistar rats. *Food Chem. Toxicol. 26*(2): 159–67.

Mallari, V. C., Jr., K. Fukuda, N. Morohoshi, and T. Haraguchi. 1989. Biodegradation of particle board. I. Decay resistance of chemically modified wood and qualities of particle board. *Mokuzai Gakkaishi 35*(9): 832–38.

NFPA. 1986. *Fire Protection Guide on Hazardous Materials*, 9th ed. Quincy, MA: National Fire Protection Association.

NIOSH. 1984. *Manual of Analytical Methods*, 3rd ed. Cincinnati, OH: National Institute for Occupational Safety and Health.

NIOSH. 1986. *Registry of Toxic Effects of Chemical Substances*, ed. D. V. Sweet. Washington, DC: U.S. Government Printing Office.

Rose, V. E. 1983. Epoxy compound. In *An Encyclopedia of Health and Occupational Safety*, 3rd ed., ed. Luigi Parmeggiani, Vol. 1, pp. 770–73. Geneva: International Labor Office.

Sikov, M. R., W. C. Cannon, D. B. Carr, R. A. Miller, R. W. Niemeier, and B. D. Hardin. 1986. Reproductive toxicology of inhaled styrene oxide in rats and rabbits. *J. Appl. Toxicol. 6*(3): 155–64.

Thiess, A. M., H. Schwegler, and W. G. Stocker. 1981. *J. Occup. Med. 23:* 343.

17

ESTERS

17.1 GENERAL DISCUSSION

Esters are a wide class of organic compounds with the general formula RCOOR′ and produced by esterification of carboxylic acids with alcohols.

$$R-\overset{\overset{\displaystyle O}{\|}}{C}-OH + R'OH \xrightarrow{\text{acid}}$$
$$R-\overset{\overset{\displaystyle O}{\|}}{C}-O-R' + H_2O$$

where R and R′ are alkyl, cycloalkyl, or aryl group(s). Unsaturated esters may be obtained from alkenyl or alkynyl groups. The esterification reaction shown above is catalyzed by mineral acids. These compounds are also formed by the reactions of alcohols with (1) acid anhydrides, or (2) acyl halides. Lower esters have a pleasant fruity odor. The lower aliphatic esters find wide applications as solvents for polymers, as fumigants, and as flavoring agents. Phthalic acid esters are used as plasticizers.

The acute toxicity of esters is generally of low order. These are narcotics at high concentrations. Vapors are an irritant to the eyes and mucous membranes. The toxicity increases with an increase in the alkyl chain length. Thus *n*-amyl acetate is more toxic than ethyl acetate or *n*-propyl acetate.

Lower aliphatic esters are flammable liquids. Some of these have low flash points and may readily propagate or flash back the flame from a nearby source of ignition to the open bottles containing the liquids. The vapors form explosive mixtures with air. The reactivity of esters is low. There is no report of any serious explosion caused by esters.

Analysis

The carbonyl group (C=O) gives infrared absorption bands in 1750–1735 cm^{-1}. The C−H stretching vibrations produce bands in 1300–1180 cm^{-1}. Aliphatic and aromatic esters may conveniently be analyzed by GC using flame ionization detector. A column such as 10% SP-1000 or 5% FFAP on Chromosorb W-HP is suitable for the purpose.

Acetic acid esters and ethyl acrylate in air may be analyzed by NIOSH (1984) Method 1450. One to 10 L of air at a flow rate of 10–200 mL/min is passed over coconut shell charcoal, desorbed with carbon disulfide, and injected onto GC-FID.

Phthalate esters listed as EPA priority pollutants may be analyzed in wastewaters by U.S. EPA (1984) Methods 606 (GC), 625 (GC/MS), and 1625 (isotope dilution). Soils, sediments, groundwater, and hazardous wastes may be analyzed by GC using FID or ECD (Method 8060) or by GC/MS (Method 8270 or 8250) (U.S. EPA 1986). The primary and secondary characteristic ions to identify these compounds by GC/MS (electron-impact ionization) are as follows: dimethyl phthalate, 149, 177, and 150; di-*n*-butyl phthalate, 149, 150, and 104; butyl benzyl phthalate, 149, 167, and 279; and di-*n*-octyl phthalate, 149, 167, and 43. Other phthalate esters may be analyzed by GC or GC/MS techniques. A fused-silica capillary column is suitable for the purpose.

ALIPHATIC ESTERS

17.2   METHYL FORMATE

DOT Label: Flammable Liquid, UN 1243
Formula $C_2H_4O_2$; MW 60.06; CAS [107-31-3]
Structure:

$$H-C{<}^{O}_{OCH_3}$$

first member of a homologue series of aliphatic esters.

Synonyms: methyl methanoate; formic acid methyl ester

Uses and Exposure Risk

Methyl formate is used as a fumigant, as a larvacide for food crops, and as a solvent for cellulose acetate.

Physical Properties

Colorless liquid with a pleasant odor; density 0.987 at 15°C; boils at 31.5°C; solidifies at −100°C; miscible with water, alcohol, and ether.

Health Hazard

Methyl formate is a moderately toxic compound affecting eyes, respiratory tract, and central nervous system. It is an irritant to the eyes, nose, and lungs. Exposure to high concentrations of its vapors in air may produce visual disturbances, irritations, narcotic effects, and respiratory distress in humans. Such effects may be manifested at a 1-hour exposure to about 10,000-ppm concentration. Cats died of pulmonary edema from 2-hour exposure to this concentration.

The acute oral toxicity of methyl formate was low in test subjects. The symptoms were narcosis, visual disturbances, and dyspnea. An oral LD_{50} value in rabbit is in the range 1600 mg/kg.

Exposure Limits

TLV-TWA 100 ppm ($\sim$250 mg/m^3) (ACGIH, MSHA, and OSHA); TLV-STEL 150 ppm ($\sim$375 mg/m^3) (ACGIH); IDLH 5000 ppm (NIOSH).

Fire and Explosion Hazard

Flammable liquid; flash point (closed cup) −19°C (−2°F); vapor pressure 476 torr at 20°C; vapor density 2.1 (air = 1), the vapor is heavier than air and can travel a considerable distance to a source of ignition and flash back; autoignition temperature 456°C (853°F); fire-extinguishing agent: dry chemical, CO_2, or "alcohol" foam; use a water spray to keep the fire-exposed containers cool and to flush or dilute any spill.

Methyl formate vapors form explosive mixtures with air, the LEL and UEL values are 5% and 23% by volume in air, respectively. Its reactions with strong oxidizers are vigorous.

17.3   ETHYL FORMATE

DOT Label: Flammable Liquid, UN 1190

Formula C_3H_6O; MW 74.09; CAS [109-94-4]

Structure:

$$H-\underset{\underset{O}{\|}}{C}-O-CH_2-CH_3$$

Synonyms: formic acid ethyl ester; ethyl methanoate

Uses and Exposure Risk

Ethyl formate is used as a solvent; as a flavor for lemonades and essences; and as a fungicide and larvicide for cereals, dry fruits, tobacco, and so on.

Physical Properties

Colorless liquid with a pleasant fruity odor; density 0.917 at 20°C; boils at 54°C; freezes at −80°C; moderately soluble in water (13.5% at 20°C), soluble in alcohol and ether.

Health Hazard

The irritant action of ethyl formate in the eyes, nose, and mucous membranes is milder than that of methyl formate. However it is more narcotic than the methyl ester. Cats exposed to 10,000 ppm died after 90 minutes, after deep narcosis. A 4-hour exposure to 8000 ppm was lethal to rats. Inhalation of 5000 ppm for a short period produces eye and nasal irritation and salivation in rats. The toxic effects from ingestion include somnolence, narcosis, gastritis, and dyspnea. The oral LD_{50} values in various test animals range between 1000 and 2000 mg/kg.

Exposure Limits

TLV-TWA 100 ppm (∼300 mg/m^3) (ACGIH, MSHA, and OSHA); IDLH 8000 ppm (NIOSH).

Fire and Explosion Hazard

Flammable liquid; flash point (closed cup) −20°C (−4°F); vapor pressure 194 torr at 20°C; vapor density 2.55 (air = 1); the vapor is heavier than air and can travel a considerable distance to a source of ignition and flash back; autoignition temperature 455°C (851°F); fire-extinguishing agent: dry chemical, CO_2, or "alcohol" foam; use a water spray to keep fire-exposed containers cool and to flush or dilute any spill.

Ethyl formate vapors form explosive mixtures in air; the LEL and UEL values are 2.8 and 16.0% by volume in air. Its reactions with strong oxidizers, alkalies, and acids can be vigorous to violent.

17.4   METHYL ACETATE

DOT Label: Flammable Liquid, UN 1231
Formula C_3H_6O; MW 74.09; CAS [79-20-9]
Structure:

$$CH_3-\underset{\underset{O}{\|}}{\overset{\overset{O}{\|}}{C}}-OCH_3$$

Synonyms: methyl ethanoate; acetic acid methyl ester

Uses and Exposure Risk

Methyl acetate is used as a solvent for lacquers, resins, oils, and nitrocellulose; in paint removers; as a flavoring agent; and in the manufacture of artificial leather.

Physical Properties

Colorless liquid with a pleasant odor; density 0.934 at 20°C; boils at 57°C; freezes at −98°C; soluble in water (24.5% at 20°C) and most organic solvents.

Health Hazard

The toxic effects from exposure to methyl acetate include inflammation of the eyes, visual and nervous disturbances, tightness of the chest, drowsiness, and narcosis. It hydrolyzes in body to methanol, which probably produces the atrophy of the optic nerve. A 4-hour exposure to 32,000 ppm was lethal to rats. Oral and dermal toxicities of this compound are low. An oral LD_{50} value in rats is on the order of 5000 mg/kg.

Exposure Limits

TLV-TWA 200 ppm ($\sim$610 mg/m^3) (ACGIH, MSHA, and OSHA); TLV-STEL 250 ppm ($\sim$760 mg/m^3) (ACGIH); IDLH 10,000 ppm (NIOSH).

Fire and Explosion Hazard

Flammable liquid; flash point (closed cup) $-10°C$ (14°F); vapor pressure 173 torr at 20°C; vapor density 2.55 (air $= 1$); the vapor is heavier than air and can travel a considerable distance to a source of ignition and flash back; autoignition temperature 454°C (850°F); fire-extinguishing agent: dry chemical, CO_2, or "alcohol" foam; use water to flush and dilute any spill and to keep fire-exposed containers cool.

Methyl acetate forms explosive mixtures with air within the range 3.1–16.0% by volume in air. Its reactions with strong oxidizers, alkalies, and acids are vigorous.

17.5   ETHYL ACETATE

EPA Designated Toxic Waste, RCRA Waste Number U112; DOT Label: Flammable Liquid, UN 1173

Formula: C_4H_8O; MW 88.12; CAS [141-78-6]

Structure:

$$CH_3-\overset{\overset{\displaystyle O}{\|}}{C}-O-CH_2-CH_3$$

Synonyms: ethyl ethanoate; acetoxyethane; acetic acid ethyl ester

Uses and Exposure Risk

Ethyl acetate is used as a solvent for varnishes, lacquers, and nitrocellulose; as an artificial fruit flavor; in cleaning textiles; and in the manufacture of artificial silk and leather, perfumes, and photographic films and plates (Merck 1996).

Physical Properties

Colorless volatile liquid; pleasant fruity odor; pleasant taste when diluted; density 0.902 at 20°C; bp 77°C; mp $-83°C$; moderately soluble in water ($\sim$10% at 20°C), miscible with most organic solvents.

Health Hazard

The acute toxicity of ethyl acetate is low in test animals. It is less toxic than methyl acetate. Inhalation of its vapors can cause irritation of the eyes, nose, and throat. Exposure to a concentration of 400–500 ppm in air may produce mild eye and nose irritation in humans. Its odor threshold is 50 ppm.

Exposure Limits

TLV-TWA 400 ppm ($\sim$1400 mg/m^3) (ACGIH, MSHA, and OSHA); IDLH 10,000 ppm (NIOSH).

Fire and Explosion Hazard

Flammable liquid; flash point (closed cup) $-4.5°C$ (24°F), (open cup) 7.2°C (45°F); vapor pressure 76 torr at 20°C; vapor density 3.01 (air $= 1$); the vapor is heavier than air and presents a flashback fire hazard when bottles are left open near an ignition source; autoignition temperature 426°C (800°F); fire-extinguishing agent: dry chemical, CO_2, or "alcohol" foam; use water to keep fire-exposed containers cool and to flush or dilute any spill.

Ethyl acetate vapor forms explosive mixtures with air in the range 2.0–11.5% by volume in air. It ignites when reacts with potassium *tert*-butoxide (MCA 1973). An explosion was reported when methyl acetate was added dropwise to the residue after 3-chloro-2-methylfuran was subjected to reductive dechlorination with lithium aluminum hydride (NFPA 1986). Its reactions with strong acids or alkalies are exothermic. Reactions with strong oxidizers can be vigorous to violent.

17.6 *n*-PROPYL ACETATE

DOT Label: Flammable Liquid, UN 1276
Formula $C_5H_{10}O_2$; MW 102.15; CAS [109-60-4]
Structure:

$$CH_3-\underset{\underset{O}{\|}}{C}-O-CH_2-CH_2-CH_3$$

Synonyms: 1-propyl acetate; acetoxypropane; acetic acid *n*-propyl ester

Uses and Exposure Risk

n-Propyl acetate is used as a solvent for cellulose derivatives, plastics, and resins; in flavors and perfumes; and in organic synthesis.

Physical Properties

Colorless liquid with a mild odor of pears; density 0.887 at 20°C; boils at 101.6°C; solidifies at −92.5°C; slightly soluble in water (2% at 20°C), miscible with most organic solvents.

Health Hazard

The acute toxicity of *n*-propyl acetate is low in test animals. The toxicity, however, is slightly greater than ethyl acetate and isopropyl acetate. Exposure to its vapors produces irritation of the eyes, nose, and throat and narcotic effects. A 5-hour exposure to 9000- and 6000-ppm concentrations produced narcotic symptoms in cats and mice, respectively (Flury and Wirth 1933). A 4-hour exposure to 8000 ppm was lethal to rats. Ingestion of the liquid can cause narcotic action. A high dose can cause death. A dose of 3000 mg/kg by subcutaneous administration was lethal to cats. The liquid may cause mild irritation upon contact with skin.

LD_{50} value, oral (mice): 8300 mg/kg

Exposure Limits

TLV-TWA 200 ppm (~840 mg/m^3) (ACGIH, MSHA, and OSHA); TLV-STEL 250 ppm (~1050 mg/m^3) (ACGIH); IDLH 8000 ppm (NIOSH).

Fire and Explosion Hazard

Flammable liquid; flash point (closed cup) 14°C (58°F); vapor pressure 25 torr at 20°C; vapor density 3.5 (air = 1); autoignition temperature 450°C (842°F). The vapor forms explosive mixtures with air in the range 2.0–8.0% by volume in air. It reacts vigorously with strong oxidizers, acids, and alkalies. Reaction with potassium *tert*-butoxide may cause ignition.

17.7 ISOPROPYL ACETATE

DOT Label: Flammable Liquid, UN 1220
Formula $C_5H_{10}O_2$; MW 102.15; CAS [108-21-4]
Structure:

$$CH_3-\underset{\underset{O}{\|}}{C}-O-CH\overset{\diagup CH_3}{\diagdown CH_3}$$

Synonyms: 2-propyl acetate; 2-acetoxypropane; acetic acid isopropyl ester; acetic acid 1-methylethyl ester

Uses and Exposure Risk

Isopropyl acetate is used as a solvent for nitrocellulose, plastics, oils, and fats, and as a flavoring agent.

Physical Properties

Colorless liquid with a fruity odor; density 0.870 at 20°C; boils at 89.5°C; solidifies at −73.5°C; low solubility in water (~3% at 20°C), miscible with most organic solvents.

Health Hazard

Isopropyl acetate is an irritant to the eyes, nose, and throat. The acute toxicity in laboratory animals was low. Exposure to high concentrations in air or ingestion can produce narcotic effects. A 4-hour exposure to a concentration of 32,000 ppm in air was fatal to rats (ACGIH 1986). The oral LD_{50} value in rats is in the range 6000 mg/kg.

Exposure Limits

TLV-TWA 250 ppm (~950 mg/m^3) (ACGIH, MSHA, and OSHA); TLV-STEL 310 ppm (~1185 mg/m^3) (ACGIH); IDLH 16,000 ppm (NIOSH).

Fire and Explosion Hazard

Flammable liquid; flash point (closed cup) 2°C (36°F), (open cup) 4°C (40°F); vapor pressure 43 torr at 20°C; vapor density 3.5 (air = 1); autoignition temperature 460°C (860°F). Its vapors form explosive mixtures with air within the range 1.8–8.0% by volume in air. Its reactions with strong oxidizers, alkalies, and acids can be vigorous.

17.8 *n*-BUTYL ACETATE

DOT Label: Flammable Liquid, UN 1123
Formula $C_6H_{12}O_2$; MW 116.18; CAS [123-86-4]

Structure:

$$CH_3-\underset{\underset{O}{\|}}{C}-O-CH_2-CH_2-CH_2-CH_3$$

Synonyms: 1-butyl acetate; acetoxybutane; acetic acid *n*-butyl ester

Uses and Exposure Risk

n-Butyl acetate is used in the manufacture of lacquers, plastics, photographic films, and artificial leathers.

Physical Properties

Colorless liquid with a fruity odor; density 0.883 at 20°C; boils at 126°C; freezes at −77°C; slightly soluble in water (0.83% at 20°C), miscible with most organic solvents.

Health Hazard

The narcotic effects of *n*-butyl acetate is greater than the lower alkyl esters of acetic acid. Also, the toxicities and irritant actions are somewhat greater than *n*-propyl, isopropyl, and ethyl acetates. Exposure to its vapors at about 2000 ppm caused mild irritation of the eyes and salivation in test animals. A 4-hour exposure to 14,000 ppm was fatal to guinea pigs. In humans, inhalation of 300–400 ppm of *n*-butyl acetate may produce moderate irritation of the eyes and throat, and headache.

Exposure Limits

TLV-TWA 150 ppm (~710 mg/m^3) (ACGIH, MSHA, and OSHA); TLV-STEL 200 ppm (~950 mg/m^3); IDLH 10,000 ppm (NIOSH).

Fire and Explosion Hazard

Flammable liquid; flash point (closed cup) 22°C (72°F), (open cup) 33°C (92°F); vapor pressure 10 torr at 20°C; vapor density 4.0 (air = 1); autoignition temperature 425°C

(797°F). The vapor forms explosive mixtures with air in the range 1.7–7.6% by volume in air. Its reactions with strong oxidizers, acids, and alkalies can be vigorous.

17.9 ISOBUTYL ACETATE

DOT Label: Flammable liquid, UN 1213
Formula $C_6H_{12}O_2$; MW 116.18; CAS [110-19-0]
Structure:

$$CH_3-\underset{\underset{O}{\|}}{C}-O-CH_2-CH\underset{CH_3}{\overset{CH_3}{<}}$$

Synonyms: 2-methylpropyl acetate; acetic acid isobutyl ester; acetic acid 2-methyl-propyl ester; β-methylpropyl ethanoate

Uses and Exposure Risk

Isobutyl acetate is used as a solvent and as a flavoring agent.

Physical Properties

Colorless liquid with a fruity smell; density 0.871 at 20°C; boils at 118°C; freezes at −99°C; slightly soluble in water (0.5% at 20°C), miscible with most organic solvents.

Health Hazard

Isobutyl acetate is more toxic but less of an irritant than *n*-butyl acetate. The toxic symptoms include headache, drowsiness, irritation of upper respiratory tract, and anesthesia. A 4-hour exposure to 8000 ppm was lethal to rats. It produced mild to moderate irritation on rabbits' skin. The irritation in eyes was also mild to moderate. The LD_{50} oral value in rabbit is within the range 4800 mg/kg.

Exposure Limits

TLV-TWA 150 ppm (~700 mg/m³) (ACGIH, MSHA, and OSHA); IDLH 7500 ppm (NIOSH).

Fire and Explosion Hazard

Flammable liquid; flash point (closed cup) 18°C (64°F); vapor pressure 13 torr at 20°C; vapor density 4.0 (air = 1); autoignition temperature 421°C (790°F). The vapor forms explosive mixtures with air, LEL values 1.3% (NFPA 1986), and 2.4% (NIOSH 1986) and UEL value 10.5% by volume in air, respectively. Vigorous reactions may occur when isobutyl acetate is mixed with strong oxidizers, alkalies, and acids.

17.10 *n*-AMYL ACETATE

DOT Label: Flammable Liquid, UN 1104
Formula $C_7H_{14}O_2$; MW 130.21; CAS [628-63-7]
Structure:

$$CH_3-\underset{\underset{O}{\|}}{C}-O-CH_2-CH_2-CH_2-CH_2-CH_3$$

Synonyms: 1-pentyl acetate; acetic acid amyl ester; acetic acid pentyl ester; 1-pentanol acetate; amyl acetic ester; pear oil

Uses and Exposure Risk

n-Amyl acetate is used as a solvent for lacquers and paints; in fabrics' printing; in nail polish; and as a flavoring agent.

Physical Properties

Colorless liquid with an odor of banana oil; boils at 149°C; solidifies at −70°C; density 0.879 at 20°C; very slightly soluble in water (~0.2%), miscible with most organic solvents.

Health Hazard

n-Amyl acetate is a narcotic, an irritant to the eyes and respiratory passage, and at high concentrations, an anesthesia. Exposure to about 300 ppm in air for 30 minutes may produce eye irritation in humans. Higher concentrations (>1000 ppm) may produce headache, somnolence, and narcotic effects. Exposure to 5200 ppm for 8 hours was lethal to rats. It is more toxic than the lower aliphatic esters. An LD_{50} value in rats is within the range 6000 mg/kg.

Exposure Limits

TLV-TWA 100 ppm ($\sim$525 mg/m^3) (ACGIH, MSHA, and OSHA); IDLH 4000 ppm.

Fire and Explosion Hazard

Flammable liquid; flash point (closed cup) 23°C (75°F); vapor pressure 4 torr at 20°C; vapor density 5.3 (air = 1); autoignition temperature 379°C (714°F). Its vapors form explosive mixtures with air within the range 1.1–7.5% by volume in air. Its reactions with strong oxidizers, alkalies, and acids can be vigorous to violent.

17.11 ISOAMYL ACETATE

Formula $C_7H_{14}O_2$; MW 130.21; CAS [123-92-2]

Structure:

$$CH_3-\overset{\displaystyle O}{\overset{\displaystyle \|}{C}}-O-CH_2-CH_2-CH\overset{\textstyle CH_3}{\underset{\textstyle CH_3}{<}}$$

Synonyms: isopentyl acetate; 3-methyl-1-butyl acetate; isoamyl ethanoate; acetic acid isopentyl ester; pear oil; banana oil

Uses and Exposure Risk

Isoamyl acetate is used to impart pear flavor to mineral waters and syrups, in perfumes, in the manufacture of artificial silk or leather, in photographic films, in dyeing textiles, and as a solvent.

Physical Properties

Colorless liquid with a pear-like odor and taste; density 0.876 at 15°C; bp 142°C; mp −78.5°C; very slightly soluble in water (0.2% at 20°C), miscible with organic solvents.

Health Hazard

Isoamyl acetate exhibits low toxicity; the toxic effects are comparable to those of n-amyl acetate. The toxic symptoms include irritation of the eyes, nose, and throat; fatigue; increased pulse rate; and narcosis. Inhalation of its vapors at 1000 ppm for 30 minutes may cause irritation, fatigue, and respiratory distress in humans. It is more narcotic than are the lower acetic esters. The LD_{50} value in rabbits is on the order of 7000 mg/kg.

Exposure Limits

TLV-TWA 100 ppm ($\sim$530 mg/m^3) (ACGIH, MSHA, and OSHA); TLV-STEL 125 ppm ($\sim$655 mg/m^3); IDLH 3000 ppm (NIOSH).

Fire and Explosion Hazard

Flammable liquid; flash point (closed cup) 25°C (77°F) (NFPA 1986), 33°C (92°F) (Merck 1989); vapor pressure 4 torr at 20°C; autoignition temperature 360°C (680°F). It forms explosive mixtures with air; the LEL and UEL values are 1.0% (at 100°C) and 7.5% by volume in air, respectively.

17.12 METHYL ACRYLATE

DOT Label: Flammable Liquid, UN 1919

Formula $C_4H_6O_2$; MW 86.10; CAS [96-33-3]

Structure:

$$CH_2{=}CH{-}\overset{\overset{\textstyle O}{\|}}{C}{-}O{-}CH_3$$

Synonyms: methyl-2-propenoate; methyl propenate; 2-propenoic acid methyl ester; acrylic acid methyl ester

Uses and Exposure Risk

Methyl acrylate is a monomer used in the manufacture of plastic films, textiles, paper coatings, and other acrylate ester resins. It is also used in amphoteric surfactants.

Physical Properties

Colorless, volatile liquid with a sharp fruity odor; density 0.956 at 20°C; bp 80°C; mp −76.5°C; low solubility in water (6 g/100 mL at 20°C), soluble in alcohol and ether; polymerizes to a transparent, elastic substance.

Health Hazard

The liquid is a strong irritant, and prolonged contact with the eyes or skin may cause severe damage. Inhalation of its vapors can cause lacrimation, irritation of respiratory tract, lethargy, and at high concentrations, convulsions. One-hour exposure to a concentration of 700–750 ppm in air caused death to rabbits. The oral toxicity of methyl acrylate in animals varied from low to moderate, depending on species, the LD_{50} values ranging between 250 and 850 mg/kg. The liquid may be absorbed through the skin, producing mild toxic effects.

Exposure Limits

TLV-TWA 10 ppm (~35 mg/m^3) (ACGIH and MSHA), TLV-TWA skin 10 ppm (~35 mg/m^3) (OSHA); IDLH 1000 ppm (NIOSH).

Fire and Explosion Hazard

Flammable liquid; flash point (closed cup) −4°C (25°F), (open cup) −3°C (27°F); vapor pressure 68 torr at 20°C; vapor density 3.0 (air = 1); the vapor is heavier than air and can travel a considerable distance to a source of ignition and flashback; autoignition temperature not established; fire-extinguishing agent: dry chemical, CO_2, or "alcohol" foam; use water to keep the fire-exposed containers cool and to flush or dilute any spill; the vapors may polymerize and block the vents.

The vapors of methyl acrylate form explosive mixtures with air, over a relatively wide range; the LEL and UEL values are 2.8% and 25.0% by volume in air, respectively. Methyl acrylate undergoes self-polymerization at 25°C. The polymerization reaction proceeds with evolution of heat and the increased pressure can cause rupture of closed containers. The reaction rate is accelerated by heat, light, or peroxides. Vigorous to violent reaction may occur when mixed with strong oxidizers (especially nitrates and peroxides) and strong alkalies.

Storage and Shipping

Methyl acrylate is stored in a flammable-materials storage room or cabinet below 20°C, separated from oxidizing substances. It is inhibited with 200 ppm of hydroquinone monomethyl ether to prevent self-polymerization. It is shipped in bottles, cans, drums, or tank cars.

17.13 ETHYL ACRYLATE

EPA Designated Toxic Waste, RCRA Waste Number U113; DOT Label: Flammable Liquid UN 1917

Formula $C_5H_8O_2$; MW 100.13; CAS [140-88-5]

Structure:

$$CH_2{=}CH{-}\underset{\underset{\textstyle O}{\|}}{C}{-}O{-}CH_2{-}CH_3$$

Synonyms: ethyl-2-propenoate; ethyl prope-nate; 2-propenoic acid ethyl ester; acrylic acid ethyl ester

Uses and Exposure Risk

Ethyl acrylate is used in the manufacture of acrylic resins, acrylic fibers, textile and paper coatings, adhesives, and leather finish resins; and as a flavoring agent.

Physical Properties

Colorless liquid with a sharp acrid odor; density 0.9405 at 20°C; boils at 99.4°C; solidifies at −71°C; slightly soluble in water (2 g/100 mL at 20°C), soluble in alcohol and ether; polymerizes to a transparent elastic substance.

Health Hazard

Ethyl acrylate is a strong irritant to the eyes, skin, and mucous membranes. The liquid or its concentrated solutions can produce skin sensitization upon contact. It is toxic by all routes of exposure. The toxicity is low in rats and mice and moderate in rabbits. The toxic effects from inhalation noted in animals were congestion of lungs and degenerative changes in the heart, liver, and kidney. Monkey exposed to 272 ppm for 28 days showed lethargy and weight loss; while exposure to 1024 ppm caused death to the animals after 2.2 days (Treon et al. 1949). By compari-son, guinea pigs died of exposure to about 1200 ppm for 7 hours. Ingestion of the liquid may result in irritation of gastrointestinal tracts, nausea, lethargy, and convulsions.

The LD_{50} values varied significantly in different species of animals. The oral LD_{50} values in rabbits, rats, and mice are in the range 400, 800, and 1800 mg/kg, respectively. Animals administered ethyl acrylate showed increased incidence of tumors in forestomach. However, there is no evidence of carcinogenicity caused by this compound in humans.

Exposure Limits

TLV-TWA 5 ppm (∼20 mg/m^3) (ACGIH), 25 ppm (∼100 mg/m^3 (MSHA, NIOSH), TWA skin 25 ppm (100 mg/m^3) (OSHA); IDLH 2000 ppm (NIOSH).

Fire and Explosion Hazard

Flammable liquid; flash point (open cup) 15°C (60°F); vapor pressure 29.5 torr at 20°C; vapor density 3.45 (air = 1); the vapor is heavier than air and presents the danger of propagation of fire in the presence of a source of ignition; autoignition temperature 372°C (702°F). The vapor forms explosive mixtures with air; LEL value 1.8% by volume in air, UEL value not established. It polymerizes readily at elevated temperature, fire condi-tions, in the presence of peroxides, or by light. The reaction is exothermic and may cause violent rupture of closed containers. Reactions with strong oxidizers and alkalies are vigorous. Polymerization is inhibited by adding hydroquinone or its methyl ether to ethyl acrylate.

17.14 *n*-BUTYL ACRYLATE

DOT Label: Flammable Liquid, UN 2348
Formula $C_7H_{12}O_2$; MW 128.19; CAS [141-32-2]
Structure:

$$CH_2{=}CH{-}\underset{\underset{O}{\|}}{C}{-}O{-}CH_2{-}CH_2{-}CH_2{-}CH_3$$

Synonyms: butyl-2-propenoate; 2-propenoic acid butyl ester; acrylic acid *n*-butyl ester

Uses and Exposure Risk

n-Butyl acrylate is used to make polymers that are used as resins for textile and leather finishes, and in paints.

Physical Properties

Colorless liquid; density 0.899 at 20°C; boils at 145°C; solidifies at −64°C; very slightly soluble in water (0.14% at 20°C), soluble in alcohol and ether; polymerizes.

Health Hazard

n-Butyl acrylate is moderately irritating to skin. Its vapor is an irritant to mucous membranes. The liquid caused corneal necrosis when instilled into rabbit eyes. The toxic and irritant properties of this compound are similar to those of methyl acrylate. The adverse health effects, however, are somewhat less than those of methyl and ethyl acrylates. The LD_{50} and LC_{50} values reported in the literature show significant variations. Also, these data varied widely between mice and rats. The dermal LD_{50} value in rabbits is 2000 mg/kg.

Exposure Limit

TLV-TWA 10 ppm (∼55 mg/m^3) (ACGIH).

Fire and Explosion Hazard

Combustible liquid; flash point (open cup) 48°C (118°F); vapor pressure 3.2 torr at 20°C; autoignition temperature 293°C (559°F); vapor forms explosive mixtures with air within the range 1.5–9.9% by volume in air. *n*-Butyl acrylate may polymerize at elevated temperature. The reaction is exothermic and may cause pressure buildup and violent rupture of closed containers. The polymerization may be prevented by 40–60 ppm of hydroquinone or its monomethyl ether. Its reactions with strong oxidizers can be vigorous to violent.

17.15 MISCELLANEOUS ALIPHATIC ESTERS

The toxicity and flammability data for miscellaneous aliphatic esters are presented in Table 17.1. The toxic symptoms are quite similar to those of the esters described earlier. These compounds are a mild irritant to the skin. The vapors are an irritant to the eyes and mucous membranes. The inhalation and oral toxicity are of low order. A narcotic effect is noted at high concentrations. Esters of low molecular weights are flammable liquids. The flash point increases with increase in the alkyl chain length. These compounds may react vigorously with strong oxidizers.

AROMATIC ESTERS

17.16 METHYL BENZOATE

DOT Label: Poison B, UN 2938
Formula $C_8H_8O_2$; MW 136.16; CAS [93-58-3]
Structure:

Synonyms: benzoic acid methyl ester; methyl benzenecarboxylate; Niobe oil

Uses and Exposure Risk

Methyl benzoate is used in perfumes.

Physical Properties

Colorless liquid with a pleasant odor; density 1.094 at 15°C; bp 198°C; mp −15°C; insoluble in water, soluble in alcohol and ether.

Health Hazard

Methyl benzoate is a mild skin irritant. The acute oral toxicity in test animals was of low order. The toxic symptoms in animals from oral administration of this compound were tremor, excitement, and somnolence. The LD_{50} value varies with species. The oral LD_{50} values in mice and rats are 3330 and 1350 mg/kg, respectively.

TABLE 17.1 Toxicity and Flammability of Miscellaneous Aliphatic Esters

Compound/Synonyms/CAS No.	Formula/MW/Structure	Toxicity	Flammability
Allyl formate (formic acid allyl ester, formic acid 2-propenyl ester) [1838-59-1]	$C_4H_6O_2$ 86.10 H—C(=O)—O—CH_2—CH=CH_2	Moderate to highly toxic; toxicity greater than saturate esters of formic acid; ingestion produces behavioral effects; vapors irritant to eyes and mucous membranes; LC_{50} inhalation (mice): 4000 ppm/2 hr; LD_{50} (mice): 136 mg/kg	Flammable liquid; flash point data not available; vapor forms explosive mixtures with air, explosive range not established; DOT Label: Flammable Liquid, UN 2336
n-Propyl formate (propyl methanoate, formic acid propyl ester) [110-74-7]	$C_4H_8O_2$ 88.12 H—C(=O)—O—CH_2—CH_2—CH_3	Low toxicity; mild irritant to skin; vapors can cause irritation of eyes, nose, and lungs; ingestion can produce somnolence and narcotic symptoms; LD_{50} oral (mice): 3400 mg/kg	Flammable liquid; flash point (closed cup) −3°C (26°F); vapor forms explosive mixtures with air, LEL and UEL data not available; DOT Label: Flammable Liquid, UN 1281
n-Butyl formate, (butyl methanoate, formic acid butyl ester) [592-84-7]	$C_5H_{10}O_2$ 102.5 H—C(=O)—O—CH_2—CH_2—CH_2—CH_3	Inhalation of its vapors at about 10,000 ppm concentration in air may cause conjunctivitis, lung irritation, and muscle contraction in humans; exposure to such concentration for 1-hr caused death in test animals; ingestion produces narcotic symptoms; LD_{50} oral (rabbits): 2650 mg/kg	Flammable liquid; flash point (closed cup) 13°C (57°F); autoignition temperature 322°C (612°F); vapor forms explosive mixtures with air in the range 1.7–8.2% by volume in air; DOT Label: Flammable Liquid, UN 1128
n-Amyl formate, (pentyl formate, formic acid pentyl ester) [638-49-3]	$C_6H_{12}O_2$ 116.18 H—C(=O)—O—$(CH_2)_4$—CH_3	Mild skin irritant; low toxicity; narcotic at high concentrations	Flammable liquid; flash point (closed cup) 29.5°C (85°F); vapor forms explosive mixtures with air, explosive limits not established; DOT Label: Flammable Liquid, UN 1109

348

Name (synonyms) [CAS]	Formula, MW	Structure	Toxicity	Fire and explosion hazard
Isobutyl formate (tetryl formate, formic acid isobutyl ester) [542-55-2]	$C_5H_{10}O_2$ 102.15		Mild skin irritant; vapor is irritant to eyes, nose, and lungs; low oral toxicity; less toxic than its n-isomer; narcotic effects at high doses; LD_{50} oral (rabbits): 3064 mg/kg	Flammable liquid; flash point (closed cup) 10°C (50°F); vapor forms explosive mixtures with air, the explosive limits not reported; DOT Label: Flammable Liquid, UN 2393
sec-Butyl acetate (2-butyl acetate, 2-butanol acetate, acetic acid sec-butyl ester) [105-46-4]	$C_6H_{12}O_2$ 116.18		Vapor is irritant to eyes and respiratory passages; the irritant action and narcotic effects are lower than n-butyl acetate; TLV-TWA air 200 ppm ($\sim$950 mg/m^3) (ACGIH, MSHA, and OSHA); IDLH 10,000 ppm (NIOSH)	Flammable liquid; flash point (closed cup) 16.5°C (62°F), (open cup) 31°C (88°F); vapor pressure 10 torr at 20°C; vapor forms explosive mixtures with air, LEL 1.7%, UEL 9.8%; reacts vigorously with strong oxidizers, acids, and alkalies; DOT Label: Flammable Liquid, UN 1123
tert-Butyl acetate (acetic acid tert-butyl ester, acetic acid 1,1-dimethylethyl ester) [540-88-5]	$C_6H_{12}O_2$ 116.18		Least toxic of the four isomers of acetic butyl ester; vapors irritant to eyes and respiratory tract; narcotic at high concentrations; TLV-TWA air 200 ppm ($\sim$950 mg/m^3) (ACGIH, MSHA, and OSHA); IDLH 10,000 ppm (NIOSH)	Flammable liquid; flash point (closed cup) 15°C (60°F); vapor forms explosive mixtures with air, LEL 1.5%, UEL not reported; reacts vigorously with strong oxidizers, acids, and alkalies; DOT Label: Flammable Liquid, UN 1123
sec-Amyl acetate (2-pentyl acetate, 2-pentanol acetate, 1-methylbutyl acetate, acetic acid 2-pentyl ester) [626-38-0]	$C_7H_{14}O_2$ 130.21		Irritant to eyes, nose, and lungs; narcotic; less toxic than n-, and iso-isomers; 5-hr exposure to 10,000 ppm was lethal to guinea pigs; TLV-TWA air 125 ppm ($\sim$660 mg/m) (ACGIH, MSHA, and OSHA), IDLH 9000 ppm (NIOSH)	Flammable liquid; flash point (closed cup) 31.7°C (89°F); vapor forms explosive mixtures with air, LEL 1%, UEL not reported; its reactions with strong oxidizing substances and strong acids and alkalies can be vigorous; DOT: Flammable Liquid, UN 1104

(continued)

TABLE 17.1 (Continued)

Compound/Synonyms/CAS No.	Formula/MW/Structure	Toxicity	Flammability
Cyclohexyl acetate (cyclohexanol acetate, acetic acid cyclohexyl ester) [622-45-7]	$C_8H_{14}O_2$ 142.22 $CH_3-C(=O)-O-$ (cyclohexyl)	Exposure to 500 ppm for 1 hr caused irritation of eyes and respiratory tract in humans, narcotic at high concentrations; LD_{50} oral in rats is in the range 6700 mg/kg	Combustible liquid; flash point (closed cup) 57°C (136°F); vapor forms explosive mixtures with air, explosive range not established. DOT Label: Flammable Liquid, UN 2243
Ethylcyclohexyl acetate (cyclohexylethyl acetate, cyclohexane, ethanol acetate, acetic acid cyclohexyl ethyl ester) [21722-83-8]	$C_{10}H_{18}O_2$ 170.28 $CH_3-C(=O)-O-CH_2-$ (cyclohexyl)	Narcotic; toxicity is greater than cyclohexyl acetate; LD_{50} oral (rats): 3200 mg/kg; LD_{50} skin (rabbits): 5000 mg/kg	Combustible liquid; flash point (closed cup) 80°C (176°F)
Phenyl acetate (acetic acid phenyl ester, acetyl phenol) [122-79-2]	$C_8H_8O_2$ 136.16 $CH_3-C(=O)-O-$ (phenyl)	Oral toxicity greater than that of the aliphatic esters of acetic acid; information on toxicity data in the published literature is too scant; LD_{50} oral (rats): 1630 mg/kg	Combustible liquid; flash point (closed cup) 76°C (170°F)
Ethyl phenylacetate (ethyl phenacetate, ethyl benzeneacetate, phenylacetic acid ethyl ester, α-toluic acid ethyl ester) [101-97-3]	$C_{10}H_{12}O_2$ 164.22 $CH_3-C(=O)-O-CH_2-$ (phenyl)	oral LD_{50} value in rats is 3300 mg/kg	Noncombustible liquid
Methyl propylate (methyl propionate, methyl propanoate, propionic acid methyl ester) [554-12-1]	$C_4H_8O_2$ 88.12 $CH_3-CH_2-C(=O)-O-CH_3$	Irritant to skin; low oral toxicity; lethal dose in rabbits 2550 mg/kg; oral LD_{50} values in mice and rats are in the range of 3500 and 5000 mg/kg, respectively	Flammable liquid; flash point (closed cup) −2°C (28°F); autoignition temperature 469°C (876°F); vapor forms explosive mixtures with air, LEL 2.5%, UEL 13.0% DOT Label: Flammable Liquid, UN 1248

Compound (synonyms) [CAS]	Formula MW / Structure	Health hazard	Fire hazard
Ethyl propylate (ethyl propionate, ethyl propanoate, propionic acid ethyl ester) [105-37-3]	$C_5H_{10}O_2$ 102.15 $CH_3—CH_2—\overset{\displaystyle O}{\overset{\|}{C}}—O—CH_2—CH_3$	Irritant to skin; low oral toxicity; LD_{50} oral (rabbits): 3500 mg/kg	Flammable liquid; flash point (closed cup) 12°C (54°F); autoignition temperature 440°C (824°F); vapor forms explosive mixtures with air in the range 1.9–11.0% by volume in air; DOT Label: Flammable Liquid, UN 1195
Methyl butyrate (methyl n-butanoate, butyric acid methyl ester) [623-42-7]	$C_5H_{10}O_2$ 102.15 $CH_3—CH_2—CH_2—\overset{\displaystyle O}{\overset{\|}{C}}—O—CH_3$	Skin irritant; vapors produce irritation of eyes and mucous membranes; low toxicity; narcotic at high concentrations; oral LD_{50} value in rabbits is in the range 3400 mg/kg	Flammable liquid; flash point (closed cup) 14°C (57°F); vapor forms explosive mixtures with air, the LEL and UEL data not available. DOT Label: Flammable Liquid, UN 1237
Ethyl butyrate (ethyl n-butanoate, butyric acid ethyl ester) [105-54-4]	$C_6H_{12}O_2$ 116.18 $CH_3—CH_2—CH_2—\overset{\displaystyle O}{\overset{\|}{C}}—O—CH_2—CH_3$	Skin irritant; low toxicity; narcotic at high concentrations; vapors are irritating to eyes and respiratory tract; oral LD_{50} value in rabbits is 5200 mg/kg	Flammable liquid; flash point (closed cup) 24°C (75°F); autoignition temperature 463°C (865°F); vapor forms explosive mixtures with air, explosive range not reported; DOT Label: Flammable Liquid, UN 1180

Fire and Explosion Hazard

Combustible liquid; flash point (closed cup) 82°C 181°F).

17.17 METHYL SALICYLATE

Formula $C_8H_8O_3$; MW 152.16; CAS [119-36-8]

Structure:

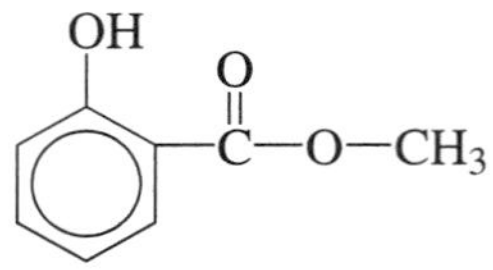

Synonyms: 2-hydroxybenzoic acid methyl ester; *o*-hydroxy methyl benzoate; salicylic acid methyl ester; wintergreen oil; betula oil; gaultheria oil; sweet birch oil

Uses and Exposure Risk

Methyl salicylate occurs in the leaves of *Gaultheria procumbens* L. and in the bark of Betulaceae. It is produced by esterification of salicylic acid with methanol. It is used in perfumery and as a flavoring agent.

Physical Properties

Colorless to yellowish oily liquid having odor and taste of gaultheria; density 1.184 at 25°C; bp 220–224°C; mp −8.6°C; very slightly soluble in water (0.07% at 20°C), soluble in organic solvents.

Health Hazard

Methyl salicylate is a highly toxic compound. The toxic symptoms in humans include nausea, vomiting, gastritis, diarrhea, respiratory stimulation, labored breathing, pulmonary edema, convulsions, and coma. Ingestion of 15 to 25 mL of this compound may be fatal to humans. Application of the liquid on the skin and eyes produced severe irritation in rabbits. Oral, subcutaneous, or dermal administration of methyl salicylate in test animals produced specific developmental abnormalities affecting the eyes, ears, and central nervous system.

Toxicity of this compound is relatively more severe in humans than in many common laboratory animals. The oral LD_{50} values in test animals were within the range 800–1300 mg/kg.

Fire and Explosion Hazard

Noncombustible liquid.

PHTHALATE ESTERS

Phthalic esters are the esters of phthalic acid and are prepared industrially from phthalic anhydride and alcohol:

These esters are noncombustible liquids.

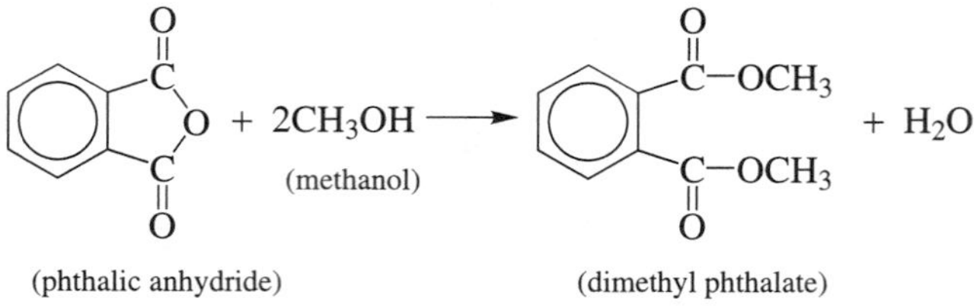

(phthalic anhydride)   (dimethyl phthalate)

The chief commercial uses of these substances are as solvents and as plasticizers of synthetic polymers such as polyvinyl chloride and cellulose acetate. Some of the lower aliphatic phthalates are used in the manufacture of varnishes, dopes, and insecticides.

Some of the phthalate esters are EPA-listed priority pollutants. These compounds occur in the environment and are often found in trace quantities in wastewaters, soils, and hazardous wastes. These substances may leach out from PVC bags and plastic containers into the liquid stored in these containers on prolonged storage.

The acute toxicity of phthalate esters is very low. High doses may, however, produce somnolence, weight loss, dyspnea, and cyanosis. Animals administered large doses manifested weak developmental toxicity. Carcinogenicity was not observed in animal studies. The pure liquids are mild irritants to skin. The toxicity of some of these esters of commercial importance and listed as priority pollutants is discussed in brief in the following sections. It may be noted that phthalate esters are relatively harmless and are among the least toxic organic industrial products.

Phthalates are destroyed by incineration. The esters are dissolved in a combustible solvent and burned in a chemical incinerator equipped with an afterburner and scrubber. Phthalates undergo anaerobic biodegradation, producing carbon dioxide and methane (Battersby and Wilson 1989; O'Connor et al. 1989). Acclimation took several days before biodegradation commenced. Sundstrom and co-workers (1989) have reported the destruction of phthalates and other aromatic pollutants in wastewaters by oxidation by hydrogen peroxide catalyzed by UV light. The reaction rates were slow for phthalates in comparison with other aromatics.

17.18 DIMETHYL PHTHALATE

EPA Priority Pollutant; RCRA Toxic Waste, Number U109

Formula $C_{10}H_{10}O_4$; MW 194.20; CAS [131-11-3]

Structure:

Synonyms: dimethyl 1,2-benzenedicarboxylate; dimethyl benzeneorthodicarboxylate; phthalic acid dimethyl ester

Health Hazard

The acute toxicity of dimethyl phthalate in test animals was found to be very low. Ingestion may produce irritation of the gastrointestinal tract, somnolence, hypotension, and coma. The oral LD_{50} value in mice is 6800 mg/kg. Animal studies have shown that administration of this compound caused developmental toxicity, having effects on fertility. Maternal toxicity in Sprague-Dawley rats was observed at a dietary treatment level of 5% (Field 1989). There was no effect on average litter size, fetal body weight, or the incidence of skeletal or visceral malformations.

Exposure Limit

TLV-TWA air 5 mg/m^3 (ACGIH, MSHA, and OSHA).

17.19 DIETHYL PHTHALATE

EPA Priority Pollutant, RCRA Toxic Waste Number U088

Formula $C_{12}H_{14}O_4$; MW 222.26; CAS [84-66-2]

Structure:

Synonyms: 1,2-benzenedicarboxylic acid diethyl ester; phthalic acid diethyl ester

Health Hazard

Diethyl phthalate exhibited low to very low acute toxicity in laboratory animals. Ingestion produced somnolence and hypotension. Inhalation of its vapors may result in lacrimation, coughing, and irritation of the throat in humans. The oral LD_{50} value in mice is 6170 mg/kg. Diethyl phthalate administered to pregnant rats at 5% concentration in the feed showed no adverse effect upon embryo or fetal growth, viability, or the incidence of malformations (Price et al. 1988).

Exposure Limit

TLV-TWA air 5 mg/m^3 (ACGIH).

17.20   DIBUTYL PHTHALATE

EPA Priority Pollutant, RCRA Toxic Waste Number U069

Formula $C_{16}H_{22}O_4$; MW 278.38; CAS [84-74-2]

Structure:

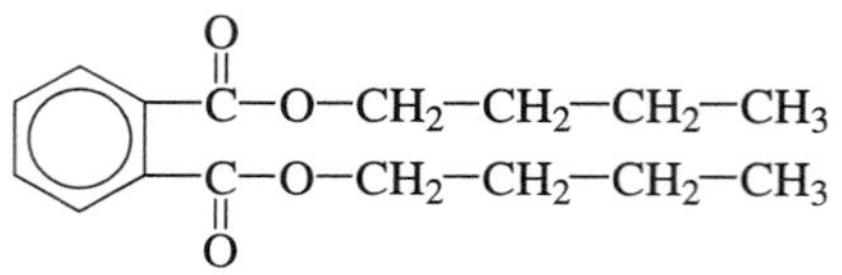

Synonyms: 1,2-benzenedicarboxylic acid dibutyl ester; dibutyl 1,2-benzenedicarboxylate; phthalic acid dibutyl ester; n-butyl phthalate

Health Hazard

The toxicity of this compound is very low. In humans, oral intake of dibutyl phthalate at a dose level of 150 mg/kg may cause nausea, vomiting, dizziness, hallucination, distorted vision, lacrimation, and conjunctivities with prompt recovery. It metabolizes to monobutyl ester and phthalic acid and is excreted in urine. The inhalation toxicity should be insignificant because of its negligible low vapor pressure (<0.1 torr at 20°C). However, exposure to its mist or aerosol can cause irritation of eyes and mucous membranes.

LD$_{50}$ value, oral (mice): 5300 mg/kg

Exposure Limit

TLV-TWA air 5 mg/m^3 (ACGIH, MSHA, and OSHA).

17.21   DI-n-OCTYL PHTHALATE

EPA Priority Pollutant; RCRA Toxic Waste Number U107

Formula $C_{24}H_{38}O_4$; MW 390.62; CAS [117-84-0]

Structure:

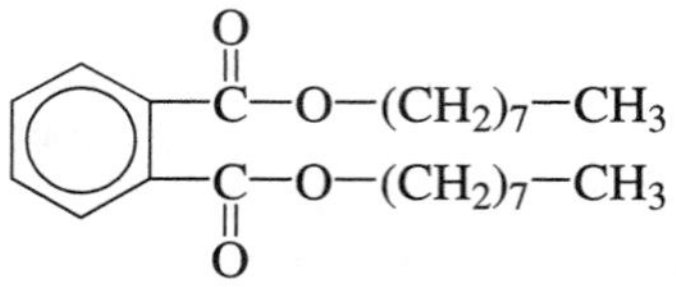

Synonyms: 1,2-benzenedicarboxylic acid dioctyl ester; dioctyl o-benzenedicarboxylate; phthalic acid dioctyl ester

Health Hazard

The acute oral toxicity is very low. Ingestion may result in nausea, somnolence, hallucination, and lacrimation. In humans, such effects may be noted at a dose level of 150–200 mg/kg. The recovery is prompt. The oral LD$_{50}$ value in mice is in the range 6500 mg/kg. Its irritant action on the skin and the eyes of rabbits was mild. Di-n-octyl phthalate fed to mice at the 5% level in diet showed no reproductive toxicity.

17.22   BUTYL BENZYL PHTHALATE

EPA Priority Pollutant

Formula $C_{19}H_{20}O_4$; MW 312.39; CAS [85-68-7]

Structure:

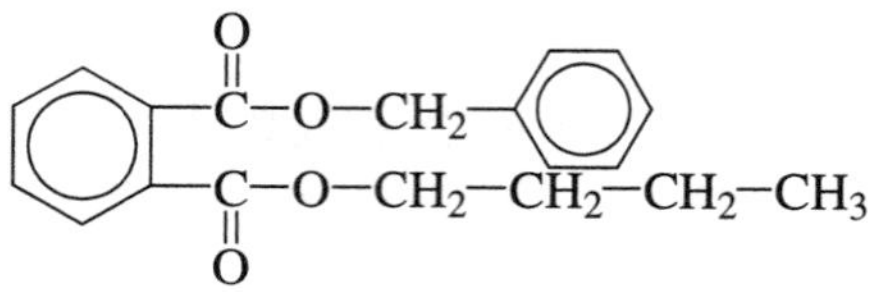

Synonyms: benzyl butyl phthalate; 1,2-benzenedicarboxylic acid butyl phenylmethyl ester; phthalic acid benzyl butyl ester;

butyl phenylmethyl 1,2-benzenedicar-boxylate

Health Hazard

The acute oral toxicity of this compound is low. It is, however, more toxic, than the phthalic acid dialkyl esters discussed in the preceding sections. The toxic symptoms include nausea, dizziness, somnolence, and hallucination. The oral LD_{50} value in mice is within the range of 4200 mg/kg. Oral administration produced reproductive toxicity in male mice (paternal effects). At a dose of 2% in diet, it caused maternal and developmental toxicity and an increased incidence of malformations in Sprague-Dawley rats (Research Triangle Institute 1989). There is evidence of cancer-causing effects observed in animals. Oral administration of butyl benzyl phthalate induced tumors in blood tissues in rats. There is no evidence of carcinogenicity in humans.

17.23 DIETHYLHEXYL PHTHALATE

EPA Priority Pollutant, RCRA Toxic Waste Number U028

Formula $C_{24}H_{38}O_4$; MW 390.62; CAS [117-81-7]

Structure:

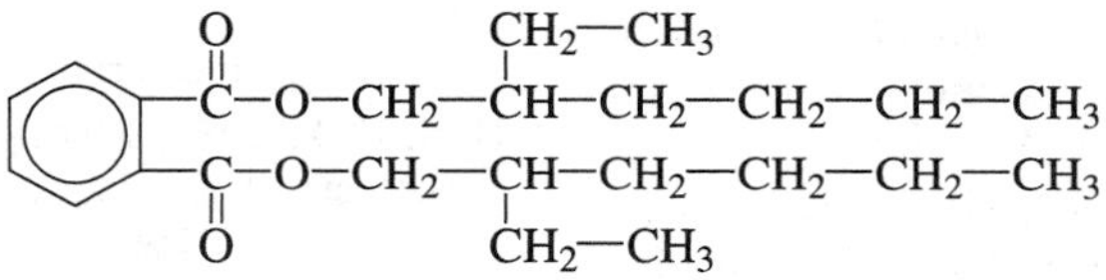

Synonyms: bis(2-ethylhexyl)phthalate; di-*sec*-octyl phthalate; bis(2-ethylhexyl)-1,2-benzenedicarboxylate; 1,2-benzenedicarboxylic acid bis(2-ethylhexyl) ester; phthalic acid bis(2-ethylhexyl) ester

Health Hazard

The acute oral toxicity is of extremely low order. The oral LD_{50} value in mice is in the range 30,000 mg/kg. Ingestion of about 10 mL of the liquid may produce gastrointestinal pain, hypermotility, and diarrhea in humans. Chronic administration by oral route did not produce any adverse effect in animals. A dose of 100 mg/kg/day produced increased liver and kidney weight, as well as retardation of growth in dogs. Mice given diets containing 12,000 ppm of diethylhexyl phthalate for 40 weeks developed renal hyperplasia (Ward et al. 1988). There was no evidence of renal carcinogenicity. Oral administration, however, produced tumors in livers in the laboratory animals. It is a suspected cancer-causing agent, and there is sufficient evidence of its carcinogenicity in animals. Such evidence in humans, however, is lacking.

Rock and co-workers (1988) reported the interaction of diethylhexyl phthalate with blood during storage in PVC plastic containers. The plasticizer is leached into the plasma and converted to mono(2-ethylhexyl) phthalate by a plasma enzyme. At a level of >125 mg/mL in the blood mono(2-ethylhexyl) phthalate produced a 50% decrease in heart rate and blood pressure in rats.

REFERENCES

ACGIH. 1986. *Documentation of the Threshold Limit Values and Biological Exposure Indices*, 5th ed. Cincinnati, OH: American Conference of Governmental Industrial Hygienists.

Battersby, N. S., and V. Wilson. 1989. Survey of the anerobic biodegradation potential of organic chemicals in digesting sludge. *Appl. Environ. Micobiol.* 55(2): 433–39.

Field, E. A. 1989. Developmental toxicity evaluation of dimethyl phthalate administered to CD rats on gestational days 6 through 15. *Govt. Rep. Announce. Index (U.S.)* 89(12), Abstr. No. 933, p. 227; cited in *Chem. Abstr. CA 111*(17): 148626q.

Flury, F., and W. Wirth. 1933. *Arch. Gewerbepathol. Gewerbehyg.* 5:1; cited in *Documentation of the Threshold Limit Values and Biological Exposure Indices*, 5th ed. 1986. Cincinnati, OH: American Conference of Governmental Industrial Hygienists.

MCA. 1973. *Case Histories of Accidents in the Chemical Industry.* Washington, DC: Manufacturing Chemists' Association.

Merck. 1996. *The Merck Index*, 12th ed. Rahway, NJ: Merck & Co.

NFPA. 1986. *Fire Protection Guide on Hazardous Materials*, 9th ed. Quincy, MA: National Fire Protection Association.

NIOSH. 1984. *Manual of Analytical Methods*, 3rd ed. Cincinnati, OH: National Institute for Occupational Safety and Health.

NIOSH. 1986. *Registry of Toxic Effects of Chemical Substances*, ed. D. V. Sweet, Washington, DC: U.S. Government Printing Office.

O'Connor, O. A., M. D. Rivera, and L. Y. Young. 1989. Toxicity and biodegradation of phthalic acid esters under methanogenic conditions. *Environ. Toxicol. Chem.* 8(7): 569–76.

Price, C. J., R. B. Sleet, J. D. George, M. C. Marr, and B. A. Schwetz. 1988. Developmental toxicity evaluation of diethyl phthalate administered to CD rats on gestational days 6 through 15. *Govt. Rep. Announce. Index (U.S.)* 89(8), Abstr. No. 919, p. 310; cited in *Chem. Abstr. CA 111*(7): 52275y.

Research Triangle Institute. 1989. Developmental toxicity evaluation of butyl benzyl phthalate

administered in feed to CD rats on gestational days 6 to 15. *Govt. Rep. Announce. Index (U.S.)* 90(3), Abstr. No. 006, p. 325; cited in *Chem. Abstr. CA 112*(23): 210322j.

Rock, G., R. S. Labow, C. Franklin, R. Burnett, and M. Tocchi. 1988. Plasticizer interaction with stored blood produces a toxic metabolite, mono(2-ethylhexyl) phthalate. *Symposium Proceedings of the Material Research Society*, Volume Date 1987, *110:* 767–72; cited in *Chem. Abstr. CA 111*(26): 239356t.

Sundstrom, D. W., B. A. Weir, and H. E. Klei. 1989. Destruction of aromatic pollutants by UV light catalyzed oxidation with hydrogen peroxide. *Environ. Prog.* 8(1): 6–11.

Treon, J., et al. 1949. *J. Ind. Hyg. 37:* 317; cited in *Documentation of the Threshold Limit Values and Biological Exposure Indices*, 5th ed. 1986. Cincinnati, OH: American Conference of Governmental Industrial Hygienists.

U. S. EPA. 1984. Methods for organic chemical analysis of municipal and industrial wastewater. 40 CFR, Part 136, Appendix A. *Fed. Register.* 49(209), Oct. 26.

U. S. EPA. 1986. *Test Methods for Evaluating Solid Waste*, 3rd ed. Washington DC: Office of Solid Waste and Emergency Response.

Ward, J. M., A. Hagiwara, L. M. Anderson, K. Lindsay, and B. A. Diwan. 1988. The chronic hepatic or renal toxicity of di(2-ethylhexyl) phthalate, acetaminophen, sodium barbital and phenobarbitol in male B6C3F1 mice: autoradiographic, immunohistochemical, and biochemical evidence for levels of DNA synthesis not associated with carcinogenesis or tumor promotion. *Toxicol. Appl. Pharmacol.* 96(3): 494–506.

18

ETHERS

18.1 GENERAL DISCUSSION

An ether is an organic compound containing an oxygen atom, bridging two alkyl or aryl groups. The general chemical formula for this class of compounds is $R-O-R'$, where R and R' can be alkyl, alkenyl, alkynyl, cycloalkyl, or aryl groups. Some of the low-molecular-weight aliphatic ethers, such as ethyl and isopropyl ethers, are widely used as solvents.

The hazards due to ethers arise primarily for two reasons: (1) high degree of flammability, and (2) formation of unstable peroxides that can explode spontaneously or on heating. Such fire and/or explosion hazards are characteristics of lower aliphatic ethers containing two to eight carbon atoms. With an increase in the carbon chain of R, the flash point decreases. Lower aliphatic ethers are some of the most flammable organic compounds (e.g., ethyl ether, vinyl ether, ethyl vinyl ether, etc., have a flash point of less than $-30°C$). Many such flammable ethers can be ignited by static electricity or lightning. Methyl ether and ethyl methyl ether are flammable gases, and other ethers are volatile liquids with low boiling points. The vapor densities being greater than 1, the vapors of these liquids are heavier than air. These vapors can travel considerable distances to an open flame or a source of ignition and pose a flashback fire hazard. Most of these ethers have relatively low autoignition temperatures, below 200°C. These compounds form explosive mixtures with air. By contrast, aromatic ethers are noncombustible liquids or solids and do not exhibit the flammable characteristics common to aliphatic ethers.

Ethers react with oxygen to form unstable peroxides:

$$R-O-R_1 \longrightarrow R-O-O-R_1$$

This reaction is catalyzed by sunlight. Thus ethers in the presence of oxygen or stored for a long period may contain peroxides. When evaporated to dryness, the concentrations of such peroxides increase, resulting in violent explosions. Tests for peroxides and the method of their removal from ethers are described in Section 18.3.

The toxicity of ethers is low to very low. At high concentrations, these compounds exhibit anesthetic effects. A concentration above 5% by volume in air can be dangerous to humans. Inhalation of larger volumes may cause unconsciousness. However, at ordinary conditions of handling or use, the health hazards from ethers are very low.

18.2 METHYL ETHER

DOT Label: Flammable Gas, UN 1033

Formula $(CH_3)_2O$; MW 46.08; CAS [115-10-6]

Structure and functional group: $H_3C-O-CH_3$ first member of a homologous series of aliphatic ethers, containing two methyl groups linked via an oxygen atom

Synonyms: dimethyl ether; wood ether; dimethyl oxide

Uses and Exposure Risk

Methyl ether is used as an aerosol propellant and in refrigeration.

Physical Properties

Colorless gas with ethereal smell; liquefies at $-23.6°C$; density 1.617 (air = 1); soluble in water (3700 mL of gas dissolves in 100 mL of water) and alcohol.

Health Hazard

Methyl ether produced low inhalation toxicity in rats. Caprino and Togna (1975) reported a 30-minute LC_{50} value of 396 ppm for rats. In lethal doses it caused sedation, a gradual depression of motor activity, loss of the sighting reflex, hypopnea, coma, and death in mice. Exposure to a 40% mixture of methyl ether in air resulted in an initial slight increase in heart rate in rabbits, which was followed by depression of arterial blood pressure. Death occurred in 45 minutes. The arterial and venous partial oxygen pressure was found to decrease while the venous CO_2 pressure and the blood pH increased.

Reuzel et al. (1981) reported that subchronic inhalation of methyl ether in rats did not cause significant adverse effects. No noticeable effect on organ and body weights and no treatment-related changes were observed. In humans, adverse health effect from inhalation of this compound should be minimal. However, inhalation of excessive quantities can produce intoxication and loss of consciousness.

Fire and Explosion Hazard

Highly flammable gas; ignition temperature 350°C (662°F); forms explosive mixtures with air, with LEL and UEL values of 3.4% and 27% by volume of air, respectively. It is, however, less flammable than propane–butane mixtures, as shown by closed drug and flame extension tests. The flame extension of compositions containing 45% methyl ether is 0–15 cm (20–35 cm for propane–butane mixture) (Bohnenn 1982); fire-extinguishing method: stop flow of gas; use a water spray to keep fire-exposed containers cool.

On long standing or exposure to oxygen or sunlight, methyl ether may form unstable peroxides that may explode spontaneously. Distillation of methyl ether with aluminum hydride or lithium aluminum hydride may result in explosion if the ether is not pure.

Storage and Shipping

Methyl ether is stored in a cool and well-ventilated location away from sources of ignition. It is shipped in steel cylinders.

18.3 ETHYL ETHER

EPA Classified Hazardous Waste, RCRA Waste Number U117; DOT Label: Flammable Liquid 1155

Formula $(C_2H_5)_2O$; MW 74.14; CAS [60-29-7]

Structure and functional group: $H_3C-CH_2-O-CH_2-CH_3$, aliphatic ether containing two ethyl groups

Synonyms: ether, diethyl ether; solvent ether; 1,1'-oxybis(ethane); diethyl oxide

Uses and Exposure Risk

Ethyl ether is used as a solvent for fats, oils, waxes, gums, perfumes, and nitrocellulose; in making gun powder; as an anesthetic; and in organic synthesis.

Physical Properties

Pungent odor; sweet burning taste; boils at 34.6°C; mp −116°C; density 0.7134 at 20°C; soluble in alcohol, benzene, chloroform, acetone, and many oils; mixes with concentrated HCl, low solubility in water (6.05 g/100 mL water at 25°C); forms azeotrope with water (1.3%).

Health Hazard

Ethyl ether is a narcotic substance and a mild irritant to the skin, eyes, and nose; at low concentrations, <200 ppm in air, exposure to this compound does not produce noticeable effects in humans. Eye and nasal irritation may become intolerable at 250–300 ppm. Repeated exposure can cause drying and cracking of skin, due to extraction of oils.

Inhalation of its vapors at high concentrations, above 1% (by volume in air), could be hazardous to human health. A concentration of 3.5–6.5% could produce an anesthetic effect; respiratory arrest may occur above this concentration (Hake and Rowe 1963). Inhalation of 10% ethyl ether by volume in air can cause death (ACGIH 1986). Repeated exposure to this compound exhibited the symptoms of exhaustion, loss of appetite, sleepiness, and dizziness.

Acute oral toxicity of ethyl ether was found to be low to moderate, varying with species. Ingestion of 300–350 mL can be fatal to humans.

LC$_{50}$ value, inhalation (mice): 6500 ppm/ 100 min

LD$_{50}$ value, oral (rats): 1215 mg/kg

In a comparison with other anesthetic agents, diethyl ether was reported to be less toxic than methoxyfluorane [76-38-0], halothane [151-67-7], and isoflurane [26675-46-7] on test animals upon repeated exposures at subanesthetic concentrations (Chenoweth et al. 1972; Stevens et al. 1975). At 2000 ppm it did not cause hepatotoxic responses. Matt et al. (1983) reported that ether exposure for 6 minutes induced significant and variable elevations of serum prolactin [900-62-4] in female golden hamsters.

In contrast to volatile hydrocarbons, the respiratory arrest caused by ethyl ether was reversible (Swann et al. 1974). Such reversibility, however, was observed at a lower concentration, about 105 ppm for a 5-minute exposure period in mice. There is no report of its carcinogenicity in animals or humans.

Exposure Limit

TLV-TWA 1200 mg/m^3 (400 ppm) (ACGIH and OSHA); STEL 1500 mg/m^3 (500 ppm) (ACGIH).

Fire and Explosion Hazard

Extremely flammable; flash point (closed cup) −45°C (−49°F); vapor pressure 439 torr at 20°C; vapor density 2.6 (air = 1); the vapor is heavier than air and may travel a considerable distance to a source of ignition and flashback dangerously; ignition temperature 180°C (356°F); fire-extinguishing agent: dry chemical, CO$_2$, or foam; use a water spray to keep fire-exposed containers cool, to flush the spills, and to disperse the vapors if the spill has not ignited.

Ethyl ether forms explosive mixtures with air, with flammable limits of 1.85% and 48% (by volume of air). Absolute dry ether when shaken vigorously can generate enough static electricity to start a fire. Radiant energy emerging from a fiber optic system can ignite ether vapor. Tortoishell (1985) reported that a dust particle 50 μm in diameter in an ether atmosphere absorbed 6 mW of energy from light from a fiber, igniting the vapors and starting a fire.

On long standing or in the presence of oxygen, ether forms unstable peroxides;

$$(C_2H_5)_2O \longrightarrow H_5C_2-O-O-C_2H_5$$

The reaction is catalyzed by light. Bottles exposed to sunlight are susceptible to peroxide formation. Unstable peroxides may explode spontaneously or when heated.

Ethyl ether presents a fire and/or explosion hazard when it comes in contact with strong oxidizers such as ozone (due to ethyl peroxide), permanganate–concentrated H_2SO_4, perchloric acid and perchlorates, many organic peroxides, sodium and potassium peroxides (catches fire), chromic acid, liquid air, and chlorine (NFPA 1986). Explosions have occurred when ethyl ether is distilled with lithium aluminum hydride or aluminum chloride. The presence of carbon dioxide in ether has been attributed to such explosions.

Storage and Shipping

Ethyl ether is stored in a flammable-liquids storage room or cabinet isolated from combustible oxidizing materials. It should be protected against direct sunlight, lightning, and static electricity. Ether left in opened bottles should be discarded after 30 days, and unopened bottles should not be used after a year. Naphthols, aminophenols, or aromatic amines may be added in trace amounts to stabilize ether. Ethers sold commercially are inhibited with ~0.0001% butylated hydroxytoluene. Ethyl ether is shipped in amber glass bottles, steel drums, or tanks packaged under nitrogen.

Tests and Removal of Peroxides

A simple qualitative test involves oxidation of iodide ion to iodine by peroxides. One milliliter of ether is added to a freshly prepared solution of sodium or potassium iodide in glacial acetic acid (100 mg/mL). Low to high concentrations of peroxides are indicated by yellow to brown colorations of the solutions (National Research Council 1983). Dialkyl peroxides that are reduced with difficulty can be tested by using an iodide–acetic acid reagent in 37% HCl. Organic peroxides in ether can also be detected by peroxidase-coated test papers which are sold commercially.

Peroxides are removed by passing ether through a column of basic activated alumina or refluxing the contaminated solvent with 4-Å molecular sieve pellets for a few hours. Alternatively, the solvent is shaken vigorously with acidified ferrous sulfate solution ($FeSO_4$–H_2SO_4 mixture).

Disposal/Destruction

Ethyl ether is mixed with an excess of higher-boiling solvent and burned in a chemical incinerator equipped with an afterburner and scrubber. Small amounts of ether free of peroxides can be evaporated in a hood in the absence of an open flame or source of ignition nearby.

Analysis

Ethyl ether is analyzed by GC-FID or GC/MS. Air analysis is done by passing air over coconut shell charcoal. The absorbed ether is desorbed with ethyl acetate and injected into a GC equipped with an FID (NIOSH 1984, Suppl. 1985, Method 1610). A packed Porapak column or a fused-silica capillary column (DB-1 or equivalent) is suitable for the purpose.

18.4   ISOPROPYL ETHER

DOT Label: Flammable Liquid, UN 1159
Formula $(C_3H_7)_2O$; MW 102.20; CAS [108-20-3]

Structure and functional group:

$$H_3C \diagdown CH-O-CH \diagup CH_3$$
$$H_3C \diagup \qquad \diagdown CH_3$$

an aliphatic ether containing two isopropyl groups held by ethereal linkage

Synonyms: diisopropyl ether; diisopropyl oxide; 2-isopropoxypropane; 2,2'-oxybis-(propane)

Uses and Exposure Risks

Isopropyl ether is used as a solvent for oils, waxes, resins, and dyes; and as a varnish remover.

Physical Properties

Colorless liquid with an unpleasant odor; bp 68.5°C; mp −60°C; density 0.7258 at 20°C; mixes readily with common organic solvents; very low solubility in water (0.2% at 20°C).

Health Hazard

Isopropyl ether is a narcotic and an irritant to the skin and mucous membranes. It is more toxic than ethyl ether and the toxic symptoms are similar to the latter compound. Inhalation of its vapors can produce anesthetic effects. Exposure to high concentrations can cause intoxication, respiratory arrest, and death. Exposure to 3.14% and 2.9% of isopropyl ether in air (by volume) produced the symptoms of somnolence, change in motor activity, and muscle contractions in mice and rabbits, respectively, causing the death of 50% of test animals. Exposure to 7–10% by volume concentration in air can be fatal to humans.

Acute oral toxicity of isopropyl ether is low. The liquid is irritating to the mucous membranes. Skin contact may cause mild irritation, and repeated exposure may cause dermatitis. The irritation effect of isopropyl ether on eyes is mild. In humans, exposure to 800 ppm in air for a few minutes caused irritation of the eyes and nose. There is no report of its carcinogenic action in animals or humans.

Exposure Limits

TLV-TWA 1045 mg/m^3 (250 ppm) (ACGIH), 2090 mg/m^3 (500 ppm) (OSHA); STEL 1300 mg/m^3 (310 ppm) (ACGIH).

Fire and Explosion Hazard

Highly flammable; flash point (closed cup) −27.8°C (−18°F), (open cup) −9°C (15°F); vapor pressure 119 torr at 20°C; vapor density 3.52 (air = 1); the vapor is heavier than air and can travel a considerable distance to a source of ignition and flash back; ignition temperature 443°C (830°F); fire-extinguishing agent: dry chemical, "alcohol" foam, or CO$_2$; use a water spray to keep fire-exposed containers cool and disperse the vapors.

Isopropyl ether forms explosive mixtures with air; the LEL and UEL values are 1.4% and 7.9% by volume in air, respectively. It forms unstable peroxides on long standing or in contact with air. These peroxides may explode on mechanical shock or on heating. Peroxides can be destroyed with sodium sulfite or acidified ferrous sulfate solutions.

Mixing isopropyl ether with concentrated nitric and sulfuric acids can generate heat. Mixing with acid chlorides can be violent. Violent exothermic reaction occurred when propionyl chloride was mixed with isopropyl ether in the presence of trace zinc chloride or ferric chloride (Koenst 1981).

Storage and Shipping

Isopropyl ether is stored in a flammable-liquids storage room, isolated from combustible and oxidizing materials. It should also be protected from direct sunlight, static electricity, and lightning. It is stabilized with 0.01% hydroquinone or p-benzylaminphenol.

The ether is shipped in amber glass bottles, steel cans, and drums.

Disposal/Destruction

Isopropyl ether is burned in a chemical incinerator equipped with an afterburner and scrubber. A small amount of ether, if free of peroxides, can be evaporated in a fume hood in the absence of any open flame or source of ignition nearby.

Analysis

Isopropyl ether is analyzed by GC using an FID. Other instrumental techniques are GC/MS and FTIR. Air analysis can be performed by charcoal adsorption, desorption of the ether with ethyl acetate, followed by the analysis of the eluant by GC or GC/MS.

18.5   VINYL ETHER

DOT Label: Flammable Liquid, UN 1167

Formula $(C_2H_3)_2O$; MW 70.10; CAS [109-93-3]

Structure and functional group: $H_2C=CH-O-CH=CH_2$, an aliphatic ether containing two olefinic double bonds

Synonyms: divinyl ether; divinyl oxide; 1,1'-oxybis(ethene); ethenyloxyethene

Uses and Exposure Risk

Vinyl ether is used as an anesthetic and in organic synthesis.

Physical Properties

Clear, colorless liquid with a characteristic odor; boils at 28.5°C; density 0.774 (at 20°C); soluble in most organic solvents, slightly soluble in water (0.53 g/dL at 37°C).

Health Hazard

Vinyl ether exhibits low inhalation toxicity. Like ethyl ether, inhalation of its vapors produces an anesthetic effect. At concentrations above 5% (by volume in air), it can cause unconsciousness in humans. Exposure to a 5.1% concentration of its vapor (in air) was lethal to mice.

Divinyl ether exhibited mutagenicity in microsomal assay and microbial mutation tests. There is no report on its carcinogenicity in animals or humans.

Fire and Explosion Hazard

Highly flammable; flash point (closed cup) $<$ $-30°C$ ($<-22°F$); vapor density 2.4 (air = 1); the vapor is heavier than air and can travel a considerable distance to a source of ignition and flash back; autoignition temperature 360°C (680°F); can be ignited by lightning or static electricity; fire-extinguishing agent: dry chemical, foam, or CO_2; use a water spray to keep fire-exposed containers cool and to disperse the vapors.

Vinyl ether forms explosive mixtures with air, with LEL and UEL values of 1.7% and 27% by volume of air, respectively. It forms unstable peroxides on long standing or in the presence of oxygen. These peroxides can explode spontaneously or when heated. It decomposes to acetaldehyde (toxic and highly flammable) when exposed to light.

Storage and Shipping

Vinyl ether is stored in a flammable liquids storage room or cabinet separated from oxidizing and combustible materials. It should be protected from physical damage and from static electricity or lightning. It is shipped in amber glass bottles or metal drums.

Disposal/Destruction

Vinyl ether is mixed with an excess of higher-boiling solvent and burned in a chemical incinerator equipped with an afterburner and scrubber. Small amounts, free of peroxides, can be evaporated in a fume hood away from a source of ignition.

TABLE 18.1 Toxicity and Flammability of Miscellaneous Ethers

Compound/Synonyms/CAS No.	Formula/MW/Structure	Toxicity	Flammability
Ethyl methyl ether (methyl ethyl ether, methoxythane, ethoxymethane) [540-67-0]	$C_2H_5OCH_3$ 60.11 $H_3C-CH_2-O-CH_3$	Strong anesthetic effect; low inhalation toxicity at concentrations below 1000 ppm in air; exposure to high concentrations (>5% by volume in air) may cause respiratory arrest in humans	Highly flammable liquid (or gas, liquefies at 10.8°C); flash point (closed cup) −37°C (−35°F); vapor pressure 712 torr at 5°C (calculated), vapor density 2.1 (air = 1); vapor heavier than air and can travel some distance to an ignition source and flash back; autoignition temperature 190°C (374°F); can be ignited by lightning and static electricity; forms explosive mixtures with air within the range 2–10% by volume of air; may form unstable peroxide on long standing, in the presence of oxygen, or when exposed to sunlight, which may explode spontaneously or when heated; fire-extinguishing agent: dry chemical, CO_2, or "alcohol" foam; DOT Label: Flammable Gas, UN 1039
Ethyl vinyl ether (vinyl ethyl ether, ethoxyethene) [109-92-2]	C_4H_8O 72.12 $H_3C-CH_2-O-CH=CH_2$	Mild skin irritant; low inhalation toxicity; low to very low LD_{50} oral (rats): 6153 mg/kg	Highly flammable liquid; bp 36°C (96°F); flash point (closed cup) <−46°C (<−50°F) autoignition temperature. 202°C (395°F); vapor density 2.5 (air = 1), vapor heavier than air and can travel a considerable distance to a source of ignition and flash back; vapor forms explosive mixtures with air, LEL and UEL 1.7–28% by volume of air, respectively; can be ignited by lightning or static electricity; forms unstable peroxides in presence of oxygen or on long standing, which can explode spontaneously or on heating; fire-extinguishing agent: dry chemical, CO_2, or "alcohol" foam; DOT Label: Flammable Liquid, UN 1302

(continued)

TABLE 18.1 (Continued)

Compound/Synonyms/ CAS No.	Formula/MW/Structure	Toxicity	Flammability
Ethyl propyl ether (propyl ethyl ether, 1-ethoxypropane) [628-32-0]	$C_5H_{12}O$ 88.17 $H_3C-CH_2-CH_2-O-CH_2-CH_3$	Low inhalation toxicity; anesthetic effect; human exposure to concentrations in air >7% by volume of air may cause respiratory arrest	Highly flammable; flash point <−20°C (<−4°F); vapor forms explosive mixtures with air, explosive limits 1.7–9% by volume of air; may form unstable peroxides in presence of air, on long standing or when exposed to sunlight; DOT Label: Flammable Liquid, UN 2615
Methyl propyl ether (propyl methyl ether, 1-methoxypropane) [557-17-5]	$C_4H_{10}O$ 74.14 $H_3C-CH_2-CH_2-O-CH_3$	Inhalation toxicity low; vapor produces anesthetic effect at high concentrations; concentration in air above 5% by volume of air may cause respiratory arrest in humans	Highly flammable liquid; bp 39°C (102°F); flash point <−20°C (<−4°F), can be ignited by static electricity; forms explosive mixtures with air; may form unstable peroxides in the presence of oxygen, on long standing, or when exposed to sunlight; DOT Label: Flammable Liquid, UN 2612
Allyl ethyl ether (ethyl allyl ether, 3-ethoxy-1-propene) [557-31-3]	$C_5H_{10}O$ 86.15 $CH_2=CH-CH_2-O-CH_2-CH_3$	Moderately toxic by inhalation and ingestion; irritant to skin, eyes, and mucous membranes; toxicity data not available; DOT Label: Poison, UN 2335	Highly flammable; flash point (closed cup) −20°C (−5°F); vapor density 3.0 (air = 1); flashback fire hazard; forms explosive mixtures with air; susceptible to form explosive peroxides; DOT Label: Flammable Liquid and Poison, UN 2335
Ethyl propenyl ether (ethyl-1-propenyl ether) [928-55-2]	C_3H_8O 60.11 $H_3C-CH_2-O-CH=CH-CH_3$	Skin irritant, mild on rabbit skin; low inhalation toxicity exposure to 0.8% concentration in air by volume for 4 hr was lethal to rats	Flammable liquid; flash point (open cup) −6°C (21°F); flashback fire hazard; forms explosive mixtures with air
Methyl vinyl ether (vinyl methyl ether) [107-25-5]	C_3H_6O 58.02 $CH_2=CH-O-CH_3$	Low inhalation and oral toxicity; toxic characteristics similar to those of low aliphatic ethers; LD_{50} oral (rats): 4900 mg/kg	Flammable gas; liquefies at 6°C (43°F); autoignition temperature 287°C (549°F); forms explosive mixtures with air; in the event of fire, shut off the flow of gas; use CO_2 or dry chemical to extinguish the flame and a water spray to keep the surroundings cool

Name [CAS]	Formula / Structure	Toxicity	Hazard
Isopropyl vinyl ether (vinyl isopropyl ether) [926-65-8]	$C_5H_{10}O$ 86.16 H_3C—$\overset{\text{}}{CH}$—O—CH=CH$_2$ with H_3C	Low inhalation toxicity; anesthetic effect on humans at concentrations above 1% in air by volume	Highly flammable; flash point (closed cup) −32°C (−26°F); vapor density 3 (air = 1); flashback fire hazard; autoignition temperature 272°C (522°F); may form unstable peroxides in the presence of oxygen or on long standing
Allyl vinyl ether (vinyl allyl ether) [3917-15-5]	C_5H_8O 84.13 CH$_2$=CH—CH$_2$—O—CH=CH$_2$	Low inhalation toxicity; narcotic effect; inhalation of 0.8% vapor in air by volume for 4 hr was fatal to rats; 5% concentration in air can cause unconsciousness in humans; low oral toxicity; LD$_{50}$ oral (rats): 550 mg/kg	Flammable liquid; vapor density 2.9 (air = 1); flashback fire hazard; forms explosive mixtures with air; forms explosive peroxides; degree of reactivity due to olefinic unsaturation in the molecule; certain additive reactions can be exothermic
Allyl propyl ether (propyl allyl ether) [1471-03-0]	$C_6H_{12}O$ 100.16 CH$_2$=CH—CH$_2$—O—CH$_2$—CH$_2$—CH$_3$	Mild irritant to skin and eyes; low inhalation toxicity (lower than that of low aliphatic ethers)	Flammable liquid; flash point −5°C (23°F); vapor density 3.45 (air = 1); flashback fire hazard; forms explosive mixtures with air; may form explosive peroxides
n-Propyl ether [dipropyl ether, dipropyl oxide, 1,1′-oxybis(propane)] [111-43-3]	$(C_3H_7)_2O$ 102.20 H$_3$C—CH$_2$—CH$_2$—O—CH$_2$—CH$_2$—CH$_3$	Low toxicity; anesthetic effect; inhalation toxicity lower than that of isopropyl ether; LD$_{50}$ intravenous (mice): 204 mg/kg	Highly flammable; flash point (closed cup) 21°C (70°F); ignition temperature 188°C (377°F); vapor pressure (calculated) 49 torr at 20°C; vapor density 3.53 (air = 1); can cause flashback fires; fire-extinguishing agent: dry chemical, CO$_2$, or "alcohol" foam; forms explosive mixtures with air in the range 1.3–7.0% by volume; forms explosive peroxide; DOT Label: Flammable Liquid, UN 2384

(continued)

TABLE 18.1 (Continued)

Compound/Synonyms/CAS No.	Formula/MW/Structure	Toxicity	Flammability
Allyl ether [diallyl ether, propenyl ether; 3,3′-oxybis(1-propene)] [557-40-4]	$(C_3H_5)_2O$ 98.16 $CH_2=CH-CH_2-O-CH_2-$ $CH=CH_2$	Skin and eye irritant, irritating effect mild on rabbits; moderately toxic by oral route; LD_{50} oral (rats); 320 mg/kg; can be absorbed through the skin; DOT Label: Poison	Flammable liquid; flash point (open cup) −6.5°C (20°F); vapor density 3.38 (air = 1); vapor heavier than air and can present flashback fire hazard; forms explosive mixtures with air; may form unstable peroxides on standing or in the presence of oxygen; DOT Label: Flammable Liquid and Poison, UN 2360
Methyl-2-propynyl ether (methyl propargyl ether) [627-41-8]	C_4H_6O 70.09 $CH≡C-CH_2-O-CH_3$	Skin and eye irritant; toxicity data not available	Flammable liquid; flash point −18°C (−1°F); vapor density 2.4 (air = 1); flashback fire hazard; vapor forms explosive mixtures with air
n-Butyl ether [n-dibutyl ether, 1-butoxybutane, dibutyl oxide, 1,1′-oxybis(butane)] [142-96-1]	$(C_4H_9)_2O$ 130.26 $H_3C-CH_2-CH_2-CH_2-O-$ $CH_2-CH_2-CH_2-CH_3$	Irritant to eyes, skin, and nose; exposure to 200 ppm for 15 min was irritating to human eyes; pure liquid caused mild skin irritation; low toxicity—narcotic inhalation of 4000 ppm/4 hr was fatal to mice; LD_{50} oral (rats): 7400 mg/kg	Flammable liquid; flash point (closed cup) 25°C (77°F); vapor pressure (calculated) 218 torr at 100°C; vapor density 1.1 (air = 1) at 100°F; flashback fire hazard; autoignition temperature 194°C; vapor forms explosive mixtures with air in the range 1.5–7.6% by volume of air; forms explosive peroxides when anhydrous; DOT Label: Flammable Liquid, UN 1149
Butyl methyl ether (methyl butyl ether, 1-methoxybutane) [628-28-4]	$C_5H_{12}O$ 88.17 $H_3C-CH_2-CH_2-CH_2-$ $O-CH_3$	Toxicity data not available; narcotic properties and low irritating effects expected	Flammable liquid; flash point −10°C (14°F); vapor density 3.04 (air = 1); flashback fire hazard; forms explosive mixtures with air; forms explosive peroxides; DOT Label: Flammable Liquid, UN 2350
Butyl vinyl ether [vinyl butyl ether, 1-(ethenyloxy)butane] [111-34-2]	$C_6H_{12}O$ 100.18 $CH_2=CH-O-CH_2-CH_2-$ CH_2-CH_3	Mild skin and eye irritant; low inhalation toxicity LC_{50} inhalation (mice): 15,000 ppm/2 hr	Flammable liquid; flash point (open cup) −9°C (15°F); vapor density 3.45 (air = 1); flashback fire hazard; autoignition temperature 225°C (437°F); forms explosive mixtures with air; forms explosive peroxides; DOT Label: Flammable Liquid, UN 2352

Compound [CAS No.]	Formula/Structure	Toxicity	Properties
Butyl ethyl ether (ethyl butyl ether) [628-81-9]	$C_6H_{14}O$ 102.20 $H_3C-CH_2-CH_2-CH_2-O-CH_2-CH_3$	Mild irritating action on skin and eyes; low toxicity; LD_{50} oral (rats): 1870 mg/kg	Flammable liquid; flash point (closed cup) $-5°C$ (22°F); vapor density 3.52 (air = 1); flashback fire hazard; vapor pressure (calculated) 113 torr at 40°C; vapor forms explosive peroxides when anhydrous; DOT Label: Flammable Liquid, UN 1179
Pentyl ether [n-pentyl ether, n-amyl ether, diamyl ether, dipentyl ether, 1,1'-oxybis pentane)] [693-65-2]	$(C_5H_{11})_2O$ 158.32 $H_3C-CH_2-CH_2-CH_2-CH_2-O-CH_2-CH_2-CH_2-CH_2-CH_3$	Very low oral toxicity; LD_{50} intravenous (mice): 146 mg/kg	Combustible liquid; flash point (open cup) 57°C (135°F); autoignition temperature 170°C (338°F); may form unstable peroxides
Phenyl ether (diphenyl ether, biphenyl oxide, oxydiphenyl, phenoxybenzene) [101-84-8]	$(C_6H_5)_2O$ 170.22 (an aromatic ether)	Mild skin irritant; acute oral toxicity low to very low; the lethal oral doses varied between 3000 and 4000 mg/kg, depending on the species; ingestion could cause injury to liver, kidney, gastrointestinal tract, and thyroid; LD_{50} oral (rats): 3370 mg/kg; no signs of toxicity were observed in animals inhaling vapors; irritation of eyes and nose occurred in rabbits exposed to a concentration of 10 ppm in air; exposure limits: TLV-TWA 6.96 mg/m^3 (1 ppm) in air (ACGIH and OSHA); STEL 13.9 mg/m^3 (2 ppm) (ACGIH) — based on its disagreeable odor	Noncombustible liquid (solid); melts at 28°C; boils at 257°C; flash point (open cup) >100°C (>212°F)

Analysis

The analysis can be performed by GC, GC/MS, FTIR, and other instrumental techniques. Air analysis can be done by charcoal adsorption, desorbed with ethyl acetate and injected into a GC equipped with an FID (see Section 18.3).

18.6 MISCELLANEOUS ETHERS

See Table 18.1.

REFERENCES

ACGIH. 1986. *Documentation of the Threshold Limit Values and Biological Exposure Indices*, 5th ed. Cincinnati, OH: American Conference of Governmental Industrial Hygienists.

Bohnenn, L. J. M. 1982. Evaluation of flammability of dimethyl ether versus propane–butane mixtures. *Aerosol Rep. 21* (11): 494–98.

Caprino, L., and G. Togna. 1975. Toxicological aspects of dimethyl ether. *Eur. J. Toxicol. Environ. Hyg. 8* (5): 287–90.

Chenoweth, M. B., B. K. J. Leong, G. L. Sparschu, and T. R. Torkelson. 1972. Toxicities of methoxyflurane, halothane, and diethyl ether in laboratory animals on repeated inhalation at subanesthetic concentrations. *Proceedings of the Symposium on Cellular Biology, Toxicity and Anesthetics*. ed. R. B. Fink, pp. 275–85. Baltimore: Williams & Wilkins.

Hake, C. L., and V. K. Rowe. 1963. Industrial *Hygiene and Toxicology*, 2nd ed., Vol. 2, p. 1656. New York: Interscience.

Koenst, W. M. B. 1981. The hazard of a propionyl chloride–diisopropyl ether mixture. *J. Hazard. Mater. 4* (3): 291–98.

Matt, K. S., M. J. Soares, F. Talamantes, and A. Bartke. 1983. Effects of handling and ether anesthesia on serum prolactin levels in the golden hamster. *Proc. Soc. Exp. Biol. Med. 173* (4): 463–66.

National Research Council. 1983. *Prudent Practices for Disposal of Chemicals from Laboratories*. Washington, DC: National Academy Press.

NFPA. 1986. *Fire Protection Guide on Hazardous Materials*, 9th ed. Quincy, MA: National Fire Protection Association.

NIOSH. 1984. *Manual of Analytical Methods* 3rd ed. Cincinnati, OH: National Institute for Occupational Safety and Health.

NIOSH. 1986. *Registry of Toxic Effects of Chemical Substances*, ed. D. V. Sweet. Washington, DC: U.S. Government Printing Office.

Reuzel, P. G. J., J. P. Bruyntjes, and R. B. Beems. 1981. Subchronic (13-week) inhalation toxicity study with dimethyl ether in rats. Report R 5717. *Aerosol Rep. 20* (1): 23–28.

Stevens, W. C., E. Eger, A. White, M. J. Halsey, W. Munger, R. D. Gibbons, W. Golan, and R. Shargel. 1975. Comparative toxicity of halothane, isoflurane, and diethyl ether at subanesthetic concentrations in laboratory animals. *Anesthesiology 42* (4): 408–19.

Swann, H. E., Jr., B. K. Kwon, G. K. Hogan, and W. M. Snellings. 1974. Acute inhalation toxicology of volatile hydrocarbons. *J. Am. Ind. Hyg. Assoc. 35* (9): 511–18.

Tortoishell, G. 1985. Safety of fiber optic systems in flammable atmospheres. *Proc. SPIE Int. Soc. Opt. Eng. 522 (Fibre Opt.)*: 132–41.

19

GASES, COMMON TOXIC AND FLAMMABLE

19.1 HYDROGEN

Symbol H; elemental state H_2; at. wt. 1.0079; at. no. 1; isotopes: 2 (deuterium) and 3 (tritium); CAS [1333-74-0]; synonym: protium.

Hydrogen is a colorless, odorless, and tasteless gas, lighter than air. It liquefies at $-252.7°C$. It occurs in the earth's atmosphere in only trace amounts ($\sim 0.00005\%$). It is produced by reactions of metals with acids; reactions of alkali metals with water; hydrolysis of metal hydrides; electrolysis of water; or from coke and steam. It is also produced from methane.

Toxicity

Hydrogen is a nontoxic gas, a simple asphyxiant in high concentrations. Liquid hydrogen can cause severe frostbite.

Fire and Explosion Hazard

Flammable gas, burns with oxygen, forming water. It forms explosive mixtures with air within a wide range, 4–74% by volume in air. Autoignition temperature 500°C 932°F).

At ordinary temperature, a hydrogen–oxygen mixture can explode in the presence of finely divided platinum or other metals. Hydrogen combines with halogens explosively. Its reaction with fluorine can result in severe explosion. It forms explosive mixtures with chlorine within the concentration range 5–95% (NFPA 1986). A hydrogen–chlorine mixture can explode readily by heat, sunlight, or spark. It also reacts explosively with bromine, but with less violence. Explosive reactions occur with many interhalogen compounds, such as chlorine trifluoride, and with halogen oxides such as chlorine monoxide or oxygen difluoride. A violent explosion occurs upon heating a mixture of magnesium and calcium carbonate in a current of hydrogen (Mellor 1946–47).

Hydrogen gas reacts with selenium or sulfur in the vapor phase to form toxic hydrogen selenide or hydrogen sulfide.

In the event of a hydrogen fire, shut off the flow of gas. Small fires may be extinguished by dry chemical, CO_2, or Halon extinguishers. Use water to keep fire-exposed containers cool.

19.2 CARBON MONOXIDE

DOT Label: Flammable Gas and Poison Gas, UN 1016

Formula CO; MW 28.01; CAS [630-08-0]

Synonym: carbonic oxide

Uses and Exposure Risk

Carbon monoxide is used in the oxo process or Fischer–Tropsch process in the production of synthetic fuel gas (producer gas, water gas, etc.); as a reducing agent in the Monod process for the recovery of nickel; in carbonylation reactions; and in the production of metal carbonyls and complexes. It is produced by incomplete combustion of organic materials. Risk of exposure to this gas arises under fire conditions; from a burning stove or from burning wood or candles in a closed room; in the exhausts of internal combustion engines; in a closed garage with the autoengine on; and from oil or gas burners, improperly adjusted.

Physical Properties

Colorless, odorless, tasteless gas; liquefies at $-191.5°C$; freezes at $-205°C$; slightly soluble in water, 2.3% at 20°C (Merck 1989), soluble in alcohol, ethyl acetate, acetic acid, and chloroform; absorbed by acidic solution of cuprous chloride.

Health Hazard

Carbon monoxide is a highly poisonous gas. The acute toxic symptoms include headache, tachypnea, nausea, dizziness, weakness, confusion, depression, hallucination, loss of muscular control, and an increase and then a decrease in pulse and respiratory rate. If the dose is high, these symptoms progress to collapse, unconsciousness, and death. The severity of toxic effects depends on the concentration of carbon monoxide and the duration of exposure. Prolonged exposure to a concentration of 50 ppm does not result in adverse health effects in humans, but a 6-hour exposure to 100 ppm may produce perceptible effects. A 10-minute exposure to 5000 ppm is lethal to humans.

LC_{50} value, inhalation (rats): 1800 ppm/4 hr

The biochemical action of carbon monoxide involves its reaction with hemoglobin (Hb) in the blood. It enters into the bloodstream through the lungs and combines with hemoglobin to form carboxyhemoglobin (COHb). Hemoglobin is essential for the transportation of oxygen into the tissues. The affinity of carbon monoxide to combine with hemoglobin is about 300 times greater than that of oxygen (Meyer 1989). Thus, the CO molecule readily displaces oxygen from the oxyhemoglobin (O_2Hb) to form the more stable adduct, carboxyhemoglobin:

$$O_2Hb + CO \longrightarrow COHb + O_2$$

Thus the hemoglobin is tied up. It cannot, therefore, supply oxygen to the tissues, thus resulting in hypoxia and death.

Therapy involves artificial respiration using a compression chamber at about 2 atmospheres of oxygen, or a 95% O_2/5% CO_2 mixture, thus supplying more oxygen to compete for the hemoglobin and increase the solubility of oxygen in the blood plasma.

Exposure Limits

TLV-TWA 50 ppm ($\sim$55 mg/m^3) (ACGIH, MSHA, and OSHA); STEL 400 ppm (ACGIH); IDLH 1500 ppm (NIOSH).

Fire and Explosion Hazard

Flammable gas; burns in air with a bright blue flame; autoignition temperature 906°C (1128°F); forms explosive mixtures with air within the wide range 12.5–74% by volume of air.

Carbon monoxide reacts with many metals to form metal carbonyls, some of which explode upon heating. Reactions with alkali metals yield the corresponding carbonyls, which explode on heating. Lithium carbonyl detonates when mixed with water, igniting gaseous products (Mellor 1946, Suppl. 1961). It undergoes violent reactions with many interhalogen compounds of fluorine, such as chlorine trifluoride, bromine pentafluoride, bromine trifluoride, and iodine heptafluoride. Reactions with fluorides of other nonmetals are also violent; however, these occur under more drastic conditions. Mixtures of carbon monoxide and nitrogen trifluoride explode upon heating (Lawless and Smith 1968). Carbon monoxide is a reducing agent. It reacts violently with strong oxidizers. It reduces many metal oxides exothermically.

19.3 CARBON DIOXIDE

Formula CO_2; MW 44.01; CAS [124-38-9]
Synonyms: carbonic acid anhydride; dry ice (solid carbon dioxide)

Carbon dioxide occurs in the atmosphere at 0.033%. It is produced by burning carbonaceous materials. It is used in the carbonation of beverages; in fire extinguishers, in the manufacture of carbonates, as dry ice for refrigeration, and as a propellant for aerosols.

Physical Properties

Colorless, odorless gas with faint acid taste; gas density 1.527 (air = 1); freezes at $-78.5°C$ (sublimes); soluble in water (88 mL/dL) at 20°C, less soluble in alcohol and ether; absorbed by alkalies to form carbonates.

Health Hazard

Carbon dioxide is an asphyxiant. Exposure to about 9–10% concentration can cause unconsciousness in 5 minutes. Inhalation of 3% CO_2 can produce weak narcotic effects. Exposure to 2% concentration for several hours can produce headache, increased blood pressure, and deep respiration.

Exposure Limits

TLV-TWA   5000 ppm   (~9000 mg/m³) (ACGIH, MSHA, and OSHA); STEL 30,000 ppm (ACGIH).

19.4 NITRIC OXIDE

EPA Designated Toxic Waste, RCRA Waste Number P076; DOT Label: Poison Gas, UN 1660
Formula NO; MW 30.01; CAS [10102-43-9]
Synonym: nitrogen monoxide

Uses and Exposure Risk

Nitric oxide is used as an intermediate in the manufacture of nitric acid, in the preparation of metal nitrosyls, in bleaching of rayon, and in incandescent lamps. It is produced by heating air at high temperatures.

Physical Properties

Colorless gas with a sharp and sweet odor; deep blue in liquid form; converts readily to nitrogen dioxide and nitrogen tetroxide on contact with air; density of gas 1.04 (air = 1); liquefies at $-151.7°C$; solidifies at $-163.6°C$; solubility in water 4.6 mL NO/dL water at 20°C.

Health Hazard

Nitric oxide is an irritant to the eyes, nose, and throat. Inhalation of this gas causes methemoglobinemia. Its actions are somewhat similar to those of carbon monoxide. It binds with hemoglobin (Hb) in blood to form metheglobin (NOHb), affecting the transportation of oxygen to body tissues and organs:

$$Hb(aq) + NO(g) \rightarrow NOHb(aq)$$

Animal experiments indicate nitric oxide to be much less toxic than nitrogen dioxide. However, because of its spontaneous oxidation to highly toxic nitrogen dioxide, nitric oxide should be viewed as a severe health hazard.

Exposure Limit

TLV-TWA 25 ppm (~30 mg/m³) (ACGIH, MSHA, and OSHA).

Fire and Explosion Hazard

Noncombustible gas; burns with fuels, hydrocarbons, or when heated with hydrogen. Nitric oxide reacts violently with carbon disulfide vapors, producing green luminous flame; with fluorine, it produces a pale yellow flame. It explodes when mixed with ozone, chlorine monoxide, or a nitrogen trihalide. Reactions with many pyrophoric metals produces incandescence. Reaction with amorphous boron produces brilliant flashes.

19.5   NITROGEN DIOXIDE

EPA Designated Toxic Waste, RCRA Waste Number P078; DOT Label: Poison Gas and Oxidizer, UN 1067

Formula NO_2; MW 46.01; CAS [10102-44-0]

Synonyms: nitrogen peroxide; dinitrogen tetroxide; nitrogen tetroxide (NTO)

Uses and Exposure Risk

Nitrogen dioxide is produced by the reaction of nitric acid with metals or other reducing agents; decomposition of nitrates; when air is heated to high temperatures; and during fire. It occurs in the exhausts of internal combustion engines and in cigarette smoke. It is used as an intermediate in the production of nitric and sulfuric acids, in rocket fuels, as a nitrating and oxidizing agent, and in bleaching flour.

Physical Properties

Reddish-brown gas with an acrid, pungent odor; liquefies under pressure to a brown fuming liquid, which is a mixture of nitrogen dioxide and colorless nitrogen tetroxide, N_2O_4; density of gas 1.58 (air = 1), liquid 1.448 at 20°C; 1 L of gas weighs 3.3 g at 21°C; liquefies at 21°C, solidifies at −9.3°C; reacts with water to form nitric acid and nitric oxide, reacts with alkalies to form nitrates and nitrites; soluble in nitric and sulfuric acids.

Health Hazard

Nitrogen dioxide is a highly toxic gas. It is an irritant to the eyes, nose, and throat and to the respiratory system. The toxic symptoms are cough, frothy sputum, chest pain, dyspnea, congestion, and inflammation of lungs and cyanosis. Even a short exposure can cause hemorrhage and lung injury. Death may result within a few days after exposure. Toxic symptoms may be noted in humans following a 10-minute exposure to a 10 ppm concentration in air. One or two minutes of exposure to 200 ppm can be lethal to humans.

Exposure Limits

TLV-TWA 3 ppm (∼6 mg/m^3) (ACGIH), ceiling in air 5 ppm (MSHA and OSHA); STEL 5 ppm (ACGIH); IDLH 50 ppm (NIOSH).

Fire and Explosion Hazard

Noncombustible gas (or liquid), but supports the combustion of several types of substances. Nitrogen tetroxide is an extremely strong oxidizer and can set fire to clothes, paper, and many readily combustible substances. Nitrogen tetroxide or its mixture with nitrogen dioxide reacts with explosive violence with a number of compounds. These include alcohols, aldehydes, hydrocarbons (alkanes, alkenes, and aromatics), nitrobenzene, acetic anhydride, and incompletely halogenated hydrocarbons. Alkali metals ignite in nitrogen dioxide. Reactions with most other metals are vigorous but not violent. Nitrogen dioxide reacts violently with ozone (producing light and/or explosion) (NFPA 1986), liquid ammonia, and phosphine.

19.6   NITROUS OXIDE

Formula N_2O; MW 44.02; CAS [10024-97-2]

Synonyms: hyponitrous acid anhydride; dinitrogen monoxide; laughing gas

Uses and Exposure Risk

Nitrous oxide is used in the production of nitrites, in rocket fuel, as an inhalation anesthesia and analgesic agent.

Physical Properties

Colorless gas with mild sweet odor and taste; liquefies at $-88.5°C$; freezes at $-91°C$; dissociates around $320°C$; soluble in water, alcohol, ether, oils, and sulfuric acid.

Health Hazard

Toxicity and irritant effects of nitrous oxide in humans are very low. It is an anesthetic. Inhalation of this gas at high concentrations can produce depression of the central nervous system, decrease in body temperature, and fall in blood pressure. The LC_{50} value of a 4-hour exposure in mice is in the range 600 ppm.

Fire and Explosion Hazard

Nonflammable gas; supports the combustion of many compounds, including carbon, sulfur, and phosphorus. Nitrous oxide decomposes explosively at high temperatures. It forms explosive mixtures with hydrogen, ammonia, silane, and phosphine. Reactions with hydrazine and alkali metal hydrides produce flame. It is a strong oxidizer at elevated temperatures, at about $300°C$, when its dissociation begins.

19.7 AMMONIA (ANHYDROUS)

Formula NH_3; MW 17.04; CAS [7664-41-7]

Uses and Exposure Risk

Ammonia is used in the manufacture of nitric acid, hydrazine hydrate, and acrylonitrile, as well as fertilizers, explosives, and synthetic fibers. It is also used in refrigeration.

Physical Properties

Colorless gas with an intense suffocating odor; density 0.597 (air = 1), 1 L of gas weighs 0.771 g; easily liquefied by pressure; bp $-3.5°C$; freezes at $-77.5°C$; highly soluble in water, alcohol, and ether; aqueous solution highly alkaline.

Health Hazard

Ammonia is intensely irritating to the eyes, nose, and respiratory tract. Toxic effects include lachrymation, respiratory distress, chest pain, and pulmonary edema. A concentration of 10 ppm may be detected by odor; irritation of eyes and nose is perceptible at about 200 ppm. A few minutes of exposure to 3000 ppm can be intolerable, causing serious blistering of the skin, lung edema, and asphyxia, leading to death. It is corrosive to skin because it reacts with moisture to form caustic ammonium hydroxide. Long exposure may result in destruction of tissues.

LC_{50} value, inhalation (mice): 4200 ppm/hr

Exposure Limits

TLV-TWA 25 ppm ($\sim$18 mg/m^3) (ACGIH and MSHA), 50 ppm (OSHA); STEL 35 ppm; IDLH 500 ppm (NIOSH).

Fire and Explosion Hazard

Flammable gas; forms explosive mixtures with air in the range 16–25% by volume of air; autoignition temperature $651°C$ ($1204°F$).

Ammonia reacts violently with halogens. At ordinary temperatures, a mixture of ammonia–fluorine bursts into flame. On heating, the products, nitrogen trihalides resulting from reactions with halogens, explode violently. Violent reactions occur with many interhalogen compounds of fluorine, as well as with many inorganic chlorine compounds. The latter include chlorites, hypochlorites, chlorine monoxide, and certain metal chlorides, such as silver chloride or mercuric chloride, which form shock-sensitive nitrides.

Liquid ammonia reacts with potassium and sodium nitrate. Violent reaction occurs with potassium chlorate and potassium ferricyanide. Ammonia reacts with gold, silver, or mercury to form fulminate-type compounds, which can explode violently. Reaction with sulfur produces sulfur nitride, which explodes. Ignition occurs when ammonia reacts with nitrogen dioxide or phosphorus trioxide. Among organics, low-molecular-weight aldehydes (e.g., acetaldehyde), epoxide (e.g., ethylene oxide), and olefin halides (e.g., ethylene dichloride) react violently with ammonia. Ammonia catalyzes the polymerization of acrolein and other unsaturates, causing an increase in temperatures and pressures, which may rupture closed containers.

19.8 HYDROGEN SULFIDE

DOT Label: Poison Gas and Flammable Gas, UN 1053

Formula H_2S; MW 34.08; CAS [7783-06-4]

Synonyms: sulfureted hydrogen; sulfur hydride

Uses and Exposure Risk

Hydrogen sulfide is used as an analytical reagent and in the manufacture of heavy water. It occurs in natural gas and sewer gas. It is formed by the reaction of a metal sulfide with dilute mineral acid, and in petroleum refining.

Physical Properties

Colorless gas with an odor of rotten eggs; sweetish taste; fumes in air; liquefies at $-60.2°C$; freezes at $-85.5°C$; slightly soluble in water (0.4% at 20°C); pH of a saturated aqueous solution is 4.5; aqueous solution unstable, absorbs oxygen, decomposes to sulfur, and the solution turns turbid.

Health Hazard

Hydrogen sulfide is a highly toxic gas. Exposure to high concentrations can result in unconsciousness and respiratory paralysis. A 5-minute exposure to a concentration of 1000 ppm can be lethal to humans. Prolonged exposure to concentrations between 250 and 500 ppm can cause respiratory irritation, congestion of the lung, and bronchial pneumonia. Toxic symptoms that have been noted from occupational exposure to hydrogen sulfide in a heavy water plant are headache, nausea, cough, nervousness, and insomnia (ACGIH 1986). In addition, it is an irritant to the eyes. Conjunctivities may result from exposure to 20–30 ppm.

Fire and Explosion Hazard

Flammable gas, burns with a pale blue flame; autoignition temperature 260°C (500°F); forms explosive mixtures with air within the wide range of 4.3–45% by volume in air.

Hydrogen sulfide reacts violently with soda-lime (sodium hydroxide and calcium oxide). The reaction progresses to incandescence in air, and causes explosion in oxygen (Mellor 1946). It reacts violently with many metal oxides, such as lead dioxide, nickel oxide, chromium oxide, and iron oxide, producing incandescence. Reactions with fuming nitric acid and other strong oxidizers can result in incandescence. Reaction with fluorine, interhalogen compounds, and finely divided metals can cause incandescence.

19.9 SULFUR DIOXIDE

Formula SO_2; MW 64.06; CAS [7446-09-5]

Synonyms: sulfurous anhydride; sulfurous oxide

Uses and Exposure Risk

Sulfur dioxide is used as a bleaching and fumigating agent; as a disinfectant, for treating wood pulp for manufacturing paper, in metal refining, for preserving food and vegetables, and as a reducing agent. It is a major air pollutant and is produced when soft coal, oils, or other sulfur-containing substances are burned. Automobile exhaust gases also contribute to air pollution. Sulfur dioxide in the atmosphere reacts with moisture to form sulfurous acid, or is oxidized to sulfur trioxide, which forms sulfuric acid, causing acid rain.

Physical Properties

Colorless gas with a strong suffocating odor; liquefies at $-10°C$; solidifies at $-72°C$; soluble in water (8.5% at 25°C, 12% at 15°C), alcohol, ether, and chloroform.

Health Hazard

Exposure to sulfur dioxide can cause severe irritation of the skin, eyes, mucous membranes, and respiratory system. The effects are coughing, choking, or suffocation; bronchoconstriction; and skin burn. A 10-minute exposure to 1000 ppm can be fatal to humans.

LC_{50} value, inhalation (mice): 3000 ppm/30 min (NIOSH 1986)

Exposure Limits

TLV-TWA 2 ppm ($\sim$5 mg/m^3) (ACGIH), 5 ppm (OSHA and MSHA); IDLH 100 ppm (NIOSH).

Fire and Explosion Hazard

Noncombustible gas. Sulfur dioxide reacts violently with alkali metals at their melting points. Reactions with finely divided metals produce incandescence. Explosion occurs when it is mixed with fluorine or interhalogen compounds of fluorine. Incandescence occurs when carbides of alkali metals are placed in a sulfur dioxide atmosphere or by heating metal oxides with sulfur dioxide. An alcoholic or ethereal solution of sulfur dioxide explodes when mixed with powdered potassium chlorate (Mellor 1946). The dry gas reacts with chlorates to form chlorine dioxide, which ignites and explodes on heating.

19.10 HALOGENS AND HYDROGEN HALIDES

Among the halogens, fluorine and chlorine are gases at ordinary temperatures, bromine is a liquid, and iodine is a solid. These gases, as well as bromine vapors, are highly toxic. These are intensely irritating to the respiratory tract and can cause damage to tissues. A brief exposure to high concentrations in air can cause death.

Fluorine is the most electronegative and reactive among all the elements. Chlorine is the second most electronegative element and highly reactive. These gases react explosively with a large number of chemical substances. Halogens are discussed at length in Chapter 23.

Hydrogen fluoride, hydrogen chloride, and hydrogen bromide are intensely pungent gases that cause severe irritation to the eyes, nose, and respiratory tract. Exposure to these gases can cause choking, coughing, eye burn, and lung injury, and at high concentrations, death. Their aqueous solutions are strong acids that are highly corrosive. Among these, hydrogen fluoride is extremely corrosive, which can produce severe skin injury and painful burn. The toxicity, and the fire and explosion hazard of hydrogen halides (as acids), are discussed in detail in Chapter 2.

19.11 HYDROCARBON GASES

The common hydrocarbon gases of commercial use are the alkanes: methane, ethane, propane, butane, and isobutane; alkenes:

ethylene, propene, 1-butene, and 2-butene; and acetylene, which is an alkyne. Natural gas consists of over 85% methane.

These gases are flammable in air and form explosive mixtures with air within a relative wide range. They exhibit very little toxic action and are all simple asphyxiants. A detailed discussion of their hazardous properties is presented in Chapter 25.

19.12   INERT GASES

Inert gases placed in the Group 0 elements in the periodic table are also known as noble gases. These are helium [7440-59-7], neon [7440-01-9], argon [7440-37-1], krypton [7439-90-9], xenon [7440-63-3], and radon [10043-92-2]. Radon is a radioactive gas, discussed separately in Chapter 54. By virtue of their stable electronic configurations, inert gases exhibit no chemical reactivity. These gases are nontoxic. However, at high concentrations, they are simple asphyxiants.

19.13   ANALYSIS

Most of the common gases can be analyzed by various instrumental techniques, including GC, GC/MS, ion chromatography, IR, and visible absorption spectroscopy. The detector used in GC for gas analysis is mostly TCD. By using an appropriate column at low or ambient temperatures (cooling), analysis of most of these gases can be performed effectively. GC/MS is a confirmatory test. A capillary column at ambient temperature or a packed column under cryogenic conditions may be suitable for analyzing common toxic gases.

Nitrogen dioxide in air may be analyzed by a visible absorption spectrophotometer at 540-nm wavelength. The sampler consists of a Palmer tube with three triethanolamine-treated screens (NIOSH 1984, Method 6700). Nitrous oxide in ambient air may be analyzed

by a portable IR spectrophotometer (wavelength 4.48 μm) (NIOSH 1984, Method 6600). Sulfur dioxide in air is absorbed by KOH in a filter; the sulfite and sulfate ions are desorbed by an $NaHCO_3-Na_2CO_3$ mixture and analyzed by an ion chromatograph using a conductivity detector (NIOSH 1984, Method 6004). Analysis for ammonia may be performed similarly by an ion chromatograph using a conductivity detector. Ammonia is absorbed by 0.01 N H_2SO_4 to form NH_4^+ ion in the aqueous solution prior to its estimation (NIOSH 1984, Method 6701).

Removal of SO_x and NO_x

Sulfur oxides and nitrogen oxides may be removed from combustion gases by various processes. One of the cost-effective and efficient methods involve dry scrubbing of SO_x and NO_x over lanthanide–oxygen–sulfur compounds (Jalan and Desai 1992). Cerium sulfate is found to be an effective catalyst toward the reduction of NO_x by ammonia. A combined removal of NO_x and SO_x has been achieved using cerium oxide doped with strontium oxide, lanthanum oxide, calcium oxide, or cerium sulfate.

REFERENCES

ACGIH. 1986. *Documentation of the Threshold Limit Values and Biological Exposure Indices*, 5th ed. Cincinnati, OH: American Conference of Governmental Industrial Hygienists.

Jalan, V., and M. Desai. 1992. Dry scrubbing of SO_x and NO_x over lanthanide-oxygen-sulfur compounds. Abstracts of Small Business Innovative Research Program. Washington, DC: U.S. EPA.

Lawless, E. W., and I. C. Smith. 1968. *High Energy Oxidizer*. New York: Marcel Dekker.

Mellor, J. W. 1946. *A Comprehensive Treatise on Inorganic and Theoretical Chemistry*. London: Longmans, Green & Co.

Merck. 1989. *The Merck Index*, 11th ed. Rahway, NJ: Merck & Co.

Meyer, E. 1989. *Chemistry of Hazardous Materials*, 2nd ed. Englewood Cliffs, NJ: Prentice-Hall.

NFPA. 1986. *Fire Protection Guide on Hazardous Materials*, 9th ed. Quincy, MA: National Fire Protection Association.

NIOSH. 1984. *Manual of Analytical Methods*, 3rd ed. Cincinnati, OH: National Institute for Occupational Safety and Health.

NIOSH. 1986. *Registry of Toxic Effects of Chemical Substances*, ed. D. V. Sweet. Washington, DC: U.S. Government Printing Office.

20

GLYCOL ETHERS

20.1 GENERAL DISCUSSION

Glycol ethers are the ether derivatives of gly-cols or dialcohols, that is, compounds containing two −OH groups. These compounds are formed by substitution of alcoholic −OH group(s) in the glycols with −OR groups, where R is an alkyl or aryl radical. Thus monomethyl or monophenyl ethers of ethylene glycol have the following structures:

$$HO-CH_2-CH_2-OCH_3$$
$$HO-CH_2-CH_2-OC_6H_5$$

The second alcoholic −OH group of the glycol may undergo further replacement with another −OR group to form dimethyl or diphenyl ethers. In the present discussion, the term *glycol ethers* refers to organic compounds containing one alcoholic −OH group and one or more ethereal oxygen atom(s).

Many glycol ethers have found extensive industrial applications as solvents. These are also known by the trade name Cellosolve. These compounds have unique solvent properties and are used as solvents for cellulose esters, resins, acrylics, oils, greases, dyes, paints, varnishes, and many other materials.

Toxicity of glycol ethers is low in test animals. In humans the toxic effects should be mild. However, moderate to severe poisoning can occur from excessive dosage. The routes of exposure are inhalation, ingestion, and absorption through the skin. Compounds with high molecular weights and low vapor pressures do not manifest an inhalation hazard. The metabolites of glycol ethers are primarily the oxidation products, aldehydes and the carboxylic acids. In the case of propylene glycol monomethyl ether, propylene glycol was reported to be the metabolite, formed by dealkylation of the ether-bound methyl group (Miller et al. 1984). C_1 to C_4 alkyl ethers of ethylene glycol exhibit reproductive toxicity in test species.

Testicular atrophy was the common reproductive toxic effect observed with these compounds. No teratogenicity was observed with glycol ethers of high molecular weight.

The low-molecular-weight alkyl ethers are flammable or combustible liquids forming explosive mixtures with air. The ethers of diethylene glycol are noncombustible liquids. The former compounds may form peroxides upon exposure to air. The reactivity of glycol ethers is low. There is no report of any violent explosive reactions.

Individual compounds in air, water, or wastes may be analyzed by common instrumental techniques such as gas chromatography using a flame ionization detector. Other techniques, such as HPLC and GC/MS, are equally suitable. Any column that is efficient for the separation of alcohols may be used to analyze glycol ethers. Analysis of some of these compounds is discussed in greater detail in the sections on individual compounds.

20.2 ETHYLENE GLYCOL MONOMETHYL ETHER

DOT Label: Flammable Liquid, UN 1188
  Formula $C_3H_4O_2$; MW 76.11; CAS [109-86-4]

Structure and functional group: $HO-CH_2-CH_2-O-CH_3$, a primary $-OH$ group

Synonyms: 2-methoxyethanol; glycolmethyl ether; methyl glycol; Methyl Cellosolve; methyl oxitol

Uses and Exposure Risk

The primary use of this compound is as a solvent for cellulose acetate, certain synthetic and natural resins, and dyes. Other applications are in jet fuel deicing, sealing moisture-proof cellophane, dyeing leather, and use in nail polishes, varnishes, and enamels.

Physical Properties

Colorless liquid; bp 124°C; freezes at −85°C; density 0.965; miscible with water, alcohol, ether, and acetone.

Health Hazard

Ethylene glycol monomethyl ether (EGME) is a teratogen and a chronic inhalation toxicant. The target organs are blood, kidney, and the central nervous system. In addition to inhalation, the other routes of exposure are absorption through the skin, and

ingestion. Animal studies indicated that overexposure to this compound produced anemia, hematuria, and damage to the testes. In humans, inhalation of EGME vapors can cause headache, drowsiness, weakness, irritation of the eyes, ataxia, and tremor. The acute inhalation toxicity, however, is low and any toxic effect may be felt at a concentration of about 25–30 ppm in air.

The oral and dermal toxicities of this compound in test animals were found to be lower than the inhalation toxicity. Ingestion can produce an anesthetic effect and in a large dosage can be fatal. An oral intake of about 200 mL may cause death to humans.

LC_{50} value (mice): 1480 ppm/7 hr
LD_{50} value (rabbits): 890 mg/kg

EGME is a teratogen exhibiting fetotoxicity, affecting the fertility and the litter size, and causing developmental abnormalities in the urogenital and musculoskeletal systems in test animals.

Toxicology

Despite its low toxicity, there have been extensive toxicological studies on EGME. The subacute percutaneous administration of this compound in male rats dosed 5 days/week for 4 weeks resulted in testicular and bone marrow damage at doses of 1000 mg/kg/day (Fairhurst et al. 1989). HPLC and isotope dilution analysis indicated that the major metabolites were methoxyacetic acid [625-45-6] and methoxyacetyl glycin (Moss et al. 1985). Foster et al. (1986) proposed that 2-methoxyacetaldehyde [10312-83-1] was a possible metabolite that played an important role in EGME-induced testicular toxicity. The former produced specific cellular toxicity to pachytene spermatocytes in mixed testicular cell cultures. Subchronic vapor inhalation studies in rats and rabbits showed that overexposure to EGME produced adverse effects on testes, bone marrow, and lymphoid tissues (Miller et al. 1984). The compound oxidized primarily to methoxyacetic acid in the body. A single exposure to 600 ppm for 4 hours produced testicular atrophy in rats, and 100 ppm showed teratogenic potential (Doe 1984). Such effects were observed in mice and hamsters when the compound was administered by oral gavage (Nagano et al. 1984).

Methoxyacetic acid is produced in the body by in vivo bioactivation of EGME, the toxicological properties of which are similar to the former (Miller et al. 1982). EGME vapor was toxic to male rats at 50 ppm concentration for a 2-week exposure period (Savolainen 1980).

Doe et al. (1983) studied the reproductive toxicology of EGME by inhalation in rats. A concentration of 300 ppm significantly reduced white and red blood cell counts, hemoglobin concentration, and hematocrit. Eisses (1989) investigated alcohol dehydrogenase activity in relation to toxicity and teratogenicity of EGME in *Drosophila melanogaster*. EGME was oxidized to methoxyacetic acid; the former was much more toxic than its oxidation product, and a teratogenic compound by itself.

Several substances have been reported to exhibit protective action against EGME toxicity. Pretreatment with an inhibitor of alcohol dehydrogenase offered complete protection against a testicular toxic dose of 500 mg/kg of EGME (Foster et al. 1984). Moss et al. (1985) reported protective action of pyrazole pretreatment (400 mg/kg, intraperitoneal) 1 hour before EGME dosing. Certain physiological compounds, including serine, acetate, and sarcosine, were reported to suppress EGME testicular toxicity by virtue of their ability to donate one-carbon units to purine-base biosynthesis, thus affecting the DNA and RNA synthesis (Mebus and Welsch 1989; Mebus et al. 1989).

Exposure Limits

TLV-TWA skin 5 ppm (15.5 mg/m^3) (ACGIH), 25 ppm (77.5 mg/m^3) (OSHA).

Fire and Explosion Hazard

Combustible liquid; flash point: (closed cup) 42°C (107°F), (open cup) 46°C (115°F); vapor pressure 6 torr at 20°C; vapor density 2.6 (air = 1); autoignition temperature 233°C (455°F); fire-extinguishing agent: "alcohol" foam or water spray; use a water spray to dilute the spill and to keep fire-exposed containers cool.

EGME forms explosive mixtures with air; the lower and upper limits are 2.5% and 19.8% by volume of air. It may form peroxides, which can explode. EGME increases the flammability of jet fuel. At a 0.15% concentration in jet fuel it reduces the flash point of the latter by 7°F and increases the flammability index by 11% (Affens and McLaren 1972). This property has been exploited commercially, making use of EGME as an icing inhibitor.

Disposal/Destruction

EGME is burned in a chemical incinerator equipped with an afterburner and scrubber. Small amounts may be flushed down the drain with a large volume of water.

Analysis

EGME can be analyzed by common instrumental techniques such as GC-FID, HPLC, and GC/MS. Air analysis may be performed by passing 1–10 L of air over coconut shell charcoal at a rate of 10–50 mL/min. The analyte is desorbed in 1 mL of 5% methanol in methylene chloride and injected into a GC equipped with an FID (NIOSH 1984, Method 1403). A column such as 10% SP-1000 on Chromosorb W-HP or any other equivalent column may be used for the purpose.

20.3 ETHYLENE GLYCOL MONOETHYL ETHER

EPA Classified Toxic Waste, RCRA Waste Number U359; DOT Label: Combustible Liquid UN 1171

Formula $C_4H_{10}O_2$; MW 90.12; CAS [110-80-5]

Structure: $HO-CH_2-CH_2-O-CH_2-CH_3$, primary $-OH$ group

Synonyms: 2-ethoxyethanol; glycol ethyl ether; hydroxy ether; Cellosolve, Ethyl Cellosolve, oxitol

Uses and Exposure Risk

Ethylene glycol monoethyl ether (EGEE) is used as a solvent for nitrocellulose, lacquers, and varnishes; in dye baths and cleansing solutions; and as an emulsion stabilizer.

Physical Properties

Colorless and odorless liquid; bp 135°C; freezes at -70°C; density 0.931; miscible with water and most organic solvents.

Health Hazard

EGEE is a teratogen and at high concentration a toxic substance. The target organs are the lungs, kidney, liver, and spleen. Animal experiments indicated that inhalation of its vapors at 2000 ppm for several hours could produce toxic effect. Death resulted from kidney injury when the test species were subjected to 24-hour exposure. It produced kidney injury, hematuria, and microscopic lesions of both the liver and kidney. EGEE may be absorbed through the skin. When inserted into the eyes, it produced corneal irritation. The recovery occurred within 24 hours.

Investigating the subchronic inhalation toxicology of EGEE in the rat and rabbit, Barbee et al. (1984) reported no biological significant effect of this compound in these animals at an exposure level of 400 and 100 ppm, respectively. Chronic treatment of rats with EGEE at 0.5–1.0 g/kg in an oral dose caused enlargement of adrenal gland in male rats and affected the development of spontaneous lesions of the spleen (males and females), pituitary (males and females), and testis (males) (Melnick 1984).

LC_{50} value (mice): 1820 ppm/7 hr (NIOSH 1986)

LD_{50} value (rats): 3000 mg/kg (NIOSH 1986)

In humans there is no report of any severe poisoning case. The toxic effect from inhaling its vapors at 1000 ppm may be less than noticeable. EGEE is less toxic than ethylene glycol monomethyl ether (EGME). When administered orally to young male rats, EGEE produced testicular atrophy similar to that of EGME (Nagano et al. 1984). However, a fivefold dose, 250–1000 mg/kg/day, was required to elicit equivalent severity (Foster et al. 1984).

Reproductive toxicity of EGEE has been investigated extensively (Lamb et al. 1984; Hardin et al. 1984; Oudiz et al. 1984). Testicular atrophy, decline in sperm count, and increased abnormal sperm were observed in treated male rats, but no specific anomalies were noted in the females. Wier et al. (1987) investigated postnatal growth and survival. EGEE produced embryo lethality and malformations and decreased fetal weight. Prenatal exposure to EGEE produced kinked tails in pups. Ethanol caused potentiation of reproductive toxicity of EGEE (Nelson et al. 1984).

Exposure Limit

TLV-TWA skin 5 ppm (18.5 mg/m^3) (ACGIH).

Fire and Explosion Hazard

Combustible liquid; flash point (closed cup) 44°C (112°F), (open cup) 49°C (120°F); vapor pressure 3.2 torr at 20°C; vapor density 3.1 (air = 1); autoignition temperature 235°C (455°F); fire-extinguishing agent: "alcohol" foam; a water spray may be effective in extinguishing the flame and should be used to dilute the spill and keep fire-exposed containers cool.

EGEE forms explosive mixtures with air; the LEL and UEL values are 2.9% and 18% by volume in air, respectively. Peroxide formation may occur on prolonged exposure to air.

Disposal/Destruction

EGEE may be burned in a chemical incinerator equipped with an afterburner and scrubber. Small amounts may be drained down the sink with an excess of water.

Analysis

EGEE is analyzed by GC techniques using either an FID or TCD. Alternatively, it can be analyzed by HPLC or GC/MS. Air analysis may be done by NIOSH (1984) Method 1403 (see Section 21.2). Esposito and Jamison (1971) described a GC-TCD method for the analysis of EGEE and other compounds in hydraulic brake fluids. The GC column was made up of 20 wt% Carbowax M or diethylene glycol succinate on a Chromosorb W support.

20.4   ETHYLENE GLYCOL MONOBUTYL ETHER

DOT Label: Poison B and Combustible Liquid, UN 2369

Formula $C_6H_{14}O_2$; MW 118.20; CAS [111-76-2]

Structure: $HO-CH_2-CH_2-O-CH_2-CH_2-CH_2CH_3$

Synonyms: 2-butoxyethanol; n-butoxyethanol; glycol monobutyl ether; monobutyl glycol ether; 3-oxa-1-heptanol; Butyl Cellosolve; butyl glycol; butyl oxitol

Uses and Exposure Risk

Ethylene glycol monobutyl ether (EGBE) is used as a solvent for nitrocellulose, resins, oil, and grease, and in dry cleaning.

Physical Properties

Colorless liquid with a mild odor of ether; boils at 172°C; freezes at −70°C; density 0.902 at 20°C; miscible with water and most organic solvents.

Health Hazard

EGBE exhibited mild to moderate toxicity in test animals. The toxic symptoms are similar to those of ethylene glycol monomethyl ether (EGME). It is an irritant to the eyes and skin. The toxic routes of exposure are inhalation, ingestion, and absorption through the skin. In animals, prolonged exposure to high concentrations or high oral intake caused hematuria, kidney damage, and increased osmotic fragility of the blood cells. Such effects, however, were noted only at a high level of exposures (i.e., 700 ppm for 7 hours in mice, the minimal lethal concentration) (Werner et al. 1943). The other toxic effects noted were respiratory distress, change in motor activity, and lung, kidney, and liver changes. EGBE is absorbed rapidly through the skin. A 3-minute contact with 0.56 mL/kg over a 4.5% skin area produced the increased red blood cell fragility within an hour in rabbits (Carpenter et al. 1956; ACGIH 1986). Rabbits treated percutaneously with 0.08 to 0.25 mL/kg EGBE developed prostration, hypothermia, hemoglobinuria, spleen congestion, and kidney enlargement (Duprat and Gradiski 1979). Some rabbits survived and did not show the foregoing histopathological signs. The percutaneous toxicity of EGBE was found to be greater than that of the industrial solvents carbon tetrachloride and dimethyl formamide but lower than that of 2-chloroethanol and 1,1,2-trichloroethane (Wahlberg and Boman 1979). Chronic exposures at lower concentrations produced mild hemolytic anemia and thymic atrophy in rats (Grant et al. 1985). Subchronic oral doses of undiluted EGBE produced a significant dose-dependent decrease in the blood hemoglobin concentration (Krasavage 1986). Such an effect was manifested at the moderately high dose of 222–885 mg/kg/day for 5 days/week over a 6-week period. No adverse effects on the testes, bone marrow, thymus, or white blood cells were observed.

LC_{50} value, (rats): 450 ppm/4 hr
LD_{50} value, oral (rats): 530 mg/kg

In humans the toxic effects of EGBE are generally low. Exposure to 200 ppm for 8 hours may produce nausea, vomiting, and headache. In a study on male volunteers exposed to 20 ppm of EGBE for 2 hours during light physical exercise, Johansson and co-workers (1986) determined that the respiratory uptake of EGBE was about 57% of the inspired amount; its concentration in the blood reached a plateau level of 0.87 mg/L; the elimination half-time and the mean residence time were 40 and 42 minutes, respectively. The amount excreted in the urine was <0.03% of the total intake, whereas that of butoxyacetic acid ranged from 17% to 55%.

No reproductive toxicity has been reported in male animals. No tetratogenic effects have been reported in females. However, when pregnant animals received high oral or inhalation doses of this material, signs of embryo-fetal toxicity were reported but only at doses that were maternally toxic. Unlike EGME, EGBE caused no testicular atrophy (Doe 1984; Nagano et al. 1984).

Toxicology

In an acute toxicity study on F344 male rats, Ghanayem and co-workers (1987b) observed that EGBE produced severe acute hemolytic anemia, causing significant increases in free plasma hemoglobin concentration. Hemoglobinuria and histopathological changes in the liver and kidney were also noted. All these effects were dose, time, and age dependent. The degradation of the metabolite butoxyacetic acid to CO_2 was greater in young rats than in the adult animals. In another study, Ghaneyem and co-workers (1987a) reported the protective actions of pyrazole (an alcohol dehydrogenase inhibitor) and cyanamide (an aldehyde dehydrogenase inhibitor) against the hematotoxicity of EGBE. Pretreatment of rats with pyrazole and cynamide inhibited the conversion of EGBE to butoxyacetic

acid and butoxyacetaldehyde, respectively. Butoxyacetaldehyde is a metabolic intermediate formed from EGBE, converting to the ultimate metabolite, butoxyacetic acid.

Exposure Limits

TLV-TWA skin 25 ppm (121 mg/m^3) (ACGIH), 50 ppm (242 mg/m^3) (OSHA); STEL 75 ppm (363 mg/m^3) (ACGIH); IDLH 700 ppm (NIOSH).

Fire and Explosion Hazard

Combustible liquid; flash point (closed cup) 60°C (141°F); vapor pressure 0.76 torr at 20°C; vapor density 4.07 (air = 1) autoignition temperature 244.5°C (472°F); fire-extinguishing agent: "alcohol" foam or water spray; use a water spray to dilute the spill and keep fire-exposed containers cool.

EGBE forms explosive mixtures with air in the range 1.1–10.6% by volume of air. Comas et al. (1974) reported the hazards associated with EGBE–perchloric acid solutions used for electropolishing alloys. Solutions containing 50–95% acid constitute explosion danger at room temperature. Acid concentration up to 30% is safe in the absence of an electric spark, solvent evaporation, or heating up to the flash point, 60°C of the solvent.

Disposal/Destruction

EGBE is destroyed by burning in an incinerator. In the laboratory, small amounts may be disposed of in the sink with a large volume of water.

Analysis

Gas chromatography is the most common analytical technique. FID or TCD may be used as the detector. Alternative analytical methods use GC/MS or HPLC. Air analysis may be performed by NIOSH (1984)

Method 1403 (see Section 21.2). EGBE or its metabolites in the blood or urine can be analyzed by HPLC techniques. Blood or urine samples may also be analyzed by derivatizing EGBE with pentafluorobenzoyl chloride and then analyzing by GC with an electron capture detector (Johansson et al. 1986).

20.5 ETHYLENE GLYCOL MONOISOPROPYL ETHER

Formula $C_5H_{12}O_2$; MW 104.17; CAS [109-59-1]

Structure:

$$HO-CH_2-CH_2-O-CH\begin{smallmatrix}CH_3\\CH_3\end{smallmatrix}$$

Synonyms: 2-isopropoxyethanol; β-hydroxy-ethyl isopropyl ether; isopropyl glycol; Isopropyl Cellosolve

Uses and Exposure Risk

Ethylene glycol monoisopropyl ether (EGIE) is used as a solvent for resins, dyes, and cellulose esters, and in coatings.

Physical Properties

Colorless liquid; bp 142°C; density 0.903; soluble in water and organic solvents.

Health Hazard

EGIE shows low toxicity from inhalation, ingestion, and skin absorption. The toxic symptoms are similar to those produced by the butyl ether of the ethylene glycol. The effects noted in rats from exposure to 1000 ppm for 6 hours were hemoglobinuria, anemia, and lung congestion (Gage 1970). Other effects observed were increases in the osmotic fragility of the red blood cells. Such effects were manifested at EGIE concentrations of >200 ppm. A 4-hour exposure to 4000 ppm was lethal to rats. In human exposure to this compound can result in anemia.

LD_{50} value, oral (rats): 5660 mg/kg

No testicular toxicity was observed in rats from a single exposure to 3500 ppm of EGIE in air (Doe 1984).

Exposure Limit

TLV-TWA skin 25 ppm (106 mg/m^3) (ACGIH).

Fire and Explosion Hazard

Combustible liquid; flash point (closed cup) 45°C (114°F); vapor pressure 2.6 torr at 20°C; vapor density 3.6 (air = 1). It forms explosive mixtures with air; LEL and UEL values not available.

Disposal/Destruction

Chemical incineration is the most suitable method of destruction. Drain disposal may be done for small amounts of EGIE.

Analysis

GC-FID, GC-TCD, HPLC, and GC/MS are suitable techniques for analysis.

20.6 PROPYLENE GLYCOL MONOMETHYL ETHER

Formula $C_4H_{10}O_2$; MW 90.14; CAS [107-98-2]

Structure:

$$\begin{smallmatrix}&&&OH&\\&&&|&\\CH_3-O-CH_2-CH-CH_3\end{smallmatrix}$$

Synonyms: 1-methoxy-2-propanol; α-propylene glycol monomethyl ether

Uses and Exposure Risk

Propylene glycol monomethyl ether (PGME) is used as a solvent for cellulose, acrylics, dyes, inks, and cellophane.

Physical Properties

Colorless liquid; bp 120°C; mp −95°C; density 0.931; soluble in water and organic solvents.

Health Hazard

PGME is a mild toxicant. The toxicity is lower than that of the methyl, ethyl, and butyl ethers of ethylene glycol. The toxic symptoms from inhaling high concentrations are nausea, vomiting, and general anesthetic effects. In humans, toxic effects may be felt at exposure to a level of 3000–4000 ppm.

The oral and dermal toxicities in test animals were low. The effects were mild depression of the central nervous system and a slight change in liver and kidney. The recovery was rapid. Irritant actions on the skin and eyes of rabbits were low.

LC$_{50}$ value (rats): 7000 ppm /6 hr

LD$_{50}$ value (rats): 5660 mg/kg

Exposure Limit

TLV-TWA 100 ppm (370 mg/m^3) (ACGIH); STEL 150 ppm (555 mg/m^3) (ACGIH).

Fire and Explosion Hazard

Flammable liquid; flash point (closed cup) 36°C (97°F); (open cup) 38°C (100°F); vapor density 3.11 (air = 1); fire-extinguishing agent: use a water spray.

PGME forms explosive mixtures with air; LEL and UEL values are not available. Reactions with strong oxidizers may be violent.

Disposal/Destruction

PGME is destroyed by burning in a chemical incinerator equipped with an afterburner and scrubber.

Analysis

GC techniques using FID or TCD are suitable for analysis. HPLC and GC/MS methods may be applied to analyze this compound.

20.7 DIETHYLENE GLYCOL MONOMETHYL ETHER

Formula C$_5$H$_{12}$O$_3$; MW 120.15; CAS [111-77-3]

Structure: HO−CH$_2$−CH$_2$−O−CH$_2$−CH$_2$−O−CH$_3$

Synonyms: 2-(2-methoxyethoxy)ethanol; methoxydiglycol; β-methoxy-β-hydroxy-diethyl ether; methyl carbitol

Uses and Exposure Risk

Diethylene glycol monomethyl ether (DGME) is used as a thermally stable solvent for many substances, such as nitrocellulose, lacquers, varnishes, and dyes.

Physical Properties

Colorless liquid; hygroscopic; bp 193°C; density 1.035; miscible with water and common organic solvents.

Health Hazard

DGME is a mild to moderate toxicant via ingestion or absorption through the skin. High doses produced lowering of hemoglobin levels and increased relative kidney weight. Renal damage may occur near the lethal dose. Eye contact of the liquid can result in mild to moderate irritation.

LD$_{50}$ value, oral (guinea pigs): 4160 mg/kg

Preliminary developmental toxicity test in pregnant mice dosed with DGME indicated teratogenicity of this compound (Hardin et al. 1987; Cheever et al. 1988). Earlier, Doe (1984) reported no teratogenic property of DGME when administered subcutaneously in rats up to 100 mL/kg. In comparison, ethylene glycol monomethyl ether produced effects at 40 mL/kg.

Fire and Explosion Hazard

Noncombustible liquid; flash point (open cup) 93°C (200°F); very low reactivity.

Disposal/Destruction

DGME is mixed with a combustible solvent and burned in a chemical incinerator.

Analysis

DGME is analyzed by GC-FID using a column containing 10% SP-1000 on Chromosorb W-HP. Alternatively, it may be analyzed by HPLC techniques.

20.8   DIETHYLENE GLYCOL MONOETHYL ETHER

Formula $C_6H_{14}O_3$; MW 134.17; CAS [111-90-1]

Structure: $HO-CH_2-CH_2-O-CH_2-CH_2-O-CH_2-CH_3$

Synonyms: 2-(2-ethoxyethoxy)ethanol; carbitol; Carbitol Cellosolve; ethoxydiglycol

Uses and Exposure Risk

Diethylene glycol monoethyl ether (DGEE) is used as a solvent for cellulose esters, lacquers, varnishes, enamels, and wood stains.

Physical Properties

Colorless liquid; hygroscopic; bp 196°C; density 1.027; miscible with water and most organic solvents.

Health Hazard

DGEE exhibited low to moderate oral and dermal toxicity in test animals. When fed to male mice at dietary levels of 5.4 mg/kg for 90 days, DGEE produced renal damage (Gaunt et al. 1968). The other symptoms noted in rats and mice were decreased food intake and growth, lowering of hemoglobin levels, increased relative kidney weight, and enlarged liver cells.

LD_{50} value, oral (rats): 5500 mg/kg

DGEE is a mild irritant to the skin and eyes. It showed very weak mutagenic activity in *Salmonella typhimurium* and *Saccharomyces cerevisiae* (Berte et al. 1986).

Fire and Explosion Hazard

Noncombustible liquid; flash point (tag open cup) 96°C (207°F). The reactivity of this compound is very low.

Disposal/Destruction

DGEE is mixed with a combustible solvent and burned in a chemical incinerator equipped with an afterburner and scrubber.

Analysis

GC-FID, GC/MS, and HPLC are suitable instrumental techniques to analyze this compound.

20.9   DIETHYLENE GLYCOL MONOBUTYL ETHER

Formula $C_8H_{18}O_3$; MW 166.22; CAS [112-34-5]

Structure: $HO-CH_2-CH_2-O-CH_2-CH_2-O-CH_2-CH_2-CH_2-CH_3$

Synonyms: 2-(2-butoxyethoxy)ethanol; butoxydiglycol; butoxydiethylene glycol; butyl carbitol

Uses and Exposure Risk

Diethylene glycol monobutyl ether (DGBE) is used as a solvent for cellulose ester, lacquers, varnishes, and dyes; as a primary component of the aqueous film-forming foam

TABLE 20.1   Toxicity of Miscellaneous Glycol Ethers

Compound/Synonyms CAS No.	Formula/MW/Structure	Toxicity
Ethylene glycol monopropyl ether (2-propoxyethanol, Propyl Cellosolve) [2807-30-9]	$C_5H_{12}O_2$ 104.17 $HO-CH_2-CH_3-O-CH_3-CH_2-CH_3$	Low toxicity via inhalation, ingestion, and skin absorption; toxic symptoms in animals from high exposures were tremor, labored breathing, and hematuria; eye irritant; teratogen, causing stillbirth and developmental abnormalities in musculoskeletal system; toxic effects in humans should be low; LD_{50} oral (rats): ~3000 mg/kg
Ethylene glycol mono-*sec*-butyl ether (2-*sec*-butoxyethanol) [7795-91-7]	$C_6H_{14}O_2$ 118.20 $HO-CH_2-CH_2-O-\overset{\overset{\textstyle CH_3}{\mid}}{CH}-CH_2-CH_3$	Low toxicity via oral, dermal, and inhalation routes; eye irritant; toxicity data not available
Ethylene glycol mono-*tert*-butyl ether (2-*tert*-butoxyethanol) [7580-85-0]	$C_6H_{14}O_2$ 118.20 $HO-CH_2-CH_2-O-\overset{\overset{\textstyle CH_3}{\mid}}{\underset{\underset{\textstyle CH_3}{\mid}}{C}}-CH_3$	Low toxicity in rats; a 5-hr exposure to 2400 ppm was lethal to rats; eye irritant
Ethylene glycol monoisobutyl ether (2-isobutoxyethanol, Isobutyl Cellosolve) [4439-24-1]	$C_6H_{14}O_2$ 118.20 $HO-CH_2-CH_2-O-CH_2-CH\overset{\textstyle CH_3}{\underset{\textstyle CH_3}{}}$	Low to moderate toxicity in test species; toxicity slightly greater than that of the *sec*- and *tert*-butyl isomers; toxicity data not available; eye irritant; teratogenicity not reported
Ethylene glycol mono-2-methylpentyl ether [2-((2-methylpentyl)oxy)ethanol, 2-Methylpentyl Cellosolve] [10137-96-9]	$C_8H_{18}O_2$ 146.26 $HO-CH_2-CH_2-O-CH_2-\overset{\overset{\textstyle CH_3}{\mid}}{CH}-CH_2-CH_2-CH_3$	Skin and eye irritant; low toxicity; toxic effects are similar to those of other monoalkyl glycol ethers; teratogenic effects not reported; LD_{50} oral (rats): 3730 mg/kg

(*continued*)

TABLE 20.1 *(Continued)*

Compound/Synonyms CAS No.	Formula/MW/Structure	Toxicity
Ethylene glycol monohexyl ether (2-(hexyloxy)ethanol, *n*-Hexyl Cellosolve) [112-25-4]	$C_8H_{18}O_2$ 146.26 $HO-CH_2-CH_2-O-CH_2-CH_2-CH_2-CH_2-CH_2-CH_3$	Skin and eye irritant; low toxicity; the toxic effects of this straight-chain glycol ether are greater than those of its 2-methylpentyl isomer; no report on its teratogenic effect; LD_{50} oral (rats): 1480 mg/kg
Diethylene glycol monohexyl ether [*n*-hexoxyethoxy ethanol, 2-((2-hexyloxy)ethoxy) ethanol, hexyl carbitol] [112-59-4]	$C_{10}H_{22}O_3$ 190.32 $HO-CH_2-CH_2-O-CH_2-CH_2-O-CH_2-CH_2-CH_2-CH_2-CH_2-CH_3$	Irritant to skin and eyes; toxicity lower than that of the hexyl ether of monoethylene glycol containing one ethereal oxygen atom; LD_{50} oral (rats): 4920 mg/kg
Ethylene glycol monophenyl ether (2-phenoxyethanol, 1-hydroxy-2-phenoxy ethane, *β*-hydroxyethyl phenyl ether, Phenyl Cellosolve) [122-99-6]	$C_8H_{10}O_2$ 138.18 $HO-CH_2-CH_2-O-$⟨benzene ring⟩	Irritant to skin and eyes; low toxicity; route of entry — ingestion and absorption through the skin; oral toxicity is somewhat greater than that of the aliphatic glycol ethers; LD_{50} oral (rats): 1260 mg/kg; weak mutagen; teratogenicity not reported
Diethylene glycol monophenyl ether (2-2-(phenoxyethoxy) ethanol, phenyl carbitol [104-68-7]	$C_{10}H_{14}O_3$ 182.24 $HO-CH_2-CH_2-O-CH_2-CH_2-O-$⟨benzene ring⟩	Irritant to skin and eyes; toxic via oral and dermal routes; low toxicity; LD_{50} oral (rats): 2140 mg/kg; teratogenic effects not reported
β-Propylene glycol monomethyl ether (3-methoxy-1-propanol) [1589-49-7]	$C_4H_{10}O_2$ 90.14 $HO-CH_2-CH_2-CH_2-O-CH_3$	Mild skin irritant; low oral and dermal toxicity in rats; toxic effects similar to those of the *α*-isomer (*see* Propylene glycol monomethyl ether); LD_{50} oral (rats): 5710 mg/kg

TABLE 20.1   *(Continued)*

Compound/Synonyms CAS No.	Formula/MW/Structure	Toxicity
β-Propylene glycol monoethyl ether (3-ethoxy-1-propanol) [111-35-3]	$C_5H_{12}O_2$ 104.17 $HO-CH_2-CH_2-CH_2-O-CH_2-CH_3$	Low toxicity in test animals dosed by oral and dermal routes; LD_{50} skin (rabbits): 2558 mg/kg; caused high maternal mortality and decrease in the birth weights of the pups (Environmental Health Research and Testing 1987); mild irritation on skin
β-Propylene glycol monoisopropyl ether (3-isopropoxy-1-propanol) [110-48-5]	$C_6H_{14}O_2$ 118.20 $HO-CH_2-CH_2-CH_2-O-CH\big\langle\!\begin{smallmatrix}CH_3\\CH_3\end{smallmatrix}$	Low oral toxicity in rats; LD_{50} oral (rats): 4000 mg/kg
α-Propylene glycol monoisobutyl ether (1-isobutoxy-2-propanol) [23436-19-3]	$C_7H_{16}O_2$ 132.23 $\begin{smallmatrix}H_3C\\H_3C\end{smallmatrix}\!\big\rangle CH-CH_2-O-CH_2-\overset{OH}{\underset{}{CH}}-CH_3$	Low oral toxicity in rats; LD_{50} oral (rats): 4300 mg/kg
Propylene glycol-n-propyl ether (1-propoxy-2-propanol) [1569-01-3]	$C_6H_{14}O_2$ 118.20 $CH_3-CH_2-CH_2-O-CH_2-\underset{OH}{CH}-CH_3$	Low toxicity observed in test animals from ingestion and absorption through the skin; LD_{50} oral (rats): 3250 mg/kg
1,2-Propanediol monomethyl ether (2-methoxy-1-propanol) [1320-67-8]	$C_4H_{10}O_2$ 90.14 $HO-CH_2-\underset{OCH_3}{CH}-CH_3$	Low toxicity; overexposure to its vapors caused increased liver weights and depression of central nervous system in rats and rabbits (Miller et al. 1984); propylene glycol has been identified as the metabolite; thus the toxicological properties are different from those of ethylene glycol monomethyl ether; showed no evidence of testicular atrophy in mice when administered by oral gavage (Doe et al. 1983; Nagano et al. 1984); irritant to eyes

that is used by the U.S. Navy in shipboard firefighting systems (Hobson et al. 1987).

Physical Properties

Colorless and odorless liquid; bp 234.5°C; mp −68°C; density 0.954; soluble in water, oil, and organic solvents.

Health Hazard

DGBE showed low toxicity in test species. Toxic symptoms are similar to those of other glycol ethers containing two ethereal oxygen atoms. The routes of entry into the body are ingestion and absorption through the skin. Hobson and co-workers (1987) investigated the subchronic oral toxicity of DGBE in rats. The toxic effects observed were lowering of food consumption, elevated liver and spleen weights, lowered red blood cells counts and lymphocyte counts, and a dose-related decrease in corpuscular hemoglobin concentration. The high doses caused pulmonary congestion. No renal damage was reported.

LD_{50} value, oral (guinea pigs): 2000 mg/kg

DGBE is an eye irritant. There is no report on teratogenicity of this compound.

Fire and Explosion Hazard

Noncombustible liquid; flash point 110°C (230°F); very low reactivity.

Disposal/Destruction

DGBE is mixed with a combustible solvent and burned in a chemical incinerator. Small amounts may be disposed down the drain with large amounts of water.

Analysis

DGBE may be analyzed by GC-FID.

20.10 MISCELLANEOUS GLYCOL ETHERS

The toxicity of glycol ethers is generally of low order. The possible health hazards caused from exposure to these compounds are presented in Table 20.1. Listed in the table are the compounds that are of commercial use. Dialkyl glycol ethers having both the alcoholic −OH groups converted to −OR (where R is an alkyl or aryl group) are excluded from the discussion. Such types of compounds do not form acid or aldehyde metabolites, and their toxicology pattern is essentially different from that of the glycol ethers containing one −OH group and one or more ethereal oxygen atom(s). Only compounds of the latter type are discussed.

As these compounds are nonreactive, nonvolatiles and those with high molecular weights are noncombustible; no flammability or explosivity data are presented.

REFERENCES

ACGIH. 1986. *Documentation of the Threshold Limit Values and Biological Exposure Indices*, 5th ed. Cincinnati, OH: American Conference of Governmental Industrial Hygienists.

Affens, W. A., and G. W. McLaren. 1972. Effect of icing inhibitor and copper passivator additives on the flammability properties of hydrocarbon fuels. *Govt. Rep. Announce. (U.S.)* 72(20): 223; cited in *Chem. Abstr. CA Index* 78(14): 86840u.

Barbee, S. J., J. B. Terrill, D. J. DeSousa, and C. C. Conaway. 1984. Subchronic inhalation toxicology of ethylene glycol monoethyl ether in the rat and rabbit. *Environ. Health Perspect.* 57: 157–63.

Berte, F., A. Bianchi, C. Gregotti, L. Bianchi, and F. Tateo. 1986. In vivo and in vitro toxicity of carbitol. *Boll. Chim. Farm.* 125(11): 401–3; cited in *Chem. Abstr. CA 107*(3): 19235e.

Carpenter, C. P., U. C. Pozzani, and C. S. Weil. 1956. *Arch. Ind. Health 14*: 114.

Cheever, K. L., D. E. Richards, W. W. Weigel, J. B. Lal, A. M. Dinsmore, and B. F. Daniel. 1988. Metabolism of bis(2-methoxyethyl)ether in the adult male rat: evaluation of the principal metabolite as a testicular toxicant. *Toxicol. Appl. Pharmacol.* 94(1): 150–59.

Comas, S. M., R. Gonzalez Palacin, and D. Vassallo. 1974. Hazards associated with perchloric acid–Butyl Cellosolve polishing solutions. *Metallography* 7(1): 45–57.

Doe, J. E. 1984. Further studies on the toxicology of glycol ethers with emphasis on rapid screening and hazard assessment. *Environ. Health Perspect.* 57: 199–206.

Doe, J. E., D. M. Samuels, D. J. Tinston, and G. A. de Silva Wickramratne. 1983. Comparative aspects of the reproductive toxicology by inhalation in rats of ethylene glycol monomethyl ether and propylene glycol monomethyl ether. *Toxicol. Appl. Pharmacol.* 69(1): 43–47.

Duprat, P., and D. Gradiski. 1979. Percutaneous toxicity of Butyl Cellosolve (ethylene glycol monobutyl ether). *IRCS Med. Sci.* 7(1): 26; cited in *Chem. Abstr. CA* 90(15): 116113b.

Eisses, K. T. 1989. Teratogenicity and toxicity of ethylene glycol monomethyl ether (2-methoxyethanol) in *Drosophila melanogaster*: involvement of alcohol dehydrogenase activity. *Teratog. Carcinog. Mutagen.* 9(5): 315–25.

Environmental Health Research and Testing, Inc. 1987. Screening of priority chemicals for reproductive hazards. Benzethonium chloride; 3-ethoxy-1-propanol; acetone. *Govt. Rep. Announce. Index (U.S.)* 89(9), Abstr. No. 922, p. 789.

Esposito, G. G., and R. G. Jamison. 1971. Rapid analysis of brake fluids uses gas–liquid chromatography. *Soc. Automot. Eng. J.* 79(1): 40–42; cited in *Chem. Abstr. CA* 74(12): 60694s.

Fairhurst, S., R. Knight, T. C. Marrs, J. W. Scawin, M. S. Spurlock, and D. W. Swanston. 1989. Percutaneous toxicity of ethylene glycol monomethyl ether and of dipropylene glycol monomethyl ether in the rat. *Toxicology* 57(2): 209–15.

Foster, P. M. D., D. M. Creasy, J. R. Foster, and T. J. B. Gray. 1984. Testicular toxicity produced by ethylene glycol monomethyl ether and monoethyl ethers in the rat. *Environ. Health Perspect.* 57: 207–17.

Foster, P. M. D., D. M. Blackburn, R. B. Moore, and S. C. Lloyd. 1986. Testicular toxicity of 2-methoxyacetaldehyde, a possible metabolite of

ethylene glycol monomethyl ether in the rat. *Toxicol. Lett.* 32(1–2): 73–80.

Gage, J. C. 1970. *Brit. J. Ind. Med.* 27: 1.

Gaunt, I. F., J. Colley, P. Grasso, A. B. G. Lansdown, and S. D. Gangolli. 1968. Short-term toxicity of diethylene glycol monoethyl ether in the rat, mouse and pig. *Food Cosmet. Toxicol.* 6(6): 689–705; cited in *Chem. Abstr. CA* 71(5): 20547c.

Ghanayem, B. I., L. T. Burka, and H. B. Matthews. 1987a. Metabolic basis of ethylene glycol monobutyl ether (2-butoxyethanol) toxicity: role of alcohol and aldehyde dehydrogenases. *J. Pharmacol. Exp. Ther.* 242(1): 222–31.

Ghanayem, B. I., P. C. Blair, M. B. Thompson, R. R. Maroupot, and H. B. Matthews. 1987b. Effect of age on the toxicity and metabolism of ethylene glycol monobutyl ether (2-butoxyethanol) in rats. *Toxicol. Appl. Pharmacol.* 91(2): 222–34.

Grant, D., S. Sulsh, H. B. Jones, S. D. Gangolli, and W. H. Butler. 1985. Acute toxicity and recovery in the hemopoietic system of rats after treatment with ethylene glycol monomethyl and monobutyl ethers. *Toxicol. Appl. Pharmacol.* 77(2): 187–200.

Hardin, B. D., P. T. Goad, and J. R. Burg. 1984. Developmental toxicity of four glycol ethers applied cutaneously to rats. *Environ. Health Perspect.* 57: 69–74.

Hardin, B. D., R. L. Schuler, J. R. Burg, G. M. Booth, K. P. Hazelden, K. M. MacKenzie, V. J. Piccirillo, and K. N. Smith. 1987. Evaluation of 60 chemicals in preliminary developmental toxicity test. *Teratog. Carcinog. Mutagen* 7(1): 29–48.

Hobson, D. W., J. F. Wyman, L. H. Lee, R. H. Bruner, and D. E. Uddin. 1987. Subchronic toxicity of diethylene glycol monobutyl ether administered orally to rats. *Govt. Rep. Announce. Index (U.S.)* 88(14), Abstr. No. 836, p. 751.

Johansson, G., H. Kronberg, P. H. Naeslund, and M. B. Nordqvist. 1986. Toxicokinetics of inhaled 2-butoxyethanol (ethylene glycol monobutyl ether) in man. *Scand. J. Work. Environ. Health* 12(6): 594–602; cited in *Chem. Abstr. CA* 106(15): 114796b.

Krasavage, W. J. 1986. Subchronic oral toxicity of ethylene glycol monobutyl ether in male rats. *Fundam. Appl. Toxicol.* 6(2): 349–55.

Lamb, J. C., IV, D. K. Gulati, V. S. Russel, L. Hommel, and P. S. Sabharwal. 1984. Reproductive toxicity of ethylene glycol monoethyl ether tested by continuous breeding of CD-1 mice. *Environ. Health Perspect.* 57: 85–90.

Mebus, C. A., and F. Welsch. 1989. The possible role of one-carbon moieties in 2-methoxyethanol and 2-methoxyacetic acid-induced developmental toxicity. *Toxicol. Appl. Pharmacol.* 99(1): 98–109.

Mebus, C. A., F. Welsch, and P. K. Working. 1989. Attenuation of 2-methoxyethanol-induced testicular toxicity in the rat by simple physiological compounds. *Toxicol. Appl. Pharmacol.* 99(1): 110–21.

Melnick, R. L. 1984. Toxicities of ethylene glycol and ethylene glycol monoethyl ether in Fischer 344/N rats and B6C3F mice. *Environ. Health Perspect.* 57: 147–55.

Miller, R. R., R. E. Carreon, J. T. Young, and M. J. McKenna. 1982. Toxicity of methoxyacetic acid in rats. *Fundam. Appl. Toxicol.* 2(4): 158–60.

Miller, R. R., E. A. Hermann, J. T. Young, T. D. Landry, and L. L. Calhoun. 1984. Ethylene glycol monomethyl ether and propylene glycol monomethyl ether: metabolism, disposition, and subchronic inhalation toxicity studies. *Environ. Health Perspect.* 57: 233–39.

Moss, E. J., L. V. Thomas, M. W. Cook, D. G. Walters, P. M. D. Foster, D. M. Creasy, and T. J. B. Gray. 1985. The role of metabolism in 2-methoxyethanol-induced testicular toxicity. *Toxicol. Appl. Pharmacol.* 79(3): 480–89.

Nagano, K., E. Nakayama, H. Obayashi, T. Nishizawa, H. Okuda, and K. Yamazaki. 1984. Experimental studies on toxicity of ethylene glycol alkyl ethers in Japan. *Environ. Health Perspect.* 57: 75–84.

Nelson, B. K., W. S. Brightwell, J. V. Setzer, and T. L. O'Donohue. 1984. Reproductive toxicity of the industrial solvent 2-ethoxyethanol in rats and interactive effects of ethanol. *Environ. Health Perspect.* 57: 255–59.

NIOSH. 1984. *Manual of Analytical Methods*, 3rd ed. Cincinnati, OH: National Institute for Occupational Safety and Health.

NIOSH. 1986. *Registry of Toxic Effects of Chemical Substances*, ed. D. V. Sweet. Washington, DC: U.S. Government Printing Office.

Oudiz, D. J., H. Zenick, R. J. Niewenhuis, and P. M. McGinnis. 1984. Male reproductive toxicity and recovery associated with acute ethoxyethanol exposure in rats. *J. Toxicol. Environ. Health* 13(4–6): 763–75.

Savolainen, H. 1980. Glial cell toxicity of ethylene glycol-monomethyl ether vapor. *Environ. Res.* 22(2): 423–30.

Wahlberg, J. E., and A. Boman. 1979. Comparative percutaneous toxicity of ten industrial solvents in the guinea pig. *Scand. J. Work. Environ. Health* 5(4): 345–51; cited in *Chem. Abstr. CA* 92(23): 192183t.

Werner, H. W., J. L. Mitchell, J. W. Miller, and W. F. von Oettingen. 1943. *J. Ind. Hyg. Toxicol.* 25: 157.

Wier, P. J., S. C. Lewis, and K. A. Taul. 1987. A comparison of developmental toxicity evident at term to postnatal growth and survival using ethylene glycol monoethyl ether, ethylene glycol monobutyl ether, and ethanol. *Teratog. Carcinog. Mutagen.* 7(1): 55–64.

21

HALOETHERS

21.1 GENERAL DISCUSSION

Haloethers are ethers containing halogen atoms. These compounds contain halogen-substituted alkyl, alkenyl, cycloalkyl, and/or aryl groups linked by a bridging oxygen atom. There can be one or more halogen atoms in the molecule, and they can be present on either or both sides of the oxygen bridge. Some examples are

$$ClCH_2—CH_2—O—CH_2—CH_2Cl$$

[bis(2-chloroethyl)ether]

(2-chloro-1,1,2-trifluoroethyl difluoromethyl ether)

Low aliphatic chloroethers are useful solvents for many purposes. The fluoroethers are used as clinical anesthetics.

Flammability

The flammable characteristics of haloethers are very much different from those of the ether class of compounds. Halogen substitutions make ether molecules less flammable or nonflammable. Among C_2 and C_3 chloroethers, an increase of one chlorine atom increases the flash point by 20°C; one extra —CH_2 unit contributes an additional 4 to 5°C (Table 21.1). Such a pattern, however, varies markedly depending on the carbon chain length, unsaturation, and position of the chlorine atom in the molecule. Among the haloethers, bis(chloromethyl)-ether, 2-chloroethyl vinyl ether, bis(2-chloroethyl) ether, chloromethyl methyl ether, chloromethyl ethyl ether, and a few others are flammable liquids with flash point (open cup) <37.7°C (100°F). Other C_4 and C_5 haloethers are combustible liquids. Those with higher carbon numbers are noncombustible liquids or solids. C_4 haloethers containing six or more halogen atoms are almost noncombustible.

The explosion hazards of low aliphatic ethers caused by peroxide formation are not manifested by the haloethers. The halogens inhibit the ether oxidation to peroxides.

Health Hazard

Aliphatic chloro- and bromoethers exhibit toxicity apparently different from that of the fluoroethers. Inhalation of the latter compounds can produce anesthesia similar to

TABLE 21.1   Effect of Increase in Chlorine Substitution and Carbon Number on Flammability of C_2–C_4 Chloroethers

C_2–C_4 Chloroether/ CAS No.	Structure	Flash Point (Open Cup) (°C)
Chloromethyl methyl ether [107-30-2]	$Cl-CH_2-O-CH_3$	15
Chloromethyl ethyl ether [3188-13-4]	$Cl-CH_2-O-CH_2-CH_3$	19
Bis(chloromethyl)ether [542-88-1]	$Cl-CH_2-O-CH_2-Cl$	35
2-Chloroethyl vinyl ether [110-75-8]	$Cl-CH_2-CH_2-O-CH=CH_2$	27
Bis(2-chloroethyl)ether [111-44-4]	$Cl-CH_2-CH_2-O-CH_2-CH_2-Cl$	55

that of the lower aliphatic ethers. Such anesthetic effects of fluoroethers on humans are manifested at high concentrations in air, on the order of several thousand ppm (1.5–2% by volume). On the other hand, anesthetic actions of ethers in humans occur at higher concentrations (4–6% by volume of air).

Lower aliphatic chloro- and bromoethers can be injurious to lungs. Many of these are cancer-causing to lungs in animals or humans. It is the carcinogenicity that makes some of these haloethers compounds of primary health concern.

Aromatic chloroethers are toxic by inhalation, ingestion, and skin absorption only at high dosages. The toxic effects should be attributed primarily to the chlorine content and to a lesser extent on the aromaticity of the molecule. The role of ethereal oxygen atoms is not clear. C—O bond scission at the ether bridge can produce fragmented moieties with active sites. Table 21.2 presents the chlorine contents of some aromatic chloroethers and their lethal doses in guinea pigs. A crude pattern of increase in chlorine contents associated with a decrease in the LD_{50} values may be observed. The inconsistency at certain numbers may be due to error in the experimental data or the presence of isomers and/or trace polymeric contaminants.

TABLE 21.2   Chlorine Contents and Toxicity of Aromatic Chloroethers

Compound/ CAS No.	Chlorine Content (%)	LD_{50} Oral (Guinea Pigs) (mg/kg)
Diphenyl ether [101-84-8]	0	3370[a]
Monochlorodiphenyl ether [55398-86-2]	17.35	600
Dichlorodiphenyl ether [28675-08-3]	29.28	1000
Tetrachlorodiphenyl ether [31242-94-1]	46.11	50
Pentachlorodiphenyl ether [42279-29-8]	51.84	100
Hexachlorodiphenyl ether [55720-99-3]	56.52	50

[a]LD_{50} oral (rats)

21.2   BIS(CHLOROMETHYL)ETHER

EPA Classified Hazardous Waste; RCRA Waste Number P016; DOT Label: Poison and Flammable Liquid, UN 2249

Formula $(CH_2Cl)_2O$; MW 114.96; CAS [542-88-1]

Structure and functional group: $Cl-CH_2-O-CH_2-Cl$, a chloro-substituted aliphatic ether, the first member of a homologous series of haloethers

Synonyms: chlorodimethyl ether; dimethyl-1:1-dichloroether; oxybis(chloromethane); chloro(chloromethoxy)methane

Uses and Exposure Risk

Bis(chloromethyl)ether (BCME) is used as an intermediate in anion exchange quaternary resins. Its use as a chloromethylation reagent in industry is being discontinued because of its high carcinogenic properties. Exposure risks associated with this compound can arise during the use or production of chloromethyl methyl ether when the latter compound comes in contact with traces of water in the presence of hydrogen or hydroxyl ions. It may occur in trace amounts in chloromethyl methyl ether.

Physical Properties

Colorless liquid with a suffocating odor; bp 106°C; mp −41.5°C; density 1.315 at 20°C; soluble in most organic solvents; decomposes in water to HCl and formaldehyde.

Health Hazard

BCME is a highly toxic and carcinogenic compound. The inhalation toxicity and carcinogenicity of this compound are greatest among the haloethers.

Exposure to its vapors can cause irritation of the eyes, nose, and throat in humans. The primary target organ is the lungs. Inhalation of 100 ppm of this compound in air for a few minutes can cause death to humans. Irritation of the eyes can be moderate to severe and conjunctival. The acute oral and dermal toxicity of this compound, however, is moderate and comparable to that of bis(2-chloroethyl)ether:

LC_{50} value, inhalation (rats): 33 mg (7 ppm)/m^3/7 hr

LD_{50} value, oral (rats): 210 mg/kg (NIOSH 1986)

Tests on animals have confirmed the carcinogenic action of BCME. It is carcinogenic by inhalation and by subcutaneous and skin applications. In humans it can produce lung cancer — a fact that is now well established. Tests on rats indicate that exposure to 0.1-ppm concentrations in air for 6 hours per day for 6 months produced tumors in the nose and lungs. Subcutaneous and skin applications produced tumors at the site of application.

Exposure Limits

TLV-TWA 0.0047 mg (0.001 ppm)/m^3 (Human Carcinogen) (ACGIH); Human and Animal Carcinogen-Sufficient Evidence (IARC); Carcinogen (OSHA).

Fire and Explosion Hazard

Flammable liquid; flash point (closed cup) 35°C (95°F) (calculated); vapor forms explosive mixtures with air; the LEL and UEL values not reported; fire-extinguishing agent: dry chemical, CO_2, or foam. BCME reacts with water, decomposing to HCl and formaldehyde. Decomposition also occurs with moist air.

Disposal/Destruction

Chemical incineration is the most effective method of its destruction (Aldrich 1997). In the laboratory it may be converted to nontoxic products by treatment with a concentrated solution of ammonia for several minutes.

Analysis

BCME may be analyzed by GC using an FID. A fused-silica capillary column such as DB-1 may be suitable. It can be analyzed by GC/MS techniques.

21.3   BIS(2-CHLOROETHYL)ETHER

EPA Priority Pollutant; RCRA Toxic Waste, Number U025; DOT Label: Poison, UN 1916

Formula $(C_2H_4Cl)_2O$; MW 143.02; CAS [111-44-4]

Structure and functional group:

$$Cl-CH_2-CH_2-O-CH_2-CH_2-Cl$$

chloro-substituted aliphatic ether

Synonyms: 2,2'-dichloroethyl ether; di(2-chloroethyl)ether; dichloroethyl oxide; 1, 1'-oxybis(2-chloroethane)

Uses and Exposure Risk

Bis(2-chloroethyl)ether (BCEE) is used as a scouring agent for textiles; as a dewaxing agent for lubricating oils; as a soil fumigant; as a solvent for resins, oils, and lacquers; and in organic synthesis.

Physical Properties

Colorless, clear liquid with a pungent smell; boils at 122°C; freezes at −50°C; density 1.22; soluble in most organic solvents, insoluble in water.

Health Hazard

BCEE is an acute toxic compound with cancer-causing properties. The toxic routes are inhalation, ingestion, and absorption through the skin. Exposure to its vapors caused irritation of the eyes, nose, and respiratory tracts in test animals. Exposure to a concentration of 250 ppm in air for 4 hours proved lethal to rats (ACGIH 1986). At a lower concentration, it caused delayed death following damage to lungs. Other organs affected to a lesser degree were the liver, kidneys, and brain.

The irritant action of BCEE on the eyes varied from mild to severe. In humans, contact of pure liquid can cause conjunctival irritation and injury to the cornea. On the skin, its irritation effect is mild. However, on prolonged contact, the liquid may be absorbed through the skin and manifest toxic effects.

Ingestion of this compound in small amounts may produce nausea and vomiting. An oral dose of 50–75 mL is expected to be fatal to humans.

LC_{50} value, inhalation (rats): 330 mg (56 ppm)/m^3/4 hr

LD_{50} value, oral (rats): 75 mg/kg

LD_{50} value, skin (guinea pigs): 300 mg/kg

BCEE is a cancer-causing compound. There is sufficient evidence of its carcinogenicity in animals. A dose of 33 g/kg given orally to mice over 79 weeks produced tumors in liver and blood. When applied subcutaneously, a smaller dose, 2.4 g/kg/60 weeks, produced tumors at the site of application.

Exposure Limits

TLV-TWA 30 mg/m^3 (5 ppm) (ACGIH); STEL 60 mg/m^3 (10 ppm) (ACGIH); ceiling 90 mg/m^3 (15 ppm) (OSHA); IDLH 250 ppm (NIOSH); Carcinogen — Animal Sufficient Evidence (IARC).

Fire and Explosion Hazard

Combustible liquid; flash point (open cup) 55°C (131°F); vapor pressure 0.4 torr at 20°C; autoignition temperature 369°C (696°F); the vapor forms explosive mixtures with air; LEL value 2.7% by volume of air, UEL value not reported; fire-extinguishing agent: dry chemical, foam, or CO_2; use a water spray to keep fire-exposed containers cool, to disperse the vapor, and to blanket the fire. The reactivity of this compound toward most chemicals is low.

Disposal/Destruction

BCEE is burned in a chemical incinerator equipped with an afterburner and scrubber. In the laboratory BCEE in the wastes can be destroyed by treatment with concentrated ammonia solution for several minutes. Treatment with sodium phenate and sodium methoxide is reported to be effective (Castegnaro et al. 1984). Biodegradability of this compound was reported by Dojlido (1979).

Analysis

GC and GC/MS are the most widely used techniques for the analysis of bis(2-chloroethyl)ether. Methods for analysis of this compound in wastewaters and soils and solid wastes along with other haloethers are described in EPA procedures (U.S. EPA 1992, Methods 611, 625; 1986, Methods 8010 and 8270).

Air analysis is performed by passing 2–15 L of air over coconut shell charcoal. The analyte is desorbed with carbon disulfide and injected onto a GC equipped with an FID (NIOSH 1984, Method 1004). Columns suitable for the purpose are 10% FFAP on 80/100-mesh DMCS Chromosorb W-AW, SP-2100 with 0.1% Carbowax 1500, 10% SP-1000 on 80/100-mesh Supelcoport, or a DB-1 fused-silica capillary column.

21.4 CHLOROMETHYL METHYL ETHER

EPA Classified Toxic Waste, RCRA Waste Number U046; DOT Label: Flammable Liquid and Poison, UN 1239

Formula C_2H_5ClO; MW 80.52; CAS [107-30-2]

Structure and functional group: $Cl-CH_2-O-CH_3$, an aliphatic monochloro ether

Synonyms: monochlorodimethyl ether; methyl chloromethyl ether; chloromethoxymethane

Uses and Exposure Risk

Chloromethyl methyl ether (CMME) is used as a methylating agent in the synthesis of chloromethylated compounds.

Physical Properties

Colorless liquid; boils at 59.5°C; melts at −103.5°C; density 1.0605 at 20°C; soluble in alcohol, ether, and other organic solvents; decomposes with water and hot alcohol.

Health Hazard

CMME is a moderately toxic compound with cancer-causing action. The inhalation toxicity of this compound, like that of other lower aliphatic haloethers, is greater than the acute oral or dermal toxicity. It is less toxic than dichloroethers. Exposure to this compound can cause irritation of the eyes, nose, and throat. At high concentrations, lung injury can occur.

LC_{50} value, inhalation (rats): 180 mg (55 ppm)/m^3/7 hr

LD_{50} value, oral (rats): 817 mg/kg

CMME was found to be carcinogenic in test animals by inhalation and subcutaneous applications. It produced lung and endocrine tumors. In humans its exposure can cause lung cancer.

Exposure Limits

TLV-TWA — none assigned, A2-Suspected Human Carcinogen (ACGIH 1988); Carcinogen–Human Limited Evidence (IARC); Carcinogen (OSHA).

Fire and Explosion Hazard

Flammable liquid; flash point (open cup) 15°C (60°F); vapor density 2.77 (air = 1); vapor forms explosive mixtures with air; the LEL and UEL values not reported; fire-extinguishing agent: dry chemical, CO_2, or foam. CMME may react with water, decomposing to HCl.

Disposal/Destruction

CMME may be burned in a chemical incinerator equipped with an afterburner and scrubber. Wastes containing chloroether residues may be treated with concentrated ammonia solution.

Analysis

CMME is analyzed by GC and GC/MS techniques similar to those used for bis(2-chloroethyl)ether.

21.5   2-CHLOROETHYL VINYL ETHER

EPA Priority Pollutant; RCRA Toxic Waste, Number U042

Formula C_4H_7ClO; MW 106.56; CAS [110-75-8]

Structure and functional group: $Cl-CH_2-CH_2-O-CH=CH_2$, aliphatic haloether containing an olefinic double bond

Synonyms: vinyl 2-chloroethyl ether; (2-chloroethoxy)ethene

Uses and Exposure Risk

2-Chloroethyl vinyl ether is used to produce sedatives, anesthetics, and cellulose ethers.

Physical Properties

Colorless liquid; bp 109°C; density 1.0525 at 15°C; soluble in alcohols, ethers, and most organic solvents, insoluble in water.

Health Hazard

2-Chloroethyl vinyl ether is moderately toxic to humans by inhalation and ingestion. Exposure to its vapors can produce irritation of the eyes, nose, and lungs. Rats exposed to its vapors at 250 ppm concentration in air died 4 hours after exposure. Pure liquid is an irritant to the skin. The liquid may be absorbed through the skin. The dermal toxicity, however, is very low.

LD_{50} value, oral (rats): 250 mg/kg

The carcinogenicity of this compound is not documented. However, drawing a similarity with other low aliphatic haloethers, this compound at high dosage may exhibit carcinogenicity to animals.

Exposure Limit

No exposure limit is set for this compound.

Fire and Explosion Hazard

Flammable liquid; flash point (open cup) 27°C (80°F); vapor density 3.7 (air = 1); vapors may form explosive mixtures with air; fire-extinguishing agent: dry chemical, foam, or CO_2.

It hydrolyzes with dilute acids to form acetaldehyde and ethylene chlorohydrin. Reactions with concentrated acids can become violent.

Disposal/Destruction

Burning in a chemical incinerator is the most suitable method of destruction.

Analysis

2-Chloroethyl vinyl ether in potable and nonpotable waters and solid and hazardous wastes may be analyzed by EPA Methods based on GC and GC/MS instrumentation (U.S. EPA 1992; 1997: Methods 611, 625, 8010, 8270) using a purge and trap or thermal desorption technique. Characteristic masses to identify this compound by GC/MS using electron-impact ionization are 106, 63, and 65.

21.6   BIS(2-CHLOROISOPROPYL) ETHER

EPA Priority Pollutant; RCRA Toxic Waste, Number U027; DOT Label: Poison B and Corrosive Material, UN 2490

Formula $(C_3H_6Cl)_2O$; MW 171.08; CAS [108-60-1]

Structure and functional group:

$$CH_3 \quad CH_3$$
$$ClCH_2-CH-O-CH-CH_2Cl$$

an aliphatic haloether containing two chloroisopropyl groups

Synonyms: 2,2'-dichloroisopropyl ether; dichlorodiisopropyl ether; bis(β-chloroisopropyl) ether; bis(2-chloro-1-methylethyl) ether

Uses and Exposure Risk

Bis(2-chloroisopropyl)ether is used as a solvent for resins, waxes, and oils, and in organic synthesis.

Physical Properties

Colorless liquid with characteristic odor; bp 187°C; density 1.4505; soluble in alcohols, ethers, and most organic solvents, insoluble in water.

Health Hazard

Bis(2-chloroisopropyl)ether exhibited moderately toxic and carcinogenic actions in test animals. The acute inhalation toxicity of this compound is considerably lower than those of bis(chloromethyl)ether and bis (chloroethyl)ether. Exposure to its vapors can cause irritation of the eyes and upper respiratory tract. Inhalation of 700 ppm of this compound in air for 5 hours proved fatal to rats.

The oral toxicity of bis(2-chloroisopropyl) ether in rats was found to be moderate, with an LD_{50} value of 240 mg/kg.

The compound is carcinogenic to animals. Although there is no evidence of its carcinogenicity in humans, exposure may cause lung cancer.

Exposure Limit

No exposure limit is set for this compound. Carcinogen — Animal Limited Evidence (IARC).

Fire and Explosion Hazard

Combustible liquid; flash point (open cup) 85°C (185°F); vapor density 5.9 (air = 1); low reactivity.

Disposal/Destruction

The compound is mixed with a combustible solvent and burned in a chemical incinerator.

Analysis

Bis(2-chloroisopropyl)ether in wastewater and soil and sludges can be analyzed by EPA Methods 611, 625, 8010, and 8270 (U.S. EPA 1992, 1997). These methods are based on GC or GC/MS techniques. In the latter technique the characteristic ions to identify this compound have the *m/z* values 45, 77, and 121 (electron-impact ionization mode).

21.7 ETHRANE

Formula $C_3H_2ClF_5O$; MW 184.50; CAS [13838-16-9]

Structure and functional group:

$$F \quad F \qquad F$$
$$Cl-C-C-O-C-H$$
$$H \quad F \qquad F$$

exhibits low reactivity due to a greater degree of halogen substitution

Synonyms: Enflurane; Efrane; methylflurether; 2-chloro-1,1,2-trifluoroethyl difluoromethyl ether; 2-chloro-1-(difluoromethoxy)-1,1,2-trifluoroethane

Uses and Exposure Risk

Ethrane is widely used clinically as an anesthesia (by inhalation). Workers in operating rooms are susceptible to inhaling this compound at low concentrations.

Physical Properties

Clear and colorless liquid with a pleasant odor; bp 56.5°C; density 1.5167 at 25°C; soluble in most organic solvents, mixes with oils and fats, insoluble in water.

Health Hazard

The acute toxicity of ethrane by inhalation, ingestion, or intraperitoneal or subcutaneous applications in rodents was found to be low to very low at concentrations of <1000 ppm in air or dosages below 1000 mg/kg. In humans it causes anesthesia at 1.5–2% concentrations (by volume of air). Exposure to concentrations above this level can be dangerous. The target organs are the central nervous system, cardiovascular system, kidney, and bladder. The symptoms are anesthesia, respiratory depression, and seizure. Hypotension can occur due to its action on the cardiovascular system. Exposure to a 1% concentration in air for 6 hours caused anuria in humans.

LC_{50} value, inhalation (mice): 8100 ppm/
  3 hr
LD_{50} value, oral (mice): 5000 mg/kg

The pure liquid can cause mild to moderate irritation of the eyes.

Inhalation of ethrane vapors produced teratogenic effects in mice and rats. These effects pertained to specific developmental abnormalities in the urogenital, musculoskeletal, and central nervous systems. These reproductive effects, however, were manifested at high exposure levels in the range of LC_{50} concentrations.

Under the conditions of its use, the concentrations of ethrane in air should be too low to produce any adverse health effects on humans. However, it should be borne in mind that this compound is highly volatile (vapor pressure 175 torr at 20°C) and its concentration in air can go up from a spill or improper handling in confined space.

Inhalation of its vapors resulted in lung and liver tumors in mice. There is no evidence of its carcinogenic actions in humans.

Exposure Limit

TLV-TWA 570 mg/m^3 (75 ppm) (ACGIH).

Fire and Explosion Hazard

Noncombustible liquid; flash point >94°C (200°F); low reactivity. Pressure buildup in a closed bottle may occur at elevated temperatures.

Disposal/Destruction

Ethrane is mixed with a combustible solvent and burned in a chemical incinerator.

Analysis

Suitable analytical techniques are GC and GC/MS. Air analysis can be performed by charcoal adsorption, desorption with carbon disulfide, and injection onto a GC column. Alternatively, thermal desorption of charcoal instead of solvent elution may be carried out.

21.8 METHOXYFLURANE

Formula $C_3H_4Cl_2F_2O$; MW 164.97; CAS [76-38-0]

Structure and functional group:

$$Cl-\overset{\overset{\displaystyle H}{|}}{\underset{\underset{\displaystyle Cl}{|}}{C}}-\overset{\overset{\displaystyle F}{|}}{\underset{\underset{\displaystyle F}{|}}{C}}-O-CH_3$$

low reactivity due to high degree of halo-substitution

Synonyms: 2,2-dichloro-1,1-difluoroethyl methyl ether; 2,2-dichloro-1,1-difluoro-1-methoxyethane; 1,1-difluoro-2,2-dichloroethyl methyl ether; Penthrane; Pentrane; Methoxane

TABLE 21.3 Toxicity and Flammability of Miscellaneous Haloethers

Compound/Synonyms/CAS No.	Formula/MW/Structure	Toxicity	Flammability
Chloromethyl ethyl ether [ethyl chloromethyl ether, monochlorodimethyl ether, (chloromethoxy)ethane, ethoxychloromethane, ethoxymethyl chloride] [3188-13-4]	C_3H_7ClO 94.55 $H_3C-CH_2-O-CH_2-Cl$	Acute toxicity from inhalation and ingestion low; toxicity data not available; LC_{50} (mice) for 2-hr exposure is expected to be ~400 ppm; no report on any carcinogenic studies; expected to cause lung cancer; DOT Label: Poison, UN 2254	Flammable liquid; flash point (open cup) 19°C (67°F); vapor density 3.26 (air = 1); vapor forms explosive mixtures with air, flammable limits not reported; fire-extinguishing agent: dry chemical, foam, or CO_2; DOT Label: Flammable Liquid, UN 2254
Bis(1-chloroethyl) ether [1,1′-oxybis(1-chloroethane] [6986-48-7]	$C_4H_8Cl_2O$ 143.02 $H_3C-\underset{Cl}{CH}-O-\underset{Cl}{CH}-CH_3$	Low inhalation toxicity; toxicity data not available; expected to cause anesthesia at concentrations >2% in air by volume; caused cancers in mice at the site of application when given subcutaneously	Combustible liquid; flash point data not available
Bis(4-chlorobutyl) ether [4,4′-dichlorodibutyl ether; 1,1′-oxydi-4-chlorobutane; oxybis(4-chlorobutane)] [6334-96-9]	$C_8H_{16}Cl_2O$ 199.14 $Cl-CH_2-CH_2-CH_2-CH_2-$ $O-CH_2-CH_2-CH_2-CH_2-Cl$	Low oral toxicity in mice; target organs—central nervous system, liver, and gastrointestinal tract; caused muscle contraction and ulceration; LD_{50} oral (mice): 1250 mg/kg	Noncombustible liquid
Bis(pentabromophenyl)ether [decabromediphenyl ether; decabromobiphenyl oxide; 1,1′-oxybis(2,3,4,5,6-pentabromobenzene); Saytex 102] [1163-19-5]	$C_{12}Br_{10}O$ 959.22 	Teratogenic: caused postimplantation mortality in rats; some evidence of carcinogenicity in test animals, oral feeding resulted in liver tumors in rats	Noncombustible solid

(continued)

TABLE 21.3 (Continued)

Compound/Synonyms/CAS No.	Formula/MW/Structure	Toxicity	Flammability
Bis(pentafluoro)ethyl ether (pentafluoroether) [358-21-4]	$C_4F_{10}O$ 254.04	Low to very low toxicity; however, at high dosages inhalation or ingestion can be harmful; a 2-hr exposure to its vapors at 17,000 ppm in air caused hemorrhage in rats; injury to gastrointestinal tract occurred when rats were fed very high dosages, >10,000 mg/kg; health hazard to humans should be very low	Noncombustible
Dichlorophenyl ether (dichlorodiphenyl oxide) [28675-08-3]	$C_{12}H_8Cl_2O$ 239.10	Acute toxicity low to moderate in test animals; toxic routes — inhalation and ingestion; an oral dose of 1000 mg/kg was fatal to guinea pigs; exposure limits: TLV-TWA in air 0.5 mg/m³ (OSHA)	Noncombustible liquid
1-Chloro-2,2,2-trifluoroethyl difluoromethyl ether (2-chloro-2-(difluoromethoxy)-1,1,1-trifluoroethane, isoflurane, forane) [26675-46-7]	$C_3H_2ClF_5O$ 184.50	Exposure to high concentrations can produce anesthetic effect; inhalation exceeding 3% concentration in air by volume is expected to cause unconsciousness in humans; toxic to fetus in rats at an exposure level of 10,000 ppm for 6 hr	Noncombustible liquid
2-Chloro-1,1,2-trifluoroethyl methyl ether [425-87-6]	$C_3H_4ClF_3O$ 148.52	Eye and skin irritant; low to moderately toxic via absorption through skin; LD_{50} skin (rabbits): 200 mg/kg	Noncombustible liquid

Name (synonym) [CAS]	Formula	Properties	
Monochlorophenyl ether (monochlorodiphenyl oxide) [55398-86-2]	$C_{12}H_9ClO$ 204.66 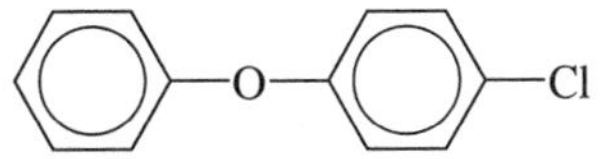	Low to moderate toxicity in test species; an oral dose of 600 mg/kg was lethal to guinea pigs	Noncombustible liquid
Pentachlorophenyl ether (pentachlorodiphenyl oxide) [42279-29-8]	$C_{12}H_5Cl_5O$ 342.42 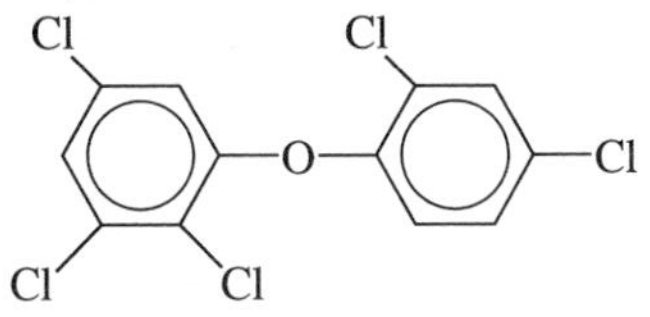 (mixture of isomers)	Moderate oral toxicity; lethal dose on guinea pigs by ingestion, 100 mg/kg	Noncombustible solid
Hexachlorophenyl ether (hexachlorodiphenyl oxide) [55720-99-5]	$C_{12}H_4Cl_6O$ 376.86 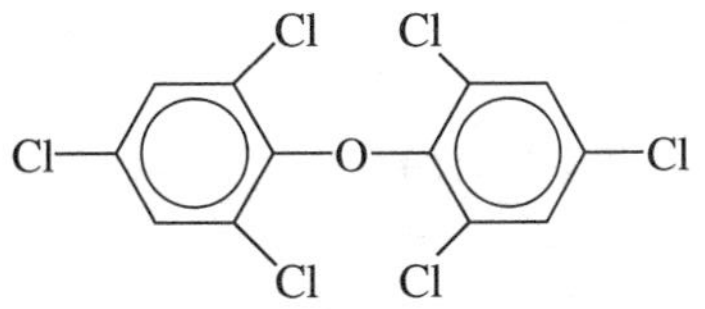	Moderate to high oral toxicity; lethal dose in guinea pigs 50 mg/kg; in humans, skin contact can cause an acne-type dermatitis and itching; cumulative effects from prolonged exposure can lead to liver damage; TLV-TWA 0.5 mg/m^3 (ACGIH)	Noncombustible solid
Tetrachlorophenyl ether (tetrachlorodiphenyl oxide) [31242-94-1]	$C_{12}H_6Cl_4O$ 307.98 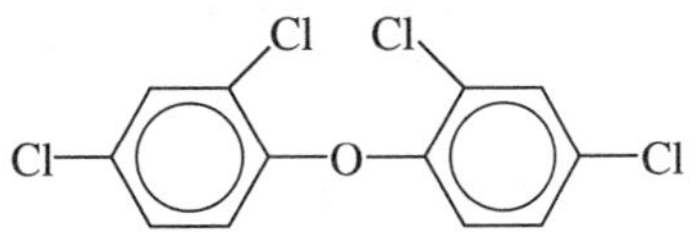	Moderate to high oral toxicity; ingestion of 50 mg/kg was lethal in guinea pigs	Noncombustible solid

(continued)

403

TABLE 21.3 (Continued)

Compound/Synonyms/CAS No.	Formula/MW/Structure	Toxicity	Flammability
2,2,2-Trifluoroethyl vinyl ether [(2,2,2-trifluoroethoxy) ethene, fluroxene] [406-90-6]	$C_4H_5F_3O$ 126.09 $F-\underset{\underset{F}{\vert}}{\overset{\overset{F}{\vert}}{C}}-CH_2-O-CH=CH_2$	Exhibits inhalation toxicity only at high exposure levels; inhalation of its vapors produces anesthetic effects; exposure to 4% concentrations in air for 1 hr was lethal to mice; in humans, inhalation of 2% concentration for 1.5 hr may impair liver, showing jaundice-like symptoms; damaging effects on embryo were observed in animals from a short exposure to unusually high concentrations of 8% in air; ingestion of 10–12 mL may be fatal to humans; tested positive in histidine reversion–Ames test for mutagenicity	Noncombustible liquid; may explode on heating; boils at 45°C

Uses and Exposure Risk

Methoxyflurane is used as a clinical anesthesia (inhalation).

Physical Properties

Clear and colorless liquid with a pleasant odor; bp 105°C; mp −35°C; density 1.4226 at 20°C; miscible with organic solvents, insoluble in water.

Health Hazard

Methoxyflurane exhibited low to very low acute toxicity via inhalation, slightly lower than that of ethrane. Oral toxicity was low to moderate depending on the species. Inhalation of its vapors at 1.5–2% by volume concentrations in air can cause anesthesia in humans. The toxic symptoms are similar to those of ethrane, and the target organs are primarily the central nervous system, kidney, and liver. At subanesthetic concentrations of 0.3–0.5% by volume in air, its exposure to humans for 1 hour resulted in the onset of low toxicity. The sites of biological effects were in the kidney.

LC_{50} value, inhalation (mice): 17500 ppm/ 2 hr

LD_{50} value, oral (mammals): 3600 mg/kg

The liquid may be an irritant to the eyes. The teratomeric properties of this compound were observed in rats and mice. The symptoms were embryo deaths and developmental abnormalities in the urogenital and musculoskeletal systems.

No carcinogenic actions in animals or humans have been reported. The histidine reversion–Ames test for mutagenicity was inconclusive.

Exposure Limit

No exposure limit is set. Based on comparison with related compounds, a TLV-TWA of 675 mg/m^3 (100 ppm) is recommended.

21.9   MISCELLANEOUS HALOETHERS

See Table 21.3.

REFERENCES

ACGIH. 1986. *Documentation of the Threshold Limit Values and Biological Exposure Indices*, 5th ed. Cincinnati, OH: American Conference of Governmental Industrial Hygienists.

Aldrich. 1997. *Aldrich Catalog*. Milwaukee: Aldrich Chemical Company.

Castegnaro, M., M. Alvarez, M. Iovu, E. B. Sansone, G. M. Telling, and D. T. Williams. 1984. Laboratory decontamination and destruction of carcinogens in laboratory wastes: some haloethers. *IARC Sci. Publ. 61*: 1–52.

Dojlido, J. R. 1979. Investigations of biodegradability and toxicity of organic compounds. *Rep. EPA-600/2-79-163*, NTIS Order No. PB80-179336, p. 118; cited in *Chem. Abstr. CA 94*(14): 108594b.

NIOSH. 1984. *Manual of Analytical Methods*, 3rd ed. Cincinnati, OH: National Institute for Occupational Safety and Health.

NIOSH. 1986. *Registry of Toxic Effects of Chemical Substances*, ed. D. V. Sweet. Washington, DC: U.S. Government Printing Office.

U.S. EPA. 1992. Analysis for organic priority pollutants in wastewaters. 40 CFR, Part 136, Appendix A, *Fed. Register*. Oct. 24.

U.S. EPA. 1997. *Test Methods for Evaluating Solid Wastes*, SW 846. Washington, DC: Office of Solid Waste and Emergency Response.

22

HALOGENATED HYDROCARBONS

22.1 GENERAL DISCUSSION

Halocarbons are halogen-substituted hydrocarbons that contain carbon, hydrogen, and halogen atoms in the molecules. These compounds are formed when hydrogen atoms of hydrocarbons are partially or fully replaced by halogen atoms: fluorine, chlorine, bromine, and iodine. Some of the widely used halogenated hydrocarbons are chloroform, $CHCl_3$; methylene chloride, CH_2Cl_2; methyl iodide, CH_3I; 1,1,1-trichloroethane, CH_3CCl_3; and vinyl chloride, $CH_2=CHCl$. Carbon tetrachloride, CCl_4; carbontetrabromide, CBr_4; tetrachloroethylene, $Cl_2C=CCl_2$; 1,1,2-trichloro-1,2,2-trifluoroethane, Cl_2FCCF_2Cl; and hexachlorobenzene, C_6Cl_6 are examples of fully substituted halocarbons that do not contain hydrogen atoms.

Although most of the halogenated hydrocarbons that are commonly used are liquids at ambient temperature and pressure, the low-molecular-weight compounds such as 1,2-difluoroethane, methyl bromide, or vinyl chloride are gases, whereas those of higher molecular weight, such as iodoform or hexachloronaphthalene, are solids. Halogenated hydrocarbons are used as solvents, in extractions, as intermediates, and in the manufacture of numerous chemicals of wide commercial use. Many fluorocarbons are used as refrigerants and in fire extinguishers. Because of the health hazard, many anesthetic halocarbons used earlier as clinical anesthesia have been replaced by less toxic compounds.

Flammability

The flammability of halogenated hydrocarbons shows wide variation. Whereas compounds such as carbon tetrachloride and chloroform are noncombustible liquids, there are substances such as Halons which are used for fire-extinguishing purposes. Several low-molecular-weight halocarbons are flammable gases or liquids. Many C_1 and C_2 fluorocarbons are flammable gases. These include trifluoromethane [75-46-7], 1,1-difluoroethane [75-37-6], chlorodifluoroethane [25497-29-4], chlorotrifluoroethylene [79-38-9], 1-chloro-1,1-difluoroethane [75-68-3], bromotrifluoroethylene [598-73-2], and tetrafluoroethylene [116-14-3]. Not all fluorocarbon gases are flammable. Hexafluoroethane [67-72-1], for example, is a nonflammable gas. Among the chlorocarbons, methyl chloride [74-87-3], ethyl chloride [75-00-3], and vinyl chloride [75-01-4] are flammable gases.

Methyl bromide [74-83-9] is a C_1 bromo compound that is a flammable gas.

Among the fluorocarbons that are flammable liquids at ambient temperatures, fluorobenzenes are notable. These include fluorobenzene [462-06-6]; 1,2-difluorobenzene [367-11-3]; 1,3-difluorobenzene [372-18-9]; 1,2,3-trifluorobenzene [1489-53-8]; 1,2,4-trifluorobenzene [367-23-7]; 1,3,5-trifluorobenzene [372-38-3]; the tetrafluoro isomers, 1,2,3,4-[551-62-2], 1,2,3,5-[2367-82-0], and 1,2,4,5 [327-38-3]; pentafluorobenzene [363-72-4]; and hexafluorobenzene [392-56-3], which are all flammable liquids having flash points 100°F (38°C) (Table 22.1).

Aliphatic hydrocarbons containing one chlorine atom in the molecules are flammable liquids up to carbon number 6 and combustible to carbon number 10. Thus *n*-propyl chloride [540-54-5], *n*-butyl chloride [109-69-3], *sec*-butyl chloride [78-86-4], isopropyl chloride [75-29-6], and hexyl chloride

[544-10-5] are all flammable liquids. The same holds true for monobromo hydrocarbons. However, the bromo compounds are less flammable than their chloro counterparts. The difference in flammability, however, is not great. Whereas bromopentanes and iodopentanes are flammable liquids, the compounds between carbon numbers 6 and 9 are combustible liquids.

An increase in the atomic radii of the halogens and an increase in the halo substitution in the molecules decreases the flammability of the compounds. On the other hand, the presence of a double bond may enhance the flammability.

With an increase in chloro or bromo substitution from 1 to 2, the molecules lose their flammability to a significant extent. Such a decrease in flammability is greater in bromo compounds than in fluoro- or chlorocarbons. Whereas 1,2-dichloroethane [107-06-2] is a flammable liquid, 1,2-dibromoethane [106-93-4] is nonflammable. In halogenated hydrocarbons, an increase in the halosubstitutions in the molecule increases the flash point. This is highlighted in Table 22.1 for fluorobenzenes. However, among fluorobenzenes such a pattern is less significant.

The flammable halocarbon gases and the vapors of the flammable liquids form explosive mixtures with air. The range of explosive limits is wide for vinyl chloride, methyl chloride, and ethyl chloride.

Explosion Hazard

Halogenated hydrocarbons are stable compounds with low reactivity. The explosion hazard is low and reported cases are rare. These compounds, however, may react violently with alkali metals such as sodium or potassium and their alloys, or with finely divided magnesium, calcium, aluminum, or zinc. Explosions may occur when mixtures are either heated or subjected to impact. Violent reactions may occur with powerful oxidizers, especially upon heating. Chloroform and carbon tetrachloride may react violently

TABLE 22.1 Flammability of Fluorobenzenes in Relation to Degrees of Fluoro Substitution

Fluorobenzene/ CAS No.	F Atoms on the Ring	Flash Point [°C (°F)]
Fluorobenzene [462-06-6]	1	−12 (9)
1,2-Difluorobenzene [367-11-3]	2	2 (36)
1,3-Difluorobenzene [372-18-9]	2	2 (36)
1,2,3-Trifluorobenzene [1489-53-8]	3	−3 (25)
1,2,4-Trifluorobenzene [367-23-7]	3	4 (40)
1,2,3,4-Tetrafluorobenzene [551-62-2]	4	4 (40)
1,2,3,5-Tetrafluorobenzene [2367-82-0]	4	4 (40)
Pentafluorobenzene [363-72-4]	4	13 (57)
Hexafluorobenzene [392-56-3]	6	10 (50)

with dimethylformamide or potassium-*tert*-butoxide on heating. Reactions with silicon hydrides can be violent.

Volatile halocarbons may rupture glass containers due to simple pressure buildup (see Section 22.5) or to exothermic polymerization in a closed vessel (e.g., vinyl chloride).

Explosive decomposition of benzyl fluoride [350-50-5] has been reported (Szucs 1990). The reaction is catalyzed on glass surfaces, proceeding with evolution of hydrogen fluoride with explosive violence. The compound is stable in nonglass containers.

Toxicity

Halogenated hydrocarbons, in general, exhibit low acute toxicity. Inhalation toxicity is greater for gaseous or volatile liquid compounds. The health hazard from exposure to these compounds may be due to their anesthetic actions; damaging effects on liver and kidney; and in case of certain compounds, carcinogenicity. The toxic symptoms are drowsiness, incoordination, anesthesia, hepatitis, and necrosis of liver. Vapors may cause irritation of the eyes and respiratory tract. Anesthetic effects in animals or humans occur at high levels of exposures, which may go up to several percentage volume concentrations in air. Death may result from cardiac arrest due to prolonged exposure to high concentrations. Ingestion can produce nausea, vomiting, and liver injury. Fatal doses in humans vary with the compound. Fluorocarbons are less toxic than the chloro, bromo, and iodo compounds. The toxicity increases with increase in the mass number of the halogen atoms, as shown in Table 22.2 for the monohalogen derivatives of benzene. Some of the halogenated hydrocarbons cause cancer in humans. These include vinyl chloride [75-01-4], vinyl bromide [593-60-2], methyl iodide [74-88-4], ethylene dibromide [106-93-4], chloroform [67-66-3], carbon tetrachloride [56-23-5], and hexachlorobutadiene [87-68-3]. In addition, many other compounds of commercial

TABLE 22.2 Toxicity of Monohalogenated Benzenes in Relation to the Increase in Mass of Halogen Atoms

Benzene Halide/CAS No.	Range of LD_{50} Oral (Rats) (mg/kg)
Fluorobenzene [462-06-6]	4400
Chlorobenzene [108-90-7]	2900
Bromobenzene [108-86-1]	2700
Iodobenzene [591-50-4]	1800

use, such as methylene chloride [75-09-2], 1,1,2-trichloroethane, trichloroethylene [79-01-6], tetrachloroethylene [127-18-4], 1,1-dichloroethylene [75-35-4], tetrafluoroethylene polymers [9002-84-0], *p,p'*-dichlorodiphenyl dichloroethylene [72-55-9], chlorobutadiene [126-99-8], ethylene dichloride [107-06-2], hexachloroethane [67-72-1], *n*-propyl iodide [107-08-4], and chlorofluoromethane [593-70-4], are animal carcinogens. The EPA has listed more than 50 compounds as priority pollutants in the environment.

The toxicity of individual compounds is discussed in detail in the following sections.

Analysis

Halogenated hydrocarbons are commonly analyzed by GC techniques. The Hall electrolytic conductivity detector is highly sensitive for the purpose. Although ECD is sensitive, too, it is not convenient for analyzing low-molecular-weight gaseous halocarbons. GC-FID is applicable for ppm-level detection. Halocarbons that are EPA listed priority pollutants may be analyzed by EPA Methods 601 (for wastewaters), 502 (for potable waters), and 8010 (for soils, sediments, and hazardous wastes) (U.S. EPA 1992, 1997). These are all purge and trap GC methods, in which the aqueous samples or an aqueous solution of a soil sample is purged with an inert gas, trapped over adsorbent columns, desorbed with a carrier gas onto a capillary or packed column, and analyzed by GC using a Hall detector. Several columns are commercially available for such analysis.

Gaseous substances may be analyzed either at low temperatures under cryogenic conditions or by using a capillary column 60 m or greater in length. In general, any volatile halocarbon can be analyzed by the foregoing technique. Instead of the purge and trap method, the sample may be heated and the halocarbons may be thermally desorbed onto a column.

Alternatively, halogenated hydrocarbons may be analyzed by GC/MS using a purge and trap technique (EPA Methods 624, 524, and 8240) or a thermal desorption method. GC/MS analysis is a confirmatory test in which the primary and secondary ions characterizing the compounds can be identified and quantified.

Halogenated hydrocarbons in aqueous samples are stable at low temperature and reduced pH. Acidification with HCl prevents degradation and allows prolonged storage of samples. Sodium bisulfite and ascorbic acid are also effective preservatives. Samples preserved with either the bisulfite or ascorbic acid were stable over a 112-day experimental period (Maskarinec et al. 1990).

Halogenated hydrocarbons in air may be analyzed by various NIOSH methods (NIOSH 1984). A known volume of air is passed over coconut shell charcoal. The adsorbed halocarbons are desorbed with a suitable solvent such as carbon disulfide, propanol, benzene, or toluene and are analyzed by GC using mostly FID, or for certain compounds, ECD as detectors. Table 22.3 lists the NIOSH Method numbers and the analytical techniques used for the analysis of halocarbons in air.

Disposal/Destruction

Halogenated hydrocarbons may be destroyed by incineration, hydrolysis with caustic alkalies, and biodegradation. Ultrahigh destruction of volatile organic compounds to sub-ppt level in fumes or air stream has been reported using catalytically stabilized thermal incineration process (Pfefferle 1992). Methylene

chloride at an inlet concentration of 50 ppm burned in a residence time of 17 msec to a level below the detection limit of 2 ppt. The catalyst used in the process is a base metal in honeycomb structure form.

HALOGEN-SUBSTITUTED ALIPHATIC HYDROCARBONS

22.2   METHYL CHLORIDE

EPA Priority Pollutant, RCRA Toxic Waste Number U045; DOT Label: Flammable Gas, UN 1063

Formula CH_3Cl; MW 50.49; CAS [74-87-3]

Structure:

$$H-\overset{\overset{\displaystyle H}{|}}{\underset{\underset{\displaystyle H}{|}}{C}}-Cl$$

Synonyms: chloromethane; monochloromethane

Uses and Exposure Risk

Methyl chloride is used as a refrigerant, as a local anesthetic, as a blowing agent for polystyrene foams, and as a methylating agent in the synthesis of a number of chemicals of commercial application.

Physical Properties

Colorless gas with a faint sweet odor; gas density 1.74 (air = 1) at 20°C; liquefies at −23.7°C; solidifies at −97°C; slightly soluble in water, miscible with organic solvents.

Health Hazard

Inhalation of methyl chloride can produce headache, dizziness, drowsiness, nausea, vomiting, convulsions, coma, and respiratory failure. It is narcotic at high concentrations. Repeated exposures can produce liver and

TABLE 22.3   Analysis of Halogenated Hydrocarbons in Air by NIOSH Methods

Compound	Desorbing Solvent	GC Detector	NIOSH Method
Allyl chloride	Benzene	FID	1000
Bromotrifluoromethane	CH_2Cl_2	FID	1017
Dibromodifluoromethane	Propanol	FID	1012
Dichlorodifluoromethane 1,2-Dichlorotetrafluoroethane	CH_2Cl_2	FID	1018
Dichlorofluoromethane	CS_2	FID	2516
1,2-Dichloropropane	Acetone/cyclohexane (15 : 85)	Hall	1013
Ethyl bromide	2-Propanol	FID	1011
Ethyl chloride	CS_2	FID	2519
Ethylene dibromide	Benzene/methanol (99 : 1)	ECD	1008
Hexachloro-1,3-cyclopentadiene	—	ECD	2518
Chloroform, bromoform, chlorobenzene, carbon tetrachloride, benzyl chloride, chlorobromomethane, 1,1-dichloroethane, 1,2-dichloroethane, 1,1,1-trichloroethane, 1,1,2-trichloroethane, ethylene dichloride, 1,2,3-trichloropropane, tetrachloroethylene, hexachloroethane, benzyl chloride, o- and p-Dichlorobenzene	CS_2	FID	1003
Methyl bromide	CS_2	FID	2520
Methyl chloride	CH_2Cl_2	FID	1001
Methylene chloride	CS_2	FID	1005
Methyl iodide	Toluene	FID	1014
Pentachloroethane	Hexane	ECD	2517
1,1,2,2-Tetrabromoethane	Tetrahydrofuran	FID	2003
1,1,2,2-Tetrachloroethane	CS_2	FID	1019
1,1,2,2-Tetrachloro-2,2-difluoroethane, 1,1,2,2-tetrachloro-1,2-difluoroethane	CS_2	FID	1016
Trichloroethylene	CS_2	FID/PID[a]	1022/3701
Trichlorofluoromethane	CS_2	FID	1006
1,1,2-Trichloro-1,2,2-trifluoroethane	CS_2	FID	1020
Vinyl bromide	Ethanol	FID	1009
Vinyl chloride	CS_2	FID	1007
Vinylidene chloride	CS_2	FID	1015

[a]Air is injected directly into a portable GC equipped with a PID.

kidney injury and depression of bone marrow activity (Patty 1963). Severe exposure can affect the central nervous system. Other symptoms that have been noted in humans are a staggering gait, weakness, tremor, difficulty in speech, and blurred vision. The ocular symptoms include mistiness, visual disturbances, and difficulty in accommodation (Fairhall 1969). Massive exposure can cause death. A 2-hour exposure to methyl chloride at a concentration of 2% in air can be fatal to humans.

Methyl chloride caused adverse reproductive effects in test animals. These include embryo toxicity, fetal death, developmental abnormalities, and paternal effects in rats and mice. It tested positive to the histidine reversion–Ames test for mutagenicity. The carcinogenic properties of this compound have not been established. The evidence in animals and humans is inadequate.

Exposure Limits

TLV-TWA 50 ppm (~105 mg/m^3) (ACGIH), 100 ppm (~210 mg/m^3) (OSHA); ceiling 100 ppm (MSHA), 200 ppm (OSHA); TLV-STEL 100 ppm (ACGIH); carcinogenicity: Animal Inadequate Evidence, Human Inadequate Evidence (IARC).

Fire and Explosion Hazard

Flammable gas, burns with a smoky flame; autoignition temperature 632°C (1170°F). Methyl chloride forms explosive mixtures with air within the range 7.6–19.0% by volume in air. It reacts explosively with alkali metals, potassium, sodium, or lithium; sodium–potassium alloy; and with magnesium, aluminum, or zinc in powder form.

22.3 METHYL BROMIDE

EPA Priority Pollutant, RCRA Waste Number U029; DOT Label: Poison B, UN 1062

Formula CH$_3$Br; MW 94.95; CAS [74-83-9]

Structure:

$$\begin{array}{c} H \\ | \\ H-C-Br \\ | \\ H \end{array}$$

Synonyms: bromomethane; monobromomethane

Uses and Exposure Risk

Methyl bromide is used as a fumigant for pest control, for degreasing wool, and as a methylating agent. Its use as a refrigerant and in fire extinguishers is restrained due to its health hazards.

Physical Properties

Colorless gas with a chloroform-like odor at high concentrations; 1 L of gas weighs 3.974 g at 20°C; liquefies at 3.5°C; solidifies at −93.6°C; slightly soluble in water, miscible with organic solvents.

Health Hazard

The acute poisoning effects from inhaling methyl bromide are headache, weakness, nausea, vomiting, loss of coordination, visual disturbance, pulmonary edema, tremor, convulsions, hyperthermia, and coma. Massive exposure may cause death from respiratory paralysis. The toxicity of this compound is comparable to that of methyl chloride. The lethal concentration in humans has not been measured accurately. The LC$_{50}$ value in rats is in the range 300 ppm after an 8-hour exposure. Chronic exposure can cause injury to the kidney and depression of the central nervous system.

The liquid, as well as the gas, may be absorbed through the skin. Contact with the liquid can cause burns. Oral administration of the liquid caused gastrointestinal tumors in rats. Its carcinogenicity in humans is not known.

Exposure Limits

TLV-TWA 5 ppm (~20 mg/m^3) (ACGIH), 15 ppm (~60 mg/m^3) (MSHA); ceiling 20 ppm (~80 mg/m^3) (OSHA); carcinogenicity: Animal Limited Evidence, Human Inadequate Evidence (IARC).

Fire and Explosion Hazard

Noncombustible gas in air; however, it will burn in oxygen or in the presence of a high-energy source of ignition or within its narrow flammability range. Autoignition temperature 538°C (1000°F). It forms explosive mixtures within a narrow range, of 13.5–14.5% by volume in air. An explosion occurred when methyl bromide was reacted with dimethyl sulfoxide (Scaros and Serauskas 1973). It reacts with aluminum to form methyl aluminum bromide, which may ignite in air.

22.4 METHYL IODIDE

EPA Designated Toxic Waste, RCRA Waste Number U138; DOT Label: Poison B, UN 2644

Formula CH_3I; MW 142.94; CAS [74-88-4]

Structure:

$$H-\underset{\underset{H}{|}}{\overset{\overset{H}{|}}{C}}-I$$

Synonym: iodomethane

Uses and Exposure Risk

Methyl iodide is used in the analysis of pyridine; microscopy; as an embedding material for examining diatoms (Merck 1996); and as a methylating agent.

Physical Properties

Colorless liquid that turns yellow or brown on exposure to light or moisture; decomposes at 270°C; density 2.280; boils at 42.5°C; freezes at −66.5°C; low solubility in water (2%), soluble in alcohol and ether; vapor pressure 375 torr at 20°C.

Health Hazard

The acute oral toxicity and inhalation toxicity of methyl iodide is moderate in test animals. It is more toxic than methyl bromide. The toxic symptoms are nausea, vomiting, diarrhea, ataxia, drowsiness, slurred speech, visual disturbances, and tremor. Pulmonary edema, coma, and death can result from massive exposures. The vapors are an irritant to the eyes. Repeated exposures may cause depression of the central nervous system. Prolonged contact with the liquid can cause skin burn and dermatitis. The reported values of LD_{50}, as well as LC_{50}, for this compound as published in the literature show variations. The fatal doses by inhalation and ingestion are 900 ppm/hr in mice and 150 mg/kg in rats, respectively (Buckell 1950).

Methyl iodide exhibited carcinogenic properties in test animals. Administration of this compound produced tumors in lungs and colon. ACGIH (1986) lists it as a suspected human carcinogen.

Exposure Limits

TLV-TWA 2 ppm ($\sim$11 mg/m^3) (ACGIH), 5 ppm (MSHA and OSHA); carcinogenicity: Animal Limited Evidence (IARC), Suspected Human Carcinogen.

Fire and Explosion Hazard

Noncombustible liquid. No violent or explosive reactions of this compound are reported. Its reactions with alkali metals or finely divided alkaline–earth metals can be vigorous to violent.

Disposal/Destruction

Methyl iodide is mixed with an excess of combustible solvent and burned in a chemical incinerator equipped with an afterburner and scrubber. Small amounts may be destroyed in the laboratory by hydrolyzing with an aqueous or ethanolic solution of caustic potash. The reaction is

$$CH_3I + KOH \longrightarrow CH_3OH + KI$$

22.5 METHYLENE CHLORIDE

EPA Priority Pollutant; RCRA Toxic Waste Number U080; DOT Label: None, Number UN 1593

Formula CH_2Cl_2; MW 84.93; CAS [75-09-2]
Structure:

$$Cl-\overset{\displaystyle H}{\underset{\displaystyle H}{C}}-Cl$$

Synonyms: dichloromethane; methylene dichloride

Uses and Exposure Risk

Methylene chloride is widely used as a solvent, as a degreasing and cleaning reagent, in paint removers, and in extractions of organic compounds from water for analyses.

Physical Properties

Colorless liquid; density 1.3255 at 20°C; boils at 39.7°C; freezes at −95°C; slightly soluble in water (1.3%), miscible with organic solvents; vapor pressure 350 torr at 20°C.

Health Hazard

Methylene chloride is a low to moderately toxic compound, the toxicity varying with the animal species. It is less toxic in small animals than in humans. The toxic routes of exposure are inhalation of its vapors, ingestion, and absorption through the skin. It may be detected from its odor at a concentration of 300 ppm. Acute toxic symptoms include fatigue, weakness, headache, lightheadedness, euphoria, nausea, and sleep. High concentrations may produce narcosis. Rabbits exposed to 10,000 ppm for 7 hours died from exposure. The LC_{50} value in mice is 14,400 ppm/7 hr (NIOSH 1986). Mild effects may be felt in humans from an 8-hour exposure to 500 ppm of methylene chloride vapors. Oral intake of 15–20 mL of the liquid may be lethal to humans. Chronic exposure to this compound can lead to liver injury. Contact of the liquid with skin or eyes can cause irritation.

Methylene chloride metabolizes in body to carbon monoxide, which forms carboxyhemoglobin in blood. The concentration of the latter is related to the vapor concentration and the duration of exposure.

Methylene chloride is carcinogenic to animals. Rats inhaling its vapors at concentrations of 2000–3500 ppm, 5–6 hours per day for 2 years developed lung and endocrine tumors. It is a suspected human carcinogen. The evidence of carcinogenicity in humans is inadequate, however.

Exposure Limits

TLV-TWA 50 ppm ($\sim$175 mg/m^3) (ACGIH); carcinogenicity: Suspected Human Carcinogen (ACGIH), Animal Sufficient Evidence, Human Inadequate Evidence (IARC).

Fire and Explosion Hazard

Methylene chloride is a noncombustible liquid. It reacts explosively with alkali metals and their alloys, and with powdered magnesium or aluminum. Ignition occurs with potassium *tert*-butoxide. It forms explosive mixtures with nitrogen tetroxide (Turley 1964). There are reports of ruptures of separatory funnels or other glass containers during extractions because of pressure buildup (Kelling 1990; Bean et al. 1990). Prerinsing the separatory funnel with methylene chloride before adding the sample to the funnel may prevent excess pressure buildup (Gibbons 1990).

Disposal/Destruction

Methylene chloride is mixed with an excess of combustible solvent and burned in a chemical incinerator equipped with an afterburner and scrubber. Alternatively, it is mixed with vermiculite and dry caustics (sodium carbonate and slated lime), wrapped in paper, and incinerated (Aldrich 1997). It may be disposed of in a landfill in lab packs. Small amounts may be destroyed in the laboratory by hydrolyzing with ethanolic solution of potassium hydroxide (National Research Council 1995). Woods and associates (1988) have reported the development of a biofilm reactor using a gas-permeable membrane for the growth of bacterial film. Methylene chloride and other chlorinated hydrocarbons containing one or two carbon atoms were reported to be biodegraded by the technique described above.

22.6 CHLOROFORM

EPA Priority Pollutant, RCRA Toxic Waste Number U044; DOT Label: Poison, UN 1888

Formula $CHCl_3$; MW 119.37; CAS [67-66-3]
Structure:

$$\begin{array}{c} Cl \\ | \\ H-C-Cl \\ | \\ Cl \end{array}$$

Synonyms: trichloromethane; methenyl trichloride; formyl trichloride

Uses and Exposure Risk

Chloroform is used in industry as a solvent, as a cleansing agent, in the manufacture of refrigerant, and in fire extinguishers. In the past it was used as an anesthetic.

Physical Properties

Colorless liquid with a pleasant sweet odor; sweet taste; volatile, vapor pressure 158 torr at 20°C; density 1.484 at 20°C; boils at 61.2°C; solidifies at −63.5°C; slightly soluble in water (0.82 mL/100 mL water at 20°C), miscible with organic solvents; sensitive to light.

Health Hazard

The inhalation toxicity of chloroform in animals and humans is low. It is moderately toxic by ingestion. Toxic symptoms include dizziness, lightheadedness, dullness,

hallucination, nausea, headache, fatigue, and anesthesia. In humans the toxic effects may be felt from a 5- to 10-minute exposure to a concentration of 1000 ppm in air. Clinical anesthesia is produced from inhalation of 10,000-ppm vapors (Morris 1963), while a 5-minute exposure to 25,000 ppm can be fatal to humans. By an oral route, the lethal dose may be 6–8 mL of the liquid. The target organs are the liver, kidney, central nervous system, and heart. Chronic effects from high dosage involve digestive disturbances, necrosis of the liver, and kidney damage.

Chloroform produced embryo toxicity in experimental animals, causing a high incidence of fetal resorption and retarded fetal growth. Animal experiments indicate this compound to be carcinogenic, causing cancers in the kidney, liver, and thyroid. It is a suspected human carcinogen for which there is limited evidence.

Exposure Limits

TLV-TWA 10 ppm (~48 mg/m^3) (ACGIH); ceiling 50 ppm (~240 mg/m^3) (MSHA and OSHA), 2 ppm/60 min (NIOSH); carcinogenicity: Suspected Human Carcinogen (ACGIH), Animal Sufficient Evidence, Human Limited Evidence (IARC).

Fire and Explosion Hazard

Chloroform is a noncombustible liquid. Mixing chloroform with certain substances can produce explosions. This includes alkali metals such as sodium, potassium, or lithium; finely divided magnesium or aluminum; sodium–potassium alloy; nitrogen tetroxide (Turley 1964); and perchloric acid–phosphorus pentoxide (NFPA 1986). Contact with disilane may produce incandescence (Mellor 1946–47), and with potassium *tert*-butoxide or sodium methoxide can result in ignition (MCA 1961, 1972).

Disposal/Destruction

Chloroform may be destroyed by being mixed with vermiculite, sodium carbonate, and slated lime, wrapped in paper, and burned in a chemical incinerator equipped with an afterburner and scrubber (Aldrich 1997). Chloroform in drinking water may be aerobically biodegraded to carbon dioxide (Speitel et al. 1989). Bacterial cultures from contaminated sites produced efficient degradation of chlorinated hydrocarbons in laboratory-scale continuous-flow reactors (Kaestner 1989). Woods and co-workers (1988) have described an aerated biofilm reactor containing a gas-permeable membrane support for the growth of bacterial films. The biofilm degraded methylene chloride, chloroform, and 1,2-dichloroethane.

22.7 CARBON TETRACHLORIDE

EPA Priority Pollutant; RCRA Toxic Waste Number U211; DOT Identification Number UN 1846, Label: None

Formula CCl_4; MW 153.81; CAS [56-23-5]

Structure:

$$Cl-\underset{\underset{Cl}{|}}{\overset{\overset{Cl}{|}}{C}}-Cl$$

Synonyms: tetrachloromethane; tetrachlorocarbon; tetraform

Uses and Exposure Risk

Carbon tetrachloride is used as a solvent, in fire extinguishers, in dry cleaning, and in the manufacture of fluorocarbon propellents.

Physical Properties

Colorless, heavy liquid with characteristic odor; density 1.589 at 25°C; bp 76.7°C; mp −23°C; insoluble in water (0.05 mL dissolves in 100 mL of water), miscible with organic solvents; vapor pressure 89.5 torr at 20°C.

Health Hazard

Carbon tetrachloride exhibits low acute toxicity by all routes of exposure. The acute poisoning effects include dizziness, fatigue, headache, nervousness, stupor, nausea, vomiting, diarrhea, renal damage, and liver injury. The dosages that produce toxic actions in animals vary with the species. The oral LD_{50} values in rats, rabbits, and mice are 2800, 5760, and 8263 mg/kg, respectively (NIOSH 1986).

Ingestion of carbon tetrachloride can be fatal to humans, death resulting from acute liver or kidney necrosis. Chronic exposure may cause liver and kidney damage. Exposure to a 10-ppm concentration for several weeks produced accumulation of fat in the livers of experimental animals (ACGIH 1986). Substances such as ethanol and barbiturates cause potentiation of toxicity of carbon tetrachloride. Skin contact can cause dermatitis.

Azri and co-workers (1990) have investigated carbon tetrachloride–induced hepatotoxicity in rat liver slices. Liver slices from male rats were incubated and exposed to carbon tetrachloride vapors, and the degree of injury to cellular tissue was determined. Covalent binding of CCl_4 radical to proteins and lipid molecules in a slice caused the cellular injury. The toxicity depended on the vapor concentration and the time of exposure. Azri and co-workers reported further that rats pretreated with phenobarbital were more rapidly intoxicated even at a lower concentration of carbon tetrachloride vapors. On the other hand, pretreatment with allylisopropylacetamide inhibited the toxicity of carbon tetrachloride.

Carbon tetrachloride is a suspected human carcinogen. Oral and subcutaneous administration of this compound in rats caused liver and thyroid cancers in the animals.

Exposure Limits

TLV-TWA skin 5 ppm (~32.5 mg/m^3) (ACGIH), 10 ppm (~65 mg/m^3) (MSHA and OSHA); ceiling 2 ppm/60 min (NIOSH); carcinogenicity: Suspected Human Carcinogen (ACGIH), Animal Sufficient Evidence, Human Limited Evidence (IARC).

Fire and Explosion Hazard

Carbon tetrachloride is a noncombustible liquid. Explosion may occur when this compound is mixed with alkali metals such as sodium, potassium, lithium, or their alloys; or finely divided aluminum, magnesium, calcium, barium, beryllium, and other metals on heating or impact. Reactions with hydrides of boron or silicon, such as diborane, disilane, trisilane, or tetrasilane, can be explosively violent. When mixed with dimethyl formamide and heated, carbon tetrachloride may explode (Kittila 1967). Its mixture with potassium *tert*-butoxide may ignite. Its reaction with fluorine or a halogen fluoride is generally vigorous and may become violent on heating. A violent reaction occurs with hypochlorites.

Disposal/Destruction

Carbon tetrachloride wastes may be destroyed by mixing with vermiculite, caustic soda, and slaked lime, followed by burning in a chemical incinerator equipped with an afterburner and scrubber. Incomplete combustion may produce chloroform (Thurnau 1988). It may be mixed with an excess of nonhalogenated solvent (combustible) and incinerated. Carbon tetrachloride wastes in laboratories may be disposed of in a landfill in lab packs.

22.8   ETHYL CHLORIDE

EPA Priority Pollutant; DOT Label: Flammable Gas/Liquid, UN 1037

Formula C_2H_5Cl; MW 64.52; CAS [75-00-3]

Structure:

$$
\begin{array}{ccc}
 & H & H \\
 & | & | \\
H - & C - & C - Cl \\
 & | & | \\
 & H & H
\end{array}
$$

Synonyms: chloroethane; monochloroethane; hydrochloric ether; muriatic ether; chlorethyl

Uses and Exposure Risk

Ethyl chloride is used as a refrigerant, as a solvent, in the manufacture of tetraethyl lead, and as an alkylating agent. It is also used as a topical anesthetic.

Physical Properties

Colorless gas at ordinary temperature and pressure; volatile liquid at low temperature or increased pressure; pungent ether-like odor; vapor density 2.22 (air = 1), liquid density 0.921 at 0°C; liquefies at 12.3°C; solidifies at −138.7°C; slightly soluble in water (0.57% at 20°C), miscible with organic solvents.

Health Hazard

Exposure to high levels of ethyl chloride can cause stupor, eye irritation, incoordination, abdominal cramps, anesthetic effects, cardiac arrest, and unconsciousness. No toxic effects were noted at a concentration of 10,000 ppm. A 45-minute exposure to a 4% concentration of ethyl chloride in air was lethal to guinea pigs. A brief exposure for 5 to 10 minutes to a concentration of 10% of the gas was not fatal to the test animals but caused kidney and liver damage. In humans narcotic effects may occur after a few inhalations of 5–10% concentrations of the gas. Irritant effects on the eyes, skin, and respiratory tract are mild. Skin contact with the liquid can cause frostbite due to cooling by rapid evaporation.

LC_{50} value, inhalation (rats): 60,000 ppm/2 hr

Exposure Limits

TLV-TWA 1000 ppm ($\sim$2600 mg/m^3) (ACGIH, MSHA, NIOSH, and OSHA); IDLH 20,000 ppm (NIOSH).

Fire and Explosion Hazard

Flammable gas; burns with a smoky, greenish flame, forming hydrogen chloride; flash point (closed cup) −50°C (−58°F), (open cup) $-43°C$ ($-45°F$); the vapor is heavier than air and can travel a considerable distance to a source of ignition and flash back; vapor pressure 1064 torr at 20°C; autoignition temperature 519°C (966°F); fire-extinguishing measure: stop the flow of gas; use CO_2 to extinguish the fire; water may be used to keep fire-exposed containers cool.

Ethyl chloride forms explosive mixtures with air, the LEL and UEL values being 3.8 and 15.4% by volume of air, respectively. Its reactions with alkali metals or powdered calcium, magnesium, or aluminum can be violent.

22.9 ETHYL BROMIDE

DOT Label: Poison B, UN 1891
Formula C_2H_5Br; MW 108.98; CAS [74-96-4]
Structure:

$$H-\overset{\overset{\displaystyle H}{|}}{\underset{\underset{\displaystyle H}{|}}{C}}-\overset{\overset{\displaystyle H}{|}}{\underset{\underset{\displaystyle H}{|}}{C}}-Br$$

Synonyms: bromoethane; monobromoethane; hydrobromic ether; Halon 2001

Uses and Exposure Risk

Ethyl bromide is used as a refrigerant and as an ethylating agent. It was formerly used as a topical anesthetic.

Physical Properties

Colorless, volatile liquid turning yellowish on exposure to air or light; burning taste and ether-like odor; density 1.461 at 20°C; boils at 38.4°C; freezes at $-119°C$; slightly soluble in water (0.9%), miscible with organic solvents.

Health Hazard

Ethyl bromide is a depressant of the central nervous system, causing narcosis. The health hazard is greater than with ethyl chloride.

In addition to the narcotic effects that occur at exposure to high concentrations, other toxic symptoms include irritation of the eyes and respiratory tract, pulmonary edema, fatty degeneration of the liver and renal tissue, and damage to the liver, kidney, and intestine. A 15-minute exposure to a 15% concentration of vapor in air was lethal to rats. The oral LD_{50} value in rats is 1350 mg (NIOSH 1986).

Exposure Limits

TLV-TWA 200 ppm ($\sim$890 mg/m^3) (ACGIH, MSHA, OSHA, and NIOSH); TLV-STEL 250 ppm ($\sim$110 mg/m^3) (ACGIH); IDLH 3500 ppm (NIOSH).

Fire and Explosion Hazard

Flammable liquid; flash point (closed cup) $-23°C$ ($-10°F$) (Aldrich 1997); vapor pressure 375 torr at 20°C; vapor density 3.76 (air = 1); the vapor is heavier than air and can travel a considerable distance to a source of ignition and flash back; autoignition temperature 511°C (952°F).

Ethyl bromide vapors form explosive mixtures with air within the range 6.75–11.25% by volume in air. Violent reactions may occur when mixed with alkali metals or powdered magnesium, aluminum, or zinc.

Disposal/Destruction

Ethyl bromide is mixed with a combustible solvent and burned in a chemical incinerator equipped with an afterburner and scrubber.

22.10 VINYL CHLORIDE

EPA Priority Pollutant, RCRA Toxic Waste Number U043; DOT Label: Flammable Gas, UN 1086
Formula C_2H_3Cl; MW 62.50; CAS [75-01-4]
Structure:

$$H-\overset{\overset{\displaystyle H}{|}}{C}=\overset{\overset{\displaystyle H}{|}}{C}-Cl$$

Synonyms: chloroethylene; chloroethene; monochloroethylene

Uses and Exposure Risk

Vinyl chloride is used as a monomer in the manufacture of polyvinyl chloride resins and plastics, as a refrigerant, and in organic synthesis.

Physical Properties

Colorless gas with an ethereal odor; liquefies at −13.4°C; solidifies at −153.8°C; polymerizes on exposure to light or in the presence of a catalyst; slightly soluble in water, miscible with organic solvents.

Health Hazard

The acute inhalation toxicity is of low order. Since it is a gas, the route of exposure is primarily inhalation. The target organs are the liver, central nervous system, respiratory system, and blood. Exposure to high concentrations can produce narcosis. A 30-minute exposure to 30% vinyl chloride in air was fatal to experimental animals. Chronic exposure produced minor injury to the liver and kidneys. Such effects were noted at a 7-hour exposure daily to 200 ppm for 6 months.

Vinyl chloride is an animal and human carcinogen. Rats subjected to 12 months' inhalation developed tumors of the lungs, skin, and bones. Occupational exposure to this compound demonstrated an increased incidence of liver cancer. Tabershaw and Gaffey (1974) conducted epidemiological studies on workers who had at least 1 year of occupational exposure to vinyl chloride. The study indicated that cancers of the digestive system, respiratory system, and brain, as well as lymphomas, were greater among people who had the greatest estimated exposure to vinyl chloride.

Exposure Limits

TLV-TWA 5 ppm (∼12.5 mg/m³) (ACGIH), 1 ppm (OSHA), 200 ppm (MSHA), Lowest Detection Limit (NIOSH); ceiling 5 ppm/15 min (OSHA); carcinogenicity: Recognized Human Carcinogen (ACGIH), Animal Sufficient Evidence, Human Sufficient Evidence (IARC), Cancer Suspect Agent (OSHA).

Fire and Explosion Hazard

Flammable gas; heavier than air, density 2.2 (air = 1), flame propagation and flashback fire hazard if the container is placed near a source of ignition; autoignition temperature 472°C (882°F); polymerization may occur at elevated temperatures, which may cause possible rupture of containers; fire-extinguishing measure: stop the flow of gas; water may be used to keep fire-exposed containers cool. Vinyl chloride may decompose under fire conditions, producing the toxic gases carbon monoxide and hydrogen chloride.

Vinyl chloride forms explosive mixtures with air in a wide range; the LEL and UEL values are 3.6% and 33.0% by volume in air, respectively. It may undergo oxidation by atmospheric oxygen, producing an unstable polyperoxide that may explode (MCA 1969). Such a reaction is catalyzed by a variety of contaminants.

22.11 VINYL BROMIDE

DOT Label: Flammable Gas, UN 1085
Formula C_2H_3Br; MW 106.96; CAS [593-60-2]
Structure:

$$H-\overset{\overset{\displaystyle H}{|}}{C}=\overset{\overset{\displaystyle H}{|}}{C}-Br$$

Synonyms: bromoethylene; bromoethene

Uses and Exposure Risk

Vinyl bromide is used as a fire retardant in plastics.

Physical Properties

Colorless gas with a characteristic odor; liquefies at 16.8°C; solidifies at −139.5°C; density of liquid 1.493; insoluble in water, miscible with organic solvents.

Health Hazard

Inhalation of this gas at high concentrations can produce anesthesia and kidney damage. Exposure to 2.5% concentration in air produced anesthetic effects in rats. A 15-minute exposure to 10% concentration was fatal to rats (ACGIH 1986). Kidney injury occurred at a 5% exposure level. Human toxicity data for this compound are not available. The liquid is mild to moderately irritating to the eyes.

Vinyl bromide manifested carcinogenic properties in laboratory animals. It produced liver and blood tumors in rats and mice. Its cancer-causing effects in humans are not known. However, based on its similarity to vinyl chloride, it is suspected to be a human carcinogen.

Exposure Limit

TLV-TWA 5 ppm ($\sim$22 mg/m^3) (ACGIH); Lowest Detectable Level (NIOSH); carcinogenicity: Suspected Human Carcinogen (ACGIH), Animal Sufficient Evidence (IARC).

Fire and Explosion Hazard

Flammable gas; heavier than air, density 3.7 (air = 1), the gas can travel a considerable distance to a source of ignition and flash back; autoignition temperature 530°C (986°F). Vinyl bromide forms explosive mixtures with air; LEL and UEL values are 9.0% and 15.0% by volume in air, respectively. It forms polyperoxide by reaction with atmospheric oxygen on prolonged contact. It decomposes to toxic carbon monoxide and hydrogen bromide under fire conditions.

22.12   1,2-DICHLOROETHANE

EPA Priority Pollutant, RCRA Toxic Waste Number U077; DOT Label: Flammable Liquid, UN 1184

Formula $C_2H_4Cl_2$; MW 98.96; CAS [107-06-2]

Structure:

$$\text{Cl}-\overset{\displaystyle H}{\underset{\displaystyle H}{C}}-\overset{\displaystyle H}{\underset{\displaystyle H}{C}}-\text{Cl}$$

Synonyms: ethylene dichloride; dichloroethylene; ethylene chloride; ethane dichloride

Uses and Exposure Risk

1,2-Dichloroethane is used in the manufacture of acetyl cellulose and vinyl chloride; in paint removers; as a fumigant; as a degreaser; as a wetting agent; and as a solvent for oils, waxes, gums, resins, and rubber.

Physical Properties

Colorless liquid with a pleasant odor; sweet taste; density 1.257 at 20°C; boils at 83.5°C; freezes at −35.5°C; slightly soluble in water (0.8%), miscible with organic solvents.

Health Hazard

The toxic symptoms from exposure to 1,2-dichloroethane include depression of the central nervous system, irritation of the eyes, corneal opacity, nausea, vomiting, diarrhea, ulceration, somnolence, cyanosis, pulmonary edema, and coma. Repeated exposure may produce injury to the kidney and liver. Ingestion of the liquid can cause death. A fatal dose in humans may range between 30 and 50 mL. The liquid is an irritant to the skin and damaging to the eyes.

LC$_{50}$ value, inhalation (rats): 1000 ppm/7 hr

LD$_{50}$ value, oral (rabbits): 860 mg/kg

1,2-Dichloroethane tested positive to the histidine reversion–Ames test and other mutagenic tests. The compound is carcinogenic to animals. Inhalation or oral administration caused lung, gastrointestinal, and skin cancers in mice and rats.

Exposure Limits

TLV-TWA 10 ppm ($\sim$40 mg/m^3) (ACGIH), 1 ppm (NIOSH), 50 ppm (MSHA and OSHA); ceiling 2 ppm/15 min (NIOSH); carcinogenicity: Animal Sufficient Evidence, Human Limited Evidence (IARC).

Fire and Explosion Hazard

Flammable liquid; burns with a smoky flame; flash point (closed cup) 13°C (56°F), (open cup) 18°C (65°F); vapor pressure 62 torr at 20°C; the vapor is heavier than air and can travel a considerable distance to a source of ignition and flash back; autoignition temperature 413°C (775°F); fire-extinguishing agent: dry chemical, CO$_2$, or foam; water may be used to keep fire-exposed containers cool and to disperse the vapors and flush away any spill.

1,2-Dichloroethane forms explosive mixtures with air, with LEL and UEL values of 6.2% and 16.0% by volume in air, respectively. Its reactions with alkali metals, powdered aluminum, or magnesium can be violent. It forms explosive mixtures with nitrogen tetroxide.

Disposal/Destruction

1,2-Dichloroethane is burned in a chemical incinerator equipped with an afterburner and scrubber. It may be buried in an approved landfill site in a lab pack.

22.13   1,2-DIBROMOETHANE

EPA Priority Pollutant, RCRA Toxic Waste Number U067; DOT Identification Number, UN 1605

Formula C$_2$H$_4$Br$_2$; MW 187.88; CAS [106-93-4]

Structure:

$$\begin{array}{ccc} & H & H \\ & | & | \\ Br-&C-&C-Br \\ & | & | \\ & H & H \end{array}$$

Synonyms: ethylene dibromide; ethylene bromide

Uses and Exposure Risk

1,2-Dibromoethane (EDB) is used as a fumigant for grains, in antiknock gasolines, as a solvent, and in organic synthesis.

Physical Properties

Colorless, heavy liquid with a mild sweet odor; density 2.172 at 25°C; boils at 131°C; freezes at 9°C; slightly soluble in water (0.4%), miscible with most organic solvents; vapor pressure 11 torr at 20°C.

Health Hazard

1,2-Dibromoethane is toxic by inhalation, ingestion, or skin contact. The acute toxic symptoms are depression of the central nervous system, irritation and congestion of lungs, hepatitis, and renal damage. Chronic exposure can produce conjunctivitis, bronchial irritation, headache, depression, loss of appetite, and loss of weight. Recovery occurs after cessation of exposure. Prolonged or repeated exposures to high concentrations can be fatal to animals and humans. Lethal concentration for a 2-hour exposure period is 400 ppm in rats.

1,2-Dibromoethane is moderate to highly toxic by ingestion. Its toxicity is far greater than that of 1,2-dichloroethane. An oral intake of 5 to 10 mL of the liquid can be fatal to humans. Death occurs from necrosis of the liver and kidney damage. The oral LD$_{50}$ values varied between 50 and 125 mg/kg for different species of laboratory animals.

Vapors are irritant to the eyes. Contact with the liquid can damage vision. Skin contact may produce severe irritation and blistering.

Mutagenic tests were positive, while the histidine reversion–Ames test gave inconclusive results (NIOSH 1986). 1,2-Dibromoethane is carcinogenic to animals and is suspected to cause cancer in humans. Inhalation of this compound produced tumors in the lungs and nose in mice and rats. Oral administration caused cancers in the liver and gastrointestinal tract.

Exposure Limits

TLV-TWA 20 ppm (145 mg/m^3) (MSHA and OSHA), none (ACGIH); Ceiling 30 ppm (OSHA); carcinogenicity: Suspected Human Carcinogen (ACGIH), Animal Sufficient Evidence (IARC).

Fire and Explosion Hazard

Noncombustible liquid. EDB reacts vigorously with alkali metals, magnesium, aluminum; caustic alkalies; and strong oxidizing substances.

Disposal/Destruction

EDB is mixed with a combustible solvent and burned in a chemical incinerator equipped with an afterburner and scrubber.

22.14   1,1-DICHLOROETHYLENE

EPA Designated Toxic Waste, RCRA Waste Number U078

Formula C$_2$H$_2$Cl$_2$; MW 96.94; CAS [75-35-4]

Structure:

$$\underset{Cl}{\overset{Cl}{\diagdown}}C=CH_2$$

Synonyms: 1,1-dichloroethylene; vinylidene chloride

Uses and Exposure Risk

1,1-Dichloroethylene (1,1-DCE) is used to produce vinylidene copolymers for films and coatings.

Physical Properties

Colorless liquid with a mild chloroform-like odor; density 1.213 at 20°C; boils at 31.7°C; freezes at −122.5°C; slightly soluble in water, miscible with organic solvents.

Health Hazard

1,1-DCE exhibits low acute toxicity. Vapors are irritant to the mucous membranes. At high concentrations it produces narcotic effects. Chronic exposure to a 50-ppm concentration for 8 hours/daily, 5 days/week for 6 months resulted in liver and kidney injury in experimental animals. The liquid in contact with the eyes causes irritation. The LC$_{50}$ value in rats is within the range 6300 ppm for a 4-hour exposure period. The oral toxicity is low. A lethal dose by subcutaneous administration is 3700 mg/kg in rabbits. Ingestion can cause nausea and vomiting.

Tests on laboratory animals indicate that 1,1-DCE is cancer causing. Rats and mice subjected to 12 months' inhalation of this compound developed tumors of the liver, kidney, skin, and blood. Carcinogenicity in humans is not reported.

Exposure Limits

TLV-TWA 5 ppm (~20 mg/m^3) (ACGIH); TLV-STEL 20 ppm (ACGIH); carcinogenicity: Animal Limited Evidence, Human Inadequate Evidence (IARC).

Fire and Explosion Hazard

Flammable liquid; flash point (closed cup) −18°C (0°F) (flash point data reported in the literature differ); vapor pressure 500 torr at 20°C; vapor density 3.34 (air = 1); the vapor is heavier than air and can travel a considerable distance to a source of ignition and flash back; autoignition temperature 570°C (1058°F); fire-extinguishing agent: dry chemical, CO_2, or foam; use water to keep fire-exposed containers cool and to flush any spill.

1,1-DCE vapors form explosive mixtures with air within the range 7.3–16.0% by volume in air. It polymerizes at elevated temperatures. If polymerization occurs in a closed container, the container may rupture violently. Polymerization is inhibited in the presence of 200 ppm of hydroquinone monomethyl ether (Aldrich 1997). It forms a white deposit of peroxide on long standing which may explode. It decomposes when involved in fire, producing toxic hydrogen chloride. Reactions with concentrated mineral acids are exothermic.

Disposal/Destruction

1,1-DCE is burned in a chemical incinerator equipped with an afterburner and scrubber.

22.15 1,1,1-TRICHLOROETHANE

EPA Priority Pollutant, RCRA Toxic Waste Number U226; DOT Label: none, Number UN 2831

Formula $C_2H_3Cl_3$; MW 133.40; CAS [71-55-6]

Structure:

$$H-\overset{\overset{\displaystyle H}{|}}{\underset{\underset{\displaystyle H}{|}}{C}}-\overset{\overset{\displaystyle Cl}{|}}{\underset{\underset{\displaystyle Cl}{|}}{C}}-Cl$$

Synonyms: methylchloroform; strobane

Uses and Exposure Risk

1,1,1-Trichloroethane is used as a cleaning solvent for cleaning metals and plastic molds.

Physical Properties

Colorless liquid with a mild chloroform-like odor; density 1.338 at 20°C; bp 74°C;

mp −32.5°C; very slightly soluble in water (0.07%), soluble in organic solvents; vapor pressure 100 torr at 20°C.

Health Hazard

The oral and inhalation toxicity of 1,1,1-trichloroethane is of low order in animals and humans. It is an anesthetic at high concentrations. Exposure to its vapors at a 1.5% concentration in air may be lethal to humans. Death may result from anesthesia and/or cardiac sensitization. Prolonged skin contact may cause defatting and reddening of eyes. Vapors are irritant to the eyes and mucous membranes.

The acute oral toxicity is low in test animals. The oral LD_{50} values in rabbits and guinea pigs are 5660 and 9470 mg/kg, respectively (NIOSH 1986). The carcinogenicity of this compound in animals and humans is not known.

Exposure Limits

TLV-TWA   350 ppm   ($\sim$1900 mg/m^3) (ACGIH, MSHA, and OSHA); TLV-STEL 450 ppm ($\sim$2450 mg/m^3) (ACGIH); IDLH 1000 ppm (NIOSH).

Fire and Explosion Hazard

Noncombustible liquid at room temperature, but moderately flammable at higher temperatures; autoignition temperature 537°C (998°F). Vapors of 1,1,1-trichloroethane form explosive mixtures with air within the narrow range 8.0–10.5% by volume in air. Its reactions with sodium, potassium, sodium–potassium alloy, powdered magnesium, or powdered aluminum may cause explosions. It forms explosive mixtures with nitrogen tetraoxide (NFPA 1997).

Disposal/Destruction

It is mixed with an excess of combustible solvent and burned in a chemical incinerator equipped with an afterburner and scrubber. It may be disposed of in an approved landfill in lab packs.

22.16 TRICHLOROETHYLENE

EPA Priority Pollutant, RCRA Toxic Waste Number U228; DOT Identification Number UN 1710

Formula C_2HCl_3; MW 131.38; CAS [79-01-6]

Structure:

$$\underset{Cl}{\overset{Cl}{\diagdown}}C=C\underset{Cl}{\overset{H_1}{\diagup}}$$

Synonyms: trichloroethene; ethylene trichloride

Uses and Exposure Risk

Trichloroethylene is used as a solvent, in dry cleaning, in degreasing, and in limited use as a surgical anesthetic.

Physical Properties

Colorless liquid with a chloroform-like odor; density 1.465 at 20°C; boils at 87°C; solidifies at −85°C; very slightly soluble in water (0.1%), miscible with organic solvents; vapor pressure 58 torr at 20°C.

Health Hazard

The toxic effects manifested in humans from inhaling trichloroethylene vapors are headache, dizziness, drowsiness, fatigue, and visual disturbances. A 2-hour exposure to a 1000-ppm concentration affected the visual perception. Higher concentrations can produce narcotic effects. Heavy exposures may cause death due to respiratory failure or cardiac arrest. A 4-hour exposure to 8000 ppm was lethal to rats. Chronic exposure caused increase in kidney and liver weights in test animals.

The symptoms of poisoning from oral intake of trichloroethylene are nausea, vomiting, diarrhea, and gastric disturbances. The acute oral toxicity, however, is low. The oral LD_{50} value in mice is in the range 2500 mg/kg. Trichloroethylene metabolizes to trichloroacetic acid, which is excreted in the urine.

Although trichloroethylene exhibits low toxicity, its metabolite trichloroethanol, and oxidative degradation products phosgene, $COCl_2$, and chlorine, can cause severe unexpected health hazards. Kawakami and associates (1988) reported a case of Steven–Johnson syndrome in a worker in a printing factory. In another case, fire on a stove in a metal-degreasing workplace produced phosgene and chlorine inhalation, which caused dyspnea, fever, and fatigue.

Trichloroethylene exhibited evidence of carcinogenicity in laboratory animals. Oral administration produced liver tumors, while inhalation caused lung and blood tumors in mice and rats.

Exposure Limits

TLV-TWA 50 ppm (~270 mg/m³) (ACGIH), 100 ppm (MSHA and OSHA); TLV-STEL 200 ppm (ACGIH); ceiling 200 ppm (OSHA); carcinogenicity: Animal Limited Evidence, Human Inadequate Evidence (IARC).

Fire and Explosion Hazard

Noncombustible liquid; moderately flammable at higher temperatures; autoignition temperature 410°C (770°F); vapor forms explosive mixtures with air, with LEL and UEL values of 8.0% and 10.5% by volume in air, respectively (NFPA 1997).

Trichloroethylene reacts explosively with sodium, potassium, or lithium. Alkaline–earth metals in a finely divided state may ignite the liquid upon mixing. A violent reaction may occur with caustic soda or caustic potash. Trichloroethylene forms explosive mixtures with nitrogen tetroxide.

Disposal/Destruction

Trichloroethylene is mixed with an excess of combustible solvent and burned in a chemical incinerator equipped with an afterburner and scrubber. It may be destroyed in aqueous waste streams or groundwater by UV peroxidation, involving treatment with hydrogen peroxide in the presence of UV light (Yost 1989; Sundstrom et al. 1990). Oku and Kimura (1990) have reported reductive dechlorination using sodium naphthalenide in tetrahydrofuran at 0°C for 10 minutes. Chlorine is removed as sodium chloride to the extent of 97–100%.

22.17   TETRACHLOROETHYLENE

EPA Priority Pollutant, RCRA Toxic Waste Number U210; DOT Identification Number UN 1897

Formula C_2Cl_4; MW 165.82; CAS [127-18-4]

Structure:

$$\begin{array}{c} Cl \\ \diagdown \\ Cl \diagup \end{array} C = C \begin{array}{c} Cl \\ \diagup \\ \diagdown Cl \end{array}$$

Synonyms: tetrachloroethene; ethylene tetrachloride; perchloroethylene

Uses and Exposure Risk

Tetrachloroethylene is used as a solvent, in drycleaning, and in metal degreasing.

Physical Properties

Colorless liquid with an odor of ether; density 1.623 at 20°C; boils at 121°C; solidifies at −22°C; insoluble in water, miscible with organic solvents; vapor pressure 19 torr at 25°C.

Health Hazard

Exposure to tetrachloroethylene can produce headache, dizziness, drowsiness, incoordination, irritation of eyes, nose, and throat, and flushing of neck and face. Exposure to high concentrations can produce narcotic effects. The primary target organs are the central nervous system, mucous membranes, eyes, and skin. The kidneys, liver, and lungs are affected to a lesser extent. Symptoms of depression of the central nervous system are manifested in humans from repeated exposure to 200 ppm for 7 hours/day. Chronic exposure to concentrations ranging from 200 to 1600 ppm caused drowsiness, depression, and enlargement of the kidneys and livers in rats and guinea pigs. A 4-hour exposure to 4000 ppm of vapor in air was lethal to rats.

Ingestion of tetrachloroethylene may produce toxic effects ranging from nausea and vomiting to somnolence, tremor, and ataxia. The oral toxicity, is low, however, with LD_{50} ranging between 3000 and 9000 mg/kg in animals. Skin contact with the liquid may cause defatting and dermatitis of skin.

Evidence of carcinogenicity of this compound has been noted in test animals subjected to inhalation or oral administration. It caused tumors in the blood, liver, and kidney in rats and mice. Carcinogenicity in humans is not reported.

Exposure Limits

TLV-TWA 50 ppm (~325 mg/m^3) (ACGIH), 100 ppm (MSHA and OSHA); TLV-STEL 200 ppm (ACGIH); carcinogenicity: Animal Limited Evidence.

Fire and Explosion Hazard

Noncombustible liquid. When mixed with alkali- or powdered alkaline–earth metals, it may explode or produce spark upon heavy impact. The reactivity of this compound at ambient temperatures is very low.

Disposal/Destruction

Tetrachloroethylene is mixed with an excess of combustible solvent and burned in a chemical incinerator equipped with an afterburner and scrubber. Other methods of destruction of tetrachloroethylene involve reductive dechlorination using sodium naphthalenide, and destructive oxidation in the aqueous stream by ultraviolet peroxidation (see Section 22.16).

22.18 1,1,2-TRICHLORO-1,2,2-TRIFLUOROETHANE

Formula $C_2Cl_3F_3$; MW 187.37; CAS [76-13-1]

Structure:

$$Cl-\underset{\underset{F}{|}}{\overset{\overset{Cl}{|}}{C}}-\underset{\underset{F}{|}}{\overset{\overset{Cl}{|}}{C}}-F$$

Synonyms: fluorocarbon-113; Freon 113

Uses and Exposure Risk

1,1,2-Trichloro-1,2,2-trifluoroethane is used as a refrigerant and as a drycleaning solvent. It is also used as an extraction solvent for analyzing petroleum hydrocarbons, oils, and greases.

Physical Properties

Colorless liquid with a characteristic odor; density 1.563 at 25°C; boils at 47.6°C; freezes at −35°C; insoluble in water, miscible with organic solvents; vapor pressure 284 torr at 20°C.

Health Hazard

The acute oral toxicity of this compound is very low. The oral LD_{50} value in rats is 43,000 mg/kg. The inhalation toxicity is also low. Exposure to high concentrations can produce a weak narcotic effect, cardiac sensitization, and irritation of respiratory passage. Chronic exposure caused liver enlargement in rats. A 6-hour exposure to 87,000 ppm of 1,1,2-trichloro-1,2,2-trifluoroethane was lethal to rats. Ingestion of the liquid may cause nausea, lethargy,

nervousness, and tremor. The irritant action of the liquid was mild on rabbits' skin.

Exposure Limits

TLV-TWA   1000 ppm   ($\sim$7600 mg/m^3) (ACGIH, MSHA, and OSHA); TLV-STEL 1250 ppm (ACGIH).

Fire and Explosion Hazard

Noncombustible liquid.

Disposal/Destruction

The compound is buried in a landfill site approved for the disposal of chemical waste.

22.19 HALOTHANE

Formula $C_2HBrClF_3$; MW 197.39; CAS [151-67-7]

Structure:

$$F-\underset{\underset{F}{|}}{\overset{\overset{F}{|}}{C}}-\underset{\underset{Cl}{|}}{\overset{\overset{Br}{|}}{C}}-H$$

Synonyms: 2-bromo-2-chloro-1,1,1-trifluoroethane; bromochlorotrifluoroethane; 2,2,2-trifluoro-1-chloro-1-bromoethane; fluothane

Uses and Exposure Risk

Halothane is used as a clinical anesthetic.

Physical Properties

Colorless volatile liquid with a sweetish odor; density 1.871 at 20°C; boils at 50.2°C; vapor pressure 243 torr at 20°C; slightly soluble in water (0.345%), soluble in organic solvents.

Health Hazard

Health hazard may arise from exposure to high concentrations of vapors. The target sites are the central nervous system, cardiovascular system, and liver. The toxic effects include depression of the central nervous system, nausea, vomiting, increased body temperature, excitability, arrhythmias, vasodilatation, hepatitis, and liver necrosis. An anesthetic effect in humans is noted at an exposure level of 10,000 ppm. Repeated exposure to halothane at anesthetic concentrations may result in hepatic lesions and necrosis. Contact of liquid with the eyes may result in severe irritation. The lethal concentration in humans for a 3-hour exposure period is in the range of 7000 ppm.

LC$_{50}$ value, inhalation (mice): 22,000 ppm/ 10 min

LD$_{50}$ value, oral (guinea pigs): 6000 mg/kg

No mutagenic effect was observed. Carcinogenicity in animals has not been reported.

Exposure Limit

TLV-TWA 50 ppm ($\sim$400 mg/m^3) (ACGIH).

Fire and Explosion Hazard

Noncombustible liquid.

Disposal/Destruction

The compound is mixed with a combustible solvent and burned in a chemical incinerator equipped with an afterburner and scrubber.

22.20   1,2,3-TRICHLOROPROPANE

Formula C$_3$H$_5$Cl$_3$; MW 147.43; CAS [96-18-4]

Structure:

$$\text{Cl} - \text{CH}_2 - \overset{\overset{\displaystyle \text{Cl}}{|}}{\text{CH}} - \text{CH}_2 - \text{Cl}$$

Synonyms: allyl trichloride; glycerol trichlorohydrin; glyceryl trichlorohydrin; trichlorohydrin

Uses and Exposure Risk

1,2,3-Trichloropropane is used as a solvent and as an intermediate in organic synthesis.

Physical Properties

Colorless to yellowish liquid with a chloroform-like odor; density 1.389 at 20°C; boils at 156°C; freezes at −14.7°C; slightly soluble in water (0.2%), miscible with organic solvents.

Health Hazard

Inhalation of its vapors can produce depression of the central nervous system, which can progress to narcosis and convulsion as the concentration increases. A 30-minute exposure to a 5000-ppm concentration caused convulsions in rats. Acute as well as chronic exposure to high concentrations can cause liver damage. 1,2,3-Trichloropropane is more toxic than its 1,1,1-isomer. Acute oral toxicity is moderate, with LD$_{50}$ values ranging between 300 and 550 mg/kg in different species of experimental animals. The liquid is a strong irritant to the eyes.

Exposure Limits

TLV-TWA 10 ppm ($\sim$60 mg/m^3) (ACGIH), 50 ppm (MSHA, OSHA, and NIOSH); IDLH 1000 ppm (NIOSH).

Fire and Explosion Hazard

Combustible liquid; flash point (closed cup) 73°C (164°F), (open cup) 82°C (180°F). Vapors of 1,2,3-trichloropropane form explosive mixtures with air, with LEL and UEL values of 3.2% and 12.6% by volume in air, respectively. The compound reacts vigorously with alkali metals, powdered magnesium, or aluminum; caustic alkalies; and oxidizers.

Disposal/Destruction

The compound is burned in a chemical incinerator equipped with an afterburner and scrubber.

22.21 HEXACHLOROETHANE

EPA Designated Toxic Waste, RCRA Waste Number U131

Formula C_2Cl_6; MW 236.72; CAS [67-72-1]

Structure:

Synonyms: carbon hexachloride; ethane hexachloride; hexachloroethylene; ethylene hexachloride; perchloroethane

Uses and Exposure Risks

Hexachloroethane is used as a solvent, in fireworks and smoke devices; in explosives, in celluloid, as an insecticide, and as a rubber vulcanizing accelerator. Earlier it was used as an anthelmintic for livestock.

Physical Properties

Colorless crystalline solid with camphor-like odor; sublimes at 187°C; practically insoluble in water, soluble in organic solvents.

Health Hazard

Vapors of hexachloroethane are an irritant to the eyes and mucous membranes. Oral doses of 1000 mg/kg produced weakness, staggering gait, and twitching muscles in dogs. Rabbits fed 1000 mg/kg for 12 days developed necrosis; a lower amount, 320 mg/kg, caused liver degeneration; no effects were observed at a dose level of 100 mg/kg (Weeks 1979).

Acute inhalation toxicity is of a low order in animals. Subacute toxic effects in dogs exposed to 260-ppm vapors of hexachloroethane for 6 hours per day, 5 days a week for 6 weeks were tremors, ataxia, hypersalivation, head bobbling, and facial muscular fasciculations (Weeks 1979). The lethal concentration in rats is 5900 ppm from an 8-hour exposure.

LD_{50} value, oral (rats): 4460 mg/kg

Tests for mutagenicity and teratogenicity were negative. The carcinogenic potential of hexachloroethane was noted in test animals only at extremely heavy dosages given continuously for a long period of time (ACGIH 1986). It caused liver tumors in mice.

Exposure Limits

TLV-TWA 10 ppm ($\sim$100 mg/m^3) (ACGIH), 1 ppm (MSHA and OSHA), Lowest Feasible Limit (NIOSH); carcinogenicity: Animal Limited Evidence (IARC).

Fire and Explosion Hazard

Noncombustible solid; vapor pressure 0.22 torr at 20°C; very low reactivity.

Disposal/Destruction

Hexachloroethane is dissolved in a combustible solvent and burned in an incinerator equipped with an afterburner and scrubber.

22.22 BENZYL CHLORIDE

EPA Designated Acute Hazardous Waste, RCRA Waste Number P028; DOT Label: Corrosive Material, UN 1738

Formula C_7H_7Cl; MW 126.59; CAS [100-44-7]

Structure:

Synonyms: tolyl chloride; α-chlorotoluene; chloromethylbenzene; chlorophenylmethane

Uses and Exposure Risk

Benzyl chloride is used in the manufacture of dyes, artificial resins, tanning agents, pharmaceuticals, plasticizers, perfumes, lubricants, and miscellaneous benyl compounds.

Physical Properties

Colorless to slightly yellow liquid with a pungent aromatic odor; density 1.100 at 20°C; boils at 179°C; solidifies at −43°C; insoluble in water, miscible with most organic solvents.

Health Hazard

Benzyl chloride is a corrosive liquid. Contact with the eyes can cause corneal injury. Exposure to its vapors can produce intense irritation of the eyes, nose, and throat. High concentrations may cause lung edema and depression of the central nervous system. Flury and Zernik (1931) stated that exposure to 16 ppm for 1 minute was intolerable to humans. The LC_{50} values for a 2-hour exposure in mice and rats are 80 and 150 ppm, respectively. The subcutaneous LD_{50} value in rats is 1000 mg/kg (NIOSH 1986).

Benzyl chloride tested positive to the histidine reversion–Ames test for mutagenicity. Subcutaneous administration of this compound in laboratory animals caused tumors at the site of application.

Exposure Limits

TLV-TWA 1 ppm (~5 mg/m^3) (ACGIH, MSHA, and OSHA); IDLH 10 ppm (NIOSH); carcinogenicity: Animal Limited Evidence, Human Inadequate Evidence (IARC).

HALOGEN-SUBSTITUTED AROMATICS

22.23   CHLOROBENZENE

EPA Priority Pollutant; DOT Label: Flammable Liquid, UN 1134

Formula C_6H_5Cl; MW 112.56; CAS [108-90-7]
Structure:

Synonyms: monochlorobenzene; chlorobenzol; phenyl chloride; benzene chloride

Uses and Exposure Risk

Chlorobenzene is used as a solvent for paints, as a heat transfer medium, and in the manufacture of phenol and aniline.

Physical Properties

Colorless liquid with a faint almond-like odor; density 1.107 at 20°C; boils at 131°C; solidifies at −55°C; insoluble in water, miscible with organic solvents.

Health Hazard

The acute inhalation and oral toxicity is low in test animals. The target organs are the respiratory system, central nervous system, liver, skin, and eyes. The toxic symptoms from inhalation of its vapors are drowsiness, incoordination, and liver injury. Repeated exposures at 1000 ppm caused changes in the lungs, liver, and kidney; no effects were observed at 200 ppm (Patty 1963). Exposure above 1000 ppm produced narcosis in animals. The vapors are an irritant to the eyes and nose.

Ingestion of chlorobenzene produced ataxia, respiratory distress, and liver damage in rats. The oral LD_{50} values in rabbits and rats are in the range 2200 and 2900 mg/kg, respectively. This compound may be carcinogenic to animals. There is some evidence manifesting its carcinogenic potential in mice, which, however, is not fully confirmed.

Exposure Limits

TLV-TWA 75 ppm ($\sim$345 mg/m^3) (ACGIH, MSHA, OSHA, and NIOSH); IDLH 2400 ppm.

Fire and Explosion Hazard

Flammable liquid; flash point (closed cup) 29°C (84°F); vapor pressure 8.8 torr at 20°C; autoignition temperature 638°C (1180°F).

Chlorobenzene vapors form explosive mixtures with air within the range 1.3–7.1% by volume in air. Dimethyl sulfoxide decomposes violently in contact with chlorobenzene (NFPA 1997). Many metal perchlorates, such as those of silver and mercury, may form shock-sensitive solvated perchlorates that may explode on impact.

Disposal/Destruction

Chlorobenzene is burned in a chemical incinerator equipped with an afterburner and scrubber. Waste solvent may be buried in an approved hazardous waste landfill site in lab packs.

22.24 1,2-DICHLOROBENZENE

EPA Priority Pollutant, RCRA Toxic Waste Number U070; DOT Identification Number, UN 1591

Formula C$_6$H$_4$Cl$_2$; MW 147.00; CAS [95-50-1]

Structure:

Synonyms: *o*-dichlorobenzene; orthodichlorobenzol

Uses and Exposure Risk

1,2-Dichlorobenzene is used as a solvent; as a fumigant; as an insecticide for termites; as a degreasing agent for metals, wool, and leather; and as a heat transfer medium.

Physical Properties

Colorless to pale yellow liquid with a faint aromatic odor; density 1.306 at 30°C; boils at 180.5°C; freezes at −17°C; insoluble in water, miscible with organic solvents.

Health Hazard

1,2-Dichlorobenzene exhibits low acute toxicity by inhalation, ingestion, and skin absorption. It is more toxic than chlorobenzene. The symptoms are lacrimation, depression of central nervous system, anesthesia, and liver damage. Lethal concentration in rats for a 7-hour exposure period is in the range of 800 ppm. The oral LD$_{50}$ value in rabbits is 500 mg/kg. There is no evidence of carcinogenicity in animals.

Exposure Limits

Ceiling 50 ppm ($\sim$300 mg/m^3) (MSHA, OSHA, and NIOSH); IDLH 1700 ppm (NIOSH).

Fire and Explosion Hazard

Combustible liquid; flash point (closed cup) 66°C (151°F); vapor pressure 1.2 torr at 20°C; autoignition temperature 648°C (1198°F). Vapors of 1,2-dichlorobenzene form explosive mixtures with air within the range 2.2–9.2% by volume in air.

Disposal/Destruction

1,2-Dichlorobenzene is burned in a chemical incinerator equipped with an afterburner and scrubber. It may be buried in an approved landfill in a lab pack, an open-head steel drum filled with small containers of wastes securely packed with vermiculite.

22.25 1,4-DICHLOROBENZENE

EPA Priority Pollutant, RCRA Toxic Waste Number U072

Formula $C_6H_4Cl_2$; MW 147.00; CAS [106-46-7]

Structure:

Synonyms: *p*-dichlorobenzene; paradichlorobenzol; Parazene

Uses and Exposure Risk

1,4-Dichlorobenzene is used as a fumigant and as an insecticide.

Physical Properties

Colorless crystalline solid with a characteristic odor; sublimes at ambient temperature; melts at 54°C; boils at 174°C; insoluble in water, soluble in organic solvents.

Health Hazard

Toxic symptoms are headache, weakness, dizziness, nausea, vomiting, diarrhea, loss of weight, and injury to liver and kidney. These symptoms occur from repeated inhalation of high concentrations of vapors or from ingestion. The vapors are an irritant to the eyes, throat, and skin. Chronic exposure may cause jaundice and cirrhosis. The oral LD_{50} value in mice is in the range 3000 mg/kg. The fatal oral dose in humans is estimated to be 40–50 g. Carcinogenic studies on animals have not produced adequate evidence of any cancer-causing action.

Exposure Limits

TLV-TWA 75 ppm (~450 mg/m^3) (MSHA, OSHA, and NIOSH); IDLH 1000 ppm (NIOSH).

Fire and Explosion Hazard

Combustible solid; flash point (closed cup) 65.5°C (150°F). The vapors of 1,4-dichlorobenzene form explosive mixtures with air, with an LEL value of 2.5%; UEL data not available.

Disposal/Destruction

1,4-Dichlorobenzene is dissolved in a combustible solvent and burned in a chemical incinerator equipped with an afterburner and scrubber. It may be mixed with vermiculite and then with dry sodium carbonate and slaked lime, wrapped in paper, and burned in a chemical incinerator equipped with an afterburner and scrubber (Aldrich 1997).

22.26 HEXACHLOROBENZENE

EPA Designated Toxic Waste, RCRA Waste Number U127; DOT Label: Poison B, UN 2729

Formula C_6Cl_6; MW 284.76; CAS [118-74-1]

Structure:

Synonyms: perchlorobenzene; phenylperchloryl

Uses and Exposure Risk

Hexachlorobenzene is used as a fungicide and as an intermediate in organic synthesis.

Physical Properties

Crystalline solid; melts at 231°C; boils at 325°C; insoluble in water, sparingly soluble in cold alcohol, dissolves readily in benzene, ether, and chloroform.

Health Hazard

The acute oral and inhalation toxicity of hexachlorobenzene is low in test animals. Repeated ingestion of this compound may produce porphyria hepatica (increased formation and excretion of porphyrin) caused by disturbances in liver metabolism. The oral LD_{50} value in rabbits is 2600 mg/kg; the inhalation LC_{50} value from a single exposure is 1800 mg/m^3 (NIOSH 1986). The occupational health hazard from inhalation should be very low because of its very low vapor pressure (0.00001 torr).

Hexachlorobenzene causes cancer in animals. Oral administration of this compound for 18 weeks to 2 years caused tumors in the liver, kidney, thyroid, and blood in rats, mice, and hamsters. It is a suspected human carcinogen, evidence of which occurs to a limited extent.

Exposure Limits

No exposure limit has been set for this compound. Carcinogenicity: Animal Sufficient Evidence, Human Limited Evidence (IARC).

Fire and Explosion Hazard

Noncombustible solid; very low reactivity. Reaction with dimethyl formamide is reported to be violent at temperatures above 65°C (NFPA 1997).

Disposal/Destruction

Hexachlorobenzene is dissolved in a combustible solvent and burned in a chemical incinerator equipped with an afterburner and scrubber.

22.27 HEXACHLORONAPHTHALENE

Formula $C_{10}H_2Cl_6$; MW 334.74; CAS [1335-87-1]

Structure:

Synonym: Halowax 1014

Uses and Exposure Risk

Hexachloronaphthalene is used in electric wire insulation and as an additive to lubricants.

Physical Properties

Light yellow solid with an aromatic odor; melts at 137°C; boils between 343 and 387°C; insoluble in water, soluble in organic solvents.

Health Hazard

The symptoms of acute toxicity are chloracne or an acneform of dermatitis, nausea, weakness, confusion, and jaundice. The target organs are mainly the liver and skin. Mixtures of penta- and hexachloronaphthalene at concentrations between 1 and 2 mg/m^3 have been attributed to fatal cases of yellow atrophy of the liver (Elkins 1959). The LD_{50} values for this compound are not reported.

Exposure Limits

TLV-TWA (skin) 0.2 mg/m^3 (ACGIH, MSHA, OSHA, and NIOSH); IDLH 2 mg/m^3 (NIOSH).

Fire and Explosion Hazard

Noncombustible solid; vapor pressure <1 torr at 20°C.

Disposal/Destruction

Hexachloronaphthalene is dissolved in a combustible solvent and burned in a chemical incinerator equipped with an afterburner and scrubber.

22.28 MISCELLANEOUS HALOCARBONS

Most halogenated hydrocarbons exhibit a low order of toxicity. A discussion on the general toxicity of these compounds is highlighted in Section 22.1. Presented in Table 22.4 are the toxicity data for miscellaneous individual compounds. The flammability and explosive reactions of these compounds are of little significance and are therefore omitted from the table. As mentioned earlier (Section 22.1), only the low-molecular-weight fluorocarbons and some chloro compounds are flammable. Most halogenated hydrocarbons are noncombustible substances.

TABLE 22.4   Toxicity of Miscellaneous Halogenated Hydrocarbons

Compound/Synonyms/ CAS No.	MS/Formula/ Structure	Toxicity
1,1-Difluoroethane (ethylidene fluoride, ethylidene difluoride, Freon-152, Halocarbon-152A) [75-37-6]	$C_2H_4F_2$ 66.06	Toxicity of this gas is very low; causes anesthesia and congestion of lungs only at very high concentrations; a 10-min exposure to a 50% concentration was lethal to rats; DOT Label: Flammable Gas, UN 1030
1,1-Dichloroethane (ethylidene dichloride) [75-34-3]	$C_2H_4Cl_2$ 98.96	EPA Priority Pollutant, RCRA Toxic Waste Number U076; low oral toxicity; LD_{50} oral (rats): 725 mg/kg; carcinogenic to animals, caused uterine tumor in mice; exposure limits: TLV-TWA 100 ppm ($\sim$400 mg/m^3) (OSHA), 200 ppm (ACGIH and MSHA); STEL 250 ppm (ACGIH); DOT Label: Flammable Liquid, UN 2362
1,1-Dibromoethane (ethylidene dibromide) [557-91-5]	$C_2H_4Br_2$ 187.88	Less toxic than its 1,2-isomer; LD_{50} rectal (rabbits) 1250 mg/kg; human toxicity data not available
trans-1,2-Dichloroethylene (*trans*-acetylene dichloride) [156-60-5]	$C_2H_2Cl_2$ 96.94 $Cl-CH=CH-Cl$	Low acute toxicity; inhalation of vapors may cause somnolence and ataxia; narcotic at high concentrations; a 10-min exposure to 5000 ppm may produce such effects; a lethal concentration in mice for a 2-hr exposure period is 7.5%; EPA Designated Toxic Waste, RCRA Waste Number U079
cis-1,2-Dichloroethylene (*cis*-acetylene dichloride) [156-59-2]	$C_2H_2Cl_2$ 96.94	Low acute toxicity; narcotic at high concentrations; 2-hr exposure to 6.5% vapor was lethal to mice

TABLE 22.4 *(Continued)*

Compound/Synonyms/ CAS No.	MS/Formula/ Structure	Toxicity
1,1,2-Trichloroethane (ethane trichloride, β-trichloro-ethane) [79-00-5]	$C_2H_3Cl_3$ 133.40 Cl Cl H—C—C—H Cl H	EPA Priority Pollutant, RCRA Toxic Waste Number U227; moderately toxic by inhalation and ingestion; more toxic than its 1,1,1-isomer; vapor is irritant to eyes and mucous membranes; the toxic symptoms from ingestion of this compound are somnolence, nausea, vomiting, ulceration, hepatitis, and necrosis; inhalation of 500 ppm for 8 hr was lethal to rats: LD_{50} oral (rats): 580 mg/kg; exhibited evidence of carcinogenicity in test animals, caused liver and endocrine tumors in mice; exposure limits: TLV-TWA 10 ppm ($\sim$45 mg/m^3) (ACGIH, MSHA, and OSHA); carcinogenicity: Animal Limited Evidence (IARC)
1,1,1-Trichloropropane [7789-89-1]	$C_3H_5Cl_3$ 147.43 Cl Cl—C—CH$_2$—CH$_3$ Cl	Acute toxicity by inhalation and ingestion is low; causes CNS depression, narcosis, and liver damage at high concentrations; lethal concentration for a 4-hr exposure is 8000 ppm in rats; LD_{50} oral (rats): 7460 mg/kg; irritant to skin (mild) and eyes (severe); a 15-min exposure to 100 ppm may cause irritation of eyes in humans
1,1,2-Trichloropropane [598-77-6]	$C_3H_5Cl_3$ 147.43 Cl Cl CH—CH—CH$_3$ Cl	Low toxicity; more toxic than 1,1,1-isomer but less toxic than 1,2,3-isomer; toxic symptoms are CNS depression, narcosis, and liver damage; the liquid is highly irritating to rabbit eyes; mild irritant action on rabbits' skin; LC_{50} inhalation (rats): 2000 ppm/4 hr; LD_{50} oral (rats): 1230 mg/kg
1,2,2-Trichloropropane [3175-23-3]	$C_3H_5Cl_3$ 147.43 Cl Cl—CH$_2$—C—CH$_3$ Cl	Low toxicity; the toxic symptoms and the doses required to produce such effects are similar to those of 1,1,2-isomer; the effects are CNS depression, narcosis, and liver injury; LD_{50} oral (rats) in the range 1200 mg/kg
Bromoform (tribromomethane, methenyl tribromide) [75-25-2]	CHBr$_3$ 252.75 Br H—C—Br Br	EPA Priority Pollutant, RCRA Waste Number U225; DOT Label: Poison B, UN 2515); vapor is irritant to respiratory passages; causes lacrimation, salivation, and narcosis in test animals; lethal concentration in rats for a 4-hr exposure period is 4480 ppm; causes liver injury;

(continued)

TABLE 22.4   *(Continued)*

Compound/Synonyms/ CAS No.	MS/Formula/ Structure	Toxicity
		ingestion of 3–4 mL of the liquid can be fatal to humans; intraperitoneal administration caused lung tumors in mice; evidence of carcinogenicity in animals, however, is insufficient; exposure limits: TLV-TWA 0.5 ppm (5 mg/m^3) (ACGIH, MSHA, OSHA, and NIOSH)
Iodoform (triiodomethane, methenyl triiodide) [75-47-8]	CHI_3 393.72 H—C—I with I above and I below	Moderately toxic; toxicity is somewhat greater than that of bromoform and similar to methyl iodide; few data on human toxicity; LD_{50} oral (rabbits) 450 mg/kg; no evidence of carcinogenicity in animals; exposure limits: 0.6 ppm (~10 mg/m^3) (ACGIH)
Carbon tetrabromide (tetrabromomethane) [558-13-4]	CBr_4 331.65	Moderately toxic; eye irritant; subcutaneous LD_{50} value in mice is in the range 300 mg/kg; inhalation of its vapors causes irritation of upper respiratory tract and damage to lungs, liver, and kidneys, while chronic exposure may cause liver injury (Torkelson and Rowe 1981); TLV-TWA 0.1 ppm (~1.4 mg/m^3), TLV-STEL 0.3 ppm (~4.0 mg/m^3) (ACGIH); DOT Label: Poison B, UN 2516
Methylene bromide (dibromomethane, methylene dibromide) [74-95-3]	CH_2Br_2 173.85 Br—C—C—Br with H H above and H H below	Low toxicity; vapors cause irritation of eyes and respiratory tract; subcutaneous LD_{50} value in mice is in the range 3700 mg/kg; EPA Designated Toxic Waste, RCRA Waste Number U068; DOT Label: Poison B, UN 2664
Bromotrifluoromethane (trifluorobromomethane, Halon-1301, Freon-1381) [75-63-8]	$CBrF_3$ 148.92 Br—C—F with F above and F below	Toxic only at very high concentrations in air; human exposure to 10% for 3 min caused lightheadedness and paresthesia; chronic exposure to 2–3% did not show toxic signs in animals; a 15-min exposure to 85% concentrated gas in the air was lethal to rats; TLV-TWA 1000 ppm (~6100 mg/m^3) (ACGIH, MSHA, and OSHA)
Dichlorodifluoromethane (Fluorocarbon-12, Freon-12) [75-75-8]	CCl_2F_2 120.91 Cl—C—F with F above and Cl below	Very low toxicity; inhalation of this gas at 20% concentration in air for 30 min can cause irritation of eyes and mucous membranes, headache, and lightheadedness in humans; LC_{50} value for 30-min exposure in rats, rabbits, and guinea pigs is 80%; TLV-TWA 1000 ppm (4950 mg/m^3) (ACGIH, MSHA, and OSHA)

TABLE 22.4   *(Continued)*

Compound/Synonyms/ CAS No.	MS/Formula/ Structure	Toxicity
Dichlorofluoromethane (fluorodichloromethane, Freon-21) [75-43-4]	$CHCl_2F$ 102.92 $$\begin{array}{c} Cl \\ | \\ H-C-F \\ | \\ Cl \end{array}$$	Low to very low inhalation toxicity; toxic symptoms are similar to those of other fluorocarbon refrigerants; a 4-hr LC_{50} value in rats is about 5%; TLV-TWA 1000 ppm (4200 mg/m^3) (ACGIH, MSHA, and OSHA)
Dibromodifluoromethane (difluorodibromomethane, Halon-1202, Freon-12-B2) [75-61-6]	CBr_2F_2 209.83 $$\begin{array}{c} Br \\ | \\ F-C-Br \\ | \\ F \end{array}$$	Low inhalation toxicity; may cause headache, drowsiness, and excitement; a 15-min exposure to 6400- and 8000-ppm concentrations was fatal to mice and rats, respectively; TLV-TWA 100 ppm (860 mg/m^3) (ACGIH, MSHA, and OSHA)
2-Chlorotoluene (*ortho*-chlorotoluene,2- methylchlorobenzene,2- chloro-1-methylbenzene, *o*-tolyl chloride) [95-49-8]	C_7H_7Cl 126.59 CH_3 (ring with Cl)	Acute oral or inhalation toxicity is low to very low; the symptoms are ataxia, somnolence, excitement, muscle contraction, and respiratory depression; exposure to 17.5% concentration in air was fatal to rats; TLV-TWA 50 ppm (~250 mg/m^3) (ACGIH); DOT Label: Combustible Liquid, UN 2238
4-Chlorotoluene (*para*-chlorotoluene, 4-methylchlorobenzene, 4-chloro-1-methylbenzene, *p*-tolyl chloride) [106-43-4]	C_7H_7Cl 126.59 CH_3 (ring with Cl)	Low acute toxicity; toxic symptoms are similar to those of its ortho-isomer, the effects are somnolence, excitement, muscle contraction, and respiratory depression; LD_{50} oral (rats): 3600 mg/kg; LC_{50} inhalation (mice): 6500 ppm/2 hr; DOT Label: Combustible Liquid, UN 2238

REFERENCES

ACGIH. 1986. *Documentation of the Threshold Limit Values and Biological Exposure Indices*, 5th ed. Cincinnati, OH: American Conference of Governmental Industrial Hygienists.

Aldrich. 1997. *Catalog Handbook of Fine Chemicals*. Milwaukee, WI: Aldrich Chemical Company.

Azri, S., J. A. Gandolfi, and K. Brendel. 1990. Carbon tetrachloride toxicity in precision-cut rat liver slices. *In Vitro Toxicol. 3* (2): 127–38.

Bean, K., B. Lincoln, W. Clapp, S. Vicente, M. Loza, and B. Church. 1990. Methylene chloride extraction. *Chem. Eng. News* July 9.

Buckell, M. 1950. *Br. J. Ind. Med. 7*: 122; cited in *Documentation of Threshold Limit Values and Biological Exposure Indices*, 5th ed. 1986. Cincinnati, OH: American Conference of Governmental Industrial Hygienists.

Elkins, H. B. 1959. *The Chemistry of Industrial Toxicology*, 2nd ed., pp. 151–52. New York: Wiley.

Fairhall, L. T. 1969. *Industrial Toxicology*, 2nd ed. p. 284. New York: Hefner.

Flury, G., and F. Zernik. 1931. *Schadliche Gase*, p. 339. Berlin: J. Springer.

Gibbons, D. W. 1990. Solvent extraction hazard. *Chem. Eng. News*, June 18.

Kaestner, M. 1989. Biodegradation of volatile chlorinated hydrocarbons. *Chem. Abstr. CA 113* (20): 177895y.

Kawakami, T., T. Takano, Y. Kumagai, H. Hirose, and K. Nakaaki. 1988. Unexpected health hazards in trichloroethylene workers. *Chem. Abstr. CA 112*(24): 222551a.

Kelling, R. E. 1990. Chemical safety. *Chem. Eng. News*, Apr. 30.

Kittila, R. S. 1967. In *Fire Protection Guide on Hazardous Materials*, 9th ed., p. 491 M-81. 1986. Quincy, MA: National Fire Protection Association.

Maskarinec, M. P., L. H. Johnson, S. K. Holladay, R. L. Moody, C. K. Bayne, and R. A. Jenkins. 1990. Stability of volatile organic compounds in environmental water samples during transport and storage. *Environ. Sci. Technol. 24*(11): 1665–70.

MCA. 1969. *Case Histories of Accidents in the Chemical Industry*. Washington, DC: Manufacturing Chemists' Association.

MCA. 1972. *Guide for Safety in the Chemical Laboratory*, 2nd ed. New York: Van Nostrand.

Mellor, J. W. 1946. *A Comprehensive Treatise on Inorganic and Theoretical Chemistry*. London: Longmans, Green & Co.

Merck, 1996. *The Merck Index*, 12th ed. Rahway, NJ: Merck & Co.

Morris, L. E. 1963. *Clinical Anesthesia: Halogenated Anesthetics*, pp. 24–41. Philadelphia: F. A. Davis.

National Research Council. 1995. *Prudent Practices for Disposal of Chemicals from Laboratories*. Washington, DC: National Academy Press.

NFPA. 1997. *Fire Protection Guide on Hazardous Materials*, 12th ed. Quincy, MA: National Fire Protection Association.

NIOSH. 1984. *Manual of Analytical Methods*, 3rd ed. Cincinnati, OH: National Institute for Occupational Safety and Health.

NIOSH. 1986. *Registry of Toxic Effects of Chemical Substances*, ed. D. V. Sweet. Washington, DC: U.S. Government Printing Office.

Oku, A., and K. Kimura. 1990. Complete destruction of tetra- and trichloroethylene by reductive dechlorination using sodium naphthalenide. *Chem. Express. 5*(3): 181–84; cited in *Chem. Abstr. CA 112*(20): 185220u.

Patty, F. A. 1963. *Industrial Hygiene and Toxicology*, 2nd ed., Vol. II, p. 1115. New York: Interscience.

Pferfferle, W. 1992. Catalytically stabilized thermal incineration of volatile organic compounds.

Abstracts of Small Business Innovative Research Program. Washington, DC; U.S. EPA.

Scaros, M. G., and J. A. Serauskas. 1973. Cited in *Fire Protection Guide on Hazardous Materials*, 9th ed. 1986. Quincy, MA: National Fire Protection Association.

Speitel, G. E., Jr., A. M. Patterson, C. J. Lu, and R. C. Thompson. 1989. Aerobic biodegradation of chloroform and trichloroethylene in drinking water treatment. *American Water Works Association Annual Conference Proceedings*, Part 2, pp. 1443–49; cited in *Chem. Abstr. CA 113*(22): 197501v.

Sundstrom, D. W., B. A. Weir, and K. A. Redig. 1990. Destruction of mixtures of pollutants by UV-catalyzed oxidation with hydrogen peroxide. *ACS Symp. Ser.*, Vol. Date 1989, *422*: 67–76; cited in *Chem. Abstr. CA 113*(18): 157994b.

Szucs, S. S. 1990. Decomposition of benzyl fluoride. *Chem. Eng. News*, vol. 68, p. 4. Aug. 20, 1990.

Tabershaw, I. R., and W. R. Gaffey. 1974. *J. Occup. Med. 16*(3): 508; cited in *Documentation of the Threshold Limit Values and Biological Exposure Indices*, 5th ed. 1986. Cincinnati, OH: American Conference of Governmental Industrial Hygienists.

Thurnau, R. C. 1988. The incomplete combustion of carbon tetrachloride during normal/abnormal hazardous waste incineration. *U.S. EPA Res. Dev. Rep. EPA-600/9-88/021*; cited in *Chem. Abstr. CA 112*(24): 222757x.

Torkelson, T. R., and V. K. Rowe. 1981. *Patty's Industrial Hygiene and Toxicology*, 3rd ed, Vol. 28, *Toxicology*, pp. 3478–3480. New York: John Wiley & Sons.

Turley, R. E. 1964. Safety study of halogenated hydrocarbon–nitrogen tetroxide detonations, pp. M64–M171. Denver: Martin Co.; cited in *Fire Protection Guide on Hazardous Materials*, 9th ed. 1986. Quincy, MA: National Fire Protection Association.

U.S. EPA. 1992. Method for organic chemical analysis of municipal and industrial wastewater. 40 CFR, Part 136, Appendix A. *Fed. Register 49*(209), Oct. 26.

U.S. EPA. 1997. *Test Methods for Evaluating Solid Waste*, 3rd ed update. Washington, DC: Office of Solid Waste and Emergency Response.

Weeks, M. H. 1979. The toxicity of hexachloro-ethane in laboratory animals. *Am. Ind. Hyg. Assoc. J. 40*: 187.

Woods, S., K. Williamson, S. Strand, K. Ryan, J. Polonsky, K. Gardener, and P. Defarges. 1988. Development of a novel support aerated biofilm reactor for the biodegradation of toxic organic compounds. U.S. EPA Res. Report, 600/9-88/021; cited in *Chem. Abstr., CA 112* (24): 222643g.

Yost, K. W. 1989. Ultraviolet peroxidation: an alternative treatment method for organic contamination destruction in aqueous waste streams. *Proceedings of the Industrial Waste Conference*, Volume Date 1988: 441–47; cited in *Chem. Abstr. CA 112* (22): 204092x.

23

HALOGENS, HALOGEN OXIDES, AND INTERHALOGEN COMPOUNDS

23.1 GENERAL DISCUSSION

Halogens are the elements in group VIIA of the periodic table and are five in number: fluorine, chlorine, bromine, iodine, and astatine, with atomic numbers, 9, 17, 35, 53, and 85, respectively. Astatine, a radioactive halogen found in uranium ores, has a very short half-life (8.3 hours for ^{210}At). The other four halogens are among the earliest discovered elements. Although they do not occur in nature in the free elemental state, the innumerable organic and inorganic compounds of these elements are well known. Fluorine occurs in the ores fluorspar, cryolite, and fluorapatite; chlorine, bromine, and iodine occur in igneous rock and in seawater. Common salt, sodium chloride, constitutes about 60% by weight of chlorine.

The electronic configuration of halogens shows that there are seven electrons in the outermost shell of these atoms. Thus the most common valence state is 1. Although uncommon, halogens exhibit higher valencies, too, especially in interhalogen compounds. For example, in bromine pentafluoride and iodine heptafluoride, bromine and iodine exhibit valencies of 5 and 7, respectively.

Reactivity and Explosion Hazard

Fluorine is the most electronegative element in the periodic table. It is the most reactive nonmetal and it is dangerously reactive. Chlorine is the second most reactive element and combines with almost all elements except nitrogen and inert gases (except xenon). Bromine and iodine are less reactive. Halogens are strong oxidizing agents and react violently with many easily oxidizable substances. These are noncombustible substances but cause ignition. Reactions with hydrogen, acetylene, ammonium hydroxide, and azides are explosive. In addition, violent reactions occur when halogens react with many common organic substances, including lower aldehydes, ethers, carbon disulfide, ethylene imine, and lower alcohols. Violent reactions, usually accompanied by flame, occur when halogens are mixed with metal hydrides, boranes, silanes, phosphine, arsine, or stibine. Reactions with finely divided metals, metal carbides, and silicides occur with incandescence. The violence of reaction decreases from fluorine to iodine.

Interhalogen compounds and halogen oxides are unique in their reactivities and are very different from all other halogen-containing compounds. The explosion hazard is greater for these substances than for the halogens. Bromine pentafluoride, chlorine pentafluoride, and chlorine trifluoride are dangerously reactive with many substances. Violent reactions occur with water. Several organic and inorganic compounds explode or burst into flame in contact with these compounds. The fire and explosion hazards caused by halogens, halogen oxides, and interhalogen compounds are presented in the following sections.

Toxicity

Halogens are severe irritants to the eyes, skin, and mucous membranes. Fluorine and chlorine gases and the vapors of bromine and iodine are suffocatingly irritating and can cause pulmonary edema and death. Chlorine monoxide and dioxide and the vapors of most interhalogen compounds exhibit similar irritant action. The latter compounds, as liquids, can destroy tissues on contact. A detailed discussion is presented below for each compound.

Analysis

There is no NIOSH Method for the analysis of halogens in the air. Airborne fluorine ion, hydrogen fluoride, and hydrogen chloride may be analyzed by the atomizing trace gas monitoring system (TGA analyzer), which uses an air ejector pump and ion-specific electrodes (Dharmarajan and Brouwers 1987). Beswick and Pitt (1988) have described optical detection of chlorine using fluorescent porphyrin Langmuir–Blodgett films. The film consists of tetraphenylporphine (TPP) mixed with arachidic acid and pentacosa-10,12-diynoic acid. The fluorescence of TPP is quenched in the presence of very low concentrations (1–10 ppm) of chlorine, HCl, or NO_2 gas.

Hegedues and co-workers (1987) have reported a method based on total-reflectance x-ray fluorescence spectrometry for low-level detection of iodine.

Chlorine in water may be analyzed by several wet chemical methods, which include iodometric, amperometric, and N,N-diethyl-p-phenylenediamine (DPD) methods based on titrations. It may also be analyzed by colorimetric methods using DPD, syringaldazine, or $4,4',4''$-methylidynetris(N,N-dimethylaniline(leucocrystal violet)).

Iodine in water may be analyzed by amperometric titration or by a colorimetric procedure using leucocrystal violet. In the latter method, mercuric chloride is added to the potable water sample, hydrolyzing iodine to hypoiodous acid. The latter reacts instantaneously with leucocrystal violet, forming a crystal violet dye. The absorbance is measured at a wavelength of 592 nm.

HALOGENS

23.2 FLUORINE

EPA Designated Acute Hazardous Waste, RCRA Waste Number P056; DOT Label: Poison and Oxidizer, UN 1045

Symbol F; at. wt. 18.998; at. no. 1

Formula F_2; MW 37.996; CAS [7782-41-4]

Uses and Exposure Risk

Fluorine is used in the manufacture of various fluorocarbons and fluorides, as a rocket propellant, and in many inorganic and organic syntheses.

Physical Properties

Pale yellow gas with an irritating pungent odor; liquefies at $-188.1°C$; solidifies at $-219.6°C$; reacts with water.

Health Hazard

Fluorine is a severe irritant to the eyes, skin, and mucous membranes. In humans its irritant effect on the eyes can be felt at a level of 5 ppm in air. The acute toxicity of fluorine was found to be moderate in animals. Exposure to this gas can cause respiratory distress and pulmonary edema. Chronic exposure can produce mottled enamel of the teeth, calcification of ligaments, and injury to the lungs, liver, and kidney. The latter effects, however, were observed in animals at high concentrations. The LC_{50} value in mice is 150 ppm for an exposure period of 1 hour. Human toxicity data on fluorine are very limited.

Exposure Limits

TLV-TWA 1 ppm ($\sim$2 mg/m^3) (ACGIH and MSHA), 0.1 ppm (OSHA); IDLH 25 ppm (NIOSH).

Fire and Explosion Hazard

Fluorine is a nonflammable gas that produces a flame when reacted with many substances. It is highly reactive and reacts vigorously or violently with a large number of compounds. It reacts violently with water even at low temperature, forming ozone and hydrofluoric acid:

$$3F_2 + 3H_2O \longrightarrow O_3 + 6HF$$

Reactions of fluorine with hydrogen; acetylene; a large number of organics, including many hydrocarbons and halogenated organics; graphite; ammonium hydroxide; chlorine dioxide; and sulfur dioxide cause explosions. Fluorine reacts with nitric acid, sodium nitrate, or potassium nitrate to form fluorine nitrate, which is highly shock sensitive and explodes on slight impact. Reaction with sodium acetate produces diacetyl peroxide, which explodes violently. Fluorine reacts with perchloric acid, or potassium or other metal perchlorates, forming fluorine perchlorate, which is unstable and explodes. Reaction with hydrogen azide produces fluorine azide, an explosive compound.

Fluorine reacts with other halogens, chlorine, bromine, and iodine, causing luminous flame. The reaction with gaseous or aqueous hydrochloric, hydrobromic, and hydroiodic acid is accompanied by flame (Mellor 1946). Reactions with halides are vigorous in cold, and violent upon heating. Powdered metals react with incandescence in fluorine. In nonpowdered state, metals react with incandescence at elevated temperatures. Sodium and potassium burn spontaneously in dry fluorine. Metal hydrides are spontaneously flammable in fluorine. Carbides, carbonates, silicides, cyanides, silicates, and phosphides of many metals burn with incandescence in fluorine. Phosphorus (yellow and red), phosphorus pentachloride, and phosphorus trichloride react with fluorine at ordinary temperatures, causing incandescence. Other inorganic compounds that ignite with fluorine at ordinary temperatures include sulfur, hydrogen sulfide, nitric oxide, hydrazine, calcium oxide, and nickel monoxide.

Polymeric materials such as nylon, polyethylene, neoprene, rubber, acrylonitrile–butadiene copolymer, polyvinyl chloride–acetate, and polyurethane foam burn or react violently with fluorine (Schmidt and Harper 1967).

In case of fire involving fluorine, shut off the flow of gas if possible. Allow the fire to burn. Use water from an explosion-resistant location to keep the fire-exposed containers cool. Do not direct water onto fluorine leaks, as water reacts violently with fluorine.

23.3 CHLORINE

DOT Label: Nonflammable Gas and Poison, UN 1017

Symbol Cl; at. wt. 35.453; at. no. 17

Formula Cl$_2$; MW 71.906; CAS [7782-50-5]; natural isotope: masses 35 and 37

Uses and Exposure Risk

Chlorine is used as a disinfectant; for purifying water; in the manufacture of bleaching powder, chlorinated hydrocarbons, synthetic rubber, and plastics; and as a chlorinating and oxidizing agent. It was used as a poison gas in the war under the name Bertholite.

Physical Properties

Greenish yellow gas with a suffocating odor; occurs as a diatomic gas, monoatomic Cl is unstable; liquefies at −34°C; solidifies at −101°C; slightly soluble in water (0.7%).

Health Hazard

Chlorine is a severely irritating gas, causing irritation of the eyes, nose, and throat. Exposure also causes burning of the mouth, coughing, choking, nausea, vomiting, headache, dizziness, pneumonia, muscle weakness, respiratory distress, and pulmonary edema. A 30-minute exposure to 500–800 ppm can be fatal to humans. Chronic exposure to concentrations around 5 ppm have produced corrosion of the teeth, inflammation of the mucous membranes, respiratory ailments, and increased susceptibility to tuberculosis among workers (Patty 1963).

Klonne and associates (1987) have reported a 1-year inhalation toxicity study of chlorine in rhesus monkeys. Exposure to 2.3 ppm chlorine caused ocular irritation during the daily exposures. Histopathological changes were observed in the respiratory epithelium of the nasal passages and trachea. These changes, however, were mild at the foregoing level of exposures. Monkeys were less sensitive to chlorine toxicity than were rats.

Zwart and Woutersen (1988) studied the acute inhalation toxicity of chlorine in rats and mice and have proposed a time–concentration–mortality relationship. The relationship between any LC value, concentration, and time of exposure could be described by the probit (P) equations, as follows:

$$\text{For rats:} \quad P = -16.67 + 1.33 \ln C - 4.31 \ln \times T + 1.01 \ln C \times \ln T$$

$$\text{For mice:} \quad P = -33.74 + 4.05 \ln C + 2.72 \ln T$$

where C and T are concentration and time, respectively. Zwart and Woutersen also observed that there was rapid shallow breathing in animals after the exposure began. Some animals exhibited the formation of pulmonary edema near the end of exposure.

Exposure Limits

TLV-TWA 1 ppm (∼3 mg/m^3) (ACGIH and MSHA); ceiling 1 ppm (OSHA), 0.5 ppm/15 min (NIOSH); IDLH 30 ppm (NIOSH).

Fire and Explosion Hazard

Chlorine is a noncombustible gas but supports combustion, similar to oxygen. Most combustible substances would burn in a chlorine atmosphere. Flammable gases and vapors form explosive mixtures with chlorine.

Chlorine is a strong oxidizer. It explodes when mixed with hydrogen in the presence of sunlight, heat, or a spark. The explosive range of a hydrogen–chlorine mixture is 5–95% by volume. Chlorine reacts explosively with acetylene at ordinary temperatures. It explodes with ethylene, methane, ethane, propane, butane, propene, and benzene vapors in the presence of sunlight, ultraviolet light, or a catalyst such as oxide of mercury or silver. In general, chlorine reacts vigorously to violently with most hydrocarbons. Its mixture with hydrocarbon vapors can ignite, especially in the presence of a catalyst such as mercuric oxide or UV light. It reacts with ammonia on heating, forming nitrogen trichloride, which is highly shock sensitive and explodes. A similar reaction can occur with excess chlorine without heating (Mellor 1946). It reacts with ethyleneimine, forming an explosive compound, 1-chloroethyleneimine (NFPA 1997). Its reaction with fluorine produces flame; the mixture can explode in the presence of a spark. Explosion occurs when chlorine is warmed with oxygen difluoride and heated with bromine trifluoride or bromine pentafluoride. Liquid chlorine reacts violently with iodine.

With alcohols, chlorine produces alkyl hypochlorites, which explode on exposure to sunlight or heat. Reaction of chlorine with formaldehyde, acetaldehyde, or diethyl ether is violent. Liquid chlorine reacts explosively with carbon disulfide, glycerol, dialkyl phthalates, polypropylene, wax, rubber, and linseed oil.

Metals in a finely divided state burn spontaneously in chlorine. Solid metals (nonpowdered form) burn spontaneously at elevated temperatures. Metal carbides react with chlorine with incandescence. Metal hydrides burn spontaneously in chlorine. Diborane explodes with chlorine at ordinary temperature; silane, phosphine, arsine, or stibine produce a flame. The latter compounds react explosively with chlorine at elevated temperatures. Metal sulfides, silicides, nitrides, phosphides, and oxides burn in chlorine. Phosphorus undergoes a highly exothermic reaction with chlorine. White phosphorus explodes with liquid chlorine; white phosphorus and finely divided red phosphorus burn spontaneously in chlorine gas with a pale green light (NFPA 1997).

In case of fire involving chlorine, stop the flow of gas. Use a water spray to keep the fire-exposed containers cool.

Storage and Shipping

Chlorine is stored in a well-ventilated and detached area, preferably outdoors and separated from readily oxidizable substances, hydrogen, acetylene, hydrocarbons, ether, ammonia, and finely divided metals. It is shipped in steel pressure cylinders and tank cars.

23.4 BROMINE

DOT Label: Corrosive Material, UN 1744

Symbol Br; at. wt. 79.904; at. no. 35

Formula Br_2; MW 159.808; CAS [7226-95-6]; isotopes: two stable isotopes of mass 79 and 81

Uses and Exposure Risk

Bromine is used for bleaching fibers and silk, as a disinfectant for purifying water, in the manufacture of bromo compounds for dyes and pharmaceutical uses, as an analytical reagent, and in organic synthesis. It occurs in igneous rock and seawater.

Physical Properties

Dark reddish brown liquid; suffocating odor; volatile; density 3.102 at 25°C; boils at 59.5°C; solidifies at −7.25°C.

Health Hazard

Bromine is a corrosive liquid and a moderately toxic substance. The target organs are the respiratory system, eyes, and the central nervous system; the routes of exposure are inhalation of its vapors, ingestion of the liquid, and skin contact. Bromine is an irritant to respiratory passages and can cause injury to the lungs. A concentration in the range of 50 ppm of bromine vapor in air is highly irritating to humans, and a short exposure to 1000 ppm for 15 minutes can be fatal. Other symptoms are dizziness, headache, coughing, and lacrimation, manifested at a short exposure to 10–20 ppm of vapor. Ingestion of the liquid can cause nausea, abdominal pain, and diarrhea. It is corrosive to the skin and eyes, causing burns. Its odor can be detected at a level of 3 ppm in air.

Exposure Limits

TLV-TWA 0.1 ppm (0.7 mg/m^3) (ACGIH, MSHA, NIOSH, and OSHA); TLV-STEL 0.3 ppm (ACGIH); IDLH 10 ppm (NIOSH).

Fire and Explosion Hazard

Noncombustible liquid. Bromine is a strong oxidizer and reacts violently with readily oxidizable substances. Flaming or explosion can occur when bromine is mixed with combustible organic compounds. A violent explosive reaction occurs when bromine is combined with methanol, acetaldehyde, acrolein, acrylonitrile, or dimethyl formamide. Its mixture with hydrogen or acetylene is explosive. Heating liquid bromine with ammonia solution produces shock-sensitive nitrogen tribromide. Reactions with many metals at ordinary temperatures or on heating can cause spontaneous flaming or incandescence. This includes aluminum, antimony, germanium, titanium, mercury, and tin. Among the alkali metals, cesium and potassium explode in contact with liquid bromine; sodium or lithium explode when subjected to mechanical shock. Bromine reacts violently with metal hydrides, nitrides, azides, carbides, phosphides, metal carbonyls, and many silicides. Ignition occurs with phosphine, phosphorus trioxide, and red phosphorus. Mixing with yellow phosphorus can result in ignition as well as explosion. Reaction with fluorine produces bromine trifluoride, accompanied by a luminous flame. Heating with oxygen difluoride may result in explosion. Reaction with ozone can produce a violent explosion (Mellor 1946, Suppl. 1956).

Bowman and co-workers (1990) have reported a violent explosion in preparing a bromine–methanol solution useful in semiconductor etching. The mixture is exothermic, showing initial increases in temperature, which then decreases at varying rates. The authors have reported that a second temperature increase occurred at higher bromine concentrations, causing explosion at a 50% volume mixture.

Spillage

If there is a small spill of liquid bromine, pour sodium thiosulfate or lime water over it. Anhydrous ammonia vapor should be used from a safe distance to neutralize large quantities of bromine vapors.

Storage and Shipping

Bromine is stored in a cool dry place separated from organic, combustible, and readily oxidizable substances. It should be kept out of direct sunlight. It should be stored above its freezing temperature.

Bromine should be shipped in quart glass bottles, nickel or Monel drums or in lead-lined tank cars.

Destruction/Disposal

Cautiously acidify to pH 2 with sulfuric acid. Add a 50% excess of sodium bisulfite solution gradually with stirring, at room temperature. Reaction should occur with an increase in temperature. If no increase in temperature occurs, add 10% excess of bisulfite solution followed by cautious addition of more sulfuric acid. Neutralize to pH 7 and flush the solution down the drain (Aldrich 1997).

Bromine may be added slowly to sodium hydroxide solution with stirring. This results in the formation of mixtures of sodium bromide and bromate, which are water soluble and nontoxic and can be flushed down the drain with an excess of water.

23.5 IODINE

Symbol I; at. wt. 126.90; at. no. 53; CAS [7553-56-2]; valency 1, may occur in valence states from 1 to 7; natural isotope 127

Uses and Exposure Risk

Iodine is used in the manufacture of many iodine compounds; in photographic materials; as an antiseptic, disinfectant, and germicide; and as a reagent in analytical chemistry. It occurs in traces in seawater and in igneous rocks.

Physical Properties

Bluish black solid; volatilizes at ordinary temperature to violet vapors; sharp characteristic odor; acrid sharp taste; melts at 113.6°C; boils at 185.2°C; very slightly soluble in water (0.03%), soluble in aqueous iodide solutions, soluble in alcohol, ether, benzene, and carbon disulfide; vapor pressure 0.3 torr at 25°C.

Health Hazard

The vapors are an irritant to the eyes, nose, skin, and mucous membranes. The toxic routes of exposure are inhalation of vapors, and ingestion and absorption through skin. Inhalation of its vapors may produce irritation, headache, tightness of the chest and congestion of the lungs. A concentration of 1 ppm of iodine vapor in air is highly irritating to humans.

Ingestion can cause burning of mouth, vomiting, abdominal pain, and diarrhea. Oral intake of about 2 g of the solid can be fatal to humans. The toxicity is greater in humans than in experimental animals. Iodine is corrosive to the skin. Contact with the skin may cause burn and rash.

Exposure Limits

Ceiling 0.1 ppm (~1 mg/m^3) (ACGIH, MSHA, OSHA, and NIOSH); IDLH 10 ppm (NIOSH).

Fire and Explosion Hazard

Iodine is a noncombustible solid. It reacts explosively with several compounds. Alkali metals, their carbides, and hydrides burn with incandescence with liquid iodine and iodine vapors (Mellor 1946). Ignition with hydrides occurs at about 100°C. Powdered magnesium or aluminum, white phosphorus, and many metal carbides burn similarly with iodine, causing incandescence. Mixtures of iodine vapors with fluorine produce luminous flame. Violent reaction can occur with liquid chlorine. Mixture with acetylene or ammonia can explode. Concentrated solutions of ammonium hydroxide react with excess iodine, forming nitrogen triiodide, which detonates on drying. Silver azide reacts with a cold ethereal solution of iodine, forming iodoazide, which explodes (Meller 1946). Heating mixtures of iodine with acrylonitrile, oxygen difluoride, formaldehyde, or

acetaldehyde can produce a violent reaction. The reaction with bromine pentafluoride or dimethylformamide is exothermic.

Denes (1987) reported a violent explosion during synthesis of 500 mg of titanium tetraiodide from a solid-state reaction between iodine and titanium under a nitrogen atmosphere in a sealed copper tube immersed in a water bath. An explosion occurred when the mixture was heated to 90°C.

HALOGEN OXIDES

23.6 CHLORINE MONOXIDE

Formula Cl$_2$O; MW 86.91; CAS [7791-21-1]
Synonyms: dichlorine monoxide; dichloroxide; hypochlorous anhydride

Uses and Exposure Risk

Chlorine monoxide is used as a strong and selective chlorinating agent. It is stored below −80°C as a liquid or solid.

Physical Properties

Yellowish brown gas with a disagreeable and suffocating odor; unstable at ordinary temperatures; liquefies at 2.2°C; solidifies at −120.6°C; highly soluble in water, forming hypochlorous acid, soluble in carbon tetrachloride.

Health Hazard

Chlorine monoxide is severely irritating to the eyes, skin, and mucous membranes. Exposure can cause lung damage. LC$_{50}$ data are not available for this compound. A short exposure to 100 ppm concentration can cause death to humans.

Fire and Explosion Hazard

Nonflammable gas. Chlorine monoxide is a highly reactive compound, exploding by itself when rapidly heated. Chlorine monoxide explodes with organic compounds, charcoal, metals, metal sulfides, sulfur, phosphorus, ammonia, nitric oxide, and carbon disulfide.

23.7 CHLORINE DIOXIDE

DOT Label: Oxidizer and Poison (for the frozen, hydrated compound); the anhydrated compound may not be transported; NA 9191

Formula ClO_2; MW 67.45; CAS [10049-04-4]

Synonyms: chlorine peroxide; chloroperoxyl; Alcide

Uses and Exposure Risk

Chlorine dioxide is used for several purposes, including its applications as a bleaching agent to bleach fats, oils, textiles, cellulose, paper pulp, flour, and leather. It is also used for purifying water; as an oxidizing agent; as an antiseptic; and in the manufacture of many chlorite salts.

Physical Properties

Yellow to orange gas at ordinary temperatures; pungent, irritating odor somewhat similar to chlorine; liquefies at 11°C to a reddish brown liquid; solidifies at −59°C to yellowish red crystals; slightly soluble in water, forming chlorous and chloric acids, soluble in acidic and alkaline solutions.

Health Hazard

Chlorine dioxide is highly irritating to the eyes, nose, and throat. Inhalation can cause coughing, wheezing, respiratory distress, and congestion in the lungs. Its toxicity in humans is moderate to high. Its irritant effects in humans can be intense at a concentration level of 5 ppm in air. A concentration of 19 ppm of the gas inside a bleach tank caused the death of one worker (Elkins

1959). The chronic toxicity signs are mainly dyspnea and asthmatic bronchitis, and in certain cases irritation of the gastrointestinal tract. Ingestion of the liquid may cause somnolence and respiratory stimulation.

Exposure Limits

TLV-TWA 0.1 ppm (0.3 mg/m^3); (ACGIH, MSHA, OSHA, and NIOSH); TLV-STEL 0.3 ppm (ACGIH); IDLH 10 ppm (NIOSH).

Fire and Explosion Hazard

Nonflammable gas; however, it is highly reactive and a strong oxidizing agent. Chlorine dioxide explodes violently upon heating, exposure to sunlight, contact with dust, or when subjected to a spark. Detonation occurs at concentrations above 10% in air in the presence of an energy source or catalyst. It undergoes violent reactions with organic matter; explosion occurs when the mixture is subjected to shock or a spark. It reacts spontaneously with sulfur or phosphorus, causing ignition and/or explosion. Liquid chlorine dioxide may explode violently when mixed with mercury, caustic potash, caustic soda, or many metal hydrides. The gas reacts explosively with fluorine and with difluoroamine (Lawless and Smith 1968).

INTERHALOGEN COMPOUNDS

23.8 CHLORINE TRIFLUORIDE

DOT Label: Oxidizer and Poison, UN 1749

Formula ClF_3; MW 92.45; CAS [7790-91-2]

Synonym: chlorotrifluoride

Uses and Exposure Risk

Chlorine trifluoride is used as a fluorinating agent, as a rocket propellant, in processing of nuclear reactor fuel, and in incendiaries. It is also used as an inhibitor of pyrolysis of fluorocarbon polymers.

Physical Properties

Colorless gas or a greenish yellow liquid with a sweet but suffocating odor; liquefies at 11.75°C; solidifies to a white solid at −76.3°C; reacts violently with water.

Health Hazard

Chlorine trifluoride is a severe irritant to the skin, eyes, and mucous membranes. Exposure to this gas can cause lung damage. A 30-minute exposure to 400 ppm was lethal to rats. It decomposes in the presence of moisture to chlorine, chlorine dioxide, and hydrogen fluoride, all of which are highly toxic. Chronic inhalation study on animals for a period of 6 months (6 hours/day, 5 days/week) indicated that at an exposure level of nearly 1 ppm the early symptoms were sneezing, salivation, and expulsion of frothy fluid from the mouth and nose (ACGIH 1986). This progressed to muscle weakness, pneumonia, and lung damage. Some animals died.

In humans, exposure to this gas can produce severe injury to the eyes, skin, and respiratory tract, and pulmonary edema. The liquid is severely corrosive to the skin and eyes. Skin contact can cause painful burns.

Exposure Limits

Ceiling 0.1 ppm (~0.4 mg/m^3) (ACGIH, MSHA, NIOSH, and OSHA); IDLH 20 ppm (NIOSH).

Fire and Explosion Hazard

Nonflammable gas; dangerously reactive. Chlorine trifluoride reacts explosively with water, forming hydrogen fluoride and chlorine. It reacts violently with most elements and common substances. Paper, cloth, wood, glass, wool, charcoal, and graphite burst into flame in contact with the liquid. The vapors, even when diluted, can set fire to organic compounds. Reactions with most metals are vigorous to violent, often causing a fire. It catches fire when mixed with phosphorus, arsenic, antimony, silicon, sulfur, selenium, tellurium, tungsten, osmium, and rhodium (Mellor 1946, Suppl. 1956). Among the alkali- and alkaline–earth metals, reaction is violent with potassium at ordinary temperatures, and with sodium, calcium, or magnesium it reacts violently at elevated temperatures. Violent reaction occurs with oxides, sulfides, halides, and carbides of metals, causing flames. Chlorine trifluoride attacks sand, glass, and asbestos. Prolonged contact can ignite glass. Explosive reactions occur with many common gases, including hydrogen, lower hydrocarbons, carbon monoxide, ammonia, hydrogen sulfide, and sulfur dioxide. Reactions with mineral acids and alkalies are violent.

In case of a small fire involving chlorine trifluoride, use a dry chemical or water spray in large amounts (NFPA 1997). Allow large fires to burn. Avoid contact of chlorine trifluoride with the body or with protective clothing.

Storage and Shipping

Chlorine trifluoride is stored and shipped in special steel cylinders. It is stored in moisture-free, cool, and isolated areas separated from other chemicals. The cylinders are kept upright, covered, and protected against physical damage.

23.9 CHLORINE PENTAFLUORIDE

DOT Label: Poison Gas, Oxidizer, Corrosive; UN 2548

Formula ClF$_5$; MW 130.45; CAS [13637-63-3]

Uses and Exposure Risk

Chlorine pentafluoride does not have any significant commercial application. It is used as a fluorinating and oxidizing agent.

Physical Properties

Colorless gas with a suffocating odor; bp and mp data not available; violent reaction with water.

Health Hazard

Chlorine pentafluoride is a highly toxic gas. It is a severe irritant to the eyes, skin, and mucous membranes. Exposure can cause lacrimation, corneal damage, skin burn, and lung damage. Other symptoms are nausea, vomiting, and dyspnea. The liquid is highly corrosive to skin, causing painful burns.

LC_{50} value, inhalation (rats): 122 ppm/hr

Exposure Limit

TLV-TWA 2.5 mg(F)/m^3 (ACGIH, MSHA, and OSHA).

Fire and Explosion Hazard

Nonflammable gas. Chlorine pentafluoride is a highly reactive substance. It reacts explosively with water. Paper, cloth, wood, and other organic matter would burst into flames upon contact with the liquid or vapor of chlorine pentafluoride. Vigorous to violent reaction occurs with metals. Reactions with oxides, sulfides, halides, and carbides of metals are violent. It forms explosive mixtures with hydrogen, carbon monoxide, hydrocarbon gases, ammonia, phosphine, sulfur dioxide, and hydrogen sulfide. It reacts violently with sulfur, phosphorus, silicon compounds, charcoal, and mineral acids.

23.10 BROMINE TRIFLUORIDE

DOT Label: Oxidizer and Poison, UN 1746
Formula BrF$_3$; MW 136.91; CAS [7787-71-5]

Uses and Exposure Risk

The commercial uses of this compound are very limited. It is used as a solvent for fluorides.

Physical Properties

Colorless to pale yellow liquid; density 2.803 at 25°C; solidifies at 8.8°C; boils at 125.7°C; fumes in air; miscible with water (reacts violently).

Health Hazard

The vapors of bromine trifluoride are highly irritating to the eyes, skin, and mucous membranes. Upon contact with the skin, the liquid can cause severe burns. The toxicity data for this compound are not available.

Exposure Limit

TLV-TWA 2.5 mg(F)/m^3 (ACGIH, MSHA, and OSHA).

Fire and Explosion Hazard

Noncombustible liquid. Bromine trifluoride reacts violently with water. It reacts with incandescence when combined with molybdenum, vanadium, tungsten, niobium, tantalum, titanium, tin, arsenic, antimony, or boron. Such an incandescent reaction occurs when the metal is in powdered form. In general, it reacts vigorously to violently with any metal. When mixed with sulfur, phosphorus, bromine, or iodine, the mixture burns with incandescence. Bromine trifluoride reacts violently with a number of metal chlorides, including those of alkali- and alkaline–earth metals. An explosive reaction occurs with ammonium halides. It undergoes vigorous reactions with many metal oxides. A violent reaction occurs with antimony trioxide. It reacts explosively with carbon monoxide at high temperatures. It reacts violently with organic matter.

23.11 BROMINE PENTAFLUORIDE

DOT Label: Oxidizer, UN 1745
Formula BrF$_5$; MW 174.91; CAS [7789-30-2]

Uses and Exposure Risk

There is very little commercial application of this compound. It is sometimes used as a fluorinating agent and an oxidizer.

Physical Properties

Colorless to pale yellow liquid; fumes in air; density 2.460 at 25°C; boils at 40.8°C; freezes at −60.5°C; reacts with water.

Health Hazard

Bromine pentafluoride is more active and toxic than elemental fluorine or bromine trifluoride. The liquid is severely corrosive to the skin. The vapors are highly irritating to the eyes, skin, and mucous membranes. Exposure to 500 ppm vapor caused gasping, swelling of eyelids, cloudiness of the cornea, lacrimation, salivation, and respiratory distress in test animals (ACGIH 1986). A few minutes' exposure to 100 ppm was lethal to most experimental animals. Chronic exposure can cause nephrosis and hepatosis. Ingestion of a few drops can cause severe corrosion and burn the mouth.

Exposure Limits

TLV-TWA 0.1 ppm (~0.7 mg/m^3) (ACGIH and MSHA), 2.5 mg(F)/m^3 (OSHA).

Fire and Explosion Hazard

Noncombustible liquid; stable to heat, shock, and electric spark. However, bromine pentafluoride is a highly reactive substance. It reacts explosively with water, producing toxic and corrosive fumes. It decomposes in contact with acids, producing toxic fumes of bromine and fluorine. The reaction is violent, especially with concentrated nitric or sulfuric acid.

Bromine pentafluoride reacts violently with organic materials, such as carboxylic acids, alcohols, ethers, hydrocarbons, grease, wax, and cellulose. Spontaneous flaming and/or explosion can occur when mixed with these or any other organic compound. Reactions with metals in powder form and/or upon warming can result in a violent explosion. Ignition occurs under cold condition. Violent reactions occur with metal halides, oxides, and sulfides. Bromide pentafluoride ignites when mixed with iodine and explodes with chlorine on heating. Mixtures of bromine pentafluoride with sulfur, phosphorus, and carbon ignite.

Avoid using water in the case of fire, since the compound explodes with water. However, water may be used from a safe distance to extinguish a fire involving large amounts of combustible materials, and to keep the fire-exposed containers cool. Fires involving small amounts of combustible material may be extinguished by CO_2 or a dry chemical (NFPA 1997).

23.12 IODINE MONOCHLORIDE

DOT Label: Corrosive Material, UN 1792
  Formula ICl; MW 162.35; CAS [7790-99-0]
Synonyms: iodine chloride; Wijs' chloride

Uses and Exposure Risk

Iodine monochloride is used to estimate the iodine values of fats and oils and as a topical anti-infective (Merck 1996).

Physical Properties

Black crystalline solid or brown liquid; density of liquid 3.10 at 29°C; the crystals occur in a α-form (mp 27.2°C), or β-form (mp 14°C, bp 97°C); soluble in water, alcohol, ether, and carbon disulfide.

Health Hazard

Iodine monochloride is highly corrosive to the skin. Contact with the skin causes burns and dark patches. Upon contact, wash immediately with 15–20% HCl. Vapors are irritating to the skin, eyes, and mucous membranes. The compound is moderate to highly toxic by an oral route. The lethal dose in rats is 59 mg/kg (NIOSH 1986).

Fire and Explosion Hazard

Noncombustible solid or liquid. Iodine monochloride explodes in contact with potassium, and burns with a bluish white flame upon prolonged contact with aluminum foil. The reaction with sodium is slow. Violent reaction occurs on heating with phosphorus, and a violent to vigorous reaction occurs upon mixing with phosphorus trichloride. Iodine monochloride reacts vigorously with cork, rubber, and organic material.

23.13 IODINE MONOBROMIDE

Formula IBr; MW 206.84; CAS [7789-33-5]
  Iodine monobromide does not have much commercial use.

Physical Properties

Brownish black crystalline solid; melts at 40°C; boils at 116°C (decomposes); soluble in water, alcohol, ether, and carbon disulfide.

Health Hazard

Vapors of iodine monobromide are irritating to the skin, eyes, and mucous membranes. Its concentrated solutions are corrosive to the skin. Toxicity data for this compound are not available.

Fire and Explosion Hazard

Noncombustible solid. The molten salt explodes when mixed with potassium, and ignites with finely divided aluminum. Its mixture with sodium explodes on impact. The reaction of phosphorus with molten salt is violent.

23.14 IODINE TRICHLORIDE

Formula ICl_3; MW 233.39; CAS [865-44-1]
Synonym: trichloroiodine

Uses and Exposure Risk

Iodine trichloride is used as a chlorinating and oxidizing agent.

Physical Properties

Yellow crystalline solid or powder; irritating pungent odor; melts at 33°C; volatile; soluble in water, alcohol, and ether.

Health Hazard

Iodine trichloride is a highly corrosive substance. Skin contact can cause burn. Vapors are irritating to the skin, eyes, and mucous membranes. When heated at 77°C, it decomposes to chlorine and iodine monochloride.

Fire and Explosion Hazard

Noncombustible solid. Violent reaction occurs with potassium. It reacts violently when heated with sodium, aluminum, phosphorus, phosphorus trichloride, or organic matter.

23.15 IODINE PENTAFLUORIDE

DOT Label: Oxidizer and Poison, UN 2495
  Formula IF_5; MW 221.90; CAS [7783-66-6]
Synonym: pentafluoroiodine

Uses and Exposure Risk

Iodine pentafluoride has very limited use. It finds application as a mild fluorinating agent.

Physical Properties

Colorless viscous liquid; density 3.190 at 25°C; boils at 100.5°C; solidifies at 9.4°C; miscible with water (reacts violently).

Health Hazard

The toxicity data for this compound in the published literature are very limited. Iodine pentafluoride is highly corrosive. Contact with the skin can cause severe burns. The vapors are highly irritating to the skin, eyes, and mucous membranes.

Exposure Limit

TLV-TWA 2.5 mg(F)/m^3 (ACGIH, MSHA, and OSHA).

Fire and Explosion Hazard

Noncombustible liquid. Iodine pentafluoride is a highly reactive substance and undergoes violent reactions with many substances. It fumes in air, reacts violently with water, and the hot liquid attacks glass. Explosion occurs when it is mixed with potassium or molten sodium. It reacts spontaneously with arsenic, bismuth, tungsten, silicon, sulfur, or phosphorus with incandescence (Sidgwick 1950). Heating with concentrated mineral acids, or organic compounds in general, can be violent.

23.16 IODINE HEPTAFLUORIDE

Formula IF$_7$; MW 259.91; CAS [16921-96-3]
Synonym: heptafluoroiodine

Uses and Exposure Risk

Iodine heptafluoride is used as a fluorinating agent.

Physical Properties

Colorless gas at ordinary temperature; acrid odor; mp 6.45°C; sublimes at 4.77°C; density of liquid 2.80 at 6°C; soluble in water, absorbed by caustic soda solution.

Health Hazard

Iodine heptafluoride is highly irritating to the skin and mucous membranes. Toxicity data on this compound are not available.

Fire and Explosion Hazard

Nonflammable gas. It reacts violently with ammonium halides (chloride, bromide, or iodide) and with organic matter (Mellor 1946, Suppl. 1956). It burns with carbon monoxide.

REFERENCES

ACGIH. 1986. *Documentation of the Threshold Limit Values and Biological Exposure Indices*, 5th ed. Cincinnati, OH: American Conference of Governmental Industrial Hygienists.

Aldrich. 1997. *Catalog Handbook of Fine Chemicals*. Milwaukee, WI: Aldrich Chemical Company.

Beswick, R. B., and C. W. Pitt. 1988. Optical detection of toxic gases using fluorescent porphyrin Langmuir–Blodgett films. *J. Colloid Interface Sci. 124*(1): 146–55.

Bowman, P. T., E. I. Ko, and P. J. Sides. 1990. A potential hazard in preparing bromine–methanol solutions. *J. Electrochem. Soc. 147*(4): 1309–11.

Denes, G. 1987. Explosion hazard in the synthesis of titanium iodides. *Chem. Eng. News 65*(7): 2.

Dharmarajan, V., and H. J. Brouwers. 1987. Advances in continuous toxic gas analyzers for process and environmental applications. *Chem. Abstr. CA 108*(16): 136891x.

Elkins, H. B. 1959. *The Chemistry of Industrial Toxicology*, 2nd ed. New York: John Wiley & Sons.

Hegedues, F., P. Winkler, P. Wobrauschek, and C. Streli. 1987. Low level iodine detection by TXRF in a reactor safety simulation experiment. *Adv. X-Ray Anal. 30*: 85–88.

Klonne, D. R., C. E. Ulrich, M. G. Riley, T. E. Hamm, Jr., K. T. Morgan, and C. S. Barrow. 1987. One-year inhalation toxicity study of chlorine in rhesus monkeys *(Macaca mulatta)*. *Fundam. Appl. Toxicol.* *9*(3): 557–72.

Lawless, E. W., and I. C. Smith. 1968. *High Energy Oxidizers*. New York: Marcel Dekker.

Mellor, J. W. 1946. *A Comprehensive Treatise on Inorganic and Theoretical Chemistry*. London: Longmans, Green & Co.

Merck, 1996. *The Merck Index*, 11th ed. Rahway, NJ: Merck & Co.

NFPA. 1997. *Fire Protection Guide on Hazardous Materials*, 12th ed. Quincy, MA: National Fire Protection Association.

NIOSH. 1986. *Registry of Toxic Effects of Chemical Substances*, ed. D. V. Sweet. Washington, DC: U.S. Government Printing Office.

Patty, F. A. 1963. *Industrial Hygiene and Toxicology*, Vol. 2, p. 847. New York: Interscience.

Schmidt, H. W., and J. T. Harper. 1967. Handling and use of fluorine and fluorine–oxygen mixture in rocket systems. *NASA SP-3037*. Washington, DC: National Aeronautics and Space Administration.

Sidgwick, N. V. 1950. *The Chemical Elements and Their Compounds*. New York: Oxford University Press.

Zwart, A., and R. A. Woutersen. 1988. Acute inhalation toxicity of chlorine in rats and mice: time–concentration–mortality relationships and effects on respiration. *J. Hazard. Mater. 19*(2): 195–208.

24

HETEROCYCLIC COMPOUNDS

24.1 GENERAL DISCUSSION

Heterocyclic organics are cyclic ring compounds containing one or more hetero-atom(s), that is, atoms other than carbon, in the rings. Heterocyclic compounds usually contain nitrogen, oxygen, or sulfur atoms(s) in the ring. Five- or six-membered cyclic compounds containing a nitrogen atom in the ring are among the most common heterocyclic compounds. Pyridine, pyrrole, indole, and so on, are some examples of nitrogen-containing rings; tetrahydrofuran, tetrahydropyran, and p-dioxane are common examples of five- and six-membered oxygen-containing heterocyclic compounds; and thiophene and tetrahydrothiopyran are examples of sulfur-containing ring compounds. Compounds that are of common laboratory use are only a few in number and include pyridine, piperidine, tetrahydrofuran, and 1,4-dioxane. Ethylene oxide and its alkyl homologues, three-membered cyclic compounds containing an oxygen atom in the ring, are discussed separately in Chapter 17. Many alkaloids containing five- or six-membered nitrogen heterocyclic units are discussed in Chapter 7. Cyclic compounds containing S atoms are highlighted in Chapter 55.

It may be noted that the chemical properties and reactivity of nitrogen ring compounds are different from those of oxygen-containing rings. Heterocyclic compounds of nitrogen are basic, and some are strong bases.

Toxicity

The toxicity data available from animal studies are too inadequate to generalize the toxicological pattern for either nitrogen or oxygen ring compounds. Human toxicity data are very limited, but animal studies indicate that the toxicity of most compounds of these two classes are low to very low. Piperidine, pyrrole, pyrrolidine, quinoline, and isoquinoline are moderately toxic in the decreasing order above, based on acute oral LD_{50} values in rats and mice. The health hazard from inhalation of vapors is insignificant for a large number of nitrogen compounds, such as indole and imidazole, which are solid at ambient temperature, having negligible vapor pressure.

The heterocyclic compounds of oxygen that are largely used as solvents, tetrahydrofuran and 1,4-dioxane, are an irritant to the eyes and respiratory tract, a depressant to the central nervous system, and exhibit a low order of toxicity. Exposure to high concentrations, however, may cause injury to the kidney and liver in humans.

Flammability

Piperidine, pyrrolidine, and pyridine are among the common heterocyclic compounds of nitrogen that are flammable liquids. Most five- or six-membered heterocyclic compounds of oxygen, containing one or two oxygen atoms in the rings, are also flammable liquids. The vapors of these substances form explosive mixtures with air. These compounds tend to form explosive peroxides on prolonged storage or exposure to air.

Analysis

Heterocyclic compounds of nitrogen and oxygen can be analyzed by GC techniques using a flame ionization detector. Alternatively, these substances may be analyzed by GC/MS or FTIR spectroscopy. Analysis of pyridine in groundwater, soil, and hazardous wastes may be performed by GC/MS in accordance to EPA Method 8270 (U.S. EPA 1986). A fused-silica capillary column such as DB-5 (J&W Scientific) or equivalent is suitable for the purpose.

Pyridine in air may be analyzed by NIOSH Method 1613. A known volume of air between 18 and 150 L at a flow rate of 10–1000 mL/min is passed through a solid sorbent tube containing coconut shell charcoal. The analyte is desorbed with methylene chloride and injected into a GC equipped with FID. A DB-5 capillary column or 5% Carbowax 20M on 80/100-mesh acid-washed Chromosorb W support is suitable for analysis. Tetrahydrofuran may be analyzed by NIOSH (1984) Method 1609. About 1–9 L of air (at a flow rate of 10–200 mL/min) is passed over coconut shell charcoal. The adsorbed tetrahydrofuran is eluted with carbon disulfide and analyzed by GC-FID using a Porapak Q or a fused-silica capillary column. Air analysis for 1,4-dioxane may be performed in a similar way (NIOSH 1984, Method 1602). Adsorption over charcoal, desorption into CS_2, and analysis by GC-FID are the sequential steps in this method. Ten percent FFAP on 80/100-mesh Chromosorb W-HP or a fused-silica capillary coated with DB-5 column may be used for analysis. Other heterocyclic compounds may be analyzed by methods similar to those stated above, using GC or GC/MS techniques.

HETEROCYCLIC COMPOUNDS OF NITROGEN

24.2 PYRIDINE

EPA Designated Toxic Waste, RCRA Waste Number U196; DOT Label: Flammable Liquid, UN 1282

Formula C_5H_5N; MW 79.11; CAS [110-86-1]

Structure:

Synonyms: azine; azabenzene

Uses and Exposure Risk

Pyridine is used as a solvent in paint and rubber industries; as an intermediate in dyes and pharmaceuticals; for denaturing alcohol; and as a reagent for cyanide analysis. It occurs in coal tar.

Physical Properties

Colorless liquid with a characteristic disagreeable odor, the odor is detectable below 1 ppm; sharp taste; density 0.983 at 20°C; bp 115°C; mp −41.5°C; miscible with water and most organic solvents; weak base, pH of 0.2 molar solution 8.5, forms salts with strong acids.

Health Hazard

The toxic effects of pyridine include headache, dizziness, nervousness, nausea, insomnia, frequent urination, and abdominal pain. The symptoms were transient, occurred in people from subacute exposure to pyridine vapors at about 125 ppm for 4 hours a day for 1–2 weeks (Reinhardt and Brittelli 1981). The target organs to pyridine toxicity are the central nervous system, liver, kidneys, gastrointestinal tract, and skin. The routes of exposure are inhalation of vapors, and ingestion and absorption of the liquid through the skin. Serious health hazards may arise from chronic inhalation, which may cause kidney and liver damage, and stimulation of bone marrow to increase the production of blood platelets. Low-level exposure to 10 ppm may produce chronic poisoning effects on the central nervous system. Ingestion of the liquid may produce the same symptoms as those stated above. Skin contact can cause dermatitis. Vapor is an irritant to the eyes, nose, and lungs. Because of its strong disagreeable odor, there is always a sufficient warning against any overexposure. A concentration of 10 ppm is objectionable to humans.

LC$_{LO}$ value, inhalation (rats): 4000 ppm/4 hr

LD$_{50}$ value, oral (mice): 1500 mg/kg

Huh and co-workers (1986) have investigated the effect of glycyrrhetinic acid [471-53-4] on pyridine toxicity in mice. Pretreatment with glycerrhetinic acid decreased depression of the central nervous system and mortality in animals induced by pyridine. Such pretreatment markedly decreased the activity of the enzyme serum transaminase [9031-66-7], and increased the activity of hepatic microsomal aniline hydroxylase [9012-90-0], a pyridine-metabolizing enzyme.

Exposure Limits

TLV-TWA 5 ppm (~15 mg/m^3) (ACGIH, MSHA, and OSHA); STEL 10 ppm (ACGIH), IDLH 3600 ppm (NIOSH).

Fire and Explosion Hazard

Flammable liquid; flash point (closed cup) 20°C (68°F); vapor pressure 18 torr at 20°C; vapor density 2.7 (air = 1), vapor heavier than air and can travel a considerable distance to a source of ignition and flash back; autoignition temperature 482°C (900°F); fire-extinguishing agent: dry chemical, "alcohol" foam, or CO$_2$; water may be used to keep fire-exposed containers cool, to disperse the vapors, and to flush and dilute the spill.

Pyridine forms explosive mixtures with air, with LEL and UEL values of 1.8% and 12.4% by volume in air, respectively. Its reactions with strong oxidizers can be violent. There is a report of an explosion with chromium trioxide during the preparation of a chromium trioxide–pyridine complex (MCA 1967). Many metal perchlorates, such as silver perchlorate, form solvated salts with pyridine, which can explode when struck. It undergoes exothermic reactions with concentrated acids.

Storage and Shipping

It is stored in a flammable liquid cabinet or preferably outside, separated from strong oxidizing substances and acids. It is shipped in bottles, drums, and tank cars.

Disposal/Destruction

Pyridine is burned in a chemical incinerator equipped with an afterburner and scrubber. It may be disposed of in a secure landfill in lab packs.

24.3 PIPERIDINE

DOT Label: Flammable Liquid, UN 2401

Formula C$_5$H$_{11}$N; MW 85.17; CAS [110-89-4]

Structure:

Synonyms: hexahydropyridine; azacyclohexane; pentamethyleneimine; cyclopentimine

Uses and Exposure Risk

It is used in organic synthesis, especially in the preparation of many crystalline derivatives of aromatic nitro compounds.

Physical Properties

Colorless liquid with characteristic odor; density 0.862 at 20°C; bp 106°C; mp −13°C; soluble in water and organic solvents; strong base.

Health Hazard

Piperidine is a highly toxic compound. The acute oral toxicity is high in many species of test animals. The oral LD_{50} values in mice and rabbits are 30 and 145 mg/kg, respectively (NIOSH 1986). The liquid is moderately toxic by skin absorption. Inhalation toxicity in experimental animals was low, however. A 4-hour exposure to 4000 ppm was lethal to rats. Piperidine is corrosive to skin. Contact with eyes can produce severe irritation.

Fire and Explosion Hazard

Flammable liquid; flash point (closed cup) 4°C (40°F); vapor forms explosive mixtures with air; the LEL and UEL values are not established. Vigorous to violent reaction may happen when combined with strong oxidizers.

24.4 PIPERAZINE

DOT Label: Corrosive Material, UN 2685
Formula $C_4H_{10}N$; MW 86.16; CAS [110-85-0]

Structure:

Synonyms: hexahydropyrazine; hexahydro-1,4-diazine; *N,N*-diethylene diamine; dispermine

Uses and Exposure Risk

Piperazine is used as an intermediate in the manufacture of dyes, pharmaceuticals, polymers, surfactants, and rubber accelerators.

Physical Properties

Crystalline solid with a salty taste; darkens on exposure to air and light; mp 106°C; bp 146°C; soluble in water and alcohol, insoluble in ether; aqueous solution strongly alkaline.

Health Hazard

Piperazine is a corrosive substance. The solid and its concentrated aqueous solutions are irritants to the skin and eyes. The irritant effect in rabbits' eyes was severe.

The toxic symptoms from ingestion of piperazine include nausea, vomiting, excitement, change in motor activity, somnolence, and muscle contraction. The toxicity of this compound is low, however. The oral LD_{50} value in rats is 1900 mg/kg. The inhalation toxicity is very low. The inhalation LC_{50} value in mice is 5400 mg/m³/2 hr.

Fire and Explosion Hazard

Combustible liquid; flash point (open cup) 81°C (178°F).

24.5 PYRROLE

Formula C_4H_5N; MW 67.10; CAS [109-97-7]

Structure:

Synonyms: 1H-pyrrole; azole; imidole; divinylenimine

Uses and Exposure Risk

Commercial applications of this compound are very limited. It is used in organic synthesis. Pyrrole is formed by heating albumin or by pyrolysis of gelatin.

Physical Properties

Colorless liquid with chloroform-like odor; darkens on exposure to air or light; density 0.969 at 20°C; boils at 130°C; solidifies at −23°C; sparingly soluble in water, readily miscible with organic solvents, insoluble in aqueous alkalies; decomposes with dilute acids; polymerizes in the presence of acids.

Health Hazard

The toxicity data on pyrrole are scant. It is moderately toxic on test animals. The routes of exposure are inhalation of vapors, ingestion, and skin absorption. Vapors are an irritant to the eyes and respiratory tract. The lethal doses in rabbits by oral and dermal routes are with in the range 150 and 250 mg/kg, respectively.

Fire and Explosion Hazard

Combustible liquid; flash point (closed cup) 39°C (102°F); vapor forms explosive mixtures with air; LEL and UEL values are not available. Heating with strong oxidizers can be violent.

24.6   PYRROLIDINE

DOT Label: Flammable Liquid, UN 1922
Formula C_4H_9N; MW 71.14; CAS [123-75-1]

Structure:

Synonyms: tetrahydropyrrole; azacyclopentane; tetramethylenimine; prolamine

Uses and Exposure Risk

Uses of this compound in any major scale have not been made. It occurs in tobacco and carrot leaves. It is formed by reduction of pyrrole.

Physical Properties

Colorless liquid with ammonia-like odor; fumes in air; density 0.892 at 20°C; boils at 89°C; soluble in water and organic solvents; aqueous solution is strongly basic.

Health Hazard

The acute toxicity of pyrrolidine is moderate on test animals. It is somewhat less toxic than pyrrole. The vapors are an irritant to the eyes and respiratory tract. The liquid is corrosive to the skin. Contact with the eyes can cause damage. The oral LD_{50} value in rats is 300 mg/kg, while the inhalation LC_{50} value in mice is 1300 mg/m^3/2 hr (NIOSH 1986).

Fire and Explosion Hazard

Flammable liquid; flash point (closed cup) 3°C (37°F). Vapor forms explosive mixtures with air; the LEL and UEL data are not available. Vigorous to violent reactions can occur with powerful oxidizers. It reacts exothermically with concentrated acids.

24.7   PYRAZOLE

Formula $C_3H_4N_2$; MW 68.01; CAS [288-13-1]

Structure:

Synonym: 1,2-diazole

Uses and Exposure Risk

Pyrazole is used in organic synthesis and as a chelating agent.

Physical Properties

Crystalline solid with pyridine-like odor; bitter taste; melts at 70°C; boils at 187°C; soluble in water and organic solvents.

Health Hazard

The acute toxic symptoms from oral administration of pyrazole in experimental animals were ataxia, muscle weakness, and respiratory depression. It is less toxic than pyrrole and pyrrolidine. The oral LD_{50} value in mice is within the range 1450 mg/kg.

Pyrazole exhibited reproductive toxicity in rats and mice when administered by oral or intraperitoneal route. The effects include fetal death, developmental abnormalities pertaining to the urogenital system, and postimplantation mortality.

Fire and Explosion Hazard

Noncombustible solid.

24.8   IMIDAZOLE

Formula $C_3H_4N_2$; MW 68.09; CAS [288-32-4]
Structure:

Synonyms: 1,3-diaza-2,4-cyclopentadiene; 1,3-diazole; glyoxaline

Uses and Exposure Risk

It is used in organic synthesis and as an anti-irradiation agent.

Physical Properties

Colorless solid; melts at 90°C; boils at 257°C; soluble in water, alcohol, and ether, slightly soluble in benzene and petroleum ether; weak base.

Health Hazard

It is less toxic relative to pyrrole and other five-membered heterocyclic compounds of nitrogen. Intraperitoneal administration of imidazole caused somnolence, muscle contractions, and convulsions in mice. The oral LD_{50} value in mice is in the range 900 mg/kg.

Fire and Explosion Hazard

Noncombustible solid.

24.9   INDOLE

Formula C_8H_7N; MW 117.16; CAS [120-72-9]
Structure:

Synonyms: 1-azaindene; 1-benzazole; benzopyrrole; 2,3-benzopyrrole

Uses and Exposure Risk

Indole occurs in coal tar. It is used, under high dilution, in perfumery, and as an intermediate in organic synthesis.

Physical Properties

Crystalline solid with a strong fecal odor; mp 52°C; bp 253°C; soluble in hot water and most organic solvents.

Health Hazard

Low to moderate toxicity was observed in experimental animals resulting from oral or subcutaneous administration of indole. The oral LD_{50} value in rats is 1000 mg/kg. It is an animal carcinogen. It caused tumors in blood and lungs in mice subjected to subcutaneous administration.

Fire and Explosion Hazard

Noncombustible solid.

24.10   QUINOLINE

Formula C_9H_7N; MW 129.17; CAS [91-22-5]
Structure:

Synonyms: 1-azanaphthalene; 1-benzazine; benzo[b]pyridine; chinoline

Uses and Exposure Risk

Quinoline is used in the manufacture of dyes and hydroxyquinoline salts; as a solvent for resins and terpenes; and therapeutically as an antimalarial agent. It occurs in coal tar in small amounts.

Physical Properties

Colorless liquid with a characteristic odor; hygroscopic; darkens on storage; bp 237.5°C; mp −15°C; soluble in water, miscible with organic solvents, and dissolves sulfur and phosphorus; weakly basic.

Health Hazard

There is little information in the published literature on the toxic properties of quinoline. The acute toxicity is moderate in rodents from oral and dermal administration. The reported oral LD_{50} values in rats show inconsistent values ranging between 300 and 500 mg/kg. Its irritant action was mild on rabbits' skin and severe in the animals' eyes.

Quinoline exhibited carcinogenicity in rats and mice, causing liver cancer. There is no evidence of its carcinogenicity in humans. It tested positive to the histidine reversion–Ames test for mutagenicity.

Fire and Explosion Hazard

Noncombustible liquid; autoignition temperature 480°C (896°F). There is no report of any explosion or fire by quinoline. When mixed with powerful oxidizing substances, quinoline may undergo vigorous to violent reactions.

24.11   ISOQUINOLINE

Formula C_9H_7N; MW 129.17; CAS [119-65-3]
Structure:

Synonyms: 2-azanaphthalene; 2-benzazine; benzo[c]pyridine

Uses and Exposure Risk

It is used in the manufacture of dyes, pharmaceuticals, and insecticides, and as an antimalarial agent.

Physical Properties

Colorless liquid (or solid) with a pungent odor; hygroscopic; mp 26.5°C; bp

242°C; insoluble in water, soluble in organic solvents.

Health Hazard

The toxic properties of this compound are similar to those of quinoline. It is moderately toxic in rats and rabbits by oral route and skin absorption. The oral LD_{50} value in rats is 360 mg/kg. The irritation effects on skin and eyes in rabbits were moderate to severe. Carcinogenicity due to isoquinoline in animals or humans is not known. The histidine reversion–Ames test for mutagenicity was inconclusive.

Fire and Explosion Hazard

Noncombustible liquid (solid).

HETEROCYCLIC COMPOUNDS OF OXYGEN

24.12   FURAN

EPA Designated Hazardous Waste, RCRA Waste Number U124; DOT Label: Flammable Liquid, UN 2389

Formula C_4H_4O; MW 68.08; CAS [110-00-9]

Structure:

Synonyms: divinylene oxide; oxacyclopentadiene; oxole; tetrole; furfuran

Uses and Exposure Risk

Furan is used as an intermediate in organic synthesis.

Physical Properties

Colorless liquid; density 0.937 at 20°C; boils at 31.5°C; insoluble in water, soluble in alcohols and ether.

Health Hazard

Furan is a highly toxic compound. Inhalation of its vapors can cause acute pulmonary edema and lung damage. The intraperitoneal LD_{50} value in rodents is between 5 and 7 mg/kg. The inhalation LC_{50} value in mice for a 1-hour exposure is 120 mg/m^3.

Fire and Explosion Hazard

Highly flammable liquid; flash point (closed cup) −35°C (−32°F); vapor density 2.34 (air = 1); the vapor is heavier than air and can travel a considerable distance to a source of ignition and flash back.

The vapors form explosive mixtures with air; the LEL and UEL values are 2.3% and 14.3% by volume in air, respectively. Reactions with strong oxidizers can be vigorous to violent. Polymerization can occur in the presence of mineral acids; the reaction can be violent in a closed container.

24.13   TETRAHYDROFURAN

EPA Designated Hazardous Waste, RCRA Waste Number U213; DOT Label: Flammable Liquid, UN 2056

Formula C_4H_8O; MW 72.12; CAS [109-99-9]

Structure:

Synonyms: oxolane; oxacyclopentane; diethylene oxide; butylene oxide; cyclotetramethylene oxide; 1,4-epoxybutane

Uses and Exposure Risk

Tetrahydrofuran is used as a solvent for resins, vinyls, and high polymers; as a Grignard reaction medium for organometallic, and metal hydride reactions; and in the synthesis of succinic acid and butyrolactone.

Physical Properties

Colorless liquid with an ethereal odor; density 0.889 at 20°C; boils at 66°C; solidifies at −108.5°C; miscible with water and most organic solvents.

Health Hazard

The toxicity of tetrahydrofuran is of low order in animals and humans. The target organs are primarily the respiratory system and central nervous system. It is an irritant to the upper respiratory tract and eyes. At high concentrations it exhibits anesthetic properties similar to those of many lower aliphatic ethers. Exposure to concentrations above 25,000 ppm in air can cause anesthesia in humans. Other effects noted were strong respiratory stimulation and fall in blood pressure (ACGIH 1986). Kidney and liver injuries occurred in experimental animals exposed to 3000 ppm for 8 hours/day for 20 days (Lehman and Flury 1943). Inhalation of high concentrations of vapors or ingestion of the liquid also causes nausea, vomiting, and severe headache. The acute oral toxicity is low; the LD_{50} value in rats is in the range of 2800 mg/kg. The inhalation LC_{50} value in rats is 21,000 ppm/3 hr.

Exposure Limits

TLV-TWA 200 ppm (590 mg/m^3) (ACGIH, MSHA, and OSHA); STEL 250 ppm (ACGIH); IDLH 20,000 ppm (NIOSH).

Fire and Explosion Hazard

Highly flammable liquid; flash point (closed cup) −14.4°C (6°F) (NFPA 1986), −17°C (1°F) (Merck 1989; Aldrich 1990); vapor pressure 145 torr at 20°C; vapor density 2.5 (air = 1); the vapor is heavier than air and can travel a considerable distance to a source of ignition and flash back; autoignition temperature 321°C (610°F); fire-extinguishing agent: dry chemical or CO_2; a water spray may be used to keep the fire-exposed containers cool and to dilute spills.

Tetrahydrofuran forms explosive mixtures with air within the range 2.0–11.8% by volume in air. It is susceptible to form organic peroxides when exposed to air or light. Severe explosion can occur during distillation, purification, or use of impure tetrahydrofuran. Solvent containing peroxide can explode when dried with caustic soda or caustic potash, or when evaporated; and may catch fire in contact with lithium aluminum hydride or other metal hydrides. Violent reactions can occur when combined with strong oxidizers.

Storage and Shipping

It is stored in a cool, dark, and well-ventilated area, separated from heat sources and oxidizing materials. An inhibitor such as butylated hydroxytoluene (~0.025%) is added to inhibit peroxide formation. Discard the unused solvent 2 months after opening the container. It is shipped in cans, drums, and tank cars.

Disposal/Destruction

Tetrahydrofuran waste solvent is burned in a chemical incinerator equipped with an afterburner and scrubber. It may be disposed of in a secure landfill in lab packs.

24.14 1,4-DIOXANE

EPA Designated Toxic Waste, RCRA Waste Number U108; DOT Label: Flammable Liquid, UN 1165

Formula $C_4H_8O_2$; MW 88.12; CAS [123-91-1]

Structure:

Synonyms: diethylene ether; 1,4-diethylene dioxide; 1,4-dioxacyclohexane; glycol ethylene ether; *p*-dioxane

Uses and Exposure Risk

1,4-Dioxane is used as a solvent for cellulose esters, oils, waxes, resins, and numerous organic and inorganic substances. It is also used in coatings and as a stabilizer in chlorinated solvents.

Physical Properties

Colorless liquid with a faint ethereal odor; density 1.033 at 20°C; boils at 101°C; solidifies between 10° and 12°C; miscible with water and most organic solvents.

Health Hazard

The toxicity of 1,4-dioxane is low in test animals by all routes of exposure. However, in humans the toxicity of this compound is severe. The target organs are the liver, kidneys, lungs, skin, and eyes. Exposure to its vapors as well as the absorption through the skin or ingestion can cause poisoning, the symptoms of which include drowsiness, headache, respiratory distress, nausea, and vomiting. It causes depression of central nervous system. There are reports of human deaths from subacute and chronic exposures to dioxane vapors at concentration levels ranging between 500 and 1000 ppm. Serious health hazards may arise from its injurious effects on the liver, kidneys, and brain. Rabbits died of kidney injury resulting from repeated inhalation of 1,4-dioxane vapors for 30 days (Smyth 1956). It is an irritant to the eyes, nose, skin, and lungs. In humans, a 1-minute exposure to 5000-ppm vapors can cause lacrimation.

LC_{50} value, inhalation (rats): 13,000 ppm/2 hr
LD_{50} value, oral (mice): 5700 mg/kg

1,4-Dioxane is an animal carcinogen of low potential. Ingestion of high concentrations of this compound at a level of 7000–18,000 ppm in drinking water for 14–23 months caused nasal and liver tumors in rats (ACGIH 1986). Guinea pigs developed lung tumors.

Exposure Limits

TLV-TWA 25 ppm ($\sim$90 mg/m^3) (ACGIH), 100 ppm (MSHA and OSHA); carcinogenicity: Animal Sufficient Evidence (IARC).

Fire and Explosion Hazard

Flammable liquid; flash point (closed cup) 12°C (54°F), (open cup) 18°C (65°F); vapor pressure 29 torr at 20°C; vapor density 3.0 (air = 1); the vapor is heavier than air and can travel a considerable distance to a source of ignition and flash back; autoignition temperature 180°C (356°F); fire-extinguishing agent: dry chemical, CO_2, or "alcohol" foam; use water to keep fire-exposed containers cool and to flush and dilute spills.

1,4-Dioxane vapors form explosive mixtures with air within a relatively wide range. The LEL and UEL values are 2.0 and 22.0% by volume in air, respectively. Under certain conditions, 1,4-dioxane can form explosive peroxides, which may cause explosion during distillation, drying, or evaporation.

Vigorous to violent reactions may occur when mixed with strong oxidizers. 1,4-Dioxane forms explosive complexes with perchlorates of many metals, such as silver and mercury. Reaction with hydrogen and nickel above 210°C is explosive (NFPA 1997).

Storage and Shipping

1,4-Dioxane is stored in a flammable-liquids cabinet or storage room isolated from combustible or oxidizing substances and sources of ignition. It is shipped in bottles, cans, and metal drums.

Disposal/Destruction

It is burned in a chemical incinerator equipped with an afterburner and scrubber.

Waste solvent may be buried in a secure landfill in lab packs.

24.15 1,3-DIOXOLANE

Formula $C_3H_6O_2$; MW 74.09; CAS [646-06-0]

Structure:

Synonyms: 1,3-dioxacyclopentane; ethylene glycol formal; formal glycol

Uses and Exposure Risk

1,3-Dioxolane is used as an intermediate in organic synthesis.

Physical Properties

Colorless liquid; density 1.061 at 20°C; boils at 74°C; solidifies at −95°C; miscible with water and organic solvents.

Health Hazard

The acute inhalation and oral toxicity of 1,3-dioxolane is low in test animals. The vapor is irritant to eyes and respiratory tract. Application of the liquid produced severe irritation in rabbits' eyes and mild action on the animals' skin. The information on the toxicity of this compound in humans is not known.

The inhalation LC_{50} value of 4-hour exposure in rats is in the range of 20,000 mg/m^3, and the oral LD_{50} is 3000 mg/kg (NIOSH 1986).

Fire and Explosion Hazard

Highly flammable liquid; flash point (open cup) 2°C (35°F); vapor density 2.55 (air = 1); the vapor is heavier than air and can travel a considerable distance to a source of ignition and flash back; fire-extinguishing agent: dry chemical or CO_2; use water to keep fire-exposed containers cool and to flush and dilute the spills.

1,3-Dioxolane vapors form explosive mixtures with air; the flammability range has not been established. It is capable of forming peroxides on prolonged storage or exposure to air. Such peroxide formation may be inhibited with trace amounts of triethylamine.

REFERENCES

ACGIH. 1986. *Documentation of the Threshold Limit Values and Biological Exposure Indices*, 5th ed. Cincinnati, OH: American Conference of Governmental Industrial Hygienists.

Aldrich. 1997. *Catalog Handbook of Fine Chemicals*. Milwaukee, WI: Aldrich Chemical Company.

Huh, K., S. I. Lee, J. M. Park, and J. R. Chung. 1986. Effect of glycyrrhetinic acid on pyridine toxicity in mouse. *Korean J. Toxicol.* 2(1): 31–36; cited in *Chem. Abstr. CA 105*(21): 185455e.

Lehman, K. B., and F. Flury. 1943. *Toxicology of Industrial Solvents*, p. 269. Baltimore: Williams & Wilkins.

MCA. 1967. *Case Histories of Accidents in the Chemical Industry*. Washington, DC: Manufacturing Chemists' Association.

Merck. 1996. *The Merck Index*, 12th ed. Rahway: Merck & Co.

NFPA. 1997. *Fire Protection Guide on Hazardous Materials*, 12th ed. Quincy, MA: National Fire Protection Association.

NIOSH. 1984. *Manual of Analytical Methods*, 3rd ed. Cincinnati, OH: National Institute for Occupational Safety and Health.

NIOSH. 1986. *Registry of Toxic Effects of Chemical Substances*, ed. D. V. Sweet. Washington, DC: U.S. Government Printing Office.

Reinhardt, C. F., and M. R. Brittelli. 1981. *Patty's Industrial Hygiene and Toxicology*, 3rd ed., Vol. 2A, Toxicology, p. 2730. New York: Wiley-Interscience.

Smyth, H. F., Jr. 1956. *Am. Ind. Hyg. Assoc. Q. 17*: 129; cited in *Documentation of the*

Threshold Limit Values and Biological Exposure Indices, 5th ed. 1986. Cincinnati, OH: American Conference of Governmental Industrial Hygienists.

U.S. EPA. 1997. *Test Methods for Evaluating Solid Waste*, 3rd ed., Washington, DC: Office of Solid Waste and Emergency Response.

25

HYDROCARBONS, ALIPHATIC AND ALICYCLIC

25.1 GENERAL DISCUSSION

Hydrocarbons are an important class of organic compounds composed solely of carbon and hydrogen and are widely distributed in nature. These substances occur in petroleum products, oil and grease, natural gases, coal and coal tars, and deep earth gases. Hydrocarbons may be classified into three broad categories: (1) open-chain aliphatic compounds, (2) cyclic or alicyclic compounds of naphthene type, and (3) aromatic ring compounds.

Open-chain aliphatic hydrocarbons constitute alkanes, alkenes, alkynes, and their isomers. Alkanes have the general formula C_nH_{2n+2}, where n is the number of carbon atoms in the molecules, such as methane, propane, n-pentane, and isooctane. Alkenes or olefins are unsaturated compounds, characterized by one or more double bonds between the carbon atoms. Their general formula is C_nH_{2n}. Examples are ethylene, 1-butene, and 1-octene. Alkynes or acetylenic hydrocarbons contain a triple bond in the molecule and are highly unsaturated. The molecular formula for an alkyne is C_nH_{2n-2}. An alicyclic hydrocarbon is a cyclic ring compound of three or more carbon atoms.

Typical examples are cyclopropane, cyclobutane, cyclopentane, and cyclohexane, which are three-, four-, five-, and six-membered rings, respectively. Aromatics are ring compounds, too; but these are of different types — characterized by six-carbon atoms unsaturated benzenoid rings. Aromatics are discussed in detail in Chapter 26.

The toxicities of aliphatic and alicyclic hydrocarbons in humans and animals are very low. The gaseous compounds are all nontoxic and are simple asphyxiants. The presence of a double or triple bond in the molecule or cyclization does not enhance the toxicity of the hydrocarbons, especially the lower ones. Only at very high concentrations in air are these gases narcotics. The higher hydrocarbons, which are liquid at ambient conditions, are also of low toxicity, producing anesthetic effects only at high doses. 1,3-Butadiene is the only open-chain hydrocarbon of commercial application, which has been recorded as cancer causing.

Lower hydrocarbons are highly flammable substances. Compounds containing one to four carbon atoms are gases at room temperature and 760 torr pressure. Aliphatic and alicyclic hydrocarbons containing five to nine carbon atoms are mostly flammable liquids with a flash point of <100°F (38°C); and longer-chain molecules of 12 or 13 carbon atoms are combustible liquids. With an increase in the carbon number, combustibility decreases, and the hydrocarbons become noncombustible solids. Lower hydrocarbons form explosive mixtures with air. Acetylene is among the most flammable organic compounds, with a very wide range of flammability in air. Thus it is primarily the flammable properties that characterize the hazardous nature of hydrocarbons.

The reactivity of alkanes and cycloalkanes are very low. Alkenes and alkynes containing double and triple bonds are reactive. The addition reactions follow Markovnikov rule. Conjugated dienes undergo Diels–Alder reactions. The substances that may react violently with unsaturated hydrocarbons are halogens, strong oxidizers, and nitrogen dioxide. Alkynes may form acetylides with certain metals.

The hazardous properties of individual hydrocarbons are discussed below in detail.

25.2 METHANE

DOT Label: Flammable Gas, UN 1971
Formula CH_4; MW 16.04; CAS [74-82-8]
Structure: tetrahedral, simplest hydrocarbon
Synonyms: methyl hydride; marsh gas

Uses and Exposure Risk

Methane is widely distributed in nature. As a deep earth gas, it is outgassing from earth's crust. It is also present in the atmosphere (0.00022% by volume). It is the prime constituent of natural gas (85–95% concentration). It is formed from petroleum cracking and decay of animal and plant remains. It is found in marshy pools and muds. Methane is used as a common heating fuel in natural gas; in the production of hydrogen, acetylene, ammonia, and formaldehyde; and as a carrier gas in GC analysis.

Physical Properties

Colorless and odorless gas; density 0.554 (air = 1), weighing 0.7168 g/L at 25°C; liquefies at −161.4°C; solidifies at −182.6°C; soluble in organic solvents, slightly soluble in water, 3.5 mL/100 mL water (Merck 1989).

Health Hazard

Methane is a nonpoisonous gas. It is an asphyxiate. Thus exposure to its atmosphere can cause suffocation.

Fire and Explosion Hazard

Flammable gas; burns with a pale luminous flame; heat of combustion 213 kcal/mol; autoignition temperature 537°C (999°F) (NFPA 1997). Fire extinguishing measure: stop the flow of gas; allow the flame to continue while keeping the surrounding cool with water spray; CO_2 or a dry chemical may be used to extinguish fire.

Methane forms explosive mixtures with air; the LEL and UEL values are 5.5% and 14% by volume of air, respectively. The explosion is loudest at 10% volume concentration. The combustion of methane in air forming CO_2 involves chain reactions, propagated by free radicals, $CH_3\cdot$, $HO\cdot$, $H\cdot$, and $O\cdot$ (Meyer 1989). Being a saturated molecule, the reactivity of methane toward most substances is very low. However, under catalytic conditions, violent reactions have been reported with halogens and certain fluorides of nonmetals. This includes reactions with chlorine at room temperature over yellow mercuric oxide, bromine pentafluoride, oxygen difluoride in the presence of a spark (Mellor 1946, Suppl. 1956), and nitrogen trifluoride (Lawless and Smith 1968).

Landfill Gas

Methane is a major component of landfill gas, occurring as well in the leachate from the landfill. It is formed by anaerobic fermentation of municipality landfill contents (Baubeau et al. 1990; Herrera et al. 1988). Wang et al. (1989) have reported that *o*-cresol containing wastewater, subjected to ozone treatment, biodegraded to methane by anaerobic bacteria. Fermentation progressed to a greater extent in basic solutions than in acidic conditions. Methane concentrations in the landfill gas increased by injecting lime into the landfill, apparently due to neutralization of CO_2 by lime. Lime and fly ash mixtures injected into the municipal solid waste landfill, however, ceased methane formation (Kinman et al. 1989). Methane consumption by methane-oxidizing bacteria was reported to be inhibited by trace amounts of trichloroethylene or 1,1,1-trichloroethane (Broholm et al. 1990).

Analysis

Methane is analyzed by GC using a thermal conductivity detector. It may be analyzed by GC/MS using a capillary column under cryogenic conditions.

25.3 ETHANE

DOT Label: Flammable Gas, UN 1035, UN 1961

Formula C_2H_6; MW 30.08; CAS [74-84-0]

Structure: H_3C-CH_3, second member in the homologous series of alkanes

Synonyms: bimethyl, dimethyl, ethyl hydride; methylmethane

Uses and Exposure Risk

Ethane is the second major component in natural gas. It is formed by petroleum cracking. It is used as a fuel gas, in the manufacture of chloro derivatives, and as a refrigerant.

Physical Properties

Colorless and odorless gas; weighs 1.356 g/L, slightly heavier than air, 1.049 (air = 1);

liquefies at $-88°C$; solidifies at $-172°C$; slightly soluble in water (4.7 mL/100 mL at 20°C), more soluble in organic solvents.

Health Hazard

Like methane, ethane is a nonpoisonous gas. It is a simple asphyxiate. At high concentrations it may exhibit narcotic effects.

Fire and Explosion Hazard

Flammable gas; burns with a faintly luminous flame; heat of combustion 1727 Btu/ft^3 at 25°C, or 1 lb (0.45 kg) gives 20,420 Btu; autoignition temperature 472°C (882°F) (NFPA 1997), 530°C (986°F) (Merck 1989); fire-extinguishing measure: shut off the flow of gas; CO_2 or a dry chemical may be used to extinguish the fire; use a water spray to keep the surroundings cool.

Ethane forms explosive mixtures with air within the range 3.2–12.5% by volume of air. Its reactions with chlorine in the presence of mercuric oxide and with fluorides of nonmetals may cause explosions.

Analysis

Ethane is analyzed by GC using a TCD.

25.4 ETHYLENE

DOT Label: Flammable Gas, UN 1038, UN 1962 (refrigerated liquid)

Formula C_2H_4; MW 28.06; CAS [74-85-1]

Structure:

$$\begin{array}{c} H \diagdown \quad \diagup H \\ C = C \\ H \diagup \quad \diagdown H \end{array}$$

first member of the alkene class of compounds

Synonyms: ethene; olefiant gas

Uses and Exposure Risk

Ethylene occurs in petroleum gases, in illuminating gas, and in ripening fruits. It is made by dehydrating alcohol. It is used in oxyethylene flame for welding and cutting metals; in the manufacture of polyethylene, polystyrene, and other plastics; in making ethylene oxide; and as an inhalation anesthetic.

Physical Properties

Colorless gas; liquefies at $-169.4°C$; solidifies at $-181°C$; gas density 0.978 (air $= 1$), 1 L of gas at 25°C weighs 1.147 g; slightly soluble in water (0.0127 g/100 g at 25°C) and insoluble in alcohol and ether, soluble in acetone and benzene.

Health Hazard

Exposure to ethylene atmosphere can cause asphyxiation. At high concentrations it is a narcotic and can cause unconsciousness.

Fire and Explosion Hazard

Flammable gas; burns with a luminous flame; autoignition temperature 490°C (914°F) (NFPA 1997), 543°C (1009°F) (Merck 1996); fire-extinguishing measure: shut off the flow of gas; use a water spray to keep fire-exposed containers cool.

Ethylene forms explosive mixtures in air; the LEL and UEL values are 2.7% and 36% by volume of air, respectively. Its reaction with fluorine is explosively violent ($\Delta H = -112$ kcal/mol), and violent with chlorine ($\Delta H = -36$ kcal/mol). In the presence of sunlight or UV light, an ethylene–chlorine mixture will explode spontaneously. The reaction is explosive at room temperature over the oxides of mercury or silver (Mellor 1946, Suppl. 1956). Ethylene reacts vigorously with oxidizing substances. It reacts with ozone to form ethylene ozonide, $H_2C(O_3)CH_2$, which is unstable and explodes on mechanical shock. Acid-catalyzed addition of hydrogen peroxide may produce ethyl hydroperoxide, which is unstable and explodes on heat or shock:

$$CH_2{=}CH_2 + H_2O_2 \longrightarrow$$

$$CH_3{-}CH_2{-}O{-}OH$$

Explosions have been reported with carbon tetrachloride (under pressure in the presence of organic peroxide catalysts) (Bolt 1947; Joyce 1947), bromotrichloromethane, nitromethane and aluminum chloride (NFPA 1997), and nitrogen dioxide.

Ethylene undergoes a variety of reactions to form highly toxic and/or flammable substances. It catalytically oxidizes to ethylene oxide (highly toxic and explosive gas); reacts with chlorine and bromine to form ethylene dichloride and ethylene dibromide (toxic and carcinogens); reacts with hypochlorite to form ethylene chlorohydrin (poisonous); reacts with chlorine in the presence of HCl and light or chlorides of copper, iron, or calcium to form ethyl chloride (flammable gas, narcotic); and hydrates in the presence of H_2SO_4 to form diethyl ether (highly flammable).

Storage and Shipping

Ethylene is stored in a cool, well-ventilated area isolated from oxygen, chlorine, and flammable and oxidizing substances. It is protected against lightning, statical electricity, heat, and physical damage. It is shipped in steel pressure cylinders and tank barges.

Analysis

Ethylene is analyzed by gas chromatography using a thermal conductivity detector. Any column that is suitable for hydrocarbons analysis may be used. Air analysis is done by direct injection into the GC. A capillary column under cryogenic conditions should be more efficient.

25.5 ACETYLENE

DOT Label: Flammable Gas, UN 1001

Formula C_2H_2; MW 26.04; CAS [74-86-2]

Structure: HC≡CH, first member of the alkyne homolog series

Synonyms: ethyne; ethine

Uses and Exposure Risk

Acetylene is used in oxyacetylene flame for welding and cutting metals; as an illuminant; as a fuel; for purifying copper, silver, and other metals; and in the manufacture of acetic acid, acetaldehyde, and acetylides. It is formed when calcium carbide reacts with water. It is also obtained from cracking of petroleum naphtha fractions.

Physical Properties

Colorless gas; the presence of trace contaminants such as phosphine renders a garlic-like odor; slightly lighter than air, density 0.9 (air = 1), 1 L of gas weighs 1.063 g (at 25°C and 760 torr); solidifies at −81°C (sublimes); liquefies at 0°C at 21.5 atm pressure; slightly soluble in water (0.106 g/100 g at 25°C); the solubility is 4, 6, 25, and 33 times greater in benzene, ethanol, acetone, and methyl sulfoxide, respectively.

Health Hazard

Acetylene is an asphyxiate like other hydrocarbon gases. Exposure to its atmosphere can cause death from suffocation or asphyxiation by exclusion of oxygen. It is nontoxic but narcotic at high concentrations. An exposure to 20% concentration in air may produce headache and dyspnea in humans; while inhaling a 50% acetylene in air for 5 minutes may be fatal (NIOSH 1986).

Fire and Explosion Hazard

Highly flammable gas; burns brilliantly in air, producing sooty flame; autoignition temperature 305°C. The ignition temperature varies with the composition of the mixture, pressure, moisture content, and initial temperature (NFPA 1997). The minimum temperature at which ignition occurs is 300°C; fire-extinguishing measures: stop the flow of gas; use water to keep fire-exposed containers cool.

Acetylene is an extremely hazardous gas. Its hazardous properties are attributed to the following factors:

1. It is thermodynamically unstable and has a high positive Gibbs free energy of formation.

2. It forms oligomers and polymers in the absence of air. Such a polymerization process is exothermic, causing acetylene to deflagrate.

3. It forms explosive mixtures with air. It is one of the few substances known to show such an exceedingly high explosive range, 2–80% by volume.

4. Acetylene being unstable, as mentioned above, is sensitive to shock and pressure. When compressed under great pressure, it explodes. Therefore, special methods are applied in making special types of steel cylinders to hold this gas. Also, it is less stable in wide-bore transportation piping.

5. The triple bond in acetylene makes the molecule highly reactive to many addition-type reactions.

6. Because of the greater proportion of s character in the sp σ orbitals of the carbon atoms than that in sp^2 hybrids (alkenes) or sp^3 hybrids (alkanes), any bonding pair of electrons can come closer to carbon nuclei in acetylene than in alkenes or alkanes. Hence hydrogen is released as a proton more easily in the acetylenic carbon atoms. The increased acidity of hydrogen attached to a carbon atom with a triple bond results in the formation of a class of carbides with certain metals, called acetylides. These acetylides are unstable and highly sensitive to shock and heat, exploding violently (see Chapter 11).

Acetylene reacts with fluorine with explosive violence. Reactions with other halogens, chlorine, bromine, and iodine are violent, too. Acetylene chloride, which is formed from addition reaction with chlorine, explodes spontaneously. It forms unstable acetylides with metals such as copper, silver, gold, and mercury, and their alloys and salts.

Among the hazardous products are the vinyl ethers (flammable) formed from reactions with alcohols in the presence of caustic potash at 150°C, acetaldehyde (flammable) formed by catalytic addition of water, and trinitromethane (melts at 15°C and explodes in the liquid state) formed when acetylene reacts with concentrated nitric acid.

Storage and Shipping

Cylinders used to store acetylene are made up of steel, containing a porous material and acetone. Acetone dissolves 25 volumes of acetylene at 1 atm and 300 volumes at 12 atm. Porous materials such as asbestos are used as a filler inside the cylinder. Acetone is absorbed into it; then the gas is filled (Meyer 1989). The cylinder pressure should not exceed 250 psi at 21°C (70°F). The distribution or transportation through hose, pipe, or supply line should be at a pressure below 15 psi. Cylinders are stored upright.

Acetylene is stored in a cool and well-ventilated area protected from heat, lightning, statical electricity, and oxidizing and combustible substances. It is shipped in steel cylinders of 10–300 standard cubic feet capacity. In a closed system avoid using more than 1 L of gaseous acetylene. It is stabilized by adding methane or acetone. Acetylene is purified by scrubbing the gas through concentrated H_2SO_4 and NaOH or activated alumina. Activated carbon should not be used, as the heat of absorption may decompose the gas (National Research Council 1983).

Analysis

GC-TCD at low-temperature column conditions should be used to analyze acetylene.

25.6 PROPANE

DOT Label: Flammable Gas, UN 1075, UN 1978

Formula C_3H_8; MW 44.11; CAS [74-98-6]
Structure: $CH_3-CH_2-CH_3$
Synonyms: propyl hydride; dimethylmethane

Uses and Exposure Risk

Propane is used as a fuel gas, as a refrigerant, and in organic synthesis.

Physical Properties

Colorless gas; odorless when pure; density 1.832 g/L at 25°C and 760 torr; liquefies at −42°C; solidifies at −187.7°C; slightly soluble in water (6.5 mL/100 mL water, solubility increases in organic solvents as turpentine > benzene > chloroform > ether > ethanol).

Health Hazard

Propane is a nontoxic gas. It is an asphyxiate. At high concentrations it shows narcotic effects.

Exposure Limit

TLV-TWA 1000 ppm (OSHA).

Fire and Explosion Hazard

Flammable gas; burns with a luminous smoky flame; heat of combustion 553 cal/mol; autoignition temperature 450°C (842°F) (NFPA 1997), 468°C (874°F) (Meyer 1989). Fire-extinguishing agent: stop the flow of gas; use a water spray to keep the fire-exposed containers and the surrounding cool.

Propane forms explosive mixtures with air, the LEL and UEL values are 2.2–9.5% by volume of air, respectively. Propane's reactivity is very low toward most compounds.

Analysis

Propane is analyzed by GC equipped with a TCD.

25.7 PROPYLENE

DOT Label: Flammable Gas, UN 1075, UN 1077

Formula C_3H_6; MW 42.09; CAS [115-07-1]

Structure: $CH_3-CH=CH$

Synonyms: 1-propene; propene; methyl-ethylene

Uses and Exposure Risk

Propylene is obtained from refining of gasoline and thermal or catalytic cracking of hydrocarbons. It is used to produce polypropylene (plastic) and in the manufacture of acetone, isopropanol, cumene, and propylene oxide.

Physical Properties

Colorless gas; gas density 1.46 (air = 1), 1 L of gas weighs 1.717 g (at 25°C and 760 torr); liquefies at −48°C; solidifies at −185°C; insoluble in water, dissolves in acetone and benzene.

Health Hazard

Propylene is an asphyxiate and at high concentrations a mild anesthetic. Exposure to high concentrations can cause narcosis and unconsciousness. Contact with the liquefied gas can cause burns.

Fire and Explosion Hazard

Flammable gas; burns with yellow sooty flame; autoignition temperature 455°C. Being heavier than air, the gas can travel a considerable distance to a source of ignition and flash back; fire-extinguishing measure: shut off the flow of gas; use a water spray to keep fire-exposed containers cool.

Propylene forms explosive mixtures with air within the range 2.4–10.3% by volume. Violent explosion may occur by mixing propylene with fluorine or nitrogen dioxide. Reactions with ozone or hydrogen peroxide in the presence of H_2SO_4 give unstable products that may explode.

Analysis

The analysis may be performed by GC at low temperature using a TCD.

25.8 n-BUTANE

DOT Label: Flammable Gas, UN 1011

Formula C_4H_{10}; MW 58.14; CAS [106-97-8]

Structure: $CH_3-CH_2-CH_2-CH_3$

Synonyms: butane; diethyl; methylethyl-methane

Uses and Exposure Risk

n-Butane occurs in petroleum, natural gas, and in refinery cracking products. It is used as a liquid fuel, often called liquefied petroleum gas, in a mixture with propane. It is also used as a propellant for aerosols, a raw material for motor fuels, in the production of synthetic rubber, and in organic synthesis.

Physical Properties

Colorless gas; density of gas 2.04 (air = 1); liquefies at −0.5°C; solidifies at −138.4°C; soluble in water (15 mL/100 mL water) and very soluble in alcohol, ether, chloroform, and other organic solvents.

Health Hazard

n-Butane is a nontoxic gas. Exposure to its atmosphere can result in asphyxia. At high concentrations it produces narcosis. Exposure to 1% concentration in air for 10 minutes may cause drowsiness. Its odor is detectable at a concentration of 5000 ppm.

Exposure Limits

TLV-TWA 800 ppm (~1920 mg/m^3) (ACGIH), 500 ppm (1200 mg/m^3) (MSHA).

Fire and Explosion Hazard

Flammable gas; flash point (closed cup) $-60°C$ ($-76°F$); autoignition temperature $405°C$ ($761°F$) (Meyer 1989); fire-extinguishing measure: stop the flow of gas; use a water spray to keep fire-exposed containers cool and to disperse the gas.

n-Butane forms explosive mixtures with air within the range 1.9–8.5% by volume of air.

Analysis

n-Butane may be analyzed by GC, using a TCD.

25.9   ISOBUTANE

DOT Label: Flammable Gas, UN 1075, UN 1969 (for isobutane mixtures)
Formula C_4H_{10}; MW 58.14; CAS [75-28-5]
Structure:

$$\begin{array}{c} H_3C \\ \\ H_3C \end{array} CH-CH_3$$

Synonym: 2-methyl propane

Uses and Exposure Risk

Isobutane occurs in petroleum, natural gas, and petroleum cracking products. It is used as a fuel gas or a liquefied petroleum gas. It is also used in organic synthesis.

Physical Properties

Colorless gas; density 2.04 (air = 1) at $20°C$; liquefies at $-11.7°C$; soluble in water, very soluble in organic solvents.

Health Hazard

Isobutane, like other saturated aliphatic alkanes, is nontoxic. It is an asphyxiate. Exposure to high concentrations of 1% in air may cause narcosis and drowsiness. Other than this, there is no report of any adverse health effect from exposure to this gas.

Fire and Explosion Hazard

Flammable gas; autoignition temperature $460°C$ ($860°F$); fire-extinguishing measure: shut off the flow of gas and use a water spray to disperse the gas and keep fire-exposed containers cool. Isobutane forms explosive mixtures with air; the LEL and UEL values are 1.9% and 8.5% by volume of air, respectively.

Analysis

Isobutane may be analyzed by GC-TCD or GC-FID.

25.10   1,3-BUTADIENE

Formula C_4H_6; MW 54.10; CAS [106-99-0]
Structure: $CH_2=CH-CH=CH_2$
Synonyms: biethylene; bivinyl, vinylethylene, buta-1,3-diene, erythrene, α,γ-butadiene.

Uses and Exposure Risk

1,3-Butadiene is a petroleum product obtained by catalytic cracking of naphtha or light oil or by dehydrogenation of butene or butane. It is used to produce butadiene–styrene elastomer (for tires), synthetic rubber, thermoplastic elastomers, food wrapping materials, and in the manufacture of adiponitrile. It is also used for the synthesis of organics by Diels–Alder condensation.

Physical Properties

Colorless gas with a mild aromatic odor; heavier than air, gas density 1.865 (air = 1), 1 L of gas weighs 2.212 g (at $25°C$ and 760 torr); liquefies at $-4.5°C$; solidifies at

−109°C; slightly soluble in water (∼0.05%), dissolves in organic solvents.

Health Hazard

The toxicity of 1,3-butadiene has been found to be very low in humans and animals. It is an asphyxiant. In humans, low toxic effects may be observed at exposure to 2000 ppm for 7 hours. The symptoms may be hallucinations, distorted perception, and irritation of eyes, nose, and throat. Higher concentrations may result in drowsiness, lightheadedness, and narcosis. High dosages of 1,3-butadiene were toxic to animals by inhalation and skin contact. General anesthetic effects and respiratory depression were noted. Concentrations of 25–30% may be lethal to rats and rabbits. Contact with the liquefied gas can cause burn and frostbite.

Exposure to 1,3-butadiene caused cancers in the stomach, lungs, and blood in rats and mice. It is suspected to be a human carcinogen. It is a mutagen and a teratogen.

Exposure Limits

TLV-TWA 10 ppm (∼22 mg/m^3) (ACGIH), 1000 ppm (OSHA and NIOSH); IDLH 20,000 ppm (NIOSH); A2–Suspected Human Carcinogen (ACGIH).

Fire and Explosion Hazard

Flammable gas; vapor pressure 910 torr; being heavier than air, it can travel a considerable distance to a source of ignition and flash back; autoignition temperature 429°C (804°F); fire-extinguishing measure: stop the flow of gas; use a water spray to keep the surrounding area and fire-exposed containers cool. Butadiene vapors may polymerize and block the vents (NFPA 1997).

1,3-Butadiene forms explosive mixtures with air within the range 2–12% by volume. On heating under pressure, it decomposes violently. It polymerizes at elevated temperatures. The reaction is exothermic, causing a pressure rise and rupture of containers. It forms peroxides on prolonged contact with air. These peroxides are highly sensitive to heat and shock. There is a report of an explosion during Diels–Alder reaction between 1,3-butadiene and crotonaldehyde under pressure (Greenlee 1948).

Storage and Shipping

1,3-Butadiene is stored in a cool and well-ventilated location separated from combustible and oxidizing substances. Small amounts of stabilizers, such as *o*-dihydroxybenzene, *p-tert*-butylcatechol, or aliphatic mercaptans, are added to prevent its polymerization or peroxides formation. The cylinders are stored vertically and protected against physical damage.

Analysis

1,3-Butadiene is analyzed by GC using a thermal conductivity or flame ionization detector. Air analysis is performed by NIOSH Method 1024 (NIOSH 1984). About 3–25 L of air is passed over 400 and 200 mg of coconut charcoal in separate tubes at a flow rate of 10–500 mL/min. The analyte is desorbed with methylene chloride and injected into a GC equipped with an FID. A fused-silica porous-layer open tubular column coated with Al_2O_3/KCl with or without a back-flushable precolumn may be used as a GC column. Alternatively, a 10% FFAP on a 80/100-mesh Chromosorb W column may be used.

1,3-Butadiene may be identified by UV spectroscopy. It shows a strong absorption band at 210 μm.

25.11 *n*-PENTANE

DOT Label: Flammable Liquid, UN 1265
Formula C_5H_{12}; MW 72.17; CAS [109-66-0]
Structure: $CH_3-CH_2-CH_2-CH_2-CH_3$
Synonyms: pentane; amyl hydride

Uses and Exposure Risk

n-Pentane occurs in volatile petroleum fractions (gasoline) and as a constituent of petroleum ether. It is used as a solvent, in the manufacture of low-temperature thermometers, and as a blowing agent for plastics.

Physical Properties

Colorless, volatile liquid with a gasoline-like smell; exists in three isomeric forms: *n*-pentane, isopentane, and neopentane; density 0.6264 at 20°C; bp 36.1°C; mp −129.7°C; slightly soluble in water (0.036% at 16°C), miscible with organic solvents.

Health Hazard

n-Pentane did not exhibit any marked toxicity in animals. However, inhalation of its vapors at high concentrations can cause narcosis and irritation of the respiratory passages. Such effects may be observed within the range 5–10% concentration in air. In humans, inhalation of 5000 ppm for 10 minutes did not cause respiratory tract irritation or other symptoms (Patty and Yant 1929).

There is no report in the literature indicating any adverse effects from pentane other than narcosis and irritation. An intravenous LD_{50} value in mouse is recorded as 446 mg/kg (NIOSH 1986).

Exposure Limits

TLV-TWA 600 ppm (~1800 mg/m^3) (ACGIH), 1000 ppm (~3000 mg/m^3) (OSHA), 500 ppm (~1500 mg/m^3) (MSHA); STEL 750 ppm (~2250 mg/m^3) (ACGIH).

Fire and Explosion Hazard

Highly flammable liquid; flash point (closed cup) −49.5°C (−57°F); vapor pressure 500 torr at 25°C; autoignition temperature 260°C (403°F) (NFPA 1997), 309°C (588°F) (Merck 1996); fire-extinguishing measures: stop the flow of gas (vapor); use a dry chemical, CO_2, or foam; use a water spray to keep fire-exposed containers cool and to disperse the vapors.

Analysis

n-Pentane may be analyzed by GC techniques using either a flame ionization or thermal conductivity detector.

25.12 CYCLOPENTANE

DOT Label: Flammable Liquid, UN 1146
Formula C_5H_{10}; MW 70.15; CAS [287-92-3]
Structure:

a five-membered ring compound
Synonym: pentamethylene

Uses and Exposure Risk

Cyclopentane is a petroleum product. It is formed from high-temperature catalytic cracking of cyclohexane or by reduction of cyclopentadiene. It occurs in petroleum ether fractions and in many commercial solvents. It is used as a solvent for paint, in extractions of wax and fat, and in the shoe industry.

Physical Properties

Colorless liquid; density 0.746 at 20°C; bp 49.3°C; mp −94.4°C; insoluble in water, miscible with organic solvents.

Health Hazard

Cyclopentane is a low-acute toxicant. Its exposure at high concentrations may produce depression of the central nervous system with symptoms of excitability, loss of equilibrium, stupor, and coma. Respiratory failure may occur in rats from 30–60 minutes' exposure to 100,000–120,000 ppm in air. It is an irritant to the upper respiratory tract, skin, and eyes. No information is available in the literature on the chronic effects from prolonged exposure to cyclopentane.

Exposure Limit

TLV-TWA   600 ppm   (~1720 mg/m^3) (ACGIH).

Fire and Explosion Hazard

Highly flammable liquid; flash point (open cup) −7°C (18°F); autoignition temperature 361°C 682°F; fire-extinguishing agent: dry chemical, foam, or CO_2; use a water spray to keep fire-exposed containers cool.

Cyclopentane forms explosive mixtures with air. The LEL value is 1.5% by volume of air; UEL value not available.

Analysis

Cyclopentane may be analyzed by GC-FID using any column that is suitable for general hydrocarbons.

25.13   CYCLOPENTADIENE

Formula C_5H_6; MW 66.11; CAS [542-92-7]
Structure:

$$\begin{array}{c} CH \\ H_2C \quad\quad CH \\ HC = CH \end{array}$$

Synonyms: 1,3-cyclopentadiene; pentole; pyropentylene

Uses and Exposure Risk

Cyclopentadiene is used in the manufacture of resins, in the synthesis of sesquiterpenes and camphors, and as a ligand in the preparation of metal complexes.

Physical Properties

Colorless, volatile liquid; dimerizes to dicyclopentadiene on standing; density 0.802 at 20°C; bp 42°C; mp −85°C; immiscible with water, soluble in organic solvents.

Health Hazard

Cyclopentadiene exhibited low toxicity in animals. Inhalation produced irritation of the eyes and nose. A 3-mL amount injected subcutaneously into rabbits resulted in narcosis, convulsions, and death (von Oettingen 1940). A dose of ≤1 mL was nontoxic. Repeated exposure to 500 ppm caused liver and kidney injuries in rats; but longer repeated exposures to 250 ppm produced no such effects in test animals (ACGIH 1986). An oral LD_{50} value in rats for the dimeric form has been recorded as 820 mg/kg (Smyth 1954).

Exposure Limits

TLV-TWA 75 ppm (~202 mg/m^3) (ACGIH, NIOSH, and OSHA); IDLH 2000 ppm (NIOSH).

Fire and Explosion Hazard

Flammable liquid; flash point (open cup) 32°C (90°F); fire-extinguishing agent: dry chemical, foam, or CO_2; a water spray may be used to cool the surroundings. Prolonged exposure to air may cause peroxide formation.

Analysis

Cyclopentadiene may be analyzed by GC-FID. Air analysis may be performed by NIOSH Method 2523 (NIOSH 1984, Suppl. 1985). About 1–5 L of air at a flow rate of 10–50 mL/min is passed through a solid sorbent tube containing maleic anhydride on Chromosorb 104. The analyte forms an adduct and is desorbed with ethyl acetate and injected into GC-FID with a column 5% OV-17 on Chromosorb W-HP.

25.14   *n*-HEXANE

DOT Label: Flammable Liquid; UN 1208
Formula C_6H_{14}; MW 86.20; CAS [110-54-3]
Structure: $CH_3-CH_2-CH_2-CH_2-CH_2-CH_3$
Synonyms: hexane; hexyl hydride

Uses and Exposure Risk

n-Hexane is a chief constituent of petroleum ether, gasoline, and rubber solvent. It is used as a solvent for adhesives, vegetable oils, and in organic analysis, and for denaturing alcohol.

Physical Properties

Colorless liquid with a faint odor, highly volatile; density 0.66 g/mL at 20°C; boils at 69°C; solidifies at −95° to −100°C; insoluble in water, miscible with organic solvents.

Health Hazard

n-Hexane is a respiratory tract irritant and at high concentrations a narcotic. Its acute toxicity is greater than that of *n*-pentane. Exposure to a concentration of 40,000 ppm for an hour caused convulsions and death in mice. In humans a 10-minute exposure to about 5000 ppm may produce hallucination, distorted vision, headache, dizziness, nausea, and irritation of eyes and throat. Chronic exposure to *n*-hexane may cause polyneuritis.

The metabolites of *n*-hexane injected in guinea pigs were reported as 2,5-hexanedione and 5-hydroxy-2-hexanone, which are also metabolites of methyl butyl ketone (DiVincenzo et al. 1976). Thus methyl butyl ketone and *n*-hexane should have similar toxicities. The neurotoxic metabolite, 2,5-hexanedione, however, is produced considerably less in *n*-hexane. However, in the case of hexane, the neuorotoxic metabolite 2,5-hexanedione is produced to a much lesser extent. Continuous exposure to 250 ppm *n*-hexane produced neurotoxic effects in animals. Occupational exposure to 500 ppm may cause polyneuropathy (ACGIH 1986).

Inhalation of *n*-hexane vapors have shown reproductive effects in rats and mice.

Exposure Limits

TLV-TWA 50 ppm (~175 mg/m^3) (ACGIH), 500 ppm (~1750 mg/m^3) (OSHA); IDLH 5000 ppm (NIOSH).

Fire and Explosion Hazard

Highly flammable liquid; flash point (closed cup) 30.5°C (−23°F) (ACGIH 1986), −22°C (−7°F) (NFPA 1997), (open cup) −23°C (−10°F) (Aldrich 1997); vapor pressure 124 torr at 20°C; autoignition temperature 223°C (433°F) (NFPA 1997); fire-extinguishing agent: dry chemical, foam, or CO_2; use a water spray to keep fire-exposed containers cool.

n-Hexane forms explosive mixtures with air; the LEL and UEL values are 1.2 and 8.0% by volume of air, respectively.

Disposal/Destruction

n-Hexane may be destroyed by chemical incineration. Small quantities packaged in lab packs can be put into a secure landfill.

Analysis

n-Hexane may be analyzed by GC-FID. Air analysis may be performed in accordance with NIOSH Method 1500 (NIOSH 1984). *n*-Hexane in air is adsorbed over coconut shell charcoal. The analyte is desorbed with carbon disulfide and injected into a GC equipped with a FID. A glass or capillary column containing 20% SP-2100 on 80/100-mesh Supelcoport or any equivalent column may be used for the purpose.

25.15   CYCLOHEXANE

EPA Classified Hazardous Waste; RCRA Waste Number U056; DOT Label: Flammable Liquid, UN 1145

Formula C_6H_{12}; MW 84.18; CAS [110-82-7]
Structure:

$$\begin{array}{ccc}
& CH_2 & \\
H_2C & & CH_2 \\
| & & | \\
H_2C & & CH_2 \\
& CH_2 &
\end{array}$$

a six-membered alicyclic hydrocarbon which exists in two conformations, boat and chair forms, which are interconvertible

Synonyms: hexahydrobenzene; hexamethylene; hexanaphthene; benzene hexahydride

Uses and Exposure Risk

Cyclohexane is a petroleum product obtained by distilling C_4-400°F boiling range naphthas, followed by fractionation and superfractionation; also formed by catalytic hydrogenation of benzene. It is used extensively as a solvent for lacquers and resins, as a paint and varnish remover, and in the manufacture of adipic acid, benzene, cyclohexanol, and cyclohexanone.

Physical Properties

Colorless liquid with a mild solvent odor, the odor may be pungent in presence of impurities; density 0.779 at 20°C; boils at 80.7°C; freezes at 6.4°C; insoluble in water (solubility in water at room temperature may be only about 50 ppm), readily mixes with alcohol, ether, chloroform, benzene, and acetone.

Health Hazard

Cyclohexane is an acute toxicant of low order. It is an irritant to the eyes and respiratory system. Exposure to a 1–2% concentration in air caused lethargy, drowsiness, and narcosis in test animals. The lethal concentration for a 1-hour exposure in mice is estimated at around 30,000 ppm. Ingestion of cyclohexane exhibited low toxic effects in test species. The LD_{50} values in the literature show a wide variation. It may be detected from its odor at 300 ppm concentration.

Exposure Limits

TLV-TWA 300 ppm (~1050 mg/m³) (ACGIH, OSHA, and NIOSH); IDLH 10,000 ppm (NIOSH).

Fire and Explosion Hazard

Flammable liquid; flash point (closed cup) −18°C (1°F) (Merck 1996), −20°C (−4°F) (NFPA 1997), −1°C (30°F) for 98% grade (ACGIH 1986); autoignition temperature 245°C (473°F); fire-extinguishing measure: use dry chemical, foam, or CO_2; a water spray may be used to keep fire-exposed containers cool.

Cyclohexane forms explosive mixtures with air within the range 1.3–8.4% by volume in air. Its reactivity is very low. An explosion has been in record when liquid nitrogen dioxide was mixed accidentally with hot cyclohexane in a nitration column (MCA 1962).

Analysis

Cyclohexane may be analyzed by GC-FID, HPLC, and FTIR techniques. Air analysis may be performed by NIOSH Method 1500 (see Section 25.14).

25.16 CYCLOHEXENE

DOT Label: Flammable Liquid, UN 2256
Formula C_6H_{10}; MW 82.16; CAS [110-83-8]
Structure:

$$\begin{array}{ccc}
& CH_2 & \\
H_2C & & CH \\
| & & \| \\
H_2C & & CH \\
& CH_2 &
\end{array}$$

The molecule displays olefinic reactivity because of the unsaturation.

Synonyms: tetrahydrobenzene; benzene tetra-hydride; hexanaphthylene

Uses and Exposure Risk

It occurs in coal tar and is obtained by catalytic dehydration of cyclohexanol and dehydrogenation of cyclohexane. Cyclohexene is used in making adipic acid, maleic acid, and butadiene; in oil extraction; as a stabilizer for high-octane gasoline; and in organic synthesis.

Physical Properties

Colorless liquid; bp 83°C; mp −103.5°C; density 0.8098 at 20°C; insoluble in water, soluble in organic solvents.

Health Hazard

There are very few toxicological data available on cyclohexene. Inhalation produces irritation of the eyes and respiratory tract. It is also an irritant to skin. Its acute toxicity is low; the toxic effects are similar to those of cyclohexane. Exposure to high concentrations or ingestion may cause drowsiness. Single exposures to 15,000 ppm of cyclohexene could be lethal to rats.

Exposure Limits

TLV-TWA 300 ppm (~1015 mg/m^3) (ACGIH, OSHA, and NIOSH); IDLH 10,000 ppm (NIOSH).

Fire and Explosion Hazard

Flammable liquid; flash point (closed cup) −11.6°C (11°F); vapor pressure 67 torr at 20°C; autoignition temperature 244°C (471°F); fire-extinguishing agent: dry chemical, foam, or CO$_2$; a water spray may be used for cooling purposes. Cyclohexene forms explosive mixtures with air; the LEL and UEL values are 1 and 5% by volume of air, respectively, at 100°C. It is susceptible to forming peroxide on long exposure to air.

Analysis

Cyclohexene is analyzed by GC-FID, HPLC, and FTIR techniques. Air analysis may be performed by NIOSH Method 1500 (see Section 25.14).

25.17 *n*-OCTANE

DOT Label: Flammable Liquid, UN 1262
Formula C$_8$H$_{18}$; MW 114.26; CAS [111-65-9]
Structure: CH$_3$−CH$_2$−CH$_2$−CH$_2$−CH$_2$−CH$_2$−CH$_2$−CH$_3$
Synonyms: octane; octyl hydride

Uses and Exposure Risk

n-Octane occurs in petroleum cracking products, gasoline, petroleum ether, and petroleum naphtha. It is used as a solvent and in organic synthesis.

Physical Properties

Colorless liquid; density 0.703 at 20°C; boils at 125.6°C; solidifies at −56.8°C; insoluble in water, slightly soluble in alcohol, miscible with ether, benzene, and toluene.

Health Hazard

The toxic properties of *n*-octane are similar to those of other paraffinic hydrocarbons. It is an irritant to mucous membranes, and at high concentrations it shows narcotic actions. The narcotic concentrations in mice were reported to be 8000–10,000 ppm (Patty and Yant 1929) and the fatal concentration was 13,500 ppm (Flury and Zernick 1931). Death occurred from respiratory arrest. The acute toxicity of *n*-octane is somewhat greater than that of *n*-heptane.

Exposure Limits

TLV-TWA 300 ppm (~1450 mg/m^3) (ACGIH and NIOSH), 500 ppm (~2420 mg/m^3) (OSHA); STEL 375 ppm (~1800 mg/m^3).

TABLE 25.1 Toxicity and Flammability of Miscellaneous Aliphatic and Cycloparaffins

Compound/Synonyms/ CAS No.	Formula/MW/Structure	Toxicity	Flammability
Cyclopropane (trimethylene) [75-19-4]	C_3H_6 42.09 H_2C-CH_2 $\quad\ \ CH_2$	Nontoxic; asphyxiant; narcotic at very high concentrations in air; contact with liquefied gas can cause frostbite	Highly flammable gas; heavier than air, gas density 1.45 (air = 1); flashback fire hazard; autoignition temperature 498°C; forms explosive mixtures with air, LEL 2.4%, UEL 10.4% by volume; DOT Label: Flammable Gas, UN 1027
Cyclobutane (tetramethylene) [287-23-0]	C_4H_8 56.12 H_2C-CH_2 H_2C-CH_2	Nontoxic; asphyxiant; causes narcosis at very high concentrations	Flammable gas; burns with a luminous flame; density 1.94 (air = 1), gas heavier than air, causing flashback fire hazard in the presence of an ignition source; forms explosive mixtures with air, LEL 1.8% by volume, UEL data not available; DOT Label: Flammable Gas, UN 2601
1-Butene (α-butylene, ethylethylene) [106-98-9]	C_4H_8 56.10 $CH_3-CH_2-CH=CH_2$	Nontoxic; asphyxiant; narcotic at high concentrations	Flammable gas; density 1.93 (air = 1), heavier than air, can cause flashback fire in the presence of an ignition source; autoignition temperature 385°C; forms explosive mixtures with air, LEL and UEL 1.6% and 10.0% by volume, respectively
cis-2-Butene (cis-2-butylene) [590-18-1]	C_4H_8 56.10 $H_3C\quad\ \ CH_3$ $\quad CH=CH$	Nontoxic, asphyxiant; narcotic at high concentrations	Flammable gas; flashback fire hazard in the presence of ignition source; autoignition temperature 325°C (NFPA 1986); forms explosive mixtures with air within the range 1.7–9.0% by volume of air

Compound	Formula / Structure	Hazard / Toxicity	Flammability
trans-2-Butene (*trans*-2-butylene, β-butylene) [624-64-6]	C_4H_8 56.10 $H_3C-\underset{\textstyle CH=CH}{}-CH_3$	Nonpoisonous gas; asphyxiant; narcotic at high concentrations	Flammable gas; flashback fire hazard in the presence of an ignition source; autoignition temperature 324°C (NFPA 1986); forms explosive mixtures with air within the range 1.8–9.7% by volume
Isobutene (isobutylene, α-butylene, 2-methyl propene) [115-11-7]	C_4H_8 56.10 $CH_3-\underset{\textstyle CH_3}{C}=CH_2$	Nontoxic gas; asphyxiant; an exposure to 20–30% concentration for 4 hr may be lethal to mice	Flammable gas; heavier than air; the gas can travel a considerable distance to a source of ignition and flash back; autoignition temperature 465°C (NFPA 1986); forms explosive mixtures with air within the range 1.8–9.6% by volume; DOT Label: Flammable Gas, UN 1055, UN 1075
Isopentane (2-methylbutane, ethyl dimethyl methane) [78-78-4]	C_5H_{12} 72.15 $H_3C-\underset{\textstyle H_3C}{}CH-CH_2-CH_3$	Nontoxic; vapors or gas at very high concentrations may produce narcotic effects	Highly flammable liquid (or gas); bp 28°C; flash point <−50°C; vapors can cause flashback fire hazard in the presence of an ignition source; autoignition temperature 420°C; forms explosive mixtures with air, LEL 1.4%, UEL 6.9% by volume of air
Isohexane (2-methylpentane) [107-83-5]	C_6H_{14} 86.20 $CH_3-\underset{\textstyle CH_3}{CH}-CH_2-CH_2-CH_3$	Low toxicity; toxic symptoms quite similar to those of *n*-hexane; irritant to respiratory tract, TLV-TWA 500 ppm, STEL 1000 ppm (ACGIH)	Highly flammable liquid; flash point −23°C; autoignition temperature 264°C (for mixed isomers); forms explosive mixtures with air within the range 1–7% by volume; DOT Label: Flammable Liquid, UN 1208, UN 2462

(continued)

TABLE 25.1 (Continued)

Compound/Synonyms/CAS No.	Formula/MW/Structure	Toxicity	Flammability
3-Methylpentane [96-14-0]	C_6H_{14} 86.20 $CH_3-CH_2-CH-CH_2-CH_3$ $\hspace{2.2cm}\vert$ $\hspace{2.4cm}CH_3$	Low toxicity; an irritant to respiratory tract and a narcotic at very high concentrations	Highly flammable liquid; flash point −6°C (20°F); autoignition temperature (for mixed isomers) 264°C; forms explosive mixtures with air, LEL 1%, UEL 7% by volume
n-Heptane [142-82-5]	C_7H_{16} 86.20 $CH_3-CH_2-CH_2-CH_2-CH_2-$ CH_2-CH_3	Low toxicity; at 1–15% concentration in air an exposure for 30–60 min was narcotic to mice; 2% concentration may cause convulsions and death; inhaling 5000 ppm for a few minutes may produce nausea, dizziness, and a gasoline taste in humans; neurotoxic; LD_{50} intravenous (mice): 222 mg/kg (NIOSH 1986); TLV TWA 400 ppm (~1600 mg/m³) (ACGIH), 500 ppm (~2000 mg/m³) (OSHA); STEL 500 ppm (ACGIH)	Flammable liquid; flash point (open cup) −1°C (30°F), (closed cup) −4°C (25°F); vapor pressure 47.7 torr at 25°C; autoignition temperature 204°C; vapor forms explosive mixture with air within the range 1–6.7% by volume; DOT Label: Flammable Liquid, UN 1206
Methylcyclohexane (cyclohexylmethane, hexahydroxytoluene) [108-87-2]	C_7H_{14} 98.19 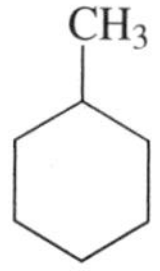	Low toxicity; vapors are irritating to mucous membranes; toxic effects similar to those of cyclohexane; exposure to 1% concentration in air for 2 hr was fatal to mice	Flammable liquid; flash point (closed cup) −4°C (25°F); vapor pressure 40 torr at 22°C; autoignition temperature 250°C; vapor forms explosive mixtures with air within the range 1.2–6.7% by volume

480

Methylcyclopentane (cyclopentylmethane) [96-37-7]

C_6H_{12}
84.16

CH$_3$ (cyclopentyl structure)

Low toxicity; vapors irritant to respiratory tract; inhalation of 10% concentration in air was fatal to mice

Highly flammable liquid; flash point −23°C (−11°F); autoignition temperature 258°C; vapor forms explosive mixtures with air within range 1–8.35% by volume of air; DOT Label: Flammable Liquid, UN 2298

Cyclopentene [142-29-0]

C_5H_8
68.13

(cyclopentene structure)

Low toxicity; toxicity data not available; LD$_{50}$ oral (rats): 1656 mg/kg; LD$_{50}$ skin (rabbits): 1231 mg/kg (NIOSH 1986)

Highly flammable liquid; flash point (closed cup) −29°C (−29°F); autoignition temperature 395°C; vapor forms explosive mixtures with air; LEL and UEL data not available

---

Methylcyclopentadiene/methyl cyclopentadiene dimer [26472-00-4]

C_6H_8 (monomer)
80.13

$C_{12}H_{16}$ (dimer)
160.23

Low toxicity; toxicity data not available; cancer-suspect agent (dimer)

Monomer: combustible liquid; flash point (closed cup) 49°C (120°F) (NFPA 1986); autoignition temperature 445°C; explosive limits in air; 1.3–7.6% by volume at 100°C; dimer: combustible liquid; flash point 26°C (80°F) (Aldrich 1990)

Methyl acetylene (propyne, propine, allylene) [74-99-7]

C_3H_4
40.07

$CH_3-C{\equiv}CH$

This colorless gas with sweet odor is an asphyxiant and toxic at high concentrations; the toxicity, however, is low; symptoms in test animals from chronic exposures at 2.87% in air were excitement, ataxia, salivation, and tremors; TLV-TWA 1000 ppm (~1650 mg/m^3) (ACGIH)

Highly flammable gas; forms explosive mixtures with air, LEL 1.7% by volume, UEL data not available, decomposes violently under pressure at about 5 atm; reactions with halogens and nitrogen dioxide may be violent; may form unstable acetylides with metals (e.g., Cu, Ag)

Fire and Explosion Hazard

Flammable liquid; flash point (closed cup) 13.3°C (56°F), (open cup) 22°C (72°F); autoignition temperature 206°C (403°F); fire-extinguishing agent: dry chemical, foam, or CO_2; use a water spray to keep fire-exposed containers cool.

n-Octane forms explosive mixtures in air; the LEL and UEL values are 0.96% and 4.66% by volume of air, respectively.

Analysis

n-Octane is analyzed by GC-FID. Air analysis may be done by NIOSH Method 1500 (see Section 25.14).

25.18   ISOOCTANE

DOT Label: Flammable Liquid, UN 1262

Formula C_8H_{18}; MW 114.26; CAS [540-84-1]

Structure:

$$CH_3-\underset{\underset{CH_3}{|}}{\overset{\overset{CH_3}{|}}{C}}-CH_2-\underset{}{\overset{\overset{CH_3}{|}}{CH}}-CH_3$$

Synonyms: 2,2,4-trimethylpentane; isobutyl-trimethylmethane

Uses and Exposure Risk

Isooctane is a petroleum product, produced by the refining of petroleum. It is used as the standard in determining the octane numbers of fuels (its antiknock octane number is 100) and as a solvent in chemical analysis.

Physical Properties

Colorless mobile liquid with odor of gasoline; density 0.692 at 20°C; boils at 99.3°C; mp −107.5°C; immiscible with water, sparingly soluble in alcohol and mixes readily with ether, chloroform, benzene, toluene, carbon disulfide, and oils.

Health Hazard

The acute toxicity of isooctane is very low and is similar to *n*-octane. Exposure to high concentrations may produce irritation of respiratory tract. Exposure to 30,000 ppm for an hour may be lethal to mice. There is no report of any other adverse effects from exposure to isooctane.

Fire and Explosion Hazard

Highly flammable liquid; flash point (closed cup) −12.2°C (10°F); autoignition temperature 418°C (784°F) (NFPA 1997); fire-extinguishing agent: dry chemical, foam, or CO_2; use a water spray to keep fire-exposed containers cool. Isooctane forms explosive mixtures with air within the range 1–4.6% by volume of air.

Analysis

It is analyzed by GC-FID. Air analysis may be done by NIOSH Method 1500 (see Section 25.14).

25.19   MISCELLANEOUS HYDROCARBONS

The flammability, explosivity, and toxicity of miscellaneous alkanes, alkenes, alkynes, and cycloparaffins are listed in Table 25.1.

REFERENCES

ACGIH. 1986. *Documentation of the Threshold Limit Values and Biological Exposure Indices*, 5th ed. Cincinnati, OH: American Conference of Governmental Industrial Hygienists.

Aldrich. 1997. *Catalog Handbook of Fine Chemicals*. Milwaukee, WI: Aldrich Chemical Company.

Baubeau, P. L., M. Heguy, and J. Gitton. 1990. Biogas from municipal landfills. *Energy Biomass Wastes 13*: 1209–50.

Bolt, R. O. 1947. *Chem. Eng. News 25*: 1866.

Broholm, K., B. K. Jensen, T. H. Christensen, and L. Olsen. 1990. Toxicity of 1,1,1-trichloroethane and trichloroethene on a mixed culture of methane-oxidizing bacteria. *Appl. Environ. Microbiol. 56* (8): 2488–93.

DiVincenzo, G. D. 1976. *Toxicol. Appl. Pharmacol. 36*: 511; cited in *Documentation of the Threshold Limit Values and Biological Exposure Indices*, 5th ed. 1986. Cincinnati, OH: American Conference of Governmental Industrial Hygienists.

Flury, F., and F. Zernik. 1931. *Schadliche Gase*, pp. 257–264. Berlin: J. Springer; cited in Documentation of the *Threshold Limit Values and Biological Exposure Indices*, 5th ed. 1986. Cincinnati, OH: American Conference of Governmental Industrial Hygienists.

Greenlee, K. W. 1948. *Chem. Eng. News 26*: 1985.

Herrera, T. A., R. Lang, and G. Tchobanoglous. 1988. A study of the emissions of volatile organic compounds found in landfill gas. *Proceedings of the Industrial Waste Conference*, Vol. 43(1988): 229–38; cited in *Chem. Abstr. CA 112* (16): 144775k.

Joyce, R. M. 1947. *Chem. Eng. News 25*: 1866–67.

Kinman, R. N., J. Rickabaugh, M. Lambert, and D. L. Nutini. 1989. Control of methane from municipal solid waste landfills by injection of lime and flyash. *Proceedings of the Industrial Waste Conference*, Vol. 43(1988): 239–50; cited in *Chem. Abstr. CA 112* (24): 222754u.

Lawless, E. W., and I. C. Smith. 1968. *High Energy Oxidizers*. New York: Marcel Dekker.

MCA. 1962. *Case Histories of Accidents in the Chemical Industry*. Washington, DC: Manufacturing Chemists' Associations.

Mellor, J. W. 1946. *A Comprehensive Treatise on Inorganic and Theoretical Chemistry*. London: Longmans, Green & Co.

Merck. 1996. *The Merck Index*, 12th ed. Rahway, NJ: Merck & Co.

Meyer, E. 1989. *Chemistry of Hazardous Materials*. Englewood Cliffs, NJ: Prentice-Hall.

National Research Council. 1981. *Prudent Practices for Handling Hazardous Chemicals in Laboratories*. Washington, DC: National Academy Press.

NFPA. 1997. *Fire Protection Guide on Hazardous Materials*, 12th ed. Quincy, MA: National Fire Protection Association.

NIOSH. 1984. *Manual of Analytical Methods*, 3rd ed. Cincinnati, OH: National Institute of Occupational Safety and Health.

NIOSH. 1986. *Registry of Toxic Effects of Chemical Substances*, ed. D. V. Sweet. Washington, DC: U.S. Government Printing Office.

Patty, F. A., and W. P. Yant. 1929. *U.S. Bureau of Mines Rep. 2979*; cited in *Documentation of the Threshold Limit Values and Biological Exposure Indices*, 5th ed. 1986. Cincinnati, OH: American Conference of Governmental Industrial Hygienists.

Smyth, H. F., Jr. 1954. *Arch. Ind. Hyg. Occup. Med. 10*: 61.

von Oettingen, W. F. 1940. Toxicity and potential dangers of aliphatic and aromatic hydrocarbons. A critical review of the literature. *U.S. Public Health Bull. 255*: 40–41.

Wang, Y. T., P. C. Pai, and J. L. Latchaw. 1989. Evaluating anaerobic biodegradability and toxicity of ozonation products of resistant phenolic compounds. *Proceedings of the International Conference on Physicochemical and Biological Detoxification of Hazardous Wastes*, 1988, Vol. 2, ed. Y. C. Wu, pp. 759–71. Lancaster, PA: Technomic.

26

HYDROCARBONS, AROMATIC

26.1 GENERAL DISCUSSION

Aromatics are a class of hydrocarbons having benzene-ring structures. Mononuclear aromatics such as toluene, cumene, and xylenes are single-ring compounds, while polynuclear aromatics such as anthracene, fluoranthrene, or pyrene are fused-ring systems containing more than one benzenoid ring. The structures and synonyms of hazardous compounds of this class are presented later in this chapter.

Aromatics are an important class of organic compounds. These hydrocarbons occur in petroleum and its products as well as in coal and coal tar. Whereas the mononuclear compounds benzene and alkylbenzenes are widely used as solvents and as raw materials for the manufacture of several chemicals, the polynuclear compounds are of little commercial applications. Many polyaromatics are carcinogens. Their ubiquitous presence in the environment is a matter of concern. Benzene forms a large number of derivatives, many of which are useful raw materials. The hazardous characteristics of each class of such derivatives may be significantly different from others and are discussed separately (see Chapters 9, 17, and 51). Only

the aromatic hydrocarbons and their alkyl, alkeny, and alkynyl substitution products that are hazardous are discussed in detail in the following sections.

Toxicity

The acute toxicity of mononuclear aromatics is low. Inhalation of vapors at high concentrations in air may cause narcosis with symptoms of hallucination, excitement, euphoria, distorted perception, and headache. Under severe intoxication these may progress to depression, stupor, and coma. The narcotic effects of alkylbenzenes are somewhat greater than the alkanes with the same number of carbon atoms. Benzene is the only mononuclear aromatic with possible human carcinogenicity and other severe chronic effects. Many polynuclear aromatic hydrocarbons (PAH) may cause cancers, affecting a variety of tissues. In the PAH class, 16 compounds in potable water and wastewater and 22 compounds in soil and solid wastes are listed under priority pollutants by the EPA. However, only benzo[*a*]pyrene is a potent human carcinogen, while benzo[*a*]anthracene, benzo[*b*]fluoranthene, benzo[*j*]fluoranthene, chrysene, dibenz[*a,h*]anthracene, and indeno(1,2,3-*cd*)pyrene have shown sufficient evidence of carcinogenicity in animals. The oral toxicity of PAHs is low to very low. The information in the published literature on the acute toxicity of this class of compounds is very little.

Flammability

Benzene and alkylbenzenes are flammable liquids. With greater degree of substitutions in the benzene ring and/or increase in the carbon chain length of the alkyl substituents, the flammability decreases. Polynuclear aromatics are noncombustible solids.

Disposal/Destruction

Hazardous aromatics are destroyed by thermal processes that include incineration of the liquid hydrocarbons or PAH solutions in combustible solvents. Anaerobic microbial degradation in the presence of high concentrations of dissolved oxygen in wastewater and oxidation with hydrogen peroxide or ozone in the presence of UV light are some of the methods reported in the literature to destroy mononuclear aromatics such as benzene and toluene.

Analysis

Gas and liquid chromatography, mass spectrometry, and UV and IR spectrophotometry are among the common instrumental techniques applied to analyze aromatic hydrocarbons. Benzene and alkylbenzenes pollutants in water, soils, and solid wastes may be analyzed by various GC or GC/MS methods as specified by the EPA (1984, Methods 602 and 624; 1986, Methods 8020, 8024). In general, any mononuclear aromatic in any matrix may be analyzed in a similar way. Analysis of these substances in air may be performed by NIOSH Methods involving adsorption over coconut charcoal, desorption with carbon disulfide, and analysis by GC-FID.

The priority pollutants of PAH type may be analyzed by HPLC, GC, and GC/MS by EPA Methods 610, 8100, 8270, and 8320 (U.S. EPA 1992, 1997). GC techniques involve the use of a flame ionization detector and a fused-silica capillary column or a packed column such as 3% OV-17 on Chromosorb W-AW or equivalent. Certain PAH compounds may coelute on a GC column. Liquid chromatography and GC/MS are more suitable methods. In the latter technique, a combination of characteristic ions along with the retention times should be used to identify these hydrocarbons, as many compounds of this class form the same primary and secondary ions. HPLC analysis using UV and fluorescence detectors and a column, reversed-phase HC-ODS Sil-X having a 5-μm particle size, does better separation of PAH mixture than that done by GC. Analysis of PAH in air may be done by NIOSH Methods 5506 and 5515. About 200 to 1000 L of air at a flow rate of 2 L/min is passed over sorbent filter consisting of PTFE and washed XAD-2. The analytes are extracted with a solvent and analyzed by GC or by HPLC using UV absorbence at 254 nm, or fluorescence excitation at 340 nm or emission at 425 nm.

26.2 BENZENE

EPA Priority Pollutant; also designated as Hazardous Waste, RCRA Waste Number U019; DOT Label: Flammable Liquid, UN 1114

Formula C_6H_6; MW 78.12; CAS [71-43-2]

Structure: , first member of the aromatic class of compounds

Synonyms: benzol; cyclohexatriene; phenyl hydride

Uses and Exposure Risk

Benzene occurs in coal and coal-tar distillation products and in petroleum products such as gasoline. It is also found in the gases and leachates of landfills for industrial wastes, construction debris, and landscaping refuse (Oak Ridge National Laboratory 1989). Trace amounts of benzene, toluene, xylenes, and other volatile organics have been found in the soils and groundwaters near many sanitary landfills (U.S. EPA 1989a,b). Kramer (1989) has assessed the level of exposures to benzene during removal, cleaning, pumping, and testing of underground gasoline storage tanks. The average human exposures were 0.43–3.84 ppm (in 1.5–6 hours) and the highest short-term (15–minute) exposure was 9.14 ppm. Benzene also occurs in the tobacco smoke (Hoffmann et al. 1989); thus the risk of its exposure may enhance from inhaling such smoke.

Benzene is used as a solvent for waxes, resins, and oils; as a paint remover; as a diluent for lacquers; in the manufacture of dyes, pharmaceuticals, varnishes, and linoleum; and as a raw material to produce a number of organic compounds.

Physical Properties

Colorless liquid with a characteristic odor; density 0.878 at 20°C; bp 80.1°C; freezes at 5.5°C; slightly soluble in water (0.188 g/ 100 g water at 24°C), readily miscible with organic solvents.

Health Hazard

Benzene is an acute as well as a chronic toxicant. The acute toxic effects from inhalation, ingestion, and skin contact are low to moderate. The symptoms in humans are hallucination, distorted perception, euphoria, somnolence, nausea, vomiting, and headache. The narcotic effects in humans may occur from inhaling benzene in air at a concentration of 200 ppm. High concentrations may cause convulsions. A 5- to 10-minute exposure to 2% benzene in air may be fatal. Death may result from respiratory failure.

Benzene is an irritant to the eyes, nose, and respiratory tract. The chronic poisoning from benzene is much more severe than its acute toxicity. The target organs to acute and chronic poisoning are the blood, bone marrow, central nervous system, respiratory system, eyes, and skin. Heavy occupational exposures to benzene can cause bone marrow depression and anemia, and in rare cases, leukemia. Leukemia may develop several years after the exposure ceases. Deaths from leukemia, attributed to occupational exposure to benzene in the workplace, which may be on the order of 200 ppm concentration, have been documented (ACGIH 1986). Benzene is listed as a suspected human carcinogen. In addition to leukemia, malignant lymphoma, and myeloma, lung cancer in subjects exposed to benzene has been reported (Aksoy 1989).

Absorption of liquid benzene through the skin may be harmful. The main elimination pathway for benzene absorbed through inhalation or skin contact is metabolism. Hydroxyl radicals play an important role in the process of metabolism. Khan and co-workers (1990) have reported the formation of formaldehyde and degradation of deoxyribose, suggesting the generation of hydroxyl radicals during benzene toxicity to the bone marrow S-9 fraction. The hydroxyl radicals react with benzene to form phenols and dihydroxyphenols, which are excreted rapidly in urine. About one-third of the retained benzene is excreted as phenols in the urine. The remaining two-thirds may be further degraded and attached onto the tissue or oxidized and exhaled as CO_2.

Kalf and associates (1989) have investigated the action of prostaglandin H synthase in benzene toxicity and prevention of benzene-induced myelo- and genotoxicity by nonsteroidal anti-inflammatory drugs (NSAIDs). Indomethacin, a prostaglandin H synthase inhibitor prevented the dose-dependent bone marrow depression and increase in marrow prostaglandin E level in mice intravenously dosed with benzene. Indomethacin, aspirin, or meclofenamate prevented the decrease in cellularity and increase in micronucleated polychromatic erythrocytes in peripheral blood, caused by intravenous injection of benzene (100–1000 mg/kg) in mice.

Exposure Limits

TLV-TWA 10 ppm ($\sim$32 mg/m^3) (ACGIH and OSHA); ceiling 25 ppm ($\sim$80 mg/m^3) (OSHA and MSHA); peak 50 ppm ($\sim$160 mg/m^3)/10 min/8 hr (OSHA); carcinogenicity: Suspected Human Carcinogen (ACGIH), Human Sufficient Evidence (IARC).

Fire and Explosion Hazard

Flammable liquid; flash point (closed cup) $-11°C$ (12°F); vapor pressure 75 torr at 20°C; vapor density 2.69 (air $= 1$); the vapor is heavier than air and may travel a considerable distance to an ignition source and flash back; autoignition temperature 562°C (1044°F); fire-extinguishing agent: dry chemical, foam, or CO_2; use a water spray to keep fire-exposed containers cool and to disperse the vapors.

Benzene forms explosive mixtures with air; the LEL and UEL values are 1.3% and 7.1% by volume, respectively. Benzene reacts violently with halogens. The reaction is explosive with fluorine, as well as with chlorine in the presence of light and to a lesser degree with bromine in the presence of a catalyst. It may react violently with other oxidizing substances, such as perchlorates, ozone, permanganate-sulfuric acid, fluorides of chlorine and bromine, and chromic anhydride. Explosions could result from inadvertent mixing. One of the reaction products with ozone is ozobenzene, a white gelatinous shock-sensitive compound which may explode (Mellor 1946). Benzene is flammable with potassium and sodium peroxides. Benzene vapor forms flammable mixtures with nitrous oxide (Jacobs 1989).

Disposal/Destruction

Benzene is burned in a chemical incinerator equipped with an afterburner and a scrubber. Benzene and many other organic compounds in water and wastes may be destroyed by UV-catalyzed oxidation with hydrogen peroxide (Sundstrom 1989). Aerobic biodegradation of benzene in gasoline contaminated subsurface sites has been reported (Chiang et al. 1989). The degradation was dependent on the concentration of the dissolved oxygen in the shallow aquifer beneath the field site. Use of hydrogen peroxide as an oxygen source in the microbial degradation has been reported (Thomas et al. 1990).

Benzene and chlorophenol contaminants may be removed from aqueous solutions by absorption with modified sodium smecticide clay (Martin et al. 1989).

Analysis

Benzene may be analyzed by several techniques, which include GC, MS, IR spectroscopy, and hydrocarbon analyzer. Its analysis at the ppb level in wastewater, groundwater, and potable water may be performed by a GC equipped with a photoionization detector and using the purge and trap or thermal desorption techniques. GC-FID may be employed to analyze benzene at ppm range. GC/MS is an excellent confirmatory test. Benzene may be identified by its primary characteristic ion with an m/z ratio of 78. The EPA has set several analytical methods for its analysis in water, soils, and solid wastes (Methods 501, 502, 524, 602, 624, 1624, 8010, and 8240).

Air analysis for benzene may be performed by NIOSH Methods 3700, 1500, 1501 (NIOSH 1984). An air bag (Tedlar bag) is filled to about 80% capacity at a flow rate of 20 to 50 mL/min. The air is injected within 4 hours after collection into a portable GC equipped with a photoionization detector. Alternatively, the air is passed over coconut shell charcoal; the adsorbed benzene is desorbed into CS_2 and injected into GC-FID, using 20% SP-2100 on Supelcoport or 10% OV-275 on Chromosorb W or an equivalent column.

26.3 TOLUENE

EPA Priority Pollutant; designated Toxic Waste, RCRA Waste Number, U220; DOT Label: Flammable Liquid, UN 1294

Formula C_7H_8; MW 92.15; CAS [108-88-3]

Structure:

Synonyms: methylbenzene; phenylmethane; toluol; methylbenzol

Uses and Exposure Risk

Toluene is derived from coal tar as well as petroleum. It occurs in gasoline and many petroleum solvents. Toluene is used to produce TNT, toluene diisocyanate, and benzene; as an ingredient for dyes, drugs, and detergents; and as an industrial solvent for rubbers, paints, coatings, and oils.

Physical Properties

Colorless liquid with a characteristic aromatic odor; density 0.866 at 20°C; boils at 110.7°C; freezes at −95°C; slightly soluble in water (0.063 g/100 g at 25°C), readily miscible with organic solvents.

Health Hazard

The acute toxicity of toluene is similar to that of benzene. The exposure routes are inhalation, ingestion, and absorption through the skin; and the organs affected from its exposure are the central nervous system, liver, kidneys, and skin. At high concentrations it is a narcotic. In humans, acute exposure can produce excitement, euphoria, hallucination, distorted perceptions, confusion, headache, and dizziness. Such effects may be perceptible at an exposure level of 200 ppm in air. Higher concentrations can produce depression, drowsiness, and stupor. Inhalation of 10,000 ppm may cause death to humans from respiratory failure.

Toluene is metabolized to benzoic acid and finally, to hippuric acid and benzoyl-glucuronide. The latter two are excreted in urine along with small amounts of cresols, formed by direct hydroxylation of toluene. Chronic exposure may cause some accumulation of toluene in fatty tissues, which may be eliminated over a period of time. The chronic effects of toluene are much less severe to benzene. It is not known to cause bone marrow depression or anemia. Animal tests showed no carcinogenic effects.

Exposure Limits

TLV-TWA 100 ppm ($\sim$375 mg/m^3) (ACGIH, NIOSH, and MSHA), 200 ppm ($\sim$750 mg/m^3) OSHA; ceiling 300 ppm, peak 500 ppm/15 min (OSHA); STEL 150 ppm (ACGIH).

Fire and Explosion Hazard

Flammable liquid; flash point (closed cup) 4.4°C (40°F); vapor pressure 28 torr at 25°C; vapor density 3.17 (air = 1); the vapor is heavier than air and may cause a flashback fire hazard in the presence of a source of ignition; autoignition temperature 536°C (997°F); fire-extinguishing agent: dry chemical, foam, or CO$_2$; use a water spray to keep fire-exposed containers cool.

Toluene forms explosive mixtures with air within the range 1.4–6.7% by volume in air. It may react violently with strong oxidizers. There is a report of an explosion when toluene was mixed with nitrogen tetroxide (Mellor 1946, Suppl. 1967). Metal perchlorates form solvated salts with hydrocarbon solvents, including toluene, which are shock sensitive and may explode on being struck (NFPA 1986). Toluene reacts with concentrated nitric acid in the presence of concentrated sulfuric acid to form nitrotoluenes, many of which are explosives. This includes nitrocresols, which on further nitration under uncontrolled conditions, decompose violently. Trinitrotoluene, which is a powerful explosive, however, does not detonate without a high-velocity initiator. Some of the nitrotoluenes are highly toxic.

Analysis

Analysis of toluene may be performed by GC using photoionization or flame ionization detectors or by GC/MS. The analysis of wastewaters, potable waters, soils, and hazardous wastes may be done by the same EPA Methods as those used for benzene (see Section 26.2). In GC/MS analysis the primary characteristic ion for toluene is m/z 91. Air analysis may be done by charcoal adsorption, followed by desorption with carbon disulfide and injecting into GC-FID (see Section 26.2 and NIOSH Methods 1500 and 1501).

26.4 ETHYL BENZENE

DOT Label: Flammable Liquid, UN 1175
Formula C_8H_{10}; MW 106.18; CAS [100-41-4]
Structure:

Synonyms: ethyl benzol; phenylethane

Uses and Exposure Risk

Ethyl benzene is used as a solvent and as an intermediate to produce styrene monomer.

Physical Properties

Colorless liquid with a characteristic aromatic odor; density 0.8626 at 25°C; boils at 136.2°C; freezes at −95°C; practically insoluble in water (0.015 g/100 g), readily miscible with organic solvents.

Health Hazard

The acute toxicity of ethyl benzene is low. At high concentrations its exposure produces narcotic effects similar to benzene and toluene. A 4-hour exposure to a concentration of 4000 ppm proved fatal to rats. The lethal dose varies with species. Deaths resulted from intense congestion and edema of the lungs.

Other than the narcotic effects, ethyl benzene exhibits irritant properties that are somewhat greater than those of benzene or toluene. It is an irritant to the skin, eyes, and nose. Repeated contact with the liquid may cause reddening of the skin and blistering. The vapors at 200 ppm may cause mild irritation of the eyes in humans, which may become severe and lacrimating at 2000–3000 ppm.

The oral toxicity in animals was found to be low to very low. An LD_{50} value of 3500 mg/kg for rats has been documented (NIOSH 1986). No adverse effects were noted in animals subjected to chronic inhalation exposure at below 400 ppm. At higher dosages only the liver was affected (ACGIH 1986). Ethyl benzene is eliminated from the body by metabolic excretion. The urinary metabolites in humans are mainly mandelic acid, $C_6H_5CH(OH)COOH$, and benzoylformic acid, $C_6H_5COCOOH$.

Exposure Limits

TLV-TWA 100 ppm (~433 mg/m³) (ACGIH, NIOSH, MSHA, and OSHA); STEL 125 ppm (541 mg/m³) (ACGIH); IDLH 2000 ppm (NIOSH).

Fire and Explosion Hazard

Flammable liquid; flash point (closed cup) 18°C (64°F); vapor pressure 7.1 torr at 21°C;

vapor density 3.65 (air = 1); the vapor is heavier than air and can travel a considerable distance to a source of ignition and flash back; autoignition temperature 432°C (810°F); fire-extinguishing agent: dry chemical, foam, or CO_2; use a water spray to keep fire-exposed containers cool and to disperse the vapors.

Ethyl benzene forms explosive mixtures with air within the range 1.0–6.7% by volume in air. Violent reactions may occur with fluorine and nitrogen tetroxide. It may form shock-sensitive products with metal perchlorates.

Analysis

The methods for analyzing ethyl benzene are the same as those for benzene or toluene. GC-PID, GC-FID, and GC/MS are the most suitable techniques. Water, soil, sludges, and air can be analyzed by EPA and NIOSH methods as specified in Section 26.2.

26.5 XYLENE

EPA Priority Pollutant; RCRA Hazardous Waste, Number U239; DOT Label: Flammable Liquid, UN 1306

Formula C_8H_{10}; MW 106.18; CAS [1330-20-7] ([95-47-6] (*ortho*-isomer), [108-38-3] (*meta*-isomer), and [106-42-3] (*para*-isomer)

Structures: Xylenes occur in three isomeric forms: *ortho, meta*, and *para* forms, having the following structures:

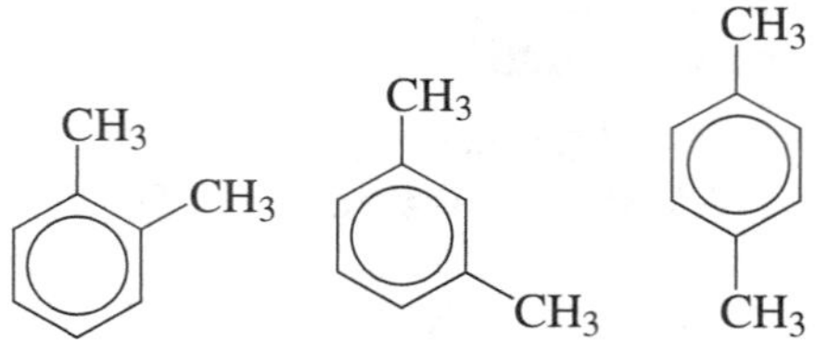

(*o*-xylene)   (*m*-xylene)   (*p*-xylene)

The positions of methyl groups as seen above are in the 1,2-, 1,3-, and 1,4-positions in the benzene ring.

Uses and Exposure Risk

Xylene occurs in petroleum solvents and gasoline. The widest applications of xylene are as solvents in paints, coatings, and rubber. Xylene isomers are used in the manufacture of dyes, drugs, pesticides, and in many organic intermediates, such as terephthalic acid and pthalic anhydride.

Health Hazard

The toxic properties of xylene isomers are similar to toluene or ethylbenzene. The target organs are the central nervous system, eyes, gastrointestinal tract, kidneys, liver, blood, and skin, which, however, are affected only at high levels of exposure. In humans its exposure may cause irritation of the eyes, nose, and throat, headache, dizziness, excitement, drowsiness, nausea, vomiting, abdominal pain, and dermatitis. The irritation effects in humans may be felt at a concentration of 200 ppm in air, while exposure to 10,000 ppm for 6–8 hours may be fatal.

The oral toxicity of xylene is low. Ingestion of a high dose, however, can cause depression of the central nervous system, dizziness, nausea, and vomiting and abdominal pain. The oral LD_{50} values in rats for xylene isomers are within the range of 5000 mg/kg.

The major route of absorption of xylene is inhalation. Another significant route is skin absorption of the liquid. About 5% of absorbed xylene is excreted unchanged in expired air within a few hours, while less than 2% is hydroxylated to xylenols. Over 90% of absorbed xylenes are metabolized to *o*-, *m*-, and *p*-isomers of methyl benzoic acid and excreted in urine as methyl hippuric acids (ACGIH 1986). Small amounts of xylenes may remain stored in adipose tissue. Repeated exposures may cause accumulation in the blood.

Exposure Limits

TLV-TWA 100 ppm (~434 mg/m^3) (ACGIH, MSHA, and OSHA); STEL 150 ppm

($\sim$651 mg/m^3) (ACGIH); ceiling 200 ppm/ 10 min (NIOSH); IDLH 1000 ppm (NIOSH).

Fire and Explosion Hazard

Flammable liquid; flash point (closed cup): *o*-xylene 32°C (90°F), *m*-xylene 27°C (81°F), *p*-xylene 27°C (81°F); vapor pressure 7–9 torr at 20°C; vapor density 3.7 (air = 1); the vapor is heavier than air and can travel a considerable distance to a source of ignition and flash back; autoignition temperature: *o*-xylene 463°C (867°F), *m*-xylene 527°C (982°F) and *p*-xylene 528°C (984°F); fire-extinguishing agent: dry chemical, CO_2, or foam; a water spray may be used to cool the fire-exposed containers and disperse the vapors.

Xylene vapors form explosive mixtures with air within the range 1.0–7.0% by volume in air. It may react vigorously with strong oxidizers.

Analysis

Xylenes in drinking water, wastewaters, soils, and hazardous wastes may be analyzed by EPA analytical procedures (Methods 501, 602, 524, 624, 1624, 8020, and 8240) (U.S. EPA 1992; 1997). These methods involve concentration of the analytes by purging and trapping over suitable adsorbent columns before their GC or GC/MS analyses. The primary characteristic ion for GC/MS identification (by electron-impact ionization) is 106, which characterizes ethylbenzene as well. Photoionization and flame ionization detectors are, in general, suitable for ppb- and ppm-level GC analysis, respectively. Air analysis may be done by NIOSH Method 1501 (see Section 26.2).

Absorption of xylene into the body may be estimated by analyzing the metabolites, methylhippuric acids in the urine by HPLC, colorimetry, or GC. For GC analysis methylhippuric acids are derivatized with silane or diazomethane and injected onto the GC column. The HPLC technique is more accurate.

26.6 CUMENE

EPA Designated Hazardous Waste, RCRA Waste Number U055; DOT Label: Flammable or Combustible Liquid, UN 1918

Formula C_9H_{12}; MW 120.21; CAS [98-82-8]

Structure:

Synonyms: isopropylbenzene; (1-methylethyl)benzene; cumol; 2-phenylpropane

Uses and Exposure Risk

It is used as a solvent and in the production of phenol, acetone, and acetophenone.

Physical Properties

Colorless liquid with an aromatic odor; density 0.862 at 20°C; bp 152.5°C; mp −96°C; insoluble in water, miscible with organic solvents.

Health Hazard

Cumene is an irritant to the eyes, skin, and upper respiratory system, and a low acute toxicant. It is narcotic at high concentrations. The narcotic effect is induced slowly and is of longer duration relative to benzene and toluene (ACGIH 1986). Although the toxicity may be of same order, the hazard from inhalation is low due to its high boiling point and low vapor pressure. An exposure to 8000 ppm for 4 hours was lethal to rats. The oral toxicity of cumene was determined to be low in animals. In addition to narcosis, it caused gastritis. An LD_{50} value documented for mice is 1400 mg/kg (NIOSH 1986).

Chronic inhalation toxicity of cumene was very low in animals. Repeated exposures caused congestion in the lungs, liver, and kidney and an increase in the kidney weight. A major portion of cumene absorbed into the body is metabolized in the liver and excreted. The urinary metabolites constituted conjugated alcohols or acids.

Exposure Limits

TLV-TWA 50 ppm ($\sim$245 mg/m^3) (ACGIH, OSHA, MSHA, and NIOSH); IDLH 9000 ppm (NIOSH).

Fire and Explosion Hazard

Flammable liquid; flash point (closed cup) 36°C (97°F) (NFPA 1986), 39°C (102°F) (Merck 1996), 35.5°C (96°F) (Meyer 1989); vapor pressure 8 torr at 20°C; vapor density 4.1 (air = 1); the vapor is heavier than air and may travel a considerable distance to a nearby ignition source and flash back; autoignition temperature 425°C (797°F); fire-extinguishing agent: dry chemical, foam, or CO$_2$; use a water spray to keep fire-exposed containers cool and to disperse the vapors.

Cumene forms explosive mixtures in the air within the range 0.9–6.5% by volume in air. Cumene may form peroxide on prolonged exposure to air. It should be tested for peroxides before it is subjected to distillation or evaporation.

(cumene)   (cumene hydroperoxide)

Cumene reacts exothermically with concentrated nitric or sulfuric acids.

Analysis

Cumene may be analyzed by GC using a PID or FID or by GC/MS. Air analysis may be done by NIOSH Method 1501 (see Section 26.2).

26.7 STYRENE

DOT Label: Flammable or Combustible Liquid, UN 2055

Formula C$_8$H$_8$; MW 104.16; CAS [100-42-5]
Structure:

Synonyms: ethenylbenzene; vinylbenzene; phenylethylene; cinnamene; cinnamol; phenethylene; styrene monomer

Uses and Exposure Risk

Styrene polymers and copolymers are used extensively in making polystyrene plastics, polyesters, protective coatings, resins, and synthetic rubber (styrene–butadiene rubber).

Physical Properties

Colorless to yellowish oily liquid; sharp aromatic odor; density 0.906 at 20°C; bp 145.5°C; mp −30.6°C; slightly soluble in water (0.03 g/100 g at 25°C), miscible with alcohol, ether, acetone, benzene, and carbon disulfide.

Health Hazard

Like all other aromatic hydrocarbons, styrene is an irritant to skin, eyes, and mucous membranes and is narcotic at high concentrations. Exposure to its vapors may cause drowsiness, nausea, headache, fatigues, and dizziness in humans (Hamilton and Hardy 1974). Inhalation of 10,000 ppm for 30–60 minutes may be fatal to humans.

Absorption of styrene by inhalation is the major path of absorption into the body. Skin absorption of the liquid is also significant. According to an estimate, contact with styrene-saturated water for an hour or brief contact with the liquid may result in absorption equivalent to 8 hours of inhalation of 12 ppm (Dutkiewicz and Tyras 1968). It may accumulate in the body due to its high solubility in fat. This would happen when the metabolic pathway becomes saturated at exposure concentrations of 200 ppm

(ACGIH 1986). Mandelic acid and benzoyl-formic acid are the major urinary metabolites. However, the excretion of mandelic acid was less when styrene was absorbed through the skin.

Styrene tested positive in an EPA mutagenicity study. It tested positive in a histidine reversion–Ames test, *Saccharomyces cerevisiae* gene conversion, in vitro human lymphocyte micronucleus, and *Drosophila melanogaster* sex-linked lethal tests (NIOSH 1986). Carcinogenicity of styrene in humans is not known. There is limited evidence of carcinogenicity in animals for both the monomer and the polymer.

Exposure Limits

TLV-TWA 50 ppm ($\sim$212 mg/m^3) (ACGIH and NIOSH), 100 ppm ($\sim$425 mg/m^3) (OSHA and MSHA); ceiling 200 ppm, peak 600 ppm/5 min/3 hr (OSHA); STEL 100 ppm ($\sim$425 mg/m^3) (ACGIH).

Fire and Explosion Hazard

Flammable liquid; flash point (closed cup) 31°C (88°F); vapor pressure 4.3 torr at 15°C; vapor density 3.6 (air = 1); the vapor is heavier than air and may travel a considerable distance to a source of ignition and flash back; autoignition temperature 490°C (914°F); fire-extinguishing agent: use dry chemical, foam, or CO_2; a water spray may be used to cool fire-exposed containers, disperse the vapor, and dilute the spill. Styrene vapor may form polymers in the vents, stopping the airflow.

Styrene forms explosive mixtures with air within the range 1.1–6.1% by volume in air. Styrene may form peroxides on exposure to light and air. Its reactions with concentrated acids are exothermic. Polymerization is exothermic. Reactions with strong oxidizers may be violent.

Storage and Shipping

Styrene is stored in a flammable liquid storage cabinet, separated from oxidizing substances. An inhibitor such as 4-*tert*-butylcatechol in trace amounts is added to the monomer to prevent polymerization. It is shipped in glass bottles, metal cans, drums, and tank cars.

Analysis

Styrene may be analyzed by GC, using a flame ionization detector. Air analysis may be performed by charcoal adsorption, followed by desorption of the analyte with carbon disulfide and injection of the eluant into GC-FID (NIOSH Method 1501; see Section 26.2). Styrene intake in the body may be estimated by analyzing mandelic acid in the urine by liquid chromatography, polarography, or GC. However, the presence of other aromatics may interfere, as these compounds also generate the same urinary metabolite. Styrene in exhaled air may be analyzed by absorption over ethanol or charcoal followed by GC, UV, or IR analysis.

POLYNUCLEAR AROMATICS

26.8 NAPHTHALENE

EPA Priority Pollutant; RCRA Toxic Waste
 Number U165; DOT Label: Flammable
 Solid, UN 1334, 2304
Formula $C_{10}H_8$; MW 128.18; CAS [91-20-3]
Structure:

Synonyms: naphthene; tar camphor; white tar; mothballs

Uses and Exposure Risk

Naphthalene is used as a moth repellent; in scintillation counters; and as a raw material in the manufacture of naphthol, phthalic anhydride, and halogenated naphthalenes. It is also used in dyes, explosives and lubricants, and in breaking emulsion.

Physical Properties

White volatile crystalline flakes with a strong aromatic odor; mp 80.2°C; bp 218°C; insoluble in water, dissolves in most organic solvents.

Health Hazard

Inhalation of naphthalene vapor may cause irritation of the eyes, skin, and respiratory tract, and injury to the cornea. Other symptoms are headache, nausea, confusion, and excitability. The routes of exposure of this compound into the body are inhalation, ingestion, and absorption through the skin; and the organs that may be affected are the eyes, liver, kidney, blood, skin, and central nervous system.

The most severe toxic effects from naphthalene, however, may come from oral intake of large doses of this compound. In animals, as well as in humans, ingestion of large amounts may cause acute hemolytic anemia and hemoglobinuria. Other symptoms are gastrointestinal pain and kidney damage. The LD_{50} values reported in the literature show variation among different species. In mice, an oral LD_{50} value may be on the order of 600 mg/kg. Symptoms of respiratory depression and ataxia were noted.

Chronic exposure to naphthalene vapor may affect the eyes, causing opacities of the lens and optical neuritis.

Exposure Limits

TLV-TWA 10 ppm ($\sim$50 mg/m^3) (ACGIH, MSHA, and OSHA); STEL 15 ppm ($\sim$75 mg/m^3) (ACGIH); IDLH 500 ppm.

Fire and Explosion Hazard

Solid volatilizes on heating, producing flammable vapors; vapor pressure 0.087 torr at 25°C; flash point 79°C (174°F); autoignition temperature 526°C (979°F); the vapor forms explosive mixtures in the air within the range 0.9–5.9% by volume. Naphthalene may react violently with strong oxidizers.

26.9 BENZO[a]PYRENE

EPA Priority Pollutant; Designated Toxic Waste, RCRA Waste Number U022

Formula $C_{20}H_{12}$; MW 252.32; CAS [50-32-8]

Structure:

Synonyms: 3,4-benzpyrene; 3,4-benzopyrene

Physical Properties

Yellowish plates or needles; orthorhombic or monoclinic crystals; mp 179°C; bp 311°C; insoluble in water, slightly soluble in alcohols, dissolves in benzene, toluene, xylenes, and methylene chloride.

Health Hazard

The acute oral toxicity of benzo[a]pyrene is low. This may be due to the poor absorption of this compound by the gastrointestinal tract. The lethal dose in mice from intraperitoneal administration is reported as 500 mg/kg (NIOSH 1986).

Animal studies show sufficient evidence of its carcinogenicity by all routes of exposure affecting a variety of tissues, which include the lungs, skin, liver, kidney, and blood.

Benz[a]pyrene exhibited teratogenic effects in test species. It is a mutagen. It showed positive in a histidine reversion–Ames test, cell transform mouse embryo test, and in in vitro sister chromatid exchange (SCE)–human lymphocytes.

26.10 BENZO[*e*]PYRENE

Formula $C_{20}H_{12}$; MW 252.32; CAS [192-97-2]

Structure:

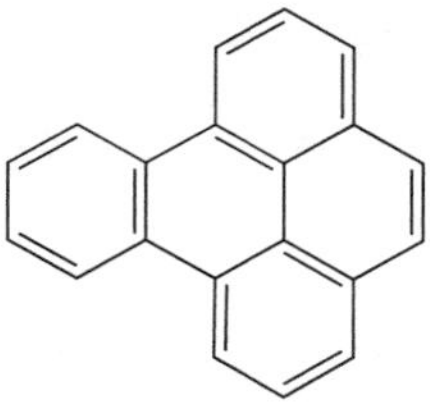

Synonyms: 1,2-benzpyrene; 1,2-benzopyrene

Physical Properties

Crystalline solid, prisms, or plates from benzene; mp 178–179°C; insoluble in water, slightly soluble in methanol, dissolves in benzene, toluene, and methylene chloride.

Health Hazard

There is very little information available on its toxicity. The oral toxicity is expected to be low. Its carcinogenic potential is lower than that of benz[*a*]pyrene. Animal studies gave inconclusive results. Oral administration may produce tumors in the stomach. However, the evidence of carcinogenicity of this compound in animals is inadequate. Benz[*e*]pyrene is a mutagen, testing positive to the histidine reversion–Ames test and in vitro UDS–human fibroblast.

26.11 BENZO[*g,h,i*]PERYLENE

EPA Priority Pollutant

Formula $C_{22}H_{12}$; MW 276.34; CAS [191-24-2]

Structure:

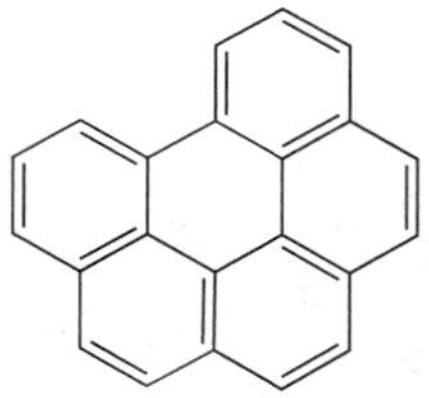

Synonym: 1,12-benzoperylene

Health Hazard

There is very little information available in the literature on the toxicity of this compound. Benzo[*g,h,i*]perylene has low oral toxicity. Based on its structural similarities with other carcinogenic polynuclear aromatics, this compound is expected to show carcinogenic properties. Such evidence, however, is inadequate at the moment. A histidine reversion–Ames test for mutagenicity gave inconclusive results.

26.12 BENZO[*a*]ANTHRACENE

EPA Priority Pollutant; RCRA Toxic Waste Number U018

Formula $C_{18}H_{12}$; MW 228.30; CAS [56-55-3]

Structure:

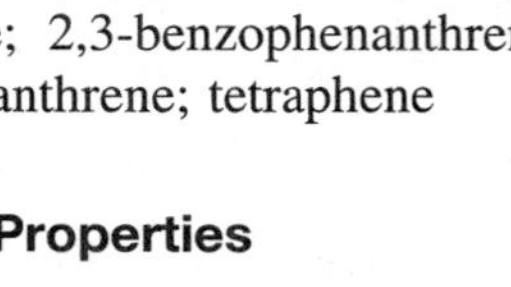

Synonyms: benz[*a*]anthracene; 1,2-benzo-anthracene; 1,2-benzanthracene; naphthan-thracene; 2,3-benzophenanthrene; benzo-[*a*]phenanthrene; tetraphene

Physical Properties

Crystallizes as plates from glacial acetic acid or alcohol; produces greenish yellow fluorescence; melts at 160°C (sublimes); insoluble in water, slightly soluble in alcohol, dissolves in most organic solvents.

Health Hazard

There is no report on its oral toxicity. However, it may be highly toxic by intravenous administration. A lethal dose in mice is reported as 10 mg/kg. Its carcinogenic actions in animals is well established. Subcutaneous administration of this compound in mice resulted in tumors at the sites of application.

Flesher and Myers (1990) have correlated carcinogenic activity of benzo[a]anthracene to its bioalkylation at the site of injection. Male rats were dosed subcutaneously and the tissue in contact with the hydrocarbon was visualized after 24 hours under UV light. Bioalkylation or the biochemical introduction of an alkyl group occurred at the mesoanthracenic centers, which are the most reactive sites in the molecule.

26.13 BENZO[b]FLUORANTHENE

EPA Priority Pollutant
Formula $C_{20}H_{12}$; MW 252.32; CAS [205-99-2]
Structure:

Synonyms: benzo[e]fluoranthene; 3,4-benzofluoranthene; benz[e]acephenanthrylene; 2,3-benzfluoranthene

Health Hazard

Acute oral toxicity data are not available. There is sufficient evidence on the carcinogenicity of this compound in animals. It produced tumors at the site of

application. Cancers in lungs and skin have been observed in animals.

26.14 BENZO[j]FLUORANTHENE

Formula $C_{20}H_{12}$; MW 252.32; CAS [205-82-3]
Structure:

Synonyms: 10,11-benzofluoranthene; benz-[j]fluoranthene

Health Hazard

Benzo[j]fluoranthene is a cancer suspect agent. Animal studies present sufficient evidence on its carcinogenicity. It may cause tumor in lungs and skin.

26.15 BENZO[k]FLUORANTHENE

EPA Priority Pollutant
Formula $C_{20}H_{12}$; MW 252.32; CAS [207-08-9]
Structure:

Synonyms: 11,12-benzofluoranthene; 8.9-benzofluoranthene; 2,3,1′,8′-binaphthylene

Health Hazard

Benzo[k]fluoranthene caused lungs and skin cancers in animals. It produced tumors at the site of application. Its carcinogenicity in humans is not known.

26.16 ACENAPHTHENE

EPA Priority Pollutant

Formula $C_{12}H_{10}$; MW 154.22; CAS [83-32-9]

Structure:

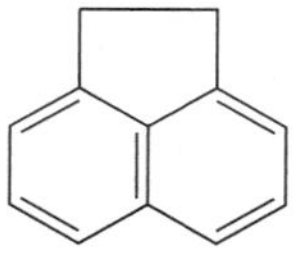

Synonyms: 1,2-dihydroacenaphthylene; 1,2-ethylenenaphthalene; naphthyleneethylene

Health Hazard

Carcinogenicity of acenaphthene in animals is not established. Tests for mutagenicity have given inconclusive results.

26.17 ACENAPHTHYLENE

EPA Priority Pollutant

Formula $C_{12}H_8$; MW 152.20; CAS [208-96-8]

Structure:

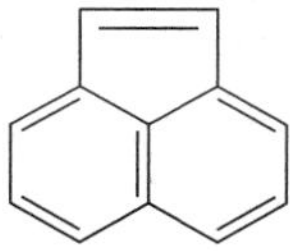

Synonym: cyclopenta[*d*,*e*]naphthalene

Health Hazard

Carcinogenic properties of this compound in animals or humans are not known. Toxicity data are not available.

26.18 ANTHRACENE

EPA Priority Pollutant

Formula $C_{14}H_{10}$; MW 178.24; CAS [120-12-7]

Structure:

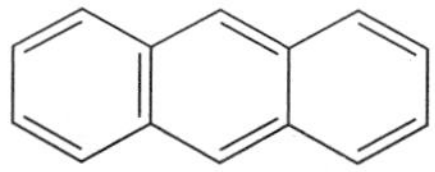

Synonym: green oil

Health Hazard

Carcinogenicity of anthracene is not known. Its toxicity is very low. An intraperitoneal LD_{50} in mice is recorded at 430 mg/kg (NIOSH 1986).

26.19 9,10-DIMETHYLANTHRACENE

Formula $C_{16}H_{14}$; MW 206.30; CAS [781-43-1]

Structure:

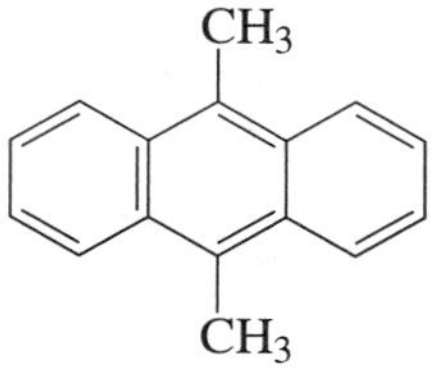

Health Hazard

This compound caused lungs and skin tumor in mice at the site of application. Carcinogenicity in humans is unknown. It tested positive in the histidine reversion–Ames test; the mammalian micronucleus test was inconclusive.

26.20 PHENANTHRENE

EPA Priority Pollutant

Formula $C_{14}H_{10}$; MW 178.24; CAS [85-10-8]

Structure:

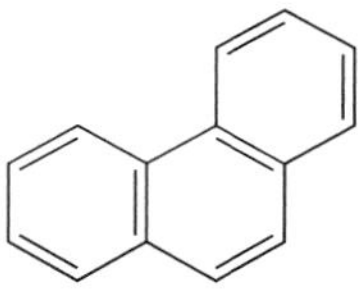

Physical Properties

Monoclinic plates crystallized from alcohol; isomeric with anthracene; exhibits blue fluorescence in solutions; mp 100°C; bp 340°C, sublimes in vacuum; insoluble in water, moderately soluble in alcohol, dissolves readily in benzene, toluene, chloroform, carbon disulfide, and anhydrous ether.

Health Hazard

The acute oral toxicity of phenanthrene is low. It is more toxic than anthracene. An oral LD_{50} value in mice is reported at 700 mg/kg. It may cause tumor in skin at the site of application. The evidence of carcinogenicity in animals, however, is inadequate.

26.21   FLUORANTHENE

EPA Priority Pollutant; RCRA Toxic Waste Number, U120
Formula $C_{16}H_{10}$; MW 202.26; CAS [206-44-0]
Structure:

Synonyms: 1,2-benzacenaphthene; benzo[j, k]fluorene; 1,2-(1,8-naphthylene)benzene

Physical Properties

Plates crystallized from alcohol; mp 110°C; sublimes; insoluble in water, moderately soluble in alcohol, dissolves in most other organic solvents.

Health Hazard

Fluoranthene exhibited mild oral and dermal toxicity in animals. The acute toxicity is lower than that of phenanthrene. An oral

LD_{50} value in rats is reported as 2000 mg/kg. It may cause skin tumor at the site of application. However, any carcinogenic action from this compound in animals is unknown.

26.22   FLUORENE

EPA Priority Pollutant
Formula $C_{13}H_{10}$; MW 166.23; CAS [86-73-7]
Structure:

Synonyms: 9H-fluorene;   diphenylenemethane; 2,2′-methylenebiphenyl

Physical Properties

White leaflets or flakes from alcohol; mp 116°C; sublimes in vacuum; insoluble in water, moderately soluble in hot alcohol, dissolves readily in most other organic solvents.

Health Hazard

Acute toxicity in animals is very low. An LD_{50} (intraperitoneal) in mice is 2000 mg/kg. Carcinogenicity of this compound in animals is not well established. It tested negative in a histidine reversion–Ames test.

26.23   CHRYSENE

EPA Priority Pollutant; RCRA Toxic Waste Number U050
Formula $C_{18}H_{12}$; MW 228.30; CAS [218-01-9]
Structure:

Synonyms: 1,2-benzophenanthrene; benz[*a*]-phenanthrene; 1,2,5,6-dibenzonaphthalene

Physical Properties

Orthorhombic bipyramidal plates crystallized from benzene; mp 254°C; bp 448°C; insoluble in water, slightly soluble in alcohol, slightly soluble in cold organic solvents, moderately soluble in these solvents when hot.

Health Hazard

There is very little information published on the acute toxicity of chrysene. The oral toxicity is expected to be low. Animal studies show sufficient evidence of carcinogenicity. It produced skin cancer in animals. Subcutaneous administration of chrysene in mice caused tumors at the site of application. Cancer-causing evidence in humans is not known. A histidine reversion–Ames test for mutagenicity showed positive.

26.24 PYRENE

EPA Priority Pollutant
Formula $C_{16}H_{10}$; MW 202.26; CAS [129-00-0]
Structure:

Synonyms: benzo[*d,e,f*]phenanthrene; *β*-pyrene

Physical Properties

Colorless monoclinic prisms crystallized from alcohol, yellowish due to the presence of tetracene; produces slight blue fluorescence; mp 156°C; bp 404°C; insoluble in water, soluble in organic solvents.

Health Hazard

Inhalation of its vapors or ingestion caused irritation of the eyes, excitement, and muscle contraction in rats and mice. An oral LD_{50} value in mice has been reported as 800 mg/kg. Studies on experimental animals do not give evidence of carcinogenicity. Skin tumors, however, have been reported in mice (NIOSH 1986). Pyrene tested negative to a histidine reversion–Ames test and other mutagenic tests.

26.25 DIBENZ[*a, h*]ANTHRACENE

EPA Priority Pollutant; RCRA Toxic Waste Number U063
Formula $C_{22}H_{14}$; MW 278.36; CAS [53-70-3]
Structure:

Synonyms: 1,2:5,6-dibenzanthracene; 1,2:5, 6-dibenzoanthracene

Physical Properties

Monoclinic or orthorhombic crystals, forming as plates or leaflets from acetic acid; mp 266°C; sublimes; insoluble in water, slightly soluble in alcohol and ether, soluble in most other organic solvents.

Health Hazard

The toxicity of dibenz[*a,h*]anthracene is on the same order as that of benz[*a*]anthracene. A lethal dose in mice by intravenous route is 10 mg/kg. There is no report on its oral toxicity. It is a mutagen. Its carcinogenicity in animals is well established, causing cancers in the lungs, liver, kidney, and skin.

26.26   DIBENZ[*a*, *j*]ANTHRACENE

Formula C$_{22}$H$_{14}$; MW 278.36; CAS [224-41-9]

Structure:

Synonyms: 3,4:5,6-dibenzanthracene; 1,2:7, 8-dibenzanthracene; dibenz[*a*,*j*]-anthracene

Health Hazard

Dibenz[*a*,*j*]anthracene has shown to cause skin cancer in animals, producing tumor at the site of application. There is no report of its toxicity. It tested positive in a histidine reversion–Ames test.

26.27   INDENO(1,2,3-*cd*)PYRENE

EPA Priority Pollutant; RCRA Toxic Waste Number U137

Formula C$_{22}$H$_{12}$; MW 276.34; CAS [193-39-5]

Structure:

Synonyms: *o*-phenylenepyrene; 2,3-phenylenepyrene; 1,10-(*o*-phenylene)pyrene; 1,10-(1,2-phenylene)pyrene

Health Hazard

Indeno(1,2,3-*cd*)pyrene causes lung cancer in animals. There is sufficient evidence of its carcinogenic actions in animals. However, the cancer-causing effects of this compound in humans are un known. There is no report on its toxicity.

REFERENCES

ACGIH. 1986. *Documentation of the Threshold Limit Values and Biological Exposure Indices*, 5th ed. Cincinnati, OH: American Conference of Governmental Industrial Hygienists.

Aksoy, M. 1989. Leukemogenic and carcinogenic effects of benzene. *Adv. Mod. Environ. Toxicol. 16:* 87–98; cited in *Chem. Abstr. CA 113*(2): 11293a.

Chiang, C. Y., J. P. Salanitro, E. Y. Chai, J. D. Colthart, and C. L. Klein. 1989. Aerobic biodegradation of benzene, toluene, and xylene in a sandy aquifer-data analysis and computer modeling. *Ground Water 27*(6): 823–34.

Dutkiewicz, T., and H. Tyras. 1968. *Br. J. Ind. Med. 25:* 253; cited in *Documentation of the Threshold Limit Values and Biological Exposure Indices*, 5th ed., p. 539. 1986. Cincinnati, OH: American Conference of Governmental Industrial Hygienists.

Flesher, J. W., and S. R. Myers. 1990. Bioalkylation of benz(a)anthracene as a biochemical probe for carcinogenic activity. Lack of bioalkylation in a series of six noncarcinogenic polynuclear aromatic hydrocarbons. *Drug Metab. Dispos. 18*(2): 163–67; cited in *Chem. Abstr. CA 113*(3): 19245g.

Hamilton, A., and H. L. Hardy. 1974. *Industrial Toxicology*, 3rd ed., p. 335. Acton, MA: Publishing Sciences Group.

Hoffmann, D., K. D. Brunnemann, and I. Hoffmann. 1989. Significance of benzene in tobacco carcinogenesis. *Adv. Mod. Environ. Toxicol. 16:* 99–112; cited in *Chem. Abstr. CA 113*(1): 1602m.

Jacobs, R. A. 1989. Benzene/nitrous oxide flammability in the precipitate hydrolysis process. *Energy Res. Abstr. 1990 15*(1): Abstr. No. 121.

Kalf, G. F., M. J. Schlosser, J. F. Renz, and S. J. Pirozzi. 1989. Prevention of benzene-induced myelo- and genotoxicity by nonsteroidal anti-inflammatory drug: a possible role for prostaglandin H synthase in benzene toxicity. *Adv. Mod. Environ. Toxicol. 16:* 1–17.

Khan, S., R. Krishnamurthy, and K. P. Pandya. 1990. Generation of hydroxyl radicals during benzene toxicity. *Biochem. Pharmacol. 39*(8): 1393–95.

Kramer, W. H. 1989. Benzene exposure assessment of underground storage tank contractors. *Appl. Ind. Hyg. 4*(11): 269–72.

Martin, J. F., M. J. Stutsman, and E. Grossman III. 1989. Hazardous waste volume reduction research in the U.S. Environmental Protection Agency. *ASTM Spec. Tech. Publ. 1043:* 163–71.

Mellor, J. W. 1946. *A Comprehensive Treatise on Inorganic and Theoretical Chemistry.* London: Longmans, Green & Company.

Merck. 1996. *The Merck Index*, 12th ed. Rahway, NJ: Merck & Co.

Meyer, E. 1989. *Chemistry of Hazardous Materials*, 2nd ed. Englewood Cliffs, NJ: Prentice-Hall.

NFPA. 1986. *Fire Protection Guide on Hazardous Materials*, 9th ed. Quincy, MA: National Fire Protection Association.

NIOSH. 1984. *Manual of Analytical Methods*, 3rd ed. Cincinnati, OH: National Institute for Occupational Safety and Health.

NIOSH. 1986. *Registry of Toxic Effects of Chemical Substances*, ed. D. V. Sweet. Washington, DC: U.S. Government Printing Office.

Oak Ridge National Laboratory. 1989. Health assessment for Midway Landfill, Seattle, Washington, Region 10. *Govt. Rep. Announce. Index (U.S.) 1990 90*(3): Abstr. No. 005, p. 974.

Sundstrom, D. W., B. A. Weir, and K. A. Redig. 1989. Destruction of mixtures of pollutants by UV-catalyzed oxidation with hydrogen peroxide. *ACS Symp. Ser. 422:* 67–76.

Thomas, J. M., V. R. Gordy, S. Fiorenza, and C. H. Ward. 1990. Biodegradation of BTEX in subsurface materials contaminated with gasoline: Granger, Indiana. *Water Sci. Technol. 22*(6): 53–62.

U.S. EPA. 1992. Guideline establishing test procedures for the analysis of pollutants under the Clean Water Act, 40 CFR Part 136. *Fed. Regist. 49*(209), Oct. 26.

U.S. EPA. 1997. *Test Methods for Evaluating Solid Waste*, 3rd ed. up date, Vol. 1B. Washington, DC: Office of Solid Waste and Emergency Response.

U.S. EPA. 1989a. Superfund record of decision (EPA region 10): Commencement Bay — south Tacoma channel, Tacoma landfill site, Tacoma, Washington (final remedial action) March 1988. *Govt. Report Announce. Index (U.S.) 1989 89*(19): Abstr. No. 951, p. 467.

U.S. EPA. 1989b. Superfund record of decision (EPA region 5): Kummer sanitary landfill, Beltrami county, Minnesota (second remedial action) September 1988. *Govt. Report Announce. Index (U.S.) 1989 89*(17): Abstr. No. 947, p. 265.

27

INDUSTRIAL SOLVENTS

27.1 GENERAL DISCUSSION

A large number of organic substances have found applications as solvents for a wide range of compounds. Although the toxic effects of most of the solvents are of low order, chronic exposures or large doses can produce moderate to severe poisoning. Most organic solvents are flammable or combustible liquids, the vapors of which form explosive mixtures with air. In addition to toxicity and flammability, the hazardous qualities manifested by many common solvents also include (1) flashback of vapors from an ignition source to containers; (2) peroxide formations on prolonged storage, especially in compounds containing an ether functional group, such as ethyl ether or tetrahydrofuran; and (3) formations of shock-sensitive solvated complexes with many metal perchlorates.

The toxic symptoms, median lethal doses, carcinogenic potentials if any, flammability, flash points, lower and upper explosive limits, violent reactions, and section numbers in this book where these compounds are described at length are highlighted below in condensed form for 63 common solvents. Full discussions of most of these substances

may be found throughout the book. Only the compounds not discussed elsewhere in the book are presented in detail in this chapter. Solvents are listed in alphabetical order.

27.2 ACETAL

Formula $C_6H_{14}O_2$; MW 118.20; CAS [105-57-7]

Structure:

$$CH_3{-}CH\begin{matrix} O{-}CH_2{-}CH_3 \\ \\ O{-}CH_2{-}CH_3 \end{matrix}$$

Synonyms: 1,1-diethoxyethane; ethylidene diethyl ether; acetaldehyde diethyl acetal

Physical Properties

Colorless liquid; density 0.8254 at 20°C; bp 102.5°C; moderately soluble in water (5 g/100 mL at 20°C), miscible with organic solvents.

Health Hazard

Mild irritant to skin and eyes; acute toxicity of low order; narcotic at high concentrations;

4-hour exposure to 4000 ppm lethal to mice; the oral LD_{50} value for mice is 3500 mg/kg.

Fire and Explosion Hazard

Highly flammable; flash point (closed cup) $-21°C$ ($-6°F$); vapor density 4.1 (air = 1), vapor heavier than air and can travel some distance to a source of ignition and flash back; autoignition temperature 230°C (446°F); vapor forms explosive mixtures with air, LEL and UEL values are 1.6% and 10.4% by volume in air, respectively (DOT Label: Flammable Liquid, UN 1088).

Analysis

GC-FID (GC column: fused-silica capillary column or Porapak Q); GC/MS; FTIR.

27.3 ACETAMIDE

Formula C_2H_5NO; MW 59.08; CAS [60-35-5]
Structure:

$$CH_3{-}\underset{\underset{O}{\|}}{C}{-}NH_2$$

Synonyms: ethanamide; acetic acid amide; methanecarboxamide

Physical Properties

Hexagonal crystals; deliquescent; mousy odor when impure; mp 81°C; bp 222°C; highly soluble in water, alcohol, and pyridine, soluble in most organic solvents.

Health Hazard

Mild irritant; acute oral toxicity in animals very low; oral LD_{50} value (rats): 7000 mg/kg; carcinogenic to animals; oral administration caused blood and liver tumors in mice and rats; carcinogenicity: animal limited evidence, no evidence in humans.

Fire and Explosion Hazard

Noncombustible liquid.

27.4 ACETIC ACID

Formula CH_3COOH; MW 60.06; CAS [64-19-7].

Glacial acetic acid highly corrosive; toxic; oral LD_{50} value (rats) 3530 mg/kg; combustible liquid; flash point (closed cup) 39°C (102°F); LEL and UEL values 4% and 16% by volume, respectively; explosive reactions with fluorides of chlorine and bromine, strong oxidizers, and phosphorus isocyanate.
See Section 1.3 for a detailed discussion.

27.5 ACETONE

Formula CH_3COCH_3; MW 58.08; CAS [67-64-1].

Vapors irritant to the eyes, nose, and throat; low toxicity; oral LD_{50} value (rats): 5800 mg/kg; highly flammable; flash point (closed cup) $-18°C$ (0°F); LEL and UEL values 2.6% and 12.8% by volume, respectively; explodes when mixed with chromic acid, hydrogen peroxide, or Caro's acid; ignites with potassium *tert*-butoxide.
See Section 29.3 for a detailed discussion.

27.6 ACETONITRILE

Formula CH_3CN; MW 41.05; CAS [75-05-8].

Causes nausea, vomiting, gastrointestinal pain, and stupor on inhalation or ingestion; oral LD_{50} value (mice): $\sim$270 mg/kg; flammable liquid; flash point (open cup) 5.5°C (42°F); LEL and UEL values 4.4% and 16.0% by volume, respectively; forms shock-sensitive solvated complexes with many metal perchlorates.
See Section 13.3 for a detailed discussion.

27.7 ACETOPHENONE

Formula $C_6H_5COCH_3$; MW 120.16; CAS [98-86-2].

Eye irritant; low toxicity; intraperitoneal LD_{50} value (mice): 200 mg/kg; combustible liquid; flash point (closed cup) 82°C (180°F).
See Section 29.22 for a detailed discussion.

27.8 ACETYL ACETONE

Formula $C_5H_8O_2$; MW 100.12; CAS [123-54-6].

Vapors an irritant to the eyes and mucous membranes; oral LD_{50} value (rat): 1000 mg/kg; flammable liquid; flash point (closed cup) 34°C (93°F).
See Section 29.5 for a detailed discussion.

27.9 ANILINE

Formula $C_6H_5NH_2$; MW 93.14; CAS [62-53-3].

Severe irritant to the eyes; causes cyanosis; oral LD_{50} value (mice): ~500 mg/kg; combustible liquid; flash point (closed cup) 70°C (~158°F); reacts violently with chloromelanamines or ozone.
See Section 9.2 for a detailed discussion.

27.10 BENZENE

Formula C_6H_6; MW 78.12; CAS [71-43-2].

Narcotic; causes somnolence, nausea, vomiting, and headache; irritant; chronic toxicant; occupational exposure can cause bone marrow depression and anemia; carcinogenic to humans; flammable liquid; flash point (closed cup) −11°C (12°F); LEL and UEL values 1.3% and 7.1% by volume, respectively; forms explosive product with ozone; violent reactions with other oxidizers.
See Section 26.2 for a detailed discussion.

27.11 BROMINE TRIFLUORIDE

Formula BrF_3; MW 136.91; CAS [7787-71-5].

Causes severe skin burn; vapors highly irritating; reacts violently with water, organic matter, and ammonium halides.
See Section 23.10 for a detailed discussion.

27.12 BROMOBENZENE

DOT Label: Combustible Liquid, UN 2514
Formula C_6H_5Br; MW 157.02; CAS [108-86-1]
Structure:

Synonyms: phenyl bromide; monobromobenzene

Physical Properties

Colorless liquid with aromatic odor; density 1.495 at 20°C; bp 156°C; mp −30.5°C; immiscible with water, miscible with most organic solvents.

Health Hazard

The acute toxicity of bromobenzene is low in test animals. The toxic symptoms include somnolence, respiratory stimulation, and muscle contraction. The oral LD_{50} value in rats is 2700 mg/kg.

Fire and Explosion Hazard

Combustible liquid; flash point (closed cup) 51°C (124°F); autoignition temperature 565°C (1050°F).

Analysis

GC-FID, GC-PID, GC-HECD, or GC/MS. A fused-silica capillary column is suitable for chromatographic analysis. Characteristic ions for GC/MS identification (electron-impact ionization) are 157, 78, and 79.

27.13   1-BUTANOL

Formula C_4H_9OH; MW 74.10; CAS [71-36-3].

Vapors an irritant to the eyes and respiratory tract; chronic exposure causes photophobia and lacrimation; oral LD_{50} value (rats): 790 mg/kg; flammable liquid; flash point (closed cup) 35°C (95°F); LEL and UEL values 1.4% and 11.2% by volume, respectively.
See Section 4.8 for a detailed discussion.

27.14   2-BUTANOL

Formula C_4H_9OH; MW 74.10; CAS [78-92-2].

Vapors an irritant to the eyes; defatting action on skin; narcotic at high concentrations; oral LD_{50} value (rats): 6480 mg/kg; flammable liquid; flash point (closed cup) 24.5°C (76°F); LEL and UEL values 1.7% and 9.8% by volume, respectively.
See Section 4.9 for a detailed discussion.

27.15   *n*-BUTYL ACETATE

Formula $C_6H_{12}O_2$; MW 116.18; CAS [123-86-4].

Irritant; narcotic; may cause degeneration of fatty liver and muscle weakness; oral LD_{50} value (guinea pigs): 4700 mg/kg; flammable liquid; flash point (closed cup) 22°C (72°F); LEL and UEL values 1.7% and 7.6% by volume, respectively.
See Section 17.8 for a detailed discussion.

27.16   CARBON DISULFIDE

Formula CS_2; MW 76.13; CAS [75-15-0].

Causes dizziness, headache, nausea, and central nervous system depression; chronic effects: psychic disturbances; oral LD_{50} value (rats): ~3200 mg/kg; lethal dose (oral) in humans: ~200 mg/kg; flammable liquid; flash point (closed cup) −30°C (−22°F); LEL and UEL values 1.3% and 50.0% by volume, respectively; vapors may readily catch fire.
See Section 56.2 for a detailed discussion.

27.17   CARBON TETRACHLORIDE

Formula CCl_4; MW 153.81; CAS [56-23-5].

Narcotic; vapors an irritant to the eyes and lungs; affects liver and kidneys; oral LD_{50} value (rats): 2800 mg/kg; carcinogenic; violent reactions with finely divided metals or strong oxidizers.
See Section 22.7 for a detailed discussion.

27.18   CHLOROFORM

Formula $CHCl_3$; MW 119.39; CAS [67-66-3].

Narcotic; hypotensive; oral LD_{50} value (rats): 900 mg/kg; cancer causing; explosive reactions with many finely divided metals or strong oxidizers.
See Section 22.6 for a detailed discussion.

27.19   CYCLOHEXANE

Formula C_6H_{12}; MW 84.18; CAS [110-82-7].

Vapors an irritant to the eyes and respiratory tract; narcotic at high concentrations; causes drowsiness; flammable liquid;

flash point (closed cup) $-20°C$ ($-4°F$); LEL and UEL values 1.3% and 8.4% by volume, respectively.

See Section 25.15 for a detailed discussion.

27.20 CYCLOHEXANOL

Formula $C_6H_{13}OH$; MW 100.16; CAS [108-93-0].

Vapors an irritant to the eyes, nose, and throat; chronic exposure may injure the kidney; oral LD_{50} value (rats): 2060 mg/kg; combustible liquid; flash point (closed cup) $67°C$ ($154°F$).

See Section 4.13 for a detailed discussion.

27.21 CYCLOHEXENE

Formula C_6H_{10}; MW 82.16; CAS [110-83-8].

Eye and respiratory tract irritant; narcotic; flammable liquid; flash point (closed cup) $-11.5°C$ ($11°F$); LEL and UEL values 1.0% and 5.0% by volume, respectively.

See Section 25.16 for a detailed discussion.

27.22 CYCLOHEXYLAMINE

Formula $C_6H_{11}NH_2$; MW 99.20; CAS [108-91-8].

Causes eye injury, skin burn, irritation of respiratory passage, and degenerative changes in the liver and kidney; narcotic; oral LD_{50} value (rats): 150 mg/kg; flammable liquid; flash point (closed cup) $32°C$ ($90°F$).

See Section 8.7 for a detailed discussion.

27.23 CYCLOPENTANE

Formula C_5H_{10}; MW 70.15; CAS [287-92-3].

Acute toxicity of very low order; central nervous system depressant; irritant to the upper respiratory tract, eyes, and skin; flammable liquid; flash point (open cup) $-7°C$ ($18°F$); LEL value 1.5% by volume.

See Section 25.12 for a detailed discussion.

27.24 DIETHYL PHTHALATE

Formula $C_{12}H_{14}O_4$; MW 222.26; CAS [84-66-2].

Vapors an irritant to the eyes and throat; causes somnolence and hypotension; oral LD_{50} value (mice): 6170 mg/kg.

See Section 17.19 for a detailed discussion.

27.25 DIMETHYL PHTHALATE

Formula $C_{10}H_{10}O$; MW 194.20; CAS [131-11-3].

Irritant to the eyes and throat; ingestion causes somnolence and hypotension; oral LD_{50} value (mice): 6800 mg/kg.

See Section 17.18 for a detailed discussion.

27.26 1,4-DIOXANE

Formula $C_4H_8O_2$; MW 88.12; CAS [123-91-1].

Central nervous system depressant; injurious to the kidneys, liver, and lungs on chronic exposure; oral LD_{50} value (mice): 5700 mg/kg; weak carcinogen; flammable liquid; flash point (closed cup) $12°C$ ($54°F$); LEL and UEL values 2.0% and 22.0% by volume, respectively; may react violently with strong oxidizers.

See Section 24.14 for a detailed discussion.

27.27   ETHANOL

Formula C_2H_5OH; MW 46.00; CAS [64-17-5].

Large doses may produce excitation, intoxication, stupor, hypoglycemia, and coma; hypotensive; chronic consumption may impair liver; flammable liquid; flash point (closed cup) 12.7°C (55°F); LEL and UEL values are 4.3% and 19.0% by volume, respectively; may explode when combined with strong oxidizers.
See Section 4.3 for a detailed discussion.

27.28   ETHYL ACETATE

Formula $C_4H_8O_2$; MW 88.12; CAS [141-78-6].

Causes eyes and nasal irritation; narcotic; oral LD_{50} value (rats): 5600 mg/kg; flammable liquid; flash point (closed cup) −4.5°C (24°F); LEL and UEL values 2.0% and 11.5% by volume, respectively.
See Section 17.5 for a detailed discussion.

27.29   ETHYL BENZENE

Formula C_8H_{10}; MW 106.18; CAS [100-41-4].

Narcotic at high concentrations; irritant; oral LD_{50} value (rats): 3500 mg/kg; flammable liquid; flash point (closed cup) 18°C (64°F); LEL and UEL values 1.0% and 6.7% by volume, respectively.
See Section 26.4 for a detailed discussion.

27.30   ETHYLENEDIAMINE

Formula $C_2H_8N_2$; MW 60.12; CAS [107-15-3].

Can cause eye injury and blistering on the skin; vapors damaging to the lung, liver, and kidney on repeated exposure; oral LD_{50} value (rats): 500 mg/kg; flammable liquid; flash point (closed cup) 34°C (93°F); LEL and UEL values 5.8% and 11.1% by volume, respectively; explodes with perchlorates.
See Section 8.11 for a detailed discussion.

27.31   ETHYLENE GLYCOL MONOBUTYL ETHER

Formula $C_6H_{14}O_2$; MW 118.20; CAS [111-76-2].

Irritant to the eyes and skin; causes nausea, vomiting, and headache; prolonged exposure to high concentrations affects kidney, liver, and red blood cells; oral LD_{50} value (rats): 530 mg/kg; teratogenic; combustible liquid; flash point (closed cup) 60°C (141°F); LEL and UEL values 1.1% and 10.6% by volume, respectively.
See Section 20.4 for a detailed discussion.

27.32   ETHYLENE GLYCOL MONOETHYL ETHER

Formula $C_4H_{10}O_2$; MW 90.12; CAS [110-80-5].

High doses may cause injury to the kidney, liver, and spleen; oral LD_{50} value (rats): 3000 mg/kg; combustible liquid; flash point (closed cup) 44°C (112°F); LEL and UEL values 2.8% and 18.0% by volume, respectively; forms peroxides.
See Section 20.3 for a detailed discussion.

27.33   ETHYLENE GLYCOL MONOISOPROPYL ETHER

$C_5H_{12}O_2$; MW 104.17; CAS [109-59-1].

Symptoms of anemia, hemoglobinuria, and pulmonary edema in animals from high exposure; oral LD_{50} value (rats): 5660 mg/kg; combustible liquid; flash point (closed cup) 45°C (115°F).
See Section 20.5 for a detailed discussion.

27.34 ETHYLENE GLYCOL MONOMETHYL ETHER

$C_3H_8O_2$; MW 76.11; CAS [109-86-4].

Affects the blood, kidney, and central nervous system; anesthetic effect from large doses; testicular atrophy and teratogenic; oral LD_{50} value (rabbits): 890 mg/kg; combustible liquid; flash point (closed cup) 42°C (107°F); LEL and UEL values 2.5% and 19.8% by volume, respectively; forms peroxides.

See Section 20.2 for a detailed discussion.

27.35 ETHYL ETHER

Formula $C_4H_{10}O$; MW 74.14; CAS [60-29-7].

Narcotic; anesthetic effect in humans at 3.5 to 6.5% in air; flammable liquid; flash point (closed cup) −45°C (−49°F); LEL and UEL values 1.85% and 48.0% by volume, respectively; forms peroxides; may catch fire or explode when combined with ozone, chromic acid, chlorine, perchlorates, and other oxidizers.

See Section 18.3 for a detailed discussion.

27.36 ETHYL FORMATE

Formula $C_3H_6O_2$; MW 74.09; CAS [109-94-4].

Vapors an irritant to the eyes, nose, and mucous membranes; narcotic; lethal concentration in rats 8000 ppm (4 hours); flammable liquid; flash point (closed cup) −20°C (−4°F); LEL and UEL values 2.8% and 16.0% by volume, respectively.

See Section 17.3 for detailed discussion.

27.37 *n*-HEPTANE

Formula C_7H_{16}; MW 100.20; CAS [142-82-5].

Vapors an irritant to the respiratory tract; narcotic at high concentrations; flammable liquid; flash point (closed cup) −4°C (25°F); autoignition temperature 204°C (399°F); LEL and UEL values 1.05% and 6.7% by volume, respectively.

27.38 *n*-HEXANE

Formula C_6H_{14}; MW 86.20; CAS [110-54-3].

Vapors an irritant to the eyes and throat; narcotic at high concentrations; symptoms of headache, dizziness, nausea, and hallucination; polyneuritis from chronic exposure; flammable liquid; flash point (closed cup) −22°C (−7°F); LEL and UEL values 1.2% and 8.0% by volume, respectively.

See Section 25.14 for detailed discussion.

27.39 *n*-HEXANOL

DOT Label: Combustible Liquid, UN 2282

Formula $C_6H_{13}OH$; MW 102.20; CAS [111-27-3]

Structure: $CH_3-CH_2-CH_2-CH_2-CH_2-CH_2-OH$

Synonyms: hexyl alcohol; pentylcarbinol; amylcarbinol; caproyl alcohol; 1-hydroxyhexane

Physical Properties

Colorless liquid; density 0.815 at 25°C; boils at 157°C; solidifies at −51.5°C; slightly soluble in water, miscible with organic solvents.

Health Hazard

Vapors of *n*-hexanol are irritant to the eyes and respiratory tract. Application of the liquid produced severe irritation in rabbits' eyes. It exhibits narcotic effects at high concentrations. The oral LD_{50} value in mice is in the range 2000 mg/kg.

Fire and Explosion Hazard

Combustible liquid; flash point (closed cup) 63°C (145°F); vapor forms explosive mixtures with air; flammability range not established.

27.40 ISOBUTANOL

Formula C_4H_9OH; MW 74.1; CAS [78-83-1].

Vapors an irritant to the eyes and throat; causes central nervous system depression; flammable liquid; flash point (closed cup) 27.5°C (81.5°F); LEL and UEL values 1.2% and 10.9% by volume, respectively.
See Section 4.11 for a detailed discussion.

27.41 ISOOCTANE

Formula C_8H_{18}; MW 114.26; CAS [540-84-1].

An irritant to the respiratory tract at high concentrations; lethal concentration in rats 30,000 ppm (1 hour); flammable liquid; flash point (closed cup) −12°C (10°F); LEL and UEL values 1.0% and 4.6% by volume, respectively.
See Section 25.18 for a detailed discussion.

27.42 ISOPROPYL ACETATE

Formula $C_5H_{10}O_2$; MW 102.15; CAS [108-21-4].

Vapors an irritant to the eyes and nose; narcotic; oral LD_{50} value (rats): ~6000 mg/kg; flammable liquid; flash point (closed cup) 2°C (36°F); LEL and UEL values 1.8% and 8.0% by volume, respectively.
See Section 17.7 for a detailed discussion.

27.43 ISOPROPYL ETHER

Formula $C_6H_{14}O$; MW 102.20; CAS [108-20-3].

Narcotic; irritant to mucous membranes; flammable liquid; flash point (closed cup) −27.8°C (−18°F); LEL and UEL values 1.4% and 7.9% by volume, respectively; forms peroxides; reacts violently with strong oxidizers.
See Section 18.4 for a detailed discussion.

27.44 LIGROIN

DOT Label: Flammable Liquid, UN 1271

A composition of hydrocarbons containing 55.4% paraffins, 30.3% naphthenes, 11.7% alkyl benzenes, and 2.4% dicycloparaffins; MW 112.0; CAS [8032-32-4]

Synonyms: VM&P naphtha; varnish makers' and printers' naphtha; petroleum ether; refined solvent naphtha; benzoline

Physical Properties

Clear, mobile liquid; density 0.85–0.87 at 15°C; boiling range 118–179°C; immiscible with water.

Health Hazard

The vapors of VM&P solvent are irritant to the eyes and upper respiratory tract. Such irritant action in humans may be manifested at an exposure level of 1000 ppm for 15 minutes. It is narcotic at high concentrations. The symptoms noted in rats of acute exposure were loss of coordination as well as convulsions. The inhalation LC_{50} value for 4-hour exposure in rats is 3400 ppm (NIOSH 1986).

Exposure Limits

TLV-TWA 300 ppm (~1350 mg/m³) (ACGIH), 75 ppm (~350 mg/m³) (NIOSH); TLV-STEL 400 ppm (ACGIH).

Fire and Explosion Hazard

Flammable liquid; flash point (closed cup) 14°C (57°F); vapors form explosive mixtures with air within the range 1.2–6.0% by volume in air.

27.45   METHANOL

Formula CH_3OH; MW 32.04; CAS [67-56-1].

Highly toxic; ingestion can cause acidosis and blindness; flammable liquid; flash point (closed cup) 12°C (52°F); LEL and UEL values 6% and 36% volume, respectively.
See Section 4.2 for a detailed discussion.

27.46   METHYL ACETATE

Formula $C_3H_6O_2$; MW 74.09; CAS [79-20-9].

Causes inflammation of the eyes; narcotic; oral LD_{50} value (rats): ~5000 mg/kg; flammable liquid; flash point (closed cup) −10°C (14°F); LEL and UEL values 3.1% and 16.0% by volume, respectively.
See Section 17.4 for a detailed discussion.

27.47   METHYL BUTYL KETONE

Formula $C_6H_{12}O$; MW 100.0; CAS [591-78-6].

Neurotoxic; disorder of peripheral nervous system from chronic exposure; irritant; narcotic at high concentrations; oral LD_{50} value (rats): 2600 mg/kg; flammable liquid; flash point (closed cup) 25°C (77°F); LEL and UEL values 1.2% and 8.0% by volume, respectively.
See Section 29.11 for a detailed discussion.

27.48   METHYLENE CHLORIDE

Formula CH_2Cl_2; MW 84.93; CAS [75-09-2].

Narcotic; causes intestine ulceration; oral LD_{50} value (rats): 2800 mg/kg; carcinogenic; reacts violently with finely divided metals or strong oxidizers.
See Section 22.5 for a detailed discussion.

27.49   METHYL ETHYL KETONE

Formula C_4H_8O; MW 72.11; CAS [78-93-3].

Vapors an irritant to the eyes and respiratory tract; narcotic at high concentrations; oral LD_{50} value (rats): 5500 mg/kg; flammable liquid; flash point (closed cup) −6°C (21°F); LEL and UEL values 1.8% and 12.0% by volume, respectively; may explode with strong oxidizers; ignites with potassium *tert*-butoxide.
See Section 29.4 for a detailed discussion.

27.50   METHYL FORMATE

Formula $C_2H_4O_2$; MW 60.06; CAS [107-31-3].

An irritant to the eyes, nose, and lungs; causes dyspnea; narcotic; oral LD_{50} value (rabbits): 1600 mg/kg; flammable liquid; flash point (closed cup): −19°C (−2°F); LEL and UEL values 5.0% and 23.0% by volume, respectively.
See Section 17.2 for a detailed discussion.

27.51   METHYL ISOBUTYL KETONE

Formula $C_6H_{12}O$; MW 100.16; CAS [108-10-1].

Vapors an irritant to the eyes and mucous membranes; defatting of skin on prolonged contact; narcotic at high concentrations; flammable liquid; flash point (closed cup) 18°C (64°F); LEL and UEL values 1.4% and 7.5%, respectively.

See Section 29.10 for a detailed discussion.

27.52   METHYL PROPYL KETONE

Formula $C_5H_{10}O$; MW 86.13; CAS [107-87-9].

Vapors an irritant to the eyes and respiratory tract; narcotic at high concentrations; fatal dose in rats 2000 ppm/4 hr; flammable liquid; flash point (closed cup) 7°C (45°F); LEL and UEL values 1.5% and 8.2% by volume, respectively.

See Section 29.6 for a detailed discussion.

27.53   MORPHOLINE

DOT Label: Flammable Liquid, UN 2054
Formula C_4H_9NO; MW 87.14; CAS [110-91-8]
Structure:

a heterocyclic compound containing an atom of oxygen and nitrogen in the ring

Physical Properties

Colorless liquid with a characteristic amine odor; hygroscopic; density 1.007 at 20°C; bp 129°C; mp −5°C; miscible with water and most organic solvents.

Health Hazard

Morpholine is an irritant to the eyes, skin, and mucous membranes. The irritant actions in rabbit eyes and skin were severe. In humans the inhalation of its vapors can cause visual disturbance, nasal irritation, and coughing. High concentrations can produce respiratory distress.

LD_{50} value, oral (mice): 525 mg/kg
LC_{50} value, inhalation (mice): 1320 mg/m^3/ 2 hr

Exposure Limits

TLV-TWA 20 ppm (~70 mg/m^3) (ACGIH, MSHA, and OSHA); STEL skin 30 ppm (ACGIH); IDLH 8000 ppm.

Fire and Explosion Hazard

Flammable liquid; flash point (open cup) 37°C (98.5°F); vapor pressure 7 torr at 20°C; autoignition temperature 290°C (555°F). Morpholine vapors form explosive mixtures with air within the range 1.8–11.0% by volume in air.

27.54   NITROBENZENE

EPA Priority Pollutant, Designated Toxic Waste, RCRA Waste Number U169; DOT Label: Poison B, UN 1662
Formula $C_6H_5NO_2$; MW 123.12; CAS [98-95-3]
Structure:

Synonyms: essence of myrbane; mirbane oil; oil of myrbane; nitrobenzol

Physical Properties

Oily liquid pale yellow to brown in color; characteristic odor; density 1.205 at 15°C; boils at 210°C; solidifies at 6°C; slightly soluble in water (0.2%), readily miscible in most organic solvents.

Health Hazard

The routes of entry of nitrobenzene into the body are the inhalation of its vapors or absorption of the liquid or the vapor through the skin and, to a much lesser extent, ingestion. The target organs are the blood, liver, kidneys, and cardiovascular system. Piotrowski (1967) estimated that in an exposure period of 6 hours to a concentration of 5 mg/m^3, 18 mg of nitrobenzene was absorbed through the lungs and 7 mg through the skin in humans. Furthermore, about 80% of inhaled vapor is retained in the respiratory tract. The dermal absorption rate at this concentration level is reported as 1 mg/hr, while the subcutaneous absorption of the liquid is between 0.2 and 0.3 mg/cm^3/hr (ACGIH 1986).

The symptoms of acute toxicity are headache, dizziness, nausea, vomiting, and dyspnea. Subacute and chronic exposure can cause anemia. Nitrobenzene effects the conversion of hemoglobin to methemoglobin. It is metabolized to aminophenols and nitrophenols to about 30%, which are excreted.

Exposure Limits

TLV-TWA 1 ppm ($\sim$5 mg/m^3) (ACGIH, MSHA, and OSHA); IDLH 200 ppm (NIOSH).

Fire and Explosion Hazard

Combustible liquid; flash point (closed cup) 88°C (190°F); vapor pressure <1 torr at 20°C; autoignition temperature 482°C (900°F); vapors form explosive mixtures with air, LEL 1.8% by volume in air, UEL not determined.

Nitrobenzene reacts explosively with nitric acid. Reaction with nitrogen tetroxide produces explosive products. Violent reactions may occur when combined with finely divided metals. Many metal perchlorates, such as silver or mercury salt, form shock-sensitive adducts with nitrobenzene.

An explosion occurred when aluminum chloride was mixed with nitrobenzene containing 5% phenol (NFPA 1986).

Disposal/Destruction

Nitrobenzene is burned in a chemical incinerator equipped with an afterburner and scrubber. It may be disposed of in a landfill in lab packs.

Analysis

Nitrobenzene in aqueous samples may be analyzed by GC and GC/MS techniques. Wastewaters and solid and hazardous wastes may be analyzed by EPA Methods 609 (GC) or 625 and 8250 or 8270, respectively, using GC/MS (U.S. EPA 1992, 1997). The characteristic ions to identify nitrobenzene by mass spectroscopy (electron-impact ionization) are 77, 123, and 65. The GC techniques involve the use of FID and NPD (in the nitrogen mode) as detectors. The former is less sensitive than the latter.

Air analysis may be performed in accordance with NIOSH (1984) Method 2005. Between 10 and 150 L of air is passed over a solid sorbent tube containing silica gel at a flow rate of 10–1000 mL/min. Nitrobenzene adsorbed over silica gel is desorbed with methanol in an ultrasonic bath for 1 hour and analyzed by GC-FID using the column 10% FFAP on Chromosorb W-HP support.

Biological monitoring of exposure to nitrobenzene may be performed by determining the total p-nitrophenol (a metabolite of nitrobenzene) in the urine. p-Nitrophenol is reduced to p-aminophenol by treatment with zinc and HCl. p-Aminophenol is further treated with phenol to convert to indophenol, the absorbance of which is determined by a spectrophotometer. The amount of nitrobenzene absorbed, E, may be calculated as follows (Piotrowski 1967):

$$E = \text{TLV}(\text{mg/m}^3) \times T(VR + \alpha)$$

where TLV is the threshold limit value in mg/m^3, T the period of exposure in hours, V the pulmonary ventilation (0.42 m^3/hr at rest), R the pulmonary retention (0.8 at rest), and α the coefficient of skin absorption (experimentally determined as 0.25 mg/hr).

The amount of nitrobenzene absorbed, E (in mg), and the urinary excretion rate, I (in mg/hr), at the end of a single exposure and at the end of a week are $I = 3.6E$ and $I = 10E$, respectively (Salmowa et al. 1963).

GC and HPLC methods, not fully investigated, may be used for qualitative monitoring. It may be noted that exposure to certain chemicals, such as parathion, may also produce p-nitrophenol as a urinary metabolite. A screening test for nitrobenzene exposure may be performed by monitoring methemoglobin in the blood.

27.55 *n*-OCTANE

Formula C$_8$H$_{18}$; MW 114.26; CAS [111-65-9].

Vapors an irritant to the respiratory tract; narcotic at high concentrations; flammable liquid; flash point (closed cup) 13°C (56°F); LEL and UEL values 0.96% and 4.66% by volume, respectively.

See Section 25.17 for a detailed discussion.

27.56 *n*-PENTANE

Formula C$_5$H$_{12}$; MW 72.17; CAS [109-66-0].

Very low toxicity; an irritant to the respiratory passage; narcotic at high concentrations; highly flammable liquid; flash point (closed cup) −49.5°C (−57°F); LEL and UEL values 1.5% and 7.8% by volume, respectively.

See Section 25.11 for a detailed discussion.

27.57 1-PROPANOL

Formula C$_3$H$_7$OH; MW 60.09; CAS [71-23-8].

Narcotic at high concentrations; oral LD$_{50}$ value (rats): 5400 mg/kg; flammable liquid; flash point (open cup) 29°C (59°F); LEL and UEL values 2.0% and 14.0% by volume, respectively; forms explosive product with concentrated HNO$_3$/H$_2$SO$_4$ mixture.

See Section 4.4 for a detailed discussion.

27.58 2-PROPANOL

Formula C$_3$H$_7$OH; MW 60.09; CAS [67-63-0].

Narcotic; flammable liquid; flash point (closed cup) 11.7°C (33.5°F); LEL and UEL values 2.02% and 7.99% by volume, respectively.

See Section 4.5 for a detailed discussion.

27.59 *n*-PROPYL ACETATE

Formula C$_5$H$_{10}$O$_2$; MW 102.15; CAS [109-60-4].

Irritant; narcotic; lethal concentration in rats 8000 ppm (4 hours); flammable liquid; flash point (closed cup) 14°C (58°F); LEL and UEL values 2.0% and 8.0% by volume, respectively.

See Section 17.6 for a detailed discussion.

27.60 PYRIDINE

Formula C$_5$H$_9$N; MW 79.11; CAS [110-86-1].

Causes headache, dizziness, nausea, and abdominal pain; kidney and liver damage from chronic exposure; oral LD$_{50}$ value (mice): 1500 mg/kg; flammable liquid; flash point (closed cup) 20°C (68°F); LEL and UEL values 1.8% and 12.4% by volume, respectively; reacts violently with strong oxidizers.

See Section 24.2 for a detailed discussion.

27.61 STODDARD SOLVENT

Formula C_9H_{20}; MW 128.29; CAS [8052-41-3]

A mixture of straight- and branched-chain alkanes, cycloalkanes, and aromatic hydrocarbons

Synonyms: mineral spirits; dry cleaning safety solvent

Physical Properties

Colorless liquid with a kerosene-like odor; density 0.79 at 20°C; boils at 154°–202°C; insoluble in water, miscible with most organic solvents.

Health Hazard

No serious health hazard has been reported as resulting from exposure to Stoddard solvent. Vapors are an irritant to the eyes, nose, and throat. Skin contact can cause defatting and irritation. A 7-hour exposure to 1700 ppm was lethal to cats.

Exposure Limits

TLV-TWA 100 ppm (~525 mg/m^3) (ACGIH), 500 ppm (OSHA), 350 mg/m^3/ 10 hr (NIOSH); ceiling 1800 mg/m^3/15 min (NIOSH).

Fire and Explosion Hazard

Combustible liquid; flash point (closed cup) 39–60°C (102–140°F); vapors form explosive mixtures with air, LEL value 0.8% by volume; UEL not determined. Nitric acid, perchlorates, and other strong oxidizers or nitrogen tetroxide react violently with Stoddard solvent.

Analysis

Air analysis may be performed by passing air over coconut charcoal, desorbing the hydrocarbon components adsorbed over charcoal into carbon disulfide, and injecting the eluant onto GC-FID.

27.62 TETRAHYDROFURAN

Formula C_4H_8O; MW 72.12; CAS [109-99-9].

Vapors an irritant to the eyes and upper respiratory tract; anesthetic at high concentrations; injurious to kidneys and liver on chronic exposure; oral LD$_{50}$ value (rats): 2800 mg/kg; flammable liquid; flash point (closed cup): −14.4°C (6°F); LEL and UEL values 2.0% and 11.8% by volume, respectively; forms peroxide.

See Section 24.13 for a detailed discussion.

27.63 TOLUENE

C_7H_8; MW 92.15; CAS [108-88-3].

Narcotic at high concentrations; flammable liquid; flash point (closed cup) 4.5°C (40°F); LEL and UEL values 1.4% and 6.7% by volume, respectively.

See Section 26.3 for a detailed discussion.

27.64 XYLENES

Formula C_8H_{10}; MW 106.18; CAS [1330-20-7]; [95-47-6], [108-38-3] and [106-42-3] for o-, m-, and p-isomers.

Central nervous system depressant; irritant; oral LD$_{50}$ value (rats): within the range

5000 mg/kg; flammable liquids; flash point 27–32°C (81–90°F) for all isomers; LEL and UEL values ~1.0 and ~7.0% by volume, respectively.

See Section 26.5 for a detailed discussion.

References

ACGIH. 1986. *Documentation of the Threshold Limit Values and Biological Exposure Indices*, 5th ed. Cincinnati, OH: American Conference of Governmental Industrial Hygienists.

NFPA. 1986. *Fire Protection Guide on Hazardous Materials*, 9th ed. Quincy, MA: National Fire Protection Association.

NIOSH. 1992. *Manual of Analytical Methods*, 3rd ed. Cincinnati, OH: National Institute for Occupational Safety and Health.

NIOSH. 1986. *Registry of Toxic Effects of Chemical Substances*, ed. D. V. Sweet. Washington, DC: U.S. Government Printing Office.

Piotrowski, J. 1967. Further investigations on the evaluation of exposure to nitrobenzene. *Br. J. Ind. Med. 24:* 60–65.

Salmowa, J., J. Piotrowski, and U. Neuhorn. 1963. Evaluation of exposure to nitrobenzene. *Br. J. Ind. Med. 20:* 241–46.

U.S. EPA. 1992. Method for organic chemical analysis of municipal and industrial wastewater. 40 CFR, Part 136, Appendix A. *Fed. Register 49* (209), Oct. 26.

U.S. EPA. 1997. *Test Methods for Evaluating Solid Waste*, 3rd ed. (update) Washington, DC: Office of Solid Waste and Emergency Response.

28

ISOCYANATES, ORGANIC

28.1 GENERAL DISCUSSION

Organic isocyanates are compounds containing the isocyanate group $-N=C=O$ attached to an organic group. They are esters of isocyanic acid HNCO. They form polymers, commonly called polyurethanes, which are of great commercial application.

These compounds are highly reactive due to the high unsaturation in the isocyanate functional group. They undergo addition reactions to the carbon–nitrogen double bond. Isocyanates react readily with alcohols to form urethanes:

$$RNCO + R'OH \longrightarrow RNH\overset{\overset{\displaystyle O}{\|}}{C}OR'$$

Urethane

Halogen acids react with isocyanates to form carbamyl halides:

$$RNCO + HCl \rightleftharpoons RNHCOCl$$

Isocyanates hydrolyze with water to give substituted urea. Disubstituted ureas are formed with amines. Isocyanates in general are highly reactive toward compounds containing active hydrogen atoms (Chadwick and Cleveland 1981).

Health Hazard

Most isocyanates are hazardous to health. They are lachrymators and irritants to the skin and mucous membranes. Skin contact can cause itching, eczema, and mild tanning. Inhalation of isocyanate vapors can produce asthma-like allergic reaction, with symptoms from difficulty in breathing to acute attacks and sudden loss of consciousness.

The toxicities of isocyanates vary widely. Whereas methyl isocyanate is highly toxic, dicyclohexylmethane diisocyanate displays moderate action. In addition, health hazards differ significantly on the route of exposure. The toxicities of these isocyanates are discussed individually below.

Although methyl isocyanate presents a severe health hazard, other aliphatic isocyanates are relatively less dangerous and their acute inhalation toxicity is much lower than that of the vapors and particulates of aromatic diisocyanates. Also, the latter class of compounds exhibits greater inhalation toxicity than that of the aromatic monoisocyanates. This may be attributed to twice the concentration of highly reactive isocyanate, $-N=C=O$, moieties present per mole of diisocyanates.

TABLE 28.1   LD$_{50}$ Values of Fluoroalkyl Isocyanates

Compound/CAS No.	Carbon Chain	LD$_{50}$ Intraperitoneal (Mice) (mg/kg)
2-Fluoroethyl isocyanate [505-12-4]	FCH_2CH_2-	17.0
3-Fluoropropyl isocyanate [407-99-8]	$FCH_2CH_2CH_2-$	10.0
4-Fluorobutyl isocyanate [353-16-2]	$FCH_2CH_2CH_2CH_2-$	4.7

On the other hand, aliphatic isocyanates exhibit greater acute oral toxicity than that of the aromatic diisocyanates. The isocyanate class as a whole manifests toxicity primarily through the inhalation route. Although toxicity decreases with the increase in the carbon chain length, this is not essentially true, as many anomalies can be found with substituted aliphatic isocyanates, as illustrated in Table 28.1. As highlighted in the table, the LD$_{50}$ values showed a pattern of decreasing from fluorobutyl to fluoroethyl isocyanate. Even the nonsubstituted *n*-butyl isocyanate has a lower intravenous LD$_{50}$ value in mice than that of ethyl isocyanate (1 and 56 mg/kg, respectively).

Fire and Explosion Hazard

Most isocyanates have high flash points. The fire hazard from these compounds is therefore low. However, closed containers can rupture due to the pressure built up from CO_2, which is formed from reaction with moisture:

$$RNCO + H_2O \longrightarrow RNHCOOH$$
$$\longrightarrow RNH_2 + CO_2$$

CO_2 may also form from reactions with carboxylic acids. Such reactions in closed containers may rupture the containers.

$$RNCO + R'COOH \longrightarrow RNHCOOCOR'$$
$$\longrightarrow RNHCOR' + CO_2$$

Safety and Handling

Isocyanates that have high vapor pressure should be handled carefully with a proper exhaust system. There should be adequate ventilation. Protective clothing, goggles, gloves, and respiratory masks should be worn when handling highly toxic isocyanates. Use self-contained breathing apparatus when handling high concentrations.

In the case of poisoning, use of a bronchodilator such as the Ventolin (Hadengue and Philbert 1983) is recommended. Oxygen should be given in acute attacks. Wash the affected skin with copious volumes of water.

28.2   METHYL ISOCYANATE

EPA Classified Acute Hazardous Waste, RCRA Waste Number P064; DOT Label: Flammable Liquid and Poison UN 2480

Formula CH_3NCO; MW 57.05; CAS [624-83-9]

Structure and functional group:

$$H_3C-N=C=O$$

highly reactive due to the presence of unsaturated isocyanate functional group

Synonyms: MIC; isocyanic acid methyl ester; isocyanatomethane

Uses and Exposure Risk

Methyl isocyanate is used primarily in making pesticides and herbicides such as carbaryl and aldicarb. It is also used to a lesser extent to produce plastics and polyurethane foams. Methyl isocyanate was attributed to the deaths of more than 3000 people in Bhopal, India, in a very tragic industrial accident during the 1980s.

Physical Properties

Colorless liquid with a sharp, unpleasant smell; highly volatile; bp 39°C; mp −80°C; density 0.96 at 20°C; soluble in hexane, isooctane, chloroform, methylene chloride, and other hydrocarbons or halogenated hydrocarbons; it decomposes in water.

Health Hazard

Methyl isocyanate is a dangerously toxic compound. Inhalation, oral intake, and absorption through skin can seriously affect the lungs, eyes, and skin.

Exposure to its vapor can cause lacrimation, irritation of nose and throat, and respiratory difficulty. In humans, exposure to 2 ppm for 5 minutes caused irritation and lacrymation, which became unbearable at 20 ppm. Odor was perceived at 4–5 ppm concentration. However, irritation of the nose and throat, occurring to a lesser extent than lacrymation, was evident in humans at 2 ppm (5 minutes exposure). The recovery from low exposure was fast — all effects disappeared within 10–20 minutes.

Methyl isocyanate is a severe acute toxicant. Penetration through the skin caused hemorrhage and edema. Acute symptoms from oral intake or absorption through skin were asthma, chest pain, dyspnea, pulmonary edema, difficulty in breathing, and death. Direct contact into eyes can damage vision.

LC_{50} value, inhalation (rats): 6.1 ppm/6 hr
LD_{50} value, oral (rats): 69 mg/kg
LD_{50} value, skin (rabbits): 213 mg/kg

In an acute and subacute toxicity study of inhaled methyl isocyanate in rats, Sethi et al. (1989) observed that the animals had congestion in the eyes, lacrimation, nasal secretion, dyspnea, progressively increased ataxia, and immobility following a single exposure to 3.52- and 33.32-ppm doses for 10 minutes each. The higher concentration dose produced severe effects. The same clinical signs were observed in subacute toxicity experiments at doses of 0.21–0.35 ppm for 30 minutes daily for 6 days by inhalation. Uncoordinated movements were also observed during this exposure period, which indicated that the nervous system was injured. A 2-hour exposure to 65 ppm was lethal to rats — the death resulting from pulmonary edema (Salmon et al. 1985). There was erosion of the corneal epithelium in eyes, which was most severe at intermediate levels of exposure. Congestion of the blood vessels, edematous changes, and cellular reaction persisted several days after a single exposure to 3.2 mg/L for 8 minutes (Dutta et al. 1988). The recovery was slow and gradual. Exposure to higher concentrations resulted in necrotizing or corrosive action, lung injury, and death of all exposed rats.

Acute exposure of mice and rats to sublethal and lethal concentrations for 2 hours produced restlessness, lacrimation, and a reddish discharge from the nose and mouth (Bucher et al. 1987a,b). The lethal concentrations were 20–30 ppm. The deaths occurred within 15–18 hours after severe respiratory distress. Males were more prone to early death than females. Animals exposed to 10 ppm died after 8–10 days. The study indicated that there was no cyanide in the blood. Thus at these doses methyl isocyanate toxicity was not due to cyanide. Repeated inhalation of 3 and 6 ppm for 6 hours/day for 4 consecutive days was lethal for rats and mice, respectively. The respiratory system was the primary site of injury. There was little evidence of direct effects on nonrespiratory tissues.

Methyl isocyanate administered subcutaneously in rabbits resulted in a fall in arterial blood pressure and respiratory inhibition

(Jeevarathinam et al. 1988). The LD_{50} values for the pure compound for male and female mice were 81.9 and 85.3 mg/kg, respectively (Vijayaraghavan and Kaushik 1987). Rats exhibited greater resistance.

Methyl isocyanate was not mutagenic in the *Salmonella* reversion assay but cytotoxic at relatively higher concentrations (Meshram and Rao 1987). McConnell et al. (1987) observed that it was genotoxic in cultured mammalian cells and showed marginal evidence of chromosomal damage in the bone marrow of mice.

Bhopal Study

Several studies have been performed on the toxic effects of methyl isocyanate on the survivors of the Bhopal disaster. The results of a clinical study (Mishra et al. 1988) indicated that eye and respiratory tract irritation were the most important clinical features, causing keratitis and pneumonia in patients. The report presents the fact that 55% of the patients fainted after inhaling the toxic gas and remained unconscious for varying periods. Hyperreflexia and encephalopathy were detected in a few cases of prolonged unconsciousness. High levels of blood urea were commonly found.

Systematic follow-up studies have been reported on after effects of exposure to the toxic gas on the population of Bhopal (Gupta et al. 1988; Rastogi et al. 1988; Srivastava et al. 1989; Saxena et al. 1988). The symptoms pertained to respiratory, gastrointestinal, cardiovascular, and musculoskeletal systems. The respiratory impairment involved bronchial obstruction, pulmonary defect, and ventilatory disorder. Behavior studies showed that visual perceptual response was severely affected. Among the biochemical parameters, there was an increase in hemoglobin values, lymphocyte and eosinophil counts, serum ceruloplasmin, and creatinine content in urine, while a decrease in blood glutathione was observed in the exposed people.

Prasad and Pandey (1985) observed that in the vicinity of the accident there was damage to the vegetation. Defoliation was manifested by blackening of chlorophyllic lamina and partial burning of the fallen leaves.

Antidote

Cyanide antidotes such as sodium nitrite and sodium thiosulfate were not effective in combating methyl isocyanate poisoning (Bucher et al. 1987a,b; Vijayaraghavan and Kaushik 1987). Atropine and ethanol were also ineffective.

Methyl isocyanate exhibited toxicity in two phases; 40 ppm exposure caused death of a majority of animals within 1–2 days or between 7 and 21 days after exposure. Varma et al. (1988) observed that injection of 2 mg dexamethasone/kg before exposure inhibited the toxicity within the first 6–7 days. Administration of this compound after exposure, however, was not effective. Twenty-four or 48 hours of starvation before exposure exhibited some antidotal effects, probably due to an increase in serum corticosterone levels, which may suppress the inflammatory response to methyl isocyanate.

Exposure Limits

TLV-TWA skin 0.047 mg/m³ (0.02 ppm) (ACGIH and OSHA); IDLH 47.4 mg/m³ (20 ppm) (NIOSH).

Fire and Explosion Hazard

Highly flammable; flash point (closed cup) −7°C (19°F); vapor density 1.97 (air = 1); vapor heavier than air and can travel to a source of ignition and flashback; ignition temperature 534°C (994°F). Fire-extinguishing agent: dry chemical, CO_2, and "alcohol" foam.

Methyl isocyanate forms an explosive mixture with air, the LEL and UEL values of 5.3% and 26% by volume of air, respectively.

Due to the highly unsaturated isocyanate functional group, methyl isocyanate exhibits high reactivity. It decomposes with water and reacts vigorously with acids, bases, amines, and metals and their salts.

Disposal/Destruction

Burning in a chemical incinerator equipped with an afterburner and scrubber is a suitable method of destruction. In a laboratory small amounts can be destroyed by slowly adding an aqueous solution of sodium bisulfite to the isocyanate. The stable water-soluble derivative may be drained down with large volumes of water.

Analysis

Methyl isocyanate can be analyzed by several instrumental techniques. These include GC, TLC, HPLC, and gel-permeable chromatography. Air samples can be analyzed by collecting the isocyanate in a solution of 1-(2-methoxyphenyl)piperazine in toluene in an impinger. The solution is acetylated with acetic anhydride and evaporated to dryness. The residue is extracted with methanol and analyzed by HPLC, using an UV detector at 254 nm [NIOSH (1984) Method for Isocyanate Group, Method 5505 1985].

28.3   TOLUENE-2,4-DIISOCYANATE

EPA Classified Toxic Waste, RCRA Waste Number U223

Formula $CH_3C_6H_3(NCO)_2$; MW 174.16; CAS [584-84-9]

Structure and functional group:

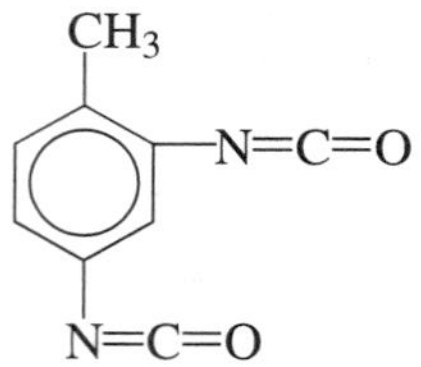

reactive sites — highly unsaturated isocyanate sites, steric hindrance may partially inhibit addition reactions, substitution can occur on the ring

Synonyms: 2,4-tolylene diisocyanate; 2,4-diisocyanatotoluene; toluene diisocyanate (TDI); 4-methyl phenylene diisocyanate; isocyanic acid 4-methyl-*m*-phenylene ester

Uses and Exposure Risk

Toluene-2,4-diisocyanate is one of the most extensively used isocyanates. It is used in the production of rigid and flexible urethane foams, elastomers, and coatings. In addition to its use as a pure compound, it is commercially available as a mixture of 2,4- and 2,6-isomers (80:20% and 65:35% ratios, respectively).

Physical Properties

Colorless liquid becoming straw on standing and dark on exposure to sunlight; fruity pungent smell; bp 238°C (250°C for 80:20 mixture); mp 20.5°C (13°C for 80:20 mixture); density 1.22; soluble in most organic solvents; decomposes with water and alcohol.

Health Hazard

Toluene-2,4-diisocyanate is a highly toxic compound by inhalation, a skin and eye irritant, and a carcinogenic substance. Exposure to its vapors can cause tracheobronchitis, pulmonary edema, hemorrhage, and death. The target organs are the respiratory system and skin. The toxic effects were also noted in the liver, kidney, and gastrointestinal tract.

In humans, exposure to low concentrations, 0.1–0.2 ppm, can result in irritation of the eyes, nose, and mucous membranes. Acute exposure to higher concentrations can cause bronchitis, pneumonitis, headache, sleeplessness, pulmonary edema, and sometimes an asthma-like syndrome. Chronic exposure can result in wheezing, coughing, shortness of breath, and chest congestion. Such effects may be manifested from inhalation of 0.02–0.05 ppm of the diisocyanate over a period of time.

Acute oral toxicity of this compound, however, is low. Symptoms may be coughing, vomiting, and gastrointestinal pain. Absorption through skin can produce toxic effects similar to those of inhalation toxicity: bronchitis, pulmonary edema, and asthma. In addition, the acute toxic symptoms can be nausea, vomiting, abdominal pain, dermatitis, and skin sensitization. Contact with eyes can cause burning, lacrimation, pricklingtype sensation, and injury to vision.

LC_{50} value, (rats): 14 ppm/4 hr
LD_{50} value, oral (rats): 5800 mg/kg
LD_{50} value, intravenous (mice): 56 mg/kg

Toluene-2,4-diisocyanate caused tumor in the liver and pancreas in test animals. There is sufficient evidence of its carcinogenicity in animals. Its cancer-causing activity in humans has not been established.

Exposure Limits

TLV-TWA 0.0355 mg/m^3 (0.005 ppm) (ACGIH and NIOSH); STEL or ceiling/10 min 0.142 mg/m^3 (0.02 ppm) (ACGIH, NIOSH, and OSHA); IDLH 71 mg/m^3 (10 ppm) (NIOSH).

Fire and Explosion Hazard

Noncombustible; flash point (open cup) 130°C (266°F); vapor pressure 0.03 torr; LEL and UEL values 0.9% and 9.5% by volume of air, respectively; fire-extinguishing agent: CO_2 or water spray.

Contact with strong oxidizers, acids, bases, and amines can produce vigorous to violent reactions with foaming and spattering. It reacts with water, forming CO_2.

Storage and Shipping

Toluene-2,4-diisocyanate is stored in a cool, dry, well-ventilated area separated from oxidizing substances. If stored in tanks, it should be blanketed with nitrogen, dry air, or an inert gas (NFPA 1997). It is shipped in steel drums and tank cars.

Disposal/Destruction

Toluene-2,4-diisocyanate is dissolved in a combustible solvent and burned in a chemical incinerator equipped with an afterburner and scrubber.

Analysis

Toluene-2,4-diisocyanate can be analyzed by GC, TLC, GC/MS, and HPLC. Air analysis is performed by HPLC using an UV detector at 254 nm. Methanol solution of its acetylated derivative of 1-(2-methoxyphenyl)piperazine is analyzed (NIOSH 1984, Method 5505, see Section 28.2).

28.4 HEXAMETHYLENE DIISOCYANATE

DOT Label: Poison B, UN 2281
Formula $(CH_2)_6(NCO)_2$; MW 168.22; CAS [822-06-0]
Structure and functional group:

$$O=C=N-CH_2-CH_2-CH_2-CH_2-CH_2-$$
$$CH_2-N=C=O$$

aliphatic isocyanate, reactive sites are the highly unsaturated $-N=C=O$ groups
Synonyms: 1,6-diisocyanatohexane, 1,6-hexanediol diisocyanate, isocyanic acid hexamethylene ester

Uses and Exposure Risk

Hexamethylene diisocyanate (HDI) is used for the production of polyurethane foam and exceptionally high-quality coatings.

Physical Properties

Colorless liquid; bp 213°C; freezes at −67°C; density 1.05 at 20°C; soluble in most organic solvents, decomposes with water.

Health Hazard

HDI is moderately toxic by inhalation. In humans the acute toxic symptoms could be wheezing, dyspnea, sweating, coughing, difficulty in breathing, and insomnia. In addition, this compound can produce irritation of the skin, eyes, nose, and respiratory tract. Chronic exposure may cause obstruction of airways and asthma.

The lethal concentration for rats from inhalation of this compound for 4 hours was 60 mg/m^3. The oral toxicity of this compound was found to be low in test animals. The toxicity order was much higher when given intravenously.

LD$_{50}$ value, oral (mice): 350 mg/kg
LD$_{50}$ value, intravenous (mice): 5.6 mg/kg

There is no report of any carcinogenic or teratogenic study for this compound.

Exposure Limit

TLV-TWA 0.0343 mg/m^3 (0.005 ppm) (ACGIH).

Fire and Explosion Hazard

Noncombustible; flash point (open cup) 140°C; vapor pressure 0.05 torr at 25°C.

HDI may react violently with strong oxidizers, acids, and bases. It reacts with water, forming CO_2. Pressure buildup may occur in closed containers due to moisture.

Disposal/Destruction

Disposal is by chemical incineration of HDI solution in a combustible solvent.

Analysis

HPLC (UV detector), GC-FID, and GC/MS are suitable instrumental techniques for analysis.

28.5 DIPHENYLMETHANE-4,4'-DIISOCYANATE

DOT Label: Poison B, UN 2489
Formula $(C_6H_4)_2CH_2(NCO)_2$; MW 250.27; CAS [101-68-8]
Structure and functional group:

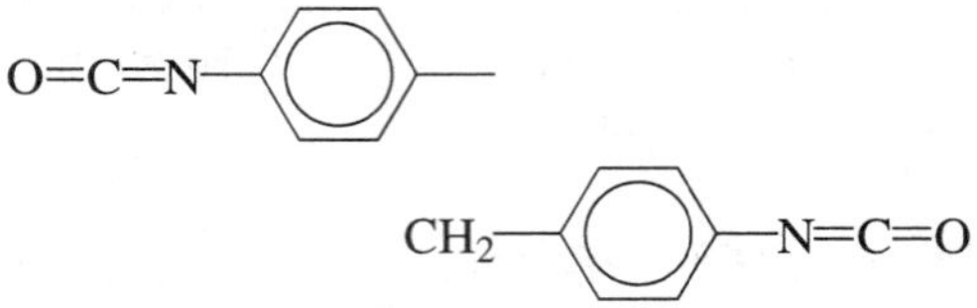

reactive sites: the unsaturated isocyanate functional groups and the aromatic rings where substitution can occur
Synonyms: methylenebis(4-phenylene isocyanate); bis(*para*-isocyanatophenyl)methane; bis(1,4-isocyanatophenyl)methane; 1,1'-methylenebis(4-isocyanotobenzene); methylenedi-*para*-phenylene diisocyanate; isocyanic acid methylene di-*para*-phenylene ester

Uses and Exposure Risk

Diphenylmethane-4,4'-diisocyanate (MDI) is widely used in the production of rigid urethane foam products, coatings, and elastomers.

Physical Properties

Light yellow to white crystal; mp 37.2°C; bp 172°C; density 1.197 at 7°C; slightly soluble in water (0.2 g/100 mL).

Health Hazard

MDI can present a moderate to severe health hazard, due to respiration of its vapors and particulates. It can contaminate the environment during foam application; concentrations in air were measured as high as 5 mg/m^3, mostly as particulates (ACGIH 1986).

The toxic route is primarily inhalation. The vapor pressure of this compound at ambient temperature is very low, 0.00014 torr at 25°C. However, when heated to about 75°C, the acute health hazard is greatly enhanced (Hadengue and Philbert 1983). The acute toxic symptoms were found to be similar to those of toluene-2,4-diisocyanate and other aromatic isocyanates. Inhalation of its vapors or particulates can cause bronchitis, coughing, fever, and an asthma-like syndrome. Other symptoms were nausea, shortness of breath, chest pain, insomnia, and irritation of the eyes, nose, and throat. The immunologic response, however, varied among humans. Exposure to 0.1–0.2 ppm for 30 minutes is likely to manifest the acute toxic effects in humans.

MDI is an eye and skin irritant. Contact with skin can produce eczema. The acute oral toxicity of this compound is very low, considerably lower than that of toluene-2,4-diisocyanate. The lethal dose for rats was 31.7 g/kg.

MDI showed positive in mutagenic testing on *Salmonella typhimurium*. There is no report that indicates its carcinogenicity.

Exposure Limits

TLV-TWA 0.051 mg/m^3 (0.005 ppm) (ACGIH and NIOSH); ceiling (air) 0.204 mg/m^3 (0.02 ppm)/10 min (NIOSH and OSHA); IDLH 102 mg/m^3 (10 ppm).

Fire and Explosion Hazard

Noncombustible; flash point (open cup) 202°C. MDI reactions with strong oxidizers, acids, and bases can be vigorous.

Disposal/Destruction

MDI is dissolved in a combustible solvent and burned in a chemical incinerator equipped with an afterburner and scrubber.

Analysis

MDI may be analyzed by GC, GC/MS, TLC, and HPLC. The latter technique is followed

for air analysis using an UV detector. Air is passed through an impinger that contains 1-(2-methoxyphenyl)piperazine solution. The derivative is treated further with acetic anhydride and analyzed (NIOSH 1984, Method 5505).

28.6 METHYLENE BIS(4-CYCLOHEXYLISOCYANATE)

Formula $(C_6H_{10})_2CH_2(NCO)_2$; MW 262.39; CAS [5124-30-1]

Structure and functional group:

reactive sites: the unsaturated isocyanate functional groups

Synonyms: bis(4-isocyanatocyclohexyl) methane; isocyanic acid methylenedi-4,1-cyclohexylene ester

Uses and Exposure Risk

Methylene bis(4-cyclohexylisocyanate) is used to produce urethane foam with color stability.

Physical Properties

Colorless liquid; freezes below −10°C; density 1.07; soluble in most organic solvents, decomposes with water and ethanol.

Health Hazard

Studies on test animals indicate that methylene bis(4-cyclohexylisocyanate) is highly toxic by an inhalation route and can cause severe skin reaction. Exposure to 20 ppm for 5 hours produced irritation of the respiratory tract, tremor, convulsion, congestion of lungs, and edema in rats. The symptoms at lower levels were decreased as to respiration rate and pulmonary irritation.

Contact with the skin can produce severe irritation, erythema, and edema. The oral toxicity of this compound is very low.

LD_{50} value, oral (rats): 9900 mg/kg

There is no report of the compound's mutagenicity or carcinogenicity.

Exposure Limit

TLV-TWA 0.0535 mg/m^3 (0.005 ppm) (ACGIH).

Fire and Explosion Hazard

Noncombustible; flash point 100°C; vapor pressure 0.4 torr at 150°C. Methylene bis(4-cyclohexylisocyanate) can react vigorously with strong oxidizers, acids, and bases.

Disposal/Destruction

Methylene bis(4-cyclohexylisocyanate) is dissolved in acetone or any other combustible solvent and burned in a chemical incinerator equipped with an afterburner and scrubber.

Analysis

Methylene bis(4-cyclohexylisocyanate) can be analyzed by GC, GC/MS, and HPLC (UV detector). Air analysis can be performed according to NIOSH (1984) Method 5505 (see Section 28.3).

28.7 *n*-BUTYL ISOCYANATE

DOT Label: Flammable Liquid and Poison, UN 2485

Formula C_4H_9CNO; MW 99.15; CAS [111-36-4]

Structure and functional group:

$$H_3C-CH_2-CH_2-CH_2-N=C=O$$

aliphatic   isocyanate,   reactive   site —N=C=O group

Synonyms: isocyanatobutane; isocyanic acid butyl ester

Uses and Exposure Risk

n-Butyl isocyanate is used as an acylating agent in the Friedel–Crafts reaction to produce amide.

Physical Properties

Colorless liquid with mild odor; bp 46°C; freezes below −70°C; density 1.4064 at 20°C; soluble in most organic solvents, decomposes with water and alcohol.

Health Hazard

n-Butyl isocyanate exhibits low inhalation toxicity and relatively higher oral toxicity. This is in contrast to the aromatic isocyanates. The toxic effects are nausea, dyspnea, insomnia, coughing, and chest pain. Such symptoms, however, are much less marked than those of methyl isocyanate.

LC_{50} value, inhalation (mice): 680 mg/m^3
LD_{50} value, oral (mice): 150 mg/kg

There is no report of any carcinogenic or teratogenic study of this compound.

Exposure Limit

No exposure limit is set for this compound. Based on its low inhalation toxicity, a TLV-TWA value of 8 mg/m^3 (2 ppm) can be safely ascribed.

Fire and Explosion Hazard

Flammable liquid; flash point (open cup) 20°C (68°F); vapor density 3.4 (air = 1);

forms explosive mixture with air; LEL and UEL values not available. Fire-extinguishing agent: CO_2, dry chemical, "alcohol" foam.

Reactive with compounds containing active hydrogen atoms. It can undergo brisk to violent reactions with strong oxidizers, acids, and bases.

Disposal/Destruction

n-Butyl isocyanate may be carefully burned in a chemical incinerator equipped with an afterburner and scrubber.

Analysis

HPLC (UV detector), GC-FID, TLC, and GC/MS are the most suitable techniques. Its presence in air can be detected and estimated by passing the air through an impinger, where the isocyanate is converted to an amine derivative, acylated, and analyzed by HPLC at 254 nm.

28.8 ISOPHORONE DIISOCYANATE

DOT Label: Poison B, UN 2290
Formula $C_{10}H_{18}(NCO)_2$; MW 222.3; CAS [4098-71-9]
Structure and functional group:

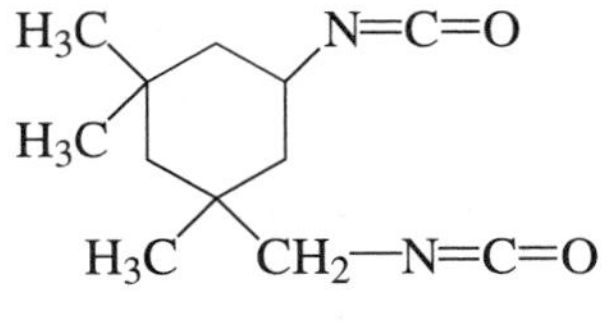

reactive sites: the unsaturated isocyanate functional groups, which readily react with compounds containing active hydrogen atoms

Synonyms: 3-isocyanato methyl-3,5,5-trimethyl cyclohexylisocyanate; 5-isocyanato-1-(isocyanatomethyl)-1,3,3-trimethylcyclohexane; isocyanic acid methylene(3,5,5-trimethyl-3,1-cyclohexylene)ester

Uses and Exposure Risk

Isophorone diisocyanate (IPDI) is used in the production of high-quality coatings, polyurethane paints, and varnishes and as an elastomer for casting compounds.

Physical Properties

Colorless to slightly yellow liquid; bp 158°C at 10 torr; freezes at −60°C; density 1.062 at 20°C; soluble in most organic solvents, decomposes with water and alcohol.

Health Hazard

Like most other isocyanates, IPDI exhibits moderate toxicity via inhalation. The acute toxic symptoms are somewhat similar to those of toluene-2,4-diisocyanate and diphenylmethane-4,4′-diisocyanate. Thus the toxicity of such types is a characteristic of the −N=C=O functional group and to a great extent is independent of the nature of the ring.

Inhalation of its vapor can cause bronchitis, asthma, tightness of chest, and dyspnea in humans. Recovery from these effects may occur in a short period from a low-concentration exposure.

IPDI is an irritant to the skin and eyes. Exposure to this compound produces skin sensitization and eczema. Its oral toxicity is very low.

LC_{50} value, inhalation (rats): 123 mg/m³/4 hr
LD_{50} value, skin (rats): 1060 mg/m³

There is no report of its carcinogenic or reproductive effects in animals or humans.

Exposure Limits

TLV-TWA 0.0454 mg/m³ (0.005 ppm) (ACGIH and NIOSH); ceiling 0.181 mg/m³ (0.02 ppm)/10 min (NIOSH).

TABLE 28.2 Toxicity and Flammability of Miscellaneous Isocyanates

Compound/CAS No.	Carbon Chain	LD$_{50}$ Intraperitoneal (Mice) (mg/kg)	
Ethyl isocyanate (isocyanatoethane, isocyanic acid ethyl ester) [109-90-0]	C$_2$H$_5$NCO 71.09 H$_3$C–CH$_2$–N=C=O	Moderately toxic by inhalation, ingestion, and skin absorption; LD$_{50}$ intravenous (mice): 56 mg/kg; DOT Label: Poison, UN 2481	Flammable liquid, flash point (open cup) −5°C (23°F); forms explosive mixture with air; highly reactive with strong oxidizers, acids and bases; decomposes with water, forming CO$_2$; DOT Label: Flammable Liquid, UN 2481
Propyl isocyanate (isocyanatopropane, isocyanic acid propyl ester) [110-78-1]	C$_3$H$_7$NCO 85.12 H$_3$C–CH$_2$–CH$_2$–N=C=O	Moderately toxic; poisoning effect similar to that of ethyl isocyanate; LD$_{50}$ intravenous (mice): 56 mg/kg; DOT Label: Poison, UN 2482	Flammable liquid, flash point (open cup) 22°C (72°F); highly reactive with strong oxidizers, acids, and bases; forms CO$_2$ with water; DOT Label: Flammable Liquid, UN 2482
Isopropyl isocyanate (2-isocyanatopropane, isocyanic acid isopropyl ester) [1795-48-8]	C$_3$H$_7$NCO 85.12 CH$_3$ \ CH–N=C=O / CH$_3$	Moderately toxic; toxic symptoms similar to those of lower aliphatic isocyanates; LD$_{50}$ data not reported; DOT Label: Poison, UN 2483	Flammable liquid, flash point −1°C (30°F); vapors form explosive mixture with air; highly reactive with strong oxidizers, acids, and bases; DOT Label: Flammable Liquid, UN 2483
Allyl isocyanate (isocyanic acid allyl ester) [1476-23-9]	C$_3$H$_5$NCO 83.12 CH$_2$=CH–CH$_2$–N=C=O	Lachrymator; skin irritant; moderately toxic; LD$_{50}$ intravenous (mice): 18 mg/kg	Combustible liquid, flash point 43°C (110°F); highly reactive, undergoing many addition-type reactions
Phenyl isocyanate (carbanil, phenyl carbonimide, isocyanatobenzene; isocyanic acid phenyl ester, Mondur P) [103-71-9]	C$_6$H$_5$NCO 119.13 (benzene ring)–N=C=O	Eye irritant, can cause lachrymation; low acute inhalation and oral toxicity; LD$_{50}$ oral (rats): 950 mg/kg; DOT Label: Poison B, UN 2487	Combustible liquid, flash point 55°C (132°F) (Aldrich 1989); DOT Label: Flammable Liquid, UN 2487

Benzoyl isocyanate (isocyanic acid benzoyl ester) [4461-33-0]	C_6H_5CONCO 147.13	Eye irritant; toxicity data not available	Combustible liquid, flash point 71°C (160°F); decomposes with water
Durene isocyanate (3-isocyanato-1,2,4,5-tetramethyl benzene) [58149-28-3]	$C_6H_2(CH_3)_4NCO$ 175.25	Exhibited low to moderate toxicity in test animals; LD_{50} intraperitoneal (mice): 83 mg/kg	Noncombustible
Benzene-1,3-diisocyanate (m-phenylene diisocyanate, 1,3-diisocyanatobenzene, isocyanic acid m-phenylene ester) [123-61-5]	$C_6H_4(NCO)_2$ 160.14	Moderate to high toxicity; toxic route—inhalation; LD_{50} data not available; in humans, acute toxic symptoms can be bronchitis, wheezing, congestion in chest, and pulmonary edema—similar to other aromatic diisocyanates; low oral toxicity; LD_{50} intravenous (mice): 5.6 mg/kg exposure limits: TLV-TWA (relative to diisocyanate) 0.0327 mg/m³ (0.005 ppm), ceiling 0.13 mg/m³ (0.02 ppm)/10 min (NIOSH)	Noncombustible

(continued)

TABLE 28.2 *(Continued)*

Compound/Synonyms/CAS No.	Formula/MW/Structure	Toxicity	Flammability
Cyclohexyl isocyanate (isocyanatocyclohexane, isocyanic acid cyclohexyl ester) [3173-53-3]	$C_6H_{11}NCO$ 125.19 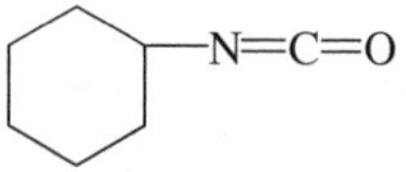 N=C=O (cyclohexyl)	Moderate to high toxicity in test animals when administered intravenously and intraperitoneally; no data on inhalation toxicity; LD_{50} intraperitoneal (mice): 13 mg/kg; DOT Label: Poison B, UN 2488	Combustible liquid, flash point 48°C (120°F); DOT Label: Flammable Liquid, UN 2488
p-Chlorophenyl isocyanate (isocyanic acid p-chlorophenyl ester) [104-12-1]	C_6H_4ClNCO 153.57 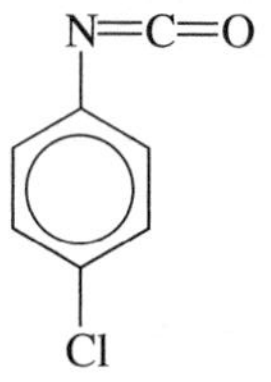 N=C=O (phenyl–Cl)	Highly poisonous when vapors or particulates inhaled; in humans, toxic symptoms characteristic of aromatic isocyanates seen at 1-min exposure to 1 mg/m³ (0.16 ppm); LC_{LO} (mice): 40 mg/m³; low oral toxicity; skin and eye contact can produce severe irritation; no exposure limit is set, a TLV-TWA of 0.005 ppm (0.031 mg/m³) is recommended	Noncombustible solid/liquid; flash point 110°C (230°F)
3,4-Dichlorophenyl isocyanate (isocyanic acid 3,4-dichlorophenyl ester) [102-36-3]	$C_6H_3Cl_2NCO$ 188.01 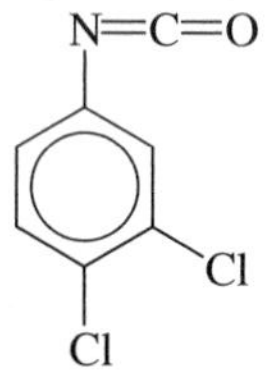 N=C=O (phenyl–Cl,Cl)	Moderately toxic in test animals via inhalation route; exposure to 18 ppm (140 mg/m³) for 2 hr was fatal to mice following respiratory tract irritation and lung injury; no report on human toxicity	Noncombustible solid

528

Name (CAS)	Formula / Structure	Toxicity	Properties
2-Fluoroethyl isocyanate (isocyanic acid 2-fluoroethyl ester) [505-12-4]	C₂H₄FNCO 89.08 $F–CH_2–CH_2–N=C=O$	Exhibited moderate to high toxicity in mice when given intraperitoneally; LD_{50} intraperitoneal (mice): 17 mg/kg	Combustible liquid
3-Fluoropropyl isocyanate (isocyanic acid 3-fluoropropyl ester) [407-99-8]	C₃H₆FNCO 103.11 $F–CH_2–CH_2–CH_2–N=C=O$	More toxic than the fluoroethyl isocyanate; LD_{50} intraperitoneal (mice): 10 mg/kg	Combustible liquid
4-Fluoropropyl isocyanate (isocyanic acid 4-fluorobutyl ester) [353-16-2]	C₄H₈FNCO 117.14 $F–CH_2–CH_2–CH_2–CH_2–N=C=O$	Highly toxic when administered intraperitoneally in mice; LD_{50} intraperitoneal (mice): 4.7 mg/kg	Combustible liquid
4-Fluorophenyl isocyanate (isocyanic acid 4-fluorophenyl ester) [1195-45-5]	C₆H₄FNCO 137.11	Eye irritant; toxicity data not available	Combustible liquid; flash point 52°C (127°F) (Aldrich 1989)

Fire and Explosion Hazard

Noncombustible; flash point 100°C; vapor pressure 0.0003 torr at 20°C. It reacts briskly with strong oxidizers, acids, bases, phenols, amines, and so on.

Disposal/Destruction

Disposal is by chemical incineration of IPDI solution in a combustible solvent.

Analysis

HPLC (UV detection) of an IPDI derivative is suitable for air analysis. Other instrumental methods include GC-FID and GC/MS.

28.9 MISCELLANEOUS ISOCYANATES

See Table 28.2.

REFERENCES

ACGIH. 1986. *Documentation of the Threshold Limit Values and Biological Exposure Indices*, 5th ed. Cincinnati, OH: American Conference of Governmental Industrial Hygienists.

Aldrich. 1989. *Aldrich Catalog*. Milwaukee, WI: Aldrich Chemical Company.

Bucher, J. R., B. N. Gupta, B. Adkins, Jr., M. Thompson, C. W. Jameson, J. E. Thigpen, and B. A. Schwetz. 1987a. Toxicity of inhaled methyl isocyanate in F344/N rats and B6C3F1 mice. I. Acute exposure and recovery studies. *Environ. Health Perspect.* 72: 53–61.

Bucher, J. R., B. N. Gupta, M. Thompson, B. Adkins, Jr., and B. A. Schwetz. 1987b. Toxicity of inhaled methyl isocyanate. II. Repeated exposure and recovery studies. *Environ. Health Perspect.* 72: 133–38.

Chadwick, D. H., and T. H. Cleveland. 1981. Isocyanates, organic, in *Kirk-Othmer Encyclopedia of Chemical Technology*, Vol. 13, pp. 798–818. New York: Wiley.

Dutta, K. K., G. S. D. Gupta, A. Mishra, G. S. Tandon, and P. K. Ray. 1988. Inhalation toxicity studies of methyl isocyanate (MIC) in rats. I. Pulmonary pathology and genotoxicity evaluation. *Indian J.Exp. Biol.* 26(3): 177–82.

Gupta, B. N., S. K. Rastogi, H. Chandra, A. K. Mathur, N. Mathur, P. N. Mahendra, B. S. Pangtey, S. Kumar, and P. Kumar. 1988. Effect of exposure to toxic gas on the population of Bhopal. I. Epidemiological, clinical, radiological and behavioral studies. *Indian J. Exp. Biol.* 26(3): 149–60.

Hadengue, P., and M. Philbert. 1983. Isocyanates, in *Encyclopedia of Health and Occupational Safety*, 3rd ed., Vol. 2, ed. L. Parmeggiani. pp. 1161–62. Geneva: International Labor Office.

Jeevarathinam, K., W. Seevamurthy, U. S. Ray, S. Mukhopadhyay, and L. Thakur. 1988. Acute toxicity of methyl isocyanate, administered subcutaneously in rabbits: changes in physiological, clinico-chemical and histological parameters. *Toxicology* 51(2–3): 223–40.

McConnell, E. E., J. R. Bucher, B. A. Schwetz, B. N. Gupta, M. D. Shelby, M. I. Luster, A. R. Brody, G. A. Boorman, and C. Richter. 1987. Toxicity of methyl isocyanate. *Environ. Sci. Technol.* 21(2): 188–93.

Meshram, G. P., and K. M. Rao. 1987. Mutagenic and toxic effects of methylisocyanate (MIC) in *Salmonella typhimurium*. *Indian J. Exp. Biol.* 28(8): 548–50.

Mishra, U. K., D. P. Nath, W. A. Khan, B. N. Gupta, and P. K. Ray. 1988. A clinical study of toxic gas poisoning in Bhopal, India. *Indian J. Exp. Biol.* 26(3): 201–4.

NFPA. 1997. *Fire Protection Guide on Hazardous Materials*, 12th ed. Quincy, MA: National Fire Protection Association.

NIOSH. 1984. *Manual of Analytical Methods*, 3rd ed. Cincinnati: National Institute for Occupational Safety and Health.

NIOSH. 1986. *Registry of Toxic Effects of Chemical Substances*, ed. D. V. Sweet. Washington, DC: U.S. Government Printing Office.

Prasad, R., and R. K. Pandey. 1985. Methyl isocyanate hazard to the vegetation of Bhopal. *J. Trop. For.* 1(1): 40–50.

Rastogi, S. K., B. N. Gupta, T. Hussain, A. Kumar, S. Chandra, and P. K. Ray. 1988. Effect of exposure to toxic gas on the population of Bhopal. II. Respiratory impairment. *Indian J. Exp. Biol.* 26(3): 161–64.

Salmon, A. G., M. K. Muir, and N. Anderson. 1985. Acute toxicity of methyl isocyanate: a preliminary study of the dose response for eye and other effects. *Br. J. Ind. Med. 42* (12): 795–98.

Saxena, A. K., K. P. Singh, S. L. Nagle, B. N. Gupta, P. K. Ray, R. K. Srivastav, S. P. Tewari, and R. Singh. 1988. Effect of exposure to toxic gas on the population of Bhopal. IV. Immunological and chromosomal studies. *Indian J. Exp. Biol. 26* (3): 173–76.

Sethi, N., R. Dayal, and R. K. Singh. 1989. Acute and subacute toxicity study of inhaled methyl isocyanate in Charles Foster rats. *Ecotoxicol. Environ. Saf. 18* (1): 68–74.

Srivastava, R. C., B. N. Gupta, M. Athar, J. R. Behari, R. S. Dwivedi, A. Singh, M. Mishra, and P. K. Ray. 1988. Effect of exposure to toxic gas on the population of Bhopal. III. Assessment of toxic manifestation in humans: hematological and biochemical studies. *Indian J. Exp. Biol. 26* (3): 165–72.

Varma, D. R., J. S. Ferguson, and Y. Alarie. 1988. Inhibition of methyl isocyanate toxicity in mice by starvation and dexamethasone but not by sodium thiosulfate, atropine and ethanol. *J. Toxicol. Environ. Health 24* (1): 93–101.

Vijayaraghavan, R., and M. P. Kaushik. 1987. Acute toxicity of methyl isocyanate and ineffectiveness of sodium thiosulfate in preventing its toxicity. *Indian J. Exp. Biol. 25* (8): 531–34.

29

KETONES

29.1 GENERAL DISCUSSION

Ketones are a class of organic compounds containing a carbonyl group ($\diagdown$ C=O ; an oxygen atom doubly bound to a single carbon atom) where the carbon atom of the carbonyl group is bound only to alkyl, alkenyl, acyl, and/or aryl groups. Thus the structure of a ketone is

$$\begin{array}{c} R \diagdown \\ C{=}O \\ R' \diagup \end{array}$$

where R and R′ may be alkyl, alkenyl, acyl, and/or aryl groups. That is, the ketone must contain a carbonyl functional group similar to an aldehyde, the only difference being that in aldehydes the carbonyl group is bound to a hydrogen atom.

Toxicity

In general, toxicity caused by the carbonyl keto group is much lower than that of other functional groups, such as cyanides or amines. Unlike aldehydes and alcohols, some of the simplest ketones are less toxic than higher ones. Thus an increase in toxicity is

observed from acetone to methyl isobutyl ketone among saturated ketones. Beyond C_6–C_7 the toxicity goes down and the higher ones are almost nontoxic. Substitution or unsaturation in the molecule can alter the toxicity significantly. An increase in toxicity would depend on the nature of the substituent or degree of unsaturation.

Ketones containing other functional groups sometimes exhibit toxicity, which may be attributed to the presence of such functional groups.

Ketene, which is unsaturated and unstable, is the most hazardous of all ketones. The toxicity data on aromatic ketones are too scant to postulate a structure–toxicity relationship. Among the aromatic ketones, benzoquinones have relatively higher toxicity.

Flammability

Simplest ketones are highly flammable. The flammability decreases with increase in carbon number. Ketones containing seven to nine carbon atoms are combustible, having flash points ranging between 100 (38°C) and 200°F (93°C). Higher ones are noncombustible. Aromatic ketones are solids, and most are noncombustible.

Disposal/Destruction

Incineration is cited exclusively as a method of destruction, applicable to neat compounds or waste solvents. Other thermal methods, such as molten metal salt treatment, which involves intimate contact with a molten salt, such as Al_2O_3 (Shultz 1985), are suitable. Chemical processes that may be effective are wet air oxidation, electrochemical oxidation, and catalytic destruction. Ketones in aqueous wastes can be altered to innocuous gases by heating at 300–460°C and 150–400 atm pressure with or without catalyst. Ni and Fe_2O_3 were found to be effective catalysts in such thermal treatments (Baker and Sealock 1988).

Hazardous Reaction Products

Halogenation of ketones may produce haloketones, many of which are toxic. Molecular chlorine, bromine, and occasionally iodine in alkaline or acidic solution react to form haloketones. In a strongly acidic solution (60% H_2SO_4) the halogenation reaction is fast.

Ketones having an α-methylene group, that is, $-CH_2$ bound to the carbonyl group, can react with nitrous acid, nitrite esters, or nitrosyl chloride to form nitroso compounds and oximes. These reactions are catalyzed by acids or bases.

Ketones can be reduced to secondary alcohols by H and a catalyst or a complex metal hydride or by hydrogen transfer reagents. Deoxygenation with a strong base at 150°C can form olefin. Ketones react with thiols to form thioacetals. Reaction with anhydrous hydrazine may be explosive.

Some of the ketones may react violently with strong oxidizers. Explosion hazards are discussed below for individual ketones.

Analysis

Qualitative and quantitative determinations of ketones can be performed by several instrumental techniques, such as GC, LC, and MS. A capillary column GC with FID or HPLC with a UV detector at 254 nm absorbance were suitable for ketone analyses (Nomeir and Abou-Donia 1985). Among other methods, IR and UV spectroscopy and ^{1}H- and ^{13}C-NMR spectroscopy are useful for the determination of carbonyl functional groups and other structural features.

C=O stretching frequency in the IR spectrum is characterized by a strong peak at about the 1720-cm^{-1} region. The presence of electronegative groups on the α-carbon atom, aromatic rings, or conjugate unsaturation in the ketone molecule can change the position of the peak.

Very strong UV absorption in the wavelength region 215–275 nm and weak absorption at 270–320 nm are characteristic of conjugated unsaturated ketones. In ^{13}C-NMR, the carbonyl group gives a strong signal at 200–220 ppm, downfield from tetramethyl silane.

29.2 KETENE

Formula C_2H_2O; MW 42.04; CAS [463-51-4]

Structure and functional group: CH_2=C=O, simplest unsaturated ketone, reactive site: olefinic double bond and the carbonyl ($\diagdown$C=O) group, polymerizes readily

Synonyms: ethenone; carbomethene; ketoethylene

Uses and Exposure Risk

Ketene is used as an acetylating agent in the production of cellulose acetate, aspirin, acetic anhydride, and in various organic syntheses.

Physical Properties

Colorless gas with a sharp penetrating odor; liquefies at −56°C; solidifies at −151°C;

soluble in alcohol and acetone, decomposed by water.

Health Hazard

Ketene is a highly toxic gas. It causes severe irritation to the eyes, nose, throat, and skin. Exposure to 10–15 ppm for several minutes can injure the respiratory tract. It causes pulmonary edema. A 30-minute exposure to 23 ppm was lethal to mice and a 10-minute exposure to 200 and 750 ppm caused death to monkeys and cats.

Exposure Limits

TLV-TWA 0.9 mg/m^3 (0.5 ppm); STEL 3.0 mg/m^3 (1.5 ppm) (ACGIH); IDLH 50 ppm (NIOSH).

Fire and Explosion Hazard

Ketene in its gaseous state should be flammable and explosive in air. The pure compound, however, polymerizes readily and cannot be stored as a gas. Its flash point and LEL and UEL values are not reported. It can react violently with oxidizers and many organic compounds. Its small size and the olefinic unsaturation imparts further reactivity to the molecule.

Analysis

The gas is bubbled through a bubbler containing hydroxyl ammonium chloride and analyzed by colorimetry.

29.3 ACETONE

EPA Classified Priority Pollutant, RCRA Hazardous Waste Number U002; DOT Label: Flammable Liquid UN1090

Formula C_2H_6CO; MW 58.08; CAS [67-64-1]

Structure and functional group:

$$H_3C-\underset{\underset{O}{\|}}{C}-CH_3$$

contains a keto group; the simplest aliphatic ketone

Synonyms: 2-propanone; dimethyl ketone; ketone propane; β-ketopropane; dimethylketal

Uses and Exposure Risk

Acetone is used in the manufacture of a large number of compounds, such as acetic acid, chloroform, mesityl oxide, and methyl isobutyl ketone; in the manufacture of rayon, photographic films, and explosives; as a common solvent; in paint and varnish removers; and for purifying paraffins.

Physical Properties

Colorless liquid with a characteristic pungent odor; sweetish taste; bp 56.5°C; mp −94°C; density 0.788 (at 25°C); mixes readily in water, alcohol, chloroform, and ether.

Health Hazard

The acute toxicity of acetone in test animals was found to be low. However, in large dosages it could be moderately toxic. Inhalation can cause irritation of the eyes, nose, and throat. A short exposure for 5 minutes to 300–500 ppm can be slightly irritating to humans. The odor threshold is 200–400 ppm. In high concentrations it can produce dryness in the mouth, fatigue, headache, nausea, dizziness, muscle weakness, loss of coordinated speech, and drowsiness. Ingestion can cause headache, dizziness, and dermatitis. Swallowing 200–250 mL can result in vomiting, stupor, and coma in humans. A dose of 25 mg/kg may be fatal to dogs.

LD$_{50}$ value, oral (rats): 5800 mg/kg

No mutagenic or carcinogenic effects have been reported.

Exposure Limits

TLV-TWA 1780 mg/m^3 (750 ppm), STEL 2375 mg/m^3 (ACGIH); 10 hr–TWA 590 mg/m^3 (250 ppm); IDLH 20,000 ppm (NIOSH).

Fire and Explosion Hazard

Highly flammable liquid; flash point (closed cup) −17.8°C (0°F); vapor density 2 (air = 1); vapor pressure 180 torr at 20°C; the vapor can travel some distance to a source of ignition and flash back; autoignition temperature 465°C (869°F); fire-extinguishing agent: dry chemical, "alcohol" foam, or CO$_2$; water may be used to flush and dilute the spill. It forms an explosive mixture with air within the range 2.6–12.8% by volume of air. It explodes when mixed with a nitric–sulfuric acid mixture in a narrow container. It may catch fire and/or explode with the following compounds: chromium trioxide; sulfuric acid–potassium dichromate; chloroform and a base, such as caustic soda; hexachloromelamine; hydrogen peroxide; nitrosyl chloride in the presence of Pt catalyst; and nitrosyl perchlorate. A violent explosion may occur when acetone is mixed with Caro's acid (permonosulfuric acid); or thiodiglycol–H$_2$O$_2$. It ignites with potassium *tert*-butoxide. An explosion was reported when acetone was reacted with sodium hypobromite (haloform reaction) (NFPA 1997).

Disposal/Destruction

Acetone is burned in a chemical incinerator equipped with an afterburner and scrubber. Small amounts may be disposed down the drain with an excess of water. Acetone may be destroyed by oxidizing with potassium permanganate (National Research Council 1995). A 20% excess of permanganate solution is slowly added to an aqueous solution of acetone; the mixture is refluxed. When the solution decolorizes, a small amount of permanganate is further added and the mixture is heated. It is then cooled and a 6 *N* solution of H$_2$SO$_4$ is added cautiously, followed by enough solid sodium sulfite to reduce the manganese to the divalent state. This would be indicated by the loss of purple color. The nontoxic products formed are washed down the drain. For land disposal, the concentration of acetone in wastewater should be below 0.05 ppm, and less than 0.59 ppm in the TCLP extract of spent solvent waste. The waste treatment technique is steam stripping (U.S. EPA 1988).

Analysis

Acetone is analyzed by GC-FID. A Carbowax 20 M, Carbopack, fused-silica capillary column, or any equivalent column may be suitable. It may be analyzed by GC/MS using a purge and trap technique or by direct injection. Characteristic ions are 43 and 58 (electron-impact ionization) (U.S. EPA 1986, Method 8240). Acetone in air is adsorbed over coconut shell charcoal, desorbed with CS$_2$, and analyzed by GC-FID using a 10% SP-2100 or DB-1 fused-silica capillary column (NIOSH 1984, Method 1300). Airflow rate: 10 to 200 mL/min; volume: 0.5–3 L of air. Trace acetone in water can be determined by the fast HPLC method discussed in Chapter 5.

29.4 METHYL ETHYL KETONE

EPA Classified Toxic Waste, RCRA Waste Number U159; DOT Label: Flammable Liquid UN 1193

Formula C$_4$H$_8$O: MW 72.11; CAS [78-93-3]

Structure and functional group:

$$CH_3\!-\!\underset{\underset{O}{\|}}{C}\!-\!CH_2\!-\!CH_3$$

reactive site: carbonyl keto group ($\overset{\diagdown}{\underset{\diagup}{C}}\!=\!O$)

Synonyms: 2-butanone; butan-2-one; ethyl methyl ketone

Uses and Exposure Risk

Methyl ethyl ketone (MEK) is used in the manufacture of smokeless powder and colorless synthetic resins, as a solvent, and in surface coating.

Physical Properties

Colorless liquid with acetone-like smell; bp 79.6°C; mp −86°C; density 0.8054 at 20°C; soluble in water, alcohol, ether, acetone, and benzene.

Health Hazard

Inhalation of MEK can cause irritation of the eyes and nose and headache. Exposure to 300 ppm for several hours may have a mildly irritating effect on humans. At high concentrations it is narcotic. Ingestion can cause dizziness and vomiting. Serious ill effects from poisoning is low.

LD$_{50}$ value, oral (rats): 5500 mg/kg

Odor threshold detection: 10 ppm.

Exposure Limits

TLV-TWA 590 mg/m^3 (200 ppm); STEL 885 mg/m^3 (300 ppm) (ACGIH); IDLH 3000 ppm (NIOSH).

Fire and Explosion Hazard

Flammable liquid; flash point (closed cup) −6°C (21°F); vapor density 2.5 (air = 1); vapor pressure 77.5 torr at 20°C; at this vapor pressure and temperature it is saturated in air to a concentration of 102,000 ppm; the vapor can travel some distance to an ignition source and flash back; autoignition temperature 515.6°C (960°F); fire-extinguishing agent: "alcohol" foam; a water spray may be effective in absorbing the heat and diluting the material. MEK forms an explosive mixture with air; LEL and UEL values 1.8% and 12% by volume in air. It may react explosively with strong oxidizers. It ignites when reacted with potassium *tert*-butoxide.

Disposal/Destruction

MEK is burned in a chemical incinerator equipped with an afterburner and scrubber. Land disposal: concentration of MEK in wastewater should not exceed 0.05 ppm, and that in TCLP extract of spent solvent wastes, 0.75 ppm; the waste treatment is by steam stripping (U.S. EPA 1988).

29.5   ACETYL ACETONE

DOT Label: Flammable Liquid UN 2310
Formula C$_5$H$_8$O$_2$; MW 100.12; CAS [123-54-6]
Structure and functional group:

$$CH_3-\underset{\underset{O}{\|}}{C}-CH_2-\underset{\underset{O}{\|}}{C}-CH_3$$

reactive sites: two carbonyl keto (C=O) groups
Synonyms: 2,4-pentanedione; diacetyl methane; acetoacetone

Uses and Exposure Risk

Acetyl acetone is used as a reagent for organic synthesis and as a transition metal chelating agent. Its organometallic complexes are used as additives for gasoline and lubricants, and in varnishes, color, ink, and fungicides.

Physical Properties

Colorless liquid with a pleasant smell; bp 140.5°C freezes at −23°C; density 0.972 at 25°C; soluble in alcohol, ether, acetone, and chloroform, moderately soluble in water (12.5 g/100 g water).

Health Hazard

Exposure to the vapors of acetyl acetone can cause irritation of the eyes, mucous membrane, and skin. In rabbits 4.76 mg produced severe eye irritation; the effect on skin was mild. Other than these, the health hazards from this compound have not been reported. However, based on its structure and the fact that it has two reactive carbonyl groups in the molecule, this compound should exhibit low to moderate toxicity at high concentrations, which should be greater than that of the C_5-monoketones.

LD_{50} value, intraperitoneal (mice): 750 mg/kg

LD_{50} value, oral (rats): 1000 mg/kg

There is no report on its carcinogenicity in animals or humans.

Exposure Limit

No exposure limit has been set.

Fire and Explosion Hazard

Flammable liquid; flash point (closed cup) 34°C (93°F); vapor density 3.45 (air = 1); fire-extinguishing agent: "alcohol" foam; a water spray may be used to flush and dilute the spill. It may form explosive mixtures with air; LEL and UEL values are not reported. It may react violently with strong oxidizers.

Disposal/Destruction

Incineration and molten metal treatment are effective ways to destroy this compound. Potassium permanganate oxidation is a suitable method of destruction in the laboratory.

Analysis

GC-FID, GC/MS, and HPLC are the suitable instrumental techniques for acetyl acetone analysis.

29.6 METHYL PROPYL KETONE

DOT Label: Flammable Liquid UN 1249

Formula $C_5H_{10}O$; MW 86.13, CAS [107-87-9]

Structure and functional group:

$$CH_3-\underset{\underset{O}{\|}}{C}-CH_2-CH_2-CH_3$$

reactive site: carbonyl keto group ($\diagdown C{=}O$)

Synonyms: 2-pentanone; ethyl acetone

Uses and Exposure Risk

Methyl propyl ketone (MPK) is used as a solvent, in organic synthesis, and as a flavoring agent.

Physical Properties

Colorless liquid with a characteristic pungent odor; bp 102.2°C; freezes at −78°C; density 0.809 at 20°C; soluble in alcohol and ether, slightly soluble in water.

Health Hazard

Inhalation of MPK vapors can cause narcosis and irritation of the eyes and respiratory tract. Chronic poisoning from this compound is not known. Exposure to 1500 ppm was severely irritating to humans and 2000 ppm for 4 hours was fatal to rats. In guinea pigs 5000 ppm produced coma.

Exposure Limits

TLV-TWA 700 mg/m^3 (200 ppm); STEL 875 mg/m^3 (250 ppm) (ACGIH).

Fire and Explosion Hazard

Flammable liquid; flash point (closed cup) 7°C (45°F); vapor density 3 (air = 1); vapor pressure 27 torr at 20°C; the vapor can travel a considerable distance to an ignition source and flash back; autoignition temperature 452°C (846°F); fire-extinguishing agent: "alcohol" foam; a water spray may be used to absorb the heat and flush the spill away from exposures. MPK forms an explosive mixture with air in the range 1.5–8.2% by volume of air. Heating with oxidizers can cause an explosion (U.S. EPA 1988).

Disposal/Destruction

MPK is burned in a chemical incinerator equipped with an afterburner and scrubber. Waste treatment is by steam stripping.

Analysis

GC-FID, GC/MS, and HPLC (see Section 29.10).

29.7 METHYL ISOPROPYL KETONE

DOT Label: Flammable Liquid UN 2397
Formula $C_5H_{10}O$; MW 86.14; CAS [563-80-4]
Structure and functional group:

$$CH_3-\underset{\underset{O}{\|}}{C}-\underset{\underset{CH_3}{|}}{CH}-CH_3$$

reactive site: carbonyl keto group ($\diagdown C=O$)

Synonyms: 3-methyl-2-butanone

Uses and Exposure Risk

Methyl isopropyl ketone (MIPK) is used as a solvent for lacquers.

Physical Properties

Colorless liquid; bp 94°C; mp −92°C; density 0.8051 at 20°C; soluble in alcohol, ether, and acetone, slightly soluble in water.

Health Hazard

The acute toxic effects from exposure to the vapors of methyl isopropyl ketone are narcosis and irritation of the eyes, skin, and respiratory passages. In rabbits the irritation effect of 100 mg for a 24-hour exposure was mild in the eyes, while 500 mg was moderate on the skin. The toxicity data on this compound are too scant. In general, its toxicity is low and narcosis symptoms in humans may be felt only at very high concentrations. In guinea pigs exposure to 5000 ppm caused coma.

Exposure Limits

TLV-TWA 700 mg/m^3 (200 ppm); STEL 875 mg/m^3 (250 ppm) (ACGIH).

Fire and Explosion Hazard

Flammable liquid; flash point (closed cup) 7°C (45°F); vapor density 3 (air = 1); vapor pressure 27 torr at 20°C; vapor can travel some distance to nearby ignition source and flash back; fire-extinguishing agent: dry chemical, "alcohol" foam, or CO$_2$; a water spray can be used in diluting and flushing the spill away from exposure. MIPK forms an explosive mixture with air within the range 1.6–8.2% by volume in air.

Disposal/Destruction

MIPK is burned in a chemical incinerator equipped with an afterburner and scrubber.

Analysis

GC-FID, GC/MS, and HPLC (see Section 29.10).

29.8 DIETHYL KETONE

DOT Label: Flammable Liquid UN 1156
Formula $C_5H_{10}O$; MW 86.13; CAS [96-22-0]

Structure and functional group:

$$CH_3-CH_2-\underset{\underset{O}{\|}}{C}-CH_2-CH_3$$

reactive group: carbonyl keto group ($\diagup C=O$)

Synonyms: 3-pentanone; propione; dimethyl acetone; methacetone; pentan-3-one

Uses and Exposure Risk

Diethyl ketone is used as a solvent, in medicine, and in organic synthesis.

Physical Properties

Colorless liquid with a characteristic pungent odor; bp 101.7°C; mp −39.8°C; density 0.814 at 20°C; soluble in alcohol and acetone, slightly soluble in water.

Health Hazard

Diethyl ketone is a mild narcotic compound as well as an irritant. Its acute toxicity is less than that of methyl propyl ketone. Exposure to 80,000 ppm for 4 hours was fatal to rats.

LD$_{50}$ value, oral (rats): 2140 mg/kg

Exposure Limit

TLV-TWA 705 mg/m^3 (200 ppm) (ACGIH).

Fire and Explosion Hazard

Flammable liquid; flash point (open cup) 13°C (55°F); vapor density 3 (air = 1); the vapor can travel some distance to an ignition source and flashback; autoignition temperature 450°C (842°F); fire-extinguishing agent: "alcohol" foam; a water spray can be used to flush and dilute the spill. Diethyl ketone forms explosive mixtures with air; the LEL value is 1.6% by volume of air, the UEL value is not reported.

Disposal/Destruction

Incineration; molten salt treatment.

Analysis

GC-FID, GC/MS, and HPLC (see Section 29.4).

29.9 MESITYL OXIDE

DOT Label: Flammable Liquid UN 1229
Formula C$_6$H$_{10}$O: MW 98.14; CAS [141-79-7]
Structure and functional group:

$$\underset{H_3C}{\overset{H_3C}{\diagup}}C=CH-\underset{\underset{O}{\|}}{C}-CH$$

reactive site: carbonyl keto (C=O) group and the olefinic double bond

Synonyms: methyl isobutenyl ketone; isobutenyl methyl ketone; 4-methyl-3-pentene-2-one; isopropyledene acetone

Uses and Exposure Risk

Mesityl oxide is used as a solvent for resins, gums, nitrocellulose, oils, lacquers, and inks; as an insect repellant; and in ore flotation.

Physical Properties

Colorless oily liquid with sweet odor; bp 129.7°C; mp −52.8°C; density 0.8653 at 20°C; soluble in alcohol, ether, and acetone, slightly soluble in water (3.3%).

Health Hazard

Mesityl oxide is a moderately toxic substance, more toxic than the saturated ketones. Inhalation of its vapors can cause irritation to the eyes, skin, and mucous membranes. In humans, the irritation effect on the eyes and nose are reported to be 25 and 50 ppm. At high concentrations narcosis can result. Other than narcosis, prolonged exposure to high concentrations can injure the lungs, liver, and kidney. A concentration of 2500 ppm was lethal to rats.

LD_{50} value, oral (rats): 1120 mg/kg

Exposure Limits

TLV-TWA 60 mg/m^3 (15 ppm) (ACGIH), 10-hr TWA 40 mg/m^3 (10 ppm) (NIOSH); STEL 100 mg/m^3 (25 ppm); IDLH 5000 ppm.

Fire and Explosion Hazard

Flammable liquid; flash point (closed cup) 31°C (87°F); vapor pressure 8 torr at 20°C; vapor density 3.4 (air = 1); autoignition temperature 344°C (652°F); the vapor can travel some distance to a nearby ignition source and flashback; fire-extinguishing agent: "alcohol" foam, CO_2, or dry chemical. Mesityl oxide forms an explosive mixture with air within the range 1.4–7.25% by volume of air. It may react explosively with concentrated nitric and sulfuric acids, strong alkalies, and strong oxidizers.

Disposal/Destruction

Mesityl oxide is burned in a chemical incinerator equipped with an afterburner and scrubber. Waste treatment is by steam stripping, catalytic hydrogenation, or molten salt treatment.

Analysis

GC-FID; GC/MS, and HPLC (see Section 29.7). Air samples are analyzed by NIOSH Method 1301 (see Section 29.15).

29.10 METHYL ISOBUTYL KETONE

EPA Classified Hazardous Waste, RCRA Waste Number U161; DOT Label: Flammable Liquid UN 1245

Formula $C_6H_{12}O$; MW 100.16; CAS [108-10-1]
Structure and functional group:

$$CH_3-\underset{\underset{O}{\|}}{C}-CH_2-\underset{\underset{CH_3}{|}}{CH}-CH_3$$

active site: carbonyl group

Synonyms: hexone; isobutyl methyl ketone; 4-methyl-2-pentanone; isopropyl acetone

Uses and Exposure Risk

Methyl isobutyl ketone (MIBK) is used as a solvent for gums, resins, nitrocellulose and cellulose ethers, and various fats, oils, and waxes.

Physical Properties

Colorless liquid with faint camphor odor; bp 117°C; mp −87.4°C; density 0.801 at 20°C; soluble in alcohol, ether, and benzene, solubility in water is low (1.91%).

Health Hazard

MIBK exhibits low to moderate toxicity. It is more toxic than acetone. Exposure to 200 ppm can cause irritation of the eyes, mucous membranes, and skin. Prolonged skin contact can leach out fat from the skin. Exposure to high concentrations can cause nausea, headache, and narcosis. Animal studies indicate that this compound could probably cause kidney damage, with symptoms of a heavier kidney, higher kidney-to-body weight ratio, and tubular necrosis. An increase in liver weight was noted, too, associated with its exposure in animal subjects. In male rats the effect was observed at 2000 ppm on 2 weeks' exposure (6 hours/day) (Phillips et al. 1987). Other than for the male rat kidney effect, the levels up to 1000 ppm for 14 weeks had no significant toxicological effect. In another study, exposure to 3000 ppm in rats and mice was found to cause increased liver and kidney weights, decrease in food consumption, incidence of dead fetuses, and reduced fetal body weight (Tyl et al. 1987).

Ingestion of MIBK can result in narcosis and coma. A genetic toxicology study of MIBK showed a negative response in the bacterial mutation assays and the yeast mitotic gene conversion assay (Brooks et al. 1988).

Exposure Limits

TLV-TWA 205 mg/m³ (50 ppm); STEL 300 mg/m³ (75 ppm) (ACGIH); IDLH 3000 ppm (NIOSH).

Fire and Explosion Hazard

Flammable liquid; flash point (closed cup) 18°C (64°F); vapor pressure 7.5 torr at 25°C; vapor density 3.4 (air = 1); the vapor can travel a considerable distance to a source of ignition and flashback; autoignition temperature 448°C (840°F); fire-extinguishing agent: dry chemical, "alcohol" foam, or CO₂; a water spray may be used to flush and dilute the spills.

MIBK forms explosive mixtures with air within the range 1.4–7.5% (by volume of air). It ignites when reacted with potassium *tert*-butoxide.

Disposal/Destruction

MIBK is burned in a chemical incinerator equipped with an afterburner and scrubber. It can be destroyed by catalytic hydrogenation and molten salt treatment. The former process is applicable to wastewater. A nickel catalyst at 300°C and 200–300 atm has been reported to be effective (Baker and Sealock 1988). In the laboratory, MIBK can be destroyed by oxidation with potassium permanganate (see Section 29.3).

Land disposal: concentration in wastewater should be less than 0.05 ppm, and that in the TCLP extract of spent solvent waste should be less than 0.33 ppm; waste treatment is by steam stripping (U.S. EPA 1988).

Analysis

MIBK can be analyzed by GC-FID using Carbowax, SP-2100, or a DB-1 fused-silica capillary column. It can be analyzed by HPLC using a UV detector (absorbance at 254 nm) and by GC/MS (purge and trap or direct injection). Air analysis is done by NIOSH Method 1300 (see Section 29.3).

29.11 METHYL BUTYL KETONE

Formula $C_6H_{12}O$; MW 100; CAS [591-78-6]
Structure and functional group:

$$CH_3-\underset{\underset{O}{\|}}{C}-CH_2-CH_2-CH_2-CH_3$$

active site: carbonyl keto group ($\diagdown C{=}O$)

Synonyms: butyl methyl ketone; 2-hexanone; methyl-*n*-butyl ketone

Uses and Exposure Risk

Methyl butyl ketone (MBK) is used as a solvent for nitrocellulose, resins, lacquers, oils, fats, and waxes.

Physical Properties

Colorless liquid with pungent odor; bp 127°C; freezes at −59.6°C; density 0.821 at 20°C; soluble in alcohol, ether, and benzene, slightly soluble in water (1.4%).

Health Hazard

Chronic exposure to MBK via inhalation or skin absorption can cause disorders of the peripheral nervous system and neuropathic diseases. The symptoms in humans, depending on the severity of exposure, were parasthesias in the hands or feet, muscle weaknesses, weakness in ankles and hand, and difficulty in grasping heavy objects. Animal studies indicate that prolonged exposure to this compound at a concentration of 200–1300 ppm for 4–12 weeks can result in neuropathy. Neurotoxicity is attributed to its primary metabolite, 2,5-hexanedione. Improvement in conditions can occur slowly over months after cessation of exposure. Acute exposure to high concentrations, 1000 ppm, for several minutes can cause irritation of the eyes and nose in humans. Inhalation to 20,000 ppm for 30 minutes caused narcosis in guinea pigs and prolonged exposure for 70 minutes was lethal. The acute oral toxicity in rats was low.

LD_{50} value, oral (rats): 2600 mg/kg

Fire and Explosion Hazard

Flammable liquid; flash point (closed cup) 25°C (77°F); vapor density 3.4 (air = 1); vapor pressure 3.8 torr at 25°C; the vapor can travel some distance to an ignition source and flash back; autoignition temperature 423°C (795°F); fire-extinguishing agent: dry chemical, "alcohol" foam, or CO_2; a water spray may be used to flush and dilute the spill. It forms an explosive mixture with air within the range 1.2–8.0% by volume of air. Its reactions with strong oxidizers can be violent.

Disposal/Destruction

MBK is burned in a chemical incinerator equipped with an afterburner and scrubber. In the laboratory it can be destroyed by potassium permanganate oxidation (see Section 29.3).

Analysis

MBK can be analyzed by GC-FID, HPLC, or GC/MS techniques. Air analysis may be performed by NIOSH (1984) Method 1301 (see Section 29.15).

29.12 4-HYDROXY-4-METHYL-2-PENTANONE

DOT Label: Flammable Liquid, UN 1148

Formula $C_6H_{12}O_2$; MW 116.16; CAS [123-42-2]

Structure and functional group:

molecule has a carbonyl, $\diagdown C{=}O$ and a tertiary —OH group

Synonyms: diacetone alcohol; 2-methyl-2-pentanol-4-one; tyranton; 4-methyl-2-pentanon-4-ol

Uses and Exposure Risk

4-Hydroxy-4-methyl-2-pentanone is used as a solvent for nitrocellulose, cellulose acetate, resins, fats, oils, and waxes; and in hydraulic fluids and antifreeze solutions.

Physical Properties

Colorless liquid with a faint pleasant odor; bp 164°C; freezes at −44°C; density 0.9387 at 20°C; soluble in water, alcohol, and ether.

Health Hazard

4-Hydroxy-4-methyl-2-pentanone is a mild irritant and a strong narcotic. It can cause irritation in the eyes, nose, throat, and skin. The effect on humans, however, is mild at 100 ppm concentration.

Animal experiments indicated that it could produce sleep after a period of restlessness and excitement. The symptoms of its toxicity are a marked decrease in breathing and blood pressure, and relaxation of the muscles. Ingestion of this compound in high doses can damage corneal tissue and liver. The oral toxicity in rats was very low, with a LD_{50} value of 4000 mg/kg.

Exposure Limits

TLV-TWA 240 mg/m^3 (50 ppm) (ACGIH); IDLH 2100 ppm (NIOSH).

Fire and Explosion Hazard

Combustible liquid; flash point (closed cup) 58°C (136°F); vapor pressure 1 torr at 22°C; vapor density 4 (air = 1); autoignition temperature 643°C (1190°F); fire-extinguishing agent: "alcohol" foam; water may be used to dilute and flush the spill.

Disposal/Destruction

4-Hydroxy-4-methyl-2-pentanone is burned in a chemical incinerator equipped with an afterburner and scrubber.

Analysis

GC-FID, HPLC, and GC/MS. Air analysis can be done by NIOSH (1984) Method 1301 (see Section 29.15).

29.13   METHYL ISOAMYL KETONE

DOT Label: Flammable Liquid UN 2302
Formula C$_7$H$_{14}$O; MW 114.21; CAS [110-12-3]
Structure and functional group:

$$CH_3-\underset{\underset{O}{\|}}{C}-CH_2-CH_2-CH\underset{CH_3}{\overset{CH_3}{<}}$$

reactive site: carbonyl keto group

Synonyms: 5-methyl-2-hexanone; 2-methyl-5-hexanone; 5-methyl hexan-2-one

Uses and Exposure Risk

Methyl isoamyl ketone (MIAK) is used as a solvent for polymers, cellulose esters, and acrylics.

Physical Properties

Colorless liquid with a sweet odor; bp 144°C; mp −74°C; density 0.888 at 20°C; refractive index 1.4062; soluble in alcohol, ether, acetone, and benzene, slightly soluble in water.

Health Hazard

Exposure to MIAK can produce a strong narcotic effect and irritation of the eyes and respiratory tract. The oral toxicity, however, was found to be low in rats: 1670 mg/kg.

Exposure Limit

TLV-TWA 240 mg/m^3 (50 ppm) (ACGIH).

Fire and Explosion Hazard

Flammable liquid; flash point (closed cup) 36°C (96°F); vapor density 3.9 (air = 1); vapor pressure 4.5 torr at 20°C; autoignition temperature 425°C; fire-extinguishing agent: "alcohol" foam or water spray. MIAK forms an explosive mixture with air at an elevated temperature (93°C) within the range 1–8.2% by volume of air.

Analysis

GC-FID, GC/MS, and HPLC (see Section 29.15).

29.14   DIPROPYL KETONE

DOT Label: Flammable Liquid UN 2710
Formula C$_7$H$_{14}$O; MW 114.19; CAS [123-19-3]
Structure and functional group:

$$CH_3-CH_2-CH_2-\underset{\underset{O}{\|}}{C}-CH_2-CH_2-CH_3$$

reactive site: carbonyl keto group (C=O)

Synonyms: butyrone; 4-heptanone; heptan-4-one; propyl ketone

Uses and Exposure Risk

Dipropyl ketone (DPK) is used as a solvent in oils, resins, nitrocellulose, and polymers; and as a flavoring compound.

Physical Properties

Colorless liquid with a pungent odor; bp 144°C; mp −33°C; density 0.8174 at 20°C; soluble in alcohol and ether, insoluble in water.

Health Hazard

Inhalation of DPK vapors can cause narcosis and irritation of the eyes and respiratory tract. Exposure to 4000 ppm for 4 hours was fatal to rats. This compound exhibited low to very low oral toxicity in test animals.

LD_{50} value, oral (rats): 3730 mg/kg

Exposure Limit

TLV-TWA 235 mg/m^3 (50 ppm) (NIOSH).

Fire and Explosion Hazard

Combustible liquid; flash point (closed cup) 49°C (120°F); vapor density 3.9 (air = 1); vapor pressure 5.2 torr at 20°C; fire-extinguishing agent: "alcohol" foam or water spray. DPK forms explosive mixtures with air; LEL and UEL values have not been reported. It is incompatible with strong acids, alkalies, and oxidizers.

Disposal/Destruction

DPK is burned in a chemical incinerator equipped with an afterburner and scrubber.

Analysis

GC-FID, GC/MS, and HPLC (see Section 29.10).

29.15   ETHYL BUTYL KETONE

Formula $C_7H_{14}O$; MW 114.19; CAS [106-35-4]

Structure and functional group:

$$CH_3—CH_2—\underset{\underset{O}{\|}}{C}—CH_2—CH_2—CH_2—CH_3$$

reactive site: carbonyl ($\overset{\diagdown}{\underset{\diagup}{C}}=O$) group

Synonyms: butyl ethyl ketone; 3-heptanone; heptan-3-one

Uses and Exposure Risk

Ethyl butyl ketone is used as a solvent for nitrocellulose and polyvinyl resins, and as an intermediate in organic synthesis.

Physical Properties

Clear colorless liquid with a mild fruity odor; bp 147°C; freezes at −39°C; density 0.8183 at 20°C; soluble in alcohol and ether, insoluble in water.

Health Hazard

Inhalation of the vapor of ethyl butyl ketone can cause irritation to the eyes, skin, and mucous membranes. Its irritation effect was mild on rabbit skin and eyes. Prolonged skin contact can cause dermatitis. Exposure to 4000 ppm for 4 hours proved fatal to rats. Ingestion can cause headache and narcosis, and in large doses coma can occur.

LD_{50} value, oral (rats): 2760 mg/kg

Exposure Limits

TLV-TWA 230 mg/m^3 (50 ppm) (ACGIH); IDLH 3000 ppm (NIOSH).

Fire and Explosion Hazard

Combustible liquid; flash point (open cup) 46°C (115°F); vapor density 4.0 (air =

1); vapor pressure 1.4 torr at 25°C; fire-extinguishing agent: dry chemical, "alcohol" foam, CO_2, or water spray; the latter can also disperse the vapor and flush spills away from exposures. Ethyl butyl ketone forms explosive mixtures with air within the range 1.4–8.8% by volume. It can react violently with strong oxidizers.

Disposal/Destruction

Ethyl butyl ketone is burned in a chemical incinerator equipped with an afterburner and scrubber. Waste treatment methods involve steam stripping, high-temperature catalytic hydrogenation, or molten metal destruction.

Analysis

Ethyl butyl ketone is analyzed by GC-FID, GC/MS, and HPLC (see Section 29.10). Air analysis may be performed by NIOSH (1984) Method 1301. One to 25 L of air is flowed over coconut charcoal; the ketone is desorbed with a CS_2–methanol mixture (99 : 1) and analyzed by GC-FID using a column, 10% FFAP on 80/100 Chromosorb W-AW or an alternative column, such as 10% SP-2100 or Carbowax 1500 on Supelcoport, or DB-1 fused silica.

29.16   METHYL AMYL KETONE

DOT Label: Combustible Liquid UN 1110

Formula $C_7H_{14}O$; MW 114.18; CAS [110-43-0]

Structure and functional group:

$$CH_3-\overset{\overset{\text{O}}{\|}}{C}-CH_2-CH_2-CH_2-CH_2-CH_3$$

active site: carbonyl keto group ($\diagdown C = O$)

Synonyms: 2-heptanone; amyl methyl ketone; methyl-*n*-amyl ketone

Uses and Exposure Risk

Methyl amyl ketone is used as a flavoring agent and as a solvent in lacquers and synthetic resins.

Physical Properties

Colorless liquid with a fruity smell; bp 150.6°C; mp −26.9°C; density 0.8166 at 20°C; soluble in alcohol, ether, and benzene, very slight soluble in water.

Health Hazard

Exposure to methyl amyl ketone caused irritation of mucous membranes, mild to moderate congestion of the lungs, and narcosis in test animals. A 4-hour exposure to a 4000-ppm concentration in air was lethal to rats; 1500–2000 ppm produced lung irritation and narcosis. The concentration at which it produces similar symptoms in humans is not known.

The oral toxicity of this compound is low. Its irritant action on skin should be low to very low.

LD_{50} value, oral (mice): 730 mg/kg

Exposure Limits

TLV-TWA 235 mg/m^3 (50 ppm) (ACGIH), 465 mg/m^3 (100 ppm) (NIOSH).

Fire and Explosion Hazard

Combustible liquid, flash point (closed cup) 39.9°C (102°F), (open cup) 48.9°C (12°F); vapor density 3.9 (air = 1) vapor pressure 2.6 torr at 20°C; autoignition temperature 393°C (740°F); fire-extinguishing agent: "alcohol" foam; a water spray may be used to cool below its flash point.

Methyl amyl ketone forms an explosive mixture with air in the range 1.1% (at 66°C) to 7.9% (at 121°C) by volume. It can react explosively with strong acids, alkalies, and oxidizing agents.

Disposal/Destruction

Methyl amyl ketone is burned in a chemical incinerator equipped with an afterburner and scrubber. Waste treatment is by steam stripping; for wastewater, catalytic hydrogenation may be effective (see Section 29.10).

Analysis

GC-FID, HPLC, and GC/MS (see Section 29.10). Air analysis can be performed by NIOSH (1984) Method 1301 (see Section 29.15).

29.17   ETHYL AMYL KETONE

Formula $C_8H_{16}O$; MW 128.21; CAS [541-85-5]

Structure and functional group:

$$CH_3-CH_2-\underset{\underset{O}{\|}}{C}-CH_2-\underset{\underset{CH_3}{|}}{CH}-CH_2-CH_2-CH_3$$

reactive site: carbonyl keto group; the alkyl group attached to a secondary C atom

Synonyms: amyl ethyl ketone; ethyl *sec*-amyl ketone; 5-methyl-3-heptanone; 3-methyl-5-heptanone

Uses and Exposure Risk

Ethyl amyl ketone is used as a solvent for vinyl resins and nitrocellulose resins.

Physical Properties

Colorless liquid with a mild fruity odor; bp 160°C; density 0.8184 at 20°C; mixes readily with alcohols, ethers, ketones, and other organic solvents, slightly soluble in water.

Health Hazard

Exposure to ethyl amyl ketone can cause irritation of the eyes, nose, and skin. At high concentrations its exposure can lead to ataxia, prostration, respiratory pain, and narcosis. A 5-minute exposure to 50 ppm may produce a mild irritating effect on the eyes and nose in humans. Inhalation of 3000 ppm of ethyl amyl ketone for 4 hours was highly toxic to mice, and 6000 ppm for 8 hours was lethal to rats. The odor threshold is 6 ppm.

Exposure Limit

TLV-TWA 130 mg/m^3 (25 ppm) (ACGIH).

Fire and Explosion Hazard

Combustible liquid; flash point (open cup) 59°C (138°F); vapor density 2 torr at 25°C; fire-extinguishing agent: "alcohol" foam, dry chemical, or CO_2; a water spray may be effective to cool it below its flash point. It forms explosive mixtures with air at elevated temperatures; the LEL and UEL values are not reported.

Disposal/Destruction

Ethyl amyl ketone wastes can be destroyed by incineration or by treatment with molten metal salts.

Analysis

GC-FID, GC-MS, and HPLC are suitable analytical methods.

29.18   DIISOBUTYL KETONE

DOT Label: Combustible Liquid UN 1157
Formula $C_9H_{18}O$; MW 142.27; CAS [108-83-8]

Structure and functional group:

$$CH_3-\underset{\underset{CH_3}{|}}{CH}-CH_2-\underset{\underset{O}{\|}}{C}-CH_2-\underset{\underset{CH_3}{|}}{CH}-CH_3$$

reactive site: carbonyl keto group ($C{=}O$)

Synonyms: 2,6-dimethyl-4-heptanone; iso-valerone; valerone

Uses and Exposure Risk

Diisobutyl ketone is used as a solvent for nitrocellulose, lacquers, and synthetic resins; in organic syntheses.

Physical Properties

Colorless liquid with mild ether-like smell; bp 168°C; density 0.8053 at 20°C, soluble in alcohol, ether, and chloroform, insoluble in water.

Health Hazard

Inhalation of the vapors of diisobutyl ketone can produce irritation of the eyes, nose, and throat. At 25 ppm its odor was unpleasant, but the irritation effect on humans was insignificant. At 50 ppm the irritation was mild. A 7-hour exposure to 125 ppm had no adverse effect on rats; however, at 250 ppm, female rats developed increased liver and kidney weights. An 8-hour exposure to 2000 ppm was lethal. Ingestion of this compound can cause the symptoms of headache, dizziness, and dermatitis.

LD_{50} value, oral (rats): 5.8 g/kg

Exposure Limits

TLV-TWA   150 mg/m^3 (25 ppm);   IDLH 1000 ppm.

Fire and Explosion Hazard

Combustible; flash point (closed cup) 60°C (140°F); vapor density 4.9; vapor pressure 1.7 torr; fire-extinguishing agent: dry chemical, CO_2, or "alcohol" foam; a water spray may be used to disperse the vapors and bring the temperature down below the flash point.

Diisobutyl ketone may form explosive mixtures with air at elevated temperatures within the range 0.8–7.1% by volume.

Disposal/Destruction

Incineration, molten metal salt destruction.

Analysis

GC-FID, GC/MS, and HPLC (see Section 29.15).

29.19   ISOPHORONE

EPA Priority Pollutant

Formula $C_9H_{14}O$:   MW   138.20;   CAS [78-59-1]

Structure and functional group:

reactive site: the carbonyl group and the double bond in the ring

Synonyms: 1,1,3-trimethyl-3-cyclohexene-5-one; 3,5,5-trimethyl-2-cyclohexene-1-one; isoacetophorone

Uses and Exposure Risk

Isophorone is used as a solvent for vinyl resins and cellulose esters, and in pesticides.

Physical Properties

Colorless liquid with a camphor-like smell; bp 215°C; mp −8°C; density 0.923 at 20°C; soluble in alcohol, ether, and acetone, slightly soluble in water.

Health Hazard

Isophorone is an irritant, moderately toxic at high concentrations, mutagenic and possibly carcinogenic. Inhalation of its vapors can cause mild irritation of the eyes, nose, and throat. Exposure to 840 ppm for 4 hours resulted in severe eye irritation in guinea pigs. Its irritation effect on human eyes may be felt at 25–40 ppm. At concentrations above 200 ppm, it may cause irritation of the throat, headache, nausea, dizziness, and a feeling of suffocation (ACGIH 1986). In rats, exposure to 1840 ppm for 4 hours was fatal. Ingestion of isophorone can cause narcosis, dermatitis, headache, and dizziness.

LD_{50} value, oral (rats): 2330 mg/kg

Isophorone is mutagenic and when fed to rats orally, 258 g/kg for 2 years, it caused kidney tumor. Its carcinogenicity on humans is not reported.

Exposure Limits

TLV-TWA 25 mg/m^3 (5 ppm); IDLH 800 ppm.

Fire and Explosion Hazard

Combustible liquid; flash point (closed cup) 84.5°C (184°F); vapor pressure 0.2 torr at 20°C; vapor density 4.75 (air = 1); autoignition temperature 460°C (860°F); fire-extinguishing agent: water spray, CO_2, dry chemical, or "alcohol" foam. It forms an explosive mixture with air in the range 0.8–3.8% by volume in air. Its reaction with strong oxidizers may be violent.

Disposal/Destruction

Incineration.

Analysis

GC-FID, GC/MS, and HPLC.

29.20   CYCLOHEXANONE

EPA Classified Toxic Waste, RCRA Waste Number U057; DOT Label: Flammable Liquid UN 1915

Formula $C_8H_{16}O$: MW 98.14; CAS [108-94-1]

Structure and functional group:

an acyclic ketone, reactive site: carbonyl keto group ($C=O$)

Synonyms: cyclohexyl ketone; pimelic ketone; ketohexamethylene; Nadone; Anone; Hytrol O; Sextone

Uses and Exposure Risk

Cyclohexanone is used in the production of adipic acid for making nylon; in the preparation of cyclohexanone resins; and as a solvent for nitrocellulose, cellulose acetate, resins, fats, waxes, shellac, rubber, and DDT.

Physical Properties

Pale yellow oily liquid with an odor of acetone and peppermint; bp 155.6°C; mp −16.4°C; density 0.9478 at 20°C; soluble in alcohol, ether, acetone, benzene, and chloroform, sparingly soluble in water (50 g/L at 30°C).

Health Hazard

The toxicity of cyclohexanone in test species was found to be low to moderate. Exposure to its vapors can produce irritation in the eyes and throat. Splashing into the eyes can damage the cornea. Throat irritation in humans may occur from 3–5 minute exposure to a 50-ppm concentration in air. The symptoms of chronic toxicity in animals from its inhalation were liver and kidney damage, as well as weight loss. However, its acute toxicity was low below 3000 ppm. The symptoms in guinea pigs were lacrimation, salivation, lowering of heart rate, and narcosis. Exposure to 4000 ppm for 4–6 hours was lethal to rats and guinea pigs.

The oral toxicity of this compound was low. Ingestion may cause narcosis and depression of the central nervous system. It can be absorbed through the skin.

LD_{50} value, dermal (rabbits): 1000 mg/kg
LD_{50} value, intraperitoneal (rats): 1130 mg/kg

Exposure Limits

TLV-TWA 100 mg/m^3 (25 ppm) (ACGIH); IDLH 5000 ppm (NIOSH).

Fire and Explosion Hazard

Combustible liquid; flash point 44°C (111°F); vapor density 3.4 (air = 1); vapor pressure 2 torr at 20°C; autoignition temperature 420°C (788°F); fire-extinguishing agent: water spray, dry chemical, "alcohol" foam, or CO_2. It forms explosive mixtures with air only at a high temperature, >100 C. The LEL value at this temperature is 1.1% by volume; the upper limit is not reported. Violent reactions can occur with nitric acid and strong oxidizers.

Disposal/Destruction

Cyclohexanone is burned in a chemical incinerator equipped with an afterburner and scrubber. For land disposal: concentrations in wastewater should be less than 0.125 ppm; and concentrations in the TCLP extract of spent solvent waste, less than 0.75 ppm. Waste treatment is by steam stripping.

Analysis

GC-FID, GC/MS, or HPLC.

29.21 *o*-METHYL CYCLOHEXANONE

Formula C$_7$H$_{12}$O; MW 112.17; CAS [583-60-8]

Structure and functional group:

methyl group in *ortho*-position, reactive site: carbonyl group ($\overset{|}{\underset{|}{C}}$=O)

Synonym: 2-methyl cyclohexanone

Uses and Exposure Risk

o-Methyl cyclohexanone is used as a solvent for varnishes, lacquers, and plastics; and as a rust remover.

Physical Properties

Colorless viscous liquid with acetone-like odor; bp 165°C; mp −19°C; density 0.925 at 20°C; soluble in alcohol and ether, insoluble in water.

Health Hazard

o-Methyl cyclohexanone is a mildly toxic compound. Any health hazard from inhaling its vapors is low. Its vapors at high concentrations in air can be irritating to the eyes, nose, and throat. Ingestion of the liquid can cause sleepiness. Toxicity studies on animals suggest that methyl cyclohexanone may cause an adverse effect on the lungs, kidney, and liver. An intravenous administration of 270 mg/kg was lethal to mice.

LD_{50} value, intraperitoneal (mice): 200 mg/kg

Exposure Limits

TLV-TWA 230 mg/m^3 (50 ppm) (ACGIH); STEL 345 mg/m^3 (75 ppm); IDLH 2500 ppm (NIOSH).

Fire and Explosion Hazard

Combustible liquid; flash point (closed cup) 48°C (118°F); vapor pressure 10 torr at 55°C; vapor density 3.9 (air = 1); fire-extinguishing agent: dry chemical or CO_2. o-Methyl cyclohexanone can react explosively with strong oxidizers at elevated temperatures.

Disposal/Destruction

o-Methyl cyclohexanone is burned in a chemical incinerator equipped with an after-burner and scrubber.

Analysis

GC-FID, GC/MS, or HPLC.

29.22 ACETOPHENONE

EPA Classified Toxic Waste, RCRA Waste Number U004

Formula C_8H_8O; MW 120.16; CAS [98-86-2]

Structure and functional group:

reactive site: carbonyl group ($C=O$) and positions in the ring

Synonyms: acetyl benzene; 1-phenyl ethanone; benzoyl methide; phenyl methyl ketone; methyl phenyl ketone; Hypnone; Dimex

Uses and Exposure Risk

Acetophenone is used in perfumery, as a photosensitizer in organic synthesis, and as a catalyst in olefin polymerization.

Physical Properties

Colorless liquid; bp 202°C; mp 20°C; density 1.033 at 15°C; soluble in alcohol, ether, and chloroform; slightly soluble in water.

Health Hazard

Acetophenone is an irritant, mutagen, and a mildly toxic compound. In rabbits 0.77 mg produced severe eye irritation, but the action on skin was mild. In mice, subcutaneous administration of this compound produced sleep; a dose of 330 mg/kg was lethal.

LD_{50} value, intraperitoneal (mice): 200 mg/kg

No symptoms of severe toxicity, nor its carcinogenicity in humans, has been reported.

Exposure Limit

No exposure limits are set. The health hazard from exposure to this compound should be low, due to its low vapor pressure and low toxicity.

Fire and Explosion Hazard

Combustible liquid; flash point (closed cup) 82°C (180°F); vapor pressure 1 torr at 37°C; vapor density 4.1 (air = 1); autoignition temperature 570°C (1058°F); fire-extinguishing agent: dry chemical, foam, or CO_2; water may cause frothing, but it can be used to flush and dilute the spill. Its reaction with strong oxidizers may be violent.

Disposal/Destruction

Incineration or molten salt treatment are suitable methods for its destruction.

Analysis

Acetophenone can be analyzed by GC-FID, GC-PID, HPLC, or GC/MS. A silicone-coated fused-silica capillary column is suitable for GC/MS analysis. Characteristic ions for its identification are 105, 71, 51, and 120 (electron-impact ionization). Acetophenone in solid wastes, soil, and sludges can be analyzed by GC/MS by EPA Method 8270 (U.S. EPA 1986).

29.23 1,4-BENZOQUINONE

EPA Classified Toxic Waste, RCRA Waste Number U197; DOT Label: Poison UN 2587

Formula $C_6H_4O_2$; MW 108.09; CAS [106-51-4]

Structure and functional group:

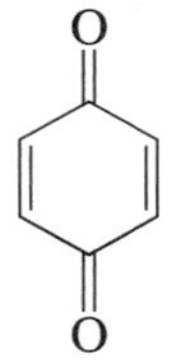

the molecule has two carbonyl keto groups ($C=O$) which are the reactive sites; in addition, substitution can occur on the ring.

Synonyms: *p*-benzoquinone; quinone; *p*-quinone; 1,4-cyclohexadienedione

Uses and Exposure Risk

1,4-Benzoquinone is used in the manufacture of dyes, fungicide, and hydroquinone; for tanning hides; as an oxidizing agent; and in photography.

Physical Properties

Pale yellow crystalline solid; acrid odor like that of chlorine; mp 116°C (sublimes); density 1.318 at 20°C; soluble in alcohol, ether, and alkalies, slightly soluble in water (1.5 g/100 mL).

Health Hazard

1,4-Benzoquinone is moderately toxic via ingestion and skin contact. It is a mutagen and may cause cancer. Because of its low vapor pressure, 0.1 torr (at 25°C), the health hazard due to inhalation of its vapor is low. However, prolonged exposure may produce eye irritation, and its contact with the eyes can injure the cornea. Contact with the skin can lead to irritation, ulceration, and necrosis.

The toxicity of benzoquinone is similar to that of hydroquinone and benzenetriol. Repeated intraperitoneal administration of 2 mg/kg/day to rats for 6 weeks produced significant decreases in red blood cell, bone marrow counts, and hemoglobin content (Rao et al. 1988). In addition, relative changes in organ weights and injuries to the liver, thymus, kidney, and spleen were observed. Lau et al. (1988) investigated the correlation of toxicity with increased glutathione substitution in 1,4-benzoquinone. With the exception of the fully substituted isomer, increased substitution resulted in enhanced nephrotoxicity. Although the conjugates were more stable to oxidation, the toxicity increased. The oral and intravenous toxicities of this compound in rats are as follow:

LD$_{50}$ value, oral (rats): 130 mg/kg

LD$_{50}$ value, intravenous (rats): 25 mg/kg

The carcinogenicity of 1,4-benzoquinone in humans is not reported. However, it is a mutagen. It produced tumors in the lungs and skin of mice.

Exposure Limits

TLV-TWA 0.4 mg/m^3 (0.1 ppm); STEL 1.2 mg/m^3 (0.3 ppm) (ACGIH); IDLH 75 ppm (NIOSH).

Fire and Explosion Hazard

Noncombustible solid; ignition can occur after only moderate heating, autoignition temperature 560°C (1040°F); fire-extinguishing agent: water spray. 1,4-Benzoquinone may react violently with strong oxidizers, especially at elevated temperatures.

Disposal

1,4-Benzoquinone is dissolved in a combustible solvent and burned in a chemical incinerator equipped with an afterburner and scrubber.

Analysis

1,4-Benzoquinone can be analyzed by GC-FID and GC/MS. A fused-silica capillary column such as DB-1 may be used. It can be analyzed by HPLC using an electrochemical detector.

29.24 *dl*-CAMPHOR

DOT Label: Flammable Solid UN 2717

Formula $C_{10}H_{16}O$; MW 152.23; CAS [76-22-2]

Structure and chemical nature:

terpene class, active site: carbonyl keto (C=O) group, occurs as optical isomeric forms

Synonyms: 1,7,7-trimethyl bicyclo[2,2,1]-2-heptanone; 2-bornanone; 2-oxobornane; 2-camphanone; gum camphor; synthetic camphor

Uses and Exposure Risk

dl-Camphor is used as a plasticizer for cellulose esters and ethers; in the manufacture of plastics and cymene; in cosmetics, lacquers, medicine, explosives, and pyrotechnics; and as a moth repellent.

Physical Properties

White crystalline solid with an aromatic odor; mp 178.8°C; sublimes at 204°C; density 0.992 at 25°C; soluble in alcohol, ether, acetone, and benzene, insoluble in water.

Health Hazard

Vapors of camphor can irritate the eyes, nose, and throat. In humans, such irritation may be felt at >3 ppm concentration. Prolonged exposure can cause headache, dizziness, and loss of sense of smell. Ingestion can cause headache, nausea, vomiting, and diarrhea, and at high dosages can lead to convulsion, dyspnea, and coma. High dosages can be harmful to gastrointestinal tracts, kidney, and brain.

LD$_{50}$ value, intraperitoneal (mice): 3000 mg/kg

Exposure Limits

TLV-TWA 12 mg/m^3 (2 ppm), STEL 18 mg/m^3 (3 ppm) (ACGIH); IDLH 200 mg/m^3 (NIOSH).

Fire and Explosion Hazard

Flammable solid; flash point (closed cup) 65.6°C (150°F); vapor pressure 0.18 torr at 20°C; vapor density 5.2; autoignition temperature 466°C (871°F); fire-extinguishing agent: none reported; water or foam may cause frothing; water spray may be applied carefully on the surface to blanket and extinguish a fire. *dl*-Camphor forms explosive mixtures with air; the LEL and UEL values are 0.6% and 3.5% by volume of air, respectively. It may react violently with chromium trioxide and other strong oxidizers.

Disposal

dl-Camphor is dissolved in a combustible solvent and burned in a chemical incinerator equipped with an afterburner and scrubber.

Analysis

GC-FID, HPLC, and GC/MS; air samples using the NIOSH Method 1301 (see Section 29.15).

29.25 MISCELLANEOUS KETONES

See Table 29.1.

TABLE 29.1 Toxicity and Flammability of Miscellaneous Ketones

Compound/ Synonyms/ CAS No.	Formula/MW/ Structure	Toxicity	Flammability
Benzophenone (benzoyl benzene, diphenyl ketone, diphenyl methanone, α-oxodiphenyl methanone) [119-61-9]	$C_{13}H_{10}O$ 182.23	Mild irritant; low toxicity; LD_{50} intraperitoneal (mice): 727 mg/kg; LD_{50} oral (mice): 2900 mg/kg (NIOSH 1986)	Noncombustible solid
Benzyl acetone (4-phenyl-2-butanone, methyl phenethyl ketone, β-phenyl-ethyl methyl ketone) [2550-26-7]	$C_{10}H_{12}O$ 148.22	Skin irritant, 500 mg/24 hr (rabbits) severe; mild eye irritant	Noncombustible liquid; flash point 98°C (209°F)
Acetonyl acetone (2,5-hexanedione, 2,5-diketohexane, 1,2-diacetylethane) [110-13-4]	$C_6H_{10}O_2$ 114.16	Eye irritant; teratogen (paternal effect); low toxicity— LD_{50} oral (rats): 2700 mg/kg	Combustible liquid; flash point 78°C (174°F)
Cyclopentanone (adipic ketone, ketocyclopentane, ketopentamethy-lene) [120-92-3]	C_5H_8O 84.13	Irritant: mild on skin, could be severe to eyes (rabbits 100 mg) mild to moderate toxicity: LD_{50} intraperitoneal (mice): 1950 mg/kg (NIOSH 1986)	Flammable liquid; flash point (closed cup) 26°C (79°F); vapor density 2.3 (air = 1); fire-extinguishing agent: "alcohol" foam; DOT Label: Flammable Liquid, UN 2245
4-Cyclopentene-1,3-dione [930-60-9]	$C_5H_4O_2$ 96.09	Irritant	Combustible liquid; flash point (closed cup) 83°C (183°F)
1,3-Cyclohexane-dione (dihydro-resorcinol, cyclo-hexane-1,3-dione) [504-02-9]	$C_6H_8O_2$ 112.14	Low toxicity; inadequate data; LD_{50} intraperi-toneal (mice) 64 mg/kg	Noncombustible solid

(continued)

TABLE 29.1   (*Continued*)

Compound/ Synonyms/ CAS No.	Formula/MW/ Structure	Toxicity	Flammability
3-Methylcyclo-hexanone [591-24-2]	$C_7H_{11}O$ 112.19	Irritant to eyes, nose, and throat; toxic: LD_{50} intravenous (dogs) 310 mg/kg	Flammable liquid; flash point (closed cup) 53°C (127°F); vapor density 3.9 (air = 1); fire-extinguishing agent: dry chemical or CO_2
4-Methylcyclo-hexanone [589-92-4]	$C_7H_{11}O$ 112.19	Irritant; toxicity similar to that of *meta*-isomer; LD_{50} intra-venous (dogs) 370 mg/kg (NIOSH 1986)	Flammable liquid; flash point (closed cup) 42°C (108°F); explosion occurred when added gradually to nitric acid at 70°C

REFERENCES

ACGIH. 1986. *Documentation of the Threshold Limit Values and Biological Exposure Indices*, 5th ed. Cincinnati, OH: American Conference of Governmental Industrial Hygienists.

Baker, E. G., and L. J. Sealock, Jr. 1988. Catalytic destruction of hazardous organics in aqueous solutions. *Report PNL-6491-2*, NTIS Order No. DE 88009535. Richland, WA: Pacific Northwest Laboratory.

Brooks, T. M., A. L. Meyer, and D. H. Hutson. 1988. The genetic toxicology of some hydrocarbon and oxygenated solvents. *Mutagenesis* 3(3): 227–32.

Lau, S. S., B. A. Hill, R. J. Highet, and T. J. Monks. 1988. Sequential oxidation and gluta-thione addition to 1,4-benzoquinone: correlation of toxicity with increased glutathione substitution. *Mol. Pharmacol.* 34(6): 829–36.

Merck. 1996. *The Merck Index*, 12th ed. Rahway, NJ: Merck & Co.

National Research Council. 1995. *Prudent Practices for Handling and Disposal of micals from Laboratories*. Washington, DC: National Academy Press.

NFPA. 1997. *Fire Protection Guide on Hazardous Materials*, 12th ed. Quincy, MA: National Fire Protection Association.

NIOSH. 1984. *Manual of Analytical Methods*, 3rd ed. Cincinnati, OH: National Institute for Occupational Safety and Health.

NIOSH. 1986. *Registry of Toxic Effects of Chemical Substances*, ed. D. V. Sweet. Washington, DC: National Institute for Occupational Safety and Health.

Nomeir, A. A., and M. B. Abou-Donia. 1985. Analysis of n-hexane, 2-hexanone, 2,5-hexanedione, and related chemicals by capillary gas chromatography and high-performance liquid chromatography. *Anal. Biochem.* 151(2): 381–88.

Phillips, R. D., E. J. Moran, D. E. Dodd, E. H. Fowler, C. D. Kary, and J. O'Donoghue. 1987. A 14-week vapor inhalation toxicity study of methyl isobutyl ketone. *Fundam. Appl. Toxicol.* 9(3): 380–88.

Rao, G. S., S. M. Siddiqui, K. P. Pandya, and R. Shanker. 1988. Relative toxicity of metabolites of benzene in mice. *Vet. Hum. Toxicol.* 30(6): 517–20.

Schultz, C. G. 1985. Destruction of organic hazardous wastes. U.S. Patent 4,552,667A,

Nov. 12; cited in *Chem. Abstr. CA 104*(6): 39264e.

Tyl, R. W., K. A. France, L. C. Fisher, I. M. Pritts, T. R. Tyler, R. D. Phillips, and E. J. Moran. 1987. Developmental toxicity evaluation of inhaled methyl isobutyl ketone in Fischer 344 rats and CD-1 mice. *Fundam. Appl. Toxicol. 8*(3): 310–27.

U.S. EPA. 1988. *Code of Federal Regulations: Protection of Environment.* Title 40, Parts 190-299. Washington DC: Office of the Federal Register, National Archives and Records Administration.

U.S. EPA. 1986. *Test Methods for Evaluating Solid Waste*, 3rd ed. Washington, DC: Office of Solid Waste and Emergency Response.

30

METAL ACETYLIDES AND FULMINATES

30.1 GENERAL DISCUSSION

Acetylides and fulminates form highly explosive shock- and heat-sensitive salts with many metals. Acetylides are the metal derivatives of acetylene. Hydrogen attached to carbon atoms bearing a triple bond is acidic in nature. It can be substituted by a metal ion to form acetylides, with the general structure $M-C\equiv C-M$. These substances are made from acetylene, $HC\equiv CH$, or alkyl acetylene, $RC\equiv CH$, by passing acetylene gas or alkyl acetylene vapors over the aqueous solutions of ammoniacal metal salts, as shown below in the following reactions:

$$HC\equiv CH + 2Ag(NH_3)_2NO_3 \longrightarrow$$

$$\overset{+}{Ag}\,\overset{-}{C}\equiv\overset{-}{C}\,\overset{+}{Ag} + 2NH_4NO_3 + 2NH_3$$

(silver acetylide)

$$HC\equiv CH + 2Cu(NH_3)_2OH \longrightarrow$$

$$CuC\equiv CCu + 4NH_3 + 2H_2O$$

(cuprous acetylide)

An alkyne of the type $RC\equiv CH$ would yield a monocopper derivative. The acetylide formation is specific to acetylenic hydrogen only and does not occur with alkanes, alkenes, or even with a dialkyl alkynes of the type $RC\equiv R'C$. Acetylide ion exhibits a strong nucleophilic property. These compounds are used in organic synthesis, in the purification of acetylene to identify a $-C\equiv H$ unit, and in explosives. Acetylides of heavy metals are extremely shock sensitive when dry, whereas the salts of alkali metals such as sodium acetylide are fairly stable. Sodium acetylide is stable up to 400°C and calcium carbide melts at 2300°C without decomposition.

Many metals are known to form fulminates. These compounds contain the $-CNO$ functional group bound to the metal atoms by $C-$metal bonds. Fulminates are formed by mixing nitric acid solutions of metal nitrates with ethanol. Even in the absence of nitric acid, mixing mercuric nitrate with ethanol can form mercury fulminate. Fulminates, like azides and acetylides, are thermodynamically unstable, having high positive Gibbs free energy (or heat of formation). While the fulminates of alkali metals are highly sensitive to impact and friction, those of the heavy metals are powerful explosives.

Hazardous properties of some selected acetylides and fulminates are presented in the following sections.

30.2 CUPROUS ACETYLIDE

Formula Cu_2C_2; MW 151.11; CAS [1117-94-8]

Structure and reactivity: $\overset{+}{Cu}\,\overset{-}{C}{\equiv}\overset{-}{C}\,\overset{+}{Cu}$, copper salt of weakly acidic acetylene, strong nucleophile

Synonyms: cuprous carbide; copper(I) acetylide

Uses and Exposure Risk

Cuprous acetylide is used as a diagnostic test to identify the $\equiv CH$ unit; to purify acetylene; in the preparation of pure copper powder; and as a catalyst in the manufacture of 2-propyn-1-ol and acrylonitrile.

Physical Properties

Red amorphous powder; explodes on heating; insoluble in water, soluble in acid.

Health Hazard

There are no toxicity data in the published literature on cuprous acetylide. As with many copper salts, inhalation of its dust can cause irritation of the respiratory tract and ulceration of nasal septum.

Exposure Limit

TLV-TWA in air 1 mg(Cu)/m^3 (ACGIH).

Fire and Explosion Hazard

Cuprous acetylide in the dry state is highly sensitive to shock, causing explosion on impact. It explodes in contact with acetylene after being warmed in air or oxygen for several hours (Mellor 1946). It forms a highly explosive mixture containing silver acetylide when mixed with silver nitrate. It ignites spontaneously with chlorine, bromine, and iodine vapors and with fine crystals of iodine. Reactions with dilute acids form acetylene, which is flammable and explosive in air. It explodes when heated above 100°C. It forms black copper(II)acetylide when exposed to air or oxygen.

Disposal/Destruction

Cuprous acetylide may be destroyed by slowly adding dilute hydrochloric acid to it in a three-necked flask under nitrogen flow. Acetylene formed is vented away with nitrogen.

30.3 CUPRIC ACETYLIDE

Formula CuC_2; CAS[12540-13-5]
Structure:
$$(-Cu{\equiv}C-)_n$$

Synonyms: copper(II)acetylide; cupric carbide

Uses and Exposure Risk

Cupric acetylide is used as a detonator. Its applications are very limited, however, because of its high sensitivity to impact or friction. It is susceptible to form and build up upon prolong contact of copper metal with organic vapors.

Physical Properties

Brownish black powder; insoluble in water.

Fire and Explosion Hazard

Cupric acetylide is much more sensitive to impact and friction than the cuprous salt. Friction heating or mild impact can result in violent explosion. In dry state, its sensitivity is much greater to impact and is flammable.

30.4 SILVER ACETYLIDE

Formula Ag_2C_2; MW 239.76; CAS [7659-31-6]; Structure $Ag-C{\equiv}C-Ag$
Synonym: silver carbide

Uses and Exposure Risk

Its limited use is only as a detonator. Risk of its formation may arise when silver metal or its salt solution comes in contact with acetylene.

Physical Properties

White precipitated solid; explodes on heating; insoluble in water, slightly soluble in alcohol, soluble in acid.

Fire and Explosion Hazard

It is extremely sensitive to shock when fully dried and can explode at a slight touch. It is, however, stable under moist conditions. It is somewhat more sensitive than cuprous acetylide. It explodes violently when heated to 120°C. Reactions with dilute acids produce acetylene which is flammable and explosive in air. It reacts with silver nitrate, forming a complex $C_2Ag_2 \bullet AgNO_3$, which can explode when dry by heat or spark.

30.5 MERCURY ACETYLIDE

Mercury forms two types of acetylides; one is a polymeric compound $(C_2Hg)_n$ having a structure $(-HgC \equiv C-)_n$ and CAS No. [37297-87-3] and the other acetylide probably has the structure

$$HgHC_2(-HgC \equiv CH)$$

CAS [68833-55-6]. Both the acetylides of mercury are highly sensitive to heat or mechanical shock.

30.6 CALCIUM ACETYLIDE

Formula: CaC_2; MW 64.10; CAS [75-20-7]
Structure:

$$(-CaC \equiv C-)_n$$

Synonym: calcium carbide, calcium ethynide

Uses and Exposure Risk

It is used to produce acetylene gas in numerous applications. Also, it is used as a reducing agent to produce metals from their salts.

Physical Properties

Colorless tetragonal crystal; density 2.22 g/cm^3; decomposes in water; melting point reported in the literature show wide discrepancy.

Health Hazard

It is a corrosive solid. Because it is highly water-reactive, skin contact can cause burn.

Fire and Explosion Hazard

Hazards associated with calcium acetylide can be attributed primarily to its three types of reactions: (1) it decomposes to acetylene, (2) it is itself a strong reducing agent, and (3) it forms acetylides with many metals. It decomposes in water producing acetylene, a highly flammable gas that also forms explosive mixture in air. The reaction is exothermic, which in closed containers can cause excessive pressure buildup and explosion. One kg of the solid can generate about 350 L of acetylene. Reactions with alcohols can be vigorous to violent.

Being a strong reducing agent, it reduces many metal oxides and other salts into the metals which are flammable in their finely divided states. The reactions are often accompanied with incandescence (depending on the nature of the metal and heating.) Incandescence occurs with halogen gases and vapors too, however, under heating above 200°C (incandesce with chlorine at 245°C and bromine vapor at 350°C). Violent to explosive reactions can occur upon mixing with strong oxidizing agents. It is susceptible to form the precipitates acetylides of copper, silver, gold and other metals when mixed with their solutions. These acetylides are highly sensitive to shock and heat.

30.7 ACETYLIDES OF ALKALI METALS

All alkali metals are known to form acetylides. Their formulas and CAS Registration No. are presented below in Table 30.1. The explosivity of alkali metal acetylide, or its sensitivity to shock or impact, is relatively lower than the acetylide of transition metals. However, all alkali metal acetylides ignite when warmed or heated in air. Contact with halogens produces similar ignition. Reactions with fluorine and chlorine occur at ambient temperature, while with bromine or iodine vapors, ignition usually occurs upon warming. The acetylides of cesium and rubidium are more reactive than those of lithium and sodium. All these acetylides incandesce in nonmetal oxides, including carbon dioxide, nitrogen pentoxide, and sulfur dioxide. When warmed or heated with metal oxides or metal salts of oxoacids such as sulfate or nitrate, acetylides ignite.

Heating with nonmetals such as phosphorus or sulfur, or contact with their vapors, can produce incandescence. Most of the reactions of alkali metal acetylides are accompanied with burning or incandescence, especially under heating or warming. Such reactions are susceptible to cause explosion when their mixtures are grinded. Mixing with strong oxidizing acids such as concentrate nitric or perchloric acids can cause explosion. Combination with limited amounts of water can generate acetylene, which can result in ignition and/or explosion. Such a hazard may not arise in the presence of excess water.

30.8 MERCURY FULMINATE

EPA Classified Acute Hazardous Waste, RCRA Waste Number P065; DOT Label (wetted with not less than 20% by weight of water): Class A Explosive, UN 0135.

Formula $Hg(CNO)_2$; MW 284.63; CAS [628-86-4]

Synonym: Fulminate of mercury

TABLE 30.1 Alkali Metal Acetylides

Compounds	Formula	CAS No.
Lithium acetylide	$LiC\equiv CLi$	[1070-75-3]
Sodium acetylide	$NaC\equiv CNa$	[2881-62-1]
Potassium acetylide	$KC\equiv CK$	[22754-97-8]
Rubidium acetylide	$RbC\equiv CRb$	[22754-97-8]
Cesium acetylide	$CsC\equiv CCs$	[22750-56-7]

Uses and Exposure Risk

Mercury fulminate is used as a primary explosive to initiate boosters.

Physical Properties

White cubic crystals explodes on heating; density 3.6 g/cm^3; slightly soluble in water.

Health Hazard

Mercury fulminate is a highly toxic compound, exhibiting the toxicity symptoms of mercury. No toxicity data are available on this compound.

Fire and Explosion Hazard

Mercury fulminate is a powerful explosive. It is highly sensitive to impact and heat. It is moderately endothermic, $\Delta H_f(s) + 267.7$ kJ/mol (64.0 kcal/mol). Detonation temperature is 180°C (356°F), and the detonation velocity is 4.7 km/s (Meyer 1989), and 5.4 km/s at a density $4.29/cm^3$. The heat of detonation is 510 kJ (122 kcal)/mol producing 90L of gas at STP. It can detonate when struck by a 2-kg weight falling only 5 cm (Fieser and Fieser 1944). It is somewhat more shock sensitive than lead azide. In the dry state, it can be initiated by intense radiation. It reacts explosively with sulfuric acid.

30.9 COPPER(II)FULMINATE

Formula: $CuC_2N_2O_2$; MW 147.58; CAS [22620-90-2];

Structure:

$$Cu(C\equiv N \longrightarrow O)_2;$$

Uses and Exposure Risk

It is used as a detonator. Its applications, however, are limited.

Fire and Explosion Hazard

Copper(II)fulminate is a powerful explosive detonating with a brisance stronger than that of mercury fulminate. It is moderately sensitive to heat and impact. It is susceptible to decompose explosively when combined with sulfuric acid.

30.10 SILVER FULMINATE

Formula: $Ag_2C_2N_2O_2$; MW 299.77; CAS [5610-59-3];
Structure:

$$(AgC\equiv N \longrightarrow O)_2$$

forms dimeric salt.

Uses and Exposure Risk

It is used as a detonator. A potential risk of its formation may arise when silver metal, silver nitrate or any other silver salt is mixed with nitric acid and ethanol.

Physical Properties

White crystals of needle shape; explodes on heating; very slightly soluble in water (approximately 750 mg/L) (CRC Handbook 1996); soluble in NH_4OH.

Fire and Explosion Hazard

It is one of the most powerful detonating fulminates. Its detonating power and sensitivity to heat and impact are greater than for mercury fulminate. The heat of formation, ΔH_f is +361.5 kJ (+86.4 kcal)/mol for the dimer (Collins 1978). Reaction with hydrogen sulfide at ambient temperature could proceed to explosive violence (Bretherick 1995). Although soluble in ammonia, however, at a pH of >13, it is likely to produce an explosive precipitate.

30.11 SODIUM FULMINATE

Formula NaCNO; MW 79.02; CAS [15736-98-8];
Structure:

$$NaC\equiv N \longrightarrow O$$

Any commercial application of sodium or other alkali metal fulminates is unknown. They are formed by the reactions of alkali metal salts with nitric acid and ethanol. Sodium fulminate is highly sensitive to shock and heat. It explodes when warmed or lightly touched with a spatula or a glass rod. It forms a double salt with mercuric fulminate which is unstable and explodes readily.

30.12 THALLIUM FULMINATE

Formula: TlCNO; MW 372.45; CAS {20991-79-1]
Structure:

$$Tl(C\equiv N \longrightarrow O)_3$$

No practical application is known for this compound. It is highly sensitive to heat and shock, exploding on light impact or warming.

REFERENCES

Bretherick, L. 1995. *Handbook of Reactive Chemical Hazards*, 5th ed. P. G. Urben. ed. Oxford, UK: Butterworth-Heinemann.

Collins, P. H. 1978. Propellants, *Explosives 3*, 159–162.

CRC Press. 1995. *Handbook of Chemistry and Physics*, 76th ed. D. R. Lide ed. Boca Raton, FL: CRC Press.

Fieser, L. F. and M. Fieser. 1944. *Organic Chemistry*. Boston: D. C. Heath.

Mellor, J. W. 1947–71. *Comprehensive Treatise on Inorganic and Theoretical Chemistry*, London: Longmans, Green & Co.

Meyer, E. 1989. *Chemistry of Hazardous Materials*, 2nd ed. Englewood Cliffs, NJ: Prentice-Hall.

31

METAL ALKOXIDES

31.1 GENERAL DISCUSSION

Metal alkoxides, also known as alcoholates, constitute a class of compounds in which the metal atom is attached to one or more alkyl groups by an oxygen atom. These substances have the general formula, $M-O-R$, where M is the metal atom and R, an alkyl group. Many metals in the Periodic Table are known to form alkoxides. However, only a few of them have commercial applications. These include the alkali and alkaline earth metals, aluminum, titanium, and zirconium. Metal alkoxides have found applications as catalysts, additives for paints and adhesives, and hardening agents for synthetic products. They are also used in organic synthesis.

Metal alkoxides can be prepared by several methods, which primarily involve the reaction of an alcohol with either the metal or its compound, such as oxide, hydroxide, halide, or amide. A few preparative reactions are illustrated in the following examples:

$$Li + C_2H_5OH \xrightarrow{\text{ether}} LiOC_2H_5 + \tfrac{1}{2}H_2$$

$$KOH + CH_3OH \longrightarrow KOCH_3 + H_2O$$

$$Cr[N(CH_3)_2]_4 + 4C_2H_5OH \longrightarrow$$
$$Cr(OC_2H_5)_4 + 4(CH_3)_2NH$$

Hazardous Reactions

The alkoxides of metals, in general, do not present any serious fire or explosion hazards, as metal alkyls do. Only a few are flammable compounds undergoing violent to vigorous explosive reactions. The single most hazardous property attributed to metal alkoxides is their exothermic reaction with water. An alcohol and a metal hydroxide are generated. The reaction is as follows:

$$MOR + H_2O \longrightarrow MOH + ROH$$

The heat of hydrolysis can ignite the solid alkoxides and can present a serious fire hazard. Any contact with water or exposure of these substances to moist air should therefore be avoided. Among all the alkoxides, the compounds of alkali metals, especially the potassium derivatives, are the most reactive. Among the alkyl groups, the lower and branched alkyl groups form most reactive derivatives. A few heavy metal alkoxides, such as uranium hexa-*tert*-butoxide, does not react with water.

Thermal decomposition can produce flammable products, such as diethyl ether, lower olefin, and alcohols, and, in some cases, metals. The relative ease of decomposition increases with branching in the alkyl group and longer carbon chain. Rapid mixing of secondary and tertiary alkoxides of alkali metals with many common solvents, such as chloroform, acetone, ethanol, or ethyl acetate, can proceed to uncontrollable exothermic reactions that may result in ignition or explosion, or both.

Health Hazard

Most alkoxides are highly basic. These moisture-sensitive substances are corrosive and caustic. Skin contact can cause irritation. However, highly exothermic water-reactive compounds such as potassium *tert*-butoxide, are strongly corrosive and can cause burns.

Analysis

The metal alkoxides are analyzed by hydrolyzing the sample with dilute acid and determining the particular metal in the solution by atomic absorption or emission spectrometry. Alcohols liberated in hydrolysis may be distilled out from the solution and analyzed by GC/MS for the identification of the alkoxy group and its quantification. The liberated alcohol may, alternatively, be determined by oxidation with a standard solution of potassium dichromate and measuring the excess of standard oxidant by iodometric titration. Higher alkoxy groups may be determined by standard elemental analysis for carbon and hydrogen. Also, the hydrolysis product, alcohol or the oxidation product, carboxylic acid, can be measured to identify and quantitate the alkoxy groups.

31.2 POTASSIUM-*tert*-BUTOXIDE

Formula: $(CH_3)_3COK$; MW 112.2; CAS [865-47-4]

Synonym: potassium *tert*-butylate

Uses and Exposure Route

It is used as a catalyst and finds application in many organic syntheses.

Physical Properties

White crystalline powder; hygroscopic; melts at 257°C (decomposes); powder density 0.50 g/mL; soluble in tetrahydrofuran and *tert*-butanol (20% at 20°C), low solubility in hexane, benzene, and ether.

Fire and Explosion Hazard

Potassium *tert*-butoxide is a flammable solid. It ignites on heating. Being very strongly basic, its reactions with acids are highly exothermic. Contact of solid powder with drops of sulfuric acid and vapors of acetic acid caused ignition after an induction period of 0.5 and 3 minutes, respectively (Manwaring 1973). Ignition occurs upon reactions with many common solvents of the type ketone, lower alcohols, esters, and halogenated hydrocarbons. Such solvents include acetone, methyl ethyl ketone, methyl isobutyl ketone, methanol, ethanol, *n*-propanol, isopropanol, ethyl acetate, *n*-propyl formate, *n*-butyl acetate, chloroform, methylene chloride, carbon tetrachloride, epichlorohydrin, dimethyl carbonate, and diethyl sulfate (NFPA 1997). Such ignition may arise from accidental contact of the alkoxide with these solvents and may be attributed to sudden release of energy from exothermic reactions. However, slow mixing of the powder with excess solvent will dissipate the heat. It reacts violently with water, producing *tert*-butanol and potassium hydroxide, as follows:

$$K-O-C(C_4H_9)_3 + H_2O \longrightarrow$$
$$tert\text{-}C_4H_9OH + KOH$$

The addition of potassium *tert*-butoxide to the solvent dimethyl sulfoxide can cause ignition of the latter (Bretherick 1995).

31.3 SODIUM METHOXIDE

Formula: CH_3ONa; MW 54.03; CAS [124-41-4]

Synonym: sodium methylate

Uses and Exposure Risk

It is used as a catalyst for treatment of edible fats and oils, as an intermediate in many synthetic reactions, to prepare sodium cellulosate; and as a reagent in chemical analysis.

Physical Properties

White free-flowing powder; hygroscopic; powder density 0.45 g/mL; soluble in methanol (33% at 20°C); insoluble in hydrocarbons and many common organic solvents; decomposes in water.

Fire and Explosion Hazard

Sodium methoxide is a flammable solid, igniting spontaneously in air when heated at 70–80°C. Also, it may ignite in moist air at ambient temperature, undergoing exothermic decomposition. The heat of decomposition at the temperature range 410–460°C has been measured as 0.77 kJ/g (Bretherick 1995), or 9.9 kcal/mol. Rapid addition of this compound to chloroform–methanol mixture have resulted in explosion. It is likely to explode when its solution in a halogenated hydrocarbon solvent is heated.

Disposal

Metal alkoxides, especially those compounds that are susceptible to react violently with water, should be dissolved or mixed in a combustible solvent (e.g., hexane, toluene) and burned in a chemical incinerator equipped with an afterburner and scrubber.

31.4 POTASSIUM METHOXIDE

Formula: CH_3KO; MW 70.14; CAS [865-33-8]
Synonym: potassium methylate

Uses and Exposure Route

It is used as a catalyst and an intermediate in organic synthesis.

Physical Properties

Yellowish white, free-flowing powder; hygroscopic; powder density 1.0 g/mL; soluble in alcohol, insoluble in hydrocarbons.

Fire and Explosion Hazard

Flammable solid, igniting in moist air. Rapid addition in bulk amount into halogenated solvents may cause an explosion. Reaction with arsenic pentafluoride in benzene can result in explosion (Kolditz 1965).

31.5 ALUMINUM ISOPROPOXIDE

Formula: $[(CH_3)_2CHO]$; MW 204.33; CAS [555-31-7]
Synonym: aluminum isopropylate

Uses and Exposure Risk

It is used in a number of organic synthetic reactions and in the manufacture of many types of products, including aluminum soaps, paints, and waterproofing finishes.

Physical Properties

White solid; hygroscopic; melts around 120°C; density 1.035 g/mL at 20°C; soluble in chloroform, benzene and ethanol.

Fire and Explosion Hazard

The flash point of this compound is 26°C (79°F) (Bretzinger and Josten). It is less flammable than the *sec*- and *tert*-alkoxides of alkali metals. Ignition may occur when this compound is heated in moist air. It

TABLE 31.1   Miscellaneous Metal Alkoxides

Compounds/CAS No.	Formula/MW	Hazardous Properties
Sodium ethoxide [141-52-6]	$NaOC_2H_5$ 68.05	Flammable solid; ignites in moist air; also ignites on heating
Potassium ethoxide [917-58-8]	KOC_2H_5	Flammable solid; ignites in contact with moist air at room temperature, often after air induction period; reaction with water highly exothermic, heat of solution being 13 kcal/mol
Magnesium methoxide [109-88-6]	$Mg(OCH_3)_2$	Flammable solid; ignites in moist air; ignites upon heating; reacts exothermically with water
Magnesium ethoxide [2414-98-4]	$Mg(OC_2H_5)_2$ 114.44	Flammable solid; sensitive to moisture
Aluminum *sec*-butoxide [2269-22-9]	$[C_2H_5CH(CH_3)O]_3Al$	Flammable liquid; flash point 27°C (82°F) (Aldrich 1995); corrosive
Titanium(III)methoxide [7245-18-3]	$Ti(OCH_3)_3$	Flammable solid; ignites in air; moisture sensitive
Titanium(IV)ethoxide [3087-36-3]	$Ti(OC_2H_5)_4$ 228.15	Flammable liquid; flash point 29°C (84°F); moisture sensitive
Titanium(IV)isopropoxide [546-68-9]	$Ti[OCH(CH_3)_2]$ 284.26	Flammable liquid; flash point 22°C (72°F); sensitive to moisture
Zirconium tetra-*n*-butoxide [1071-76-7]	$Zr(OC_4H_9)_4$ 383.70	Flammable liquid; flash point 21°C (70°F); moisture sensitive

decomposes in water. The reaction is exothermic, producing isopropanol. It may decompose when heated to 250°C, producing highly flammable isopropyl ether.

REFERENCES

Aldrich Chemical 1995. *Catalog Handbook of Fine Chemicals*. Milwaukee, WI: Aldrich Chemical Company.

Bretherick, L. 1995. *Handbook of Reactive Chemical Hazards*, 5th. edition. P. G. Urben, ed. Oxford, UK: Butterworth-Heinemann.

Kolditz, L. 1965. *Z. Anorg. Chem. 341*: 88–92.

Manwaring, R. 1973. *Chem & Ind.* 172.

NFPA. 1997. *Fire Protection Guide on Hazardous Materials*, 12th ed. Quincy, MA: National Fire Protection Association.

32

METAL ALKYLS

32.1 GENERAL DISCUSSION

Metal alkyls constitute one of the most important classes of organometallics. These compounds have the general formula R_nM, where M is the metal atom, n its valency, and R an alkyl group, usually methyl, ethyl, vinyl, allyl, butyl, or isobutyl group. These substances are formed by covalent binding of the single electron of the C atom of the alkyl radicals with metal atoms. Thus, the number of alkyl radicals bound to the metal atom is equal to the valency of the metal. Most metals listed in the Periodic Table, especially the normal elements, are known to form such alkyl compounds. Examples of such compounds, formed by elements of each group of metals are presented below in Table 32.1. The transition metal alkyls are fewer in number than the alkyls of normal elements. No alkyl compound has been reported for any metal in group IIIB or IVB.

Metal alkyls are prepared by reaction of the metal with the alkyl halide or from a Grignard reagent (RMgX) in ether, as shown below in the following equations:

$$Zn(s) + CH_3I(l) \xrightarrow{\text{ether}} CH_3ZnI(l)$$

$$2CH_3ZnI(l) \xrightarrow{\text{ether}} Zn(CH_3)_2^{(l)} + ZnI_2(s)$$

The alkyls formed may be determined by NMR or FTIR techniques. More stable compounds, such as tetraethyl lead, may be identified by GC/MS as well. Many metal alkyls are used as alkylating or reducing agents in organic or inorganic syntheses. Aluminum alkyls are employed as catalysts in many polymerization and hydrogenation processes.

Fire and Explosion Hazard

The single most hazardous property of metal alkyls is their high flammability and detonating reactions. All metal alkyls are pyrophoric, spontaneously igniting in air. Some compounds, however, such as trimethylaluminum, are extremely pyrophoric. The latter types of compounds are formed by some of the alkali metals, magnesium and aluminum, attached to lower alkyl groups with one or two C atoms. Also, the branched alkyl compounds are more pyrophoric than the straight-chain alkyl metals containing the same number of C atoms. Such a pattern of higher reactivity with branching is found to be more conspicuous among the trialkylaluminums. All such

TABLE 32.1   Examples of Metal Alkyls of Elements from Each Group of Metals

Group	Elements	Metal Alkyls/CAS No.
I A	Lithium	Butyllithium [109-72-8]
II A	Magnesium	Dibutylmagnesium [81065-77-2]
III A	Aluminum	Triethylaluminum [97-93-8]
IV A	Tin	Tetramethyltin [594-27-4]
V A	Antimony	Triethylantimony
VI A	Tellurium	Tetramethyltellurium (IV)
I B	Copper	Propylcopper (I)
II B	Zinc	Diethylzinc [557-20-0]
V B	Tantalum	Pentamethyltantalum
VI B	Chromium	Tetraisopropylchromium
VII B	Manganese	Dimethylmanganese
VIII	Platinum	Tetramethylplatinum

pyrophoric reactions produce the corresponding metal oxides, as shown in the following equations:

$$Mg(C_2H_5)_2(l) + 7O_2(g) \longrightarrow MgO(s)$$
$$+ 4CO_2(g) + 5H_2O(l)$$

$$2Pb(CH_3)_4(l) + 15O_2(g) \longrightarrow 2PbO(s)$$
$$+ 8CO_2(g) + 12H_2O(l)$$

All metal alkyls are moisture sensitive, reacting violently with water. The reaction produces a metal hydroxide as shown below in the following examples:

$$Zn(C_2H_5)_2(l) + 2H_2O(l) \longrightarrow Zn(OH)_2(s)$$
$$+ 2C_2H_6(g)$$

$$Al(C_4H_9)_3(l) + 3H_2O(l) \longrightarrow Al(OH)_3(s)$$
$$+ C_4H_{10}(g)$$

The lower alkyls of lithium, sodium, potassium, rubidium, beryllium, magnesium, and aluminum decompose in water, with detonation. All metal alkyls react violently with halogens and other oxidizing substances. The alkyls of reactive metals are susceptible to ignition in contact with any compound that contains oxygen, even CO_2.

For example, diethylmagnesium [557-18-6] ignites and catches fire in CO_2.

Fires from these substances should be extinguished with bicarbonate or other dry chemical powder (NRC 1995). Sand, graphite, or other inert substances may be used to smother and put out fires. Water or CO_2 must not be applied to extinguish such fires.

Toxicity

Some metal alkyls, especially those of heavy metals, such as dimethylmercury, tetraethyltin, or tetraethyl lead, are highly toxic. The toxicity of these alkyls are attributed to the toxicity of the metals in the compounds. Since most metal alkyls are pyrophoric and react violently with water, any contact of these substances with the moisture in the tissues of the skin can cause sever burn. Similarly, contact with eyes can result in permanent injury. Some lower alkyls of lighter elements have low but significant vapor pressure to present inhalation risk. Their vapors can damage the respiratory tract. The hazardous properties of some selected metal alkyls of commercial applications are discussed in detail under their titled entries. Miscellaneous metal alkyls are presented in Table 32.2.

TABLE 32.2 Miscellaneous Metal Alkyls

Compound/CAS No.	Formula/MW	Hazardous Properties
Ethyllithium [811-49-4]	LiCH$_3$ 21.976	Highly pyrophoric; ignites spontaneously in air; reacts explosively with water, alcohols, acids, halogenated hydrocarbons, and oxidizers
n-Butyllithium [109-72-8] *sec*-Butyllithium [598-30-1] *tert*-Butyllithium [594-19-4]	LiC$_4$H$_9$ 64.057	All the three isomers ignite spontaneously in air; tert-isomer being most pyrophoric; solutions of butyllithium at concentrations of >50% in hydrocarbon solvents also ignite immediately on exposure to air (NRC 1995); reacts explosively with water producing butane; ignite on contact with CO$_2$; violent reactions with acids, halogenated hydrocarbons, lower alcohols, and oxidizers
Ethylsodium [676-54-0]	NaC$_2$H$_5$ 52.052	Highly pyrophoric powder material, igniting spontaneously in air; reacts explosively with water; violent reactions with acids, lower alcohols, halogenated hydrocarbons, and oxidizers
Methylpotassium [17814-73-2]	KCH$_3$ 54.133	Highly pyrophoric, ignites spontaneously in air; reacts explosively with water, alcohols, acids, halogenated hydrocarbons, and oxidizers
Dimethylzinc [544-97-8]	Zn(CH$_3$)$_2$ 95.460	Pyrophoric; ignites spontaneously in air; explodes in oxygen; reacts explosively with water, lower alcohols, halogenated hydrocarbons, and oxidizers; violent reactions can occur with halogens, SO$_2$ and nitroorganics; susceptible to ignition when mixed with chlorides of phosphorus, arsenic, and antimony
Dimethylmagnesium [2999-74-8]	Mg(CH$_3$)$_2$ 54.375	Highly pyrophoric, igniting spontaneously in air; ignites in N$_2$O and CO$_2$; reacts explosively with water; reacts violently with lower alcohols and ammonia; soluble in ether; decomposes when heated under vacuum
Dimthylmanganese [33212-68-9]	Mn(CH$_3$)$_2$ 85.001	Polymeric powder; ignites in air; reacts violently with water
Dimethylberyllium [506-63-8]	Be(CH$_3$)$_2$ 39.082	Ignites spontaneously in air; ignites in CO$_2$; ethereal solution also ignites spontaneously in air; reacts explosively with water; reactions with lower alcohols, halogenated hydrocarbons, ammonia, and oxidizing substances expected to be violent to explosive
Dimethylcadmium [506-82-1]	Cd(CH$_3$)$_2$ 142.48	Ignites in air (slow oxidation); forms peroxide which explodes on friction (Davies 1961); reacts violently with water
Trimethylantimony [594-10-5]	Sb(CH$_3$)$_3$ 166.86	Ignites spontaneously in air; reacts violently with halogens and other oxidizing substances
Trimethylgallium [1445-79-0]	Ga(CH$_3$)$_3$ 114.83	Highly pyrophoric; ignites in air even at −76°C (Sidgwick, 1950); violent reactions with water, lower alcohols and many organics; reacts violently with oxidizers; likely to form explosive complex with diethyl ether
Trimethylthallium [3003-15-4]	Tl(CH$_3$)$_3$ 249.49	Ignites spontaneously in air; explodes on heating; explosion reported in contact with diethyl ether at 0°C (Bretherick 1995)
Trimethylindium [3385-78-2]	In(CH$_3$)$_3$ 159.92	Ignites in air; forms explosive complex with diethyl ether

TABLE 32.2 *(Continued)*

Compound/CAS No.	Formula/MW	Hazardous Properties
Trimethylbismuth [593-91-9]	$Bi(CH_3)_3$ 254.08	Ignites spontaneously in air; explodes on heating above 110°C (Bretherick 1995); stable towards water (Coates 1967) and alcohol
Triethylgallium [1115-99-7]	$Ga(C_2H_5)_3$ 156.91	Ignites spontaneously in air; reacts violently with water and alcohol; forms explosive complex with diethyl ether
Tetramethylgermanium [865-52-1]	$Ge(CH_3)_4$ 132.75	Highly flammable liquid; flash point −39°F; boils at 45°C; reacts violently with strong oxidizers; irritant
Tetraethylgermanium [597-63-7]	$Ge(C_2H_5)_4$ 188.86	Combustible liquid; flash point 95°F; boils at 163°C; susceptible to react violently with strong oxidizers; irritant
Tetraethylsilane [631-36-7]	$Si(C_2H_5)_4$ 144.33	Flammable liquid; flash point 79°F; boils at 153°C; the flammability and ease of oxidation much lower than the mono-, di-, and trialkylsilanes
Tetramethylsilane [75-76-3]	$Si(CH_3)_4$ 88.225	Highly volatile and flammable liquid; flash point — 17°F (−27°C); boils at 27°C; hygroscopic; reacts violently with strong oxidizers; explosive reaction with chlorine at 100°C catalyzed by $SbCl_3$
Tetramethylplatinum [22295-11-0]	$Pt(CH_3)_4$ 255.23	Stable at ambient temperature; however, explodes weakly on heating; vigorous to violent reactions with strong oxidizers
Tetramethyllead [75-74-1]	$Pb(CH_3)_4$ 267.35	Flammable liquid; flash point 100°F (Sax 1996); stable at room temperature but explodes when exposed to flame or heated to >90°C; LEL 1.8%; explodes with tetrachlorotrifluoromethylphosphorane under vacuum at ambient temperature; highly toxic by ingestion; moderately toxic by skin contact; toxic effects are those of lead; oral LD_{50} (rat): 105 mg/kg; TLV-TWA 0.15 mg (Pb)/m^3 (skin) (ACGIH); PEL-TWA 0.075 mg (Pb)/m^3 (skin) (OSHA)
Tetramethyltellurium (IV) [123311-08-0]	$Te(CH_3)_4$ 187.74	Ignites spontaneously in air; explodes on warming or in contact with oxidizers
Pentamethyltantalum [53378-72-6]	$Ta(CH_3)_5$ 256.12	Extremely pyrophoric; reacts explosively with water; a dangerous heat and shock sensitive substance; warming of frozen material or shaking could cause violent explosion
Hexamethylrhenium [56090-02-9]	$Re(CH_3)_6$ 276.42	Dangerously sensitive to heat and shock; warming of frozen sample or transferring the material can cause violent explosion (Bretherick 1995); unstable above −20°C; likely to explode even in absence of oxygen under vacuum; can ignite and explode in contact with oxygen and water

Storage

Because of their pyrophoric nature, all metal alkyls are packed and stored under dry nitrogen, hydrogen, or an inert gas. No oxygen or moisture should be present in these gases or in the solvents employed in the synthetic applications. Hexane, heptane, ether, toluene, and tetrahydrofuran are some of the solvents that may be used to store these substances.

Disposal/Destruction

In laboratory, small amounts of metal alkyls may be destroyed by diluting the pure compounds or its more concentrated solutions to a concentration below 5% with a hydrocarbon solvent, such as hexane or toluene. Alternatively, a water-miscible solvent, such as ethanol or *tert*-butanol may be used. Small volumes of such solutions are then slowly and cautiously added to water in wide-mouthed containers in a hood and swirled gently. The metal alkyls are converted into their oxides or hydroxides. The organic solvent, if immiscible in water, is separated and evaporated in a hood. The entire content may, alternatively, be placed in waste containers and labeled for disposal. The toxic oxides or hydroxides of the metals formed are disposed for landfill burial; while the nontoxic metal oxides or hydroxides could be flushed down the drain. Alternatively, the metal alkyl solution or its waste may be diluted to a concentration below 5% with toluene or heptane. The diluted solution is then placed in a labeled container under argon for waste disposal.

32.2   TRIMETHYLALUMINUM

Formula: $Al(CH_3)_3$; MW 72.08; readily forms a dimer $[Al(CH_3)_3]_2$; CAS [85-24-1]

Synonym: aluminum trimethyl

Uses and Exposure Risk

Trimethyl aluminum is a highly reactive reducing and alkylating agent. It is used in a Ziegler-Natta catalyst for polymerization and hydrogenation.

Physical Properties

Colorless liquid; corrosive odor; boils at 125°C; vapor pressure 12 torr at 25°C; vapor decomposes above 150°C; solidifies at 15.4°C; density 0.75 g/mL; soluble in ether, toluene, hexane; reacts explosively with water.

Fire and Explosion Hazard

Trimethylaluminum is most reactive and volatile among all organoaluminum compounds. It is extremely pyrophoric, igniting spontaneously in air, even as a frozen solid. Its flash point is −18°C (NRC 1995). It decomposes and detonates when combined with water. It reacts violently with lower alcohols, halogenated hydrocarbons, and oxidizing substances. Reactions with low-molecular-weight aldehydes, ketones and amides could be violent to explosive (especially upon warming or slight heating).

Health Hazard

As it is pyrophoric and reacts explosively with moisture, skin contact can cause a dangerous burn. Contact with eyes can cause blindness. Because of its significant volatility, the risk of inhalation of this compound is higher than with most other alkyls. Inhalation of its vapors can severely damage the respiratory tract.

TLV-TWA: 2 mg(Al)/m^3 (ACGIH)

32.3   TRIETHYLALUMINUM

Formula: $Al(C_2H_5)_3$; MW 114.17; CAS [97-93-8]

Synonym: aluminum triethyl

Uses and Exposure Risk

Triethylaluminum, in combination with many transition metal complexes, is used as

Ziegler-Natta polymerization and hydrogenation catalyst. Also, it is used as intermediate in organic syntheses.

Physical Properties

Colorless liquid; occurs as a dimer; boils at 194°C; density 0.83 g/mL at 25°C; soluble in ether, hexane, isooctane; reacts violently with water, alcohol, chloroform, carbon tetrachloride, and many other solvents.

Fire and Explosion Hazard

Triethylaluminum is extremely pyrophoric, igniting spontaneously in air. It reacts violently with water, alcohol, halogenated hydrocarbons, and oxidizing substances. Among the alcohols, the lower alcohols, methanol, ethanol, n-propanol, and isopropyl alcohol, react explosively with triethylaluminum. Reactions with lower aldehydes, ketones and amides can be vigorous to violent. It may explode on contact with halocarbons in excess molar ratios or upon slight warming. When heated to 200°C, it decomposes, liberating ethylene and hydrogen.

Health Hazard

The health hazard from exposure to this compound is attributed to its violent reactions with many substances, including air and water. Because of its violent reaction with moisture, skin contact can cause a dangerous burn. Contact with eyes can damage vision.

32.4   TRIISOBUTYLALUMINUM

Formula: $[(CH_3)_2CHCH_2]_3Al$ MW 198.33; CAS [100-99-2]
Synonym: aluminum triisobutyl

Uses and Exposure Risk

Triisobutylaluminum is used as a reducing agent. It is also used in combination with transition metal compounds as a Ziegler-Natta catalyst in polymerization and hydrogenation reactions. A dilute solution of the compound is employed in commercial applications.

Physical Properties

Colorless liquid; exists as a monomer; decomposes above 50°C; solidifies at 6°C; density 0.78 g/mL at 25°C; soluble in ether, hexane and toluene; reacts explosively with water.

Fire and Explosion Hazard

It is a highly pyrophoric compound, igniting spontaneously in air. The flash point is measured to be $-1°F$ ($-18°C$) (Aldrich 1996). A 1.0 M solution in hexane or toluene is pyrophoric too. It decomposes explosively with water. Reactions with lower alcohols, halogenated hydrocarbons, halogens, and common oxidizing substances can be violent or explosive. Triisobutylaluminum is thermally less stable than triethylaluminum, decomposing above 50°C, producing isobutene and hydrogen.

Health Hazard

Being moisture sensitive, the pure liquid or its concentration solution can cause server burns to the skin.

32.5   DIETHYLMAGNESIUM

Formula: $Mg(C_2H_5)_2$; MW 82.44; CAS [557-18-6];
Synonym: magnesium diethyl

Uses and Exposure Risk

Diethylmagnesium is used as intermediate in organic synthesis.

Physical Properties

Colorless liquid; forms solvated crystals from ether, $(C_2H_5)_2Mg \cdot (C_2H_5)_2O$; loses its ether of crystallization when heated under vacuum; freezes at $0°C$; decomposes at $250°$ under high vacuum; soluble in ether.

Fire and Explosion Hazard

Diethylmagnesium catches fire spontaneously in air. It is susceptible to glow and can catch fire in other gases, as well, that contain oxygen atoms in the molecules, such as N_2O and even CO_2. It explodes with water. The ether solution of the compound is also susceptible to ignition, upon contact with water (Bretherick 1995). It reacts violently with lower alcohols and ammonia (Sidgwick 1950).

Health Hazard

Skin contact can cause severe burns.

32.6   DIETHYLZINC

Formula: $ZN(C_2H_5)_2$; MW 123.50; CAS [544-97-8];

Synonym: zinc diethyl

Uses and Exposure Risk

Diethyl zinc is used in organic synthesis. It is also used in preservation of archival papers.

Physical Properties

Colorless liquid; boils at $118°C$; freezes at $-28°C$; density 1.21 g/mL; soluble in ether, hexane, and benzene; reacts violently with water.

Fire and Explosion Hazard

Diethylzinc ignites spontaneously in air, burning with a blue flame. Reactions with water and lower alcohols can be violent. Violent reactions can occur with halogens,

halogenated hydrocarbons, nitroorganics, oxidizers, sulfur dioxide, and chlorides of phosphorus, arsenic, and antimony. With the latter compounds, diethylzinc forms pyrophoric triethylphosphine, triethyl arsine, and triethylstibine, respectively.

Health Hazard

Being moisture sensitive, any accidental contact of the pure liquid or its concentrated solution with the skin can cause a severe burn.

32.7   DIETHYLBERYLLIUM

Formula: $Be(C_2H_5)_2$; MW 67.13; CAS [542-63-2];

Synonym: beryllium diethyl

Uses and Exposure Risk

Diethylberyllium is used as an intermediate in organic synthesis.

Physical Properties

Colorless liquid; decomposes at $85°C$; soluble in ether, hexane, and benzene; reacts violently with water.

Fire and Explosion Hazard

Diethylberyllium ignites spontaneously in air, producing dense white fumes of beryllium oxide. Ether solution of this compound is also highly flammable, igniting spontaneously in air. It reacts explosively with water. Reactions with lower alcohols, halogens, halogenated hydrocarbons, ammonia, and oxidizers may proceed to explosive violence.

Health Hazard

An accidental spill of the pure liquid or its concentrated solution on the skin can cause a severe burn.

32.8 TRIETHYLSILANE

Formula: $(C_2H_5)_3SiH$; MW 116.28; CAS [617-86-7];

Synonyms: silicon triethyl hydride, triethyl-silicon hydride

Uses and Exposure Risk

Triethylsilane is used as a reducing agent in many organic synthetic reactions.

Physical Properties

Colorless liquid; boils at 109°C; density 0.73 g/mL; insoluble in water

Fire and Explosion Hazard

Highly flammable; flash point −4°C (25°F) (Aldrich 1996).

It does not ignite spontaneously in air, but it can explode on heating. The ease of oxidation of this compound is relatively lower than mono- and dialkylsilanes. Reaction with oxidizing substances, however, can be violent. Reaction with boron trichloride could be explosive at room temperature. Even at −78°C, mixing these reagents caused pressure buildup and combustion (Matteson 1990).

32.9 TRIETHYLBORANE

Formula: $(C_2H_5)_3B$; MW 98.00; CAS [97-94-9];

Synonym: boron triethyl

Uses and Exposure Risk

It is used in mixtures with triethylaluminum as hypergolic igniters in rocket propulsion systems.

Physical Properties

Colorless liquid; boils at 95°C; solidifies at −93°C; density 0.70 g/mL at 23°C; soluble in ether, alcohol, hexane and tetrahydrofuran; reacts violently with water.

Fire and Explosion Hazard

Triethylborane is a pyrophoric substance, igniting spontaneously on exposure to air, chlorine, or bromine. Explosion may result when mixed with oxygen. Contact with ozone, peroxides and other oxidants can cause explosion. Its concentrated solutions can be pyrophoric. Flash point of 1 M solution in hexane is −23°C (−9°F), while that in tetrahydrofuran is −17°C (1°F) (Aldrich 1996). It decomposes explosively when mixed with water.

Health Hazard

Triethylboron is a toxic substance. The toxic symptoms from chronic exposure to this compound are not reported. Skin contact can burn tissues.

32.10 TETRAMETHYLTIN

Formula: $(CH_3)_4Sn$; MW 178.85; CAS [594-27-4]

Synonym: tetramethylstannane, tin tetramethyl

Uses and Exposure Risk

Tetramethyltin is used in many organic synthetic reactions.

Physical Properties

Colorless liquid; boils at 78°C; highly volatile; solidifies at −55°C; thermally stable to 400°C (Meyer 1989); density 1.314 g/mL; insoluble in water; soluble in carbon disulfide, and other organic solvents.

Fire and Explosion Hazard

The ease of oxidation, however, is low in comparison with most other metal alkyls. It is relatively stable to air and does not ignite spontaneously as other alkyls do. Also, it is stable in water. The compound, however, is highly flammable, the flash point being (−13°C) 9°F (Aldrich 1996). It explodes violently in contact with dinitrogen tetraoxide even, at −80°C (−112°F) (Bailar 1973). Reaction with strong oxidizers is expected to be vigorous to violent.

Health Hazard

Tetramethyltin is moderately toxic by all routes of exposure. The symptoms include headache, muscle weakness, and paralysis. Exposure limits:

TLV-TWA: 0.1 mg (Sn)/m^3 (skin) (ACGIH)
PEL-TWA: 0.1 mg (Sn)/m^3 (skin) (OSHA)

32.11 DIMETHYLMERCURY

Formula: $Hg(CH_3)_2$; MW 230.66; CAS [593-74-8]
Synonyms: methyl mercury, mercury dimethyl

Uses and Exposure Risk

Dimethylmercury is used as a reagent in inorganic synthesis, and as a reference standard for mercury nuclear magnetic resonance (Hg NMR). It is an environmental pollutant found in bottom sediments and also in the bodies of fishes and birds (Merck 1996). It occurs in fishes and birds along with monomethylmercury.

Physical Properties

Colorless liquid with faint sweet odor, boils at 94°C, freezes at −43°C, density 3.19 g/mL at 20°C, soluble in ether and alcohols; insoluble in water.

Fire and Explosion Hazard

It is a flammable liquid; flash point 101°F. The flammability of this compound, its ease of oxidation and the energy of decomposition is relatively lower than the alkyls of lighter metals. It is mildly endothermic. The heat of formation, ΔH_f° is +75.3 kJ/mol (Bretherick 1995). Unlike most other metal alkyls formed by elements of lower atomic numbers, dimethylmercury does not pose any serious fire or explosion hazard. Although it does not ignite in air, the compound is easily inflammable. It dissolves in lower alcohols without any violent decomposition. Heating with oxidizing substances can cause explosion. Violent explosion is reported with diboron tetrachloride at −63°C under vacuum (Wartik et al. 1971).

Health Hazard

All alkylmercury compounds are highly toxic by all routes of exposure. There are many serious cases of human poisoning from methylmercury (Lu 1966). Outbreaks of mass poisoning from consumption of contaminated fish occurred in Japan during the 1950s, causing a severe neurological disease, so-called "Minamata disease," which resulted in hundreds of deaths. A similar outbreak of food poisoning from contaminated wheat caused several hundred deaths in Iraq in 1972. Recently there was a tragic death from a single acute transdermal exposure to dimethylmercury (estimated between 0.1 to 0.5 ml) that penetrated into the skin through disposable latex gloves (Blayney et al. 1997; The New York Times, June 11, 1997). The symptoms reported were episodes of nausea and vomiting occurring three months after the exposure followed by onset of ataxia, slurred speech (dysarthia), and loss of vision and hearing two months after that. The death occurred in about six months after the accident.

Methylmercury can concentrate in certain fetal organs, such as the brain. The target organs are the brain and the central nervous system. It can cause death, miscarriage, and deformed fetuses. Unlike inorganic mercury compounds, it can penetrate through the membrane barrier of the erythrocyte, accumulating at about 10 times greater concentration than that in the plasma (WHO 1976). Its rate of excretion on the blood level is very slow. It gradually accumulates in the blood. Such accumulation was found to reach 60% equilibrium at about 90 days, culminated after 270 days (Munro and Willes, 1978). Skin absorption exhibits the symptoms of mercury poisoning. The toxic threshold concentration of mercury in the whole blood is usually in the range 40 to 50 µg/L, while the normal range should be below 10 µg/L.

Exposure Limit

TLV-TWA: 0.01 mg (Hg)/M^3 (ACGIH)
PEL-TWA: 0.01 mg (Hg)/M^3 (OSHA)
STEL: 0.03 mg (Hg)/M^3 (ACGIH)

32.12 TETRAETHYLLEAD

EPA Designated Toxic Waste, RCRA Waste Number P110; DOT Label: Poison, Flammable Liquid

Formula: $(C_2H_5)_4Pb$; MW 323.45; CAS [78-00-0]

Synonyms: lead tetraethyl, tetraethylplumbane

Uses and Exposure Risk

Tetraethyllead is used as an additive to gasoline to prevent knocking in motors. However, because of its high toxicity and the pollution problem, its use in gasoline has been drastically curtailed.

Physical Properties

Colorless liquid; boils at 200°C; decomposes at 228°C; density 1.65 g/mL at 20°C; insoluble in water; miscible with benzene, toluene, petroleum ether, and gasoline.

Fire and Explosion Hazard

Combustible liquid; flash point 163°F. It is least combustible among the heavy metal alkyls. The fire hazard is significantly low. The rate and spontaneity of air oxidation is also low. However, because of its moderate endothermicity (ΔH_f° within the range +215 kcal/mol), it should be reasonably unstable and therefore susceptible to violent decomposition. Heating, rigorous agitation or prolonged contact with oxidants and nonmetal halides should, therefore, be considered a potential hazard. It is liable to explode if exposed to air for several days.

Health Hazard

Tetraethyllead is highly toxic by oral route. The LD$_{50}$ values for rats, mice, and rabbits were all found to be <15 mg/kg when administered by the oral, intravenous, subcutaneous, intraperitoneal, and parenteral routes. It may be absorbed through the skin causing lead poisoning. It is toxic to the central nervous system. The toxicity, however, is low to moderate by dermal route. It is an acute as well as chronic toxicant. The toxic effects are insomnia, hypotension, tremor, hypothermia, pallor, weight loss, hallucination, nausea, convulsion, and coma. The toxicity of this compound by inhalation route is also low to moderate. Because of its low vapor pressure, 0.2 torr at 20°C, inhalation hazard is relatively low.

LD$_{50}$ oral (rat): 12 mg/kg
Oral lethal dose (rabbit): 30 mg/kg
LC$_{50}$ inhalation (rat): 850 mg/m^3/1 hr.

Exposure Limit

TLV-TWA 0.1 mg (Pb)/m^3 (skin) (ACGIH)
PEL-TWA 0.075 mg (Pb)/m^3 (skin) (OSHA)

32.13 MISCELLANEOUS METAL ALKYLS

Hazardous properties, including toxicity, flammability, and explosive reactions of miscellaneous metal alkyls are presented in Table 32.2.

References

Aldrich 1996. *Catalog Handbook of Fine Chemicals*. Milwaukee, WI: Aldrich Chemical Company.

Bretherick, L. 1995. *Handbook of Reactive Chemical Hazards*, 5th edition. P.G. Urben, ed. Oxford: Butterworth-Heinemann.

Bailar, J. C., H. J. Emelus, and R. S. Nyholm. 1973. *Comprehensive Inorganic Chemistry*. A. F. Trotman-Dickenson, ed. Vol. 2, p. 355. Oxford: Pergamon.

Blayney, M. B., J. S. Winn and D. W. Nierenberg. 1997. Handling diamethylmercury. *Chem. Engg. News*, 75(19), May 12, 1997.

Croates, G. E., M. L. H. Green, and K. Wade. 1967–68. *Organometallic Compounds*, Vols. 1 and 2. London: Methuen.

Davies, S. G. 1961. *Organic Peroxides*. London: Butterworths.

Lewis R. J. 1996. *Sax's Dangerous Properties of Industrial Materials*, 9th ed. New York: Van Nostrand Reinhold.

Lu, F. C. 1996. *Basic Toxicology. Fundamentals, Target Organs and Risk Assessment*. 3rd ed. Washington, DC: Taylor & Francis.

Matteson, D. S. 1990. *J. Org. Chem. 55*: 2274.

Merck. 1996. *The Merck Index*, 12th ed. Rahway, NJ: Merck & Co.

Meyer, E. 1989. *Chemistry of Hazardous Materials*, 2nd ed. p. 254. Englewood Cliffs, NJ: Prentice-Hall.

Munro, I. O., and A. F. Willes. 1978. Reproductive toxicity and the problems of in vitro exposure, in *Chemical Toxicology of Food*, A. Galli, R. Paoletti, and G. Veterazzi, eds. pp. 133–145. Amsterdam: Elsevier.

National Research Council. 1995. *Prudent Practices in the Laboratory: Handling and Disposal of Chemicals*. Washington DC: National Academy Press.

Sidgwick, N. V. 1950. *The Chemical Elements and Their Compounds*. Oxford: Oxford University Press.

The New York Times, New York. June 11, 1997. News report, Exposure to mercury kills professor at Dartmouth.

World Health Organization. 1986. *Principles of Toxicokinetic Studies. Environmental Health Criteria 57*. Geneva: WHO.

33

METAL AZIDES

33.1 GENERAL DISCUSSION

Azides form highly explosive shock- and heat-sensitive salts with many metals. They are derivatives of hydrazoic acid, HN_3, containing a $-N$ moiety attached to the metal, by ionic or covalent bond. The azide group has a linear structure that is stabilized by resonance as follows:

$$RN\overset{+}{=}N\overset{-}{=}\overset{..}{\underset{..}{N}}: \longleftrightarrow R\overset{-}{\underset{..}{N}}-N\overset{+}{\equiv}N:$$

Azide anion is a good nucleophile. Many azide reactions proceed via formation of an active intermediate, nitrene, analogous to carbene. This results from the photolysis of hydrazoic acid:

$$H-\overset{-}{\underset{..}{N}}-N\overset{+}{\equiv}N: \xrightarrow{h\nu} H-\overset{..}{\underset{..}{N}} + N_2$$

(hydrazoic acid)   (nitrene)

Alkali metal azides are prepared by passing nitrous oxide into the molten metal amide. Azides of other metals are made by reacting sodium azide with the corresponding metal or alkyl salts. Whereas alkali metal azides are inert to shock, the salts of copper, silver, lead, mercury, and most heavy metals are dangerously shock sensitive.

Azides react explosively or form other explosive azides when they come in contact with a number of substances. With acids, almost all metal azides react to form hydrazoic acid which is dangerously sensitive to heat, friction or impact. Azides can react with salt solutions of many metals forming azides of those metals, some of which — especially, the heavy metal azides — are highly sensitive to friction and impact. The rates and yields of such reaction products would depend on the equilibrium constants and solubility products. For example, soluble alkali metal azides can readily form lead or cadmium azide when mixed with a salt solution of lead or cadmium. The hazardous properties of some of the compounds of this class are discussed below.

The metal derivatives of all azides (especially the heavy metal derivatives) are thermodynamically unstable. These endothermic compounds have high positive Gibbs free energy and can explode on impact or heating. The explosivity of these compounds is a function of their thermodynamic unstability and decreases in the following order for some of these azides: mercuric azide > cadmium azide > lead azide > hydrazoic acid > silver azide > zinc azide > cupric azide >

577

TABLE 33.1 Heats of Formation of Selected Azides

Azide[a]	Formula	Heat of Formation $\Delta H_f^\circ(s)$ (kcal/mol)
Mercuric azide	$Hg(N_3)_2$	+133.0
Cadmium azide	$Cd(N_3)_2$	+107.8
Lead azide	$Pb(N_3)_2$	+104.3
Hydrazoic acid	HN_3	+70.3
Silver azide	AgN_3	+66.8
Cuprous azide	$Cu_2(N_3)_2$	+60.5
Strontium azide	$Sr(N_3)_2$	+48.9
Sodium azide	NaN_3	+16.8
Calcium azide	$Ca(N_3)_2$	+11.0

[a]Pure solid in crystalline form.

cuprous azide > strontium azide > calcium azide.

The decreasing pattern above is of little practical interest, however, as all the heavy metal azides detonate violently upon heating and mechanical impact. Table 33.1 lists the heat of formation $\Delta H_f^\circ(s)$ for some azides. It may be seen that explosivity decreases with a decrease of ΔH_f° and at a low value of +16.8 kcal/mol, sodium azide is nonexplosive. Discussed below are individual compounds of commercial interest or those presenting severe explosion hazard. The explosive properties of additional compounds are highlighted in Table 33.2.

33.2 HYDRAZOIC ACID

Formula HN_3; MW 43.04; CAS [7782-79-8]
Structure:

$$H-\overset{..}{\underset{..}{N}}-\overset{+}{N}\equiv N:$$

Synonyms: hydrogen azide; azoimide; triazoic acid; hydronitric acid

Uses and Exposure Risk

Hydrazoic acid is used in making heavy metal azides for detonators. It forms readily when sodium azide reacts with acid or hydrazine is mixed with nitrous acid.

Physical Properties

Colorless mobile liquid with a strong pungent odor; bp 37°C; mp −80°C; soluble in water and organic solvents.

Health Hazard

The acute toxicity of hydrazoic acid through inhalation and other routes of exposure has been found to be high to very high. The symptoms and the intensity of poisoning are similar to sodium azide. It is, however, less toxic than hydrogen cyanide. In humans, inhalation of its vapors can produce irritation of eyes and respiratory tract, bronchitis, headache, dizziness, weakness, and decreased blood pressure (Matheson 1983). Prolonged exposure to high concentrations can result in collapse, convulsion, and death. An exposure to 1100 ppm for 1 hour was lethal to rats. Chronic exposure to a low level of this compound in air may produce hypotension.

Animals given intraperitoneal dosages of hydrazoic acid showed the symptoms of heavy breathing, convulsions, depression, and fall in blood pressure. It affected

TABLE 33.2   Miscellaneous Metal Azides

Compound/CAS No.	Formula/MW	Explosive Properties
Lithium azide [19597-69-4]	Li(N$_3$) 48.96	Colorless crystalline solid; inert to shock and stable at elevated temperature; reactions with lead, copper, silver, cadmium zinc, and mercury salts can form explosive azides of these metals; reactions with chlorine, bromine vapors and carbon disulfide can be violent
Magnesium azide [39108-12-8]	Mg(N$_3$)$_2$ 108.35	Crystalline solid; insensitive to shock; explodes on heating; reacts with acids forming hydrazoic acid; reacts with a number of heavy metal salts in solutions forming their explosive azides; susceptible to form unstable and explosive azodithioformate when its solution is mixed with carbon disulfide; forms explosive bromine azide with bromine vapors
Potassium azide [20762-60-1]	K(N$_3$) 81.118	Insensitive to shock or impact; rapid heating to high temperature can cause explosion; it melts first and then decomposes above melting point; evolves nitrogen on decomposition; reacts with acids forming explosive hydrazoic acid; forms explosive potassium azidodiathioformate when its aqueous solution is mixed with carbon disulfide; reacts with bromine vapor, forming explosive bromine azide; reaction with methylene chloride, chloroform, and other halogenated hydrocarbon solvents can produce diazidoalkanes; upon distillation or concentration, may explode; reactions with a number of salts of lead, copper, silver, mercury, cadmium, and other heavy metals can produce explosive heavy metal azides
Ammonium azide [12164-94-2]	NH$_4$(N$_3$) 60.059	Explodes on rapid heating, friction, and impact; reacts with acids and metals to form hydrazoic acid and metal azides, which explode readily
Calcium azide [19465-88-8]	Ca(N$_3$)$_2$ 124.12	Colorless crystalline solid; mildly endothermic, $\Delta H_f^\circ(s) + 46$ kJ/mol; not sensitive to shock; explodes on heating at 150°C; reactions with acids can form hydrazoic acid (explosive); susceptible to the formation of many heavy metal azides when mixed with their salts solutions
Strontium azide [19465-89-5]	Sr(N$_3$)$_2$ 171.66	More endothermic than calcium azide; $\Delta H_f^\circ(s) + 205$ kJ/mol; moderately sensitive to impact; explodes on heating to 170°C; susceptible to form heavy metal azides; and hydrazoic acid on treatment with acids and heavy metal salt solutions
Barium azide [18810-58-7]	Ba(N$_3$)$_2$ 221.37	Crystalline solid; one of the metal azides that is not an endothermic compound, $\Delta H_f^\circ(s) - 22.2$ kJ/mol; however, highly sensitive to shock or impact when dry; when damp or moistened with a solvent or in solution, it is relatively insensitive to impact and safe to handle; presence of metal ions as impurities can enhance the danger of explosion; explodes when the anhydrous salt or the nonhydrate is heated to 200°C, forming nitride; forms explosive products, hydrazoic acid and azides of lead,

(*continued*)

TABLE 33.2 (*Continued*)

Compound/CAS No.	Formula/MW	Explosive Properties
		silver, and copper in contact with acids or the corresponding salts of these metals, respectively; DOT restricted material, Label: Class A Explosive and Poison; UN 0224 (when containing less than 50% water); Flammable Solid and Poison, UN 1571 (when containing 50% or more water)
Zinc azide [14215-28-2]	$Zn(N_3)_2$ 149.42	Colorless crystalline solid; explodes easily; highly sensitive to heat, shock, and static charge; forms explosive azides with heavy metal salts; forms explosive hydrazoic acid an reactions with acids
Mercury (II) azide [14215-33-9]	$Hg(N_3)_2$ 284.63	Dangerously explosive; highly endothermic salt, $\Delta H_f^\circ + 556.5$ kJ/mol (Bretherick 1995); explodes when subjected to heat or shock; can explode even on mild scratching of large crystals
Mercury (I) azide [38232-63-2]	$Hg_2(N_3)_2$ 485.06	Less sensitive than the Hg(II) salt, and other heavy metal azides; explodes on heating to above 270°C; forms explosive azides of heavy metals when mixed with their salt solutions; forms explosive hydrazoic acid with acids
Cadmium azide [14215-28-3]	$Cd(N_3)_2$ 196.45	Dangerously explosive similar to the azides of Hg(II), Pb and Ag; highly endothermic, $\Delta H_f^\circ(s)$ 451 kJ/mol (Bretherick 1995); sensitive to heat, shock, and friction; explodes on heating; slight friction or shock can cause explosion when the salt is in dry state; its solutions susceptible to explode spontaneously on long standing, especially when saturated, or upon crystallization or in presence of impurities; reacts with heavy metal salts solutions or acids, proceeding to explosive violence
Silicon tetrazide [27890-58-0]	$Si(N_3)_4$ 196.17	Dangerously explosive; explodes spontaneously when pure (Mellor 1967); highly sensitive to heat and shock; crystallization and purification should be handled very carefully; explodes on reactions with acids; susceptible to react with heavy metal salts forming their azides
Cobalt azide [14215-31-7]	$Co(N_3)_2$ 142.97	Explodes on heating at 200°C; susceptible to form hydrazoic acid with acids and heavy metal azides in contact with heavy metals or their salts
Nickel azide [59865-91-7]	$Ni(N_3)_2$ 142.70	Explodes at 200°C; no data available on shock or impact sensitivity; forms hydrazoic acid with acids
Palladium azide [13718-25-7]	$Pd(N_3)_2$ 190.44	Crystalline solid; explodes on heating; sensitive to impact when dry; forms hydrazoic acid with acids
Copper amine azide	$Cu(NH_3)_2(N_3)_2$ 181.65	Dark green crystals; explodes when heated to 105°C; reacts with acids forming hydrazoic acid; DOT classified forbidden substance; Hazard Label: Forbidden

the central nervous system, but no damage was observed in liver or kidney.

LD_{50} value, intraperitoneal (mice): 22 mg/kg

Exposure Limit

Ceiling 0.1 ppm vapor (ACGIH).

Fire and Explosion Hazard

In pure form or highly concentrated solution, hydrazoic acid is a dangerous explosive compound. It is unstable and sensitive to heat and shock. The explosion hazard decreases significantly with more dilute solutions.

It forms shock-sensitive metal azides when react with metal salts, and fluorine azide with fluorine (Lawless and Smith 1968) and susceptible to form chlorine azide and bromine azide with chlorine gas and bromine vapor. All these products can explode violently on impact. With carbon disulfide it forms a violently explosive salt (Mellor 1946; NFPA 1997).

Disposal/Destruction

Hydrazoic acid may be destroyed by converting it to sodium azide. The latter is decomposed with nitrous acid in a hood (National Research Council 1995). The following method is used. It is diluted in water to a strength below 5%; or its solution in organic solvents that is immiscible in water is shaken vigorously with water in a separatory funnel. The aqueous solution containing hydrazoic acid is neutralized with sodium hydroxide and separated from any organic layer. Sodium azide, so formed, is destroyed by reacting the aqueous solution with an excess of sodium nitrite followed by 20% sulfuric acid until the solution is acidic. The reaction is carried out in a three-necked flask equipped with a stirrer, a dropping funnel, and a gas outlet line to vent out nitric oxide. The reaction mixture is flushed down the drain.

33.3 SODIUM AZIDE

EPA Classified Acute Hazardous Waste, RCRA Waste Number P105
DOT Label: Poison B, UN 1687
Formula NaN_3; MW 65.02; CAS [26628-22-8]
Structure:

$$\overset{+}{Na} : \overset{-}{\underset{\cdot\cdot}{N}} = \overset{+}{N} = \overset{-}{\underset{\cdot\cdot}{N}} :$$

ionic bond between sodium ion and azide anion, hexagonal crystals.
Synonyms: azium; smite

Uses and Exposure Risk

Sodium azide is used in making other metal azides, therapeutically to control blood pressure, as a propellant for automotive safety bags, as a preservative for laboratory reagents, as an analytical reagent, and in organic synthesis. It is also used as an antifading reagent for immunofluorescence (Boeck et al. 1985).

Physical Properties

Colorless crystalline solid; decomposes at 300°C; density 1.846; readily soluble in water (41.7 g/100 mL at 17°C), soluble in alcohol (0.314 g/100 mL at 16°C) insoluble in ether.

Health Hazard

Sodium azide is a highly toxic compound; the order of toxicity is the same as that of hydrazoic acid. It is converted to hydrazoic acid in water. The aqueous solutions of sodium azide contains hydrazoic acid, which escapes at 37°C, presenting a danger of inhalation toxicity.

Sodium azide, by itself, is a severe acute toxicant causing hypotension, headache, tachpnea, hypothermia, and convulsion.

The toxic symptoms from ingestion of 100–200 mg in humans may result in headache, respiratory distress, hypermotility, and diarrhea. An oral intake of 10–20 g may be fatal to humans. The target organs are primarily the central nervous system and brain. There are reports on azide poisoning of brain and nerve tissue (ACGIH 1986; Mettler 1972). Owing to the availability of electron pairs in azide ion for coordinate bonding, azide forms strong complexes with hemoglobin, which blocks oxygen transport to the blood (Alben and Fager 1972).

LD$_{50}$ value, oral (rats): 27 mg/kg

Sodium azide is strongly mutagenic in the Ames test. However, it shows a very weak mutagenic effect on *Saccharomyces cerevisiae* C658-K42 (Morita et al. 1989). Macor et al. (1985) found that light decreased the mutagenic effect of sodium azide to induce hereditary bleaching of *Euglena gracilis*. Carcinogenicity of this compound on animals or humans has not yet been fully established, although skin and endocrine tumors in rats have been observer (NIOSH 1986).

Exposure Limit

Ceiling 0.3 mg/m^3 in air (ACGIH).

Fire and Explosion Hazard

Although metal azides are known explosives, sodium azide is inert to shock. However, violent decomposition may occur when heated to 275°C. It forms lead, copper, and other metal azides, which are highly detonating when it comes in contact with the corresponding metal salts. Therefore, pouring sodium azide solution into lead or copper drain must be avoided.

Reactions with chlorine or bromine vapor produce chloroazide or bromoazide, which can explode spontaneously. Explosions can occur when sodium azide reacts with carbon disulfide, chromyl chloride (Mellor 1946,

Suppl. 1967), and dibromomalononitrile (MCA 1962). Such violent explosions may be attributed to the formation of highly unstable reaction products that are extremely sensitive to shock and/or light.

Disposal/Destruction

Sodium azide may be destroyed by nitrous acid treatment (National Research Council 1995):

$$2NaN_3 + 2HNO_2 \longrightarrow 3N_2$$
$$+ 2NO + NaOH$$

The reaction is carried out in a three-necked flask equipped with a stirrer, a dropping funnel, and an outlet for toxic nitric oxide to carry the gas to an exhaust system. A 20% aqueous solution of sodium nitrite containing about 1.5 g of the salt per 1 g of sodium azide is added onto a 5% aqueous solution of sodium azide. This is followed by slow addition of 20% sulfuric acid, until the mixture is acidic. The decomposition is complete when starch–iodide paper turns blue, indicating the presence of excess nitrite. The reaction mixture is washed down the sink with a large volume of water.

Analysis

Sodium azide may be analyzed by reacting the aqueous solution with iodine. The reaction is induced by 2-mercaptopyrimidine (Kurzawa 1987). The excess of iodine is back-titrated using potassium iodide, sodium thiosulfate, and starch indicator. The azide functional group may be identified by FTIR.

33.4   LEAD (II) AZIDE

DOT Label: Class A Explosive. Forbidden (dry); when wetted with not less than 20% by weight water or mixed with mixture of water and alcohol, it is labeled Explosive 1.1A. UN 0129

Formula: PbN_6; MW 291.25; CAS [13424-46-9]

Structure: $Pb(N_3)_2$

Uses and Exposure Risk

Lead azide is used as a primary explosive in detonators and fuses to initiate the booster or bursting charge. Generally, it is used in destrinated form. Lead azide is also used in shells, cartridges, and percussion caps.

Physical Properties

Colorless needles or white powder; bp 350°C (explodes); density about 4.0 g/cm^3; slightly soluble in water (0.023% at 18°C and 0.09% at 70°C), soluble in acetic acid.

Health Hazard

Toxicity data for lead azide are not available. Its aqueous solution is toxic, exhibiting poisoning effect of lead.

Fire and Explosion Hazard

Lead azide is a primary explosive. Its detonation temperature is 350°C, and the detonation velocity is 5.1 km/s (Meyer 1989). Its heat of combustion and heat of detonation are 631 and 368 cal/g, respectively (or 184 and 107 kcal/mol). The released gas volume is 308 cm^3/g at STP. It forms highly shock-sensitive copper and zinc azides when mixed with the solutions of copper and zinc salts. Its contact with these metals or their alloys over a period of time results in the formation of their azides, too. Reaction with carbon disulfide is violently explosive. There is a report of an explosion resulting from the addition of calcium stearate in a lead azide preparation (MCA 1962).

Disposal/Destruction

Lead azide is decomposed by treatment with nitrous acid or ceric ammonium nitrate (Wear 1981).

33.5 LEAD (IV) AZIDE

Formula PbN_{12}; MW 375.28; CAS [73513-16-3]

Structure: $Pb(N_3)_4$

Synonym: lead tetrazide

Uses and Exposure Risk

The tetrazide salt, unlike the diazide has no commercial use because it is too unstable.

Health Hazard

No data are available on the toxicity of this unstable substance. It is expected to be highly toxic. The toxic effects should be those of lead.

Fire and Explosion Hazard

The tetrazide salt is unstable, decomposing spontaneously. It is highly sensitive to heat, shock and static charge. The decomposition can occur explosively (Mellor 1967).

33.6 SILVER AZIDE

DOT Label: Hazard: forbidden

Formula: AgN_3; MW: 149.80; CAS [13863-88-2];

Structure: $Ag(N_3)$

Uses and Exposure Risk

It is used as a primary explosive, often as a replacement for lead azide. It may be used in smaller quantities than lead azide as a initiator.

Physical Properties

White rhombic prism; crystal density 5.1 g/cm^3; melts at 252°C; insoluble in water.

Health Hazard

Silver azide is a highly toxic substance. The toxic effects have not been reported.

Fire and Explosion Hazard

It is a primary explosive. It explodes violently upon thermal and mechanical shock. It requires lesser energy for initiation than lead azide and also fires with a shorter time delay. The heats of combustion and detonation are 1037 and 454 cal/g, respectively (i.e., 156 and 68 kcal/mol, respectively). The detonation velocity is 6.8 km/sec (at the crystal density 5.1 g/cm^3). The pure compound explodes at 340°C (Mellor 1967). The detonation can occur at much lower temperatures in an electric field when initiated by irradiation. Also, the presence of impurities can lower down the temperature of detonation. Such impurities include oxides, sulfides, and selenides of copper and other metals.

It reacts with halogens forming halogen azides, often producing explosions. It reacts explosively with chlorine dioxide, most interhalogen compounds, and many dyes. It reacts with acids forming heat and shock sensitive hydrazoic acid. When mixed with salt solutions of heavy metals, their azides are formed, which are sensitive to heat, impact and electric charge.

33.7   COPPER AZIDE

Copper forms two azides: cuprous azide or copper (I) azide, and cupric azide or copper (II) azide. Cuprous azide, $Cu_2(N_3)_2$, MW 211.14, CAS [14336-80-2] is a colorless crystalline solid. It is highly sensitive to impact. It explodes on heating. It decomposes at 205°C. The heat of formation [$\Delta H_f^\circ(s)$] is reported as 253.1 kJ/mol (60 kcal/mol) (Bretherick 1995).

Cupric azide, $Cu(N_3)_2$, MW 147.59, CAS [14212-30-6] forms brownish red crystals.

Violent explosion can occur when this substance is subjected to shock or friction. Scratching the crystals out from the containers may result in explosion. It can detonate when dry. It may even explode when moist.

Both azides can readily form explosive products when combined with acids, halogens, halogen oxides, ammonia, or heavy metal salt solutions. Cupric azide has limited use as a powerful initiator. It is a DOT-classified forbidden substance; Hazard Label: Forbidden.

REFERENCES

ACGIH. 1986. *Documentation of the Threshold Limit Values and Biological Exposure Indices*, 5th ed. Cincinnati, OH: American Conference of Governmental Industrial Hygienists.

Alben, J. O., and L. Y. Fager. 1972. Infrared studies of azide bound to myoglobin and hemoglobin temperature dependence of ionicity. *Biochemistry 11*: 842.

Boeck, G., M. Hilchenback, K. Schauenstein, and G. Wick. 1985. Photometric analysis of antifading reagents for immunofluorescence with laser and conventional illumination sources. *J. Histochem. Cytochem. 33*(7): 699–705.

Bretherick, L. 1995. *Handbook of Reactive Chemical Hazards*, 5th ed. P. G. Urben, ed. Oxford, UK: Butterworth-Heinemann.

Fieser, L. F., and M. Fieser. 1944. *Organic Chemistry*. Boston: D.C. Heath.

Kurzawa, J. 1987. The iodine–azide reaction induced by mercaptopyrimidines and its application in chemical analysis. *Chem. Anal. 32*(6): 875–90; cited in *Chem. Abstr. Ca 111*(20): 186650a.

Lawless, E. W., and I. C. Smith. 1968. *High Energy Oxidizers*. New York: Marcel Dekker.

Macor, M., Ebringer, L. and P. Siekel. 1985. Hyperthemia and other factors increasing sensitivity of *Euglena* to mutagens and carcinogens. *Teratog. Carcinog. Mutagen. 5*(5): 329–37; cited in *Chem Abstr. CA 103*(21): 173797p.

Matheson, D. 1983. Hydrazoic acid and azide, in *Encyclopedia of Health and Occupational Safety*, 3rd ed. Vol. 1, L. Parmeggiani, ed. Geneva: International Labor Office.

MCA. 1962. *Case Histories of Accidents in the Chemical Industry*. Washington, DC: Manufacturing Chemists' Association.

Mellor, J. W. 1946–71. *A Comprehensive Treatise on Inorganic and Theoretical Chemistry*. London: Longmans, Green & Co.

Merck. 1996. *The Merck Index* 12th ed. Rahway, NJ: Merck & Co.

Mettler, F. A. 1972. Neuropathological effects of sodium azide administration in primates. *Chem. Absr. CA 77*(24): 160676t.

Meyer, E. 1989. *Chemistry of Hazardous Materials* 2nd ed. Englewood Cliffs, NJ: Prentice-Hall.

Morita, T., Y. Iwamoto, T. Shimizu, T. Masuzawa. and Y. Yanagihara. 1989. Mutagenicity tests with permeable mutant of yeast on carcinogens showing false-negative in *Salmonella* assay. *Chem. Pharm. Bull 37*(2): 407–9; cited in *Chem. Abstr. CA 110*(21): 187387m.

National Research Council. 1995. *Prudent Practices for Handling and Disposal of Chemicals from Laboratories*. Washington, DC: National Academy Press.

NFPA. 1997. *Fire Protection Guide on Hazardous Materials*, 12th ed. Quincy, MA: National Fire Protection Association.

NIOSH. 1986. *Registry of Toxic Effects of Chemical Substances*, D. B. Sweet, ed. Washington, DC: U.S. Government Printing Office.

Wear, J. O. 1981. *Safety Chem. Lab. 4*: 77

34

METAL CARBONYLS

34.1 GENERAL DISCUSSION

A number of transition metals in the periodic table form complexes with carbon monoxide. The general formula of metal carbonyls is $M_x(CO)_y$, where M is a metal and x and y are integers (the number of metal atoms and CO units, respectively, in the carbonyl molecule). The oxidation state of the metal in the carbonyl is zero. The number of CO molecules bound to the metal atom is equal to the number of electron pairs needed to be donated to the metal to achieve the nearest inert gas configuration. For example, five CO molecules have to be coordinated to iron atom (atomic number 26). These CO ligands would totally contribute ten electrons, providing a 36-electron stable electronic configuration of the inert gas, krypton. The complex formed is iron pentacarbonyl, $Fe(CO)_5$. Metals with an odd number of electrons, such as cobalt, can achieve such inert gas configuation by forming a covalent metal–metal bond and CO bridging between two metal atoms, with each atom metal receiving one electron from each CO molecule (Wagner 1978). All the Group VIII noble metals in the periodic table are known to form metal carbonyls. Other transition metals that form carbonyls include chromium, molybdenum, tungsten, vanadium, manganese, technecium, and rhenium. These carbonyls may be either mononuclear or polynuclear, containing one or more metal atoms, respectively, in their structures. The only stable four-coordinate carbonyl is nickel tetracarbonyl, which has a tetrahedral structure. The pentacarbonyls are five-coordinate trigonal bipyramidal compounds, which include the carbonyls of iron, ruthenium, and osmium. Chromium, molybdenum, tungsten, and vanadium form six-coordinate mononuclear neutral complexes. Dicobalt octacarbonyl, $Co_2(CO)_8$ and dimanganese decacarbonyl, $Mn_2(CO)_8$ are the only important binuclear carbonyls. Tri-, poly-, and heteronuclear metal carbonyls of Group VIII metals and manganese are known. Such compounds have complex structures and perhaps very little or no commercial applications.

Metal carbonyls are used as catalysts in a number of organic syntheses. Some of these industrial processes include carbonylation of olefins to produce aldehydes and alcohols, carboxylation reactions, and isomerization of olefins, as well as various coupling, cyclization, and hydrogenation reactions. Also, metal carbonyls are often used as sources to obtain pure metals, produced by the thermal decomposition of carbonyls. Some metal carbonyls find applications in the preparation of ketones and olefins.

Health Hazard

All carbonyls are highly toxic by all routes of exposure: the volatile liquid compounds present additional inhalation risk. Toxicity may be attributed to their decomposition products, the metals and carbon monoxide, essentially the toxic metabolites of all metal carbonyls. Therefore, the unstable carbonyls, especially the mononuclear complexes, which are readily susceptible to breakdown, should be treated as dangerous poisons. The polynuclear carbonyls of heavy metals may bioactivate and exhibit severe delayed effects. Oral intake of such substances can be fatal. The health hazards associated with individual compounds are discussed in the following sections. Toxicity data for most compounds of this class are not available.

Fire and Explosion Hazard

The mononuclear carbonyls of nickel and iron that are liquid at the ambient temperatures are extremely flammable. Their vapors present flashback fire hazard. They also form explosive mixtures with air. Also, most mononuclear carbonyls that are solids at ambient conditions are pyrophoric. These include the carbonyls of vanadium and tungsten. All carbonyl complexes are air sensitive. Some of them ignite on prolonged exposure to air or may catch fire when opened to air after long storage. The dimeric and trimeric derivatives, however, are less pyrophoric than the mononuclear complexes. All carbonyl metal complexes are susceptible to explode on heating. They react violently with strong oxidants.

Chemical Analysis

Metal carbonyls may be analyzed by various instrumental techniques, including GC/MS,

AA, and IR spectroscopy. Metal analysis may be performed by atomic absorption or emission spectroscopy and their stoichiometric amounts in the carbonyl complexes may be determined. Samples may be digested carefully and cautiously with nitric acid in a fume hood before the determination of metal contents. For mass spectrometric determination the carbonyl complex should be dissolved in an organic solvent, injected onto a GC column and identified from the characteristic mass ions. The carbonyl ligands in the complex may be determined by thermal decomposition of the metal carbonyl followed by analysis of carbon monoxide on a GC-TCD.

The hazardous properties of some selected metal carbonyls are presented in the following sections.

34.2 NICKEL TETRACARBONYL

Formula: $Ni(CO)_4$; MW 170.75; CAS [13463-39-3]

Structure:

Synonyms: tetracarbonyl nickel, nickelcarbonyl, UN 1259.

Uses and Exposure Risk

Nickel tetracarbonyl is used in the manufacture of nickel powder and nickel-coated metals, and as a catalyst in carboxylation, coupling, and cyclization reactions.

Physical Properties

Colorless, volatile liquid; characteristic sooty odor; boils at 43°C; solidifies at −19.3°C;

vapor pressure 400 torr at 26°C; density 1.318 at 17°C (Merck 1996); practically insoluble in water (solubility 0.02%); soluble in most organic solvents.

Health Hazard

Nickel tetracarbonyl is an extremely toxic substance by all routes of exposure exhibiting both immediate and delayed effects. The delayed effects may manifest in a few hours to days after exposure. Exposure to its vapors can cause dizziness, giddiness, headache, weakness, and increased body temperature. Vapors are irritating to eyes, nose, and throat. Prolonged exposure or inhalation of its vapors at a further increased level of concentration may produce rapid breathing, followed by congestion of the lungs. The respiration will initially be rapid with non-productive cough, progressing to pain and tightness in the chest (U.S. EPA 1995). High exposure can cause convulsion, hemorrhage, and death. Other symptoms from inhalation of vapors or ingestion of the liquid include hallucinations, delirium, nausea, vomiting, diarrhea, and liver and brain injury.

In humans, a 30-minute exposure to a 30-ppm concentration in air could be fatal. A few whiffs of the vapors of the liquid can cause death. Similarly, swallowing 5–10 mL of the liquid can be fatal.

Nickel tetracarbonyl can be absorbed through the skin. While the skin contact with a dilute solution can cause dermatitis and itching, that from a concentrated solution or the pure liquid can produce a burn. Absorption of the liquid through the skin may result in death. The subcutaneous and intravenous LD_{50} values in rats are 60–70 mg/kg.

LC_{50} (mouse): 0.067 mg/L/30 min (RTEC 1986)

Evidence of carcinogenicity observed in experimental animals dosed with nickel tetracarbonyl is limited. It caused tumors in the lungs and liver. The compound is also teratogenic, causing birth defects.

Fire and Explosion Hazard

Highly flammable; burns with a yellow flame (Clayton and Clayton 1982) flash point (closed cup) below −18°C (−4°F) (NFPA 1997); vapor density 5.89 (air = 1). The vapors form explosive mixtures with air; the LEL value is 2% by volume, the UEL has not been determined. The vapor is heavier than air and may travel a considerable distance to a source of heat and flash back. The liquid may explode when heated in a closed container (NFPA 1997). Sewer disposal, therefore, may create a hazardous situation for fire and explosion.

It explodes in air or in oxygen after an induction period. The oxidation reaction can proceed to explosive violence in the presence of a catalyst without an induction period. This can happen when the carbonyl is dry and is vigorously shaken with oxygen in the presence of mercury or its oxide (Mellor 1946). With atmospheric oxygen, it produces a deposit which forms a peroxide that may ignite (Bretherick 1995). Reactions with strong oxidants such as permanganate, perchlorate, or persulfate in combination with sulfuric acid, liquid bromine, or dinitrogen tetraoxide can be explosive. Violent reactions can occur with concentrated acids or acid fumes.

Exposure Limit

TLV-TWA: 0.05 ppm (0.35 mg as Ni/M^3) (ACGIH)

PEL: 0.001 ppm (0.007 mg Ni/M^3) (OSHA, MSHA and NIOSH)

IDLH: 0.001 ppm (NIOSH, OSHA)

34.3 IRON PENTACARBONYL

DOT Label: Flammable Liquid and Poison, UN 1994

Formula: $Fe(CO)_5$; MW 195.90; CAS [13463-40-6]

Structure:

trigonal bypyramidal

Synonym: pentacarbonyl iron

Uses and Exposure Risk

It is used to produce carbonyl iron (finely divided iron) for high frequency coils and also for radio and television (Merck 1996). It is also used as a catalyst in many organic syntheses and as an antiknock agent in motor fuels.

Physical Properties

Colorless oily liquid that turns yellow; freezes at $-20°C$; boils at $103°C$; density 1.50 at $20°C$; insoluble in water, slightly soluble in alcohol, and readily dissolves in most organic solvents.

Health Hazard

Iron pentacarbonyl is moderately toxic, the acute toxicity is considerably lower than that of nickel tetracarbonyl. The toxic symptoms, however, are nearly the same. Being highly volatile (the vapor pressure being 30 torr at $20°C$), this compound presents a serious risk of inhalation to its vapors. Furthermore, it evolves toxic carbon monoxide when exposed to light. The reaction is as follows:

$$2Fe(CO)_5 \xrightarrow{light} Fe_2(CO)_9 + CO$$

Therefore, all handling and operations must be carried out in fume hoods or under adequate ventilation. Inhalation of its vapor can cause headache, dizziness, and somnolence. Other symptoms are fever, coughing and cyanosis, which may manifest several hours after exposure. The vapor is an irritant to the lung. Chronic exposure may cause injury to liver and kidneys. The inhalation LC_{50} value in rats is 5 ppm for a 10-minute exposure period. The oral LD_{50} value in rats is 40 mg/kg. Any cancer-causing effect in animals or humans has not been reported. Sodium salt of EDTA or dithiocarb salts of sodium or calcium are antidotes against iron pentacarbonyl poisoning.

Fire and Explosion Hazard

Highly flammable liquid; flash point (closed up) $-15°C$ ($5°F$); vapor pressure 40 torr at $30.3°C$ (CRC Handbook 1996). It ignites in air. It reacts violently with strong oxidants including ozone and hydrogen peroxide. Explosion has been reported when heated rapidly with nitric oxide above $50°C$ in an autoclave (Bretherick 1995). Reactions with other oxides of nitrogen under similar conditions are expected to be explosive.

Exposure Limits

TLV-TWA: 0.1 ppm ($\sim$0.8 mg(Fe)/m^3) (ACGIH), 0.01 ppm (MSHA)

TLV-STEL: 0.2 ppm ($\sim$1.6 mg(Fe)/m^3) (ACGIH)

34.4 DICOBALT OCTACARBONYL

Formula: $Co_2(CO)_8$; MW 341.94; CAS [10210-68-1]

Structure:

Synonyms: cobalt carbonyl; cobalt octacarbonyl

Physical Properties

Orange crystalline solid; melts at 51°C; decomposes above 52°C on exposure to air; insoluble in water; soluble in organic solvents.

Uses and Exposure Risk

It is used as a catalyst in many organic conversion reactions, which include hydrogenation, isomerization, hydroformylation, polymerization, and carbonylation.

Health Hazard

Dicobalt octacarbonyl exhibits moderate toxicity by inhalation route and somewhat lower toxicity by intraperitoneal and oral routes. However, it is much less toxic than nickel tetracarbonyl or iron pentacarbonyl. A 2-hour LC_{50} value in mice is reported as 27 mg/m^3 (Lewis 1996). An oral LD_{50} value in rats is within the range of 750–800 mg/kg. It decomposes, evolving toxic carbon monoxide.

Exposure Limits

TLV-TWA: 0.1 mg/m^3 as Co (ACGIH)
PEL-TWA: 0.1 mg/m^3 as Co (OSHA)

34.5 COBALT HYDROCARBONYL

Formula: $HCO(CO)_4$; MW 171.98; CAS [16842-03-8]

Uses and Exposure Risk

It is used as a catalyst in certain organic synthesis.

Physical Properties

Light yellow gas or liquid; unstable, decomposes rapidly at ambient temperature; solidifies at −26°C ; boils at 10°C; slightly soluble in water (0.05%), dissolves in organic solvents.

Health Hazard

The toxic effects are similar to those of nickel tetracarbonyl or iron pentacarbonyl. The acute toxicity, however, is lower than that of these two carbonyls. Inhalation of the gas can cause dizziness, giddiness, and headache. It readily decomposes at room temperature producing toxic carbon monoxide. A 30-minute LC_{50} in rats is 165 mg/m^3 (Palmes et al. 1959).

Exposure Limits

TLV-TWA 0.1 mg/m^3 as Co (ACGIH)
PEL-TWA 0.1 mg/m^3 as Co (OSHA)

Fire and Explosion Hazard

Flammable gas; the liquid form can explode when heated in a closed container due to rapid decomposition and heavy pressure buildup.

34.6 CHROMIUM HEXACARBONYL

Formula: $Cr(CO)_6$; MW 220.07; CAS [13007-92-6]
Structure:

octahedral

Synonym: hexacarbonyl chromium, chromium carbonyl

Uses and Exposure Risk

It is used as a catalyst for polymerization and isomerization of olefins. It is also used as an additive to gasoline, to increase the octane number.

Physical Properties

White crystalline solid; sublimes at room temperature; decomposes at 130°C; vapor pressure 1 torr at 48°C (Merck 1989); insoluble in water and alcohol; soluble in most other organic solvents.

Exposure Limit

TLV-TWA: 0.05 mg (Cr)/m^3, confirmed human carcinogen (ACGIH)
PEL: 0.1 mg (CrO3)/m^3 (ceiling) (OSHA)

Health Hazard

Chromium hexacarbonyl is a highly toxic substance by all routes of exposure. The toxic effects are similar to those of molybdenum and tungsten carbonyls. The symptoms are headache, dizziness, nausea, vomiting, and fever. The oral LD$_{50}$ in mice is 150 mg/kg. As a hexavalent compound of chromium, it is a carcinogenic substance.

Fire and Explosion Hazard

It explodes when heated at 210°C. Its ethereal solution is susceptible to explode after prolonged storage.

34.7 MOLYBDENUM HEXACARBONYL

Formula: Mo(CO)$_6$; WE 264.00; CAS [13939-06-5]
Structure:

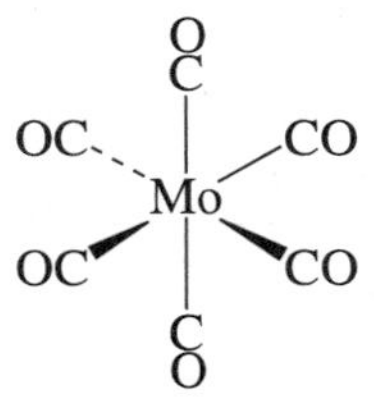

octahedral

Synonyms: molybdenum carbonyl, hexacarbonylmolybdenum

Uses and Exposure Risk

It is used as a catalyst in many organic synthetic reactions.

Physical Properties

White crystalline solid; melts at 150°C (decomposes); density 1.96 g/cm^3; insoluble in water, soluble in most organic solvents.

Health Hazard

Molybdenum hexacarbonyl is a highly toxic substance. It is listed in Toxic Substance Control Act. Acute symptoms include headache, dizziness, nausea, vomiting, and fever. Because of its low vapor pressure, the risk of exposure of this compound, however, by inhalation is lower than that from the volatile tetra- or pentacoordinated metal carbonyls. Ingestion of the compound can cause death. Absorption of its solution through skin may cause severe poisoning, manifesting the same effects as those from other routes of exposure.

Fire and Explosion Hazard

It reacts violently with strong oxidizers. Contact with concentrated acids is expected to be violent. Its solutions are susceptible to detonate spontaneously (Owen 1950). Ethereal solutions of this compound can explode on extended storage (Bretherick 1995).

34.8 VANADIUM HEXACARBONYL

Formula: V(CO)$_6$; MW 291.01; CAS [14024-00-1]
Structure:

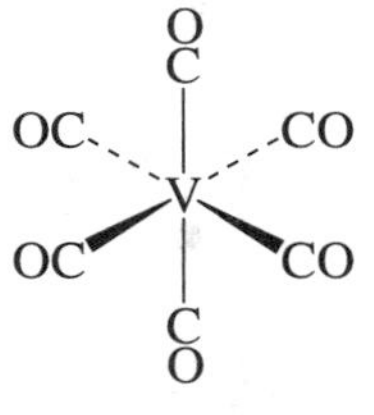

octahedral

Synonyms: vanadium carbonyl, hexacarbonyl vanadium

Uses and Exposure Risk

It is used as a catalyst in isomerization and
hydrogenation reactions.

Physical Properties

Bluish green crystalline solid; decomposes
on heating at 70°C; insoluble in water
and alcohol; low solubility in hexane and
other saturated hydrocarbons, but soluble in
most other organic solvents (solution turns
yellow).

Health Hazard

There is no report on its toxicity in the published literature. However, this substance,
like most other metal carbonyls, should
be treated as a potential health hazard as
it can produce highly toxic decomposition
products, carbon monoxide, and vanadium
oxides. Skin contact can cause irritation.

Fire and Explosion Hazard

Vanadium hexacarbonyl is a pyrophoric compound. It can explode on heating.

34.9 TUNGSTEN HEXACARBONYL

Formula W$(CO)_6$; MW 351.91; CAS [14040-
11-0]
Structure:

octahedral

Uses and Exposure Risk

It is used as a catalyst in many organic
synthetic reactions.

Physical Properties

White crystalline solid; melts at 170°C
(decomposes); sublimes; boils at 175°C at
766 torr (CRC Handbook 1996); density
2.65 g/mL; insoluble in water and alcohol; soluble in fuming nitric acid and most
organic solvents.

Health Hazard

Toxicity of this compound is not reported.
Although air stable at ambient temperature, upon heating it emits toxic carbon
monoxide. Chronic exposure to its vapors
or dusts can cause bronchitis. Ingestion is
likely to produce the toxic effect of tungsten
oxides.

Fire and Explosion Hazard

It explodes when heated to high temperature.
Solution in ether can explode if stored for a
long time.

34.10 DIMANGANESE DECACARBONYL

Formula: $Mn_2(CO)_{10}$; MW 389.99; CAS
[10170-69-1]
Structure:

Synonyms: decacarbonyl dimanganese, manganese carbonyl

Uses and Exposure Risk

It is used as a catalyst and a fuel additive to increase octane number.

Physical Properties

Golden yellow crystalline solid; melts at 155°C; decomposition begins at 110°C; density 1.75 g/mL at 25°C; insoluble in water, soluble in most organic solvents.

Health Hazard

Toxicity data are not available for this compound. Chronic exposure to its dusts has the potential to produce damaging effects on the pulmonary system and the central nervous system. Skin or eye contact can cause irritation. When heated, it emits toxic carbon monoxide.

Fire and Explosion Hazard

There is no report of any fire or explosion hazard. However, based on its structure and analogy with dirhenium decacarbonyl [14285-68-8], it may be anticipated to ignite when heated above 150°C. The fine powder is susceptible to produce pyrophoric tetramer.

REFERENCES

ACGIH. 1986. *Documentation of the Threshold Limit Values and Biological Exposure Indices,* 5th ed. Cincinnati, OH: American Conference of Governmental Industrial Hygienists.

Bretherick, L. 1995. *Handbook of Reactive Chemical Hazard,* 5th edition, ed. P.G. Urben, Oxford, UK: Buterworth-Heinemann.

CRC Press. 1996. *Handbook of Chemistry and Physics,* 76th edition, D. R. Lide, ed. Boca Raton, FL: CRC Press.

Clayton, G. D., and F. E. Clayton (eds.). 1982. Patty's Industrial Hygiene and Toxicology, 3rd. ed. New York; John Wiley & Sons.

Lewis R. J. 1996. *Sax's Dangerous Properties of Industrial Materials,* 9th ed. New York: Van Nostrand Reinhold.

Mellor, J. W. 1947–71. *Comprehensive Treatise on Inorganic and Theoretical Chemistry,* London: Longmans & Green.

Merck. 1996. *The Merck Index,* 12th ed. Rahway, N.J.: Merck & Co.

Owens, B. B. 1950. Molybdenum hexacarbonyl. *Inorg. Synth. 3*; 158

NFPA. 1986. *Fire Protection Guide on Hazardous Material,* 9th ed. Quincy, MA: National Fire Protection Association.

NIOSH. 1986. *Registry of Toxic Effects of Chemical Substances,* D. V. Sweet, ed. Washington, DC: US Government Printing Office.

Palmes, E. D., N. Nelson, S. Laskin, and M. Kuschner. 1959. *Am. Ind. Hyg. Assoc. J. 20:* 453

US EPA. 1995. *Extremely Hazardous Substances: Superfund Chemical Profiles.* Park Ridge, NJ: Noyes Data Corp.

Wagner, F. S. 1978. Carbonyls, in *Kirk-Othmer Encyclopedia of Chemical Technology,* 3rd ed., Vol. 4, pp. 794–814. New York: Wiley-Interscience.

35

METAL HYDRIDES

35.1 GENERAL DISCUSSION

Hydrides are compounds containing hydrogen in a reduced state, bound to a metal by ionic or covalent bond. Structurally, these compounds can be grouped into three types: (1) ionic hydrides, (2) covalent hydrides, and (3) complex hydrides. In ionic hydrides, a negatively charged hydride anion is bound to a metal ion by electrostatic force of attraction. Such simple binary compounds are formed by the alkali- and alkaline-earth metals, which are more electropositive than hydrogen.

In covalent hydrides, hydrogen and the metal are linked by a covalent bond. Aluminum, silicon, germanium, arsenic, and tin are some of the metals whose covalent hydride structures have been studied extensively. Some ionic hydrides, such as LiH or MgH_2, exhibit partial covalent character. The complex hydrides, such as lithium aluminum hydride and sodium borohydride, contain two different metal atoms, usually an alkali metal cation bound to a complex hydrido anion. The general formula for these compounds is $M(M'H_4)_n$, where the tetrahedral $M'H_4$ contains a group IIIA metal such as boron, aluminum, or gallium, and n is the valence of metal M.

The single most hazardous property of hydrides is their high reactivity toward water. The reaction with water is violent and can be explosive with liberation of hydrogen:

$$NaH + H_2O \longrightarrow NaOH + H_2$$

Many hydrides are flammable solids that may ignite spontaneously on exposure to moist air. Many ionic hydrides are strongly basic; their reactions with acids are violent and exothermic, which can cause ignition. Hydrides are also powerful reducing agents. They react violently with strong oxidizing substances, causing explosion.

Covalent volatile hydrides, such as arsine, silane, or germane, are highly toxic. Ionic alkali metal hydrides are corrosive to skin, as they form caustic alkalies readily with moisture.

The hazardous properties of individual compounds are discussed below in detail. Only the metal hydrides, which are highly reactive or toxic, and the commercially important hydrides are discussed. Excluded from the foregoing discussion are the hydrides of nonmetals. Phosphine is presented in Chapter 52.

35.2 SODIUM HYDRIDE

DOT Flammable Solid, Label: Flammable
  Solid and Danger When Wet, UN 1427
Formula NaH; MW 24.00; CAS [7646-69-7]
Structure: cubic lattice similar to the corre-
  sponding metal halide

Uses and Exposure Risk

Sodium hydride is used as a reducing agent
in organic synthesis, for reduction of oxide
scale for metals and high-alloy steel, and as
a hydrogenation catalyst.

Physical Properties

White crystalline powder or silver needles;
decomposes at 420°C without melting; density 1.36 g/cm^3; soluble in fused salt mixtures, fused caustic soda, and caustic potash;
insoluble in organic solvents; reacts violently
with water.

Health Hazard

The health hazard from sodium hydride may
arise due to its intense corrosive action on
skin. It is a strong nasal irritant and corrosive
on ingestion. It readily hydrolyzes to sodium
hydroxide, which is also highly corrosive.
Other than this, there is no report of health
hazards in animals or humans.

Fire and Explosion Hazard

Flammable solid; ignites spontaneously on
contact with moist air. Sodium hydride reacts
violently with water, liberating hydrogen,
which is flammable and which also forms
explosive mixtures with air. It can form dust
clouds, which may explode in contact with
flame or oxidizing substances.

Sodium hydride ignites with oxygen at
230°C and reacts vigorously with sulfur
vapors (Mellor 1946). It ignites spontaneously with fluorine and with chlorine in the
presence of moisture (Mellor 1946). It burns
with bromine and iodine to incandescence at
high temperature. Reactions between sodium
hydride and dimethyl sulfoxide (DMSO)
have resulted in explosions (NFPA 1986).

Sodium hydride fire should be extinguished
by lime, limestone, or dry graphite. Water,
carbon dioxide, dry chemical, and the Halon
agents must not be used to extinguish the fire.

Disposal/Destruction

Sodium hydride is decomposed in the laboratory by mixing it with an excess of a
dry hydrocarbon solvent such as n-hexane or
heptane. When its concentration is less than
5%, an excess of *tert*-butyl alcohol is added
dropwise under nitrogen or an inert atmosphere with stirring. This is followed by addition of cold water gradually. The two layers
are separated. The aqueous layer is drained
down the sink with an excess of water. The
organic layer is incinerated or sent to a landfill (National Research Council 1983).

35.3 POTASSIUM HYDRIDE

Formula: KH; MW 40.11; CAS [7693-26-7]
Structure: cubic lattice similar to potassium
  chloride

Uses and Exposure Risk

It is used as a strong reducing agent and
in making super bases RNHK and ROK
(Sullivan and Wade 1980). It is sold as
35 wt% dispersion in mineral oil.

Physical Properties

White crystalline powder or needles (in the
pure form); decomposes to potassium and
hydrogen; density 1.43 g/cm^3; reacts violently with water; insoluble in most organic
solvents, soluble in fused hydroxides and salt
mixtures.

Health Hazard

The toxicity data on potassium hydride are
not reported in the literature. In the pure

state, this compound should be highly corrosive by inhalation, ingestion, and skin contact. It yields potassium hydroxide, which is also very corrosive, when reacted with moisture.

Fire and Explosion Hazard

Flammable solid; ignites spontaneously in moist air. It reacts violently with water, liberating hydrogen. Reactions with oxygen, CO, and lower alcohols can be vigorous to violent. It is strongly basic, and its reactions with acids can be violent. Violent explosive reactions may take place when the pure compound is mixed with strong oxidizers. It ignites spontaneously with oxygen, fluorine, and chlorine. The oil dispersions of potassium hydride are quite safe to handle, however, as the oil phase provides a barrier to moisture and air.

Disposal/Destruction

An excess of *tert*-butyl alcohol is added dropwise with stirring to potassium hydride in heptane under a nitrogen atmosphere. The organic phase is sent to a landfill or incinerated, and the aqueous phase is flushed down the drain (see Section 35.2).

35.4 LITHIUM HYDRIDE

DOT Flammable Solid, Label: Flammable Solid and Danger When Wet, UN 1414 and UN 2805

Formula: LiH; MW 7.95; CAS [7580-67-8]

Structure: ionic salt containing Li^+ ion and hydrogen anion, cubic lattice structure

Uses and Exposure Risk

Lithium hydride is used in the manufacture of lithium aluminum hydride and silane, as a powerful reducing agent, as a condensation agent in organic synthesis, as a portable source of hydrogen, and as a lightweight nuclear shielding material. It is now being used for storing thermal energy for space power systems (Morris et al. 1988).

Physical Properties

White crystalline solid, darkens on exposure to light; mp 688°C, thermally stable, melts without decomposition; density 0.77 g/cm^3; insoluble in most organic solvents; decomposes rapidly in water.

Health Hazard

The health hazard due to lithium hydride may be attributed to the following properties: (1) corrosivity of the hydride, (2) its hydrolysis to strongly basic lithium hydroxide, and (3) toxicity of the lithium metal. However, the latter property, which may arise because of the formation of lithium resulting from the decomposition of lithium hydride and the metabolic role of lithium, is not yet established.

This compound is highly corrosive to skin. Contact with eyes can produce severe irritation and possible injury. It can hydrolyze with body fluid, forming lithium hydroxide, which is also corrosive to the skin and harmful to the eyes. Animal tests indicated that exposure to its dust or vapor at a level exceeding 10 mg/m^3 eroded the body fur and skin, caused severe inflammation of the eyes, and led to the destruction of external nasal septum (ACGIH 1986). No chronic effects were observed.

Exposure Limit

TLV-TWA 0.025 mg/m^3 (ACGIH).

Fire and Explosion Hazard

Flammable solid; reacts vigorously with water, liberating hydrogen. Hydrogen is also liberated when lithium hydride reacts with chlorine, ammonia, carboxylic acids, and lower aliphatic alcohols at high temperatures.

It reacts violently with strong oxidizing agents.

Disposal/Destruction

Lithium hydride is added to an excess of *n*-heptane. Onto this solution, dry *n*-butanol is added cautiously under an inert atmosphere. The resulting solution is neutralized with acid. The aqueous solution is separated and drained down the sink with an excess of water. The organic layer is separated and burned in a chemical incinerator equipped with an afterburner and scrubber.

35.5 LITHIUM ALUMINUM HYDRIDE

DOT Flammable Solid, Label: Flammable Solid and Danger When Wet, UN 1410; Flammable Liquid (its ethereal solution), UN 1411

Formula: $LiAlH_4$; MW 37.96; CAS [16853-85-3]

Structure: complex hydride containing ionic bond between Li^+ and AlH_4^- ions

Synonyms: lithium tetrahydroaluminate; aluminum lithium hydride; lithium aluminum tetrahydride

Uses and Exposure Risk

It is used as a powerful reducing agent in organic synthesis. Except for olefinic double bonds, almost all organic functional groups are reduced by lithium aluminum hydride (Sullivan and Wade 1980). It is used extensively in pharmaceutical synthesis and in catalytic hydrogenation.

Physical Properties

White crystalline solid; hygroscopic; decomposes slowly in moist air or at elevated temperature; mp 190°C (decomposes); density 0.917 g/cm^3; soluble in ether (~40 g/100 g at 25°C) and tetrahydrofuran (13 g/100 g at 25°C), reacts vigorously with water.

Health Hazard

The toxicological information on lithium aluminum hydride is scant. Because it is highly moisture sensitive, it causes skin burn on contact with moist skin. Its hydrolysis product lithium hyroxide is strongly alkaline and corrosive. Ingestion could result in gas embolism due to hydrogen, which forms readily on reaction with water.

Exposure Limit

TLV-TWA 2 mg(Al)/m^3 (ACGIH).

Fire and Explosion Hazard

Flammable solid; ignites in moist or heated air. It undergoes vigorous exothermic reaction with water evolving hydrogen [2.36 L(H$_2$)/g of LiAlH$_4$]:

$$LiAlH_4 + 2H_2O \longrightarrow LiAlO_2 + 4H_2$$

Its reactions with alcohols and carboxylic acids are vigorous and exothermic. Several explosions have been reported during distillation of various ethers with lithium aluminum hydride (MCA 1966, 1968, 1972; NFPA 1986). These ethers include alkyl ethers, glycol ethers, dimethoxyalkanes, and cyclic ether such as tetrahydrofuran. The presence of traces of dissolved carbon dioxide in ethers, small amounts of peroxides, or the addition of water drops have been attributed to these explosions.

Lithium aluminum hydride reacts with metal halides of silicon, germanium, tin, arsenic, and antimony to form hydrides, which are flammable and toxic:

$$GeCl_4 + LiAlH_4 \longrightarrow GeH_4 + LiCl + AlCl_3$$

Addition of lithium aluminum hydride to boron trifluoride etherate in ether at liquid nitrogen temperature resulted in an explosion with orange flame when the flask accidentally jarred (NFPA 1986). Cellulose materials may ignite in contact with lithium aluminum hydride.

Disposal/Destruction

It is destroyed by dropwise adding 95% ethanol to its solution in ether or tetrahydrofuran under nitrogen atmosphere. When the decomposition is complete (after the addition of enough methanol and when no hydrogen evolution occurs), an equal volume of water is added gradually. The organic layer (if toluene or ether is used as solvent) is separated and sent for incineration, while the aqueous layer is drained down with a large volume of water.

Alternatively, lithium aluminum hydride may be destroyed by treating with ethyl acetate (National Research Council 1995):

$$LiAlH_4 + 2CH_3COOC_2H_5 \longrightarrow LiOC_2H_5$$
$$+ Al(OC_2H_5)_3$$

It may be noted that no hydrogen is formed in the reaction above. Ethyl acetate is added gradually with stirring. This is followed by the addition of a saturated aqueous solution of ammonium chloride. The upper organic layer is separated and sent for incineration or landfill disposal. The aqueous phase containing the inert inorganic products is flushed down the drain with a large volume of water.

35.6 SODIUM BOROHYDRIDE

DOT Label: Flammable Solid, UN 1426
Formula: $NaBH_4$; MW 37.83; CAS [16940-66-2]
Structure: complex hydride containing ionic bond between Na^+ and BH_4^- ions
Synonyms: sodium tetrahydroborate; sodium boron tetrahydride

Uses and Exposure Risk

It is used as an effective reducing agent in many organic synthetic reactions. As a reducing agent, its action is milder to lithium aluminum hydride, and it can be applied in the aqueous phase.

Physical Properties

White solid; melts at 400°C (decomposes); density 1.074

Health Hazard

It is mildly corrosive to skin. Oral intake or intravenous administration of the solid or its solution produced high toxicity in animals. Ingestion of 160-mg/kg dose was lethal to rats (NIOSH 1986).

Fire and Explosion Hazard

Flammable solid; however, less flammable that lithium aluminum hydride. It burns quietly in air. Reaction with concentrated sulfuric, phosphoric, and fluorophosphoric acids is highly exothermic, forming β-diborane. Rapid mixing with these acids can cause dangerous explosion. Violent reactions may occur if its solution is mixed with formic or acetic acid. With alkalies it decomposes, liberating hydrogen. Reaction is highly exothermic, and pressure buildup may rupture glass apparatus. It reacts with ruthenium and other noble metal salts in solution, forming solid hydrides that are shock sensitive and water reactive. Mixture with charcoal ignites on exposure to air.

Disposal/Destruction

It may be destroyed in several ways. One method is as follows (Aldrich 1995). The solid or its solution is dissolved or diluted in large volume of water. Diluted acetic acid or acetone is then slowly added to this solution in a well-ventilated area. Hydrogen generated from decomposition of borohydride should be carefully vented out. The pH is adjusted to 1. The solution is then allowed to stand for several hours. It is then neutralized to 7, and the solution is then evaporated to dryness. The residue is then buried in a landfill site approved for hazardous waste disposal. Sodium borohydride may be destroyed in the laboratory by alternative methods mentioned for other hydrides.

35.7 SILANE

DOT: Flammable Gas, UN 2203

Formula: SiH_4; MW 32.13; CAS [7803-62-5]

Structure: Similar to that of methane; the Si—H bond, however, much more reactive than the C—H bond, forms higher homologues, such as disilane (Si_2H_6), trisilane (Si_3H_8), and hexasilane (Si_6H_{14})

Synonyms: silicon tetrahydride; monosilane; silicane

Uses and Exposure Risk

It is used for doping of solid-state devices and for preparing semiconducting silicon for the electronic industry.

Physical Properties

Colorless gas with a disagreeable odor; liquefies at $-120°C$; freezes at $-185°C$; liquid density 0.68 at $-185°C$, gas density at $20°C$ 1.44 g/L; insoluble in common organic solvents, decomposes slowly in water.

Health Hazard

Very little information has been published on the toxicology of this compound. The acute toxicity of silane is much less than that of germane. Among the hydrides of group IVB elements, the toxicity of silane falls between the nontoxic methane and moderately toxic germane. Inhalation of the gas can cause respiratory tract irritation. A 4-hour exposure to a concentration of about 10,000 ppm in air proved fatal to rats.

Exposure Limit

TLV-TWA 5 ppm (6.5 mg/m^3) (ACGIH).

Fire and Explosion Hazard

Flammable gas; ignites spontaneously in air and chlorine. It reacts with water under alkaline conditions, evolving hydrogen; fire-extinguishing steps: shut off the flow of gas; use a water spray to keep fire-exposed containers cool; do not use a Halon extinguisher. Compounds containing a strained Si—Si bond undergo spontaneous air oxidation. Therefore, higher homologues of silane should ignite in air, readily oxidizing to siloxanes. Bubbling the gas through solutions of strong oxidizers can cause violent reactions. Silane decomposes to silicon and hydrogen at $400°C$. Its reaction with nitrous oxide produces an unstable mixture that can be dangerously explosive. Three people were killed in an explosion resulting from a tainted silane cylinder contaminated with 50% nitrous oxide (The Courier News 1988; S. Wirth 1992, personal communication). The cylinder blasted when the gas was being vented out.

Disposal/Destruction

Silane can be destroyed by burning under controlled conditions by a suitable method. Small quantities can be decomposed by bubbling the gas through potassium hydroxide solution in a fume hood.

Analysis

Silane can be analyzed by GC/MS techniques under cryogenic conditions. A GC method using thermal conductivity detector may be suitable.

35.8 GERMANE

DOT Label: Poison A and Flammable Gas, UN 2192

Formula: GeH_4; MW 76.63; CAS [7782-65-2]

Structure: covalent hydride, tetrahedral structure; forms higher homologues

Synonyms: monogermane; germanium hydride; germanium tetrahydride

Uses and Exposure Risk

It is used to produce high-purity germanium metal and as a doping substance for electronic components.

Physical Properties

Colorless gas; liquefies at $-90°C$; freezes at $-165°C$; liquid density 1.523 at $-142°C$, gas density 3.43 g/L at $0°C$; insoluble in water and organic solvents, slightly soluble in HCl, soluble in liquid ammonia, decomposed by nitric acid.

Health Hazard

Germane is a moderately toxic gas. It exhibits acute toxicity, lower than that of stannane, but much greater than that of silane. By contrast, its poisoning effects are somewhat similar to the group VB metal hydrides, arsine, and stibine, while being much less toxic than the latter two compounds. Exposure to this gas can cause injury to the kidney and liver. A 1-hour exposure to a concentration of 150–200 ppm in air was fatal to test animals, including mice, guinea pigs, and rabbits. Inhalation of the gas can also cause irritation of the respiratory tract.

Exposure Limit

TLV-TWA 0.62 mg/m^3 (0.3 ppm) (ACGIH).

Fire and Explosion Hazard

Flammable gas; ignites spontaneously in air. Its higher analogues also burn in air with flame.

Disposal/Destruction

It is destroyed by burning it by a suitable method under controlled conditions.

Analysis

It is converted into a suitable derivative and analyzed by an atomic absorption spectrophotometer. Germane can be analyzed by GC/MS techniques either in the gaseous state or as a derivative.

35.9 ARSINE

DOT: Poison A and Flammable Gas, UN 2188

Formula: AsH$_3$; MW 77.95; CAS [7784-42-1]

Structure: covalent hydride

Synonyms: arsenic hydride; arsenic trihydride; arsenous hydride; hydrogen arsenide

Uses and Exposure Risk

It is used as a doping agent for solid-state electronic components and as a military poison gas. Risk of exposure may arise when an arsenic compound reacts with an acid or a strong alkali.

Physical Properties

Colorless gas with a disagreeable garlic smell; liquefies at $-62°C$; freezes at $-117°C$; liquid density 1.604 at $-64°C$, gas density 2.695 g/L at atmospheric pressure; slightly soluble in water (700 mg/L).

Health Hazard

Arsine is a highly poisonous gas showing severe acute and chronic toxicity in test animals as well as in humans. Exposure to a concentration of 5–10 ppm in air for several minutes may be hazardous to human health. There are conflicting reports on the concentrations at which arsine is fatal to humans. However, an exposure to 100 ppm in air for 1 hour should be lethal. The LC$_{50}$ values determined on animals varied according to the species. The symptoms of acute toxicity are headache, weakness, dizziness, vomiting, abdominal pain, and dyspnea. Arsenic is excreted in urine following exposure. In severe cases of acute poisoning, death can result from renal failure and pulmonary edema. Chronic poisoning can lead to jaundice, hemoglobinuria, hemolysis, and renal damage. The target organs are the blood, kidneys, and liver. Arsine is a cancer-causing gas. There is sufficient evidence of its carcinogenicity in humans.

Exposure Limits

TLV-TWA 0.2 mg/m^3 (0.05 ppm) (ACGIH and OSHA); 0.002 mg(As)/m^3/15 min; ceiling 0.005 ppm(As)/15 min (NIOSH).

Fire and Explosion Hazard

Flammable gas; ignites in chlorine; reacts vigorously with potassium in liquid ammonia at $-78°C$ (Mellor 1946, Suppl. 1963); reacts violently with fuming nitric acid. Reactions with strong oxidizers can be violent.

Disposal/Destruction

Arsine may be destroyed by oxidation with aqueous copper sulfate under nitrogen atmosphere. The reaction is carried out in a three-necked flask equipped with a stirrer, gas inlet, and nitrogen inlet (National Research Council 1995).

Analysis

Arsine in air is absorbed over coconut charcoal; the latter is digested with nitric acid, and arsenic is analyzed by flameless atomic absorption in a high-temperature graphite.

35.10   STIBINE

DOT Label: Poison A and Flammable Gas, UN 2676

Formula: SbH_3; MW 124.78; CAS [7803-52-3]

Structure: covalent hydride

Synonyms: antimony hydride; antimony trihydride; hydrogen antimonide

Uses and Exposure Risk

Stibine is used as a fumigating agent. Although the commercial application of this compound is limited, an exposure risk may arise when it is formed as a result of mixing an antimony compound with a strong reducing agent in the presence of acid.

Physical Properties

Colorless gas with a disagreeable odor; decomposes slowly at ambient temperature but rapidly at $200°C$; liquefies at $-18°C$; freezes at $-18°C$; gas density 5.515 g/L at $20°C$; slightly soluble in water, dissolves readily in alcohol, ether, carbon disulfide, and other organic solvents.

Health Hazard

Stibine is a highly toxic gas; the acute and chronic effects are similar to those of arsine. Exposure to 100 ppm in air for 1 hour was lethal to mice and guinea pigs, causing delayed death within $1-2$ days. The lethal concentration in air for humans is unknown. Like arsine, stibine is a hemolytic agent, causing injury to the kidney and liver. The toxicity is somewhat lower than that of arsine. In severe poisoning, death can result from renal failure and pulmonary edema.

Stibine is an irritant to the lung. Other toxic symptoms from inhaling this gas are headache, weakness, nausea, and abdominal pain. There is no report on its carcinogenicity in animals or humans. Because of its similarity to arsine in chemical and toxicological properties, stibine is expected to exhibit carcinogenic action.

Exposure Limit

TLV-TWA 0.5 mg/m^3 (0.1 ppm) (ACGIH and OSHA).

Fire and Explosion Hazard

Flammable gas; explodes when heated with ammonia, chlorine, or ozone at $90°C$, or treated with concentrated nitric acid (Mellor 1946). Reactions with strong oxidizers may cause explosions. Stibine decomposes to hydrogen and antimony metal at $200°C$.

Disposal/Destruction

Stibine may be destroyed by oxidation with aqueous copper sulfate solution under nitrogen (see Section 35.8).

Analysis

Stibine in air may be analyzed by charcoal adsorption, digestion with nitric acid, followed by atomic absorption spectrophotometric analysis. Since its reaction with concentrated HNO_3 may be violent, as mentioned above, diluted acid may be used.

35.11 DIBORANE

DOT Label: Poison and Flammable Gas, UN 1911

Formula: B_2H_6; MW 27.69; CAS [19287-45-7]

Structure:

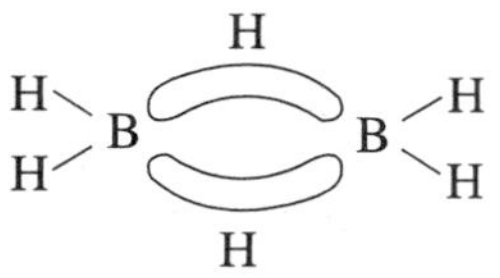

Both $-BH_2$ units are bound by three-center/two-electron hydride bridge structures; thus, the molecule contains four terminal and two bridge H atoms bound to boron atoms; first member of boron hydride class of compounds.

Synonyms: diboron hexahydride; boron hydride; boroethane

Uses and Exposure Risk

Diborane is used as a rocket propellant, in the vulcanization of rubber, as a polymerization catalyst, as a reducing agent, in the synthesis of trialkyl boranes, and as a doping agent (Merck 1996).

Physical Properties

Colorless gas with a repulsive smell; liquefies at $-92.5°C$; freezes at $-165°C$; liquid density 0.447 at $-112°C$; gas density 1.15 g/L; decomposes in water, soluble in carbon disulfide.

Health Hazard

Animal studies indicate that exposure to diborane results in irritation and possible infection in respiratory passage. In addition to the acute poisoning of the lungs, this gas may cause intoxication of the central nervous system. A 4-hour exposure to 60 ppm may be lethal to mice, resulting in death from pulmonary edema.

Exposure Limits

TLV-TWA 0.11 mg/m^3 (0.1 ppm) (ACGIH and OSHA); IDLH 40 ppm (NIOSH).

Fire and Explosion Hazard

Flammable gas; ignites with moist air at room temperature; vapor density 0.96 (air = 1); autoignition temperature of $40-50°C$ ($105-122°F$); fire-extinguishing step: stop the flow of gas; use a water spray to keep fire-exposed containers cool; do not use a Halon extinguishing agent. Diborane forms explosive mixtures with air over a wide range; the LEL and UEL values are 0.8% and 88% by volume of air, respectively. It explodes in contact with fluorine or chlorine, producing boron halide; halogenated hydrocarbons, such as carbon tetrachloride or chloroform; and oxygenated surfaces. It ignites with fuming nitric acid (Mellor 1946, Suppl. 1971). Violent explosions have been reported when liquid diborane at low temperature reacted with nitrogen trifluoride (Lawless and Smith 1968). Diborane reacts with the metals lithium and aluminum to form metal hydrides; and with phosphorus trifluoride, it forms an adduct, borane–phosphorus trifluoride (Mellor 1946, Suppl. 1971). These reaction products ignite spontaneously in air. Diborane produces hydrogen when reacted with water or on heating.

Storage and Shipping

It is stored in a cold, dry, well-ventilated place separated from halogenated hydrocarbons and oxidizing substances. It should be protected against physical damage, spark, and heat. It is shipped in high-pressure steel cylinders. The containers should be free from oxygen.

Disposal/Destruction

Diborane may be burned by any suitable method, under controlled conditions. It may be oxidized by aqueous copper sulfate under nitrogen in a three-necked flask equipped with a stirrer, a gas inlet, and a nitrogen inlet (National Research Council 1995).

Analysis

Analysis for diborane in air may be performed by NIOSH Method 6006 (NIOSH 1984). Air is passed through a PTFE filter and oxidizer-impregnated charcoal at a flow of 0.5–1 L/min. Diborane is oxidized to boron, which is desorbed with 3% H_2O_2 and analyzed by plasma emission spectrometry. Alternatively, boron may be analyzed by inductively coupled plasma atomic emission spectrometry (NIOSH 1984, Method 7300).

35.12   DECABORANE

DOT Label: Flammable Solid and Poison, UN 1868

Formula $B_{10}H_{14}$; MW 122.24; CAS [17702-41-9]

Structure: three-center/two-electron hydride bridge linking boron atoms

Synonym: decaboron tetradecahydride

Uses and Exposure Risk

It is used in rocket propellants and as a catalyst in olefin polymerization.

Physical Properties

Colorless orthorhombic crystals; pungent smell; mp 99.6°C; bp 213°C; density 0.94 g/cm^3 at 25°C; slightly soluble in cold water, decomposes in hot water, dissolves in carbon disulfide, alcohol, benzene, ethyl acetate, and acetic acid.

Health Hazard

Decaborane is a highly toxic compound by all routes of administration. Its toxicity is somewhat greater than that of diborane. The acute toxic symptoms in humans from inhalation of its vapors could be headache, nausea, vomiting, dizziness, and lightheadedness. In severe poisoning, muscle spasm and convulsion may occur. Symptoms of toxicity may appear 1 or 2 days after exposure, and the recovery is slow. An LC_{50} value for mice from a 40-hour exposure was 12 ppm.

Ingestion can cause spasm, tremor, and convulsion. It can be absorbed through the skin, producing drowsiness and loss of coordination. Toxic effects from skin absorption, however, are relatively moderate.

LD_{50} value, oral (mice): 41 mg/kg

LD_{50} value, skin (rats): 740 mg/kg

Exposure Limits

TLV-TWA skin 0.05 ppm (0.25 mg/m^3) (ACGIH and OSHA); STEL skin 0.15 ppm (0.75 mg/m^3) (ACGIH); IDLH 20 ppm (NIOSH).

Fire and Explosion Hazard

Flammable solid; vapor pressure 0.05 torr at 25°C; flash point (closed cup) 80°C (176°F); fire-extinguishing agent: dry chemical, CO_2, or water spray; do not use halogenated extinguishing agents. Decaborane ignites spontaneously in oxygen. Reactions with halogen compounds yield products that may explode on impact. Reactions with strong oxidizers could be violent. Decaborane produces hydrogen when it hydrolyzes with water or on decomposition at 300°C.

TABLE 35.1 Toxicity and Flammability of Miscellaneous Hydrides

Compound/Synonyms/ CAS No.	Formula/ MW	Toxicity	Flammability
Calcium hydride [7789-78-8]	CaH_2 42.10	No toxic effect reported; can cause skin burn	Flammable solid; ignites in air on heating; explodes violently when mixed with chlorates, bromates, or perchlorates and rubbed in a mortar (Mellor 1946); reactions with other strong oxidizers can be violent; reacts exothermically with water and to incandescence with silver fluoride; evolves hydrogen when reacted with water, lower alcohols, and carboxylic acids
Beryllium hydride [13597-97-2]	BeH_2 11.03	Toxic polymeric solid; toxicity data not available	Hydrolyzed by acid, liberating hydrogen; stable in water
Magnesium hydride [7693-27-8]	MgH_2 26.34	No toxic effect reported; can cause skin burn	Flammable solid; ignites spontaneously in air; reacts violently with water, liberating hydrogen; liberates hydrogen when reacted with methanol or dissociates at 280°C; DOT label: Flammable Solid and Danger When Wet, UN 2010
Aluminum hydride (aluminum trihydride) [7784-21-6]	AlH_3 29.99	No toxic effect has been reported	Flammable solid; ignites spontaneously in air; reacts violently with water; explosion may occur when distilled with ether, due to the presence of trace CO_2; DOT Label: Flammable Solid and Dangerous When Wet, UN 2463
Stannane (tin hydride, stannic hydride) [2406-52-2]	SnH_4 122.72	Poisonous gas; exposure can cause irritation of respiratory tract and injury to kidney and liver; 1-hr exposure to 150 ppm can be lethal to mice; no exposure limit is set for this gas; recommended TLV-TWA: 0.2 ppm (1 mg/m^3)	Flammable gas; reacts violently with strong oxidizers
Titanium hydride [7704-98-5]	TiH_2 49.92	No toxic effects reported	Stable in air; burns quietly when ignited; violent reactions occur when burned with oxidizers

TABLE 35.1   *(Continued)*

Compound/Synonyms/ CAS No.	Formula/ MW	Toxicity	Flammability
Copper(I)hydride [13517-00-5]	Cu_2H_2	No toxic effect reported	The dry solid ignites in air; ignites with halogens; reacts violently with oxidizers
Zinc hydride [14018-82-7]	Z_nH_2 67.40	No toxic effect reported	Reacts violently with aqueous acid liberating hydrogen; air oxidation slow; violent reactions with oxidizers
Lanthanum hydride [13864-01-2]	LaH_3 141.93	No toxic effect reported; may cause skin burn	Ignites in air; Reacts violently with aqueous acid liberating hydrogen; violent reactions with oxidizers.
Sodium aluminum hydride (sodium tetrahydroaluminate, aluminum sodium hydride) [13770-96-2]	$NaAlH_4$ 54.01	Moisture sensitivity similar to that of lithium aluminum hydride; can cause skin burn on contact with moist skin	Flammable solid; ignites in air; ignites or explodes in contact with water; a violent explosion has been reported during its preparation from sodium and aluminum in tetrahydrofuran (NFPA 1997); DOT Label: Flammable Solid and Danger When Wet, UN 2835

Disposal/Destruction

Decaborane may be destroyed by carefully dissolving with water followed by treating with 1 M sulfuric acid to a pH value of 1 (Aldrich 1995). Hydrogen, which may evolve vigorously after acid addition, should be vented. The solution is evaporated to dryness and the residue is buried in a landfill site for hazardous waste disposal. Decaborane spill should be decontaminated by flushing with a 3% aqueous ammonia solution.

Analysis

It may be analyzed by FTIR techniques. It may be oxidized to boron and analyzed by atomic absorption spectrophotometry.

35.13   MISCELLANEOUS HYDRIDES

See Table 35.1.

REFERENCES

ACGIH. 1986. *Documentation of the Threshold Limit Values and Biological Exposure Indices*, 5th ed. Cincinnati, OH: American Conference of Governmental Industrial Hygienists.

Aldrich. 1995. *Aldrich Catalog*. Milwaukee, WI: Aldrich Chemical Company. *The Courier News*, Bridgewater, NJ. March 17–24, 1988.

Lawless, E. W., and I. C. Smith. 1968. *High Energy Oxidizers*. New York: Marcel Dekker.

MCA. 1966. *Case Histories of Accidents in Chemical Industry*. Washington, DC: Maufacturing Chemists' Association.

MCA. 1968. *Case Histories of Accidents in Chemical Industry*. Washington, DC: Manufacturing Chemists' Association.

MCA. 1972. *Guide for Safety in the Chemical Laboratory*, 2nd ed. Manufacturing Chemists' Association. New York: Van Nostrand.

Mellor, J. W. 1946. *A Comprehensive Treatise on Inorganic and Theoretical Chemistry*. London: Longmans, Green.

Merck. 1996. *The Merck Index*, 12th ed. Rahway, NJ: Merck & Co.

Morris, D. G., J. P. Foote, and M. Olszewski. 1988. *Development of encapsulated lithium hydride thermal energy storage for space power system. Energy Res. Abstr.* 13(7): Abstr. No. 14709.

NFPA. 1997. *Fire Protection Guide on Hazardous Materials*, 12th ed. Quincy, MA: National Fire Protection Association.

National Research Council. 1995. *Prudent Practices for Handling and Disposal of Chemicals from Laboratories*. Washington, DC: National Academy Press.

NIOSH. 1984. *Manual of Analytical Methods*, 3rd ed. Cincinnati, OH: National Institute for Occupational Safety and Health.

NIOSH. 1986. *Registry of Toxic Effects of Chemical Substances*, ed. D. V. Sweet. Washington, DC: U.S. Government Printing Office.

Sullivan, E. A., and R. C. Wade. 1980. *Hydrides*. In *Kirk-Othmer Encyclopedia of Chemical Technology*, Vol. 12, pp. 772–92. New York: Wiley-Interscience.

36

METALS, REACTIVE

36.1 GENERAL DISCUSSION

Alkali metals constitute group IA of the periodic table. These metals are lithium, sodium, potassium, rubidium, cesium, and francium. Among these, only lithium, sodium, and potassium are of industrial use. Rubidium and cesium are of very little commercial application. Francium is a radioactive element.

These elements have electronic configurations, containing one extra electron over the stable inert-gas configurations. Thus, the elements tend to lose this extra electron to attain a geometrically stable configuration. The ionization potentials of alkali metals are therefore low, as the energy required to pull this unpaired electron out of the shell is low. As a result, the alkali metals are highly active, forming univalent ionic salts. By contrast, alkaline-earth metals would need to lose two electrons to attain a stable inert-gas configuration. This would require a higher ionization potential. Thus, the group IIA metals, also known as alkaline-earth metals, exhibit valency 2 and are less active than the alkali metals. As we go farther down in a period, metals tend to become less and less electropositive.

Among the three alkali metals of commercial application, potassium is most reactive, followed by sodium and lithium. Other reactive metals that undergo several violent reactions include magnesium, calcium, aluminum, zinc, and titanium. The reactivity of these metals is lower than that of alkali metals. Several other metals are flammable, too, but only in a very finely divided state. In the powdered form, aluminum is highly reactive, either bursting into flame or exploding, or both.

Fire-Extinguishing Agents

Water, carbon dioxide, and Halon extinguishers are ineffective against alkali metal fires. These substances, including carbon tetrachloride, which is used in Halon extinguishers, react violently with alkali metals. Sodium and potassium fires should be extinguished using a dry-powder fire extinguisher containing dry graphite. For lithium fires, lithium chloride or graphite should be used to smother the fire. It may be noted that when graphite is used, the metal may form a carbide. The metal carbide reacts with moisture to form acetylene, which is flammable and may rekindle the fire (Meyer 1989).

Therefore, the graphite residues from fire should be reacted with water under controlled conditions. Other substances that may be used effectively to extinguish sodium or potassium fire are soda ash and dry sodium chloride (in powder form). The same fire-extinguishing agents can be used to extinguish fires involving other metals.

Storage and Shipping

Alkali metals and other reactive metals are stored under nitrogen or kerosene. Care should be taken to keep these metals away from water, halogenated hydrocarbons, acids, oxidizers, and combustible materials. These compounds are shipped in hermetically sealed cans and drums.

36.2 LITHIUM

DOT Label: Flammable Solid and Dangerous When Wet, UN 1415

Symbol Li; at. wt. 6.941; at. no. 3; CAS [7439-93-2]; isotopes: 7, 6 (natural), 5, 8, 9 (artificial radioactive isotopes)

Uses and Exposure Risk

Lithium is used in making alloys, in the manufacture of lithium salts, and in vacuum tubes.

Physical Properties

Silvery white metal, turns grayish on exposure to moist air; melts at 181°C; bp 1340°C; dissolves in liquid ammonia, forming a blue solution.

Health Hazard

Lithium can react with moisture on the skin to produce corrosive hydroxide. Thus, contact of this metal with the skin or eyes can cause burn. The fumes are irritating to the skin, eyes, and mucous membranes. Ingestion of lithium can cause kidney injury,

especially when sodium intake is limited (Merck 1996).

Fire and Explosion Hazard

Lithium is less active than sodium or potassium. Finely divided metal ignites spontaneously in air. The ignition of the bulk metal occurs when heated to its melting point. It burns with a carmine-red flame. Burning evolves dense white and opaque fumes.

Vigorous reaction occurs when the metal is mixed with water. The heat of reaction, if not dissipated, can ignite or explode hydrogen that is liberated.

Violent explosive reactions occur with carbon tetrachloride; carbon tetrabromide; chloroform, bromoform, or iodoform (on heating); carbon monoxide in the presence of water; phosphorus (on heating); arsenic (on heating); and sulfur (molten). Among the substances that constitute high explosion hazards, the halogenated hydrocarbons are most significant. A number of compounds of this class in addition to those mentioned above form impact-sensitive products that can detonate on heating or impact.

Heating with nitric acid can cause fires. Lithium reacts with nitrogen at elevated temperatures to form lithium nitride, which can ignite on heating.

36.3 SODIUM

DOT Label: Flammable Solid and Danger When Wet; UN 1428

Symbol [7440-23-5]; isotopes: 23 (natural), 20, 21, 22, 24, 25, 26 (radioactive)

Synonym: natrium

Uses and Exposure Risk

Sodium is used in the manufacture of many highly reactive sodium compounds and tetraethyllead, as a reducing agent in organic synthesis, and as a catalyst in the production of synthetic rubber. It is also used in making sodium lamp and photovoltaic cells.

Physical Properties

Light silvery white metal when freshly cut; becomes gray and tarnished when exposed to air; soft and waxy at room temperature; melts at 98°C; boils at 881°C; dissolves in liquid ammonia.

Health Hazard

Sodium is highly corrosive and can cause burns on the skin. Its fumes are highly irritating to the skin, eyes, and mucous membranes.

Fire and Explosion Hazard

Sodium reacts violently with water, liberating hydrogen and forming caustic soda. Much heat is released in this reaction, which may ignite or explode hydrogen that is liberated. It also reacts spontaneously in air, burning with a golden yellow flame. The product is sodium oxide. The reaction occurs only in the presence of moisture and not in absolute dry air. Sodium vapors ignite at room temperature, while the droplets ignite at about 120°C. The ignition temperature in air depends on the surface area of sodium exposed.

Sodium reacts with explosive violence when mixed with mineral acids, especially hydrofluoric, sulfuric, and hydrochloric acids. It burns spontaneously in fluorine, chlorine, and bromine; and a luminous reaction occurs with iodine. Sodium added to liquid bromine can explode when subjected to shock (Mellor 1946). Violent explosions may result when a mixture of sodium and a metal halide is subjected to impact. Shock-sensitive mixtures are formed upon combination with halogenated hydrocarbons, such as methyl chloride, methylene chloride, or chloroform. It reacts with incandescence when combined with many metal oxides, such as lead oxide or mercurous oxide.

Reaction of sodium with carbon dioxide is vigorous at elevated temperatures. Sodium in liquid ammonia reacts with carbon monoxide to form sodium carbonyl, which explodes on heating. Its mixture with sulfur can explode upon heating or impact; with phosphorus, the reaction is vigorous. Reaction with sulfur dioxide is violent at about 100°C; with sulfur dichloride, a shock-sensitive product is formed that explodes upon strong impact (NFPA 1997).

36.4 POTASSIUM

DOT Label: Flammable Solid and Danger When Wet, UN 2257

Symbol K; at. wt. 39.098; at. no. 19; CAS [7440-09-7]; isotopes: 39 (natural), 40 (radioactive)

Synonym: kalium

Uses and Exposure Risk

Potassium is used in the manufacture of many reactive potassium salts, in organic synthesis, and as a heat exchange fluid when alloyed with sodium.

Physical Properties

Silvery white metal, loses its luster when exposed to air; soft at ordinary temperature, becomes brittle at low temperatures; melts at 63°C; bp 765.5°C; soluble in liquid ammonia, ethylenediamine, and aniline.

Health Hazard

Potassium reacts with the moisture on the skin, causing severe burns. Contact with the solid metal can produce skin and eye burns. The metal fumes are highly irritating to the skin, eyes, and mucous membranes.

Fire and Explosion Hazard

Potassium is one of the most reactive metals. It ignites in air or oxygen at room temperature. It burns with a characteristic purple flame, forming potassium oxide, K_2O. Violent reaction occurs when mixed with water, producing hydrogen and caustic potash. Sufficient heat is released to ignite and explode hydrogen, liberated in the reaction.

The violent reactions of potassium are quite similar to those of sodium. However, it is more reactive than sodium. Numerous metal salts form explosive mixtures with potassium, which can explode on impact. These include metal halides, oxychlorides, and sulfates. When combined with potassium, a large number of metal oxides burn with incandescence either at room temperature or on heating. Contact of potassium metal with potassium peroxide, potassium superoxide, or potassium ozonide can result in explosions.

Many interhalogen compounds explode on contact with potassium. These include chlorine trifluoride, iodine monochloride, iodine monobromide, iodine pentafluoride, and others. A violent explosion occurs when chlorine monoxide comes in contact with potassium. Potassium burns spontaneously with fluorine, chlorine, and bromine vapor. Its mixture with liquid bromine can explode on impact, and mixture with iodine ignites.

Potassium forms explosive mixtures with halogenated hydrocarbons, carbon tetrachloride, carbon dioxide (dry ice), carbon monoxide, and carbon disulfide, resulting in explosions when subjected to shock. With carbon monoxide, it forms potassium carbonyl, which explodes on exposure to air. It ignites in nitrogen dioxide, sulfur dioxide, and phosphine.

36.5 CALCIUM

DOT Label: Flammable Solid and Danger When Wet

Symbol Ca; at. wt. 40.08; at. no. 20; CAS [7440-70-2]

Uses and Exposure Risk

Calcium is used as a deoxidizer for copper, steel, and beryllium in metallurgy; to harden lead for bearing; and in making alloys.

Physical Properties

Silvery white metal, becoming bluish gray on exposure to moisture; melts at 850°C; boils at 1440°C; soluble in liquid ammonia, forming a blue solution.

Health Hazard

Contact of dusts with skin or eyes can cause burns. It reacts with moisture on skin to form caustic hydroxide, which may cause such a burn. Fumes from burning calcium are strongly irritant to the eyes, skin, and mucous membranes.

Fire and Explosion Hazard

Finely divided metal ignites in air. It burns with a crimson flame. It ignites at ordinary temperature only in finely divided form and when the air is moist. It reacts with water to form calcium hydroxide and hydrogen. The reaction is much less violent than with sodium.

Calcium burns spontaneously in fluorine or chlorine at ordinary temperatures. In chlorine ignition occurs at room temperature when the metal is in the finely divided state. Calcium reacts explosively when heated with sulfur, alkali hydroxides, alkali carbonates, and many interhalogen compounds. It reacts violently with acids.

36.6 MAGNESIUM

DOT Label: Flammable Solid and Danger When Wet

Symbol Mg; at. wt. 24.31; at. no. 12; CAS [7439-95-4]

Uses and Exposure Risk

Magnesium is used in the manufacture of alloys, optical mirrors, and precision instruments; in pyrotechnics; as a deoxidizing and desulfurizing agent in metallurgy; in signal lights, flash bulbs, and dry batteries; and in Grignard reagent.

Physical Properties

Silvery white metal, obtained in the form of powder, pellets, wire turnings, bars, and ribbons; melts at 651°C; boils at 1100°C; reacts with water.

Health Hazard

Inhalation of magnesium dust can produce irritation of the eyes and mucous membranes. Magnesium may react with water in the bronchial passage to form magnesium hydroxide, which is caustic and may cause adverse effects on lungs. The fumes can cause metal fever.

Fire and Explosion Hazard

Magnesium is a combustible metal. It burns with a brilliant white flame, producing intense heat. Magnesium powder forms explosive mixtures with air. Such mixtures can be readily ignited by a spark. Its reaction with water is slow. Finely divided metal reacts with water, liberating hydrogen, which can ignite or explode.

Magnesium reacts explosively with many compounds or classes of substances. These include metal oxides, metal carbonates, metal sulfates, and many metal phosphates. An explosion resulted when adding bromobenzyl trifluoride to magnesium turnings in dried ether (MCA 1972).

Many cyanide salts burn with incandescence when heated with magnesium. Magnesium catches fire when mixed with moist fluorine or chlorine. Reactions with methyl chloride, chloroform, or other halogenated hydrocarbons can be explosive. Heating magnesium with perchlorates, peroxides, nitric acid, or other oxidizers can cause explosion. Magnesium catches fire when combined with moist silver.

36.7 ALUMINUM

DOT Label: Flammable Solid, UN 1309

Symbol Al; at. wt. 26.98; at. no. 13; CAS [7429-90-5]; isotopes: 27 (natural), 26 (radioactive); five other radioactive isotopes are also known.

Uses and Exposure Risk

Aluminum finds wide applications for industrial and domestic purposes. Fine powder is used in explosives, in fireworks, as flashlights in photography, and in aluminum paints. It is commonly used in alloys with other metals and is nonhazardous as alloys.

Physical Properties

White malleable metal with bluish tint; produced as bars, sheets, wire, or powder; forms oxide in moist air; melts at 660°C; boils at 2327°C.

Health Hazard

Repeated inhalation of high concentrations of aluminum may result in aluminosis or lung fibrosis. Accumulation of aluminum particles in the kidney may occur especially among dialysis patients, as the dialyzing solution may contain high concentrations of this metal (Hodgson et al. 1988). This may result in impaired memory, dementia, ataxia, and convulsions. Accumulation in brain has been suggested as a factor in Alzheimer disease. The evidence is not definitive, however.

Exposure Limits

TLV-TWA 10 mg/m^3 (Al dust), 5 mg/m^3 (pyrophoric Al powder and welding fumes), 2 mg/m^3 (soluble Al salts and alkyls) (ACGIH).

Fire and Explosion Hazard

Aluminum in the form of finely divided dusts is flammable. The dusts form explosive mixtures with air. Aluminum powder is a component of many explosives. It is used in many explosive compositions, combined with ammonium nitrate or nitroguanidine, for example (see Chapter 40). When combined with aluminum, many deflagrating substances attain detonation brisance.

Finely divided aluminum in combination with chlorates, bromates, iodates, or perchlorates of alkali- and alkaline-earth metals (in finely powdered form) can explode on impact, heat, or friction. Aluminum is a powerful reducing substance. It reduces sulfates, persulfates, and oxides on heating. Its reaction with copper oxide, lead oxide, or silver oxide on heating can progress to violent explosion. Reaction with powdered iron(III) oxide, known as a thermite reaction, can cause fires. Chlorinated hydrocarbons as liquids or vapors can explode when coming in contact with finely divided aluminum. Such halogenated hydrocarbons include chloroform, methyl chloride, methylene chloride, carbon tetrachloride, 1,4-dichlorobenzene, dichlorodifluoromethane, ethylene dichloride, and fluorochloro lubricants. Aluminum powder and halogens in contact will burst into flame. Similar ignition can occur when aluminum is combined with many interhalogen compounds, diborane, carbon disulfide, sulfur dioxide, sulfur dichloride, oxides of nitrogen, and chromic anhydride. Aluminum reacts vigorously with acids and alkalies producing hydrogen.

36.8 ZINC

DOT Label: Flammable Solid and Danger When Wet

Symbol Zn; at. wt. 65.37; at. no. 30; CAS [7440-66-6]

Uses and Exposure Risk

Zinc is an ingredient in alloys such as brass, bronze, and German silver. It is used as a protective coating to prevent corrosion of other metals, for galvanizing sheet iron, in gold extraction, in utensils, in dry cell batteries, and as a reducing agent in organic synthesis.

Physical Properties

Bluish white metal with luster; exposure to moist air produces a white carbonate coating; stable in dry air; melts at 420°C; boils at 908°C; malleable at 100–150°C; becomes brittle at 200°C; reacts with water.

Health Hazard

Exposure to zinc dust can cause irritation, coughing, sweating, and dyspnea. A 1-hour exposure to a concentration of 100 mg/m^3 in air may manifest the foregoing symptoms in humans. Toxic effects from inhalation of its fumes include weakness, dryness of throat, chills, aching, fever, nausea, and vomiting. Many zinc salts, such as zinc chloride [7646-85-7] and zinc oxide [1314-13-2], can produce metal fume fever when the fumes are inhaled. The oral toxicity of zinc chloride in experimental animals is moderate, the LD_{50} value in rats being 350 mg/kg (NIOSH 1986). Oral administration of this salt caused colon tumors in hamsters.

Fire and Explosion Hazard

Zinc dusts form explosive mixtures with air. In the presence of moisture the dusts may heat spontaneously and ignite in air. Reaction with water produces hydrogen. The heat of reaction may ignite the liberated hydrogen. Much vigorous reaction occurs with acids, with brisk evolution of hydrogen.

When combined with oxidizers such as chlorates, bromates, peroxides, persulfates, and chromium trioxide and subjected to impact, percussion, or heating, powdered zinc explodes. Explosion may result when the powder metal is heated with manganese chloride, hydroxylamine, ammonium nitrate (an oxidizer), potassium nitrate (an oxidizer), sulfur, or interhalogen compounds. Zinc burns in fluorine and chlorine (moist), and reacts with incandescence when mixed with carbon disulfide.

36.9 TITANIUM

DOT Label: Flammable Solid, UN 2546 (dry metal powder), UN 2878 (wet with less than 20% water)

Symbol Ti; at. wt. 47.88; at. no. 22; CAS [7440-32-6]

Uses and Exposure Risk

Titanium is added to steel and aluminum to enhance their tensile strength and acid resistance. It is alloyed with copper and iron in titanium bronze.

Physical Properties

Dark gray metal; melts at 1677°C; boils at 3277°C; brittle when cold; strength of the metal increases by traces of oxygen or nitrogen.

Health Hazard

Inhalation of metal powder may cause coughing, irritation of the respiratory tract, and dyspnea. Intramuscular administration of titanium in rats caused tumors in blood. Animal carcinogenicity is not fully established. Human carcinogenicity is not known.

Fire and Explosion Hazard

Titanium metal in a very finely divided state is flammable. It explodes when combined with liquid oxygen or red fuming nitric acid (NFPA 1997). Heating the metal with other oxidizing substances, such as potassium permanganate, potassium nitrate, or potassium chlorate, can cause explosion. Heating with cupric oxide, silver oxide, or lead oxide can produce a violent reaction. Fluorine or interhalogen compounds of fluorine react on powdered titanium with incandescence.

REFERENCES

Hodgson, E. H., R. B. Mailman, and J. E. Chambers. 1988. *Dictionary of Toxicology*. New York: Van Nostrand Reinhold.

Manufacturing Chemists' Association. 1972. *Guide for Safety in the Chemical Laboratory*, 2nd ed. New York: Van Nostrand.

Mellor, J. W. 1946. *A Comprehensive Treatise on Inorganic and Theoretical Chemistry*. London: Longmans, Greene.

Merck. 1996. *The Merck Index*, 12th ed. Rahway, NJ: Merck & Co.

Meyer, E. 1989. *Chemistry of Hazardous Materials*, 2nd ed. Englewood Cliffs, NJ: Prentice-Hall.

NFPA. 1997. *Fire Protection Guide on Hazardous Materials*, 12th ed. Quincy, MA: National Fire Protection Association.

NIOSH. 1986. *Registry of Toxic Effects of Chemical Substances*, ed. D. V. Sweet. Washington, DC: U.S. Government Printing Office.

37

METALS, TOXIC

37.1 GENERAL DISCUSSION

More than two-thirds of the elements in the Periodic Table are metals. Many of these are distributed widely but unevenly in nature — in rocks, soils, sediments, waters, and air. The uses of some of the metals by mankind are enormous, finding extensive applications in industry, agriculture, utensils, weaponry, medicine, and jewelry. Anthropogenic activities, including mining, smelting, production, and waste generation, have enhanced the concentrations of many metals in the environment, thereby increasing the level of human and animal exposure to such metals.

The physical, chemical, and toxicological properties of metals show wide variation. Not all metals are highly toxic, nor different compounds of the same metals exhibit the same degree of toxicity. The toxicity of many meal salts may be fully attributed to the metal ions, as the salts dissociate into their ions. However, the same is not true for organometallic substances. The latter may exhibit toxic properties that are different from the inorganic compounds of the same metals. For example, the toxic properties of mercury salts are different from alkyl mercury compounds. Thus, the degree of ionization and the covalent character in the compound play a significant role. Elements in the same group may differ significantly in their toxic properties. Not only heavier metals in a group are usually more toxic than the lighter elements, but also their toxicological properties may be quite different. Also, the nature and degree of toxicity may widely differ with the physical state of the metal or its compounds and the routes of exposure. For example, chronic inhalation of finely powdered dusts of practically any metal can produce serious lung diseases. On the other hand, ingestion of aqueous solutions of soluble salts of the same metals may affect gastrointestinal tract or other target organs, producing entirely different toxic effects. It is therefore not very prudent to classify the toxic metals under one heading, as there may be little or no correlation at all in their toxic properties. However, for the sake of convenience, and in a broader perspective, these are groups together in this chapter. The physical properties, exposure risks, toxic properties, and chemical analyses of a few selected metals are discussed below. The flammable and explosive characteristics of these and other metals are presented separately in Chapter 36.

It may be noted that many toxic metals are also essential for the body, at trace levels.

Their absence from the diet can produce various deficiency syndromes and adverse health effects. Such essential metals include selenium, copper, cobalt, zinc, and iron. On the other hand, excessive intake can produce serious adverse reactions. Also, a number of metals, such as aluminum, bismuth, lithium, gold, platinum, and thallium, have been used in medicine. Despite their beneficial effects, excessive intake of these metals and their salts can cause serious poisoning.

37.2 TOXICOLOGY

The toxicological features that may be common to many metals are outlined below:

Enzymes: Toxicity of metals may arise from their actions on enzymes. Many metals may inhibit enzymes by (1) interacting with the SH group of the enzyme; (2) displacing an essential metal cofactor of the enzyme, and (3) inhibiting the synthesis of enzyme.

An example of the latter is nickel, which can inhibit the synthesis of an enzyme, δ-aminolevulinic acid synthetase (ALAS). (Maines and Kappas, 1977) ALAS catalyzes synthesis of heme, an important component of hemoglobin and cytochrome. Thus, such metals as nickel or platinum, may affect the synthesis of heme. Also actions of metals may differ widely, as well as, the degree of susceptibility of enzymes to metals. For example, lead can inhibit several enzymes and therefore interfere with the synthesis of heme to a greater extent than nickel or platinum.

Organelles: Subcellular organelles are other sites of reactions of metals. A metal may enter the cell and react with intracellular components.

37.3 LEAD

Symbol: Pb; At. Wt. 207.2; At. No. 82; valences 2, 4; CAS [7439-92-1]

Uses and Exposure Risk

Lead has been known to humankind since ancient times. It is a major component of many alloys, such as bronze and solder. It is used for tank linings, piping, and building construction; in the manufacture of pigments for paints, tetraethyllead, and many organic and inorganic compounds; in storage batteries; and in ceramics. Lead levels in many soils have been range from 5 to 25 mg/kg and in groundwaters from 1 to 50 µg/L. These concentrations may vary significantly.

Physical Properties

Silvery gray metal; lustrous when freshly cut, loses its shine when exposed to air; soft; resistant to corrosion; opacity to gamma and x-rays; mp 327.5°C; bp 1740°C; density 11.35 at 20°C; reacts with hot concentrated nitric, hydrochloric, and sulfuric acids; resistant to hydrofluoric acid and brine.

Health Hazard

Toxic routes of exposure to lead are food, water, and air. It is an acute as well as a chronic toxicant. The toxic effects depend on the dose and the nature of the lead salt. Ingestion of lead paint chips is a common cause of lead poisoning among children. Chronic toxic effects may arise from occupational exposure.

Acute toxic symptoms include ataxia, repeated vomiting, headache, stupor, hallucinations, tremors, convulsions, and coma. Such symptoms are manifested by the encephalopathic syndrome. Chronic exposure can cause weight loss, central nervous system effects, anemia, and damage to the kidney. Lead can severely affect the nervous system. Chronic lead poisoning adversely affects the central and peripheral nervous systems, causing restlessness, irritability, and memory loss. At lead concentrations of >80 µg/dL, encephalopathy can occur. Cerebral edema neuronal degenerationa and glial proliferation can occur. The clinical symptoms are ataxia, stupor, convulsion, and coma. Permanent brain damage has been noted among children from lead poisoning. Kidney damage arising from short-term ingestion of lead is reversible: while a longer-term effect may develop to general degradation of the kidney, causing glomular atrophy, interstitial fibrosis, and sclerosis of vessels (Manahan 1989). Inhalation of lead justs can cause gastritis and changes in the liver. Lead is significantly bioaccumulated in bones and teeth, where it is stored and released. It binds to a number of cellular ligands, interfering with some calcium-regulated functions. Lead has an affinity for sulfhydryl groups (−SH), which are present in many enzymes. Thus it inhibits enzymatic activity. One such effect is the inhibition of δ-amino-levulinic acid dehydratas (ALAD) an enzyme required for the biosynthesis of heme, an iron(II)−porphyrin complex in hemoglobin and cytochrome. Another enzyme which is also highly susceptible to the inhibitory effect of lead is heme synthetase. The impaired heme synthesis may cause anemia. The clinical anemia is perceptible at a blood lead level of 50 μg/dL. Concentrations of lead in the blood at levels of 10 μg/dL can cause ALAD inhibition. Carcinogenicity of lead has not been observed in humans; the evidence in animals is inadequate.

Exposure Limits

TLV-TWA 0.15 mg/m^3 as Pb (ACGIH and MSHA), 0.05 mg (Pb)/m^3 (OSHA); 10-hr TWA 0.1 mg(inorganic lead)/m^3 (NIOSH).

37.4 CADMIUM

Symbol Cd; At. Wt. 112.4; At. No. 48; valence 2; CAS [7440-43-9]; a group IIB element in the Periodic Table.

Uses and Exposure Risk

Cadmium is used in electroplating, in nickel-cadmium storage batteries, as a coating for other metals, in bearing and low-melting alloys, and as control rods in nuclear reactors. Cadmium compounds have numerous applications, including dyeing and printing textiles, as TV phosphors, as pigments and enamels, and in semiconductors and solar cells.

Physical Properties

Bluish white metal; malleable; density 8.64 at 20°C; mp 321°C; bp 767°C; vapor pressure 0.095 torr at 321°C; vapor pressure of solid at room temperature produces 0.12 mg/m^3 of Cd; soluble in acids.

Health Hazard

There are several reports of cadmium poisoning and human death. Cadmium can enter the body by inhalation of its dusts or fumes, or by ingestion. In humans the acute toxic symptoms are nausea, vomiting, diarrhea, headache, abdominal pain, muscular ache, salivation, and shock. In addition, inhalation of its fumes or dusts can cause cough, tightness of chest, respiratory distress, congestion in lungs, and bronchopneumonia. A 30-minute exposure to about 50 mg/m^3 of its fumes or dusts can be fatal to humans. The oral LD$_{50}$ value in rats is within the range of 250 mg/kg.

Cadmium is a poison that is accumulated in the liver and kidneys. Thus, chronic poisoning leads to liver and kidney damage. It is very slowly excreted. Its biological half-life in humans is estimated at about 20–30 years (Manahan 1989). Cadmium level in the kidney at 200 μg/g, can damage proximal tubules, resulting in their inability to reabsorb small-molecule proteins, such as β_2-microglobulin (Lu 1996). Cigarette smoking and calcium-deficient diet enhance its toxicity. The absorption of this metal through the gastrointestinal tract is low. Cadmium is also known to produce the so-called *itai-itai* disease, which is a chronic renal disease, producing bone deformity and kidney malfunction. Cadmium, similar to other heavy meals, binds to the sulfhydryl (−SH) groups in enzymes, thus inhibiting enzymatic acitivity. Intramuscular administration of cadmium produced tumors in the lungs and blood rats. There is sufficient evidence of its carcinogenicity in animals.

Exposure Limits

TLV-TWA 0.05 mg/m^3 (for dusts and salts) (ACGIH), 0.2 mg/m^3 (MSHA), 0.1 mg/m^3 (OSHA), lowest feasible level in air (NIOSH); ceiling 0.3 mg/m^3 (OSHA).

37.5 MERCURY

Symbol: Hg; At. Wt. 200.59; At. No. 80; valences 1, 2; CAS [7439-97-6]; a group IIB element

Synonyms: quicksilver; hydrargyrum

Uses and Exposure Risk

Mercury is used in mercury arc and fluorescent lamps; in thermometers, barometers, and hydrometers; to extract gold and silver form ores; and as amalgams with many metals.

Physical Properties

Silvery white, heavy liquid; mobile; density 13.53 at 25°C; solidifies at −39°C; boils at 356.7°C; does not oxidize at ambient temperatures; immiscible with water; reacts with nitric acid and with hot concentrated sulfuric acid.

Health Hazard

Elemental mercury and its inorganic salts, as well as organomercury compounds, are all highly toxic substances. The element has a vapor pressure of 0.0018 torr at 25°C, which is high enough to make it a severe inhalation hazard. Exposure to mercury vapors at high concentrations for a short period can cause bronchitis, pneumonitis, coughing, chest pain, respiratory distress, salivation, and diarrhea. The toxic symptoms due to its effects on the central nervous system include tremor, insomnia, depression, and irritability. A 4-hour exposure to mercury vapors at a concentration of about 30 mg/m^3 in air produced damage in the kidneys, liver, lungs, and brain in rabbits (ACGIH 1986). Elemental mercury is rapidly oxidized to Hg(II) in red blood cells. Prior to its oxidized and accumulates (Manahan 1989). Mercury(II) accumulates in the kidneys.

The toxicities of mercury compounds show significant variation with solubility. Less soluble mercurous compounds are relatively less toxic than the more soluble mercuric salts. The latter compounds are highly toxic ingestion. Chronic effects due to Hg(II) ions are inflammation of the muth, salivation, loose teeth, muscle tremors, jerky gait, depression, irritability, and nervousness (Hodgson et al. 1988). The Hg^{2+} ions have an affinity for sulfhydryl groups (−SH) in proteins, enzymes, serum albumin, and hemoglobin to form complexes, thereby causing enzyme inhibition. Cycteine, penicillamine, and 2,3-dimercapto-1-propanol are effective antidotes against mercury(II) poisoning. Also reported are the other sulfur antidotes; unithiol [4076-02-2], BAL [59-67-5], and D-penicillamine [52-67-5] (Softova et al. 1984). These substances can form chelates with mercury and excrete out in the urine, thus exhibiting a protective effect against mercury-induced renal damage. *N*-Benzyl-D-glucamine is another chelating agent that has shown protective action against renal toxicity of inorganic mercury in rats (Kojima et al. 1989).

Exposure Limits

TLV-TWA 0.05 mg/m^3 for Hg vapor, and 0.10 mg/m^3, as Hg for alkyl mercury and inorganic compounds (ACGIH); ceiling 0.1 mg/m^3 (OSHA); IDLH 28 mg/m^3 (NIOSH).

37.6 ARSENIC

Symbol: As; At. Wt. 74.92; At. No. 33; valences 3, 5; CAS [7440-38-2]; a group IIIA metal

Uses and Exposure Risk

Arsenic is used for hardening metals such as copper and lead and as a doping agent in solid-state products of silicon and germanium. Its salts are used in making herbicides and rodenticides, in semiconductors, and in pyrotechnics.

Physical Properties

Shiny graph metals; tarnish when exposed to air; exist n several allotropic modifications; mg 650°C; bp 1380°C; sublimes.

Health Hazard

One of the allotropic forms, yellow arsenic, is a severe human poison. The fatal dose in humans is 1–2 mg/kg body weight. All arsenic compounds are toxic, the toxicity varying with the oxidation state of the metal and the solubility. Thus, the trivalent inorganic compounds of arsenic, such as arsenic trichloride, arsenic trioxide, and arsine, are highly toxic — more poisonous than the metal and its pentavalent salts. The organic arsenic compound Lewisite is a severe blistering agent that can penetrate the skin and cause damage at the point of exposure. Lewisite was used as a poison gas in World War I. Less soluble arsenic sulfide exhibits a lower acute toxicity.

Arsenic is absorbed into the body through a gastrointestinal route and inhalation. The acute symptoms include fever, gastrointestinal disturbances, irritation of the respiratory tract, ulceration of the nasal septum, and dermatitis. Chronic exposure can produce pigmentation of the skin, peripheral neuropathy, and degeneration of liver and kidneys.

The toxic effects of arsenic are attributed to its binding properties with sulfur. It forms complexes with coenzymes. This inhibits the production of adenosine triphosphate (ATP), which is essential for energy in body metabolism. 2,3-Dimercapto-1-propanol is an antidote against acute intoxication. 2,3-Dithioerythritol is reported to be a more effective and less toxic antidote (Boyd et al. 1989). Arsenic is carcinogenic to humans. Ingestion by an oral route has caused an increased incidence of tumors in the liver, blood, and lungs.

Exposure Limits

TLV-TWA 0.2 mg(As)/m^3 (ACGIH), 0.5 mg (As)/m^3 (MSHA), 0.01 mg(As)m^3 (OSHA); ceiling 0.002 mg(As)/m^3/15 min (NIOSH); carcinogenicity: Human Sufficient Evidence (IARC).

37.7 CHROMIUM

Symbol: Cr; At. Wt. 52.00; At No. 24; CAS [7440-47-3]; a transition metal, contains partially filled d-orbitals; exhibits valences from 1 to 6, commonly occurs in the +3 and +6 oxidation states.

Uses and Exposure Risk

Chromium is used in the manufacture of its alloys, such as chrome-steel or chrome-nickel-steel. It is also used for chromeplating of other metals, for tanning leather, and in catalysts. It occurs in chromite ores (FeO·Cr$_2$O$_3$).

Physical Properties

Grayish, hard, lustrous metal; density 7.14; melts at 1900°C; boils at 2642°C; reacts with dilute HCl and H$_2$SO$_4$; attacked by alkalies.

Health Hazard

The toxicity of chromium alloys and compounds varies significantly. Chromium metal does not exhibit toxicity. Divalent and trivalent compounds of chromium have a low order of toxicity. Exposure to the dusts of chromite or ferrochrome alloys may cause lung diseases, including pneumoconiosis and pulmonary fibrosis.

Among all chromium compounds only the hexavalent salts are a prime health hazard. Cr^{6+} is more readily taken up by cells, than any other valence state of the metal. Occupational exposure to these compounds can produce skin ulceration, dermatitis, perforation of the nasal septa, and kidney damage. It on induce hypersensitivity reactions of the skin and renal tubular necrosis. Examples of hexavalent salts are the chromates and dichromates of sodium, potassium, and other metals. The water-soluble hexavalent chromium salts are absorbed into the bloodstream through inhalation. Many chromium(VI) compounds are carcinogenic, causing lung cancers in animals and humans. The carcinogenicity may be attributed to intracellular conversion of Cr^{6+} to Cr^{3+}, which is biologically more active. The trivalent Cr ion can bind with nucleic acid and thus initiate carcinogenesis.

Exposure Limits

TLV-TWA: chromium metal 0.5 mg/m^3 (ACGIH and MSHA), 1 mg/m^3 (OSHA); Cr(II) and Cr(III) compounds 0.5 mg/m^3 (ACGIH); Cr(VI) compounds, water soluble and certain water insoluble, 0.05 mg/m^3 (ACGIH).

37.8 SELENIUM

Symbol Se; At. Wt. 78.96; At. No. 34; valences 2, 4, and 6; CAS [7782-49-2]; a group VIA metal

Uses and Exposure Risk

Selenium is used in the manufacture of colored glass, in photocells, in semiconductors, as a rectifier in radio and television sets, and as a vulcanizing agent in the manufacture of rubber.

Physical Properties

Dark red to bluish black amorphous solid, or dark red or gray crystals; exists in several allotropic forms; density 4.26–4.28 for amorphous, 4.26–4.81 for crystals; mp 170–217°C (crystals); bp 685°C; amorphous form becomes elastic at 70°C.

Health Hazard

The toxicity of selenium and its compounds varies substantially. Sodium selenite is highly toxic; many sulfur compounds of selenium are much less toxic. The target organs are the respiratory tract, liver, kidneys, blood, skin, and eyes. The sign of acute poisoning is a garlic-like odor in the breath and sweat. The other symptoms are headache, fever, chill, sore throat, and bronchitis. Chronic intoxication can cause loss of hair, teeth, and nails, depression, nervousness, giddiness, gastrointestinal disturbances dermititis, blurred vision, and a metallic taste. Although inhalation toxicity is severe in test animals, oral toxicity is of low order. The LD$_{50}$ values for selenium compounds vary with the compounds.

Paul and co-workers (1989) have investigated the antidotal actions of several compounds on the acute toxicity of selenium in rats. Male Wistar rats were injected sodium [^{75}Se]selenite subcutaneously in this study. Intraperitoneal administration of diethyldithiocarbamate or treatment with citrate salt of bismuth, antimony, or germanium, administered subcutaneously, reduced selenium-induced loss of body weight in the animals. Germanium citrate and bis(carboxyethyl)germanium sesquioxide promoted increases in the 24-hour urinary excretion of selenium when administered 15 minutes after sodium selenite.

Exposure Limits

TLV-TWA 0.2 mg(Se)/m^3 (ACGIH, MSHA, and OSHA); IDLH 100 mg/m^3.

37.9 SILVER

Symbol: Ag; At. Wt. 107.87; At. No. 47; valences 1, 2; CAS [7440-22-4]; a group 1B metal

Uses and Exposure Risk

Silver is a precious metal, used in jewelry and ornaments Other applications include its use in photography, electroplating, dental alloys, high-capacity batteries, printed circuits, coins, and mirrors.

Physical Properties

Brilliant, white, hard, lustrous metal; very good conductor of heat and electricity; density 10.5; mp 960°C; boils above 2000°C; does not rust, inert to water or atmospheric oxygen; most salts are photosensitive; soluble in fused-alkali hydroxides.

Health Hazard

Silver dusts are irritating to the eyes, nose, and respiratory tract. The toxicity of this metal is low by all routes of exposure. Prolonged skin contact with fine particles of silver or inhalation of dusts of silver compounds can result in argyria, a blue-gray discoloration of the skin. Argyria may occur in the conjunctiva of the eye, gum tissue, or nasal septum; it may result from the accumulation of silver in the body.

Exposure Limits

TLV-TWA (metal dusts and fumes) 0.1 mg/m^3 (ACGIH), 0.01 mg/m^3 (MSHA and OSHA), soluble compounds 0.01 mg/m^3 (AIGIH).

37.10 BARIUM

Symbol: Ba; At. Wt. 147.34; At. No. 56; valence 2; CAS [7440-39-3]; a group IIA metal (alkaline-earth metal)

Uses and Exposure Risk

Barium is used in electronic tubes and as a carrier for radium. Many barium salts are used for various purposes, such as in paints, ceramics, lubricating oils, and analytical work.

Physical Properties

Yellowish white, malleable metal; readily oxidized: density 3.6; mp 710°C; bp 1600°C.

Health Hazard

Inhalation of barium dusts can cause irritation of the nose and upper respiratory tract. All soluble salts of barium are acute poisons. Barium ion is toxic to muscle. Ingestion can cause severe hypokalemia. The toxicity of barium salts depends on their solubility. Ingestion of about 1 g of barium chloride or 4 g of barium carbonate can be lethal to humans.

Exposure Limits

TLV-TWA 0.5 mg/m^3 (for soluble compounds) (ACGIH and MSHA); IDLH (for soluble compounds) 250 mg/m^3 (NIOSH).

37.11 BERYLLIUM

Symbol: Be: At. Wt. 9.012; At. No. 4; valence 2; CAS [7440-41-7]; a group IIA metal (alkaline-earth metal)

Uses and Exposure Rish

Beryllium is used in beryllium−copper alloys, as a neutron moderator in nuclear reactors, and in radio tubes.

Physical Properties

Gray metal; density 1.848; mp 1287°C; bp 2500°C; resistant to acid attack.

Health Hazard

Beryllium is a highly toxic metal. Inhalation of its dusts can cause irritation of the eyes, skin, and respiratory system. Other toxic symptoms are weakness, fatigue, and weight loss. The main acute effect from a single exposure to beryllium dusts is pneumonitis. Chronic exposure can cause berylliosis, a pulmonary disease. The disease is characterized by granulomas which develop into fibrotic tissues. Dermal contact can produce hypersensitivity reactions of the skin, causing delayed, cell-mediated type IV allergy. Beryllium is carcinogenic to animals and humans, causing lung cancer. Animal studies show sufficient evidence of its carcinogenicity. In humans, such evidence is limited.

Exposure Limit

TLV-TWA 0.002 mg/m^3 (ACGIH, MSHA, and OSHA).

37.12 THALLIUM

Symbol: Tl; At. Wt. 204.38; At. No. 81; valences 1 and 3; CAS [7440-28-0]; a group IIIA heavy metal

Uses and Exposure Risk

Thallium is used in photoelectric cells, in semiconductor studies, and in low-range glass thermonmeters. It is alloyed with many metals. Many of its salts are used as rodent poisons.

Physical Properties

Bluish white metal; very soft and easily fusible; density 11.85; mp 303.5°C; bp 1457°C; volatilizes around 174°C; oxidizes in air, forming a coating of thallium oxide, Tl$_2$O; reacts with sulfuric and nitric acids.

Health Hazard

Thallium and its soluble compounds are highly toxic in experimental animals. The acute toxic symptoms in humans are nausea, vomiting, diarrhea, polyneuritis, convulsion, and coma. Ingestion of 0.5 g can be fatal to humans. Severe chronic toxicity can lead to kidney and liver damage, deafness, and loss of vision. Other signs of toxicity from chronic exposure include reddening of the skin, abdominal pain, polyneuritis, loss of hair, pain in legs, and occasionally cataracts. Ingestion of thallium salts in children has caused neurological abnormalities, mental retardation, and psychoses.

Exposure Limits

TLV-TWA 0.1 mg/m^3 (thallium and its soluble salts) (ACGIH, MSHA, and OSHA); IDHL 10/mg/m^3

37.13 NICKEL

Symbol: Ni: At. Wt. 58.69; At. No. 28; valence 2, CAS [7440-02-0]; a group VIII transition metal

Uses and Exposure Risk

Nickel is used in various alloys, such as German silver, Monel, and nickel–chrome; for coins; in storage batteries; in spark plugs; and as a hydrogenation catalyst.

Physical Properties

White, lustrous, hard metal; ferromagnetic; density 8.90; mp 1555°C; bp ~2800°C; not attacked by water or fused alkalies; reacts slowly with dilute acids; forms NiO at high temperatures.

Health Hazard

Ingestion of nickel can cause hyperglycemia, depression of the central nervous system, myocardial weakness, and kidney damage. A subcutaneous lethal dose in rabbits is in the range 10 mg/kg. The oral toxicity of the metal, however, is very low. Skin contact can lead to dermatitis and "nickel itch," a chronic eczema, caused by dermal hypersensitivity reactions. Nickel itch may result from wearing pierced earrings. Inhalation of metal dusts can produce irritation of the nose and respiratory tract. Nickel and some of its compounds have been reported to cause lung cancer in experimental animals. It may also induce cancer in nose, stomach, and possibly the kidney. The experimental data on the latter, are not fully confirmative. Nickel refinery flue dust, nickel sulfide (Ni_3S_2) [12035-72-2], and nickel oxide (NiO) [1313-99-1] produced localized tumors in experimental animals when injected intramuscularly. IARC has classified nickel and its compounds as carcinogenic to humans (IARC 1990). Inhalation of metal dusts can produce lung and sinus cancers in humans, with a latent period of about 25 years.

Exposure Limits

TLA-TWA (metal) 1 mg/m^3 (ACGIH, MSHA, and OSHA); (soluble inorganic compounds) 0.1 mg(Ni)/m^3 (ACGIH) 0.015 mg (Ni)/m^3 (NIOSH); (insoluble inorganic compounds) 1 mg/m^3 (ACGIH).

37.14 COPPER

Symbol: Cu; At. Wt. 63.546; At. No. 29; valences 1 and 2; CAS [7440-50-8]; a group IB metal (coinage metal)

Uses and Exposure Risk

Copper has been known since early times. It is used to make utensils, electrical conductors, and alloys such as bronze, brass, and other copper alloys.

Physical Properties

Reddish, lustrous, malleable metal; tarnishes on exposure to air; density 8.94; mp 1083°C; bp 2595°C; dissolves slowly in ammonia water.

Health Hazard

The toxicity of metallic copper is very low. However, inhalation of its dusts, fumes, or mists or its salts can cause adverse health effects. Inhalation causes irritation of the eyes and mucous membranes, nasal perforation, cough, dry throat, muscle ache, chills, and metal fever. Skin contact can result in dermatitis. Many copper(II) salts are toxic.

Exposure Limits

TLV-TWA 1 mg(Cu)/m^3 (dusts and mists) (ACGIH and MSHA); 0.2 mg/m^3 (fumes) (ACGIH).

37.15 ANTIMONY

Symbol: Sb; At. Wt. 121.75; At. No. 51; valences 3 and 5; CAS [7440-36-0]; a group VA element Synonym: stibium

Uses and Exposure Risk

Antimony is used to make alloys such as Babbit metal, white metal, and hard lead; in bullets and fireworks; and for coating metals.

Physical Properties

Silvery white metal; hard and brittle: tarnishes slowly in moist air; density 6.68: mp 630°C; bp 1635°C; attacked by hot concentrated acid.

Health Hazard

The toxicity of antimony is of low order, much less poisonous than arsenic. The symptoms of acute poisoning include weight loss,

loss of hair, eosinophilia, and congestion of heart, liver, and kidney. The toxic routes are primarily inhalation of its dusts or fumes, and skin absorption.

37.16 MANGANESE

Symbol: Nn; At. Wt. 54.938; At. No. 25; valence, shows several oxidation states; CAS [7439-96-5]; a group VIIB metal

Uses and Exposure Risk

Manganese is a constituent of many alloys. It is used in the manufacture of steel.

Physical Properties

Steel-gray lustrous metal; mp 1244°C; bp 2095°C; density 7.21; soluble in dilute mineral acids.

Health Hazard

Although manganese is a cofactor in many enzyme systems, and, thus, an essential element, chronic exposure to this metal can cause encephalopathy. In the elemental form, oral toxicity is low. Ingestion of its salts solutions may induce disturbance in gastrointestinal tract. Oral intake of large doses of metal produced fatty degeneration of liver and kidney. Such effects result from over exposure to the metal. Short-term acute exposure to Mn dust and fume can cause pneumonitis. Some of the symptoms of manganese poisoning include sleepiness, weakness, emotional disturbance and spastic gait (ACGIH 1989). There are conflicting reports in the literature on the toxicity data. This may be attributed to a generalized and oversimplified picture of grouping the metal and many inorganic salts of manganese together. The degree of toxicity can vary from compound to compound.

Exposure Limit

Ceiling: 5 mg(Mn)/m^3 (ACGIH and OSHA)
TWA: 1 mg(Mn)/m^3 (NIOSH)

37.17 COBALT

Symbol: Co; At. Wt. 58.933; At. No. 22; valence 2 and 3; CAS [7440-48-4]

Uses and Exposure Risk

Cobalt is used in steel alloys, cemented carbide abrasives and jet engines.

Physical Properties

Silvery-gray hard metal; mp 1493°C; bp about 3150°C; density 8.92.

Health Hazard

Cobalt is an essential element. Its deficiency can result in pernicious anemia. It is present in vitamin B$_{12}$. Excessive intake of this element may result in polycythemia or overproduction of erythrocytes and heart lesions. Exposure to its dusts can produce cough and respiratory irritation. Chronic inhalation of its dusts or fumes can decrease pulmonary functions and may cause diffuse nodular fibrosis and other pulmonary diseases. Skin contact may induce dermal hypersensitivity reactions, producing an allergy-type dermatitis.

Exposure Limit

TLV-TWA 0.05 mg as Co/m^3 (ACGIH)
PEL-TWA: 0.05 mg as Co/m^3 (NIOSH, OSHA)
TLV-STEL 0.1 mg as Co/m^3 (ACGIH)
IDLH 20 mg as Co/m^3 (NIOSH)

37.18 ZINC

Symbol: Zn; At. Wt. 65.37; At. No. 30; valence 2; CAS [7440-66-6]; a group IIB metal

Uses and Exposure Risk

Zinc is a constituent of many common alloys, including brass, bronze, Babbit metal, and German Silver. It is used to make household utensils, castings, printing plates, building materials, electrical apparatus, dry-cell batteries and many zinc salts. It is also used to galvanize sheet iron, bleaching bone glue and as a reducing agent in many organic reactions.

Physical Properties

Bluish white metal with lusterl; mp 420°C; bp about 908°C; density 7.14; stable in dry air.

Health Hazard

Zinc is another essential metal, a cofactor to many metalloenzymes. Its deficiency can induce effects on liver, nervous system, eye, skin and testis. Excessive intake of this metal, however, may produce adverse effects. The toxicity of the metal from ingestion is relatively low as it is readily excreted. Chronic exposure to the fume, however, may lead to "metal fume fever," which could probably be attributed to the bivalent zinc ion, Zn^{2+}. Although the metal in its zero-valent state exhibits little inhalation toxicity, in its oxidation state or as oxide it can present a serious health hazard. Inhalation of Zn^{2+} or metal oxide fume can produce a sweet metallic taste, cough, chills, fever, dry throat, nausea, vomiting, blurred vision, ache, weakness, and other symptoms. The nontoxic fume of the metal is susceptible to oxidize in the air in the presence of moisture and in contact with many other substances in air. The metal powder or its dust is a skin irritant.

37.19   ZIRCONIUM

Symbol: Zr; At. Wt. 91.224; At. No. 40; CAS [7440-67-7]

Uses and Exposure Risk

It is used in lamp filaments; flash bulbs; vacuum tubes and in explosives. It is also used as a structural material for nuclear reactors (Merck 1996)

Physical Properties

Grayish white lustrous metal or bluish black amorphous powder; mp 1857°C; bp 3575°C; density 6.50.

Health Hazard

The toxicity of zirconium and its compounds has been found to be of low order. Lethal dose in rabbits when administered intravenously is reported as 150 mg/kg (Lewis(Sr) 1996). Inhalation of dust of the metal or its compounds can form skin and pulmonary granulomas that may be attributed to reaction of sensitized T cells with antigen. X-ray studies in animals indicate retention of the metal in the lungs. Inhalation may produce irritation of mucous membrane. Skin contact can cause irritation.

Exposure Limit

TLV-TWA 0.05 mg (Zr)/m^3 (ACGIH)
PEL-TWA: 0.05 mg (Zr)/m^3 (OSHA)
STEL 10 mg (Zr)/m^3 (NIOSH and OSHA)
IDLH 500 mg Zr/m^3 (NIOSH)

37.20   TELLURIUM

Symbol: Te; At. Wt. 127.60; At. No. 52; CAS [13494-80-9]; a group VIA element

Uses and Exposure Risk

The metal is used in vulcanizing rubber, in storage batteries, and as a coloring agent in ceramics. It is also used as an additive to iron, steel, and copper. Many tellurium salts find application on semiconductors.

Physical Properties

Silvery-white metal; mp 450°C; bp about 990°C; density 6.24; reacts with concentrated mineral acids

Health Hazard

Although tellurium in elemental form has low toxicity, ingestion can produce nausea, vomiting, tremors, convulsions, and central nervous system depression. In addition, exposure to the metal or to its compounds can generate garlic-like odor in breath, sweat, and urine. Such odor is imparted by dimethyl telluride that is formed in the body. Oral intake of large doses of the metal or its compounds can be lethal. Clinical symptoms are similar for most tellurium salts, which include headache, drowsiness, loss of appetite, nausea, tremors, and convulsions. High exposure can produce metallic taste, dry throat, chill and other symptoms. Inhalation of dust or fume of the metal can cause irritation of the respiratory tract. Chronic exposure can produce bronchitis and pneumonia.

Exposure Limit

TLV-TWA 0.1 mg (Te)/m^3 (ACGIH)
PEL-TWA: 0.1 mg (Te)/m^3 (OSHA)
TWA 0.1 mg (Te)/m^3 (NIOSH)

REFERENCES

ACGIH. 1986. *Documentation of the Threshold Limit Values and Biological Exposure Indices*, 5th ed. and updates. Cincinnati, OH: American Conference of Governmental Industrial Hygienists.

Boyd, V. L., J. W. Harbell, R. J. O'Connor and E. L. McGown. 1989. 2,3-Dithioerythritol, a possible new arsenic antidote. *Chem. Res. Toxicol.* 2(5): 301–6.

Hodgson, E., R. B. Mailman and J. E. Chambers. 1988. *Dictionary of Toxicology*. New York: Van Nostrand Reinhold.

International Agency for Research on Cancer. 1990. Nickel Compounds. Monograph vol. 49. Geneva, Switzerland: World Health Organization(WHO).

Kojima, S., H. Shimada and M. Kiyozumi. 1989. Comparative effects of chelating agents on distribution, excretion and renal toxicity of inorganic mercury in rats. *Res. Commun. Chem. Pathol. Pharmacol.* 64(3): 471–84.

Lewis(Sr), R. J. 1996. *Sax's Dangerous Properties of Industrial Materials*, 9th ed. New York: Van Nostrand Reinhold.

Lu, F. C. 1996. *Basic Toxicology: Fundamentals, Target Organs, and Risk Assessment*, 3rd ed. Washington, DC: Taylor & Francis.

Manahan, S. E. 1989. *Toxicological Chemistry*. Chelsea, MI: Lewis Publishers.

Merck. 1996. *The Merck Index*, 12th ed. Rahway, NJ: Merck & Co.

Paul, M., R. Mason and R. Edwards. 1989. Effect of potential antidotes on the acute toxicity, tissue deposition and elimination of selenium in rats. *Res. Commun. Chem. Pathol. Pharmacol.* 66(3): 441–50.

Softova, E., A. Belcheva and M. Mangarova. 1984. Comparative study of the influence of mono- and dithiol antidotes upon renal structural changes and urea level in acute mercury intoxication. *Scr. Sci. Med.* 21: 31–7; cited in *Chem. Abstr.* CA 102(17): 144380W.

38

MUSTARD GAS AND SULFUR MUSTARDS

38.1 GENERAL DISCUSSION

Sulfur mustards have become one of the most lethal weapons in chemical warfare. These substances were developed between World Wars I and II, along with some other lethal agents, such as nerve gases, nitrogen mustards and phosgene. Sulfur mustards rank among the worst toxic military poisons, second to the nerve agents in terms of potential for mass casualties.

The common structural feature of all sulfur mustards is that their molecules are composed of two chloroethyl groups attached to a sulfur atom. Some compounds may have additional S and/or O atoms bridging $-CH_2-CH_2-$ units. It may be noted that in all these compounds, both the chlorine atoms are in the terminal positions. Some of these structures are shown below:

$$Cl-CH_2-CH_2-S-CH_2-CH_2-Cl$$

(Mustard Gas)

$$Cl-CH_2-CH_2-S-CH_2-CH_2-S-$$
$$CH_2-CH_2-Cl$$

(Chemical Agent (Q) or Sesquimustard (Q))

$$Cl-CH_2-CH_2-S-CH_2-CH_2-O-CH_2-$$
$$CH_2-S-CH_2-CH_2-Cl$$

(Chemical Agent (T) or O-Mustard (T))

$$Cl-CH_2-CH_2-S-CH_2-CH_2-S-CH_2-$$
$$CH_2-S-CH_2-CH_2-Cl$$

(1,2-Bis(2-chloroethylthioethyl)thioether)

These compounds are made by the reaction of hydroxyderivative of the sulfide with HCl gas. This is shown in the following equation:

$$HO-CH_2-CH_2-S-CH_2-CH_2-OH + 2HCl \longrightarrow Cl-CH_2-CH_2-S-CH_2-CH_2-Cl$$

(2,2-dihydroxyethyl sulfide)

(mustard gas)

$$+ 2H_2O$$

Mustard gas may also be prepared by the reaction of ethylene with sulfur chloride.

These substances exhibit moderate to high reactivity towards a variety of functional groups, including carboxyl, sulfhydryl, and primary, secondary, and tertiary amino nitrogen atoms. Such reactivity may be attributed to the formation of a highly reactive intermediate, a cyclic sulfonium ion in water, as shown below:

$$Cl-CH_2-CH_2-S-CH_2-CH_2-Cl \Longleftrightarrow Cl-CH_2-CH_2-\overset{+}{S}\underset{CH_2}{\overset{CH_2}{\diagup}} \, Cl^-$$

Sulfur mustards are oxidized with strong oxidants and are hydrolyzed with alkali, generally producing reaction products that are less toxic. Reactions with chlorine or chlorinating agents such as NaOCl or Ca(OCl)$_2$ yield nontoxic products. This reaction is suitable to inactivate and decontaminate sulfur mustards. The physical, chemical, and toxic properties of mustard gas and a few selective compounds of this class are outline below.

38.2 CHEMICAL ANALYSIS

Sulfur mustards at trace levels may be separated on a GC column and detected by an ECD or other halogen specific detectors. A better method, however, would be the GC/MS analysis, where these compounds may be identified from their characteristic mass spectra. Detection and quantitation may be performed by colorimetry. Reaction with *p*-nitrobenzylpyridine followed by alkali treatment produces a blue color, the absorbance of which can be measured by a spectrophotometer. Mustards at trace levels, as little as 0.1 μg, can be measured by this method.

38.3 MUSTARD GAS

Formula: $C_4H_8Cl_2S$; MW 159.08; CAS [505-60-2];

Structure: $Cl-CH_2-CH_2-S-CH_2-CH_2-Cl$;

Synonyms: bis(2-chloroethyl)sulfide; 2,2′-dichlorodiethyl sulfide; β,β-dichloroethyl sulfide; 1,1-thiobis(2-chloroethane); 1-chloro-2-(β-chloroethylthio)ethane; ypperite; HD

Physical Properties

Colorless, oily liquid when pure; faint sweet odor; melts at 13.5°C; boils at 216°C; density 1.274 at 20°C; soluble in common organic solvents, fats and lipids; slightly soluble in water (0.68 g/L)

Chemical Reactions

Mustard gas can be oxidized with strong oxidizing agents, such as H_2O_2, $K_2MnO_4-H_2SO_4$, and $K_2Cr_2O-H_2SO_4$, forming sulfone. Oxygen atoms attach to S atom, oxidizing the latter to +6 oxidation state. The reaction is as follows:

$$Cl-CH_2-CH_2-S-CH_2-CH_2-Cl \xrightarrow[\text{H}_2\text{O}_2/\text{K}_2\text{Cr}_2\text{O}_7-\text{H}_2\text{SO}_4]{\text{Oxidation}} Cl-CH_2-CH_2-\overset{\overset{O}{\|}}{\underset{\underset{O}{\|}}{S}}-CH_2-CH_2-Cl$$

(bis(2-chloroethyl)sulfone)

The product sulfone is also a vesicant, however, less toxic than the mustard gas.

It reacts readily with chlorine or chlorine-releasing substances, such as hypochlorites.

Alkaline hydrolysis produces hydroxy derivative, as follows:

$$Cl-CH_2-CH_2-S-CH_2-CH_2-Cl \xrightarrow[\text{KOH}]{\text{hydrolysis}} HO-CH_2-CH_2-S-CH_2-CH_2-OH$$

(2,2′-dihydroxyethyl sulfide)

Toxicity

Mustard gas is a deadly vesicant and a blistering poison. Because of its high penetrative action, this substance can cause severe local injury to most organs in the body, including skin, eyes, nose, throat, and lung. It can penetrate the skin and destroy the tissues and blood vessels. This can result in severe inflammation of tissues and painful blistering. The skin becomes red and itching and is often infected. The effect on the eye, can be severe, ranging from conjunctivitis and swollen eyelids to blindness. A 1-minute exposure to 30 ppm in air can cause severe eye injury in humans. Inhalation of its vapors can produce inflammation of the nose, throat, bronchi, and lung tissues, resulting in cough, edema, ulceration, and necrosis of the respiratory tract. Death can occur from pulmonary lesions (Manahan 1989). The median lethal dosage for inhalation is 1500 mg/m^3 per minute (US EPA 1988). At low concentrations, mustard gas exhibits delayed effect with symptoms appearing a few hours after exposure. However, at higher concentrations, the onset of such toxic symptoms may occur after shorter delays. While the nonchloro precursor of mustard gas, diethyl sulfide ($CH_3-CH_2-S-CH_2-CH_3$), is relatively far less toxic, the incorporation of two terminal Cl atoms in the molecule enormously enhances and alters the toxic properties of the chloroderivative. The rate of detoxication of this toxicant in the body is very slow. As a result, repeated exposures can be cumulative. The process of healing is very slow. The therapy available for treatment is very little.

At ambient temperature, mustard gas is a liquid. Its volatility at 25°C is reported to be 925 mg/m^3 (Harris et al. 1984) Such volatility is sufficiently high to enable its application in warm weather. A few minutes exposure to 25–50 ppm concentration in the air can produce the delayed fatal effect in humans.

Ingestion of materials contaminated with mustard gas can cause nausea and vomiting (Merck 1996).

LD$_{50}$ (intravenous)(rats): 3.3 mg/kg
LD$_{50}$ (intravenous)(mice): 8.6 mg/kg
LD$_{50}$ (skin) rat: 5 mg/kg

Mustard gas is an experimental mutagen (Levi 1987) and a confirmed human carcinogen (Lewis 1996). Carcinogenicity may be attributed to the highly reactive sulfonium ion which this compound forms in the solution. Such bioactivated reactive intermediate alkylates functional groups of macromolecules.

38.4   O-MUSTARD(T)

Formula: $C_8H_{16}Cl_2OS_2$; MW 263.26; CAS [63918-89-8];

Structure: $Cl-CH_2-CH_2-S-CH_2-CH_2-O-CH_2-CH_2-S-CH_2-CH_2-O-Cl$

Synonyms: 1,1′-oxybis(2-2-chloroethyl)thioethane; 2,2′-di(3-chloroethylthio)diethyl ether; Chemical Agent (T); bis(2-chloroethylthioethyl)ether

Physical Properties

Colorless liquid; solidifies at 10°C; b.p. 120°C; density 1.24 at 25°C; readily dissolves in common organic solvents; slightly soluble in water.

Toxicity

O-Mustard(T) is a dangerous vesicant. Its toxic properties are similar to mustard gas. However, it is less volatile than mustard gas; the volatility at 25°C being 2.8 mg/m^3 (Harris et al. 1984). The inhalation hazard from its vapors is therefore lower than that of mustard gas. Contact with the body can result in severe local injury. It can penetrate through skin, causing inflammation and blistering, which are difficult to heal. Like mustard gas, this substance also exhibits delayed clinical symptoms at low concentrations. There is no report of carcinogenicity. In chemical warfare, it is used in combination with mustard gas. Such a combined mixture is known as HT. The toxicity and vesicant action of HT are greater than those of its components. The mixture solidifies at a lower temperature and is more persistent than the mustard gas alone.

38.5 SESQUIMUSTARD

Formula: $C_6H_{12}Cl_2S_2$; MW 219.20; CAS [3563-36-8];

Structure: $Cl-CH_2-CH_2-S-CH_2-CH_2-S-CH_2-CH_2-Cl$

Synonyms: 1,2-bis(2-chloroethylthio)ethane; 1,2-bis(β-chloroethylthio)ethane; sesquimustard (Q); Chemical Agent (Q)

Physical Properties

Solid at room temperature; melts at 56°C; low volatility, 0.4 mg/m^3 at 25°C; readily dissolves in common organic solvents; very slightly soluble in water.

Toxicity

The toxic effects are similar to those associated with mustard gas. It is a dangerous vesicant and a highly toxic substance, exhibiting delayed symptoms. Although sesquimustard is less volatile than O-Mustard(T), its toxicity in humans via inhalation route is somewhat greater than that of the latter. The subcutaneous lethal dose in mice is in the range 10 mg/kg.

LC$_{50}$ (mouse): 6 mg/m^3/10 min.

There is no report of its carcinogenicity. Its mixture with mustard gas, known as HQ, is more toxic than either of the component and is used as a war agent. HQ is a deadly vesicant like HT and is more persistent than mustard alone.

REFERENCES

Harris, B. L., F. Shanty, and W. J. Wiseman. 1984. Chemicals in War. In *Kirk-Othmer Encyclopedia of Chemical Technology*, 3rd ed. pp. 395–397. New York: Wiley Interscience.

Levi, P. E. 1987. Toxic action. In *Modern Toxicology*, ed. E. Hodgson and P. E. Levi, pp 134–84. New York: Elsevier.

Manahan, S. E. 1988. *Toxicological Chemistry*, pp 265–66. Chelsea, MI: Lewis Publishers.

Merck. 1996. *The Merck Index*, 12th ed. Rahway, NJ: Merck & Co.

Lewis (Sr.), R. J. 1996. *Sax's Dangerous Properties of Industrial Materials*, 9th ed. New York: Van Nostrand Reinhold.

US EPA 1988. *Extremely Hazardous Substances; Superfund Chemical Profiles*, Park Ridge, NJ: Noyes Data Corporation.

39

NERVE GASES

39.1 GENERAL DISCUSSION

Nerve gases are extremely toxic organophosphates developed for use as military poisons. These substances, developed during World War II as possible chemical warfare agents and used in recent years in terrorist attacks, are among the most toxic substances ever made. At ambient temperatures, these compounds are volatile liquids. The best known nerve agents are tabun (GA), sarin (GB), soman (GD), and VX. The general structure of this class of compounds is as follows:

$$\begin{array}{c} R \\ \diagdown \\ Y \end{array} \overset{\displaystyle \overset{O}{\|}}{P}-X$$

where R is an alkyl group, generally methyl or ethyl group; Y is RO or R_2N; and X is a leaving group, such as, CN^-, F^-, or NO_2^-, that breaks away from the molecule during hydrolysis.

The nerve gases are powerful inhibitors of the enzyme acetylcholinesterase (AChE), which causes rapid hydrolysis of acetylcholine. Acetylcholine is the chemical that is released to transmit nerve impulses in the central nervous system and also at several peripheral locations. Once the impulse is

transmitted, acetylcholine must be removed instantaneously for proper functioning of the nervous system. Such removal is done by the enzyme AChE, which is found extensively in nervous system and many nonnervous tissues. Nerve gases and other organophosphorus compounds bind AChE by phosphorylation via the active "oxon" (P=O) portion of the molecule forming P—O—AChE. Such binding of AChE inactivates its action to hydrolyze acetylcholine. Thus the latter accumulates at the synapse, disrupting the normal responses to discrete nerve impulses.

The biological action of nerve agents is similar to that of organophosphorus insecticides, that is, binding and inactivation of the enzyme AChE in nerve tissues. However, the degree of toxicity of the former is much greater than that of the latter. The difference in toxicity may be attributed to the rate of chemical binding of AChE to the oxon, which becomes irreversible over a period of time. This process is referred to as aging. It occurs at different rates depending on the structure of the molecules and other factors, which are not well understood. It apparently involves hydrolysis of an alkoxy group on the phosphorus atom. In sarin and DFP, the isopropoxy group ages within 1–2 hours, while the trimethylpropoxy group in soman ages within minutes. Among the four nerve gases, the lethality of soman is somewhat greater than that of sarin and VX. The latter two showed a slightly greater difference in lethality over tabun in rats. The toxic response was found to vary among species. The lethality of these compounds is relatively greater in humans than in monkeys, rats, and rabbits.

These substances are absorbed through the lungs, skin, eyes, and gastrointestinal tract. A short exposure may be lethal, causing death within minutes. The toxic symptoms are runny nose, miosis, tightness of chest, dimness of vision, salivation, nausea, diarrhea, difficulty in breathing, convulsions, coma, respiratory paralysis, and death. Death occurs from asphyxia. Atropine and oximes are antidotes against such poisoning.

Presented in the following sections are the general chemistry, toxicology, antidotes, analysis, and methods of destruction for each of these compounds. Organophosphorus compounds that are not used as nerve gases but show comparable high toxicity are mentioned briefly.

39.2 SARIN

Formula, $C_4H_{10}FPO_2$; MW 140.09; CAS [107-44-8]

Structure:

$$(CH_3)_2CH-O \underset{H_3C}{\overset{O}{\diagdown P}} -F$$

Synonyms: methylphosphonofluoridic acid isopropyl ester; isopropylmethylphosphonofluoridate; isopropylmethylfluorophosphonate; isopropoxymethylphosphoryl fluoride; fluoroisopropoxymethyl phosphine oxide; isopropyl ester of methylfluorophosphoric acid; GB

Physical and Chemical Properties

Colorless liquid; boils at 147°C; freezes at −57°C; density 1.10 at 20°C; miscible

with organic solvents and water (hydrolyzes). Hydrolysis with water occurs rapidly with removal of the fluorine atom, forming isopropoxymethyl phosphoric acid:

$$(CH_3)_2CH-O \underset{H_3C}{\overset{O}{\diagdown P}} -F \xrightarrow{H_2O}$$
$$(CH_3)_2CH-O \underset{H_3C}{\overset{O}{\diagdown P}} -OH$$

Aqueous alkaline solutions of caustic soda, caustic potash, and sodium carbonate hydrolyze sarin rapidly (Merck 1989).

Health Hazard

Sarin is one of the most toxic compounds ever synthesized. It is a very powerful inhibitor of acetylcholinesterase enzyme. It exhibits systemic toxicity to the central nervous system. The liquid can be absorbed readily through the skin. One drop so absorbed can be fatal to humans (Manahan 1989). It is extremely toxic by inhalation and oral intake. The toxic symptoms are similar to those of the insecticide parathion but more severe than the latter. The poisoning effects arise essentially from the deactivation of cholinesterase and the symptoms are described in Section 39.1.

In humans the toxic effects from sarin may be manifested at an oral dose as low as 0.002–0.005 mg/kg with symptoms of nausea, vomiting, muscle weakness, bronchiolar constriction, and asthma. A lethal dose may be as low as 0.01 mg/kg. A LC_{50} value from 10-minute exposure may be 100–150 mg/m^3 for most animals; and the oral LD_{50} value for rats should be within the range of 0.5 mg/kg.

In a study on the toxicity of sarin, soman, and tabun in brain acetylcholinesterase activity in mice, Tripathi and Dewey (1989) observed differential potencies between lethality and inhibition of acetylcholinesterase. These investigators conclude that there were more factors involved than simply the inhibition of acetylcholinesterase within the brain, causing lethality of these nerve agents. An LD_{50} value by intravenous administration for sarin was reported as 0.109 mg/kg. The acute toxicity of sarin based on behavioral test models indicated that certain species, such as marmosets, showed greater sensitivity to the lethal action of sarin than rodents and rabbits (D'Mello and Duffey 1985). Doses in the range 33–55% LD_{50} produced 88% or more inhibition of erythrocyte acetylcholinesterase [9000-81-1] and disrupted the performance of a food-reinforced visually guided reaching response.

Acute neurotoxic investigation using perfused canine brain indicated that 400 μg of sarin caused seizure in 5.6 minutes after stimulation of cerebral metabolic activity (Drewes and Singh 1987). Inhalation of sarin and soman in amounts of 30 and 13.14 μg/kg, respectively, caused irregular pulse rate, development of apnea, and a significant decrease in mean blood pressure in baboons (Anzueto et al. 1990). These effects were reversed with atropine.

Goldman et al. (1987) evaluated mutagenic potential of sarin and soman in vitro and in vivo. Both these nerve agents gave no significant evidence of mutagenicity.

Antidote

Several compounds and their combinations exhibiting protective action against lethality of sarin have been reported. Among these, atropine [51-55-8] is probably the most effective.

Methylthioobidoxime [99761-22-5], methylthio-TMB 4 [99761-21-4], and related compounds have been reported to be reactivators of acetylcholinesterase [9000-81-1] inhibited by sarin and VX (61–84%) but not by soman or tabun (Bevandic et al. 1985). These methylthio analogs were more effective than obidoxime [7683-36-5], TMB 4 [56-97-3], or PAM 2 [94188-80-4]. The chemical structure of methylthioobidoxime is

These compounds given together with atropine (15–30 μmol/kg +10 mg/kg) were effective against sarin and VX poisoning in mice.

Schoene and co-workers (1985) have reported the protective action of atropine combined with either toxogonin [114-90-9] or HI 6 [34433-31-3] against sarin, soman, and DFP. Pretreatment with atropine combined with toxogonin or HI 6, 10 and 13 mg/kg, respectively, 10 minutes before exposure increased the LCt_{50} (median lethal products of concentration × time) by a factor of 2.5 and 14, respectively, in rats.

Gray and co-workers (1988) have reported the potential antidote actions of (naphthylvinyl)pyridine derivatives against nerve agent poisoning. These compounds are analogs of (E)-4-(1-naphthylvinyl)pyridine methiodide, which is a potent inhibitor of cholineacetyltransferase. Given below are the structures of some of these derivatives that produced significant protection against sarin and soman poisoning in mouse and guinea pig, respectively.

Compound III, (E)-1-methyl-4-(1-naphthylvinyl)piperidinium hydrochloride, was found to be most effective among the (naphthylvinyl)pyridine derivatives. It is reported further that (III) showed no inhibition of cholineacetyltransferase; the protection action was therefore not related to inhibition of the latter.

Disposal/Destruction

It is dissolved in a combustible solvent and burned in a chemical incinerator equipped with an afterburner and scrubber. Sarin may also be destroyed by the Shultz process of molten metal reduction (Shultz 1987). Molten aluminum, aluminum alloys, recovered scrap metal, or eutectic melts may be used at 780–1000°C. Sarin is reduced to phosphorus, alkenes, and hydrogen. The hydrocarbon products may be used in preheating the feed.

Worley (1989) reported decomposition of sarin, soman, VX, and other chemical warfare agents by oxidizing with 1,3-dibromo-4,4,5,5-tetramethyl-2-imidazolidinone or other N,N'-dihalo-2-imidazolidinone. The reaction is carried out in an aqueous emulsion containing tetrachloroethylene or a similar organic solvent.

Sarin and other nerve agents may be removed from cleaning organic solvents (trichlorotrifluoroethane and its mixtures) by such adsorbents as Fuller's earth, activated alumina, silica gel, and silica gel impregnated with a metal salt (Fowler and McIlvaine 1989). Hydrolysis with water or dilute alkalies should yield products of low toxicity.

Analysis

Sarin in water may be analyzed by gas chromatography after extraction with chloroform. Using an internal standard linearity was determined to fall in the range 10–1000 ppb with flame photometric detector (Shih and Ellin 1986). Flame photometric or a nitrogen-phosphorus detector should be operated in the phosphorus specific mode. A fused-silica capillary GC column should be efficient in separation. Sarin may be analyzed by GC/MS.

39.3 SOMAN

Formula: $C_7H_{16}FO_2P$; MW 182.19; CAS 96-64-0]

Structure:

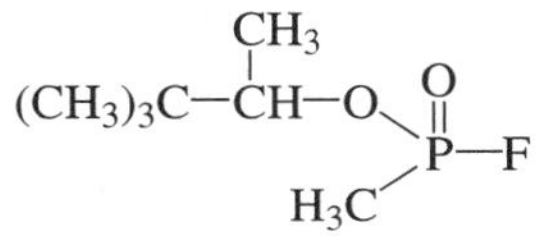

Synonyms: methylphosphonofluoridic acid 1,2,2-trimethylpropyl ester; 3,3-dimethyl-n-but-2-yl methylphosphonofluoridate; methyl pinacolyloxy phosphorylfluoride; pinacolyl methylfluorophosphonate; fluoromethyl(1,2,2-trimethylpropoxy)phosphine oxide; GD

Physical and Chemical Properties

Colorless liquid; forms four stereoisomers; density ~1.15; miscible with organic solvents and water (hydrolyzes). The stability of soman and its stereoisomers in aqueous solution has been studied by Broomfield and co-workers (1986). One millimole of soman in saline was stable for 5 months when stored at −90°C. When stored above freezing it hydrolyzed to 50% in 5 months. At room temperature, hydrolysis to the same extent occurred in 5 days. The rate was the same for all four stereoisomers. In buffered solutions at pH 8, the hydrolysis to 50% occurred in 3.2 hours at 27°C.

Health Hazard

Soman is the most toxic of all the nerve agents. It is extremely toxic by all routes of exposure. The symptoms of toxic effects are those of organophosphate insecticides, but the severity of poisoning is much greater.

An oral dose of 0.01 mg/kg in humans could be fatal. In animals, soman toxicity varied among species; the LD_{50} values by subcutaneous administration were 20, 28, and 126 µg/kg for rabbits, guinea pigs, and rats, respectively (Maxwell et al. 1988). Exposure to a concentration of 21 mg/m^3 soman caused a large inhibition of the activities of the enzyme carboxylesterase [9016-08-5] in bronchi, lungs, and blood tissues in rats (Aas et al. 1985). There was an increase in soman toxicity by 70% following subcutaneous pretreatment with tri-o-cresyl phosphate, a carboxylesterase inhibitor. This study indicates that carboxylesterase is important as a detoxifying enzyme.

Jimmerson and co-workers (1989a) reported the 24-hour subcutaneous LD_{50} value in rats as 118.2 µg/kg. Soman inhibited carboxylesterase activity in plasma and cholinesterase activity in brain regions in a dose-related manner. Such cholinesterase inhibition and elevation of acetylcholine in the brain are very similar to those caused by sarin, DFP, paraoxon, and other organophosphates. However, organophosphates are dissimilar in their effects on choline levels, neuronal activity, and phospholipase A activity. These differential effects are attributed to the differences in the neurotoxicity of soman and other organophosphates (Wecker 1986). Intravenous injection of soman in rats (by six times the LD_{50} amount) followed by isolation of diaphragms 1 or 2 hours after the injection showed detectable amounts of soman P($-$)-isomer in diaphragm tissue (Van Dongen et al. 1986). Pretreatment of the rats with

pinacolyl dimethylphosphinate [92411-69-3] prevented the storage of soman in diaphragm tissue.

A relation between soman toxicity and the aging process has been suggested by many investigators. Sterri and co-workers (1985) measured the activity of the enzymes carboxylesterase and cholinesterase in the plasma, liver, and lung of young rats 5–31 days old. Soman was six- to sevenfold higher in toxicity in 5-day-old rats than in 30-day-old animals. The decrease in toxicity was attributed to the increase in plasma carboxylesterase. Plasma and brain regional cholinesterase activity profiles have been investigated by Shih and co-workers (1987) in four groups of male rats of 30, 60, 120, and 240 days old. The calculated 24-hour intramuscular LD_{50} values were 110.0, 87.2, 66.1, and 48.6 µg/kg, respectively. Young rats showed a less severe initial weight loss and a more rapid and sustained recovery of growth than older animals. These data indicate the relationship between the toxicity of soman to age-related changes of cholinesterase in certain brain areas. In a latter paper, Shih and co-workers (1990) reported that survivors from the two oldest groups of the studied animals did not recover to baseline body weights by the end of the observation period. The activity of plasma cholinesterase did not change significantly with age, while brain regional cholinesterase showed distinct patterns of age dependence. These data further correlate between soman toxicity and the aging process. However, no definite relationship could be established from these studies between the toxicity and the cholinesterase activity in the blood and brain of the test animals.

Pretreatment with certain substances showed potentiation in soman toxicity in test animals. Pretreating rats with cresylbenzodioxaphosphorin oxide (CBDP) by 1 mg/kg reduced the 24-hour subcutaneous LD_{50} value by approximately sixfold (Jimmerson et al. 1989b). CBDP blocks tissue carboxylesterase sites that serve to detoxify soman. This enhances soman-induced inhibition of cholinesterase in the central nervous system, potentiating its lethality. Similar potentiating effects from CBDP pretreatment were reported earlier in other animals, such as mice, guinea pigs, and rabbits (Maxwell et al. 1987). Pretreatment with tri-*o*-cresyl phosphate, another inhibitor, decreased the LD_{50} dose of soman in rats (Tekvani and Srivastava 1989).

Wheeler (1989) observed that the toxicity of soman in rats increased during exposure of the species to either cold or hot environments and after removal from the cold temperatures. Such an increase in toxicity under cold environmental temperatures was attributed to a generalized adrenocortical stress response.

Antidote

Antidote properties of several substances toward the toxicity of soman and other nerve gases have been well investigated (see Sections 39.2, 39.4, and 39.5). These include atropine, phenobarbitol, certain oximes and carbamates, clonidine, pinacolyl dimethylphosphinate, and certain enzymes and their combinations with receptors. Some of these studies are discussed briefly below.

The effects of atropine, 6 mg/kg, and clonidine, 0.2 mg/kg, on the lethality and behavioral changes induced by soman in rats have been investigated (Buccafusco et al. 1988a). Over a concentration range of 50–200 µg/kg injected subcutaneously, soman produced increases in tremor, salivation, hind limb extension, convulsions, and chewing behavior, which depended on time and dose. Pretreatment with clonidine and atropine protected against a soman dose of 200 µg/kg. Both were equally effective and the protective effects were synergistic. Clonidine blocks acetylcholine release and ligand binding to cortical muscarinic receptors in vitro (Buccafusco and Aronstam 1986), and protects acetylcholinesterase from permanent inactivation by soman (Aronstam et al. 1986). Unlike atropine, it also protects

against the chronic behavioral toxic effects of soman (Buccafusco et al. 1989). Pargyline, a monoamine oxidase inhibitor, that causes elevation of brain catecholamines showed similar protection against soman toxicity (Buccafusco et al. 1988b). Its action was additive with that of clonidine.

Certain oximes, such as bispyridinium oxime HI 6 and 2-PAM, have been reported to protect against soman toxicity. Pretreatment with HI 6 (50 mg/kg) together with atropine (10 mg/kg) increased the LC_{50} ($LC_{50} \times$ time) in rats by a factor of 7 (Schoene et al. 1985). HI 6 is an effective reactivator of soman-inhibited acetylcholinesterase. Its protective action was found to be greater in mice than in guinea pigs (Maxwell and Koplovitz 1990). In addition to reactivation of the enzyme acetylcholinesterase inhibited by soman, HI 6 produced an effect on carboxylesterase that increased its therapeutic effect in test animals. The antidotal actions of oximes may be based on mechanisms involving attenuation of soman-induced lesions unrelated to acetylcholinesterase reactivation (Dekleva et al. 1989). Among other oxime antidotes that have been reported to combat soman toxicity in test animals are 2-PAM (Albuquerque et al. 1988), 2-PAM-atropine, and their combinations with *N*-hydroxyethylnaphthylvinylpyridine or *N*-allyl-3-quinuclidinol (Sterling et al. 1988).

Certain carbamates combined with atropine have shown protection against soman toxicity in all species. Injection of physostigmine or pyridostigmine combined with atropine 25 minutes before subcutaneous administration of soman increased the LD_{50} values in rats, guinea pigs, and rabbits (Maxwell et al. 1988). Diazepam or prodiazepam (2-benzoyl-4-chloro-*N*-methyl-*N*-lysilglycinanilide) added to pyridostigmine–atropine showed a further increase in protection against soman in guinea pigs (Karlsson et al. 1990). Physostigmine alone without atropine when given by acute administration failed to protect against soman lethality (Lim et al. 1988). However, continuous administration of physostigmine alone or in addition with scopolamine via an implanted miniosmotic pump before exposure to soman protected from soman-induced toxicity in guinea pigs. In contrast, subchronic treatment of pyridostigmine in the absence of atropine or oxime enhanced soman toxicity in rats (Shiloff and Clement 1986). The LD_{50} value lowered from 104 µg/kg in control animals to 82 µg/kg in treated rats along with the time to onset of soman-induced convulsions.

Rats pretreated with diazepam [439-14-5] or benactyzine [302-40-9] 10 minutes before soman injection prevented the convulsive activity of soman (Pazdernik et al. 1986). By comparison, pretreatment with atropine only reduced the duration of convulsive action of soman after exposure. Diazepam, pralidoxime, and atropine showed antidotal action against soman-induced cardiovascular toxicity in rats in postpoisoning treatment (Bataillard et al. 1990).

Atropine methyl nitrate, phenytoin, and ketamine have been reported to prevent neuromuscular toxicity of soman and diisopropylphosphorofluoridate (DFP) in rats (Clinton et al. 1988).

Among other substances that have been reported to inhibit soman toxicity are phenobarbital and 3-methylcholanthrene, which are microsomal enzyme inducers (Tekvani and Srivastara 1989), pinacolyl dimethylphosphinate (Van Dongen and De Lange 1987), and fetal bovine serum acetylcholinesterase enzyme in conjunction with atropine and 2[(hydroxyimino)methyl]-1-methylpyridinum chloride (Wolfe et al. 1987). The latter antidote system protected mice from multiple doses of soman, VX, and other nerve gases.

Membranes that can inactivate the toxic actions of organophosphorus compounds have been reported (Taylor 1989). These are made by binding two separate proteins onto them. One is a receptor to bind the toxic compound, while the other is an enzyme to detoxify it. Enzymes prepared from rat liver or from squid that can hydrolyze organophosphorus compounds are immobilized along with acetylcholinesterase upon the surfaces of polyvinyl chloride or bovine serum albumin. These membranes were incorporated into bandages to protect wounds on test animals against organophosphorus compounds.

Disposal/Destruction

Among the thermal processes, chemical incineration and molten metal reduction can efficiently destroy soman (see Section 39.2). It may be decomposed by oxidizing with N,N'-dihalo-2-imidazolidinone in an aqueous emulsion containing tetrachloroethylene (Worley 1989). Hydrolysis with dilute alkalies should form products of low toxicity. Adsorbents such as Fuller's earth, activated carbon, alumina, or silica gel have been reported to remove soman from cleaning organic solvents (Fowler and McIlvaine 1989).

Analysis

Soman may be analyzed by GC or GC/MS techniques. In GC methods, flame photometric or nitrogen–phosphorus detectors operated in the phosphorus-specific mode should be used. A wide-bore fused-silica capillary column may be suitable for the purpose. Waste samples or waters may be extracted with methylene chloride, exchanged with hexane, and injected into the GC. Soman at high concentrations between 10 and 1000 ppm may be analyzed by GC-FID. GC/MS using a selective-ion monitoring technique may be applied for the identification of soman.

39.4   VX

Formula: $C_{11}H_{26}NO_2S$; MW 236.44; CAS [50782-69-9]

Structure:

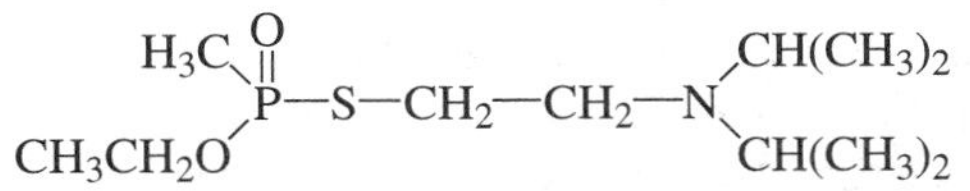

Synonyms: methylphosphonothioic acid *S*-(2-(diisopropylamino)ethyl)-*O*-ethyl ester; ethyl *S*-dimethylaminoethyl methylphosphonothiolate; ethyl *S*-diisopropylaminoethyl methylthiophosphonate

Physical Properties

Colorless and odorless liquid, soluble in organic solvents and water.

Health Hazard

VX is an extremely toxic organophosphorus compound. Like soman and sarin, it is a highly potent cholinesterase inhibitor, a property that results in severe toxicity. Its toxic effects in humans may become moderate to severe from oral intake of about 5 µg/kg. The symptoms may be nausea, vomiting, hypermotility, and diarrhea. Higher dosages can cause difficulty in breathing, bronchial constriction, tremor, convulsions, and death.

Intramuscular injections of VX at 2–20 µg/kg in hens increased the levels of plasma enzymes such as creatine kinase, causing tissue damage (Wilson et al. 1988). There was depression of plasma acetylcholinesterase but not butyrylcholinesterase 2 hours after injections. Goldman and co-workers (1988) performed in vitro and in vivo tests to determine any possible genotoxic, teratogenic, and reproductive effects of exposure to VX. Mutagenicity was tested using the Ames *Salmonella* assay. There were no mutations induced in any of these assays. Exposure to concentrations up to 100 µg VX/mL failed to increase the recombinant activity in the yeast *Saccharomyces cerevisiae*.

Wilson and co-workers (1988) reported the antidote properties of atropine and 2-pralidoxime against VX, paraoxon, and DFP in hens. The birds tolerated 150 µg/kg (five times the LD$_{50}$) dose when atropine and 2-pralidoxime were given before and immediately after subcutaneous injections of VX. Methylthiooobidoxime and methylthio-TMB 4 given together with atropine showed protection against VX and sarin poisoning in mice (see Section 39.2) (Bevandic et al. 1985). An effective antidote against VX and other nerve agents is the enzyme fetal bovine serum acetylcholinesterase. This enzyme has been reported to protect mice from multiple LD$_{50}$ doses (Wolfe et al. 1987).

Disposal/Destruction

VX may be destroyed by chemical incineration or molten metal reduction (see Section 39.2).

Analysis

VX may be analyzed by GC techniques using a flame photometric or a nitrogen–phosphorus detector operated in the phosphorus-specific mode. A fused-silica capillary column may be used for GC analysis. It may be identified and analyzed by GC/MS using a selective-ion monitoring technique.

Hoke and Shih (1987) have described a solid-phase extraction procedure for concentration and storage of nerve agents from natural waters. VX and soman were stable at room temperature on the solid-phase extraction column for 7 days.

39.5 TABUN

Formula: C$_5$H$_{11}$N$_2$O$_2$P; MW 162.12; CAS [77-81-6]

Structure:

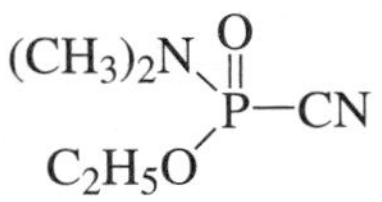

Synonyms: dimethylphosphoramidocyanidic acid ethyl ester; ethyl dimethylphosphoramidocyanidate; ethyl *N*,*N*-dimethylamino cyanophosphate; ethyl dimethylamidocyanophosphate; dimethylamidoethoxyphosphoryl cyanide; GA

Physical and Chemical Properties

Colorless liquid with a fruity smell of bitter almond; boils at 240°C; freezes at −50°C; density 1.077 at 20°C; miscible with organic solvents and water (hydrolyzes); decomposed by bleaching powder forming cyanogen chloride (Merck 1989); synthesized by reacting sodium cyanide with dimethylamidophosphoryl dichloride in the presence of ethanol.

Health Hazard

Like soman and sarin, tabun is a highly potent cholinesterase inhibitor. It is extremely toxic by inhalation, ingestion, and absorption through the skin and eyes. The toxic effects are characteristic of deactivation of acetylcholinesterase enzyme and the symptoms are outlined in Section 39.1. Small doses of tabun can produce severe poisoning, causing constriction of the pupils of the eye, respiratory distress, bronchial constriction, tremor, convulsions, and death. A fatal dose in humans may be about 0.01 mg/kg.

The LD_{50} value in mice from intravenous administration of tabun has been reported as 0.287 mg/kg (Tripathi and Dewey 1989). Gupta and co-workers (1987) investigated acute toxicity of tabun and its biochemical consequences in the brain of rats. An acute nonlethal dose of 200 μg/kg was injected subcutaneously. Within 0.5–1 hour, the toxicity was maximal; it persisted for 6 hours, accompanied by a sharp decline in acetylcholinesterase activity. The prolonged inhibition of this enzyme in muscle and brain may be due to storage and delayed release of tabun from nonenzymic sites. In addition,

cyanide released from a tabun molecule could cause further delay in recovery from its toxic effects. Atropine and its combination with various compounds may offer protection against tabun (see Sections 39.2 and 39.3).

Disposal/Destruction

Tabun may be removed from cleaning organic solvents by adsorbing over Fuller's earth, activated carbon, activated alumina, or silica gel (Fowler and McIlvaine 1989). Incineration and molten metal reduction (Schultz 1987) are efficient thermal processes. Chemical decomposition can be achieved by oxidation with *N*,*N*′-dihalo-2-imidazolidinone in tetrachloroethylene (Worley 1989). It may be fully decomposed by bleaching powder. One of the products, however, is cyanogen chloride, which is toxic and undesirable. Tabun may be hydrolyzed by water, forming less toxic substances.

Analysis

GC using FID or FPD may be the most convenient method for analyzing tabun. FPD is suitable when the analyte level is low in the ppb range, while the FID technique is useful at ppm concentration. A nitrogen–phosphorus detector in the phosphorus-specific mode may alternatively be used with GC. Tabun may be analyzed by GC/MS. It may be extracted from natural waters using a solid-phase extraction column (see Section 39.4).

MISCELLANEOUS ORGANOPHOSPHATES OF HIGH TOXICITY

39.6 DFP

EPA Classified Acute Hazardous Waste; RCRA Waste Number P040

Formula: $C_6H_{14}FO_3P$; MW 184.15; CAS [55-91-4]

Structure:

$$(CH_3)_2CHO \diagdown \underset{\overset{\|}{P}-F}{\overset{O}{}} (CH_3)_2CHO \diagup$$

Synonyms: *O,O*-diisopropyl fluorophosphate; diisopropyl fluorophosphate; diisopropylphosphonofluoridate, diisopropoxyphosphoryl fluoride; fluorophosphoric acid diisopropyl ester; fluoropryl; isofluorophate

Uses and Exposure Risk

DFP served as a basis for developing nerve gases during World War II. It is not used as a nerve gas; it is employed as a pharmacology tool to investigate the toxic properties of nerve gases. It is applied as a miotic and in ophthalmic use.

Physical Properties

Colorless oily liquid; bp 183°C (Merck 1989); mp −82°C; density 1.055 at 20°C; soluble in water and vegetable oil; decomposes in acidic solution at pH 2.5; forms HF in the presence of moisture.

Health Hazard

DFP is a highly toxic organophosphate. Its chemical structure is similar to that of sarin and soman. It is a potent inhibitor of acetylcholinesterase, and its toxic actions are lower but similar to those of sarin and soman. Subcutaneous administration of 1.5 mg/kg caused cholinergic symptoms in rats, inducing a progressive dose-related necrosis (Dettbarn 1984). DFP administered to rats produced muscle fiber discharges from the peripheral nerves as well as from the central nervous system (Clinton et al. 1988). This was significantly reduced when the animals were pretreated with atropine methyl nitrate, ketamine, or phenytoin.

In humans, exposure to its vapors at 1.5 ppm concentration in air for 10 minutes may produce headache and constriction of the pupil of the eye.

LD_{50} value, oral (rabbits): 10 mg/kg (NIOSH 1986)

LD_{50} value, subcutaneous (rats): 2 mg/kg (NIOSH 1986)

Disposal/Destruction

It is dissolved in a combustible solvent and burned in a chemical incinerator. It may be destroyed by treating with molten metal at 800–1000°C. It may be decomposed by treatment with dilute acid. Care should be taken, as highly toxic HF may be generated.

Analysis

GC techniques using FPD or NPD in the phosphorus-specific mode should be employed to analyze DFP.

39.7 GF

Formula: $C_7H_{14}FO_2P$; MW 180.18; CAS [329-99-7]

Structure:

Synonyms: methylcyclohexyl fluorophosphonate; cyclohexyl methylphosphonofluoridate (CMPF)

Health Hazard

GF is a highly toxic nerve agent. It is a potent inhibitor of acetylcholinesterase and a neurotoxicant. The toxic symptoms are characteristics of sarin and other similar organophosphates. The toxicity is lower than GA, GB, GD, and VX.

LD_{50} value, subcutaneous (guinea pigs): 0.1 mg/kg (NIOSH 1986)

LD$_{50}$ value, subcutaneous (rats): 0.225 mg/kg (NIOSH 1986)

39.8 DIMETHYL FLUOROPHOSPHATE

Formula: $C_2H_6FO_3P$; MW 128.05; CAS [5954-50-7]

Structure:

$$\begin{array}{c} CH_3O \\ \diagdown \\ \diagup \\ CH_3O \end{array} \overset{\overset{\textstyle O}{\|}}{P} - F$$

Synonyms: dimethyl phosphorofluoridate; phosphorofluoridic acid dimethyl ester; dimethoxyphosphoryl fluoride

Health Hazard

Dimethyl fluorophosphate is a strong acetylcholinesterase inhibitor. It is not used as a military poison. Its toxic actions are similar to those of DFP; the poisoning effects may be slightly greater than those of DFP.

LD$_{50}$ value, skin (mice): 36 mg/kg
LC$_{50}$ value, (mice): 290 mg/m^3/10 min

REFERENCES

Aas, P., S. H. Sterri, H. P. Hjermstad, and F. Fonnum. 1985. A method for generating toxic vapors of soman: toxicity of soman by inhalation in rats. *Toxicol. Appl. Pharmacol.* *80*(3): 437–45.

Albuquerque, E. X., M. Alkondon, S. S. Deshpande, W. M. Cintra, Y. Aracave, and A. Brossi. 1988. The role of carbamates and oximes in reversing toxicity of organophosphorus compounds: a perspective into mechanisms. *Int. Cong. Ser. Excerpt Med. 832*: 349–73; cited in *Chem. Abstr. CA 110* (23): 207350z.

Anzueto, A., R. A. DeLemos, J. Seidenfeld, G. Moore, H. Hamil, D. Johnson, and G. S. Jenkinson. 1990. Acute inhalation toxicity of soman and sarin in baboons. *Fundam. Appl. Toxicol. 14*(4): 676–87.

Aronstam, R. S., M. D. Smith, and J. J. Buccafusco. 1986. Clonidine protection from soman and echothiophate toxicity in mice. *Life Sci. 39*(22): 2097–2102.

Bataillard, A., F. Sannajust, D. Yoccoz, G. Blanchet, H. Sentenac-Roumanou, and J. Sassard. 1990. Cardiovascular consequences of organophosphorus poisoning and of antidotes in conscious unrestrained rats. *Pharmacol. Toxicol. (Copenhagen) 67*(1): 27–35; cited in *Chem. Abstr. CA 113*(19): 167157v.

Bevandic, Z., A. Deljac, M. Maksimovic, B. Boksovic, and Z. Binenfeld. 1985. Methylthio analogs of PAM-2, TMB-4 and obidoxime as antidotes in organophosphate poisonings. *Acta Pharm. Jugosl. 35*(3): 213–18; cited in *Chem. Abstr. CA 104*(7): 46784c.

Broomfield, C. A., D. E. Lenz, and B. MacIver. 1986. The stability of soman and its stereoisomers in aqueous solution: toxicological considerations. *Arch. Toxicol. 59*(4): 261–65.

Buccafusco, J. J., and R. S. Aronstam. 1986. Clonidine protection from the toxicity of soman, an organophosphate acetylcholinesterase inhibitor in the mouse. *J. Pharmacol. Exp. Ther. 239*(1): 43–47.

Buccafusco, J. J., J. H. Graham, and R. S. Aronstam. 1988a. Behavioral effects of toxic doses of soman, an organophosphate cholinesterase inhibitor, in the rat: protection afforded by clonidine. *Pharmacol. Biochem. Behav. 29*(2): 309–13.

Buccafusco, J. J., R. S. Aronstam, and J. H. Graham. 1988b. Role of central biogenic amines on the protection afforded by clonidine against the toxicity of soman, an irreversible cholinesterase inhibitor. *Toxicol. Lett. 42*(3): 291–99.

Buccafusco, J. J., J. H. Graham, J. Van Lingen, and R. S. Aronstam. 1989. Protection afforded by clonidine from the acute and chronic behavioral toxicity produced by the cholinesterase inhibitor soman. *Neurotoxicol. Teratol. 11*(1): 39–44.

Clinton, M. E., K. E. Misulis, and W. D. Dettbarn. 1988. Effects of phenytoin, ketamine, and atropine methyl nitrate in preventing neuromuscular toxicity of acetylcholinesterase inhibitors soman and diisopropylphosphorofluoridate. *J. Toxicol. Environ. Health 24*(4): 439–49.

Dekleva, A., D. Sket, J. Sketelji, and M. Brzin. 1989. Attenuation of soman-induced lesions of skeletal muscle by acetylcholinesterase reactivating and nonreactivating antidotes. *Acta Neuropathol. 79*(2): 183–89; cited in *Chem. Abstr. CA 112*(1): 93434p.

Dettbarn, W. D. 1984. Nerve agent toxicity and its prevention at the neuromuscular junction; an analysis of acute and delayed toxic effects in extraocular and skeletal muscle. *Govt. Rep. Announce. Index (U.S.) 84*(25): 55; cited in *Chem. Abstr. CA 102*(17): 144447y.

D'Mello, G. D., and E. A. M. Duffy. 1985. The acute toxicity of sarin in marmosets (*Callithrix jacchus*): a behavioral analysis. *Fundam. Appl. Toxicol. 5*(6, part 2): S169–S174.

Drewes, L. R., and A. K. Singh. 1987. Cerebral metabolism and blood brain transport: toxicity of organophosphorus compounds. *Govt. Rep. Announce. Index (U.S.) 88*(22), Abstr. No. 856, p. 413; cited in *Chem. Abstr. CA 111*(1): 2296k.

Fowler, D. E., and E. T. McIlvaine. 1989. Removal of toxic agents from trichlorotrifluoroethane cleaning solvents. U.S. Patent 4,842,746, June 27; cited in *Chem. Abstr. CA 111*(18): 159675f.

Goldman, M., A. K. Klein, T. G. Kawakami, and L. D. Rosenblatt. 1987. Toxicity studies on Agents GB (sarine types I and II) and GD (soman). *Govt. Rep. Announce. Index (U.S.) 88*(10), Abstr. No. 825, p. 196; cited in *Chem. Abstr. CA 110*(1): 2633x.

Goldman, M., L. S. Rosenblatt, B. W. Wilson, T. G. Kawakami, and M. R. Culbertson. 1988. Toxicity studies on agent VX. *Govt. Rep. Announce. Index (U.S.) 89*(8), Abstr. No. 919, p. 324; cited in *Chem. Abstr. CA 111*(7): 52276z.

Gray, A. P., R. D. Platz, T. R. Henderson, T. C. P. Chang, K. Takahashi, and K. L. Dretchen. 1988. Approaches to protection against nerve agent poisoning. (Naphthylvinyl) pyridine derivatives as potential antidotes. *J. Med. Chem. 31*(4): 807–14.

Gupta, R. C., G. T. Patterson, and W. D. Dettbarn. 1987. Acute Tabun toxicity; biochemical and histochemical consequences in brain and skeletal muscles of rat. *Toxicology 46*(3): 329–41.

Hoke, S. H., and M. L. Shih. 1987. Solid phase extraction for concentration and storage of nerve agents. *Govt. Rep. Announce. Index. (U.S.) 88*(14), Abstr. No. 836, p. 777; cited in *Chem. Abstr. CA 110*(10): 82155f.

Jimmerson, V. R., T. M. Shih, and R. B. Mailman. 1989a. Variability of soman toxicity in the rat: correlation with biochemical and behavioral measures. *Toxicology 57*(3): 241–54.

Jimmerson, V. R., T. M. Shih, D. M. Maxwell, and R. B. Mailman. 1989b. Cresylbenzodioxaphosphorin oxide pretreatment alters soman-induced toxicity and inhibition of tissue cholinesterase activity of the rat. *Toxicol. Lett. 48*(1): 93–103.

Karlsson, B., B. Lindgren, E. Millquist, M. Sandberg, and A. Sellstroem. 1990. On the use of diazepam (2-benzoyl-4-chloro-*N*-methyl-*N*-lysilglycin anilide), as adjunct antidotes in the treatment of organophosphorus intoxication in the guinea pig. *J. Pharm. Pharmacol. 42*(4): 247–51.

Lim, D. K., Y. Ito, Z. J. Yu, B. Hoskins, and I. K. Ho. 1988. Prevention of soman toxicity after the continuous administration of physostigmine. *Pharmacol. Biochem. Behav. 31*(3): 633–39.

Manahan, S. E. 1989. *Toxicological Chemistry*. Chelsea, MI: Lewis Publishers.

Maxwell, D. M., and I. Koplovitz. 1990. Effect of endogenous carboxylesterase on HI 6 protection against soman toxicity. *J. Pharmacol. Exp. Ther. 254*(2): 440–44.

Maxwell, D. M., K. M. Brecht, and B. L. O'Neill. 1987. The effect of carboxylesterase inhibition on interspecies differences in soman toxicity. *Toxicol. Lett. 39*(1): 35–42.

Maxwell, D. M., K. M. Brecht, D. E. Lenz, and B. L. O'Neill. 1988. Effect of carboxylesterase inhibition on carbamate protection against soman toxicity. *J. Pharmacol. Exp. Ther. 246*(3): 986–91.

Merck. 1989. *The Merck Index*, 11th ed. Rahway, NJ: Merck & Co.

NIOSH. 1986. *Registry of Toxic Effects of Chemical Substances*, ed. D. V. Sweet. Washington, DC: U.S. Government Printing Office.

Pazdernik, T. L., S. R. Nelson, R. Cross, L. Churchill, M. Giesler, and F. E. Samson. 1986. Effects of antidotes on soman-induced brain changes. *Arch. Toxicol. Suppl. 9*: 333–36; cited in *Chem. Abstr. CA 106*(9): 62533s.

Schoene, K., D. Hochrainer, H. Oldiges, M. Kruegel, N. Franzes, and H. J. Bruckert. 1985. The protective effect of oxime pretreatment upon the inhalative toxicity of sarin and soman in rats. *Fundam. Appl. Toxicol.* 5 (6, Part 2): S84–S88.

Shih, M. L., and R. I. Ellin. 1986. Determination of toxic organophosphorus compounds by specific and nonspecific detectors. *Anal. Lett.* 19 (23–24): 2197–2205.

Shih, T. M. A., D. M. Penetar, J. H. McDonough, J. A. Romano, and J. M. King. 1987. Age-related changes in cholinesterase activity and soman toxicity in the rat. *Govt. Rep. Announce. Index (U.S.)* 87, Abstr. No. 743, p. 794; cited in *Chem. Abstr. CA 108* (15): 126103c.

Shih, T. M., D. M. Penetar, J. H. McDonough, Jr., J. A. Romano, and J. M. King. 1990. Age-related differences in soman toxicity and in blood and brain regional cholinesterase activity. *Brain Res. Bull.* 24 (3): 429–36; cited in *Chem. Abstr. CA 112* (25): 230931q.

Shiloff, J. D., and J. G. Clement. 1986. Effects of subchronic pyridostigmine pretreatment on the toxicity of soman. *Can. J. Physiol. Pharmacol.* 64 (7): 1047–49.

Shultz, C. G. 1987. Destruction of nerve gases and other cholinesterase inhibitors by molten metal reduction. U.S. Patent 4,666,696, May 19; cited in *Chem. Abstr. CA 107* (14): 120536d.

Sterling, G. H., P. H. Doukas, R. J. Sheldon, and J. J. O'Neill. 1988. In vivo protection against soman toxicity by known inhibitors of acetylcholine synthesis in vitro. *Biochem. Pharmacol.* 37 (3): 379–84.

Sterri, S. H., G. Berge, and F. Fonnum. 1985. Esterase activities and soman toxicity in developing rat. *Acta Pharmacol. Toxicol.* 57 (2): 136–40.

Taylor, R. F. 1989. Enzymes and receptors on membranes for the inactivation of toxic materials. U.S. Patent 8,902,920, Apr. 6; cited in *Chem. Abstr. CA 111* (24): 219332d.

Tekvani, P., and R. Srivastava. 1989. Role of carboxylesterase in protection against soman toxicity. *Pharmacology 38* (5): 319–26.

Tripathi, H. L., and W. L. Dewey. 1989. Comparison of the effects of diisopropylfluorophate, sarin, soman, and tabun on toxicity and brain acetylcholinesterase activity in mice. *J. Toxicol. Environ. Health 26* (4): 437–46.

Van Dongen, C. J., and J. De Lange. 1987. Influence of pinacolyl dimethylphosphinate on soman storage in rats. *J. Pharm. Pharmacol. 39* (8): 609–13.

Van Dongen, C. J., H. P. M. Van Helden, and O. L. Wolthius. 1986. Further evidence for the effect of pinacolyl dimethylphosphinate on soman storage in muscle tissue. *Eur. J. Pharmacol. 127* (1–2): 135–38.

Wecker, L. 1986. Relationship between organophosphate toxicity and choline metabolism. *Govt. Rep. Announce. Index (U.S.) 89* (11), Abstr. No. 930, p. 597; cited in *Chem. Abstr. CA 111* (17): 148458m.

Wheeler, T. G. 1989. Soman toxicity during and after exposure to different environmental temperatures. *J. Toxicol. Environ. Health 26* (3): 349–60.

Wilson, B. W., J. D. Henderson, E. Chow, J. Schreider, M. Goldman, R. Culbertson, and J. C. Dacre. 1988. Toxicity of an acute dose of agent VX and other organophosphorus esters in the chicken. *J. Toxicol. Environ. Health 23* (1): 103–13.

Wolfe, A. D., R. S. Rush, B. P. Doctor, I. Koplovitz, and D. Jones. 1987. Acetylcholinesterase prophylaxis against organophosphate toxicity. *Fundam. Appl. Toxicol. 9* (2): 266–70.

Worley, S. D. 1989. Method for decontamination of toxic chemical agents. U.S. Patent 4,874,532, Oct. 17; cited in *Chem. Abstr. CA 112* (20): 185258n.

40

NITRO EXPLOSIVES

40.1 GENERAL DISCUSSION

Nitrated organics and inorganics constitute the largest class of chemical substances that are known for their explosive characteristics. Explosives that are used in bombs, actuating cartridges, boosters, bursters, torpedoes, igniter cords, rifle powder, blasting devices (for mining, etc.), propellants, signal flares, fireworks, and for many other military and commercial purposes contain primarily nitrated organics and inorganic substances in their compositions. Among the inorganic nitrates, ammonium nitrate (used in dynamite and other detonators) and potassium nitrate (used in black gunpowder together with sulfur and charcoal in the molar ratio $32:3:16$,

respectively, are most prominent. These are discussed in Chapter 41.

Organic nitro explosives may be broadly classified under two structural categories: (1) nitro-substituted hydrocarbons and (2) nitrate esters. Nitromethane, trinitrobenzene, trinitrotoluene, and pentanitroaniline are examples of the former class, in which the nitro group ($-NO_2$) is attached to the carbon atom(s) of the alkane or the aromatic ring. Nitroglycerine, ethylene glycol dinitrate, and pentaerythritol tetranitrate are nitrate esters in which the $-NO_2$ group is attached to an oxygen atom in the molecule. In addition, $-NO_2$ may be attached to an N atom, as in cyclonite, shown below. The structures typifying both these classes are as follows:

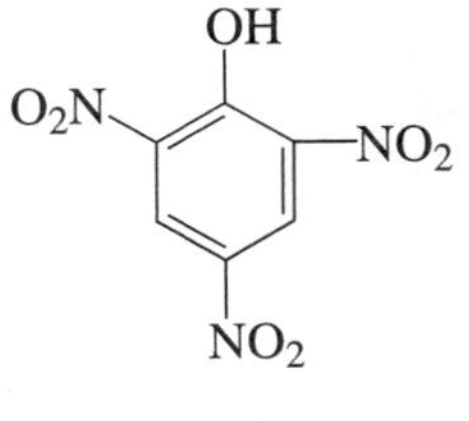

(picric acid, in which all the three $-NO_2$ groups are attached to the C atoms)

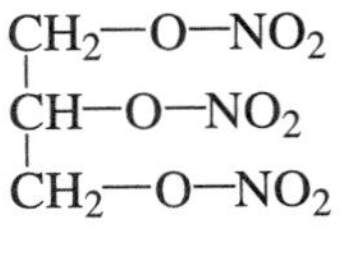

(nitroglycerine, $-NO_2$ groups attached to O atoms)

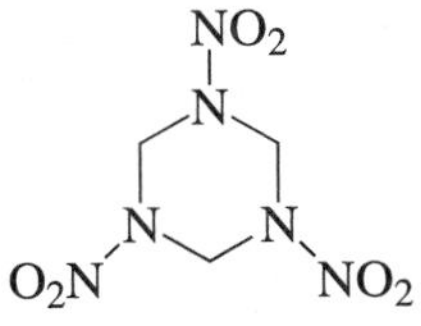

(cyclonite, $-NO_2$ groups attached to N atoms)

Whether it is a $-C-NO_2$, $-O-NO_2$, or $-N-NO_2$ link, all these structural features yield substances that are high explosives. Nitroglycerine, cyclonite, pentaerythrytol tetranitrate, tetryl, trinitrobenzene, trinitrotoluene, and pentanitroaniline are examples of high explosives. The terms *high explosives* and *low explosives* are used for qualitative comparisons only. The explosive power of chemicals or their compositions is determined by their brisance and detonation velocity (see Section IV in Part A).

With a greater degree of nitro substitution in the ring or in the molecule, there is an increase in the energy of detonation. In other words, the more nitro groups, the more the molecule tends to be explosive. For example, mononitrobenzene does not exhibit explosive characteristics. But the addition of two more $-NO_2$ groups makes trinitrobenzene a powerful explosive. The explosive characteristic of a molecule can be evaluated from its energy content and oxygen balance. For composite explosives, the heat of detonation can be calculated from the heat of formation of products and the components of charges.

Sulimov and co-workers (1988) investigated two possible mechanisms of transition from deflagration to detonation in high-porosity explosives. Granular nitrocellulose in steel tubes was used in the study and the process was monitored simultaneously by optical and piezometric techniques. The detonation wave, according to the mechanism, rises upstream from the convective flame front in the unburned explosive column. Precompaction of the explosive ahead of the flame front stabilizes flame propagation, causing explosive particle movement and decreases the rate of pressure rise. Increase of the particle velocity to 200 m/sec caused the low-velocity detonation of granular nitrocellulose. Such an onset of detonation depended on the initial particle diameter. According to an alternative mechanism, transition from deflagration to detonation in the explosives occurs when a strong secondary compression wave is formed in the combustion zone downstream of the flame front. When this secondary wave overtakes the leading flame front, it results in detonation.

Major ingredients of propellants used in rifles and large-caliber guns are nitrocellulose, nitroguanidine, and nitroglycerine. The major products of combustion occurring in exhaust are carbon monoxide, carbon dioxide, hydrogen, water vapor, and nitrogen. Minor species include ammonia, oxides of nitrogen, sulfur dioxide, methane, polycyclic aromatics of unknown origin, and metal particulates (Ross et al. 1988). When TNT or NH_4NO_3 containing nitroguanidine was detonated, the composition of detonation gases was found to depend on the grain size of spherical nitroguanidine (Volk 1986).

Disposal of Explosive Wastes

The reactivity of explosive-containing wastes must be known as a prerequisite for disposal of such waste materials. Such explosive reactivity can be determined by the gap and internal ignition tests developed by U.S. Bureau of Mines (Bajpayee and Mainiero 1988). These tests determine the sensitivity to shock and heat, respectively.

Explosives should be detonated carefully by trained personnel in a safe and remote area under controlled conditions. Alternatively, such materials should be desensitized by diluting with a combustible solvent and fed carefully into an incinerator in small amounts (National Research Council 1983).

Explosive wastes should never be sent for landfill burial. It should either be detonated under control conditions or desensitized with a combustible solvent for incineration. Picric acid waste should always remain wettened with plenty of water. Trace amount of metals can form highly shock sensitive metal picrates. Jayawant (1989) has reported decomposition of polynitrophenols and their salts in wastewater. The method involves treatment with H_2O_2 at a molar ratio greater than 2 in the presence of 0.002–0.7 mol of an iron salt, such as ferrous sulfate or ferrous ammonium sulfate. The treatment is performed above 65°C at a pH below 4. Bockrath and Kirksey (1987) described a process for destruction of nitrophenols in wastewaters. Finely divided Fe particles at 0.2–7 molar equivalent was used at pH 1.3–3.0. Urea was added to reduce NO_x evolution. Fernando and associates (1990) have reported biodegradation of TNT by the white rot fungus Phanerochaete chrysosporium. Decontamination of TNT-contaminated sites in the environment may be achieved by this fungus.

Important Organic Nitro Explosives

Some commercially important or highly hazardous nitro explosives are listed below.

1-Bromo-2-nitrobenzene [577-19-5]
1-Bromo-3-nitrobenzene [585-79-5]
1-Bromo-4-nitrobenzene [586-78-7]
Dinitroaniline [26471-56-7]
Dinitrobenzene [25154-54-5]
Dinitrochlorobenzene [25567-67-3]
1,1-Dinitroethane [600-40-8]
1,2-Dinitroethane [7570-26-5]
1,3-Dinitroglycerine [623-87-0]
Dinitroglycol [628-96-6]
Dinitromethane [625-76-3]
Dinitrophenol [25550-58-7]
2,4-Dinitrophenylhydrazine (dry) [119-26-6]
Dinitropropylene glycol [6423-43-4]
Dinitroresorcinol [35860-51-6]
2,2-Dinitrostilbene [6275-02-1]
Ethanol amine dinitrate
Ethyl 4,4-dinitropentanoate
Galactsan trinitrate
Glycerol monogluconate trinitrate
Glycerol monolactate trinitrate
Hexamethylol benzene hexanitrate
Hexanitro dihydroxazobenzene
Hexanitroazoxy benzene
Hexanitrodiphenyl ether

Hexanitrodiphenylamine (dry)
Hexanitrodiphenylurea
Hexanitroethane [918-37-6]
Hexanitrooxanilide
Hexanitrostilbene
Inositol hexanitrate (dry)
Inulin trinitrate (dry)
Mannitan tetranitrate
Methyl nitrate [598-58-3]
Methyl picric acid
Methylamine dinitramine
Methylene glycol dinitrate
o-Nitroaniline [29757-24-2]
N-Nitroaniline [645-55-6]
5-Nitro 1H-benzotriazole [2338-12-7]
Nitrocellulose [9004-70-0]
Nitroethane [79-24-3]
Nitroethyl nitrate [4528-34-1]
Nitroglycerine [55-63-0]
Nitroglycide
Nitroguanidine [556-88-7]
1-Nitrohydantoin
Nitromannite [15825-70-4]
Nitromethane [75-52-5]
1-Nitropropane [108-03-2]
2-Nitropropane [79-46-9]
Nitrotoluene [1321-12-6]
N-Nitrourea [556-89-8]
Octogen [2691-41-0] (Cyclotetramethylene tetranitramine)
Pentaerythritol tetranitrate [78-11-5]
Pentanitroaniline
Pentolite [8066-33-9] [2,2-bis(nitrooxy) methyl)-1,3-propanediol dinitrate mixed with 2-methyl-1,3,5-trinitrobenzene]
Picramic acid (dry) [96-91-3] (4,6-Dinitro-2-aminophenol)
Picric acid [88-89-1] (2,4,6-Trinitrophenol)
Styphnic acid [82-71-3] (2,4,6-Trinitroresorcinol)
Tetranitroaniline [53014-37-2]
Tetranitrocarbazole

Tetranitrochrysazin [517-92-0] (1,8-Dihy-droxy-2,4,5,7-tetranitroanthraquinone)

Trinitroacetic acid

Trinitroacetonitrile [630-72-8]

Trinitroaniline [26952-42-1]

2,4,6-Trinitroanisole [606-35-9]

Trinitrobenzene [99-35-4]

2,4,6-Trinitrobenzenesulfonic acid [2508-19-2]

Trinitrobenzoic acid [129-66-8]

2,4,6-Trinitrochlorobenzene [28260-61-9]

2,4,6-Trinitro-m-cresol [602-99-3]

1,1,1-Trinitroethane [595-86-8]

2,2,2-Trinitroethanol [918-54-7]

Trinitrofluorenone [25322-14-9]

Trinitromethane [517-25-9]

Trinitronaphthalene

Trinitrophenetol

2,4,6-Trinitrophenyl-N-methylnitramine [479-45-8] (Tetryl)

1,3,5-Trinitro-1,3,5-triazacyclohexane [121-82-4] (Cyclonite)

2,4,6-Trinitro-1,3,5-triazidobenzene [29306-57-8]

2,4,6-Trinitrotoluene [118-96-7]

Tritonal [54413-15-9] (2-Methyl-1,3,5-tri-nitrobenzene mixed with aluminum)

40.2 NITROGLYCERINE

EPA Designated Hazardous Waste, RCRA Waste Number P081; DOT Hazard: Class A Explosive (desensitized nitroglycerine), forbidden for transportation when not desensitized

Formula $C_3H_5N_5O_9$; MW 227.11; CAS [55-63-0]

Structure:

$$CH_2-O-NO_2$$
$$CH-O-NO_2$$
$$CH_2-O-NO_2$$

Synonyms: glycerol trinitrate; nitroglycerin; nitroglycerol; trinitroglycerol; glycerol nitric acid triester; 1,2,3-propanetriol trinitrate; blasting gelatin

Preparation

Nitroglycerine is prepared by mixing glycerol (trihydroxy alcohol) into a mixture of nitric and sulfuric acids at a cold temperature.

$$
\begin{array}{ccc}
CH_2OH & & CH_2O-NO_2 \\
CHOH & + \ 3HNO_3 \longrightarrow & CHO-NO_2 \\
CH_2OH & & CH_2O-NO_2
\end{array}
$$

Physical Properties

Pale yellow oily liquid; burning sweet taste; density 1.596 at 20°C; solidifies at 13.2°C; decomposition begins at ~60°C, volatile at 100°C; slightly soluble in water (1.25 g/L at 20°C) (Merck 1996), miscible with alcohol, ether, acetone, ethyl acetate, glacial acetic acid, benzene, chloroform, and pyridine; slightly soluble in carbon disulfide, glycerol, and petroleum ether.

Health Hazard

Severe acute poisoning may result from ingestion of nitroglycerine or inhalation of its dust. The acute toxic symptoms include headache, nausea, vomiting, abdominal pain, tremor, dyspnea, paralysis, and convulsions. In addition, methemoglobinemia and cyanosis may occur. Ingestion of a relatively small amount, 1.5–2.0 g, could be fatal to humans. Inhalation of its vapors or dust at 0.3 mg/m³ concentration in air produced an immediate fall in blood pressure and headache in human volunteers (ACGIH 1986). Chronic poisoning may produce headache and hallucination.

LD_{50} value, oral (rats): 105 mg/kg

Exposure Limits

TLV-TWA skin 0.05 ppm (0.5 mg/m^3) (ACGIH), 0.2 ppm (MSHA, OSHA, and NIOSH).

Fire and Explosion Hazard

Nitroglycerine in the pure form is a dangerously explosive compound. It detonates when subjected to shock. It explodes when mixed with ozone. Spontaneous detonation occurs when it is heated to approximately 180°C. It may hydrolyze to glycerol and nitric acid upon exposure to moisture. Such a mixture of nitric acid and glycerol can explode readily. Nitroglycerine detonates with a brisance which is roughly three times that of an equivalent quantity of gunpowder and proceeds 25 times faster (Meyer 1989). The detonation velocity is 7.8 km/sec. The products of detonation, however, are harmless gases, carbon dioxide, water vapor, nitrogen, and oxygen. The decomposition reaction is as follows:

$$4C_3H_5(ONO_2)_3 \longrightarrow 12CO_2 + 10H_2O$$
$$+ 6N_2 + O_2$$

Because of its high sensitivity, nitroglycerine in the pure form is rarely used as an explosive. It is mostly desensitized for its commercial applications, as in dynamite (see Section 40.3). Bosch and Pereira (1988) reported the burning behavior of nitroglycerine — solvent mixtures. The risk of detonation resulting from burning of a mixture of nitroglycerine and ethyl acetate could be decreased by adding a liquid of high boiling point as a desensitizer. A water-spray system was effective to extinguish fires involving up to 24 tons of propellant.

40.3 DYNAMITE

Dynamite, discovered by Alfred Nobel, is a composition of nitroglycerine and silicious earth. Nitroglycerine absorbed into a porous material makes it much safer to handle than the direct liquid. Dynamite is a high explosive and can be transported without the danger of spontaneous decomposition. A detonating cap is used to explode it. A typical composition of dynamite is 75% nitroglycerine, 24.5% porous adsorbent (diatomaceous earth), and 0.5% sodium carbonate. Adsorbents such as wood pulps, sawdust, and other carbonaceous substances are used to make dynamite. Sodium carbonate or calcium carbonate is added to the composition to neutralize nitric acid, which forms by spontaneous decomposition.

Various forms of dynamites are manufactured, containing nitroglycerine concentrations usually between 20 and 60%. Certain types of dynamites have greater nitroglycerine content. Oxidizers such as sodium nitrate are also added. Gelatin dynamite consists of nitroglycerine gelatinized in nitrocellulose (1% by weight). Ammonia dynamite is composed of nitroglycerine (at concentrations depending on the grade) and ammonium nitrate, a carbonaceous adsorbent, sulfur, sodium carbonate, and moisture. The mixture is packed in cylindrical cartridges of varying size, made of waxed paper.

Nitrocellulose or ammonium nitrate in dynamite produces additional brisance to the explosive. The detonation velocity ranges between 0.75 and 6.0 km/sec, depending on the type of dynamite and its composition.

40.4 CYCLONITE

DOT Label: Class A Explosive, Corrosive, UN 0072, UN 0118

Formula $C_3H_6N_6O_6$; MW 222.15; CAS [121-82-4]

Structure:

Synonyms: cyclotrimethylenetrinitramine; trinitrocyclotrimethylene triamine; hexa-hydro-1,3,5-trinitro-1,3,5-triazine; 1,3,5-trinitro-1,3,5-triazacyclohexane; hexolite; hexogen; RDX

Uses and Exposure Risk

Cyclonite is a high explosive and principal constituent of plastic bombs. It is prepared by treating hexamethylenetetramine with nitric acid, ammonium nitrate, and acetic anhydride:

It is also used as a rat poison.

Physical Properties

White solid, orthorhombic crystals from acetone; density 1.82 at 20°C; melts at 205°C (decomposes); insoluble in water, soluble in acetone (4 g/100 mL at 20°C), slightly soluble in alcohol, ether, and glacial acetic acid.

Health Hazard

Cyclonite is a highly toxic substance. Repeated exposure can cause nausea, vomiting, and convulsions.

LD_{50} value, oral (rats): 100 mg/kg

Exposure Limit

TLV-TWA skin 1.5 mg/m^3 (ACGIH and MSHA).

Fire and Explosion Hazard

Cyclonite is a highly powerful explosive with a brisance greater than that of TNT. The detonation velocity is between 7 and 8 km/sec. The pure compound is highly sensitive to shock, decomposing explosively. The products of decomposition are, however, harmless gases, carbon dioxide, oxygen, nitrogen, and water vapor.

The sensitivity of cyclonite can be greatly reduced by combining with beeswax. Such desensitization yields high stability for its safe handling and transportation. Commercial cyclonite has two types of compositions: (1) 91% cyclonite and 9% beeswax and (2) 60% cyclonite, 39% TNT, and 1% beeswax. Both compositions are used in artillery shells. Dry cyclonite is forbidden by the U.S. Department of Transportation for transportation. Wet explosive containing not less than 15% water by mass and desensitized cyclonite are regulated substances.

40.5 NITROCELLULOSE

DOT Label: Flammable Solid, Class A Explosive (dry solid); CAS [9004-70-0]

Structure: polymeric substance, varying with degree of nitration; the explosive material is more nitrated than other forms of nitrocellulose; part of the structure is as follows:

Synonyms: nitrocotton; guncotton; collodion; cellulose nitrate; pyroxylin

Uses and Exposure Risk

Nitrocellulose is used as a propellant in artillery ammunition, in small-arms ammunition, in chemical explosives, and in smokeless powder. It is made by reacting cotton with nitric acid.

Fire and Explosion Hazard

Nitrocellulose is a white fibrous solid or amorphous powder. It is wetted with water, alcohol, or other solvent for handling and storage. It may be made to various forms, gel, flake, granular, or powder. Dry material is a low explosive and often used in combination with another explosive, such as nitroglycerine, to obtain more brisance for the composition. Dry nitrocellulose does not detonate but deflagrates. When wetted with water or alcohol, its sensitivity is considerably reduced.

Nitrocellulose presents three types of hazards. As mentioned earlier, it is an explosive compound. It explodes upon burning or friction. It is a flammable solid having a flash point of 13°C (55°F). It can therefore ignite at ambient temperatures, thus presenting a severe fire and explosion hazard. It burns at a very rapid rate. The combustion products consist of extremely toxic gases: notably, hydrogen cyanide, carbon monoxide, and oxides of nitrogen.

Fires involving nitrocellulose should be fought with extreme caution. Unmanned fixed turrets and hose nozzles should be used. Since nitrocellulose produces oxygen on decomposition, a large volume of water should be applied through spray nozzles to cool the material and wet the entire surface. Self-contained breathing apparatus must be worn by firefighters for protection against highly toxic gases.

Storage and Shipping

Nitrocellulose should be stored as a wetted substance and never allowed to go dry. Storage should be in a cool, well-ventilated location isolated from all heat sources. Shipping should be in steel drums or barrels wet with 25–35% alcohol, water, or other solvent.

40.6   2,4,6-TRINITROTOLUENE

DOT Label: Class A Explosive, UN 0209 (dry or containing less than 30% water by weight), UN 1356 (wetted with not less than 30% water)

Formula $C_7H_5N_3O_6$; MW 227.15; CAS [118-96-7]

Structure:

Synonyms: 2-methyl-1,3,5-trinitrobenzene; 1-methyl-2,4,6-trinitrobenzene; 2,4,6-trinitrotoluol; trotyl; TNT

Uses and Exposure Risk

2,4,6-Trinitrotoluene (TNT) is used as a high explosive in mining and in military. It is produced by nitration of toluene with a mixture of nitric and sulfuric acids.

Physical Properties

Colorless or light yellow solid in the form of crystals, pellets, or flakes; melts at 80°C; very slightly soluble in water (0.01% at 25°C), soluble in acetone and benzene.

Health Hazard

The toxic effect of TNT are dermatitis, cyanosis, gastritis, yellow atrophy of the liver, somnolence, tremor, convulsions, and aplastic anemia. Sneezing, sore throat, and muscular pain have also been noted in people exposed to this compound. It is an irritant to skin, respiratory tract, and urinary tract. Prolonged exposure may produce liver damage. The oral LD_{50} value in rats is in the range 800 mg/kg.

Levine and co-workers (1990) have conducted a 6-month oral toxicity test of TNT in beagle dogs. The major toxic effects observed were hemolytic anemia, methemoglobinemia, liver injury, splenomegaly, and death. A dose of 32 mg/kg/day was lethal to the dogs.

TNT tested nongenotoxic to the bone marrow of mice and the liver of rats (Ashby et al. 1985). It produced a negative response at dose levels up to 1000 mg/kg in the liver assay. High levels of hemoglobin and red-colored urine in the TNT-treated rats suggest a possible carcinogenic hazard to the hemopoietic and urinary tissues of animals at toxic levels on chronic exposure (Ashby et al. 1985).

Exposure Limits

TLV-TWA (skin) 0.5 mg/m^3 (ACGIH and MSHA), 1.50 mg/m^3 (OSHA).

Fire and Explosion Hazard

TNT is a high explosive. In comparison to many other high explosives, it is insensitive to heat, shock, or friction. Small amounts may burn quietly without detonation. However, when heated rapidly or subjected to strong shock, it detonates. Its detonation temperature is 470°C and its velocity is between 5.1 and 6.9 km/sec. In combination with other explosives, TNT is widely used as a military and industrial explosive. Amatol, cyclonite, and tetrytol are some of the examples of such explosive combinations. Amatol is a composition of 80% ammonium nitrate and 20% TNT by mass. TNT itself has a very high brisance.

Products from the detonation of 1.5–2.0 kg of TNT in air- and oxygen-deficient atmospheres consisted of low-molecular-weight gases and high-molecular-weight polycyclic aromatic hydrocarbons (Johnson et al. 1988). Greiner and associates (1988) examined the soots produced from the detonation of cast composites of TNT mixed with nitroguanidine or RDX in 1 atmosphere of argon. The soot contained 25 wt% diamond 4–7 nm in diameter, the IR spectrum and particle size of which resembled those from meteorites.

Storage and Shipping

TNT is stored in a permanent magazine, separated from combustible and oxidizable materials, initiators, and heat sources. It is shipped in amounts not exceeding 60 lb (27 kg) in weight in metal containers enclosed in wooden or fiberboard boxes.

40.7   TRINITROBENZENE

Trinitrobenzene wetted (containing between 10–30% water) is designated by EPA as Hazardous Waste, RCRA Waste Number U234; DOT Label: Explosive A, UN 0214 (for dry solid or material containing not more than 30% water), Flammable Solid, UN 1354 (for material wetted with not less than 30% water)

Formula: $C_6H_3N_3O_6$; MW 213.12; CAS [99-35-4]

Structure:

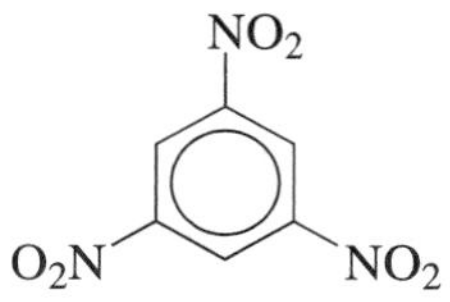

Synonyms: trinitrobenzol; 1,3,5-trinitro-benzene; benzite

Uses and Exposure Risk

Trinitrobenzene is used as an explosive. It is obtained by oxidation of trinitrotoluene.

Physical Properties

Pale yellow crystalline solid; melts at 122.5°C; sublimed by careful heating; very slightly soluble in water (350 mg/L at 20°C), slightly soluble in carbon disulfide and petroleum ether, moderately soluble in benzene and methanol, freely soluble in sodium sulfite solution.

Health Hazard

The toxic effects from ingestion of the solid or inhalation of its dusts include irritation of respiratory tract, headache, dyspnea, and cyanosis. Other effects noted in animals were degenerative changes in the brain. The oral LD_{50} values in rats and mice are 450 and 570 mg/kg, respectively.

Fire and Explosion Hazard

Trinitrobenzene is a high explosive, similar to trinitrotoluene. However, it is less sensitive to impact than the latter. Its brisance and power are higher than those of TNT. It detonates when heated rapidly. The dry material is highly sensitive to shock and heat. Slow and careful heating of a small amount of material does not cause detonation. Trinitrobenzene is a flammable solid. It reacts vigorously with reducing substances. It emits highly toxic oxides of nitrogen on decomposition.

Storage and Shipping

Storage and shipping are similar to those used for TNT and other high explosives. Trinitrobenzene is stored in a permanent magazine well separated from initiator explosive, combustible and oxidizing materials, and heat sources. It is shipped in metal containers enclosed in wooden boxes or strong siftproof cloth or paper bags in amounts not exceeding 60 lb net weight (NFPA 1997).

40.8 TETRYL

DOT Label: Class A Explosive, UN 0208
Formula $C_7H_5N_5O_8$; MW 287.17; CAS [479-45-8]
Structure:

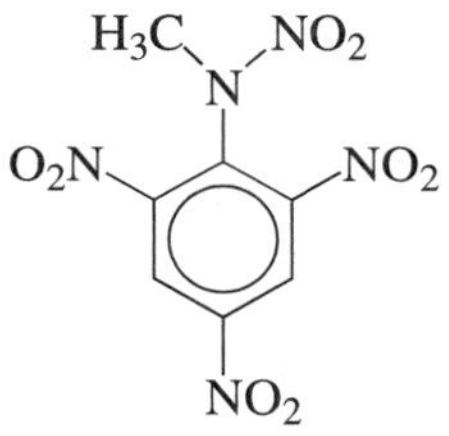

Synonyms: *N*-methyl-*N*,2,4,6-tetranitro-aniline; picrylmethylnitramine; picrylnitromethylamine; 2,4,6-trinitrophenyl-*N*-methylnitramine; Tetralite; nitramine

Uses and Exposure Risk

Tetryl is used as an initiator for many less sensitive explosives. It is used as the booster in artillery ammunition. A combination of Tetryl with trinitrotoluene and a small amount of graphite is known as Tetrytol, an explosive used as the bursting charge in artillery ammunition. It is also used as an indicator.

Tetryl is made by reacting 2,4-dinitrochlorobenzene with methylamine and nitrating the product with nitric acid and sulfuric acid. The reaction steps are as follows:

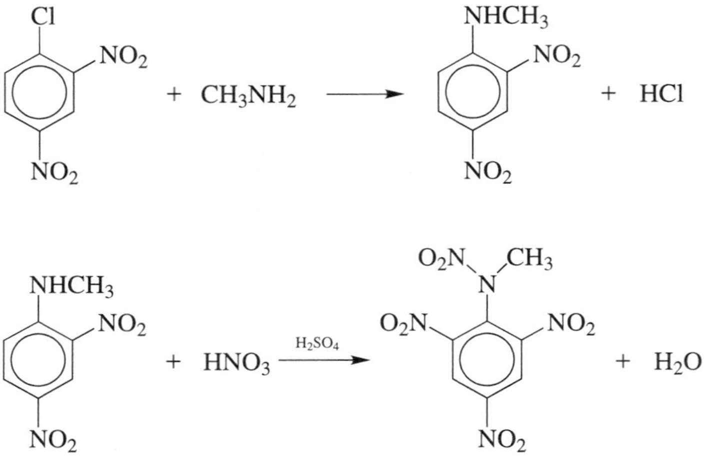

Physical Properties

Yellow crystalline solid; alcoholic solution colorless at pH 10.8 and reddish brown at pH 13.0; melts at 130°C; insoluble in water, soluble in alcohol, ether, benzene, and glacial acetic acid.

Health Hazard

There is very little information on the human toxicity of this compound. A dose of 5000 mg/kg given subcutaneously was lethal to dogs (NIOSH 1986).

Fire and Explosion Hazard

Tetryl is a high explosive, very sensitive to shock, friction, and heat. It detonates at 260°C. Detonation velocity is between 7.0 and 7.5 km/sec. It produces highly toxic oxides of nitrogen on decomposition.

40.9 PENTAERYTHRITOL TETRANITRATE

DOT Label: Class A Explosive, Forbidden (dry); Class A Explosive (wet or desensitized), UN 0150, UN 0411

Formula $C_5H_8N_4O_{12}$; MW 316.17; CAS [78-11-5]

Structure:

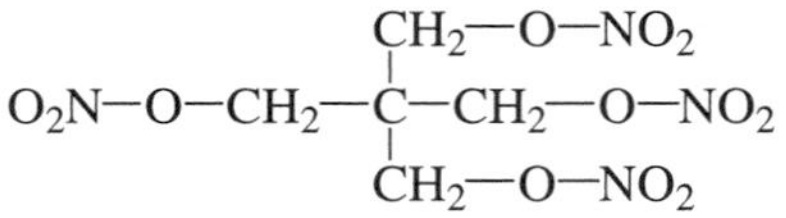

Synonyms: tetranitropentaerythrite; 2,2-bis (hydroxymethyl)-1,3-propanediol tetranitrate, 1,3-dinitrato-2,2-bis(nitratomethyl) propane; pentaerythrite tetranitrate

Uses and Exposure Risk

Pentaerythritol tetranitrate (PETN) is a high explosive, often used as the booster in artillery ammunition. It is frequently used as a detonation fuse, which is a form of primacord in which PETN powder is wrapped in waterproof fabric. PETN is produced by nitration of pentaerythritol:

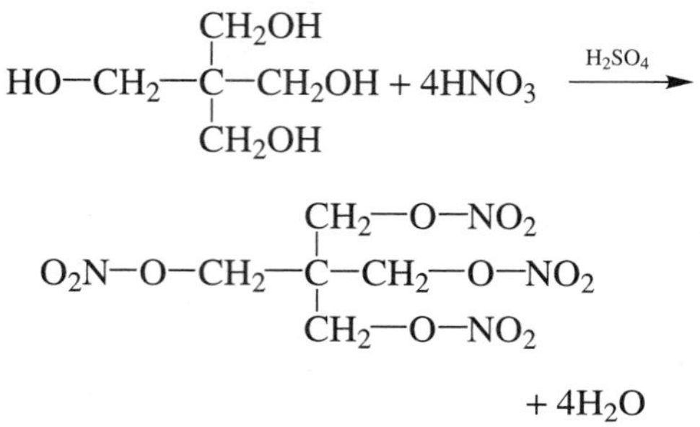

Health Hazard

Skin contact can cause dermatitis. There is no report of any major adverse effects in humans. However, drawing a parallelism with other nitroorganics of similar structures, PETN is expected to cause poisoning with symptoms of headache, nausea, abdominal pain, and dyspnea.

Fire and Explosion Hazard

PETN is a high explosive, as powerful as cyclonite. It is more sensitive than TNT to shock. It explodes on percussion or heating. The detonating temperature is 210°C. The detonation velocity is 7.9 km/sec.

40.10   PICRIC ACID

DOT Label: Class A Explosive, UN 0154 (dry or wetted with not more than 30% water), UN 1344 (wetted with not less than 30% water)

Formula: $C_6H_3N_3O_7$; MW 229.12; CAS [88-89-1];

Structure:

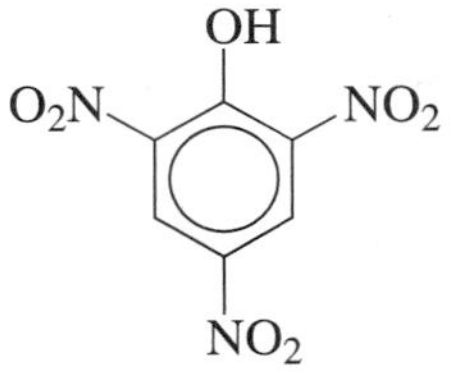

Synonyms: 2,4,6-trinitrophenol; 2-hydroxy-1,3,5-trinitrobenzene; nitroxanthic acid; carbozotic acid, picronitric acid; Melinite

Uses and Exposure Risk

Picric acid and its metal salts are used as explosives. It is also used in making matches; electric batteries, colored glass; in etching copper; and for dyeing textiles.

It is prepared by nitration of phenol:

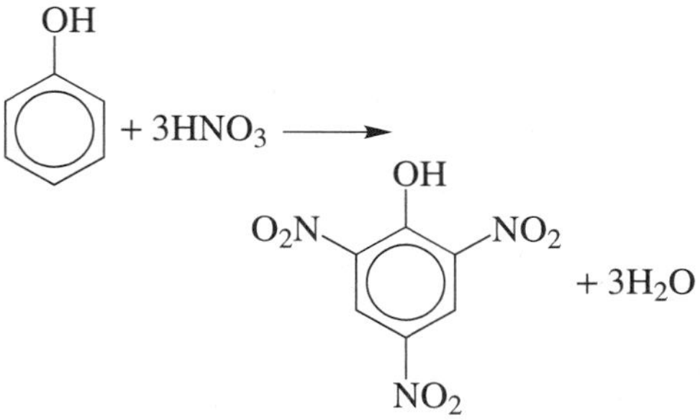

Physical Properties

Yellow crystalline solid; bitter in taste; melts at 122°C; slightly soluble in water (1.28 g/dL at 20°C), moderately soluble in benzene, alcohol, and boiling water.

Health Hazard

Picric acid is a highly toxic substance. Ingestion can cause severe poisoning in humans. The toxic symptoms are headache, nausea, vomiting, abdominal pain, and yellow coloration of the skin. High doses can cause destruction of erythrocytes, gastroenteritis, nephritis, hepatitis, and hematuria. Contact with the eyes can produce irritation and corneal injury. Skin contact with powder can result in sensitization dermatitis. The lethal dose (oral) in rabbits is 120 mg/kg.

Exposure Limit

TLV-TWA skin 0.1 mg/m^3 (ACGIH, OSHA, and MSHA).

Fire and Explosion Hazard

Picric acid is a high explosive. It is as stable as TNT and about as sensitive to explosive decomposition as TNT. It reacts with many metals and bases, readily forming metal picrates, which are highly sensitive explosive compounds. Metal picrates of iron(III), copper(II), and lead in the dry state are as sensitive as PETN. Its sensitivity is reduced by wetting with water. It explodes when heated above 300°C.

40.11   NITROGUANIDINE

DOT Label: Class A Explosive, UN 0282 (for dry material or containing less than 20% water), Flammable Solid, UN 1336 (wet with not less than 20% water)

Formula: $CH_4N_4O_2$; MW 104.08; CAS [556-88-7]

Structure:

$$H_2N-\underset{\underset{NH}{\|}}{C}-NH-NO_2$$

Structure and Exposure Risk

Nitroguanidine is used as an explosive and as an intermediate in the synthesis of pharmaceuticals.

Physical Properties

Crystalline solid; occurs in two forms, α and β; the former is more stable; melts at 239°C (decomposes); moderately soluble in cold water (4.4 g/dL at 20°C), readily dissolves in hot water, soluble in concentrated acids and cold alkalies, insoluble in alcohol and ether.

Health Hazard

Toxicity of nitroguanidine is very low in test animals. It is nontoxic in rats when given by oral administration at doses as high as 1000 mg/kg/day for 14 days (Morgan et al. 1988). The LD_{50} value, as determined by the oral gavage single-dose limit test, is >5000 mg/kg in both male and female rats (Brown et al. 1988).

Fire and Explosion Hazard

Nitroguanidine is an explosive of moderate power. Its brisance is less than that of TNT or trinitrobenzene. It requires an initiator to explode. On the other hand, nitroguanidine acts as a sensitizer, causing many nonexploding fuel–air systems detonable. Tulis (1984) investigated the detonation of unconfined clouds of aluminum powder in air sensitized with a small amount of nitroguanidine. Aluminum powder dispersed in air becomes more readily detonable when mixed with nitroguanidine. The latter added to a mixture of aluminum powder and potassium chlorate causes transition of the deflagrating mixture into detonable (Tulis et al. 1984). A similarly insensitive explosive, ammonium chlorate, becomes readily detonable when combined with nitroguanidine or nitroguanidine and aluminum powder. Aluminum powder added to a detonable composition of nitroguanidine, and NH_4ClO_4, caused degradation of the detonation velocity. Moist nitroguanidine containing about 25% water is a flammable solid.

REFERENCES

ACGIH. 1986. *Documentation of the Threshold Limit Values and Biological Exposure Indices*, 5th ed. Cincinnati, OH: American Conference of Governmental Industrial Hygienists.

Ashby, J., B. Burlinson, P. A. Lefevre, and J. Topham. 1985. Non-genotoxicity of 2,4,6-trinitrotoluene (TNT) to the mouse bone marrow and the rat liver: implications for its carcinogenicity. *Arch. Toxicol.* 58(1): 14–19.

Bajpayee, T. S., and R. J. Mainiero. 1988. Methods of evaluating explosive reactivity of explosive-contaminated solid waste substances. *Bur. Mines Rep. Invest. RI 9217*; cited in *Chem. Abstr. CA 110*(16): 140915s.

Bockrath, R. E., and K. Kirksey. 1987. Destruction of nitrophenol byproducts in wastewaters from dinitrobenzene manufacture. U.S. Patent 4,708,806, Nov. 24; cited in *Chem. Abstr. CA 108*(6): 43406c.

Bosch, L., and J. R. Pereira. 1988. Deflagration to detonation transition of small arms propellants and nitroglycerine/solvent mixtures. *Chem. Abstr. CA 111*(20): 177413y.

Brown, L. D., C. R. Wheeler, and D. W. Korte. 1988. Acute oral toxicity of nitroguanidine in male and female rats. *Govt. Rep. Announce. Index (U.S.) 88*(17), Abstr. No 845, p. 237; cited in *Chem. Abstr. CA 110*(13): 109671k.

Fernando, T., J. A. Bumpus, and S. D. Aust. 1990. Biodegradation of TNT (2,4,6-trinitrotoluene) by *Phanerochaete chrysosporium. Appl. Environ. Microbiol. 56*(6): 1666–81.

Greiner, R. N., D. S. Phillips, J. D. Johnson, and F. Volk. 1988. Diamonds in detonation soot. *Nature 333*(6173): 440–42.

Jayawant, M. D. 1989. Destruction of nitrophenols. U.S. Patent 4,804,480, Feb. 14; cited in *Chem. Abstr. CA 110*(24): 218489c.

Johnson, J. H., E. D. Erickson, R. S. Smith, D. J. Knight, D. A. Fine, and C. A. Heller. 1988. Products from the detonation of trinitrotoluene and some other Navy explosives in air and nitrogen. *J. Hazard. Mater. 18*(2): 145–70.

Levine, B. S., J. H. Rust, J. J. Barklay, E. M. Furedi, and P. M. Lish. 1990. Six month oral toxicity study of trinitrotoluene in beagle dogs. *Toxicology 63*(2): 233–44.

Merck. 1996. *The Merck Index*, 12th ed. Rahway, NJ: Merck & Co.

Meyer, E. 1989. *Chemistry of Hazardous Materials*, 9th ed. Quincy, MA: National Fire Protection Association.

Morgan, E. W., L. D. Brown, C. M. Lewis, R. R. Dahlgren, and D. W. Korte. 1988. Fourteen-day subchronic oral toxicity study of nitroguanidine in rats. *Govt. Rep. Announce. Index (U.S.) 88*(23), Abstr. No. 859, p. 176; cited in *Chem. Abstr. CA 111*(3): 19019v.

National Research Council. 1983. *Prudent Practices for Disposal of Chemicals from Laboratories*. Washington, DC: National Academy Press.

NFPA. 1997. *Fire Protection Guide on Hazardous Materials*, 12th ed. Quincy, MA: National Fire protection Association.

NIOSH. 1986. *Registry of Toxic Effects of Chemical Substances*, ed. D. V. Sweet. Washington, DC: U.S. Government Printing Office.

Ross, R. H., B. C. Pal, R. S. Ramsey, R. A. Jenkins, and S. Lock. 1988. Problem definition study on techniques and methodologies for evaluating the chemical and toxicological properties of combustion products of gun systems. *Govt. Rep. Announce. Index (U.S.) 88*(15), Abstr. No. 838, p. 331; cited in *Chem. Abstr. CA 110*(10): 78784t.

Sulimov, A. A., B. S. Ermolaev, and V. E. Khrapovskii, 1988. Mechanism of deflagration-to-detonation transition in high porosity explosives. *Prog. Astronaut. Aeronaut. 114*: 322–30; cited in *Chem. Abstr. CA 110*(14): 117764h.

Tulis, A. J. 1984. On the detonation of sensitized unconfined clouds of aluminum powder in air. *Chem. Phys. Processes Combust.* 51/1–51/4; cited in *Chem. Abstr. CA 103*(2): 8547z.

Tulis, A. J., D. E. Baker, and D. J. Hrdina. 1984. Further studies on the effects of inert and reactive additives on the detonation characteristics of insensitive explosives. *Proceedings of the 9th International Pyrotechnics Seminar*, pp. 961–69; cited in *Chem. Abstr. CA 102*(16): 134435d.

Volk, F. 1986. Detonation gases and residues of composite explosives. *J. Energy Mater. 4*(1–4): 93–113.

41

OXIDIZERS

41.1 GENERAL DISCUSSION

Oxidizers include certain classes of inorganic compounds that are strong oxidizing agents, evolving oxygen on decomposition. These substances include ozone, hydrogen peroxide and metal peroxides, perchlorates, chlorates, chlorites, hypochlorites, periodates, iodates, perbromates, bromates, permanganates, dichromates, chromates, nitrates, nitrites, and persulfates. All these substances are rich in oxygen and decompose violently on heating. The explosion hazard arises when these substances come into contact with easily oxidizable compounds, such as organics, metals, or metal hydrides. When the solid substances are finely divided and combined, the risk of explosion is enhanced. The unstable intermediate products, so formed, are sensitive to heat, shock, and percussion. The degree of explosivity may vary with the anion and, not so significantly, on the metal ions in the molecules. Ammonium salts of these anions, especially ammonium nitrate, have high explosive power when thermally decomposed.

The oxidizers mentioned above are noncombustible substances. However, contact with comustible materials such as alcohols, amines, ethers, hydrocarbons, aldehydes, ketones, and other organics may produce flame (sometimes, with explosion). For example, cotton in contact with 90% hydrogen peroxide can burst into flame spontaneously.

The health hazard from thes substances arises due to their strong corrosive action on the skin and eyes. Ozone and hydrogen peroxide vapors can cause injury to mucous membranes. High concentrations of the former can be fatal to humans. The toxicity of oxidizers has little to do with their oxidizing properties but depends on the metal ions in the molecules. For example, all barium and mercury salts are poisons. Ingestion of sodium and potassium salts (nitrates) in large amounts may cause gastroenteritis, anemia, and methemoglobinemia. Hydrogen peroxide and sodium nitrate have produced tumors in animals. The explosion, fire, and health hazards of individual and different classes of oxidizers are presented in the following sections.

41.2 OZONE

Formula O_3; MW 48.00; CAS [1028-15-6]

Uses and Exposure Risk

Ozone is used as an oxidizing compound, as a disinfectant for air and water, for bleaching waxes and oil, and in organic synthesis. It occurs in the atmosphere at sea level to about 0.05 ppm. It is produced by the action of ultraviolet (UV) radiation on oxygen in air.

Physical Properties

Bluish gas with a characteristic pleasant odor at low concentrations (<2 ppm); liquefies at $-112°C$; solidifies at $-193°C$; insoluble in water (0.00003% at 20°C).

Health Hazard

Ozone is a highly toxic gas. Inhalation of this gas at a relatively low concentration can cause death in a short period of time. The target organs are lungs and eyes. The toxic symptoms include lacrimation, cough, labored breathing, headache, dryness of throat, lowering pulse rate and blood pressure, congestion of lungs, edema, and hemorrhage. A 1-hour exposure to 100 ppm can be lethal to humans, while a concentration of 1 ppm in air can produce toxic symptoms in humans. The LC_{50} value in rats is within the range 50 ppm of a 4-hour exposure period. Chronic exposure may result in pulmonary disease. Continual daily exposures to ozone, like ionizing radiation, can cause premature aging (ACGIH 1986).

Exposure Limits

TLV-TWA 0.1 ppm ($\sim$0.2 mg/m^3) (ACGIH, NIOSH, and MSHA), 0.2 ppm ($\sim$0.4 mg/m^3) (OSHA); IDHL 10 ppm (NIOSH).

Fire and Explosion Hazard

Nonflammable by itself but an explosive gas. It is highly endothermic, the heat of formation, ΔH_f°(s) being +34.1 kcal/mol (142 kJ/mol). The liquefied gas or concentrated solutions of ozone explode on warming. This a powerful oxidizing agent reacts explosively with readily oxidizable substances, even well below ambient temperatures. Reaction with olefins or other unsaturated compounds produces ozonides, $R_2C(O_3)CR_2$, which are cyclic or polymeric peroxides, in which both the σ and π bonds have been broken, with formation of bonds between carbon and oxygen. Low-molecular-weight ozonides are highly unstable and explode. Its reactions with acetylene and other alkynes can be dangerously explosive. Both ozone and acetylene are endothermic substances. Violent reactions can occur even at low concentrations.

Ozone reacts with benzene, aniline, and other aromatics, one of the products being ozobenzene, a white gelatinous explosive compound. Reactions with oxides of nitrogen produce flame and explosions. It explodes when mixed with hydrogen, carbon monoxide, ammonia, or hydrogen sulfide. All such explosive reactions can occur at temperatures below 0°C. Violent explosions occur when ozone is mixed with bromine or hydrogen bromide (Mellor 1946, Suppl. 1956). Ozone forms explosive mixtures with nitrogen halides. It forms dialkyl peroxides with dialyl ethers. The peroxides are sensitive to shock and explode. It reacts explosively when combined with reducing agents such as hydrogen iodide, sodium borohydride, lithium aluminum hydride, hydrazine, stibine, phosphine, diborane, silane, or arsine. Liquid ozone may explode when mixed with oxidizable organic compounds.

Analysis

Ozone produces an intense UV absorbance band that begins at 290 nm. Air is passed through an impinger containing potassium iodide/caustic soda or phosphoric and sulfamic acids. Colorimetric analysis may be performed using a spectrophotometer.

41.3 HYDROGEN PEROXIDE

DOT Label: Oxidizer, Identification Numbers, UN 2015 for solution greater than 52%

strength, UN 2014 for 20–52% solution. UN 2984 for 8–20% solution.

Formula: H_2O_2; MW 34.02; CAS [7722-84-1]

Synonyms: hydrogen dioxide; perone

Uses and Exposure Risk

Hydrogen peroxide is used for bleaching silk, fabrics, feathers, and hairs; in refining oils and fats; for cleaning metals surfaces; as an antiseptic; and in rocket propulsion (90% solution). It is marketed as an aqueous solution of 3–90% by weight.

Physical Properties

Colorless liquid; unstable (pure compound or a 90% solution); bitter in taste; density 1.463 at 0°C; boils at 152°C; solidifies at −0.43°C; miscible in water, soluble in ether, decomposed by many organic solvents.

Health Hazard

Concentrated solutions of hydrogen peroxide is very caustic and can cause burns of skin and mucous membranes. Exposure to its vapors can produce body irritation, lacrimation, sneezing, and bleaching of hair. A dose of 500 mg/kg by dermal route caused convulsions and deaths in rabbits. The oral LD_{50} value for 90% peroxide solution in mice is 2000 mg/kg.

Oral administration of hydrogen peroxide produced tumors in gastrointestinal tract in mice. There is limited evidence of carcinogenicity in animals. Cancer-causing effects of hydrogen peroxide in humans are unknown. Padma and co-workers (1989) reported the promoting effect of hydrogen peroxide on tobacco-specific N-nitrosoamines in inducing tumors in the lung, liver, stomach, and cheek pouch in Syrian golden hamsters and mice. The incidence of cheek pouch tumors increased when peroxide was administered concurrently or applied for a long period after a low initiator dose of N-nitrosamines.

Exposure Limits

TLV-TWA 1 ppm (∼1.5 mg/m^3) (ACGIH, MSHA, and OSHA), IDLH 75 ppm (NIOSH).

Fire and Explosion Hazard

Hydrogen peroxide is a powerful oxidizer. Concentrated solutions can decompose violently if trace impurities are present. Commercially available product is inhibited by addition of trace acetanilide against decomposition.

Contact with readily oxidizable substances may produce spontaneous combustion. Concentrated solutions can detonate on heating. The rate of decomposition increases by 1.5 times for every 10°F increase in temperature (NFPA 1997). Mixing with organic compounds may cause violent explosions. This includes alcohols, acetone and other ketones, aldehydes, carboxylic acids, and their anhydrides. With acetic acid and acetic anhydride, peroxyacetic acid is formed. In the presence of excessive anhydride, diacetyl peroxide is formed. Both the peroxy acids and the organic peroxides are shock-sensitive explosive compounds. Spontaneous ignition occurs when added to cotton (cellulose).

Violent explosions may occur when hydrogen peroxide is brought in contact with metals such as sodium, potassium, magnesium, copper, iron, nickel, chromium, manganese, lead, silver, gold, platinum; metal alloys such as brass or bronze; metal oxides such as lead oxides, mercury oxides, or manganese dioxide; and many metal salts (which include certain oxidizing substances such as potassium permanganate or sodium iodate). It forms unstable mixtures with concentrated mineral acids. When organic compounds come in contact with these solutions, violent explosions may result. Storing hydrogen peroxide in a closed container that does not have a vent may rupture the vessel caused by the pressure buildup by released oxygen.

41.4 PEROXIDES, METAL

Several metals form peroxides. Some of the common inorganic peroxides are sodium peroxide, Na_2O_2 [1313-60-6]; potassium peroxide, K_2O_2 [17014-71-0]; magnesium peroxide, MgO_2 [1335-26-8]; calcium peroxide, CaO_2 [1305-79-9]; and barium peroxide, BaO_2 [1304-29-6]. Peroxides are used for bleaching fibers, waxes, and woods; for dyeing and printing textiles; and for purifying air. Peroxides of alkali metals, calcium, and magnesium are nontoxic. These compounds, however, are corrosive to the skin. Barium peroxide is poisonous. The subcutaneous LD_{50} value in mice is 50 mg/kg.

Peroxides are powerful oxidizing compounds. Potassium peroxide is highly reactive, which can decompose violently when exposed to air or water. Combustible organic compounds catch fire spontaneously when combined with sodium or potassium peroxide. Such organic compounds include benzene, aniline, toluene, alkanes, cycloalkanes, ethers, glycerine, paper, and wood. Peroxides oxidize metals with incandescence. The reactions may become explosive at high temperatures and in the presence of moisture. Peroxides ignite charcoal, calcium carbide, boron nitride, hydrogen sulfide, or sulfur monochloride (Mellor 1946, Suppl. 1961). Ignition and/or explosion occurs when mixed with phosphorus.

41.5 PERCHLORATES

Perchlorates are the metal salts of perchloric acid, $HClO_4$. Perchlorates that have commercial applications are potassium perchlorate, $KClO_4$ [7778-74-7]; sodium perchlorate, $NaClO_4$ [7601-89-01]; ammonium perchlorate, NH_4ClO_4 [7790-98-9]; lithium perchlorate, $LiClO_4$ [7791-03-9]; magnesium perchlorate, $Mg(ClO_4)_2$ [10034-81-8]; and barium perchlorate, $Ba(ClO_4)_2$ [13465-95-7]. These compounds are used in explosives, pyrotechnics, photography, as desiccants, or in analytical chemistry.

Perchlorates are low to moderately soluble in water and decompose at around 400°C. Health hazard is low. These compounds are an irritant to the skin and eyes. The dusts cause irritation of respiratory passage. Rats treated with perchlorates shows increase in levels of chlolesteral, phspholipids, and triglycerides in lipoprotein fractions, increasing the risk of cardiac disease (Vijaya Lakshmi and Motlag 1989).

Perchlorates are powerful oxidizing substances. These compounds explode when mixed with combustible, organic, or other easily oxidizable compounds and subjected to heat or friction. Perchlorates explode violently at ambient temperatures when mixed with mineral acids, finely divided metals, phosphorus, trimethylphosphite, ammonia, or ethylenediamine. Explosions may occur when perchlorates are mixed with sulfur, or hydride of calcium, strontium, or barium and are subjected to impact or ground in a mortar. Perchlorates react with fluorine to form fluorine perchlorate, an unstable gas that explodes spontaneously. Heating perchlorates to about 200°C with charcoal or hydrocarbons can produce violent explosions. Metal perchlorates from complexes with many organic solvents, which include benzene, toluene, xylenes, aniline, diozane, pyridine, and acetonitrile. These complexes are unstable and explode when dry. Many metal perchlorates explode spontaneously when recrystallized from ethanol. Saturated solution of lead perchlorate in mathanol is shock sensitive.

The explosive reactions stated above may occur with perchlorates of most metals and are not confined solely to those of alkali- or alkaline-earth metals. The explosivity of these salts is attributed to the perchlorate anions, while metal cations do not play a very significant role. Ethyl perchlorate, which is not a metal salt, is a dangerous explosive.

Hydrated metal perchlorates are stable. Dehydrated and lower hydrated salts are endothermic and unstable. When mixed with concentrated sulfuric acid, they form anhydrous perchloric acid, which is unstable and explodes.

41.6 PERIODATES

Periodates are the metal salts of periodic acid [10450-60-9]. The general formula of these compounds is $M(IO_4)_x$, where M is a metal ion having the valency x. Periodic acids occur in two forms; the *meta* acid, HIO_4, and the *ortho* acid, H_5IO_6, which is the dihydrate of meta acid. Periodic acid is a much weaker acid than perchloric acid and is reduced to iodic acid by nitrous and sulfurous acids. Potassium periodate [7790-21-8] is the only metal periodate of commercial use. It is used as an oxidizing agent and for colorimetric analysis of manganese.

Potassium periodate and sodium periodate [7790-28-5] exhibit low toxicity. An intraperitoneal LD_{50} value for the sodium salt in mice is reported to be 58 mg/kg (NIOSH 1986). These compounds are highly irritating to skin, eyes, and mucous membranes.

Periodates are powerful oxidizers. They explode when heated. Ammonium periodate, NH_4IO_4, explodes when touched by a spatula, or upon mild shock.

Decomposition evolves oxygen. Periodates are noncombustible substances but ignite when mixed with combustible materials. Fire and/or explosion may occur when periodates are mixed with organics or other readily oxidizable compounds. Reactions with finely divided metals, hydrides, mineral acids, charcoal, phosphorus, alkyl phosphites, amines, sulfur, and a number of inorganic and organic compounds can be violently explosive.

41.7 CHLORATES

Chlorates are metal or ammonium salts of chloric acid, having the general formula $M(ClO_3)_x$, where M is the metal ion with valency x or amonium ion. The formula of chloric acid [7790-93-4] is $HClO_3$. Some of the chlorates that have commercial uses

are potassium chlorate. $KClO_3$ [3811-04-9]; sodium chlorate, $NaClO_3$ [7775-09-9]; magnesium chlorate, $Mg(ClO_3)_2$ [10326-21-3]; and calcium chlorate, $Ca(ClO_3)_2$ [10137-74-3]. barium chlorate, $Ba(ClO_3)_2$ [13477-00-4]; ammonium chlorate, NH_4ClO_3 [10192-29-7]; cupric chlorate, $Cu(ClO_3)_2$ [14721-21-2]; mercurous chlorate, $Hg_2(ClO_3)_2$ [10294-44-7]; silver chlorate, $AgClO_3$ [7783-92-8]; strontium chlorate, $Sr(ClO_3)_2$ [7791-10-8]; and lead chlorate, $Pb(ClO_3)_2$ [10294-47-0] are known, but do not have commercial applications. Potassium and sodium chlorates are used in explosives, fireworks, matches, in dyeing cotton, and in chemical analysis. Calcium chlorate is used as a disinfectant and in herbicides and insecticides.

Sodium and potassium chlorates are low to moderately toxic in test animals. Oral administration produced irritation of the gastrointestinal tract, anemia, and methemoglobinemia. The oral LD_{50} values for sodium, potassium, calcium, and magnesium salts in rats are within the range 1200, 1800, 2500, and 6300 mg/kg, respectively. The toxicity data for other metal chlorates are not reported.

Chlorates are powerful oxidizers and react explosively with organic or other readily oxidizable substances. They explode violently when mixed with concentrated sulfuric acid. In general, chlorates would react with concentrated mineral acids at ordinary temperature or with dibasic organic acids on heating, forming chlorine dioxide, which is an explosive gas. Mixtures of chlorates with powdered metals may ignite or explode by friction or percussion. Similarly, reactions with metal sulfides, cyanides, or thiocyanates initiated by heat, shock, or friction may cause explosions. Chlorates mixed with calcium hydride or stontium hydride and rubbed in a mortar can explode violently (Mellor 1946). Similar explosive reactions are likely to occur with other metal hydrides. Reactions with phosphorus, hypophosphite, phosphonium iodide, sulfur, sulfur dioxide, sulfites, carbon, or manganese dioxide are violent, producing flame and/or explosions. Ammonium chlorate decomposes explosively upon shock or impact and is listed by the U.S. Department of Transportation (DOT) as a forbidden material for transportation.

41.8 BROMATES

Bromates have the general formula $M(BrO_3)_x$, where M is a metal ion with valency x or the ammonium radical. These are salts of bromic acid, $HBrO_3$ [7789-31-3]. Alkali metal bromates are used in analytical work for volumetric analysis, in extraction of gold, and as oxidizing agents.

Potassium bromate, $KBrO_3$ [7558-01-2], and sodium bromate, $NaBrO_3$ [7789-38-0], are moderately toxic compounds. Ingestion causes nausea, vomiting, diarrhea, respiratory stimulation, decrease in body temperature, methemoglobinemia, and renal injury. The oral LD_{50} value for potassium bromate in rats is in the range 300 mg/kg. Experiments on laboratory animals present sufficient evidence of carcinogenicity of potassium bromate. It caused kidney and thyroid tumors in rats. Carcinogenicity in humans is not known. Barium bromate is a strong poison. The DOT Label for this compound is "Oxidizer and Poison," UN 2719. The TLV-TWA value in air is 0.5 mg(Ba)/m^3 (ACGIH, MSHA, and OSHA).

Bromates are powerful oxidizers. On heating, these compounds decompose explosively, producing oxygen. Ammonium bromate [13843-59-9] explodes upon mild impact. DOT lists it under substances forbidden for transportation; while sodium, potassium, and magnesium bromates are labeled as oxidizers.

Bromates react violently with combustible, organic, and other readily oxidizable substances. Finely divided bromates when combined with finely divided organic compounds can explode by heat, percussion, or friction (MCA 1963). When mixed with hydrides of calcium or strontium and rubbed in a mortar, a violent explosion results (Mellor

1946, Suppl. 67). Laboratory fires have been reported, caused by contact of moisture with a potassium bromate, malonic acid, and cerium ammonium nitrate mixture, setting a high exothermic reaction (Bartmess et al. 1998). Bromates ignite when mixed with concentrated mineral acids, lead acetate, or phosphonium iodide, PH_4I. Finely divided mixtures of bromates with finely divided metals, phosphorus, sulfur, or metal sulfides can explode when heated or subjected to friction (Mellor 1946).

41.9 IODATES

Iodates are compounds having the structure $M(IO_3)_x$ or an ammonium ion. These are salts of iodic acid, HIO_3 [7782-68-5]. Only potassium iodate, KIO_3 [7758-05-6], and sodium iodate, $NaIO_3$ [7681-55-2], are of commercial use. These are used as oxidizing agents in volumetric analysis, as dough conditioners, and as antiseptics. Calcium iodate, $Ca(IO_3)_2$ [7789-80-2], is used as a nutritional source of iodine in food.

Sodium and potassium iodates are mild skin irritants but moderately toxic. Ingestion causes nausea, vomiting, and diarrhea. An oral dose of 400 mg/kg is lethal to guinea pigs. The hazardous properties of iodates are similar to those of chlorates and bromates. Iodates are noncombustible substances but strong oxidizers. When these compounds decompose, oxygen is evolved. They ignite combustible compounds. Violent reactions occur when iodates are mixed with organic or other oxidizable substances and are subjected to heat, percussion, or shock.

41.10 CHLORITES

Chlorites have the general formula, $M(ClO_2)_x$, where M is a metal ion with valency x or the ammonium ion. These are the salts of chlorous acid. Only sodium chlorite, $NaClO_2$ [7758-19-2], is of commercial use. It is used as a bleaching agent for textiles and paper pulp, in water purification, and to produce chlorine dioxide.

Sodium chlorite is a moderately toxic compound. Ingestion produces adverse effects on liver and kidney. It causes jaundice and interstitial nephritis. Developmental toxicity in animals has been noted. The oral LD_{50} values in rats and mice are 165 and 350 mg/kg, respectively. The toxicity data for other chlorites have not been reported.

Chlorites are strong oxidizing substances. They, however, are less stable than the chlorates. They are sensitive to heat or impact. Most metal chlorites explode upon heating at $100°-110°C$. Sodium chlorite does or explode on percussion. Mixtures of chlorites with finely divided metals or organic compounds may burst into flame on friction. Ammonium chlorite explodes when subjected to heat or shock, as does silver chlorite, $AgClO_2$ [7783-91-7].

Dry silver chlorite is forbidden for transportation by DOT. Reaction of a chlorite with acids produces explosive chlorine dioxide gas. Ignition occurs with sulfur or phosphorus (red) under moist conditions.

41.11 HYPOCHLORITES

Hypochlorites are the salts of hypochlorous acid, HOCl. The general formula of this class of compounds is $M(OCl)_x$, where M is a metal ion having the valency x. Among the hypochlorites, calcium and sodium salts are of commercial importance. Calcium hypochlorite or the bleaching powder [7778-54-3] is used as a bleaching agent, fungicide, deodorant, and disinfectant, and as an oxidizing agent. Sodium hypochlorite [7681-52-9] solution is used in laundry bleach and as a deodorant and disinfectant.

Hypochlorites are irritant to skin. Acute oral toxicity is low for the calcium salt. The oral LD_{50} value in rats is 850 mg/kg. Ingestion of sodium hypochlorite may cause gastric perforation and corrosion of mucous membranes. Toxicity of metal hypochlorites is attributed to the toxicity of the metal cations present in the salts.

Hypochlorites undergo violent reactions with many compounds, causing fire or explosion. Reactions with primary amines form chloroamines, which are explosives. Violent explosions may occur when hypochlorites are mixed with ethanol, carbon tetrachloride, urea (forming explosive nitrogen trichloride), mercaptans, or ethyleneimine (forming an explosive, 1-chloroethylene imine). Metal oxide catalysts decompose hypochlorites, evolving oxygen and often causing explosions. Reactions with glyceral, methyl carbitol, and many organic sulfides produce fire. With the latter compounds, explosions may occur, too. Hypochlorites may react violently with nitromethane, many ammonium salts, and oil and grease. Explosions may occur when heated with finely divided charcoal (at 1 : 1 ratio) in a closed container (NFPA 1986); or with oil at 135°C. Hypochlorites produce mild flames when heated with wood at 180°C. Reaction with sulfur in the presence of moisture produces a crimson flash.

41.12 PERMANGANATES

Permanganates have the general formula $M(MnO_4)_x$, where M can be ammonium or a metal ion with valency x. These are salts of permanganic acid, $HMnO_4$. The oxidation state of manganese in permanganate ion is +7. Manganeses also forms anion (manganate) at a lower oxidation state, +6. Among the permanganates, potassium permanganate, $KMnO_4$ [7722-64-7], has the greatest commercial use. It is used for bleaching resins, waxes, oils, silk, cotton, and other fibers; in purifying water; in tanning leathers; in photography; in organic synthesis; and as an analytical reagent. Permanganates of sodium, $NaMnO_4$ [10101-50-5]; ammonium, NH_4MnO_4 [13446-10-1]; and barium, $Ba(MnO_4)_2$ [7787-36-2], are known but have few commercial applications.

Potassium permanganate is a moderately toxic compound. In humans the toxic effects from ingestion are nausea, vomiting, and dyspnea. Ingestion of about 10 g of salt can be fatal to humans. The oral LD_{50} value in rats is within the range 1100 mg/kg. Its concentrated solutions are caustic to the skin and eyes.

Permanganates are powerful oxidizers. Explosion occurred when concentrated sulfuric acid was mixed with crystalline potassium permanganate in the presence of moisture. The reaction produces manganese heptoxide, Mn_2O_7, which explodes at 70°C (Archer 1948). Potassium permanganate/sulfuric acid mixture does not explode at low temperature. However, the solution oxidizes most organic substances. Concentrated solutions of this mixture containing permanganyl sulfate in contact with benzene, alcohols, ethers, acetone, carbon disulfide, petroleum, or other organic matter may react with explosive violence. In the absence of sulfuric acid, finely divided solid permanganates may explode, as well, when they come in contact with organic substances. The contact of finely divided metals, powdered sulfur (on heating), or phosphorus (on grinding) may cause an explosion. Reaction with acetic acid or acetic anhydride above room temperature is explosive. Contact with concentrated hydrogen peroxide can cause explosion. Reaction with diethyl sulfoxide or hydroxyl amine may produce flame.

Ammonium permanganate is shock sensitive at 60°C, decomposing violently when heated at higher temperatures. Permanganic acid decomposes explosively above 3°C. Aliphatic or aromatic hydrocarbons, cycloalkanes, alcohols, ethers, amines, and amides burst into flame when mixed with permanganic acid. Explosion may occur with ethanol. Silver permanganate, $AgMnO_4$ [7783-98-4], reacts with ammonium hydroxide, forming a shock-sensitive complex of the formula $[Ag(NH_3)_2]MnO_4$ (NFPA 1997).

41.13 PERSULFATES

Persulfate anion has the formula $S_2O_7^{2-}$. Persulfates of sodium, $Na_2S_2O_7$ [7775-27-1];

potassium, $K_2S_2O_7$ [7727-54-0], are known. Persulfates are used for bleaching fabrics, in photography, and as a reagent in chemical analysis. The toxicity of these compounds is low. The oral LD_{50} value of ammonium persulfate in rats is 820 mg/kg. The lethal dose of sodium salt in rabbits by intravenous administration is 178 mg/kg (Merck 1989). These compounds are strong irritants to the skin and mucous membranes.

Persulfates are strong oxidizers; they decompose violently upon heating, releasing oxygen. Reactions with organic and other easily oxidizable compounds can be violent. A mixture of ammonium persulfate powder with aluminum powder and water explodes (Pieters and Creyghton 1957). Ammonium persulfate and sodium peroxide mixture explodes when crushed in a mortar or heated (Mellor 1946).

41.14 DICHROMATES

Dichromate anion has the formula $Cr_2O_7^{2-}$, in which chromium ion occurs in the +6 oxidation state. It also occurs in the same valence state (+6) oxidation state. It also occurs in the same valence state (+6) in chromates, CrO_4^{2-}. Sodium and potassium dichromates of formula $Na_2Cr_2O_7$ and $K_2Cr_2O_7$ and CAS numbers [10588-01-9] and [7778-50-9], respectively, are of commercial importance. These compounds are used for bleaching resins, oils, and waxes; in chrome tanning of hides; in the manufacture of chrome pigments; in pyrotechniques and safety matches; and as corrosion inhibitors.

Dichromates of sodium, potassium, and ammonium, $NH_4Cr_2O_7$ [7789-09-5], are moderately toxic compounds and irritants to the skin. These are corrosive substances causing ulceration of hand and injury to mucous membranes. The toxic effects are similar for all three compounds, and the lethal doses in guinea pigs are comparable: 23.0, 29.4, and 25.0 mg/kg, respectively, by subcutaneous administration. Ingestion of 5 g can be lethal to humans. Dichromates of sodium and potassium caused lung tumors in rats. However, the evidence of carcinogenicity in animals or humans is inadequate at the moment. Exposure limit: TLV-TWA air 0.05 mg $(Cr)/m^3$ (ACGIH), 0.5 mg$(Cr)/m^3$ (MSHA), 0.025 mg$(Cr^{6+})/m^3$ (NIOSH); ceiling 0.1 mg $(CrO_3)/m^3$ (OSHA), 0.05 mg $(Cr^{6+})/m^3/15$ min (NIOSH).

Dichromates are powerful oxidizing materials. These substances decompose on heating, evolving oxygen. Decomposition of ammonium dichromate begins at 180°C, becoming self-sustaining at 225°C with luminescence (NFPA 1997). Nitrogen gas is liberated. Closed containers may rupture as a result of pressure buildup.

Dichromates can ignite readily oxidizable substances. Combustion can be violent with finely divided oxidizable compounds. A dichromate/sulfuric acid mixture may explode violently in contact with acetone. Dichromates react explosively with hydrazine; in the powder form, these substances produce violent explosions when mixed with anhydrous hydroxyl amine (Mellor 1946).

41.15 NITRATES

All metals in the periodic table form nitrates, $M(NO_3)_x$. Nitrate is a common anion Many nitrates occur in nature. These are widely used in industry, and many are common laboratory reagents. Nitrates are salts of nitric acid. Their hazardous properties vary widely. The toxicity of nitrates depends exclusively on the metal cation part of the molecule.

Explosivity also differs significantly among nitrates. Nitrates are oxidizing substances that decompose by heat, evolving oxygen. Ammonium nitrate, NH_4NO_3 [6484-52-2], is a dangerous explosive. Potassium nitrate, KNO_3 [7757-79-1], and sodium nitrate, $NaNO_3$ [7631-99-4], have wide commercial applications. These are used in fireworks, matches, gun powder, freezing mixtures, and for pickling meats. Ingestion of these compounds in large quantities can produce

serious adverse effects, especially gastroenteritis. Prolonged exposure to potassium nitrate may cause anemia, methemoglobinemia, and nephritis (Merck 1996). The oral LD_{50} value in rats is 3750 mg/kg (NIOSH 1986). Ingestion of about 8 g of sodium nitrate may be lethal to humans. Oral administration of sodium nitrate in rats caused tumors in the liver, skin, and testes. Carcinogenicity in humans is not known.

Alkali metal nitrates on heating decompose to nitrates, evolving oxygen:

$$2KNO_3 \longrightarrow 2KNO_2 + O_2$$

It may be noted that 0.5 mol O_2 is released from decomposition of one mol potassium and other alkali metal nitrates. This is much lower in comparison with 3 or 3.5 mol gaseous products produced from equimolar amount of NH_4NO_3, causing extremely high internal pressure buildup. However, when mixed with easily oxidizable materials and heated, ignition and/or explosion can occur. Nitrates explode on reacting with esters, phosphorus, or stannous chloride (Pieters and Creyghton 1957). Explosion occurs when these are mixed and heated with hypophosphites, cyanides, or thiocyanates. Unlike alkali metal salts, heavy metal nitrates do not form any nitrites on decomposition. Metal oxides are obtained with evolution of oxygen and toxic nitrogen dioxide:

$$2AgNO_3(s) \longrightarrow 2Ag(s)$$
$$+ 2NO_2(g) + O_2(g)$$

The explosive reactions are similar to those mentioned above.

Ammonium Nitrate

Among the inorganic nitrates, ammonium nitrate is by far the most dangerous explosive. At a lower temperature, 80–90°C, ammonium nitrate decomposes endothermically to ammonia and nitric acid:

$$NH_4NO_3(s) \longrightarrow NH_3(g) + HNO_3(g)$$

At a temperature above 210°C, the decomposition becomes exothermic, producing nitrous oxide and water:

$$NH_4NO_3(s) \longrightarrow N_2O(g) + 2H_2O(g)$$

It melts at 167°C with slow decomposition. In a close confinement, heating the molten mass causes a severe pressure buildup, resulting in explosion. Furthermore, evolved nitrous oxide supports combustion. Thus, if ammonium nitrate is containment with oil, charcoal, or other combustible substances, the entire mass may explode.

The decomposition accelerates as the temperature increases. At 300°C, there is rapid evolution of brow fumes of nitrogen oxides, which are highly toxic gases. The decomposition is explosive. In the presence of easily oxidizable substances or finely divided metals, a self-sustained ignition may take place. At a further elevated temperature or under fire conditions the decomposition produces nitrogen, oxygen, and water vapors:

$$2NH_4NO_3(s) \longrightarrow 2N_2(g)$$
$$+ 4H_2O(g) + O_2(g)$$

This equation indicates that 2 mol of solid ammonium nitrate produces 7 mol of gaseous products. Thus the internal pressure buildup is extremely high, causing a disastrous explosion. There are several cases of ammonium nitrate explosions, resulting in the loss of human lives. Thousands of people have died worldwide in some of the worst industrial accidents, ship fires and terrorist bombings. The worst one was an explosion of a cargo vessel in 1947 in Galvenston Bay near Texas City, carrying 2200 tons of ammonium nitrate (Thayer 1997; Meyer 1989). The ship caught fire exploding in one hour later. The devastative blast caused a tidal wave that washed 150 feet inland. More than 600 people died in the disaster. The ship also carried cotton, a readily combustible material for fire. In an earlier disaster, a warehouse full of ammonium nitrate exploded in a chemical plant in Germany, killing thousands. A mixture of ammonium nitrate fertilizer and fuel oil was a tool in the Oklahoma City bombing in 1995. Such combinations are common industrial explosives. As mentioned earlier, the presence of flammable or combustible material can accelerate the fire and sustain it because of nitrous oxide and/or oxygen released during the exothermic decomposition of ammonium nitrate.

Among the metals, sodium reduces ammonium nitrate to disodium nitrite, a yellow explosive substance (Mellor 1946, Suppl. 1964). Potassium or sodium–potassium alloy mixed with ammonium nitrate and ammonium sulfate results in explosion (NFPA 1986). Violent reactions may occur when a metal such as aluminum, magnesium, copper, cadmium, zinc, cobalt, nickel, lead, chromium, bismuth, or antimony in powdered form is mixed with fused ammonium nitrate. An explosion may occur when the mixture above is subjected to shock. A mixture with white phosphorus or sulfur explodes by percussion or shock. It explodes when heated with carbon. Mixture with concentrated acetic acid ignites on warming. Many metal salts, especially the chromates, dichromates, and chlorides, can lower the decompostion temperature of ammonium nitrate. For example, presence of 0.1% $CaCl_2$, NH_4Cl, $AlCl_3$, or $FeCl_3$ can cause explosive decomposition at 175°C. Also, the presence of acid can further catalyze the decomposition of ammonium nitrate in presence of metal sulfides.

41.16 NITRITES

Nitrites have the general formula $M(NO_2)_x$, containing nitrogen in a +3 oxidation state. Nitrites act as oxidizing agents as well as reducing substances. With strong oxidizers such as peroxides, nitrites are oxidized to nitrates. Reactions with reducing agents reduce nitrites to nitric oxide. These compounds decompose at relatively high temperatures as follows:

$$2KNO_2(s) \longrightarrow NO_2(g) + NO(g) + K_2O(s)$$

The explosion hazard is lower than that for nitrates. Heating with ammonium salts, finely divided metals, or easily oxidizable substances can be violent or explosive. Evaporating a mixture of sodium nitrite, $NaNO_2$ [7632-00-0], with sodium thiosulfate to dryness caused a violent explosion (Mellor 1946). Molten potassium nitrite, KNO_2 [7758-09-0], reacts violently with boron. A mixture of nitrite and cyanide explodes when heated to 45°C. Ammonium nitrite, NH_4NO_2 [13446-48-5], is an unstable compound, decomposing explosively on mild heating to 60°C. Also, its concentrated aqueous solutions can explode on heating. Presence of acids or impurities can lower the temperature of its decomposition. DOT forbids its transportation.

REFERENCES

ACGIH. 1986. *Documentation of the Threshold Limit Values and Biological Exposure Indices*, 5th ed. Cincinnati, OH; American Conference of Governmental Industrial Hygienists.

Archer, J. R., 1948. *Chem. Eng. News* 26:205

Bartmess, J., C. Feigerle, G. Schweitzer, W. Fellers, and P. Smith. 1998. *Chem. Eng. News,* 76(26): 4.

MCA. 1963. *Case Histories of Accidents in the Chemical Industry*. Washington, DC: Manufacturing Chemists' Association.

Mellor, J. W. 1946. *A Comprehensive Treatise on Inorganic and Theoretical Chemistry*. London: Longmans, Green & Co.

Merck. 1996. *The Merck Index*. 12th ed. Rahway, NJ: Merck & Co.

Meyer, E. 1989. *Chemistry of Hazardous Materials*, 2nd ed. Englewood Cliffs, NJ: Prentice-Hall.

NFPA. 1997. *Fire Protection Guide on Hazardous Materials*, 12th ed. Quincy, MA: National Fire Protection Association.

NIOSH, 1986. *Registry of Toxic Effects of Chemical Substances*. D. V. Sweet, ed. Washington, DC: U.S. Government Printing Office.

Padma, P. R., V. S. Lalitha, A. J. Amonkar, and S. V. Bhide. 1989. Carcinogenicity studies on the two tobacco-specific *N*-nitrosamines, *N*-nitrosonornicotine and 4-(methylnitrosoamino)-1-(3-pyridyl)-1- butanone. *Carcinogenesis* 10(11): 1997–2002.

Pieters, H. A. J., and J. W. Creyghton. 1957. *Safety in the Chemical Laboratory*, 2nd ed. New York: Academic Press.

Thayer, A. 1997. *Chem. Eng. News,* 75(16): page no. 11.

Vijayalakshmi, K., and D. B. Motlag. 1989. Lipoprotein profile during perchlorate toxicity. *Indian J. Biochem. Biophys.* 26: 273–74.

42

PARTICULATES

42.1 GENERAL DISCUSSION

Particulates are air pollutants consisting of various types of fine solid particles or liquid droplets suspended in air. Small particulate matter of diameter of <1 μm are known as fumes. Liquid particles of diameters less than 2 μm are termed "mist." The term *aerosol* generally refers to liquid or solid particles of diameter less than 0.1 μm suspended in air or another gas (Hodgson et al. 1988). Dusts are particles produced directly from a substance under use and are the largest particulate matter, with a diameter of about 100 μm. Cement, sawdust, and ash are examples of dust particles contributing to air pollution. Smoke is small particulate matter of diameter 0.05–1.0 μm, resulting from the incomplete burning of fossil fuels or other organic material. Thus, the general term *particulates* refer to a wide array of fine particles of various diameters suspended in air. These substances may differ widely in their toxic properties. Particles of quartz, mica, asbestos, talc, coal dust, coaltar pitch volatiles, lead oxide fumes, cotton dust, sulfuric acid mists, and oil mists are some of the examples of particulate matters.

Health hazards associated with inhalation of particulates involve either respiratory problems or irritation of eyes and skin (e.g., acid mist). Particulates can be classified into two categories: (1) nuisance or (2) proliferative or fibrogenic dusts. While the former are "inert" dusts, the latter are harmful. Proliferative dusts accumulate in the lung, causing fibrotic hardening and irreversible lung damage, known as pneumoconiosis (ILO 1977). These dusts are not easily removed by phagocytosis or other pulmonary defense mechanisms. Dusts of quartz, coal, and asbestos are examples of fibrogenic dusts.

Particulates may be emitted from processing equipments, such as reactors, furnaces, kilns, and dryers; from industrial sources; or from incineration of municipal and industrial wastes. These may also be generated during the crushing and grinding of materials. Spink (1988) has described the use of a Waterloo scrubber for removal of particulates from waste gases. Water droplets 15–20 times the diameter of the particles were injected into the gas stream and a fan provided turbulent mixing to effect interactions between particles and droplets and cause complete removal of resulting agglomerates. A 98.5% particle removal efficiency has been reported.

667

42.2 NUISANCE DUSTS

Nuisance particulates have very little adverse effect on the lung unless inhaled in excessive amounts. Occupational exposures do not produce any pathological irreversible changes in the lung tissues. The toxic effects resulting from exposures to these dusts are insignificant. Deposits in the eyes and nasal passages may cause irritation but no injury. ACGIH (1986) recommends a TLV-TWA of 10 mg/m^3 for nuisance particulates. Total and respirable nuisance dusts in air may be determined by NIOSH (1984) Methods 0500 and 0600.

Some common nuisance particles are as follows (when toxic impurities are not present):

Alundum (Al$_2$O$_3$)

Calcium carbonate

Cellulose (paper fibers)

Corundum (Al$_2$O$_3$)

Graphite (synthetic)

Gypsum

Kaolin

Limestone

Magnesite

Plaster of Paris

Portland cement

Silicon carbide

Starch

Titanium dioxide

Vegetable oil mist

42.3 SILICA, CRYSTALLINE

CAS Registry No. [14464-46-1] (cristobalite), [14808-60-7] (quartz), [1317-95-9] (tripoli), [15468-32-3] (tridymite).

There are four main varieties of crystalline silica, SiO$_2$: cristobalite, quartz, tripoli, and tridymite. All these forms are found in association with each other. Amorphous silica or diatomaceous earth on calcination converts into crystalline silica. Silica is soluble in hydrofluoric acid. Exposure risk to crystalline silica particulates may arise during its use or manufacture. The main uses of all the forms above include the production of water glass, refractories, abrasives, enamels, and ceramics; decolorization and purification of oils; and in metal polishes and paint filler.

Chronic exposure to crystalline silica particulates may cause silicosis, which is a chronic lung disease with symptoms of scattered rounded nodules of scar tissue in the lungs. Very heavy exposures to quartz dusts of very small particle size can cause acute silicosis. This disease is rapidly progressive and can develop within a few months of initial exposure. It is often associated with tuberculosis. Death can occur within 1 or 2 years.

Among the various forms of crystalline silica, fibrogenicity is highest in tridymite. The decreasing order of fibrogenic potential is tridymite > cristobalite > quartz $\sim$ tripoli $\sim$ coesite.

Exposure limits (TLV-TWA):

ACGIH: 0.05 mg/m^3 (tridymite as respirable dust)
0.05 mg/m^3 (cristobalite as respirable dust)
0.1 mg/m^3 (quartz as respirable dust)
0.1 mg/m^3 (tripoli as respirable dust)

NIOSH: 0.05 mg/m^3 as respirable free silica (10-hr TWA)

OSHA: 10 mg/m^3/2 (% respirable dust +2)
30 mg/m^3/2 (% tridymite, cristobalite, or tripoli as total dust +2)
30 mg/m^3/2 (% quartz as total dust +3)

Crystalline silica in the air may be determined by x-ray diffraction, colorimetric or IR spectrophotometric techniques (NIOSH 1984, Methods 7500, 7601, and 7602, respectively).

42.4 FUSED SILICA

CAS [60676-86-0]; amorphous; formed by heating amorphous silica or quartz; insoluble in water or acids, soluble in HF; used in rockets as an ablative material, used for reinforcing plastics and in GC capillary columns.

Fused-silica dusts are fibrogenic, impairing the functioning of lung. TLV-TWA 0.1 mg/m^3 as respirable dust (ACGIH). Air analysis may be performed by X-ray diffraction method.

42.5 DIATOMACEOUS EARTH

CAS [68855-54-9]; an amorphous form of silica; highly porous, absorbs water; used to purify liquids, in the manufacture of firebrick and heat insulators, and in metal polishes.

There is no adverse effect on the lungs. Under normal conditions of occupational exposures, any fibrogenic or toxic effect is insignificant. Silicosis observed in animals from intratracheal instillation or inhalation is attributed to the presence of crystalline quartz in diatomaceous earth. TLV-TWA 10 mg/m^3 as total dust (ACGIH).

42.6 GRAPHITE, NATURAL

CAS [7782-42-5]; a crystalline form of carbon; used to make lead pencils, crucibles, recarburizing steel, electrodes, and in electrical equipment.

Exposure to graphite particulates in high concentrations can cause pneumoconiosis and anthracosilicosis. Graphite dusts containing a small amount of silica produced a fibrogenic change in animals. ACGIH recommends a TLV-TWA of 2.5 mg/m^3.

42.7 GRAPHITE, SYNTHETIC

CAS [7440-44-0]; crystalline form; used as an adsorbent, and in water purification; synonyms are activated carbon, Filtrasorb, and Norit.

Inhalation causes cough, dyspnea, black sputum, and fibrosis. TLV-TWA 2.5 mg/m^3 (respirable dust), 5.0 mg/m^3 (total dust) (ACGIH), 15 mppcf (OSHA).

42.8 GRAIN DUST

(These include dusts of the grains, wheat, barley, and oats.) Inhalation of airborne grain dusts at high concentrations can produce coughing, wheezing, breathlessness, dyspnea, and bronchial asthma. Also, symptoms of bronchitis, chest tightness, and "grain" fever have been observed in workers from chronic exposures to grain dusts. TLV-TWA 4 mg/m^3, as total particulate (ACGIH).

42.9 COAL DUST

(Carbon in amorphous form; used primarily as a fuel, and to produce coal gas, water gas, coke, coal tar, synthetic rubber, and fertilizers.) Chronic exposure to high concentrations may cause pneumoconiosis, bronchitis, and impair the function of the lungs. TLV-TWA 2 mg/m^3 as respirable dust (ACGIH).

42.10 COAL TAR PITCH VOLATILES

CAS [65996-93-2] (the volatile components of dark brown amorphous residue obtained after distillation of coal-tar pitch; composed of polycyclic aromatics up to 10%, including benzopyrene to about 1.4%).

Occupational exposure may cause bronchitis and dermatitis. Exposure to particulates caused tumors in the lungs and kidneys of animals. In humans, lung and kidney cancers may develop years after the cessation of exposure. TLV-TWA 0.2 mg/m^3, as a benzene-soluble fraction (ACGIH).

42.11 MICA

CAS [12001-26-2]; formula K_2Al_4 $(Al_2Si_6O_{20})(OH)_4$ for muscovite or white mica, a hydrated aluminum potassium silicate; the other major form of mica is phlogopite or amber mica, an aluminum potassium magnesium silicate; nonfibrous; insoluble in water.

Cases of pneumoconiosis have been observed in workers exposed to mica dust at concentrations greater than 10 mppcf for several years. There have also been high incidences of silicosis. Other symptoms are coughing, respiratory distress, weakness, and weight loss. TLV-TWA 3 mg/m^3 as respiratory dust (ACGIH), 20 mppcf (MSHA).

42.12   MINERAL OIL MIST

CAS 8012-95-1; mist of white mineral petroleum oil with an odor of burned lube oil. There are very few adverse effects from acute or chronic exposure to mist of mineral oil. The particulates may cause mild irritation of respiratory tract. ACGIH specifies a TLV-TWA of 5 mg/m^3, which is low, keeping a safety margin, as certain light oils may contain volatile hydrocarbons in the mist.

42.13   RAW COTTON FIBER

(Untreated cotton fibers before any processing.) Exposure to cotton dusts can cause byssinosis, bronchitis, and chest tightness. TLV-TWA 0.2 mg/m^3 (fibers <15 μm in length) (ACGIH).

42.14   PORTLAND CEMENT

CAS [65997-15-1]; gray powder containing less than 1% crystalline silica; composed of dicalcium silicate, tricalcium silicate, and small amounts of alumina, iron oxide, and tricalcium aluminate; insoluble in water.

Portland cement is classified as a nuisance particulate. It does not cause fibrosis or lung damage. Exposure to its dusts may cause irritation of the eyes and nose, coughing, wheezing, bronchitis, and dermatitis. TLV-TWA 10 mg/m^3 (ACGIH), 50 mppcf (OSHA and MSHA).

REFERENCES

ACGIH. 1986. *Documentation of the Threshold Limit Values and Biological Exposure Indices*, 5th ed. Cincinnati, OH: American Conference of Governmental Industrial Hygienists.

Hodgson, E., R. B. Mailman, and J. E. Chambers. 1988. *Dictionary of Toxicology*. New York: Van Nostrand Reinhold.

ILO. 1977. *Occupational Exposure Limits for Airborne Toxic Substances*. Geneva: International Labor Office.

NIOSH. 1984. *Manual of Analytical Methods*, 3rd ed. Cincinnati, OH: National Institute for Occupational Safety and Health.

Spink, D. R. 1988. Handling mists and dusts. *Chemtech* 18(6): 364–68.

43

PEROXIDES, ORGANIC

43.1 GENERAL DISCUSSION

Organic peroxides are compounds containing peroxide functional group, $-O-O-$, bonded to organic groups. These could be considered as derivatives of hydrogen peroxide, $H-O-O-H$, having one or both hydrogen atoms replaced by alkyl or other organic groups. Thus the structures of peroxides could be as follows:

$R-O-O-H$ (hydroperoxides)

$R-O-O-R'$ (dialkyl peroxides)

where R and R′ could be alkyl, cycloalkyl, aryl, or other organic groups. There may be more than one peroxide functional group in the compound, or the peroxy linkage may be part of a ring or polymeric system. Apart from the dialkyl and the hydroperoxides, this class of compounds include acyl and aliphatic ketone peroxides, peroxyesters, peroxydicarbonates, and peroxyacids. The structures of these compounds are as follows:

$$R'-\overset{\overset{\displaystyle O}{\|}}{C}-O-O-\overset{\overset{\displaystyle O}{\|}}{C}-R'' \qquad \text{acyl peroxide}$$

$$R'-\overset{\overset{\displaystyle O}{\|}}{C}-O-O-R'' \qquad \text{peroxyester}$$

$$R'-O-\overset{\overset{\displaystyle O}{\|}}{C}-O-O-R'' \qquad \text{peroxyester}$$

$$R'-O-O-\overset{\overset{\displaystyle O}{\|}}{C}-O-O-R'' \qquad \text{peroxyester}$$

$$R'-O-\overset{\overset{\displaystyle O}{\|}}{C}-O-O-\overset{\overset{\displaystyle O}{\|}}{C}-O-R'' \qquad \text{peroxydicarbonate}$$

Ketone peroxides are mixtures of various isomers, and no definite structure can be ascribed to these compounds. Nitrogen- and sulfur-containing peroxides are well known, some of which are hazardous. Peroxyacids are discussed separately in Chapter 3.

Health Hazards

In general, peroxides exhibit low toxicity. However, some are moderately toxic. Their irritant action on skin and eyes vary widely. For example, methyl ethyl ketone peroxide is a severe eye irritant, whereas lauroyl peroxide is innocuous to eyes and skin and is nontoxic. As peroxides are a wide class of compounds with several possible organic moieties held by peroxy linkage, the relationship of toxicity with structure cannot be generalized. Also the toxicity of solvents or diluents should be borne in mind when handling commercial peroxide solutions.

Fire and Explosion Hazard

Peroxides are indeed a hazardous class of compounds, some of which are extremely dangerous to handle. The dangerous ones are highly reactive, powerful oxidizers, highly flammable, and often form decomposition products which are more flammable.

The oxygen–oxygen bond in the peroxide is weak and can cleave due to heat or light to produce free radicals. This is shown as follows:

$$R'OOR'' \xrightarrow[hv]{heat} R'O \cdot + \cdot OR''$$

The heat of reaction, which depends on the nature of R' and R'', is

$$\Delta H = -30 \text{ kcal to } -44 \text{ kcal/mol}$$

(Sheppard and Mageli 1982)

The free radicals, which have a very short lifetime of milliseconds, are highly reactive, quickly undergoing a variety of reactions.

Many organic peroxides, as discussed in the following sections, can explode violently due to one or a combination of the following factors: (1) mechanical shock, such as impact, jarring, or friction; (2) heat; and (3) chemical contact. Shock- and heat-sensitive peroxides can absorb energy from shock and/or heat, which cleaves the O—O bond, liberating energy to accelerate further decomposition. Compounds such as diisopropyl peroxydicarbonate or dibenzoyl peroxide are extremely sensitive to shock and heat. Death, injury, and property damage due to violent explosions from peroxides are known (Meidl 1970).

Short-chain alkyl and acyl peroxides, hydroperoxides, peroxyesters, and peroxydicarbonates with low carbon numbers are of much greater hazard than the long-chain peroxy compounds. The active oxygen content of peroxides is measured as the amount of active oxygen (from peroxide functional group) per 100 g of the substance. The greater the percent of active oxygen in a formulation, the higher is its reactivity. An active oxygen content exceeding 9% is too dangerous for handling and shipping. However, this is not the only guideline when estimating peroxide danger. Another important factor that indicates the explosive decomposition is the self-accelerating decomposition temperature (SADT). This is the minimum storage temperature at which the decomposition of the highest commercial package of a specific peroxide becomes spontaneous, self-accelerating, and rapid. The degree of hazard may further be evaluated from other experimental tests, such as rupturing of pressure vessel disks, lead pipe deformation, impact shock sensitivity, and bullet impact.

Some peroxides present great fire hazard even with moderate flammability. This happens because of the formation of highly flammable products from their decomposition. The ignition is supported by the oxygen generated from peroxide decomposition. Peroxides burn vigorously, and firefighting is often difficult.

Diluents such as dimethyl phthalate, dibutyl phthalate, tricresyl phosphate, silicone fluid, benzene, toluene, xylene, cyclohexane, dichloroethane, mineral spirit, water and fire-retardant pastes, and so on, are used which greatly lower the shock and heat sensitivity of peroxides. Care should be taken to maintain the storage temperature within the recommended temperature range. Too much cooling can result in crystallizations of peroxides from their diluent solvents and can cause danger.

Analysis

Instrumental methods for analysis depend on the nature of peroxide. Gas chromatography is applicable where peroxides are thermally stable. It may also be applied as a pyrolytic decomposition technique where the decomposition products are separated and quantified (Swern 1986). Liquid and column chromatography, thin-layer chromatography, and paper chromatography have been used for analyses of peroxides.

Peroxides can be reduced by excess iodide ion in acetic acid or isopropanol solvents, and the liberated iodine may be titrated with standard thiosulfate solution. Iodide ion, titanium(IV), ferrous ion, and *N,N*-dimethyl-*p*-phenylenediamine are some of the reducing reagents used for colorimetric analyses for trace peroxides.

Peroxides that are reduced irreversibly at the mercury electrode can be analyzed by polarography. IR, UV, NMR, and GC/MS are useful in structure characterization. IR absorption bands at 800–900 cm^{-1} and in the region of carbonyl group are characteristics of diacyl peroxides, peroxydicarbonates, and peroxyesters.

43.2 DIACETYL PEROXIDE

DOT Label: Forbidden for solid and solutions with strength over 25%; DOT UN 2084 for solutions less than 25%

Formula: $C_4H_6O_4$; MW 118.10; CAS [110-22-5]

Structure and functional group:

$$H_3C-\overset{\overset{\textstyle O}{\|}}{C}-O-O-\overset{\overset{\textstyle O}{\|}}{C}-CH_3$$

two acetyl groups bound by peroxy linkage

Synonyms: acetyl peroxide; ethanoyl peroxide

Uses and Exposure Risk

Diacetyl peroxide diluted solutions with <25% strength are used as free-radical sources to initiate polymerizations and in organic syntheses.

Physical Properties

A 25% solution in dimethyl phthalate is a colorless liquid with pungent odor; bp 63°C at 21 torr; soluble in alcohol and ether, slightly soluble in water.

Health Hazard

The concentrated compound is a severe eye hazard. An amount of 60 mg in a 1-minute rinse caused severe irritation in rabbit's eyes. Skin contact of its solution may cause irritation. Toxicity of this compound should be of low order. Diacetyl peroxide caused lung and blood tumors in rats. Its carcinogenic action on humans is not reported.

Fire and Explosion Hazard

Pure diacetyl peroxide is extremely shock sensitive and therefore stored as a 25% solution in dimethyl phthalate. It is a flammable compound, but it is the shock sensitivity that makes it a highly dangerous substance. It forms crystals from its solution below −5°C (17°F) which are shock sensitive. Kunh (1948) reported an explosion of diacetyl peroxide while it was being removed from an ice chest. Evaporation of its solution should never be carried out. It is sensitive to heat; self-accelerating decomposition temperature of 25% solution being 49°C (120°F); the decomposition is sudden with release of force and smoke.

Diacetyl peroxide reacts violently with easily oxidizable, organic, and flammable compounds. Contact of solid peroxide with ether or any volatile solvent has resulted in violent explosions (Kunh 1948; Shanley 1949). Firefighting should be conducted from an explosion-resistant location using water from a sprinkler or fog nozzle.

Storage and Shipping

Diacetyl peroxide is stored and shipped as a 25% solution in dimethyl pthalate; the storage temperature must not go down below −5°C or above 32°C; it is stored in a cool and well-ventilated area isolated from other chemicals and protected from physical damage. Concentration, maximum to 25%, can be shipped in amber bottles or polyethylene carboy containers with vent caps not exceeding 10 and 45 lb.

Spillage

If there is a spill, absorb the material with vermiculite, perlite, or other noncombustible absorbent. Paper, wood, or spark-generating metals should never be used for sweeping up or in handling operations. The absorbed materials should be placed in a plastic container and sent for disposal.

Disposal/Destruction

It is ignited on the ground in a remote outdoor area using a long torch. Rinse the empty containers with 5–10% caustic soda or caustic potash solution.

Analysis

Diacetyl peroxide can be analyzed by polarography and HPLC techniques. Wet method involves iodide reduction using sodium iodide in acetic acid and measuring the excess iodine with sodium thiosulfate using starch indicator. Carbonyl groups can be identified by IR absorption at the 800–900 cm^{-1} region.

43.3 *tert*-BUTYL HYDROPEROXIDE

DOT Label: Organic Peroxide UN2093 for <72% and UN 2094 for 72 to 90% solutions in water; Forbidden above 90%

Formula: $C_4H_{10}O_2$; MW 90.12; CAS [75-91-2]

Structure and functional group:

$$H_3C-\underset{\underset{CH_3}{|}}{\overset{\overset{CH_3}{|}}{C}}-O-O-H$$

a tertiary C atom bound to peroxide functional group

Synonyms: 1,1-dimethylethyl hydroperoxide; Perbutyl H; Cardox TBH

Uses and Exposure Risk

It is used to initiate polymerization reactions and in organic syntheses to introduce peroxy groups into the molecule.

Physical Properties

Colorless liquid; bp 89°C (decomposes); freezes at 6°C; density 0.896 at 20°C; soluble in organic solvents, moderately soluble in water.

Health Hazard

tert-Butyl hydroperoxide is a strong irritant. Floyd and Stockinger (1958) observed that direct cutaneous application in rats did not cause immediate discomfort, but the delayed action was severe. The symptoms were erythema and edema within 2–3 days. Exposure to 500 mg in 24 hours produced a severe effect on rabbit skin, while a rinse of 150 mg/min was severe to eyes.

It is moderately toxic; the effects are somewhat similar to those of methyl ethyl ketone peroxide. Symptoms from oral administration in rats were weakness, shivering, and prostration.

LD_{50} value, intraperitoneal (rats): 87 mg/kg
LD_{50} value, oral (rats): 406 mg/kg

Inhalation of its vapors in high concentrations may be injurious to the lungs. Toxicity in humans should generally be low. This compound is known to exhibit mutagenicity. Any carcinogenic effect on animals or humans has not been observed.

Exposure Limit

No exposure limit is set. Based on its irritant properties a ceiling limit of 1.2 mg/m^3 (0.3 ppm) is recommended.

Fire and Explosion Hazard

Highly reactive, oxidizing, and flammable liquid; flash point varies from <27°C (80°F) to 54°C (130°F) for commercially available products (NFPA 1986); vapor density 3.1 (air = 1); autoignition temperature not reported.

t-Butyl hydroperoxide forms explosive mixtures with air; UEL and LEL values have not been reported. *t*-Butyl hydroperoxide is sensitive to shock and heat. Its shock sensitivity is lower than benzoyl and methyl ethyl ketone peroxides. When exposed to heat or flame, it explodes; self-accelerating decomposition temperature 88–93°C (190–200°F).

Mixing *t*-butyl hydroperoxide with readily oxidizable, organic, or flammable substances may cause ignition and/or explosion. Fire-extinguishing agent: water from a sprinkler or a hose with fog nozzle; use water to keep the containers cool; fight the fire from a safe location.

Storage and Shipping

See Section 43.7.

Disposable/Destruction

See Section 43.6.

Analysis

Polarography (see Section 43.6). HPLC and GC-FID analyses may be applicable.

43.4 *tert*-BUTYL PEROXYACETATE

DOT Label: Organic Peroxide UN 2095 for 52–76% solution and UN 2096 for <52% solution. Forbidden above 76%

Formula: $C_6H_{12}O_3$; MW 132.18; CAS [107-71-1]

Structure and functional group:

$$CH_3-\overset{\overset{\displaystyle O}{\|}}{C}-O-O-\overset{\overset{\displaystyle CH_3}{|}}{\underset{\underset{\displaystyle CH_3}{|}}{C}}-CH_3$$

acetyl and *tert*-butyl groups are bound by peroxy linkage

Synonyms: *tert*-butyl peracetate; peroxyacetic acid *tert*-butyl ester; Lupersol 70

Uses and Exposure Risk

It is used to initiate polymerization, and in organic syntheses.

Health Hazard

Mild skin and eye irritant. Its toxicity is low, both via inhalation and ingestion routes.

LD_{50} value, oral (rats): 675 mg/kg

Fire and Explosion Hazard

The pure compound is highly reactive and oxidizing, and sensitive to heat and shock.

A 75% solution in benzene or mineral spirits is the maximum assay that is sold commercially. Its benzene solution at this concentration is reactive, oxidizing, and flammable. The flash point varies depending on the solvent; the autoignition temperature not reported; the self-accelerating decomposition temperature is 93°C (200°F).

t-Butyl peroxyacetate forms an explosive mixture with air, explosive range not reported. It can ignite or explode when in contact with combustible organic compounds. Fire-extinguishing agent: water from a sprinkler from an explosion-resistant location; keep the containers cool.

Storage and Shipping

Store in a cool and well-ventilated place isolated from other chemicals; protect from physical damage. It is shipped in glass and earthenware containers not exceeding 7 lb, inside a wooden or fiberboard box.

Spillage and Disposal/Destruction

See Section 43.7.

Analysis

Polarography, HPLC, iodide, or other reduction method.

43.5 DIISOPROPYL PEROXYDICARBONATE

DOT Label: Organic Peroxide UN 2133, and UN 2134 for a 52% solution.

Formula $C_8H_{14}O_6$; MW 206.18; CAS [105-64-6]

Structure and functional group:

$$H_3C \underset{H_3C}{\diagdown} CH - O - \overset{\overset{O}{\|}}{C} - O - O - \overset{\overset{O}{\|}}{C} - O - CH \underset{CH_3}{\overset{CH_3}{\diagup}}$$

peroxyester containing two carbonyl groups, and secondary alkyl groups

Synonyms: isopropyl peroxydicarbonate; peroxydicarbonic acid diisopropyl ester; diisopropyl perdicarbonate; isopropyl percarbonate

Uses and Exposure Risk

It is used as a catalyst to initiate polymerization reactions to produce polymers from ethylene, styrene, vinyl acetate, vinyl chloride, and vinyl esters. Polymers of increased linearity and better properties can be obtained by optimizing its concentration, temperature, and pressure conditions.

Physical Properties

White crystalline solid with a mild characteristic odor; mp 8–10°C (45–50°F); density 1.080 at 15°C; slightly soluble in water, 0.04% at 25°C, soluble in ether, chloroform, benzene, hexane, and isooctane.

Health Hazard

Diisopropyl peroxydicarbonate is a moderate skin irritant. The irritation can be severe on rabbit skin. Eye contact can cause conjunctivitis and corneal ulcerations, which, however, are fully recoverable in 1–2 weeks. Its toxicity is very low. There is no report of its ill effect on humans. It may cause dermatitis on sensitive skin.

LD_{50} value, oral (rats): 21,410 mg/kg

Exposure Limit

No exposure limit is set. Because of low toxicity and vapor pressure, the health hazard from its exposure does not arise.

Fire and Explosion Hazard

This compound is a highly reactive oxidizing and combustible substance. It explodes on heating. It can also decompose violently on its own below room temperature, the self-accelerating decomposition temperature being 12°C (53°F). It is sensitive to shock and friction, its shock sensitivity increasing with concentration in solutions. It does not detonate but decomposes with explosive violence. In a 100% solid form it partially decomposes at 0–10°C when subjected to a bullet impact test (Armitage and Strauss 1964). In the frozen state, it is less sensitive to impact than are benzoyl and some other peroxides.

Diisopropyl peroxydicarbonate can present a greater fire hazard than explosion danger. It has an active oxygen content of 7.8%. When warmed above its melting point to 12–14°C, initially the decomposition takes place slowly, but in the final phase, it goes very rapidly (Strong 1964). Highly flammable decomposition products, such as acetaldehyde, ethane, acetone, and isopropyl alcohol, can cause fire if the heat of reaction is not dissipated. The fire hazard may be enhanced in the presence of solvents that are susceptible to hydrogen abstraction. This may be due to the decreased evolution of CO_2, a major decomposition product, in the presence of such solvents. Alkyl carbonate radicals react with solvent moieties to form carbonate esters.

It is highly reactive because of its peroxide functional group. When mixed with flammable or readily oxidizable compounds, ignition or explosion can occur.

Firefighting should be performed from an explosion-resistant area. Temperature control is essential. Use water to fight flames.

Storage, Handling, and Shipping

Diisopropyl peroxydicarbonate should be stored in a deep freeze below −18°C (0°F), if possible in small amounts. The same storage temperature should be maintained for solutions. Experiments by Strong (1964) suggest that 10% solutions are safe up to 40°C in the case of refrigeration failure. Aluminum, plastic, and earthenware containers are suitable. Shipping should be done in quantities not exceeding 2 gallons in wooden boxes having glass or earthenware inside containers maintained at −18°C (0°F).

Disposal/Destruction

Larger quantities can be disposed of by scattering in a disposal area, where it melts and gradually decomposes. Small quantities are destroyed by adding slowly to 5% alcoholic KOH. Solutions in small amount may be disposed of in porous sandy soils.

Analysis

Analysis techniques are HPLC and pyrolysis followed by estimation of decomposition products by GC. Reduction with excess potassium iodide and titration of liberated iodine with sodium thiosulfate may be suitable for its determination.

43.6   METHYL ETHYL KETONE PEROXIDE

RCRA Hazardous Waste U160; DOT Label: Forbidden (pure compound), DOT Organic Peroxide UN 2550 (<9% active oxygen), UN 2127 (<60% concentration)

Methyl ethyl ketone (MEK) peroxide is a mixture of monomeric and polymeric acyclic as well as cyclic products of peroxidic structure. No single specific structure can be assigned to this compound. The two predominant structures have molecular formula $C_8H_{16}O_4$ and $C_8H_{18}O_6$; CAS [1338-23-4]. Commercial MEK peroxide is a colorless liquid mixture containing 60% peroxide and 40% diluent. The diluents are dimethyl or diethyl phthalate or cyclohexanone peroxide.

Synonyms: 2-butanone peroxide; Lupersol; Thermacure

Uses and Exposure Risk

MEK peroxide it is used to initiate the polymerization of ethylene, styrene, vinyl chloride, and other monomers; and for room temperature curing of polyester resins.

Health Hazard

MEK peroxide is a strong skin and eye irritant with moderate acute and subchronic toxicity. Its toxicity is greater than di-*t*-butyl peroxide and benzoyl peroxide. Inhalation of its vapors can cause injury to lungs with symptoms of gross hemorrhages and hyperemia (Floyd and Stockinger 1958). Exposure to high concentrations may have damaging effects on the liver and kidney. In humans, ingestion of 30–40 g can be toxic, which may cause gastrointestinal pain, nausea, and vomiting.

LC_{50} value, inhalation (mice): 170 ppm/4 hr
LD_{50} value, intraperitoneal (rats): 65 mg/kg
LD_{50} value, oral (mice): 470 mg/kg

Its carcinogenicity is not yet fully established. It is reported (NIOSH 1986) to cause tumor in mice.

Exposure Limit

ACGIH (1986) recommends a ceiling limit of 1.5 mg/m^3 (0.2 ppm). This concentration in air is based on its irritant properties.

Fire and Explosion Hazard

MEK peroxide is highly reactive, oxidizing, and flammable. It is extremely shock sensitive and can explode in pure form. It can decompose explosively when subjected to heat or in contact with easily oxidizable or flammable compounds. The active oxygen concentration is 7.6–9.1% (corresponding to formulas $C_8H_{18}O_6$ and $C_8H_{16}O_4$), but this varies according to the composition, depending on the presence of adducts, polymeric forms, and diluents.

MEK peroxide is highly sensitive to contaminants. Trace amounts of strong acids, bases, metals, metal salts, amines, sulfur compounds, and reducing agents can cause violent decomposition. The violence depends on the amount and type of contaminants, the rate of temperature rise, and the degree of confinement (Woodcock 1983). Precipitation of peroxides is dangerously shock sensitive. Such precipitations can take place if MEK peroxide is contaminated with acetone.

Fire-extinguishing agent: water from a sprinkler system or hose with a fog nozzle from a safe distance; if diluted in a low-density flammable solvent, use foam.

Storage and Shipping

MEK peroxide is diluted in a solvent or dispersed in a plasticizer, which greatly reduces its shock sensitivity. It is stored in a refrigerator in a well-ventilated area and well separated from other chemicals. Peroxide with an active oxygen content >9% may not be shipped. Diluted material is shipped in metal drums with polyethylene liners or polyethylene-lined paper bags in wooden boxes.

Spillage and Disposal/Destruction

See Section 43.11.

Analysis

MEK peroxide can be analyzed by polarographic method using a dropping mercury electrode assembly and a saturated calomel electrode (Floyd and Stockinger 1958).

43.7   Di-*t*-BUTYL PEROXIDE

DOT Label: Organic Peroxide UN 2102
Formula: $C_8H_{18}O_2$; MW 146.22; CAS [110-05-4]

Structure and functional group:

$$H_3C-\underset{\underset{CH_3}{|}}{\overset{\overset{CH_3}{|}}{C}}-O-O-\underset{\underset{CH_3}{|}}{\overset{\overset{CH_3}{|}}{C}}-CH_3$$

two tertiary butyl groups are bound by a peroxide functional group

Synonyms: bis(1,1'-dimethyl ethyl)peroxide, *tert*-butyl peroxide; Trigonox B

Uses and Exposure Risk

Di-*t*-butyl peroxide (DTBP) is used as a polymerization catalyst.

Physical Properties

Colorless liquid; bp 111°C; freezes at −40°C; density 0.794 at 20°C; soluble in most organic solvents, slightly soluble in water to about 0.01%.

Health Hazard

DTBP is slightly toxic by inhalation and in general exhibits low to very low toxicity by other routes. However, toxic effects are observed only at very high concentrations. Rats exposed to 4103-ppm vapor developed head and neck tremor after 10 minutes of exposure (Floyd and Stockinger 1958). Other symptoms were weakness, hyperactivity, and labored breathing. However, the animals recovered fully in 1 hour.

LD$_{50}$ value, intraperitoneal rats): 3210 mg/kg

DTBP is nonirritating to the skin and mild on the eyes. It is reported to cause lung and blood tumors in mice (NIOSH 1986). However, its carcinogenicity is not yet fully established.

Fire and Explosion Hazard

Highly flammable and reactive; flash point 18°C (65°F); vapor pressure 19.5 torr at 20°C; vapor density 5.03. Its decomposition products are ethane and acetone, which enhance the fire hazard. Use a water spray to fight fire and to keep the containers cool.

DTBP forms an explosive mixture with air. The explosive range is not reported. Its decomposition products may explode above its boiling point, 111°C. However, as it is thermally stable and shock insensitive, its explosion hazard is much lower. It may, however, react with explosive violence when in contact with easily oxidizable substances.

Storage and Shipping

Store in a cool and well-ventilated area isolated from easily oxidizable materials. Protect against physical damage. Shipping containers are amber glass and polyethylene bottles or steel drums not exceeding 100-lb capacity.

Spillage

If DTBP spills, absorb the material with a noncombustible absorbent such as vermiculite. Do not use paper, wood, or spark-generating metals for sweeping and handling. Place the absorbed peroxide in a plastic container for disposal.

Disposal/Destruction

DTBP is disposed on the ground in a remote area and ignited with a long torch. 10% NaOH may be used to wash empty containers.

Analysis

DTBP can be analyzed by polarography (see Section 43.6). GC-FID and HPLC may be applicable.

43.8   *tert*-BUTYL PEROXYPIVALATE

DOT Label: Forbidden, pure compound; DOT Organic Peroxide UN 2110, not more than 77% in solution

Formula $C_9H_{18}O_3$; MW 174.27; CAS [927-07-1]

Structure and functional group:

H₃C, O CH₃
H₃C—C—O—O—C—C—CH₃
H₃C CH₃

peroxy linkage binding a *tert*-butyl group to an acyl moiety containing another *tert*-butyl group

Synonyms: *tert*-butyl perpivalate; *tert*-butyl trimethyl peroxyacetate; Esperox 31M; peroxypivalic acid *tert*-butyl ester

Uses and Exposure Risk

t-Butyl peroxypivalate is used as a polymerization initiator to make acrylic polymers.

Physical Properties

Oily liquid; soluble in mineral spirits, benzene, and other organic solvents, insoluble in water; its 75% solution in mineral spirits is a colorless clear liquid that solidifies at $-19°C$ ($-2°F$).

Health Hazard

t-Butyl peroxypivalate is a mild irritant to the eyes and skin with low toxicity.

LD_{50} value, oral (rats): 4300 mg/kg

Fire and Explosion Hazard

The pure compound is dangerously shock sensitive. It is sold commercially up to 75% maximum concentration in mineral spirits. This solution is reactive, oxidizing, and combustible; flash point (open cup) $68°C$ ($155°F$). Its burning is vigorous, which is difficult to extinguish (NFPA 1997). It is sensitive to shock and heat; self-accelerating decomposition temperature $29°C$ ($84°F$); decomposition of the 75% concentrated solution may be explosive when heated. It may form flammable decomposition products, which can enhance the fire hazard. It may ignite and/or explode in contact with accelerators, acids, and combustible and readily oxidizable substances.

Fight fires from a safe and explosion-resistant location. Use water from a sprinkler or a fog nozzle. Since it decomposes little above room temperature, giving off flammable volatile products, care should be taken for proper venting and to keep the containers and vicinity cool.

Storage and Shipping

t-Butyl peroxypivalate is stored in a deep-freeze box with a free-opening cover at -18 to $-1°C$ ($0-30°F$) in a well-ventilated and unheated area, isolated from other chemicals. It is shipped in 5-gallon polyethylene containers with vented caps and outer fiberboard covering, maintained at -18 to $-1°C$.

Spillage and Disposal/Destruction

Absorb the spilled material with vermiculite or other noncombustible material. Then sweep up and place in a plastic container for disposal. It is burned in a shallow pit in a remote area by igniting with a long torch.

Analysis

t-Butyl peroxypivalate may be analyzed by estimating its volatile decomposition products. It may be derivatized carefully and analyzed by colorimetric or HPLC techniques.

43.9   CUMENE HYDROPEROXIDE

EPA Classified Hazardous Waste, RCRA Waste Number U096; DOT Label: Organic Peroxide UN 2116

Formula $C_9H_{12}O_2$; MW 152.21; CAS [80-15-9]

Structure and functional group:

peroxide functional group is bound to isopropyl benzene via the tertiary C atom

Synonyms: α,α-dimethylbenzyl hydroperoxide; α-cumyl hydroperoxide; cumenyl hydroperoxide; isopropylbenzene hydroperoxide

Uses and Exposure Risks

Cumene hydroperoxide is used for the manufacture of acetone and phenols; for studying the mechanism of NADPH-dependent lipid peroxidation; and in organic syntheses.

Physical Properties

Colorless liquid; bp 65°C (at 0.18 torr) density 1.048; soluble in organic solvents, insoluble in water.

Health Hazard

Cumene hydroperoxide is a mild to moderate skin irritant on rabbits. Subcutaneous application exhibited a strong delayed reaction with symptoms of erythema and edema (Floyd and Stockinger 1958). Strong solutions can irritate the eyes severely, affecting the cornea and iris.

Its toxicity is comparable to that of *tert*-butyl hydroperoxide. The toxic routes are ingestion and inhalation. The acute toxicity symptoms in rats and mice were muscle weakness, shivering, and prostration. Oral administration of 400 mg/kg resulted in excessive urinary bleeding in rats.

LD$_{50}$ value, oral (rats): 382 mg/kg

LD$_{50}$ value, intraperitoneal (rats): 95 mg/kg

Although cumene hydroperoxide is toxic, its pretreatment may be effective against the toxicity of hydrogen peroxide. In humans, its toxicity is low.

Cumene hydroperoxide is mutagenic and tumorigenic (NIOSH 1986). It may cause tumors at the site of application. In mice, skin and blood tumors have been observed. Its cancer-causing effects on humans are not known.

Exposure Limit

No exposure limit is set. On the basis of its irritant properties, a ceiling limit of 2 mg/m^3 (0.3 ppm) is recommended.

Fire and Explosion Hazard

Flammable; highly reactive and oxidizing. Flash point 79°C (175°F); vapor density 5.2 (air = 1); autoignition temperature not reported; self-accelerating decomposition temperature 93°C (200°F).

When exposed to heat or flame, it may ignite and/or explode. A 91–95% concentration of cumene hydroperoxide decomposes violently at 150°C (NFPA 1986). Duswalt and Hood (1990) reported violent decomposition when this compound mixed accidentally with a 2-propanol solution of sodium iodide.

It forms an explosive mixture with air. The explosive concentration range is not reported. Hazardous when mixed with easily oxidizable compounds. Fire-extinguishing agent: water from a sprinkler or fog nozzle from an explosion-resistant location.

Storage and Shipping

Cumene hydroperoxide is stored in a cool, dry and well-ventilated area isolated from other chemicals. It should be protected against physical damage. It may be shipped in wooden boxes with inside glass or earthenware containers or in 55-gallon metal drums.

Spillage and Disposal/Destruction

See Section 43.7.

Analysis

Polarography (see Section 43.6). HPLC and GC-FID may be applicable.

43.10 *tert*-BUTYL PEROXYBENZOATE

DOT Label: Organic Peroxide UN 2097 for technical pure or concentrated >75% UN 2098 for concentrated <75% in solution; UN 2890 for <50% with inorganic solid

Formula: $C_{11}H_{14}O_3$; MW 194.2; CAS [614-45-9]

Structure and functional group:

peroxy ester, benzoyl, and *tert*-butyl groups are bound by peroxy linkage

Synonyms: *tert*-butyl perbenzoate; peroxybenzoic acid *tert*-butyl ester; Novox; Esperox 10

Uses and Exposure Risk

t-Butyl peroxybenzoate is used for elevated-temperature curing of polyesters and to initiate polymerization reactions.

Physical Properties

A light yellow liquid with a mild aromatic smell; bp 112°C (explodes); freezes at 9°C; density 1.023 at 20°C; soluble in organic solvents, insoluble in water.

Health Hazard

t-Butyl peroxybenzoate is a mild skin and eye irritant. Exposure to 500 mg/day caused mild irritation in rabbit eyes and skin. Toxicity data on animals show a low order of toxicity.

LD_{50} value, oral (mice): 914 mg/kg

It has been reported to cause tumors (blood) in mice. Its carcinogenic actions on humans are unknown.

Fire and Explosion Hazard

Highly reactive and oxidizing compound with moderate flammability; flash point (open cup) 107–110°C (225–230°F); autoignition temperature not reported.

It forms an explosive mixture with air; the explosive range not reported. It is not sensitive to shock but is heat sensitive. It explodes on heating; self-accelerating decomposition temperature 60°C (140°F).

It may react explosively when mixed with readily oxidizable, organic, and flammable substances. Fire-extinguishing agent: water from a sprinkler; use water to keep the containers cool.

Storage and Shipping

It should be stored in a well-ventilated place at a temperature between 10 and 27°C (50–80°F), isolated from oxidizable, flammable, organic materials and accelerators. It should be shipped in glass, polyethylene, and earthenware containers of up to 5-gallon capacity placed inside wooden or fiberboard boxes.

Spillage and Disposal/Destruction

See Section 43.8.

Analysis

Polarography, HPLC, and various reduction techniques.

43.11 BENZOYL PEROXIDE

DOT Label: Organic Peroxide UN 2085, 2086, 2087, 2088, 2089

Formula: $(C_6H_5CO)_2 O_2$; MW 242.22; CAS [94-36-0]

Structure and functional group:

two benzoyl carbonyl groups are bound by peroxy linkage

Synonyms: dibenzoyl peroxide; benzoyl; acetoxyl; Benzac; Oxylite; Superox; Lucidol

Uses and Exposure Risk

Benzoyl peroxide is used as a source of free radicals in many organic syntheses and to initiate polymerizations of styrene, vinyl chloride, vinyl acetate, and acrylics; to cure thermoset polyester resins and silicone rubbers; in medicine for treating acne; and for bleaching vegetable oil, cheese, flour, and fats.

Physical Properties

White granular powder, or crystals with faint odor; mp 103–106°C, decomposes explosively; density 1.334 at 25°C; sparingly soluble in water and alcohol; dissolves readily in benzene, chloroform, and ether.

Health Hazard

The health hazard from benzoyl peroxide is low. It can cause irritation of the skin, mucous membranes, and eyes. An intraperitoneal injection of 250 mg/kg was lethal to adult mice. Systemic toxicity in humans is not known. It may be mild to moderately toxic on an acute basis. The oral LD_{50} value in rats is 7710 mg/kg (NIOSH 1986). Its toxicity from inhalation is low; an LC_{50} value of 700 ppm in mice is suggested (ACGIH 1986).

Benzoyl peroxide may cause gene damage and DNA inhibition. It has been found to cause skin tumor. The evidence of its carcinogenicity in animals and humans is inadequate.

Exposure Limits

TLV-TWA 5 mg/m^3; IDLH 7000 mg/m^3.

Fire and Explosion Hazard

Benzoyl peroxide can cause a major fire and explosion hazard. It is highly flammable and a strong oxidizer; autoignition temperature 80°C (176°F). It ignites instantly. The rate and violence of decomposition and the potential ease of such ignition or decomposition have been experimentally measured by Noller et al. (1964). Lead pipe deformation (LPD), pressure vessel test (PVT), and self-accelerating decomposition test (SADT) have been performed to measure these explosive characteristics. Heating 5 g of benzoyl peroxide in an aluminum tester containing an aperture vent and 6-atm rupture disk, caused the disk to blow up in 95 seconds when the aperture vent area was less than 174.7 mm^2. Redried material was more violent. The decomposition hazard was greatly reduced with wet and diluted benzoyl peroxide.

Noller et al. (1964) measured the SADT temperature at 82.2°C (180°F), above which the decomposition was self-accelerating, sudden, and produced smoke.

Benzoyl peroxide is a deflagrant, posing a severe explosion hazard. The compound is sensitive to heavy shock, such as impact or blows, as well as to friction and heat. Especially in the dry state, it is highly dangerous.

A water sprinkler should be used to extinguish fires. Water should be used to keep the containers cool.

Hazardous Reactions

Benzoyl peroxide reacts exothermically with strong acids, strong bases, amines, reducing agents, and sulfur compounds. Explosions have been reported when it reacted with carbon tetrachloride and ethylene (Bolt and Joyce 1947), lithium aluminum hydride (Sutton 1951), N,N-dimethyl aniline (Horner and Betzel 1953), hot chloroform (NFPA

1986), and methyl methacrylate (NFPA 1986). Lappin (1948) reported an explosion when a bottle was opened. Organic matter entrapped in the threads of the bottle probably reacted explodingly with benzoyl peroxide.

Storage and Shipping

Benzoyl peroxide should be stored in a cool and well-ventilated area, isolated from other chemicals and free of heating and electrical installations. Dry compound may be shipped in polyethylene-lined paper bags or fiber containers packed in wooden boxes or fiberboards. Metal drums or barrels with polyethylene liners are recommended.

Spillage

In the event of a spill, mix the spilled material with water-wetted vermiculite or sand or any other inert moist diluent and place it in a plastic container for immediate disposal. Do not use paper, wood, or cellulose material or spark-generating tools.

Disposal/Destruction

Benzoyl peroxide is diluted in water and disposed of in a large flowing stream of water. Another common method of disposal is to spread the material over a large area on the downwind side of a pit. The waste solvent is fed to the pit and ignited carefully from a remote position (Armitage and Strauss 1964). The solvent will burn and decompose the dispersed peroxide.

Waste containing benzoyl peroxide can be destroyed by mixing it slowly with 10 times its weight of 10% caustic soda solution. The solution is stirred at room temperature vigorously for 3 hours (Woodcock 1983) and flushed down the drain.

Analysis

Benzoyl peroxide can be analyzed by HPLC, polarography, and chemical reduction methods. In the latter method, excess sodium iodide in acetic acid is treated with benzoyl peroxide. The iodine liberated is titrated with standard sodium thiosulfate solution. Air analysis may be performed by NIOSH Method 5000 (NIOSH 1984). Forty to 400 L of air is passed over a 0.8-mm cellulose ester membrane at a rate of 1 to 3 L/min. Adsorbed benzoyl peroxide is extracted with ethyl ether and analyzed by HPLC using UV detection at 254 nm. A stainless steel pressure column, Spherisorb ODS or equivalent, may be used.

43.12 LAUROYL PEROXIDE

DOT Label: Organic Peroxide UN 2124 (pure compound), and UN 2893, (<42% solution)

Formula: $C_{24}H_{46}O_4$; MW 398.70; CAS [105-74-8]

Structure and functional group:

$$CH_3-(CH_2)_{10}-\overset{\overset{\displaystyle O}{\|}}{C}-O-O-\overset{\overset{\displaystyle O}{\|}}{C}-(CH_2)_{10}-CH_3$$

peroxide linkage between two lauroyl groups

Synonyms: dodecanoyl peroxide; dilauroyl peroxide; dilauryl peroxide; bis(1-oxododecyl)peroxide; Laurox, Laurydol; Alperox C

Uses and Exposure Risk

Lauroyl peroxide is used as an initiator for free-radical polymerization in making polyvinyl chloride. Lauroyl peroxide constitutes about 4% of all organic peroxides consumption in the United States.

Physical Properties

White powder or flakes with a faint odor; tasteless; mp 53–55°C; density 0.91; insoluble in water, slightly soluble in alcohols, mixes readily with most other organic solvents.

TABLE 43.1 Toxicity and Flammability of Miscellaneous Organic Peroxides

Compound/Synonyms/ CAS No.	Formula/MW/Structure	Toxicity	Fire and Explosion Hazard
Methyl hydroperoxide (methyl hydrogen peroxide) [3031-73-0]	CH_4O_2 48.08 $H_3C{-}O{-}O{-}H$	Toxicity not reported	Pure liquid is extremely shock and heat sensitive; explodes violently with heating or jarring; DOT Forbidden for shipping
Ethyl hydroperoxide (ethyl hydrogen peroxide) [3031-74-1]	$C_2H_6O_2$ 62.08 $CH_3{-}CH_2{-}O{-}O{-}H$	Nontoxic	Extremely shock and heat sensitive
Hydroperoxymethanol (hydroxymethyl hydroperoxide) [15932-89-5]	CH_4O_3 64.08 $HO{-}CH_2{-}O{-}O{-}H$ with an OH group	Toxicity not reported	The oily liquid in the pure state decomposes violently on heating and/or in contact with metals
Hydroperoxyethanol	$C_2H_7O_3$ 79.08 $HO{-}CH_2{-}CH_2{-}O{-}O{-}H$ with OH OH groups	Toxicity not reported	Decomposes violently on heating and/or in contact with metals
Acetyl acetone peroxide (2,4-pentanedione peroxide) [37187-22-7]	A C_{10}-ketone peroxide with no single definite structure, coexists with isomers	Toxicity not reported	Pure solid violently decomposes; sensitive to heat, shock, and contaminants (transition metals, etc.); DOT Forbidden for shipping, when active oxygen content >9%; DOT Label: Organic Peroxide, UN 2080
1,1-Peroxydicyclo-hexanol [bis-(1-hydroxycyclohexyl) peroxide; 1,1-di-oxydicyclohexanol] [2407-94-5]	$C_{12}H_{22}O_4$ 230.34	Nontoxic	Pure solid explodes on heating; can ignite and/or explode in contact with reducing agents

(continued)

TABLE 43.1 (Continued)

Compound/Synonyms/ CAS No.	Formula/MW/Structure	Toxicity	Fire and Explosion Hazard
Methyldioxymethanol [40116-50-5]	$C_2H_6O_3$ 78.067 $HO-CH_2-O-O-CH_3$	Toxicity not reported	Liquid in the pure state is unstable and explosive
Diisobutyryl peroxide [isobutyroyl peroxide; bis(2-methyl-1-oxopropyl)peroxide] [3437-84-1]	$C_8H_{14}O_4$ 174.22 (structure)	Nontoxic	Pure solid is shock and heat sensitive, explodes; DOT maximum concentration for shipping 52%; DOT Label: Organic Peroxide, UN 2182
Dipropionyl peroxide [propionyl peroxide, bis(1-oxopropyl)peroxide] [3248-28-0]	$C_6H_{10}O_4$ 146.16 (structure)	Low to moderately toxic via inhalation, 100 ppm was lethal to rats	Oily liquid, highly shock and heat sensitive, decomposes violently in contact with readily oxidizable organics and flammable substances; concentration >28% in solution is forbidden by DOT for shipping, UN 2132 (<28% in solution)
Acetyl propionyl peroxide [13043-82-8]	$C_5H_8O_4$ 132.11 (structure)	Toxicity not reported	Liquid in the pure state is heat and shock sensitive, may decompose violently when mixed with reducing agents and organics
Acetyl benzoyl peroxide [644-31-5]	$C_9H_8O_4$ 180.17 (structure)	Nontoxic	Solid with a mp 37–39°C is highly sensitive to shock and heat; explodes; decomposes violently with readily oxidizable and organic contaminants; the pure compound or its solution above 40% strength is DOT Forbidden; DOT Label: Organic Peroxide, UN 2081 (<40% solution)
Octanoyl peroxide [caprolyl peroxide; capryl peroxide; dioctanoyl peroxide; bis(1-oxo octyl)peroxide] [762-16-3]	$C_{16}H_{30}O_4$ 286.46 $CH_3-(CH_2)_6-C(=O)-O-O-C(=O)-(CH_2)_6-CH_3$	Nontoxic	Less hazardous than the acyl peroxides of lower carbon numbers; ignites slowly and burns vigorously

Name	Formula / MW	Structure	Toxicity	Hazards
Dicumyl peroxide (cumyl peroxide; cumene peroxide, diisopropylbenzene peroxide; bis(α,α-dimethylbenzyl)peroxide [80-43-3]	$C_{18}H_{22}O_2$ 270.40		Low toxicity; LD_{50} oral (rats): 4100 mg/kg	Less hazardous than lower alkyl peroxides; insensitive to shock, ignites slowly, burns vigorously
Bis(p-chlorobenzoyl) peroxide[p,p'-dichlorobenzoyl peroxide; di-(4-chlorobenzoyl)peroxide] [94-17-7]	$C_{14}H_8Cl_2O_4$ 311.12		Low to moderate toxicity; an intraperitoneal dose of 500 mg/kg was lethal to mice	Pure compound presents severe decomposition hazard; aqueous solution above 75% is dangerous for handling and shipping
Bis(2,4-dichlorobenzoyl)peroxide (dichlorobenzoyl peroxide) [133-14-2]	$C_{14}H_6Cl_4O_4$ 380.00		Low toxicity; data not reported	Pure compound decomposes violently
tert-Butyl peroxycarbamate [18389-96-3]	$C_5H_{11}O_3N$ 133.15		Toxicity not reported	Heat sensitive, decomposes violently above 80°C

(continued)

TABLE 43.1 *(Continued)*

Compound/Synonyms/CAS No.	Formula/MW/Structure	Toxicity	Fire and Explosion Hazard
tert-Butyl-*p*-chloroperoxybenzene sulfonate [77482-48-5]	$C_{10}H_{13}O_4SCl$ 264.70	Toxicity not reported	Heat sensitive, explodes at 30–35°C (melting point)
tert-Butyl-*p*-methoxyperoxybenzene sulfonate [77482-50-9]	$C_{11}H_{16}O_5S$ 250.29	Toxicity not reported	Heat sensitive, explodes at its mp (47°C)
Diethyl peroxydicarbonate (peroxydicarbonic acid diethyl ester) [14666-78-5]	$C_6H_{10}O_6$ 178.16	Nontoxic	Dangerous; highly shock and heat sensitive, highly reactive; explosion hazard; shipping of >27% solution forbidden by DOT; DOT Label: Organic Peroxide, UN 2175 (for solutions not exceeding 27% concentration)
Di-*n*-propyl peroxydicarbonate (*n*-propyl percarbonate, peroxydicarbonic acid dipropyl ester) [16066-38-9]	$C_8H_{14}O_6$ 206.22	Skin irritant; low toxicity; LD50 oral (rats): 3400 mg/kg	Can decompose violently on heating or impact; highly reactive, can react violently with organics and reducing agents

Name (CAS)	Formula / MW	Structure	Toxicity	Hazards
tert-Butyl peroxyisobutyrate (peroxyisobutyric acid *tert*-butyl ester) [109-13-7]	$C_8H_{16}O_3$ 160.24 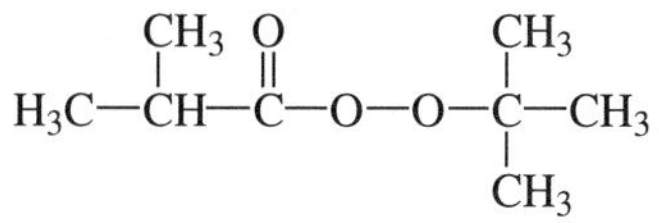	$H_3C-CH-C-O-O-C-CH_3$ with CH_3 and O (double bond) and CH_3, CH_3 groups	Nontoxic	Highly reactive; sensitive to heat and shock; reacts violently with organics and reducing agents; shipping of >77% solution forbidden by DOT; DOT Label: UN 2562 (<52% solution), UN 2142 (52–77% solution)
Di-*n*-butyl peroxydicarbonate (butyl peroxydicarbonate; peroxydicarbonic acid dibutyl ester) [16215-49-9]	$C_{10}H_{18}O_6$ 234.28 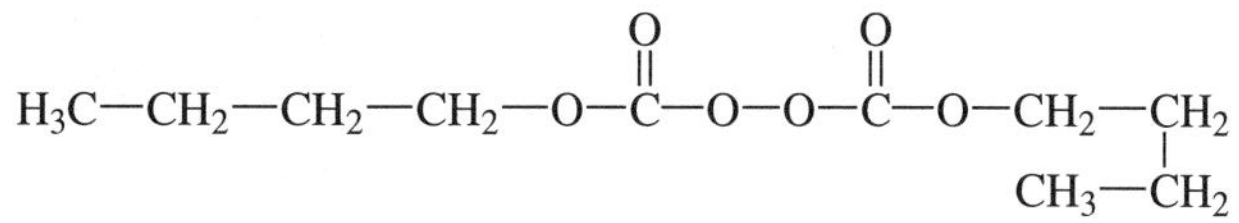	$H_3C-CH_2-CH_2-CH_2-O-C-O-O-C-O-CH_2-CH_2$ with O (double bonds) and CH_3-CH_2	Nontoxic	Pure compound highly sensitive to shock, heat, and contaminants; decomposes explosively; shipping of >52% solutions. DOT prohibited; DOT Label: Organic Peroxide, UN 2169 (27–52% solution), UN 2170 (<27% solution)
Di-*sec*-butyl peroxydicarbonate (*sec*-butyl peroxydicarbonate, peroxydicarbonic acid di-*sec*-butyl ester) [19910-65-7]	$C_{10}H_{18}O_6$ 234.28 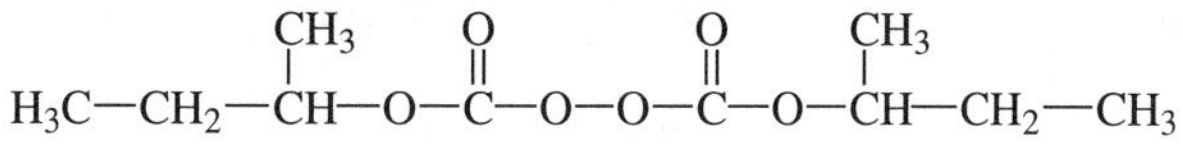	$H_3C-CH_2-CH-O-C-O-O-C-O-CH-CH_2-CH_3$ with CH_3 and O (double bonds) and CH_3	Low toxicity; LD$_{50}$ skin (rabbits): 1200 mg/kg	Pure compound highly sensitive to shock, heat, and contaminants

(continued)

689

TABLE 43.1 (Continued)

Compound/Synonyms/CAS No.	Formula/MW/Structure	Toxicity	Fire and Explosion Hazard
Dibenzyl peroxydicarbonate (peroxydicarbonic acid dibenzyl ester) [2144-45-8]	$C_{16}H_{14}O_6$ 302.30	Nontoxic	Pure compound decomposes explosively; sensitive to heat, shock, and contaminants; more than 87% aqueous solution DOT forbidden; DOT Label: Organic Peroxide, UN 2149 (<87% solution)
Dicyclohexyl peroxydicarbonate (dicyclohexyl peroxide carbonate, peroxydicarbonic acid dicyclohexyl ester) [1561-49-5]	$C_{14}H_{22}O_6$ 286.36	Nontoxic	Can decompose violently; sensitive to heat and contaminants; DOT forbidden for shipping when pure or above 91% solution in water, DOT Label: Organic Peroxide, UN 2152 (<91% solution)

Health Hazard

Lauroyl peroxide is a mild eye irritant; the irritation from 500 mg/day in rabbits' eyes was mild. It is nontoxic. Prolonged exposure to laboratory animals caused tumors at the site of application. However, the evidence of carcinogenicity in animals is inadequate to date.

Fire and Explosion Hazard

Lauroyl peroxide presents a much smaller hazard than do most other diacyl peroxides. It is combustible in a dry state and shock sensitive only at elevated temperatures. Its self-accelerating decomposition temperature is 49°C (120°F).

It may ignite and explode when mixed with chemical accelerators, readily oxidizable, flammable, and organic compounds. Fire-extinguishing agent: use a water spray from a safe location.

Storage and Shipping

Store in a well-ventilated, cool area, isolated from other chemicals. It is shipped in fiber drums not exceeding 100 lb. Small amounts may be shipped in 1-lb fiberboard boxes.

Spillage and Disposal/Destruction

Although lauroyl peroxide is relatively less hazardous, it is recommended that to handle spills and disposal, the same safety measures be followed as those for other hazardous organic peroxides.

Analysis

Iodide or other reduction method, polarography, GC-FID, or HPLC.

43.13 MISCELLANEOUS ORGANIC PEROXIDES

See Table 43.1.

REFERENCES

ACGIH. 1986. *Documentation of the Threshold Limit Values and Biological Exposure Indices*, 5th ed. Cincinnati, OH: American Conference of Governmental Industrial Hygienists.

Armitage, J. B., and H. W. Strauss. 1964. Safety considerations in industrial use of organic peroxides. *Ind. Eng. Chem. 56* (12): 28–32.

Bolt, R. O., and R. M. Joyce. 1947. *Chem. Eng. News 25*: 1866.

Duswalt, A. A., and H. E. Hood. 1990. Potential hydroperoxide hazard. *Chem. Eng. News 68* (6): 2.

Floyd, E. P., and H. E. Stockinger. 1958. Toxicity studies of certain organic peroxides and hydroperoxides. *Ind. Hyg. J.*, June, pp. 205–212.

Horner, L., and C. Betzel. 1953. *Chem. Ber. 86*: 1071–1072.

Kuhn, L. P. 1948. *Chem. Eng. News 26*: 3197.

Lappin, G. R. 1948. *Chem. Eng. News 26*: 3518.

Meidl, J. H. 1970. *Explosive and Toxic Hazardous Materials*. Beverly Hills, CA: Glencoe Press.

NFPA. 1997. *Fire Protection Guide on Hazardous Materials*, 12th ed. Quincy, MA: National Fire Protection Association.

NIOSH. 1984. *Manual of Analytical Methods*, 3rd ed. Cincinnati, OH: National Institute for Occupational Safety and Health.

NIOSH. 1986. *Registry of Toxic Effects of Chemical Substances*, ed. D. V. Sweet. Washington, DC: US Government Printing Office.

Noller, D. C., S. J. Mazurowski, G. F. Linden, F. J. G. De Leeuw, and O. L. Magelli. 1964. A relative hazard classification of organic peroxides. *Ind. Eng. Chem. 56* (12): 18–27.

Shanley, E. S. 1949. *Chem. Eng. News 27*: 175.

Sheppard, C. S., and O. L. Mageli. 1982. Peroxides and peroxy compounds, organic. In *Kirk-Othmer Encyclopedia of Chemical Technology*, 3rd ed., pp. 27–90. New York: Wiley.

Strong, W. A. 1964. Organic peroxides: diisopropyl peroxydicarbonate. *Ind. Eng. Chem.* *56*(12): 33–38.

Sutton, D. A. 1951. *Chem. Ind. 1951*: 272.

Swern, D. 1986. Peroxides. In *Comprehensive Organic Chemistry: The Synthesis and Reactions of Organic Compounds*, ed. J. F. Stoddart, pp. 909–939. New York: Pergamon Press.

Woodcock, R. C. 1983. Peroxides, Organic. In *Encyclopedia of Health and Occupational Safety*, 3rd ed. ed. L. Parmeggiani. Vol. 2, pp. 1611–1613. Geneva: International Labor Office.

44

PESTICIDES AND HERBICIDES: CLASSIFICATION, STRUCTURE, AND ANALYSIS

44.1 GENERAL

Pesticides are chemical agents used for pest control, that is, substances used by humans to kill or control undesired organisms. The term excludes pathogenic organisms such as bacteria, viruses, and protozoa. Pesticides may be classified according to their functions. Such a classification includes some common terms, such as the following:

Insecticide: kills insects

Herbicide: kills unwanted vegetation (weed control)

Rodenticide: kills rodents

Miticide: kills mites

Avicide: kills birds

Piscicide: controls fish

Fungicide: kills fungi

Slimicide: controls slime

Algicide: kills algae

Molluscicide: kills snails, clams, and oysters

Ovicide: destroys eggs

Repellant: repels insects

Defoiliant: causes leaves to fall

More than 1500 chemicals are used as active ingredients in pesticides. These compounds make up several thousands of formulations or products that are marketed all over the world.

Although pesticides are applied to kill specific target organisms, they are not highly selective, often causing adverse effects on nontarget species. Since all pesticides are generally toxic to a certain degree on species other than the target organisms, their residues in the environment can contaminate crops, foods, air, and groundwater and produce ill effects on humans and animals.

44.2 STRUCTURAL CLASSIFICATION

Pesticides may be classified according to their chemical structures. Such classification may fall under the following structural patterns:

1. *Organochlorine insecticides*, such as DDT, aldrin, lindane, or chlordane — The term usually refers to a broad range of chlorine-containing organics used for pest control. No single structural feature or specific functional groups(s) may be assigned to pesticides of this class.

2. *Organophosphates*, such as parathion, phorate, mevinphos, or demeton — The organophosphates have the following general structure:

$$\begin{array}{c} RO \\ \backslash \\ RO \diagup \end{array} \overset{\displaystyle \overset{O}{\parallel}}{P} - O \text{ (or S)} - \text{(leaving group)}$$

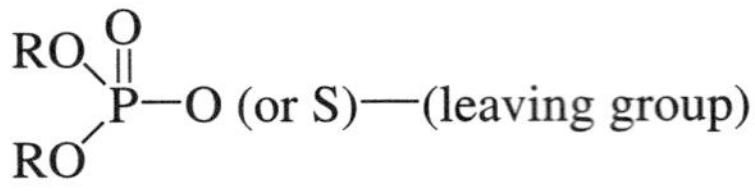

where R is most often a methyl or ethyl group attached to the phosphorus atom via ethereal linkage. Phosphorus is pentavalent. The other three bonds constitute a double bond with an oxygen or sulfur atom and a single bond attachment. An oxygen or sulfur atom bound to a leaving group.

3. *Carbamates*, such as aldicarb, carbaryl, methomyl, or terbucarb — Carbamates have the following general structure:

$$\begin{array}{c} R \\ \backslash \\ R \diagup \end{array} N - \overset{\displaystyle \overset{O}{\parallel}}{C} - O - \text{(leaving group)}$$

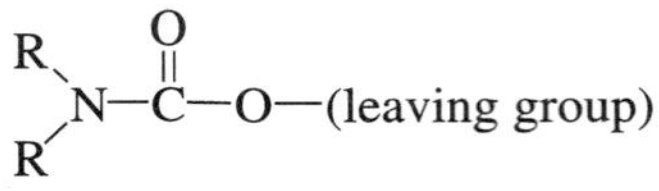

where R is a lower alkyl or aryl group or hydrogen atom. Pesticides that contain a methyl group attached to the N atom are also known as *N*-methylcarbamates. Sulfur substitution of one or both the oxygen atoms in the above structure produce similar type compounds, known as thiocarbamates.

4. *Urea-type pesticides*, such as, fenuron, siduron, monuron, or linuron — The structure is similar to carbamate except that the terminal oxygen atom is substituted with a nitrogen atom, as follows:

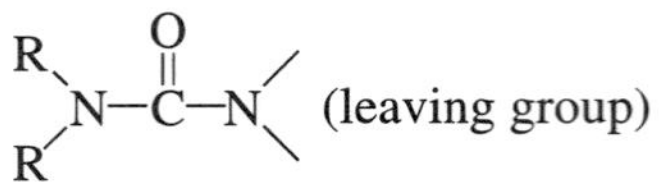

$$\begin{array}{c} R \\ \backslash \\ R \diagup \end{array} N - \overset{\displaystyle \overset{O}{\parallel}}{C} - N \begin{array}{c} \diagup \\ \backslash \end{array} \text{(leaving group)}$$

5. *Triazines and triazole herbicides*, such as, atrazine, simazine, metribuzin, or amitrole — These substances contain a heterocyclic triazine ring.

6. *Chlorophenoxy acid herbicides*, such as, 2,4-D, silvex, or 2,4,5-T — In these compounds, the alkyl group of a lower carboxylic acid usually acetic or propionic acid is attached to a chlorophenoxy moeity as follows:

Cl substitution in the aromatic ring

7. *Bipyridyl herbicides*, such as paraquat or diquat dibromide — These compounds, relatively very few with practical application, have bipyridyl structure containing two quaternary N atoms.

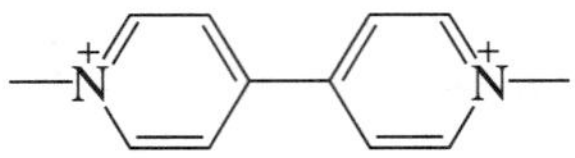

In addition to those listed above, many other classes of compounds are also used in several pesticide formulations: dithiocarbamates, chlorophenols, nitrophenols, and various phthalimides. While the former three classes of substances — organochlorine pesticides, organophosphates, and carbamates — are among the best known pesticides (e.g., insecticides, rodenticides), triazines, chlorophenoxy acids, and bipyridyls are used in making herbicides. There are also many pesticides that do not fall under any specific class of structures. These are discussed separately.

44.3 ANALYSIS

Methods of analyses of pesticides and herbicides at trace concentrations in environmental matrices of potable and nonpotable waters, soils, sediments, solid and hazardous wastes, and ambient air have been described in US EPA, NIOSH, ASTM, and WEF literature. The subject is fully reviewed (Patnaik 1997).

Organochlorine pesticides may be determined by GC and GC/MS techniques. ECD and HECD are suitable GC detectors for such analyses. Samples are extracted with hexane and the extracts injected into a GC equipped with an ECD. Alternatively, the solvent extracts (either in hexane or methylene chloride) may be analyzed by GC/MS (a packed or a capillary column) employing electron impact or chemical ionization. Analysis of these substances in potable and wastewaters and solid wastes may be performed by EPA Methods 505, 508, 515, 608, 625, 8080, 8250, and 8270. Of these, Methods 625, 8250, and 8270 are based on GC/MS (US EPA 1984, 1986).

Organophosphorus insecticides may be analyzed by GC using a detector, which may be either an NPD or an FPD in the phosphorus mode. If the phosphate contains halogen, HECD may be used for the GC analysis. Qualitative identifications may be performed by GC/MS. Chemical ionization is preferred over electron-impact ionization for mass spectroscopic identification. An electron impact process causes extensive fragmentation. Organophosphate pesticides in soils, solid wastes, and ground waters may be analyzed by GC and GC/MS by US EPA Methods 8141 and 8270, respectively (U.S. EPA 1986). Table 44.1 presents the characteristic primary and secondary ions for identification of pesticides by GC/MS techniques using electron-impact ionization at 70 eV. Compounds producing same characteristic ions may be identified from their retention times.

Carbamates, urea, and triazine type compounds can be analyzed on an HPLC. For carbamates postcolumn derivatization technique may be applied. Carbamates separated on a C−18 column are hydrolyzed with NaOH and the products amines are then derivatized with σ-phthalaldehyde and 2-mercaptoethanol to form highly fluorescent derivatives. These carbamate derivatives are then detected by a fluorescence detector. Urea pesticides, structurally similar to carbamates may be determined by an LC/MS technique, using a reverse-phase HPLC column interfaced to a mass spectrometer with a particle beam interface (US EPA Method 553). Aqueous samples may be extracted by liquid–liquid extraction using methylene chloride or by solid-phase extraction. Triazine herbicides may be analyzed by all major instrumental techniques, namely, GC, GC/MS, and HPLC. The solvent extracts may be analyzed by GC on NPD (in N-mode) or FID detector or by HPLC using an ultraviolet (UV) detector.

Chlorophenoxy acid herbicides, such as acids, or as their esters or salts, are first converted into acids, which are then extracted with solvent ether. The solvent extract is treated with a methylating agent, then converted to their methyl esters. The acids are converted into their methyl esters. The esters are extracted with toluene or benzene and determined by a GC, using an ECD, or by a GC/MS.

Pesticides in air may be analyzed by collecting these compounds over various filters, followed by desorption with suitable solvents and analysis by GC, HPLC, or colorimetric techniques. NIOSH Methods for the determination of some of the pesticides are highlighted below.

Aldrin and lindane in air are collected over a filter bubbler, dissolved in isooctane, and injected into a GC-HECD equipped with a column containing 5% SE-30 on acid-washed DMCC Chromosorb W or equivalent (NIOSH 1984, Method 5502). The recommended airflow is 200–1000 mL/240L. Endrin is collected over an 0.8-μm cellulose ester membrane and Chromosorb 102; extracted with toluene and analyzed by GC-ECD (NIOSH 1984, Method 5519). A column suitable for the purpose is 3% OV-1 on 100/120-mesh Chromosorb Q.

Mevinphos, TEPP, and ronnel are trapped over a Chromosorb 102 filter, extracted with

TABLE 44.1 Characteristic Ions for GC/MS Identification of Selected Pesticides

Pesticide/CAS No.	Primary Ions	Secondary Ions
Aldrin [309-00-2]	66	263, 220
Azinphos-methyl [86-50-0]	160	132, 93, 104, 105
α-BHC [319-84-6]	183	181, 109
β-BHC [319-85-7]	181	183, 109
δ-BHC [319-86-8]	183	181, 109
γ-BHC (Lindane) [58-89-9]	183	181, 109
Carbaryl [63-25-2]	144	115, 116, 201
Chlordane [57-74-9]	373	375, 377
Chlorfenvinphos [470-90-6]	267	269, 323, 295, 325
Coumaphos [56-72-4]	362	226, 210, 364, 97
Crotoxyphos [7700-17-6]	127	105, 193, 166
4,4'-DDD [72-54-8]	235	237, 165
4,4'-DDE [72-55-9]	246	246, 176
4,4'-DDT [50-29-3]	235	237, 165
Demeton-O [298-03-3]	88	89, 60, 61, 115
Demeton-S [126-75-0]	88	60, 81, 89, 114
Dichlorvos [62-73-7]	109	185, 79, 145
Dicrotophos [141-66-2]	127	67, 72, 109, 193
Dieldrin [60-57-1]	79	263, 279
Dimethoate [60-51-5]	87	93, 125, 143, 229
Dinocap [39300-45-3]	69	41, 39
Disulfoton [298-04-4]	88	97, 89, 142, 186
Endosulfan I [959-98-8]	195	339, 341
Endosulfan II [33212-65-9]	337	339, 341
Endosulfan sulfate [1031-07-8]	272	387, 422
Endrin [72-20-8]	263	82, 81
Endrin aldehyde [7421-93-4]	67	345, 250
Endrin ketone	317	67, 319
EPN [2104-64-5]	157	169, 185, 141, 323
Ethion [593-12-2]	231	97, 153, 121, 125
Ethyl carbamate [51-79-6]	62	44, 45, 74
Fensulfothion [115-90-2]	293	97, 308, 125, 292
Fenthion [55-38-9]	278	125, 127, 93, 158
Heptachlor [76-44-8]	100	272, 274
Heptachlor epoxide [1024-57-3]	353	355, 351
Kepone [143-50-0]	272	274, 237, 178, 143
Leptophos [21609-90-5]	171	377, 375, 77, 155
Malathion [121-75-5]	173	125, 127, 93, 158
Methoxychlor [72-43-5]	227	228, 152, 114, 274
Methyl parathion [298-00-0]	109	125, 263, 79, 93
Mevinphos [7786-34-7]	127	192, 109, 67
Mexacarbate [315-18-4]	165	150, 134, 164, 222
Mirex [2385-85-5]	272	237, 274, 270, 239
Monocrotophos [6923-22-4]	127	192, 67, 97, 109
Naled [300-76-5]	109	145, 147, 301, 79

TABLE 44.1 (*Continued*)

Pesticide/CAS No.	Primary Ions	Secondary Ions
Parathion [56-38-2]	109	97, 291, 139, 155
Phorate [298-02-2]	75	121, 97, 93, 260
Phosalone [2310-17-0]	182	184, 367, 121, 379
Phosmet [732-11-6]	160	77, 93, 317, 76
Phosphamidon [13171-21-6]	127	264, 72, 109, 138
TEPP [107-49-3]	99	155, 127, 81, 109
Terbufos [13071-79-9]	231	57, 97, 153, 103
Toxaphene [8001-35-2]	159	231, 233

toluene, and injected into Super-Pak 20M or equivalent GC column (NIOSH 1984, Methods 2503 and 2504). An FPD is used for such analysis. A glass fiber filter is used for collecting EPN, malathion, and parathion. Analysis is performed by GC-FPD using a 3% OV-1 or 1.5% OV-17 + 1.95% OV-210 on Gas Chrom Q support (NIOSH 1984, Method 1450). Demeton is analyzed in a similar manner. The sampling system consists of sorbent tubes containing a 2-μm mixed cellulose ester and XAD-2 (NIOSH 1984, Method 5514) Rotenone [83-79-4] is collected over a 1-μm PTFE membrane and analyzed by an HPLC-UV detector using a μ-Bondapak C column (NIOSH 1984, Method 5007). Carbaryl in air may be determined by visible absorption spectrophotometry. The pesticide forms a colored complex with

p-nitrobenzenediazonium fluoroborate, the absorbance of which is measured at 475 nm (NIOSH 1984, Method 5006).

REFERENCES

NIOSH. 1984. *Manual of Analytical Methods.* 3rd ed. Cincinnati, OH: National Institute for Occupational Safety and Health.

Patnaik. P. 1997. *Handbook of Environmental Analysis.* Boca Raton, FL: CRC Press.

U.S. EPA. 1986. *Test Methods for Evaluating Solid Waste*, 3rd ed. Vol. 1B. Washington, DC: Office of Solid Waste and Emergency Response.

U.S. EPA. 1990. Guideline establishing test procedures for the analysis of pollutants under the Clean Water Act, *Code of Federal Regulations*, Title 40, Part 136.

45

PESTICIDES, CARBAMATE

45.1 TOXICITY

The toxicity of carbamate insecticides is similar to that of organophosphates. Like organophosphates, carbamates are inhibitors of acetylcholinesterase. The general chemical structure of carbamate is

$$\begin{array}{c} R \\ \diagdown \\ N-C-O-\text{(leaving group)} \\ \diagup \quad \parallel \\ R \quad\;\; O \end{array}$$

where R is a lower alkyl group or H). These substances readily bind the enzyme acetylcholinesterase at the esteratic sites. Such carbamylation does not require metabolic activation as it does for sulfur-containing organophosphates (e.g., dimethoate). Carbamylation of enzyme differs from phosphorylation by the fact that the former is a rapidly reversible process. As a result, the duration of carbamate poisoning is relatively short, although the poisoning may be severe, and the symptoms are quite similar to that of phosphate esters. Carbamate pesticides, in general, have a higher toxicity than that of most thiocarbamate herbicides. The acute oral toxicities of carbamates, however, vary very widely. Aldicarb [116-

06-3], for example, is highly toxic by both oral and dermal routes. By contrast, carbaryl [63-25-2] is moderately toxic by the oral route and has a very low dermal toxicity. Many substances of this class, such as, benomyl [17804-35-2] or phenoxycarb [79127-80-3], exhibit very low toxicity. Most carbamate insecticides exhibit very low dermal toxicity. The toxicity data for some common carbamate pesticides are presented in Table 45.1. Their chemical structures are given in Table 45.1, comparing their toxicities with any additional structural features in those compounds. The toxicities and physical properties of some selected individual compounds are discussed in the following sections.

45.2 ALDICARB

Formula: $C_7H_{14}N_2O_2S$; MW 190.29; CAS [116-06-3]

Structure:

$$\begin{array}{c} \quad\quad\quad CH_3 \quad\quad O \\ \quad\quad\quad\; | \quad\quad\quad \parallel \\ CH_3-S-C-CH=N-C-O-NH-CH_3 \\ \quad\quad\quad\; | \\ \quad\quad\quad CH_3 \end{array}$$

TABLE 45.1 Comparison of Acute Toxicity of Some Common Carbamate Pesticides

Pesticides/ CAS No.	Structure	Oral and Skin LD_{50} (rat) (mg/kg)
Aldicarb [116-06-3]		~0.8 (oral) 2.5 (skin)
Isolan [119-38-0]		11 (oral) 5.6 (skin)
Carbofuran [1563-66-2]		~8 (oral) 120 (skin)
Methiocarb [2032-65-7]		20 (oral) 120 (skin)
Mexacarbate [315-18-4]		14 (oral)
Methomyl [16752-77-5]		17 (oral) >1600 (skin)
Metolcarb [1129-41-5]		LC_{50} inhalation 128 mg/m^3/hr
Aldoxycarb [1646-88-4]		20 (oral) 1000 (skin)

(*continued*)

TABLE 45.1 *(Continued)*

Pesticides/ CAS No.	Structure	Oral and Skin LD$_{50}$ (rat) (mg/kg)
Promecarb [2631-37-0]		35 (oral) 450 (skin)
Bendiocarb [22781-23-3]		40 (oral) 566 (skin)
Bufencarb [8065-36-9]		~80 (oral) —
Propoxur [114-26-1]		~90 (oral) 800 (skin)
Pirimicarb [23108-98-2]		100 (oral) >500 (skin)
Trimethacarb [2686-99-9]		178 (oral) >2000 (skin)
Carbaryl [63-25-2]		~500 (oral) 4000 (skin)

TABLE 45.1   *(Continued)*

Pesticides/ CAS No.	Structure	Oral and Skin LD$_{50}$ (rat) (mg/kg)
Barban [101-27-9]		~550 (oral) >1600 (skin)
Propham [122-42-9]		ranging between 1000–5000 (oral) >5000 (skin)
Chloropropham [101-21-3]		ranging between 1200–7500 (oral)
Asulam [3337-71-1]		2000 (oral) >10,000 (skin)
Benomyl [17804-35-2]		10,000 (oral) >1000 (skin)
Phenmedipham [13684-63-4]		>5000 (oral) >5000 (skin)
Desmedipham [13684-56-5]		~10,000 (oral) >10,000 (skin)
Phenoxycarb [79127-80-3]		>10,000 (oral)

Synonyms: 2-methyl-2-(methylthio)propion-aldehyde *O*-(methylcarbamoly)oxime; *O*-((methylamino)carbonyl)oxime-2-methyl-2-(methylthio)propanal; Temik; Carbanolate

Physical Properties

Crystalline solid; melts at 99°C; slightly soluble in water (0.6% at 25°C, soluble in most organic solvents.

Health Hazard

Probably the most toxic caramate insecticide; extremely toxic by all routes of exposure in rats, mice, ducks, chickens, and wild birds; cholinesterase inhibitor; toxic effects similar to, but more severe than, other carbamate esters; toxic symptoms include increased salivation, tearing, sweating, spontaneous urination, slow heart rate, blurred vision, twitching of muscle, tremor, and mental confusion; high exposure may produce convulsions and coma; gastrointestinal effects include nausea, vomiting, abdominal cramps, and diarrhea; ingestion of <0.5 g may cause seizure or sudden unconsciousness that may progress to death in humans; oral LD_{50} value in rodents <1 mg/kg.

LD_{50} oral (mouse): 0.3 mg/kg

LD_{50} oral (rat): 0.65 mg/kg

LD_{50} skin (rat): 2.5 mg/kg

US EPA-listed extremely hazardous substance; RCRA waste Number P070.

45.3 CARBARYL

Formula: $C_{12}H_{11}NO_2$, MW 201.22; CAS [63-25-2]

Structure:

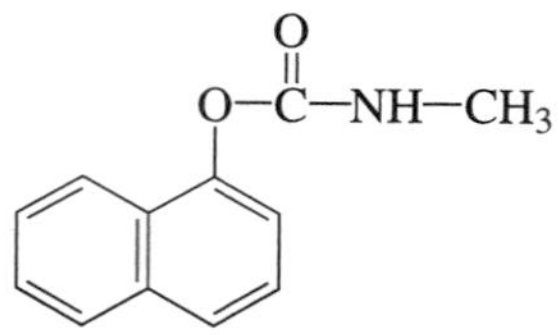

Synonyms: *N*-methyl-1-naphthyl carbamate; methylcarbamic acid 1-naphthyl carbamate; methylcarbamic acid 1-naphthyl-ester; Sevin

Physical Properties

Crystalline solid; melts at 142°C; vapor pressure 0.005 torr at 25°C; slightly soluble in water (0.012% at 30°C), moderately soluble in acetone, cyclohexanone, and dimethylformamide; hydrolyzed by alkalies.

Health Hazard

Acute oral toxicity — moderate in rats; dermal toxicity low to very low; oral LD_{50} value (rats); 250 mg/kg, skin LD_{50} value (rats); 4000 mg/kg; toxic symptoms in humans — nausea, vomiting, diarrhea, abdominal cramps, miosis, lachrimation, excessive salivation, nasal discharge, sweating, cyanosis, muscle twitching, convulsions, and coma; acetylcholinesterase inhibitor; exposure limit: TLV-TWA 5 mg/m^3 (ACGIH, OSHA, and MSHA).

45.4 PROMECARB

Formula: $C_{12}H_{17}NO_{21}$; MW 207.28; CAS [2631-37-0]

Structure:

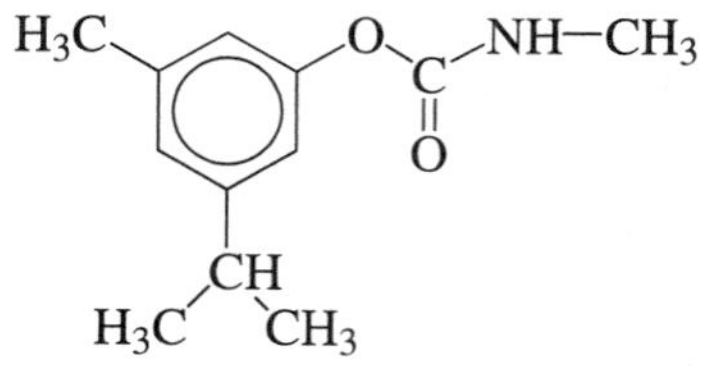

Synonyms: *N*-methyl carbamic acid 3-methyl-5-isopropylphenyl ester; 3-isopropyl-5-methylphenyl *N*-methylcarbamate; *m*-cym-5-yl methylcarbamate; Carbamult; Minacide

Physical Properties

Colorless crystals; melts at 87°C; readily dissolves in organic solvents; insoluble in water (~90 mg/L at 25°C); octanol–water partition coefficient 1545 at pH 4 (Milne 1995).

Toxicity

Highly toxic insecticide; cholinesterase inhibitor; absorbed by all routes; exhibited high toxicity in test animals when administered by oral, intravenous, intraperitoneal, and intramuscular routes; skin absorption is relatively slow, however; toxic symptoms include headache, salivation, lacrimation, blurred vision, stomach ache, and vomiting; ingestion of large dose could be fatal to human; toxicity data not available for human but, based on animal studies, the oral lethal dose in human is estimated to be in the range 2 to 3 g; LD_{50} data published in the literature vary;

LD_{50} oral (rat): 35 mg/kg (Milne 1995)
LD_{50} oral (guinea pig) 25 mg/kg
LD_{50} skin (rat): 450 mg/kg

Because of high octanol–water partition coefficient (>1500 at pH 4), its bioaccumulation potential is expected to be high.

45.5   CARBOFURAN

Formula: $C_{12}H_{15}NO_3$; MW 221.28
Structure:

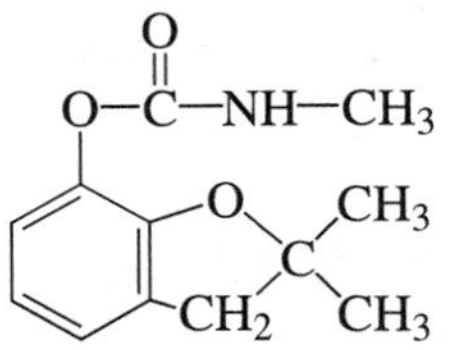

Synonyms: 2,3-dihydro-2,2-dimethyl-7-benzofuranyl-*N*-methylcarbamate; 2,2-dimethyl-7-coumaranyl-*N*-methylcarbamate; 2,2-dimethyl-2,3-dihydrobenzofuranyl 7-methylcarbamate; methylcarbamic acid 2,2-dimethyl-2,3-dihydrobenzofuran-7-yl ester; 2,3-dihydro-2,2-dimethyl-7-benzofuranol methylcarbamate; Furadan; Curaterr; BAY 70143; BAY 78537

Physical Properties

White crystalline solid; melts at 150–153°C; vapor pressure 0.0002 torr at 33°C (Worthing and Walker 1983); density 1.18; slightly soluble in water, 320 mg/L at 25°C; readily dissolves in acetone, chloroform, acetonitrile, benzene and ethanol; low solubility in xylene and petroleum ether; octanol–water partition coefficient 17–26.

Health Hazard

Extremely toxic carbamate pesticide; routes of entry — ingestion, skin absorption, and inhalation of vapors; although the vapor pressure is very low (0.00002 torr at 33°C), because of high toxicity, even trace inhalation could be harmful; can cause death if swallowed, absorbed through skin or inhaled; ingestion of about 0.5- to 2-g substance could cause death to human; manifests acute, delayed, and chronic toxicity; choline sterase inhibitor; acute exposure may cause increased salivation, lacrimation, spontaneous urination, blurred vision, tremor, confusion, muscle twitching, and convulsion; high exposure may result in coma or respiratory collapse; other effects may include gastrointestinal effects, such as nausea, vomiting, abdominal pain, and diarrhea; can cause burn on skin contact; produced adverse reproductive effects in experimental animals.

LD_{50} inhalation (guinea pig): 0.043 mg/L/4 hr (RTECS 1985)
LD_{50} oral (rat): ~10 mg/kg
LD_{50} rat (skin): 120 mg/kg

Exposure Limit

TLV-TWA: 0.1 mg/m³ (ACGIH)

45.6 METHIOCARB

Formula: $C_{11}H_{15}NO_2S$; MW 225.33; CAS [2032-65-7]

Structure:

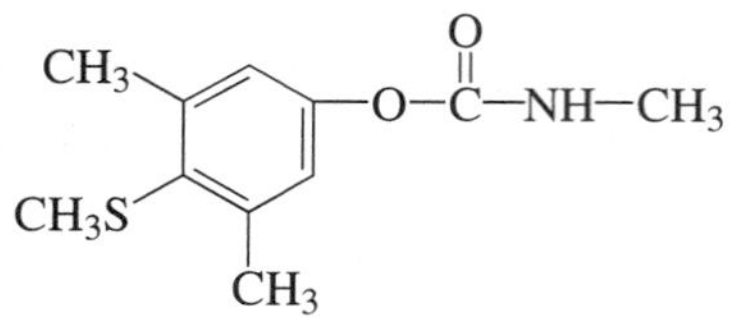

Synonyms: 4-methylthio-3,5-dimethylphenyl-N-methylcarbamate; 4-methylmercapto-3,5-xylyl-N-methylcarbamate; 4-(methylthio)-3,5-xylenol methylcarbamate; 3,5-dimethyl-4-(methylthio)phenol methylcarbamate; methylcarbamic acid 4-(methylthio)-3,5-xylyl ester; Mesurol; Bay 9026

Physical Properties

Crystalline or powder solid; mp 120°C; almost insoluble in water, 27 mg/L at 20°C; readily dissolves in methylene chloride, chloroform, toluene and isopropanol.

Health Hazard

Highly toxic cholinesterase inhibitor; exhibits acute, delayed and chronic effects; routes of entry — ingestion, skin absorption and inhalation of vapors; reversible action of short duration; toxic symptoms include salivation, lacrimation, bradycardia, blurred vision, labored breathing, headache, muscle twitching, tremor, and slight paralysis; gastrointestinal effects include nausea, vomiting, abdominal pain, and diarrhea; severe poisoning may lead to convulsions and coma; oral intake of probably 5–10 g may be fatal to adult humans.

LD_{50} oral (rat): 15–20 mg/kg
LD_{50} oral (guinea pig): 40 mg/kg
LD_{50} skin (rat): 350 mg/kg
LD_{50} skin (wild bird): 100 mg/kg

45.7 BENDIOCARB

Formula: $C_{11}H_{13}NO_4$; MW 223.25. CAS [22781-23-3];

Structure:

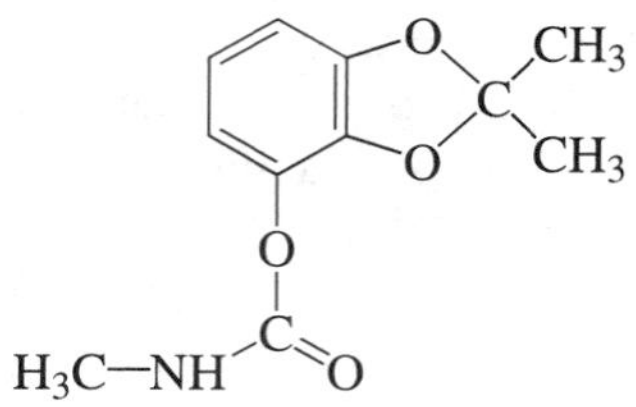

Synonyms: 2,2-dimethyl-1,3-benzdioxol-4-yl-N-methylcarbamate; methylcarbamic acid 2,3-(isopropylidenedioxy)phenyl ester; Ficam W; Garvox

Physical Properties

Colorless crystals; mp 130°C; density 1.25 at 20°C; slightly soluble in water, 40 mg/L at 20°C; readily soluble in acetone, methylene chloride, ethanol and benzene; poor solubility in hexane (280 mg/L at 20°C); octanol–water partition coefficient 50.

Health Hazard

Highly toxic by all routes of exposure; oral median lethal doses in rats, mice, guinea pigs, and rabbits range at 30–50 mg/kg; cholinesterase inhibitor; toxic symptoms include increased salivation, lacrimation, sweating, blurred vision, twitching of muscle, tremor, weakness, vomiting, abdominal cramps, and diarrhea; high doses can cause death; absorption through skin may be slow.

LD_{50} oral (rat): 40 mg/kg
LD_{50} oral (rabbit): 35 mg/kg
LD_{50} skin (rat): ~600 mg/kg

45.8 MEXACARBATE

Formula: $C_{12}H_{18}N_2O_2$; MW 222.29; CAS [315-18-4]

Structure:

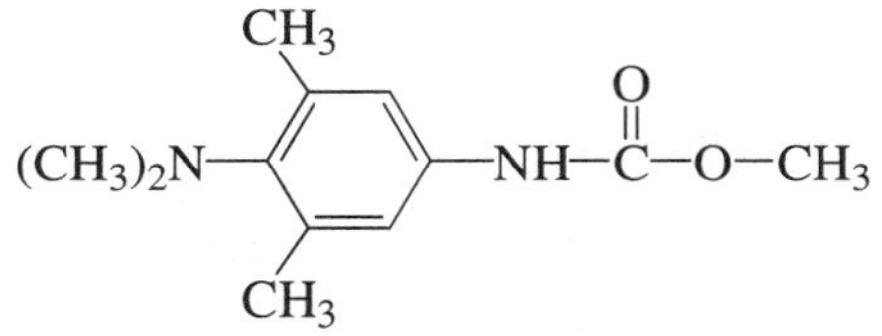

Synonyms: methyl-4-dimethylamino-3,5,-xylyl carbamate; 3,5-dimethyl-4-(dimethylamino)phenyl methylcarbamate; 4-(dimethylamino)-3,5-xylenol methylcarbamate; methylcarbamic acid 4-(dimethylamino)-3,5-dimethylphenyl ester; methylcarbamic acid 4-(dimethylamino)-3,5,xylyl ester; ENT 25766; Dowco 139; Zectran

Physical Properties

White crystalline solid; odorless; melts at 85°C; slightly soluble in water (100 mg/L at 25°C) readily soluble in organic solvents; decomposes in highly alkaline media.

Health Hazard

Extremely toxic carbamate insecticide; exhibits acute, delayed and chronic effects; toxic by all routes of exposure; cholinesterase inhibitor; symptoms of poisoning similar to other carbamate, especially carbaryl esters; symptoms include excessive salivation, lacrimation, blurred vision, twitching of muscles, loss of muscle coordination, and slurring of speech; gastrointestinal effects include nausea, vomiting, diarrhea, abdominal cramps, sweating and weakness; inhalation of vapors may cause runny nose and tightness in the chest; oral lethal dose in adult humans probably ranges at 0.5–2 g; skin and eye contact can cause burn;

LD_{50} oral (rat): 14 mg/kg (RTECS 1985)
LD_{50} oral (mouse): 12 mg/kg
LD_{50} skin (mouse): 107 mg/kg

45.9 PIRIMICARB

Formula: $C_{11}H_{18}N_4O_2$; MW 238.33; CAS [23103-98-2]

Structure:

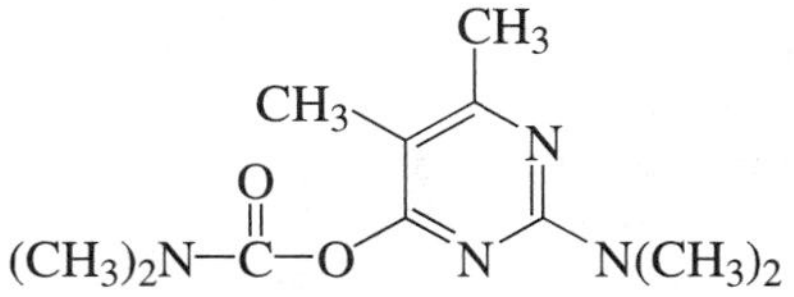

Synonyms: 2-(dimethylamino)-5,6-dimethyl-4-pyrimidinyldimethylcarbamate; 5,6-dimethyl-2-(dimethylamino)-4-pyrimidinyldimethylcarbamate; dimethylcarbamic acid 2-(dimethylamino)-5-6-dimethyl-4-primidyl ester; Aphox; Pirimor; Fernos

Physical Properties

Colorless crystalline solid; melts at 90.5°C; low solubility in water and polar organic solvents (2.7 g/L at 25°C in water); octanol–water partition coefficient 50.

Health Hazard

Highly toxic insecticide by oral and possibly other routes of entry; cholinesterase inhibitor; median oral lethal doses in experimental animals ranged at 100–150 mg/kg; toxic symptoms include excessive salivation, lacrimation, slow heartbeat, blurred vision, headache, muscle twitching, tremor and convulsion; gastrointestinal effects include vomiting, nausea, abdominal pain, and diarrhea; severe poisoning can progress to death.

LD_{50} oral (rat): ~50 mg/kg
LD_{50} oral (dog): 100 mg/kg

45.10 METHOMYL

Formula: $C_5H_{10}N_2O_2S$; MW 162.23; CAS [16752-77-5]

Structure:

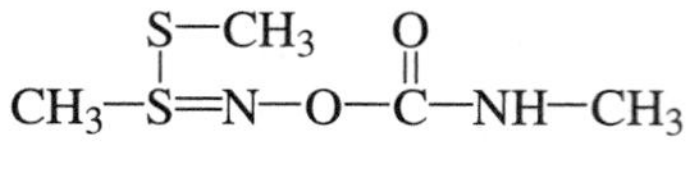

$$CH_3-S=N-O-\overset{\overset{\displaystyle S-CH_3}{|}}{\underset{}{}}\overset{\overset{\displaystyle O}{\|}}{C}-NH-CH_3$$

Synonyms: 1-(methylthio)ethylideneamino methylcarbamate; thio-*N*-((methylearbamoy)oxy)acetimidie acid methyl ester; 1-(methylthio)acetaldehyde-*O*-methylcarbamoyl oxime; methylacetimidothioic acid *N*-(methylcarbamoyl) ester; methyl *N*-((methylcarbamoyl)oxy)thioacetimidate; Lannate; Mesomule; DuPont 1179.

Physical Properties

White crystalline solid with slight sulfurous odor; melts at 78°C; vapor pressure 0.0005 at 25°C (ACGIH 1986); density 1.295 at 25°C (Merck 1989); solubility in water, low to moderate 5.8%; readily dissolves in most organic solvents; low solubility in hydrocarbons.

Health Hazard

Extremely toxic by oral route; moderately toxic by inhalation; exhibits acute, delayed, and chronic effects; cholinesterase inhibitor; symptoms are those of phosphate — or carbamate esters — the major signs of which are increased salivation, lacrimation, spontaneous urination, blurred vision, pinpoint pupils, tremor, twitching of muscle and loss of coordination; confusion, convulsions, and coma may occur as well; other signs of poisoning include nausea, vomiting, cramps, diarrhea, slow heart rate, shortness of breath, and pulmonary edema (US EPA 1988); death may result from respiratory arrest (Gosselin 1984); the probable lethal dose from ingestion in adult human could be 0.5–2 g; oral lethal dose in dog 30 mg/kg, and monkey 40 mg/kg; toxic effects from skin absorption low; listed as extremely hazardous substance by US EPA.

LD$_{50}$ oral (mouse): 10 mg/kg

LD$_{50}$ oral (rat): 17 mg/kg

LD$_{50}$ inhalation (rat): 77 ppm

Exposure Limit

TLV-TWA: 2.5 mg/m^3 (ACGIH)
PEL-TWA: 2.5 mg/m^3 (OSHA)

45.11   METOLCARB

Formula: $C_9H_{11}NO_2$; MW 165.21; CAS [1129-41-5]

Structure:

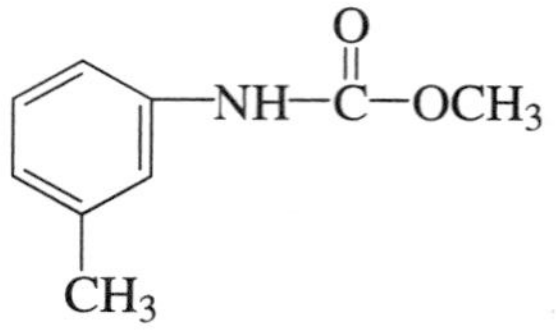

Synonyms: 3-tolyl *N*-methylcarbamate; 3-ethylphenyl *N*-methylcarbamate; *m*-cresyl methylcarbamate; *N*-methylcarbamic acid *m*-cresyl ester; methylcarbamic acid 3-tolyl ester; Tsumacide

Physical Properties

Colorless crystalline solid; no odor; melts at 76°C; sparingly soluble in water 2.6 g/L at 30°C; soluble in organic solvents.

Health Hazard

Highly toxic by ingestion and moderately toxic by inhalation and skin absorption; cholinesterase inhibitor; exhibits acute, delayed, and chronic toxicity; toxic effects are those of organophosphorus pesticides and carbamate esters; the symptoms include excessive salivation, lacrimation, blurred vision, headache, labored breathing, twitches of muscle, loss of reflexes, headache, weakness, sweating, nausea, giddiness, vomiting, cramps, diarrhea, convulsions, and coma; US EPA-listed extremely hazardous substance.

LD$_{50}$ oral (rat): 268 mg/kg

LD$_{50}$ skin (rat): 268 mg/kg

LD$_{50}$ inhalation (rat): 128 mg/m^3 hour

45.12 PROPOXUR

Formula: $C_{11}H_{15}NO_3$; MW 209.27; CAS [114-26-1]

Structure:

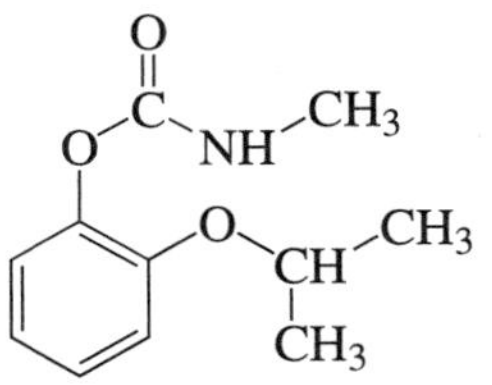

Synonyms: *o*-isopropoxyphenyl *N*-methyl-carbamate; *N*-methyl-2-isopropoxyphe-nylcarbamate; *o*-isopropoxy phenol methylcarbamate; methylcarbamic acid *o*-isopropoxyphenyl ester; Aprocarb; Baygon; Isocarb; Sendran

Physical Properties

White crystalline solid or powder; melts at 85–90°C; decomposes on further heating; density 1.12 at 20°C; slightly soluble in water (2 g/L at 20°C); readily dissolves in most common organic solvents; decomposes in highly alkaline solution.

Health Hazard

A highly toxic substance by ingestion, and possibly by most other routes of exposure; moderately toxic by inhalation and skin contact; cholinesterase inhibitor; toxic effects are similar to those of other carbamate pesticides and include excessive salivation, lacrimation, slow heart rate, blurred vision, twitching of muscle and lack of coordination, nausea, weakness, diarrhea and abdominal pain; oral intake of probably 1.5–3 g could be fatal to adult humans; a teratogenic substance, producing adverse reproductive effects in experimental animals.

LD$_{50}$ oral (rat): 70 mg/kg

LD$_{50}$ skin (rat): 800 mg/kg

LC$_{50}$ inhalation (rat): 1440 mg/m^3/1 hr

Exposure Limit

TLV-TWA: 0.5 mg/m^3 (ACGIH)

PEL-TWA: 0.5 mg/m^3 (OSHA)

45.13 BARBAN

Formula: $C_{11}H_9Cl_2NO_2$; MW 258.11; CAS [101-27-9]

Structure:

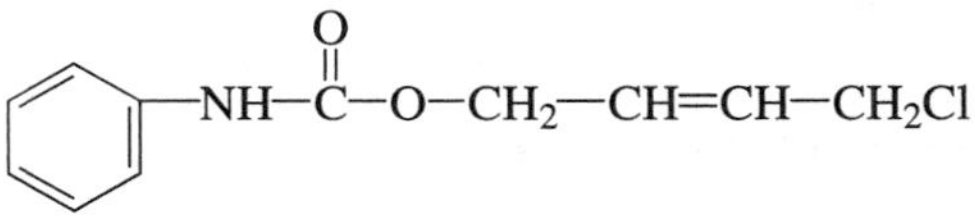

Synonyms: (4-chlorobutyn-2-yl)-*N*-(3-chlorophenyl)carbamate; chlorobutynyl chlorocarbanilate; (3-chlorophenyl) carbamic acid 4-chloro-2-butynyl ester; *m*-chlorocarbanilic acid 4-chloro-2-butynyl ester; Barbamate; Carbin; Chlorinat

Physical Properties

Crystalline solid; melts at 75°C; insoluble in water (11 mg/L at 25°C); slightly soluble in hexane; soluble in benzene and chlorinated hydrocarbons.

Health Hazard

Moderately toxic by ingestion and inhalation; absorption through skin may be very slow and, therefore, almost nontoxic by dermal route; used as a herbicide, rather than as a pesticide; toxic symptoms are those of other carbamate esters;

LD$_{50}$ oral (rat): 600 mg/kg

LD$_{50}$ skin (rabbit): >20,000 mg/kg

45.14 BENOMYL

Formula: C$_{14}$H$_{18}$N$_4$O$_3$; MW 290.36; CAS [17804-35-2]

Structure:

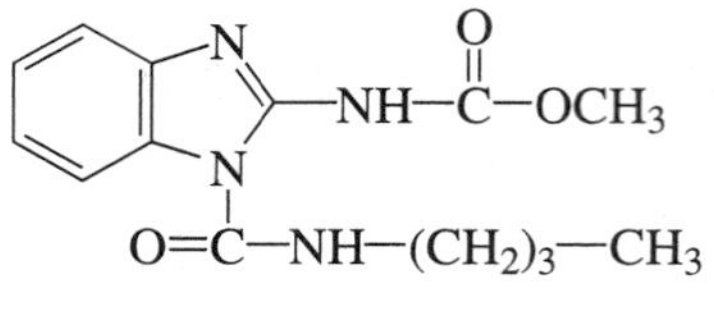

Synonyms: methyl *N*-(1-butylcarbamoyl-2-benzimidazole)carbamate; butylcarbamoylbenzimidazole-2-methyl carbamate; 1-(butylcarbamoyl)-2-benzimidazolecarbamic acid methyl ester; Benlate; Fungicide 1991

Physical Properties

Crystalline solid; decomposes on heating; insoluble in water (2 mg/L at 25°C) readily soluble in chloroform, dimethylformamide, acetone, and benzene.

Health Hazard

Mildly toxic in rodent by ingestion, inhalation, and absorption through skin; large doses can produce effects of carbamate poisoning; teratogenic and mutagenic effects reported; carcinogenic potential not known; mild skin irritant.

LD$_{50}$ oral (rat): 10,000 mg/kg

LD$_{50}$ skin (mouse): 5600 mg/kg

LD$_{50}$ skin (wild bird): 100 mg/kg

45.15 ALDOXYCARB

Formula: C$_7$H$_{14}$N$_2$O$_4$S; MW 222.29; CAS [1646-88-4]

Structure:

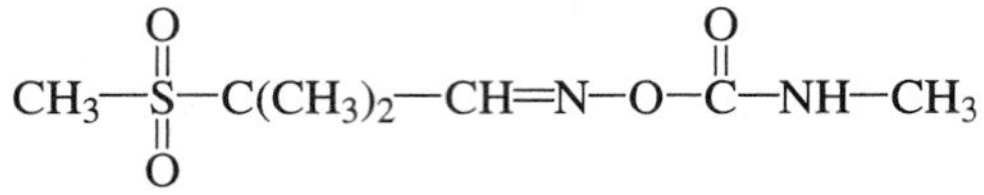

Synonyms: 2-methyl-2-(methylsulfonyl)propionaldehyde *O*-[(methylamino)carbonyl] oxime; Aldicarb sulfone; Sulfocarb; Standak; Temik sulfone

Physical Properties

Crystalline solid; melts at 140°C; slightly soluble in water; dissolves in most organic solvents.

Health Hazard

Highly toxic by all routes of exposure; cholinesterase inhibitor; ingestion of 1–3 g may cause death to adult humans.

LD$_{50}$ oral (mouse): 20 mg/kg

LD$_{50}$ inhalation (rat): 140 mg/m^3/4 hr

LD$_{50}$ skin (rabbit): 200 mg/kg

45.16 BUFENCARB

Formula: C$_{13}$H$_{19}$NO$_2$; MW; CAS [8065-36-9]

Structure:

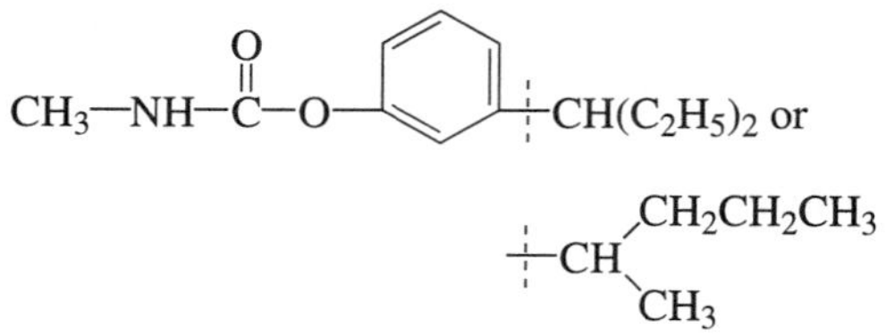

(mixture in ratio of 1 : 3)

Synonyms: mixture of *m*-(1-ethylpropyl) phenyl methylcarbamate and *m*-(1-methylbutyl)phenyl methylcarbamate; methylcarbamic acid *m*-(1-methyl)butyl)phenyl ester mixed with methylcarbamic acid *m*-(1-ethylpropyl)phenyl ester; But; Metalkamate

Physical Properties

Solid at room temperature; melts between 26–39°C; density 1.024 at 26°C; insoluble in water; soluble in methanol, benzene, xylene

Health Hazard

Moderately toxic by ingestion and skin contact; cholinesterase inhibitor; signs of poisoning include headache, weakness, salivation, tearing, blurred vision, loss of muscle coordination, vomiting, diarrhea, and difficulty in breathing.

LD_{50} oral (mouse): 85 mg/kg
LD_{50} skin (rat): 240 mg/kg
LD_{50} skin (rabbit): 400 mg/kg

45.17   TRIMETHACARB

Formula: $C_{11}H_{15}NO_2$; MW 193.27
Structure:

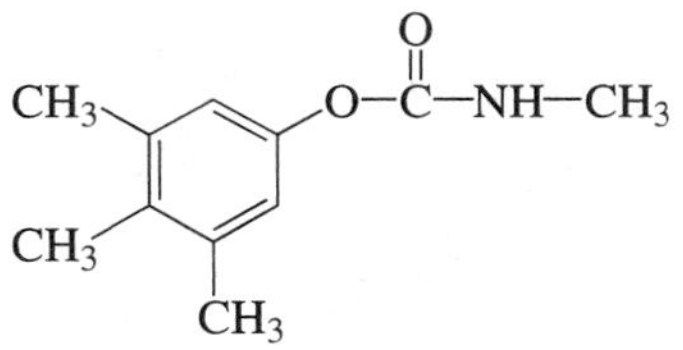

(also occurs as a 2,3,5-isomer)
CAS [2686-99-9] for 3,4,5-isomer and [3971-89-9] for 2,3,5-isomer
Synonyms: trimethylphenyl methylcarbamate; methylcarbamic acid trimethylphenyl ester; Broot

Physical Properties

Crystalline solid; melts in the temperature range 117°–123°C; slightly soluble in water; soluble in most organic solvents

Health Hazard

Moderately toxic by ingestion and possibly other routes of exposure; cholinesterase inhibitor; toxic symptoms include headache,

weakness, lacrimation blurred vision, twitching of muscle, nausea, vomiting, and convulsions.

LD_{50} oral (rat); 175–225 mg/kg

45.18   ASULAM

Formula: $C_8H_{10}N_2O_4S$; MW; CAS: [3337-71-1]
Structure:

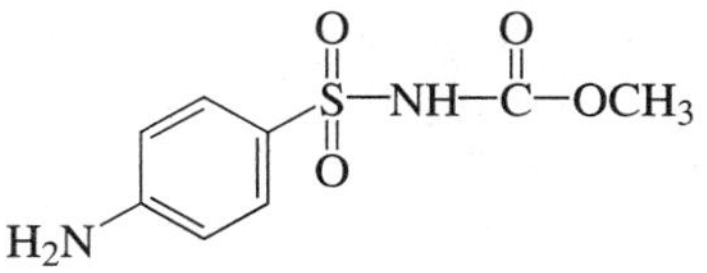

Synonyms: methyl[(4-aminophenyl)sulfonyl] carbamate; Asilan; Asulox

Physical Properties

Crystalline solid; melts at 143°C; low solubility in water (5 g/L at 20°C); highly soluble in dimethylformamide, acetone and methanol; less soluble in chloroform, methylene chloride, and hydrocarbon solvents.

Health Hazard

Low order of toxicity; no adverse effect on skin reported; ingestion of large dose could produce cholinergic effects.

LD_{50} oral (rat): 2000 mg/kg

45.19   PROPHAM

Formula: $C_{10}H_{13}NO_2$; MW 179.21; CAS [122-42-9]
Structure:

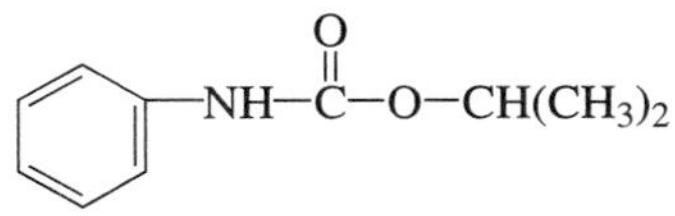

Synonyms: isopropyl phenylcarbamate; isopropyl carbanilate; Agermin; Birgin

Physical Properties

Crystalline solid; mp 87°C; sublimes on further heating; density 1.09 at 20°C; slightly soluble in water (250 mg/L at 20°C) readily dissolves in most organic solvents.

Health Hazard

Moderately toxic herbicide; exhibited low to moderate toxicity in experimental animals when administered by oral, intraperitoneal, intravenous, and subcutaneous routes; skin absorption is slow; cholinesterase inhibitor; in human ingestion can cause carbamate poisoning, which can be lethal when taken in large amount; probable lethal oral dose in adult human estimated to be larger than other carbamate insecticides within the range 35–50 g.

45.20   ISOLAN

Formula: $C_{10}H_{17}N_3O_2$; MW 211.27; CAS [119-38-0]

Structure:

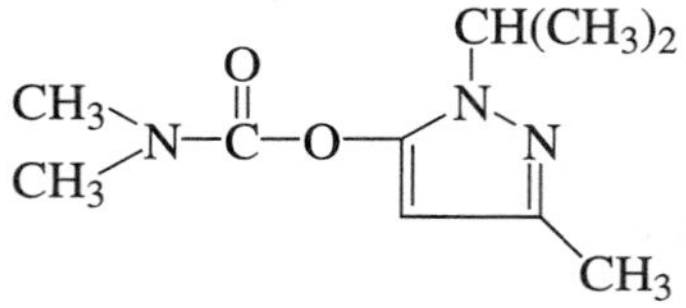

Synonyms: 1-isopropyl-3-methyl-5-pyrazolyl dimethylcarbamate; dimethylcarbamic acid 1-isopropyl-3-methylpyrazol-5-yl ester; Primin; Saolan

Physical Properties

Colorless liquid; bp 118°C at 2.5 torr; vapor pressure 0.001 torr at 20°C; specific gravity 1.07; miscible in water.

Health Hazard

Extremely toxic carbamate; exhibits acute, delayed and chronic effects similar to organophosphates; cholinesterase inhibitor; found to be more toxic to rats by dermal than oral route (Merck 1996); toxic symptoms include trembling, pinpoint pupils, lacrimation, excessive salivation, sweating, slurring of speech, jerky movements, nausea, vomiting, loss of bladder control, slight blueness of skin, lips, and nail beds, as well as convulsion, and coma (Gosselin 1984); death can occur from respiratory arrest; oral lethal dose in adult human estimated to be within the rane 0.5–2 g.

LD_{50} oral (rat): 11 mg/kg
LD_{50} skin (rat): 5.6 mg/kg

REFERENCES

ACGIH. 1986. *Documentation of the Threshold Limit Values and Biological Exposure Indices*, 5th ed. Cincinnati, OH: American Conference of Governmental Industrial Hygienists.

Gosselin, R. E., Smith, R. P., and M. C. Hodge. 1984. *Clinical Toxicology of Commercial Products*, 5th ed., Baltimore; Williams & Wilkins.

Merck. 1996. *The Merck Index*, 12th ed. Rahway, NJ: Merck & Co.

Milne, G. W. A. 1995. *CRC Handbook of Pesticides*. Boca Raton, FL: CRC Press.

NIOSH. 1986. *Registry of Toxic Effects of Chemical Substances*, D. V. Sweet, ed. Washington, DC: US Government Printing Office.

US EPA. 1988. *Extremely Hazardous Substances: Superfund Chemical Profiles*. Park Ridge, NJ: Noyes Data Corporation.

Worthing, C. R. and S. B. Walker. 1983. *The Pesicide Manual—A World Compendium*, 7th ed. Lavenham, England: The Lavenham Press Limited.

46

PESTICIDES, ORGANOCHLORINE

46.1 GENERAL DISCUSSION

Unlike organophosphorus and carbamate pesticides, the toxic properties of all organochlorine pesticides are not the same. This is because the general term '*organochlorine pesticides*' refers to many halogenated organic compounds that are being used currently or that were used in the past in pesticide formulations. Such substances may differ widely in their chemical structures and, therefore, in their physical and toxicological properties. Some of the most well-known pesticides of this class, many of which listed as US EPA priority environmental pollutants, may be structurally classified into the following five types:

1. Hexachlorooctahydronaphthalene type (e.g., Aldrin, Dieldrin, Endrin, Endrin ketone, and Endrin aldehyde)
2. Chlorinated camphene type (e.g., Heptachlor, Chlordane, Heptachlor epoxide, and Toxaphene)
3. Chlorinated diphenylethane type (e.g., DDT, DDD, Methoxychlor, and Perthane)
4. Hexachlorocyclohexane type (e.g., Lindane and its α-, β-, and δ-isomers)

5. Chlorinated pentacyclodecane type (e.g., Mirex and Kepone)

Although the toxic properties are more or less the same for structurally similar compounds, such as Heptachlor and Chlordane, the degree of toxicity may vary significantly with chlorosubstitution in the compound. For example, substitution of chlorine atoms with methoxy group in the aromatic rings in DDT decreases the latter's toxicity. Thus, the methoxy derivative, Methoxychlor, is significantly less toxic than DDT. Similarly, ethyl-substituted Perthane is much less toxic than p,p'-DDD.

The chemical structures of some organochlorine pesticides and their toxicity data as median lethal concentrations in experimental animals are presented for comparison in Table 46.1.

There are also many chlorinated pesticides such as Chlorbenside or Chlorphacinone that do not fit into the above structural classifications. Chlorinated herbicides, such as Captan, Alachlor, or Nitrofen used as herbicides rather than insecticides are excluded in this chapter. Also excluded are the chlorinated organophosphorus pesticides. It may be noted that some of these substances

TABLE 46.1 Comparison of Chemical Structures of Some Organochlorine Pesticides to Their LD$_{50}$/LC$_{50}$ Values

Pesticide/CAS No.	Chemical Structure	LD$_{50}$ (oral) (mg/kg)	LD$_{50}$ (skin) (mg/kg)	LC50 (inhalation) (mg/m^3)
Aldrin [309-00-2]		~40 (rat) 50 (rabbit)	~100 (rat) 15 (rabbit)	5.8/4 hr (rat) (LC$_{LO}$)
Dieldrin [60-57-1]		38 (rat) 45 (rabbit)	56 (rat) 250 (rabbit)	13/4 hr (rat) —
Endrin [72-20-8]	(a stereoisomer of dieldrin)	3 (rat) 7 (rabbit)	12 (rat) 60 (rabbit)	—
Lindane (γ-BHC) [58-89-9]		76 (rat) 60 (rabbit)	500 (rat) 50 (rabbit)	—
α-BHC [319-84-6]	a stereoisomer of Lindane	177 (rat)	—	—

Name [CAS]	Structure/Description			
β-BHC [319-85-7]	a stereoisomer of Lindane	6000 (rat)	—	—
Δ-BHC [319-86-8]	a stereoisomer of Lindane	1000 (rat)	—	—
Heptachlor [76-44-8]		40 (rat)	119 (rat)	—
Heptachlor epoxide [1024-57-3]		15 (rat) 39 (mouse)	—	—
Chlordane [57-74-9]		200 (rat) 100 (rabbit)	690 (rat) 780 (rabbit)	100/4 hr (cat)
α-Chlordane [5103-71-9]	A cis-isomer of Chlordane	500 (rat)	—	—
γ-Chlordane [5566-34-7]	A trans-isomer of Chlordane	500 (rat)	—	—

(continued)

TABLE 46.1 (Continued)

Pesticide/CAS No.	Chemical Structure	LD$_{50}$ (oral) (mg/kg)	LD$_{50}$ (skin) (mg/kg)	LC50 (inhalation) (mg/m^3)
Endosulfan [115-29-7]	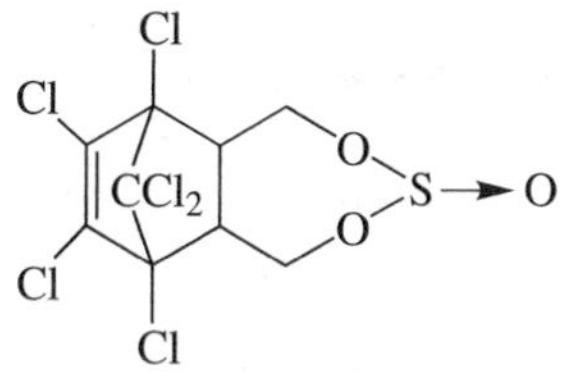	18 (rat) 28 (rabbit)	34 (rat) 90 (rabbit)	80/4 hr (rat)
Methoxychlor [72-43-5]	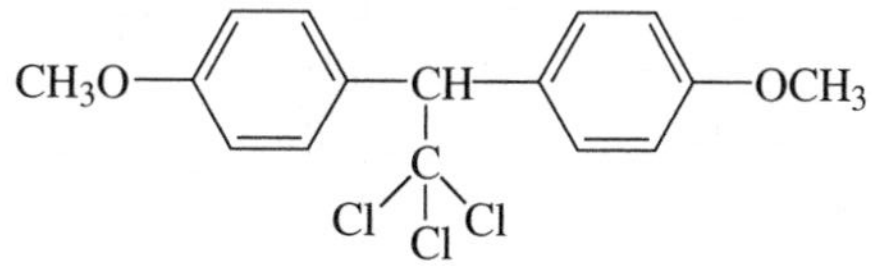	5000 (rat) 1000 (mouse)	>6000 (rat)	—
DDT [50-29-3]	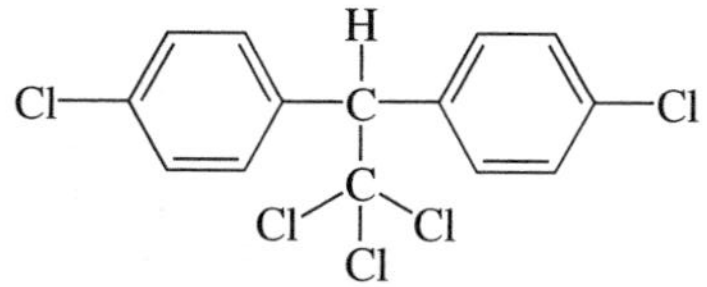	87 (rat) 135 (mouse)	1930 (rat) 300 (rabbit)	—
p,p-DDD [72-54-8]	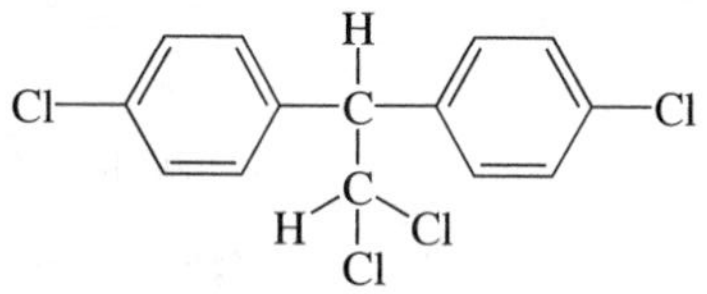	113 (rat)	1200 (rabbit)	—
p,p-DDE [72-55-9]	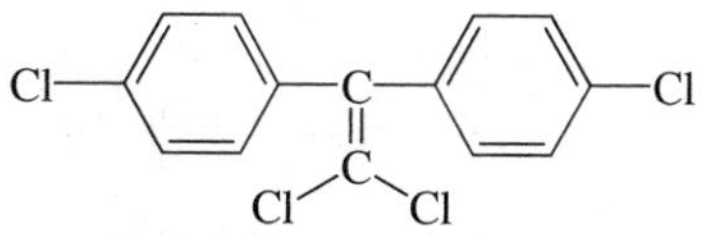	880 (rat) 700 (mouse)	—	—

714

Name [CAS]	Structure	LD50 (oral)	LD50 (dermal)	
Perthane [72-56-0]	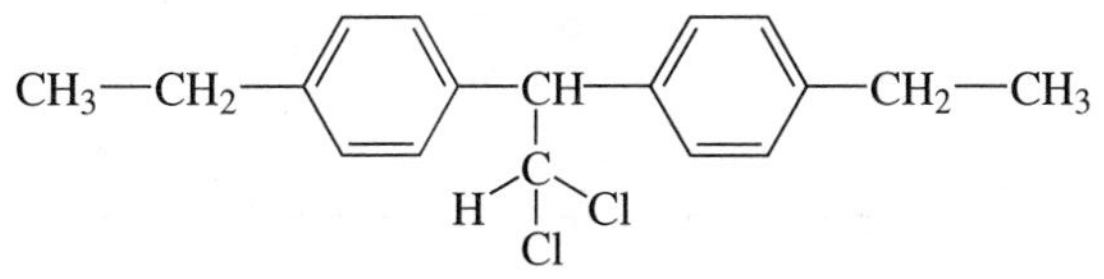 CH$_3$—CH$_2$—⟨C$_6$H$_4$⟩—CH—⟨C$_6$H$_4$⟩—CH$_2$—CH$_3$, CH—C(Cl)(Cl)(Cl), H	6600 (rat) 9000 (wild bird)	>5000 (rat)	—
Chlorobenzilate [510-15-6]	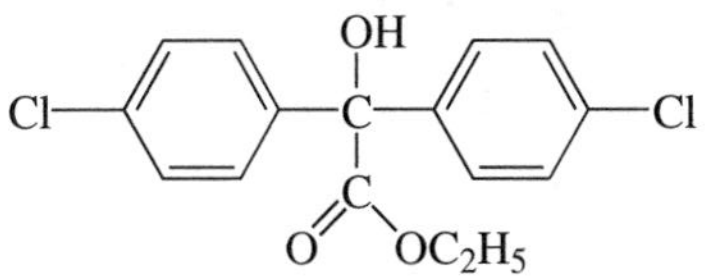 Cl—⟨C$_6$H$_4$⟩—C(OH)(⟨C$_6$H$_4$⟩—Cl)—C(=O)—OC$_2$H$_5$	700 (rat) 729 (mouse)	>5000 (rat) >5000 (rabbit)	—
Kepone [143-40-0]	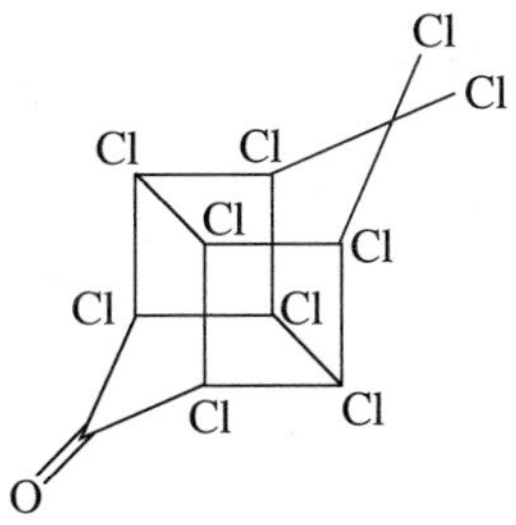 Chlorinated cage structure (Cl × 10, C=O)	95 (rat) 65 (rabbit) 250 (dog)	>2000 (rat) 345 (rabbit)	—
Mirex [2385-85-5]	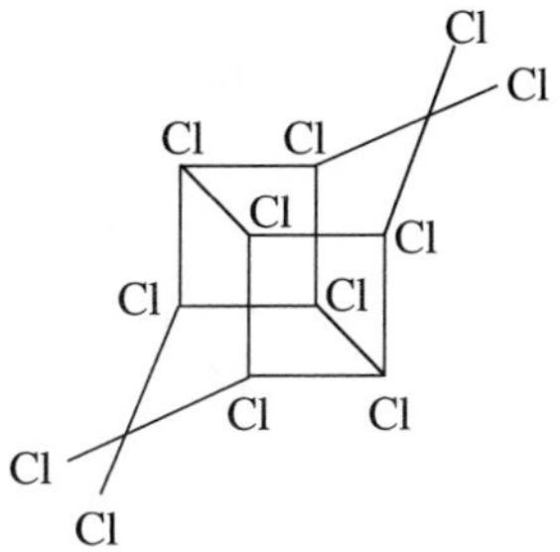 Chlorinated cage structure (Cl × 12)	235 (rat)	>2000 (rat) 800 (rabbit)	—

(continued)

715

TABLE 46.1 (Continued)

Pesticide/CAS No.	Chemical Structure	LD$_{50}$ (oral) (mg/kg)	LD$_{50}$ (skin) (mg/kg)	LC50 (inhalation) (mg/m^3)
Toxaphene [8001-35-2]	A complex mixture of chlorinated camphenes	50–80 (rat) 112 (mouse) 75 (rabbit)	600 (rat) 1025 (rabbit)	2000/2 hr (rat) (LC$_{LO+}$)
Strobane [8001-50-1]	Mixture of many terpene polychlorinates	200 (rat) 200 (dog)	—	—
Alachlor [15972-60-8]	2,6-diethylphenyl-N(CH$_2$OCH$_3$)(COCH$_2$Cl)	930 (rat) 460 (mouse)	3500 (rabbit)	—
Butachlor [23184-66-9]	2,6-diethylphenyl-N(CH$_2$OC$_4$H$_9$)(COCH$_2$Cl) (butachlor)	1740 (rat)	4080 (rabbit)	
Propachlor [1918-16-7]	phenyl-N(CH(CH$_3$)$_2$)(COCH$_2$Cl)	710 (rat) 290 (mouse) 710 (rabbit)	380 (rabbit)	—

are no longer used for pest control in the United States because of their high toxicity or carcinogenic potential, or both. These compounds, however, may have nonpesticide applications. Certain pesticides, such as DDT and its degradation products like p,p'-DDD and p,p'-DDE have very high octanol–water partition coefficients and are susceptible to bioaccummulate and persist in the environment for years.

Organochlorine pesticides are highly lipophilic. They can penetrate cell membranes rapidly and distribute between tissue lipids and extracellular water as much as within the cells. Such distribution, however, may be distorted by proteins that may bind to pesticides. Dieldrin can bind to serum proteins up to 99% in vivo (Garretson and Curly 1969). Its half-life in adipose tissue in a case study of accidental poisoning in a child is reported to be 50 days. By contrast, Hunter and Robinson (1967) obtained a half-life value of 3–4 months for this pesticide in adult humans in long-term exposure. Most chlorinated pesticides are known to accumulate in adipose and other tissues. The "storage ratio," expressed as a ratio of accumulation of residues of compounds in body fat to their concentrations in diet, can vary significantly from compound to compound. (Street and Sharma 1977). Rats fed with pesticides in their diet showed a high storage ratio of 36 and 20 for chlordane and DDT, respectively, while such ratio was much lower for toxaphene and methoxychlor ($\sim$0.2 and 0.04). The values for dieldrin and lindane were 1–2. The accumulation and elimination kinetics of many pentachloro-PCB isomers closely parallel with some chlorinated pesticides, especially dieldrin. It may be noted that the presence of stored lipophilic chemicals in adipose tissue probably does not cause injury to the tissue itself.

Certain substances are known to deplete chlorinated pesticides held in the body in the lipid phase. They induce greater biotransformation, thereby accelerating the depletion of toxicants in the lipid phase. For example, phenobarbital can induce biotransformation of dieldrin (Cook and Wilson 1971), while diphenyhydantoin may be effective against DDT residues (Davies et al. 1971). In addition, adsorbants such as activated carbon are used to treat pesticide poisoning.

Many organochlorine pesticides are neurotoxins; some of them are highly toxic. Notable among these are endrin [72-20-8], lindane [58-89-9], aldrin [309-00-2], and dieldrin [60-57-1], all of which have LD_{50} values in rats of <100 mg/kg. Structurally, aldrin, dieldrin, and endrin are quite similar. The latter two are stereoisomers. The structures of these and other pesticides are presented in the following sections. Among the hexachlorocyclohexanes, the γ-isomer, lindane, is more toxic than the other three isomers: α-BHC [319-84-6], β-BHC [319-85-7], and δ-BHC [319-86-8]. Structurally, heptachlor [76-44-8] and chlordane [57-74-9] have similar features. DDT [50-29-3] and methoxychlor [72-43-5] have similar structures. Toxaphene [8001-35-2] is a complex mixture of more than 170 chlorinated camphenes. Although the acute oral toxicity of this formulation is very low, one of its components, 9-octachlorobornane, has an acute oral LD_{50} value in mice of 3.3 mg/kg.

The acute toxic symptoms of many chlorinated pesticides include headache, dizziness, loss of coordination, nausea, vomiting, tremors, and convulsions. Many organochlorine pesticides, such as DDT, aldrin, dieldrin, endrin, mirex, and kepone were found to be teratogenic in animal studies. These pesticides are cancer causing in animals. The carcinogencity is of varying potency. There are also limited evidences of cancer-causing effects of some of these and other chlorinated pesticides in humans. Because of their high toxicity and high carcinogenic potency, several of these compounds are no longer being produced or used in the United States.

46.2 LINDANE

Formula: $C_6H_6Cl_6$; MW 290.82; CAS [58-89-9]

Structure:

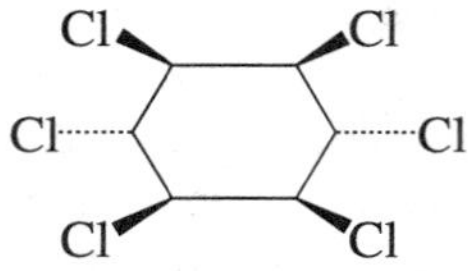

Synonyms: γ-1,2,3,4,5,6-hexachlorocyclohexane; γ-benzene hexachloride; γ-BHC; hexachlorocyclohexane γ-isomer

Physical Properties

White crystalline substance; melts at 112.5°C; vapor pressure 9.4×10^{-6} torr at 20°C (Merck 1996); forms seven other stereoisomers; insoluble in water, soluble in most organic solvents.

Health Hazard

High acute toxicity; symptoms — headache, dizziness, nausea, vomiting, diarrhea, tremor, cyanosis, epileptic convulsions; stimulant to nervous system, which can lead to violent convulsions; such convulsions may set rapidly that may either progress to recovery within 24 hours or could lead to death (Hayes 1982); ingestion of 2–10 g probably fatal to human; an irritant to eye and skin; chronic exposure causes liver injury; oral LD_{50} value (mice): 86 mg/kg; carcinogenic to animals, causing liver and lung tumors; exposure limit: TLV : TWA (skin) 0.5 mg/m^3 (ACGIH, MSHA, and OSHA); RCRA Waste Number U120.

46.3 ALDRIN

Formula: $C_{12}H_8Cl_6$; MW 364.93; CAS [309-00-2]

Structure:

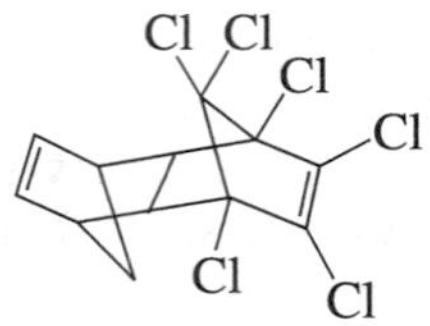

Synonyms: 1,2,3,4,10,10-hexachloro-1,4,4*a*, 5,8,8*a*-hexahydro-1,4:5,8-dimethanonaphthalene; hexachlorohexahydro-endo, exo-dimethanonaphthalene; octalene; Aldrex; Aldrite

Physical Properties

Tan to dark brown solid; melts at 104°C; vapor pressure 7.5×10^{-5} torr at 20°C; insoluble in water, highly soluble in most organic solvents.

Health Hazard

Highly toxic to humans and animals by all routes of exposure; absorbed through skin as well; toxic symptoms — headache, dizziness, nausea, vomiting, tremor, ataxia, convulsions, central nervous system depression, and respiratory failure; also causes renal damage; oral LD_{50} value 30–100 mg/kg in most test animals metabolizes to dieldrin; oral administration in rats and mice increased the incidence of liver and lung cancer, carcinogenicity: animal and human evidence inadequate; exposure limit 0.25 mg/m^3 (skin) (ACGIH); RCRA Waste Number P004.

46.4 DIELDRIN

Formula: $C_{12}H_8Cl_6O$; MW 380.93; [60-57-1]

Structure:

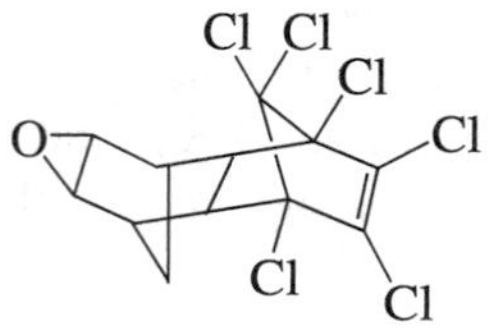

Synonyms: 1,2,3,4,10,10-hexachloro-6,7-epoxy-1,4,4a, 5,6,7,8a-octahydroendo, exo-1,4:5,8-dimethanonaphthalene; 3,4, 5,6,9,9-hexachloro-1,1a,2,2a,3,6,6a,7,7a-octahydro-2,7,3,6-dimenthanonaphth(2,3-b)oxirene; Octalox; Dieldrex

Physical Properties

Colorless to light tan solid; mild odor; melts at 176°C; vapor pressure 3.1×10^{-6} torr at 20°C; insoluble in water, moderately soluble in most organic solvents.

Health Hazard

Highly toxic; toxic symptoms similar to these of aldrin; affects central nervous system, liver, kidneys, and skin; causes headache, dizziness, nausea, vomiting, tremor, ataxia, clonic and tonic convulsions, and respiratory failure; oral LD_{50} value (mice): 38 mg/kg; causes liver cancers in animals; inadequate evidence in humans; RCRA Waste Number P037.

46.5 ENDRIN

Formula: $C_{12}H_8Cl_6O$; MW 380.90; CAS [72-20-8]

Structure:

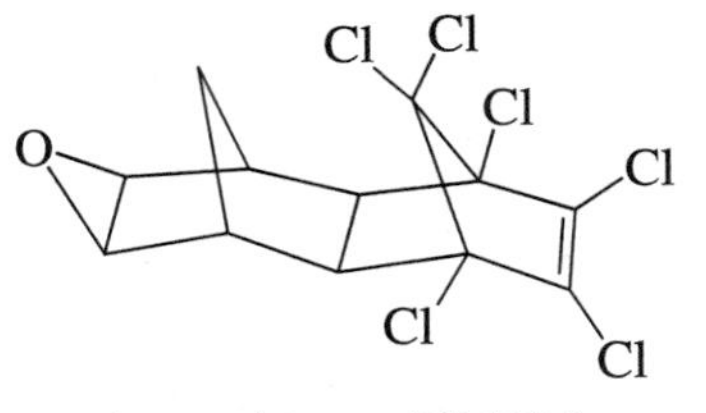

(a stereoisomer of dieldrin)

Synonyms: 1,2,3,4,10,10-hexachloro-6,7-epoxy-1,4,4a,5,6,8a-octahydroendo, endol, 4:5,8-dimethanonaphth[2,3-b]oxirene; Hexadrin; Mendrin; Endrex

Physical Properties

Colorless to tan solid with a mild odor; decomposes at 245°C; vapor pressure 2×10^{-7} torr at 25°C; insoluble in water, soluble in most organic solvents.

Health Hazard

Extremely toxic to experimental animals by all routes of exposure; also, highly toxic to humans and animals by ingestion or skin absorption; toxic effects similar to those of aldrin and dieldrin; symptoms — headache, dizziness, nausea, vomiting, abdominal pain, insomnia, confusion, stupor, convulsions, tremors, rise in blood pressure, fever, frothing of mouth, deafness, coma, and respiratory failure; stimulant to central nervous system; ingestion of 50–60 mg can produce toxic symptoms in human; can produce adverse reproductive effects; a teratogenic substance; evidence of carcinogenicity in experimental animals and humans inadequate; oral LD_{50} value (mice): ~13 mg/kg, (guinea pig): 16 mg/kg; US EPA listed extremely hazardous substance; exposure limit: PEL/TLV-TWA(skin) 0.1 mg/kg (OSHA, ACGIH)

46.6 HEPTACHLOR

Formula: $C_{10}H_5Cl_7$; MW 373.30; CAS [76-44-8]

Structure:

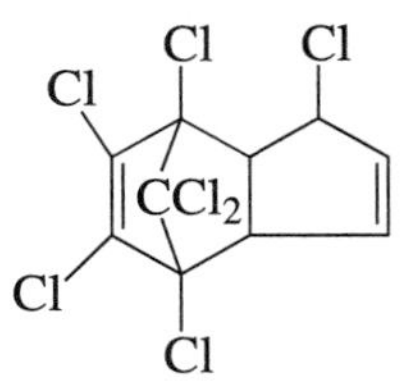

Synonyms: 1,4,5,6,7,8,8-heptachloro-3a, 4,7, 7a-tetrahydro-4,7-methanoindene; 3,4,5,6, 7,8-heptachlorodicyclopentadiene; 1,4,5,6, 7,10,10-heptachloro-4,7,8,9-tetrahydro-4, 7-endomethyleneindene: Heptamul

Physical Properties

White or light tan waxy solid with camphor-like odor; melts at 95°C; boils at 135–145°C (decomposes); vapor pressure 3×10^{-4} torr at 25°C; insoluble in water, soluble in most organic solvents.

Health Hazard

Highly toxic to animals; toxic symptoms — tremors, convulsions, and liver necrosis; also causes blood dyscrasias; oral LD_{50} value (rats): 40 mg/kg; caused liver cancers in mice; no evidence of human carcinogenicity; RCRA Waste Number P059; exposure limit: TLV-TWA (skin) 0.5 mg/m^3 (ACGIH).

46.7 CHLORDANE

Formula: $C_{10}H_6Cl_8$; MW 409.76; CS [57-74-9]

Structure:

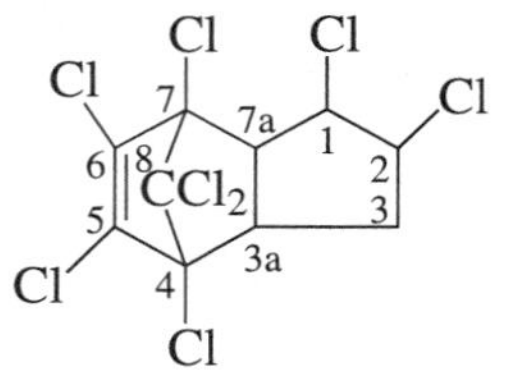

Synonyms: 1,2,4,5,6,7,8,8-octachloro-2,3,3a, 4,7,7a-hexahydro-4,7-methano-1H-indene: 1,2,4,5,6,7,8,8-octachloro-3a,4,7,7a-tetra-hydro-4,7-methanoindan; octachlor; topichlor; toxichlor; Velsicol 1068

Physical Properties

Colorless or amber, thick, viscous liquid; odorless; density 1.59–1.63 at 20°C; boils at 175°C; vapor pressure 1×10^{-5} torr; insoluble in water, miscible with most organic solvents; loses its chlorine in the presence of alkalies.

Health Hazard

Highly toxic to humans by ingestion; moderately toxic in test animals; skin absorption or inhalation of its vapors can produce poisoning effects; exhibits acute, delayed, and chronic effects; symptoms include nausea, vomiting, abdominal pain, irritation, confusion ataxia, tremor, and convulsions; delayed development of liver disease and blood disorder also reported (US EPA 1988); human death may result from ingestion of 10–20 g of pure compound or topical skin application of 50 g in 30 minutes; moderately irritating to skin; oral LD_{50} value in rats ~300 mg/kg; exposure limit 0.5 mg/kg: exposure limit 0.5 mg/m^3 (skin); RCRA Waste Number U036; US EPA listed extremely hazardous substance; LD_{50} data in literature inconsistent:

LD_{50} oral (rat): 200–600 mg/kg

LD_{50} oral (rabbit): 100 mg/kg

LD_{50} skin (rat): 690 mg/kg

46.8 ENDOSULFAN

Formula: $C_9H_6Cl_6O_3S$; MW 406.95; CAS [115-29-7]

Structure:

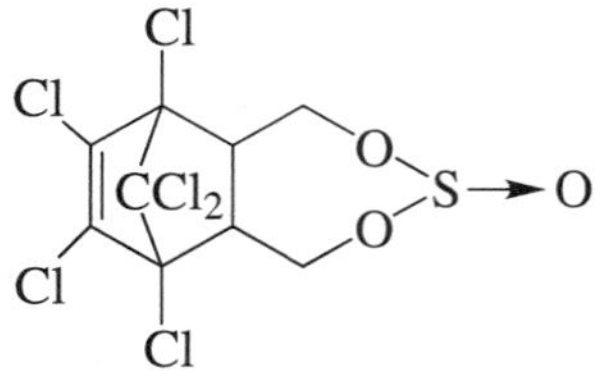

Synonyms: 1,4,5,6,7,7-hexachloro-5-norbornene-2,3-dimethanol cyclic sulfite; 1,2,3,4, 7,7-hexachlorobicyclo[2.2.1]-hepten-5-6-bioxymethylene sulfite; 6,7,8,9,10,10-hexachloro-1,5,5a,6,9,9a-hexahydro-6,9-methano-2,4,3-benzodioxathiepin-3-oxide; Chlorthoiepin; Thiosulfan; Thionex

Physical Properties

Commercial product is a mixture of α- and β-isomers; brown crystals; mp 106°C (pure compound); vapor pressure 0.00001 torr (Worthing 1979); insoluble in water, soluble in most organic solvents; hydrolyzed by alkalies.

Health Hazard

Highly toxic; exhibits acute, delayed, and chronic toxicity; symptoms include nausea, vomiting, weakness, somnolence, excitement, and muscle contraction; causes dermatitis and irritation on skin contact; a stimulant to central nervous system, producing convulsions; can cause adverse reproductive effects; no evidence of carcinogenicity in rats and mice; oral LD$_{50}$ value (mice): ~7.5 mg/kg; exposure limit: PEL/TLV-TWA (skin) 0.1 mg/m^3 (OSHA:ACGIH); RCRA Waste Number P050; US EPA listed extremely hazardous substance.

46.9 DDT

Formula: C$_{14}$H$_9$Cl$_5$; MW 354.48; CAS [50-29-3]

Structure:

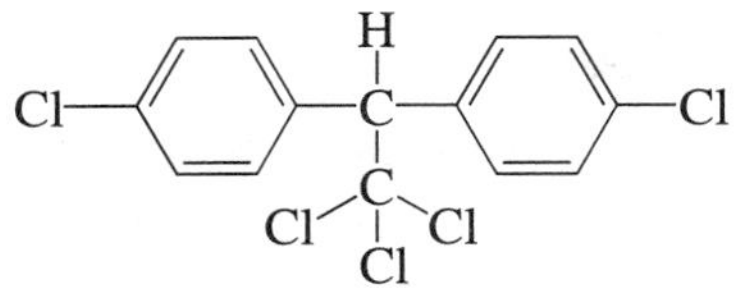

Synonyms: *p,p′*-dichlorodiphenyltrichloroethane; 2,2-bis(*p*-chlorophenyl)-1.1.1-trichloroethane; 1,1′-(2,2,2-trichloroethylidene)bis(4-chlorobenzene); Dicophane

Physical Properties

Colorless to white powder with a faint odor; melts at 109°C; decomposes at high temperatures; insoluble in water, soluble in organic solvents.

Health Hazard

Acute toxicity: low to moderate; high doses cause headache, dizziness, confusion, sweating, tremor, and convulsions; may accumulate in body tissues, causing delayed ill effects after years of low-level exposure; chronic effects include liver damage, central nervous system degeneration, dermatitis, and convulsions; oral LD$_{50}$ value (rats): ~120 mg/kg; a teratogen causing adverse effects on embroying fetal development and adverse estrogenic activity; carcinogenicity: animal sufficient evidence, causing liver cancers; human inadequate evidence; exposure limit: TLV-TWA 1 mg/m^3 (ACGIH, MSHA, and OSHA); RCRA Waste Number U061.

Bioaccumulation

DDT is highly resistant to biodegradation. It persists for a long time in the environment, and is still present in many agricultural areas, where it was last applied as a pesticide 30 years ago. It is highly lipophilic; the octanol–water partition coefficient, K$_{0w}$ being $>10^6$ and, hence, susceptible to cross, accumulatering in cell membranes easily. As a result, DDT and other toxic xenobiotics may be found at elevated levels in aquatic species (Bunce 1995).

46.10 *p,p′*-DDD

Formula: C$_{14}$H$_{10}$Cl$_4$; MW 320.05; CAS [72-54-8]

Structure:

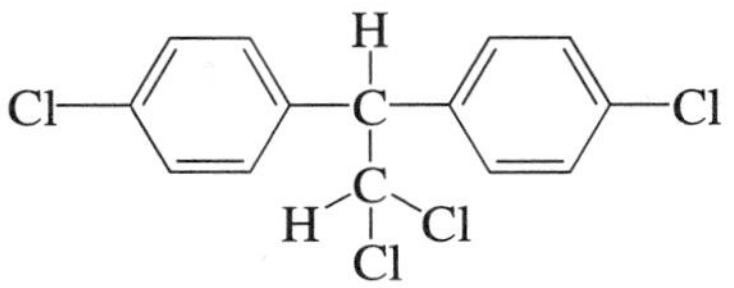

Synonyms: 1,1-dichloro-2,2-bis(*p*-chlorophenyl)ethane; 1,1′-(2,2dichloro-ethylidene)bis[bis[4-chlorobenzene]; 4,4′-(dichlorodiphenyl)dichloroethane; *p,p′*-TDE; Rhothane

Physical Properties

Crystalline solid; melts at 110°C; decomposes at high temperature; insoluble in water; soluble in most organic solvents.

Health Hazard

A derivative of DDT; toxic properties similar to DDT; systemic effects from ingestion include headache, anesthesia, cardiac arrhythmias, nausea, vomiting, sweating, and convulsions; oral lethal dose in mice 600 mg/kg; also moderately toxic by skin absorption; susceptible to accumulation in fat; sufficient evidence of carcinogenicity in experimental animals.

LD_{50} oral (rat): 113 mg/kg
LD_{50} skin (rabbit): 1200 mg/kg

46.11 *p,p'*-DDE

Formula: $C_{14}H_8Cl_4$; MW 318.02; CAS [72-55-9]
Structure:

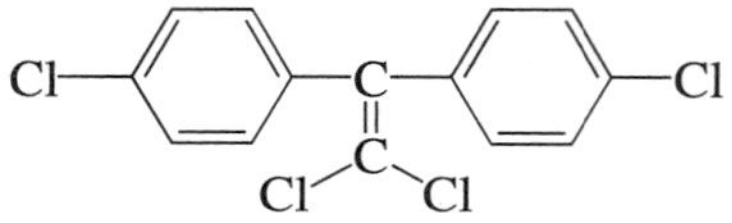

Synonyms: 2,2-bis(*p*-chlorophenyl)-1,1-dichloroethylene; *p,p'*-dichlorodiphenyl-dichloroethylene; 1,1-dichloro-2,2-bis(*p*-chlorophenyl)ethene; DDT dehydrochloride

Physical Properties

Crystalline solid; decomposes when heated to high temperature; insoluble in water; soluble in most organic solvents.

Health Hazard

A derivative of DDT; structurally and toxic properties are similar to those of DDT; however, the acute effects are somewhat milder; moderately toxic by ingestion; a teratogenic substance; adequate evidence of carcinogenicity in experimental animals

LD_{50} oral (rat): 880 mg/kg
LD_{50} oral (mouse): 700 mg/kg

46.12 METHOXYCHLOR

Formula: $C_{16}H_{15}Cl_3O_2$; MW 345.66; CAS [72-43-5]
Structure:

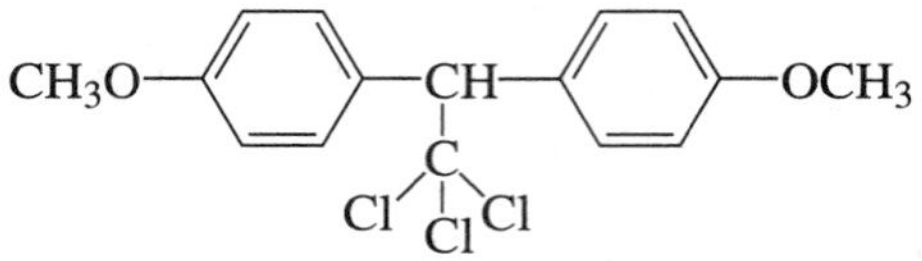

Synonyms: 1,1,1-trichloro-2,2-bis(*p*-methoxyphenyl)ethane; 2,2-bis(*p*-anisyl)-1,1,1-trichloroethane; 1,1-bis(*p*-methoxyphenyl)-2,2,2-trichloroethane; *p,p'*-dimethoxydiphenyltrichloroethane; dimethoxy-DDT

Physical Properties

Crystalline solid: melts at 78°C; insoluble in water, soluble in organic solvents.

Health Hazard

Low toxicity; acute and chronic effects are much less serious than those of the structurally similar DDT; skin absorption low; may cause kidney damage on chronic exposure; oral LD_{50} value (rats): 6000 mg/kg (ACGIH 1986); no evidence of carcinogenicity to animals or humans; exposure limit: TLV-TWA 10 mg/m^3 (ACGIH, MSHA, and OSHA); RCRA Waste Number U247.

46.13 PERTHANE

Formula: $C_{18}H_{20}Cl_2$; MW 307.28 CAS [72-56-0]

Structure:

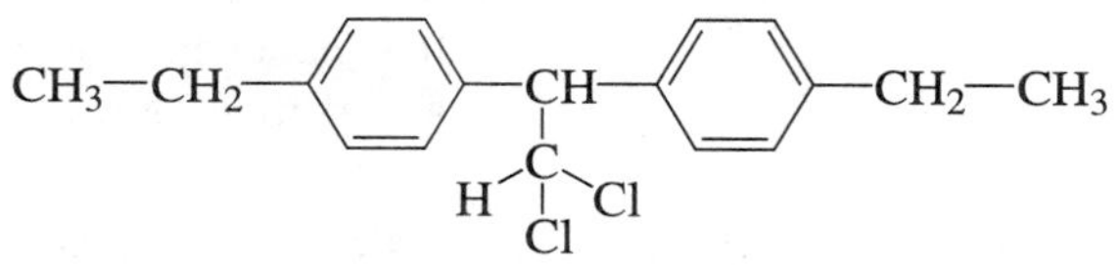

Synonyms: 1,1'-(2,2-dichloroethylidene)bis [4-ethylbenzene]; 2,2-dichloro-1,1-bis(*p*-ethylphenyl)ethane; 1,1-bis(*p*-ethylphenyl)-2,2-dichloroethane; 1,1-dichloro-2,2-bis(4-ethylphenyl)ethane; *p,p'*-ethylDDD; Ethylan

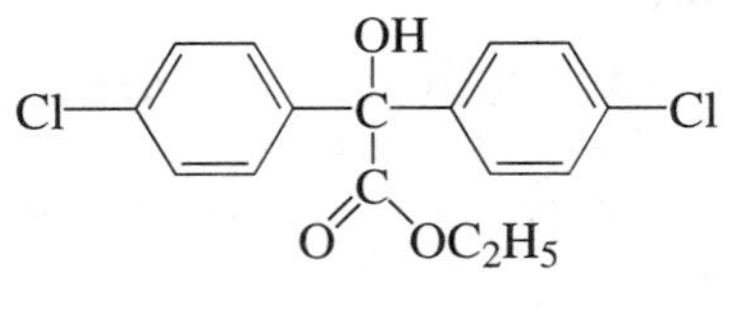

Synonyms: ethyl 2-hydroxy-2,2-bis(4-chlorophenyl)acetate; 4,4'-dichlorobenzilic acid ethyl ester; 4-chloro-α-(4-chlorophenyl)-α-hydroxybenzeneacetic acid ethyl ester; 4,4'-dichlorobenzilate

Physical Properties

Crystalline solid; mp 56°C; insoluble in water; soluble in acetone, kerosene, and diesel fuel.

Physical Properties

Yellowish viscous liquid; b.p. 146°C; vapor pressure 2.2×10^{-6} torr at 20°C; slightly soluble in water; soluble in most organic solvents; decomposes in alkaline solution.

Health Hazard

Mildly toxic by ingestion; the oral median lethal doses in all experimental animals were >5000 mg/kg; however, produced moderate to severe effects when administered intravenously; toxic properties are somewhat similar to those of DDT; susceptible to storage in fat; also bioaccumulative; a teratogenic substance; sufficient evidence of carcinogenicity in mice, but inadequate evidence in other animals; cancer-causing effects in human unknown.

LD$_{50}$ oral (rat): 6600 mg/kg
LD$_{50}$ oral (wild bird): 9000 mg/kg
LD$_{50}$ intravenous (rat): 73 mg/kg

Health Hazard

Moderately toxic by oral route; toxic symptoms similar to DDT and Perthane; ingestion of large dose can produce nausea, vomiting, tremor and convulsions; skin or eye contact can cause irritation; application of 25 mg produced moderate irritation of eye in rabbits; adequate evidence of carcinogenicity in experimental animals; produced tumors in rats and mice; evidence of carcinogenicity in humans remains unknown.

LD$_{50}$ oral (rat): 700 mg/kg
LD$_{50}$ oral (mouse): 729 mg/kg

46.14 CHLOROBENZILATE

Formula: $C_{16}H_{14}Cl_2O_3$; MW 325.20 CAS [510-15-6]
Structure:

46.15 MIREX

Formula: $C_{10}Cl_{12}$; MW 545.59; CAS [2385-85-5]

Synonyms: 1,1*a*,2,2,3,3*a*,4,5,5,5*a*,5*b*,6-dodecachlorooctahydro-1,3,4-metheno-1*H*-cyclobutal[*c,d*]pentalene; dodecachloropentacylodecane; hexachlorocyclopentadiene dimer; Dechlorane; Ferriamicide; Paramex

Structure:

Synonyms: 1,1a,3,3a,4,5,5,5a,5b,6-decachlo-
rooctahydro-1,3,4-metheno-2H-cyclobuta
[c,d]pentalen-2-one; decachlorotetrahydro-
4,7-methanoindeneone; Chlordecone

Physical Properties

Crystalline solid; decomposes at 350°C; bp
170°C at 0.05 torr; slightly soluble in water
(145 mg/L) and hydrocarbons; readily dissolves in acetone, alcohol, and acetic acid

Health Hazard

Highly toxic by oral, dermal, and possibly
other routes of exposure; toxic properties
similar to those of Mirex; however, more toxic
than the latter; symptoms include tremors,
ataxia, hyperactivity, muscle spasms, and skin
changes; highly injurious to liver, kidney,
and central nervous system; a teratogenic
substance; caused testicular atrophy, sterility,
low sperm count, and breast enlargement
in experimental animals; sufficient evidence
of carcinogenicity in experimental animal;
also, possibly carcinogenic to humans. (IARC
1987)

LD_{50} oral (rat): 95 mg/kg
LD_{50} oral (rabbit): 65 mg/kg
LD_{50} skin (rabbit): 345 mg/kg

Exposure Limit

Ceiling: 0.001 mg/m^3/15 min (NIOSH)

Physical Properties

White crystalline solid; odorless; decomposes at 485°C; insoluble in water; soluble
in most organic solvents

Health Hazard

Highly toxic by ingestion and moderately
toxic by absorption through skin and inhalation; the toxic symptoms include tremor,
ataxia, muscle spasms, and kidney and liver
damage; a teratogenic substance causing
adverse reproductive effects; adequate evidence of carcinogenicity in experimental
animals and a probable human carcinogen
(IARC 1996)

LD_{50} oral (rat): 235 mg/kg
LD_{50} oral (hamster): 125 mg/kg
LD_{50} skin (rabbit): 800 mg/kg

46.16   KEPONE

Formula: $C_{10}Cl_{10}O$; MW 490.68; CAS [143-
50-0]
Structure:

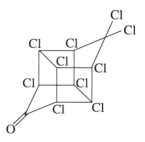

46.17   TOXAPHENE

Empirical formula $C_{10}H_{10}Cl_8$, a complex
mixture of a large number of C_{10} chloroderi-
vatives of camphene; CAS [8001-35-2]

Physical Properties

Yellow wax-like solid; mild pleasant odor;
melts at 65–90°C; insoluble in water, soluble in benzene, toluene, and other aromatic
solvents; dechlorinates in the presence of

alkalies or when heated to $>150°C$; also loses its chlorine on prolonged exposure to sunlight.

Health Hazard

Highly toxic by ingestion; moderately toxic by skin contact and inhalation; may cause skin irritation and allergic dermatitis; caused adverse reproductive effects in experimental animals; ingestion of about 1.5–3 g may be fatal to adult human; toxic effects in animals include central nervous system stimulation, tremors, convulsions, and liver injury; oral LD_{50} value (rats): ~100 mg/kg: sufficient evidence of carcinogenicity in animals, causing liver cancer; RCRA Waste Number P123.

LD_{50} oral (rat): 50 mg/kg
LD_{50} skin (rat): 600 mg/kg

46.18 STROBANE

Formula: mixture of several compounds consisting of many terpene polychlorinates including chlorinated camphene, pinene, and related compounds (~65% Cl content); CAS [8001-50-1]

Synonyms: terpene polychlorinates; Dichloricide aerosol; Dichloricide mothproofer

Physical Properties

Amber-colored viscous liquid; density 1.627 at 25°C; insoluble in water; slightly miscible in alcohol; soluble in hydrocarbon solvents.

Health Hazard

Moderately toxic by ingestion; mild skin irritant; animal experiments indicate sufficient evidence of carcinogenicity; cancer-causing effects in humans remain unknown.

LD_{50} oral (rat): 200 mg/kg
LD_{50} oral (dog): 200 mg/kg

46.19 CHLORBENSIDE

Formula: $C_{13}H_{10}Cl_2S$; MW 269.20 CAS [103-17-3]
Structure:

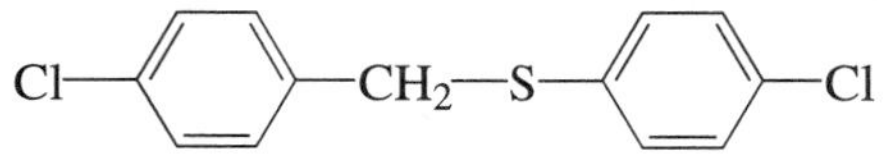

Synonyms: 1-chloro-4-[[(4-chlorophenyl) methyl]thio]benzene; *p*-chlorobenzyl *p*-chlorophenylsulfide; Chlorocide; Chloroparacide; Chlorosulfacide; Mitox

Physical Properties

Crystalline solid; almond odor (technical grade); melts at 75°C; vapor pressure 1.2×10^{-5} torr; slightly soluble in water; moderately soluble in alcohol; readily dissolves in acetone, benzene, and petroleum ether; oxidized to sulfoxide and sulfone by strong oxidizing agents.

Health Hazard

Low to moderate toxicity in experimental animals from oral administration; chronic effects may produce liver and kidney injury in humans.

LD_{50} oral (rat): 2000 mg/kg

46.20 CHLOROPHACINONE

Formula: $C_{23}H_{15}ClO_3$; MW 374.82; CAS [3691-35-8]
Structure:

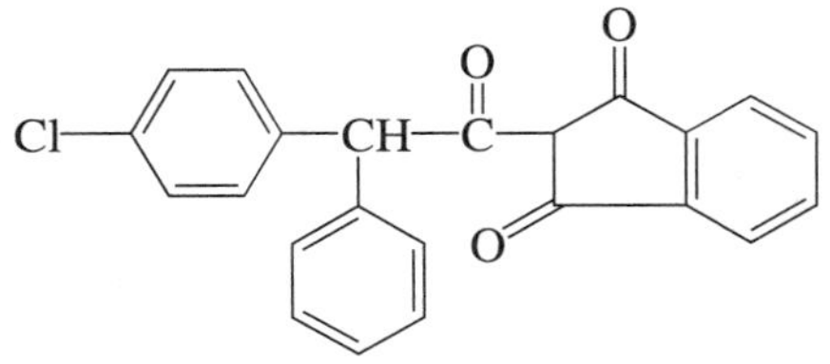

Synonyms: 2((p-chlorophenyl)phenylacetyl)-1H-indene-1,3(2H)dione; 2-((p-chlorophenyl)phenylacetyl-1,3-indandione; Liphadione; Afnor; Topitox; Ranac

Physical Properties

Yellow crystalline solid; also commercially available as oil or powder; melts at 140°C; practically insoluble in water; dissolves in most organic solvents.

Health Hazard

Highly toxic anticoagulant rodenticide; low chlorine content but highly toxic; exhibits acute, delayed, and chronic poisoning; symptoms include bleeding of nose and gum, blood in cough, urine, and stool, abdominal pain and hemorrhage; can be absorbed through skin causing systemic poisoning; ingestion of $1-2$ g is estimated to be fatal dose for adult human; US EPA-listed extremely hazardous substance.

LD_{50} oral (rat): $\sim$2 mg/kg
LD_{50} oral (mouse): $\sim$1 mg/kg
LD_{50} oral (rabbit): 50 mg/kg
LD_{50} skin (rabbit): 200 mg/kg

REFERENCES

ACGIH. 1990. *Documentation of the Threshold Limit Values and Biological Exposure Indices*, 5th ed. Cincinnati, OH: American Conference of Governmental Industrial Hygienists.

Cook, R. M. and K. A. Wilson. 1971. Removal of pesticide residues from dairy cattle. *J. Dairy Sci.*, *54*: 712–17.

Davies, J. E., Edmundson, W. F., Maceo, A., Irvin, G. L. Cassady, J. and A. Barquet. 1971. Reduction of pesticide residues in human adipose tissue with diphenylhydantoin. *Food Cosmet. Toxicol.*, *9*: 413–23.

Garretson, L. K. and A. Curley. 1969 Dieldrin. Studies on a poisoned child. *Arch. Environ. Health*, *19*: 814–22.

Hayes, W. J., Jr. 1975. *Toxicology of Pesticides*. Baltimore, MD: Williams & Wilkins.

Hunter, C. G. and J. Robinson. 1967. Pharmacodynamics of dieldrin (HEOD) I. Ingestion by human subjects for 18 months. *Arch. Environ. Health*, *15*: 614–26.

IARC. 1987. IARC Monograph. Vol. 20, Suppl. 7. Lyon, France; International Agency for Research on Cancer.

Lewis, R. J. (Sr.) 1995. *Sax's Dangerous Properties of Industrial Materials*, 9th ed. New York: Van Nostrand Reinhold.

Merck. 1996. *The Merck Index*, 12th ed. Rahway, NJ: Merck & Co.

Milne, G. W. A. 1995. *CRC Handbook of Pesticides*. Boca Raton, FL: CRC Press.

Street, J. C. and R. P. Sharma. 1977. Accumulation and release of chemicals by adipose tissue. In *Handbook of Physiology—Reactions to Environmental Agents*, Section 9, Douglas H. K. Lee, ed. pp. 483–493. Bethesda, MD: American Physiological Society.

US EPA. 1988. *Extremely Hazardous Substances: Superfund Chemical Profiles*. Park Ridge, NJ: Noyes Data Corp.

47

PESTICIDES, ORGANOPHOSPHORUS

47.1 TOXICITY

Organophosphorus insecticides have been attributed to several cases of human poisonings and deaths. The first compound of this class, or tetraethylpyrophosphate (TEPP) [107-49-3], was synthesized during World War II, which led to the development of extremely toxic nerve gases as chemical warfare agents (see Nerve Gases, Chapter 39). TEPP is unstable and rapidly hydrolyzed by moistures. It is also highly toxic to mammals. Development of other insecticides, including parathion [56-38-2] and paraoxon [311-45-5], soon followed.

The biochemical actions of organophosphates involve the inhibition of the function of the enzyme, acetylcholinesterase, which causes rapid hydrolysis of acetylcholine. The latter is the chemical transmitter of nerve impulses in several components of the nervous system. The nervous system requires almost instantaneous removal of acetylcholine once the impulse has been transmitted. The enzyme acetylcholinesterase performs that function by rapidly hydrolyzing acetylcholine to choline and acetate (Hodgson et al. 1988).

Organophosphates have the following general structure:

$$\begin{array}{c} RO \\ \diagdown \\ RO \diagup \end{array} \overset{\displaystyle O \text{ (or S)}}{\underset{\displaystyle }{\overset{\|}{P}}} - O \text{ (or S)} \text{---leaving group}$$

where R is a methyl or ethyl group. They phosphorylate or bind the enzyme at its esteractic site, thus inhibiting the action of the enzyme. The phosphorylated enzyme is hydrolyzed very slowly, a few days to weeks; thus the recovery of the enzyme is slow. In other words, the poisoning effects of organophosphorus insecticides may persist for several days. The acute toxic effects resulting from the accumulation of acetylcholine in muscarinic receptors include tightness in the chest, bronchial secretions, increased salivation and lacrimation, sweating, nausea, vomiting, abdominal cramps, diarrhea, frequent urination, and constriction of the pupils. Accumulation of acetylcholine at the ending of motor nerves produces nicotinic signs and muscular effects. The symptoms are fatigue, weakness, twitching, cramps, pallor, elevation of blood pressure, and hyperglycemia. Accumulation in the central nervous system causes headache, restlessness, insomnia, slurred speech, tremor, ataxia, convulsions, and coma (Murphy 1980). Death may result from respiratory failure. Many organophosphorus insecticides have a thiono group (=S) in which the S atom is bound to the phosphorus by a double bond. Such pesticides (e.g., parathion, azinphos methyl) in the pure form have little anticholinesterase activity. They exhibit delayed toxic actions several hours after the thiono group is converted to oxon. Such metabolic activation in which =S is replaced by =O occurs mainly in the liver.

The chronic effects from organophosphorus pesticides are very limited. These are often metabolized rapidly and excreted. Atropine exhibits protection against such toxicity. It may be noted that although organophosphates are more acutely toxic than are the organochlorine pesticides, the latter are more stable, accumulating in body tissues to produce severe ill effects after a long time. The oral and dermal LD$_{50}$ values in rats of some of the organophosphorus pesticides are presented in Table 47.1. The physical and toxic properties of some selected individual substances of this class are discussed in the following sections.

47.2 PARATHION

Formula: C$_{10}$H$_{14}$NO$_5$PS; MW 291.27; CAS [56-38-2]

Structure:

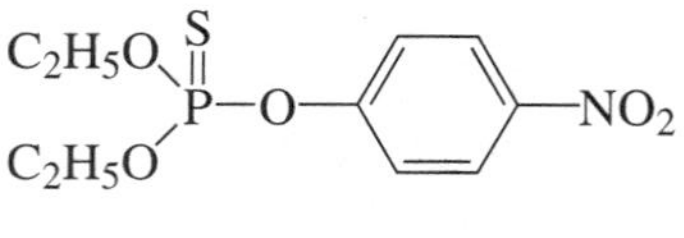

Synonyms: *O,O*-diethyl-*O*-(-*p*-nitrophenyl) thionophosphate; *O,O*-diethyl-*O*-(-*p*-nitrophenyl)phosphorothioate; phosphorothioic acid *O,O*-diethyl-*O*-(-*p*-nitrophenyl) ester; ethyl parathion; thiophos

Physical Properties

Pale yellow liquid; density 1.26 at 25°C; boils at 375°C; solidifies at 6°C; vapor pressure 3.78 × 10^{-5} torr at 20°C; insoluble in water (~20 ppm at 20°C), soluble in most organic solvents.

Health Hazard

Extremely toxic; acetylcholinesterase inhibitor; toxic symptoms include nausea, vomiting, diarrhea, excessive salivation, lacrimation, constriction of the pupils, bronchoconstriction, convulsions, coma, and respiratory failure; metabolizes to paraoxon; oral LD$_{50}$ value (rats): 2 mg/kg, LD$_{50}$ value, skin (rats): 6.8 mg/kg RCRA Waste Number P089.

47.3 MEVINPHOS

Formula: C$_7$H$_{13}$O$_6$P; MW 224.17; CAS [7786-34-7]

Structure:

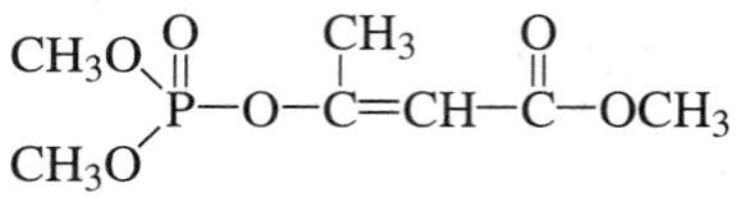

Synonyms: 2-carbomethoxy-1-methylvinyl dimethyl phosphate; dimethyl(1-methoxycarboxyproprn-2-yl)phosphate; methyl 3-(dimethoxyphosphinyloxy)crotonate; 3-hydroxycrotonic acid methyl ester dimethyl phosphate; 3-((dimethoxyphosphinyl)oxy)-2-butenoic acid methyl ester; phosdrin

Physical Properties

Colorless liquid; commercial product is a mixture of *cis*- and *trans*-isomers and is yellowish; density 1.25 at 20°C; boils at 107°C; miscible with water and most organic solvents, insoluble in hexane and petroleum ether; corrosive to iron, steel, and brass.

Health Hazard

A severely acute toxicant by all routes; cholinesterase inhibitor; absorbed through the skin, lungs, and mucous membranes; more toxic by subcutaneous or intravenous routes than by an intraperitoneal route; toxic symptoms include headache, weakness,

TABLE 47.1   Median Lethal Doses of Organophosphorus Pesticides

Pesticide/CAS No.	Structure	LD$_{50}$ (mg/kg) Oral	LD$_{50}$ (mg/kg) Skin
TEPP [107-49-3]		0.5–1.1 (rat) 3 (mouse)	2.4 (rat) 8 (mouse)
Phorate [298-02-2]		1.4* (rat) 0.6 (duck)	20 (guinea pig) 203 (duck)
Demeton [8065-48-3]		1.7 (rat)	8.2 (rat)
Fensulfothion [115-90-2]		2.0	3.0
Dimefox [115-26-4]		1.0 (rat)	—
Disulfoton [298-04-4]		2.0	6.0
Parathion [56-38-2]		2.0	6.8
Nemaphos [297-97-2]		3.5	8.0
Mevinphos [7786-34-7]		4.0 (mouse) 4.6 (duck)	12.0 (mouse) 11 (duck)
Parathion methyl [298-00-0]		6.0	63
Nemafos (Thionazin) [297-97-2]		3.5–6.5 (rat)	8 (rat) 7 (duck)

(*continued*)

TABLE 47.1 *(Continued)*

Pesticide/CAS No.	Structure	LD$_{50}$ (mg/kg) Oral	Skin
Amiton [78-53-5]		3.3	—
Fonofos (Dyfonate) [944-22-9]		3–25* (rat) 3 (dog)	147 (rat) 25 (rabbit) 278 (guinea pig)
Schradan [152-16-9]		5–10 (rat) 25 (rabbit)	15 (rat)
Metamidophos [10265-92-6]		7.5 (rat) 10 (rabbit)	50 (rat) 118 (rabbit)
EPN [2104-64-5]		7.0	25
Azinphos methyl [86-50-0]		7.0	220
Coumaphos [56-72-4]		13.0	860
Trichloronate [327-98-0]		15	64
Somonil [950-37-5]		20 (rat) 25 (mouse)	25 (rat)

TABLE 47.1 (*Continued*)

Pesticide/CAS No.	Structure	LD$_{50}$ (mg/kg)	
		Oral	Skin
Monocrotophos [6923-22-4]	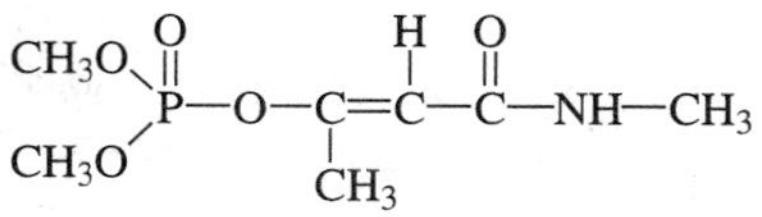	8 (rat) 0.8 (wild bird)	112 (rat) 4.2 (wild bird)
Chlorthiophos [21923-23-9]	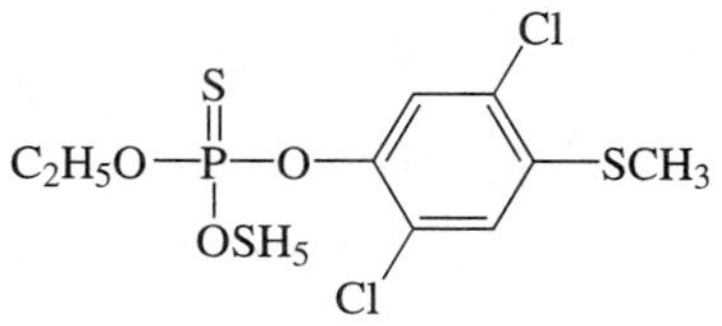	13 (rat) 20 (rabbit)	58 (rat) 31 (rabbit)
Dicrotophos [141-66-2]	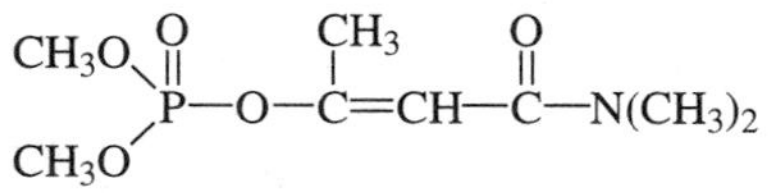	13–21 (rat) 11–15 (mouse)	42 (rat) 1.3 (wild bird)
Cyanofos [2636-26-2]	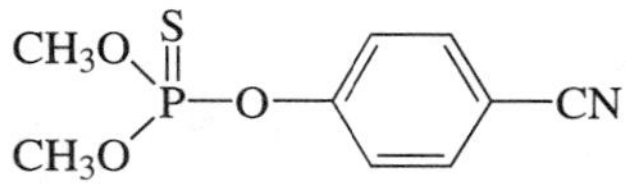	18 (rat)	—
Dialifor [10311-84-9]	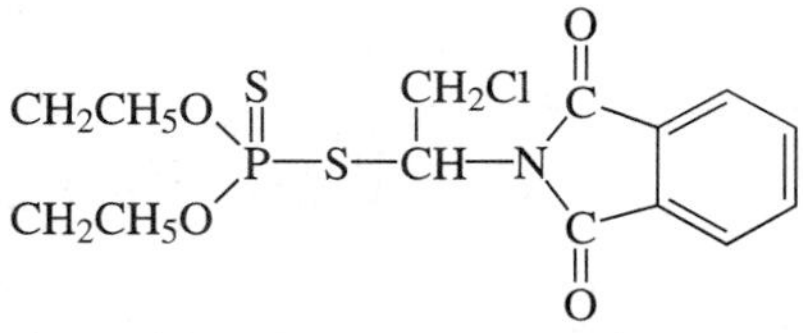	5–50* (rat) 35 (rabbit)	28 (skin) 145 (skin)
Phosphamidon (Dimecron) [13171-21-6] [297-99-4]	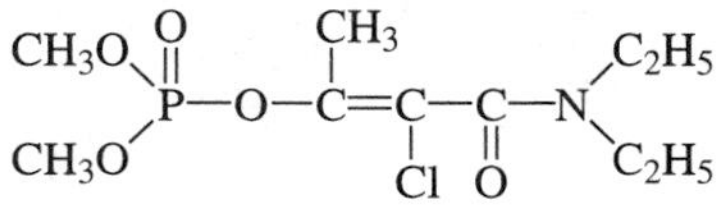	8–30* (rat) 70 (rabbit) 3.1 (duck)	125 (rat) 80 (rabbit) 26 (duck)
Fenamiphos [22224-92-6]	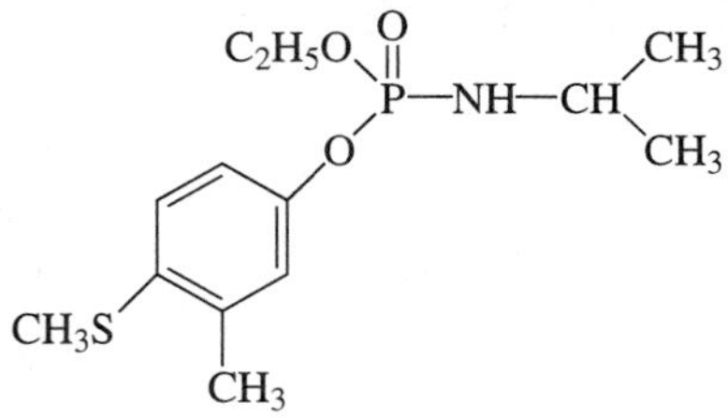	10–25* (rat) 10 (rabbit)	80 (rat) 178 (rabbit)
Chlorfenvinphos [470-90-6]	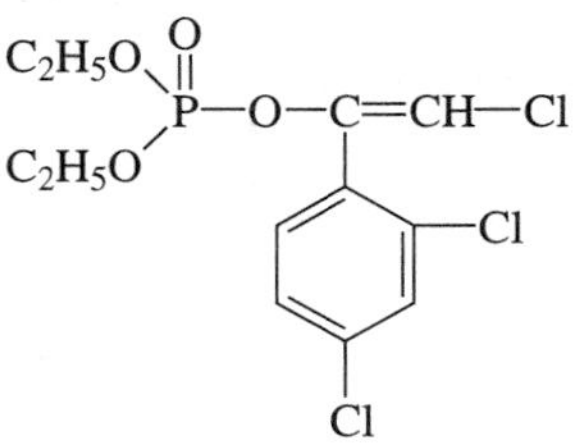	10 (rat) 65 (mouse) 300 (rabbit)	— 336 (mouse) 400 (rabbit)

(*continued*)

TABLE 47.1 (*Continued*)

Pesticide/CAS No.	Structure	LD_{50} (mg/kg) Oral	LD_{50} (mg/kg) Skin
Ethion [563-12-2]	$(C_2H_5O)\!-\!\overset{\displaystyle S}{\underset{\displaystyle OC_2H_5}{P}}\!-\!S\!-\!CH_2\!-\!S\!-\!\overset{\displaystyle S}{\underset{\displaystyle OC_2H_5}{P}}\!-\!OC_2H_5$	40–45 (mouse) 40 (guinea pig)	245 (rat) 915 (guinea pig)
Prophos [13194-48-4]	$(C_2H_5O)\!-\!P\overset{S\!-\!CH_2\!-\!CH_2\!-\!CH_3}{\underset{S\!-\!CH_2\!-\!CH_2\!-\!CH_3}{}}$	34–62* (rat) 55 (rabbit)	60 (rat) 11 (duck)
Carbofenthion [786-19-6]	$\overset{C_2H_5O}{\underset{C_2H_5O}{}}\!\overset{S}{P}\!-\!S\!-\!CH_2\!-\!S\!-\!\bigcirc\!-\!Cl$	7–30 (rat) 5.6 (wild bird)	27 (rat) —
Dichlorvos [62-73-7]	$CH_3O\!-\!\overset{\displaystyle O}{\underset{\displaystyle OCH_3}{P}}\!-\!O\!-\!CH\!=\!C\overset{Cl}{\underset{Cl}{}}$	15–80* (rat) 100 (dog)	70 (rat) 206 (mouse)
Dioxathion [78-34-2]	dioxane ring with S—P(OC$_2$H$_5$)$_2$ groups (C=S)	20–43 (rat) 10 (dog)	63–235* (rat) 85 (rabbit)
Famphur [52-85-7]	$\overset{CH_3O}{\underset{CH_3O}{}}\!\overset{S}{P}\!-\!O\!-\!\bigcirc\!-\!\overset{\displaystyle O}{\underset{\displaystyle O}{S}}\!-\!N(CH_3)_2$	35 (rat) 9.5 (mouse)	400 (rat) 1460 (rabbit)
Metasystox [919-86-8]	$\overset{CH_3O}{\underset{CH_3O}{}}\!\overset{O}{P}\!-\!S\!-\!CH_2\!-\!CH_2\!-\!S\!-\!CH_2\!-\!CH_3$	30–105* (rat)	85 (rat)
Crotoxyphos [7700-17-6]	$CH_3\!-\!\overset{\displaystyle OCH_3}{\underset{\displaystyle O}{P}}\!-\!O\!-\!\underset{CH_3}{C}\!=\!CH\!-\!\overset{\displaystyle O}{C}\!-\!O\!-\!\underset{CH_3}{CH}\!-\!\bigcirc$	35–110* (rat) 56 (wild bird)	202 (rat) 385 (rabbit)
Oxydemetonmethyl [301-12-2]	$\overset{CH_3O}{\underset{CH_3O}{}}\!\overset{O}{P}\!-\!S\!-\!CH_2\!-\!CH_2\!-\!\overset{\displaystyle O}{S}\!-\!CH_2\!-\!CH_3$	30–75* (rat) 10 (mouse)	100 (rat) —
Phosalone [2310-17-0]	$\overset{C_2H_5O}{\underset{C_2H_3O}{}}\!\overset{S}{P}\!-\!S\!-\!CH_2\!-\!N$ benzoxazolone ring with Cl	85–175* (rat) 75a–180* (mouse)	390 (rat) 1000 (rabbit)

TABLE 47.1 (Continued)

Pesticide/CAS No.	Structure	LD$_{50}$ (mg/kg)	
		Oral	Skin
Isofenphos (Oftanol) [25311-71-1]	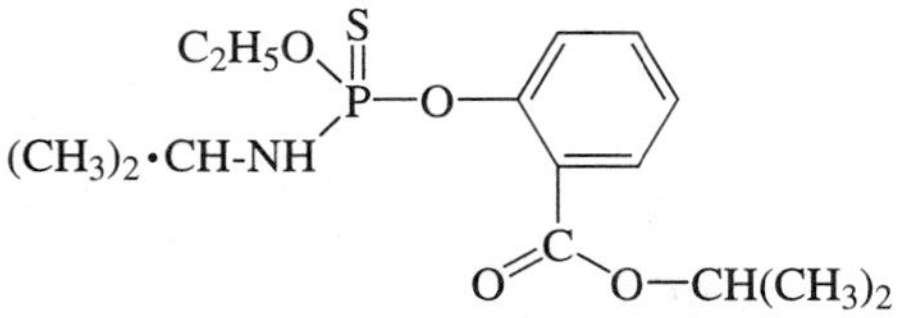	28 (rat) 91 (mouse)	188 (rat) 162 (rabbit)
Isazophos [67329-04-8]	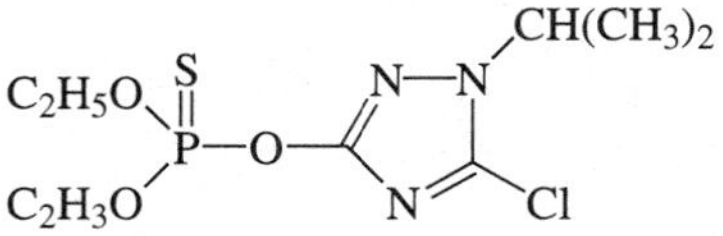	60 (rat)	—
Diazinon [333-41-5]	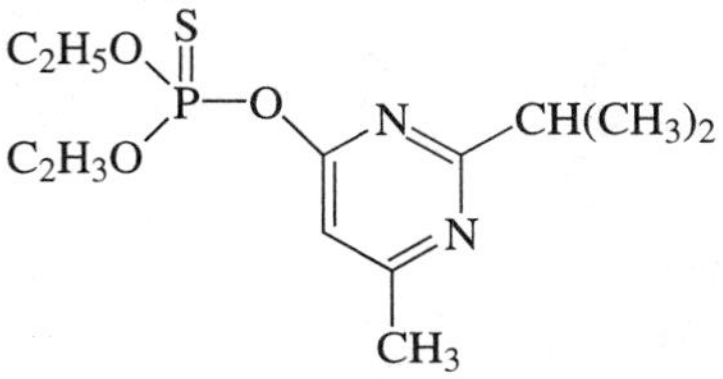	66	180
Chlorpyrifos (Dursban) [2921-88-2]	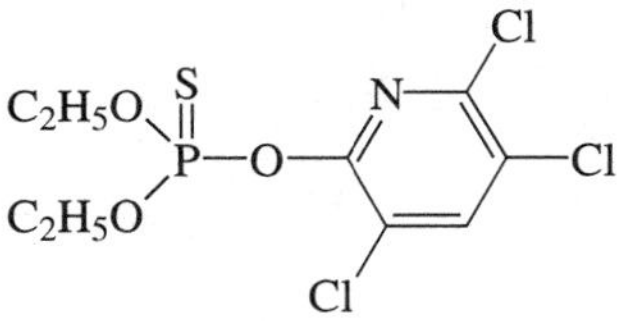	82	202
Propetamphos [31218-83-4]	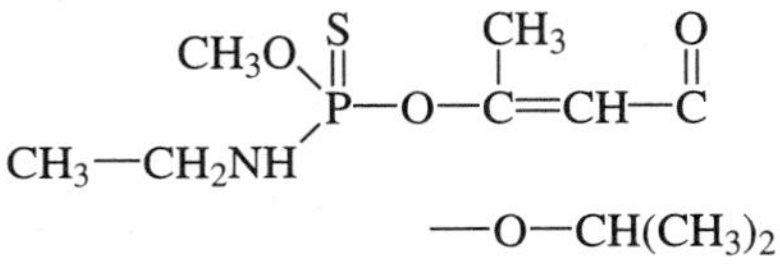	72–120 (rat)	564 (rat)
Pirimiphos-ethyl [23505-41-1]	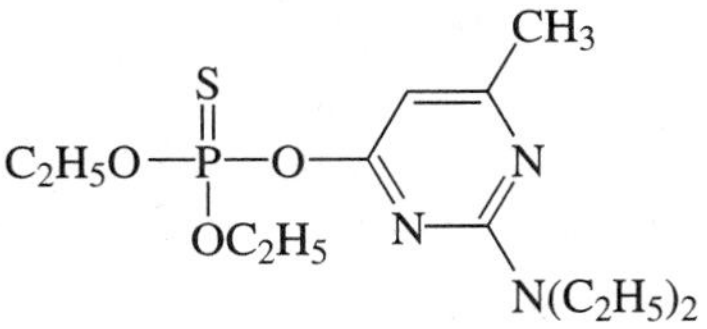	140–190 (rat) 50 (guinea pig)	1000–2000* (rat)
Dimethoate [60-51-5]	$(CH_3O)_2 \overset{\displaystyle S}{\underset{\displaystyle \|}{P}} - S - CH_2CONHCH_3$	152	353
Fenthion [55-38-9]	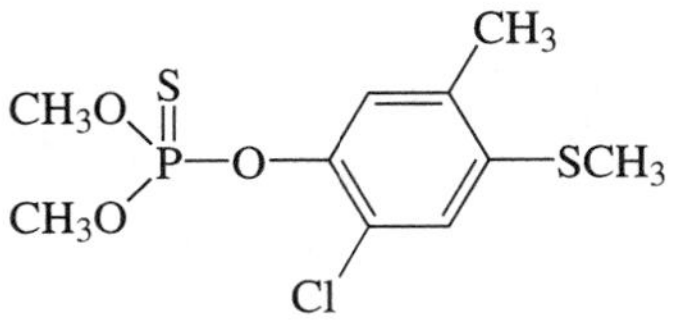	180	330

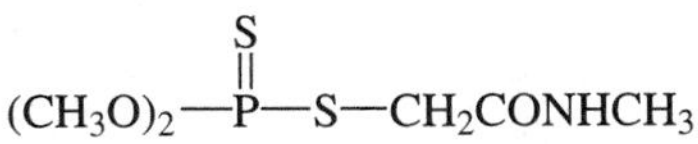

(*continued*)

TABLE 47.1   *(Continued)*

Pesticide/CAS No.	Structure	LD_{50} (mg/kg) Oral	Skin
Naled [300-76-5]		92 (rat) 222 (mouse)	800 (rat) 600 (mouse)
Butiphos [78-48-8]		—	—
Phosmet (Imidan) [732-11-6]		90–160 (rat) 200 (guinea pig)	1326 (rat) >5000 (rabbit)
Profenofos [41198-08-7]		358 (rat) 700 (rabbit)	1610 (rat) 192 (rabbit)
Metathion [122-14-5]		229 (mouse) 500 (guinea pig)	2500 (mouse) 1250 (rabbit)
Malathion [121-75-5]		370	4400
Acephate [30560-19-1]		700 (rat) 233 (mouse)	>2000 (rat) 2000 (rabbit)
Tricholorofon (Chlorofos) [52-68-6]		560–630 (rat) 160 (rabbit)	2000 (rat) 1500 (rat)
Ronnel (Fenchlorfos) [299-84-3]		420 (rabbit) 1400 (guinea pig)	1000 (rabbit) 2000 (guinea pig)

TABLE 47.1   *(Continued)*

Pesticide/CAS No.	Structure	LD$_{50}$ (mg/kg)	
		Oral	Skin
Tokuthion [34643-46-4]	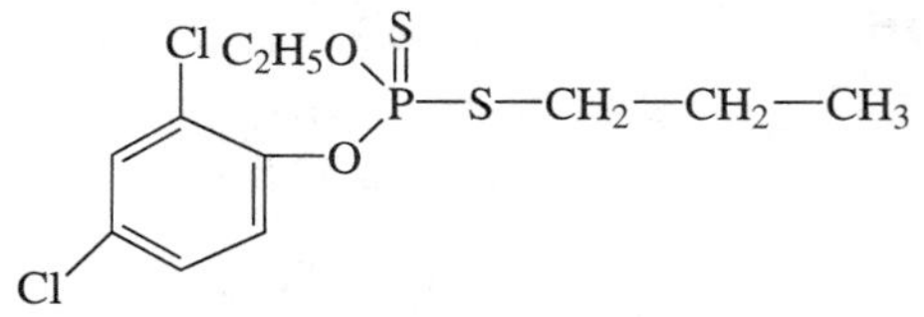	925	1600
Etrimfos [38260-54-7]	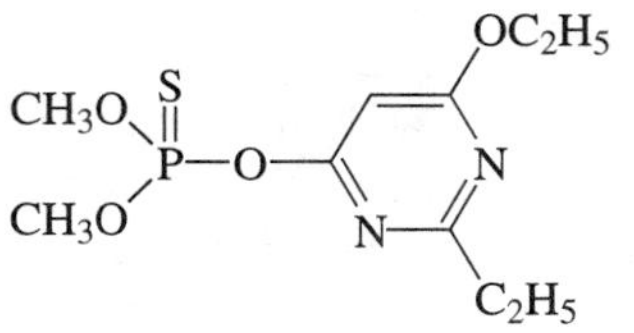	1800 (rat)	>2000 (rat)
Stirofos [961-11-5]	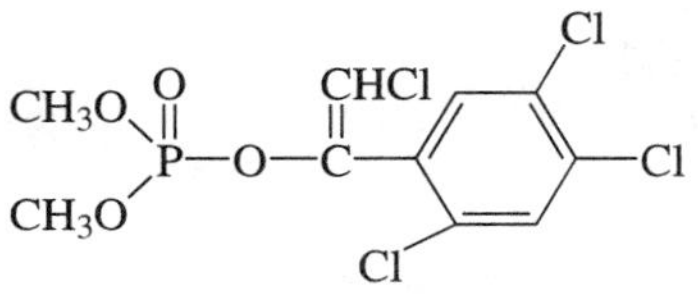	4000 (rat) 4200 (mouse)	>10,000 (rat) >7500 (mouse)
Abate [3383-96-8]	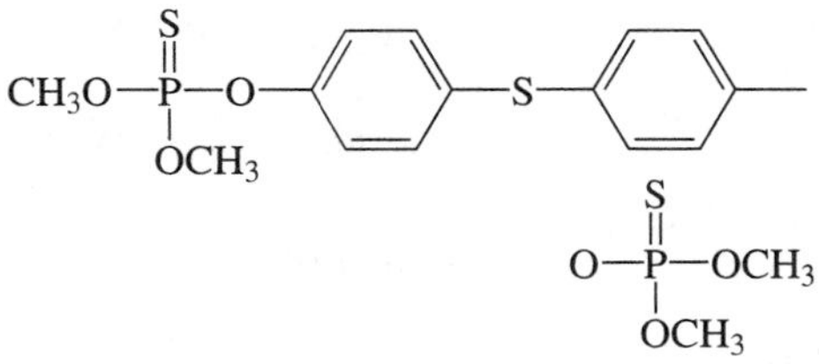	8600 —	—

blurred vision, diarrhea, and tightness in the chest; signs of severe poisoning are sweating, salivation, lacrimation, constriction of the pupils, depression, tremor, and convulsions; oral LD$_{50}$ value (mice): 4 mg/kg; skin LD$_{50}$ value (mice): 12 mg/kg; exposure limit: TLV-TWA 0.01 ($\sim$0.1 mg/m^3) (ACGIH, MSHA, and OSHA).

47.4 AZINPHOS METHYL

Formula: C$_{10}$H$_{12}$H$_3$O$_3$PS$_2$; MW 317.34; CAS [86-50-0]

Structure:

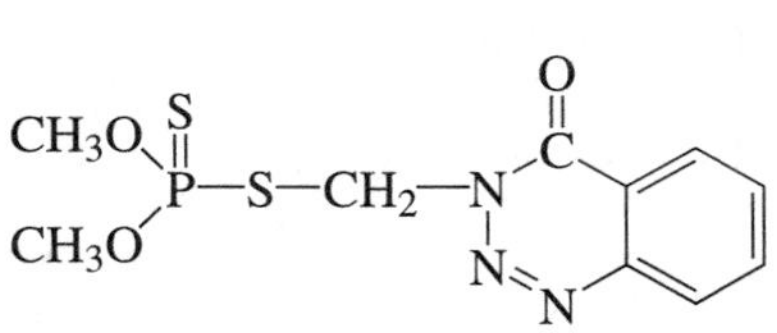

Synonyms: *O,O*-dimethyl-*S*-(3,4-dihydro-4-keto-1,2,3-benzotriazinyl-3-methyl)dithiophosphate; *O,O*-methyl-*S*-(4-oxobenzotriazino-3-methyl)phosphorodithioate; 3-(mercaptomethyl)1,2,3-benzotriazin-4(3*H*)-one; *O,O*-dimethyl phosphorodithioate *S*-ester; Guthion

Physical Properties

White crystalline solid when pure; melts at 73°C; decomposes at >200°C; very slightly soluble in waer (33 ppm at 25°C), soluble in most organic solvents; hydrolyzes in acid or alkali.

Health Hazard

Cholinesterase inhibitor; a severely acute toxicant; delayed effects may be observed after several hours; toxic symptoms similar to other organophosphorus compounds; exposure may cause headache, dizziness, blurred vision, and muscle spasms. Other toxic symptoms include vomiting, abdominal pain, diarrhea, seizures, and shortness of breath; ingestion of 5–10 g could be fatal to adult human; oral LD_{50} value (mice): 15 mg/kg, rats 7 mg/kg; LC_{50} inhalation (rat): 69 mg/m^3/1 hour

Exposure Limits

TLV-TWA: 0.2 mg/m^3 (AIGIH)
PEL-TWA: 0.2 mg/m^3 (NIOSH)
IDLH: 5 mg/m^3 (NIOSH)

47.5 AZINPHOS ETHYL

Formula: $C_{12}H_{16}N_3O_3PS_2$; MW 345.38; CAS [2642-71-9]
Structure:

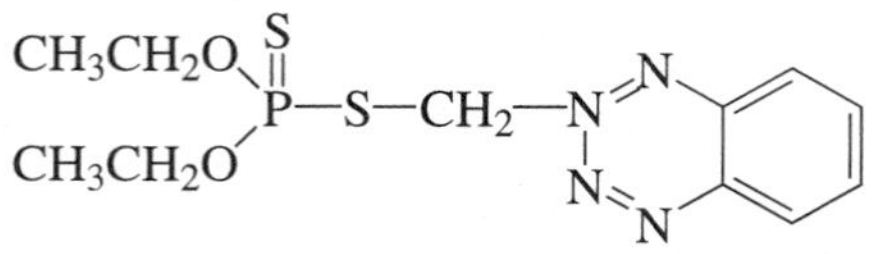

Synonyms: 3,4-dihydro-4-oxo-3-benzotria-zinylmethyl-*O,O*-diethyl phosphorodithi-oate; *O,O*-diethyl *S*-(4-oxobenzyltriazine-3-methyl)dithiophosphate; ethyl guthion; ethyl gusathion

Physical Properties

Crystalline solid; melts at 53°C; vapor pressure 2.2×10^{-7} torr at 20°C; insoluble in water; soluble in most organic solvents; hydrolyzes in acid or alkali.

Health Hazard

A highly toxic substance by all routes of entry, especially ingestion and skin contact; inhalation hazard may be low because of very low vapor pressure (2.2×10^{-7} torr at 20°C); the systemic effects are similar to parathion and azinphos methyl; the toxic effects are nausea followed by vomiting, abdominal pain, diarrhea, salivation, and secretion of excessive mucus in mouth and nose; headache, giddiness, weakness, slurring of speech, tightness in the chest, tearing, eye muscle pain and blurring of vision, breathing difficulty, convulsion, and coma (Gosselin 1976); ingestion of a small quantity can be fatal.

Oral LD_{50} (rat): 7 mg/kg

47.6 COUMAPHOS

Formula: $C_{14}H_{16}O_5ClPS$; MW 269.38; CAS [56-72-4]
Structure:

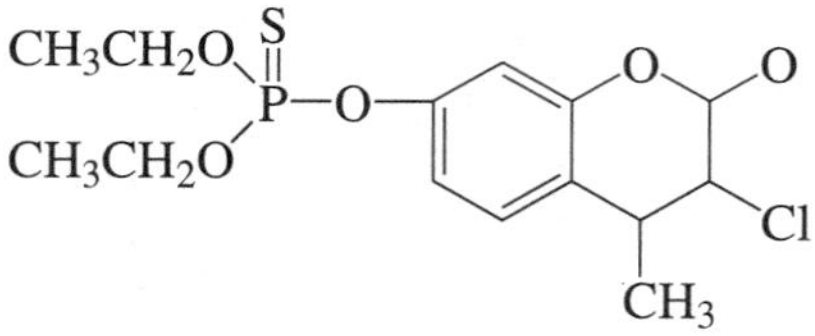

Synonyms: 3-chloro-4-methyl-7-coumarinyl diethyl phosphorothioate; 3-chloro-4-methyl-7-hydroxycoumarin diethyl thiophosphoric acid ester; diethyl-3-chloro-4-methylumbelliferyl thionophosphate; Asuntol; Muscatox; Suntol; Umbethion

Physical Properties

Crystalline solid; light brownish color; melts at 95°C; insoluble in water, soluble in most organic solvents.

Health Hazard

Highly toxic compound; choline stearase inhibitor; toxic symptoms include headache, dizziness, blurred vision, muscle spasm, vomiting, abdominal pain, diarrhea, chest pain, and shortness of breath; high dose may cause seizure, respiratory paralysis, coma and death; ingestion of 15–20 g can be fatal to adult human.

LD_{50} oral (rat): 13 mg/kg
LD_{50} skin (rat): 860 mg/kg
LC_{50} inhalation (rat): $\sim$300 mg/m^3

47.7 AMITON

Formula: $C_{10}H_{24}O_3NPS$; MW 269.38; CAS [78-53-5]
Structure:

$$CH_2H_5O \underset{CH_2H_5O}{\overset{\overset{\textstyle O}{\|}}{\diagup\!\!\!\!\searrow}} P-S-CH_2-CH_2-N(C_2H_5)_2$$

Synonyms: S-(diethylaminoethyl) O,O-diethyl phosphorothioate; (2-diethylamino) ethylphosphorothioic acid O,O-diethyl ester; O,O-diethyl S-(s-diethylaminoethyl) thiophosphate; Citram; Tetram

Physical Properties

Colorless liquid; boils at 110°C (at 2 torr); solubility data not available

Health Hazard

One of the most toxic organophosphorus insecticides; highly toxic by oral route; effects are cumulative; toxic symptoms are similar to those of other cholinestearase inhibitors and include nausea, vomiting, diarrhea, excessive salivation, blurred vision and pain in eye muscle, convulsion, coma, and respiratory failure; probable lethal dose is estimated to be 5–10 g for adult human

LD_{50} oral (rat): 3.3 mg/kg

47.8 DIMETHOATE

Formula: $C_5H_{12}NO_3PS_2$; MW 229.28; CAS [60-5-5]
Structure:

$$CH_3O \underset{CH_3O}{\overset{\overset{\textstyle S}{\|}}{\diagup\!\!\!\!\searrow}} P-S-CH_2-NH-CH_3$$

Synonyms: O,O-dimethyl S-(N-methylcarbamoylmethyl)phosporodithioate; phosphorodithioic acid, O,O-diethyl S-(2-methylamino)-2-oxoethyl)ester; O,O-dimethyldithiophosphorylacetic acid, N-monomethylamide salt; Cygon; Fosfotox; Phosphamide; Systoate

Physical Properties

White crystalline solid; camphor-like odor; melts at 52°C; vapor pressure 8.5×10^{-6} torr at 25°C; solubility in water 2 to 3 g/100 g, hydrolyzed by aqueous alkali

Health Hazard

Cholinesterase inhibitor; very toxic, exhibiting acute, delayed, and chronic toxicity; routes of entry — ingestion, skin contact, and inhalation; toxic symptoms include nausea, vomiting, diarrhea, excessive salivation, bronchoconstriction, and respiratory arrest; oral intake of 5–20 g may cause death to adult humans.

LD_{50} oral (mammal): 15 mg/kg

The oral LD_{50} in rat, however, is higher (within the range 150 mg/kg) than some other common organophosphorus insecticides, such as Nemaphos, Mevinphos, or Coumaphos.

47.9 DICROTOPHOS

Formula: $C_8H_{16}NO_5P$; MW 237.21; CAS [141-66-2]

Structure:

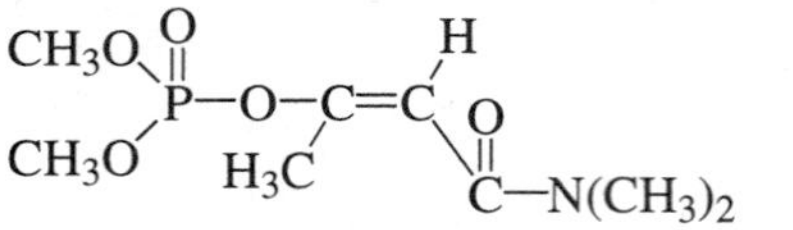

Synonyms: dimethyl *cis*-2-dimethylcarbamoyl-1-methylvinyl phosphate; 3-hydroxy-*N*,*N*-dimethyl-*cis*-crotonamide dimethyl phosphate ester; 3-hydroxy-*N*,*N*-dimethyl-*cis*-dimethyl phosphate crotonamide; Bidrin; Carbicron; Ciba 709.

Physical Properties

Yellow to brown colored liquid; mild odor; boils at 400°C; nonvolatile, vapor pressure 1.0×10^{-5} torr at 20°C; density 1.216 at 15°C; soluble in water and most organic solvents; rapidly hydrolyzes in acids and alkalies.

Health Hazard

An extremely toxic organophosphorus pesticide; cholinesterase inhibitor. Toxic effects are similar to those of monocrotophos. Toxic symptoms include headache, dizziness, muscle spasms, blurred vision, dilation of pupil, nausea, vomiting, diarrhea, abdominal pain, and seizures. Like most other organophosphorus compounds, exposure to this pesticide can dangerously affect the respiratory system, producing shortness of breath and respiratory depression, which can progress to respiratory paralysis. Dermal exposure may increase the heart rate, while oral intake may decrease the heart rate. This compound can be hypotensive and psychotic. Ingestion of a small quantity of the liquid (0.5–2 g) can be fatal to adult humans.

LD_{50} oral (rat): 15 mg/kg
LD_{50} skin (rat): 42 mg/kg

LD_{50} inhalation (rat): 0.09 mg/L/4 hr (RTEC 1985)

Exposure Limit

TLV-TWA 25 mg/m^3 (skin) (ACGIH 1986)

47.10 MONOCROTOPHOS

Formula: $C_7H_{14}NO_5P$; MW 223.16; CAS [6923-22-4]

Structure:

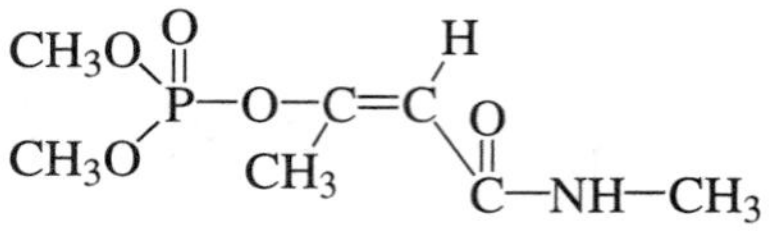

Synonyms: (E)-dimethyl-1-methyl-3-(methylamino)-3-oxo-1-propenyl phosphate; dimethyl 2-methylcarbamyl-1-methylvinyl phosphate; 3-(dimethoxyphosphinyloxy)-*N*-methyl-*cis*-crotonamide; phosphoric acid dimethyl[1-methyl-3-(methylamino)-3-oxo-1-propenyl]ester; Axodrin, Monocron

Physical Properties

Crystalline solid with ester odor; commercial product, reddish brown solid (Merck 1996); mp 54°F (25–30°C for the commercial product); density 1.33 at 20°C; readily soluble in water and most organic solvents; hydrolyzes in acid or alkali.

Health Hazard

Extremely toxic phosphate ester; however, susceptible to hydrolyze in acid or alkali; cholinesterase inhibitor; toxic properties are similar to those of dicrotophos, symptoms include headache, dizziness, pinpoint pupils, blurred vision, weakness, muscle spasms, vomiting, diarrhea, abdominal cramp, shortness of breath, and hypotension; high exposure may cause seizure, coma, and respiratory paralysis.

LD$_{50}$ oral (rat): 8 mg/kg

LD$_{50}$ oral (mouse): 15 mg/kg

LD$_{50}$ skin (rat): 112 mg/kg

LD$_{50}$ inhalation (rat): 63 mg/m^3/4 hr

47.11 DICHLORVOS

Formula: C$_4$H$_7$Cl$_2$O$_4$P; MW 220.98; CAS [62-73-7]

Structure:

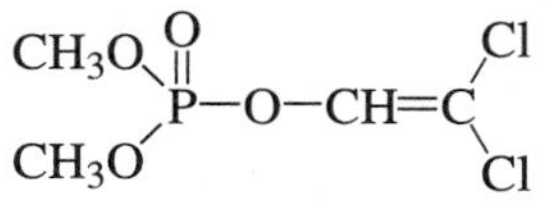

Synonyms: dimethyl-2,2-dichlorovinyl phosphate; dimethyl-2,2-dichlorovinyl phosphate; dimethyl 2,2-dichloroethenyl phosphate; dimethyl-2,2-dichlorovinyl phosphoric acid ester; phosphoric acid 2,2-dichloroethenyl dimethyl ester; Dichlorphos; Chlorvinphos; Atgard; Herkol; Vapona

Physical Properties

Colorless liquid; bp 140°C at 20 torr; vapor pressure 0.012 torr at 20°C; density 1.415 at 25°C; sparingly soluble in water (10 g/L at 25°C); readily soluble in organic solvents.

Health Hazard

Highly toxic by all routes of exposure; exhibits acute, delayed, and chronic poisoning; cholinesterase inhibitor; signs and symptoms of exposure are similar to those of other organophosphates; toxic effects include sweating, twitching of muscle, constriction of pupils, lacrimation, salivation, tightness in the chest, wheezing, slurred speech, nausea, vomiting, abdominal pain, and diarrhea; high exposure can result in coma, cessation of breathing, and death. Oral LD$_{50}$ value in rodents reported in the literature varies widely from 17 to 80 mg/kg in rat and 61–275 mg/kg in mouse.

LD$_{50}$ inhalation (mouse): 13 mg/m^3/4 hr

LD$_{50}$ skin (mouse): 206 mg/kg

LD$_{50}$ oral (wild bird): 12 mg/kg

47.12 CHLORFENVINPHOS

Formula: C$_{12}$H$_{14}$Cl$_3$O$_4$P; MW 359.56; CAS [470-90-6]

Structure:

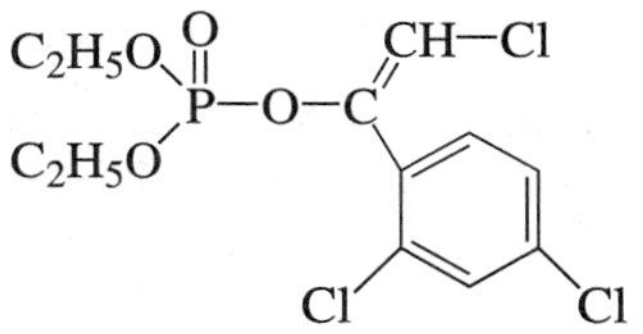

Synonyms: β-2-chloro-1-(2',4'-dichlorophenyl)vinyl diethyl phosphate; phosphoric acid 2-chloro-1-(2,4-dichlorophenyl)ethenyl diethyl ester; Birlane; Clofenvinfos; Supona

Physical Properties

Amber liquid; mild odor; solidifies between −16° to −22°C; vapor pressure 7.5 × 10^{-5} torr at 25°C; insoluble in water (~150 mg/L); hydrolyzes slowly in water; soluble in polar organic solvents.

Health Hazard

Highly toxic by all routes of exposure; however, more toxic in rat than in mouse and rabbit; cholinesterase inhibitor; exhibits acute, delayed, and chronic poisoning; toxic symptoms range from headache, twitching, salivation, blurred vision, and lacrimation to gastrointestinal effects and respiratory paralysis; in human permeation through skin at a dosage level of 10–15 mg/kg may manifest toxic effect.

LD$_{50}$ oral (rat): 10 mg/kg

LD$_{50}$ skin (rabbit): 300 mg/kg

LD$_{50}$ inhalation (rat): 50 mg/m^3/4 hr

47.13 DIALIFOR

Formula: $C_{14}H_{17}CINO_{14}PS_2$; MW 393.84; CAS [10311-84-9].

Structure:

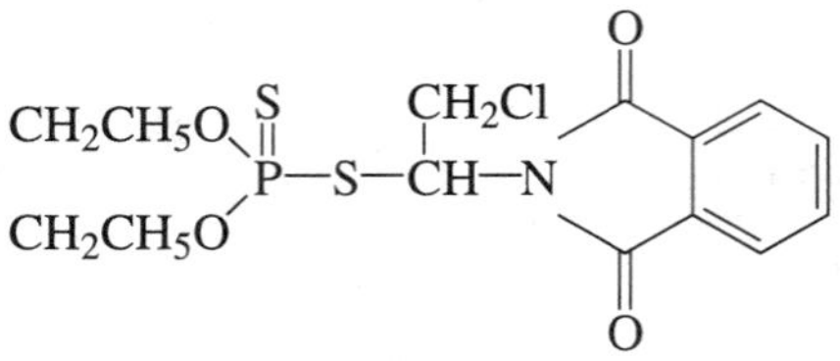

Synonyms: *O,O*-diethyl *S*-(2-chloro-1-ph-thalimidoethyl)phosphorodithioate; phosphorodithioic acid *S*-[2-chloro-1-(1,3-dihydro-1,3-dioxo-2*H*-isoindol-2-yl)ethyl]-*O,O*-diethyl ester; Dialifos; Torak

Physical Properties

White crystalline solid or colorless oily liquid; the solid melts at around 68°C; vapor pressure 0.001 torr at 35°C (Worthing, 1983); insoluble in water; hydrolyzed by concentrated alkali; low solubility in ethanol and hexane; readily dissolves in most other organic solvents.

Health Hazard

Highly toxic by all routes of exposure; exhibits acute, delayed, and chronic effect; cholinesterase inhibitor; symptoms of poisoning are similar to those of parathion, including weakness and twitching of muscle, headache, giddiness, dizziness, excessive salivation, lacrimation, tightness in chest, blurred vision, slurred speech, mental confusion, and drowsiness; gastrointestinal effects include nausea, vomiting, stomach cramps, and diarrhea; heavy exposure can lead to difficulty in breathing, convulsions, and coma; median lethal doses reported in the literature show inconsistent and varying values.

LD$_{50}$ oral (rat): 5–53 mg/kg
LD$_{50}$ oral (mouse): 39–65 mg/kg

LD$_{50}$ oral (rabbit): 35 mg/kg
LD$_{50}$ skin (rabbit): 145 mg/kg

47.14 DEMETON

Formula: $C_{18}H_{19}O_3PS_2$; MW 258.34; CAS [8065-48-3]

Structure: It is an isomeric mixture of Demeton-O and Demeton-S:

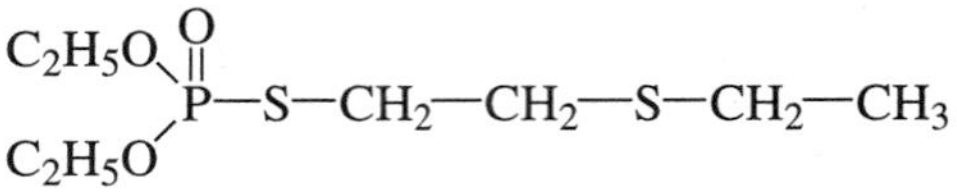

and

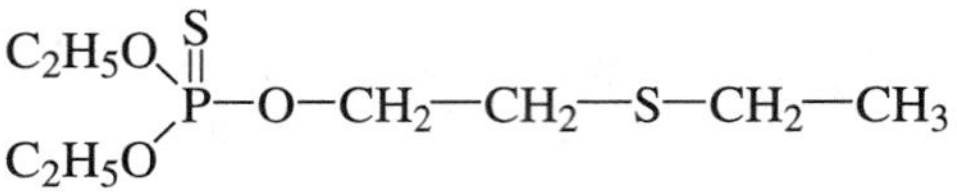

Synonyms: *O,O*-diethyl *O* (and *S*)-2-(ethylthio)ethyl phosphorothioate mixture; mixture of phosphorothioic acid *O,O*-diethyl *o*-(2-(ethylthio)ethyl)ester and phosphorothioic acid *O,O*-diethyl *S*-(2-(ethylthio)ethyl)ester; Systox; Mercaptophos

Physical Properties

Amber-colored oily liquid; mild odor; bp 134°C at 2 torr; solidifies at −25°C; specific gravity 1.183; vapor pressure 0.00026 torr at 25°C; insoluble in water (60 mg/L for Demeton-O at 25°C); soluble in most organic solvents; decomposed by alkalies and oxidizing agents.

Health Hazard

One of the extremely toxic organophosphorus pesticides; cholinesterase inhibitor; exhibits acute, delayed, and chronic effects; toxic symptoms include headache, dizziness, muscle spasms, pinpoint pupils, blurred vision, weakness, abdominal pain, vomiting, nausea, diarrhea, and seizure; high exposure may cause coma and death; other symptoms are chest pain, low or high blood pressure, shortness of breath, psychosis, respiratory depression, and respiratory paralysis; readily absorbed through skin; effects may be delayed by several hours; ingestion can cause immediate seizure or loss of consciousness; probable oral lethal dose within the range 0.4–2 g for adult humans.

LD_{50} oral (rat): 1.7 mg/kg
LD_{50} skin (rat): 8–14 mg/kg

Exposure Limit

TLV-TWA: 0.1 mg/m^3 (skin) (ACGIH)
PEL-TWA: 0.1 mg/m^3 (skin) (OSHA)
IDLH: 20 mg/m^3 (WHO)

47.15   DEMETON-S-METHYL

Formula: $C_6H_{15}O_3PS_2$; MW 230.30; CAS [919-86-8]
Structure:

$$CH_3O\underset{CH_3O}{\overset{O}{\underset{\|}{P}}}-S-CH_2-CH_2-S-CH_2-CH_3$$

Synonyms: *O,O*-dimethyl *S*-(ethylmercapto) ethyl thiophosphate; phosphorothioic acid *S*-[2-(ethylthio)ethyl]-*O,O*-dimethyl ester; Metasystox; Isometasystox; Methyl isosystox; Metasystox forte; Methyl demeton thioester; Demeton-S-methyl sulfide

Physical Properties

Pale yellow oily liquid; bp 118°C at 1 torr; vapor pressure 0.00035 torr at 20°C; specific gravity 1.207 at 20°C; sparingly soluble in water (3.3 g/L); miscible with most organic solvents; hydrolyzed by alkali.

Health Hazard

Highly toxic by all routes of exposure; cholinesterase inhibitor; toxic symptoms similar to those of Demeton and range form

headache, dizziness, blurred vision, and muscle spasms to gastrointestinal effects manifesting vomiting, diarrhea, and abdominal pain, as well as respiratory symptoms of dyspnea, respiratory depression, and paralysis; high exposures may result in onset of seizures and loss of consciousness; absorbed readily through skin.

LD_{50} oral (rat): 30–60 mg/kg
LD_{50} oral (guinea pig): 110 mg/kg
LD_{50} skin (rat): 85 mg/kg
LD_{50} inhalation (rat): 500 mg/m^3/4 hr

47.16   CYANOPHOS

Formula: $C_9H_{10}NO_3PS$; MW 243.23; CAS [2636-26-2]
Structure:

$$CH_3O\underset{CH_3O}{\overset{S}{\underset{\|}{P}}}-O-\!\!\left\langle\!\!\bigcirc\!\!\right\rangle\!\!-CN$$

Synonyms: *O,O*-dimethyl-*O*-(4-cyanophenyl) phosphorothioate; *O,O*-dimethyl *O*-(4-cyanophenyl)thionophosphate; phosphorothioic acid *O*-(4-cyanophenyl)-*O,O*-dimethyl ester; Ciafos, Sumitoma S

Physical Properties

Yellow to reddish brown transparent liquid; decomposes on heating; solidifies at 15°C; vapor pressure 0.0008 torr at 20°C; specific gravity 1.26 at 25°C; slightly soluble in water; readily dissolves in most organic solvents; decomposes when mixed with alkalies or when exposed to light.

Health Hazard

Exhibits acute, delayed, and chronic poisoning; highly toxic cholinesterase inhibitor; ingestion of small doses or skin absorption may produce delayed effects; massive oral dose can be fatal; toxic symptoms include headache, dizziness, blurred vision, pinpoint pupils, vomiting, abdominal pain and seizures; respiratory symptoms include shortness of breath, respiratory depression, and respiratory paralysis.

LD$_{50}$ oral (rat): 18 mg/kg

47.17 DIMEFOX

Formula: $C_{14}H_{12}FN_2OP$; MW 154.13; CAS [115-26-4]

Structure:

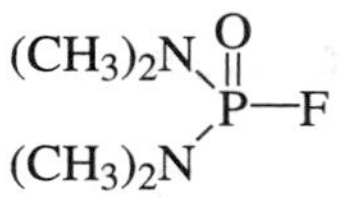

$$(CH_3)_2N\underset{(CH_3)_2N}{\overset{\overset{\displaystyle O}{\|}}{\diagdown}}P-F$$

Synonyms: bis(dimethylamido)fluorophosphate; bis(dimethylamido)phosphoryl fluoride; tetramethyldiamidophosphoric fluoride; Terrasystam

Physical Properties

Colorless liquid; fishy odor; bp 86°C at 15 torr; vapor pressure 0.36 at 25°C (Martin 1974); specific gravity 1.115 at 20°C; oxidized slowly in strong oxidizing agents; decomposes in presence of chlorine; readily soluble in water and in most organic solvents.

Health Hazard

Although structural feature in the molecule varies from most other organophosphorus pesticides, the toxic actions are similar to parathion and other phosphate esters. Extremely toxic by all routes of exposure; cholinesterase inhibitor; can present a serious inhalation hazard, if spilled, due to relatively high vapor pressure; exhibits acute, delayed, and chronic effects; symptoms of cholinergic effects similar to those of other organophosphates; death can result from respiratory arrest; ingestion of 0.3–2 g could be fatal to adult human.

LD$_{50}$ oral (rat): 1.0 mg/kg

47.18 FONOFOS

Formula: $C_{10}H_{15}OPS_2$; MW 246.32; CAS [944-22-9]

Structure:

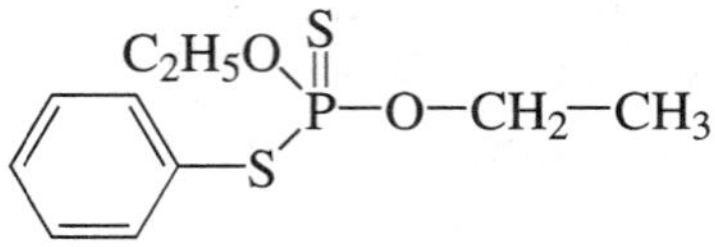

$$\underset{S}{\overset{C_2H_5O}{\diagdown}}\overset{\overset{\displaystyle S}{\|}}{P}-O-CH_2-CH_3$$

Synonyms: O-ethyl-S-phenyl ethylphosphonothiolothionate; ethylphosphonodithioic acid-O-ethyl-S-phenyl ester; Dyfonate

Physical Properties

Pale yellow liquid; bp 130°C at 0.1 torr; density 1.16 at 25°C; insoluble in water (13 mg/L); readily miscible with organic solvents; octanol–water partition coefficient 8000.

Health Hazard

Cholinesterase inhibitor; highly toxic by ingestion and skin absorption; exhibits acute, delayed, and chronic poisoning; toxic symptoms include headache, blurred vision, pinpoint pupils, salivation, tearing, muscle spasms, vomiting, abdominal pain, diarrhea, seizure, shortness of breath, and respiratory arrest; ingestion of 0.5–29 can cause death to adult human; median lethal dose (oral) in rat reported in the literature varies at 3–25 mg/kg; bioaccumulation potential of this toxicant is expected to be high because of high K_{ow}(8000):

LD$_{50}$ oral (rat): 3–25 mg/kg
LD$_{50}$ oral (dog): 3.3 mg/kg
LD$_{50}$ skin (rabbit): 25 mg/kg
LD$_{50}$ skin (rat): 147 mg

Exposure Limit

TLV-TWA: 0.1 mg/m^3 (skin) (ACGIH)
PEL-TWA: 0.1 mg/m^3 (skin) (OSHA)

47.19 NEMAFOS

Formula: $C_{18}H_{13}N_2O_3PS$; MW 248.26; CAS [297-97-2]
Structure:

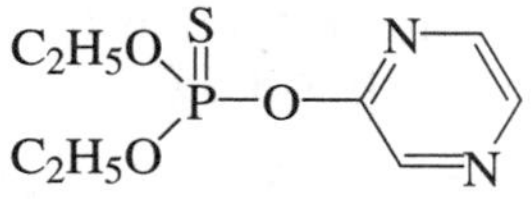

Synonyms: O,O-diethyl-O-(2-pyrazinyl) phosphorothionate; ethyl pyrazinyl phosphorothioate; phosphorothioic acid O,O-diethyl-O-pyrazinyl ester; Thionazin; Zinofos

Physical Properties

Amber liquid; boils at 80°C; solidifies at -1.7°C; vapor pressure 0.003 torr at 30°C; slightly soluble in water; readily mixes with organic solvents.

Health Hazard

Extremely toxic by ingestion and skin absorption; cholinesterase inhibitor; exhibits acute, delayed, and chronic effect; symptoms of cholinergic effects include excessive salivation, lacrimation, blurred vision, muscle spasms, headache, weakness, mental confusion, vomiting, diarrhea, stomach pain, convulsions, and coma; also causes shortness of breath, respiratory depression, and respiratory paralysis; ingestion of small quantity (0.5–1.59) could be fatal to adult humans; poisoning effects may onset several hours after exposure.

LD$_{50}$ oral (rat): 3.5–6 mg/kg
LD$_{50}$ oral (mouse): 5 mg/kg

LD$_{50}$ skin (rat): 8 mg/kg
LD$_{50}$ skin (guinea pig): 10 mg/kg

47.20 CHLORPYRIFOS

Formula $C_9H_{11}Cl_3NO_3NO_3PS$; MW 350.62; CAS [2921-88-2]
Structure:

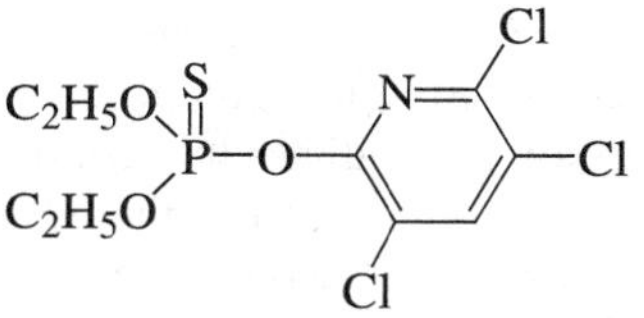

Synonyms: O,O-diethyl-O-(3,5,6-trichloro-2-pyridyl)phosphorothioate; O,O-diethyl)-(3,5,6-trichloro-2-pyridyl)thiophosphate; phosphorothioic acid O,O-diethyl)-(3,5,6-trichloro-2-pyridoinyl)ester; Dursban; Brodan; Eradex

Physical Properties

White crystal; mild odor; melts at 42°C; insoluble in water (2 mg/L at 25°C); readily soluble in organic solvents; octanol–water partition coefficient 50,000.

Health Hazard

Cholinesterase inhibitor; heavy exposure can produce acute, delayed, and chronic effect; exhibits low to moderate toxicity in experimental animals when administered by oral and dermal routes; however, severity of effects varies with species; highly toxic to birds; ingestion of 1.5–2 g would probably result in onset of cholinergic effects in adult humans.

LD$_{50}$ oral (rat): ~150 mg/kg
LD$_{50}$ oral (rabbit): 1000 mg/kg
LD$_{50}$ oral (wild bird): 5 mg/kg
LD$_{50}$ oral (chicken): 25 mg/kg
LD$_{50}$ skin (rat): ~200 mg/kg
LD$_{50}$ skin (rabbit): 2000 mg/kg

47.21 CARBOFENTHION

Formula: $C_{11}H_{16}ClO_2PS_3$; MW 342.85; CAS [786-19-6]

Structure:

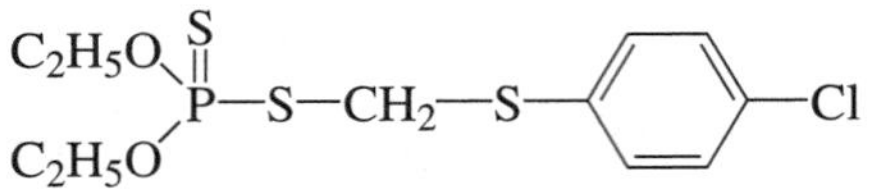

Synonyms: *O,O*-diethyl-*S*-*p*-chlorophenyl-thiomethyl dithiophosphate; *S*-(4-chloro-phenylthiomethyl)diethylphosphorothiolo-thionate; *S*-(((*p*-chlorophenyl)thio)me-thyl)-*O,O*-dietyl phosphorodithioate; phos-phorodithioic acid *S*-(((*p*-chlorophenyl) thio)methyl)-*O,O*-diethyl ester; Trithion; Hexathion

Physical Properties

Amber liquid; mild odor of sulfur; bp 82°C at 0.01 torr; density 1.271 at 25°C; insoluble in water; soluble in organic solvent.

Health Hazard

Highly toxic cholinesterase inhibitor; exhibits acute, delayed, and chronic poisoning; expo-sure risk of inhalation of vapors, how-ever, may be low because of its very low vapor pressure (3.0×10^{-7} torr at 20°C); also, severity of toxicity may vary widely from species to species; rats are more vul-nerable to its effect than are mice or rabbits; signs of toxicity in human include headache, dizziness, muscle spasms, tearing, blurred vision, salivation, vomiting, abdominal pain, diarrhea, chest pain, and convulsions; high exposure can cause coma and death; inges-tion of 0.4–1 g can be lethal to adult humans.

LD$_{50}$ oral (rat): in the range 7–10 mg/kg
LD$_{50}$ skin (rat): 27 mg/kg

47.22 DIAZINON

Formula: $C_{12}H_{21}N_2O_3PS$; MW 304.36; CAS [333-41-5]

Structure:

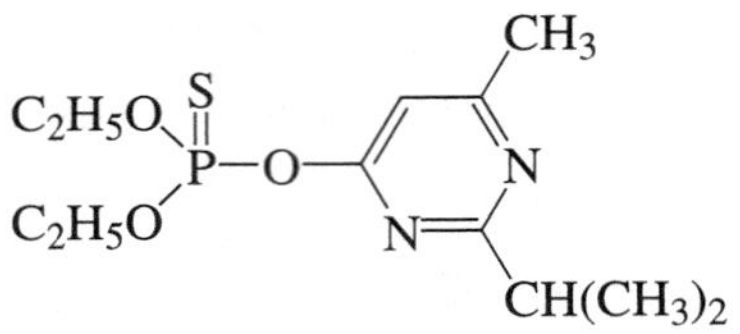

Synonyms: diethyl-2-isopropyl-4-methyl-6-pyrimidinyl phosphorothionate; diethyl 4-(2-isopropyl-6-methylpyrimidinyl)phos-phorothionate; phosphorothioic acid *O,O*-diethyl-*O*-(2-isopropyl-6-methyl-4-pyri-midinyl)ester; Basudin; Diazide

Health Hazard

Cholinesterase inhibitor; moderately toxic by ingestion and skin absorption; ingestion of approximately large quantities of 10–15 g can produce cholinergic effects in adult human; poisoning symptoms include head-ache, dizziness, weakness, blurred vision, pinpoint pupils, salivation, muscle spasms, vomiting abdominal pain, and respiratory depression; LD$_{50}$ values in experimental animals show wide variation.

LD$_{50}$ oral (rat): within the range 100–300 mg/kg
LD$_{50}$ oral (guinea pig): 250 mg/kg

47.23 MALATHION

Formula: $C_{10}H_{19}O_6PS_2$; MW 330.36; CAS [121-75-5]

Structure:

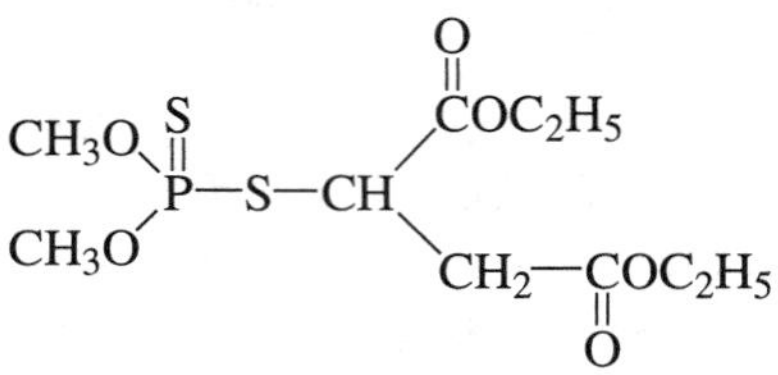

Synonyms: diethyl[dimethoxyphosphino-thioyl)thio] butanedioate; diethyl 2-(dime-thoxyphosphinothioylthio)succinate; dicarboethoxyethyl-*O,O*-dimethyl phosphorodithioate; [(dimethoxyphosphinothioyl)thio] butanedioic acid diethyl ester; dithiophosphoric acid *S*-(1,2-dicarboxyethyl)-*O,O*-dimethyl ester; Carbofos; Calmathion; Cython; Fosfothion; Carbetox

Physical Properties

Brownish yellow liquid; characteristic odor; bp 157°C at 0.7 torr; solidifies at 3°C; density 1.23 at 25°C; practically insoluble in water (145 mg/L); miscible with most organic solvents

Health Hazard

Cholinesterase inhibitor; toxic properties similar to those of parathion; however, less toxic than parathion; moderately toxic by ingestion and possibly other routes of exposure; toxic symptoms include excessive salivation, lacrimation, blurred vision, constriction of the pupils, nausea, vomiting, abdominal pain, and difficulty in breathing; also coma and death can result form large intake; ingestion of 10−25 g could be fatal to adult humans; skin contact can produce allergic sensitization reaction; also absorbed through skin, causing systemic poisoning; experimental teratogen; LD_{50} data reported in the literature show wide variation.

LD_{50} oral (rate): 300−2800 mg/kg
LD_{50} skin (mouse): 2000−3000 mg/kg

47.24 ABATE

Formula: $C_{16}H_{20}O_6P_2S_3$; MW 337.37; CAS [3383-96-8];
Structure:

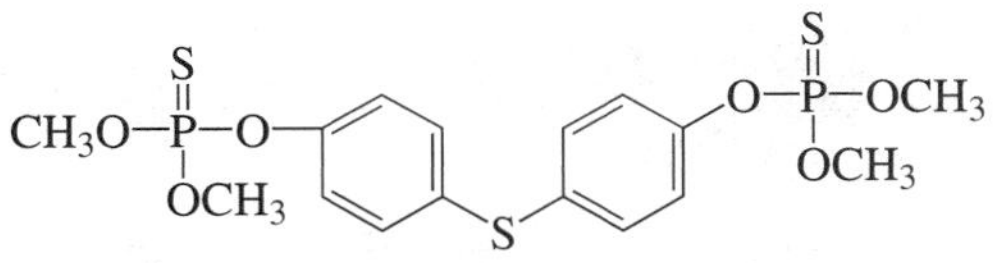

Synonyms: *O,O′*-(thiodi-4,1-phenylene)phosphorothioic acid *O,O,O′,O′*-tetramethyl ester; *O,O,O′,O′*-tetramethyl-*O,O′*-thiodi-*p*-phenylene phosphorothioate; temephos

Physical Properties

Crystalline solid; melts at 30°C; insoluble in water; soluble in acetonitrile, chloroform, ether, and toluene.

Health Hazard

Abate is less toxic in comparison to with most other organophosphorus pesticides. Among the test animals, rats showed almost 10 times greater resistance to this compound than did mice and rabbits. In humans, ingestion can cause nausea, vomiting, diarrhea, and convulsions. Intake of large amounts can result in respiratory failure.

LD_{50} value, oral (rats): 8600 mg/kg

48

HERBICIDES, CHLOROPHENOXY ACID

48.1 GENERAL DISCUSSION

Toxicity

Most herbicides of common use are of low toxicity. However, chronic exposure can produce severe health effects in humans. Chlorophenoxy acids can cause stupor, somnolence, nausea, vomiting, and convulsions in humans when ingested or when dusts are inhaled. Certain compounds of this class, notably 2,4,5-T, are known to contain trace amounts of highly toxic 2,3,7,8-tetrachlorodibenzo-p-dioxin (TCDD) as a contaminant. These herbicides inhibited the growth of freshwater bacteria, with toxicity increasing with a decrease in the phenoxy side-chain length (Milner and Goulder 1986). Signs of toxicity are similar for all phenoxy acid herbicides; the effects are usually mild in humans from single exposures. Certain substances, such as probenecid [57-66-9], were found to displace the chlorophenoxy acids from their binding sites in rat plasma protein, increasing the brain: plasma ratios in rats (Ylitalo et al. 1990). This resulted in increased toxicity. The toxicity of some of the common herbicides is presented in the following sections.

Analysis

Herbicides are conveniently analyzed by instrumental techniques: GC, HPLC, and GC/MS. For GC analysis, selection of the detector should be based on the structure of the herbicides. Chlorinated compounds, including the chlorophenoxy acid herbicides in potable water or wastewater, can be analyzed by GC-ECD. Chlorophenoxy acids (e.g., 2,4-D, 2,4,5-T, silvex.) are converted into their methyl esters by treatment with boron trifluoride–methanol or diazomethane following extraction with ether. If present in the form of their salts, these acids, may be hydrolyzed, extracted, and esterified before being injected into a GC. Alternatively, the esters may be analyzed by GC/MS, using a fused-silica capillary column. Soil, sediments, and solid wastes may be analyzed by GC-ECD techniques as discussed above (US EPA 1986, Method 8150). Liquid chromatography/particle beam mass spectrometry-based methods have been applied for detecting target herbicides in aqueous and hazardous waste samples (Brown et al. 1990). The compounds are separated by reversed-phase or anion-exchange chromatography. Triazine and other nitrogen-containing herbicides may be analyzed by GC-NPD in the nitrogen mode. GC-FID or GC/MS techniques may be applied.

Analysis of 2,4-D and 2,4,5-T in air may be performed in accordance with NIOSH (1984) Method 5001. An air volume of 15–200 L may be passed through a glass fiber filter at a flow rate of 1–3 L/min. The phenoxy acid anions are desorbed with methanol and analyzed by HPLC with UV detection. The eluants used 0.001 M solutions of $NaClO_4$ and $Na_2B_4O_7$. A stainless steel column packed with Zipax SAX (DuPont) is used for separation. The UV detector is set at 284 nm for 2,4-D and at 289 nm for 2,4,5-T.

Air analysis for paraquat may be done in a similar way (NIOSH 1984, Method 5003). The sampler consists of a 1-mm PTFE membrane filter. 1,1-Dimethyl-4,48-bipyridium cation is desorbed with water and analyzed by HPLC by UV detection (absorption at 254 nm). A stainless steel column packed with μ-Bondapak C is suitable for the purpose.

Disposal/Destruction and Biodegradation

Herbicides and herbicide-containing wastes are destroyed by incineration. Individual compounds are dissolved in a combustible solvent and are burned in a chemical incinerator equipped with an afterburner and scrubber. Landfill burial is not recommended. Borden and Yanoschak (1989) surveyed monitoring data of existing surface waters and groundwaters and found that most landfills do contaminate groundwater, the severity of which is highly variable. Contaminants detected include both inorganic pollutants and organics.

Microbial degradations of all classes of herbicides have been reported. Gibson and Suflita (1986, 1990) have reported anaerobic biodegradation of 2,4-D and 2,4,5-T in samples from a methanogenic

aquifier. The mechanism involves reductive dehalogenation, cleavage of the side chains, and cleavage of the aromatic rings in these molecules. Addition of short-chain organic acids and alcohols promoted biodegradation, while sulfate anion caused inhibitory effect. Shaler and Klecka (1986) determined the relationship between the specific growth rate of 2,4-D utilizing bacteria and the concentrations of both the organic substrate and dissolved oxygen by a modified Monod equation. For 2,4-D, the values for the maximum specific growth rate, yield, Monod coefficient for growth, and half-saturation constant for dissolved oxygen have been estimated at $0.09/hr^{-1}$ 0.14 g/g, 0.6 mg/L, and 1.2 mg/L, respectively. There is an acclimation period for the biodegradation of 2,4-D in lake water. During this period, a small population of bacteria is able to use the compound and grow sufficiently large to cause a detectable loss (Chen and Alexander 1989). Biodegradation of 2,4-D is affected strongly by the pH value of the culture. At a pH of <5.1, there is extensive accumulation of its metabolite, 2,4-dichlorophenol, which inhibits the biodegradation completely (Sinton et al. 1987). The authors noted that biodegradation ceased completely when the concentration of the metabolite reached 44 mg/L. The highest growth rates (*Pseudomonas* sp. NCIB 9340) occurred at pH 6.5–7.9, and the growth rate reached zero at 9.1 or 5.1. Cultures could be revived to resume normal growth when the pH was changed to 6.0.

Among other classes of herbicides, paraquat (a bipyridyl type) is biodegraded by the soil yeast, *Lipomyces starkeyi* (Carr et al. 1985). Kearney and co-workers (1988) have described destruction of atrazine by ozonization of subsequent biodegradation. The oxidation was rapid at pH 10 and the rate was first order for 100 mg/L ozonized solution. The mechanism involves the addition of oxygen to the alkylamino group, dealkylation, deamination, and dehalogenation. UV irradiation in the presence of ozone destroyed alachlor, atrazine, cyanazine, 2,4-D, trifluralin, and other herbicides by photooxidation (Kearney et al. 1987). The process may be used as a waste disposal pretreatment. Herbicides may be removed from wastewaters by pretreatment with oxidizing agents, activated carbon, and biological methods.

48.2 2,4-D

EPA Priority Pollutant, Designated Toxic Waste, RCRA Waste Number U240

Formula: C8H6Cl2O3; MW 221.04; CAS [94-75-7]

Structure:

$$OCH_2COOH$$

Cl

Cl

Synonyms: 2,4-dichlorophenoxyacetic acid; Herbidal; Tributol; Weedtrol

Uses and Exposure Risk

2,4-D is one of the most widely used herbicides. Exposure to this chemical may arise in the garden, farm, or plantation setting, where it is sprayed for weed control and to enhance the plant growth. Its residues have been detected in soils, sediments, and groundwater, as well as in tissues of corals (Crockett et al. 1986; Glynn et al. 1984).

Physical Properties

White to yellow crystalline powder; odorless; melts at 138°C; boils at 160°C at 0.4 torr; vapor pressure 0 torr at 20°C; insoluble in water, soluble in organic solvents.

Health Hazard

The acute toxic symptoms from ingestion, absorption through skin, and inhalation of its dusts include lethargy, stupor, weakness, incoordination, muscular twitch, nausea, vomiting, gastritis, convulsions, and coma. Death resulted in animals from ventricular fibrillation. In humans, ingestion of 4–6 g of 2,4-D can cause death. The toxic effects are nausea, vomiting, somnolence, convulsions, and coma. The oral LD_{50} value in test animals varied at 100–500 mg/kg, depending on species, which included dogs, guinea pigs, hamsters, rats, mice, and rabbits.

Gorzinski and co-workers (1987) have investigated acute, pharmacokinetic, and subchronic toxicity of 2,4-D and its esters and salts in rats and rabbits. The acute dermal LD_{50} values in rabbits for the acid, esters, and salts were >2000 mg/kg. Urinary elimination of 2,4-D was saturated in male rats given oral doses of 50 mg/kg. There was a decrease in body weight gain and increase in kidney weights in animals, resulting from subchronic dietary doses. A dose-related degenerative change was observed in the proximal renal tubules. Repeated subcutaneous dosing (150–250 mg/kg) of 2,4-D butyl ester resulted in accumulation of the compound in the brain, causing neurobehavioral toxicity (Schulze and Dougherty 1988).

Alexander and associates (1985) determined the 48-hour median lethal concentration values for 2,4-D to aquatic organisms. These values are 25, 325, 290, and 358 mg/L for *Daphnia magna*, fathead minnows, bluegill, and rainbow trout, respectively.

Exposure Limits

TLV-TWA 10 mg/m^3 (ACGIH, MSHA, and OSHA); IDLH 500 mg/m^3 (NIOSH).

48.3 2,4,5-T

EPA Priority Pollutant, Designated Toxic Waste, RCRA Waste Number U232

Formula: C$_8$H$_5$Cl$_3$O$_3$; MW 255.48; CAS [93-76-5]

Structure:

Synonyms: 2,4,5-trichlorophenoxyacetic acid; Fortex; Trioxon

Uses and Exposure Risk

2,4,5-T was formerly used as a herbicide. It is currently not used, following a ban by the EPA.

Physical Properties

Colorless to tan solid; odorless; melts at 153°C; vapor pressure 0 torr at 20°C; insoluble in water, soluble in organic solvents.

Health Hazard

The toxicity of 2,4,5-T is very similar to that of 2,4-D. The symptoms are irritation of skin, acne-like rash, somnolence, gastritis, fatty liver degeneration, and ataxia.

LD_{50} value, oral (rats): 300 mg/kg

Oral administration of this compound in hamsters and mice produced fetotoxicity, as well as developmental toxic effects on the central nervous system, eyes, ear, and prostate.

Exposure Limits

TLV-TWA 10 mg/m^3 (ACGIH, MSHA, and OSHA); IDLH 5000 mg/m^3 (NIOSH).

48.4 SILVEX

EPA Priority Pollutant, Designated Toxic Waste, RCRA Waste Number U233
Formula: $C_9H_7Cl_3O_3$; MW 269.51; CAS [93-72-1]

Structure:

Synonyms: 2-(2,4,5-trichlorophenoxy)propionic acid; 2,4,5-TP; Fenoprop

Uses and Exposure Risk

Silvex was formerly used as a herbicide. Its use in rice fields, sugarcane, and orchards has been stopped by the EPA.

Physical Properties

Crystalline solid; odorless, melts at 181.5°C; vapor pressure 0–20°C; practically insoluble in water, soluble in acetone, methanol, and ether, almost insoluble in carbon disulfide, benzene, and carbon tetrachloride.

Health Hazard

Silvex is moderately toxic to test animals. The toxic effects are comparable to those of 2,4-D and 2,4,5-T. Oral or subcutaneous administration in mice caused embryo toxicity and fetal death. The oral LD_{50} value in rats is 650 mg/kg.

48.5 2,4-DB

Formula: $C_{10}H_{10}Cl_2O_3$; MW 248.98; CAS [94-82-6]

Structure:

Synonyms: 4(2,4-dichlorophenoxy)butanoic acid; (2,4-dichlorophenoxy)butyric acid; Butyrac;

Butoxon; Embutox; Legumex

Physical Properties

Crystalline solid; melts at 118°C; practically insoluble in water (46 mg/L at 25°C); readily dissolves in acetone, ethanol, and ether; slightly soluble in benzene, toluene, and kerosene.

Health Hazard

This substance is moderately toxic by oral route. The oral LD_{50} values published in the literature vary. The LD_{50} (oral) in rat is within the range of 700 mg/kg, comparable to that of Silvex. The toxic symptoms are similar to those of other chlorophenoxy acids. Contact with skin or eye can cause irritation. The vapor is irritant to mucous membranes and upper respiratory tract.

48.6 MCPA

Formula: $C_9H_9Cl_2O_3$; MW 200.52; CAS [94-97-6]

Structure:

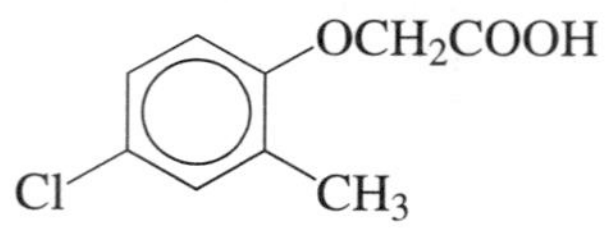

Synonyms: 2-methyl-4-chlorophenoxy acetic acid; (4-chloro-2-methyphenoxy) acetic acid; (4-chloro-*o*-tolyloxy) acetic acid; Agroxon; Chiptox; Hedonal; Herbicide M; Krezone; Leuna M

Physical Properties

Crystalline solid; melts at 119°C; slightly soluble in water (825 mg/L at 25°C); readily dissolves in most organic solvents. This compound is structurally similar to 2,4-D, except that one of the Cl atoms in the aromatic ring is substituted with a methyl group. Toxic properties are expected to be similar to 2,4-D. It is moderately toxic by ingestion, inhalation, and possibly other routes of exposure. Skin contact can cause irritation.

LD_{50} oral (rat): 700 mg/kg
LD_{50} intravenous (mouse): 28 mg/kg
LC_{50} inhalation (rat): 1370 mg/m^3/4 hr

48.7 MCPB

Formula: $C_{11}H_{13}ClO_3$; MW 228.55; CAS [94-81-5]

Structure:

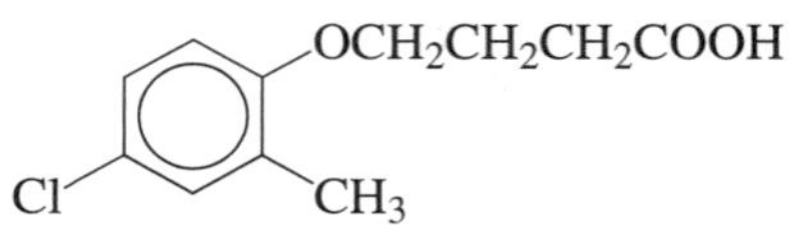

Synonyms: 4-(4-chloro-2-methylphenoxy) butanoic acid; 4-[(4-chloro-o-tolyl)oxy]butyric acid; 4-(2-methyl-4-chlorophenoxy) butyric acid; Bexone; Tritrol

Physical Properties

Colorless crystal; melts at 100°C; practically insoluble in water (44 mg/L at 25°C); readily dissolves in most organic solvents.

Health Hazard

MCPB is structurally similar to 2,4-DB, except one methyl group substituting a Cl atom in the ring. Toxic properties and the symptoms are similar to 2,4-DB. It is moderately toxic by ingestion and possibly other routes of exposure. Skin contact can cause irritation.

LD_{50} oral (rat): 680 mg/kg
LD_{50} oral (mouse): 800 mg/kg

REFERENCES

ACGIH. 1986. *Documentation of Threshold Limit Values and Biological Exposure Indices*, 5th ed. Cincinnati, OH: American Conference of Governmental Industrial Hygienists. Alexander, H. C., F. M. Gersich, and M. A. Mayes. 1985. Acute toxicity of four phenoxy herbicides to aquatic organisms. *Bull. Environ. Contam. Toxicol. 35*(3): 314–21.

Borden, R. C., and T. M. Yanoschak. 1989. North Carolina Sanitary landfills: Leachate generation, management and water quality impacts. *Chem. Abstr. CA 111*(16): 140013s.

Brown, M. A., I. S. Kim, S. Fassil I., and R. D. Stephens. 1990. Analysis of target and nontarget pollutants in aqueous and hazardous waste samples by liquid chromatography/particle beam mass spectrometry. *ACS Symp. Ser. 420*: 198–214.

Carr, R. J. G., R. F. Bilton, and T. Atkinson. 1985. Mechanism of biodegradation of paraquat by *Lipomyces starkeyi*. *Appl. Environ. Microbiol. 49*(5): 1290–94.

Chen, S., and M. Alexander. 1989. Reasons for the acclimation for 2,4-D biodegradation in lake water. *J. Environ. Qual. 18*(2): 153–56.

Crockett, A. B., A. Prop, and T. Kimes. 1986. Soil characterization study of former herbicide storage site at Johnson Island. *Govt. Rep. Announce. Index (U.S.) 87*(19), Abstr. No. 743, p. 346; *Chem. Abstr. CA 108*(17): 145271a.

Gibson, S. A., and J. M. Suflita. 1986. Extrapolation of biodegradation results to groundwater aquifers: reductive dehalogenation of aromatic compounds. *Appl. Environ. Microbiol. 52*(4): 681–88.

Gibson, S., and J. M. Suflita. 1990. Anaerobic biodegradation of 2,4,5-trichloro phenoxyacetic acid in samples from a methanogenic aquifer: stimulation by short-chain organic acids and alcohols. *Appl. Environ. Microbiol. 56*(6): 1825–32.

Glynn, P. W., L. D. Howard, E. Corcoran, and A. D. Freay. 1984. The occurrence and toxicity of herbicides in reef building corals. *Mar. Pollut. Bull. 15*(10): 370–74; cited in *Chem. Abstr. CA 103*(19): 155505n.

Gorzinski, S. J., R. J. Kociba, R. A. Campbell, F. A. Smith, R. J. Nolan, and D. L. Eisenbrandt. 1987. Acute, pharmacokinetic, and subchronic toxicological studies of 2,4-dichlorophenoxyacetic acid. *Fundam. Appl. Toxicol. 9*(3): 423–25.

Kearney, P. C., M. T. Muldoon, and C. J. Somich. 1987. UV-ozonization of eleven major pesticides as a waste disposal pretreatment. *Chemosphere 16*(10–12): 2321–30.

Kearney, P. C., M. T. Muldoon, C. J. Somich, J. M. Ruth, and D. J. Voaden. 1988. Biodegradation of ozonated atrazine as a wastewater disposal system. *J. Agric. Food Chem. 36*(6): 1301–6.

NIOSH. 1984. *Manual of Analytical Methods*, 3rd ed. Cincinnati, OH: National Institute for Occupational Safety and Health.

Schulze, G. E., and J. A. Dougherty. 1988. Neurobehavioral toxicity and tolerance to the herbicide 2,4-dichlorophenoxyacetic acid n-butyl ester (2,4-D ester). *Fundam. Appl. Toxicol. 10*(3): 413–24.

Shaler, T. A., and G. M. Klecka. 1986. Effects of dissolved oxygen concentration on biodegradation of 2,4-dichlorophenoxyacetic acid. *Appl. Environ. Microbiol. 51*(5): 950–55.

Sinton, G. L., L. E. Ericson, and L. T. Fan. 1987. The effect of pH on 2,4-D biodegradation. U.S. EPA Res. Dev. Rep. EPA/600/9-87/018F; cited in *Chem. Abstr. CA 109*(19): 166989y.

U. S. EPA. 1986. *Test Methods for Evaluating Solid Waste*, 3rd ed. Washington, DC: Office of Solid Waste and Emergency Response.

Ylitalo, P., U. Narhi, and H. A. Elo. 1990. Increase in the acute toxicity and brain concentrations of chlorophenoxyacetic acids by probenecid in rats. *Gen. Pharmacol. 21*(5): 811–14.

49

HERBICIDES, TRIAZINE

49.1 GENERAL DISCUSSION

Triazine class of herbicides are used to control broadleaf weeds and annual grasses. Atrazine and other related symmetrical triazines are used to control broadleaf and grassy weeds in tomatoes, potatoes, corn, asparagus and other vegetables. The mode of action of these compounds involve inhibition of Hill reaction in photosynthesis. All triazine herbicides are structurally related. They all contain a triazine ring in their structures. The general structure of this class of herbicides is discussed in Chapter 44.

In general, the acute oral toxicity of these substances in rats have been found to be of low to moderate order. In humans, a similar low acute toxicity may be expected. Because of the low vapor pressure of these compounds, any exposure risk from inhalation should be considered very low. The toxic symptoms in humans from overdose include abdominal pain, impaired adrenal function, nausea, diarrhea, and vomiting. Vapors are irritant to eye and mucous membrane. Skin contact may cause irritation and dermatitis. Like most other classes of pesticides and herbicides, the degree of toxicity of triazines has

been found to vary with compounds. Thus, the toxicity may not be solely attributed to the triazine ring. For example, while substances such as atrazine, metribuzin, and cyanazine are moderately toxic by all routes of exposure, compounds such as prometryn and propazine have a very low order of toxicity. However, no cases of human poisonings have been reported. A few selected triazine herbicides are discussed individually in the following sections.

Animal studies using labeled triazines indicate that the herbicides are mostly excreted out within 2–3 days of administration and not retained in body tissues or fluids. However, small amounts of labeled residues ($<10\%$) have been detected in blood, brain, heart, liver, kidney, muscle, and lung, even after a week. The nonchlorinated triazines are eliminated more rapidly than are the chlorinated compounds (ACGIH 1986). All these studies indicate that triazines do not metabolize to carbon dioxide. The mode of action of certain triazines involves carbohydrate metabolism. The chlorinated-, methoxy-, and methylthio-triazines inhibit starch accumulation by blocking the production of sugar (Gysin 1962; Gast 1958).

49.2 ATRAZINE

Formula: $C_8H_{14}ClN_5$; MW 216.06; CAS
[1912-24-9]

Structure:

CH₃CH₂NH──N──NHCH(CH₃)₂ (1,3,5-triazine ring with Cl)

Synonyms: 2-chloro-4-ethylamino-6-isopro-
pylamino-1,3,5-triazine; 1-chloro-3-ethyl-
amino-5-isopropylamino-s-triazine; Atra-
tol; Atrex; Penatrol

Physical Properties

White crystalline solid; mp 172 to 176°C;
vapor pressure 3×10^{-7} torr at 20°C; very
slightly soluble in water (70 mg/L at 25°C);
moderately soluble in ether and methanol;
readily dissolves in chloroform and dimethyl-
sulfoxide.

Health Hazard

Acute oral toxicity of atrazine in experimental animals was found to be moderate.
In human, the acute and chronic toxicity is
low. There is no reported case of poisoning. The toxic symptoms in animals include
ataxia, dyspnea, and convulsion. The oral
LD_{50} values in rats and rabbits are 672 and
750 mg/kg, respectively (NIOSH 1986).

The toxicity signs reported in rabbits
included conjuctivitis, excessive salivation,
sneezing, and muscle weakness (Salem et al
1985a,b). There was a gradual reduction
in the hemoglobin content and erythrocyte
and leukocyte count; an increase in glucose,
cholesterol, total proteins, and the enzyme
serum transminases. The oral LD_{50} value
reported is 3320 mg/kg. Rabbits fed atrazine-
treated maize for 6 months developed loss of
appetite, debility, progressive anemia, enteri-
tis, and muscle weakness. Most organs were
affected.

Rats treated orally with atrazine 12 mg/
100 g for 7 days were found to contain the
unchanged atrazine, as well as its metabolites in the liver, kidney, and brain. The
highest concentration of unchanged atrazine
was detected in the kidney, while its major
metabolite, diethylatrazine, was found in the
brain (Gojmerac and Kniewald 1989).

Donna and co-workers (1986) conducted
a 130-month study to test the carcinogenicity of atrazine in male Swiss albino mice.
Intraperitoneal administration of a total dose
of 0.26 mg/kg showed a statistically significant increase of plasma cell type and
histiocytic type of lymphomas in the animals. IARC has listed atrazine as possibly
carcinogenic to human (Group 2B Carcinogen) (IARC 1991). It produced severe eye
irritation in rabbit. Irritant action on skin
is mild.

Exposure Limit

TLV-TWA: 5 mg/m³ (ACGIH)

PEL-TWA: 5 mg/m³ (OSHA)

49.3 METRIBUZIN

Formula: $C_8H_{14}N_4OS$; MW 214.32; CAS
[21087-64-9]

Structure:

(CH₃)₃C─(1,2,4-triazin-5-one ring with SCH₃ and NH₂ substituents)

Synonyms: 4-amino-6-*tert*-butyl-3-(methylthio)-1,2,4-triazine-5-one; Sencor; Sen-
corex

Physical Properties

White crystalline solid; melts at 125 to
126°C; vapor pressure 10^{-5} torr at 20°C;
slightly soluble in water (1200 mg/L); soluble in alcohol.

Health Hazard

Low to moderate acute toxicity; single high dose may cause depression of central nervous system; repeated high doses affect thyroid and stimulate the metabolizing liver enzymes (ACGIH) 1986; excretes out rapidly; no irritant action on skin or eye; no teratogenic, mutagenic or carcinogenic effect observed in animal studies.

LD_{50} oral (rat): 1100 mg/kg
LD_{50} oral (mouse): 698 mg/kg
LD_{50} oral (guinea pig): 250 mg/kg
LD_{50} skin (rat): 2000 mg/kg

Exposure Limit

TLV-TWA: 5 mg/m^3 (ACGIH)
PEL-TWA: 5 mg/m^3 (OSHA)

49.4   CYANAZINE

Formula: $C_9H_{13}ClN_6$; MW 240.73; CAS [21725-46-2]
Structure:

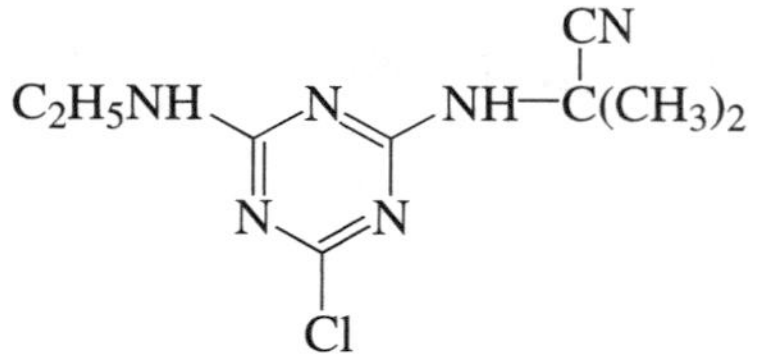

$$C_2H_5NH\text{—}\underset{\underset{Cl}{|}}{\overset{}{\text{triazine}}}\text{—}NH\text{—}\underset{CN}{\overset{}{C}}(CH_3)_2$$

Synonyms: 2-chloro-4-(ethylamino)-6-(1-cyano-1-methyl)(ethylamino)-s-triazine; 2-[[4-chloro-6-(ethylamino)-s-triazine-2-yl]amino]-2-methylpropionitrile; Bladex; Fortrol

Physical Properties

White crystalline solid; melts at 168°C; vapor pressure 1.6×10^{-9} torr at 20°C (Merck 1989); very slightly soluble in water (170 mg/L at 25°C); moderately soluble in hydrocarbon solvents; readily dissolves in chloroform and alcohol.

Health Hazard

Oral administration exhibited moderate to high toxicity in different species showing LD_{50} within the range 150–750 mg/kg; also moderately toxic by dermal absorption; such increased toxicity of this compound over other triazines may be attributed to the presence of nitrile functional group in the molecule; teratogenic effects in experimental animals; persists in the environment.

LD_{50} oral (rat): between 149 to 334 mg/kg
LD_{50} oral (rabbit): 141 mg/kg
LD_{50} skin (rat): 1200 mg/kg
LC_{50} inhalation (mouse): 2470 mg/m^3/4 hr

49.5   SYMCLOSENE

Formula: $C_3Cl_3N_3O_3$; MW 232.41; CAS [87-90-1]
Structure:

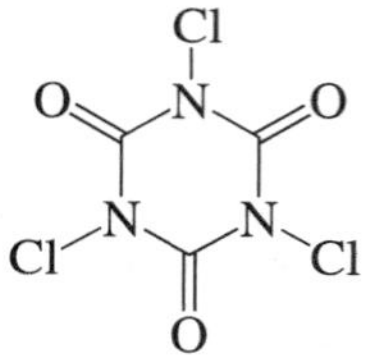

Synonyms: 1,3,5-trichloro-1,3,5-triazine-2,4,6(1H,3H,5H)-trione; 1,3,5-trichloro-s-triazine-2,4,6-trione; 1,3,5-trichloro-2,4,6-trioxohexahydro-s-triazine; 1,3,5-trichloro-s-triazinetrione; 1,3,5-trichloroisocyanuric acid; trichlorocyanuric acid; N,N',N''-trichloroisocyanuric acid

Physical Properties

White crystalline solid; odor of chlorine; melts at 246°C (decomposes); slightly soluble in water (2000 mg/L at 25°C); dissolves in chloroform, alcohol, and acetone.

Health Hazard

Moderately toxic by ingestion; causes lethargy, weakness, bleeding from stomach, and ulceration; ingestion of very large quantities, probably within the range 150–100 g, may cause delayed death; as a strong oxidizer, the toxic properties of this substance show many similarities to other powerful oxidants; injurious to liver and kidney, causing inflammation of gastrointestinal tract, liver discoloration, and kidney hyperemia as indicate from autopsy (Sax and Lewis 1995); the dermal toxicity is very low although an irritant to skin; irritation in rabbit's eye form a 50-mg dose in 24 hours was found to be severe.

LD_{50} oral (rat): 406 mg/kg

49.6 AMETRYNE

Formula: $C_9H_{17}N_5S$; MW 227.37; CAS [834-12-8]
Structure:

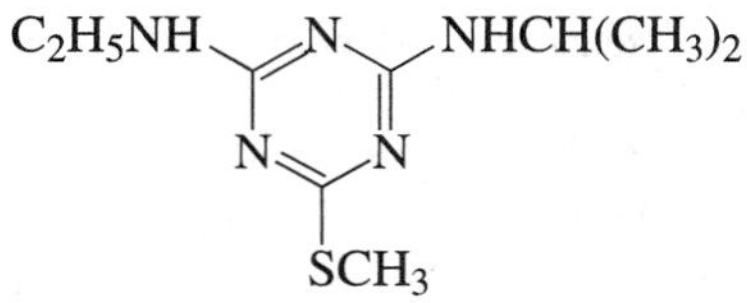

Synonyms: 2-ethylamino-4-isopropylamino-6-methylthio-1,3,5-triazine; 2-methylmer-capto-4-isopropylamino-6-ethylamino-s-triazine; 2-methylthio-4-ethylamino-6-isopropylamino-s-triazine; *N*-ethylamino-6-isopropylamino-s-triazine; *N*-ethyl-*N'*-(isopropyl)-6-(methylthio)1,2,5-triazine-2,4-diamine; Ametrex; Gesapax

Physical Properties

Crystalline solid; melts at 88–89°C; slightly soluble in water (185 mg/L at 20°C); moderately soluble in hexane; readily dissolves in methylene chloride, methanol, acetone, and toluene.

Health Hazard

Moderately toxic by ingestion; human toxicity data not available; a mild irritant to skin and eye.

LD_{50} oral (rat): 508 mg/kg
LD_{50} oral (mouse): 965 mg/kg
LD_{50} skin (rabbit): >5000 mg/kg

49.7 PROCYAZINE

Formula: $C_{10}H_{13}ClN_6$; MW 252.74; CAS [32889-48-8]
Structure:

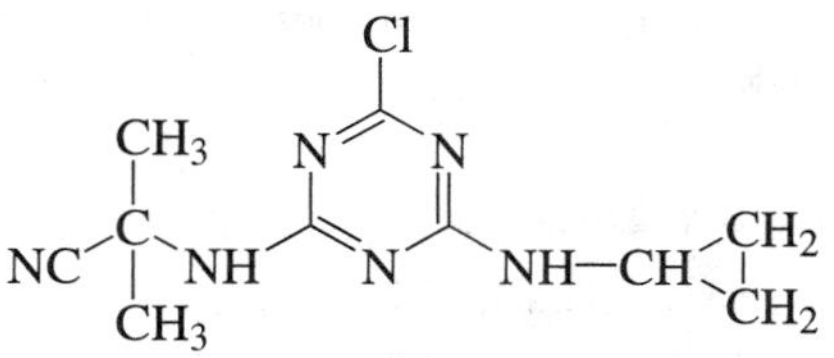

Synonyms: 2-[[4-chloro-6-(cyclopropyl-amino)-1,3,5-triain-2-yl]amino]-2-methyl-propionitrile; Cycle

Physical Properties

Crystalline solid; melts at 170°C; aqueous solubility data not available; dissolves in organic solvents

Health Hazard

Oral toxicity in rates was found to be moderate; possibly toxic by other routes of exposures; contains a nitrile functional group and is structurally related to Cyanazine; moderately toxic by skin absorption; teratogenic effect in experimental animals.

LD_{50} oral (rat): 290 mg/kg

49.8 SIMAZINE

Formula: $C_7H_{12}ClN_5$; MW 201.69; CAS [122-34-9]
Structure:

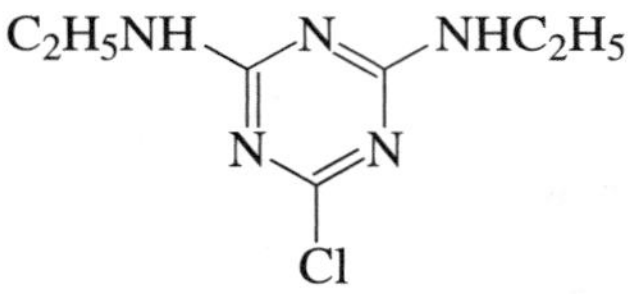

Synonyms: 2,4-bis(ethylamino)-6-chloro-*s*-triazine; 2-chloro-4,6-bis(ethylamino)-1,3,5-triazine; 6-chloro-*N*,*N'*-diethyl-1,3,5-triazine-2,4-diamine; Aquazine; Gesapun; Herbazin; Premazine.

Physical Properties

Crystalline solid; mp 226°C; insoluble in water (5 mg/L at 20°C); slightly soluble in methanol, chloroform and diethylether; octanol–water partition coefficient 91 (Milne 1995).

Health Hazard

Inconsistent data in the literature; oral LD_{50} values in rats reported as 970 and 5000 mg/L, showing a wide difference; toxicity is of low order.

REFERENCES

ACGIH. 1986. *Documentation of Threshold Limit Values and Biological Exposure Indices*, 5th ed. Cincinnati, OH: American Conference on Governmental Industrial Hygienists.

Donna, A., P, G. Betta, F. Robutti, and D. Bellingeri. 1986. Carcinogenicity testing of atrazine: preliminary report on a 13-month study on male Swiss albino mice treated by intraperitoneal administration. *G. Ital Med. Lav.*, *8* (3–4): 119–21; cited in *Chem Abstr.*, *109* (11): 87930u.

Gast, A. 1958. *Experiment, 13*: 134.

Gojemerac, T., and J. Kniewald. 1989. Atrazine biodegration in rats: A model for mammalian metabolism. *Bull. Environ. Contam. Toxicol.*, *43* (2): 199–206.

Gysin, H. 1962. *Chem. Ind.*, pp 1393.

IARC. 1991. Monograph, Vol. 53. Lyon, France: Internation Agency for Research on Cancer.

Lewis, R. J., Sr. 1995. *Sax' Dangerous Properties of Industrial Materials*, 9th ed. New York: Van Nostrand Reinhold.

Milne, G. W. A. (ed.) 1995. *CRC Handbook of Pesticides*. Boca Raton, FL: CRC Press.

NIOSH. 1986. *Registry of Toxic Effects of Chemical Substance*, D. V. Sweet, ed. Washington, DC: US Government Printing Press.

Salem, F. M. S., M, M. Lotfi, and A. H. Nounou. 1985a. Hazardous effect of application of the herbicide "Gesaprin" on Balady rabbits. *Vet. Med. J. 33* (2): 239–50.

Salem, F. M. S., M. M. Lotfi, and A. H. Nounou and H. A. M. El-Mansoury. 1985b. Acute toxicological study on the herbicide "Gesaprim" in Balady rabbits. *Vet. Med. J. 33* (2): 225–38.

50

HERBICIDES, UREA

50.1 TOXIC PROPERTIES OF UREA DERIVATIVE HERBICIDES

These substances in general exhibit low to intermediate order of toxicity in experimental animals. The severity of effects, however, depends on the nature of the species and also on the route of administration of the substance into the body. In humans, there is no reported case of poisoning. These substances have very low vapor pressure and are nonvolatile. The risk of inhalation is, therefore, very low. Also, any skin absorption or dermal route of entry of the pure material should be relatively insignificant. Accidental ingestion of a large dose, however, can be dangerous.

Structurally, all herbicides of this class contain $-NH-\overset{\overset{\textstyle O}{\|}}{C}-N\diagdown$ functional group, where the imino group ($-NH$) is mostly attached to a benzene ring having one or two chloro substitution(s). Fenuron, a common urea-type herbicide does not have any chloro-substitution nor contain any chlorine atom in the molecule and is mildly toxic by ingestion.

Many compounds of this class hydrolyze to produce chloroanilines, which can cause anemia and methemoglobinemia. Such hydrolysis may occur both under acidic and alkaline conditions, and the degree of hydrolysis depends on the water solubility. Although these substances may form nitrosamines under certain conditions, there is no report of carcinogenicity. Chemical structures, physical properties, and toxicity data for a few selected herbicides of this class are presented in the following sections.

50.2 MONURON

Formula: $C_9H_{11}ClN_2O$; MW 198.67; CAS [150-68-5]

Structure:

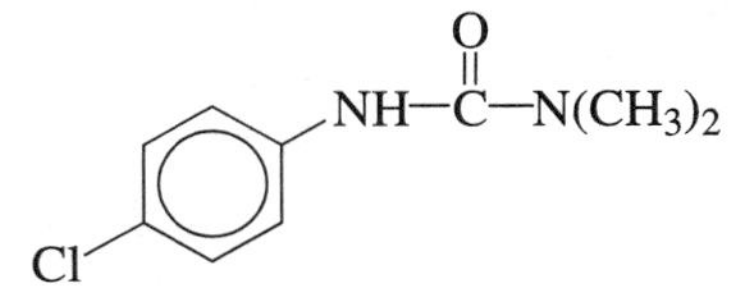

Synonyms: N-(p-chlorophenyl)-$N'N'$-dimethylurea; 1,1-dimethyl-3-(p-chlorophenyl) urea; N'-(4-chlorophenyl)-N,N-dimethylurea; Monurex; Karmex, Monuron Herbicide; Telvar

757

Physical Properties

Crystalline solid; faint odor; melts around 171°C; vapor pressure 5×10^{-7} torr at 25°C; very slightly soluble in water (230 mg/L at 25°C); moderately soluble in alcohol and acetone; practically insoluble in hydrocarbon solvents.

Health Hazard

Toxic properties are similar to Diuron; hydrolyzes under acidic or alkaline conditions to *p*-chloroaniline, which can cause anemia and methemoglobinemia; LD_{50} data published in the literature differ; acute and chronic toxicity of this herbicide is probably of low order; no reported case of human poisoning; showed clear evidence of carcinogenicity in male F344/N rats fed diets containing 750 ppm monuron for 2 years; caused cancers in the kidney and liver (National Toxicology Program 1988); female rats and male and female mice (B6C3F1) showed no evidence; induced cytomegaly of the renal epithelial cells in rats.

LD_{50} oral (rat): 3700 mg/kg (Bailey and White, 1965)

LD_{50} oral (rat): 1053 mg/kg (Lewis 1995)

50.3   DIURON

Formula: $C_9H_{10}Cl_2N_2O$; MW 233.10; CAS [330-54-1]

Structure:

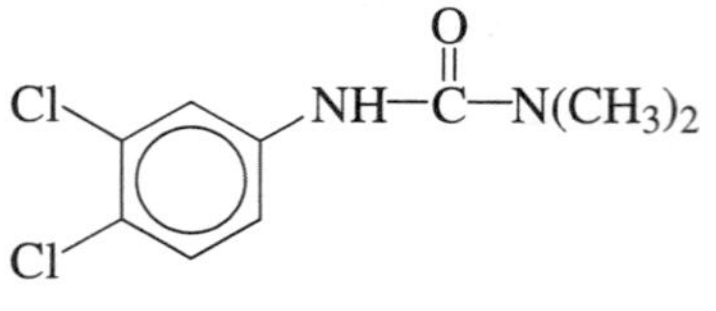

Synonyms: N'-(3,4-dichlorophenyl)-N,N-dimethylurea; 1,1-dimethyl-3-(3,4-dichlorophenyl)urea; Dinex; Duran; Diurex; Karmex; Unidron

Physical Properties

Crystalline solid; melts around 158°C; decomposes at 180°C; vapor pressure 2.0×10^{-7} torr at 30°C, and 3.1×10^{-6} torr at 50°C; practically insoluble in water (42 mg/L at 25°C) and hydrocarbons; moderately soluble in alcohol and acetone.

Health Hazard

Acute and chronic toxicity was found to be of low order in experimental animals, administered by oral route; repeated doses produced anemia in rats; the LD_{50} data, however, reported in the literature significantly differ; moderately toxic by intraperitoneal route, while the inhalation risk is low because of low vapor pressure. Susceptible to hydrolyze to dichloroaniline in vivo, which can cause methemoglobinemia; no evidence of carcinogenicity in an 18-month study in mice at 1400 ppm (Innes et al, 1969); no adverse effect observed in rats in 2-year feeding studies at dietary concentration level of 250 ppm (ACGIH 1986); adverse reproductive effects may arise from chronic exposure to high concentration level.

LD_{50} oral (rat): 1017 mg/kg (Lewis 1995)

LD_{50} oral (rat): 3400 mg/kg (Hodge et al 1968)

LD_{LO} intraperitoneal (mouse): 500 mg/kg

Exposure Level

TLV-TWA: 10 mg/m³ (ACGIH)

PEL-TWA: 10 mg/m³ (OSHA)

50.4   CHLOROXURON

Formula: $C_{15}H_{15}ClN_2O_2$; MW 290.77 CAS [1982-47-4]

Structure:

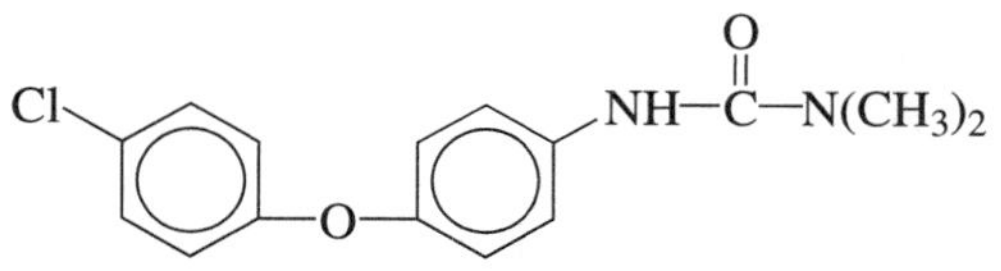

Synonyms: N'-[4-(4-chlorophenoxy)phenyl]-N,N-dimethylurea;   3-[p-(p-chlorophe-noxy)phenyl]-1,1-dimethylurea;   Norex; Tenoran

Structure:

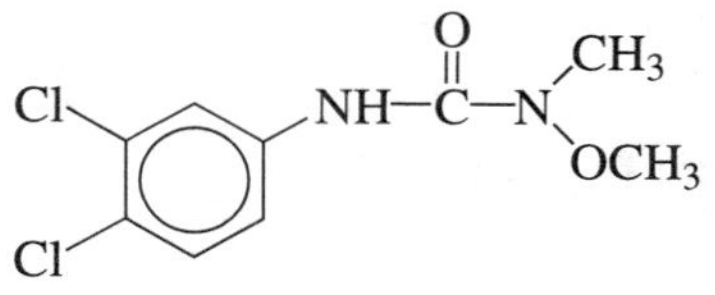

Synonyms: 3-(3,4-dichlorophenyl)-1-meth-oxymethylurea; N'-(3,4-dichlorophenyl)-N-methoxy-N-methylurea;   methoxydi-uron; Linurex; Lorox

Physical Properties

White crystal or colorless powder; odorless; melts at 151°C; vapor pressure 1.8×10^{-9} torr at 20°C (Worthing, 1983); density 1.34 at 20°C; insoluble in water (3.7 mg/L at 20°C and pH 7).

Physical Properties

Crystalline solid; melts at 94°C; practically insoluble in water (75 mg/L at 25°C); vapor pressure 1.5×10^{-5} torr at 24°C; moderately soluble in acetone and alcohol; partially soluble in benzene and other aromatic solvents.

Health Hazard

Exhibits acute, delayed, and chronic tox-icity; moderately toxic by intraperitoneal route; low to moderately toxic by inges-tion — LD_{50} values showing wide variation with species; the data reported appear to be questionable; highly toxic in dog but low oral toxicity in rat; toxic effects in humans unknown but expected to be simi-lar to other urea derivatives; susceptible to formation of dimethylnitrosamine under cer-tain conditions, which is carcinogenic; EPA-listed extremely toxic substance (US EPA 1989).

Health Hazard

Low to moderately toxic by ingestion — LD_{50} varying with experimental animals; 4-hour exposure to its vapors at 48 mg/m^3 was lethal to rats; toxic properties similar to those of Monuron.

LD_{50} oral (rat): 1146 mg/kg
LD_{50} oral (mouse): 2400 mg/kg
LD_{50} inhalation (rat): 48 mg/m^3/4 hr.

LD_{50} oral (rat): 3700 mg/kg (Lewis 1995)
LD_{50} oral (dog): 10 mg/kg (RTECS 1985)

50.6 FENURON

Formula: $C_9H_{12}N_2O$; MW 164.23; CAS [101-42-8]

Structure:

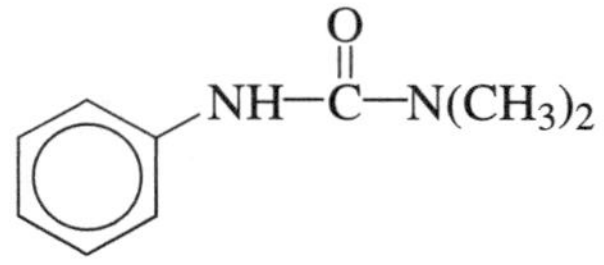

50.5 LINURON

Formula: $C_9H_{10}Cl_2N_2O_2$; MW 249.11 CAS [330-55-2]

Synonyms: 1,1-dimethyl-3-phenylurea; 1-phenyl-3,3-dimethylurea; N-phenyl-$N'N'$-dimethylurea; Beet Kleen; Dybar; Fenidin

Physical Properties

White crystals; melts around 132°C; slightly soluble in water ($\sim$3000 mg/L at 25°C)

Health Hazard

Acute and chronic toxicity of low order; moderately toxic by intraperitoneal route; in the body, it may hydrolyze to aniline, which may cause methemoglobinemia; the oral median lethal dose in experimental animals ranged between 3000 to 8000 mg/kg.

LD_{50} oral (rat): 6400 mg/kg
LD_{50} oral (rabbit): 4700 mg/kg
LD_{50} oral (guinea pig): 3200 mg/kg

50.7 NEBURON

Formula: $C_{12}H_{16}Cl_2N_2O$; MW 275.20; CAS [555-37-3]

Structure:

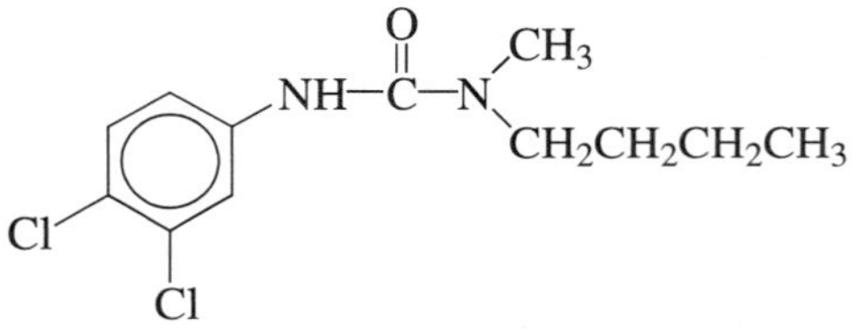

Synonyms: 1-butyl-3-(3,4-dichlorophenyl)-1-methylurea; 3-(3,4-dichlorophenyl)-1-methyl-1-butylurea; N-butyl-N'-(3,4-dichlorophenyl)-N-methylurea; Neburex; Neburea; Kloben

Physical Properties

Crystalline solid; melts around 102°C; practically insoluble in water, 48 mg/L at 24°C (Merck 1996); slightly soluble in hydrocarbon solvents.

Health Hazard

Acute oral toxicity is of very low order; moderately toxic by intravenous route.

LD_{50} oral (rat): 11,000 mg/kg
LD_{50} intravenous (mouse): 180 mg/kg

REFERENCES

ACGIH 1986. Diuron. In *Threshold Limit Values and Biological Exposure Indices*. 5th ed. Cincinnati. OH: American Conference of Governmental Industrial Hygienists.

Bailey, G. W. and J. L. White. 1965. *Residue Rev., 10*: 97

Hodge, H. C., Downs, W. L. and D. W. Smith. 1968. *Fed. Cosmet. Toxicol., 5*, 513

Innes, J. R. M. 1969. *J. Natl. Cancer Inst. 42*: 1101

Lewis, R. N. 1995. *Sax's Dangerous Properties of Industrial Materials*, 9th ed. New York: Van Nostrand Reinhold.

Merck. 1996. *The Merck Index*, 12th ed. Rahway, NJ: Merck & Co.

National Toxicology Program. 1988. Toxicology and carcinogenesis studies of monuron in F344/N rats and B6C3F1 mice (feed studies). *Chem. Abstr. CA 111*(15): 128577W

NIOSH. 1985. *Registry of Toxic Properties of Chemical Substances*, D. V. Sweet, ed. Cincinnati, OH: National Institute of Occupational Safety and Health.

US EPA. 1988. *Extremely Hazardous Substances: Superfund Chemical Profiles*. Park Ridge, NJ: Noyes Data Corp.

Worthing, C. R., and S. B. Walker. Eds. 1983. *The Pesticide Manual: A World Compendium*, 7th ed. Lavenham, Suffolk, UK: The Lavenham Press Limited.

51

PHENOLS

51.1 GENERAL DISCUSSION

Phenols are a class of organic compounds containing hydroxyl groups (−OH) attached to aromatic rings. Thus, a phenolic −OH group exhibits chemical properties that are different from an alcoholic −OH group. Phenols are weakly acidic, forming metal salts on reactions with caustic alkalies. In comparison, acid strengths of alcohols are negligibly small or several orders of magnitude lower than those of phenols.

Phenols are widely used in industry. The three major classes of phenolic compounds of commercial applications are alkyl phenols, such as cresols and pyrogallol; chlorophenols; and nitrophenols. The industrial uses of these compounds are described in this chapter, under individual headings. Because of their widespread uses, these substances occur in the environment in air, groundwater, wastewater, and soils. Pentachlorophenol produced during the incineration of municipal refuse could react with different fly ashes at 250–350°C and form dioxins (Karasek and Dickson 1987). The EPA has listed some of these substances as priority pollutants in potable and nonpotable waters and in solid wastes and has set regulations for their monitoring and control.

Toxicity

In comparison with many other classes of organic compounds, phenols show relatively greater toxicity. Most alkyl phenols, chlorophenols, and nitrophenols, although following different metabolic pathways and toxicokinetic patterns, exhibit a high degree of toxicity. The symptoms of acute toxicity are discussed under individual compounds in the following sections. The alkyl phenols, diphenols (benzenediols), and triphenols (benzenetriols) exhibit toxicities quite similar to those of phenols. The symptoms and severity of toxic effects are more or less similar.

Schultz (1987) used the ionization constant values (pK_a) for predicting the mechanism of toxicity of phenols. Two mechanisms have been proposed: polar narcosis and uncoupling of oxidative phosphorylation, based on which equations have been derived to determine the toxicities of phenols.

Chlorophenols show significant variation in their toxicities. The intraperitoneal LD_{50} values in rats are present in Table 51.1. An increase in the chlorine contents of phenols increases their toxicities. This is, however, manifested only for tetrachloro- and pentachlorophenols, not in lower chlorophenols.

TABLE 51.1 Comparison of Toxicity of Chlorophenols in Rats

Chlorophenol/CAS No.	LD_{50} Intraperitoneal (mg/kg)
o-Chlorophenol [95-57-8]	230
m-Chlorophenol [108-43-0]	355
p-Chlorophenol [106-48-9]	281
2,4-Dichlorophenol [120-83-2]	430
2,6-Dichlorophenol [87-65-0]	390
2,3,6-Trichlorophenol [933-75-5]	308
2,4,5-Trichlorophenol [95-95-4]	355
2,4,6-Trichlorophenol [88-06-2]	276
3,4,5-Trichlorophenol [609-19-8]	372
2,3,4,6-Tetrachlorophenol [58-90-2]	130
Pentachlorophenol [87-86-5]	56

Among the lower chlorophenols, 2-chlorophenol [95-57-8] and 3-chlorophenol [108-43-0] were found in mice to be considerably more toxic than the dichlorophenol series (Borzelleca et al. 1985).

2,4-Dinitrophenol is another highly toxic phenolic compound. All nitrophenols are toxic. Nitrophenols containing more than three nitro groups are unstable, the toxicity of which has not been investigated.

Destruction, Removal, and Biodegradation

Incineration is the most common method of destruction of phenolic wastes. Phenols are dissolved in combustible solvents and burned in a chemical incinerator equipped with an afterburner and scrubber. An electrically heated pyrolyzer at 2200°C has been reported to destroy chlorophenols and other highly toxic contaminants in contaminated soils and to reduce their levels below detectability (Boyd et al. 1987).

There are several reports on the destruction of phenols and other aromatics in wastewaters by ultraviolet (UV) light-catalyzed oxidation. Hydrogen peroxide and ozone have been used as oxidizing agents. McShea and co-workers (1987) have reported a decrease in the concentrations of phenol and pentachlorophenol in wastewater from a wood preservation process after treatment using oil/water separation, acidification, coagulation, and UV-ozone oxidation. Phenol concentration decreased from 30 mg/L to 0.79 mg/L; for pentachlorophenol, the decrease was from 150 mg/L to 73 mg/L. In a similar study conducted in a quartz annular reactor equipped with a low-pressure mercury lamp, Sundstrom and co-workers (1989) found that the rates of oxidation with hydrogen peroxide were fastest with 2,4,6-trichlorophenol. The mechanism of such photooxidation as proposed for phenol involves the formation of OH radicals by the irradiation of water by UV light (Ho 1987). The OH radicals hydroxylate the benzene ring, followed by photooxidation and rupture of the benzene ring to produce lower-molecular-weight carboxylic acids and aldehydes. Further photooxidation of these products forms CO_2 and water.

Garg and Ritscher (1987) have patented a process for removing phenol and other toxic organic materials from industrial wastewater. The process involves contacting the feedstock with an adsorbent, zeolitic molecular sieve (with a $SiO_2 - Al_2O_3$ molar ratio greater than 50:1). The spent molecular sieve is regenerated by contacting with an aqueous solution of hydrogen peroxide. Pillared and delaminated clays have also been used for the removal of chlorophenols from wastewaters (Zielk and Pinnavaia 1988). The binding capacity of these clays increased with high surface area, pore-size distribution, and polarity. Adsorption of chlorophenols increased as the hydrophobicity and the degree of chlorination of the phenol increased. 2,4-Dichlorophenol in water may be removed by polymerization to insoluble compounds, which may be filtered off. The polymerization is promoted by low ozone doses or treatment with enzymes (Duguet et al. 1986).

Decomposition of polynitrophenols or their salts in wastewaters can be effected at a pH below 4 and above 65°C by treatment with H_2O_2 (>2 mol) and an iron salt, such as ferrous sulfate or ferrous ammonium sulfate (0.002–0.7 mol) (Jayawant 1989). Treatment with finely divided iron particles 0.2–7 mol equivalent (based on the nitro group) at 0–100°C and pH 1.5–3 completely destroyed picric acid and styphnic acid in wastewaters containing these nitrated aromatics at 0.5–10 wt% concentrations (Bockrath and Kirksey 1987).

There are numerous reports on the biodegradation of phenolic compounds in soils, sediments, and wastewaters. In situ, biodegradation of phenol in soil has been reported to be effected by contaminated soil from another site (Damborg 1986). Aeration and addition of microorganisms further increased the biodegradation rate. Portier and Fujisaki (1986) described continuous biodegradation and detoxification of chlorinated phenols in a period of 3 months, using immobilized bacteria. Several bacterial strains from diverse aquatic environments were attached to porous solid matrix, which were immobilized by glass, cellulose, or chitin in this study. Mileski and co-workers (1988) have reported the activity of white rot fungus, *Phanerochaete chrysosporium*, toward biodegradation of pentachlorophenol ([^{14}C]PCP). Formation of water-soluble metabolites of [^{14}C]PCP was observed during degradation. In subsurface microcosms phenol degraded as rapidly as methanol. Over the range 10–30°C the average value for the temperature coefficient for biodegradation was found to be 1.04 (Hickman and Novak 1987).

Hrudey and associates (1987) have investigated the biodegradation of dichlorophenols containing an *ortho*-chlorine atom in semicontinuous cultures inoculated with 50% unacclimated anaerobic sludge. Complete mineralization to methane occurred only in cultures fed with 2,6-dichlorophenol. An *ortho*-chlorine atom was removed from the dichlorophenols in the following preferential order: 2,6- > 2,4- > 2,3-dichlorophenol. The rate of removal of the second Cl atom followed as *ortho* > *para* > *meta*. The pseudo-first-order biodegradation rates in natural waters for *p*-chlorophenol and *p*-methylphenol have been reported as 0.008 and 0.028 per hour, respectively (Vaishnav and Korthals 1988).

o-, *m*-, or *p*-Cresol was biodegraded by sulfate-reducing bacterial enrichments obtained from a shallow anoxic aquifer (Suflita et al. 1989). Bacterial strains from *Pseudomonas* and *Acinetobacter* species isolated from natural waters have been reported to biodegrade phenol at 200 mg/L concentration within 12–16 hours (D'Aquino et al. 1988). Phenol biodegradation has been carried out in granular activated-carbon columns seeded with microorganisms and subjected to ozonization (Speitel et al. 1989). *p*-Nitrophenol and dichlorophenol initially sorbed onto a granular activated-carbon bed were effectively biodegraded (Speitel et al. 1988).

Anaerobic biodegradability of 2,4-dinitrophenol using phenol-supplemented cultures is induced by ozone. Ozonization removes COD by 80%. The ozonization product is effectively biodegraded to methane (Wang et al. 1989).

Analysis

Phenolic compounds in aqueous and solid samples may be estimated as total phenols by colorimetric analysis. The method involves the reactions of phenols with 4-aminoantipyrine in the presence of potassium ferricyanide at pH 8 to form a yellow antipyrine dye (soluble in chloroform), the absorbance of which is measured by a spectrophotometer at 500 nm.

Individual phenols may be analyzed by GC, GC/MS, and HPLC techniques. GC analyses require the use of FID or ECD. If ECD is used, phenols should be converted to their haloderivatives. Although GC-ECD analysis is more cumbersome, the derivatizations of phenols eliminate interference effects. Phenols in wastewaters may be analyzed by EPA Method 604 based on GC-FID (or GC-ECD) techniques (U.S. EPA 1984). Pentafluorobenzyl bromide is used for derivatization of phenols in the extracts. The chromatographic columns suitable for the purpose are 1% SP-1240 DA on Supelcoport (GC-FID) or 5% OV-17 on Chromosorb W-AW (GC-ECD).

Identification of unknown phenols may best be achieved by GC/MS. Analyses of wastewaters and soil and solid wastes by GC/MS techniques may be performed by EPA Methods 625 (U.S. EPA 1984) and 8270 (capillary column), respectively (or 8250 for packed column) (U.S. EPA 1986). A fused-silica capillary column is suitable for the analyses. The primary and secondary ions for identification of priority pollutant phenols are given in Table 51.2. It may be noted that the characteristic ions for o-, m-, and p-cresols, as well as other phenol isomers, such as 2- and 4-nitrophenols are the same. These isomers may be distinguished by their retention times. Among the cresols, the o-, p-, and m-isomers are eluted in that order. Among nitrophenols, 2-nitrophenol is eluted before 4-nitrophenol.

Aqueous samples are acidified and extracted with a suitable solvent such as methylene chloride before their analysis.

HPLC is also widely used for analysis of phenols. Most toxicity studies in the literature refer to HPLC-UV determination of phenols.

Air analysis for phenol and cresols may be performed by NIOSH (1984) Methods

TABLE 51.2 Characteristic Ions for GC/MS Identification of Some Phenol Pollutants[a]

Compound/CAS No.	Primary Ion	Secondary Ions
Phenol [108-95-2]	94	65, 66
o-Cresol [95-48-7]	107	108, 77, 79, 90
m-Cresol [108-39-4]	107	108, 77, 79, 90
p-Cresol [106-44-5]	107	108, 77, 79, 90
2,4-Dimethylphenol [105-67-9]	122	107, 121
2-Nitrophenol [88-75-5]	139	109, 65
4-Nitrophenol [100-02-7]	139	109, 65
2,4-Dinitrophenol [51-28-5]	184	63, 154
2,6-Dinitrophenol [606-20-2]	162	164, 126, 98, 63
4,6-Dinitro-o-cresol [534-52-1]	198	51, 105
2-Chlorophenol [95-57-8]	128	64, 130
2,4-Dichlorophenol [120-83-2]	16	164, 98
2,4,6-Trichlorophenol [88-06-2]	196	198, 200
2,4,5-Trichlorophenol [95-95-4]	196	198, 97, 132, 99
2,3,4,6-Tetrachlorophenol [58-90-2]	232	131, 230, 166, 234
Pentachlorophenol [87-86-5]	266	264, 268
4-Chloro-3-methylphenol [59-50-7]	107	144, 142
Resorcinol [108-46-3]	110	81, 82, 53, 69
2-Cyclohexyl-4,6-dinitrophenol [131-89-5]	231	185, 41, 193, 266

[a]Data obtained using a fused-silica capillary column and under the conditions of electron-impact ionization at 70 eV.

3502 and 2001, respectively. For phenol, 26–240 L of air is passed through a bubbler containing 0.1 N NaOH at a flow rate of 200 to 1000 mL/min. The pH of the solution is adjusted to <4 by adding 0.1 mL of concentrated H_2SO_4 before GC injection. The detector used is FID and the column consists of 35/60 mesh Tenax (in 1.2 m × 6 mm OD stainless steel). Cresols are adsorbed over silica gel (150 mg/75 mg) by passing 5 to 20 L of air at a flow rate of 10 to 200 mL/min, desorbed with acetone, followed by injection into GC-FID. A 10% FFAP on Chromosorb W or equivalent column is suitable. Pentachlorophenol in blood or urine is analyzed by direct injection of the sample into GC-ECD using the column 4% SE-30, 6% OV-210 on 80/100-mesh silanized support (NIOSH 1984, Methods 8001 and 8303). Urine samples are stable for 40 days if stored frozen. EDTA is added to blood samples as an anticoagulant.

51.2 PHENOL

EPA Priority Pollutant; Acute Toxic Waste, RCRA Waste Number U188; DOT Label: Poison B

Formula: C_6H_6O; MW 94.12; CAS [108-95-2]

Structure:

OH

Synonyms: hydroxybenzene; carbolic acid; phenic acid; phenyl hydroxide; phenylic alcohol; benzenol

Uses and Exposure Risk

Phenol is used in the manufacture of various phenolic resins; as an intermediate in the production of many dyes and pharmaceuticals; as a disinfectant for toilets, floors, and drains; as a topical antiseptic; and as a reagent in chemical analysis. It has been detected in cigarette smoke and automobile exhaust. Smoke emitted from a burning mosquito coil (a mosquito repellent) has been found to contain submicron particles coated with phenol and other substances; a lengthy exposure can be hazardous to health (Liu et al. 1987).

Physical Properties

White or colorless crystals becoming red on exposure to light; characteristic odor; melts at 41°C; boils at 182°C; moderately soluble in water (6.6% at 20°C), dissolves in most organic solvents, insoluble in petroleum ether; aqueous solution weakly acidic, pH 6.0.

Health Hazard

Phenol is a corrosive substance with moderate to high toxicity. The acute poisoning effects are high in most animals. In humans, ingestion of 5–10 g of solid can cause death. The toxic symptoms include nausea, vomiting, weakness, muscular pain, dark urine, cyanosis, tremor, convulsions, and kidney and liver damage. In addition, it is an irritant to the eyes, nose, and throat and can cause skin burn and dermatitis. Inhalation of its vapors or absorption of the solid solution or vapor through skin can produce similar toxic effects. Phenol vapors can readily be absorbed through the skin.

Absorption of phenol through intact skin, as determined from the blood phenol levels in the test animals, depended on the surface area of applications rather than its concentrations in the test solutions (Pullin 1978). The animals (swines) showed the signs of twitching and tremors within a few minutes of applications, which were followed by salivation, nasal discharge, and labored breathing.

LD$_{50}$ value, oral (mice): 270 mg/kg
LC$_{50}$ value, inhalation (mice): 177 mg/kg

Krajnovic-Ozretic and Ozretic (1988) investigated the effects of water pollution by phenol on fish. An 8-day exposure to 7.5 mg/L of phenol produced damage to the gills, gallbladder, liver, and kidney in gray mullet. A higher concentration, above 10 mg/L, was lethal in several hours, while a lower level of 0.5 mg/L was nontoxic during an 8-day exposure.

Snails (*Indoplanorbis exustus*) secreted mucus and developed hemorrhage in a highly concentrated phenol environment. A 96-hour median lethal concentration was determined as 125.75 mg/L (Agrawal 1987).

Exposure Limits

TLV-TWA skin 5 ppm (~19 mg/m^3) (ACGIH, MSHA, and OSHA); 10-hour TWA 5.2 ppm (~20 mg/m^3) (NIOSH); ceiling 60 mg (15 minutes) (NIOSH); IDLH 250 ppm (NIOSH).

Fire and Explosion Hazard

Combustible solid; flash point (closed cup) 79°C (174°F); vapor pressure 0.35 torr at 25°C; autoignition temperature 715°C (1319°F). When heated, phenol produces flammable vapors that form explosive mixtures with air.

Vigorous to violent reactions may occur when phenol is mixed with strong oxidizers. A violent explosion occurred when aluminum chloride was added to nitrobenzene containing about 5% phenol (NFPA 1997).

Storage and Shipping

It is stored in a cool, dry, and well-ventilated area separated from heat sources. It is shipped in bottles, cans, drums, and tank cars. In case of spill, flush with plenty of water followed by caustic soda solution for neutralization.

51.3 CRESOL

EPA Priority Pollutant, Acute Toxic Waste, RCRA Waste Number U052; DOT Label: Corrosive Material, Poison B

Formula: C$_7$H$_8$O; MW 108.15; CAS [1319-77-3], [95-48-7], [108-39-4], and [106-44-5] for cresol and its *ortho- meta-*, and *para*-isomers, respectively

Structure:

(*o*-cresol)   (*m*-cresol)   (*p*-cresol)

Synonyms: hydroxytoluene; methylphenol; cresylic acid; cresylol; tricresol

Uses and Exposure Risk

Cresol is used in disinfectants and fumigants, in the manufacture of synthetic resins, in photographic developers and explosives.

Physical Properties

Colorless to yellowish or brown-yellow liquid with phenolic odor; mp 30, 11, and 35.3°C for *o*-, *m*-, and *p*-isomers; bp 191, 202, and 202°C, respectively, for *o*-, *m*-, and *p*-isomers respectively; density between 1.030 and 1.038 at 25°C; slightly soluble in water (2.0–2.5% at 20°C), miscible with most organic solvents.

Health Hazard

The toxic actions of cresol are similar to those of phenol. The *para*-isomer is somewhat more toxic than the other two isomers. The toxic symptoms are weakness, confusion, depression of the central nervous system, dyspnea, and respiratory failure. It is an irritant to the eyes and skin. Skin contact can cause burn and dermatitis. Chronic effects are gastrointestinal disorders, nervous disorders, tremor, confusion, skin eruptions, oliguria, jaundice, and liver damage. Skin LD$_{50}$ values in rats for *o*-, *m*-, and *p*-phenols are 620, 1100, and 750 mg/kg, respectively.

Dietz and Mulligan (1988a,b) have investigated the subchronic toxicity of *meta-* and *para*-cresols in Sprague-Dawley rats after 13 weeks of oral gavage. There was no adverse effects from a dose of 50 mg/kg/day. High dose levels of 150 mg/kg/day produced central nervous system depression and reduction of body weight gain. *p*-Cresol was found to be hepatotoxic and nephrotoxic, inducing a mild anemic effect. The acute toxicity of methyl phenols did not show any relationship to the number or position of methyl groups on the phenol nucleus. In an acute toxicity study on *Daphnia magna*, Devillers (1988) found cresols to be more toxic than phenol, xylenols, and trimethylphenols.

Exposure Limits

TLV-TWA (skin) for all isomers, 5 ppm ($\sim$22 mg/m^3) (ACGIH, MSHA, and OSHA), 10-hour TWA 2.3 ppm ($\sim$10 mg/m^3) (NIOSH), IDLH 250 ppm (NIOSH).

Fire and Explosion Hazard

Combustible substances; flash points (closed cup) for *o*-, *m*-, and *p*-cresols 81°C (178°F), 86°C (187°F), and 86°C (187°F), respectively; vapor pressure 0.25, 0.15, and 0.10 torr, respectively, for the three isomers at 20°C. The vapors of cresol form explosive mixtures with air; the flammable range has not been established.

51.4 RESORCINOL

EPA Designated Toxic Waste, RCRA Waste Number U201

Formula: $C_6H_6O_2$; MW 110.12; CAS [108-46-3]

Structure:

Synonyms: *m*-hydroxyphenol; 3-hydroxyphenol; 1,3-benzenediol; 1,3-dihydroxybenzene; *m*-dihydroxybenzene; resorcin

Uses and Exposure Risk

Resorcinol is used in the manufacture of resorcinol–formaldehyde resins, resin adhesives, dyes, drugs, and explosives; in tanning; in cosmetics; and in dyeing and printing textiles.

Physical Properties

White crystalline solid turning pink on exposure to air or light; sweetish taste; melts at 109–111°C; boils at 281°C; soluble in water, alcohol, and ether, slightly soluble in chloroform; aqueous solution acidic.

Health Hazard

The acute oral toxicity of resorcinol is moderate in most test animals. It is less toxic than phenol or catechol. Ingestion or skin absorption can cause methemoglobinemia, cyanosis, and convulsions. Vapors or dusts are irritant to mucous membranes. Contact with the skin or eyes can cause strong irritation. An amount of 100 mg caused severe irritation in rabbit eyes.

LD$_{50}$ value, oral (rats): 301 mg/kg (NIOSH 1986)

Exposure Limits

TLV-TWA 10 ppm ($\sim$45 mg/m^3) (ACGIH); STEL 20 ppm (ACGIH).

Fire and Explosion Hazard

Noncombustible solid; autoignition temperature 608°C (1126°F). At high temperatures ($\sim$200°C) its vapors can form explosive mixtures with air; the LEL at this temperature is 1.4% (NFPA 1986).

51.5 PYROCATECHOL

Formula: $C_6H_6O_2$; MW 110.12; CAS [120-80-9]

Structure:

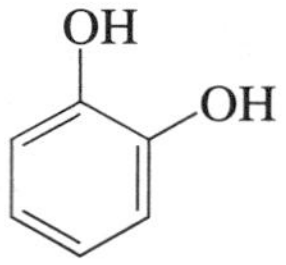

Synonyms: 1,2-benzenediol; catechol; *o*-dihydroxybenzene; 2-hydroxyphenol; *o*-diphenol; *o*-hydroquinone; *o*-phenylenediol; oxyphenic acid; pyrocatechin

Uses and Exposure Risk

Pyrocatechol is used in photography, in dyeing fur, and as a topical antiseptic.

Physical Properties

Colorless, crystalline solid; aqueous solution turns brown on exposure to air or light; sublimes; mp 105°C; bp 245.5°C; soluble in water and most organic solvents.

Health Hazard

Acute oral and percutaneous toxicity of pyrocatechol is greater than that of phenol; inhalation toxicity is less than that of phenol. The toxic symptoms include weakness, muscular pain, dark urine, tremor, dyspnea, and convulsions. Large amounts can produce degenerative changes in renal tubules. Large doses can cause death due to respiratory failure. Skin contact can cause eczematous dermatitis.

LD$_{50}$ value, oral (rats): 260 mg/kg
LD$_{50}$ value, skin (rabbits): 800 mg/kg

Exposure Limit

TLV-TWA 5 ppm (~22 mg/m^3) (ACGIH).

Fire and Explosion Hazard

Noncombustible solid.

51.6 PYROGALLOL

Formula: $C_6H_6O_3$; MW 126.12; CAS [87-66-1]

Structure:

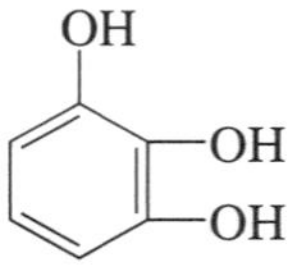

Synonyms: 1,2,3-trihydroxybenzene; 1,2,3-benzenetriol; pyrogallic acid

Uses and Exposure Risk

Pyrogallol is used in the manufacture of various dyes; in dyeing furs, hairs, and feathers; for staining leather; in engraving; as a developer in photography; and as an analytical reagent.

Physical Properties

White crystalline solid, turning grayish on exposure to air and light; melts at 131–132°C; boils at 309°C; sublimes on slow heating; highly soluble in water, alcohol, and ether; aqueous solutions become dark when exposed to air.

Health Hazard

The toxic symptoms are similar to those of phenol. It can enter the body by absorption through skin and ingestion. The poisoning effects are nausea, vomiting, gastritis, hemolysis, methemoglobinemia, kidney and liver damage, convulsions, and congestion of lungs. High doses can cause death. Ingestion of 2–3 g of solid can be fatal to humans. The LD$_{50}$ values varied widely in species. The oral LD$_{50}$ value in mice is about 300 mg/kg.

Fire and Exposure Hazard

Noncombustible solid.

51.7 2-NAPHTHOL

Formula: $C_{10}H_8O$; MW 144.18; CAS [135-19-3]

Structure:

Synonyms: β-naphthol; β-naphthyl hydroxide; 2-naphthalenol; 2-hydroxynaphthalene; CI 37500

Uses and Exposure Risk

2-Naphthol is used in the manufacture of dyes, perfumes, and medicinal organics, and in the production of antioxidants for synthetic rubber.

Physical Properties

Crystalline solid, darkens on exposure to light; slight phenolic odor; melts at 121°C; boils at 285°C; sublimes on heating; slightly soluble in water (0.1% at 20°C), soluble in organic solvents.

Health Hazard

Although the toxicity of 2-naphthol is of low order in test animals, ingestion of large amounts may result in nausea, vomiting, diarrhea, abdominal pain, convulsions, and hemolytic anemia. Death may result from respiratory failure. The oral LD_{50} value in rats is in the range 2000 mg/kg. 2-Naphthol is slightly more toxic than 1-naphthol [90-15-3], the oral LD_{50} value of which is in the range 2500 mg/kg.

Skin contact can produce peeling of the skin and pigmentation.

Fire and Explosion Hazard

Noncombustible solid.

51.8 PENTACHLOROPHENOL

EPA Priority Pollutant, Designated Toxic Waste, RCRA Waste Number U242

Formula: C_6HCl_5O; MW 266.32; CAS [87-86-5]

Structure:

Synonym: pentachlorophenate

Uses and Exposure Risk

Pentachlorophenol (PCP) is used for termite control, as a defoliant, and in the preservation of wood and wood products. It is an indoor air pollutant. It has been detected in timbers in the ppm range, causing contamination of air, surfaces, and materials in the homes. Its concentrations in blood samples have been reported in the range of sub-ppb to 110 μ/kg (Ruh et al. 1984). It has been detected in flue gas at 760–870°C exit temperature from an incinerator at a concentration of 1.033 mg/m^3 (Guinivan et al. 1985). The incinerator burned pentachlorophenol-treated wooden ammunition boxes and there was no afterburning. Methyl ethers of pentachlorophenol — pentachloroanisole [1825-21-4] and tetrachlorohydroquinone dimethyl ether [944-78-5] — formed from microbial methylation of pentachlorophenol have been identified in the pg/m^3 range in marine air samples from both the northern and southern hemispheres (Atlas et al. 1986).

Physical Properties

Colorless to light brown crystalline solid with a phenolic odor; mp 190°C; bp 310°C (decomposes); vapor pressure 0.00017 torr at 20°C; insoluble in water (solubility at 20°C 14 ppm), soluble in alcohol, ether, and benzene.

Health Hazard

Pentachlorophenol is a severe acute toxicant by ingestion and dermal penetration. The compound and its alkali salts can produce local and systemic effects. The symptoms of acute toxicity are headache, dizziness, sweating, nausea, vomiting, dyspnea, chest pain, weakness, fever, collapse, convulsions, and heart failure. Inhalation of its dusts or vapors can cause irritation of the eyes, nose, and throat, and coughing and sneezing. There is no evidence of chronic poisoning or any cumulative effects.

LD_{50} value, oral (mice): 117 mg/kg
LD_{50} value, skin (rats): 96 mg/kg

Subacute toxicity studies on rats orally administered pentachlorophenol at a dose of 0.2 mmol/kg/day for 28 days showed no effect on growth. However, this treatment induced cell alterations in liver and changes in relative liver weights (Renner et al. 1987). Fathead minnows exposed to 8–130 g/L of pentachlorophenol for 90 days experienced no adverse effects on their survival, growth, or bone development (Hamilton et al. 1986).

McKim and associates (1986) have conducted aquatic toxicokinetic studies using ^{14}C-labeled pentachlorophenol in rainbow trouts. At sublethal doses and over its 65-hour half-life period, about 50% was eliminated over the gills, 30% in the feces and bile, and 20% in the urine. It was found that pentachlorophenol and its metabolites were rapidly eliminated from the bodies of fish.

Fisher and Wadleigh (1986) have investigated the effects of pH on the acute toxicity and uptake of pentachlorophenol in the midge. The greatest toxicity was observed at pH 4 and the least at pH 9. While pentachlorophenol was fully protonated and was highly lipophilic at pH 4, it was completely ionized at pH 9. At the latter pH, the lipophilicity was low, which thus decreased the ability of pentachlorophenol to penetrate the midge. This caused a decrease in the observed toxicity of the compound.

Exposure Limits

TLV-TWA 0.5 mg/m^3 (ACGIH, MSHA, OSHA, and NIOSH); IDLH 150 mg/m^3 (NIOSH).

Fire and Explosion Hazard

Noncombustible solid.

51.9 2,4-DINITROPHENOL

EPA Priority Pollutant, Acute Hazardous Waste, RCRA Waste Number P048
Formula: $C_6H_4N_2O_5$; MW 184.12; CAS [51-28-5]
Structure:

Synonyms: 1-hydroxy-2,4-dinitrobenzene; α-dinitrophenol

Uses and Exposure Risk

Dinitrophenol is used in the manufacture of dyes, as a wood preservative, and as an indicator and analytical reagent.

Physical Properties

Yellowish crystalline solid; melts at 112°C; sublimes when heated slowly and carefully; almost insoluble in cold water, soluble in alcohol, ether, acetone, pyridine, and most organic solvents, soluble in aqueous alkaline solutions.

Health Hazard

2,4-Dinitrophenol is a severely acute toxicant, exhibiting high toxicity in animals by all routes of administration. It can be absorbed through the intact skin. The toxic effects are heavy sweating, nausea, vomiting, collapse, and death. Ingestion of 1 g of solid can be fatal to humans. A 30-minute exposure to its vapors at a concentration of 300 mg/m^3 was lethal to dogs (NIOSH 1986). Chronic effects include polyneuropathy, weight loss, cataracts, and dermatitis.

LD$_{50}$ value, oral (rats): 30 mg/kg

51.10   4,6-DINITRO-*o*-CRESOL

EPA Priority Pollutant, Acute Hazardous
    Waste, RCRA Waste Number P047
Formula: $C_7H_6N_2O_5$; MW 198.15; CAS
    [534-52-1]
Structure:

Synonyms: 3,5-dinitro-2-hydroxytoluene; 2-methyl-4,6-dinitrophenol

Uses and Exposure Risk

4,6-Dinitro-*o*-cresol is used as a selective herbicide as well as an insecticide.

Physical Properties

Yellow crystalline solid; mp 87.5°C; vapor pressure 0.00005 torr at 20°C; very slightly soluble in water (100 ppm at 20°C), readily dissolves in aqueous alkaline solution, soluble in most organic solvents.

Health Hazard

4,6-Dinitro-*o*-cresol exhibits cumulative toxicity in humans; the symptoms of poisoning are manifested when the blood levels of this compound exceeds 15–20 µg/g (ACGIH 1986). Thus chronic exposure to this compound can cause serious health hazard. The signs of toxicity in humans are headache, fever, profuse sweating, rapid pulse and respiration, cough, shortness of breath, and coma. Other symptoms noted are a decrease in hemoglobin, an increase in blood sugar, a loss of muscle tone, dyspnea, kidney and liver injury, and edema of the lung and brain.

LD$_{50}$ value, oral (mice): 47 mg/kg
LD$_{50}$ value, skin (rats): 200 mg/kg

Exposure Limits

TLV-TWA 0.2 mg/m^3 (ACGIH, MSHA, and OSHA); IDLH 5 mg/m^3 (NIOSH).

REFERENCES

ACGIH. 1986. *Documentation of the Threshold Limit Values and Biological Exposure Indices*, 5th ed. Cincinnati, OH: American Conference of Governmental Industrial Hygienists.

Agrawal, H. P. 1987. Evaluation of the toxicity of phenol and sodium pentachlorophenate to the snail *Indoplanorbis exustus* (Deshayes). *J. Animal Morphol. Physiol.* 34 (1–2): 107–12.

Atlas, E., K. Sullivan, and C. S. Giam. 1986. Widespread occurrence of polyhalogenated aromatic ethers in the marine atmosphere. *Atmos. Environ.* 20 (6): 1217–20.

Bockrath, R. E., and K. Kirksey. 1987. Destruction of nitrophenol byproducts in wastewaters from dinitrobenzene manufacture. U.S. Patent 4,708,806, Nov. 24; cited in *Chem. Abstr. CA* 108 (6): 43406c.

Borzelleca, J. F., J. R. Hayes, L. W. Condie, and J. A. Eagle, Jr. 1985. Acute toxicity of monochlorophenols, dichlorophenols and pentachlorophenol in the mouse. *Toxicol. Lett.* 29 (1): 39–42.

Boyd, J., H. D. Williams, R. W. Thomas, and T. L. Stoddart. 1987. Destruction of dioxin contamination by pyrolysis techniques. *ACS Symp. Ser. 338*: 299–310.

Damborg, A. 1986. A method for decontamination of polluted soil by biodegradation. *Chem. Abstr. CA 107*(14): 120495q.

D'Aquino, M., S. Korol, P. Santini, and J. Moretton. 1988. Biodegradation of phenolic compounds. I. Improved degradation of phenol and benzoate by indigenous strains of *Acinetobacter* and *Pseudomonas*. *Rev. Latinoam. Microbiol. 30*(3): 283–88; cited in *Chem. Abstr. CA 111*(1): 3923z.

Devillers, J. 1988. Acute toxicity of cresols, xylenols, and trimethylphenols to *Daphnia magna* Straus 1820. *Sci. Total Environ. 76*(1): 79–83.

Dietz, D., and L. T. Mulligan. 1988a. Subchronic toxicity of *para*-cresol in Sprague–Dawley rats. *Govt. Rep. Announce. Index (U.S.) 88*(14), Abstr. No. 836, p. 770; cited in *Chem. Abstr. CA 110*(9): 70771r.

Dietz, D., and L. T. Mulligan. 1988b. Subchronic toxicity of *meta*-cresol in Sprague–Dawley rats. *Govt. Rep. Announce. Index (U.S.) 88*(14), Abstr. No. 836, p. 769; cited in *Chem. Abstr. CA 110*(9): 70770q.

Duguet, J. P., B. Dussert, and J. Mallevialle. 1986. Removal of phenolic compounds by polymerization in water treatment: uses of ozone or enzymes. *Chem. Abstr. CA 108*(20): 173290w.

Fisher, W. S., and R. W. Wadleigh. 1986. Effects of pH on the acute toxicity and uptake of [^{14}C]pentachlorophenol in the midge, *Chironomus riparius*. *Ecotoxicol. Environ. Saf. 11*(1): 1–8.

Garg, D. R., and J. S. Ritscher. 1987. Process for removing toxic organic materials from weak aqueous solutions thereof. U.S. Patent 4,648,977, May 10; cited in *Chem. Abstr. CA 106*(24): 201274y.

Guinivan, T. L., T. Tiernan, M. Taylor, G. Vaness, D. Deis, J. Garrett, and H. Hodges. 1985. Emissions from and sampling of incineration of pentachlorophenol treated wood. *Chem. Abstr. CA 104*(12): 94556b.

Hamilton, S. J., L. Cleveland, L. M. Smith, J. A. Lebo, and F. L. Mayer. 1986. Toxicity of pure pentachlorophenol and chlorinated phenoxyphenol impurities to fathead minnows. *Environ. Toxicol. Chem. 5*(6): 543–52.

Hickman, G. T., and J. T. Novak. 1987. Microcosm assessment of biodegradation rates of organic compounds in soils. *Toxic Hazard. Wastes 19*: 153–62.

Ho, P. C. 1987. Evaluation of ultraviolet light/oxidizing agent as a means for the degradation of toxic organic chemicals in aqueous solutions. *Chem. Abstr. CA 109*(2): 11146x.

Hrudey, S. E., E. Knettig, P. M. Fedorak, and S. A. Daignault. 1987. Anaerobic semi-continuous culture biodegradation of dichlorophenols containing an ortho chlorine. *Water Pollut. Res. J. Can. 22*(3): 427–36.

Jayawant, M. 1989. Destruction of nitrophenols. U.S. Patent 4,804,480, Feb. 14; cited in *Chem. Abstr. CA 220*(24): 218489c.

Karasek, F. W., and L. C. Dickson. 1987. Model studies of polychlorinated dibenzo-*p*-dioxin formation during municipal refuse incineration. *Science 237*(4816): 754–56.

Krajnovic-Ozretic, M., and B. Ozretic. 1988. Toxic effects of phenol on grey mullet, *Mugilauratus Risso*. *Bull. Environ. Contam. Toxicol. 40*(1): 23–9.

Liu, W. K., M. H. Wong, and Y. L. Mui. 1987. Toxic effects of mosquito coil (a mosquito repellent) smoke on rats. I. Properties of the mosquito coil and its smoke. *Toxicol. Lett. 39*(2–3): 223–30.

McKim, J. M., P. K. Schmieder, and R. J. Ericson. 1986. Toxicokinetic modelling of [^{14}C] pentachlorophenol in the rainbow trout (*Salmo gairdneri*). *Aquat. Toxicol. 9*(1): 59–80.

McShea, L. J., M. D. Miller, and J. R. Smith. 1987. Combining UV/ozone to oxidize toxics. *Pollut. Eng. 19*(3): 58–9.

Mileski, G. J., J. A. Bumpus, M. A. Jurek, and S. D. Aust. 1988. Biodegradation of pentachlorophenol by the white rot fungus *Phanerochaete chrysosporium*. *Appl. Environ. Microbiol. 54*(12): 2885–89.

NFPA. 1997. *Fire Protection Guide on Hazardous Materials*, 12th ed. Quincy, MA: National Fire Protection Association.

NIOSH. 1984. *Manual of Analytical Methods*, 3rd ed. Cincinnati, OH: National Institute for Occupational Safety and Health.

NIOSH. 1986. *Registry of Toxic Effects of Chemical Substances*, ed. D. V. Sweet. Washington, DC: U.S. Government Printing Office.

Portier, R. J., and K. Fujisaki. 1986. Continuous biodegradation and detoxification of chlorinated phenols using immobilized bacteria. *Toxic Assess. 1* (4): 501–13.

Pullin, T. G. 1978. Decontamination of the skin of swine following phenol exposure: a comparison of the relative efficiency of water versus polyethylene glycol/industrial methylated spirits. *Toxicol. Appl. Pharmacol. 43*: 199–206.

Renner, G., C. Hopfer, J. M. Gokel, S. Braun, and W. Muecke. 1987. Subacute toxicity studies on pentachlorophenol, and the isomeric tetrachlorobenzenediols, tetrachlorohydroquinone, tetrachlorocatechol and tetrachlororesorcinol. *Toxicol. Environ. Chem. 15* (4): 301–12.

Ruh, C., I. Gebefuegi, and F. Korte. 1984. The indoor biocide pollution: occurrence of pentachlorophenol and lindane in homes. *Chem. Abstr. CA 104* (10): 74030x.

Schultz, W. T. 1987. The use of the ionization constant(pKa) in selecting models of toxicity in phenols. *Ecotoxicol. Environ. Saf. 14* (2): 178–83.

Speitel, G. E. Jr., C. J. Lu, and M. A. Turakhia. 1988. Biodegradation of synthetic organic chemicals in GAC beds. *Water Sci. Technol. 20* (11–12): 463–65.

Speitel, G. E. Jr. M. H. Turakhia, and C. J. Lu. 1989. Initiation of micropollutant biodegradation in virgin GAC columns. *J. Am. Water Works Assoc. 81* (4): 168–76.

Suflita, J. M., L. N. Liang, and A. Saxena. 1989. The anaerobic biodegradation of *o*-, *m*-, and *p*-cresol by sulfate-reducing bacterial enrichment cultures obtained from a shallow anoxic aquifer. *J. Ind. Microbiol. 4* (4): 255–66.

Sundstrom, D. W., B. A. Weir, and H. E. Klei. 1989. Destruction of aromatic pollutants by UV light catalyzed oxidation with hydrogen peroxide. *Environ. Prog. 8* (1): 6–11.

U. S. EPA. 1984. Guideline establishing test procedures for the analysis of pollutants under the Clean Water Act, 40CFR Part 136. *Fed. Regist. 49* (209), Oct. 26.

U. S. EPA. 1986. *Test Methods for Evaluating Solid Waste*, 3rd ed., Vol. 1B. Washington, DC: Office of Solid Waste and Emergency Response.

Vaishnav, D. D., and E. T. Korthals. 1988. Comparison of chemical biodegradation rates in BOD dilution and natural waters. *Bull. Environ. Contam. Toxicol. 41* (2): 291–98.

Wang, Y. T., P. C. Pai, and J. L. Latchaw. 1989. Methanogenic toxicity reduction of 2,4-dinitrophenol by ozone. *Hazard. Waste Hazard. Mater. 6* (1): 33–41.

Zielk, R. C., and T. J. Pinnavaia. 1988. Modified clays for the adsorption of environmental toxicants: binding of chlorophenols to pillared, delaminated, and hydroxy-interlayered smectites. *Clays Clay Miner. 36* (5): 403–8.

52

PHOSPHORUS AND ITS COMPOUNDS

52.1 GENERAL DISCUSSION

Phosphorus belongs to Group VB in the periodic table along with nitrogen, arsenic, antimony, and bismuth. It exhibits +3 and +5 valence states. Therefore, in most cases it forms tri- and pentavalent salts. It occurs widely in nature and many of its compounds find wide commercial applications. Phosphorus compounds that are hazardous may be classified under a few broad categories: (1) tri- and pentavalent halides, sulfides, and oxides; (2) phosphine and substituted phosphines; (3) organic phosphites or the esters of phosphorus acid; (4) organic phosphates or the esters of phosphoric acid; (5) organophosphorus pesticides and herbicides; and (6) nerve gases. Many compounds in the foregoing categories may overlap. This classification, however, may be useful for the sake of convenience and to some extent helpful in predicting the hazardous properties based on structures. In the present chapter, categories 1, 2, and 4 are discussed in detail. Pesticides and nerve gases are discussed in Chapters 47 and 39, respectively. Phosphorus acids are highlighted in Chapter 2. Many inorganic phosphates are nonhazardous. The toxic properties of metal phosphates are due to the metal ions and not to the PO_4^{3-} radical.

Toxicity

Halides, oxides, and sulfides of phosphorus are strong irritants to respiratory passage. Many of these compounds are moisture sensitive and cause acid burn on skin contact. Phosphine is one of the most toxic compounds of phosphorus. In addition to causing severe respiratory distress and lung damage, it can injure the gastrointestinal tract and produce depression of the central nervous system. Alkyl-substituted phosphines are less toxic, and there is no correlation of toxicity of these compounds to that of phosphine. Phosphorus in its trivalent state has a lone pair of electrons in its bonding orbitals, thus forming adducts with many chemicals in the body system. Some of the adducts formed contribute to the toxicity of the respective phosphorus compounds.

Organic phosphates or the phosphate esters have the structure

$$R-O-\overset{\displaystyle O}{\underset{\displaystyle O-R''}{\overset{\displaystyle \|}{P}}}-O-R'$$

where R, R', and R'' may be alkyl, cycloalkyl, or aryl groups. These compounds are cholinesterase inhibiting and neurotoxic, causing

TABLE 52.1 Comparison of LDs$_{50}$ of Some Common Phosphate Esters

Compound/CAS No.	Alkyl/Aryl Group	LD$_{50}$ oral (mice) (mg/kg)
Trimethyl phosphate [512-56-1]	Methyl	1470
Triethyl phosphate [78-40-0]	Ethyl	1400
Tributyl phosphate [126-73-8]	Butyl	1189
Dibutyl phenyl phosphate [2528-36-1]	Butyl and phenyl	1790
Triphenyl phosphate [115-86-6]	Phenyl	1320
Triorthocresyl phosphate [78-30-8]	Phenyl groups with methyl substitution in ortho position	1160[a]

[a]LD$_{50}$ value is for rats; an oral LD$_{50}$ values for mice are expected between 900 and 950 mg/kg.

muscles weakness and paralysis. Although phosphate esters exhibit low and subacute toxicity, at high dosages the poisoning may become severe. The oral LD$_{50}$ values determined on mice for some common aliphatic and aromatic esters are presented in Table 52.1. It may be noted that the values are more or less comparable for both the classes of phosphate esters. In aromatic phosphate esters, an alkyl substitution in the ortho positions of benzene rings enhances the toxicity.

Although phosphate esters show certain common toxic characteristics as discussed above, the toxic effects of many of these compounds differ markedly. For example, the sites of biological effects due to triorthocresyl phosphate are the peripheral nervous system, central nervous system, and gastrointestinal tract, whereas triphenyl phosphate causes minor changes in blood enzymes only. Trimethyl phosphate is a carcinogen. Aliphatic phosphate esters are irritant to the skin, eyes, and respiratory tract.

The toxicity of organic phosphites is low and different from the organic phosphates. The former show low toxicity and cause irritation to the eyes, skin, and the respiratory passage.

Flammability and Reactivity

In its white or yellow allotropic form, elemental phosphorus ignites spontaneously in air at room temperature. Many phosphorus compounds are pyrophoric and burn with brilliant flash. Phosphine is flammable and can explode. Replacing hydrogen atoms in phosphine with alkyl or aryl groups reduces its flammability. Substituted phosphines are flammable, too, but to a lesser extent, depending on the number and bulkiness of the substituent groups. The flammability pattern of some of the substituted phosphines are in the following decreasing order: phosphine (PH$_3$) > trimethylphosphine [(CH$_3$)$_3$P] > tributylphosphine[(C$_4$H$_9$)$_3$P] > phenyl − phosphine(C$_6$H$_5$PH$_2$) > triphenylphosphine [(C$_6$H$_5$)$_3$P]. Only the last compound is a noncombustible solid.

While halides and oxyhalides of phosphorus are noncombustible liquids or solids, the sulfides are flammable. Flammability increases with an increase in the phosphorus contents of the inorganic phosphorus−sulfur compounds. Oxides are usually noncombustible. Phosphorus trioxide ignites when heated in air.

Organophosphorus compounds are usually noncombustible. Some of the lower alkyl phosphites, such as trimethyl phosphite, are combustible liquids.

Many halides, sulfides, and oxides of phosphorus react violently with alkali metals, acids, oxidizers, halogens, sulfur, diborane, water, alcohols, and so on. Ignitions, incandescence, and violent explosions may occur if mixing is carried out under heating or grinding. These violent reactions are documented in this book under the individual compounds. These compounds are decomposed by water. The reaction with water is highly exothermic, liberating heat and causing a pressure increase. Phosphoric acid is usually one of the decomposition products.

The anions of phosphorus, such as phosphates and phosphites of metals, behave very differently and do not undergo such violent reactions as those discussed above.

Analysis

Phosphorus is analyzed by atomic absorption and ICP emission spectrometry and neutron activation techniques. The total phosphorus contents can be estimated colorimetrically by classical wet methods (American Public Health Association... 1995). Phosphorus is oxidized to orthophosphate by digesting with potassium persulfate. The solution is treated with ammonium molybdate and antimony potassium tartarate in an acid medium to form an antimony–phosphomolybdate complex that is reduced by ascorbic acid to form a deep blue coloration, the intensity of which is proportional to the concentration of phosphorus. The absorbance is measured at 650 nm by a spectrophotometer. Alternatively, it can be analyzed colorimetrically by an autoanalyzer (Technicon model).

Phosphorus particulates in air is adsorbed onto Tenax in a solid sorbent tube, desorbed by xylene. The eluant is injected into a GC equipped with a flame photometric detector using a 3% OV 101 Chromosorb W-HP column (NIOSH 1984, Suppl. 1987, Method 1905). Analysis of phosphorus in flue gases from incineration using a quartz fiber filter and different collecting impingers has been reported (Ward et al. 1988). Organic phosphorus compounds may be analyzed by GC using a nitrogen–phosphorus detector or in a more suitable way by GC/MS techniques.

52.2   PHOSPHORUS

Symbol P: at. wt. 30.9738; at. no. 15; CAS [7723-14-0]; valences 3, 5; position in the periodic table: group VB, along with nitrogen, arsenic, antimony and bismuth; natural isotope: ^{31}P; radioactive isotopes (artificial): ^{28}P, ^{29}P, ^{30}P, ^{32}P, ^{33}P, and ^{34}P; Structure and physical forms: phosphorus molecule P_4 is aggregation of four P atoms in tetrahedral form; dissociates to diatomic P_2 molecules above 800°C; in the solid state it exists in three main allotropic forms — white (or yellow), black, and red.

White and red phosphorus are used commercially for several purposes. Polymorphic black phosphorus is formed from white phosphorus under high pressure (Merck 1996). It is stable in air and the fire hazard is low. It does not catch fire spontaneously. On the other hand, white and red phosphorus are flammable and highly toxic. A detailed discussion on these is given below.

WHITE PHOSPHORUS

DOT Label: Flammable Solid, Spontaneously Combustible and Poison, UN 1381

Uses and Exposure Risk

It is used in smoke screens, in making rat poisons, and in analytical work.

Physical Properties

White or colorless solid (sometimes yellow due to impurities); wax-like transparent; turns dark on exposure to light; crystalline; density 1.88 at 20°C; mp 44°C; bp 280°C; insoluble in water (0.00033 g dissolves in 100 mL of water), low solubility in alcohol (0.25 g/dL) and chloroform (2.5 g/dL), and highly soluble in carbon disulfide (125 g/100 mL).

Health Hazard

White phosphorus is a highly poisonous substance. The toxic routes are ingestion, skin contact, and inhalation.

In humans a single oral dose of 70–100 mg can cause death. The toxic symptoms are nausea, vomiting, severe abdominal pain, diarrhea, coma, and convulsions. The other harmful effects from ingestion are liver damage and jaundice. An amount as small as 5–10 mg of white phosphorus can exhibit some of the foregoing toxic effects in humans from an oral intake. The lethal doses and symptoms for other species varied with the species. The toxic symptoms were somnolence, convulsion, and lung injury. The lethal doses ranges from 3 mg/kg for rats to 50 mg/kg for dogs.

White phosphorus can cause severe burns on skin contact. Absorption through skin can result in swelling, jaw pain, anemia, and cachexia. Eye contact can damage vision.

Inhalation of its vapors can cause irritation of respiratory tract. The chronic poisoning from inhalation (or ingestion) severely affected the lungs, kidney, and liver in test animals. The toxic symptoms were bronchopneumonia, bone changes, necrosis of the jaw ("phossy" jaw), anemia, and weight loss. Since the vapor pressure of white phosphorus is low (0.026 torr at 20°C), the acute health hazard from a short exposure to its vapors under normal conditions of its handling and uses should be low.

Exposure Limit

TLV-TWA air 0.1 mg/m^3 (ACGIH and OSHA).

Fire and Explosion Hazard

Flammable solid; catches fire spontaneously in moist air at temperature greater than 30°C (>86°F); ignites at higher temperatures in dry air. Fire-extinguishing agent: use a large volume of water, wet sand, or dry chemical. White phosphorus burns in air spontaneously, producing white fumes and flame. The dense white smoke is made up of vapors of tetraphosphorus hexoxide. P_4O_6 and/or tetraphosphorus decoxide, P_4O_{10}, are irritating to the eyes, throat, and lungs.

White phosphorus forms explosive mixtures with oxidizing agents. Its reactions with chlorates, bromates, iodates, chlorine trioxide, chromyl chloride, chlorine (gas or liquid), performic acid, peroxides, oxychlorides, chlorosulfonic acid, and lead dioxide can be explosive.

Mixing these compounds with white phosphorus, followed by warming and/or imparting shock, can be explosive. The presence of moisture may enhance the degree of violence. Among other oxidizing agents or oxygen-containing compounds, potassium permanganate, chromium trioxide, many nitrates, and mercuric oxide may cause explosion when heated and/or ground with white phosphorus.

It ignites in hot concentrated sulfuric acid or in vapors of nitric acid. With the latter, it burns with an intense white light. With aqueous hydroxides, phosphine gas is liberated, which can ignite in air or explode. It reacts vigorously with alkali metals. When white phosphorus is heated with other metals, the mixture can become incandescent.

Nitrogen bromide, sulfur, and many selenium halides and oxyhalides result in explosive reactions when mixed with white phosphorus (NFPA 1997).

Storage, Handling, and Shipping

White phosphorus is stored under water as the solid. The liquid is stored under a blanket of inert gas. This prevents its contact with atmospheric oxygen. Use forceps and wear gloves to protect against skin contact.

White phosphorus is shipped under water in drums or sealed cans. Alternatively, it is blanketed under an inert gas for shipping. It is separated from other chemicals.

Disposal/Destruction

Small quantities may be placed in a pit in an open area and allowed to burn. In general, incineration should be carried out for the destruction of large quantities. Small amounts are covered under water in bottles and placed in an incinerator.

In the laboratory, 5–10 g of white phosphorus can be destroyed by oxidizing with an excess of copper sulfate solution. The mixture is allowed to stand for days. The product copper phosphide is oxidized with NaOCl (laundry bleach) to phosphate (National Research Council 1995).

Chemical wastes containing elemental phosphorus may be disposed of by treating with oxygen under a layer of water. Phosphorus is oxidized within the water layer, liberating heat in the oxidation zone to incinerate the waste (Roberts et al. 1989).

Red Phosphorus

DOT Label: Flammable Solid, UN 1338

Uses and Exposure Risk

It is used to make safety matches, incendiary shells, and smoke bombs; in pyrotechnics; and in the manufacture of fertilizers, pesticides, phosphoric acid, and phosphorus halides.

Physical Properties

Red to violet powder; sublimes at 416°C; density 2.34; insoluble in water and organic solvents, soluble in phosphorus tribromide.

Health Hazard

Red phosphorus is much less toxic than white phosphorus. Its fumes, when burned, are highly irritating. In a subchronic inhalation toxicity study Aranyi (1986) observed that smoke of red phosphorus–butyl rubber combustion products caused loss of body weight and necrosis in rats. About 10% of animals died from the chronic exposure to 1.2 mg/L of aerosol. Poston et al. (1986) have investigated the acute toxicity of smoke screen materials to aquatic organisms. The toxic effects of white phosphorus–felt smoke and red phosphorus–butyl rubber smoke were low and similar. The toxicity was attributed to many complex oxyphosphoric acids in water formed from these phosphorus aerosols. These products acidified water and produced low toxic effects.

The determination of oral LD_{50} values of red phosphorus gave inconsistent values. The presence of white phosphorus as a contaminant can produce high toxicity.

Fire and Explosion Hazard

Flammable solid; autoignition temperature 260°C (500°F). Fire-extinguishing method: use excess water and cover the extinguished material with wet sand. At a high temperature, ~300°C, it converts to more flammable white phosphorus.

Red phosphorus is less reactive and less hazardous than white phosphorus. At ambient temperature, it does not present any explosion hazard similar to its white allotrope when in contact with oxidizers. It may ignite when reacted with chlorates, iodates, performic acid, selenium oxychloride, and chlorine. It may explode at high temperatures in contact with strong oxidizers and moistures. It catches fire when heated in air above 250°C.

When mixed with polyvinyl chloride, red phosphorus enhances the flammability and smoke formation characteristics of the latter. It manifests flame-retardant action on polystyrene while greatly enhancing smoke formation (Cullis et al. 1986).

Storage and Shipping

It is stored in a cool place, protected from physical damage and friction. It is isolated from other materials, especially strong oxidizers and alkalies. It is shipped in hermetically sealed cans in wooden boxes.

Disposal/Destruction

Small amounts of red phosphorus (5 g) may be destroyed in the laboratory by oxidizing to phosphoric acid. An aqueous solution of potassium chlorate in 1 N sulfuric acid (1.65% strength) is refluxed for several hours until phosphorus has dissolved. Excess chlorate is reduced at room temperature with sodium bisulfite and the solution is washed down the drain (National Research Council 1995).

52.3 PHOSPHINE

EPA Classified Acute Hazardous Waste, RCRA Waste Number P096; DOT Label: Poison Gas A and Flammable Gas, UN 2199

Formula PH_3; MW34.00; CAS [7803-51-2]

Synonyms: hydrogen phosphide; phosphorus trihydride

Uses and Exposure Risk

Phosphine is used as a fumigant, in the synthesis of many organophosphorus compounds, and as a doping agent for electronic components. It occurs in the waste gases from plants manufacturing semiconductors and thin-film photovoltaic cells. The presence of bound residues of phosphine in fumigated commodities has been reported (Rangaswamy and Sasikala 1986).

Physical Properties

Colorless gas with a characteristic fishy odor; liquefies at $-87°C$; solidifies at $-133°C$; slightly soluble in water (0.04% at 20°C).

Health Hazard

Phosphine is a highly poisonous gas. The symptoms of its acute toxic effects in humans can be respiratory passage irritation, cough, tightness of chest, painful breathing, a feeling of coldness, and stupor. Inhalation of high concentrations of phosphine in air can cause lung damage, convulsion, coma, and death. In addition to damaging the respiratory system, exposure to this compound can cause nausea, vomiting, diarrhea, and depression of the central nervous system. Exposure to a concentration of 1000 ppm in air for 5 min can be fatal to humans (NIOSH 1986).

LC$_{50}$ value, inhalation (rats): 11 ppm (15.3 mg/m^3)/4 hr

Chronic exposure is likely to cause phosphorus poisoning. Nutritional and toxicological studies indicated that ingestion of a phosphine-fumigated diet by rats for 2 years did not cause marked modification of growth, feed intake, functional behavior, or the incidence or type of tumors (Cabrol Telle et al. 1985).

Exposure Limits

TLV-TWA 0.42 mg/m^3 (0.3 ppm) (ACGIH and OSHA); STEL 1.4 mg/m^3 (1 ppm) (ACGIH); IDLH 200 ppm (NIOSH).

Fire and Explosion Hazard

Flammable gas; ignites spontaneously at room temperature in the presence of trace diphosphine, P_2H_3; autoignition temperature 100°C (212°F). It forms explosive mixtures with air; the LEL value is 1.6% by volume of air. The upper flammability could not be determined experimentally because of its autoignitability; it is calculated to be 100% (Ohtani et al. 1989). Phosphine oligomerizes to higher phosphines. Triphosphine and higher phosphines decompose rapidly in light at room temperature (Mellor 1946, Suppl. 1971). Solid hexaphosphine, $(P_2H_4)_3$, ignites at 160°C.

Phosphine may react violently with many nitrogen-containing inorganic compounds. It may ignite or explode by spark when mixed with nitric oxide, nitrous oxide, nitrogen trioxide, and nitrous acid. It decomposes with flame when mixed with concentrated nitric acid. Its reactions with mercuric nitrate and silver nitrate have produced explosions (Mellor 1946). With the former, the product, a yellow precipitate, exploded when heated or subjected to shock.

Phosphine ignites with halogens — fluorine, chlorine, and bromine — and explodes with chlorine monoxide and nitrogen trichloride (NFPA 1997).

Disposal/Destruction

In the laboratory phosphine can be destroyed by treating with an aqueous solution of copper sulfate (National Research Council 1995). The gas is bubbled into a copper sulfate solution under nitrogen or an inert gas. Phosphine is oxidized to phosphoric acid; the oxidation-reduction step is as follows:

$$PH_3 + 4Cu^{2+} + 4H_2O \longrightarrow H_3PO_4 \\ + 4Cu + 8H^+$$

Waste gases containing phosphine and other hydrides from semiconductor manufacturing plants are cleaned by passing them through a packed column containing pellets of copper oxide and manganese dioxide impregnated with silver oxide or carbonate (Kitahara et al. 1989). Phosphine emission in thin-film photovoltaic cell manufacturing operations may be controlled by carbon adsorption for very low concentrations and chemical scrubbing for higher concentrations (Fthenakis and Moskowitz 1986). NaOH, NaOCl/KOH, and $KMnO_4$ solutions are reported to be suitable for wet chemical scrubbing (Herman and Soden 1988).

52.4 PHOSPHORUS PENTOXIDE

DOT Label: Corrosive Material, UN 1807
Formula P_2O_5; MW 141.94; CAS [1314-56-3]

Structure: hexagonal crystals, can be modified to several crystalline and amorphous forms, dimerizes to P_4O_{10}
Synonyms: phosphorus pentaoxide; phosphoric anhydride; phosphorus (V) oxide

Uses and Exposure Risk

It is used as a dehydrating agent, in organic synthesis, and in hydrocarbon analysis.

Physical Properties

White deliquescent crystals; mp 340°C; sublimes at 360°C; density 2.3; readily absorbs moisture, mixes with water, forming phosphoric acid.

Health Hazard

Because of its dehydrating action, phosphorus pentoxide is a highly corrosive substance. It is an irritant to the eyes, skin, and mucous membranes.

Inhalation of its vapors caused chronic pulmonary edema, injury to lungs, and hemorrhage in test animals. The LC_{50} values varied significantly with the species.

LC_{50} value, inhalation (rats): 127 mg/m^3/hr

Fire and Explosion Hazard

Phosphorus pentoxide is a nonflammable compound and does not support combustion. It reacts violently with water and alcohol. Reaction with water produces phosphoric acid, H_3PO_4. When heated with caustic soda or potash or calcium oxide, the reaction can be extremely violent (Mellor 1946, Suppl. 1971). Explosion may occur when perchloric acid solution in chloroform is added to phosphorus pentoxide (NFPA 1997). Due to its dehydrating action, it can produce many dangerous substances in their anhydrous state, such as perchloric acid. Heating and the presence of sulfuric acid can enhance the explosion hazard.

Disposal/Destruction

Phosphorus pentoxide is mixed with water in an icewater bath. The acid solution is neutralized and washed down the drain.

52.5   PHOSPHORUS OXYCHLORIDE

DOT Label: Corrosive Material, UN 1810
Formula $POCl_3$; MW 153.32; CAS [10025-87-3]
  Structure:

$$O = P\begin{array}{c} \diagup Cl \\ - Cl \\ \diagdown Cl \end{array}$$

Synonyms: phosphoryl chloride; phosphorus oxytrichloride

Uses and Exposure Risk

Phosphorus oxychloride is used to produce hydraulic fluids, plasticizers, and fire-retarding agents; as a chlorinating agent; and as a solvent in cryoscopy.

Physical Properties

Colorless liquid with a pungent odor; strongly fuming; density 1.645; bp 106°C; mp 1°C; mixes with water and alcohol.

Health Hazard

Inhalation of vapors of phosphorus oxychloride produced acute and chronic toxicity in test subjects. In humans, exposure to its vapors may cause headache, dizziness, weakness, nausea, vomiting, coughing, chest pain, bronchitis, and pulmonary edema. Most of these symptoms are manifested from chronic exposure to its vapors.

LC_{50} value, inhalation (rats): 48 ppm (301 mg/m^3)/4 hr

Vapors of this compound are an irritant to the eyes and mucous membranes. The liquid is corrosive and can cause skin burns. An oral LD_{50} value for rats is documented to be 380 mg/kg (NIOSH 1986).

Exposure Limit

TLV-TWA 0.628 mg/m^3 (0.1 ppm) (ACGIH).

Fire and Explosion Hazard

Nonflammable solid. Reacts exothermically with water, alcohols, and amines. Mixing with water can result in violent heat release and splattering, forming HCl and H_3PO_4. The reaction is accelerated rapidly with pressure release and could cause an explosion. Phosphorus oxychloride reacts violently with dimethyl sulfoxide, decomposing the latter.

Disposal/Destruction

Phosphorus oxychloride is water reactive. It should not be put into lab packs for landfill disposal. It can be destroyed by hydrolyzing with sodium hydroxide solution.

$$POCl_3 + 3NaOH \longrightarrow H_3PO_4 + 3NaCl$$

Phosphorus oxychloride is added dropwise to an excess of 2.5 N caustic soda solution. The mixture is cooled to room temperature, neutralized, and washed down the drain.

52.6   PHOSPHORUS PENTACHLORIDE

DOT Label: Corrosive Material, UN 1806
Formula PCl_5; MW 208.22; CAS [10026-13-8]
Synonyms: phosphoric chloride; phosphorus perchloride; pentachlorophosphorane

Uses and Exposure Risk

Phosphorus pentachloride is used as a chlorinating agent to convert acids into acid chlorides, as a dehydrating agent, and as a catalyst.

Physical Properties

White to yellow crystalline solid; deliquescent; pungent odor; mp 148°C under pressure; sublimes at 100°C; bp 160°C; soluble in carbon tetrachloride and carbon disulfide; reacts with water and alcohol.

Health Hazard

Phosphorus pentachloride vapors are a strong irritant to the eyes and mucous membranes. Contact with skin can cause acid burns, as it reacts readily with moisture to form hydrochloric and phosphoric acids:

$$PCl_5 + 4H_2O \longrightarrow H_3PO_4 + 5HCl$$

Chronic exposure to this compound can result in bronchitis.

LC_{50} value, inhalation (rats): 205 mg (24 ppm)/m^3 (NIOSH 1986)

Exposure Limits

TLV-TWA 0.85 mg/m^3 (0.1 ppm) (ACGIH), ~1 mg/m^3 (0.1 ppm) (OSHA).

Fire and Exposion Hazard

Noncombustible solid. Phosphorus pentachloride fumes in moist air; reacts vigorously with water and alcohol. The hydrolysis products are HCl and phosphorus oxychloride, POCl$_3$. The latter reacts further with water to yield HCl and H$_3$PO$_4$. Both these reactions are exothermic. Alcohols decompose it, producing toxic alkyl chlorides.

It reacts with chlorine trioxide to form explosive chlorine monoxide. It ignites with magnesium oxide and fluorine (Mellor 1946); and its reaction with nitric acid can be violent.

Disposal/Destruction

Phosphorus pentachloride is water reactive and therefore cannot be disposed of in a landfill. It is destroyed by treatment with an excess of sodium hydroxide in an ice bath. The reaction products containing unreacted sodium hydroxide are neutralized and washed down the drain.

52.7 PHOSPHORUS TRICHLORIDE

DOT Label: Corrosive Material, UN 1809
Formula PCl$_3$; MW 137.35; CAS [7719-12-2]
Synonym: phosphorus chloride

Uses and Exposure Risk

Phosphorus trichloride is used as a chlorinating agent; as an intermediate in making gasoline additives, dyes, surfactants, and pesticides; in the manufacture of phosphorus pentachloride and phosphorus oxychloride; and as a catalyst.

Physical Properties

Colorless fuming liquid; bp 76°C; freezes at −112°C; density 1.574 at 21°C; decomposed by water and alcohol, soluble in ether, benzene, chloroform, and carbon disulfide.

Health Hazard

Phosphorus trichloride is a highly corrosive substance. Its vapors are an irritant to the upper and lower respiratory tracts. Chronic exposure to its vapors can produce coughing, bronchitis, and pneumonia.

LC_{50} value, inhalation (guinea pigs): 50 ppm (280 mg/m^3)/4 hr

The liquid is corrosive to the skin and can cause acid burns.

Exposure Limits

TLV-TWA 1.12 mg/m^3 (0.2 ppm) (ACGIH), 2.8 mg/m^3 (0.5 ppm) (OSHA).

Fire and Explosion Hazard

Noncombustible liquid. Its reaction with water can be violently exothermic, the decomposition products being phosphorus acid and hydrochloric acid:

$$PCl_3 + 3H_2O \longrightarrow H_3PO_4 + HCl$$

The heat of reaction may decompose phosphorus acid to phosphine, which may ignite spontaneously or explode. It is decomposed by alcohols, producing alkyl halides. The reactions of phosphorus trichloride with acetic acid, nitric acid, nitrous acid, and chromyl chloride can produce explosions. It may ignite or react violently with dimethyl sulfoxide, lead dioxide, fluorine, hydroxyl amine, and iodine monochloride (Mellor 1946, Suppl. 1948; NFPA 1997). Its vapors burn to incandescence when heated with alkali metals.

Disposal/Destruction

Phosphorus trichloride can be hydrolyzed with an excess of 2.5 N sodium hydroxide solution. It is added dropwise into caustic soda solution with stirring. The mixture is then cooled to room temperature, neutralized with dilute HCl or H_2SO_4, and washed down the drain. Small amounts of highly toxic and flammable phosphine may be produced during hydrolysis (National Research Council 1995).

52.8 PHOSPHORUS PENTAFLUORIDE

DOT Label: Poison Gas, UN 2198
Formula PF_5; MW 125.97; CAS [7647-19-0]
Synonyms: phosphoric fluoride; pentafluorophosphorane

Uses and Exposure Risk

Phosphorus pentafluoride is used as a catalyst in polymerization reactions.

Physical Properties

Colorless gas fuming strongly in air; liquefies at $-85°C$; solidifies at $-95°C$; density of gas 5.8 g/L; reacts with water, soluble in carbon disulfide and carbon tetrachloride.

Health Hazard

Phosphorus pentafluoride is highly irritating to the skin, eyes, and respiratory tract. Exposure to this gas can cause pulmonary edema and lung injury. The LC_{50} value for this compound is not reported. The concentration in air at which it may be lethal to mice over a 10-minute exposure period is estimated to be about 400 ppm ($\sim$2050 mg/m^3).

Exposure Limit

TLV-TWA 2.5 mg(F)/m^3 (ACGIH and OSHA).

Fire and Explosion Hazard

Nonflammable gas; high thermal stability. Sensitive to moisture, decomposed by water, forming phosphorus oxyfluoride, hydrogen fluoride, and phosphoric acid. Intermediate oxyfluorophosphates are formed, too. Reaction with water is exothermic.

Disposal/Destruction

Phosphorus pentafluoride is destroyed by bubbling the gas through a solution of caustic soda (in excess) at a cold temperature. The alkaline mixture is neutralized slowly by dilute HCl and washed down the drain.

52.9 PHOSPHORUS PENTASULFIDE

EPA Classified Toxic Waste, RCRA Waste Number U189; DOT Label: Flammable Solid and Danger When Wet
Formula P_2S_5; MW 222.24; CAS [1314-80-3]

Structure: monoclinic crystals; occurs as P_2S_5

Synonyms: phosphorus sulfide; phosphoric sulfide; phosphorus persulfide; sulfur phosphide, thiophosporic anhydride

Uses and Exposure Risk

Phosphorus pentasulfide is used in the manufacture of lubricant additives, pesticides, safety matches, and flotation agents.

Physical Properties

Light yellow to greenish yellow solid; crystalline; odor of rotten eggs; density 2.09; mp 288°C; bp 513°C; hygroscopic; soluble in carbon disulfide and aqueous solutions of caustic alkalies, decomposes violently in water.

Health Hazard

Phosphorus pentasulfide is an irritant to the skin and eyes. Inhalation of its vapors can cause irritation of the respiratory passage. Oral toxicity of this compound in rats was found to be moderate.

LD_{50} value, oral (rats): 389 mg/kg (NIOSH 1986)

Phosphorus pentasulfide can readily produce highly toxic hydrogen sulfide in the presence of moisture. Other toxic products from its combustion are sulfur dioxide and phosphorus pentoxide (corrosive).

Exposure Levels

TLV-TWA air 1 mg/m^3 (ACGIH and OSHA); STEL 3 mg/m^3 (ACGIH)

Fire and Explosion Hazard

Flammable solid; burns slowly; small flame can ignite it; autoignition temperature (liquid) 275°C (527°F), (dust) between 260 and 290°C (500–554°F).

Fire-extinguishing agent: CO_2 or dry chemical; a sodium chloride-base extinguisher suitable for metal fires is reported to be effective (NFPA 1997); do not use water.

Phosphorus pentasulfide is violently decomposed by water to form phosphoric acid and the highly toxic and flammable gas hydrogen sulfide:

$$P_2S_5 + 8H_2O \longrightarrow 5H_2S + 2H_3PO_4$$

It may ignite in the presence of moisture. Reactions with strong oxidizers may be violent.

Disposal/Destruction

Phosphorus pentasulfide is dissolved in sodium hydroxide solution. An excess amount of sodium hypochlorite (laundry bleach) is added slowly to the alkaline solution of phosphorus pentasulfide. The mixture is allowed to stand for several hours. It is neutralized to pH 7 by adding dilute sulfuric or hydrochloric acids in small amounts. The neutral solution may be washed down the drain with a large volume of water.

52.10 PHOSPHORUS SESQUISULFIDE

DOT Label: Flammable Solid and Danger When Wet, UN 1341

Formula P_4S_3; MW 220.06; CAS [1314-85-8]

Structure: rhombic crystals

Synonyms: tetraphosphorus trisulfide; trisulfurated phosphorus

Uses and Exposure Risk

Phosphorus sesquisulfide is used in making safety matches.

Physical Properties

Yellowish-green crystals; melts at 172°C; boils at 407–408°C; density 2.03 at 20°C;

soluble in carbon disulfide, benzene, and toluene, reacts with water.

Health Hazard

Phosphorus sesquisulfide exhibited low to moderate acute oral toxicity in animals. An oral dose lethal to rabbits was reported to be 100 mg/kg (NIOSH 1986). Because of its low vapor pressure, any health hazard due to inhalation of this compound in the work place should be very low. There is no report on its acute inhalation toxicity. Its vapors may produce irritation of respiratory passage. Skin contact may cause mild irritation. It produces toxic sulfur dioxide on burning.

Fire and Explosion Hazard

Highly flammable solid; autoignition temperature 100°C (212°F). It can ignite by friction. Fire-extinguishing agent: CO_2; use a flood of water to extinguish fire. Phosphorus sesquisulfide is decomposed by water. However, its reactivity toward water is not as violent as that of phosphorus pentasulfide.

Disposal/Destruction

An alkaline solution is treated with laundry bleach, allowed to stand overnight, neutralized, and washed down the drain with plenty of water.

52.11   TRIMETHYL PHOSPHITE

DOT Label: Flammable Liquid, UN 2329
Formula $(CH_3)_3PO_3$; MW 124.09; CAS [121-45-9]
Structure and functional group:

$$CH_3O-\underset{\underset{OCH_3}{|}}{P}-OCH_3$$

an organic phosphite
Synonyms: methyl phosphite; trimethoxyphosphine; phosphorus acid trimethyl ester

Uses and Exposure Risk

Trimethyl phosphite is used in the manufacture of pesticides and fire retardants.

Physical Properties

Colorless liquid with a characteristic pungent odor; bp 112°C; mp −78°C; density 1.052 at 20°C; mixes with most organic solvents, decomposed by water.

Health Hazard

Trimethyl phosphite is a skin, eye, and mucous membrane irritant with low subacute inhalation toxicity. Its irritating action on rabbits' skin was moderate to severe. The pure liquid instilled into the eyes can cause severe irritation and swelling, which can last for a few days. Chronic exposure to 300–600 ppm concentration in air produced lung inflammation and cataracts in mice. There was no acute inhalation toxicity observed in test animals. The oral and dermal toxicities were low.

LD_{50} value, oral (rats): 1600 mg/kg

Teratogenic effects showing gross abnormalities were observed in newborn rats when pregnant rats were dosed with high concentrations of trimethyl phosphite.

The odor threshold for this compound was determined to be 0.0001 ppm (ACGIH 1986). The odor is irritating and pungent at high concentrations.

Exposure Limit

TLV-TWA 10 mg/m^3 (2 ppm) (ACGIH).

Fire and Explosion Hazard

Flammable liquid; flash point (closed cup) 37.8°C (100°F); vapor pressure 24 torr at 25°C. Trimethyl phosphite in contact with a small quantity of magnesium perchlorate produced a brilliant flash and loud explosion (Allison 1968).

Disposal/Destruction

Trimethyl phosphite may be burned in a chemical incinerator equipped with an after-burner and scrubber.

52.12 TRIMETHYL PHOSPHATE

Formula $(CH_3)_3PO_4$; MW 140.08; CAS [512-56-1]

Structure and functional group:

$$\underset{\underset{\displaystyle O-CH_3}{|}}{\overset{\overset{\displaystyle O}{\|}}{H_3C-O-P-O-CH_3}}$$

the first member of a homologous series of organic phosphates, methyl ester of phosphoric acid

Synonyms: methyl phosphate; phosphoric acid trimethyl ester

Uses and Exposure Risk

Trimethyl phosphate (TMP) is used as an intermediate in organic synthesis. Its industrial applications are much fewer than those of other organic phosphates.

Physical Properties

Colorless liquid; bp 197°C; density 1.197 at 20°C; soluble in most organic solvents, low solubility in water.

Health Hazard

TMP is a mutagen, teratogen, and a cancer-causing compound. The health hazard from this compound is somewhat different from those of the higher members of the organic phosphate series. While carcinogenicity is observed only for this phosphate, with mutagenicity and the teratogenic effects being more marked, higher alkyl phosphates are more neurotoxic but noncarcinogenic.

TMP is toxic at high dosages.

LD_{50} value, oral (mice): 1470 mg/kg (NIOSH 1986)

Laboratory tests on animals indicated clear evidence of its cancer-causing actions, producing uterine and skin tumors. There is so far no report on such carcinogenic actions in humans. The compound tested positive in mutagenic tests and caused fetal deaths and birth defects in mice, rats, and hamsters when given orally or intraperitoneally.

Fire and Explosion Hazard

Noncombustible liquid. Incompatibility — none.

Disposal/Destruction

TMP is mixed with a combustible solvent and destroyed by burning in a chemical incinerator.

52.13 TRIPHENYL PHOSPHATE

Formula $(C_6H_5)_3PO_4$; MW 326.30; CAS [115-86-6]

Structure and functional group:

an aromatic phosphate, phenyl ester of phosphoric acid

Synonyms: phenyl phosphate; phosphoric acid phenyl ester

Uses and Exposure Risk

Triphenyl phosphate (TPP) is used in fire-proofing, in impregnating roofing paper, as a plasticizer in lacquers and varnishes, and as a substitute for camphor in celluloid materials to make the latter stable and fireproof.

Physical Properties

Colorless solid (crystalline needles) with a faint aromatic odor; mp 49°C; bp 370°C; soluble in most organic solvents, insoluble in water (~10 mg/L).

Health Hazard

TPP is neurotoxic, causing paralysis at high dosages. Like tri-o-cresyl phosphate, it is a cholinestearase inhibitor. The acute oral toxicity is low. The acute toxicity via subcutaneous administration is low to moderate. The toxic symptoms from high dosages in test animals were tremor, diarrhea, muscle weakness, and paralysis.

LD_{50} value, oral (mice): 1320 mg/kg

LD_{50} value, subcutaneous (cats): 100 mg/kg

Cleveland et al. (1986) investigated the acute and chronic toxicity to various species of freshwater fish of phosphate ester compounds containing triphenyl phosphate. The adverse toxic effects occurred at exposure concentrations of 0.38–1.0 mg/L.

Exposure Limit

TLV-TWA air 3 mg/m^3 (ACGIH, OSHA, and NIOSH).

Fire and Explosion Hazard

Noncombustible solid. Incompatibility — none.

Disposal/Destruction

TPP is dissolved in a combustible solvent and subjected to chemical incineration.

52.14   TRIBUTYL PHOSPHATE

Formula $(C_4H_9)PO_4$; MW 266.32; CAS [126-73-8]

Structure and functional group:

$$C_4H_9O-\overset{\overset{\displaystyle O}{\|}}{\underset{\underset{\displaystyle OC_4H_9}{|}}{P}}-OC_4H_9$$

organic phosphate, butyl ester of phosphoric acid

Synonyms: butyl phosphate; tri-n-butyl phosphate; phosphoric acid tributyl ester

Uses and Exposure Risk

Tributyl phosphate is used as a plasticizer for cellulose esters, vinyl resins, and lacquers; and in making fire retardants, biocides, defoamers, and catalysts.

Physical Properties

Colorless and odorless liquid; bp 289°C (decomposes); mp −80°C; density 0.976 at 25°C; soluble in most organic solvents, slightly soluble in water (5.9 g/L).

Health Hazard

Tributyl phosphate is a neurotoxic compound and an irritant. The toxic effects are characteristic of organic phosphates. It inhibits cholinestearase activity and causes paralysis. In addition, it can cause depression of the central nervous system, as well as irritation of the skin, eyes, and respiratory passage. Inhalation toxicity data in the literature are inconsistent.

The oral toxicity in rats was low; the LD_{50} value was reported as 1189 mg/kg (NIOSH 1986).

The pure liquid instilled into rabbits' eyes caused severe irritation but no permanent damage. The irritation effect on the skin is mild.

Tributyl phosphate exhibited teratogenic effects in rats. There is no report on its carcinogenicity.

Exposure Limits

TLV-TWA 2.18 mg/m^3 (0.2 ppm) (ACGIH), 5 mg/m^3 (OSHA and NIOSH); IDLH 1300 mg/m^3 (120 ppm) (NIOSH).

Fire and Explosion Hazard

Noncombustible solid. Incompatibility — none.

Disposal/Destruction

Tributyl phosphate is dissolved in a combustible solvent and is burned in a chemical incinerator equipped with an afterburner and scrubber.

52.15   TRI-*o*-CRESYL PHOSPHATE

Tricresyl phosphate containing >3% *o*-isomer; DOT Label: Poison, UN 2574

Formula $C_{21}H_{21}PO_4$; MW 368.39; CAS [78-30-8]

Structure and functional group:

oxygen atoms bridging aromatic rings with P

Synonyms: tri-*o*-tolyl phosphate; tris (*o*-methylphenyl) phosphate; triorthocresyl phosphate; phosphoric acid; tri-*o*-cresyl ester

Uses and Exposure Risk

Tri-*o*-cresyl phosphate (TOCP) finds wide applications in several areas. It is used as a flame retardant; as a plasticizer; as a waterproofing agent; as a synthetic lubricant; as an additive in gasoline to control preignition; and in hot extrusion molding, hydraulic fluid, and solvent mixture for resins.

Physical Properties

Colorless and oily liquid without an odor; boils at 455°C (decomposes); freezes at −33°C; density 1.1955 at 20°C; soluble in most organic solvents, exceedingly low solubility in water (0.3 mg/L).

Health Hazard

TOCP is a highly poisonous compound. Its toxicity is greater than that of the *meta*- or *para*-isomer. The toxic routes are inhalation, ingestion, and absorption through the skin; and the symptoms varied with the species and the route of admission.

Ingestion of 40–60 mL of the liquid can be fatal to humans. An oral dose of 6–7 mg/kg has produced serious paralysis in humans (Patty 1949). The toxic symptoms from oral intake can be gastrointestinal pain, diarrhea, weakness, muscle pain, kidney damage, and paralysis. The target organs are the gastrointestinal tract, kidney, central nervous system, and neuromuscular system.

LD$_{50}$ value, oral (rabbits): 100 mg/kg

Somkuti et al. (1987) reported testicular toxicity of TOCP in adult leghorn roosters. Birds dosed with 100 mg/kg/day exhibited limb paralysis in 7–10 days. Such symptoms are characteristics of delayed neurotoxicity caused by organophosphorus compounds. Analysis at the termination of 18 days indicated a significant inhibition of neurotoxic esterase activity in both brain and testes, and a decrease in sperm motility and brain acetylcholinesterase activity.

TOCP caused adverse reproductive effects in mice, such as increased maternal mortality and a decreased number of viable litters. An LD$_{50}$ value of 515 mg/kg/day is reported

(Environmental Health Research and Testing 1987).

Exposure Limits

TLV-TWA skin 0.1 mg/m^3 (ACGIH and OSHA); IDLH 40 mg/m^3 (NIOSH).

Fire and Explosion Hazard

Noncumbustible solid; vapor pressure 0.02 torr at 150°C; fire retardant.

Disposal/Destruction

TOCP is dissolved in a combustible solvent and burned in a chemical incinerator equipped with an afterburner and scrubber.

52.16 MISCELLANEOUS COMPOUNDS OF PHOSPHORUS

See Table 52.2.

TABLE 52.2 Toxicity and Flammability of Miscellaneous Phosphorus Compounds

Compound/Synonyms/ CAS No.	Formula/MW/ Structure	Toxicity	Flammability
Phosphorus trioxide (diphosphorus trioxide) [1314-24-5]	P$_2$O$_3$ 109.94	Skin irritant; vapors are irritant to respiratory passage and cause pulmonary edema; DOT Label: Corrosive Material, UN 2578	Reacts violently with hot water, producing phosphine and red phosphorus; ignites brilliantly when heated in air; ignites with chlorine (greenish flame) and bromine; reacts violently with dimethyl formamide and dimethyl sulfite (Mellor 1946, Suppl. 1971); violent reactions may occur with halides of group VB metals; may explode on mild heating with sulfur
Phosphorus trisulfide (diphosphorus trisulfide) [12165-69-4]	P$_2$S$_3$ 158.12	Skin irritant; vapors can cause lung injury	Flammable solid; ignitable by small flame; reacts violently with water; liberates H$_2$S in contact with acids; DOT Label: Flammable Solid, UN 1343
Phosphorus tribromide (phosphorus bromide, tribromophosphine) [7789-60-8]	PBr$_3$ 270.70	Corrosive liquid vapors are highly irritating; produces toxic hydrogen bromide in contact with water or acid; DOT Label: Corrosive Material, UN 1808	Nonflammable liquid; reacts vigorously with water; can explode violently when mixed with alkali metals and subjected to impact (Mellor 1946, Suppl. 1963)

(continued)

TABLE 52.2 (*Continued*)

Compound/Synonyms/ CAS No.	Formula/MW/ Structure	Toxicity	Flammability
Phosphorus trifluoride (phosphorus fluoride, trifluorophosphine) [7783-55-3]	PF_3 87.97	Irritating to eyes, skin, and respiratory tract; exposure to 200 ppm (700 mg/m^3) in air for 30 min is lethal to mice; TLV-TWA air 2.5 mg(F)/m^3 (ACGIH and OSHA)	Nonflammable gas, but explodes in oxygen; ignites when mixed with fluorine (yellow flame) or diborane
Phosphorus triiodide (phosphorus iodide, triiodophosphine) [13455-01-1]	PI_3 411.69	Irritating on skin contact	Noncombustible solid; reacts violently with oxidizing compounds
Phosphonium iodide [12125-09-6]	PH_4I 161.93	High oral toxicity; LD_{50} (rabbits) 5 mg/kg (NIOSH 1986); produces toxic phosphine on decomposition	Unstable solid; sublimes at room temperature; explodes when heated rapidly; decomposed by water or alcohol to HI and phosphine (ignites); ignites spontaneously with nitric acid and with perchloric, perbromic, and periodic acids; ignites in contact with chlorates, bromates, and iodates when dry
Phosphorus oxybromide (phosphoryl tribromide) [7789-59-5]	$POBr_3$ 286.70	Irritating to skin, eyes, and mucous membranes; DOT Label: Corrosive Material, UN 2576	Reacts exothermically with water, undergoing decomposition
Phosphorus triselenide (tetraphosphorus triselenide) [1314-86-9]	P_4Se_3 360.80	Skin irritant; vapors irritating to mucous membranes	Combustible solid; ignites when warmed in air
Phosphorus pentabromide (phosphoric bromide, pentabromophosphorane) [7789-69-7]	PBr_5 430.52	Vapors are irritating to eyes and mucous membranes; skin contact can cause burn; DOT Label: Corrosive Material, UN 2691	Noncombustible solid; decomposed by water and alcohol; reacts with water to form toxic HBr
Trimethylphosphine [594-09-2]	$(CH_3)_3P$ 76.08	Irritant; vapors cause irritation of respiratory passage; foul odor	Flammable liquid; ignites spontaneously and burns violently in air

TABLE 52.2   (*Continued*)

Compound/Synonyms/ CAS No.	Formula/MW/ Structure	Toxicity	Flammability
Triethylphosphine [554-70-1]	$(C_2H_5)_3P$ 118.16	Vapors irritant to respiratory passage	Flammable liquid; flash point −17°C (1°F); ignites in air
Phenylphosphine [638-21-1]	$(C_6H_5)PH_2$ 110.10	Moderate to high inhalation toxicity; acute toxic symptoms — shortness of breath and respiratory distress; LC_{50} inhalation (rats): 38 ppm/4 hr; chronic toxicity symptoms from inhalation were hypersensitivity to sound and touch, dermatitis, and testicular degeneration in rats at 0.6–2.2-ppm concentration level, and nausea, diarrhea, lacrimation, and tremor in dogs at the same level of concentration (ACGIH 1986); ceiling 0.25 mg/m^3 (0.05 ppm) (ACGIH 1986)	Combustible solid; ignites in air at high concentrations
Triphenyl phosphine [603-35-0]	$(C_6H_5)_3P$ 262.03	Irritant; inhalation of vapors can cause irritation of eyes and respiratory passage; LD_{50} oral (rats): 700 mg/kg	Noncombustible solid
Tributyl phosphine [998-40-0]	$(C_4H_9)_3P$ 202.36	Mild irritant: low oral toxicity: LD_{50} oral (rats): 750 mg/kg	Spontaneously flammable in air
Tributylphosphine sulfide [3084-50-2]	$(C_4H_9)_3PS$ 234.42	Skin irritant: low oral and dermal toxicity: LD_{50} oral (rats): 930 mg/kg	Noncombustible solid

REFERENCES

ACGIH. 1986. *Documentation of the Threshold Limit Values and Biological Exposure Indices*, 5th ed. Cincinnatti, OH: American Conference of Governmental Industrial Hygienists.

Allison, W. W. 1968. Communication to *Fire Protection Guide on Hazardous Materials*. Quincy, MA: National Fire Protection Association.

American Public Health Association, American Water Works Association, and Water Pollution Control Federation. 1995. *Standard Methods for the Examination of Water and Wastewater*, 19th ed. Washington, DC: APHA.

Aranyi, C. 1986. Research and development on inhalation toxicologic evaluation of red phosphorus/butyl rubber combustion products. Phase 4. *Govt. Rep. Announce. Index (U.S.) 1988 88* (12), Abstr. No. 831, p. 284; cited in *Chem. Abstr. CA 110* (7): 52482t.

Cabrol Telle, A. M., G. DeSaint Blanquat, R. Derache, E. Hollande, B. Periquet, and J. P. Thouvenot. 1985. Nutritional and toxicological effects of long-term ingestion of phosphine-fumigated diet by the rat. *Food Chem. Toxicol. 23* (11): 1001–9.

Cleveland, L., F. L. Mayer, D. R. Buckler, and D. U. Palawski. 1986. Toxicity of five-aryl phosphate ester chemicals to four species of freshwater fish. *Environ. Toxicol. Chem. 5* (3): 273–82.

Cullis, C. F., M. M. Hirschler, and Q. M. Tao. 1986. The effect of red phosphorus on the flammability and smoke-producing tendency of poly(vinyl chloride) and polystyrene. *Eur. Polym. J. 22* (2): 161–67.

Environmental Health Research and Testing, Inc. 1987. Screening of priority chemicals for reproductive hazards. Triorthocresyl phosphate; 4,4-thiobis (6-*tert*-butyl-*m*-cresol); 13-cis-retinoic acid. *Govt. Rep. Announce. Index (U.S.) 89* (9), Abstr. No. 922, p. 788; cited in *Chem. Abstr. CA 111* (17): 148326s.

Fthenakis, V. M., and P. D. Moskowitz. 1986. Manufacture of thin-film photovoltaic cells: characterization and management of phosphine hazards. *Energy Res. Abstr. 1987 12* (14), Abstr. No. 28470; cited in *Chem. Abstr. CA 108* (4): 26391b.

Herman, T., and S. Soden. 1988. Efficiently handling effluent gases through chemical scrubbing. *AIP Conference Proceedings 166 (Photovoltaic Safety)*, pp. 99–108. Napa, CA: Airproteck Inc.; cited in *Chem. Abstr. CA 109* (26): 236085b.

Kitahara, K., T. Shimada, N. Akita, H. Tadashi, and K. Sasaki. 1989. Method for cleaning gas containing toxic components, such as arsine, diborane, and monosilane from semiconductor manufacturing. European Patent, Mar. 29; cited in *Chem. Abstr. CA 111* (8): 63274y.

Mellor, J. W. 1946. *A Comprehensive Treatise on Inorganic and Theoretical Chemistry*. London: Longmans, Green & Co.

Merck. 1996. *The Merck Index*, 12th ed. Rahway, N. J.: Merck & Co. Inc.

National Research Council. 1995. *Prudent Practices for Handling and Disposal of Chemicals from Laboratories*. Washington, DC: National Academy Press.

NFPA. 1997. *Fire Protection Guide on Hazardous Materials*, 12th ed. Quincy, MA: National Fire Protection Association.

NIOSH. 1984. *Manual of Analytical Methods*, 3rd ed. Cincinnati, OH: National Institute for Occupational Health and Safety.

NIOSH. 1986. *Registry of Toxic Effects of Chemical Substances*, D. V. Sweet. ed. Washington, DC: U.S. Government Printing Office.

Ohtani, H., S. Horiguchi, Y. Urano, M. Iwasaka, K. Tokuhashi, and S. Kondo. 1989. Flammability limits of arsine and phosphine. *Combust. Flame 76* (3–4): 307–10.

Patty, F. A. 1949. *Industrial Hygiene and Toxicology*, Vol. 2. New York: Interscience.

Poston, T. M., K. M. McFadden, R. M. Bean, M. L. Clark, B. L. Thomas, B. W. Killand, L. A. Prohammer, and D. R. Kalkwarf. 1986. Acute toxicity of smoke screen materials to aquatic organisms, white phosphorus-felt, red phosphorus-butyl rubber and SCG No. 2 for oil. *Energy Res. Abstr. 11* (15), Abstr. No. 34982; cited in *Chem. Abstr. CA 106* (19): 150923z.

Rangaswamy, J. R., and V. B. Sasikala. 1986. Bound residues of phosphine and their insect toxicity in two wheat types. *Lebensum.-Wiss. Technol. 19* (5): 360–64; cited in *Chem. Abstr. CA 106* (17): 137230b.

Roberts, A. K., W. E. Trainer, D. L. Biederman, and L. C. Duffine. 1989. Method of waste disposal involving oxidation of elemental phosphorus. European Patent, Apr. 12; cited in *Chem. Abstr. CA 111* (8): 63502w.

Somkuti, S. G., D. M. Lapadula, R. E. Chapin, J. C. Lamb IV, and M. Abou-donia. 1987. Testicular toxicity following oral administration of tri-*o*-cresyl phosphate (TOCP) in roosters. *Toxicol. Lett. 37* (3): 279–90.

Ward, T. E., M. R. Midgett, G. D. Rives, N. F. Cole, and D. E. Wagoner. 1988. Evaluation of methodology for measurement of toxic metals in incinerator stack emissions. *Proceedings of the Air Pollution Control Act Annual Meeting*; cited in *Chem. Abstr. CA 111* (26): 238726.

53

POLYCHLORINATED BIPHENYLS

53.1 GENERAL DISCUSSION

Polychlorinated biphenyls (PCBs) are a group of chloro-substituted biphenyl compounds having the following general structure:

(chlorine atoms are attached to the biphenyl ring)

One to 10 chlorine atoms can attach to the rings to form 209 possible isomers: 1 decachloro, 3 each of mono- and nonachloro, 12 each of di- and octachloro, 24 each of tri- and heptachloro, 42 each of tetra- and hexachloro, and 46 pentachloro biphenyls. All are classified as EPA Priority Pollutants.

These were marketed under various trade names, among them Aroclor, Sovol, Clophen, Phenoclor, Kanechlor, and Pyralene. In the United States the trade name Aroclor (Monsanto) is perhaps best known. In the Aroclor series, the first two digits define the type of molecular structure and the last two are the weight percent of chlorine. Molecular

structural types are 12 for PCBs, 25 and 44 for blends of PCBs and PCTs (polychlorinated triphenyls), and 54 for PCTs. Thus, Aroclor 1221 would indicate a composition of PCBs with biphenyl structure having 21% chlorine by weight. The most common Aroclors are 1221, 1232, 1242, 1248, 1254, 1260, and 1262.

Uses and Exposure Risk

Because of their excellent physical and chemical properties, such as high thermal and chemical stability and high dielectric constant, boiling point, and flame resistance, PCBs were widely used in transformer oils, capacitors, hydraulic fluids, and lubricating oils. They are also used as plasticizers and adhesives, and in paints, printing inks, and fire retardants. However, due to their health hazards, current production of these compounds has been drastically reduced.

PCBs have widely contaminated the ecosystem, occurring ubiquitously in the environment. They have been detected in food packaging and animal feed and have been found in soils and sediments. The sediments of the Hudson River were found to contain very high levels of PCBs.

Physical Properties

Aroclors 1221, 1232, 1242, and 1248 are a colorless mobile oil. Aroclor 1254 is a light yellow viscous oil. Aroclors 1260 and 1262 are yellowish sticky resins, and 1268 is a white powder. Their specific gravities range from 1.182 to 1.812, increasing with chlorine contents. Their distillation temperatures range from 275 to 450°C, increasing gradually with the chlorine number.

PCBs are practically insoluble in water; the solubility decreases with increase in the degrees of chlorination. The solubility of individual chlorobiphenyls varies from about 6 ppm for monochlorobiphenyls to 0.007 ppm for octachlorobiphenyls. Their solubility is much higher in organic solvents.

PCBs are noncombustible liquids or solids with high thermal stability. The chemical reactivity of this class of compounds is low. There is no report of their undergoing violent or explosive reactions. Thus the fire or explosive hazards of PCBs are extremely low.

53.2   HEALTH HAZARD

Polychlorinated biphenyls are moderately toxic substances that have been found to cause cancers in animals and to induce birth defects. Occupational exposure to PCBs exhibit a broad range of adverse health effects on the skin, eyes, mucous membranes, and digestive and neurological systems. The severity of the health hazard depends on the concentration and chlorine content of the PCBs. Aroclors with high chlorine contents are more toxic than are the lower ones. But this is not a general rule, as toxicity reaches a peak at 54% Cl content, followed by a decrease.

Symptoms on skin and mucous membranes from chronic exposure to higher Aroclors are chloracne (an acne-type skin eruption), dermatitis, hyperpigmentation of the skin, discoloration of the finger nails, thickening of the skin, swelling of the eyelids, and burning and excessive discharge from the eyes (Wassermann and Wassermann 1983; NIOSH 1986). Digestive symptoms are nausea, vomiting, abdominal pain, and anorexia. Extreme poisoning can result in jaundice and acute yellow atrophy of the liver. Symptoms of neurotoxicity are headache, dizziness, fatigue, depression, weight loss, and occasionally, pain in the joints and muscles.

Animals exposed to PCBs developed tumors in their livers, gastrointestinal tracts, and lymphatic and leukatic tissues. Although such evidence on humans is insufficient, PCBs are likely to exhibit similar carcinogenicity in humans.

Some of the chlorinated biphenyls are teratomers on animals, causing fetal deaths and birth defects. The toxicity of coplanar polychlorinated biphenyls in avian and chick embryos has been investigated (Brunstroem and Andersson 1988; Brunstroem 1988). Pentachlorobiphenyl was found to be several times more toxic than the tetrachloro compound and 50-fold more potent than the hexachloro compound. In a 2-week toxicity study, pentachloro- and hexachlorobiphenyls were injected into the yolks of hens' eggs preincubated for 4 days. Both compounds caused embryonic death and abnormalities, including liver lesions, edema, and beak deformities. Yen et al. (1989) found that women who consumed PCB-contaminated rice oil bore children of below-average weight. This occurred especially in the first pregnancy after PCB exposure. The pregnancy data, however, were too limited and there is no other report substantiating this finding.

PCBs usually contain trace contaminants, polychlorinated dibenzofurans (PCDHs), which are highly toxic. These contaminants occur at the ppm level: Aroclor 0.8–2.0 ppm, Phenchlor A60 8.4 ppm, Phenchlor DP-6 13.6 ppm, and Kanechlors 1–17 ppm (Wassermann and Wassermann 1983). Some of the chlorinated dibenzofurans, especially the highly toxic 2,3,7,8-tetrachloro isomer, may be contributing to the health hazard from PCBs. The Yusho oil tragedy in Japan in 1968, due to the consumption of rice oil that was accidentally contaminated with 0.3% Kanechlor 400 (chlorine content 48%), contained a high level of 5 ppm polychlorinated dibenzofurans. Toxic evaluation of potentially hazardous coplanar PCBs, dibenzofurans, and dioxins in the adipose tissues of Yusho patients revealed that the principal poisoning compound was 2,3,4,7,8-pentachlorodibenzofuran, with an accountable toxic contribution from coplanar PCBs (Tanabe et al. 1989).

A comparison of the toxicity of some of the commercial-grade PCBs, along with their chlorine contents, is presented in Table 53.1. Detailed documentation is given in RTECS (NIOSH 1986).

The experimental data are too scant to derive any relationship between the toxicity and the chlorination. Aroclors 1221 and 1232, with a lower chlorine contents, are less toxic than Aroclors −1242, −1248, and −1254, the latter exhibiting the most toxicity. Aroclors 1262 and 1268 exhibit lower

toxicity. Thus a decrease in toxicity above 60% chlorine contents may be attributed to the steric hindrance exerted by too many chlorine atoms, thus shielding the rings from the approach of any reactant. In other words, the presence of too many chlorine atoms would lessen the reactivity of the PCB molecule. The scenario is different at lower chlorination. Chlorine withdraws electron density from the ring, inducing a positive charge in the ring. This makes the ring susceptible to further nucleophilic attack. Thus the reactivity is enhanced with the degree of chlorination. This reaches a maximum beyond which the steric hindrance from further chlorine addition inhibits the reactivity of the PCB molecule. Such a maximum toxicity is reached at a degree of chlorination corresponding to a 54% chlorine content in PCBs, as in Aroclor 1254.

The foregoing postulation can also explain the toxicity pattern of other classes of chlorinated aromatics, such as polychlorinated dioxins, dibenzofurans, and aromatics.

TABLE 53.1 Toxicity of Aroclors

Trade Name/CAS No.	Chlorine Content (%)	Toxicity
Aroclor 1016 [12674-11-2]	41	Teratogen; low acute toxicity
Aroclor 1221 [11104-28-2]	21	Teratogen; low acute toxicity; suspected human carcinogen
Aroclor 1232 [11141-16-5]	32	Low acute toxicity; suspected human carcinogen
Aroclor 1242 [53469-21-9]	42	Teratogen; low acute toxicity; LD_{50} subcutaneous (guinea pigs) 345 mg/kg; suspected human carcinogen
Aroclor 1248 [12672-29-6]	48	Teratogen; low acute toxicity; suspected human carcinogen
Aroclor 1254 [11097-69-1]	54	Mutagenic; teratomer; moderately toxic, LD_{50} intravenous (rats) 358 mg/kg; suspected human carcinogen
Aroclor 1260 [11096-82-5]	60	Mutagenic; teratomer; moderately toxic, LD_{50} oral (rats) 1315 mg/kg; suspected human carcinogen
Aroclor 1262 [37324-23-5]	62	Low acute toxicity, LD_{50} oral (rats) 11,300 mg/kg; suspected human carcinogen
Aroclor 1268 [11100-14-4]	68	Low acute toxicity, LD_{50} oral (rats) 10,900 mg/kg; suspected human carcinogen

Exposure Limits

ACGIH (1986) has set an exposure limit for two of the commercial PCBs:

Aroclor 1242 TLV-TWA: 1 mg/m^3
Aroclor 1254 TLV-TWA: 0.5 mg/m^3

These exposure limits are based on the injurious action of PCBs on the skin. NIOSH (1977) considers all PCBs potential carcinogens and recommends a TWA value of 0.001 mg/m^3 or less.

53.3 DISPOSAL/DESTRUCTION

Total destruction of PCBs can result from pyrolysis above 700°C. It is burned in a chemical incinerator equipped with an after-burner and scrubber. Pyrolysis at lower temperature, 300–600°C, however, can produce toxic polychlorinated dibenzofurans (Morita et al. 1978; Buser et al. 1978). Among the thermal destruction technologies, incineration by plasma arc is highly efficient (Joseph et al. 1986). Rosenthal and Wall (1988) have described an infrared incineration system. Catalytic combustion for PCB destruction in the presence of $Cr_2O_3-Al_2O_3$ or $Pt-Al_2O_3$ at 550–700°C is reported (Kolaczkowski et al. 1987).

Aerobic and anaerobic microorganisms can biodegrade PCBs. Such biodegradation depends on several factors, such as PCB concentration, chlorine content, the environment or nature of the soil, and additives. Biodegradation in soils, sediments, and water occurs more effectively when the concentration of PCBs is low to moderate. PCBs can stimulate bacterial growth by providing carbon to microorganisms. However, at a concentration level of 10,000 ppm, no biodegradation was found to occur and PCBs were toxic to microorganisms (Sawhney 1986). Biodegradation is inhibited when the chlorine content of PCBs is greater and is enhanced when an additive such as biphenyl is mixed with the substrate. Hydroxylated products generally result from biodegradation. However, under enhanced microbial conditions, PCBs decompose to methane and CO_2. The mechanisms probably involve hydroxylation, dehalogenation, formation of arene oxide, and ultimate cleavage of the aromatic rings.

McDermott et al. (1989) reported cleanup of PCB-contaminated soils using the bacteria *Pseudomonas putida*. Investigating the biodegradation kinetics in continuous cultures of a *Pseudomonas* strain, Parsons and Sijm (1988) observed that the degradation rates followed first-order kinetics and were influenced by the carbon source on which the cultures were grown. Other microorganisms that were effective were *Alcaligenes eutrophos*, which degraded PCBs in soils and surfactant solutions to chloroacetophenone under aerobic conditions (Bedard and Brennan 1989), and *Acinetobacter types* (Focht 1987).

Low-temperature oxidation is an equally suitable process for PCB destruction. The oxidizing agents reported are $RuO_4-NaClO$ solution at 64°C (Creaser et al. 1988), ozone in the presence of UV light (Carpenter and Wilson 1988), and Carbowax 6000$-K_2CO_3-$ Na_2O_2 at 85°C (Tundo 1986).

Other routes of breakdown of PCBs are photolysis and catalytic processes. High-energy UV radiation and sunlight can cause photodegradation of PCBs, which absorb energy within the range 280–300 nm. The process is accelerated by amines and TiO_2. Photochemical dechlorination is slow when the chlorine content of PCBs is low and does not produce fully dechlorinated biphenyls.

PCBs can be completely dechlorinated catalytically. Catalysts reported are nickel boride, $NaBH_4$ (Dennis et al. 1979); noble metals (Berg et al. 1972); and $LiAlH_4$ (De Kok et al. 1981). Complete destruction may be effected by treating PCBs with methanol and NaOH at 300°C and 180 atm (Yamasaki et al. 1980).

A product of caustic potash and polyethylene glycol monomethyl ether was reported to be efficient at lower temperatures, <100°C

(Kornel and Rogers 1985). Several metals in molten states or at high temperatures have been employed successfully to destroy PCBs. These include Al, Ca, Fe, Zn, Ni, Cu, Pb, and rare-earth metals (Shultz 1986; Bach and Nagel 1986; Adams 1987; Mantle 1989). The efficiencies of Al, Mg, and their oxides in the destruction of PCBs in the vapor phase at 500–800°C have been described (Ross and Lemay 1987). The metals that are most effective for dechlorination are those that have the highest standard heats of formation for their chlorides.

Adams (1987) reported the reactions of PCBs and other halogen-containing organics with sulfur in an inert atmosphere at 500–1500°C, converting into nontoxic, nonflammable, inert, refractory by-products.

Removal of PCBs from a contaminated water stream can be achieved by adsorption with chitosan [9012-76-4] and activated charcoal (Thome and Van Daele 1986). Chitosan from *Procambarus clarkii* coated on diatomaceous earth such as Cellite partially removed PCBs from waste streams (Portier 1988). Decontamination of transformer oil consists of immobilization of PCBs and halogenated organic compounds on alkali- or alkaline-metal carbonates and bicarbonates impregnated with polyethylene glycol and C_1 to C_6 alcoholates of alkali- and alkaline-earth metals (Tumiatti et al. 1989).

Taylor (1988) reported a method for in situ degradation of PCBs in structures and debris at Superfund sites. The method involves treatment of cement surfaces with alkali metal/polethylene glycol mixtures followed by shotblasting to cut away concrete surface.

53.4   ANALYSIS

PCBs in trace levels can be analyzed by GC using a halogen-sensitive detector such as a Hall electrolytic conductivity detector (HECD) or an electron capture detector (ECD). The detection range at the subnanogram level has been well achieved. Aroclors in wastewaters, groundwaters, soils, sediments, and hazardous wastes can be analyzed by EPA methods (EPA 1984; EPA 1988); 3% SP2100, OV-1, DB 5, SPB5, and other equivalent columns can be used in GC and GC/MS analyses. ECD and HECD techniques are suitable for analyzing PCBs in oils (Sonchik et al. 1984).

Aroclors produce multiple peaks. Their presence is confirmed by GC/MS tests using selective ion monitoring (SIM) or multiple-ion detection (MID) techniques in which a few ions characteristics of the PCBs are looked for selectively. Characteristic ions for Aroclors in electron-impact ionization are 190, 222, 224, 256, 292, 360, and 362. The quantitation is generally performed by GC-ECD or GC-HECD.

Samples containing high concentrations of PCBs can be analyzed by GC-FID. Concentrations in excess of 1000 ppm may be estimated by HPLC by UV detection (Brinkman et al. 1976) at 254 nm and by NMR (Wilson 1975). Cairns et al. (1986) have reviewed the analytical chemistry of PCBs.

PCBs in air may be collected on a 13-mm glass fiber filter and florisil, desorbed with *n*-hexane, and analyzed by GC-ECD (NIOSH 1984).

Hogendoorn et al. (1989) have reported using a normal-phase HPLC technique with column switching to separate PCBs from chlorinated pesticides and fats in human milk.

REFERENCES

ACGIH. 1986. *Documentation of the Threshold Limit Values and Biological Exposure Indices*, 5th ed., pp. 128–129. Cincinnati, OH: American Conference of Governmental Industrial Hygienists.

Adams, H. W. 1987. Destruction of carbonaceous material, and inert solid material formed by reaction of carbonaceous material with sulfur. European Patent 238,735, Sept. 30; cited in *Chem. Abstr. CA 107* (26): 242061r.

Bach, R. D., and C. J. Nagel. 1986. Destruction of toxic chemicals. U.S. Patent 4,574,714, Mar. 11; cited in *Chem. Abstr. CA 105*(2): 11520g.

Bedard, D. L., and M. J. Brennan, Jr. 1989. Method for biodegrading PCBs. U.S. Patent 4,876,201, Oct. 24; cited in *Chem. Abstr. CA 112*(8): 62149y.

Berg, O. W., P. L. Diosady, and G. A. V. Rees. 1972. Column chromatographic separation of polychlorinated biphenyls from chlorinated hydrocarbon pesticides, and their subsequent gas chromatographic quantitation in terms of derivatives. *Bull. Environ. Contam. Toxicol. 7*: 338.

Brinkman, U. A. Th., J. W. F. L. Seetz, and H. G. M. Reymer. 1976. *J. Chromatogr. 116*: 353.

Brunstroem, B. 1988. Toxicity of coplanar polychlorinated biphenyls in avian embryos. *Chemosphere 19*(1–6): 765–68.

Brunstroem, B., and L. Andersson. 1988. Toxicity and 7-ethoxyresorufin *O*-deethylase-inducing potency of coplanar polychlorinated biphenyls (PCBs) in chick embryos. *Arch. Toxicol. 62*(4): 263–64.

Buser, H. R., H. P. Bosshardt, and C. Rappe. 1978. Formation of polychlorinated dibenzofurans from the pyrolysis of individual PCB isomers. *Chemosphere 3*: 157–61.

Cairns, T., G. M. Doose, J. E. Froberg, R. A. Jacobson, and E. G. Siegmund. 1986. Analytical chemistry of PCBs. In *PCBs and the Environment*, J. S. Waid, ed. Vol. 1, pp. 1–45. Boca Raton, FL: CRC Press.

Carpenter, B. H., and D. L. Wilson. 1988. PCB sediment decontamination processes selection for test and evaluation. *Hazard. Waste Hazard. Mater. 5*(3): 185–202; cited in *Chem. Abstr. CA 110*(2): 13068t.

Creaser, C. S., A. R. Fernandes, and D. C. Ayres. 1988. Oxidation of aromatic substances, VIII. Oxidative destruction of polychlorinated biphenyls by ruthenium tetroxide. *Chem. Ind. 15*: 499–500.

De Kok, A., R. B., Geerdink, R. W. Frei, and U. A. Th. Brinkman. 1981. The use of dechlorination in the analysis of polychlorinated biphenyls and related class of compounds. *Int. J. Environ. Anal. Chem. 9*: 301–10.

Dennis, W. H., Jr., Y. H. Chang, and W. J. Cooper. 1979. Catalytic dechlorination of organochlorine compounds. V. Polychlorinated biphenyls: Aroclors 1254. *Bull. Environ. Contam. Toxicol. 11*: 750–754.

Focht, D. D. 1987. Analog enrichment decontamination process. European Patent 211,546, Feb. 25; cited in *Chem. Abstr. CA 107*(2): 12352g.

Hogendoorn, E. A., G. R. Van der Hoff, and P. Van Zoonen. 1989. Automated sample cleanup and fractionation of organochlorine pesticides and polychlorinated biphenyls in human milk using NP-HPLC with column-switching. *J. High Resolut. Chromatogr. 12*(12): 784–89.

Joseph, M. F., T. G. Barton, and S. C. Vorndran. 1986. Incineration of PCBs by plasma arc. In *Proceedings of the Ontario Industrial Waster Conference*, Vol. 33, pp. 201–5; cited in *Chem. Abstr. CA 106*(4): 22771m.

Kolaczkowski, S. T., B. D. Crittenden, U. Ullah, and N. Sanders. 1987. Catalytic combustion: the search for efficient commercial techniques for the destruction of polychlorinated biphenyls. In *Management of Hazardous and Toxic Wastes Process Industry*, S. T. Kolaczkowski and B. D. Crittendon, ed. pp. 489–502. London: Elsevier Applied Science.

Kornel, A., and C. Rogers. 1985. PCB destruction: a novel dehalogenation reagent. *J. Hazard. Mater. 12*(2): 161–76.

Mantle, E. C. 1989. Destruction of environmentally hazardous halogenated hydrocarbons with lead. U.K. Patent 2,210,867, June 21; cited in *Chem. Abstr. CA 111*(20): 190220p.

McDermott, J. B., R. Unterman, M. J. Brennan, R. E. Brooks, D. P. Mobley, C. C. Schwartz, and D. K. Dietrich. 1989. Two strategies for PCB soil remediation: biodegradation and surfactant extraction. *Environ. Prog. 8*(1): 46–51.

Morita, M., J. Nakagawa, and C. Rappe. 1978. Polychlorinated dibenzofuran formation from PCB mixture by heat and oxygen. *Bull. Environ. Contam. Toxicol. 19*: 665–68.

NIOSH. 1977. Criteria for a Recommended Standard: *Occupational Exposure to Polychlorinated Biphenyls*. Cincinnati, OH: National Institute for Occupational Health and Safety.

NIOSH. 1984. *Manual of Analytical Methods*, 3rd ed. Cincinnati, OH: National Institute for Occupational Health and Safety.

NIOSH. 1986. *Registry of Toxic Effects of Chemical Substances*, D. V. Sweet, ed. Washington, DC: U.S. Government Printing Office.

Parsons, J. R., and D. T. H. M. Sijm. 1988. Biodegradation kinetics of polychlorinated biphenyls in continuous cultures of a *Pseudomonas* strain. *Chemosphere 17*(9): 1755–66.

Portier, R. J. 1988. Decontamination of contaminated streams. U.S. Patent 4,775,650, Oct. 4; cited in *Chem. Abstr. CA 111*(8): 63488w.

Rosenthal, S., and H. O. Wall. 1988. *Technology Evaluation Report SITE Program Demonstration Test Shirco Infrared Incineration System.* NTIS Order No. PB89- 116024; cited in *Chem. Abstr. CA 111*(14): 120281z.

Ross, R. A., and R. Lemay. 1987. Efficiencies of aluminum, magnesium and their oxides in the destruction of vapor-phase polychlorobiphenyls. *Environ. Sci. Technol. 21*(11): 1115–18.

Sawhney, B. L. 1986. Chemistry and properties of PCBs in relation to environmental effects. In *PCBs and the Environment*, J. S. Waid, ed. pp. 47–64. Boca Raton, FL: CRC Press.

Shultz, C. G. 1986. Destruction of polychlorinated biphenyls and other hazardous halogenated hydrocarbons. European Patent 170,714, Feb. 12; cited in *Chem. Abstr. CA 105*(6): 48445n.

Sonchik, S., D. Madeleine, P. Macek, and J. Longbottom. 1984. Evaluation of sample preparation techniques for the analysis of PCBs in oil. *J. Chromatogr. Sci. 22*(July): 265–71.

Tanabe, S., N. Kannan, T. Wakimoto, R. Tatsukawa, T. Okamoto, and Y. Masuda. 1989. Isomer-specific determination and toxic evaluation of potentially hazardous coplanar PCBs, dibenzofurans, and dioxins in the tissues of "Yusho" PCB poisoning victim and in the causal oil. *Toxicol. Environ. Chem. 24*(4): 215–31.

Taylor, M. L. 1988. Decontamination of structures and debris at Superfund sites. *EPA Report 600/D-88/243*; NTIS Order No. PB89-129381; cited in *Chem. Abstr. CA 111*(10): 83508f.

Thome, J. P., and Y. Van Daele. 1986. Chitosan as a tool for the purification of waters. Adsorption of polychlorinated biphenyls on chitosan and application to decontamination of polluted stream waters. In *Proceedings of the International Conference on Chitin and Chitosan*, 3rd Meeting, R. A. A. Muzzarelli, C. Jeuniaux, and G. W. Goodway, eds. pp. 551–54, New York: Plenum Press.

Tumiatti, W., G. Nobile, and P. Tundo. 1989. Immobilized reagent for the decontamination of halogenated organic compounds. U.S. Patent 4,839,042, June 13; cited in *Chem. Abstr. CA 111*(20): 177773r.

Tundo, P. 1986. Process for the decomposition and decontamination of organic substances and halogenated toxic materials. U.S. Patent 4,632,742, Dec. 30; cited in *Chem. Abstr. CA 106*(22): 182155s.

Wassermann, M., and D. Wassermann. 1983. Polychlorinated biphenyls. In *Encyclopedia of Health and Occupational Safety*, L. Parmeggiani, ed. Vol. 2, pp. 1753–55. Geneva: International Labor Office.

Wilson, N. K. 1975. *J. Am. Chem. Soc. 97*(13): 3573.

Yamasaki, N., T. Yasui, and K. Matsuoka. 1980. Hydrothermal decomposition of polychlorinated biphenyls. *Environ. Sci. Technol. 14*: 550.

Yen, Y. L., S. J. Lan, and C. J. Chen. 1989. Follow-up study of reproductive hazards of multiparous women consuming PCBs-contaminated rice oil in Taiwan. *Bull. Environ. Contam. Toxicol. 43*(5): 647–55.

54

RADON AND RADIOACTIVE SUBSTANCES

54.1 GENERAL DISCUSSION

Heavy elements in the periodic table of atomic number 84 (polonium) and above are radioactive. The nuclei of these elements, by virtue of their high neutron/proton ratio, are unstable and tend to become stable by emitting α- and β-particles and γ-radiation. An α-particle is a positively charged helium nucleus (He^{2+}) having atomic mass 4 and two units of positive charge. A β-particle is an electron with one unit of negative charge and negligible mass. γ-particles have no charge and no mass. Naturally occurring isotopes of all the elements in the periodic table up to bismuth (atomic number 83) are nonradioactive. All the isotopes of polonium [7440-08-6] (atomic number 84) of mass numbers 193 to 210, including the naturally occurring isotope 210, are radioactive. The next two elements, astatine and radon (atomic numbers 85 and 86, respectively), are radioactive gases. Actinium (atomic number 89) followed by the next 14 elements from thorium to lawrencium (atomic numbers 90–103) constitute the actinide series. These elements have electronic configurations containing unpaired electrons in the f subshell, similar to lanthanides. While the

extranuclear electronic configuration determines the valence state and reactivity of an atom, it has little to do with the radioactivity, which is a nuclear phenomenon and depends on the intranuclear structures. Emission of radiation is independent of temperature, pressure, and other conditions.

Radioactive substances decay into stable end products with rates widely varying with the half-life periods of the elements. *Half-life* is the time period over which an element decays into half of its initial concentration. The rate of disintegration follows first-order kinetics. It is interesting to note that the half-life period of isotopes may vary enormously, from billions of years to milliseconds.

Isotopes of many lighter elements with lower mass numbers are radioactive, too. At least one radioisotope is known for every element. Most of these do not occur in nature but can be generated in reactors by nuclear reactions. Tritium (^3_1H) is a naturally occurring radioisotope of hydrogen. Artificial radioactive isotopes are known for a number of elements.

Radioactive heavy elements such as uranium, thorium, or plutonium are used as nuclear fuel; radium is used in the radiography of metals; and radon is used as a surface label to study surface reactions, as well, in the determination of radium or thorium. Among the lighter isotopes, ^{63}Ni is used in electron capture detectors for GC analysis, ^{14}C in radiocarbon dating and as a tracer, and tritium in nuclear fusion and as a tracer in the studies of reactions. Many radioactive elements are used as a source of radiation, in medicine to diagnose disease, and for treatment.

54.2   HEALTH HAZARD

Exposure to ionizing radiation can produce a health hazard, the severity of which depends on the dose, duration of exposure, and the nature of radiation. Alpha radiation is the least hazardous because of its low penetrating power. It is, however, highly energetic and upon ingestion can cause damage in localized area. β-particles can pass through the skin layer and damage living cells. γ and x-rays can produce the most severe effect. These radiations can penetrate much deeper, causing ionization of various chemical substances in the body, producing ions and reactive free radicals. Various somatic effects caused by such radiations include reduction of white blood cells or lymphocytes formed in lymphoid tissues such as spleen, tonsils, or thymus. Other effects are fatigue, nausea, vomiting, reduction of blood elements, sterilization, and reddening of the skin (Meyer 1989).

When radioactive substances pass into the body by inhalation of radioactive vapors, gases (e.g., radon) or dusts, or by ingestion, such substances may accumulate and deposit in the bone. They incorporate in the bone matter; the ionizing radiations damage cells and cause cancer. Cancer in lungs, bones, and lymphatic systems are known to occur in humans as a result of radiation exposures.

In addition to the radiation hazards, many heavy elements and their compounds are toxic. Uranium [7440-61-1] (atomic number 92) and its salts are highly toxic, causing renal damage. The toxic mechanism involves binding of uranyl ion to bicarbonate ion. The dissociation of complex produces UO_2^+

which binds to the membrane of the proximal tubule cells, causing loss of cell function. Plutonium [7440-07-5] (atomic number 94) in high doses is highly toxic, which can cause pulmonary fibrosis and radiation pneumonitis. Inhaled insoluble oxide may remain in the lung and slowly be transported out to the lymph nodes. Soluble form may cause lung, bone, and liver cancers. Inhalation, ingestion, or skin contact of radium [7440-14-4] (atomic number 88) can produce osteogenic sarcoma, osteitis, and blood dyscrasias. Exposure can cause lung cancer.

Like helium and xenon, radon [10043-92-2] (atomic number 86) is an inert gas. There are 21 isotopes, all radioactive, ranging in mass numbers from 202 to 224. ^{222}Rn, ^{220}Rn, and ^{219}Rn are common isotopes with half-life periods of 3.82 days, 55.3 seconds, and 4.0 seconds, respectively (Merck 1996). Radon is produced by α-disintegration of radium and its isotopes. The latter are formed by disintegration of uranium and thorium, which occurs in nature. Thus radon produced from disintegration of radioisotopes beneath the earth's surface migrates through groundwater or percolates through the soil and dissipates into the atmosphere. The average radon concentration in the atmosphere is estimated to be 4 picocuries per liter (pCi/L)(Meyer 1989). In trace amounts it has been detected in wells, drainpipes, and building basements. Radon-enriched water is a health concern. Cancer incidence has been found to be markedly increased in people drinking radon-contaminated well water (Mose et al. 1990).

Inhalation of radon can lead to lung cancer. Carcinogenic potential may increase by smoking.

54.3   RADIOACTIVE WASTE DISPOSAL

Disposal of radioactive waste is done in licensed radioactive waste disposal facilities. The design and constructions should address the long-term performance of geosynthetics in waste cover systems, leachate collection system, belowground vaults, and disposal cells. Highly radiating and heat-producing nuclear waste are classified for two types of repository: underground and near-surface repository and packaged in large 20-m^3 containers (Alder 1989). Bostic and co-workers (1989) have described treatment and disposal options for heavy metal wastes containing the radionuclides ^{235}U and ^{99}Tc. The process involves dilution, pH adjustment between 8.2 and 8.5, precipitation and separation of heavy metal sludge, and heavy metal filtrate by filtration. The sludge is incorporated into cement-based grout containing ground blast furnace slag to reduce the mobility of radioactive and toxic components in it. The grout waste is then disposed in a waste disposal facility. Radioactive ^{99}Tc which may be present in the solution is adsorbed over poly(4-vinylpyridine) resin and disposed of. Alternatively, the waste may be solidified with water and blast-furnace slag to a flowable slurry which is hardened with calcium sulfate (Hooykaas 1990).

Liquid wastes containing plutonium and other actinide elements may be detoxified by biological treatment using *Citrobacter* species, isolated from metal-polluted soil (Plummer and Macaskie 1990). Metals in their trivalent state were most amenable to biodegradation.

Genotoxic wastes mixed with radioactive products may be destroyed by oxidation with $KMnO_4$ and $NaOCl$ and disposed of in a radioactive disposal facility. Among the radioisotopes ^3H, ^{14}C, ^{32}P, and ^{125}I, only ^{14}C was released as a radioactive gas during treatment, constituting up to 60% of the total radioactivity (Simonnet et al. 1989).

Lamarre (1990) has described a process for removing radon and volatile organics from domestic water supplies. The process uses a perforated horizontally oriented tray designed to produce a curved (serpentine or spiral) liquid flow path. Contaminated water flows down the path into a storage tank located below the tray. Air is blown into the tank and up through the perforations, causing the water to froth. The contaminants evaporate out of the frothing water, and the air containing them is vented outside the home.

Granulated activated carbon used in the treatment of radon-contaminated water may accumulate radionuclides that can emit harmful γ-radiation. According to an estimate, maximum proximity to an activated carbon filter for 8 hr/day can produce radiation doses in humans equal to 1.172, 0.171, and 0.058 millirads per hour (mrad/hr) for occupational, public, and residential areas, respectively (Rydell et al. 1989).

54.4 MONITORING OF RADIATION

The amount of radiation to which a person is exposed in the workplace can be measured through personal monitoring equipment such as pocket dosimeters or film badges. Pocket dosimeters are calibrated to read radiation exposure at 0–250 milliroentgen. Film badges or radiation monitoring badges consist of photographic emulsions that develop upon exposure to radiation. The amount of radiation exposure is estimated by comparison with previously developed film exposed to a known amount of radiation. Among other devices, the Geiger counter is the most commonly used instrument used to measure radiation. Portable Geiger counters are often used to monitor β- and γ-radiation.

REFERENCES

Alder, J. C. 1989. Preliminary studies of packaging and disposal of decommissioning waste in Switzerland. *Nucl. Technol. 86* (2): 197–206.

Bostic, W. D., J. L. Shoemaker, P. E. Osborne, and B. Evans-Brown. 1989. Treatment and disposal options for a heavy metals waste containing soluble technetium-99. *ACS Symp. Ser.*, Volume Date 1989, 422: 345–67; cited in *Chem. Abstr. CA 113* (20): 179992b.

Hooykaas, C. W. J. 1990. Method for rendering toxic waste harmless. European Patent

Application EP 361,614, Apr. 4; cited in *Chem. Abstr. CA 113* (2): 11645s.

Lamarre, B. L. 1990. Removing hazardous contaminants from water. PCT International Application WO 9003945, Apr. 19; cited in *Chem. Abstr. CA 113* (10): 84581c.

Merck. 1996. *The Merck Index*, 12th ed. Rahway, NJ: Merck & Co.

Meyer, E. 1989. *Chemistry of Hazardous Materials*, 2nd ed., pp. 474–96. Englewood Cliffs, NJ: Prentice Hall.

Mose, D. G., G. W. Mushrus, and C. Chrosniak. 1990. Radioactive hazard of potable water in Virginia and Maryland. *Bull. Environ. Contam. Toxicol. 44* (4): 508–13

Plummer, E. J., and L. E. Macaskie. 1990. Actinide and lanthanum toxicity towards a *Citrobacter* sp.: uptake of lanthanum and a strategy for the biological treatment of liquid wastes containing plutonium. *Bull. Environ. Contam. Toxicol. 44* (2): 173–80.

Rydell, S., B. Keene, and J. Lowry. 1989. Granulated activated carbon water treatment and potential radiation hazards. *J. N. Engl. Water Works Assoc. 103* (4): 234–48.

Simonnet, F., J. C. Orts, and G. Simonnet. 1989. Destruction of genotoxic wastes mixed with radioactive products. *Health Phys. 57* (6): 885–90.

55

SULFATE ESTERS

55.1 GENERAL DISCUSSION

Sulfate esters or the alkyl sulfates are the organic esters of sulfuric acid. These compounds are produced by esterification of one or both. The $-OH$ groups in H_2SO_4 in which one or both the ionizable H atoms are replaced by alkyl groups. This is shown below in the following structures:

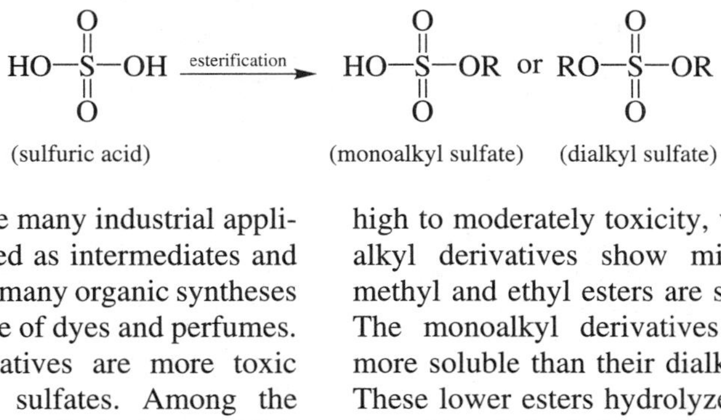

(sulfuric acid)   (monoalkyl sulfate)   (dialkyl sulfate)

These substances have many industrial applications. These are used as intermediates and alkylating reagents in many organic syntheses and in the manufacture of dyes and perfumes.

The dialkyl derivatives are more toxic than the monoalkyl sulfates. Among the former type, the lower alkyl esters exhibit high to moderately toxicity, while the higher alkyl derivatives show mild action. The methyl and ethyl esters are soluble in water. The monoalkyl derivatives are somewhat more soluble than their dialkyl counterparts. These lower esters hydrolyze forming sulfuric acid and alcohols as follows:

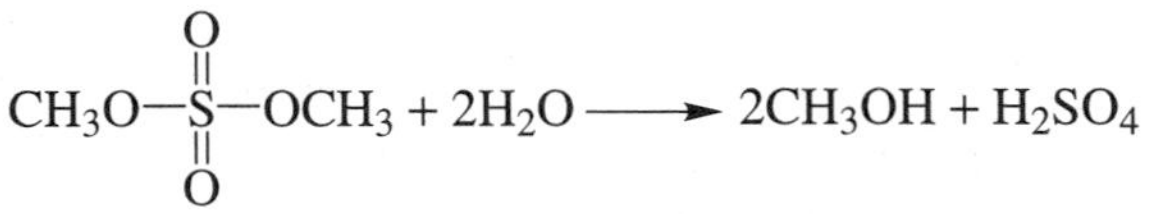

(dimethyl sulfate)

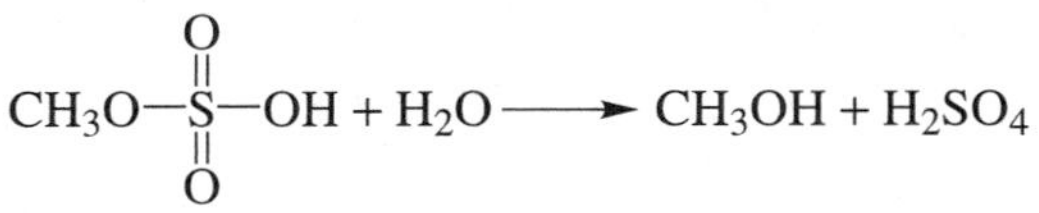

(methyl hydrogen sulfate)

The higher alkyl derivatives are insoluble in water and do not hydrolyze. All sulfate esters are noncombustible. All but dimethyl sulfate have flash points of >200°F.

Sulfate esters may be analyzed by GC and GC/MS techniques. Individual substances may be separated on a suitable column (capillary column such as DB-5 or equivalent) and determined by an FID, FPD or a mass selective detector. Alternatively, the aqueous solutions of lower esters may be distilled. The hydrolysis products alcohols are separated on distillation and determined by derivatization and/or chromatography. Aqueous solution of unhydrolyzed higher esters at trace levels may be directly injected onto the GC column for FID determination, or serially extracted with petroleum ether, the extract concentrated and analyzed by chromatographic technique.

The physical, chemical and hazardous properties of select sulfate esters are outlined in the following sections.

55.2   DIMETHYL SULFATE

Formula: $C_2H_6O_4S$; MW 126.14; CAS [77-78-1]

Structure:

$$CH_3-O-\overset{\displaystyle O}{\underset{\displaystyle O}{\overset{\|}{\underset{\|}{S}}}}-O-CH_3$$

Synonyms: sulfuric acid, dimethyl ester; DMS

Uses and Exposure Risk

It is used as a methylating agent in the manufacture of many organic compounds, such as, phenols and thiols. Also, it is used in the manufacture of dyes and perfumes, and as an intermediate for quaternary ammonium salts. It was used in the past as a military poison.

Physical Properties

Colorless, oily liquid; faint onion odor; freezes at −27°C; boils at 188°C (decomposes); vapor pressure 0.5 torr at 20°C; density 1.33 g/mL at 20°C; moderately soluble in water (28 g/L); soluble in ether, alcohol, acetone and benzene.

Health Hazard

Dimethyl sulfate is a highly toxic substance by all routes of exposure, including inhalation, ingestion, skin absorption and intravenous routes. In humans, a 10-minute inhalation of 100 ppm dimethyl sulfate in air can cause death in a few days after exposure. There are also cases of deaths from occupational exposures to this substance (ACGIH 1986)

Like many other organosulfur compounds, such as mustard gas, dimethyl sulfate manifests delayed effects too. The toxicity is lower than the sulfur mustards with latent period somewhat longer. It can damage eye, respiratory tract, skin, liver, and kidney. The effects appear 2–12 hours after contact, progressively worsening over several days or even weeks.

It affects the eye with delayed irritation and reddening. If the level of exposure is high, the inflammation can progress to photophobia. Such delayed symptoms on the eye may be attributed to its metabolites, methanol and sulfuric acid, the former impairing central vision, while the latter damaging cornea. Inhalation of vapors can cause irritation of the nose, throat, and larynx, and coughing, followed by more severe symptoms that include swelling of the tongue, hoarseness in voice, difficulty in swallowing, congestion in mucous membranes, and difficulty breathing. Other symptoms include giddiness, vomiting, and diarrhea. Ingestion of the liquid can cause lung, liver and kidney damage. It can corrode the mucous membrane. Coma and death can result in severe cases. Heavy exposure can impair liver and kidney functions, resulting in suppression of urine and the appearance of jaundice, albuminuria and hematuria (Lewis 1995).

Skin contact can produce severe blistering. It can penetrate the skin to some depth from the point of contact, resulting in necrosis and severe poisoning. Such reddening and inflammation may develop after a latent period.

LD$_{50}$ value, oral (mouse): 140 mg/kg

LD$_{50}$ value, skin (rat): 100 mg/kg

LC$_{50}$ value, inhalation (rat): 45 mg/m^3/4 hour

Studies on experimental animals have indicated its tumorigenic potency. In humans, however, its carcinogenicity has not been established. The US National Toxicology Program has classified this compound as "Reasonably anticipated to be carcinogenic to human" (NTP 1998).

Exposure Limits

TLV/PEL-TWA skin 0.1 ppm (0.52 mg/m^3) (ACGIH, OSHA, NIOSH)

IDLH 10 ppm (NIOSH)

Flammability and Explosive Reactions

Noncombustible liquid; flash point (open cup) 83°C (182°F). vapor pressure 0.5 torr at 20°C; autoignition temperature 188°C (370°F) (NFPA 1997). The fire hazard is low because of the high flash point. However, it may become flammable when exposed to heat or flame. Fire extinguishing agent: CO$_2$, dry chemical or foam.

Dimethylsulfate can react explosively when combined with concentrated ammonia solution. There are reports of violent explosions when liter quantities of these substances were mixed (Bretherick 1995). Reactions with certain trialkylamines and strong oxidizers can be violent.

Disposal/Destruction

It is burned in a chemical incinerator equipped with an afterburner and scrubber. In the laboratory, dimethyl sulfate in small amounts may be destroyed by treating it with dilute ammonia solution in a fume hood. Because the reaction is exothermic, the mixing should be done gradually in smaller quantities. Concentrated ammonia solution should not be used, as it may react explosively.

55.3   DIETHYL SULFATE

DOT Label: Poison, UN 1594

Formula: C$_4$H$_{10}$O$_4$S; MW 154.19; CAS [77-78-1];

Structure:

$$CH_3-CH_2-O-\overset{\overset{\displaystyle O}{\|}}{\underset{\underset{\displaystyle O}{\|}}{S}}-O-CH_2-CH_3$$

Synonyms: sulfuric acid, diethyl ester

Uses and Exposure Risk

Diethyl sulfate is used as an ethylating agent in many organic syntheses. It is also used as an accelerator in the sulfation of ethylene and as an intermediate in certain sulfonation reactions (Merck 1996).

Physical Properties

Colorless, oily liquid; odor of peppermint; solidifies at −24°C; boils at 208°C (decomposes); vapor pressure 1 torr at 47°C and 134 torr at 150°C; density 1.172 at 25°C; insoluble in water (decomposes gradually at ambient temperature, but rapidly in hot water); miscible with alcohol and ether.

Health Hazard

The toxicity of this compound is significantly lower than that of the corresponding methyl ester. The occupational hazard from inhalation is lower because of its lower vapor pressure. A 4-hour exposure to 250 ppm proved lethal to rats. Inhalation toxicity data in humans are unavailable. It is, however, a severe skin irritant. Such action on skin may be attributed to its gradual decomposition to sulfuric acid upon contact with moisture in skin. Contact of the liquid or the vapor on the eye can produce severe irritation. It is moderately toxic by ingestion.

LD$_{50}$ oral (mouse): ~650 mg/kg

LD$_{50}$ skin (rabbit): 600 mg/kg

There is no evidence of its carcinogenic potency in humans. However, based on its alkylating properties similar to dimethyl sulfate and other alkylating agents, this substance may be reasonably anticipated to cause cancer.

Hazardous Reactions

All hazardous reactions of this compound may be attributed to sulfuric acid, which is generated on hydrolysis, that is, reaction with water, The acid formed, based on its concentration, may readily react with metals (metal containers) and generate hydrogen. This may be hazardous if there is any buildup of pressure due to hydrogen. Also, sulfuric acid can react violently with certain nitro compounds. In the absence of any hydrolysis acid product, the parent ester does not exhibit any violent reactivity with most substances. Reactions with strong oxidants at elevated temperatures, however, can be vigorous to violent. It is a noncombustible liquid. The flash point is above 200°F (220°F)

55.4   DIISOPROPYL SULFATE

Formula: C$_6$H$_{14}$O$_4$S; MW 182.26; CAS [2973-10-6];

Structure:

Synonym: sulfuric acid, diisopropylester

Physical Properties

Colorless liquid; sharp odor; boils around 120°C at 30 torr; density 1.11; slightly soluble in water; miscible with alcohol and ether.

Health Hazard

This substance is moderately toxic by ingestion and skin contact (Lewis 1996). Its toxicity, however, is lower than the ethyl derivative. The evidence of its carcinogenic potency are inadequate. IARC has listed this substance as "possibly carcinogenic to human" (group B carcinogen) (IARC 1998).

LD$_{50}$, oral (rat): ~1100 mg/kg

LD$_{50}$, skin (rabbit): ~1400 mg/kg

55.5   DIBUTYL SULFATE

Formula: C$_8$H$_{18}$O$_4$S; MW 210.32; CAS [625-22-9];

Structure:

Synonyms: di-*n*-butyl sulfate; sulfuric acid, dibutyl ester

Physical Properties

Colorless liquid; sharp characteristic odor; boils at 115°C at 6 torr; density 1.06 at 20°C; insoluble in water; soluble in alcohol and ether.

Health Hazard

The toxicity of this substance is much lower than the methyl or the ethyl esters. It is moderately toxic by ingestion. The oral lethal dose in rabbits is within the range 200 mg/kg. Studies on experimental animals indicate a mild toxic action when administered by subcutaneous route.

LD_{50}, skin (rat): 5000 mg/kg.

55.6 ETHYL HYDROGEN SULFATE

Formula: $C_2H_6O_4S$; MW 126.13; CAS [540-82-9];
Structure:

$$CH_3-CH_2-O-\overset{\overset{\displaystyle O}{\|}}{\underset{\underset{\displaystyle O}{\|}}{S}}-OH$$

Synonyms: monoethyl sulfate; sulfuric acid, monoethyl ester; ethylsulfuric acid; ethyl sulfate; sulfovinic acid.

Uses and Exposure Risk

Ethyl hydrogen sulfate is used as an intermediate in the synthesis of ethanol from ethylene.

Physical Properties

Colorless, oily liquid; boils at 280°C (decomposes); density 1.367; highly soluble in water; hydrolyzes forming sulfuric acid.

Health Hazard

The liquid is a strong irritant to skin. Such irritant property may be attributed to sul-furic acid resulting from the reaction with the moisture in the skin. The vapors of this substance are highly irritating to the respiratory tract. It readily forms salts with group IIA metals. The barium salt, barium ethyl sulfate

$$(C_2H_5O-\overset{\overset{\displaystyle O}{\|}}{\underset{\underset{\displaystyle O}{\|}}{S}}-O)_2Ba$$

is highly toxic.

REFERENCES

ACGIH: 1986. *Documentation of the Threshold Limit Values and Biological Exposure Indices*, 5th ed. Cincinnati, OH: American Conference of Governmental Industrial Hygienists.

Bretherick, L. 1995. *Handbook of Reactive Chemical Hazards*, 5th edition, ed. P. G. Urden. Oxford, U. K: Butterworth-Heinemann.

IARC. 1998. Monograph, Vol. 71. Lyon, France: International Agency of Research on Cancer.

Lewis (Sr.), R. J. 1995. *Sax's Dangerous Properties of Industrial Materials*, 9th ed. New York: Van Nostrand Reinhold.

Merck. 1996. *Merck Index*, 12th ed. Rahway, NJ: Merck & Co.

NFPA. 1997. *Fire Protection Guide on Hazardous Materials*, 12th ed. Quicy, MA: National Fire Protection Association.

NTP. 1998. *Annual Report on Carcinogen* (8th), Washington, D. C.; National Toxicology Program, US Department of Health and Human Services.

56

SULFUR-CONTAINING ORGANICS (MISCELLANEOUS)

56.1 GENERAL DISCUSSION

Sulfur atom has six valence electrons and its electronic configuration is similar to oxygen. However, because of available $3d$ orbitals, its valency can be greater than 2. Thus it forms compounds in the $+2$, $+4$, and $+6$ oxidation states, exhibiting diversity in the types of the compounds it forms. While all oxygenated organics are bivalent, a number of S compounds have a valency of 4 and 6.

A large number of organic sulfur compounds contain sulfur atoms in various structural features. Most, but not all classes of these compounds may be paralleled as sulfur analogues of various oxygenated organics. These include thiols or thiophenols, thioethers, disulfides, thioureas, thioamides, and so on, as shown below.

thiols: —SH (sulfur analogue of alcohol —OH)

thiophenols: aromatic —SH (sulfur analogue of phenol)

thioethers: —S— (sulfur analogue of ether, —O—)

thioureas: $R_2N-\underset{\underset{S}{\|}}{C}-NR_2$ (sulfur analogue of urea, $R_2N-\underset{\underset{O}{\|}}{C}-NR_2$

thiocyanates: $R-S-C\equiv N$ (sulfur analogue of cyanates, $R-O-C\equiv N$)

isothiocyanates: $R-N=C=S$ (sulfur analogue of isocyanates, $R-N=C=O$)

thiofurans: 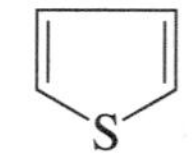   (sulfur analogue of furans, 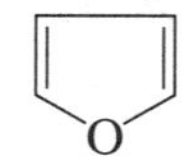)

In addition, sulfoxides, $R-\underset{\underset{O}{\|}}{S}-R$; sulfones, $R-\overset{\overset{O}{\|}}{\underset{\underset{O}{\|}}{S}}-R'$; sulfolanes, 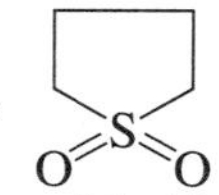 ; disulfides, $R-S-S-R$; sulfonic acids, $R-S-OH$; sulfonic esters, $R-S-O-R$; sulfate esters, $R-O-\overset{\overset{O}{\|}}{\underset{\underset{O}{\|}}{S}}-O-R$; and so on, are examples of compounds in which the valency of the S atom is 4 or 6. There are no oxygen analogues for these types of sulfur organics.

Toxicity

Most sulfur-containing organics exhibit a low order of toxicity. The toxicity, however, may enhance by substitution in the molecules. Sulfur mustards are among the most toxic organosulfur compounds. These are highly toxic and blistering military poisons. Some of the best known compounds in this group are mustard oil H, bis(2-chloroethyl)sulfide, $ClCH_2CH_2SCH_2CH_2Cl$; sesquimustard Q or 1,2-bis(2-chloroethylthio)ethane, $ClCH_2CH_2SCH_2CH_2SCH_2CH_2Cl$; and O-mustard(T) or bis(2-chloroethylthioethyl)ether, having the formula $ClCH_2CH_2SCH_2CH_2OCH_2CH_2SCH_2CH_2Cl$. A detailed discussion on the first compound of this group, namely, bis(2-chloroethyl)sulfide is presented in Chapter 38.

The toxic symptoms from exposure to thiols include headache, dizziness, nausea, muscle weakness, irritation of the nose and throat, and cyanosis. In extreme poisoning, unconsciousness, coma, and even death may result. Methanethiol, ethanethiol,

and benzenethiol are discussed in detail in Sections 56.3, 56.4, and 56.5, respectively. The median lethal doses and median lethal concentrations of some of the thiols are presented in Table 56.1. It may be noted that an $-SH$ group attached to the benzene ring imparts greater toxicity to the molecule than that attached to an alkyl group.

Alkyl thiocyanates, $R-S-C\equiv N$, of lower carbon numbers are highly toxic. These compounds liberate HCN as a result of metabolic processes and can cause death (Manahan 1989). Methyl thiocyanate [556-64-9] is highly toxic, with an intravenous LD_{50} value of 18 mg/kg in mice (NIOSH 1986). Ethyl thiocyanate [542-90-5], allyl thiocyanate [764-49-8], butyl thiocyanate [628-83-1], and amyl thiocyanate [32446-40-5] are all toxic compounds with an LD_{50} (subcutaneous) value of less than 100 mg/kg. The phenyl ester [5285-87-0] exhibits greater toxicity (intravenous LD_{50} value in mice: 6.7 mg/kg) than that of the aliphatic esters of thiocyanic acid.

The toxicity of isothiocyanates, $R-N=C=S$, varies with their carbon numbers and routes of administration. These are irritants to the eyes, skin, and mucous membranes. These substances decompose upon heating,

TABLE 56.1   Median Lethal Values for Some Common Thiols

Compound/CAS No.	LD_{50} (oral) (mg/kg)	LC_{50} Inhalation (ppm in air)
Methanethiol (methyl mercaptan) [74-93-1]		675/4 hr (rats)
Ethanethiol (ethyl mercaptan) [75-08-1]	680 (rats)	4420/4 hr (rats)
1-Propanethiol (n-propyl mercaptan) [107-03-9]	1790 (rats)	7300/4 hr (rats)
2-Methyl-1-propanethiol (isobutyl mercaptan) [513-44-0]	7170 (rats)	
2-Methyl-2-propanethiol (tert-butyl mercaptan) [75-66-1]	4700 (rats)	16,500/4 hr (rats)
1-Butanethiol (n-butyl mercaptan) [109-79-5]	1500 (rats)	4020/4 hr (rats)
1-Pentanethiol (n-amyl mercaptan) [110-66-7]		2000/4 hr (rats) (LC_{L0})
Benzenethiol (phenyl mercaptan) [108-98-5]	46 (rats)	33/4 hr (rats)
α-Toluenethiol (phenylmethyl mercaptan) [100-53-8]	490 (rats)	178/4 hr (mice)

emitting HCN and SO_2. Allyl isothiocyanate [57-06-7] is moderately toxic and an animal carcinogen, causing kidney cancer. The oral LD_{50} value in mice is in the range 300 mg/kg.

Carbon disulfide and carbon oxysulfide are volatile substances with low to moderate toxicity. The psychopathological effects of carbon disulfide arising from chronic exposure include neuritis, bizarre dreams, insomnia, irritation, and excitation. Carbon oxysulfide is an irritant and narcotic at high concentrations. Very little information is available on the human toxicity of sulfides, disulfides, and sulfones. Organic sulfur compounds are notable for their disagreeable odor. In an extremely pure state, several such compounds have a mild sweet odor. A detailed discussion of the toxicity of individual compounds of commercial interest is presented in the following sections.

Analysis

Organosulfur compounds may be analyzed by various instrumental techniques, including GC, HPLC, and GC/MS. A flame photometric or flame ionization detector is suitable for GC analysis. Akintonwa (1985) described reversed-phase HPLC separation of thiourea, thioacetamide, and phenobarbitone for screening of the purity of substances for toxicological evaluation.

Carbon disulfide in air may be analyzed by NIOSH (1984) Method 1600. Between 3 and 25 L of air at a flow rate of 10–200 mL/min is adsorbed over coconut shell charcoal, desorbed with benzene, and injected into a GC equipped with a flame photometric detector with a sulfur filter. The chromatography column consists of 5% OV-17 on an 80/100-mesh Gas Chrom Q support.

56.2 CARBON DISULFIDE

EPA Designated Acute Hazardous Waste, RCRA Waste Number PO22; DOT Label: Flammable Liquid UN 1131

Formula CS_2; MW 76.13; CAS [75-15-0]

Structure: $S=C=S$

Synonyms: carbon bisulfide; dithiocarbonic anhydride

Uses and Exposure Risk

Carbon disulfide is used in the manufacture of rayon, soil disinfectants, and electronic vacuum tubes. It is also used as a solvent and as an eluant for organics adsorbed on charcoal in air analysis.

Physical Properties

Clear colorless liquid with a strong foul odor; density 1.263 at 20°C; boils at 46.5°C; freezes at −111.6°C; slightly soluble in water (0.22% at 20°C), miscible in alcohol, ether, benzene, and chloroform.

Health Hazard

Although carbon disulfide exhibits low toxicity in most experimental animals, its toxicity is relatively greater in humans. The primary route of exposure is inhalation of vapors. It may also enter the body through skin absorption. The toxic effect from single exposure is narcosis. Repeated exposure causes headache, dizziness, fatigue, nervousness, insomnia, psychosis, irritation, tremors, loss of appetite, indigestion, and gastric disturbances. The symptoms above may be manifested in humans after a few months of 4-hour daily exposure to 150 ppm. A concentration below 30 ppm does not produce any notable toxic effects. A 15-minute exposure to 5000 ppm of carbon disulfide in air can be fatal to humans. Ingestion of 5–10 mL of the liquid may be fatal. The oral LD_{50} value in rats is in the range 3000 mg/kg.

Exposure Limits

TLV-TWA 10 ppm ($\sim$30 mg/m^3) (ACGIH), 20 ppm (MSHA and OSHA); IDLH 500 ppm (NIOSH).

Fire and Explosion Hazard

Highly flammable liquid; flash point (closed cup) $-30°C$ $(-22°F)$; vapor pressure 300 torr at $20°C$; vapor-air density 2.2 at $38°C$; vapor heavier than air and can travel a considerable distance to a source of ignition and flashback; autoignition temperature $90°C$ $(194°F)$.

Carbon disulfide is a dangerous fire hazard. Its flash point and autoignition temperature are notably low. A laboratory hot plate or even a light bulb can ignite the vapors on contact. Fire-extinguishing agent: dry chemical or CO_2; water and foam are ineffective; water may be used to keep fire-exposed containers cool.

The vapors of carbon disulfide form explosive mixtures with air within a wide range, 1.3–50.0% by volume in air. Its mixtures with fluorine or chlorine are flammable at ordinary temperatures. An explosion occurred when carbon disulfide was mixed with liquid chlorine in an iron cylinder (MCA 1964). Metals such as aluminum or zinc in a finely divided state or potassium or sodium on heating ignite with carbon disulfide vapors. Reactions with metal azides, metal fulminates, and nitrogen dioxide can progress to explosions; with nitric oxide, it burns with a greenish luminous flame.

Storage and Shipping

Carbon disulfide is stored in a flammable-liquids storage room or cabinet isolated from heat sources and protected against physical damage, lightning, or static electricity. Exposure to direct sunlight and electricity should be avoided. The containers should be kept cool. Spark-producing tools should never be used during handling or use of this liquid (NFPA 1997). Use wood sticks for measuring the contents of the tanks or drums. Carbon disulfide is shipped in glass or metal containers packed in wooden boxes.

Disposal/Destruction

Carbon disulfide is burned in a chemical incinerator equipped with an afterburner and scrubber in small amounts (less than 100 mL) and with extra care. Larger amounts pose explosion hazard. It should preferably be diluted with a higher-boiling solvent before incineration.

In the laboratory, small quantities may be evaporated in a hood in the absence of a hot plate, open flame, or nearby ignition source. Carbon disulfide can be destroyed by oxidizing with hypochlorite:

$$CS_2 + 8NaOCl + 2H_2O \longrightarrow CO_2$$
$$+ 2H_2SO_4 + 8NaCl$$

Either sodium or calcium hypochlorite is used in 25% excess for each 0.5 mol of carbon disulfide (National Research Council 1995).

56.3 METHANETHIOL

EPA Designated Hazardous Waste, RCRA Waste Number U153; DOT Label: Flammable Gas, UN 1064

Formula CH_4S; MW 48.11; CAS [74-93-1]
Structure:

$$H-\underset{\underset{H}{|}}{\overset{\overset{H}{|}}{C}}-SH$$

first member of the homologous series of thiols;

Synonyms: methyl mercaptan; thiomethanol; thiomethyl alcohol; mercaptomethane; methyl sulfhydrate

Uses and Exposure Risk

Methanethiol is used in the manufacture of pesticides and fungicides and as an intermediate in the manufacture of jet fuels (Watkins et al. 1989); it is added to natural gas to give odor; density 0.8665 at $20°C$; freezes at $5.95°C$; solidifies at $-123°C$; soluble in water, alcohol, and ether; forms a crystalline hydrate with water.

Methanethiol occurs in the landfill gas. Its concentration depends more on the stage of decomposition than on the nature of the waste. Young and Parker (1984) reported its odor in landfill gas, exceeding the odor threshold by a factor of $>10^6$.

Health Hazard

The acute toxicity of methanethiol is similar to that of hydrogen sulfide. Inhalation of this gas can cause narcosis, headache, nausea, pulmonary irritation, and convulsions in humans. Other symptoms noted are acute hemolytic anemia, methemoglobinemia, and cyanosis. Exposure to high concentrations can result in respiratory paralysis and death. The 2-hour inhalation LC_{50} value in mice is within the range 650 mg/m^3.

Exposure Limits

TLV-TWA 0.5 ppm ($\sim$1.0 mg/m^3) (ACGIH and MSHA); ceiling 10 ppm (OSHA); IDLH 400 ppm (NIOSH).

Fire and Explosion Hazard

Flammable gas; forms explosive mixtures with air; the LEL and UEL values are 3.9% and 21.8% by volume in air, respectively. It reacts violently with strong oxidizing substances.

56.4 ETHANETHIOL

Formula C$_2$H$_6$S; MW 62.14; CAS [75-08-1]; Structure: CH$_3$−CH$_2$−SH

Synonyms: ethyl mercaptan, thioethanol, thioethyl alcohol, mercaptoethane, ethyl sulfhydrate

Uses and Exposure Risk

Ethanethiol is used as an intermediate in the manufacture of insecticides, plastics, and antioxidants; and as an additive to natural gas

to give odor. It occurs in illuminating gas and in petroleum distillates.

Physical Properties

Colorless liquid with a strong disagreeable skunk-like odor; density 0.839 at 20°C; boils at 36°C; freezes at −148°C; slightly soluble in water (0.68 g/100 ml), soluble in alcohol, ether and petroleum naphtha.

Health Hazard

The inhalation toxicity of ethanethiol is very low. Intraperitoneal administration in rats at sublethal doses caused deep sedation, followed by lethargy, restlessness, lack of muscular coordination, and skeletal muscle paralysis. Higher doses produced cyanosis, kidney and liver damage, respiratory depression, coma, and death. The intraperitoneal LD_{50} value in rats was 450 mg/kg (ACGIH 1986).

Ethanethiol is metabolized to inorganic sulfate and ethyl methyl sulfone, and excreted. The oral LD_{50} value in rats is 625 mg/kg.

In humans, repeated exposures to its vapors at about 5 ppm concentration can produce irritation of the nose and throat, headache, fatigue, and nausea.

Exposure Limits

TLV-TWA 0.5 ppm ($\sim$1.3 mg/m^3) (ACGIH and MSHA); ceiling 10 ppm (OSHA); IDLH 2500 ppm (NIOSH).

Fire and Explosion Hazard

Highly flammable liquid; flash point (closed cup) −48°C (−55°F); vapor pressure 442 torr at 20°C; vapor density 2.14 (air = 1); vapor is heavier than air and can travel a considerable distance to a source of ignition and flashback; autoignition temperature 299°C (570°F).

Vapors of ethanethiol form explosive mixtures with air within the range 2.8–18.0% by volume in air. Its reactions with strong oxidizers can be violent.

56.5 THIOPHENOL

EPA Designated Acute Hazardous Waste, RCRA Waste Number PO14; DOT Label: Poison, UN 2337

Formula C_6H_6S; MW 110.18; CAS [108-98-5]

Structure:

Synonyms: benzenethiol; phenyl mercaptan

Uses and Exposure Risk

Thiophenol is used as a mosquito larvicide and as an intermediate in organic synthesis. It is effective in reducing peroxide formation in jet fuels (Watkins et al. 1989).

Physical Properties

Colorless liquid with a penetrating garlic-like odor; density 1.0728 at 25°C; bp 168°C; solidifies at −15°C; insoluble in water, soluble in alcohol, ether, benzene, toluene, and carbon disulfide; weakly acidic.

Health Hazard

Animal toxicity data show thiophenol to be highly toxic; the oral LD_{50} value in test animals is <100 mg/kg. Its irritant action on rabbits' eyes and skin is severe. Thiophenol can enter the body by ingestion, absorption of the liquid through the skin, and inhalation of vapors. In humans the toxic symptoms include restlessness, incoordination, muscle weakness, headache, dizziness, cyanosis, lethargy, sedation, respiratory depression, and coma. Death may result from high doses. Repeated exposure to thiophenol vapors caused injury to the lung, liver, and kidneys in mice.

LC_{50} value, inhalation (mice): 28 ppm/4 hr
LD_{50} value, oral (rats): 46 mg/kg

Exposure Limit

TLV-TWA 0.5 ppm ($\sim$2.5 mg/m^3) (ACGIH).

Fire and Explosion Hazard

Combustible liquid; flash point (closed cup) 56°C (132°F); vapor pressure 1 torr at 18°C.

56.6 THIOUREA

EPA Designated Acute Hazardous Waste, RCRA Waste Number, U219; DOT Label: Poison B, UN 2877

Formula: CH_4N_2S; MW 76.13; CAS [62-56-6]

Structure:

Synonyms: thiocarbamide; sulfourea

Uses and Exposure Risk

Thiourea is used in the manufacture of resins, as a vulcanization accelerator, and as a photographic fixing agent and to remove stains from negatives.

Physical Properties

Crystalline solid; melts at 176°C; density 1.405; soluble in water and alcohol.

Health Hazard

The acute oral toxicity of thiourea in most animals is of low order. The oral LD_{50} values reported in the literature show variation. Symptoms of chronic effects in rats include bone marrow depression and goiters. Administration of 32.8 mol of thiourea in chick embryos on day 17 of incubation resulted in the accumulation of parabronchial liquid in those embryos (Wittman et al. 1987). The investigators have attributed such changes to the toxic effects of thiourea, rather to than a retardation of pulmonary development.

Dedon and co-workers (1986) observed the possible protective action of thiourea against platinum toxicity. Thiourea and other sulfur-containing nucleophiles have the ability to chelate and remove platinum from biochemical sites of toxicity.

Oral administration of thiourea resulted in tumors in the liver and thyroid in rats. It is carcinogenic to animals and has shown sufficient evidence.

Fire and Explosion Hazard

Noncombustible solid. There is no report of any explosion resulting from reactions of thiourea. Small amounts of thiourea in contact with acrolein may polymerize acrolein, which is a highly exothermic reaction.

56.7 MERCAPTOACETIC ACID

DOT Label: Corrosive Material, UN 1940

Formula $C_2H_4O_2S$; MW 92.12; CAS [68-11-1]

Structure:

$$HS-CH_2-\overset{\displaystyle O}{\overset{\displaystyle \|}{C}}-OH$$

Synonyms: thioglycolic acid; thiovanic acid

Uses and Exposure Risk

Mercaptoacetic acid is used as a reagent for metals analysis; in the manufacture of thioglycolates, pharmaceuticals, and permanent wave solutions; and as a vinyl stabilizer.

Physical Properties

Colorless liquid with a disagreeable odor; density 1.325 at 20°C; boils at 108°C at 15 torr; solidifies at −16.5°C; miscible in water and most organic solvents.

Health Hazard

Mercaptoacetic acid is a highly toxic and a blistering compound. Even a 10% solution was lethal to most experimental animals by dermal absorption. The oral LD_{50} value of undiluted acid is less than 50 mg/kg (Patty 1963). The lethal dose in rabbits by skin absorption is 300 mg/kg. The acute toxic symptoms in test animals include weakness, respiratory distress, convulsions, irritation of the gastrointestinal tract, and liver damage.

Mercaptoacetic acid is a severe irritant. Contact with eyes can cause conjunctival inflammation and corneal opacity. Skin contact can result in burns and necrosis.

Exposure Limit

TLV-TWA 1 ppm (∼3.8 mg/m^3) (ACGIH).

Fire and Explosion Hazard

Noncombustible liquid.

56.8 THIOACETAMIDE

EPA Designated Acute Toxic Waste; RCRA Waste Number U218

Formula C_2H_5NS; MW 75.14; CAS [62-55-5]

Structure:

$$CH_3-\overset{\displaystyle S}{\overset{\displaystyle \|}{C}}-NH_2$$

Synonyms: ethanethioamide; acetothioamide; thiacetamide

Uses and Exposure Risk

It is used as an intermediate in organic synthesis.

Physical Properties

Crystalline solid with a slight odor of mercaptans; melts at 113°C; soluble in water and alcohol, sparingly soluble in ether.

Health Hazard

The toxicity of this compound is moderate in rats; an oral lethal dose is 200 mg/kg.

Oral administration of thioacetamide caused liver cancer in rats and mice. It is, however, a weak liver carcinogen. Malvaldi and associates (1988) investigated the mechanism of its carcinogenic activity on rat liver. Whereas the initiating ability of this compound is quite low, its promoting effect is strong. Thus thioacetamide is a very effective promoter of the liver carcinogenesis. A similar promoting activity of liver carcinogenesis has been observed with other thioamide substances, such as thiobenzamide [2227-79-4] (Malvaldi et al. 1986).

Fire and Explosion Hazard

Noncombustible solid.

REFERENCES

ACGIH. 1986. *Documentation of the Threshold Limit Values and Biological Exposure Indices*, 5th ed. Cincinnati, OH: American Conference of Governmental Industrial Hygienists.

Akintonwa, D. A. A. 1985. High-pressure liquid chromatography separation of phenobarbitone, thiourea, and thioacetamide of toxicological interest. *Ecotoxicol. Environ. Saf. 10*(2): 145–49.

Dedon, P. C., R. Qazi, and R. F. Borch. 1986. Potential mechanisms of cisplatin toxicity of diethyldithiocarbamate rescue. *Chem. Abstr. CA 105*(19): 164609t.

Manahan, S. E. 1989. *Toxicological Chemistry*. Chelsea, MI: Lewis Publishers.

Malvaldi, G., E. Chieli, and M. Saviozzi. 1986. Characterization of the promoting activity of thiobenzamide on liver carcinogenesis. *Toxicol. Pathol. 14*(3): 370–74.

Malvaldi, G., M. Saviozzi, V. Longo, and P. G. Gervasi. 1988. Studies on the mechanism of the carcinogenic activity of thioacetamide on rat liver. *Chem. Abstr. CA 109*(13): 106273a.

MCA. 1964. Case *Histories of Accidents in the Chemical Industry*, MCA Case History 971. Washington, DC: Manufacturing Chemists' Association.

National Research Council. 1995. *Prudent Practices for Handling and Disposal of Chemicals from Laboratories*. Washington, DC: National Academy Press.

NFPA. 1997. *Fire Protection Guide on Hazardous Materials*, 12th ed. Quincy, MA: National Fire Protection Association.

NIOSH. 1984. *Manual of Analytical Methods*, 3rd ed. Cincinnati, OH: National Institute for Occupational Safety and Health.

NIOSH. 1986. *Registry of Toxic Effects of Chemical Substances*, ed. D. V. Sweet. Washington, DC: U.S. Government Printing Office.

Patty, F. A. 1963. *Industrial Hygiene and Toxicology*, 2nd ed., Vol. II, p. 1807. New York: Interscience.

Watkins, J. M. Jr., G. W. Mushrush, R. N. Hazlett, and E. J. Beal. 1989. Hydroperoxide formation and reactivity in jet fuels. *Energy Fuels 3*(2): 231–36.

Wittmann, J., A. Steib, H. G. Liebach, and W. Hammel. 1987. Lung development under the influence of thiourea and L-thyroxine. Retarding and toxic effects of thiourea. *Res. Commun. Chem. Pathol. Pharmacol. 58*(2): 199–214.

Young, P., and A. Parker. 1984. Vapors, odors, and toxic gases from landfills. *ASTM Spec. Tech. Publ. 851*: 24–81.

57

MISCELLANEOUS SUBSTANCES

57.1 NITRIDES

Nitrides are the binary salts of metals and nitrogen, containing the anion N^{3-}. While the nitrides of alkali metals and some alkaline earth metals are ionic, most other metal nitrides are covalent. Nitrides of silicon and selenium are covalent and polymeric. Many complex and polynuclear metal nitrides are known. The formulas and CAS numbers for some nitrides are presented in Table 57.1.

The explosivity of metal nitrides is considerably lower than that of their corresponding azides. While some nitrides are endothermic substances, most azides have much higher positive heat of formation. The endothermic nitrides are thermodynamically unstable and decompose explosively on heating or impact or reactions with acids, bases or oxidizers. Some may even decompose violently upon mixing with water. Table 57.2 presents heats of formation, $\Delta H_f^\circ(s)$ of some selected metal nitrides. The $\Delta H_f^\circ(s)$ of some selected metal nitrides. The $\Delta H_f^\circ(s)$ values for some of the corresponding azides are presented in Table 57.2 for comparison.

It may be noted that compounds that have moderate to high negative $\Delta H_f^\circ(s)$ such as aluminum nitride are stable and do not pose

any explosion hazard from impact or heat. By contrast, substances with positive and low negative $\Delta H_f^\circ(s)$ can decompose explosively. Many nitrides with moderately negative $\Delta H_f^\circ(s)$ are pyrophoric. Nitrides of high electropositive metals are more pyrophoric than those of lower electropositive elements.

Among that alkali metal nitrides, the salts of heavier metals are more pyrophoric than that of lighter metals. For example, while the nitrides of potassium, rubidium, and cesium ignite air under ambient condition, sodium salt ignites on gentle warming an lithium salt ignites in finely divided state and in the presence of moisture. A similar pattern is manifested among group IIA nitrides. For example, whereas calcium nitride ignites in moist air when finely divided, the barium salt reacts violently with air or water. By contrast, the nitrides of heavy metals are dangerously explosive. For example, thallium(I)nitride explodes violently when subjected to shock and heat. Reactions with water or dilute acids can cause explosive decomposition. Mercury nitride can explode even below $-40^\circ C$ on slight jarring. Silver forms two nitrides, silver nitride, Ag_3N, and the trisilver tetranitride, Ag_3N_4. Both compounds are extremely sensitive to shock,

TABLE 57.1 Formulas and CAS Registry Numbers of Selected Metal Nitrides

Metal Nitrides	Formula	CAS No.
Lithium nitride	Li_3N	[26134-62-3]
Sodium nitride	Na_3N	[12135-83-3]
Potassium nitride	K_3N	[29285-24-3]
Cesium nitride	Cs_3N	[12134-29-1]
Rubidium nitride	Rb_3N	[12136-85-5]
Magnesium nitride	Mg_3N_2	[12057-71-5]
Calcium nitride	Ca_3N_2	[12057-71-5]
Barium nitride	Ba_3N_2	[12013-82-0]
Autimony(III) nitride	SbN	[12333-57-2]
Copper(I) nitride	Cu_3N	[1308-80-1]
Silver nitride	Ag_3N	[20737-02-4]
Lead nitride	Pb_3N_2	[58572-21-7]
Cadmium nitride	Cd_3N_2	[58572-21-1]
Cobalt(III) nitride	CoN	[12139-70-7]
Mercury nitride	Hg_3N_2	[12136-15-1]
Gold(III)nitride trihydrate	$Au_3N_2 \cdot 3H_2O$	—
Thallium nitride	TlN	[12033-67-9]
Chromium nitride	CrN	[24094-93-7]
Zirconium nitride	ZrN	[25658-42-8]
Trisilver tetranitride	Ag_3N_4	—
Tetraselenium dinitride	Se_4N_4	[132724-39-1]

TABLE 57.2 Heats of Formation, ΔH_f°(s) of Selected Metal Nitrides and Comparison With Some Corresponding Azides

Metal Nitrides/Azides	Formula	ΔH_f°(s) (kcal/mol)
Copper(I)nitride	Cu_3N	+17.8
Copper(I)azide	CuN_3	+66.7
Silver nitride	Ag_3N	+47.6
Silver azide	AgN_3	+73.8
Cadmium nitride	Cd_3N_2	+38.6
Cadminum azide	$Cd_3(N_3)_2$	+107.8
Nickel nitride	Ni_3N	+0.2
Nickel azide	$Ni(N_3)_2$	
Zinc nitride	Zn_3N_2	−5.4
Zinc azide	$Zn(N_3)_2$	+52.0
Strontium nitride	Sr_3N_2	−93.5
Strontium azide	$Sr(N_3)_2$	+2.1
Pentamanganese dinitride	Mn_5N_2	−48.8
Manganese azide	$Mn(N_3)_2$	+92.3
Calcium nitride	Ca_3N_2	−103
Calcium azide	$Ca(N_3)_2$	+3.5
Chromium nitride	CrN	−29.8
Molybdenum nitride	Mo_2N	−19.5
Aluminum nitride	AlN	−76.0
Lanthanum nitride	LaN	−72.5
Barium nitride	BaN_2	−41.0

heat, and moisture. The nitrides of other heavy metals, such as lead and cadmium, are unstable and decompose explosively on heating or on impact. Such heavy metal nitrides may explode when mixed with acids or alkalies.

57.2 BARBITURATES

Barbiturates are derivatives of barbituric acid [67-52-7], whose structure is as follows:

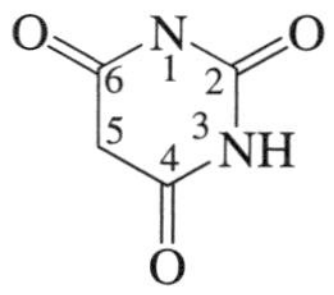

Barbituric acid itself has no sedative-hypnotic potency. However, when both the hydrogen atoms in position 5 in the ring are substituted with alkyl or aryl groups, the molecule attains sedative-hypnotic activity. An increase in the alkyl chain length increases such activity. Also, the increase in chain length in the molecule produces such sedative-hypnotic activity at a faster speed but for a shorter duration. When the number of carbon atoms in the alkyl side chains in position 5 reaches seven, the compound exhibits optimum hypnotic activity. Barbiturates containing more than eight carbon atoms in the side chains are toxic. Barbiturates are clinically used as hypnotics and anesthetics. These substances have little analagetic potency. Barbiturates are habit-forming addictive drugs. These are listed as controlled substances in the U.S. Code of Federal Regulations. Overdoses can cause respiratory depression. Structures, CAS numbers, and clinical uses of some of the barbiturates are presented in Table 57.3.

57.3 CANNABIS

Synonyms: Indian *cannabis*; Indian hemp; bhang; ganja; charas; marijuana; marihuana; hashish; pot.

The use of cannabis as a hallucinogen has been known since early time. It occurs in dried flowering tops of *Cannabis sativa* L., grown in India, Central America, and the United States. The plant produces a group of C_{21} compounds known as cannabinoids. A total of 61 specific cannabinoids have been isolated. Although cannabinol [521-35-7] and cannabidiol [13956-29-1] are the most abundant cannabinoids, γ^9-*trans*-tetrahydrocannabinol (THC) [1972-08-3] exhibits the highest psychoactivity. Cannabinol is only 5% as potent as THC, and cannabidiol is inactive. The structures of these major cannabinoids are as follows:

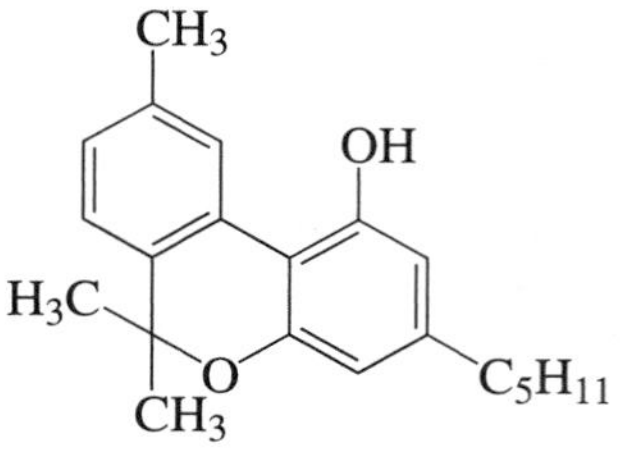

(cannabinol)

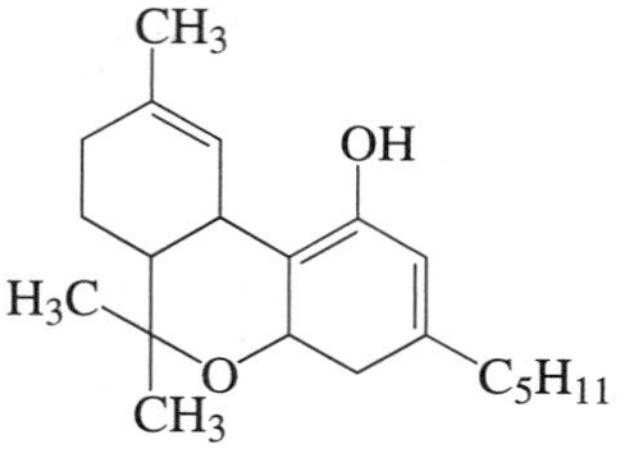

(Δ^9-THC)

THC occurs primarily in flowers, leaves, and bracts of the plants; it does not occur in seeds, roots, or stems. The THC contents of the plants determine their use. *Cannabis sativa* plants containing less than 0.5% THC are used for fiber production. The drug-type plants should have a THC content of greater than 1.0%.

Cannabis may be administered by inhalation or ingestion. The flowering tops or leaves are chopped and rolled in cigarette paper for smoking, or are added to food. The effects of intoxication vary with the dose and route of administration. The effects are

TABLE 57.3 Structures and Clinical Uses of Some Barbiturates

Compound/CAS No.	Structure	Clinical Uses
Barbital [57-44-3]		Sedative, hypnotic
Amobarbital [57-43-2]		Sedative, hypnotic
Phenobarbital [50-06-6]		Sedative, hypnotic, anticonvulsant, long-acting anesthetic
Phenobarbital sodium [57-30-7]	Monosodium salt of phenobarbital	Sedative, hypnotic, anticonvulsant
Secobarbital sodium [309-43-3]		Sedative, short-acting hypnotic
Methohexital sodium [22151-68-4]		Ultrashort-acting anesthetic
Thiogenal [730-68-7]		Sedative, hypnotic
Thiobarbital [77-32-7]		Thyroid inhibitor

euphoria, hallucinations, delirium, drowsiness, and weakness. Intoxication from oral ingestion is slow but lasts longer than that from inhalation of smoke. *Cannabis* is a complex mixture of hundreds of chemicals. THC, the major psychoactive constituent, has a biological half-life ranging between 24 and 36 hours. It is excreted in urine and feces to the extent of 15% and 35% of the dose, respectively. Among its metabolites, only 11-hydroxy-γ^9-THC is as potent as the parent compound, THC. The oral and intravenous LD_{50} values of THC in rats are in the range 660 and 30 mg/kg, respectively.

57.4 HERBICIDES MISCELLANEOUS

57.4.1 Paraquat

Formula $C_{12}H_{14}N_2$; MW 186.28; CAS [4685-14-7]

Structure:

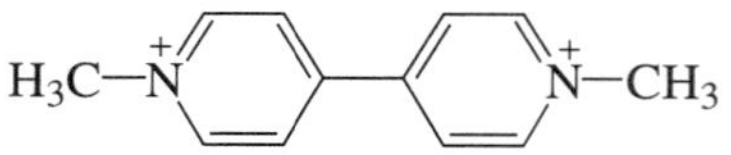

The paraquat cation shown above forms dichloride ([1910-42-5] $C_{12}H_{14}Cl_2N_2$, MW 257.18) and sulfate (bismethyl sulfate, [2074-50-2], $C_{14}H_{20}N_2O_8S_2$, MW 408.48), both of which are used as herbicides; the chloride salt is used as a biological oxidation-reduction indicator.

Synonym: 1,18-dimethyl-4,48-bipyridinum

Health Hazard

Paraquat salts show moderate to high acute toxicity among species, the oral LD_{50} values ranging between 25 and 300 mg/kg. The LD_{50} values in dogs, cats, guinea pigs, and rats are 25, 35, 30, and 100 mg/kg, respectively, for the sulfate salts. The toxic symptoms from ingestion and inhalation of dusts include headache, nausea, vomiting, diarrhea, ulceration, dyspnea, and lung injury. Single exposure to paraquat aerosols 3–5 mm in diameter at 1 mg/m^3 concentration for 6 hours caused death to rats (ACGIH 1986). Inhalation of nonrespirable size dusts caused intense irritation and nose bleeding. Repeated exposures can lead to severe pulmonary edema and fibrosis. The dusts are an irritant to the respiratory tract.

Dey and associates (1990) investigated paraquat pharmacokinetics in rats, using a subcutaneous toxic low dose (72 mmol/kg) of [$^{14}CH_3$]paraquat, which would produce lung disease but no renal damage. Paraquat was rapidly absorbed. Peak blood concentrations were 58 nmol/mL at 20 minutes, while its peak concentrations in the lung and kidney were 65 and 359 nmol/g, respectively, at 40 minutes. About 85% of dose was eliminated in the urine by 7 days. Of the remaining radioactivity, 79% remained in the body and 21% remained in the lung, causing progressive lung disease. Excretion of the retained amount was slow and prolonged. Chui and co-workers (1988) investigated the toxicokinetics of paraquat and the effects of different routes of administration. A single dose of 11.4 mg/kg of [$^{14}CH_3$]paraquat was administered in rats by intravenous, intragastric, dermal, and pulmonary (exposure by aqueous or liquid aerosols) routes. The major excretion routes were urine and feces.

The radioactivity absorbed into the systemic circulation of the rat was about 27.5, 23.8, 8.5, and 1.5 nmol for inhalation through a tracheal cannula, nose-only exposure, intragastric injection, and dermal absorption, respectively. The bulk of the herbicide administered by the inhalation and dermal routes remained at the sites of the administration.

In humans, cardiovascular collapse resulting from acute paraquat poisoning is associated with the distribution phase; the late occurrence of death-related pulmonary fibrosis is associated with the elimination phase (Houze et al. 1990). Toxicokinetics studies conducted by these investigators in acute human poisoning cases indicated a concentration of paraquat in blood that had a mean distribution half-life of 5 hours and a mean elimination half-life of 84 hours, respectively. It was excreted in the urine. It was retained in the muscle for several weeks after poisoning. Electron microscopy studies (TEM and SEM) on the lungs obtained from a dog after 7 days of intravenous administration of paraquat (12 mg/kg) indicate detachment of alveolar epithelial cells and alveolar macrophage, which plays a significant role in paraquat-induced pulmonary fibrosis (Hampson and Pond 1988). Administration of paraquat to adult rat pulmonary alveolar macrophages in primary culture caused cell death that was dependent on the dose and time. The cell death was potentiated by hyperoxia (95% O) and extracellular production of an active oxygen species, the superoxide anion radical (Wong and Stevens 1985). Although diquat can enter the cells to a greater extent, than can paraquat, the latter is about twice as potent as diquat (Wong and Stevens 1986).

Paraquat is a urinary metabolite of 4,48-bipyridyl when the latter is administered intraperitoneally to guinea pigs. Godin and Crooks (1989) detected 2.9% N-methyl-4,48-bipyridinium ion in the urine of animals treated with 4,48-bipyridyl, thus indicating the formation of such toxic metabolites through the N-methylation pathway.

Certain substances have been reported to potentiate the toxicity of paraquat. These include transition metal ions such as copper (Kohen and Chevion 1985) and ethanol (Kuo and Nanikawa 1990). Blood paraquat levels showed significant elevation in rabbits, and the mortality rates increased when the animals were orally administered paraquat combined with ethanol in amounts of 2.0 and 3.8 g/kg. Continuous breathing of high oxygen concentrations 12–24 hours after administration of paraquat caused severe and extensive pulmonary lesions and interstitial fibrosis (Selman et al. 1985). On the other hand, a reverse sequence of treatment — inhalation of high oxygen concentrations followed by paraquat administration — caused no mortality and pulmonary lesions.

The effect of light on toxicity of paraquat has been reported (Barabas et al. 1986). A 72-hour exposure to illumination increased the lethality of paraquat in mice. Changes in the activity of the enzymes, catalase [9001-05-2], superoxide dismutase [9054-89-1], and glutathione peroxidase [9013-66-5], were also noted after exposure to light.

Antidote actions of a few substances against paraquat toxicity have been reported. These include several sodium sugar sulfates, including dextran sulfate, cellulose sulfate, chondroitin sulfate, sucrose sulfate, and glucose sulfate (Tsuchiya et al. 1989; Ukai et al. 1987). Sugar sulfates 2000 mg/kg, given orally immediately after paraquat ingestion (200 mg/kg), protected mice against the acute toxicity of the herbicide. Thiols, cystein [52-90-4], d-penicillamine [52-67-5], and GSH [70-18-8] were found to protect mice against a LD_{50} dose of paraquat (Szabo et al. 1986). The protecting action of these compounds decreased in the order noted above.

57.4.2 Diquat dibromide

Formula: $C_{12}H_{12}N_2^\circ 2Br$; MW 344.08; CAS [85-00-7]

Structure:

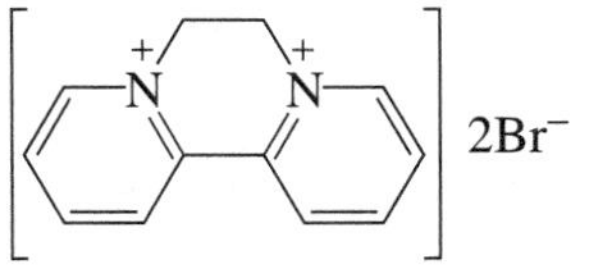

Synonyms: 1,18-ethylene-2,28-bipyridylium dibromide; ethylene dipyridylium dibromide; 6,7-dihydrodipyrido(1,2-a:28,18-c) pyrazinediium dibromide; Reglone; Aquacide

Physical Properties

Pale yellow crystals; forms monohydrate; mp 320°C (decomposes); readily soluble in water, insoluble in organic solvents; stable in acids or neutral solution.

Health Hazard

The acute toxicity of diquat dibromide is moderate to high in most species. In domestic animals, its toxicity is greater than that in small laboratory animals. The oral LD_{50} value in cows, dogs, rabbits, and mice is 30, 187, 188, and 233 mg/kg, respectively. The symptoms of acute toxicity are somnolence, lethargy, pupillary dilation, and respiratory distress. Prolonged exposure to this compound produced cataracts in experimental animals. Intratracheal administration of diquat dibromide in rats showed toxic effects in the lung and caused lung damage (Manabe and Ogata 1986). But when administered by oral or intravenous routes, there was no toxic effect on the lung.

Exposure Limit

TLV-TWA 0.5 mg/m^3 (ACGIH and MSHA).

57.4.3 Dinoseb

EPA Designated Toxic Waste, RCRA Waste Number P020
Formula $C_{10}H_{12}N_2O_5$; MW 240.24; CAS [88-85-7]
Structure:

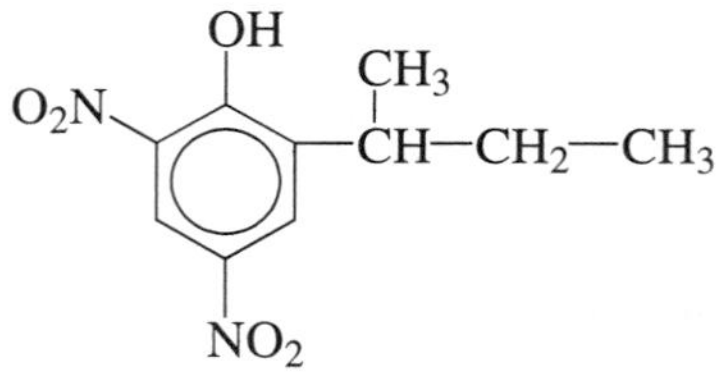

Synonyms: 2-*sec*-butyl-4,6-dinitrophenol; 2-(1-methylpropyl)-4,6-dinitrophenol; 2,4-dinitro-6-*sec*-butylphenol; Caldon; Basanite

Uses and Exposure Risk

Dinoseb is used as a herbicide and insecticide.

Physical Properties

Orange-brown viscous liquid or solid; melts between 38° and 40°C; insoluble in water, soluble in most organic solvents.

Health Hazard

Dinoseb is a highly toxic compound. The oral LD_{50} values in small laboratory animals were between 10 and 25 mg/kg. Acute toxicity tests on daphnids and fathead minnows showed high toxicity. The LC_{50} values in both these species are 0.24 and 0.17 mg/L, respectively (Gersich and Mayes 1986). Pregnant white rabbits treated with dinoseb exhibited maternal toxicity above the dose level of 1 mg/kg/day. At highly toxic dose levels, adverse effects were observed in developing fetuses (Johnson et al. 1988). Oral administration of dinoseb produced tumors in lung and liver in mice.

57.4.4 Nitrofen

Formula $C_{12}H_7Cl_2NO_3$; MW 284.10; CAS [1836-75-5]
Structure:

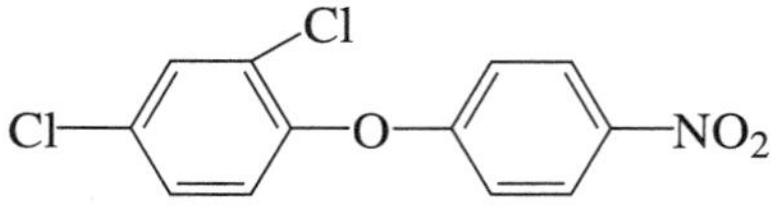

Synonyms: 2,4-dichloro-1-(4-nitrophenoxy)-benzene; 2,4-dichlorophenyl- pnitrophenyl ether; 2,4-dichloro-48-nitrodiphenyl ether; 48-nitro-2,4-dichlorodiphenyl ether; TOK; TOK E-25

Physical Properties

White crystalline solid; melts at 70°C; vapor pressure 8×10^{-6} torr at 20°C; insoluble in water, soluble in most organic solvents.

Health Hazard

Nitrofen is moderately toxic by ingestion and inhalation of dusts. The lethal doses in cats from oral administration and inhalation of dusts are 300 mg/kg and 620 mg/m^3/4 hr, respectively (NIOSH 1986). Bovine calves treated orally by 1.5 mL 25% nitrofen/kg produced toxic effects after 36–48 hours. The symptoms were increase in body temperature, depression, and progressive decrease in respiration rate and pulse rate, similar to tribulin (Gupta and Singh 1985). An increase in the activities of serum glutamic-oxaloacetic transaminase [9000-97-9] and glutamipyruvic transaminase [9000-86-6] was noted (Gupta and Singh 1984). Nitrofen has been found to cause cancer in animals. There is sufficient evidence of its carcinogenicity in animals (IARC). Oral administration in mice caused liver and lung cancers.

57.4.5 Picloram

Formula $C_6H_3Cl_3N_2O_2$; MW 241.46; CAS [1918-02-1]

Structure:

Synonyms: 4-amino-3,5,6-trichloropicolinic acid; 3,5,6-trichloro-4-aminopicolinic acid; Tordon; Borolin; Amdon

Uses and Exposure Risk

It is used as a herbicide and defoliant.

Physical Properties

White crystalline powder, melts at 218°C; vapor pressure 6.16×10^{-7} torr at 35°C; slightly soluble in water (0.43 g/dL at 25°C), soluble in acetone, alcohol, and ether.

Health Hazard

The toxic effects from ingestion or inhalation of dusts of picloram in test animals were mild. The acute oral LD_{50} values in rats and rabbits are 2900 and 2000 mg/kg, respectively. Maternal toxicity in rats was observed at a dose level of 750 mg/kg/day. Oral administration of picloram in rats and mice caused tumors in thyroid and liver.

57.4.6 Tribunil

Formula $C_{10}H_{11}N_3OS$; MW 221.30; CAS [18691-97-9]

Structure:

Synonyms: 1,3-dimethyl-3-(2-benzothiazolyl)urea; N-methyl-N-methyl-N-(2-benzothiazolyl)urea; methabenzthiazuron

Physical Properties

White crystalline solid; melts at 120°C; insoluble in water (~59 ppm at 20°C), soluble in organic solvents.

Health Hazard

Gupta and Singh (1985) investigated the toxicity of tribunil in calves. Animals treated with 500 mg/kg orally developed toxicity in 36–48 hours. The symptoms included increase in body temperature, depression, and a progressive decrease in respiration rate and pulse rate. Necropsy findings in animals that died showed congestion of the lungs. Other organs affected were liver, kidney, urinary bladder, spleen, heart, and brain. A 75-mg/kg dose produced toxicity after 2–3 weeks.

57.4.7 Barban

Formula $C_{11}H_9Cl_2NO_2$; MW 258.11; CAS [101-27-9]

Structure:

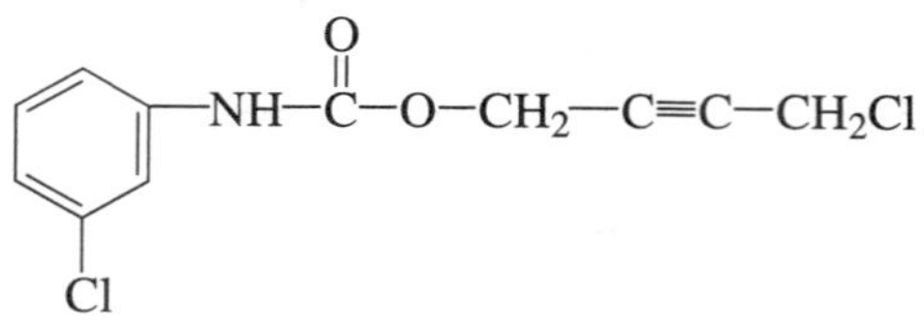

Synonyms: 2-butynyl-4-chloro-*m*-chloro-carbanilate; 4-chloro-2-butynyl *N*-(3-chlorophenyl)carbamate, (3-chlorophenyl)carbamic acid 4-chloro-2-butynyl ester

Uses and Exposure Risk

Barban is used as a selective herbicide for wild oats.

Physical Properties

White crystalline solid; melts at 75°C; insoluble in water, dissolves in benzene, toluene, and ethylene dichloride; hydrolyzed by acids and alkalies; alkaline hydrolysis liberates terminal chlorine atom.

Health Hazard

Barban is moderately toxic to test animals by oral route. The LD_{50} values range between 200 and 600 mg/kg. The oral LD_{50} value in rats is 600 mg/kg. Toxicity in humans is low. Skin contact causes irritation.

57.4.8   Alachlor

Formula $C_{14}H_{20}ClNO_2$; MW 269.80; CAS [15972-60-8]

Structure:

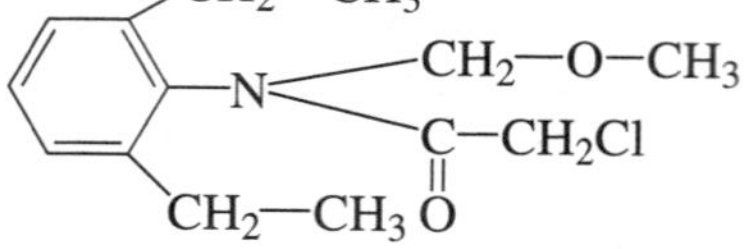

Synonyms: 2-chloro-*N*-(2,6-diethylphenyl)-*N*-(methoxymethyl)acetamide; 2-chloro-2,

6-diethyl-*N*-(methoxymethyl)acetanilide; Metachlor; Lasso; Lazo

Physical Properties

Crystalline solid; melts at 40°C; very slightly soluble in water (140 mg/L at 23°C), soluble in alcohol, ether, acetone, and benzene; hydrolyzed by strong acids or alkalies.

Health Hazard

Very little information is available on the toxic effects of alachlor in animals or humans. The oral LD_{50} value in mice is 462 mg/kg (NIOSH 1986). The dermal toxicity is very low in rabbits. It tested positive in a S. cerevisiae gene conversion test for mutagenicity.

57.5   DIAZOMETHANE

Formula CH_2N_2; MW 42.04; CAS [334-88-3]

Structure: $CH_2{=}N^+{=}N^-$

Synonym: azimethylene

Uses and Exposure Risk

It is used in organic synthesis as a methylating agent to methylate acidic compounds such as carboxylic acids and phenols. It is used in trace environmental analysis to methylate chlorophenoxy acid herbicides.

Physical Properties

Yellow gas at ambient temperature; heavier that air, gas density 1.45 (air = 1); liquefies at −23°C; solidifies at −145°C; insoluble in water; soluble in ether and benzene.

Fire and Explosion Hazard

Being a highly endothermic compound (H_f° (g) = +192.5 kJ/mol or 4.58 kJ/g) it is unstable and is susceptible to decompose explosively under a variety of conditions. The gas explodes when heated above 90°C. Even when diluted with nitrogen, it explodes at elevated temperatures (≥ 100°C). Similar explosive decomposition occurs when exposed to high-intensity light or UV radiation or in contact with rough surface. Glass stirrers and ground glass apparatus should not be used with diazomethane. The liquid diazomethane or its concentrated solutions can explode if impurities or solids are present or if exposed to strong light or radiation. Diazomethane reacts explosively with alkali metals, as well as with many metal salts. Its cyclic isomer, azirine [157-22-2] (bp -14) also explodes on heating.

Health Hazard

It is a highly toxic gas and an irritant to eye, nose, and the entire respiratory tract. Exposure can cause dizziness, weakness, chest pain, severe headache, fever, asthmatic attack and pneumonia. Exposure to trace concentrations of this substance can also produce adverse effects, causing coughing, wheezing and headache. There have been many reported cases of poisoning. Its toxicity may be attributed to its strong methylating property.

Exposure Limit

TLV-TWA 0.2 ppm (0.38 mg/m^3) (ACGIH)
PEL-TWA 0.2 ppm (0.38 mg/m^3) (OSHA)

57.6   HYDRAZINES

57.6.1   Hydrazine

EPA Designated Hazardous Waste, RCRA Waste Number U133; DOT Label: Flammable Liquid and Poison, UN 1029.

Hydrazine, H$_2$NBNH$_2$ [302-01-2], is a water-miscible oily liquid with ammonia-like penetrating odor. This highly polar solvent (dielectric constant 51.7 at 25°C) dissolves many metal salts. It reacts with inorganic acids to form salts. Hydrazine is used as a reducing agent. It is also used as a rocket fuel and in the manufacture of many organic hydrazine derivatives.

Fire and Explosion Hazard

Hydrazine (anhydrous) exhibits flammable, explosive, and toxic properties. It is a flammable liquid and burns in air with a violet flame; the closed cup flash point is 38°C (~ 100°F). The vapors of hydrazine form explosive mixtures with air over a very wide range, between 4.7% and 100% by volume in air. It ignites spontaneously when brought in contact with many porous substances, such as earth, wood, or cloth. Ignition occurs on contact with many metal oxide surfaces and on mixing with oxidizers such as nitric acid or hydrogen peroxide. It forms shock-sensitive explosive products when combined with metal perchlorates, alkali metals, or their dichromates. Spontaneous ignition occurs with fluorine, chlorine, bromine vapors, or nitrous oxide.

Health Hazard

Hydrazine vapors are highly irritating to the eyes, nose, and throat. Inhalation of its vapors or ingestion of the liquid can cause nausea, vomiting, dizziness, and convulsions. Chronic exposure can cause injury to the lung, liver, and kidney. Skin contact with the liquid may result in severe burns. Contact with the eyes can cause damage to vision.

LC$_{50}$ value, inhalation (rats): 570 ppm/4 hr
LD$_{50}$ value, oral (rats): 60 mg/kg

Hydrazine is a suspected carcinogen. Animal studies indicate sufficient evidence of its carcinogenicity. Administration of this compound by inhalation, oral, and intravenous routes caused tumors in the blood and lung in laboratory animals. There is, however, no evidence of any carcinogenic action of this compound in humans.

Exposure Limits

TLV-TWA (skin) 1 ppm (1.3 mg/m^3) (MSHA and OSHA), 0.1 ppm (ACGIH).

Analysis

Hydrazine in air may be analyzed by NIOSH Method 3503. About 10–100 L of air at a flow rate of 0.2–1 L/min is passed through a bubbler containing 0.1 M HCl. The analyte is complexed with *p*-dimethylaminobenzaldehyde and is measured by visible absorption spectrophotometry at a wavelength of 480 nm. It may also be determined by a GC method using a nitrogen–phosphorus detector in the nitrogen mode, or by a flame ionization detector.

57.6.2 1,1-Dimethylhydrazine

EPA Designated Hazardous Waste, RCRA Waste Number U098; DOT Label: Flammable Liquid and Poison

Formula C$_2$H$_8$N$_2$; MW 60.12; CAS [57-14-7]

Structure:

$$CH_3 \diagdown$$
$$N{-}NH_2$$
$$CH_3 \diagup$$

Synonyms: unsym-dimethylhydrazine; *N,N*-dimethylhydrazine

Uses and Exposure Risk

1,1-Dimethylhydrazine is used in rocket fuel.

Physical Properties

Colorless, mobile liquid, turning yellow on exposure to air; fumes in air; hygroscopic;

ammonia-like odor; density 0.782 at 25°C; boils at 64°C; freezes at −58°C; soluble in water, alcohol, and ether.

Health Hazard

1,1-Dimethylhydrazine vapor is an irritant to the eyes, nose, and throat. Inhalation causes diarrhea, hypermotility, stimulation of central nervous system, tremor, and convulsions. On contact with the skin, the liquid can cause burn.

LC$_{50}$ value, inhalation (rats): 250 ppm/4 hr
LD$_{50}$ value, oral (rats): 122 mg/kg

Tests on laboratory animals indicate its carcinogenicity. Repeated oral administration of this compound produced colon cancers in rats and mice. Its cancer-causing effect in humans is unknown.

Exposure Limits

TLV-TWA skin 0.5 ppm (1.0 mg/m^3) (ACGIH, MSHA, and OSHA); carcinogenicity:
Animal Sufficient Evidence (IARC), Suspected Carcinogen (ACGIH).

Fire and Explosion Hazard

Flammable liquid; flash point (closed cup) 15°C (59°F); vapor density 2.1 (air = 1), vapor heavier than air and can travel a considerable distance to a source of ignition and flash back; autoignition temperature 249°C (480°F). Vapors of 1,1-dimethylhydrazine form explosive mixtures with air over a wide range; the LEL and UEL values are 2% and 95% by volume of air, respectively. It ignites spontaneously in air or in contact with hydrogen peroxide, nitric acid, or other oxidizers.

1,2-Dimethylhydrazine

EPA Designated Hazardous Waste, RCRA Waste Number U099; DOT Label: UN 2382

Formula $C_2H_8N_2$; MW 60.12; CAS [540-73-8]

Structure: $H_3C-NH-NH-CH_3$

Synonyms: *N,N*-dimethylhydrazine; *sym*-dimethylhydrazine

Physical Properties

Colorless mobile liquid; turns yellow on exposure to air; odor of ammonia; hygroscopic; density of 0.827 at 20°C; bp 82–83°C; miscible with water, alcohol, ether, and hydrocarbons.

Health Hazard

The vapors of this compound are irritating to the eyes, nose, and throat. The liquid is corrosive to the skin. The toxic properties are similar to those of 1,1-dimethylhydrazine. Exposure to 280 ppm for 4 hours was fatal to rats. The oral LD_{50} value in the animal is 100 mg/kg. 1,2-Dimethylhydrazine is an animal carcinogen, similar to its unsym-isomer. Administration of this compound caused lung and colon cancers in test animals. There has been no report of cancer-causing effects in humans.

57.7   NAPALM

Napalm [8031-21-8] is an aluminum soap made from coprecipitating aluminum hydroxide, naphthenic acid, and palmitic acid. The latter is one of the fatty acids of coconut oil. Alternatively, gasoline is gelled by hydroxyaluminum-bis(2-ethylhexanoate) [30745-55-2] and a nonionic surfactant or water in making Napalm. Napalm is used in chemical warfare as fire bombs or to flame land mines. It is also used with incendiary devices to burn targets in short duration that are difficult to ignite. It has the ability to stick to its target when it burns.

57.8   NITROSOAMINES

Nitrosoamines or nitrasamines are derivatives of amines, containing *N*-nitroso groups ($>N-N=O$). Typically, a nitrosoamine is formed by the reaction of an amine with a nitrite. These substances are also produced by the action of nitrate-reducing bacteria. Nitrosoamines occur in trace quantities in tobacco smoke, processed food, meat products, and salted fish. Many nitrosoamines are used as gasoline and lubricant additives, antioxidants, stabilizers, and softeners for copolymers. These compounds are noncombustible liquids or solids at ambient temperature. The hazardous properties of nitrosoamines are different from those of their parent aliphatic or aromatic amines. Chronic exposure to these compounds can cause jaundice and liver damage. In addition, the other toxic effects caused by methyl and ethyl nitrosoamines include nausea, vomiting, ulceration, and increase in body temperature. Nitrosoamines are potent animal and human carcinogens which may cause cancers in liver, kidney, lung, bladder, and pancreas. Carcinogenicity may be attributed to their alkylating actions on DNA. A high level of nitrites in food may form nitrosoamines in the acidic medium of the stomach and can therefore be hazardous to health.

The EPA has listed nitrosoamines as priority pollutants in potable waters, industrial wastewaters, and soil and hazardous wastes. Nine such compounds, designated as priority pollutants in solid and hazardous wastes, are listed in Table 57.4. The median lethal doses

TABLE 57.4 Median Lethal Doses of Some Common Nitrosoamines

Compound/CAS No.	MW/Structure	LD$_{50}$ oral (mg/kg)	LD$_{50}$ subcutaneous (mg/kg)
N-Nitrosodimethylamine [62-75-9]	74.10 $(CH_3)_2N-N{=}O$	28 (hamsters)	28 (hamsters)
N-Nitrosomethylethylamine [10595-95-6]	88.13 $CH_3 \diagdown N-N{=}O$ $C_2H_5 \diagup$	90 (rats)	—
N-Nitrosodiethylamine [55-18-5]	102.16 $(C_2H_5)_2N-N{=}O$	280 (rats)	195 (rats)
N-Nitrosodi-n-propylamine [621-64-7]	130.22 $(C_3H_7)_2N-N{=}O$	480 (rats)	487 (rats)
N-Nitrosodi-n-butylamine [924-16-3]	158.28 $(C_4H_9)_2N-N{=}O$	1200 (rats)	1200 (rats)
N-Nitrosodiphenylamine [86-30-6]	198.24 $C_6H_5 \diagdown N-N{=}O$ $C_6H_5 \diagup$	1650 (rats)	—
N-Nitrosomorpholine [59-89-2]	116.14 (morpholine structure, N—N=O)	282 (rats)	170 (rats)
N-Nitrosopiperidine [100-75-4]	114.17 (piperidine structure, N—N=O)	200 (rats)	100 (rats)
N-Nitrosopyrrolidine [930-55-2]	100.14 (pyrrolidine structure, N—N=O)	900 (rats)	900 (rats)

of these nitrosoamines in rats by an oral route are presented in Table 57.4.

Nitrosoamines in wastewater, soils, and solid wastes may conveniently be analyzed by GC or GC/MS techniques after their extraction with a suitable solvent, such as methylene chloride under alkaline pH conditions. The detector used for GC analysis is NPD in the nitrogen mode. An FID may be used for high-range analysis.

57.9 PHOSGENE

EPA Designated Acute Toxic Waste, RCRA Waste Number P095; DOT Label: Poison Gas, UN 1076

Formula CCl_2O; MW 98.91; CAS [75-44-5]

Structure:

Synonyms: carbonyl chloride; carbon oxychloride; chloroformyl chloride; carbonic dichloride

Exposure Risk

Phosgene is used in the synthesis of dyes, pharmaceuticals, isocyanates, and pesticides; and as a war gas. Exposure risk to this gas may arise when many common solvents, such as carbon tetrachloride, chloroform, or methylene chloride, are intensely heated. Photodecomposition of these solvents in air or oxygen can also generate phosgene in small amounts. It is also formed by the reaction of carbon tetrachloride with oleum, or of carbon monoxide with chlorine or nitrosyl chloride.

Physical Properties

Colorless gas with a pungent suffocating odor at high concentrations; sweet odor of moldy hay at low concentrations in air; liquefies at 8°C; density 1.432 at 0°C; solidifies at 118°C; slightly soluble in water, reacting slowly; soluble in hexane, isooctane, benzene, toluene, and glacial acetic acid.

Health Hazard

Phosgene is a highly poisonous gas. Its effects can be treacherously dangerous, as there may not be any immediate irritation even at lethal concentrations. The initial symptoms are mild. However, severe congestion of lungs or pneumonia occurs 6–24 hours after exposure. The toxic symptoms include coughing, dry burning of throat, choking, chest pain, vomiting, foamy sputum (often containing blood), labored breathing, and cyanosis. Death results from anoxia. It hydrolyzes to HCl and CO_2 in the lungs. A 30-minute exposure to about 100 ppm of phosgene in air can be fatal to humans, causing death within a few hours of exposure. A concentration of 15–20 ppm, however, exhibits only mild effects. Chronic exposure may result in bronchitis and fibrosis. Exposure to the gas can cause eye irritation. Contact with the liquid can cause skin burns.

Exposure Limits

TLV-TWA 0.1 ppm (~0.4 mg/m^3) (ACGIH, MSHA, OSHA, and NIOSH); 0.2 ppm (15-minute ceiling) (NIOSH); IDLH 2 ppm (NIOSH).

Analysis

Phosgene may be analyzed by a GC technique using ECD or FID or by a colorimetric method. The latter method involves the reaction of phosgene with p-dimethylaminobenzaldehyde and diphenylamine. A 10% solution of equal parts of p-dimethylaminobenzaldehyde and diphenylamine in carbon tetrachloride turns from yellow to deep orange in the presence of phosgene. The test may be applied to determine the ppm-level concentration of phosgene in air.

57.10 THALIDOMIDE

Formula $C_{13}H_{10}N_2O_4$; MW 258.25; CAS [50-35-1]

Structure:

Synonyms: 2,6-dioxo-3-phthalimidopiper dine; 3-phthalimidoglutarimide; *N*-(2,6-dioxo-3-piperidyl)phthalimide

Uses and Exposure Risk

Thalidomide was formerly used as a sedative-hypnotic drug. It is used in the treatment of leprosy.

Health Hazard

Thalidomide is a strong teratogen. Exposure to this compound during the first trimester of pregnancy resulted in deformities in babies. Infants born suffered from amelia or phocomelia, the absence or severe shortening of limbs. Administration of thalidomide in experimental animals caused fetal deaths, postimplantation mortality, and specific developmental abnormalities in the eyes, ear, central nervous system, musculoskeletal system, and cardiovascular system. Several thousand children were affected. The drug has been withdrawn from the market.

57.11 WARFARIN

Formula $C_{19}H_{16}O_4$; MW 308.32; CAS [81-81-2]

Structure:

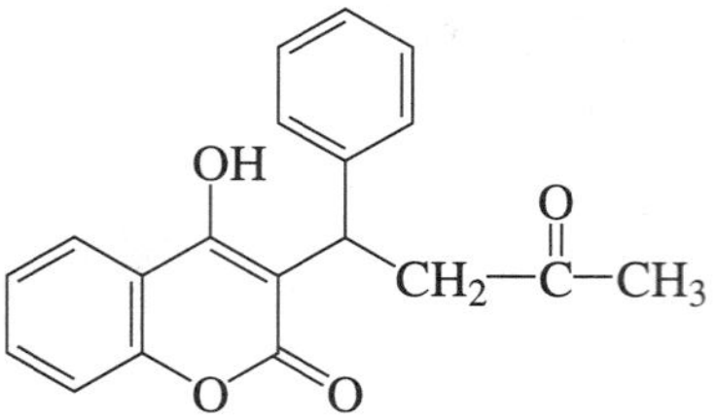

Synonyms: 4-hydroxy-3(1-phenyl-3-oxobutyl) coumarin; 3-(alpha-acetonylbenzyl)-4-hydroxycoumarin; 4-hydroxy-3-(3-oxo-1-phenylbutyl)-2H-1-benzopyran-2-one.

Physical Properties

Colorless solid, no odor; melts at 161°C; decomposes on heating; insoluble in water; soluble in alcohol, acetone, and ether.

Uses

Warfarin is used as a rodenticide.

Health Hazard

Highly toxic substance; exhibits acute, delayed and chronic effects. Ingestion of a dose of 3–15 g is thought to be fatal to adult human. It is an anticoagulant causing hemorrhage. The toxic symptoms which begins a few days or weeks after ingestion include bleeding of nose and gums, pallor and blood in the urine and feces. Another symptom may be hematomas around joints and hip. If the dose is large or lethal the delayed effects may lead to cerebral hemorrhage, paralysis and death. It exhibited teratogenic effects in laboratory animals. The LD_{50} values reported in the literature widely vary.

Exposure Limit

TLV-TWA: 0.1 mg/m^3 (ACGIH,OSHA)
TLV-STEL: 0.3 mg/m^3 (ACGIH, NIOSH)

57.12 WOOD DUST

Workers in furniture and cabinet making industries are highly likely to be exposed to wood dusts. Many published reports and case studies have linked wood dusts to serious respiratory diseases and cancer (ACGIH 1989; Gandevia and Milne 1970; Ito 1963; Sosman 1969; Mosbech and Acheson 1971). Dusts of various types of woods, including cedar wood, oak, mahogany, redwood, abiruana, kejaat, and several African woods, have been attributed to bronchial asthma, rhinitis, and acute allergic disorder of the upper respiratory tract. Chronic exposure to wood dust can cause nasal cancer. High incidence of adenocarcinoma of nasal cavity and ethmoid sinus have been found in woodworkers in the furniture industry. The IARC has classified wood dust as a human carcinogenic agent (IARC 1995). Skin contact may cause dermatitis.

Exposure Limit

TLV-TWA: 1 mg/m^3 (hard wood) and 5 mg/m^3 (soft wood) (ACGIH)

TLV-STEL: 10 mg/m^3 (soft wood) (ACGIH)

References

ACGIH, 1986. *Documentation of the Threshold Limit Values and Biological Exposure Indices*, 5th ed. Cincinnati, OH: American Conferences of Governmental Industrial Hygienists.

Barabas, K., L. Szabo, B. Matkovics, and S. I. Varga. 1986. The effect of light on the toxicity of paraquat in the mouse. *Gen. Pharmacol.* *17*(3): 359–62.

Chui, V. C., G. Poon, and F. Law. 1988. Toxicokinetics and bioavailability of paraquat in rats following different routes of administration. *Toxicol. Ind. Health 4*(2): 203–19.

Dey, M. S., R. G. Breeze, W. L. Hayton, A. H. Karara, and R. I. Krieger. 1990. Paraquat pharmacokinetics using a subcutaneous toxic low dose in the rat. *Fundam. Appl. Toxicol. 14*(1): 208–16.

Gersich, F. M., and M. A. Mayes. 1986. Acute toxicity tests with *Daphnia magna* Straus and *Pimephales promelas* Rafinesque in support of national pollutant discharge elimination permit requirements. *Water Res. 20*(7): 939–41.

Godin, S. C., and P. A. Crooks. 1989. *N*-Methylation as a toxicant route for xenobiotics. II. In vivo formation of *N*,*N*-dimethyl-4,48-bipyridyl ion (Paraquat) from 4,4-bipyridyl in the guinea pig. *Drug Metab. Dispos. 17*(2): 180–85.

Gupta, S. C., and S. P. Singh. 1984. Some biochemical changes in TOK E-25 toxicity in calves. *Int. J. Trop. Agric. 2*(2): 189–92.

Gupta, S. C., and S. P. Singh. 1985. Some observations of herbicide toxicity in calves. *Int. J. Trop. Agric. 3*(2): 132–36.

Hampson, E. C. G. M., and S. M. Pond. 1988. Ultrastructure of canine lung during the proliferative phase of paraquat toxicity. *Br. J. Exp. Pathol. 69*(1): 57–68.

Houze, P., F. J. Baud, R. Mouy, C. Bismuth, R. Bourdon, and J. M. Scherrmann. 1990. Toxicokinetics of paraquat in humans. *Hum. Exp. Toxicol. 9*(1): 5–12.

IARC, 1995. Monograph, Vol. 62, Lyon, France: International Agency for Research on Cancer.

Gandevia, B., and J. Milne, 1970. *Brit. J. Ind. Med., 27*: 235.

Ito, K., 1963. *J. Sci. Labor, 39*: 568.

Johnson, M. E., E. Bellet, M. Eugene, M. S. Christian, and A. M. Hoberman. 1988. The hazard identification and animals NOEL phases of developmental toxicity risk estimation: a case study employing dinoseb. *Adv. Mod. Environ. Toxicol. 15*: 123–32.

Kohen, R., and M. Chevion. 1985. Transition metals potentiate paraquat toxicity. *Free Radical Res. Commun. 1*(2): 79–88.

Kuo, T. L., and R. Nanikawa. 1990. Effect of ethanol on acute paraquat toxicity in rabbits. *Nippon Hoigaku Zasshi 44*(1): 12–17; cited in *Chem. Abstr. CA 113*(21): 186340r.

Mosbech, J., and E. D. Acheson, 1971. *Danish Med. Bull., 18*, 34

NIOSH. 1986. *Registry of Toxic Effects of Chemical Substances.* ed. D. V. Sweet. Washington, DC: U.S. Government Printing Office.

Selman, M., M. Montano, I. Montfort, and R. Perez-Tamayo. 1985. The duration of the pulmonary paraquat toxicity-enhancement effect of oxygen in the rat. *Exp. Mol. Pathol. 43*(3): 388–96.

Sosman, A. J. 1969. *N. Engl. J. Med. 281*: 978.

Szabo, L., B. Matkovics, K. Barabas, and G. Oroszian. 1986. Effects of various thiols on paraquat toxicity. *Chem. Abstr. CA 105*(1): 1932d.

Tsuchiya, T., K. Nagai, T. Yoshida, T. Kiho, and S. Ukai. 1989. Effectiveness of sodium sugar sulfates on acute toxicity of paraquat in mice. *J. Pharmacobio-Dyn. 12*(8): 456–60; cited in *Chem. Abstr. CA 111*(19): 169038t.

Ukai, S., K. Nagai, T. Kiho, T. Tsuchiya, and Y. Nochida. 1987. Effectiveness of dextran sulfate on acute toxicity of paraquat in mice and rats. *J. Pharmacobio-Dyn. 10*(11): 682–84; cited in *Chem. Abstr. CA 108*(9): 70369z.

Wong, R. C., and J. B. Stevens. 1985. Paraquat toxicity in vitro. I. Pulmonary alveolar macrophages. *J. Toxicol. Environ. Health* *15*(3–4): 417–29.

Wong, R. C., and J. B. Stevens. 1986. Bipyridy- lium herbicide toxicity in vitro: comparative study of the cytotoxicity of paraquat and diquat toward the pulmonary alveolar macrophage. *J. Toxicol. Environ. Health* *18*(3): 393–407.

APPENDIX A

FEDERAL REGULATIONS

Several laws have been enacted to control pollution of the nation's waters, air, and land resources and to regulate the manufacture and distribution of hazardous substances. Below are some of the major Federal laws administered by the U.S. EPA. The first five Acts are discussed in brief in the following sections.

- Resource Conservation and Recovery Act of 1976 (RCRA) authorizes the EPA to establish regulations and programs to ensure safe waste treatment and disposal.
- Clean Water Act regulates the pollution control of all waters in the nation to ensure that the waters are clean for swimming and fish cultivation.
- Safe Drinking Water Act, as amended in 1977, permits the EPA to regulate the quality of water in public drinking water systems and the disposal of waters into injection wells.
- Clean Air Act, as amended in 1977, provides the basic legal authority for the nation's air pollution control program

and is designed to enhance the air quality.

- Toxic Substances Control Act (TSCA) of 1976 provides the EPA with the authority to regulate the manufacture, import, distribution, and use of chemical substances.
- Federal Insecticide, Fungicide, and Rodenticide Act (FIFRA) authorizes EPA to regulate the manufacture, distribution, and use of pesticides and to direct the EPA to conduct research into the health and environmental effects of such pesticides.
- Marine Protection, Research, and Sanctuaries Act of 1972 provides authority to the EPA to protect the oceans from the indiscriminate dumping of wastes.
- Comprehensive Environmental Response Compensation and Liability Act of 1980 (CERCLA or "Superfund") is a program established to deal with the release of hazardous substances from inactive and abandoned disposal sites or from spills.

A1 RESOURCE CONSERVATION AND RECOVERY ACT

The Resource Conservation and Recovery Act of 1976 regulates materials and wastes that are generated, treated, stored, or disposed of by industrial facilities. The act authorizes the EPA to establish regulations and programs to ensure safe treatment of wastes and their disposal. The Act defines and lists hazardous wastes, sets guidelines for thermal processing and land disposal of solid wastes, and specifies standards applicable to generators and transporters of hazardous wastes and owners and operators of hazardous waste treatment, storage, and disposal facilities. This Act also specifies guidelines for underground storage tanks. Discussed below are some of the salient features of the Act. Owing to space limitations, only the most important aspects of the act are touched upon here.

Characteristics of Hazardous Waste

A hazardous waste has one or more of the following characteristics: (1) ignitability, (2) corrosivity, (3) reactivity, and (4) toxicity. A waste is termed *ignitable* (1) if it is a liquid other than an aqueous solution containing less than 24% alcohol by volume and has a closed cup flash point below 60°C (140°F) determined by a Pensky-Martens closed-cup tester or an equivalent test method; (2) if it is not a liquid and is capable of burning vigorously when ignited or causes fire through friction, absorption of moisture, or spontaneous chemical changes at standard temperature and pressure; (3) if it is an ignitable compressed gas; or (4) if it is an oxidizer.

A waste exhibits the characteristic of corrosivity if (1) it is aqueous and has a pH of ≤2 or ≥12.5, or (2) it is a liquid and corrodes steel at a rate >6.35 mm (0.25 inch) per year at 55°C (130°F).

A solid waste exhibits the characteristic of reactivity if (1) it is normally unstable and undergoes violent change without detonating; (2) it reacts violently with water; (3) it forms potentially explosive mixtures with water; (4) generates toxic gases, vapors, or fumes dangerous to human health when mixed with water; (5) it is a cyanide- or sulfide-bearing waste that, when exposed to pH conditions between 2 and 12.5, can generate toxic gases or vapors in a quantity sufficient to present a danger to human health; or (6) detonates or explodes when subjected to a strong initiating source or heating under confinement.

A solid waste exhibits the characteristic of toxicity if when using the extraction procedure (EP) toxicity test, the extract from a representative sample of the waste contains any of the contaminants listed in Table A1 at a concentration equal to or greater than the respective value listed in Table A1. Another test, the toxicity characteristic leaching procedure (TCLP), which is not very different from the EP toxicity test and which determines the mobility of both organic and inorganic contaminants present

TABLE A1 Maximum Concentration of Contaminants for Characteristics of EP Toxicity[a]

Contaminant	Maximum Concn. (mg/L)
Arsenic	5.0
Barium	100.0
Cadmium	1.0
Chromium	5.0
2,4-D	10.0
Endrin	0.02
Lead	5.0
Lindane	0.4
Mercury	0.2
Methoxychlor	10.4
Selenium	1.0
Silver	5.0
Toxaphene	0.5
2,4,5-TP (silvex)	1.0

[a]The maximum concentration of these parameters for groundwater protection should not exceed one-hundredth of the values above.

in liquid, solid, and multiphasic wastes, has come into effect. In addition to the old EP constituents, the following new organic compounds are now regulated under the TC rule. These are as follows:

- Benzene
- Carbon tetrachloride
- Chlordane
- Chlorobenzene
- Chloroform
- Cresol and its *o-*, *m-*, and *p*-isomers
- 1,4-Dichlorobenzene
- 1,2-Dichloroethane
- 1,1-Dichloroethylene
- 2,4-Dinitrotoluene
- Heptachlor and its hydroxide
- Hexachlorethane
- Methyl ethyl ketone
- Nitrobenzene
- Pentachlorophenol
- Pyridine
- Tetrachloroethylene
- Trichloroethylene
- 2,4,5- and 2,4,6-trichlorophenol
- Vinyl chloride

EP toxicity and TCLP tests are described in the Code of Federal Regulations, 40, Part 261, Appendix II and Part 268, Appendix I, respectively.

In addition to the four classes of hazardous wastes discussed above — ignitable, corrosive, reactive, and toxic wastes — there are 203 individual compounds and salts of some of these compounds as well as soluble cyanide salts that are listed as acute hazardous wastes, and each designated with a RCRA Hazardous Waste Number, starting with the letter P. In addition, there are 455 compounds plus the salts and isomers of some of these compounds, which are designated as toxic wastes. These substances are assigned an RCRA Waste Number, which starts with a U. Because of space limitations, a full listing of both these types of wastes is not presented in this section. However, the RCRA Waste numbers for hazardous chemicals are given under the individual compounds in Part B of the book. 40 CFR Part 261, Appendix VIII should be referred to for a full list of hazardous constituents of wastes and their corresponding RCRA Waste Numbers.

Groundwater Monitoring

The suitability of groundwater as a drinking water supply is characterized by the contaminants listed in Table A1, the maximum levels of which should not exceed one-hundredth of the concentrations for the characteristic EP toxicity given in the table. The presence of chloride, iron, manganese, phenols, sodium, and sulfate ions affects the quality of groundwater. These parameters should be used as a basis for comparison in the event that a groundwater quality assessment is required, while the groundwater contamination can be indicated from the pH, specific conductance, total organic carbon (TOC), and total organic halogen (TOX) in the water.

Potentially Incompatible Waste

When mixed with other waste, at a hazardous waste facility, many hazardous wastes can produce harmful effects, which include heat, pressure, violent reaction, fire, explosion, or liberation of toxic or flammable fumes, gases, or mists. Table A2 presents examples of such incompatible materials. This is not an exhaustive list, but it highlights incompatible wastes.

Treatment and Disposal

Incompatible wastes and materials must not be placed in the same pile unless separated or protected by means of a dike, wall, berm, or other device. A hazardous waste may be subjected to land disposal in compliance with RCRA regulations. The landfill must have a liner system to the adjacent subsurface soil or groundwater or surface water. The

TABLE A2 Examples of Some Incompatible Wastes

Incompatible Substances	Consequences

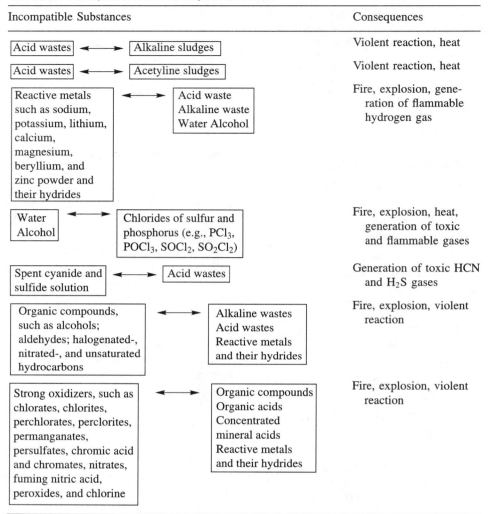

	Consequences
Acid wastes ⟷ Alkaline sludges	Violent reaction, heat
Acid wastes ⟷ Acetyline sludges	Violent reaction, heat
Reactive metals such as sodium, potassium, lithium, calcium, magnesium, beryllium, and zinc powder and their hydrides ⟷ Acid waste / Alkaline waste / Water Alcohol	Fire, explosion, generation of flammable hydrogen gas
Water Alcohol ⟷ Chlorides of sulfur and phosphorus (e.g., PCl$_3$, POCl$_3$, SOCl$_2$, SO$_2$Cl$_2$)	Fire, explosion, heat, generation of toxic and flammable gases
Spent cyanide and sulfide solution ⟷ Acid wastes	Generation of toxic HCN and H$_2$S gases
Organic compounds, such as alcohols; aldehydes; halogenated-, nitrated-, and unsaturated hydrocarbons ⟷ Alkaline wastes / Acid wastes / Reactive metals and their hydrides	Fire, explosion, violent reaction
Strong oxidizers, such as chlorates, chlorites, perchlorates, perclorites, permanganates, persulfates, chromic acid and chromates, nitrates, fuming nitric acid, peroxides, and chlorine ⟷ Organic compounds / Organic acids / Concentrated mineral acids / Reactive metals and their hydrides	Fire, explosion, violent reaction

liner must be constructed of materials with appropriate strength thickness and chemical properties to prevent physical contact and withstand pressure gradients. There should be a leachate collection system above and between the liners.

Hazardous wastes may be packaged in small drums (lab packs) and placed in a landfill if the following requirements are met:

1. The waste must be packaged in non-leaking inside containers constructed of a material that will not react dangerously with, be decomposed by, or be ignited by the contained waste. Inside containers must be sealed tightly and securely.

2. The inside containers must be overpacked in an open-head DOT specification metal shipping container of no more than 416-L (110-gall) capacity and surrounded by a sufficient quantity of absorbent material to absorb completely all the liquid contents of the inside

containers. The metal outer container must be full after packing with inside containers and absorbent material.

3. Incompatible wastes must not be placed in the same outside container.
4. Reactive wastes, other than cyanide and sulfide wastes, must be treated and rendered nonreactive before packaging.
5. The absorbent material used must not be capable of reacting dangerously or be ignited by the contents of the inside containers.

Wastes containing dioxins, dibenzofurans, chlorophenols, and their chlorophenoxy derivatives must meet additional design, operating, and monitoring requirements for landfills. An incinerator burning hazardous waste must be designed, constructed, and maintained to meet the following performance standard:

1. It must achieve a destruction and removal efficiency (DRE) of 99.99% for each principal organic hazardous constituent (POHC). It is determined from the following equation:

$$\text{DRE} = \frac{W_{in} - W_{out}}{W_{in}} \times 100$$

where W_{in} is the mass feed rate of one POHC in the waste stream feeding the incinerator, and W_{out} is the mass emission rate of the same POHC present in the exhaust emissions prior to release to the atmosphere.

2. An incinerator burning dioxins, dibenzofurans, and pentachlorophenols and their chlorophenoxy derivatives wastes must achieve a DRE value of 99.9999% for each POHC.

3. An incinerator producing stack emissions of more than 1.8 kg/hr (4 lb/hr) or HCl must control HCl emission so that the rate of emission would not be greater than the larger of either 1.8 kg/hr of 1% of the HCl in the stack

gas before entering pollution control equipment.

4. An incinerator burning hazardous waste must not emit particulate matter in excess of 180 mg per dry standard cubic meter (0.08 grain per dry standard cubic foot) when corrected to the amount of oxygen in the stack gas according to the formula

$$P_c - P_m \times \frac{12}{21 - Y}$$

where P_c is the corrected concentration of particulate matter, P_m is the measured concentration of particulate matter, and Y is the measured concentration of oxygen in the stack gas using the Orsat method for oxygen analysis of dry flue gas.

Universal Waste Rule

Universal waste rule was an amendment to the Resource Conservation and Recovery Act regulations. It is designed to reduce the amount of hazardous waste items in the municipal solid waste stream. The rule encourages the recycling of certain common hazardous wastes and their proper disposal. Another objective of this rule is to reduce paperwork and other administrative requirements and regulatory burden on businesses that generate these wastes, and to save them compliance costs.

Universal wastes include batteries, such as nickel–cadmium and lead–acid batteries, which are often found in many common items in the business and home setting. Agricultural pesticides that are obsolete, banned, or no longer needed may be found stored for long periods of time in sheds or barns are examples of universal wastes. Thermostats containing liquid mercury are also designated as universal wastes. Such wastes may be generated by individual households that, unlike small and large businesses, are not regulated under RCRA. Under this universal waste rule, EPA encourages residents to take these items to collection sites located at nearby businesses and other centers for disposal or proper recycling. Communities may establish such collection programs.

Requirements for Hazardous Waste Generators

Small quantity generators are those businesses that produce 100–1000 kg of hazardous waste in a calendar month. They are subject to federal hazardous waste requirements. Additional state requirements may apply. Businesses producing more than 1000 kg of hazardous waste per month or 1 kg of certain acute hazardous waste (that are fatal to human in low doses) would be termed as large quantity generators. They are subject to more extensive regulations. Businesses that generate <100 kg of hazardous wastes or 1 kg of acute hazardous wastes per month may be exempt from most of the federal hazardous waste requirements. However, it must be determined whether such waste is hazardous. The waste, however, should be delivered to a facility permitted, licensed, or authorized by EPA or the state to accept hazardous waste. Some states do not recognize such exemptions.

EPA identification numbers must be obtained for each site that generates small or large quantities of hazardous wastes. Also, a permit must be obtained from the EPA for storage, treatment, or disposal of the waste on site. The RCRA/Superfund hotline or an EPA regional office may be contacted for all information. The wastes may be periodically shipped off from the premises for treatment or disposal following Federal and state requirements.

Hazardous waste may be stored on site without a permit for 180 days (or 270 days if the waste is to be shipped more than 200 miles) up to an amount of 6000 kg. To store for a longer period, a permit must be obtained. An amount up to 55 gallons of hazardous waste may be accumulated in a "satellite accumulation area" — an area at or near the point of generation. When this amount is accumulated, at that point it must be moved to the hazardous waste storage area within 3 days.

Shipping of hazardous waste off the premises must be carried out only be authorized hazardous waste transporters to licensed, authorized, and permitted facilities. Small businesses, however, may send certain types of wastes, including dead automobile batteries and used oils, to recycling or reclamation establishments. Further information may be obtained from RCRA/Superfund Hotline at 800-424-9346 or 703-412-9810 or 202-382-3000. The Small Business Hotline number is 800-368-5888 or 202-557-1938. The National Solid Waste Management Association number is 202-659-4613.

A2 CLEAN WATER ACT

The Federal Water Pollution Control act was amended by the Clean Water Act of 1977. It is commonly known as the Clean Water Act. There have been many amendments since then. The objective of this Act is to restore and maintain the chemical, physical, and biological integrity of U.S. waters. The provisions of the Act are as follows:

1. To eliminate the discharge of pollutants into navigable waters
2. To achieve an interim goal of water quality for the protection and propagation of fish, shellfish, and wildlife
3. To prohibit the discharge of toxic pollutants in toxic amounts
4. To develop and implement waste treatment processes for adequate control of sources of pollutants
5. To provide federal financial assistance to construct publicly owned waste treatment works
6. To develop the technology necessary to eliminate the discharge of pollutants in navigable waters and the oceans

Section 307 of this act lists the following substances as toxic pollutants:

Inorganics

Antimony and compounds

Arsenic and compounds

Asbestos

Beryllium and compounds

Cadmium and compounds

Copper and compounds

Cyanides

Lead and compounds

Mercury and compounds

Nickel and compounds

Selenium and compounds

Silver and compounds

Thallium and compounds

Zinc and compounds

Organics

Acenaphthene

Acrolein

Acrylonitrile

Aldrin/Dieldrin

Benzene

Benzidine

Carbon tetrachloride

Chlordane

Chlorinated benzenes (other than dichlorobenzenes)

Chlorinated ethanes (including 1,2-dichloroethane, 1,1,1-trichloroethane, and hexachloroethane)

Chlorinated naphthalene

Chloroalkyl ethers (chloromethyl, chloroethyl, and mixed ethers)

Chlorophenols (2-chlorophenol, 2,4-dichlorophenol, pentachlorophenol)

Chlorophenylphenyl ether

DDT and metabolites

Dichlorobenzenes (1,2-, 1,3-, and 1,4-dichlorobenzenes)

Dichlorobenzidine

Dichloropropene

2,4-Dimethylphenol

Dinitroltoluene

Diphenylhydrazine

Endosulfan and metabolites

Endrin and metabolites

Ethylbenzene

Fluoranthene

Haloethers [chlorophenylphenyl ether, bromophenylphenyl ether, bis(dichloroisopropyl)ether, bis(chloroethoxy)methane, and polychlorinated diphenyl ethers]

Halomethanes (methylene chloride, methyl chloride, methyl bromide, bromoform, dichlorobromomethane, trichlorofluoromethane, and dichlorodifluoromethane)

Heptachlor and metabolites

Hexachlorobutadiene

Hexachlorocyclohexane (all isomers)

Hexachlorocyclopentadiene

Isophorone

Naphthalene

Nitrobenzene

Nitrophenols (2,4-dinitrophenol and dinitrocresol)

Nitrosamines

Pentachlorophenol

Phenol

Phthalate esters

Polychlorinated biphenyls (PCBs)

Polynuclear aromatic hydrocarbons (benzanthracenes, benzopyrenes, benzofluoranthene, chrysene, dibenzanthracenes, and indenopyrenes)

2,3,7,8-Tetrachlorodibenzo-p-dioxin (TCDD)

Tetrachloroethylene

Toluene

Toxaphene

Trichloroethylene

Vinyl chloride

A3 SAFE DRINKING WATER ACT

Regulations for Toxic Contaminants in Drinking Water

In accordance with the 1986 amendments to the Safe Drinking Water Act, the EPA has imposed regulations and guidelines for the maximum contaminant levels (MCLs) and

TABLE A3   Drinking Water Standards for Inorganic Chemicals

Inorganic Contaminants	MCL (mg/L)	MCLG (mg/L)
Antimony	0.006	0.006
Arsenic	0.05	—
Asbestos (fibers/L >10-μm length)	7 MFL	7 MFL
Barium	2	2
Beryllium	0.004	0.004
Bromate	zero	zero
Cadmium	0.005	0.005
Chloramine[a]	4	4
Chlorine	4	4
Chlorine dioxide	0.8	0.3
Chlorite	1	0.08
Chromium, total	0.1	0.1
Copper (at tap)	Treatment technique; 1.3 (action level)	1.3
Cyanide	0.2	0.2
Fluoride[b]	4	4
Hypochlorite (regulated as chlorine)	—	4
Hypochlorous acid (regulated as chlorine)	—	4
Lead (at tap)	Treatment technique; 0.015 (action level)	zero
Mercury (inorganic)	0.002	0.002
Nickel	0.1	0.1
Nitrate (as N)	10	10
Nitrite (as N)	1	1
Nitrate and nitrite (both as N)	10	10
Selenium	0.05	0.05
Sulfate	500	500
Thallium	0.002	0.0005

[a]Measured as free chlorine.
[b]Under review.

maximum contaminant level goals (MCLGs) of certain chemicals in the drinking water. MCL is defined as the maximum permissible level of a contaminant in water which is delivered to any user of a public water system. MCLG is a nonenforceable concentration of a drinking water contaminant that is protective of adverse human health effects and allows an adequate margin of safety. These contaminants include metals, anions, volatile organics, synthetic organic chemicals, pesticides, radionuclides, and coliforms and other bacteria that can produce adverse health effects on humans.

Table A3 lists the MCL and MCLG values of toxic substances, including regulated inorganic, organic, radionuclides, and aggregate properties. Drinking water standards for radionuclides and organics are listed in Tables A4 and A5 respectively. The MCLG values represent the desired concentration levels of contaminants at which no known or anticipated adverse health effects occur, thus allowing an adequate margin of safety. MCL and MCLG values should be as close as feasible and should be achieved by using the best available treatment techniques.

TABLE A4   Drinking Water Standards for Radionuclides

Radionuclides	MCL	MCLG
β particles and photo activity	4 mrem	zero[a]
Gross α-particle activity	15 pCi/L	zero[a]
Radium-226 and -228 (combined)	5 pCi/L	zero[a]
Radon	300 pCi/L	zero
Uranium	20 µg/L	zero

[a]Proposed in 1991, no final decision has been taken.

TABLE A5   Drinking Water Standards for Organics

Organics	MCL (mg/L)	MCLG (mg/L)
Acifluofen	—	zero
Acrylamide	Treatment technique	zero
Acrylonitrile	—	zero
Adipate (diethylhexyl)	0.4	0.4
Alachlor	0.002	zero
Aldicarb	0.007	0.007
Aldicarb sulfone	0.007	0.007
Aldicarb sulfoxide	0.007	0.007
Atrazine	0.003	0.003
Bentazon	—	0.02
Benzene	0.005	zero
Benzo(a)pyrene (PAH)	0.0002	zero
Bromodichloromethane (THM)	0.08	zero
Bromoform (THM)	0.08	zero
Carbofuran	0.04	0.04
Carbon tetrachloride	0.005	zero
Chloral hydrate	0.06[a]	0.04
Chlordane	0.001	zero
Chlorodibromomethane (THM)	0.06	0.08
Chloroform (THM)	0.08	zero
Cyanazine	—	0.1001
2,4-D	0.07	0.07
Dalapon	0.2	0.2
Di(2-ethylhexyl)adipate	0.4	0.4
Dibromochloropropane (DBCP)	0.0002	zero
Dichloroacetic acid	0.06	zero
o-Dichlorobenzene	0.6	0.6
m-Dichlorobenzene	0.6	0.6
p-Dichlorobenzene	0.075	0.075
1,2-Dichloroethane	0.005	zero
1,1-Dichloroethylene	0.007	0.007
cis-1,2-Dichloroethylene	0.07	0.07
$trans$-1.2-Dichloroethylene	0.1	0.1
Dichloromethane	0.005	zero
1,2-Dichloropropane	0.005	zero

(*continued*)

TABLE A5   *(Continued)*

Organics	MCL (mg/L)	MCLG (mg/L)
1,3-Dichloropropane	—	zero
Di(2-ethylhexyl)phthalate(PAE)	0.006	zero
Dinoseb	0.007	0.007
Diquat	0.02	0.02
Endothall	0.1	0.1
Endrin	0.002	0.002
Epichlorohydrin	Treatment technique	zero
Ethylbenzene	0.7	0.7
Ethylene dibromide (EDB)	0.00005	zero
Glyphosate	0.7	0.7
Heptachlor	0.0004	zero
Heptachlor epoxide	0.0002	zero
Hexachlorobenzene	0.001	zero
Hexachlorobutadiene	0.001	—
Hexachlorocyclopentadiene	0.05	0.05
Lindane	0.0002	0.0002
Methoxychlor	0.04	0.04
Monochlorobenzene	0.1	0.1
Oxamyl(Vydate)	0.2	0.2
Pentachlorophenol	0.001	zero
Picloram	0.5	0.5
Polychlorinated biphenyls (PCBs)	0.0005	zero
Simazine	0.004	0.004
Styrene	0.1	0.1
2,3,7,8-TCDD(Dioxin)	3×10^{-8}	zero
Tetrachloroethylene	0.005	zero
Toluene	1	1
Toxaphene	0.003	zero
2,4,5-TP	0.05	0.05
Trichloroacetic acid	0.06[a]	0.3
1,2,4-Trichlorobenzene	0.07	0.07
1,1,1-Trichloroethane	0.2	0.2
1,1,2-Trichloroethane	0.005	zero
Vinyl chloride	0.002	zero
Xylenes	10	10

[a]Total for all haloacetic acids cannot exceed 0.06 mg/L level

THM, trihalomethane; PAH, polynuclear aromatic hydrocarbons; PAE, phthalate esters.

While the MCLG for the microbial agents *Giardia lamblia, Legionella*, total coliform bacteria, viruses, and standard plate count has been set at zero, the MCL is based on the treatment technique. It is critical that no bacteria, viruses, or other microbial agents be found in the drinking water supply.

Secondary Maximum Contaminant Levels (SMCLs)

Secondary drinking water standards are unenforceable federal guidelines with regard to taste, odor, color, and other nonaesthetic effects of drinking water. Federal law does not require water systems to comply with them. States may, however, adopt their own

TABLE A6   Secondary Drinking Water Standards

Chemicals/Aggregate Properties	SMCL (mg/L)
Aluminum	0.05–0.2
Chloride	250
Color	15 color units
Copper	1.0
Corrosivity	Noncorrosive
Fluoride (under review)	2.0
Foaming agents	0.5
Iron	0.3
Manganese	0.05
Odor	3 threshold odor number
pH	6.5–8.5 SU
Silver	0.1
Sulfate	250
Total dissolved solids (TDS)	500
Zinc	5

enforceable regulations. Secondary maximum contaminant levels (SMCLs) are presented in Table A6.

Concentration levels of carcinogenic chemicals in drinking water supplies must not exceed specific values as defined under the "Health Advisories." Such "no-exceed" concentrations of cancer-causing pollutants in the water have been defined for adult (70 kg) and child (10 kg) for 1 day, 10 days, and long-term consumption, respectively. Any further discussion on the subject is beyond the scope of this book. For further information regarding drinking water regulations and "health advisories" readers may call the Safe Drinking Water Hotline at (800)-426-4791 or EPA's Office of Water at (202) 260-1332.

A4   CLEAN AIR ACT

Emission of certain toxic chemicals and particulate mattes have been regulated under this act. These pollutants include sulfur dioxide, carbon monoxide, nitrogen oxides, hydrocarbons, ozone, photochemical oxidants, and particulate materials. There are two categories of regulations for toxic substances in the air: (1) the ambient air regulations, which refer to the general atmospheric concentrations; and (2) emissions from industrial sources and automobiles.

Ambient Air Regulations

The EPA has developed standards for toxic air pollutants under the term *national ambient air quality standards* (NAAQS). There are two types of standards: primary and secondary. The former is designed to protect the public health; the latter is meant to protect public welfare, such as the effects of air pollution on vegetation, materials, and visibility. NAAQS have been established for the six sources of air pollutants presented in Table A7.

Toxic Pollutants Emissions from Mobile Sources

The Clean Air Act Amendments of 1990 set the emission standards for vehicles in two tiers. The tier I standards set the limit for NO_x tailpipe emissions at 0.6 g/mile. Automobiles must meet the standards for 100,000 miles or 10 years. Light-duty trucks must meet this standard for 75,000 miles

TABLE A7 National Ambient Air Quality Standard

Air Pollutant	Primary Standard	Secondary Standard
Particulate matter		
($<$10 µm)		
Annual mean (arithmetic)	50 µg/m^3	50 µg/m^3
24-hr average	150 µg/m^3	150 µg/m^3
Sulfur dioxide		
Annual mean (arithmetic)	0.03 ppm (80 µg/m^3)	
24-hr average[a]	0.14 ppm (365 µg/m^3)	
3-hr average[a]		0.5 ppm (1300 µg/m^3)
Carbon monoxide		
8-hr average[a]	9 ppm (10 mg/m^3)	No standard
1-hr average[a]	35 ppm (40 mg/m^3)	No standard
Nitrogen dioxide		
Annual mean (arithmetic)	0.053 ppm (100 µg/m^3)	0.053 ppm (100 µg/m^3)
Ozone		
Maximum daily 1-hr average	0.12 ppm (235 µg/m^3)	0.12 ppm (235 µg/m^3)
Lead		
Maximum quarterly average	1.5 µg/m^3	1.5 µg/m^3

[a]Not to be exceeded more than once per year.

or 7 years. The standard for hydrocarbons is set at 0.4 g/mile. Both standards have been enacted beginning 1994. Tier II tailpipe emission standards for nitrogen oxides and hydrocarbons will be set at 0.2 and 0.125 g/mile, respectively, beginning in the year 2003. The tier I standard for carbon monoxide starting is 10 g/mile. The vehicle must meet the limit for 50,000 miles or 5 years.

Emissions from Industrial Sources

Hazardous air pollutants from a stationary source involve a stationary source located within a contiguous area that emits or has the potential to emit 10 tons per year or more of any hazardous air pollutant or 25 tons or more per year of any combination of hazardous air pollutants. EPA may establish a lesser quantity for such a source on the basis of the potency of the air pollutant, persistence, potential for bioaccumulation, and other relevant factors.

Listed below are individual and classes of compounds that are air toxicants which EPA is required to regulate under the new Clear Air Act.

Compound/CAS Registry Number

Acetaldehyde [75-07-0]

Acetamide [60-35-5]

Acetonitrile [75-05-8]

Acetophenone [98-86-2]

2-Acetylaminofluorene [53-96-3]

Acrolein [107-02-8]

Acrylamide [979-06-1]

Acrylic acid [79-10-7]

Acrylonitrile [107-13-1]

Allyl chloride [107-05-1]

4-Aminobiphenyl [92-67-1]

Aniline [62-53-3]

o-Anisidine [90-04-0]

Asbestos [1332-21-4]

Benzene (including benzene from gasoline) [71-43-2]

Benzidine [92-87-5]

Benzotrichloride [98-07-7]

Benzyl chloride [100-44-7]

Biphenyl [92-52-4]

Bis(chloromethyl)ether [542-88-1]

Bis(2-ethylhexl)phthalate (DEHP) [117-81-7]

Bromoform [75-25-2]

1,3-Butadiene [106-99-0]

Calcium cyanamide [156-62-7]

Caprolactam [105-60-2]

Captan [133-06-2]

Carbaryl [63-25-2]

Carbon disulfide [75-15-0]

Carbon tetrachloride [56-23-5]

Carbonyl sulfide [463-58-1]

Catechol [120-80-9]

Chloramben [133-90-4]

Chlordane [57-74-9]

Chlorine [7782-50-5]

Chloroacetic acid [79-11-8]

2-Chloroacetophenone [532-27-4]

Chlorobenzene [108-90-7]

Chlorobenzilate [510-15-6]

Chloroform [67-66-3]

Chloromethyl methyl ether [107-30-2]

Chloroprene [126-99-8]

Cresols/cresylic acid (isomers and mixture) [1319-77-3]

m-Cresol [108-39-4]

o-Cresol [95-48-7]

p-Cresol [106-44-5]

Cumene [98-82-8]

2,4-D, salts and esters [94-75-7]

DDE [3547-04-4]

Diazomethane [334-88-3]

Dibenzofurans [132-64-9]

1,2-Dibromo-3-chloropropane [96-12-8]

Dibutylphthalate [84-74-2]

1,4-Dichlorobenzene [106-46-7]

3,3-Dichlorobenzidene [91-94-1]

Dichloroethyl ether [bis(2-chloroethyl)ether] [111-44-4]

1,3-Dichloropropene [542-75-6]

Dichloryos [62-73-7]

Diethanolamine [111-42-4]

N,*N*-Diethyl aniline (*N*,*N*-dimethylaniline) [121-69-7]

Diethyl sulfate [64-67-5]

3,3-Dimethoxybenzidine [119-90-4]

Dimethyl aminoazobenzene [60-11-7]

3,3′-Dimethyl benzidine [119-93-7]

Dimethyl carbamoyl chloride [79-44-7]

Dimethyl formamide [68-12-2]

1,1-Dimethyl hydrazine [57-14-7]

Dimethyl phthalate [131-11-3]

Dimethyl sulfate [77-78-1]

4,6-Dinitro-*o*-cresol, and salts [53-45-21]

2,4-Dinitrophenol [51-28-5]

2,4-Dinitrotoluene [121-14-2]

1,4-Dioxane (1,4-Diethyleneoxide) [123-91-1]

1,2-Diphenylhydrazine [122-66-7]

Epichlorohydrin (1-chloro-2,3-epoxypropane) [106-89-8]

1,2-Epoxybutane [106-88-7]

Ethyl acrylate [140-88-5]

Ethyl benzene [100-41-4]

Ethyl carbamate (urethane) [51-79-6]

Ethyl chloride (chloroethane) [75-00-3]

Ethylene dibromide (dibromoethane) [106-93-4]

Ethylene dichloride (1,2-dichloroethane) [107-06-2]

Ethylene glycol [107-21-1]

Ethylene imine (aziridine) [151-56-4]

Ethylene oxide [75-21-8]

Ethylene thiourea [96-45-7]

Ethylidene dichloride (1,1-dichloroethane) [75-34-4]

Formaldehyde [50-00-0]

Heptachlor [76-44-8]

Hexachlorobenzene [118-74-1]

Hexachlorobutadiene [87-68-3]

Hexachlorocyclopentadiene [77-47-4]

Hexachloroethane [67-72-1]

Hexamethylene-1,6-diisocyanate [822-06-0]

Hexamethylphosphoramide [680-31-9]

Hexane [110-54-3]

Hydrazine [302-01-2]

Hydrochloric acid [7647-01-0]

Hydrogen fluoride (hydrofluoric acid) [7664-39-3]

Hydrogen sulfide [7783-06-4]

Hydroquinone [123-31-9]

Isophorone [78-59-1]

Lindane (all isomers) [58-89-9]

Maleic anhydride [108-31-6]

Methanol [67-56-1]

Methoxychlor [72-43-5]

Methyl bromide (bromomethane) [74-83-9]

Methyl chloride (chloromethane) [74-87-3]

Methyl chloroform (1,1,1-trichloroethane) [71-55-6]

Methyl ethyl ketone (2-butanone) [78-93-3]

Methyl hydrazine [60-34-4]

Methyl iodide (iodomethane) [74-88-4]

Methyl isobutyl ketone (hexane) [108-10-1]

Methyl isocyanate [624-83-9]

Methyl methacrylate [80-62-6]

Methyl *tert*-butyl ether [1634-04-4]

4,4′-Methylene bis(2-chloroaniline) [101-14-4]

Methylene chloride (dichloromethane) [75-09-2]

Methylene diphenyl diisocyanate (MDI) [101-68-8]

4,4′-Methylenedianiline [101-77-9]

Naphthalene [91-20-3]

Nitrobenzene [98-95-3]

4-Nitrobiphenyl [92-93-3]

4-Nitrophenol [100-02-7]

2-Nitropropane [79-46-9]

N-Nitroso-*N*-methylurea [684-93-5]

N-Nitrosodimethylamine [62-75-9]

N-Nitrosomorpholine [59-89-2]

Parathion [56-38-2]

Pentachloronitrobenzene (quintobenzene) [82-68-8]

Pentachlorophenol [87-86-5]

Phenol [108-95-2]

p-Phenylenediamine [106-50-3]

Phsogene [75-44-5]

Phosphine [7803-51-2]

Phosphorus [7723-14-0]

Phthalic anhydride [85-44-9]

Polychlorinated biphenyls (Aroclors) [1336-36-3]

1,3-Propane sultone [1120-71-4]

β-Propiolactone [57-57-8]

Propionaldehyde [123-38-6]

Propoxur (Baygone) [114-26-1]

Propylene dichloride (1,2-dichloropropane) [78-87-5]

Propylene oxide [75-56-9]

1,2-Propylenimine (2-methyl aziridine) [75-55-8]

Quinoline [91-22-5]

Quinone [106-51-4]

Styrene [100-42-5]

Styrene oxide [96-09-3]

2,3,7,8-Tetrachlorodibenzo-*p*-dioxin [1746-01-6]

1,1,2,2-Tetrachloroethane [79-34-5]

Tetrachloroethylene (perchloroethylene) [127-18-4]

Titanium tetrachloride [7550-45-0]

Toluene [108-88-3]

2,4-Toluene diamine [95-80-7]

2,4-Toluene diisocyanate [584-84-9]

o-Toluidine [95-53-4]

Toxaphene (chlorinated camphene) [80001-35-2]

1,2,4-Trichlorobenzene [120-82-1]

1,1,2-Trichloroethane [79-005]

Trichloroethylene [79-01-6]

2,4,5-Trichlorophenol [95-95-4]

2,4,6-Trichlorophenol [88-06-2]

Triethylamine [121-44-8]

Trifluralin [1582-09-8]

2,2,4-Trimethylpentane [540-84-1]

Vinyl acetate [108-05-4]

Vinyl bromide [593-60-2]

Vinyl chloride [75-01-4]

Vinylidene chloride (1,1-dichloroethylene) [75-35-4]

Xylenes (isomers and misture) [1330-20-7]

m-Xylene [108-38-3]

o-Xylene [95-47-6]

p-Xylene [106-42-3]

Antimony compounds

Arsenic compounds

Beryllium compounds

Cadmium compounds

Chromium compounds

Cobalt compounds

Coke oven emissions

Cyanide compounds

Glycol ethers

Lead compounds

Manganese compounds

Mercury compounds

Fine mineral fibers

Nickel compounds

Polycyclic organic matter

Radionuclides (including radon)

Selenium compounds

A5 TOXIC SUBSTANCE CONTROL ACT

The Toxic Substance Control Act (TSCA) was enacted officially by the US Congress in 1976. This Act assigns authority and responsibility to the US EPA to (1) gather information on the toxicity of particular chemicals that someone wants to manufacture or process or import; the extent to which people and the environment would be exposed to such substances; (2) to assess any health risks to humans and pollution risks to the environment; and (3) to institute appropriate control actions after weighing and evaluating their potential risks against their benefits to the economy and social well-being.

TSCA regulation basically applies only to the newly introduced chemicals and not the "existing" chemicals. There are more than 60,000 existing chemicals in the Inventory that were already being produced or imported

before 1976, prior to TSCA or any other chemical law as enacted. Every year about 1000 new chemicals are manufactured, processed, or imported that fall under TSCA and are assessed by the EPA. About 15,000 of 70,000 chemicals on the Inventory List (which includes many "existing" chemicals) have been fully screened. A variety of approaches have been taken to reduce the risks.

The law mandates a 90-day prior notice, known as pre-manufacture notice (PMN), to be submitted to the US EPA before production or import of a new chemical substance for a non-exempt commercial purpose. Any chemical which is not listed on the Inventory of existing chemicals is considered new for the purpose of PMN. Eight product categories are exempt from TSCA's regulatory authorities. These are, pesticides, tobacco, nuclear material, firearms and ammunition, food, food additives, drugs, and cosmetics. Many of these products fall under the jurisdiction of other Federal laws. In addition, the following are excluded from PMN reporting under certain conditions: products of incidental reactions, products of end-use reactions, mixtures, byproducts, substances manufactured solely for export, nonisolated intermediates, and substances formed during the manufacture of an article.

The EPA has limited reporting requirements for new chemical substances in the following cases:

- when the substance is produced in small quantities for research and development
- when the amount of substance manufactured or imported each year is less than 10,000 kg
- when the substance is expected to have low release and low exposure
- if the substance is manufactured or imported for test marketing
- if the substance is a polymer and not chemically active or bioavailable

PMN submissions require all available data on chemical identity, production volume, byproducts, use, environmental release, disposal practices, human exposure and all possible health and environmental data.

Since 1979, EPA's New Chemicals Program has reviewed almost 30,000 new chemicals, including 22,000 PMNs, which were fully reviewed and 5500 low-volume, test market, and polymer exemptions. During this period, the Agency claims that the TSCA has prevented potential risks to people and the environment from nearly 2700 new substances.

It may be noted that the TSCA, unlike other Federal regulations, applies only to the manufacture and import of chemical substances. Other aspects of this Act, pertaining to administrative requirements, procedures, fees, and paperwork are beyond the scope of this book. Readers interested in obtaining further information may call the TSCA hotline at (202)-554-1404 or a PMN prenotice coordinator at (202)-260-1745 or (202)-260-3937.

APPENDIX B

IARC LIST OF CARCINOGENIC AGENTS

The following list contains all agents, mixtures, and exposures evaluated by the International Agency for Research on Cancer as confirmed, probable, and possible carcinogens to human. For meaning of these terms refer to Section VI, (Cancer-causing Chemicals), in Part A of this text. CAS Registry numbers, IARC Monograph volumes and the year of publication are shown in parenthesis.

GROUP 1   CARCINOGENIC TO HUMANS

Agents and groups of agents

Aflatoxins, naturally occurring [1402-68-2] (Vol. 56; 1993)

4-Aminobiphenyl [92-67-1] (Vol. 1, Suppl. 7; 1987)

Arsenic [7440-38-2] and arsenic compounds (Vol. 23, Suppl. 7; 1987)

(NB: This evaluation applies to the group of compounds as a whole and not necessarily to all individual compounds within the group)

Asbestos [1332-21-4] (Vol. 14, Suppl. 7; 1987)

Azathioprine [446-86-6] (Vol. 26, Suppl. 7; 1987)

Benzene [71-43-2] (Vol. 29, Suppl. 7; 1987)

Benzidine [92-87-5] (Vol. 29, Suppl. 7; 1987)

Beryllium [7440-41-7] and beryllium compounds (Vol. 58; 1993)

(NB: Evaluated as a group)

N,N-Bis(2-chloroethyl)-2-naphthylamine (Chlornaphazine) [494-03-1] (Vol. 4, Suppl. 7; 1987)

Bis(chloromethyl)ether [542-88-1] and chloromethyl methyl ether [107-30-2] (technical-grade) (Vol. 4, Suppl. 7; 1987)

1,4-Butanediol dimethanesulfonate (Busulphan; Myleran) [55-98-1] (Vol. 4, Suppl. 7; 1987)

Cadmium [7440-43-9] and cadmium compounds (Vol. 58; 1993)

(NB: Evaluated as a group)

Chlorambucil [305-03-3] (Vol. 26, Suppl. 7; 1987)

1-(2-Chloroethyl)-3-(4-methylcyclohexyl)-1-nitrosourea (Methyl-CCNU; Semustine) [13909-09-6] (Suppl. 7; 1987)

Chromium[VI] compounds (Vol. 49; 1990)

(NB: Evaluated as a group)

Cyclosporin [79217-60-0] (Vol. 50; 1990)

Cyclophosphamide [50-18-0] [6055-19-2] (Vol. 26, Suppl. 7; 1987)

Diethylstilboestrol [56-53-1] (Vol. 21, Suppl. 7; 1987)

Epstein-Barr virus (Vol. 70; 1997)

Erionite [66733-21-9] (Vol. 42, Suppl. 7; 1987)

Ethylene oxide [75-21-8] (Vol. 60; 1994)

(NB: Overall evaluation upgraded from 2A to 1 with supporting evidence from other data relevant to the evaluation of carcinogenicity and its mechanisms)

Helicobacter pylori (infection with) (Vol. 61; 1994)

Hepatitis B virus (chronic infection with) (Vol. 59; 1994)

Hepatitis C virus (chronic infection with) (Vol. 59; 1994)

Human immunodeficiency virus type 1 (infection with) (Vol. 67; 1996)

Human papillomavirus type 16 (Vol. 64; 1995)

Human papillomavirus type 18 (Vol. 64; 1995)

Human T-cell lymphotropic virus type I (Vol. 67; 1996)

Melphalan [148-82-3] (Vol. 9, Suppl. 7; 1987)

8-Methoxypsoralen (Methoxsalen) [298-81-7] plus ultraviolet A radiation (Vol. 24, Suppl. 7; 1987)

MOPP and other combined chemotherapy including alkylating agents (Suppl. 7; 1987)

Mustard gas (Sulfur mustard) [505-60-2] (Vol. 9, Suppl. 7; 1987)

2-Naphthylamine [91-59-8] (Vol. 4, Suppl. 7; 1987)

Nickel compounds (Vol. 49; 1990)

(NB: Evaluated as a group)

Oestrogen replacement therapy (Suppl. 7; 1987)

Estrogens, nonsteroidal (Suppl. 7; 1987)

(NB: This evaluation applies to the group of compounds as a whole and not necessarily to all individual compounds within the group)

Estrogens, steroidal (Suppl. 7; 1987)

(NB: This evaluation applies to the group of compounds as a whole and not necessarily to all individual compounds within the group)

Opisthorchis viverrini (infection with) (Vol. 61; 1994)

Oral contraceptives, combined (Suppl. 7; 1987)

(NB: There is also conclusive evidence that these agents have a protective effect against cancers of the ovary and endometrium)

Oral contraceptives, sequential (Suppl. 7; 1987)

Radon [10043-92-2] and its decay products (Vol. 43; 1988)

Schistosoma haematobium (infection with) (Vol. 61; 1994)

Silica [14808-60-7], crystalline (inhaled in the form of quartz or cristobalite from occupational sources) (Vol. 68; 1997)

Solar radiation (Vol. 55; 1992)

Talc-containing asbestiform fibres (Vol. 42, Suppl. 7; 1987)

Tamoxifen [10540-29-1] (Vol. 66; 1996)

(NB: There is also conclusive evidence that this agent (tamoxifen) reduces the risk of contralateral breast cancer)

2,3,7,8-Tetrachlorodibenzo-*para*-dioxin [1746-01-6] (Vol. 69; 1997)

(NB: Overall evaluation upgraded from 2A to 1 with supporting evidence from other data relevant to the evaluation of carcinogenicity and its mechanisms)

Thiotepa [52-24-4] (Vol. 50; 1990)

Treosulfan [299-75-2] (Vol. 26, Suppl. 7; 1987)

Vinyl chloride [75-01-4] (Vol. 19, Suppl. 7; 1987)

Mixtures

Alcoholic beverages (Vol. 44; 1988)

Analgesic mixtures containing phenacetin (Suppl. 7; 1987)

Betel quid with tobacco (Vol. 37, Suppl. 7; 1987)

Coal-tar pitches [65996-93-2] (Vol. 35, Suppl. 7; 1987)

Coal tars [8007-45-2] (Vol. 35, Suppl. 7; 1987)

Mineral oils, untreated and mildly treated (Vol. 33, Suppl. 7; 1987)

Salted fish (Chinese-style) (Vol. 56; 1993)

Shale-oils [68308-34-9] (Vol. 35, Suppl. 7; 1987)

Soots (Vol. 35, Suppl. 7; 1987)

Tobacco products, smokeless (Vol. 37, Suppl. 7; 1987)

Tobacco smoke (Vol. 38, Suppl. 7; 1987)

Wood dust (Vol. 62; 1995)

Exposure circumstances

Aluminium production (Vol. 34, Suppl. 7; 1987)

Auramine, manufacture of (Suppl. 7; 1987)

Boot and shoe manufacture and repair (Vol. 25, Suppl. 7; 1987)

Coal gasification (Vol. 34, Suppl. 7; 1987)

Coke production (Vol. 34, Suppl. 7; 1987)

Furniture and cabinet making (Vol. 25, Suppl. 7; 1987)

Hematite mining (underground) with exposure to radon (Vol. 1, Suppl. 7; 1987)

Iron and steel founding (Vol. 34, Suppl. 7; 1987)

Isopropanol manufacture (strong-acid process) (Suppl. 7; 1987)

Magenta, manufacture of (Vol. 57; 1993)

Painter (occupational exposure as a) (Vol. 47; 1989)

Rubber industry (Vol. 28, Suppl. 7; 1987)

Strong inorganic acid mists containing sulfuric acid (occupational exposure to) (Vol. 54; 1992)

Group 2A PROBABLY CARCINOGENIC TO HUMANS

Agents and groups of agents

Acrylamide [79-06-1] (Vol. 60; 1994)

(NB: Overall evaluation upgraded from 2B to 2A with supporting evidence from other data relevant to the evaluation of carcinogenicity and its mechanisms)

Adriamycin [23214-92-8] (Vol. 10, Suppl. 7; 1987)

(NB: Overall evaluation upgraded from 2B to 2A with supporting evidence from other data relevant to the evaluation of carcinogenicity and its mechanisms)

Androgenic (anabolic) steroids (Suppl. 7; 1987)

Azacitidine [320-67-2] (Vol. 50; 1990)

(NB: Overall evaluation upgraded from 2B to 2A with supporting evidence from other data relevant to the evaluation of carcinogenicity and its mechanisms)

Benz[*a*]anthracene [56-55-3] (Vol. 32, Suppl. 7; 1987)

(NB: Overall evaluation upgraded from 2B to 2A with supporting evidence from other data relevant to the evaluation of carcinogenicity and its mechanisms)

Benzidine-based dyes (Suppl. 7; 1987)

(NB: Overall evaluation upgraded from 2B to 2A with supporting evidence from other data relevant to the evaluation of carcinogenicity and its mechanisms)

Benzo[*a*]pyrene [50-32-8] (Vol. 32, Suppl. 7; 1987)

(NB: Overall evaluation upgraded from 2B to 2A with supporting evidence from other data relevant to the evaluation of carcinogenicity and its mechanisms)

Bischloroethyl nitrosourea (BCNU) [154-93-8] (Vol. 26, Suppl. 7; 1987)

1,3-Butadiene [106-99-0] (Vol. 71; 1998)

Captafol [2425-06-1] (Vol. 53; 1991)

(NB: Overall evaluation upgraded from 2B to 2A with supporting evidence from other data relevant to the evaluation of carcinogenicity and its mechanisms)

Chloramphenicol [56-75-7] (Vol. 50; 1990)

(NB: Overall evaluation upgraded from 2B to 2A with supporting evidence from other data relevant to the evaluation of carcinogenicity and its mechanisms)

α-Chlorinated toluenes (benzal chloride, benzotrichloride, benzyl chloride) and benzoyl chloride (combined exposures) (Vol. 29, Suppl. 7, Vol. 71; 1998)

1-(2-Chloroethyl)-3-cyclohexyl-1-nitrosourea (CCNU) [13010-47-4] (Vol. 26, Suppl. 7; 1987)

(NB: Overall evaluation upgraded from 2B to 2A with supporting evidence from other data relevant to the evaluation of carcinogenicity and its mechanisms)

para-Chloro-*ortho*-toluidine [95-69-2] and its strong acid salts (Vol. 48; 1990)

(NB: Evaluated as a group) Chlorozotocin [54749-90-5] (Vol. 50; 1990)

(NB: Overall evaluation upgraded from 2B to 2A with supporting evidence from other data relevant to the evaluation of carcinogenicity and its mechanisms)

Cisplatin [15663-27-1] (Vol. 26, Suppl. 7; 1987)

(NB: Overall evaluation upgraded from 2B to 2A with supporting evidence from other data relevant to the evaluation of carcinogenicity and its mechanisms)

Clonorchis sinensis (infection with) (Vol. 61; 1994)

(NB: Overall evaluation upgraded from 2B to 2A with supporting evidence from other data relevant to the evaluation of carcinogenicity and its mechanisms)

Dibenz[*a,h*]anthracene [53-70-3] (Vol. 32, Suppl. 7; 1987)

(NB: Overall evaluation upgraded from 2B to 2A with supporting evidence from other data relevant to the evaluation of carcinogenicity and its mechanisms)

Diethyl sulfate [64-67-5] (Vol. 54, Vol. 71; 1998)

Dimethylcarbamoyl chloride [79-44-7] (Vol. 12, Suppl. 7, Vol. 71; 1998)

(NB: Overall evaluation upgraded from 2B to 2A with supporting evidence from other data relevant to the evaluation of carcinogenicity and its mechanisms)

1,2-Dimethylhydrazine [540-73-8] (Vol. 4, Vol. 71; 1998)

Dimethyl sulfate [77-78-1] (Vol. 4, Suppl. 7, Vol. 71; 1998)

(NB: Overall evaluation upgraded from 2B to 2A with supporting evidence from other data relevant to the evaluation of carcinogenicity and its mechanisms)

Epichlorohydrin [106-89-8] (Vol. 11, Suppl. 7, Vol. 71; 1998)

(NB: Overall evaluation upgraded from 2B to 2A with supporting evidence from other data relevant to the evaluation of carcinogenicity and its mechanisms)

Ethylene dibromide [106-93-4] (Vol. 15, Suppl. 7, Vol. 71; 1998)

(NB: Overall evaluation upgraded from 2B to 2A with supporting evidence from other data relevant to the evaluation of carcinogenicity and its mechanisms)

N-Ethyl-*N*-nitrosourea [759-73-9] (Vol. 17, Suppl. 7; 1987)

(NB: Overall evaluation upgraded from 2B to 2A with supporting evidence from other data relevant to the evaluation of carcinogenicity and its mechanisms)

Formaldehyde [50-00-0] (Vol. 62; 1995)

Human papillomavirus type 31 (Vol. 64; 1995)

Human papillomavirus type 33 (Vol. 64; 1995)

IQ (2-Amino-3-methylimidazo[4,5-f]quinoline) [76180-96-6] (Vol. 56; 1993)

(NB: Overall evaluation upgraded from 2B to 2A with supporting evidence from other data relevant to the evaluation of carcinogenicity and its mechanisms)

Kaposi's sarcoma herpesvirus/human herpesvirus 8 (Vol. 70; 1997)

5-Methoxypsoralen [484-20-8] (Vol. 40, Suppl. 7; 1987)

(NB: Overall evaluation upgraded from 2B to 2A with supporting evidence from other data relevant to the evaluation of carcinogenicity and its mechanisms)

4,4′-Methylene bis(2-chloroaniline) (MOCA) [101-14-4] (Vol. 57; 1993)

(NB: Overall evaluation upgraded from 2B to 2A with supporting evidence from other data relevant to the evaluation of carcinogenicity and its mechanisms)

Methyl methanesulfonate [66-27-3] (Vol. 7, Vol. 71; 1998)

N-Methyl-N′-nitro-N-nitrosoguanidine (MNNG) [70-25-7] (Vol. 4, Suppl. 7; 1987)

(NB: Overall evaluation upgraded from 2B to 2A with supporting evidence from other data relevant to the evaluation of carcinogenicity and its mechanisms)

N-Methyl-N-nitrosourea [684-93-5] (Vol. 17, Suppl. 7; 1987)

(NB: Overall evaluation upgraded from 2B to 2A with supporting evidence from other data relevant to the evaluation of carcinogenicity and its mechanisms)

Nitrogen mustard [51-75-2] (Vol. 9, Suppl. 7; 1987)

N-Nitrosodiethylamine [55-18-5] (Vol. 17, Suppl. 7; 1987)

(NB: Overall evaluation upgraded from 2B to 2A with supporting evidence from other data relevant to the evaluation of carcinogenicity and its mechanisms)

N-Nitrosodimethylamine [62-75-9] (Vol. 17, Suppl. 7; 1987)

(NB: Overall evaluation upgraded from 2B to 2A with supporting evidence from other data relevant to the evaluation of carcinogenicity and its mechanisms)

Phenacetin [62-44-2] (Vol. 24, Suppl. 7; 1987)

Procarbazine hydrochloride [366-70-1] (Vol. 26, Suppl. 7; 1987)

(NB: Overall evaluation upgraded from 2B to 2A with supporting evidence from other data relevant to the evaluation of carcinogenicity and its mechanisms)

Styrene-7,8-oxide [96-09-3] (Vol. 60; 1994)

(NB: Overall evaluation upgraded from 2B to 2A with supporting evidence from other data relevant to the evaluation of carcinogenicity and its mechanisms)

Tetrachloroethylene [127-18-4] (Vol. 63; 1995)

Trichloroethylene [79-01-6] (Vol. 63; 1995)

1,2,3-Trichloropropane [96-18-4] (Vol. 63; 1995)

Tris(2,3-dibromopropyl)phosphate [126-72-7] (Vol. 20, Suppl. 7, Vol. 71; 1998)

(NB: Overall evaluation upgraded from 2B to 2A with supporting evidence from other data relevant to the evaluation of carcinogenicity and its mechanisms)

Ultraviolet radiation A (Vol. 55; 1992)

(NB: Overall evaluation upgraded from 2B to 2A with supporting evidence from other data relevant to the evaluation of carcinogenicity and its mechanisms)

Ultraviolet radiation B (Vol. 55; 1992)

(NB: Overall evaluation upgraded from 2B to 2A with supporting evidence from other data relevant to the evaluation of carcinogenicity and its mechanisms)

Ultraviolet radiation C (Vol. 55; 1992)

(NB: Overall evaluation upgraded from 2B to 2A with supporting evidence from other data relevant to the evaluation of carcinogenicity and its mechanisms)

Vinyl bromide [593-60-2] (Vol. 39, Suppl. 7, Vol. 71; 1998)

(NB: Overall evaluation upgraded from 2B to 2A with supporting evidence from other data relevant to the evaluation of carcinogenicity and its mechanisms)

Vinyl fluoride [75-02-5] (Vol. 63; 1995)

Mixtures

Creosotes [8001-58-9] (Vol. 35, Suppl. 7; 1987)

Diesel engine exhaust (Vol. 46; 1989)

Hot mate (Vol. 51; 1991)

Non arsenical insecticides (occupational exposures in spraying and application of) (Vol. 53; 1991)

Polychlorinated biphenyls [1336-36-3] (Vol. 18, Suppl. 7; 1987)

Exposure circumstances

Art glass, glass containers and pressed ware (manufacture of) (Vol. 58; 1993)

Hairdresser or barber (occupational exposure as a) (Vol. 57; 1993)

Petroleum refining (occupational exposures in) (Vol. 45; 1989)

Sunlamps and sunbeds (use of) (Vol. 55; 1992)

Group 2B: POSSIBLY CARCINOGENIC TO HUMANS

Agents and groups of agents

A-α-C (2-Amino-9H-pyrido[2,3-b]indole) [26148-68-5] (Vol. 40, Suppl. 7; 1987)

Acetaldehyde [75-07-0] (Vol. 36, Suppl. 7, Vol. 71; 1998)

Acetamide [60-35-5] (Vol. 7, Suppl. 7, Vol. 71; 1998)

Acrylonitrile [107-13-1] (Vol. 71; 1998)

AF-2 [2-(2-Furyl)-3-(5-nitro-2-furyl)acrylamide] [3688-53-7] (Vol. 31, Suppl. 7; 1987)

Aflatoxin M1 [6795-23-9] (Vol. 56; 1993)

para-Aminoazobenzene [60-09-3] (Vol. 8, Suppl. 7; 1987)

ortho-Aminoazotoluene [97-56-3] (Vol. 8, Suppl. 7; 1987)

2-Amino-5-(5-nitro-2-furyl)-1,3,4-thiadiazole [712-68-5] (Vol. 7, Suppl. 7; 1987)

Amitrole [61-82-5] (Vol. 41, Suppl. 7; 1987)

ortho-Anisidine [90-04-0] (Vol. 27, Suppl. 7; 1987)

Antimony trioxide [1309-64-4] (Vol. 47; 1989)

Aramite® [140-57-8] (Vol. 5, Suppl. 7; 1987)

Atrazine [1912-24-9] (Vol. 53; 1991)

(NB: Overall evaluation upgraded from 3 to 2B with supporting evidence from other data relevant to the evaluation of carcinogenicity and its mechanisms)

Auramine [492-80-8] (technical-grade) (Vol. 1, Suppl. 7; 1987)

Azaserine [115-02-6] (Vol. 10, Suppl. 7; 1987)

Aziridine [151-56-4] (Vol. 9, Vol. 71; 1998)

Benzo[*b*]fluoranthene [205-99-2] (Vol. 32, Suppl. 7; 1987)

Benzo[*j*]fluoranthene [205-82-3] (Vol. 32, Suppl. 7; 1987)

Benzo[*k*]fluoranthene [207-08-9] (Vol. 32, Suppl. 7; 1987)

Benzofuran [271-89-6] (Vol. 63; 1995)

Benzyl violet 4B [1694-09-3] (Vol. 16, Suppl. 7; 1987)

Bleomycins [11056-06-7] (Vol. 26, Suppl. 7; 1987)

(NB: Overall evaluation upgraded from 3 to 2B with supporting evidence from other data relevant to the evaluation of carcinogenicity and its mechanisms)

Bracken fern (Vol. 40, Suppl. 7; 1987)

Bromodichloromethane [75-27-4] (Vol. 52, Vol. 71; 1998)

Butylated hydroxyanisole (BHA) [25013-16-5] (Vol. 40, Suppl. 7; 1987)

β-Butyrolactone [3068-88-0] (Vol. 11, Suppl. 7, Vol. 71; 1998)

Caffeic acid [331-39-5] (Vol. 56; 1993)

Carbon black [1333-86-4] (Vol. 65; 1996)

Carbon tetrachloride [56-23-5] (Vol. 20, Suppl. 7, Vol. 71; 1998)

Catechol [120-80-9] (Vol. 15, Vol. 71; 1998)

Ceramic fibres (Vol. 43; 1988)

Chlordane [57-74-9] (Vol. 53; 1991)

Chlordecone (Kepone) [143-50-0] (Vol. 20, Suppl. 7; 1987)

Chlorendic acid [115-28-6] (Vol. 48; 1990)

para-Chloroaniline [106-47-8] (Vol. 57; 1993)

Chloroform [67-66-3] (Vol. 20, Suppl. 7; 1987)

1-Chloro-2-methylpropene [513-37-1] (Vol. 63; 1995)

Chlorophenoxy herbicides (Vol. 41, Suppl. 7; 1987)

4-Chloro-*ortho*-phenylenediamine [95-83-0] (Vol. 27, Suppl. 7; 1987)

Chloroprene [126-99-8] (Vol. 71; 1998)

CI Acid Red 114 [6459-94-5] (Vol. 57; 1993)

CI Basic Red 9 [569-61-9] (Vol. 57; 1993)

CI Direct Blue 15 [2429-74-5] (Vol. 57; 1993)

Citrus Red No. 2 [6358-53-8] (Vol. 8, Suppl. 7; 1987)

Cobalt [7440-48-4] and cobalt compounds (Vol. 52; 1991)

(NB: Evaluated as a group)

para-Cresidine [120-71-8] (Vol. 27, Suppl. 7; 1987)

Cycasin [14901-08-7] (Vol. 10, Suppl. 7; 1987)

Dacarbazine [4342-03-4] (Vol. 26, Suppl. 7; 1987)

Dantron (Chrysazin; 1,8-Dihydroxyanthraquinone) [117-10-2] (Vol. 50; 1990)

Daunomycin [20830-81-3] (Vol. 10, Suppl. 7; 1987)

DDT [*p,p'*-DDT, 50-29-3] (Vol. 53; 1991)

N,N'-Diacetylbenzidine [613-35-4] (Vol. 16, Suppl. 7; 1987)

2,4-Diaminoanisole [615-05-4] (Vol. 27, Suppl. 7; 1987)

4,4'-Diaminodiphenyl ether [101-80-4] (Vol. 29, Suppl. 7; 1987)

2,4-Diaminotoluene [95-80-7] (Vol. 16, Suppl. 7; 1987)

Dibenz[*a,h*]acridine [226-36-8] (Vol. 32, Suppl. 7; 1987)

Dibenz[*a,j*]acridine [224-42-0] (Vol. 32, Suppl. 7; 1987)

7H-Dibenzo[*c,g*]carbazole [194-59-2] (Vol. 32, Suppl. 7; 1987)

Dibenzo[*a,e*]pyrene [192-65-4] (Vol. 32, Suppl. 7; 1987)

Dibenzo[*a,h*]pyrene [189-64-0] (Vol. 32, Suppl. 7; 1987)

Dibenzo[*a,i*]pyrene [189-55-9] (Vol. 32, Suppl. 7; 1987)

Dibenzo[*a,l*]pyrene [191-30-0] (Vol. 32, Suppl. 7; 1987)

1,2-Dibromo-3-chloropropane [96-12-8] (Vol. 20, Suppl. 7, Vol. 71; 1998)

para-Dichlorobenzene [106-46-7] (Vol. 29, Suppl. 7; 1987)

3,3'-Dichlorobenzidine [91-94-1] (Vol. 29, Suppl. 7; 1987)

3,3'-Dichloro-4,4'-diaminodiphenyl ether [28434-86-8] (Vol. 16, Suppl. 7; 1987)

1,2-Dichloroethane [107-06-2] (Vol. 20, Vol. 71; 1998)

Dichloromethane (methylene chloride) [75-09-2] (Vol. 71; 1998)

1,3-Dichloropropene [542-75-6] (technical grade) (Vol. 41, Suppl. 7, Vol. 71; 1998)

Dichlorvos [62-73-7] (Vol. 53; 1991)

Di(2-ethylhexyl)phthalate [117-81-7] (Vol. 29, Suppl. 7; 1987)

1,2-Diethylhydrazine [1615-80-1] (Vol. 4, Vol. 71; 1998)

Diglycidyl resorcinol ether [101-90-6] (Vol. 36, Vol. 71; 1998)

Dihydrosafrole [94-58-6] (Vol. 10, Suppl. 7; 1987)

Diisopropyl sulfate [2973-10-6] (Vol. 54, Vol. 71; 1998)

3,3'-Dimethoxybenzidine (*ortho*-Dianisidine) [119-90-4] (Vol. 4, Suppl. 7; 1987)

para-Dimethylaminoazobenzene [60-11-7] (Vol. 8, Suppl. 7; 1987) *trans*-2-[(Dimethylamino)methylimino]-5-[2-(5-nitro-2-furyl)-vinyl]-1,3,4-oxadiazole [25962-77-0] (Vol. 7, Suppl. 7; 1987)

2,6-Dimethylaniline (2,6-Xylidine) [87-62-7] (Vol. 57; 1993)

3,3′-Dimethylbenzidine   (*ortho*-Tolidine) [119-93-7] (Vol. 1, Suppl. 7; 1987)

1,1-Dimethylhydrazine   [57-14-7]   (Vol. 4, Vol. 71; 1998)

3,7-Dinitrofluoranthene [105735-71-5] (Vol. 65; 1996)

3,9-Dinitrofluoranthene [22506-53-2] (Vol. 65; 1996)

1,6-Dinitropyrene   [42397-64-8]   (Vol. 46; 1989)

1,8-Dinitropyrene   [42397-65-9]   (Vol. 46; 1989)

2,4-Dinitrotoluene   [121-14-2]   (Vol. 65; 1996)

2,6-Dinitrotoluene   [606-20-2]   (Vol. 65; 1996)

1,4-Dioxane [123-91-1] (Vol. 11, Suppl. 7, Vol. 71; 1998)

Disperse Blue 1 [2475-45-8] (Vol. 48; 1990)

1,2-Epoxybutane [106-88-7] (Vol. 47, Vol. 71; 1998)

(NB: Overall evaluation upgraded from 3 to 2B with supporting evidence from other data relevant to the evaluation of carcinogenicity and its mechanisms)

Ethyl acrylate [140-88-5] (Vol. 39, Suppl. 7, Vol. 71; 1998)

Ethylene thiourea [96-45-7] (Vol. 7, Suppl. 7; 1987)

Ethyl methanesulfonate [62-50-0] (Vol. 7, Suppl. 7; 1987)

2-(2-Formylhydrazino)-4-(5-nitro-2-furyl) thiazole [3570-75-0] (Vol. 7, Suppl. 7; 1987)

Furan [110-00-9] (Vol. 63; 1995)

Glasswool (Vol. 43; 1988)

Glu-P-1   (2-Amino-6-methyldipyrido[1,2-*α*: 3′,2′-d]imidazole) [67730-11-4] (Vol. 40, Suppl. 7; 1987)

Glu-P-2 (2-Aminodipyrido[1,2-*α*:3′,2′-d]imidazole) [67730-10-3] (Vol. 40, Suppl. 7; 1987)

Glycidaldehyde [765-34-4] (Vol. 11, Vol. 71, 1998)

Griseofulvin [126-07-8] (Vol. 10, Suppl. 7; 1987)

HC Blue No. 1 [2784-94-3] (Vol. 57; 1993)

Heptachlor [76-44-8] (Vol. 53; 1991)

Hexachlorobenzene   [118-74-1]   (Vol. 20, Suppl. 7; 1987)

Hexachlorocyclohexanes (Vol. 20, Suppl. 7; 1987)

Hexamethylphosphoramide [680-31-9] (Vol. 15, Vol. 71; 1998)

Human immunodeficiency virus type 2 (infection with) (Vol. 67; 1996)

Human papillomaviruses: some types other than 16, 18, 31 and 33 (Vol. 64; 1995)

Hydrazine [302-01-2] (Vol. 4, Suppl. 7, Vol. 71; 1998)

Indeno[1,2,3-cd]pyrene [193-39-5] (Vol. 32, Suppl. 7; 1987)

Iron-dextran complex [9004-66-4] (Vol. 2, Suppl. 7; 1987)

Isoprene [78-79-5] (Vol. 60, Vol. 71; 1998)

Lasiocarpine [303-34-4] (Vol. 10, Suppl. 7; 1987)

Lead [7439-92-1] and lead compounds, inorganic (Vol. 23, Suppl. 7; 1987)

(NB: Evaluated as a group)

Magenta [632-99-5] (containing CI Basic Red 9) (Vol. 57; 1993)

MeA-*α*-C   (2-Amino-3-methyl-9H-pyrido[2, 3-b]indole) [68006-83-7] (Vol. 40, Suppl. 7; 1987)

Medroxyprogesterone acetate [71-58-9] (Vol. 21, Suppl. 7; 1987)

MeIQ (2-Amino-3,4-dimethylimidazo[4,5-f] quinoline) [77094-11-2] (Vol. 56; 1993)

MeIQx (2-Amino-3,8-dimethylimidazo[4,5-f]quinoxaline)   [77500-04-0]   (Vol. 56; 1993)

Merphalan [531-76-0]   (Vol. 9, Suppl. 7; 1987)

2-Methylaziridine (Propyleneimine) [75-55-8] (Vol. 9, Vol. 71; 1998)

Methylazoxymethanol acetate [592-62-1] (Vol. 10, Suppl. 7; 1987)

5-Methylchrysene [3697-24-3] (Vol. 32, Suppl. 7; 1987)

4,4′-Methylene bis(2-methylaniline) [838-88-0] (Vol. 4, Suppl. 7; 1987)

4,4′-Methylenedianiline [101-77-9] (Vol. 39, Suppl. 7; 1987)

Methylmercury compounds (Vol. 58; 1993) (NB: Evaluated as a group)

2-Methyl-1-nitroanthraquinone [129-15-7] (uncertain purity) (Vol. 27, Suppl. 7; 1987)

N-Methyl-N-nitrosourethane [615-53-2] (Vol. 4, Suppl. 7; 1987)

Methylthiouracil [56-04-2] (Vol. 7, Suppl. 7; 1987)

Metronidazole [443-48-1] (Vol. 13, Suppl. 7; 1987)

Mirex [2385-85-5] (Vol. 20, Suppl. 7; 1987)

Mitomycin C [50-07-7] (Vol. 10, Suppl. 7; 1987)

Monocrotaline [315-22-0] (Vol. 10, Suppl. 7; 1987)

5-(Morpholinomethyl)-3-[(5-nitrofurfurylidene)amino]-2-oxazolidinone [3795-88-8] (Vol. 7, Suppl. 7; 1987)

Nafenopin [3771-19-5] (Vol. 24, Suppl. 7; 1987)

Nickel, metallic [7440-02-0] and alloys (Vol. 49; 1990)

Niridazole [61-57-4] (Vol. 13, Suppl. 7; 1987)

Nitrilotriacetic acid [139-13-9] and its salts (Vol. 48; 1990) (NB: Evaluated as a group)

5-Nitroacenaphthene [602-87-9] (Vol. 16, Suppl. 7; 1987)

2-Nitroanisole [91-23-6] (Vol. 65; 1996)

Nitrobenzene [98-95-3] (Vol. 65; 1996)

6-Nitrochrysene [7496-02-8] (Vol. 46; 1989)

Nitrofen [1836-75-5] (technical-grade) (Vol. 30, Suppl. 7; 1987)

2-Nitrofluorene [607-57-8] (Vol. 46; 1989)

1-[(5-Nitrofurfurylidene)amino]-2-imidazolidinone [555-84-0] (Vol. 7, Suppl. 7; 1987)

N-[4-(5-Nitro-2-furyl)-2-thiazolyl]acetamide [531-82-8] (Vol. 7, Suppl. 7; 1987)

Nitrogen mustard N-oxide [126-85-2] (Vol. 9, Suppl. 7; 1987)

2-Nitropropane [79-46-9] (Vol. 29, Vol. 71; 1998)

1-Nitropyrene [5522-43-0] (Vol. 46; 1989)

4-Nitropyrene [57835-92-4] (Vol. 46; 1989)

N-Nitrosodi-n-butylamine [924-16-3] (Vol. 17, Suppl. 7; 1987)

N-Nitrosodiethanolamine [1116-54-7] (Vol. 17, Suppl. 7; 1987)

N-Nitrosodi-n-propylamine [621-64-7] (Vol. 17, Suppl. 7; 1987)

3-(N-Nitrosomethylamino)propionitrile [60153-49-3] (Vol. 37, Suppl. 7; 1987)

4-(N-Nitrosomethylamino)-1-(3-pyridyl)-1-butanone (NNK) [64091-91-4] (Vol. 37, Suppl. 7; 1987)

N-Nitrosomethylethylamine [10595-95-6] (Vol. 17, Suppl. 7; 1987)

N-Nitrosomethylvinylamine [4549-40-0] (Vol. 17, Suppl. 7; 1987)

N-Nitrosomorpholine [59-89-2] (Vol. 17, Suppl. 7; 1987)

N′-Nitrosonornicotine [16543-55-8] (Vol. 37, Suppl. 7; 1987)

N-Nitrosopiperidine [100-75-4] (Vol. 17, Suppl. 7; 1987)

N-Nitrosopyrrolidine [930-55-2] (Vol. 17, Suppl. 7; 1987)

N-Nitrososarcosine [13256-22-9] (Vol. 17, Suppl. 7; 1987)

Ochratoxin A [303-47-9] (Vol. 56; 1993)

Oil Orange SS [2646-17-5] (Vol. 8, Suppl. 7; 1987)

Oxazepam [604-75-1] (Vol. 66; 1996)

Palygorskite (attapulgite) [12174-11-7] (long fibres, >5 micrometers) (Vol. 68; 1997)

Panfuran S (containing dihydroxymethylfuratrizine [794-93-4]) (Vol. 24, Suppl. 7; 1987)

Phenazopyridine hydrochloride [136-40-3] (Vol. 24, Suppl. 7; 1987)

Phenobarbital [50-06-6] (Vol. 13, Suppl. 7; 1987)

Phenoxybenzamine hydrochloride [63-92-3] (Vol. 24, Suppl. 7; 1987)

Phenyl glycidyl ether [122-60-1] (Vol. 47, Vol. 71; 1998)

Phenytoin [57-41-0] (Vol. 66; 1996)

PhIP (2-Amino-1-methyl-6-phenylimidazo [4,5-b]pyridine) [105650-23-5] (Vol. 56; 1993)

Polychlorophenols and their sodium salts (mixed exposures) (Vol. 41, Suppl. 7, Vol. 53, Vol. 71; 1998)

Ponceau MX [3761-53-3] (Vol. 8, Suppl. 7; 1987)

Ponceau 3R [3564-09-8] (Vol. 8, Suppl. 7; 1987)

Potassium bromate [7758-01-2] (Vol. 40, Suppl. 7; 1987)

Progestins (Suppl. 7; 1987)

1,3-Propane sultone [1120-71-4] (Vol. 4, Vol. 71; 1998)

β-Propiolactone [57-57-8] (Vol. 4, Vol. 71; 1998)

Propylene oxide [75-56-9] (Vol. 60; 1994)

Propylthiouracil [51-52-5] (Vol. 7, Suppl. 7; 1987)

Rockwool (Vol. 43; 1988)

Saccharin [81-07-2] (Vol. 22, Suppl. 7; 1987)

Safrole [94-59-7] (Vol. 10, Suppl. 7; 1987)

Schistosoma japonicum (infection with) (Vol. 61; 1994)

Slagwool (Vol. 43; 1988)

Sodium *ortho*-phenylphenate [132-27-4] (Vol. 30, Suppl. 7; 1987)

Sterigmatocystin [10048-13-2] (Vol. 10, Suppl. 7; 1987)

Streptozotocin [18883-66-4] (Vol. 17, Suppl. 7; 1987)

Styrene [100-42-5] (Vol. 60; 1994)

(NB: Overall evaluation upgraded from 3 to 2B with supporting evidence from other data relevant to the evaluation of carcinogenicity and its mechanisms)

Sulfallate [95-06-7] (Vol. 30, Suppl. 7; 1987)

Tetrafluoroethylene [116-14-3] (Vol. 19, Vol. 71; 1998)

Tetranitromethane [509-14-8] (Vol. 65; 1996)

Thioacetamide [62-55-5] (Vol. 7, Suppl. 7; 1987)

4,4′-Thiodianiline [139-65-1] (Vol. 27, Suppl. 7; 1987)

Thiourea [62-56-6] (Vol. 7, Suppl. 7; 1987)

Toluene diisocyanates [26471-62-5] (Vol. 39, Vol. 71; 1998)

ortho-Toluidine [95-53-4] (Vol. 27, Suppl. 7; 1987)

Toxins derived from *Fusarium moniliforme* (Vol. 56; 1993)

Trichlormethine (Trimustine hydrochloride) [817-09-4] (Vol. 50; 1990)

Trp-P-1 (3-Amino-1,4-dimethyl-5H-pyrido [4,3-b]indole) [62450-06-0] (Vol. 31, Suppl. 7; 1987)

Trp-P-2 (3-Amino-1-methyl-5H-pyrido[4,3-b]indole) [62450-07-1] (Vol. 31, Suppl. 7; 1987)

Trypan blue [72-57-1] (Vol. 8, Suppl. 7; 1987)

Uracil mustard [66-75-1] (Vol. 9, Suppl. 7; 1987)

Urethane [51-79-6] (Vol. 7, Suppl. 7; 1987)

Vinyl acetate [108-05-4] (Vol. 63; 1995)

4-Vinylcyclohexene [100-40-3] (Vol. 60; 1994)

4-Vinylcyclohexene diepoxide [106-87-6] (Vol. 60; 1994)

Mixtures

Bitumens [8052-42-4], extracts of steam-refined and air-refined (Vol. 35, Suppl. 7; 1987)

Carrageenan [9000-07-1], degraded (Vol. 31, Suppl. 7; 1987)

Chlorinated paraffins of average carbon chain length C12 and average degree of chlorination approximately 60% (Vol. 48; 1990)

Coffee (urinary bladder) (Vol. 51; 1991)

(NB: There is some evidence of an inverse relationship between coffee drinking and cancer of the large bowel; coffee drinking could not be classified as to its carcinogenicity to other organs)

Diesel fuel, marine (Vol. 45; 1989)

(NB: Overall evaluation upgraded from 3 to 2B with supporting evidence from other data relevant to the evaluation of carcinogenicity and its mechanisms)

Engine exhaust, gasoline (Vol. 46; 1989)

Fuel oils, residual (heavy) (Vol. 45; 1989)

Gasoline (Vol. 45; 1989)

(NB: Overall evaluation upgraded from 3 to 2B with supporting evidence from other data relevant to the evaluation of carcinogenicity and its mechanisms)

Pickled vegetables (traditional in Asia) (Vol. 56; 1993)

Polybrominated biphenyls [Firemaster BP-6, 59536-65-1] (Vol. 41, Suppl. 7; 1987)

Toxaphene (polychlorinated camphenes) [8001-35-2] (Vol. 20, Suppl. 7; 1987)

Welding fumes (Vol. 49; 1990)

Exposure circumstances

Carpentry and joinery (Vol. 25, Suppl. 7; 1987)

Dry cleaning (occupational exposures in) (Vol. 63; 1995)

Printing processes (occupational exposures in) (Vol. 65; 1996)

Textile manufacturing industry (work in) (Vol. 48; 1990)

APPENDIX C

NTP LIST OF CARCINOGENS

KNOWN CARCINOGENS

For explanation of the terms "known carcinogens" and "reasonably anticipated to be carcinogens", please refer to Section VI (Cancer-causing Chemicals) in Part A of this text.

Substances or groups of substances, occupational exposures associated with a technological process, and medical treatments that are known to be carcinogenic. The following agents are listed as known carcinogens by the National Toxicology Program of the US Department of Health and Human Services in their 7th Annual Report on Carcinogens.

Aflatoxins (CAS No. 1402-68-3)

Aminobiphenyl (CAS No. 92-67-1)

Analgesic Mixtures Containing Phenacetin

Arsenic and Certain Arsenic Compounds

Asbestos (CAS No. 1332-21-4)

Azathioprine (CAS No. 446-86-6)

Benzene (CAS No. 71-43-2)

Benzidine (CAS No. 92-87-5)

Bis(Chloromethyl) Ether and Tech-grade Chloromethyl Methyl Ether (CAS Nos. 542-88-1 and 107-30-2)

1,4-Butanediol Dimethyl-sulfonate (Myleran) (CAS No. 55-98-1)

Chlorambucil (CAS No. 305-03-3)

(2-Chloroethyl)-3-(4-Methylcyclohexyl)-1-Nitrosourea (MeCCNU) CAS No. 13909-09-6)

Chromium and Certain Chromium Compounds

Conjugated Estrogens

Diethylstilbestrol (CAS No. 56-53-1)

Erionite (CAS No. 66733-21-9)

Melphalan (CAS No. 148-82-3)

Methoxsalen with Ultraviolet a Therapy (PUVA)

Mustard Gas (CAS No. 505-60-2)

2-Naphthylamine (CAS No. 91-59-8)

Radon (CAS No. 10043-92-2)

Thorium Dioxide (CAS No. 1314-20-1)

Vinyl Chloride (CAS No. 75-01-4)

Two more compounds have been nominated as known human carcinogens to be included in the NTP's 8th Annual Report on Carcinogens (May 1998). These are Cyclosporin [59865-13-3] and Thiotepa [52-24-4]

REASONABLY ANTICIPATED TO BE CARCINOGENS 7TH ANNUAL REPORT ON CARCINOGENS

The following agents are listed in the 7th Annual Report on Carcinogens of the National Toxicology Program of the US Department of Health and Human Services. For the meaning of the terms and comparison of terminologies with other cancer review panels, please refer to Chapter VI, "Cancer-causing Chemicals," in Part A of this text.

Acetaldehyde (CAS No. 75-07-0)

2-Acetylaminofluorene (CAS No. 53-96-3)

Acrylamide (CAS No. 79-06-1)

Acrylonitrile (CAS No. 107-13-1)

Adriamycin (CAS No. 23214-92-8)

2-Aminoanthraquinone (CAS No. 117-79-3)

o-Aminoazotoluene (CAS No. 97-56-3)

1-Amino-2-Methylanthraquinone (CAS No. 82-28-0)

Amitrole (CAS No. 61-82-5)

o-Anisidine Hydrochloride (CAS No. 134-29-2)

Benzotrichloride (CAS No. 98-07-7)

Beryllium and Certain Beryllium Compounds

Bischloroethyl Nitrosourea (CAS No. 154-93-8)

Bromodichloromethane (CAS No. 75-27-4)

1,3-Butadiene (CAS No. 106-99-0)

Butyulated Hydroxyanisole (CAS No. 25013-16-5)

Cadmium and Certain Cadmium Compounds

Carbon Tetrachloride (CAS No. 56-23-5)

Ceramic Fibers (respirable size)

Chlorendic Acid (CAS No. 115-28-6)

1-(2-Chloroethyl)-3-Cyclohexyl-1-Nitro-sourea (CCNU) (CAS No. 13010-47-4)

Chloroform (CAS No. 67-66-3)

3-Chloro-2-Methylpropene (CAS No. 563-47-3)

4-Chloro-o-Phenylenediamine (CAS No. 95-83-0)

C.I. Basic Red 9 Monohydrochloride (CAS No. 569-61-9)

Cisplatin (CAS No. 15663-27-1)

p-Cresidine (CAS No. 120-71-8)

Cupferron (CAS No. 135-20-6)

Dacarbazine (CAS No. 4342-03-4)

DDT (CAS No. 50-29-3)

2,4-Diaminoanisole Sulfate (CAS No. 39156-41-7)

2,4-Diaminotoluene (CAS No. 95-80-7)

1,2-Dibromo-3-Chloropropane (CAS No. 96-12-8)

1,2-Diromoethane (Ethylene Dibromide) (CAS No. 106-93-4)

1,4-Dichlorobenzene (CAS No. 106-46-7)

3,3'-Dichlorobenzidine and 3,3'-Dichloro-benzidine 2HCl (CAS No. 91-94-1 and 612-83-9)

1,2-Dichloroethane (CAS No. 107-06-2)

Dichloromethane (Methylene Chloride) (CAS No. 75-09-2)

1,3-Dichloropropene (Technical Grade) (CAS No. 542-75-6)

Diepoxybutane (CAS No. 1464-53-5)

DI(2-Ethylhexyl) Phthalate (CAS No. 117-81-7)

Diethyl Sulfate (CAS No. 64-67-5)

Diglycidyl Resorcinol Ether (CAS No. 101-90-6)

3,3'-Dimethoxybenzidine and 3,3'-Dimethoxybenzidine 2 HCl (CAS Nos. 119-90-4 and 20325-40-0)

4-Dimethylamino-Azobenzene (CAS No. 60-11-7)

3,3'-Dimethylbenzidine (CAS No. 119-93-7)

Dimethylcarbamoyl Chloride (CAS No. 79-44-7)

1,1-Dimethylhydrazine (CAS No. 57-14-7)

Dimethyl Sulfate (CAS No. 77-78-1)

Dimethylvinyl Chloride (CAS No. 513-37-1)

1,4-Dioxane (CAS No. 123-91-1)

Direct Black 38 (CAS No. 1937 -37-7)

Direct Blue 6 (CAS No. 2602-46-2)

Estrogens (not conjugated): Estradiol-17β (CAS No. 50-28-2)

Estrogens (not conjugated): Estrone: (CAS No. 53-16-7)

Estrogens (not conjugated): Ethinylestradiol: (CAS No. 57-63-6)

Estrogens (not conjugated): Mestranol: (CAS No. 72-33-3)

Ethyl Acrylate (CAS No. 140-88-5)

Ethylene Oxide (CAS No. 75-21-8)

Ethylene Thiourea (CAS No. 96-45-7)

Ethyl Methanesulfonate (CAS No. 62-50-0)

Formaldehyde (Gas) (CAS No. 50-00-0)

Glasswool (Respirable Size)

Glycidol (CAS No. 556-52-5)

Hexachlorobenzene (CAS No. 118-74-1)

Hexachloroethane (CAS No. 67-72-1)

Hexamethyl-Phosphoramide (CAS No. 680-31-9)

Hydrazine and Hydrazine Sulfate (CAS Nos. 302-01-2 and 10034-93-2)

Hydrazobenzene (CAS No. 122-66-7)

Iron Dextran Complex (CAS No. 9004-66-4)

Kepone (Chlordecone) (CAS No. 143-50-0)

Lead Acetate and Lead Phosphate (CAS Nos. 301-04-2 and 7446-27-7)

Lindane and other Hexachlorocyclohexane Isomers

2-Methylaziridine (Propyleneimine) (CAS No. 75-55-8)

4,4′-Methylenebis(2-Chloroaniline) (MBOCA) (CAS No. 101-14-4)

4,4′-Methylenebis(N,N-Dimethylbenzen-amine) (CAS No. 101-61-1)

4,4′-Methylenedianiline and its Dihydrochloride (CAS No. 101-77-9 and 13552-44-8)

Methyl Methanesulfonate (CAS No. 66-27-3)

N-Methyl-N'-Nitro-N-Nitrosoguanidine (CAS No. 70-25-7)

Metronidazole (CAS No. 443-48-1)

Michler's Ketone (CAS No. 90-94-8)

Mirex (CAS No. 2385-85-5)

Nickel and Ceratin Nickel Compounds

Nitrilotriacetic Acid (CAS No. 139-13-9)

Nitrofen (CAS No. 1836-75-5)

Nitrogen Mustard Hydrochloride (CAS No. 55-86-7)

2-Nitropropane (CAS No. 79-46-9)

N-Nitrosodi-N-Butylamine (CAS No. 924-16-3)

N-Nitrosodiethanolamine (CAS No. 1116-54-7)

N-Nitrosodiethylamine (CAS No. 55-18-5)

N-Nitrosodimethylamine (CAS No. 62-75-9)

N-Nitrosodi-N-Propylamine (CAS No. 621-64-7)

N-Nitroso-N-Ethylurea (CAS No. 759-73-9)

4-(N-Nitrosomethyl-Amino)-1-(3-Pyridyl)-1-Butanone (NNK) (CAS No. 64091-91-4)

N-Nitroso-N-Methylurea (CAS No. 684-93-5)

N-Nitrosomethyl-Vinylamine (CAS No. 4549-40-0)

N-Nitrosomorpholine (CAS No. 59-89-2)

N-Nitrosonornicotine (CAS No. 16543-55-8)

N-Nitrosopiperidine (CAS No. 100-75-4)

N-Nitrosopyrrolidine (CAS No. 930-55-2)

N-Nitrososarcosine (CAS No. 13256-22-9)

Norethisterone (CAS No. 68-22-4)

Ochratoxin A (CAS No. 303-47-9)

4,4′-Oxydianiline (CAS No. 101-80-4)

Oxymetholone (CAS No. 434-07-1)

Phenacetin (CAS No. 62-44-2)

Phenazopyridine Hydrochloride (CAS No. 136-40-3)

Phenoxybenzamine Hydrochloride (CAS No. 63-92-3)

Phenytoin (CAS No. 57-41-0)

Polybrominated Biphenyls

Polychlorinated Biphenyhls

Polycyclic Aromatic Hydrocarbons, 15 Listings

Procarbazine Hydrochloride (CAS No. 366-70-1)

Progesterone (CAS No. 57-83-0)

1,3-Propane Sultone (CAS No. 1120-71-4)

β-Propiolactone (CAS No. 57-57-8)

Propylene Oxide (CAS No. 75-56-9)

Propylthiouracil (CAS No. 51-52-5)

Reserpine (CAS No. 50-55-5)

Saccharin (CAS No. 128-44-9)

Safrole (CAS No. 94-59-7)

Selenium Sulfide (CAS No. 7446-34-6)

Silica, Crystalline (Respirable size)

Streptozotocin (CAS No. 18883-66-4)

Sulfallate (CAS No. 95-06-7)

2,3,7,8-Tetrachlorodibenzo-p-Dioxin (TCDD) (CAS No. 1746-01-6)

Tetrachloroethylene (Perchloroethylene) (CAS No. 127-18-4)

Tetranitromethane (CAS No. 509-14-8)

Thioacetamide (CAS No. 62-55-5)

Thiourea (CAS No. 62-56-6)

Toluene Diisocyanate (CAS No. 26471-62-5)

o-Toluidine and o-Toluidine Hydrochloride (CAS Nos. 95-53-4 and 636-21-5)

Toxaphene (CAS No. 8001-35-2)

2,4,6-Trichlorophenol (CAS No. 88-06-2)

Tris(1-Aziridinyl)Phosphine Sulfide (Thiotepa) (CAS No. 52-24-4)

Tris(2,3-Dibromopropyl) Phosphate (CAS No. 126-72-7)

Urethane (CAS No. 51-79-6)

4-Vinyl-1-Cyclohexene Diepoxide (CAS No. 106-87-6)

The following additional agents have been nominated as "reasonably anticipated to be human carcinogens" for NTP's 8th Annual Report on Carcinogens (May 1998).

Azactidine [320-67-2]

p-Chloro-o-Toluidine [95-69-2] and its hydrochloride

Chlorozotocin [54749-90-5]

Danthron (1,8-Dihydroxyanthraquinone) [117-10-2]

1,6-Dinitropyrene [42397-64-8]

1,8-Dinitropyrene [42397-65-9]

Disperse Blue 1 (1,4,5,8-Tetraaminoanthraquinone) [2475-45-8]

Furan [100-00-9]

o-Nitroanisole [91-23-6]

6-Nitrochrysene [7495-02-8]

1-Nitropyrene [5522-43-0]

4-Nitropyrene [57835-92-4]

1,2,3-Trichloropropane [96-18-4]

CHEMICAL SUBSTANCES–CAS REGISTRY NUMBER INDEX

Substance	CAS #	Page #
Cyclohexene oxide	[286-20-4]	322
Cyclohexyl acetate	[622-45-7]	350
Cyclohexyl alcohol	[108-93-0]	122
Cyclohexyl isocyanate	[3173-53-3]	528
Cyclohexyl ketone	[108-94-1]	548
Cyclohexyl methylphosphonofluoridate	[329-99-7]	639
Cyclohexylamine	[108-91-8]	215, 506
2-Cyclohexyl-4,6-dinitrophenol	[131-89-5]	764
Cyclohexylformaldehyde	[2043-61-0]	161
Cyclonite	[121-82-4]	646–647
Cyclopentadiene	[542-92-7]	474
1,3-Cyclopentadiene	[542-92-7]	474
Cyclopenta[d,e]naphthalene	[208-96-8]	497
Cyclopentane	[287-92-3]	473, 480, 506
Cyclopentanone	[120-92-3]	553
Cyclopentene	[142-29-0]	481
Cyclopentene oxide	[285-67-6]	317
4-Cyclopentene-1,3-dione	[930-60-9]	553
Cyclopentimine	[110-89-4]	455
Cyclophosphamide	[50-18-0]	43, 852
Cyclopropane	[75-19-4]	478
Cyclosporin	[59865-13-3]	43, 852, 862
Cyclotetramethylene	[109-99-9]	459
Cygon	[60-51-5]	737
Cythion	[121-75-5]	745
Cytisine	[485-35-8]	177, 200
2,4-D	[94-75-7]	748
Dacarbazine	[4342-03-4]	857, 863
DAM 57	[4238-84-0]	191
Dantron	[117-10-2]	857, 865
Daunomycin	[20830-81-3]	857
2,4-DB	[94-82-6]	749
p,p'-DDD	[72-54-8]	696, 714, 721
p,p'-DDE	[72-55-9]	696, 714, 722
DDT	[50-29-3]	696, 714, 721, 857, 863
DDT dehydrochloride	[72-55-9]	696, 714, 722
Decaborane	[17702-41-9]	603
Decaboron tetradecahydride	[17702-41-9]	603
Decacarbonyl dimanganese	[10170-69-1]	592
Dechlorane	[2385-85-5]	723
Demeton	[8065-48-3]	729, 740
Demeton-O	[298-03-3]	696
Demeton-S	[126-75-0]	696
Demeton-S-methyl	[919-86-8]	741
Demeton-S-methyl sulfide	[919-86-8]	741
Deserpidine	[131-01-1]	177, 195

Substance	CAS #	Page #
Glu-P-1	[67730-11-4]	858
Glu-P-2	[67730-10-3]	858
Glutaral	[111-30-8]	152
Glutaraldehyde	[111-30-8]	152
Glutaric dialdehyde	[111-30-8]	152
Glutethimide	[77-21-4]	62–63
Glycerol α,β-chlorohydrin	[616-23-9]	265
Glycerol α,β-dichlorohydrin	[96-23-1]	264
Glycerol α-monochlorohydrin	[96-24-2]	261
Glycerol trichlorohydrin	[96-18-4]	426
Glycerol trinitrate	[55-63-0]	645–647
Glyceryl trichlorohydrin	[96-18-4]	426
Glycidal	[765-34-4]	330
Glycidaldehyde	[765-34-4]	330, 858
Glycidol	[556-52-5]	332, 864
Glycidyl alcohol	[556-52-5]	332
Glycol bromohydrin	[540-51-2]	263
Glycol ethyl ether	[110-80-5]	381
Glycol ethylene ether	[123-91-1]	461
Glycol monobutyl ether	[111-76-2]	382
Glycol monochlorohydrin	[107-07-3]	258
Glycolmethyl ether	[109-86-4]	379
Glyconitrile	[107-16-4]	270
Glyoxal	[107-22-2]	152
Glyoxaline	[288-32-4]	457
Gold cyanide	[506-65-0]	300
Grain alcohol	[64-17-5]	112
Graphite, natural	[7782-42-5]	669
Graphite, synthetic	[7440-44-0]	669
Grasex	[75-87-6]	156
Griseofulvin	[126-07-6]	858
Gum camphor	[76-22-2]	552
Guncotton	[9004-70-0]	649
Guthion	[86-50-0]	735
Halocarbon-152A	[75-37-6]	432
Halon-1202	[75-61-6]	435
Halon-1301	[75-63-8]	434
Halothane	[151-67-7]	425
Halowax 1014	[1335-87-1]	431
HC blue 1	[2784-94-3]	858
Hedonal	[94-97-6]	750
Helium	[7440-59-7]	376
Heptachlor	[76-44-8]	696, 713, 719, 858
Heptachlor epoxide	[1024-57-3]	696, 713
Heptafluoroiodine	[16921-96-3]	450
Heptamul	[76-44-8]	719

Substance	CAS #	Page #
Mescaline	[54-04-6]	178, 204
Mesityl oxide	[141-79-7]	539
Mesomule	[16752-77-5]	706
Mestranol	[72-33-3]	864
Mesurol	[2032-65-7]	704
Metachlor	[15972-60-8]	826
Metaldehyde	[37273-91-9]	161
Metaldehyde II	[37273-91-9]	161
Metalkamate	[8065-36-9]	708
Metamidophos	[10265-92-6]	730
Metasystox	[919-86-8]	741
Metasystox forte	[919-86-8]	741
Metathion	[122-14-5]	734
Methabenzthiazuron	[18691-97-9]	825
Methacetone	[96-22-0]	539
Methacrolein	[78-85-3]	158
2-Methacrolein	[78-85-3]	158
Methacrylaldehyde	[78-85-3]	158
Methacrylic acid	[79-41-4]	85
Methacrylic aldehyde	[78-85-3]	158
Methacrylonitrile	[126-98-7]	282
Methaform	[57-15-8]	262
Methamphetamine	[537-46-2]	53
Methanal	[50-00-0]	140
Methanamine	[74-89-5]	211
Methane	[74-82-8]	465
Methanecarbonitrile	[75-05-8]	277
Methanecarboxamide	[60-35-5]	503
Methanecarboxylic acid	[64-19-7]	83
Methanethiol	[74-93-1]	811, 813
Methanoic acid	[64-18-6]	82
Methanol	[67-56-1]	110, 510
Methaqualone	[72-44-6]	62–63
Methenyl tribromide	[75-25-2]	433
Methenyl trichloride	[67-66-3]	414
Methenyl triiodide	[75-47-8]	434
Methiocarb	[2032-65-7]	699, 704
Methohexital sodium	[22151-68-4]	821
Methomyl	[16752-77-5]	699, 705
Methoxane	[76-38-0]	400
2-Methoxyacetaldehyde	[10312-83-1]	158
2-Methoxybenzaldehyde	[135-02-4]	156
4-Methoxybenzaldehyde	[123-11-5]	157
2-Methoxybenzenecarboxaldehyde	[135-02-4]	156
1-Methoxybutane	[628-28-4]	366
Methoxychlor	[72-43-5]	696, 714, 722

Substance	CAS #	Page #
Phosphorus iodide	[13455-01-1]	790
Phosphorus oxybromide	[7789-59-5]	790
Phosphorus oxychloride	[10025-87-3]	781
Phosphorus oxytrichloride	[10025-87-3]	781
Phosphorus pentabromide	[7789-69-7]	790
Phosphorus pentachloride	[10026-13-8]	781
Phosphorus pentafluoride	[7647-19-0]	783
Phosphorus pentaoxide	[1314-56-3]	780
Phosphorus pentasulfide	[1314-80-3]	783
Phosphorus pentoxide	[1314-56-3]	780
Phosphorus perchloride	[10026-13-8]	781
Phosphorus persulfide	[1314-80-3]	784
Phosphorus sesquisulfide	[1314-85-8]	784
Phosphorus sulfide	[1314-80-3]	784
Phosphorus tribromide	[7789-60-8]	789
Phosphorus trichloride	[7719-12-2]	782
Phosphorus trifluoride	[7783-55-3]	790
Phosphorus trihydrate	[7803-51-2]	779
Phosphorus triiodide	[13455-01-1]	790
Phosphorus trioxide	[1314-24-5]	789
Phosphorus triselenide	[1314-86-9]	790
Phosphorus trisulfide	[12165-69-4]	789
Phosphorus(V) oxide	[1314-56-3]	780
Phosphoryl chloride	[10025-87-3]	781
Phosphoryl tribromide	[7789-59-5]	790
Phthalatic acid bis(2-ethylhexyl) ester	[117-81-7]	355
Phthalic acid	[88-99-3]	88
Phthalic acid benzyl butyl ester	[85-68-7]	353
Phthalic acid dibutyl ester	[84-72-2]	353
Phthalic acid diethyl eser	[84-66-2]	353
Phthalic acid dimethyl ester	[131-11-3]	353
Phthalic acid dioctyl ester	[117-84-0]	353
Picloram	[1918-02-1]	825
Picramic acid (dry)	[96-91-3]	645
Picric acid	[88-89-1]	645, 653
Picronitric acid	[88-89-1]	653
Picrylnitromethylamine	[479-45-8]	651
Pilocarpine	[92-13-7]	206
Pimelic ketone	[108-94-1]	548
Piperazine	[110-85-0]	455
Piperidine	[110-89-4]	454
3-(2-Piperidyl)pyridine	[494-52-0]	182
Pirimicarb	[23108-98-2]	700, 705
Pirimiphos-ethyl	[23505-41-1]	733
Pirimor	[23108-98-2]	705
Pivaldehyde	[630-19-3]	159

CAS REGISTRY
NUMBER–CHEMICAL
SUBSTANCES INDEX

CAS #	Substance	Page #
[87-62-7]	2,6-Dimethylaniline	858
[87-65-0]	2,6-Dichlorophenol	762
[87-66-1]	Pyrogallol	768
[87-86-5]	Pentachlorophenol	762, 764, 769
[87-90-1]	Symclosene	754
[88-06-2]	2,4,6-Trichlorophenol	762, 764, 865
[88-75-5]	2-Nitrophenol	764
[88-85-7]	Dinoseb	824
[88-89-1]	Picric acid	645, 653
[88-99-3]	Phthalic acid	88
[90-02-8]	Salicylaldehyde	155
[90-04-0]	*o*-Anisidine	239, 856
[90-30-2]	*N*-Phenyl-α-naphthylamine	240
[90-39-1]	Sparteine	177, 200
[90-69-7]	Lobeline	182
[90-94-8]	Michler's ketone	864
[91-20-3]	Naphthalene	493
[91-22-5]	Quinoline	458
[91-23-6]	*o*-Nitroanisole	859, 865
[91-59-8]	2-Naphthylamine	44, 235, 852, 862
[91-94-1]	3,3'-Dichlorobenzidine	240, 857, 863
[92-13-7]	Pilocarpine	206
[92-44-4]	2,3-Naphthalenediol	128
[92-67-1]	Aminobiphenyl	851, 862
[92-87-5]	Benzidine	237, 851, 862
[93-53-8]	Cumene aldehyde	158
[93-58-3]	Methyl benzoate	347
[93-59-4]	Peroxybenzoic acid	104
[93-72-1]	Silvex	749
[93-76-5]	2,4,5-T	748
[94-17-7]	Bis(*p*-chlorobenzoyl)peroxide	687
[94-36-0]	Benzoyl peroxide	682
[94-58-6]	Dihhdrosafrole	857
[94-59-7]	Safrole	860, 865
[94-75-7]	2,4-D	748
[94-81-5]	MCPB	750
[94-82-6]	2,4-DB	749
[94-97-6]	MCPA	750
[95-06-7]	Sulfallate	860, 865
[95-47-6]	*o*-Xylene	514
[95-48-7]	*o*-Cresol	764, 766
[95-49-8]	2-Chlorotoluene	435
[95-50-1]	1,2-Dichlorobenzene	429
[95-53-4]	*o*-Toluidine	229, 860, 865
[95-54-5]	*o*-Phenylenediamine	231
[95-57-8]	*o*-Chlorophenol	762, 764

CAS #	Substance	Page #
[13256-22-9]	*N*-Nitrososarcosine	864
[13361-32-5]	Allyl cyanoacetate	269
[13424-46-9]	Lead(II) azide	582
[13446-10-1]	Ammonium permanganate	662
[13446-48-5]	Ammonium nitrite	666
[13455-01-1]	Phosphorus triiodide	790
[13463-39-3]	Nickel tetracarbonyl	587
[13463-40-6]	Iron pentacarbonyl	588
[13465-95-7]	Barium perchlorate	659
[13477-00-4]	Barium chlorate	660
[13494-80-9]	Tellurium	624
[13517-00-5]	Copper(I) hydride	605
[13552-44-8]	4,4'-Methylenedianiline 2HCl	864
[13597-97-2]	Beryllium hydride	604
[13637-63-3]	Chlorine pentafluoride	446
[13684-56-5]	Desmedipham	701
[13684-63-4]	Phenmedipham	701
[13718-25-7]	Palladium azide	580
[13746-66-2]	Potassium ferricyanide	300
[13768-00-8]	Actinolite	244
[13770-96-2]	Sodium aluminum hydride	605
[13838-16-9]	Ethrane	399
[13843-59-9]	Ammonium bromate	661
[13863-88-2]	Silver azide	583, 819
[13864-01-2]	Lanthanum hydride	605
[13909-09-6]	1-(2-Chloroethyl)-3-(4-methylcyclo-hexyl)-1-nitrosourea	852, 862
[13939-06-5]	Molybdenum hexacarbonyl	591
[13956-29-1]	Cannabidiol	820
[14018-82-7]	Zinc hydride	605
[14024-00-1]	Vanadium hexacarbonyl	591
[14040-11-0]	Tungsten hexacarbonyl	592
[14063-77-5]	*β,p*-Dichlorocinnamaldehyde	137, 165
[14063-78-6]	*p*-Bromo-*β*-chlorocinnamaldehyde	163
[14063-78-6]	*β*-Chloro-*p*-bromocinnamaldehyde	138
[14063-79-7]	*β*-Chloro-*p*-methoxycinnamaldehyde	138, 164
[14212-30-6]	Cupric azide	584
[14215-28-2]	Zinc azide	580, 819
[14215-28-3]	Cadmium azide	580, 819
[14215-31-7]	Cobalt azide	580
[14215-33-9]	Mercury(II) azide	580
[14336-80-2]	Cuprous azide	584
[14464-46-1]	Cristobalite	668
[14567-73-8]	Tremolite	244
[14666-78-5]	Diethyl peroxydicarbonate	688
[14721-21-2]	Cupric chlorate	660

SUBJECT INDEX